Ullmann's Encyclopedia of Industrial Chemistry

Volume B 7

Environmental Protection and Industrial Safety I

VCH

Ullmann's Encyclopedia of Industrial Chemistry

Volumes A 1 – A 28: alphabetically arranged articles
Volumes B 1 – B 8: basic knowledge

Editorial Advisory Board

Ullmann's Encyclopedia of Industrial Chemistry

Fifth, Completely Revised Edition

Editors: Barbara Elvers, Stephen Hawkins, William Russey

Volume B 7:

Environmental Protection and Industrial Safety I

Volume Editor: Eberhard Weise

Planning and Coordination:
Wolfgang Dekant, Jean-Marie Devos, Otto Fränzle,
Herwig Hulpke, Louis Jourdan, Burkhard Stüber,
John Wolf

VCH

Editorial Director: Christina Dyllick-Brenzinger
Editorial Assistants: Ilse Bedrich, Reinhilde Gutsche, Ingeborg Harders, Monika Pikart-Müller, Ulrike Winter
Copy Editor: Don Emerson

Library of Congress Card No. 84-25-829

A catalogue record for this book is available from the British Library.

Die Deutsche Bibliothek – CIP-Einheitsaufnahme

Ullmann's encyclopedia of industrial chemistry / ed.: Barbara Elvers ... [Ed. advisory board Hans-Jürgen Arpe ...]. — Weinheim ; Basel (Switzerland) ; Cambridge ; New York, NY ; Tokyo : VCH.
Teilw. executive ed.: Wolfgang Gerhartz
Bis 4. Aufl. u. d. T.: Ullmanns Encyklopädie der technischen Chemie
ISBN 3-527-20100-9

NE: Gerhartz, Wolfgang [Hrsg.]; Elvers, Barbara [Hrsg.]; Encyclopedia of industrial chemistry

Vol. B. Basic knowledge.
7. Environmental protection and industrial safety. – 1. / Vol. ed.: Eberhard Weise. – 5., completely rev. ed. – 1995
ISBN 3-527-20137-8

NE: Weise, Eberhard [Hrsg.]

Printed on acid-free and low-chlorine paper

Distribution
VCH Verlagsgesellschaft, P.O. Box 10 11 61. D-69451 Weinheim (Federal Republic of Germany)
Switzerland: VCH Verlags-AG, P.O. Box, CH-4020 Basel (Switzerland)
Great Britain and Ireland: VCH Publishers (UK) Ltd., 8 Wellington Court, Wellington Street, Cambridge CB1 1HZ (Great Britain)
USA and Canada: VCH Publishers, 220 East 23rd Street, New York NY 10010-4606 (USA)
Japan: VCH, Eikow Building, 10-9 Hongo 1-chomc, Bunkyo-ku, Tokyo 113, Japan

Production Managers: Claudia Baumann, Peter J. Biel
Cover design: Wolfgang Schmidt
Composition, printing, and bookbinding: Graphischer Betrieb Konrad Triltsch, D-97070 Würzburg
Printed in the Federal Republic of Germany

Foreword

The two *Ullmann's* volumes devoted to environmental protection and industrial safety have the purpose of examining the mutual relationship between chemical technology on the one hand and the environment and the workplace on the other. They differ from other volumes of the *Encyclopedia* above all in the fact that they not only provide information related to industrial chemistry but also discuss certain issues that could be considered political. Moreover, unlike most of the subjects treated in other volumes of the set, it would be impossible in the case of environmental protection and industrial safety to abstain from a consideration of historical developments. The various authors have also necessarily dealt with developments that can be anticipated in the future despite their obvious inability to foresee whether particular expectations will actually be met.

As with any handbook, what we attempt to present here is a comprehensive picture of the situation at a particular moment in time: the date of publication. *Ullmann's* editors have always tried to take into account differing perceptions on the part of multiple experts, but the chapters in these two volumes reflect in addition the diverse points of view associated with different sociopolitical vantage points.
Developments and production trends in industrial chemistry never evolve in a vacuum, but are instead a consequence of human actions, initiated by individuals who have their own unique interests and apprehensions.

To promote open dialogue between the various interest groups we have attempted to structure the presentation in as neutral a way as possible. This proved to be extremely difficult, especially given that the 5th Edition is expected to reflect a worldwide and transnational perspective. In many instances a parallel presentation of different viewpoints has been adopted, in order to avoid the risk of assuming power of decision between right or wrong of future developments.

Despite serious reservations on the part of the editorial staff and others in the scientific community, it was ultimately agreed that this pair of volumes should open with a chapter devoted to ethical questions. In view of the current worldwide debate over the legitimacy of various human activities, and the consequent burdens imposed on the natural environment, it seemed more fitting that we devote a judicious amount of space to the issue rather than ignore it altogether. It would be a welcome development, indeed, if such a chapter could be omitted from subsequent editions of the *Encyclopedia* on the grounds that the urgency of environmental problems had in the meantime been accepted as self-evident and therefore become an integral consideration in the everyday practice of technology.

The relationship between many topics of interest here and such other disciplines as ecology, biology, and medicine has made extensive reference to the latter increasingly indispensable, despite the difficulties this presents. We have been particularly careful never to lose sight of our primary goal, which is to provide the reader with a first-rate encyclopedia of industrial chemistry.

Fundamental and enduring aspects of environmental protection and industrial safety treated in the earlier 4th Edition of *Ullmann's* have been retained. Also carried over from the 4th Edition is a general organizational scheme that recognizes specific "compartments" within the environment as a whole. Industrial safety issues are treated both in articles focusing selectively on technological and on toxicological considerations. In addition, an introduction is presented to the important fields of ecology and ecotoxicology. This is followed by articles on human toxicology, a consideration of relevant legal matters, and a concise treatment of the subject of environmental management. The first five contributions thus constitute the basis for a general assessment of chemical technology as it relates to the environment and the workplace.

This is followed by articles that cover the subject of noise and vibration and issues related to air, soil and water. In subsequent chapters there is an examination of the fundamental distinction between *production* and *products*, together with an economic and sociological overview of factors related to raw materials and energy utilization; industrial products; the transport, handling, and storage of materials; and waste.

We have repeatedly been confronted with a problem of overlap between the various topics under consideration. For example, the subject of noise could legitimately have been treated under the heading of industrial safety—but also as an environmental protection issue. Similarly, the disposal of sewage-treatment sludge might reasonably be discussed in conjunction with either wastewater or waste in the broader sense. Duplicate coverage has been avoided whenever possible and cross-references have been used instead. Nevertheless, a certain amount of repetition was deemed desirable, if for no other reason that to spare the reader the burden of constantly paging back and forth between the two volumes.

Optimal ordering of the various chapters was a subject of considerable discussion, and that finally adopted clearly represents a compromise.

Special attention was directed to the amount of space to be allotted to the legal provisions of individual countries, the relevance of which is inevitably called into question by the worldwide distribution of the *Encyclopedia*. Assitance from members of the European Commission was especially welcome in this regard, since it permitted us to include references to national policies while at the same time pointing the way toward significant international trends anticipated in the future. An important—but at the same time challenging—task for the years ahead will be ensuring that the *Encyclopedia* remains current in this respect. It is perhaps worthwile to stress the fact that no attempt whatsoever has been made to create a substitute for the extensive specialized literature devoted to legal matters.

It would be impossible within the space of two volumes to provide in-depth treatment of unique factors associated with the manufacture and utilization of specific chemicals, apart from examples introduced to illustrate general concepts. Detailed information regarding the environmental problems posed by particular substances, including toxicological data, will be found under appropriate headings within the "A Series" of volumes. The same applies to analytical techniques, which are discussed here only insofar as they impinge upon specific aspects of environmental protection or industrial safety. A comprehensive overview of analytical methods is provided by Volumes B 5 and B 6.

At this point we wish to express our sincere appreciation to all the authors who have contributed to this effort, and especially to the coordinators of the various subject areas. Every participant was required to take cognizance of the fact that involvement in such a project entails the risk of exposing oneself to the polarity, that always surrounds the endeavors of representatives of different interest groups. Indeed, the ever-present potential for conflict was answered in several cases with extraordinary displays of civic courage. The volume editor is convinced that full publication of information reflecting the current state of the art worldwide is the best way to encourage international convergence with respect to the standards applicable to environmental protection and industrial safety. This may have been an additional motivation for the various authors, entirely consistent with our sincere wish that *Ullmann's Encyclopedia* might make a meaningful contribution to the future welfare of both mankind and the environment.

EBERHARD WEISE

Contents

Symbols and Units

Symbols and units agree with SI standards (for conversion factors see pp. VIII–IX). The following list gives the most important symbols used in the encyclopedia. Articles with many specific units and symbols have a similar list as front matter.

Symbol	Unit	Physical Quantity
a_B		activity of substance B
A_r		relative atomic mass (atomic weight)
A	m^2	area
c_B	mol/m^3, mol/L (M)	concentration of substance B
C	C/V	electric capacity
c_p, c_v	$J\,kg^{-1}K^{-1}$	specific heat capacity
d	cm, m	diameter
d		relative density (ϱ/ϱ_{water})
D	m^2/s	diffusion coefficient
D	Gy (= J/kg)	absorbed dose
e	C	elementary charge
E	J	energy
E	V/m	electric field strength
E	V	electromotive force
E_A	J	activation energy
f		activity coefficient
F	C/mol	Faraday constant
F	N	force
g	m/s^2	acceleration due to gravity
G	J	Gibbs free energy
h	m	height
h	$W \cdot s^2$	Planck constant
H	J	enthalpy
I	A	electric current
I	cd	luminous intensity
k	(variable)	rate constant of a chemical reaction
k	J/K	Boltzmann constant
K	(variable)	equilibrium constant
l	m	length
m	g, kg, t	mass
M_r		relative molecular mass (molecular weight)
n_D^{20}		refractive index (sodium D-line, 20 °C)
n	mol	amount of substance
N_A	mol^{-1}	Avogadro constant (6.023×10^{23} mol^{-1})
p	Pa, bar*	pressure
Q	J	quantity of heat
r	m	radius
R	$J\,K^{-1}mol^{-1}$	gas constant
R	Ω	electric resistance
S	J/K	entropy
t	s, min, h, d, month, a	time
t	°C	temperature
T	K	absolute temperature
u	m/s	velocity

* The official unit of pressure is the pascal (Pa).

Symbols and units (continued from p. IX)

Symbol	Unit	Physical Quantity
U	V	electric potential
U	J	internal energy
V	m^3, L, mL	volume
w		mass fraction
W	J	work
x_B		mole fraction of substance B
Z		proton number, atomic number
α		cubic expansion coefficient
α	$W\,m^{-2}K^{-1}$	heat-transfer coefficient (heat-transfer number)
α		degree of dissociation of electrolyte
$[\alpha]$	$10^{-2}deg\,cm^2g^{-1}$	specific rotation
η	Pa · s	dynamic viscosity
θ	°C	temperature
$\varkappa$		c_p/c_v
λ	$W\,m^{-1}K^{-1}$	thermal conductivity
λ	nm, m	wavelength
μ		chemical potential
ν	Hz, s^{-1}	frequency
ν	m^2/s	kinematic viscosity (η/ϱ)
π	Pa	osmotic pressure
ϱ	g/cm^3	density
σ	N/m	surface tension
τ	Pa (N/m^2)	shear stress
φ		volume fraction
χ	Pa^{-1} (m^2/N)	compressibility

Conversion Factors

SI unit	Non-SI unit	From SI to non-SI multiply by
Mass		
kg	pound (avoirdupois)	2.205
kg	ton (long)	9.842×10^{-4}
kg	ton (short)	1.102×10^{-3}
Volume		
m^3	cubic inch	6.102×10^4
m^3	cubic foot	35.315
m^3	gallon (U.S., liquid)	2.642×10^2
m^3	gallon (Imperial)	2.200×10^2
Temperature		
°C	°F	°C × 1.8 + 32
Force		
N	dyne	1.0×10^5

Conversion factors (continued from p. X)

SI unit	Non-SI unit	From SI to non-SI multiply by
Energy, Work		
J	Btu (int.)	9.480×10^{-4}
J	cal (int.)	2.389×10^{-1}
J	eV	6.242×10^{18}
J	erg	1.0×10^{7}
J	kW · h	2.778×10^{-7}
J	kp · m	1.020×10^{-1}
Pressure		
MPa	at	10.20
MPa	atm	9.869
MPa	bar	10
kPa	mbar	10
kPa	mm Hg	7.502
kPa	psi	0.145
kPa	torr	7.502

Prefixes for Powers of Ten

E (exa)	10^{18}	M (mega)	10^{6}	d (deci)	10^{-1}	n (nano)	10^{-9}
P (peta)	10^{15}	k (kilo)	10^{3}	c (centi)	10^{-2}	p (pico)	10^{-12}
T (tera)	10^{12}	h (hecto)	10^{2}	m (milli)	10^{-3}	f (femto)	10^{-15}
G (giga)	10^{9}	da (deca)	10	μ (micro)	10^{-6}	a (atto)	10^{-18}

Abbreviations

The following is a list of the abbreviations used in the text. Common terms, the names of publications and institutions, and legal agreements are included along with their full identities. Other abbreviations will be defined wherever they first occur in an article. For further abbreviations, see p. VII (Symbols and Units), p. XIV (Companies and Country Codes in Patent References). The names of periodical publications are abbreviated exactly as done by Chemical Abstracts Service.

abs.	absolute
a.c.	alternating current
ACGIH	American Conference of Governmental Industrial Hygienists
ACS	American Chemical Society
ADI	acceptable daily intake
ADN	accord européen relatif au transport international des marchandises dangereuses par voie de navigation interieure (European agreement concerning the international transportation of dangerous goods by inland waterways)
ADNR	ADN par le Rhin (regulation concerning the transportation of dangerous goods on the Rhine and all national waterways of the countries concerned)
ADP	adenosine 5′-diphosphate
ADR	accord européen relatif au transport international des marchandises dangereuses par route (European agreement concerning the international transportation of dangerous goods by road)
AEC	Atomic Energy Commission (United States)
AIChE	American Institute of Chemical Engineers

AIME	American Institute of Mining, Metallurgical, and Petroleum Engineers
AMP	adenosine 5′-monophosphate
APhA	American Pharmaceutical Association
API	American Petroleum Institute
ASTM	American Society for Testing and Materials
ATP	adenosine 5′-triphosphate
BAM	Bundesanstalt für Materialprüfung (Federal Republic of Germany)
BAT	Biologischer Arbeitsstoff-Toleranzwert (biological tolerance value for a working material, established by MAK Commission, see MAK)
Beilstein	Beilstein's Handbook of Organic Chemistry, Springer, Berlin–Heidelberg–New York
BET	Brunauer–Emmett–Teller
BGA	Bundesgesundheitsamt (Federal Republic of Germany)
BGBl.	Bundesgesetzblatt (Federal Republic of Germany)
BIOS	British Intelligence Objectives Subcommitee Report (see also FIAT)
BOD	biological oxygen demand
bp	boiling point
B.P.	British Pharmacopeia
BS	British Standard
ca.	circa
calcd.	calculated
CAS	Chemical Abstracts Service
cat.	catalyst, catalyzed
CEN	Çomité Européen de Normalisation
cf.	compare
CFR	Code of Federal Regulations (United States)
Chap.	chapter
ChemG	Chemikaliengesetz (Federal Republic of Germany)
C.I.	Colour Index
CIOS	Combined Intelligence Objectives-Subcommitee Report (see also FIAT)
CNS	central nervous system
Co.	Company
COD	chemical oxygen demand
conc.	concentrated
const.	constant
Corp.	Corporation
crit.	critical
CTFA	The Cosmetic, Toiletry and Fragrance Association (United States)
DAB 9	Deutsches Arzneibuch, 9th ed., Deutscher Apotheker-Verlag, Stuttgart 1986
d.c.	direct current
decomp.	decompose, decomposition
DFG	Deutsche Forschungsgemeinschaft (German Science Foundation)
dil.	dilute, diluted
DIN	Deutsche Industrie Norm (Federal Republic of Germany)
DMF	dimethylformamide
DNA	deoxyribonucleic acid
DOE	Department of Energy (United States)
DOT	Department of Transportation – Materials Transportation Bureau (United States)
DTA	differential thermal analysis
EC	effective concentration
EC	European Community
ed.	editor, edition, edited
e.g.	for example
emf	electromotive force
EmS	Emergency Schedule
EN	European Standard (European Community)
EPA	Environmental Protection Agency (United States)
EPR	electron paramagnetic resonance
Eq.	equation
ESCA	electron spectroscopy for chemical analysis
esp.	especially
ESR	electron spin resonance
Et	ethyl substituent ($-C_2H_5$)
et al.	and others
etc.	et cetera
EVO	Eisenbahnverkehrsordnung (Federal Republic of Germany)
exp (...)	$e^{(...)}$, mathematical exponent
FAO	Food and Agriculture Organization (United Nations)
FDA	Food and Drug Administration (United States)
FD & C	Food, Drug and Cosmetic Act (United States)
FHSA	Federal Hazardous Substances Act (United States)
FIAT	Field Information Agency, Technical (United States reports on the chemical industry in Germany, 1945)
Fig.	figure

fp	freezing point
Friedländer	P. Friedländer, Fortschritte der Teerfarbenfabrikation und verwandter Industriezweige, Vol. 1–25, Springer, Berlin 1888–1942
FT	Fourier transform
(g)	gas, gaseous
GC	gas chromatography
GefStoffV	Gefahrstoffverordnung (regulations in the Federal Republic of Germany concerning hazardous substances)
GGVE	Verordnung in der Bundesrepublik Deutschland über die Beförderung gefährlicher Güter mit der Eisenbahn (regulation in the Federal Republic of Germany concerning the transportation of dangerous goods by rail)
GGVS	Verordnung in der Bundesrepublik Deutschland über die Beförderung gefährlicher Güter auf der Straße (regulation in the Federal Republic of Germany concerning the transportation of dangerous goods by road)
GGVSee	Verordnung in der Bundesrepublik Deutschland über die Beförderung gefährlicher Güter mit Seeschiffen (regulation in the Federal Republic of Germany concerning the transportation of dangerous goods by sea-going vessels)
GLC	gas-liquid chromatography
Gmelin	Gmelin's Handbook of Inorganic Chemistry, 8th ed., Springer, Berlin–Heidelberg–New York
GRAS	generally recognized as safe
Hal	halogen substituent (–F, –Cl, –Br, –I)
Houben-Weyl	Methoden der organischen Chemie, 4th ed., Georg Thieme Verlag, Stuttgart
HPLC	high performance liquid chromatography
IAEA	International Atomic Energy Agency
IARC	International Agency for Research on Cancer, Lyon, France
IATA-DGR	International Air Transport Association, Dangerous Goods Regulations
ICAO	International Civil Aviation Organization
i.e.	that is
i.m.	intramuscular
IMDG	International Maritime Dangerous Goods Code
IMO	Inter-Governmental Maritime Consultive Organization (in the past: IMCO)
Inst.	Institute
i.p.	intraperitoneal
IR	infrared
ISO	International Organization for Standardization
IUPAC	International Union of Pure and Applied Chemistry
i.v.	intravenous
Kirk-Othmer	Encyclopedia of Chemical Technology, 3rd ed., J. Wiley & Sons, New York–Chichester–Brisbane–Toronto 1978–1984
(l)	liquid
Landolt-Börnstein	Zahlenwerte u. Funktionen aus Physik, Chemie, Astronomie, Geophysik u. Technik, Springer, Heidelberg 1950–1980; Zahlenwerte und Funktionen aus Naturwissenschaften und Technik, Neue Serie, Springer, Heidelberg, since 1961
LC_{50}	lethal concentration for 50% of the test animals
LCLo	lowest published lethal concentration
LD_{50}	lethal dose for 50% of the test animals
LDLo	lowest published lethal dose
ln	logarithm (base e)
LNG	liquefied natural gas
log	logarithm (base 10)
LPG	liquefied petroleum gas
M	mol/L
M	metal (in chemical formulas)
MAK	Maximale Arbeitsplatz-Konzentration (maximum concentration at the workplace in the Federal Republic of Germany); cf. Deutsche Forschungsgemeinschaft (ed.): Maximale Arbeits-

	platzkonzentrationen (MAK) und Biologische Arbeitsstoff-Toleranz-Werte (BAT), VCH Verlagsgesellschaft, Weinheim (published annually)
max.	maximum
MCA	Manufacturing Chemists Association (United States)
Me	methyl substituent ($-CH_3$)
Methodicum Chimicum	Methodicum Chimicum, Georg Thieme Verlag, Stuttgart
MFAG	Medical First Aid Guide for Use in Accidents Involving Dangerous Goods
MIK	maximale Immissionskonzentration (maximum immission concentration)
min.	minimum
mp	melting point
MS	mass spectrum, mass spectrometry
NAS	National Academy of Sciences (United States)
NASA	National Aeronautics and Space Administration (United States)
NBS	National Bureau of Standards (United States)
NCTC	National Collection of Type Cultures (United States)
NIH	National Institutes of Health (United States)
NIOSH	National Institute for Occupational Safety and Health (United States)
NMR	nuclear magnetic resonance
no.	number
NOEL	no observed effect level
NRC	Nuclear Regulatory Commission (United States)
NRDC	National Research Development Corporation (United States)
NSC	National Service Center (United States)
NSF	National Science Foundation (United States)
NTSB	National Transportation Safety Board (United States)
OECD	Organization for Economic Cooperation and Development
OSHA	Occupational Safety and Health Administration (United States)
p., pp.	page, pages
Patty	G. D. Clayton, F. E. Clayton (eds.): Patty's Industrial Hygiene and Toxicology, 3rd ed., Wiley Interscience, New York
PB report	Publication Board Report (U.S. Department of Commerce, Scientific and Industrial Reports)
PEL	permitted exposure limit
Ph	phenyl substituent ($-C_6H_5$)
Ph. Eur.	European Pharmacopoeia, 2nd. ed., Council of Europe, Strasbourg 1981 –
phr	part per hundred rubber (resin)
PNS	peripheral nervous system
ppm	parts per million
q. v.	which see (quod vide)
ref.	refer, reference
resp.	respectively
R_f	retention factor (TLC)
R. H.	relative humidity
RID	règlement international concernant le transport des marchandises dangereuses par chemin de fer (international convention concerning the transportation of dangerous goods by rail)
RNA	ribonucleic acid
R phrase (R-Satz)	risk phrase according to ChemG and GefStoffV (Federal Republic of Germany)
rpm	revolutions per minute
RTECS	Registry of Toxic Effects of Chemical Substances, edited by the National Institute of Occupational Safety and Health (United States)
(s)	solid
SAE	Society of Automotive Engineers (United States)
s.c.	subcutaneous
SI	International System of Units
SIMS	secondary ion mass spectrometry
S phrase (S-Satz)	safety phrase according to ChemG and GefStoffV (Federal Republic of Germany)
STEL	Short Term Exposure Limit (see TLV)
STP	standard temperature and pressure (0° C, 101.325 kPa)
T_g	glass transition temperature
TA Luft	Technische Anleitung zur Reinhaltung der Luft (clean air regulation in Federal Republic of Germany)
TA Lärm	Technische Anleitung zum Schutz gegen Lärm (low noise

	regulation in Federal Republic of Germany)
TDLo	lowest published toxic dose
THF	tetrahydrofuran
TLC	thin layer chromatography
TLV	Threshold Limit Value (TWA and STEL); published annually by the American Conference of Governmental Industrial Hygienists (ACGIH), Cincinnati, Ohio
TOD	total oxygen demand
TRK	Technische Richtkonzentration (lowest technically feasible level)
TSCA	Toxic Substances Control Act (United States)
TÜV	Technischer Überwachungsverein (Technical Control Board of the Federal Republic of Germany)
TWA	Time Weighted Average
UBA	Umweltbundesamt (Federal Environmental Agency)
Ullmann	Ullmanns Encyklopädie der Technischen Chemie, 4th ed., Verlag Chemie, Weinheim 1972–1984; 3rd ed., Urban und Schwarzenberg, München 1951–1970
USAEC	United States Atomic Energy Commission
USAN	United States Adopted Names
USD	United States Dispensatory
USDA	United States Department of Agriculture
U.S.P.	United States Pharmacopeia
UV	ultraviolet
UVV	Unfallverhütungsvorschriften der Berufsgenossenschaft (workplace safety regulations in the Federal Republic of Germany)
VbF	Verordnung in der Bundesrepublik Deutschland über die Errichtung und den Betrieb von Anlagen zur Lagerung, Abfüllung und Beförderung brennbarer Flüssigkeiten (regulation in the Federal Republic of Germany concerning the construction and operation of plants for storage, filling, and transportation of flammable liquids; classification according to the flash point of liquids, in accordance with the classification in the United States)
VDE	Verband Deutscher Elektroingenieure (Federal Republic of Germany)
VDI	Verein Deutscher Ingenieure (Federal Republic of Germany)
vol	volume
vol.	volume (of a series of books)
vs.	versus
WHO	World Health Organization (United Nations)
Winnacker-Küchler	Chemische Technologie, Carl Hanser Verlag, München
wt	weight
$	U.S. dollar, unless otherwise stated

Abbreviations for the Names of Frequently Cited Companies

Air Products	Air Products and Chemicals
Akzo	Algemene Koninklijke Zout Organon
Alcoa	Aluminum Company of America
Allied	Allied Corporation
Amer. Cyanamid	American Cyanamid Company
BASF	BASF Aktiengesellschaft
Bayer	Bayer AG
BP	British Petroleum Company
Celanese	Celanese Corporation
Daicel	Daicel Chemical Industries
Dainippon	Dainippon Ink and Chemicals Inc.
Dow Chemical	The Dow Chemical Company
DSM	Dutch Staats Mijnen
Du Pont	E.I. du Pont de Nemours & Company
Exxon	Exxon Corporation
FMC	Food Machinery & Chemical Corporation
GAF	General Aniline & Film Corporation
W.R. Grace	W.R. Grace & Company
Hoechst	Hoechst Aktiengesellschaft
IBM	International Business Machines Corporation
ICI	Imperial Chemical Industries
IFP	Institut Français du Pétrole
INCO	International Nickel Company
3M	Minnesota Mining and Manufacturing Company
Mitsubishi Chemical	Mitsubishi Chemical Industries
Monsanto	Monsanto Company
Nippon Shokubai	Nippon Shokubai Kagaku Kogyo
PCUK	Pechiney Ugine Kuhlmann
PPG	Pittsburg Plate Glass Industries
Searle	G.D. Searle & Company
SKF	Smith Kline & French Laboratories
SNAM	Societá Nazionale Metandotti
Sohio	Standard Oil of Ohio
Stauffer	Stauffer Chemical Company
Sumitomo	Sumitomo Chemical Company
Toray	Toray Industries Inc.
UCB	Union Chimique Belge
Union Carbide	Union Carbide Corporation
UOP	Universal Oil Products Company
VEBA	Vereinigte Elektrizitäts- und Bergwerks-AG
Wacker	Wacker Chemie GmbH

Country Codes

The following list contains a selection of standard country codes used in the patent references.

AT	Austria
AU	Australia
BE	Belgium
BG	Bulgaria
BR	Brazil
CA	Canada
CH	Switzerland
CS	Czechoslovakia
DD	German Democratic Republic
DE	Federal Republic of Germany (and Germany before 1949)*
DK	Denmark
ES	Spain
FI	Finland
FR	France
GB	United Kingdom
GR	Greece
HU	Hungary
ID	Indonesia
IL	Israel
IT	Italy
JP	Japan*
LU	Luxembourg
MA	Morocco
NL	Netherlands*
NO	Norway
NZ	New Zealand
PL	Poland
PT	Portugal
SE	Sweden
SU	Soviet Union
US	United States of America
YU	Yugoslavia
ZA	South Africa
EP	European Patent Office*
WO	World Intellectual Property Organization

* For Europe, Federal Republic of Germany, Japan, and the Netherlands, the type of patent is specified: EP (patent), EP-A (application), DE (patent), DE-OS (Offenlegungsschrift), DE-AS (Auslegeschrift), JP (patent), JP-Kokai (Kokai tokkyo koho), NL (patent), and NL-A (application).

Periodic Table of Elements

1A ("European" group designation according to old IUPAC recommendation)
1 (group designation according to 1985 IUPAC proposal)
IA ("American" group designation, also used by the Chemical Abstracts Service until the end of 1986)

1A 1 IA	2A 2 IIA	3A 3 IIIB	4A 4 IVB	5A 5 VB	6A 6 VIB	7A 7 VIIB	8 8 VIII	8 9 VIII	8 10 VIII	1B 11 IB	2B 12 IIB	3B 13 IIIA	4B 14 IVA	5B 15 VA	6B 16 VIA	7B 17 VIIA	0 18 VIIIA
1.0079 $_{1}$**H**																	4.0026 $_{2}$**He**
6.941 $_{3}$**Li**	9.0122 $_{4}$**Be**											10.811 $_{5}$**B**	12.011 $_{6}$**C**	14.007 $_{7}$**N**	15.9994 $_{8}$**O**	18.998 $_{9}$**F**	20.180 $_{10}$**Ne**
22.990 $_{11}$**Na**	24.305 $_{12}$**Mg**											26.982 $_{13}$**Al**	28.086 $_{14}$**Si**	30.974 $_{15}$**P**	32.066 $_{16}$**S**	35.453 $_{17}$**Cl**	39.948 $_{18}$**Ar**
39.098 $_{19}$**K**	40.078 $_{20}$**Ca**	44.956 $_{21}$**Sc**	47.88 $_{22}$**Ti**	50.942 $_{23}$**V**	51.996 $_{24}$**Cr**	54.938 $_{25}$**Mn**	55.847 $_{26}$**Fe**	58.933 $_{27}$**Co**	58.69 $_{28}$**Ni**	63.546 $_{29}$**Cu**	65.39 $_{30}$**Zn**	69.723 $_{31}$**Ga**	72.61 $_{32}$**Ge**	74.922 $_{33}$**As**	78.96 $_{34}$**Se**	79.904 $_{35}$**Br**	83.80 $_{36}$**Kr**
85.468 $_{37}$**Rb**	87.62 $_{38}$**Sr**	88.906 $_{39}$**Y**	91.224 $_{40}$**Zr**	92.906 $_{41}$**Nb**	95.94 $_{42}$**Mo**	98.906 $_{43}$**Tc** *	101.07 $_{44}$**Ru**	102.91 $_{45}$**Rh**	106.42 $_{46}$**Pd**	107.87 $_{47}$**Ag**	112.41 $_{48}$**Cd**	114.82 $_{49}$**In**	118.71 $_{50}$**Sn**	121.75 $_{51}$**Sb**	127.60 $_{52}$**Te**	126.90 $_{53}$**I**	131.29 $_{54}$**Xe**
132.91 $_{55}$**Cs**	137.33 $_{56}$**Ba**		178.49 $_{72}$**Hf**	180.95 $_{73}$**Ta**	183.85 $_{74}$**W**	186.21 $_{75}$**Re**	190.2 $_{76}$**Os**	192.22 $_{77}$**Ir**	195.08 $_{78}$**Pt**	196.97 $_{79}$**Au**	200.59 $_{80}$**Hg**	204.38 $_{81}$**Tl**	207.2 $_{82}$**Pb**	208.98 $_{83}$**Bi**	208.98 $_{84}$**Po** *	209.99 $_{85}$**At** *	222.02 $_{86}$**Rn** *
223.02 $_{87}$**Fr** *	226.03 $_{88}$**Ra** *																

138.91 $_{57}$**La**	140.12 $_{58}$**Ce**	140.91 $_{59}$**Pr**	144.24 $_{60}$**Nd**	146.92 $_{61}$**Pm** *	150.36 $_{62}$**Sm**	151.97 $_{63}$**Eu**	157.25 $_{64}$**Gd**	158.93 $_{65}$**Tb**	162.50 $_{66}$**Dy**	164.93 $_{67}$**Ho**	167.26 $_{68}$**Er**	168.93 $_{69}$**Tm**	173.04 $_{70}$**Yb**	174.97 $_{71}$**Lu**
227.03 $_{89}$**Ac** *	232.04 $_{90}$**Th** *	231.04 $_{91}$**Pa** *	238.03 $_{92}$**U** *	237.05 $_{93}$**Np** *	244.06 $_{94}$**Pu** *	243.06 $_{95}$**Am** *	247.07 $_{96}$**Cm** *	247.07 $_{97}$**Bk** *	251.08 $_{98}$**Cf** *	252.08 $_{99}$**Es** *	257.10 $_{100}$**Fm** *	258.10 $_{101}$**Md** *	259.10 $_{102}$**No** *	260.11 $_{103}$**Lr** *

* Elements with unstable isotopes; mass of most important isotope given

Ethics and Industrial Chemistry

EBERHARD WEISE, Monheim, Federal Republic of Germany, HENNING FRIEGE, Umweltdezernat der Landeshauptstadt Düsseldorf, Federal Republic of Germany (Chap. 1)

GÜNTER ALTNER, Heidelberg, Federal Republic of Germany, PHILIPP SCHMITZ, Frankfurt, Federal Republic of Germany (Chap. 2)

KLAUS K. KLOSTERMAIER, University of Manitoba, Winnipeg, Canada (Chaps. 3, 4)

The entire topic was coordinated by EBERHARD WEISE

1. Technology and Ethics

1.1. Technology and Ethics Belong Together

Technology and ethics clearly have at least one thing in common: both were born subsequent to humankind's appearance on earth. We define *technology* as the conscious, goal-oriented application of scientific knowledge, and *ethics* as a system of norms for human activity. If we now attempt to relate the two in a meaningful way it becomes clear that one cannot possibly assess technology without reference to ethics. In this sense "technology as such" is neither good nor bad, neither destructive nor beneficial. Only ethics, which stands above technology and its application, permits evaluation on the basis of moral categories.

Both technology and ethics have experienced dramatic change over the course of human existence, and both will continue to evolve in the future. Technology at any given moment is a reflection of the *status quo* of scientific knowledge, whereas ethics reflects prevailing moral attitudes. Technology has helped people improve their living conditions and chances for survival, at the same time providing the means for overcoming certain of the physical inferiorities of humans relative to other animals, as well as the detrimental effects of adverse climatic conditions. Indeed, human survival has always been

3-527-20137-8/95/$ 5.00 + .50

the paramount goal of technology. Unfortunately, humankind has always believed that war and destruction are essential for survival. Technology has thus opened the way to creating a "brave new world" catering specifically to humankind's creature comforts—and to experiences in the broadest sense—while giving clearer expression to social inequalities that have so far proven ineradicable.

Developments in ethics have not been as linear as in technology. Religions and philosophies have instead displaced one another at various times, and if we are now beginning to question the notion of "progress" in the area of technology, its validity in the field of ethics seems even less universally plausible.

1.2. Limits of Growth

For centuries humankind could afford to pursue its own ends without paying particular attention to the environment. While there are a few early examples—some even from the Middle Ages—of laws aimed at keeping the air and water clean, the lawmaker's perspective in the past rarely extended beyond some immediate localized setting. Low population densities and limited technical potential meant that, despite minor incursions, people as such never constituted a serious threat to the natural balance. Only in the twentieth century have we been confronted with the fact that fresh air, clear water, and various other resources are ultimately limited in their availability. Whereas the deforestation of vast tracts of land was once considered a pioneering achievement, it is more likely now to be regarded as a rape of the countryside. The population explosion, exponential increases in the rate of transformation of materials, and the development of new technologies whose impacts are difficult to assess—all these factors have suddenly forced humankind to become more conscious of its responsibilities. Societies in the past—excessive exploitation of specific resource supplies notwithstanding—were never equipped to provoke continental or indeed global crises, but today the situation is very different as a result of the massive flows of material associated with utilization of such resources as oil, metals, and land, as well as the hazards attributable to certain man-made materials (e.g., fluorochlorinated hydrocarbons, pesticides) and the enormous waste streams generated in both production and consumption. We must now consider from a global standpoint the consequences of human activity with respect to the technology, synthesis, and release of materials, the probable effects on future generations as well as the regenerative capacity of nature.

In other words, all technological activity must be reexamined in the light of ethical responsibility. This injunction applies first and foremost to the highly industrialized societies, and within these especially to the technical elite. Virtually no one would today question the important role of chemicals with respect to survival of the human race, nor the need to continue providing the technical services essential to an industrialized society. This means, however, that acting responsibly has become an absolute imperative.

Any inquiry into the general subject "technology and ethics" by someone working in the chemical industry might usefully start from the following basic premise: The industry regards its output as a response to orders placed by society. "Output" in this context might mean, for example, solving a particular problem through the transformation of substances, or satisfying specific customer desires. Occasionally it may transpire that society will choose to reject some solution proposed by industry, even one that has already been implemented. The human urge to explore apparently cannot be entirely dissociated from its social context and left exclusively to the discipline of voluntary self-restriction. It is precisely at this point that careful assessment of the consequences of technological activity—as difficult as that may be—takes on special significance. Once innovations have been transformed into products it becomes apparent that even research can no longer be regarded as a truly independent activity. It is driven by the marketplace.

The inherent variety and variability of chemical processes and products provide many opportunities for overcoming crises. But what sorts of values are required by the participants—researchers, technicians, and sales personnel in the materials and manufacturing industries, as well as consumers, and regulatory agencies—if the opportunities inherent in the chemical industry are to be realized?

1.3. Opportunities and Risks

Chemical manufacturers exercise free choice over their products, and thus over the processes used for transforming materials. The overall effects of such processes and products must be subjected to constant assessment, or at least estimation, if the necessary input is to be available for a discussion aimed at balancing out the various positive and negative factors. The effects in question are influenced by a combination of *retrospective* variables (including utility and attendant damage) together with such *prospective* variables as opportunity and risk. Utility and damage are rather easily assessed, but an evaluation of opportunities and risks is always based on estimates, which may be more or less accurate.

No technological development, no technical application, and thus no process or product—indeed, not even human action or inaction, or life itself—is completely free of risk. Every functioning individual is subject to risk, whether carrying out an assignment or playing the more independent role of an entrepreneur in the manufacture of a product. Those in a position to take advantage of the opportunities offered by a particular product or activity, and willing to accept a certain risk as part of the bargain, are the *beneficiaries*. But there will also be uninvolved *bystanders* who are subject only to the consequences; they derive no direct benefits or are unable to take advantage of resulting opportunities. These bystanders must also be included in any comprehensive discussion of advantages and disadvantages. In practice it is difficult to initiate the type of dialogue envisaged here, one that assures adequate consideration of justifiable risks that might be incurred by the "uninvolved," perhaps through the active participation of some of the uninvolved parties. Nature itself in both its inanimate and animate manifestations must be regarded as belonging to the "uninvolved," subject only to the consequences of entrepreneurial activity. Clearly, no assessment of risk can be left strictly in the hands of industry and such experts as industry may choose to consult; commissions representing society as a whole—perhaps the government—must necessarily play a role as well, often defined in terms of licensing procedures or even official bans on certain products. At a level below that of law, however, industry can and in fact should seize the initiative by instituting meaningful dialogue with appropriate representatives of the "uninvolved" over the issue of risks and opportunities, dialogue that also includes consumer groups and environmental protection organizations.

The pharmaceutical industry is a useful example to examine some of the problems associated with risk/opportunity assessments, in part because this is a sector already subject to extensive regulation. Assume, for example, that release of a new drug will be significantly delayed because of the requirement that all conceivable side effects first be investigated. It then becomes appropriate to consider the suffering of the individuals to whom the medication might bring swift relief. Is it possible to balance opportunities of one group against the risks of another? In the complicated licensing procedures governing pharmaceutical products, responsibility has been distributed fairly equally between industry, the regulatory authorities and lawmakers; all three groups are called upon to make essentially political decisions in the course of establishing appropriate dosage levels and reviewing side effects. In the future, considerations of this type will become increasingly important with respect not only to the pharmaceutical sector but also to other branches of the chemical industry because of the risks associated with chemicals of all types.

It is important to consider also the broader question of what activities chemical industry should or should not in fact be called upon to undertake. What should be done, for example, if the risk attributable to *inactivity* is estimated to exceed that resulting from tangible action? If industry decides to forego producing some material because the risk appears excessive, is it not obliged to show that it has also seriously considered the magnitude of the accompanying lost opportunity for solving an urgent problem?

The financial world demands that all entrepreneurs be willing to accept certain risks. But simple calculations based on the premise that the greater the potential (for fiscal profit) the greater can be the risk (of fiscal loss) are inappropriate in the case of the chemical industry; account must always be taken of safety—in terms of employees, consumers, and the environment—and not just with respect to the present generation, but also for generations to come.

1.3.1. Reduction of Risk through Adoption of the Global Ethical Imperative "Sustainable Development"

The chemical industry is faced with a unique set of challenges because damages attributable to particular substances often first become apparent only after long periods of use. This raises difficult questions regarding responsible handling of materials for which available toxic or ecotoxic data are clearly unreliable. Judicious action is particularly important if the substance in question exhibits a high level of persistence in sensitive environmental compartments such as human beings, animals, freshwater lakes, or the troposphere. The problem of responsibility acquires added dimensions from the fact that "mending" a once-damaged system may require several generations, depending on the length of the repair cycle in that particular system.

Some of the issues associated with the responsible handling of materials can be addressed with the aid of the principle of "Sustainable Development" [1], which has both technical and ethical aspects [2]. The concept of "sustainability" has guided the German forestry industry for more than 100 years. As far as environmental protection is concerned, sustainable development can be summarized in terms of two principles:

1) Utilization of a *renewable* resource must not occur faster than the rate of regeneration of that resource, whereas the rate at which a *nonrenewable* resource is consumed must be maintained at a level sufficiently low to ensure that future generations will inherit an ecologically functional earth [3]
2) The above utilization conditions must be met in such a way that the basic prerequisites for life continue to be guaranteed for people everywhere in the world

These two principles in turn lead to the following more specific guidelines for material disposition:

1) Consumption as it applies to both materials and energy must be directed primarily toward satisfying basic human needs and maintaining the technological infrastructure essential to an industrialized society.
2) Resources must be consumed with maximum efficiency, which in turn requires careful optimization of material deployment.
3) Every effort should be made to avoid filling underground disposal sites with persistent substances whose hazardous nature cannot be excluded with a high degree of probability. Any materials not expected to remain within "closed systems" should therefore be formulated explicitly with a view toward ecological safety. Persistent substances [4] should be the subject of particularly careful scrutiny, and strict control should be maintained over their whereabouts, because chemical transformations in ecosystems are often initiated or first detected only after a delay of several years.
4) Production of substances known to have the potential for causing irreversible effects [5] on the ecosystem should be suspended until it becomes clear that such threats can be reliably dealt with over a manageable period of time.

Manufacturers of basic chemicals and metals, both nonferrous and ferrous, normally do not engage in the production of consumer goods, so the special challenges to which these companies are subject relate primarily to the efficient use of resources and the ecological design of products. Their primary obligation is thus to rethink the fundamental premises associated with material flow, in both quantitative and qualitative terms. Given such currently pressing global problems as the "greenhouse effect" and simultaneous exhaustion of certain key resources, and in the face of virtually insoluble waste problems throughout the industrialized world, the much-heralded "efficiency revolution" [6] becomes of paramount importance. It is essential that material and energy expenditures attributable to production be reduced to a minimum, but without at the same time losing sight of the goal of an overall ecological optimum.

1.3.2. Risk Assessment

Assessments of risk and opportunity in the chemical industry must in principle be conducted at three levels. At the first level industry must ask itself: Is this particular activity so important that its pursuit is essential regardless of cost? Not everyone will answer the question in the same way in any given case. What one person views as crucial another may well regard as unimportant. A fundamental distinction must therefore be made between the *necessary* and the *desirable* [7]: Between those things that are absolutely indispensable, and those that are simply preferable. In any event, at this first level it is the product itself, or perhaps something to be derived from it, that represents the focus of attention.

At the second level the question is: Given an agreed set of goals, by what means should they be achieved? Here the analysis can become complicated if, for example, a process that represents no risk is dependent upon a starting material whose manufacture is hazardous. When such a starting material is in turn prepared by another company altogether any risk–benefit analysis is especially difficult. The same holds true with respect to transport problems. Situations of this type demand an *integrated analysis* involving ecobalances, product-line studies, and technology assessments; such analyses so far can be envisioned only in the distant future.

Once the decision has been made to adopt a particular process, a third level of inquiry is reached where the question to be posed is: How might residual risks associated with the planned operation be minimized? Were one to conclude at this point that, despite implementation of every conceivable technical safeguard, the residual risk cannot be reduced below a justifiable limit, then there would be no alternative but to return to the first stage of the assessment and ask: Is it still—even under these circumstances—so important and proper to engage in the proposed activity, or should the whole issue instead be reviewed later, when new insights might perhaps be available?

The complexity of such analyses becomes readily apparent in the context, for example, of the dilemmas with which modern agriculture is confronted.

1.4. The Problem of Communication

The chemical industry has a special obligation to keep the public adequately informed about its intended activities, even though this may sometimes result in society's electing to forego access to some product because the attendant risk appears too great. A decisive factor in controversies of this sort is fear. Large segments of the population, especially in the absence of detailed information, are afraid of the chemical industry. Industry for its part must learn to accept this anxiety. Fear as such is not actually bad. Indeed, fear is the only thing that can teach mankind caution, and it is as much a part of life as the desire for protection. Industry must direct all its efforts toward openly explaining company activities as the only effective way of dispelling fears that are unfounded. Residual anxieties then become a factor to be taken into account in day-to-day operation.

Manufacturers are also obliged to deal with the fact that values constantly change. In the early days of the industrial revolution, corporations and employees alike were proud of the smoke issuing from their factory chimneys, but increasing air pollution and added insight into its consequences have transformed this once positive image into a universally negative symbol.

This points to a dilemma that has plagued the chemical industry from the beginning. Chemical products should be reactive—i.e., highly effective in their intended applications—but at the same time they must not be permitted to pollute the environment. In practice this means, for example, that the ideal crop-protection agent is one that destroys a specific pest quickly and efficiently, and immediately thereafter becomes transformed into soil, thus posing no threat whatsoever to the environment. Similarly, synthetic structural materials are expected to display outstanding stability, but after years of use they should decompose rapidly upon consignment to a landfill. In practice it is of course impossible to meet such contradictory demands. Specific characteristics can be adjusted, but the fundamental dilemma remains.

If discussion is to succeed in establishing broad consensus, it must be based on a common scale of values. This means that open, fundamental dialogue directed toward implementation of the global imperative for "Sustainable Development"—such as that conducted by the German Enquête Commission under the title "Protection of Humanity and the Environment" [3]—is a key prerequisite for public confidence in the efforts of the chemical industry and its employees. Any dialogue between technical specialists and ordinary citizens is bound to be difficult, but it will be aggravated further by a climate of mutual distrust. It is therefore absolutely essential that the chemical industry openly admit the limits of present knowledge, in the spirit of the time-honored maxim: "Today's truths are often tomorrow's errors!" For its part, the public must come to realize that establishing safety standards (with respect to permissible emissions, for example, or the licensing of new technologies) is a political process, one which individual citizens can and should influence by their participation in discussions with regulatory agencies and the chemical industry. This will not always prevent conflicts between protection of the individual on one hand and economic and societal interests on the other, but openness once again represents the first step in achieving a resolution.

1.5. The Responsibility of Producers, and Self-Regulation

The chemical industry—at least in the developed countries—has already declared its intent to practice self-regulation ("Responsible Care" [8]; →Environmental Management in the Chemical Industry, Section 5.4). This move was based on the conviction that responsible handling of chemicals consists of:

1) Optimizing the utilization of raw materials
2) Conducting reactions in such a way that they generate minimal emission burdens in the form of exhaust air, wastewater, and unused byproducts
3) Manufacturing substances and products with targeted functionality and minimal risk to humankind and the environment
4) Reusing excess materials to the greatest extent possible, and practicing environmentally friendly waste management with appropriate cooperation on the part of primary producers [8]

Implicit in this agenda is the notion that suppliers acquire a "cradle-to-grave responsibility" for their products, and that manufacturers are obliged to maintain an interest in materials they produce even if other processors intervene as distribution intermediaries. Thus, it is the producers who are expected to inform consumers about risks and limits of application associated with their products, perhaps by way of pamphlets summarizing the advantages and disadvantages of particular materials, because it is the producer who is most knowledgeable regarding product characteristics. Processing companies then assume responsibility for uses not covered by the manufacturer's guidelines, subject to the same rules that apply to producers. Moreover, consumers must recognize their own obligation to acquaint themselves with production processes and their consequences (by inquiries directed to the supplier) so as to be in a position to withdraw support from activities detrimental to the health and safety of employees. Consumers likewise bear a significant responsibility with respect to appropriate use of purchased goods.

It is in the laboratories of the chemical industry—together with those of universities and government research institutes—that most scientific discoveries originate from. Whenever chemical industry in the course of its research is successful in developing a new and better (low-risk) technology it becomes obliged to pursue on its own initiative the goal of translating these findings into practice. Likewise, a corporation should be expected to withdraw voluntarily any product from the market if new information reveales the product to be more hazardous than was realized at the time the original marketing permit was issued, and the newly recognized risk cannot be justified. The same point might be expressed in a more positive way: Given the validity of the above-stated goals with respect to sustainable development, a new material should replace another already in use for the same purpose only if it poses fewer risks, and is therefore superior from the standpoint of "ecological design." The more the chemical industry regards itself as part of the service sector, with materials or chemicals representing simply the means for supplying particular services, and the more responsibility producers accept with respect to the life-cycles of their products, the more likely industry will be to invest meaningful effort in the minimization of risks. Chemical industry should take it upon itself always to seek the most efficient, least dangerous, and thus in the long run most profitable methods of rendering "chemical service."

The interpretation presented here of accountability as it applies to the chemical industry unquestionably goes far beyond what current law requires, but it often happens that moral and legal responsibilities follow diverging paths. Even so, corporations would be well advised not to wait until legal restrictions become more stringent; industry should instead take the initiative now to raise its own standards, proactively and consistently (Responsible Care). This also means, however, that companies must be permitted freely to admit that the implementation of certain improvements may take time, and that not everything inherently desirable is subject to immediate realization. In some cases this will be because of the current state of technological development, but short-term financial exigencies may also play a role.

On the other hand, the responsibilities that rest with distributors and consumers must again be stressed. For too long, consumers in the countries with the highest standards of technology have been living under the illusion that a "clean environment" would be readily achievable if only industry would make the necessary effort. The work of both the Brundtland Commission and the 1992 Rio de Janeïro Conference [1], [2] clearly demonstrates the need for a redistribution of wealth between "North" and "South." Consumer responsibility in the industrialized countries therefore implies a lowering of expectations, voluntary limits on consumption, and perhaps even a reduction in the overall standard of living.

1.6. Environmental Politics and its Goals

The chemical industry is consistently faced with a need to make long-term decisions. A company's products serve not only the well-being of humanity, but also the organization's own survival. For this reason boundary conditions affecting corporate activities must be defined on the most long-term basis possible. Just as an economy without socially defined limits would result in the exploitation of humankind in general, so the current absence of an accepted ecological framework threatens unreasonable exploitation of natural resources. Politicians face the difficult task of harmonizing the disparate interests of industry, consumers, and the environment in the form of a reasonable set of operating guidelines. It is important that industry does not function in some vaguely defined gray area; it should instead be presented with clear, quantified objectives as a basis on which to orient itself. A precise legal framework is required to point the way toward a feasible form of self-regulation. The establishment of national environmental goals would represent a useful first step in this direction, these goals having been set with the active participation of all the various groups constituting society.

It is conceivable that the social market economy itself is influenced by parameters related to environmental politics. An "ecosocial economy" would advocate environmental protection as one of its fundamental values, thereby lending support — as in the social sector — to values that could not be expected to prevail in a pure market economy.

1.6.1. Product-Integrated Environmental Protection

An essential prerequisite for evaluating the ecotoxicological effects of any substance is knowing both where it enters into the environment and where it can be expected to persist; this is termed the "use pattern." Chemical industry must not be permitted to restrict its attention to establishing use patterns only for materials that remain strictly within the confines of processing plants. It has the additional responsibility to investigate distribution pathways of consumer goods in order to ascertain the points of entry and subsequent whereabouts of such materials within the environment. In some cases it will be necessary that targeted strategies be developed for ecologically safe waste management. The new generation of industrial environmental protectionists no longer limits its activities to the disposal of production residues. Equally important in environmental planning is the handling of starting materials, intermediates, and end-products, together with the associated logistics, problems of warehousing and supervision, transport to the workplace, and relevant modes of application. Integrated environmental protection assigns a high priority to the avoidance, minimization, and reuse of residual materials as apposed to treating residues as waste. It expresses equal concern for methods of production and the products themselves, thereby furnishing a positive example of the practical application of ethical principles in technology.

1.6.2. Staff and Outsiders Sharing Responsibility

Those who work in the chemical industry are affected in a special way by the current debate over environmental protection, and for this reason they are entitled to clear and objective information regarding their own companies' activities. A corporation can be only as commendable as the attitude of its personnel permits it to be. This observation applies especially to the board of directors and corporate managers in their capacity as role models. Every employee — whether on the production line or in administration — has a stake in the manufacture of a company's products, and in-house directives related to environmental protection should guide everyone's efforts. At the same time, staff should also be given the opportunity to influence the development of such directives, a right that might, for example, be reflected in corporate bylaws, or an active employees' organization. Finally, those working in the chemical industry have important contacts with the outside world, they are in a strong position to demonstrate through personal example that environmental protection is the concern of every citizen in his or her everyday surroundings—at home as well as at the workplace.

1.7. Safety

The comments above regarding "external relations" with respect to environmental protection also apply to workplace safety, especially in the case of production staff. Workplace safety has now joined environmental protection at the very center of public debate, and companies everywhere are called upon to recognize that effective training, continuing education, and staff motivation should be regarded as obligations. Especially important is the level of trust that exists between management and employees. Only with an open atmosphere in which people are encouraged to speak their minds is it possible, for example, to address the so-called "near accident," and to learn from such an incident in order to prevent more serious situations in the future. Here again it is important that staff members be given the opportunity to engage in at least indirect contact with management via some form of employees' organization. Within the company, just as in discussions involving the public, mistakes should be freely acknowledged—indeed, honesty in this regard should be explicitly encouraged—in the interest of improving the working conditions for everyone.

1.8. An Attitude of Partnership

The interaction of people with one another has always been a central issue in ethics, and its importance is increasing as the world moves closer to democratization. In an age in which religions find it difficult to assert a claim of universal acceptance it seems essential to encourage the broad-based development of ethical principles that promote peaceful coexistence and are acceptable to diverse religions and in diverse areas of the world.

Specifically within the chemical industry, the link between technology and ethics leads inescapably to the conclusion that maximum safety for humankind and the environment depends on several factors: Constant development and improvement of technology; presentation of technological achievements to the public through a process that encourages open debate; and a willingness on the part of scientists, especially those in the chemical industry, to face criticism. Close attention to these considerations is critical to an industry striving to regain its lost credibility. On the other hand, society has certain responsibilities of its own with respect to scientists: To accept them as fellow human beings, to acknowledge their qualifications as experts, and to take them seriously as advisors.

Openness and mutual trust—behavior consistent with true partnership—can provide the foundation for communication that rises above secretiveness, insinuations, and polemics. The will to preserve our planet as a place fit for human habitation must lead to widespread acceptance of the notion that every individual shares in the responsibility.

1.9. Technology and Ethics—Two Related Disciplines

The encounter between technology and ethics represents a meeting of two intellectual disciplines with a long history of association. The Greek philosophers did not recognize an arbitrary distinction between the natural sciences and the humanities like the one we take for granted in present-day Western culture. The Greeks knew but one cosmos awaiting exploration, and for them all forms of scholarship spoke with a common tongue. How far we have strayed from this ideal in an age in which natural scientists and scholars in the humanities often find themselves unable to understand each other! If both groups would become conscious once again of their kinship, if they were to rediscover how closely their interests are related and find anew a common language for communication—at least in those realms reflecting the common good, progress, quality of life, environment and security—then the world as a whole would benefit considerably.

2. Creation, Technology, and Ethics

The word "technology" derives from the Greek expression *techne*. As the Greeks understood it, *techne* represents the *poiesis* (the creative activity) guiding the fine arts. The underlying ideal is the ability on the part of creative genius to capture that which is inherent in nature, and then enhance it. Inspired by Antiquity, the Renaissance resumed this ancient interpretation of technology and carried it forward. As the history of technology in Europe continued to unfold, however, the Greek conception of *techne* was superseded, and today it constitutes little more than an isolated episode, having given way especially during and after the Enlightenment to a spirit of inventing and conquering, with the result that creative artistic revelation was diverted in other directions.

The Greek concept of technology implies a sense of the indivisibility of the whole, presupposing a oneness of the good, the true, and the beautiful. It is particularly significant that from this point of view the divine which is inherent in nature is not subject to challenge and human disposition, but is conceived instead as something amenable to creative enhancement. One should also not forget that the Greek way of thinking assumed that processes of nature and history are determined by eternal laws that are cyclical in character, effectively ruling out any future-oriented notion of "progress."

The Judaeo–Christian tradition has been another decisive factor in the course taken by the European history of technology. The reader is reminded in this context of a much-discussed verse at the beginning of the Old Testament, Genesis 1:28, the classical translation of which reads: "Be fruitful and multiply, and replenish the earth and subdue it: have dominion over the fish of the sea, and over the fowl of the air, and over every living thing that moveth upon the earth" (Authorized King James version). Recent translations are more faithful to the original Hebrew text, replacing "subdue" and "have dominion over" with expressions like "take possession of" and "use."

CARL AMERY has suggested that this passage calls upon humans to assume total command over the earthly Creation in the name of God, but close reading of the text and its context rules out such an interpretation [9]. Nevertheless, there can be no doubt that the Sons of Abraham have taken a far more rational and secular approach than the Greeks. Earth is regarded as a space entrusted uniquely to humankind, a place in which humans are to live and assume responsibility. The terrestrial world is a part of God's Creation, but the Creator—the giver and initiator of all that exists—is entirely distinct from the Creation. Similarly, human beings, as images of God, occupy a special place in this Creation despite their natural affiliation with other creatures, inasmuch as they are cognizant of their origin and thus in a position to accept responsibility for the earthly Creation. As Genesis 2:15 affirms in the course of the second account of Creation, humankind is entitled to "till the earth" with which humans have been entrusted, but is expected also to "care for it" (New English Bible). This creative right in turn reflects something of the essence of the Creator, for which reason humankind is referred to as the "image of God."

Starting from these premises it becomes quite clear that a system of ethics based on the Judaeo–Christian tradition necessarily incorporates an ethics of responsibility for the world. It directs attention not only to constructive human interaction but also to the well-being of the Creation as a whole. Beyond that, the Biblical account of Creation expresses an element of enlightenment, which encourages us to consider in conjunction with all our plans and technological activities our responsibility toward God with respect to both our fellow human beings and our fellow creatures. The two Creation stories with which the Bible begins, but also the subsequent chapters devoted to so-called proto-history (the Fall of Man, the Expulsion from Paradise, Cain's fratricide, increasing human violence, the Flood, the Covenant with Noah, the building of the Tower of Babel, etc.) must not be misinterpreted simply as historical accounts. Rather, they are expressions of a basic reflection on the eternally ambivalent station of humankind between God and the world, both then and now. The human race is depicted as in constant danger of foundering, both on its own account and with respect to the Creation with which it has been entrusted, and thus in need of constant reminder regarding its responsibilities.

There are essentially eight elements that characterize the Biblical notion of human responsibility toward the Creation:

1) Humans represent one species among many, endowed like the others with special abilities. Exceptional human attributes are cognition and responsibility.
2) The Creation encompassing humankind and all other creatures must be regarded as a dynamic process over time, one which is still incomplete and awaiting fulfillment.
3) No component of this dynamic process is superfluous or without value; each is good in

its own way, and each has its unique place in the unfolding of Creation.

4) The goal of the Creation process is the achievement of an all-encompassing peace, as anticipated in the idea of the Sabbath (the day of rest) as the seventh day of Creation.
5) The eventful traditional history presented in the Old and New Testaments deals ultimately with nothing other than the realization of this peace, which is repeatedly frustrated by human hubris. The God of Creation is thus revealed as a sensitive and empathetic deity who, according to Christian interpretation, entered into the Creation in the form of Jesus of Nazareth as a way of confirming once and for all the promise of love toward His Creation.
6) The Creation in its various inanimate and animate aspects is in all its impulses (from the motion of elementary particles up to human consciousness) determined, driven, and borne by the will of the Creator. A systematic natural historical perspective will ultimately be forced to interpret causality within this light, taking into account the totality of natural history. However, it must in no way disturb the notion of God, which cannot be encompassed within natural dimensions.
7) Careful and reverential management of the Creation ("tilling and caring for the earth") entails respect for the God who is the precondition for the possibility of all existence. In this sense the Judaeo–Christian position is not far removed from the ancient Greek ideal of creative enhancement, even though the theological premises are quite different.
8) It is problematic to play off against each other so-called dualistic and monistic interpretations of Creation. For example, according to the American Indian way of thinking, God's reality is perceived monistically as present throughout the web of life. As American Indian culture in its totality so impressively demonstrates, this leads to an extremely sensitive approach to the exploitation of nature. But a dualistic view in which humankind has been entrusted with the Creation also involves a radical obligation to safeguard that Creation. Unfortunately, Christian churches and the European cultural history they helped to shape have yet to fulfill this important stipulation.

The associations explored above are especially significant with respect to the current debate over technology and ethics insofar as they reveal the deeper context of humankind's concern with responsibility. On one hand the idea of a Creation has the effect of relativizing every human claim to autonomy. The Creation exists on God's behalf, not on the behalf of human beings. All creatures have a share in this fundamental relationship. No single creature takes precedence over any other when it comes to the prerequisites for life. On the other hand, a human, due to its cognitive abilities and capacity for assuming responsibility, is regarded as having the potential to stand very tall or fall very low. It is in the nature of human creatureliness that what a person makes of his or her life is not entirely predestined, but reflects the conscious exercise of human responsibility. In this context action and responsibility—technology and ethics—are closely interrelated. In the stewardship function assigned to humankind in Genesis 1:28 and Genesis 2:15 ("till and care for") both elements are cited explicitly. Nothing is self-evident when it comes to transformations humankind may initiate in the course of satisfying basic needs. Everything requires careful evaluation. With respect to the Creation, humankind has received a calling to be the conferrer of form and shape.

The contemporary relevance of the Creation concept—irrespective of religious tradition—lies primarily in the fact that it can be used to support an emphatic "No" issued in response to calls for uninhibited and limitless exploitation of the earth on the part of modern utilitarian interests. This opposition is not an expression of fundamental hostility toward technology, but rather of human obligation to work toward an appropriate (from the standpoint of Creation) balance in the conditions essential to life; that is, an obligation to respect certain limits on the use of technology—limits defined by the needs of humans and other creatures.

Many points of view (religious, secular, environmental-protectionist, entrepreneurial, trade-unionist, and political) intersect in the current efforts to respond to this imperative. The chief problem that arises, notwithstanding general acceptance of the great common goal of "preserving the Creation," is that practical consequences to which this goal points are extremely diverse—and often conflicting. This is due in part to the fact that the critical potency of the concept "Creation" has been undermined and redirected through pressures exerted by various interest groups. There is another important factor as well, however. Any concrete exercise of responsibility in an advanced technological civilization represents a very complicated balancing act. Assigning relative weights to genuine human needs (including mobility, comfort, release from heavy labor) and to measures designed for environmental protection leads to disputes in which human interests all too quickly gain the upper hand.

During the past few years several attempts have been made to establish the foundation for an ethics of the common good of all creatures, based not on humankind alone but on the biosphere as a whole. Three such positions will be singled out here:

1) The environmental manifesto "Austrian Scientists for Environmental Protection" stresses as a normative value the historicity and uniqueness of life-forms [10]: "Every form of life is unique, each must be respected independent of its current utility for mankind, and all must be protected from thoughtless extermination in the spirit of an elementary cultural obligation."
2) HANS JONAS has repeatedly emphasized the inviolability and "sacredness" of the balance of life as we have inherited it [11]: "We must learn again to live in fear and trembling, and—even without God—to stand in awe before the sacred."
3) As early as the beginning of the twentieth century the renowned theologian and physician ALBERT SCHWEITZER implored humankind to revere all forms of life [12]: "I am life that desires to live in the midst of life that likewise desires to live."

SCHWEITZER's words can reasonably be interpreted as an extension of KANT's categorial imperative ("Act only according to that maxim by which you can at the same time will that it should become a universal law") [13], broadened somewhat insofar as it relativizes humanity—very much in the sense of the Biblical Creation story—within the context of a universal web of life, without at the same time depriving the human race of its special assignment (or calling). SCHWEITZER's approach essentially combines three premises:

1) What makes human beings human is the fact that they are aware of the value of life: their own lives and the lives of all other creatures as well
2) The will to live is always subject to competition, with one will to live confronting others
3) It is the specific calling of humankind to be painfully aware of this conflict, and to consider it as a challenge to achieve the maximum possible degree of balance

With these principles SCHWEITZER has formulated the conditions for a comprehensive ethics of solidarity in creation. For him it becomes a question of an unequivocal effort to establish compatibility among all forms of life, taking into account both human and non-human needs. Four far-reaching perspectives for action emerge with respect to the current situation: ecological compatibility, social compatibility, international compatibility, and generational compatibility, with an eye toward our children and grandchildren. Each of the four perspectives encompasses a multifaceted program of action. The challenge posed by this ethical system is that of achieving a judicious interlocking of the four aspects of obligation.

The central problem that arises in the attempt to develop an ethics of solidarity in creation is finding a way for justice to be done to the various non-human forms of life. Proceeding from the basic principle of equality despite indisputable diversity between the various stages of life (microorganisms, plants, animals), the discussion in recent years has produced certain criteria that have unquestionably moved the matter forward:

1) It is essential that one presuppose a right to life on the part of all living species. This right encompasses each species' existence, current nature, and biotopical interconnections. The protection of species and biotopes is a fundamental obligation that applies to anyone involved with nature. The spectrum of presently existing species must be viewed as an ethical challenge, and it should not be interfered with arbitrarily.
2) Life is always to be regarded as a dynamic process: a coming into existence and a vanishing. If humans wish to interfere with the interrelations of life—an undeniable tendency—they are obligated simultaneously to take into consideration the slow and complex dynamics of natural history in general. Due to the accelerating pace of technological progress, ecological destruction has already begun, but it must under no circumstances be permitted to accelerate. This consideration must be inti-

mately incorporated into the very organization of technology and progress. Any interference with terrestrial evolution requires an assurance that the continuity of the process and the variety of its guarantors will be preserved.

3) The principle of sustainability (see Section 1.3.1) must be the criterion applied whenever claims are made on natural processes. Production and consumption, both of which affect nature, must not be allowed to proceed at the expense of the natural potential for regeneration. Any interference with the dynamics of nature—including technology—must be perceived as a cooperative undertaking in concert with nature. Energy consumption and material flow cannot be allowed to overburden or destroy the reproductive cycles of nature. This in turn mandates development of a savings-oriented regenerative policy with respect to energy, together with a materials policy based on recycling.
4) The realization of objectives such as these will be possible only with the aid of all-inclusive processes for making decisions. Those voting in favor of the non-human aspects of Creation and those on the side of social and economic interests must enter into a dialogue in which the different vantage points are given comparable weight, a dialogue oriented toward the common obligation of survival. Only in this way will it be possible to take command of the fallow ground lying in the tense reaches between natural and human history, territory which is waiting patiently for creative development.

3. Asian and Other Noneuropean Traditions and the Contemporary Environmental Crisis [14]

3.1. Introduction

When the ecological movement began in the 1950s, thoughtful people trying to get at the roots of the ecological crisis identified one particular mind-set as the ultimate source of most of today's ecological problems. The historian LYNN WHITE JR. in his often reprinted and usually grossly misquoted article on "The Historical Roots of our Ecological Crisis" [15] suggested that a fixation on the Biblical imperative—supposedly uttered by the Creator-God—to "multiply, be fertile, and subdue the earth" (Genesis 1:28) was what made Western humanity single-mindedly exploit and dominate the earth for its own purposes. By way of an alternative, experts on Asian religions (including ALDOUS HUXLEY, ALAN WATTS, HUSTON SMITH, and others) in the 1960s proposed looking toward the East and learning from Asian traditions [16]. Taoism, Buddhism, Jainism, and Hinduism were pointed out as offering alternative, more gentle world-views, in tune with nature and capable of sustaining cultures of thousands of years duration. Flaws were soon discovered in that argument, however. It was pointed out that Asia is also showing signs of ecological stress. The ecological disaster areas on the Asian continent were—until recently—not necessarily created by modern, science-based industry, but by overgrazing, overtilling, overuse of the land, carelessness, neglect, and mismanagement. Few, if any, of the societies in Asia resisted on ideological grounds the industrialization and accompanying degradation of the environment that was pushed by their newly independent governments. In Asia, a fear began to grow instead that newly discovered ecological concern on the part of Westerners was simply one more ruse to keep Asian nations in the industrial and commercial backwaters, preserve their former state of underdevelopment, and consign them forever to the "Third World."

3.2. Asian Traditions—How Relevant are They Today?

Asia, with more than three fifths of the world's population, is the cradle of all major world religions. While the Biblical traditions have largely become "Western Religions," undergoing in the course of time cultural changes that have estranged them from their countries of origin, Hinduism, Jainism, Buddhism, Zoroastrianism, Taoism, Confucianism, and Islam have remained largely Asian, and they still claim the majority of Asians as adherents. The hold these traditions have on Asia, and the degree to which the populace identifies with them, varies greatly across the continent. On one side of the spectrum we have Islamic theocracies like Iran and Pakistan (as well as the traditionalist Gulf countries), and on the other we have the vastness of China, which through its ill-fated cultural revolution tried (to some extent successfully) to eradicate all links with the past, outlawing Confucian and Buddhist traditions and ridiculing their followers. In between there are countries like India, which is home to several major religions, with a very visible and fervent Hinduism as the majority religion that continues to maintain a strong hold on the lives of untold millions. There is also Sri Lanka, as well as the Buddhist countries of Indochina, where Buddhism is practiced by millions, and Buddhist monks exert considerable influence on politics. Then there is Japan, with a multitude of new religions side by side with both ancient traditions and modern secularism. And there are countries like Korea, Taiwan, Indonesia, Malaysia, and others, where a wide variety of religious traditions flourish and sometimes clash.

With the possible exception of China (even there, recent visitors report a renewal of interest in Buddhism and Confucianism, both private and public), Asia is still strongly influenced by indigenous traditions, and religions are major forces that motivate people in their individual and corporate lives. Events in the last few years (like the destruction of the Babri Masjid in Ayodhyā and the subsequent Hindu–Muslim riots in many large cities in India), widely reported in the media, provide ample evidence for such an assertion. Visiting these countries and living there for any length of time offers even more convincing proof.

3.3. Traditional Ecological Teachings of Asian Religions

All traditional cultures over the world have had an understanding for the need to preserve the fertility of the earth and the viability of human settlements [17]. Many ancient religions, concerned with all of earthly life and not just an abstract soul that needed to be saved for a life hereafter, had very specific ecological regulations. Hunting and gathering as well as tilling and harvesting followed a religiously sanctioned ritual routine. Care was taken not to overhunt, and not to exhaust the soil. Gods and goddesses of forests and fields were believed to oversee observance of the regulations, handed down from generation to generation and often thought to have been given to the tribal ancestor by way of higher revelation. Religion and life were closely interwoven. The daily work of providing for the necessities of life had "sacred" dimensions. Special times were set aside for religious festivals, as were special spaces for the higher powers. Harmony with nature, grounded in religious attitudes and beliefs, was achieved in many major cultural regions in Asia, and it enabled civilizations there to flourish for thousands of years over small and large areas alike. Autochthonous traditions like Hinduism and Taoism, which claim no specific founder (even if the writings of key texts were later ascribed to specific persons like LAO-TSE in the case of the Tao-te-Ching, or the Vedic ṛṣis for the Veda) preserved a great deal of possibly pre-historic wisdom regarding living in harmony with nature, a wisdom that came to be expressed in more sophisticated ways in later ages, when sages and philosophers articulated specific systems of thought. Thus, in India, the Lawbook of Manu prohibits the pollution of lakes and rivers, and threatens violators with severe punishment. The planting of trees was recommended as a religiously meritorious act in some Purāṇas. A comprehensive kṛṣiśāstra (Science of Agriculture) was also developed, which contained practical advice on agriculture within a magico-religious framework. All over Asia (including China) one finds countless little shrines and objects of worship in the fields, testifying to the continuing connection of contemporary agriculturists with the spritifual world of their ancestors, and a belief in supernatural powers considered important for the growth of the crops and the well-being of society.

While certain relevant "ecological" texts can be quoted from the major world religions ("book religions"), the majority of the followers of these traditions have on the whole ignored such passages, and instead adopted a scientific–economic, industrial–consumerist ethos. The only ones who follow religiously colored and traditional ecological wisdom seem to be "tribals," peoples with a pre-modern economy and no access to (expensive) technology [18]. But even here there have been inroads. No doubt some members (especially the youth) of these communities wish to possess the toys of modernity, and/or make life (at least initially) easier through the use of machines. One follows from the other: machines are more effective for performing certain tasks—but they are expensive and demand investment, which in turn requires that productivity be increased, often forcing nature beyond its capacity to recuperate.

3.4. Traditional Religions and the Challenge of the Contemporary Environmental Crisis

The Industrial Revolution, which continues to transform Asia with great vehemence, has created environmental problems on a scale unknown to traditional cultures, and not foreseen in their religious teachings. Traditional Asian cultures did not know the phenomenon of the megacity, which is becoming such a typical feature of the late twentieth century. They did not know the powers of modern technology, which quite literally permit the limitless exploitation of nature. They had no categories for comprehending modern theories of large-scale industrial economy, and they were not prepared for the onslaught of the modern Western mind in virtually all areas of their lives. Not surprisingly, they reacted by either withdrawing into their own past or violently rejecting what they conceived to be incompatible with their own ethos.

Internal problems of all kinds had by the nineteenth century reduced large parts of Asia to disaster areas plagued by famines, lawlessness, and social decay. Modernization, understood as Westernization, was seen as the only way to status and prosperity by a progressive elite of Asians who looked to Europe and America as their models. Indigenous traditions were believed to be worse than useless: they not only did not contribute anything to modernization, but were thought to hinder it and keep people in ignorance, superstition, and inertia. For some time it appeared that one of two great twentieth century Western ideologies—Capitalism or Communism—would transform Asia into its own likeness. But this did not happen. Together with progress, the West brought its own problems to Asia, among others the ecological crisis which is now worldwide and threatening the survival not only of a few people, but of humankind as a whole. Worldwide we are looking not only for scientific-technical resources, but also for resources of mind and spirit to cope with the disaster.

While Asian traditions themselves have hardly begun to grapple with the issue, and cannot be expected to offer ready-made solutions to a problem that took three centuries to emerge in full force, they do possess a wisdom that could be utilized globally in the service of an emerging contemporary ecological ethics [19].

3.4.1. The Human Ideal

In all those vast areas influenced by Hinduism, Buddhism, Jainism, Taoism, and Confucianism, the highest type of human was the sage, the wise person, the philosopher, the person who had gained enlightenment and achieved a state of self-contentment. Neither wealth and power (in the pursuit of which virtually all the large-scale ecological catastrophes have been caused), nor pleasure and unrestrained consumption (which has led to almost all the rest of the ecological problems) were held out as conferring the highest social status, or as constituting the essence of the good life. Enlightenment and wisdom were associated with restricting desires and demands on the environment, a sense of empathy with everything and everybody, and a sensitivity toward nature as well as toward fellow humans.

"Greed–anger–delusion" (*lobha–krodha–moha*) was called the "gate to hell," and it certainly is the cause of much of the ecological and social hell of today. The teachings of the great Asian traditions were concerned with the eradication of this triple threat. For example, Nirvāṇa, the ultimate aim of Buddhism, is a life that is free from greed, free from anger, and free from self-delusion. How much could nature and humankind benefit if such an attitude became commonly accepted?

3.4.2. One Reality

Virtually all Asian traditions assume that rebirth is a universal fact of life. Life is seen as one great cycle, comprising all stages in life and all types of life. Humans do not occupy a unique position, and their fate, both present and future, depends on relationships with other forms of life. One's own life is thought to have undergone all kinds of transformations—plant, animal, human, godlike—as have all other lives. Life as such deserves respect. Not all Asian traditions went so far as Jainism, which prescribes that its professional adherents sweep the path before their feet so as not to harm insects, and demands that a covering be worn over the mouth so as not to harm the air molecules (believed to have life as well) by breathing in or out too sharply. But all the traditions have a deep respect for life, and not harming life (ahimsā) is considered the highest ethical principle. Ethics in Asian traditions is not restricted to interpersonal relationships; it extends to all of natural (and supernatural) reality. Some, like the Vaiṣnavas of India, go so far as to consider the universe as a whole to be the Body of God, a living being deserving of worship.

3.4.3. A Universal Natural Law

The tao of the Chinese and the dharma of Indian traditions suggest a universal natural law that has not only physicomechanical but also moral-human and transcendental dimensions. The tao or the dharma encompasses all reality; it is the "nature" of things as well as the "inborn law" of which humans become aware when they are morally awakened.

Implicit in this notion is also the notion that the universe as such is a meaningful, purposeful reality, within which humans—by following its dynamics—find personal fulfillment. The cultural articulation of tao/dharma includes societal and historic moments: it must be rearticulated in every age and for each culture. It does not assume the absolute identity of all at all times, but it differentiates and specifies. The notion of tao/dharma could reconnect our scientific with our moral ethos, our concern for knowledge with our need for respecting our limits. The authors of works like "The Tao of Science" or "The Tao of Physics" [20] suggest that the natural sciences themselves may lead to the discovery of such a universal structure encompassing both nature and the human mind.

3.4.4. Discipline

All Asian traditions insist that the "Path" which leads from darkness to enlightenment, from crisis to health, demands discipline. The most systematic articulation of such discipline is found in the Indian Yoga, part of virtually all Hindu and Buddhist traditions, similar to Taoist practices, and increasingly known throughout the world. All traditional religions insist on an ethic that is non-negotiable if one is serious about a genuinely humane life. Discipline in the sense cultivated by the yogic traditions presupposes as its foundation most of what has been said above. There is no doubt that "indiscipline" in its many manifestations is causing a great deal of what we deplore today as societal breakdown and ecological degradation [21]. Discipline begins by paying careful attention to what one does and how one does it, and it progresses through the formation of character and the ordering of one's life according to reality principles, culminating in becoming aware of what Heisenberg called "the central order of things" [22], and living accordingly.

3.5. Asian Traditions: Conclusion

There is no doubt that Asia possesses traditions that have been ecologically and socially functional. Many, if not all, of these regional ecologically sound cultures have been disrupted by the intrusion of modern forms of economy and thought, and they cannot be revived in their historic shapes. However, they do contain principles that can be rearticulated and brought to bear on the contemporary situation. It will depend on the willingness as well as the ability of leaders within these traditions whether these principles will again inform the praxis of people not only in Asia but in the world as a whole. Religious leaders in the past have been surprisingly successful in bringing about large-scale movements that resulted in cultural transformations of great depth. Many people in Asia are still deeply religious, responding to their religious leaders with enthusiasm and fervor. It is obviously easier to mobilize the masses with nationalist and fundamentalist slogans than to create a large-scale ecological movement based on traditional wisdom. Nevertheless, it is not wholly unrealistic to hope that leaders will arise from within the great Asian traditions, leaders who will articulate a new deep ecology based on ancient wisdom and modern science, and will help people in both the East and the West to "save our earth"—the only earth we and our children have.

3.6. Other Traditions

Native American traditions are adding an increasingly influential voice to the current discussion of ecology within North America. Not only have Native Americans organized some of the most highly publicized anti-logging movements, they are also reviving and updating their spiritual heritage to offer a contemporary ecological spirituality, which appeals to non-Natives as well, and is soundly based on an empathy with nature.

African traditions are so numerous and diverse that it is next to impossible to offer generalizations in the context of ecology and religion. While earlier generations of Europeans described these traditions as "primitive" or "nature-religions," implying not only a certain simplicity and lack of historic consciousness but also a closeness to human origins and pristine nature, recent research has rendered any such suggestion obsolete. Certain African traditions are extremely complex and subtle, and are the result of prolonged cultural developments. Many do show great sensitivity to ecological concerns and clearly embody a "holistic" world-view; i.e., they consider human life to be part of an intricate web of relationships with other living beings and transcendent powers, overarched by notions of an agent of universal justice. Some of the traditions, suppressed not only by colonial governments but also by Christian missions, have generated active revivalist movements that include attempts to make use of the ecological wisdom of the present generation's ancestors in today's very different world of semiurbanized and war-plagued Africa.

4. The New Master Paradigm: Ecology

Ecology, understood as a concern for nature in the most comprehensive sense, is likely to be the master paradigm of the science and politics of the future. It will be the focus not only of theoretical preoccupations aimed at halting and possibly reversing the destructive tendencies inherent in modern industrial civilization, but also of sociopolitical practice: no government will be able to survive if it fails to give expression to the ecological concerns of its citizens, and this ecological focus will bring about cooperation among people crossing the boundaries of nationality, race, language, class, and caste.

Ecology will be the fulcrum of the ethics of the future, an ethics compellingly evident to individuals, put into practice by corporations, and accepted by governments and international agencies as a precondition for survival. Ecological ethics—better: an ethics based on ecology—requires contributions from everyone. It requires day-to-day diligence on the part of every human being on earth in exercising care and imaginative nurture with respect to nature. It requires the attention of those in positions of responsibility at all levels of government so as to ensure that environmental concerns are translated into enforceable and enforced policy. It requires the attention of both research scientists and applied engineers who, having found the means to exploit and despoil nature, must now apply their skills and knowledge to finding ways to restore and preserve nature. It requires also the attention of humanists, who for too long have taken nature for granted and left it to the more "practical" people in government and industry to provide for the mundane necessities of life, so that they could themselves dwell in the more refined realms of culture and art. Furthermore, it requires the attention of students of religion. Students (and often practitioners) of religions, especially those coming out of a (modern) Christian background, have often not only made a theoretical distinction between the natural and the supernatural; they have then proceeded to treat this distinction as a value judgment, and a basis for their theoretical and practical efforts. Nature has been considered not only as lower compared to super-nature, but also fallen, corrupt, the opposite to spirit, the domain of the devil, the darkness to be overcome, and the force that draws one away from God. The concerns of the truly religious person were the supernatural, a God transcendent, and a life beyond. Christianity in particular (as Judaism before), upon encountering tribal traditions in which nature or certain aspects of nature were worshipped, expressed itself in an often fiercely nature-hostile way.

Many influential religious traditions possessed and still retain a substantial environmental ethics. However, it appears important as well that a future environmental ethics connect with what has been called "conscience" in the West, and resist the notion that environmental concern is a matter of convention, aesthetics, preferences, and the like, an attitude that has characterized much of recent ethical discussion in other areas.

This line of reasoning has two consequences. The first is that those who consider themselves religious must incorporate caring for nature into their religious theory and practice. They can no longer maintain the nature/supernature dichotomy, leaving this world to the devil while saving souls for the next. A religious conscience must include a concern for nature.

The second consequence is that ecological ethics must be linked with some type of "ultimate": it is not enough to derive ecological ethics from a cost–benefit calculus, or to criminalize environmental offenses. We must go beyond pragmatism and opportunistic politicking. We need to recognize that there exists but one earth, and that if we do not preserve it there will be none. Utopian colonies will for the foreseeable future remain just that: utopias, never-never lands. At present it is utterly impossible to imagine sending even a sizable fraction of the more than 6×10^9 people living on earth into some space-world. In the sense in which this earth is for us, here and now, an ultimate and a singularity, preservation of that earth constitutes an ultimate and singular ethical imperative. Its principles are non-negotiable. It constitutes, in traditional categories, something "sacred."

Ecology also has implications for ethics in the more traditional sense, which has concerned itself exclusively with human beings and human societies. In spite of progress in many areas, permitting a much larger number of people than in the past to derive physical sustenance from a given amount of space, there is nevertheless a limit, and thus a need to allocate limited resources fairly among all.

An increasing number of religious thinkers today are devoting their attention to environmental ethics, and an impressive body of writing is emerging from within all the major religious traditions, drawing ecological conclusions from central tenets of faith. We must not set up "faith" against "reason," as has been done for so long, especially in the West; we must rather accept reason as inalienably human *and* "religious," as we must also consider faith the foundation for any human enterprise of consequence. A contemporary ecology will require both faith and reason: faith in the importance of saving the environment, faith in the ability and willingness of humans to do so, and faith in the rationality of economics as well as science, of philosophy as well as theology.

5. References and Notes

[1] World Commission on Environment and Development (WCED): *Our Common Future* (*The Brundtland Report*), Oxford 1987.

[2] United Nations Conference on Environment and Development: Proposal of the Chairman of the Preparatory Committee for UNCED on the Declaration on Environment and Development, Rio de Janeiro 1992, Agenda 21.
Commission of the European Community: *Für eine dauerhafte und umweltgerechte Entwicklung*, KOM (92) 23, Brussels 1992.

[3] Enquête-Kommission "Schutz der Menschen und der Umwelt," des Deutschen Bundestages (ed.): *Verantwortung für die Zukunftswege zum nachhaltigen Umgang mit Stoff- und Materialströmen* (*Zwischenbericht*); Economica-Verlag, Bonn 1993.
Die Industriegesellschaft gestalten (*Abschlußbericht*), Economica-Verlag, Bonn 1994.
Enquête Commission of the German Bundestag on the "Protection of Humanity and the Environment" (ed.): *Responsibility for the Future-Options for Sustainable Management of Substance Chains and Material Flows* (*Interim Report*), Economica-Verlag, Bonn 1994.
Shaping Industrial Society—Prospects for Sustainable Management of Substance Chains and Material Flows (*Summary of the Final Report*), Economica-Verlag, Bonn 1995.

[4] W. Klöpffer, *ESPR—Environ. Sci. Pollut. Res.* **1** (1994) 108–116.

[5] Meerkamp van Embden, I.C. UWSF-Z. *Umweltchem. Ökotox* ISSN 0934-3504. Coden USZOE 9.4(2)65-128 April 1992, p. 86.

[6] A. Lovins: *Soft Energy Path*, Pelican Books, Harmondsworth 1977.
Wirtschaftlicher Energieeinsatz, C. F. Müller-Verlag, Karlsruhe 1983.

[7] H. A. Pestalozzi: *Nach uns die Zukunft*, Bern 1979, p. 53.

[8] CEFIC, Brüssel, -5/1993 Dépot légal D 1993/3158/3.

[9] C. Amery: *Das Ende der Vorsehung. Die gnadenlosen Folgen des Christentums*, Rowohlt, Reinbek 1972, pp. 9.

[10] H. P. Aubauer (ed.): *Umweltmanifest*, Flugblatt, Wien 1987, p. 6.

[11] H. Jonas: "Technik, Ethik und biogenetische Kunst," in *Am Beginn des zweiten Jahrhunderts Hoechst Pharma, Internationales Symposium*, Frankfurt a.M. 1984, p. 17.

[12] R. Grabs (ed.): *Albert Schweitzer, Gesammelte Werke in fünf Bänden*, vol II, C. H. Beck, München 1971, pp. 122.

[13] R. Schmidt (ed.): *I. Kant—Die drei Kriterien in ihrem Zusammenhang mit dem Gesamtwerk*, A. Kröner, Stuttgart 1956, pp. 245.

[14] I have refrained from using the term "non-Christian religions" in this discussion of Asian traditions for various reasons. First, traditions like Buddhism, many expressions of Hinduism, Taoism, and Confucianism do not fit the definition of "religion" from a Biblical perspective. Moreover, the adherents of these traditions find it offensive to be lumped together as "non-Christian"; apart from the insult contained in any negative definition, it blurs the enormous differences between individual Asian traditions. A German Lutheran would hardly feel happy to be identified as "non-Islamic," nor would an Irish Roman Catholic wish to be called "non-Hindu."

[15] L. White, Jr., *Science* (*Washington, D.C.*) **155** (1967) 1203–1207.

[16] Ian Barbour (ed.): *Earth Might Be Fair*, Prentice Hall, Englewood Cliffs 1972.

[17] Adolf Portmann, the well-known Swiss anthropologist/zoologist, maintains that humans, like all other animals, possess an inborn ecological sense. Apparently many humans in our time have lost that sense, together with the environmental teachings of the major cultural traditions. See Lorenz: *Der Abbau des Menschlichen*, Piper, München 1976.

[18] See, e.g., W. Pereira: "The Sustainable Lifestyle of the Warlis" in: *Indigenous Vision: Peoples' of India Attitudes to the Environment,* Special issue of *India International Centre Quarterly*, 1992, 35–48.

[19] The absence of a workable environmental ethics may also be due to the fact that it was often the unworldy, impractical people who tended to take up religion as a profession, people who had neither an interest in nor the competence required for solving the types of problems an environmental ethics is to address.

[20] R. G. H. Siu: *The Tao of Science,* M.I.T. Press, Cambridge, Mass., 1957.
F. Capra: *The Tao of Physics,* Shambala, Berkeley 1975.

[21] Madhav Gadgil, an Indian Ecologist, writing on "Population, Resources, and Environment" in SEMINAR 365 (Towards 2000 AD), New Delhi, Jan. 1990, blames the "highly indisciplined fashion in which current demands are met" for much of the ecological as well as economic malaise of our world.

[22] W. Heisenberg: *Der Teil und das Ganze,* English transl. *Physics and Beyond*, Harper & Row, New York 1974.

Ecology and Ecotoxicology

OTTO FRÄNZLE, Christian-Albrechts-Universität, Kiel, Federal Republic of Germany [(Sections 1.1, 1.2, 2.1 (in part), 2.2, 2.3 (in part), 2.4, 3.1, 3.2 (in part)]

MILAN STRAŠKRABA, Akademie věd České Republiky, Biomatematická laboratoř, Entomologický institut, České Budějovice, Czech Republic (Sections 2.1.4, 2.3.5, 3.2.3)

SVEN ERIK JØRGENSEN, Danmarks Farmaceutiske Højskole, Copenhagen, Denmark (Section 3.3)

The entire topic was coordinated by OTTO FRÄNZLE

3-527-20137-8/95/$ 5.00 + .50

In addition to the symbols and abbreviations listed in the front matter of this volume the following symbols and abbreviations are used:

a species-specific maximum interception
A constant; evaporating surface
ALT(*t*) leaf senescence function
b constant; density of vegetation cover; coefficient adjusting evaporation coefficient to other substances than oxygen
B natality
BCF bioconcentration factor
C connectance of the web
C_{dis} concentration at which long-lasting disturbances of essential soil functions occur
C_{eff} concentration at which low-level adverse effects first appear
CHAID chi-square automatic interaction detection
D disturbance; Simpson's diversity index; standardized damage level
2,4-D 2,4-(dichlorophenoxy)acetic acid
DOC dissolved organic carbon
e efficiency; vapor pressure
E evenness (equitability); rate
E(*N*) photosynthetic efficiency of a particular needle age class
ECETOC European Chemical Industry Ecology and Toxicology Centre
eff specific effect constant
EFF average photosynthetic efficiency
EM ecological magnification (concentration trophic level *n*/concentration trophic level $n-1$)
$E_o(\lambda)$ scalar irradiance
ES solar radiation
ET evapotranspiration
Ex excretion coefficient
f fugacity
F amount of food uptake per day
F_m site factors
F^* rate of infiltration
F_k correction factor
f_{oc} fraction of organic carbon in the solid being considered
F_{oc} distribution coefficient for organic matter
F_s distribution coefficient
GIS geographic information system
GW groundwater
GWP global warming potential
GWR groundwater run-off
h soil water suction; vector (Eq. 52)
H Shannon's diversity index; Henry's law constant
H_a sensible heat transfer through air

HC_p protective hazard concentration
H_s heat flow in soil
i cumulative infiltration at time t; flux
INT total intake of toxic substance per day
IRPTC International Register of Potentially Toxic Chemicals
J rate of irreversible phenomena (e.g., heat flux)
k von Karman's constant; rate constant
k_b $(=\mu_m/Y)$ biodegradation constant
K hydraulic conductivity; rate constant, buffer constant; sediment–water partition coefficient (Eq. 80)
$\underline{K}$ vertical eddy diffusion coefficient
$K_{a,\lambda}$ specific absorption rate
K_G gas film mass-transfer coefficient
K_L liquid film mass-transfer coefficient; evaporation coefficient (Eq. 89)
K_{OW} *n*-octanol–water partition coefficient
K_s saturated hydraulic conductivity, half-saturation constant (Fig. 10); concentration of substrate to support half-maximum specific growth rate (Eq. 50)
L specific coefficient (e.g., of thermal conductivity or diffusion)
$L(N)$ leaf mass of particular needle age class
LAI leaf area index
$LEAF$ total leaf mass
m expectation (Eq. 52); adsorbent mass
M mortality
n number of individuals per species; number of molecules adsorbed
n_m number of molecules required for a complete unimolecular surface film
N precipitation; rate of transfer; total number of individuals in the community
$N_i(R)$ net interception
NJ number of needle age classes
N_o above-canopy precipitation
NOEC no-observed-effect concentration
O_n organisms
ODP ozone-depletion potential
p concentration of chemical in solution
P precipitation; rate of absorption (photosynthesis); equilibrium vapor pressure
PCB polychlorinated biphenyl
PCP pentachlorophenol
PEC predicted environmental concentration
p_i proportion of individuals or biomass contributed to the total of the sample
PNEC predicted no-effect concentration
POCP photochemical ozone formation potential
p_o saturation vapor pressure
p_r amount of rain required to pond the soil
q instantaneous value of water vapor concentration; water vapor content at level
QSAR quantitative structure–activity relationship
r resistance factor; increased number of organisms per unit time and unit population (Eq. 81)
r_{ET} transfer rate of latent heat due to evapotranspiration
r_p rainfall intensity
R recovery; universal gas constant
R_c long-wave counterradiation of the lower atmosphere
R_n net radiant flux
R_s surface runoff
s sorptivity; specific leaf surface
S total number of species in the community; sorptivity; nutrient concentration; saturation concentration or solubility at fixed temperature; net source – sink term (Eq. 51); entropy; standardized damage level
S^* pondage capacity
SM soil moisture, volumetric soil water content
S_s IR absorption strength in the interval 800–1200 cm^{-1}
SW perched water
T safety factor (Eq. 65)
TCO_2 percentage of total CO_2
2,4,5-T 2,4,5-(trichlorophenoxy)acetic acid
TF throughflow
TJ season
To equilibrium concentration in water
TOD theoretical oxygen demand
T_r average residence time due to reactions
u velocity
U wind speed at level
v rate of infiltration (Eq. 11)
v_z vertical flow of water at depth z (Eq. 8)
Vd vapor density
w momentary vertical wind speed
W accumulation; body weight (dry or wet matter)
$W(t)$ accumulated pollutant at time t
x amount of substance adsorbed
X thermodynamic "force" (e.g., gradient of temperature or concentration); biomass per unit volume
Y biomass produced per unit amount of substrate consumed; random function (Eq. 52)

z	depth, depth of lowest point of measurement (Eq. 8); elevation above surface (Eq. 21)
Z	fugacity capacity
α	measure of the effectiveness of Br in ozone depletion with respect to Cl; reflectivity (albedo)
β	average interaction strength
β_{ij}	effect of species *j*'s density on species *i*'s rate of increase
γ	coefficient of absorption of photosynthetically active radiation
ε_λ	molar absorptivity at wavelength λ
$\varepsilon\sigma T^4$	long-wave radiation from the earth's surface
μ	specific growth rate
μ_m	maximum specific growth rate
ϱ	constant in Hammett–Taft equation; air density
σ	Hammett constant; reflectivity
τ_s	atmospheric lifetime
Φ_d	quantum yield
Ψ	soil water potential

1. Ecology and Environment

1.1. Humans and Environment

In the course of their evolution, humans have exerted an ever-increasing influence on the terrestrial and aquatic environments that ensured their livelihood. In early Paleolithic times, *Homo erectus* was dispersed over large geographical ranges of the Old World continents and had already attempted to circumvent environmental limitations, via culture, in expressing dietary preferences different from the youngest australopithecines and in colonizing both drier and colder environments. Because of their food preferences, these hunters and gatherers probably required larger territories and thus comparatively few Acheuleans could be supported in any one area; consequently, mobility and the periodicity of seasonal activities were considerable [1], [81]. The sites occupied indicate a preference for open, grassy environments with large herds of gregarious herbivores, which means that human influence on ecosystems was by no means different from that of hunting or gathering animals.

Things changed drastically during the tenth millennium B.C., when the first agricultural societies arose. The new subsistence system, based on crop planting, livestock raising, and permanent settlements, had been developed to the point that long-term population increases became possible. Significant ecosystem changes, often associated with disturbances of the hydrological cycle and soil erosion [82], resulted when native floras and faunas began to be replaced by domesticated stock and cultigens. Thus, not only were the largely prevailing agrarian and woodland ecosystems profoundly changed in species composition, but natural recycling processes also constantly lost importance.

In the present millennium, characterized by an unprecedented population increase in the last two centuries, the formerly "isolating" woodlands or uncultivated areas around the agrarian ecosystems and settlements have been constantly reduced, which means that the ecological buffering capacity of these nonagrarian areas, which had such an important role in the intricate ecosystem interplay of the past, no longer exists in densely populated regions of the world. Consequently, both aquatic and terrestrial types of ecosystems are now open to human impact on the regional and continental levels, and present-day analytical knowledge provides ample evidence that even the most remote areas on earth bear more or less marked imprints of human activity.

Among them, chemical imprints have attracted particular attention, because their importance has grown since the middle of the last century with ever-increasing speed, and the development is not likely to change in principle, albeit sectoral differentiations are evident. A comparison of high-volume chemicals illustrates for the United States that petroleum derivatives represent 10% of the total number of entries in the inventories, but account for 55% of the total production, whereas inorganics represent both 12% of the materials and 12% of the production. Another 6.9% of the production is due to materials that are residues from the processing of ferrous metals. Saturated hydrocarbons are responsible for 6.7%. Well-defined organic substances are the most numerous, and cause major concern in testing, because they account for about 34% of the inventory sample, but only 6% of the total production. Polymers and plastics represent 24% of the number of materials and 3% of the total production [83]. Table 1 provides insight into the estimated global release of organic reference chemicals. The figures listed

Table 1. Estimated global release of organic reference chemicals in 10^3 t/a (after [84])

1,2,3-Trichlorobenzene	0.09	1,2-Dichloropropane	15
1,3,5-Trichlorobenzene	0.09	Diethyl phthalate	23.1
1,2,4,5-Tetrachlorobenzene	0.15	Pentachlorophenol	27
4-Chloroaniline	0.16	*p-p'*-DDT	28.5
2-Chloroaniline	0.17	Aniline	32.7
1,2-Dichloroethylene	0.3	Dichloroisopropyl ether	45
1,1,2,2-Tetrachloroethane	0.45	Nitrobenzene	46.5
4-Nitroaniline	0.55	Malathion	47.4
Hexachlorobutadiene	0.6	*o*-Xylene	63
N,N'-Dimethylaniline	0.72	1,4-Dichlorobenzene	66
3,4-Dichloroaniline	1.2	Hexachlorobenzene	80
4-Methylphenol	1.2	Tetrachloromethane	84
Dimethoate	1.8	*p*-Xylene	100
1,1,2-Trichloroethane	2.4	Atrazine	111
1,3-Dichlorobenzene	3	Alachlor	123
Dieldrin	3	Phenol	180
Tri-*N*-butyl phosphate	3	Styrene	210
1,2,4-Trichlorobenzene	3.3	Di-*N*-butyl phthalate	230
m-Xylene	3.6	Trichlorofluoromethane	291
Endrin	6	Trichloroethylene	326
Chlorobenzene	6.6	Di-2-ethylhexyl phthalate	402
Dimethyl phthalate	6.6	Tetrachloroethane	470
Aldrin	7.5	1,1,1-Trichloroethane	537
Heptachlor	8.1	1,2-Dichloroethane	547
1,2 Dibromoethane	9	Ethylbenzene	600
Parathion (methyl and ethyl)	9.6	Cyclohexanol	1 200
Chlorpyrifos	10	Toluene	1 800
1,1-Dichloroethane	12	Benzene	11 000
γ-HCH (lindane)	14.1		

indicate that humans, owing to the accelerated growth of the chemical industry since the 1950s, produce and use many synthetic compounds that, in localized exposure at least, no longer appear to be negligible in comparison to natural fluxes and concentrations. Thus, the ecological role of humans has changed profoundly in little more than a century of their long cultural history, because they are now capable of and responsible for introducing new substances or increasing the concentration of natural ones in the environment to an extent unthought of in the past [2].

Because the products of the chemical industry are also a most essential part of both our economic and our social life, governments and industry alike have become increasingly concerned about the potential, unintended consequences that the use of man-made chemicals could have on both human health and the environment. Therefore, a number of key chemical-producing countries have passed, or are enacting, general substance control legislation (→ Legal Aspects, Chaps. 2, 6, 7). Common to all these legal instruments is the preventive aspect (i.e., the notification of chemicals prior to marketing), which entails the presentation of data derived from laboratory investigations together with additional information to permit the evaluation of potential hazards.

1.2. Scope of Ecology and Ecotoxicology

Environments are highly complex systems whose evolution is determined by complicated networks of positive and negative feedback loops. Human interference, particularly when inadvertent, is therefore reflected in an ever-increasing number of environmental problems that cause major concern in today's societies. To better understand and resolve these problems, knowledge of the basic earth and life sciences unified under a comprehensive ecological perspective is essential, and a consideration of the bewildering number of interrelationships from a generalized system point of view appears indicated.

Ecological thinking began to emerge in the 17th and 18th centuries [3], while the very term "ecology" as derived from the Greek *oikos* (home) was coined and introduced into biological terminology by HAECKEL [4]. Paraphrasing HAECKEL, ecology can be defined as the scientific study of the interactions between organisms and their environment or, more informatively: "Ecology is the scientific study of the interactions that determine the distribution and abundance of organisms" [5]. The environment of an organism consists of all those physical and chemical (abiotic) or biotic (other organisms) factors and phenomena outside the organism that influence it.

At the *level of the organism*, ecology deals with how individuals interact with their biotic and abiotic environments (autecology). At the *population level*, ecology determines the presence or absence of particular species, their abundance or rarity, and trends and fluctuations in their numbers (demecology). Two approaches can be distinguished at this level: one deals first with the attributes of individual organisms and then considers the way in which they combine to define the characteristics of the population. The other deals directly with the characteristics of populations and relates these to relevant aspects of the environment. *Community ecology* deals with the structure of communities and with the pathways and fluxes of energy, water, nutrients, and other chemicals through them. This implies adequate consideration of human influence on nature (e.g., pollution and global warming).

Since ecology must deal explicitly with these three levels of the biological hierarchy and the overwhelming multitude of interactions between organisms and their physical, chemical, and biological environments, it is particularly subtle and complex. Yet ecology also has the distinction of being confronted with uniqueness in a very peculiar way: it must consider millions of different species, each composed of frequently countless numbers of genetically distinct individuals, all living and interacting in a highly varied and ever-changing world. Thus, "the beauty of ecology is that it challenges us to develop an understanding of very basic and apparent problems, in a way that recognizes the uniqueness and complexity of all aspects of nature but seeks patterns and predictions within this complexity rather than being swamped by it" [6].

Ecology of such a comprehensive scope cannot be an easy science but must be a complex interdisciplinary structure grouping several dozens of sciences on the basis of a hierarchically structured system of hypotheses. For instance, one of the major present-day ecological research schemes, the long-term ecosystem research project in the Bornhöved Lake District (Schleswig-Holstein) comprises no less than 25 disciplines, ranging from mathematics to social sciences, around a nucleus of physical, chemical, biological, and earth sciences [85].

Ecotoxicology may be defined as "a science which is concerned with the toxic effects of chemicals and physical agents on living organisms, especially on populations and communities within defined ecosystems; it includes the transfer pathways of those agents and their interactions with the environment" [7]. In formal analogy with the objectives of classical toxicology, which has evolved as an aut- or demecologically oriented discipline, ecotoxicology aims at defining toxic effects in terms of reactions of ecosystems or representative compartments of such systems. In view of the complicated structure of ecosystems and their tremendous variability in time and space, however, particular problems arise. While many experimental approaches exist to predict the hazards of drugs, food additives, and contaminants for humans (→ Toxicology, Chaps. 3, 5), which are based on extrapolation from animal or other data, the analogous ecotoxicological procedures are much more complicated and hitherto far less satisfactory [8].

The very crux of the ecotoxicological extrapolation problem is the difficulty that an eventual decision-making process will lead to either "false negatives or false positives" [86]. In this connection, a false negative implies that a chemical that appears acceptable on the basis of laboratory data may cause considerable damage in practical circumstances. If, for instance, the prediction of possible environmental exposure levels on the basis of standardized laboratory tests turns out to underestimate the actual levels occurring in "real-world" environments, populations of certain species will be much more at risk than expected. Thus, species with mono- or oligomaniacal feeding behaviors (i.e., food specialists), run a much higher risk of bioaccumulation effects if specific chemicals are applied against their favorite or only prey than less specialized predators.

In contrast, the false positive involves rejection of chemicals because laboratory tests indicate intolerable hazards for the environment, while in fact such undesirable effects are not likely to occur. Pesticides may be labeled toxic to fish on the basis of standardized laboratory trials, although this does not necessarily imply that they will always cause mortality in fish when applied in, or in the immediate neighborhood of, their natural aquatic habitat. The framework of tsetse fly control in West Africa amply demonstrated that the potentially fish-toxic insecticides endosulfan and deltamethrin could actually be used without having any noticeable effects on fish [87]. The effective dose rate for killing tsetse flies was minimized by careful selection of the insecticide formulation on the one hand and application methods on the other.

Thus, the hazards of pesticides usually depend on such factors as the amount applied, formulation, methods and times of application, and intensity of use, which are—in their extremely variable specific combinations—virtually impossible to simulate. Attempts must be made, therefore, to consider the ecological characteristics of target areas as well as possible nontarget effects in connection with the use pattern envisaged in order to avoid unrealistic predictions. A compound that appears unacceptable in a certain application in one place need not be so in another. Hence, basic physical–chemical data such as those described in the *Collection of Minimum Pre-Marketing Sets of Data* of the OECD Working Group on Exposure Analysis [88] have to be matched to data on the relevant properties of typified environments into which the compounds are ultimately released. From this particular perspective, the essential structural and dynamic characteristics of both chemicals and ecosystems that are relevant to ecotoxicology are described in the following chapters.

2. Principles of Ecology and Environmental Chemistry

2.1. Structure and Dynamics of Terrestrial Ecological Systems

2.1.1. Structure of Communities and Ecosystems

In a classic paper, Tansley [89] defined the term *ecosystem* as "...the whole *system* (in the sense of physics), including not only the organism-complex, but also the whole complex of physical factors forming what we call the environment of the biome—the habitat factors in the widest sense. Though the organisms may claim our primary interest, when we are trying to think fundamentally we cannot separate them from their special environment, with which they form one physical system. It is the systems so formed which, from the point of view of the ecologist, are the basic units of nature on the face of the earth.... These ecosystems, as we may call them, are of the most various kinds and sizes. They form one category of the multitudinous physical systems of the universe which range from the universe as a whole down to the atom."

From the definition of an ecosystem as an arbitrary unit with respect to both its spatial extent and the phenomena considered, ecosystems are often portrayed as food webs or graphs in which "all" the trophic relations between component species are given [9], [10], [90]. Species can also be grouped into a hierarchy of trophic levels or size spectra [11], [91], [92], which facilitates an understanding of the flows of energy and matter through a community and underscores biological differences in interactions within and between levels. Because of the impacts of pollution and environmental change on the living world [2], the concept of stability has also been germane to the argument of how best to depict ecosystems (Section 2.3.1).

Figure 1 illustrates that an ecosystem can be viewed as a composite of the lower levels of organization (i.e., communities, populations, and individuals). Coexisting individuals of a single species possess characteristics such as density, sex ratio, age–class structure, rates of natality and mortality, and immigration or emigration that are unique to *populations*. The *community* is an assemblage of species populations that occur together in space and time.

In proceeding from the level of physiological and behavioral ecology—dealing primarily with individuals—to the ecology of populations and communities, collective properties such as species diversity, community biomass, and productivity can be identified. Organisms of the same or different species interact with each other in highly diversified processes of mutualism, parasitism, predation, and physical or chemical competition. Thus, the nature of a higher organizational level is obviously more than just the sum of its constituent parts; it is their sum plus the interactions among them. Thus, *emergent properties* appear only when increased complexity (i.e., different levels of organizational structure) is the focus of attention. A primary aim of community and ecosystem ecology is to determine

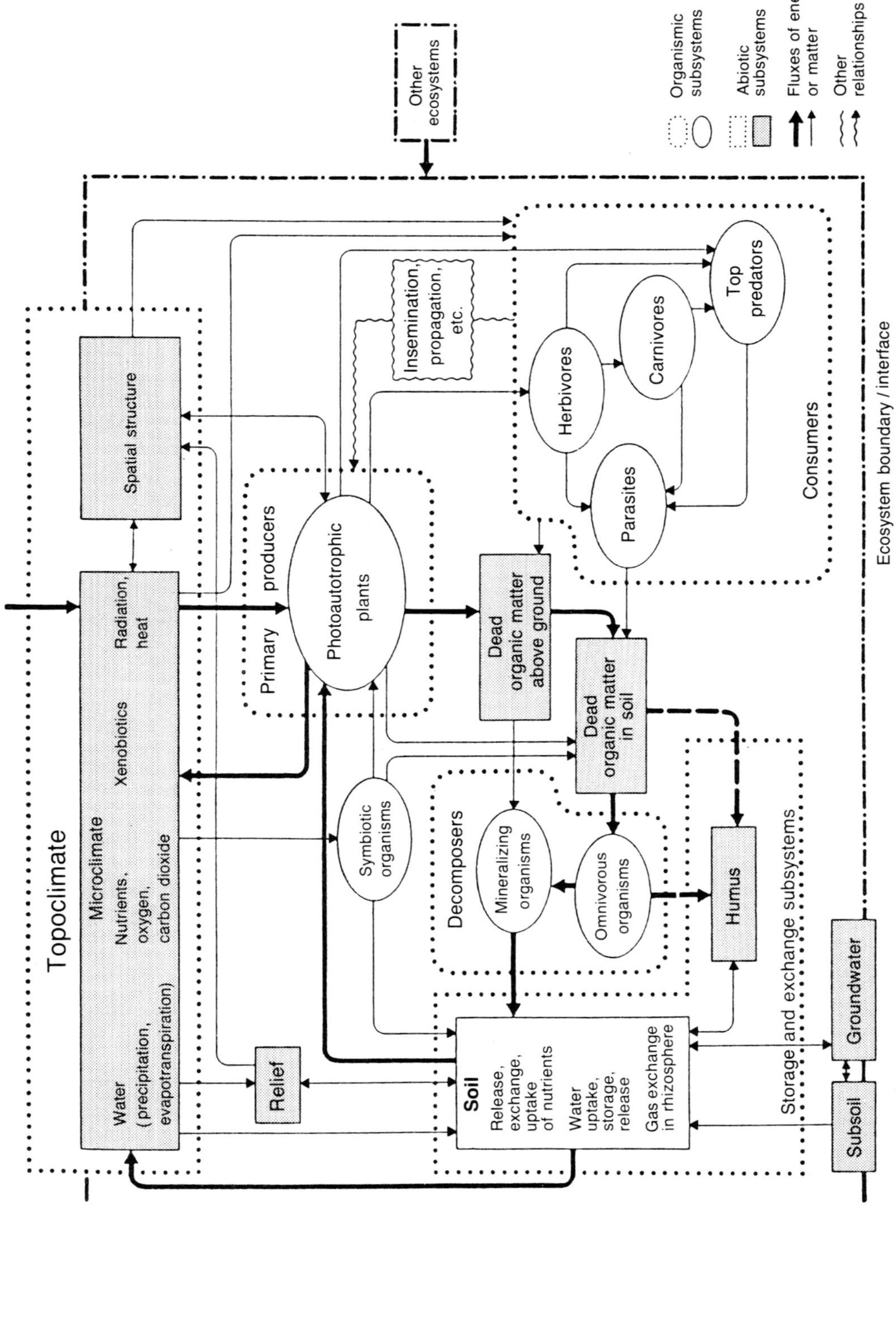

Figure 1. Graph model of an ecosystem (after [90], modified)

whether repeating patterns in collective and emergent properties exist, even when the differences in the particular species that happen to be assembled together are great. Discussions need not be segregated as ecosystem rather than community topics, since these phenomena depend explicitly on fluxes between living and nonliving components of ecosystems. Thus, for instance, detailed knowledge of the role that communities play in biogeochemical cycling is essential for understanding and adequately combating the effects of acid rain or the increasing levels of atmospheric CO_2 on forests, fields, greenlands, streams, lakes, and other communities.

2.1.1.1. Description of Community Structure

The apparently simplest way to characterize a community is to list its species, but in practice this may become surprisingly difficult, because of taxonomic problems and because usually only a subsample of the organisms in an area can be counted. Thus, the *species richness* of different communities can properly be compared only if investigations are based on the same sample sizes in terms of the area of habitat studied, the time devoted to sampling, or the number of individuals or modules included in the samples. Statistics in general, and geostatistics in particular, are essential to ensure valid sampling (see, e.g., [12], [13], [93]).

The simplest measures that take into account both species richness and abundance or biomass patterns are diversity indices. As examples, the Simpson, McIntosh, and Shannon diversity indices are given:

The Simpson index [94] is

$$D = \left(\sum_{i=1}^{S} p_i \right)^{-1} \qquad (1)$$

where D is the diversity index, S the total number of species in the community, and p_i is the proportion of individuals or biomass that it contributes to the total in the sample. An important further specification is expressed by the term evenness (or equitability, E), i.e., the proportion of the maximum possible value D would assume if individuals were completely evenly distributed among species:

$$E = D \cdot D_{\max}^{-1} \qquad (2)$$

Evenness assumes a value between 0 and 1.

The McIntosh [95] index is given as

$$D = \left[N - \left(\sum_{i=1}^{n} n_i^2 \right)^{\frac{1}{2}} \right] \cdot (N - N^{\frac{1}{2}})^{-1} \qquad (3)$$

where N is the total number of individuals in the community and n is the number of individuals per species. In Section 2.3.3, the McIntosh index is used in the framework of stability-oriented analyses of biodiversity.

Shannon's diversity index H [96] again depends on an array of p values; thus,

$$H = - \sum p_i \ln p_i \qquad (4)$$

and

$$E = H \cdot H_{\max}^{-1} = H (\ln S)^{-1} \qquad (5)$$

Since an infinite number of logarithmic systems exist, different authors use other logarithms, in particular those with bases 2 and 10, which should be specified when calculating H.

Rank–abundance diagrams provide a more complete picture of the distribution of species abundances in a community since they make use of the full array of p_i values by plotting p_i against rank. A rank–abundance diagram can be drawn for the number of individuals, for the area of ground covered by sessile species, or for the biomass contributed to a community by various species (see Fig. 2).

Other classical statistical techniques to reduce the subjectivity of community description are ordination and classification.

Ordination organizes communities on a graph so that those that are most similar in both species composition and relative abundance appear closest together. To this end, the axes of the graph are derived mathematically solely from the species composition of the various stands. Interpretation of the resultant patterns in terms of environmental variables is the second step, in which the scatter of points in the ordination is analyzed to see if the axes correspond to ecologically relevant gradients. Since the latter can be brought about by a host of interactions and a priori determination of the important ones is not possible, the same critical remarks relative to gradient analysis (see below) apply to the interpretation of ordination diagrams.

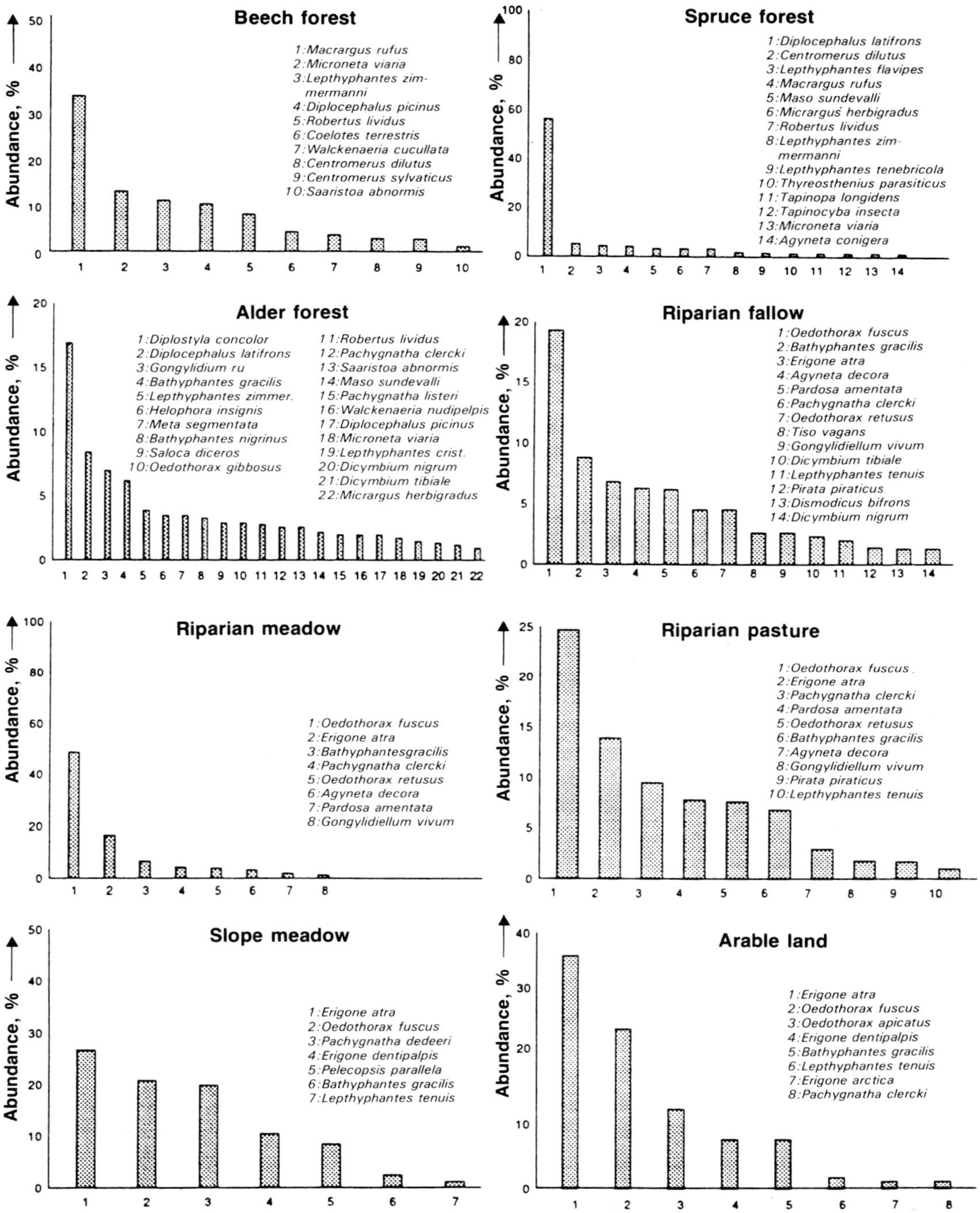

Figure 2. Rank–abundance histograms of epigeic spider communities from various ecosystems of the Bornhöved Lake District (Schleswig-Holstein) (after [85])

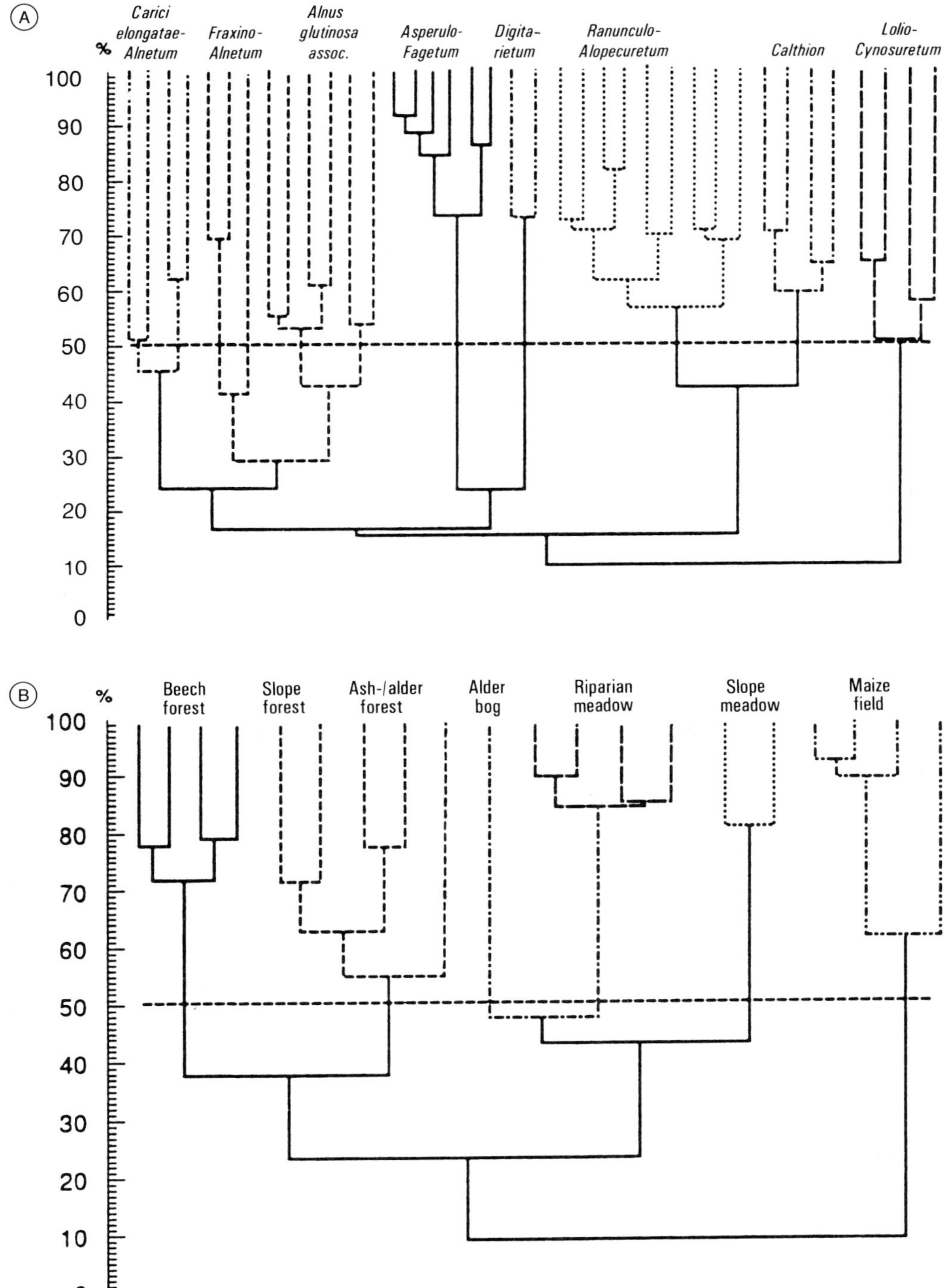

Figure 3. Cluster analysis of plant communities and related Collembola faunas in the Bornhöved Lake District (after [85])
A) Group analysis; B) Formation of subsets and clusters

Classification, in a sense opposed to ordination, produces groups of related communities by a process conceptually related to taxonomic classification. Communities with similar species compositions are grouped together in subsets, and similar subsets may be further combined in a sequential manner by means of clustering algorithms [14], [97]. Figure 3 provides an illustration of such a community analysis (Fig. 3A) and classification (Fig. 3B), in which the primary characterization of the individual phytosociological units and the related Collembola faunas was effectuated by means of Renkonen indices.

Gradient analysis and canonical correspondence analysis are uni- and multivariate approaches to detect patterns in communities. In the first case, the investigator searches (sometimes with a considerable amount of subjectivity) for some feature of the environment that appears to matter to the organism. Then, data about the species concerned are organized along a gradient of that factor, which is not necessarily a priori the most appropriate factor. The fact that individual species from a community can be arranged in a sequence along a gradient may imply only that the factor chosen is loosely correlated with whatever really matters in the lives of the species considered. Thus, gradient analysis is a minor step on the way to the objective description of communities.

Canonical Correspondence Analysis. Much better suited to unravel the intricate interplay of environmental factors and to define their relative importance more precisely are multivariate methods (for a comparative analysis, see [98]). The canonical correspondence analysis [99] compares optimum and pessimum curves of species with Gaussian distributions on the basis of multivariate data obtained in the field. Figure 4 illustrates the results of such an analysis for the Collembola fauna of several stands of the Bornhöved Lake District (Schleswig-Holstein). In analogy to biplot analysis described below, in Figure 4 the orthogonal projection of the points marking species on the various vectors (environmental factors) defines the relative importance of these factors for the distribution of the Collembola species explored.

Biplot analysis [100], [101] permits analysis of the inherent structure of comprehensive geographic data matrices, representing, for instance, organisms (O_n) in relation to metrically defined site factors (F_m).

$$[M(O_n, F_m)]_{1<n,m} \tag{6}$$

Such an $n \times m$ matrix is equivalent to a cluster in m-dimensional space, where m is the number of factors as related to n organisms considered. This equivalence means that any matrix with $n \times m$ elements may be represented by one vector for a row and another vector for each column such that the elements of the matrix are the inner products of the vectors. When the matrix is of rank 2, or if a higher-rank matrix can be closely approximated by a matrix of this rank by using singular-value decomposition procedures, the vectors may be plotted and the resulting matrix representation (i.e., the biplot), inspected visually. Thus, a biplot not only permits viewing the individual data and their differences, but further allows scanning the standardized differences between units and inspecting variances, covariances, and correlations of the variables.

Chi-Square Automatic Interaction Detection. In addition to the above analytical procedures for metric data, chi-square automatic interaction detection (CHAID) is a hierarchically disaggregative classification procedure for nominal and ordinal data [98]. As a multivariable procedure, CHAID subdivides the base set of data automatically in relation to χ^2 values and significance levels into exhaustive subsets. Because of mathematical advantages and the easy interpretability of the resultant dendrogram presentation of data, it has found wide application in forest dieback studies in Germany [102], [103]. Figure 5 illustrates the results of a comprehensive CHAID analysis of environmental factors at 895 sites in North Rhine-Westphalia that proved to be of decisive importance for forest damage.

2.1.1.2. Successions

The relative importance in the community of many organisms (particularly those with relatively short lifetimes) changes with time of year because of seasonal change. In addition, interannual variations in the abundance of species occur because individual populations respond to a multitude of environmental factors that influence their reproduction and survival rates. On still higher time scales, different processes of succession occur, which may be defined "as the non-

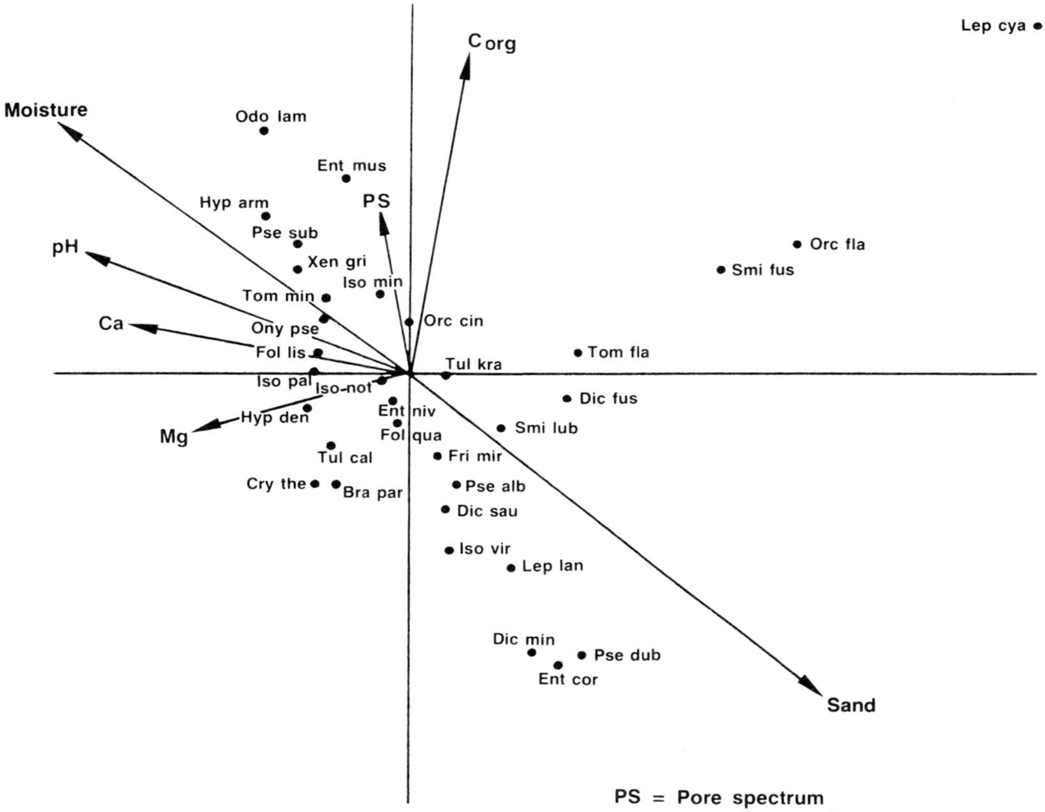

Figure 4. Canonical correspondence analysis of the Collembola faunas of different sites in the Bornhöved Lake District (after [85])

Species: *Brachystomella parvulu, Cryptopygus thermophilus, Dicyrtoma fusca, Dicyrtoma minuta, Dicyrtoma saundersi, Entomobrya corticalis, Entomobrya muscorum, Entomobrya nivalis, Folsomia listeri, Folsomia quadrioculata, Frisea mirabilis, Hypogastrura armata, Hypogastrura denticulata, Isotoma notabilis, Isotoma viridis, Isotomiella minor, Isotomurus palustris, Lepidocyrtes cyaneus, Lepidocyrtes lanuginosus, Odontella lamellata, Onychiurus pseudovanderdrifti, Orchesella cincta, Orchesella flavescens, Pseudachorutes dubius, Pseudachorutes subcrassus, Pseudosinella alba, Sminthurus fuscus, Sminthurus lubbocki, Tomocerus flavescens, Tomocerus minor, Tullbergia callipygos, Tullbergia krausbaueri, Xenylla grisea*

seasonal, directional, and continuous pattern of colonization and extinction on a site by species populations" [6]. On any time scale, successions are of particular ecotoxicological interest since the varying communities normally have different stability, resilience, and consequently sensitivity properties.

The underlying mechanisms of such serial replacements may be quite different. *Degradative successions* usually occur over a relatively short time scale of months or years [104]. Any packet of dead organic matter—for example, litter, the body of an animal, or a fecal deposit—is exploited by microorganisms and detritivores. Normally different species invade and disappear in turn, as degradation of the organic matter uses up part of the resources and makes others available. Ultimately, degradative successions come to an end because the resource is completely metabolized and (partly) mineralized [15], [16].

In addition, autogenic and allogenic successions can be distinguished. *Allogenic successions* occur as a result of changing external environmental conditions (e.g., changes in habitat quality due to erosion or deposition, modifications of soil pH). *Autogenic successions* usually occur on newly exposed landforms, and one of the principal forces driving this type of serial replacement is the change in soil conditions caused by the early colonists. These may alter the availability

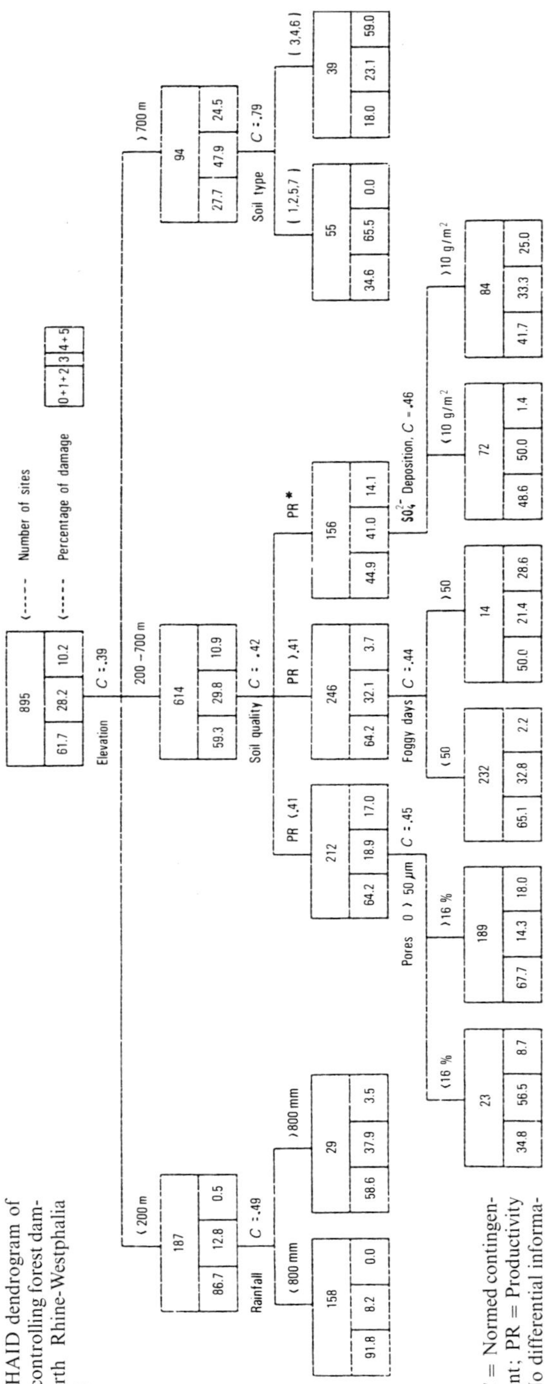

Figure 5. CHAID dendrogram of site factors controlling forest damage in North Rhine-Westphalia (after [103])

$p < 0.01$; C = Normed contingency coefficient; PR = Productivity rate; x = No differential information

of resources (e.g., nutrients) in a habitat, in such a way that the entry of new species is facilitated. The process of *facilitation* [105] may be particularly important in primary successions where conditions are initially severe [e.g., on till laid bare after the retreat of a glacier or inland ice (see Section 2.3.7)]. Another mechanism underlying autogenic succession is inhibition—specific modification of the environment by early species that makes it less suitable for recruitment of "late-successional" species.

Connell and Slatyer [105] proposed three models whose fundamental characteristics are summarized in Figure 6. Hemprich [106] modeled the changes in community composition on the basis of comparative phytosociological analyses in the Bornhöved Lakes area. Figure 7 is an example of successions in grassland communities that typically take from several decades to hundreds of years to run their course.

2.1.1.3. Patch Dynamics

Disturbances that open up gaps in woods may be caused by the death of a tree, high winds, lightning, animals, or lumberjacks. Perturbations in grassland communities may be due to frost, to grazing or burrowing animals, and to the feet and dung of grazers. The formation of such gaps of highly variable size is of importance to sensitive or sedentary species with a requirement for open space, whereas it matters comparatively little in the lives of mobile animals for whom space is not a limiting factor.

Viewing the evolution of communities or succession from such a "patch-dynamics" point of view leads to an important class of models consisting of a number of cells that are colonized at random by individuals of different species. Although species interactions within cells proceed according to Lotka–Volterra principles [107], the stochastically cellular structure of the community and the manifold opportunities in each generation for migration between cells allow species-rich communities to evolve and persist. Thus, patch-dynamics models are much more realistic than equilibrium models with resource partitioning and niche differentiation (e.g., [108]–[110]) or nonequilibrium models incorporating temporal variation in conditions (e.g., [111]), which both deal with closed systems. Patch-dynamics models including competitive interactions and dispersal are of particular interest in an ecotoxicological context. Dhôte and

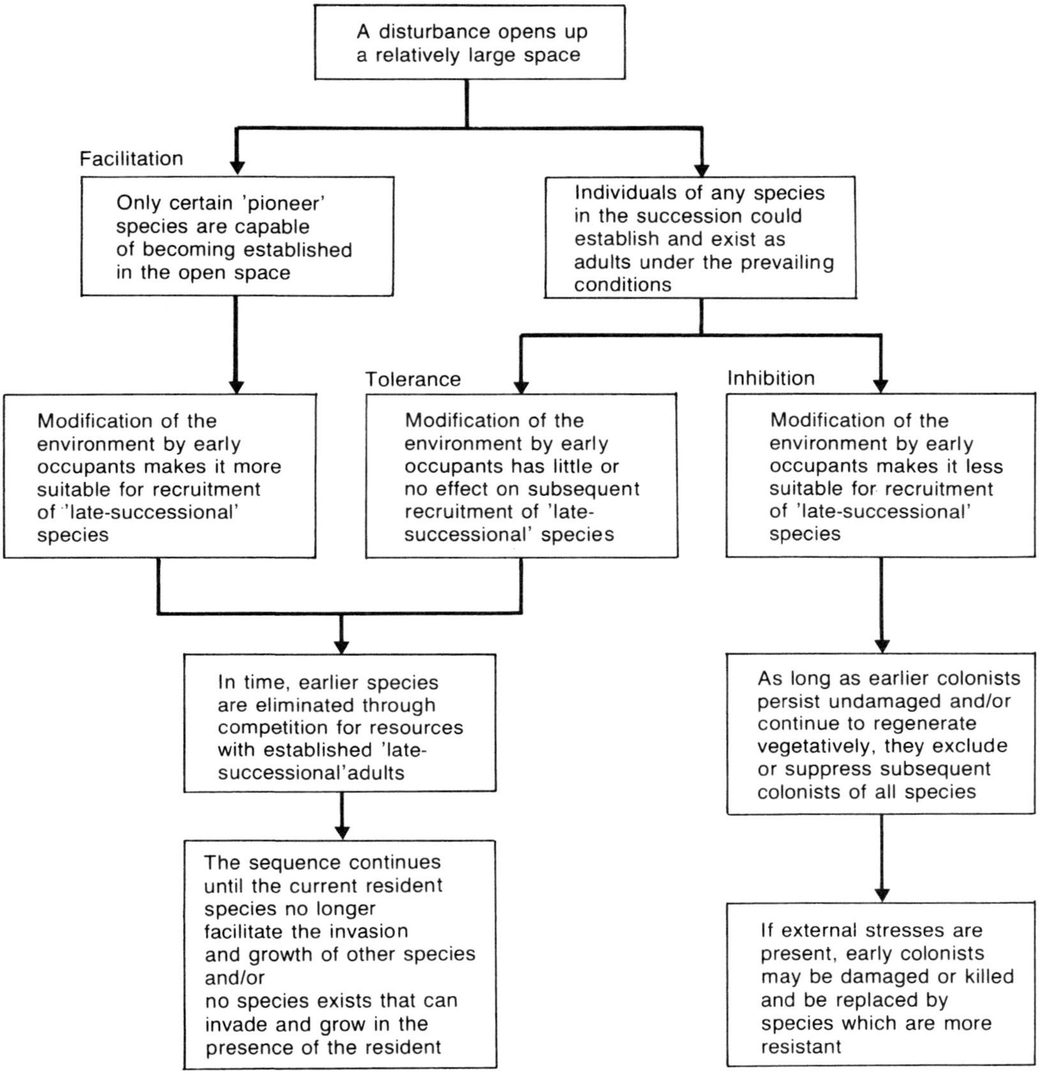

Figure 6. Essential mechanisms underlying terrestrial plant successions (after [105])

HOLLIER [112] considered the dynamics of forest stands from two complementary points of view, namely, their intrinsic biological behavior and their management. This involves explicit consideration of different spatial and temporal scales, levels of organization, and model aggregation. YODZIS [113] distinguished two different types of community organization depending on the type of competition between component species. The case in which some species are competitively superior is called "dominance controlled", while the other with all species having similar competitive abilities is "founder controlled."

Dominance-Controlled Communities. In dominance-controlled communities, disturbances that open up gaps lead to relatively predictable species sequences, where early species are good colonizers and fast growers, whereas later species can tolerate lower resource levels and grow in the presence of early species, eventually outcompeting them. Thus, the community regains the cli-

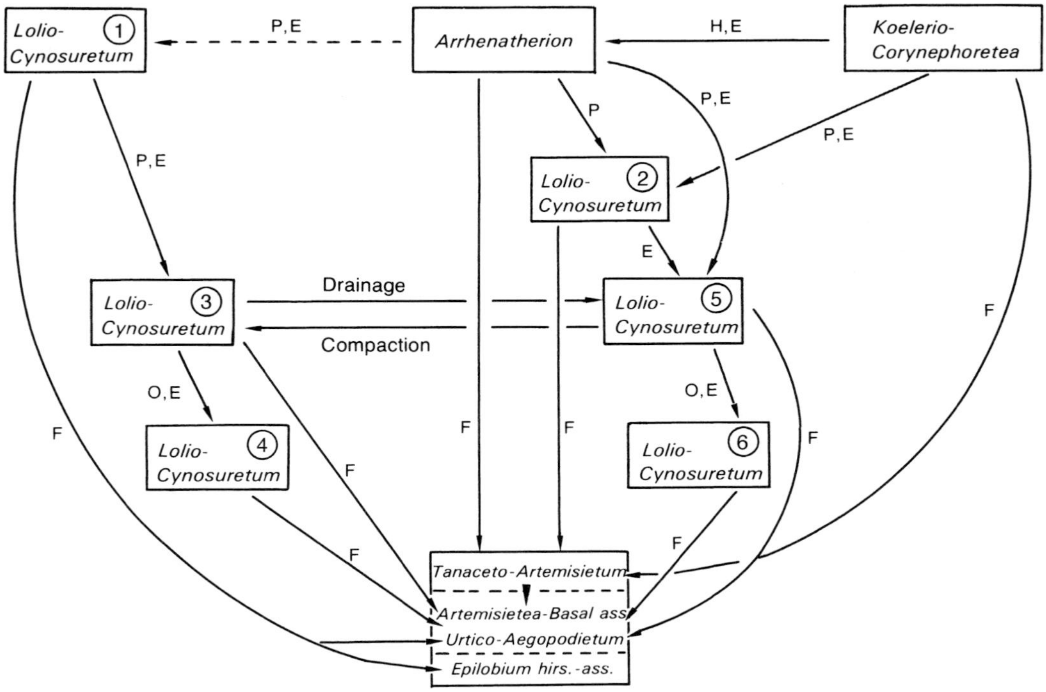

Figure 7. Successions in grassland communities of the Bornhöved Lake District (after [106], modified)
E = Eutrophication; P = Pasture; O = Overgrazing; H = Hay harvest; F = Fallow

Lolio-Cynosuretum

1: *Lotus uliginosus* sub-association
2: Typ. sub-ass., variant of *Hypochoeris radic*
3: Typ. sub-ass., variant of *Agrostis stolonife*
4: Typ. sub-ass., variant of *A. stol.*, with *Rum*
5: Typ. sub-ass., typ. variant
6: Typ. sub-ass., typ. variant, with *Rumex ob*

max stage when the most efficient competitors oust their neighbors. Disturbances may be synchronized, or phased, over extensive areas (e.g., bush or forest fires). Other disturbances are much smaller and thus produce a mosaic of patches, particularly if these disturbances are unphased. The resultant climactic macrosystem has a much higher species diversity than an extensive area undisturbed for a very long period and occupied by just a few dominant species.

Gap size and structure (defined in terms of relief, soil, climate, etc.) are important factors of community structure because they induce contrasting mechanisms of recolonization. The central parts of very large gaps are likely to be colonized by species propagules that can travel greater distances, whereas the smallest gaps may be filled by simple lateral movements of individuals around the periphery. Regrowth results from seeds, plants established prior to disturbance, and lateral growth of branches from surrounding trees. Nearer ground level, colonization is frequently due to saplings that were already present but suppressed [114]. Also, animals can be overrepresented in gaps, as Levey [115] has shown in a study of tropical rain forest in Costa Rica. Here, nectarivorous and frugivorous birds were much more abundant in treefall gaps where understory plants tend to produce distinctly more fruit over longer periods than conspecies fruiting under an undisturbed canopy.

Founder-Controlled Communities. Whereas in dominance-controlled communities colonizing ability and competitive status are inversely related, founder-controlled communities are characterized by species that are both good colonists and essentially equal competitors. If a large number of species nearly equivalent in their capacity to invade gaps, and equally tolerant of the abiotic conditions, can "defend" the gaps against all comers, then the probability of competitive exclusion is much reduced in communities where gaps open continually and randomly. On each occasion an organism dies, a gap is reopened for invasion. Thus, all conceivable replacements are possible and biodiversity will be maintained at a high level. An additional factor is the unpredictability of neighboring individuals in space and time, which contributes to reducing the likelihood of competitive exclusion of one species by another in the cause of succession. In any case, however, numerous feedback relationships exist between local and regional successional dynamics and the energy and water balances of the ecosystems affected.

2.1.2. Water Balance of Ecosystems

A number of models for plant–environment interactions, and particularly for the utilization of water and energy by plants, have been developed since the 1950s. Several of these attempt to understand plant growth and water use as related to specific physiological and environmental parameters [116], [117]. These models, however, cannot be applied to situations for which few data are available unless a number of simplifying assumptions are used. On the regional and geographical levels, other models of a predominantly qualitative character have been suggested [118]–[120]. As a consequence of this dichotomous development, attempts have been made to unify these two approaches with a view to simplifying the comprehensive models, so as to make them applicable to regional use in areas with limited data, without introducing misleading oversimplifications.

Examples of soil–water balance studies that appear to partly fill in this middle ground are those of FUCHS and STANHILL [121], SLATYER [122], BENECKE and VAN DER PLOEG [123].

The soil–water balance equation is generally written in the form:

$$P - R_S - (GWR + TF) - ET + \Delta(SM + SW + GW) = 0 \qquad (7)$$

where P denotes precipitation, R_S the surface runoff, GWR the groundwater runoff, TF the throughflow, ET evapotranspiration, SM soil moisture, SW the perched water (above a compact horizon), and GW groundwater. For the sake of greater convenience, $(GWR + TF)$ is termed deep drainage, and $\Delta(SM + SW + GW)$ is the change in soil water storage (initial minus final) during the period and for the depth of measurement. For a number of cases, deep drainage may be adequately defined as the amount of water passing beyond the root zone or, for experimental purposes, as the amount passing below the lowest point of measurement. All symbols have dimensions of length, for example, millimeters. Equation (7) can be written more specifically as

$$\int_{t_1}^{t_2} [(P - R_S) - (ET) - v_z]\,dt = \int_{t_1}^{t_2}\int_0^z \frac{\partial(SM)}{\partial t}\,dz\,dt \qquad (8)$$

where $(t_2 - t_1)$ is the time interval over which the measurements are made (s), z is the depth to the lowest point of measurement (cm), v_z is the net downward flux of water at depth z (cm/s), and SM is the soil moisture, i.e., the volumetric soil water content (cm^3 of water per cm^3 of soil). P, R_S, and ET are in units of mm/s or g $cm^{-2}\,s^{-1}$. In the absence of a water table near the surface, v_z is generally positive. Compared to aboveground measurements of estimates of the water vapor flux, the soil–water balance approach has the advantages of ease of data processing and integration, since the soil–water reservoir (SM) automatically integrates extraction rates between observations. The disadvantages are associated largely with a somewhat lower level of measurement accuracy and the difficulty of adequately assessing evapotranspiration during periods of rainy weather. Therefore, its applicability is, to a considerable degree, restricted to regions of relatively high potential evaporation rates and sufficiently well-defined alternations of rainy and dry weather.

2.1.2.1. Precipitation and Interception

The measurement of precipitation N at a site or in a region is generally considered simpler and more straightforward than that of the other terms in the water balance equation. The technical requirements are comprehensively described by VON HOYNINGEN-HUENE and NASDALACK [124] and by SEVRUK [125]. The identification of unequivocal spatial structures should be controlled by means of variogram analysis, which is particularly important in vegetation stands (see Section 3.1.4.3).

Here, marked differences in the pattern of precipitation actually reaching the ground normally develop in many plant communities because of the gross interception of precipitation by the vegetation. Subsequently, precipitation is partly transferred to the soil by channeling down the main streams ("stemflow") and partly by dripping from branches, twigs, and foliage ("canopy leaching"), or it may be lost by evaporation from the wet surface. This latter proportion constitutes the term "net interception." Further differences in the amount of precipitation reaching the ground between plants ("throughfall") are due particularly to the disturbed wind structure and are most noticeable in the case of snow.

In view of the complicated physical nature of net interception, attempts to assess it by means of measurement and by indirect estimations are quite numerous. In the framework of a comprehensive agroclimatological model, BRADEN [126] developed the following interception estimate:

$$N_i(R) = a \cdot LAI \cdot \left(1 - \frac{1}{1 + N_0 b / a\,LAI}\right) \qquad (9)$$

where (in the author's notation) $N_i(R)$ is the net interception; $a \cdot LAI$ the saturation parameter, dependent on leaf area index (LAI) and a species-specific maximum interception a; N_0 the above-canopy precipitation; and b the density of vegetation cover. The BRADEN approach has the double advantage of mathematical simplicity and physical foundation, and its validity has been widely tested [127].

With rain, drip or stemflow is usually observed after an area rain total of about 2 mm has been received. However, with freezing rain or snow under conditions favoring retention on the leaves, twigs, branches, and stems (i.e., low wind, temperatures a few degrees below freezing), several times this amount may be accumulated. Stemflow is enhanced by a smooth bark and by branches and leaves that are inclined upwards. Thus, in deciduous (beech) and evergreen (spruce) forests the amounts vary considerably, beech providing much more stemflow and much less interception loss, while oak would have an intermediate position (for greater details, see [2]).

2.1.2.2. Infiltration and Surface Runoff

Horton Overland Flow. Surficial runoff in the sense of Horton overland flow occurs whenever the rate of effective precipitation (i.e., precipitation net interception) exceeds the rate of infiltration (F^*) and the resultant accumulation of surface water exceeds the pondage capacity (S^*) at the point of measurement. The most important regulator is F^*. It is useful in applied hydrology and in relation to pollutant transport to characterize the dynamics of infiltration by a small number of parameters. PHILIP developed a simple physical model of infiltration, which is, however, closely related to more precise diffusion descriptions of infiltration [128].

$$i = S\,t^{0.5} + A\,t \qquad (10)$$

where i is the cumulative infiltration at time t, and the constants S ("sorptivity") and A have a physical meaning related to the diffusion analysis of infiltration. The first term on the right-hand side of Equation (10) describes the contribution to infiltration due to capillarity; the second term represents mainly the contribution due to gravity. The differential form of Equation (10) is Equation (11), where v is the rate of infiltration (cm/s):

$$v = {}^1\!/_2\, S\,t^{-1/2} + A \qquad (11)$$

Some values of the infiltration rate (mm/h) for particular soils in specified conditions are given below:

Sand, loess, silt	11–7
Sandy loam	7–4
Clayey loam, soils poor in organic matter	4–1
Clays, alkaline soils (solonets)	<1

These values are much lower than those determined by means of a field rain simulator in the Schönbuch Nature Reserve (Baden-Württemberg), where the use of buffered cylinder infiltrometers proved impossible because of the extremely high spatial variability of the infiltration-relevant soil characteristics. The experimental conditions were such that a constant surface runoff rate was produced by a constant amount of rainfall when the soil had reached its saturation point. Since evaporation is negligible during the short duration of the experiment in comparison to the rainfall applied (100–250 mm/h) and the increment in soil and water storage is assumed to be zero, the difference between precipitation and surface runoff may be equated to infiltration [129]. For two clay soils tested, the minimum infiltration capacity amounted to 58 and 60 mm/h; for sandy soils, 79 mm/h was measured, while the infiltration rates of loamy soils varied between 63 and 76 mm/h. These results are, on the one hand, indicative of a marked enhancement of infiltration due to organic matter and, in particular, desiccation cracks and root voids; on the other hand, they point to high throughflow rates close to the surface (e.g., piping). As a consequence, no differences in runoff characteristics could generally be found on soils in coniferous, broadleaf, or mixed forests. Variation in the runoff rates was high when the soils were initially dry, but low when the soils were wet. A comparison of these minimum infiltration rates with the maximum net precipitation rates recorded within the last hundred years leads to the conclusion that surface runoff is an exceedingly rare phenomenon in temperate forests due to preferential seepage.

Horton overland flow appears to be a common process in semiarid and arid regions, where precipitation intensities are high and the infiltration capacity of the sparsely vegetated soil is low. It is further caused or intensified by the development of a crust on the soils because the surface layer becomes compacted and the pores blocked as a result of the redistribution of soil particles following raindrop impact. Crust formation due to lateral iron translocation is a particularly widespread phenomenon in ferric luvisols of the semihumid tropics, where it largely contributes to enhance pediplanation processes.

Saturated Overland Flow. In temperate environments with normally modest precipitation rates and well-structured soils, Horton overland flow is the exception rather than the rule, except under certain conditions of cultivation and when the ground is frozen. In temperate environments, all the pore spaces may become filled with water after a period of prolonged rainfall, thus saturating the soil. At this point the water table has risen to the surface and the effective infiltration capacity is consequently reduced to zero. Subsequent rainfall runs off directly across the surface of the slope as saturated (or saturation) overland flow. This situation is likely to come about toward the base of a slope or in microtopographical depressions on a slope where both local infiltrations and throughflow received from higher up the slope contribute to soil moisture. SMITH and PARLANGE [130] describe simple relationships that enable saturating or ponding times to be estimated from values of saturated hydraulic conductivity K_s and sorptivity s. In Morocco

Table 2. Hydrometeorological characteristics of two ploughland sites in northeastern Germany for 1961–1965 (after [2])

Characteristic	Müncheberg	Bochow
Number of erosion-inducing showers	30	39
Precipitation, mm		
Median value	14.8	11.8
Mean extreme value	23.6	18.4
Absolute extreme value	34.8	25.2
Runoff, mm		
Median value	1.2	2.1
Mean extreme value	3.3	5.1
Absolute extreme value	10.2	8.1

IMESON [131] found that the amount of rain required to pond the soil p_r could be estimated reasonably well with one of these equations, namely,

$$\int_0^{t p_r} p_r \,dt = \frac{A}{K_s} \ln \frac{r_p}{r_p - K_s} \quad (12)$$

where A is 0.5 s^2 and r_p is the rainfall intensity.

Because of the number of boundary conditions operative in runoff, the latter varies considerably with the amount, intensity, and duration of precipitation, as well as with slope configuration and soil fabric, which determine the degree and extent to which pondage can occur [129], [132], [133]. In natural situations, the slope is rarely constant and, while runoff tends to reduce soil water recharge at the top of the slope and increase it at the bottom, minor changes of slope generally modify the slope–runoff interrelation. Table 2 provides some runoff figures from two sites in the Federal Republic of Germany. Bochow is representative of loamy luvisols developed from Weichselian till, while Müncheberg is characterized by a podzolized cambisol on coversand overlying Saalian till.

2.1.2.3. Deep Drainage

The deep-drainage term in balance Equation (7) comprises throughflow and groundwater flow and can be equated to a vertical flow that, in turn, may be calculated from hydraulic conductivity and soil water potential data. The normal equation for vertical flow of water v_z is

$$v_z = K + K\frac{\delta h}{\delta z} \quad (13)$$

where K is the hydraulic conductivity (cm/s) and $(\delta h/\delta z)$ is the rate of change of soil water suction (h in cm) with depth (z in cm). Soil water suction is derived from the soil water potential ψ (dyn/cm^2) by the relationship $h = -\psi/\varrho_W g$, where ϱ_W is the density of water and g the vertical acceleration due to gravity. Unless h is very small, $\delta h/\delta z$ is usually much greater than unity so the K term in Equation (13) is often negligible [122]. Under these circumstances, the deep-drainage term of the water balance equation is given by

$$GWR + TF = \int_{t_1}^{t_2} v_z \cdot dt \quad (14)$$

where $(t_2 - t_1)$ is the time between observations [134], [135].

In other situations, a net upward flux of soil water into the root zones can occur from wetter underlying soil horizons or, in particular, from a water table closer to the surface. For comprehensive reviews of methods available for the determination of deep drainage with a particular emphasis on groundwater recharge, see [136]–[140].

Under certain soil conditions, diffuse water movement through the intergranular pore spaces and voids may be supplemented by concentrated turbulent throughflow in networks of pipes. These result from large voids that exist in many soils and are enlarged by soil fauna (e.g., mice, rats, hamsters, moles, weasels, ground squirrels) and the growth and decay of roots. Frequently, soil pipes develop at the interface between organic soil and the underlying mineral soil. Discharge in completely filled pipes varies depending on pressure and gravity potentials, and in partly filled pipes in response to the gradient of the water surface. Usually, therefore, pipe flow velocity is much more rapid than that of matrix flow. Table 3 quotes estimates of flow velocities. From these figures, pipe flow may be seen to attain considerable importance in chemical transport, although the distances covered are normally small in comparison to channel flow. Piping is perhaps more strongly associated with semiarid areas than with humid regions. Drainage and slope development in badlands all over the world are frequently dominated by piping.

Table 3. Flow velocities along different routes in a catchment (after [141])

Type of flow	Flow route	Velocity, m/h
Surface	channel flow	300–10000
	overland flow	50–500
Soil flow	pipe flow	50–500
	matrix throughflow	0.005–0.3
Groundwater flow	limestone (jointed)	10–500
	sandstone	0.001–10
	shale	10^{-8}–1

2.1.2.4. Soil Water

Measurement of changes in soil water storage is conducted most accurately by the use of weighing or hydraulic lysimeters, provided they are properly designed and sited [142]. Lysimeters cannot be used, however, when the nature of the species composition, the spatial structure of the vegetation cover, the depth and ramification of the root system, or other factors make it impossible to simulate the natural environment inside the lysimeter itself. In such cases, determinations of changes in soil water storage at different points in the plant community provide the only technique for evaluating $\Delta(SM + SW + GW)$. Soil water sensing equipment may still fall short of operator requirements, although marked advances have been made since the 1980s. Probably the most commonly used techniques, at the present time, are those of neutron moderation and tensiometers or tensiographs.

2.1.3. Energy Balance of Ecosystems

The disposition of radiant energy at the surface of the earth is of prime importance for understanding soil–water balances and the related chemical transport and transformation processes. A formalized transcription of the energetic relationships yields the following form of the energy balance equation:

$$R_n + H_s + H_a + r_{ET} = 0 \quad (15)$$

In this equation, which summarizes the solar energy cascade with its numerous regulation and transformation components, any form of energy that is flowing toward the surface is considered positive and any form that is moving away from it is considered negative. R_n is the *net radiant flux*, H_s is the *heat flow* in soil at the surface, H_a is the *sensible heat* transfer through air at the surface, and r_{ET} is the transfer rate of *latent heat* due to evapotranspiration. Conventionally, all terms are expressed in joules per minute and square centimeters; 4.19 $J\,cm^{-2}\,min^{-1}$ equals ca. 1 mm of water depth evaporated per hour.

2.1.3.1. Net Radiation

The earth's surface receives short-wave and long-wave radiation from the sun—some of which is lost through reflection—and long-wave counterradiation from the atmosphere, while itself emitting long-wave radiation. Whereas photochemical reactions are controlled mostly by short-wave radiation, the diurnal variability of surface temperatures and the related energy fluxes are very much under the influence of long-wave radiation. Thus, the net radiant flux R_n may be derived from the short- and long-wave radiation balances:

$$R_n = ES(1 - \alpha) + R_c - \varepsilon \sigma T^4 \qquad (16)$$

where the solar radiation ES is the sum of direct beam radiation ES_S and diffuse radiation ES_D, each term comprising short-wave and solar long-wave components; R_c is the long-wave counterradiation of the lower atmosphere; $\varepsilon \sigma T^4$ the long-wave radiation from the earth's surface (Stefan–Boltzmann radiation); and α the reflectivity. Regionally, the net radiant flux of the earth's surface varies considerably with a general trend toward a positive balance or surplus in low latitudes and a negative balance or deficit in high latitudes (for details, see [17]). Factors such as cloudiness and atmospheric humidity, which in conjunction with aerosols affect the transmission of both solar and terrestrial radiation, and the radiative properties of the soil and vegetation canopy create the spatial variation in radiation balance over the earth's surface.

2.1.3.2. Soil Heat Flow

The rate at which heat is transferred downwards into the soil and subsurface substrate (H_s) is directly related to the nature and efficiency of the distribution mechanisms. In solids, heat is redistributed by conduction, and the flow rate depends on *thermal conductivity*. Units are watts per meter per degree Celsius ($W m^{-1}\,°C^{-1}$) or joules per centimeter per second per degree Celsius ($J\,cm^{-1}\,s^{-1}\,°C^{-1}$). Under steady-state conditions, the flow of heat ($\partial Q/\partial t$) is related to the temperature gradient ($\partial \Theta/\partial x$) and the thermal conductivity λ by

$$H_s = \frac{\partial Q}{\partial t} = \lambda \frac{\partial \Theta}{\partial x} \qquad (17)$$

A major problem in soils is that steady-state conditions are rarely achieved and thermal conductivity is a complicated function of granulometry, mineralogical composition, compaction, and water content. The conventional determination of thermal conductivity is consequently fairly difficult. An alternative parameter, *thermal diffusivity*, is used, which is given by $\lambda/(c\varrho)$, where c denotes the specific heat and ϱ the density. Units are square meters per second. For a homogeneous medium, thermal diffusivity defines the rate at which temperature changes ($\partial \Theta/\partial t$) take place:

$$\frac{\partial \Theta}{\partial t} = \frac{\lambda}{c\varrho} \cdot \frac{\partial^2 \Theta}{\partial x^2} \qquad (18)$$

2.1.3.3. Sensible Heat Flow in the Lower Atmosphere

The magnitude of the heat flow into the air at the surface H_a can be obtained by a difference when the other components of the energy balance in Equation (16) are measured, or it can be determined directly. Among the latter approaches, two methods, the aerodynamic method and the Sverdrup–Albrecht method merit particular attention.

The *aerodynamic method* is based on very precise determinations of the vertical temperature and wind profiles, and involves a horizontal homogeneity of the surface over considerable distances in the luff of the measuring station.

The *Sverdrup–Albrecht method* determines either H_a or ET as components of the energy balance in Equation (19)

$$H_a = \frac{H_a}{ET}(R_n - H_s) \cdot \left(1 + \frac{H_a}{ET}\right)^{-1} \qquad (19)$$

The ratio H_a/ET (Bowen ratio) can be estimated from measurements of the vertical gradients of temperature ($\Delta\Theta/\Delta z$) and vapor pressure ($\Delta e/\Delta z$) above the surface.

2.1.3.4. Latent Heat Flux

The latent heat flux can be estimated as a result of the vaporization or condensation of water. Consequently, five general methods are used to evaluate the water vapor flux caused by evapotranspiration.

Analogous to Equation (19), the *Sverdrup–Albrecht method* may be used to determine ET:

$$ET = (R_n - H_s)\left(1 + \frac{H_a}{ET}\right) \tag{20}$$

THORNTHWAITE and HOLZMAN [143] suggested

$$ET = \frac{\varrho k (U_2 - U_1)(q_1 - q_2)}{[\ln (z_1/z_2)]^2} \tag{21}$$

where q is the average water vapor content, ϱ the air density (g/cm³), U the average wind speed (cm/s), z the average elevation above surface (cm), and k von Karman's constant (0.4).

A *continuous record of water vapor convection by means of eddy-correlation techniques* as described by DYER [144] yields estimates of ET by

$$ET = \frac{1}{t}\int_0^t \varrho w q \, dt \tag{22}$$

where ϱ denotes air density, w the momentary vertical wind speed, and q the instantaneous value of water vapor concentration. The problem of measuring ET as the time average of the vertical flux of water vapor thus becomes one of designing instrumentation capable of measuring the turbulent air motion and structure of q. This involves equipment whose response time is sufficiently short to take account of all frequencies in the turbulent spectrum contributing significantly to the flux. This requirement depends on both the height of measurement and the stability of the air, which means, in general terms, that the sensing elements should respond adequately to signals with a period of 1 s.

If all terms of the energy budget are known except the flux due to evaporation (or evapotranspiration), the latter can be obtained by a *difference*. Finally the latent heat flux can be found by *direct measurement*, i.e., by means of weighing lysimeters on the local scale or as the difference term of the water budget of a drainage system on the regional scale.

2.1.3.5. Efficiency of Photosynthesis

In comparison to the above components of the energy balance summarized in Equation (15), the absorption of photosynthetically active radiation in the plant cover plays a quantitatively minor role. Yet, it furnishes useful information on the amounts of solar energy available for photochemical reactions at the surface of soil and vegetation under natural conditions. A quantitative description of the absorption process is provided by

$$dP = \gamma (ES)_v \, s \, dz, \tag{23}$$

where P is the rate of absorption, γ is the coefficient of absorption of photosynthetically active radiation (i.e., wavelength ranges from 400–480 nm to 640–690 nm); $(ES)_v$ is the incident direct beam and diffuse short-wave radiation; s is the specific leaf surface (i.e., the surface of leaves in a unit volume); and z is the vertical coordinate counted upward [18].

According to MONTEITH and SZEICZ [145], the daily energy rates required for gross assimilation are about 28.5, for respiration 6.7, and for net assimilation 21.3 $J\,cm^{-2}\,d^{-1}$. This means that normally the photosynthetic efficiency of the vegetation cover is in the order of magnitude of 1–2% of the total incident radiation. These findings were corroborated by numerous authors. Examples of studies from different parts of Central Europe are, for instance, [19] and [146]. From the purely quantitative budgetary point of view this may appear little, but with regard to bioproductivity and photochemical transformation reactions the situation is quite different. A number of compounds (e.g., various chlorinated aromatics) show distinctly higher conversion rates if adsorbed on particulate matter than if deposited as solids or on thin films [147]. This can be attributed to a bathochromic shift, changes in the relative extinction, the appearance of new absorption bands as a consequence of adsorption, and finally fixation of the dispersed molecules in the adsorbed phase resulting in an intensified compound–oxygen contact.

2.1.4. Nutrient Cycles and Productivity of Terrestrial and Aquatic Ecosystems

Speaking about nutrient cycles means considering how the same atoms of elements move through different components of nature or different biota, and participate in different kinds of reactions (Fig. 8). In principle, these are always chemical reactions; however, the conditions driving them vary highly if embedded in biochemical, biological, or ecological structures. A

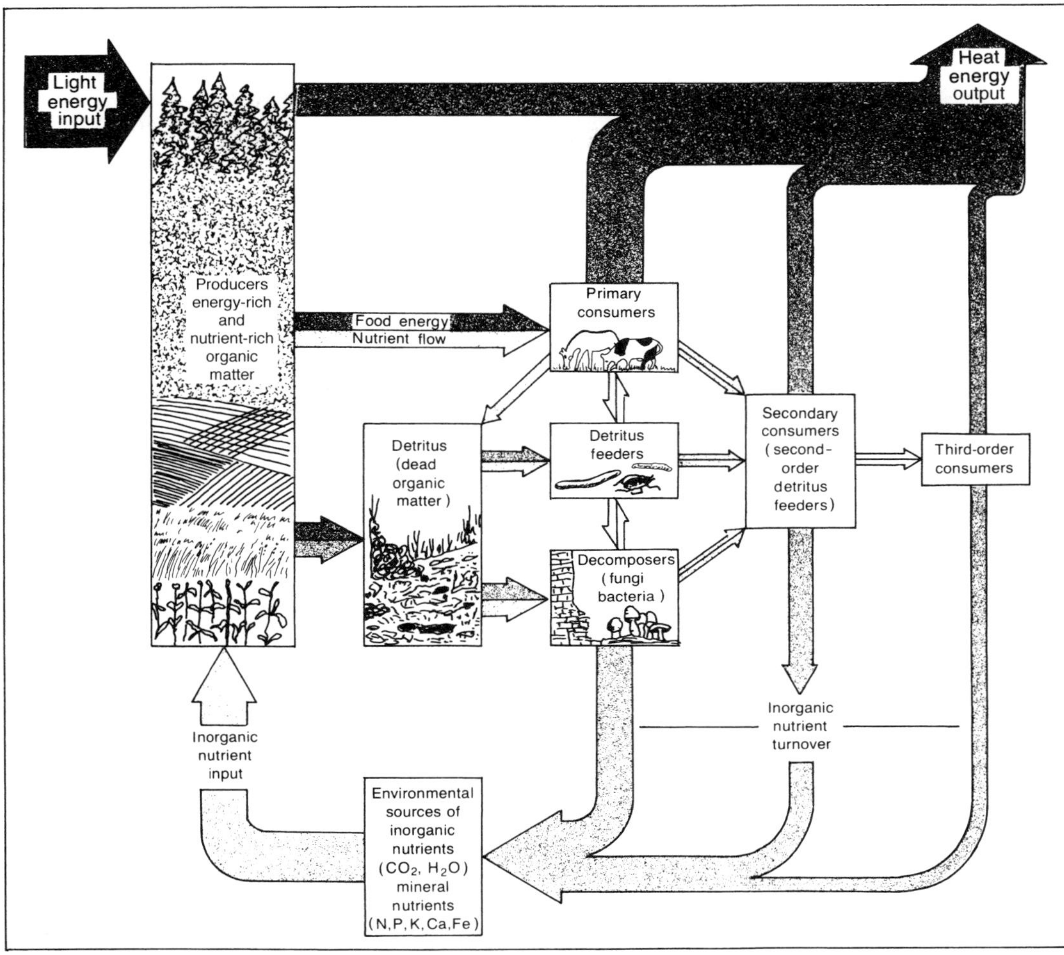

Figure 8. Nutrient cycling and productivity of ecosystems production decreases in relation to successive steps of the trophic chain (producers > herbivores > predators) (after [20], modified)

typical feature of the organic world is the hierarchy constructed of levels of increasing complexity (Fig. 9). In such a hierarchy, the control of processes is both bottom-up, from the low levels to high ones, and top-down, from the higher to lower levels. Biological reactions are controlled by several mechanical, chemical, and neurological mechanisms, which depend on the internal and external environment of biota. An example is the formation of urine in kidneys. In ecological reactions, the conditions for biological, biochemical, and chemical reactions are driven by population cycles of organisms, by mutual reactions of the environment and biota, and by interactions among biota themselves.

Biological production is based on the uptake of nutrients, but external conditions such as light, temperature, and oxygen have major effects. Production is the output of organic matter formation by plants, the primary producers, and of transformation by other groups of biota. Biomass is the organic mass of organisms, which can be expressed as fresh weight, as dry weight, as protein content,· or in energy equivalents. Biomass is the standing stock, which is not necessarily identical with production. Deciduous

World

On the globe we see only the position of land masses and of major biomes. The atmosphere extent is highly magnified - in real proportions it would represent only much less than a millimeter.

Biome

The basic vegetation cover types occupying vast areas of the global surface are called biomes. Tundra biomes represented schematically.

Ecosystem

An ecosystem is defined as a relatively isolated part of nature like a forest or a lake, with its abiotic components and biota mutually interacting.

Biocenosis

An assemblage of plants and animals living in certain specific types of environment. The example depicts the nearshore (littoral) biocenosis of a lake.

Population

One-species population consisting of different age groups which may also appear successively.

Individuum

Within the population different individua possess different capabilities.

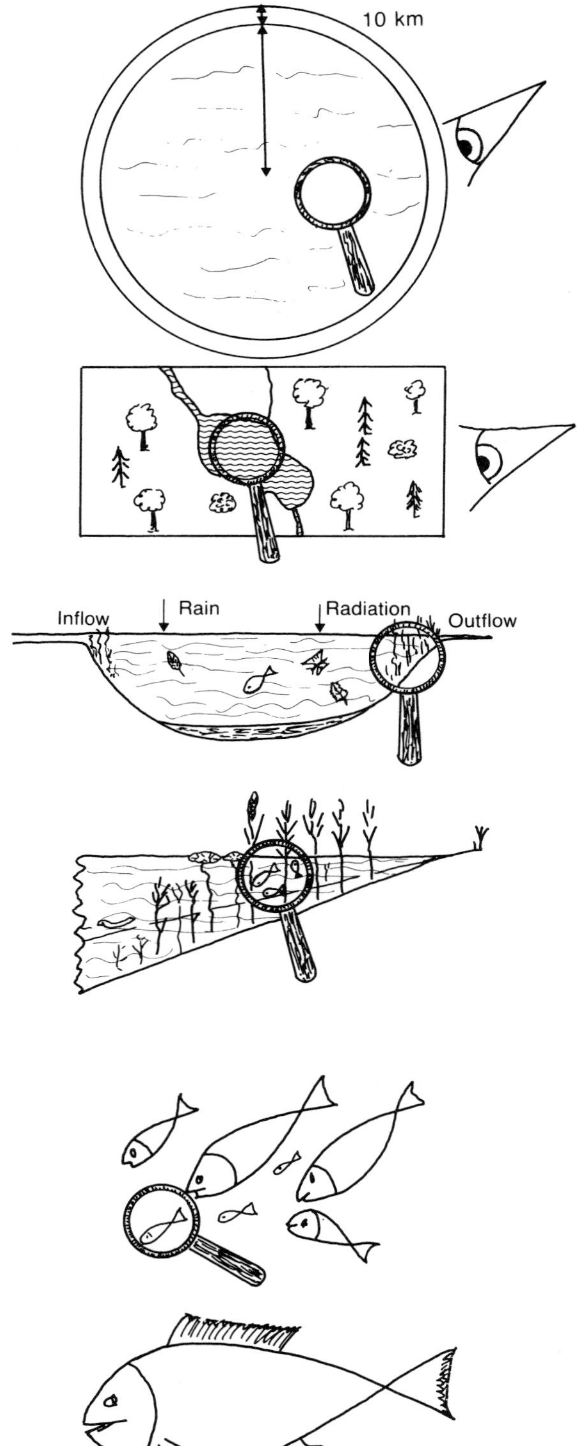

Figure 9. Hierarchy of biological organization; looking at selected objects of nature with successively larger or smaller magnification involves a simultaneous change of view

forests in temperate regions, for example, produce yearly the biomass of leaves that then decay in fall. The next year, a new leaf biomass is formed.

In the following, nutrients and biological production are described separately for the sake of simplicity. However, they are shown to be inseparably interrelated. A distinction is made between terrestrial ecosystems (those inhabiting land) and aquatic ecosystems (those inhabiting water).

The effect of human activities, particularly of industry, on the environment is enourmous. Humans are at present the dominant driving force on earth. They must take care that this force is not misused and is not driving humanity to extinction.

2.1.4.1. Nutrients

Nutrients are defined as those chemical elements that are essential for the construction of organisms. Depending on the proportion of the organism biomass that they represent, nutrients are usually divided into two classes: (1) macronutrients and (2) micronutrients or trace elements. *Macronutrients* represent major components of the biological material and are needed in large amounts. Usually, C, N, P, K, S, Ca, and Mg are considered macronutrients. *Micronutrients* are needed only in small amounts; micronutrients include Si, Mo, Cu, Zn, Mn, Fe, B, and Cl. The needs of organisms for elements belonging to the same class differ by a factor of ten. An indication of the ratio of the first three macronutrients in organisms is given by the stoichiometric ratio of the organisms' bodies. The average elemental composition (by weight) of aquatic autotrophs, for example, is about C:N:P = 106:16:1. In land plants the average ratio of C:N:P corresponds to 250 : 10:1 according to data in [21], and the entire set of macroelements will give a C:N:P:K:Mg:S ratio of 250:10:1:5:5:1:0.5. Micronutrients are present in land plants in lower amounts (Fe 100, Mn 50, B 20, Zn 20, Co 6, and Mo 0.2 μg per gram dry weight). However, their ratios can vary widely in individual plants (see, e.g. [148] for aquatic plants).

Chemical elements usually are not taken up as such by plants; rather, chemical compounds (e.g., C in the form of CO_2) are. Plants are now recognized to optimize their uptake. For example, in selection between nitrates and ammonia, the energetically more favorable ammonia is taken up first. Only some compounds of phosphorus are bioavailable [22].

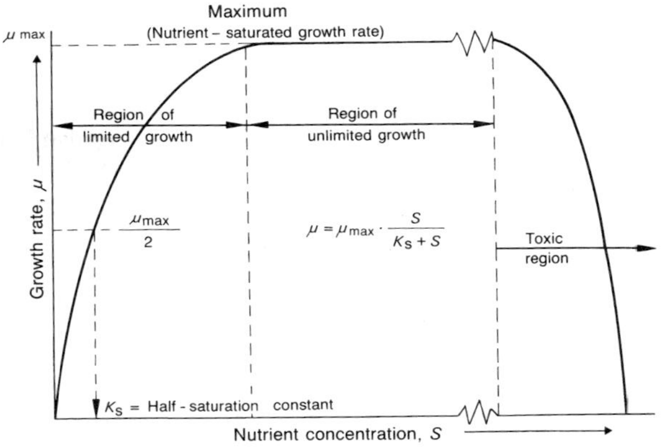

Figure 10. Kinetics of nutrient uptake by organisms
Organismic growth depends predictably on nutrient concentration, following a relationship known from enzyme kinetics.

A typical curve for the kinetics of nutrient uptake is given in Figure 10, which also shows that an element essential for the life of organisms may have a negative effect if it is present in excess quantities. In agriculture this is known as overfertilization. Plants need at least the first three macronutrients (C, N, P) simultaneously. The theory of limiting nutrient states that the nutrient in shortest supply represents the bottleneck. The uptake of all macronutrients is driven by the limiting one, the rest being taken up proportionally to the total needs.

Inorganic nutrients are taken up only by plants with photosynthetic pigments that use solar energy for the construction of highly structured organic matter from more simple inorganics. Such plants are called *autotrophs*; *heterotrophs* are organisms bound to organic matter formed primarily by autotrophs, but possibly transformed by other heterotrophs (e.g., carnivores). Bacteria, in this connection, are known as *decomposers*. They decompose matter to mineral components; bacteria belong to a larger group of destruents, degrading dead organic material, the so-called detritus.

2.1.4.2. Nutrient Cycling in Ecosystems

Two basic types of cycles can be distinguished: the short-circuit cycle in natural ecosystems (see Fig. 8) and the open cycle in agricultural ecosystems (Fig. 11). In natural ecosystems, nutrients circulate rapidly among the organisms, soil, and groundwater. Soil constantly covered

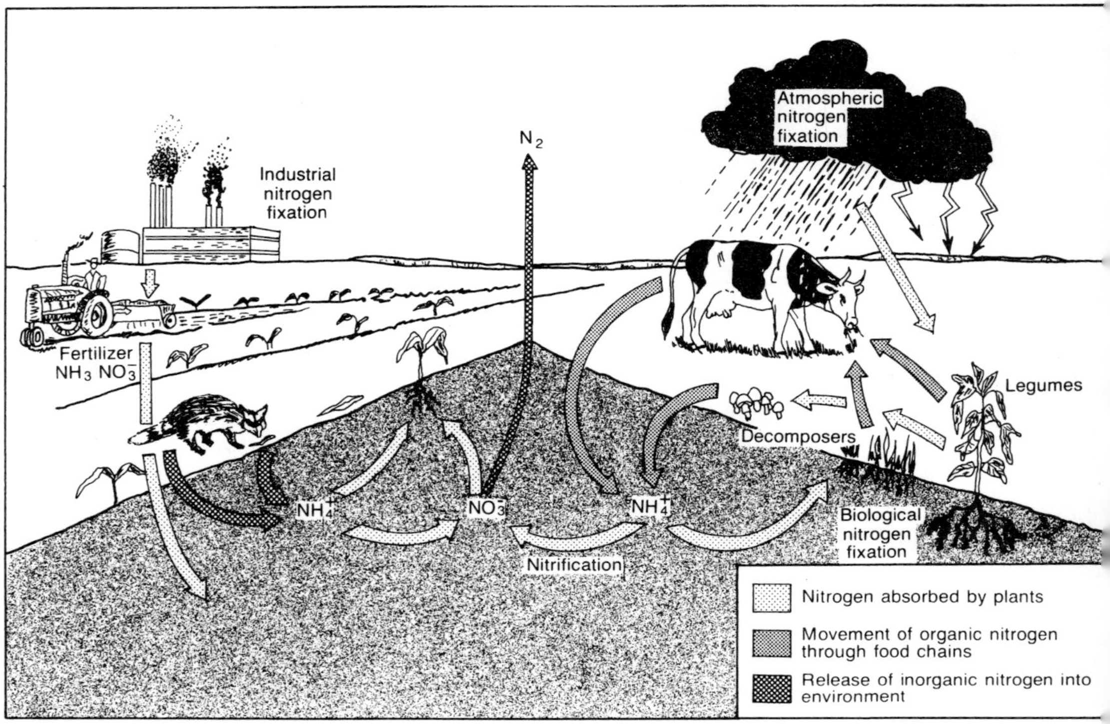

Figure 11. Nutrient (nitrogen) cycle in a man-made landscape (after [20])

by vegetation retains water with the dissolved nutrients. The vegetation cover also retains humidity. Evaporation creates backflow from the atmosphere as rain. Soil is rich in organic matter, which keeps the water capacity high. Therefore, nutrient loss from vegetated soils is low. Simultaneously, soil erosion and washout of nutrients are minimized.

In today's agricultural and inhabited landscapes, the situation is drastically different. The soil is barren or loosely covered, at least for a part of the year. Because of lack of rhizosphere, it is subject to high erosion and flushing out of nutrients to streams and rivers and, consequently, to lakes or sea. The soil is continuously deprived of organic matter, which decreases its water retention capacity. Instead of a short-circuit, an open cycle is observed, with considerable nonreversible losses. In the United States, soil erosion was estimated in 1968 and 1980 to represent 36×10^9 kg/a, about one-half of which originated from agriculture.

The opening of the cycle of organics and nutrients occurred in the Middle Ages when the increasing accumulation of people in towns caused diseases to spread. The construction of sewers disrupted the natural cycle, the original backflow of organic remnants and nutrients to the soil being diverted.

An experiment showing which major changes are produced by deforestation was obtained on the Hubbard Brook Ecosystem [4], [23]. Trees were logged within the entire watershed, and the outflow and its chemical composition were carefully measured. An increase in soil erosion and leaching of nitrogen, phosphorus, and sulfur to groundwater and streams was observed. Because water flux through the soil has increased, the soil loses its fertility. The organic forest floor rapidly decomposes and eventually disappears. The variability of flow in streams increases, with much more transport of the streambed produced during larger flows. However, the average flow decreases. Nitrogen released to the atmosphere during burning of remains has negative effects on the atmosphere. Reradiation from barren soil creates ground-level frosts on clear nights. Soil temperature is significantly increased on hot days.

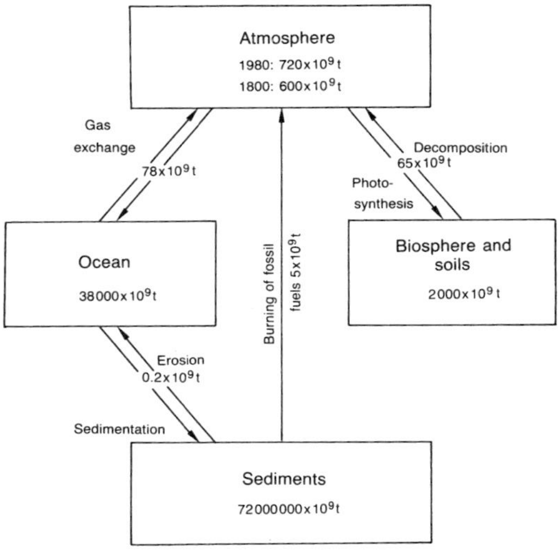

Figure 12. Major carbon reservoirs (after [24])

Carbon Cycle. The global cycling of carbon is gaining in importance, not only because of plant nutrition but mainly because of its increasing concentrations in the atmosphere and the oceans. Because of the increase of several "greenhouse" gases—among which CO_2 plays the leading role—a global change in world weather conditions is expected. Carbon is essentially exchanged among the four reservoirs shown in Figure 12, where the past (in the atmosphere) and present reserves and flows are also given. The importance of forests and oceans as balancing elements in the CO_2 budget is enormous.

Nitrogen Cycle. Basic transformation processes in the terrestrial cycle of nitrogen are shown in Figure 13. Two pathways lead out of the short-circuiting between soil, crops, and animals: to the atmosphere and to water. Losses to the atmosphere have two dominant sources: industry, which releases NO_x; and agriculture, specifically animal husbandry, which produces NH_3. Both gases contribute to the greenhouse effects. Losses of nitrogen to waters also have negative consequences. Figure 14 shows the enormous rise in concentration of nitrates in water bodies in intensively cultivated landscapes of Europe and North America. Since the pristine state, the concentration has increased by a factor of ten and is occurring in both surface and

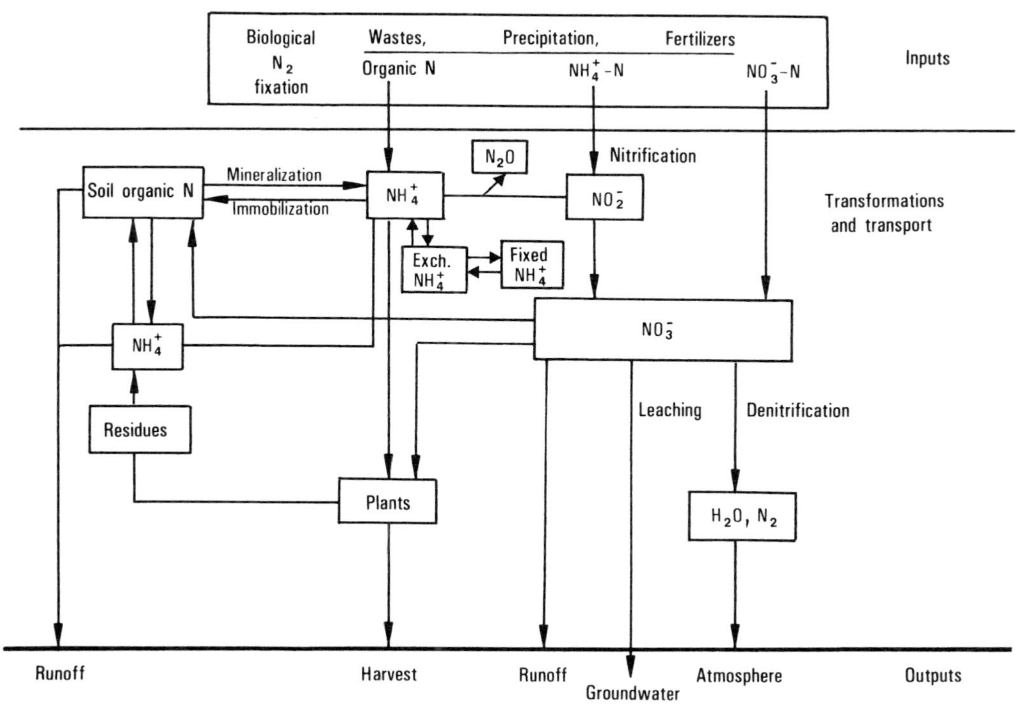

Figure 13. Soil nitrogen cycle (after [25], modified)

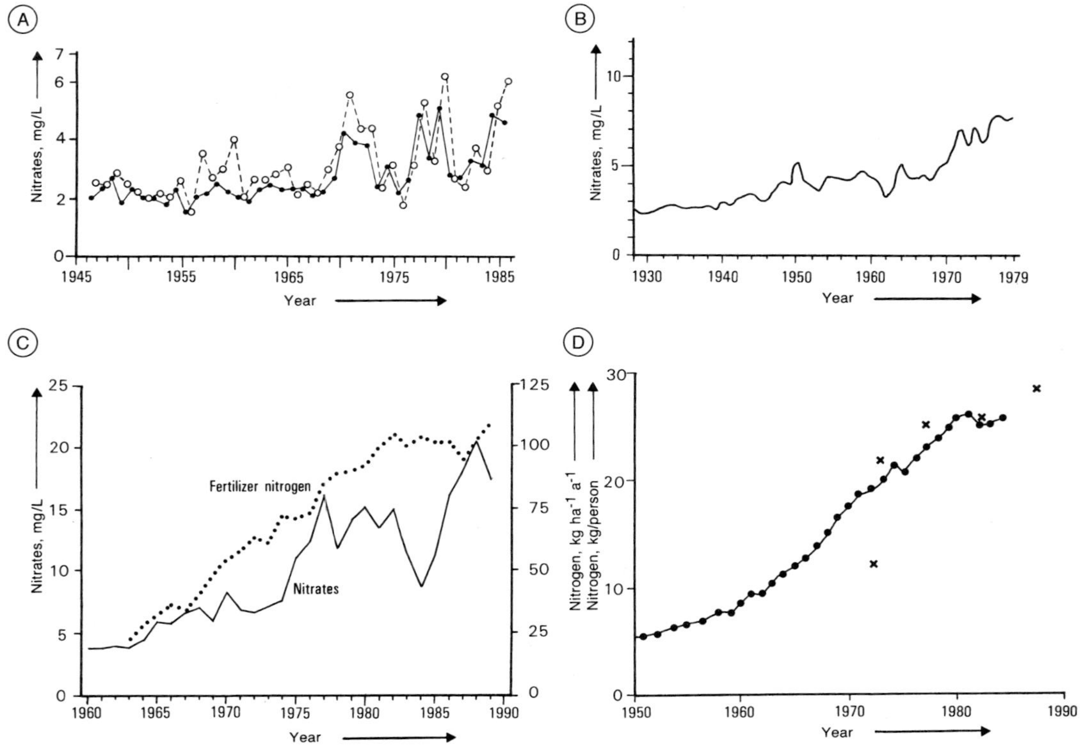

Figure 14. Trends of nitrate in freshwater and fertilizer use
A) Mean winter–spring concentrations of nitrate in Esthwaite Water and Blelham Tarn (Lake District, United Kingdom); B) Mean annual nitrate concentrations at Walton on Thames (United Kingdom), after [149]; C) Mean annual nitrate concentrations in the Slapy Reservoir (River Vltana, Czech. Republic), and related fertilizer use in the drainage basin [150]; D) Mean global fertilizer use per capita in 1950–1984 [151]; crosses represent new data [26]

groundwater. Most of this nitrogen increase is attributable to agriculture, for which it represents enourmous economic losses because of unutilized fertilizers. World fertilizer use increased from about 5 kg per capita in 1950 to 28 kg in 1985 [151]. The consequence of this fertilization of waters is an increased growth of aquatic vegetation, particularly of nuisance algae. In drinking water supply, excess nitrates result in hygienic problems, among them the deadly infant disease methemoglobinemia. For these reasons, every country has standards for the maximum permissible concentrations of nitrates in drinking water. The problem of cheap extraction of nitrogen from water is far from being solved on a global scale, although technological nitrogen-elimination processes are known. The nonsensicality of losing amounts of costly nutrients from agriculture on the one hand and eliminating them from waters at high costs on the other seems evident.

Phosphorus Cycling. A popular representation of the phosphorus cycle is given in Figure 15. Similarly to C and N, short-circuit cycling is characteristic of natural undisturbed vegetated landscapes. An open cycle with losses, mainly to waters, typifies agricultural and town environments. The difference from the previous cycles is that no phosphorus gas phase occurs. In organisms, phosphates are oxidized to produce energy in cellular respiration, and the phosphate is released in urine or similar waste back to the environment. When phosphorus compounds (particulate P, adsorbed P, dissolved orthophosphates, inorganic polyphosphates, organic phosphorus dissolved in the water phase) enter waterways, they enrich—indeed, overenrich—aquatic

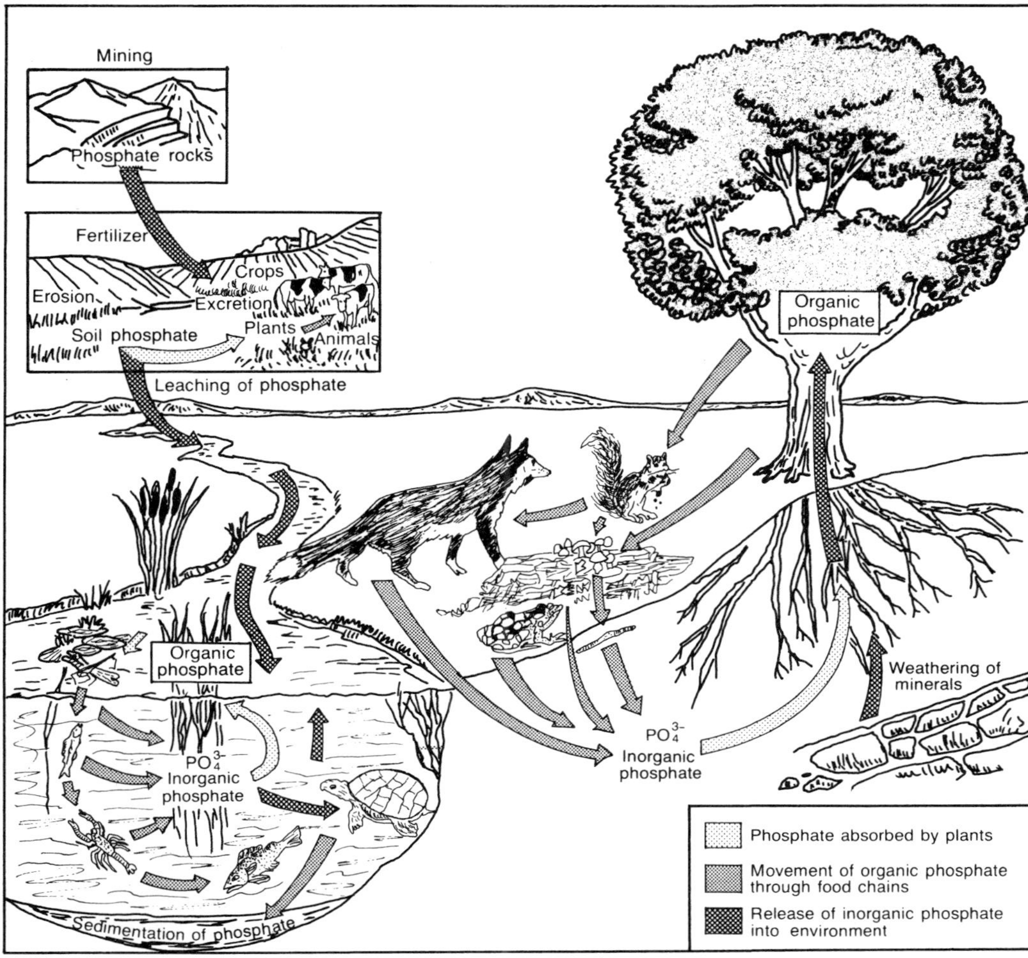

Figure 15. Phosphorus cycle (after [20])

ecosystems. In waters, the processes of phosphorus uptake by aquatic flowering plants and microflora as well as processes of P immobilization and microbial synthesis are inadequate to utilize these amounts. Besides nuisance growth, considerable amounts of P-rich sediments accumulate in estuaries, lakes, and reservoirs. From water, little natural return to the land occurs. The sources of phosphorus for water are listed below:

Natural surface load	0.07–0.12 kg/ha
Atmospheric load	0.4–1.2 kg/mm
From agricultural areas	up to 1.8 kg/ha
Livestock sewage	up to 13 kg per capita cattle
Sewers	0.85 kg per capita dweller
Septic tanks	0.14 kg per septic tank
Municipal sewage	1.1 kg per hectare of town area

Although great differentiation exists locally, two sources of phosphorus seem to dominate in the developed world: detergents and soil erosion. Phosphorus applied as fertilizer does not go into waterways in significant amounts unless some scattering from stores or spreading during aircraft application occurs.

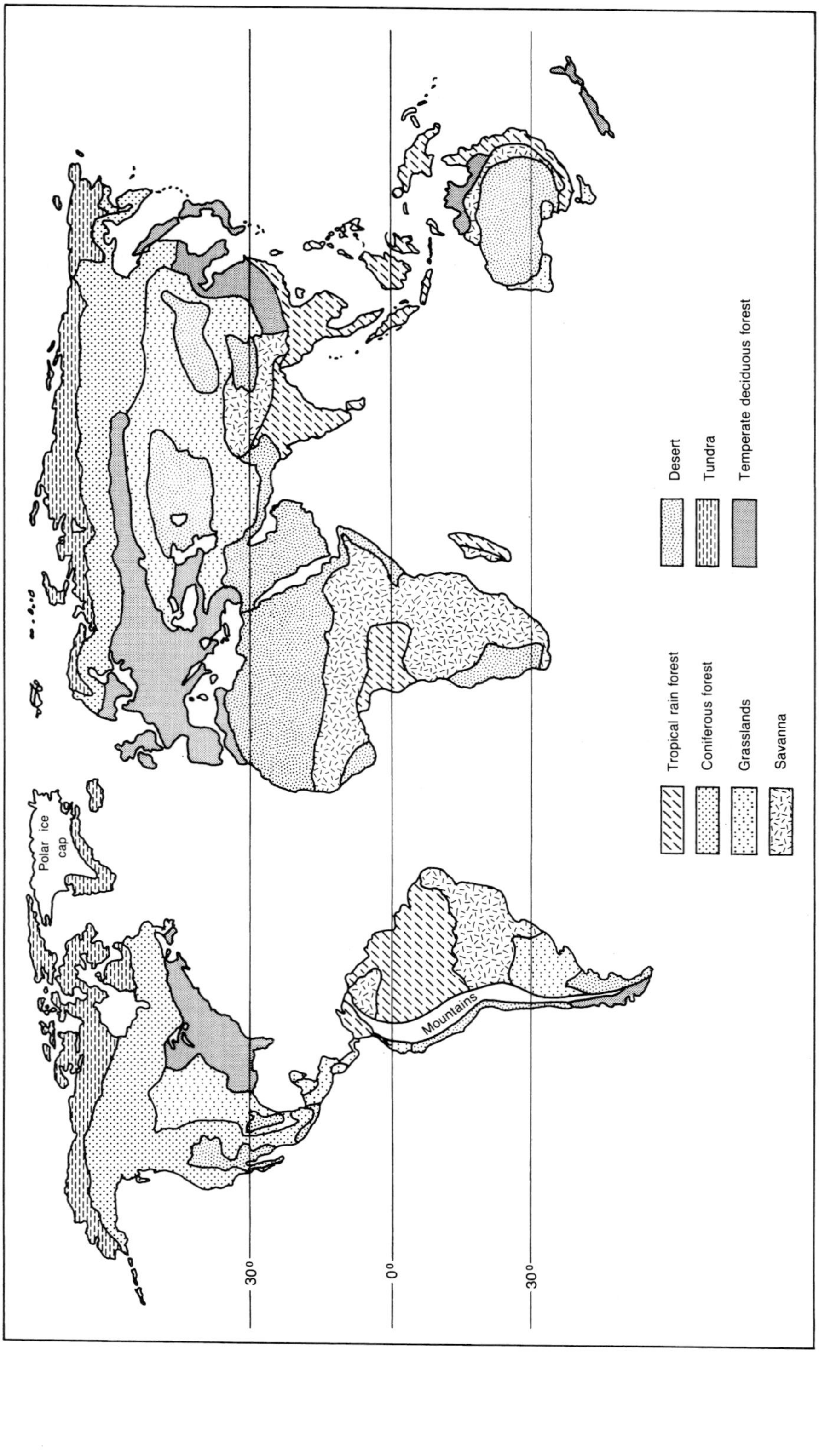

Figure 16. Major biomes of the world (after [27])

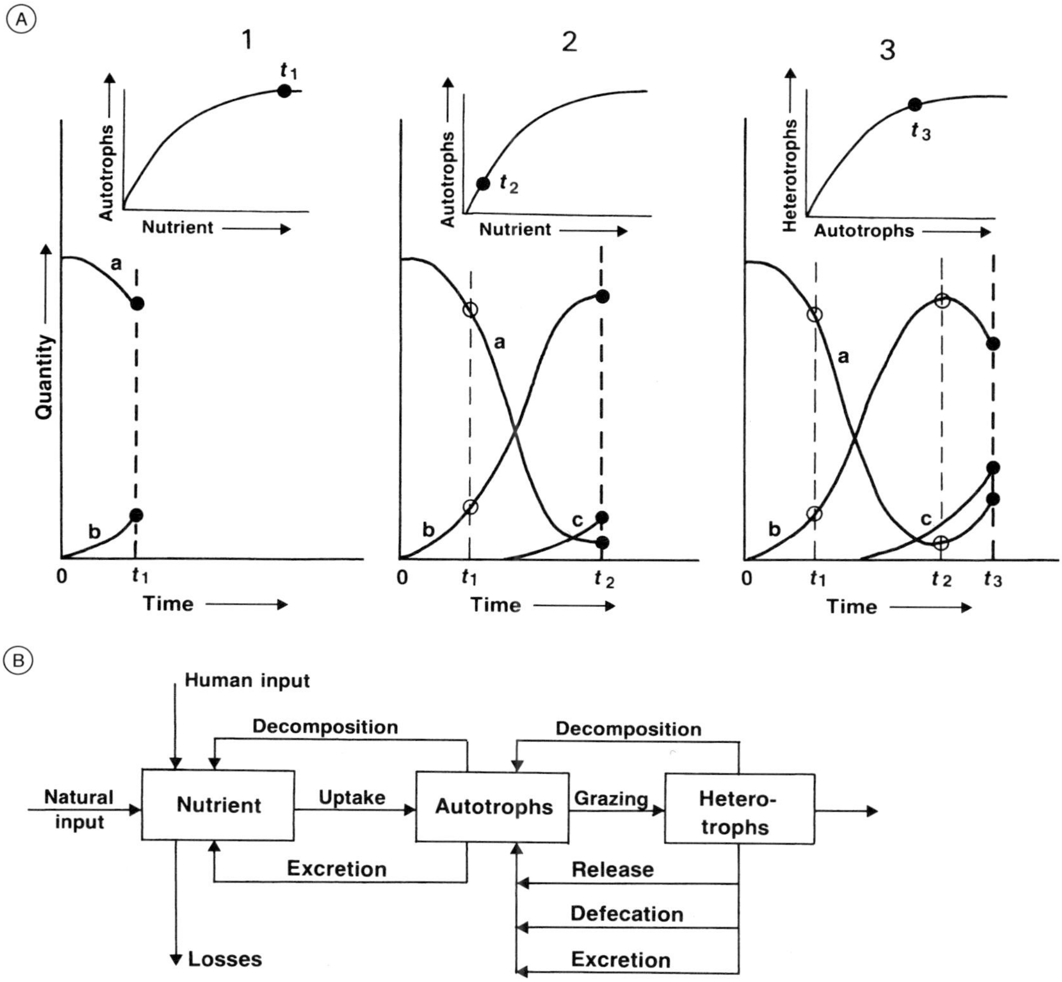

Figure 17. Essential processes of ecosystem dynamics
A) Time development of a population of autotrophs and heterotrophs in a nutrient-limited environment
1) At time t_1 (e.g., in winter) nutrients (curve a) are high and autotrophs (curve b) low, due to low temperatures. When autotrophs start to appear, they grow quickly and take up nutrients rapidly. 2) At time t_2, growth ceases, being severely limited by the decreasing quantity of nutrients (see inset). 3) Heterotrophs (curve c) appear with retardation, because their life cycles are long and at the beginning their abundance is limited by low food supply (i.e., low amounts of autotrophs on which they feed). At time t_3 autotrophs decrease, being heavily grazed on by herbivores (heterotrophs). Nutrients are regenerated both from senescent and decaying autotrophs and from excretion, defecation, and nutrient release (as in Fig. 17 B)
B) Main processes in the nutrient-productivity cycle. Losses are represented by flushing nutrients into groundwater and streams.

2.1.4.3. Productivity of Ecosystems

Major biomes of the world are represented in Figure 16. The total biomass of plants on the globe is estimated to be 1.84×10^{15} kg [28]. The value is rapidly diminishing because of forest clearing, soil salinization, and desertification. In tropical countries of America, Africa, and Asia, the rate of loss represented 37, 48, and 40 %, respectively, between 1980 and 1985 [29]. Typical feedback between the nutrients and organisms is shown in Figure 17, which illustrates the dynamics of natural systems.

However, nutrients are not the only decisive agents for biological production. As evident

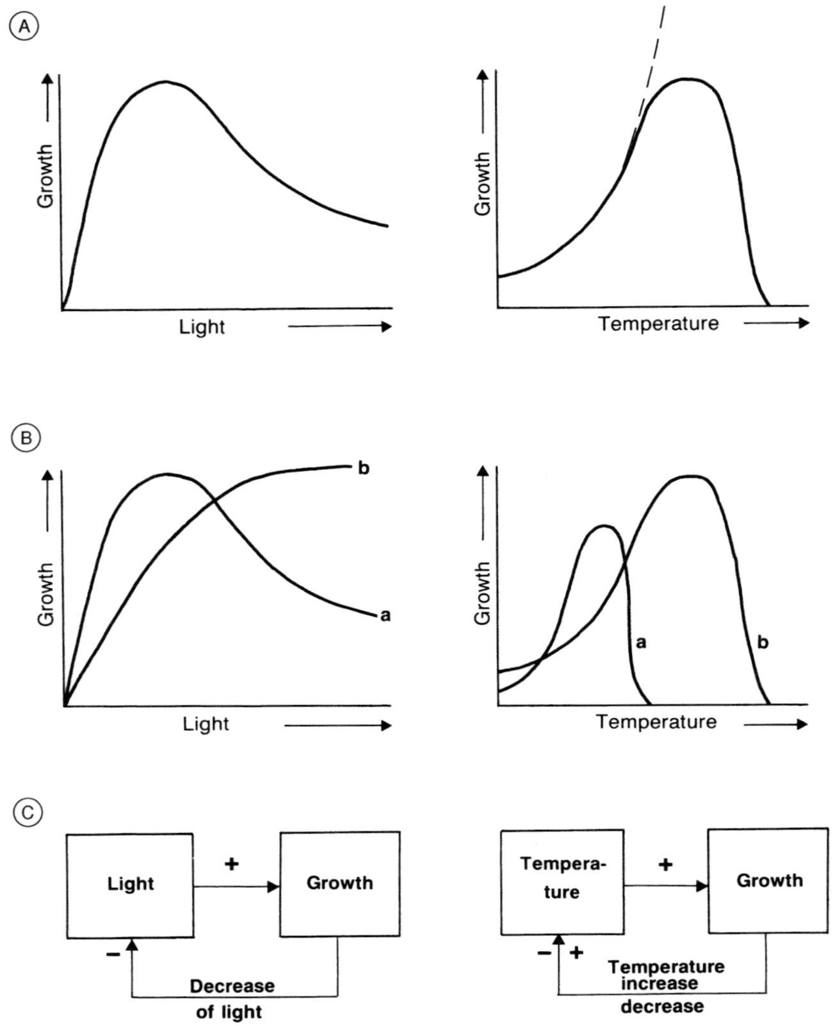

Figure 18. Light and temperature dependence of organismic growth and related feedback mechanisms
A) Typical shapes (equations not shown) of the dependence of plant growth on light and temperature. The dotted line shows the dependence typical for a chemical reaction (Arrhenius curve). B) In contrast to chemical reactions, which follow a fixed curve, the organisms are capable of adaptation to changing conditions. The two curves show changes for low (a) and high (b) light and low (a) and high (b) temperature conditions, respectively. Within the pertinent range of conditions the adapted organisms have the same high performance. However, full compensation is not always possible, as the temperature case shows. C) Feedback effect of growth on light (left) and temperature (right). In the case of light the feedback is negative; for temperature, either positive (increase of temperature, e.g., in water where more radiation is absorbed by microscopic algae) or negative feedback relationships (decrease of temperature by shading in forests) are possible

from the summary photosynthetic equation

$$6\,CO_2 + 6\,H_2O \xrightarrow{\text{Energy}} C_6H_{12}O_6 + 6\,O_2 \qquad (24)$$

energy (usually solar energy) and water are necessary. Like chemical reactions, biological reaction rates are also affected by temperature. The difference is that life is confined only to some −50 to +50 °C. Characteristic shapes of light and temperature dependence of organismic growth are given in Figure 18. For the above reasons the fate of nutrients is highly dependent on geographical position as well as seasonal and daily changes, as shown in Figure 19.

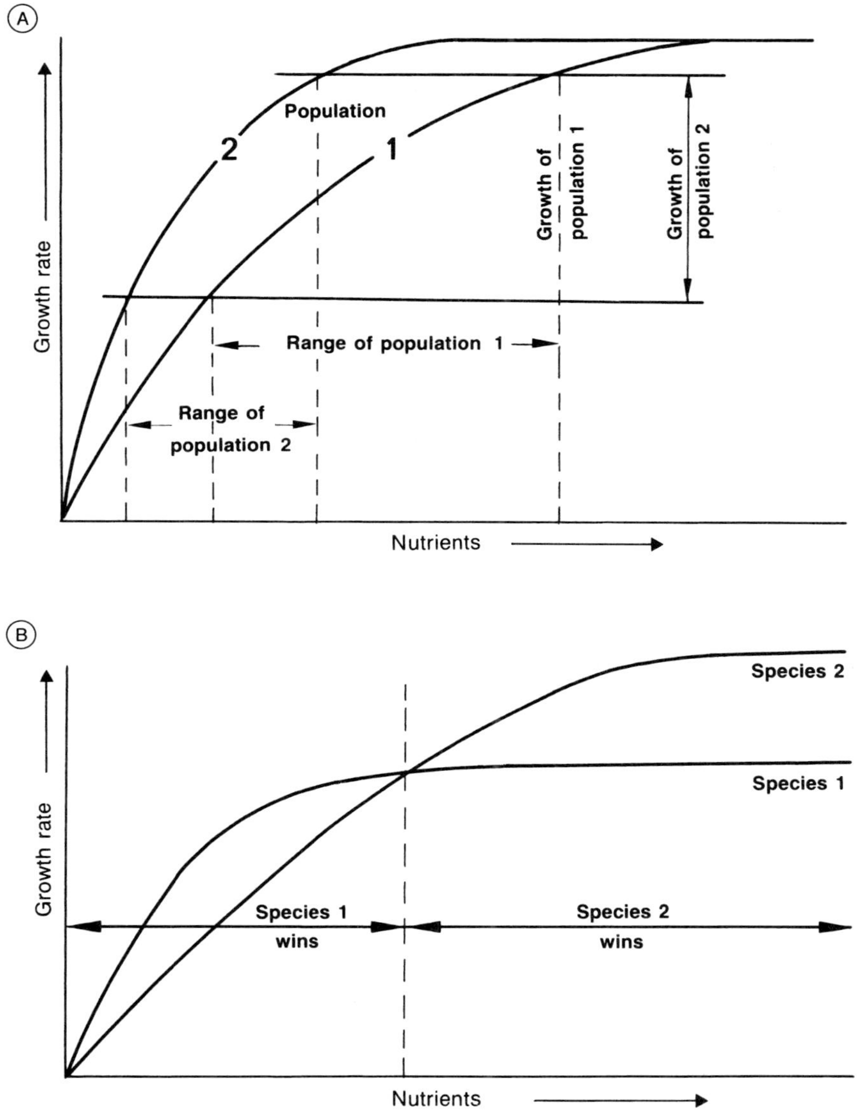

Figure 19. Adaptation of populations to nutrient supply and competition for nutrients
A) Populations adapt to low-nutrient conditions, as indicated by increasing slope of the curve. Population 1 living in high-nutrient conditions shows a lower initial slope of the curve, while population 2 living in low concentrations shows a high initial slope, enabling it to utilize low concentrations more efficiently B) Species compete for nutrients. In a certain condition range, one or the other species dominates, depending on the range in which the curve reaches the highest values.

Productivity of Terrestrial Ecosystems. A distinction is made between natural and the agricultural production, although temperate forests are now also cultivated. Quantitative data on primary production around the world were elaborated in [28], [152]; data on sites where primary and secondary production was measured simultaneously are summarized in [30], [153]. The results of these studies are given in Table 4. Such estimates are methodologically very difficult, and some of them might be highly biased. Only the net aboveground production was measured; belowground production is known to represent up to five times as much as aboveground [31]. Net

Table 4. Data on primary and secondary (herbivore) production of natural ecosystems in different biomes (after [153], modified) [a]

Ecosystem	Prim. prod. [b]	Biomass [c]	Consumption [d]	Herbivores [e]	Efficiency [f]
Desert	403	1	32 (7.9)	0.1	0.02
Tundra	257	0.5	16 (6.2)	0.2	0.07
Temperate grassland	1181	1	161 (13.6)	14.6	1.23
Temperate old field	1352	2.2	95 (7.0)	10.3	0.76
Tropical grassland	2694	8	1725 (64.0)	1.5	0.06
Temperate forest	5032	1.2	64 (1.3)	0.8	0.02
Tropical forest	5400	17	591 (10.9)		
Salt marsh	4328		345 (8.0)	46.0	1.06
Average	2276	8.1	861 (37.9)	8.0	0.35

[a] Values are given in grams of fresh weight per square meter and year. [b] Prim. prod. = net aboveground annual plant production. [c] Herbivore biomass in grams of fresh weight per square meter and year. [d] Amount of primary production consumed by herbivores. The percentage of primary production being consumed by herbivores is given in brackets. [e] Annual production by herbivores. [f] Ecological efficiency, i.e., the percentage of primary production converted to herbivore production.

production is distinguished from gross production in that it does not include organic matter produced but used immediately for maintenance (respiration, excretion). The range of values for 104 sites was estimated as 30 g fresh weight per square meter and year for a desert to 7050 $g\,m^{-2}\,a^{-1}$ for a tropical forest. The average estimates for various ecosystems range from 0–10 g dry weight per square meter and year for extreme desert through 10–250 for deserts and 400–2000 for boreal forests up to 1000–3500 for tropical rain forest and 800–6000 g dry weight per square meter and year for swamps and marshes. Table 4 shows that forests in general are the most productive ecosystems. Column 3 also shows that a high proportion of the primary production in nature is consumed by herbivores, be they insects, small rodents, or large ungulates. On average, 1.3 to 64 % of primary production is consumed in different biomes, with by far the highest proportion in tropical grassland. The primary production taken up is converted to herbivore production with an ecological efficiency (expressed as the amount of primary production converted to herbivore production) of about 0.35 %. The efficiency of herbivore utilization of primary production calculated from columns 3 and 4 results in an average value of 0.93 %, with a range between 0.09 for tropical grasslands and 13.3 % for salt marshes. Two types of ecological cycles in terrestrial ecosystems are distinguished in [31]: a short one, where most of the primary production is immediately decomposed, while in the long one, herbivores not only consume a large part of primary production but also are preyed upon by carnivores and a complex food web originates. In Table 4, column 3, only temperate forests show the short cycle, with only about 1 % of primary production being consumed by herbivores. In all other instances the primary production consumed is of the order of magnitude of 10 %. For tropical grassland, seemingly the most complex system, more than half of the primary production is consumed by herbivores. Table 5 gives examples of net aboveground primary production of agrarian ecosystems and secondary production of domestic livestock compared to game.

Surprisingly, comparison of Table 5 with Table 4 shows that natural production and agricultural production are not much different, if one ignores the extreme energy needs to achieve high agricultural yields [154]. Often the fossil energy input is greater than the energy content of the food harvested [32]. According to [33], for an eightfold increase in yield of major crops, a twentyfold increase in energy applied is needed. The idea that much more output is obtained when artificially cultivating domestic livestock is not supported by these data. The use of natural vegetation by natural livestock is far from inefficient. That was recently recognized in tropical countries and became a political issue in opposition to what was exported from the developed countries as the "green revolution." The green revolution appears to lead to the continuing poverty of poor countries. They are made dependent on Western technology and on energy

Table 5. Examples of net aboveground primary production of agricultural eosystems and production of domestic livestock per area as compared to wild animals (data from [10], [23])

Primary production, kg dry weight per square meter and year		
Crop	World average	Maximum per country
Wheat	0.4	1.3
Corn	0.7	1.3
Rice	0.6	1.5
Potatoes	2.5	4.6
Soybean	0.35	0.6
Sugarcane	0.9	3.3
Sugar beet	2.0	
Secondary production		
Type of animals	Weight gain, kg fresh weight per day	Biomass, kg/km^2
Sheep	0.05	
Cattle	0.14	
Thomson's gazelle	0.06	
Topi	0.20	
Wildebeest	0.24	
Domestic livestock		10000–27000
Domestic livestock in East Africa		10000–14000
Wild ungulates in East Africa		61000–87000
Bison		12000–17000
Black tailed deer in California		9200

sources not available at home, without adequate rewards. The simultaneous trend for sustainable development leads to major revisions of prevailing ideas concerning agricultural production.

Productivity of Aquatic Ecosystems. The primary production of fresh waters may be divided into two groups: (1) production of flowering plants living in shallow water of marshes, swamps, and along the shores of lakes and slow-flowing streams; and (2) microflora inhabiting primarily the open lake and slow river central regions. Microflora consists of several groups of microscopic algae and species of cyanobacteria (also called blue-green algae). These free-floating organisms are called phytoplankton.

The transition between land and water belongs to the most productive plant environments: no water deficit exists; nutrients are obtained both from rich aquatic soil and from water. Under comparable geographic conditions, the highest production among all vegetation is obtained in aquatic macroflora. The dependence of aquatic primary production on geographical position is due to a combination of effects such as light and temperature. For phytoplankton, conditions of mixing in lakes dependent on geographical latitude and altitude are decisive. During the annual cycle, lakes undergo periods of intensive deep mixing to the bottom, as well as periods of stratification. During summer stratification, upper layers become warmer than deeper ones, and intensive mixing is confined to several top meters. The density difference between warm top and cold bottom layers is usually sharp. At a certain depth, a jump in temperature occurs between the two basic layers called the mixing layer (thermocline) and the hypolimnion. The depth of the mixing layer affects primary production. Algae moved within this layer periodically encounter high light intensity when taken to the water surface and reduced or no light when mixed to depths. Greater lakes are usually clearer and mixed to greater depths due to higher wind stress; their primary productivity is lower than that of shallow lakes with comparable nutrients.

The most important determinant of lake productivity is nutrient input. In the 1960s, controversy existed—mainly between representatives of the soap and detergent industry and limnologists studying lakes—whether carbon, nitrogen, or phosphorus is the nutrient most often limiting aquatic primary production. The situation is now clear: depending on the mutual relation among C, N, and P any of them can be limiting, as can several micronutrients. However, in most lakes at present, phosphorus is the limiting agent, since nitrogen is usually supplied from human activities in excess. Carbon can limit in instances of an excess of both nitrogen and phosphorus; carbon deficiency is due mainly to limited exchange of CO_2 between water and atmosphere. This is the case in highly fertilized fish ponds. A systematic difference exists between fresh water on the one hand and brackish and marine waters on the other: in the latter, nitrogen limitation prevails. Instances of limitation by micronutrients are reported rarely, particularly from regions without human activities, where the lack of some elements is geologically conditioned [34].

Table 6 gives some ideas of the primary production of fresh water depending on phosphorus and nitrogen concentrations. In addition, there is a geographical differentiation of phytoplankton primary production at similar conditions of nu-

Table 6. Primary and secondary production in fresh waters, based on different sources and data obtained mainly during the International Biological Program [5] (for more detailed data see [31])

	Macroflora	
Type of water body	Aboveground primary production, kg dry weight per square meter and year	Average biomass, kg dry weight per square meter and year
Lakes and streams	0.1–3.0	0.02
Swamps and marshes	0.8–8.5	1.5
	Microflora (phytoplankton)	
Phosphorus and nitrate concentrations, µg/L	Primary production, kg dry weight per square meter and year	Average biomass, g dry weight per square meter
Low P (<5) and N (<250)	0.015–0.09	0.4–2.0
Medium P (<30), N (<500)	0.075–0.3	2–5
High P (>30), N (>500)	>0.43	>5
	Type of water body fish production,* kg fresh weight per square meter and year	
Small lakes	0.002–0.18	
Great lakes	1–7	
Tropical lakes	2–250	
European fish ponds	500–850	
Chinese fish ponds	500–7500	

* Fish production is estimated from harvested production. According to recent estimates harvested production represents in lakes about 10% of total production and in culture about 50%.

trients. In Arctic and Antarctic lakes values between 0.002 and 0.07 kg dry weight per square meter and year were observed. In alpine lakes the lowest values observed are similar to those in the Arctis but maximum values up to 0.2 are reached. For temperate and tropical lakes the ranges are 0.004–1.8 and 0.06–5.0, respectively. Some data on herbivorous production, represented by fish, are also indicated. The biomass of phytoplanktonic algae is about three factors of ten lower than in macrophytes, whereas production is similar. This is due to the short life cycle and high turnover rate of microscopic algae. Under intensive artificial cultivation, some algal species have produced up to 8 kg dry weight per square meter and year.

2.2. Terrestrial, Lotic, Lentic, and Wetland Ecotones

A boundary in nature may be created by discontinuities in the external influences that control the ecosystems or by discontinuities in the functional network of the systems themselves. Consequently, a wide spectrum of equilibrium system responses to values of environmental gradients exists.

In Figure 20 A the response of the ecosystem as characterized by one indicator variable is as smooth and linear as the changes in gradient; Figure 20 B depicts an abrupt break in response, probably due to a threshold response in one of the components of the ecosystem. In Figure 20 C the response has a twist, indicating that the system can have two stable equilibrium states. In these cases the ecosystem state and behavior with respect to driving variables depend on the history of the systems (hysteresis). Hysteresis plays a major role in catastrophe theory. In Figure 20 D the system has two stable states and can display multiple internal reactions to take on one or the other.

In more abstract or unifying terms the individual response curves of Figure 20 can be considered as a continuum of ecosystem responses to an environmental variable. Thus, the various responses grade into one another depending on internal reactions and autonomous features of the ecosystem. The environmental gradients can be continuous or discontinuous to form a variety of pattern that, together with the response of the ecosystems, control the nature and pattern of the transition zones between adjacent systems.

These transition zones or *ecotones* [155] have a set of characteristics uniquely defined by space and time scales and by the strength of interaction between adjacent ecosystems (see Section 2.3). Because ecotones usually represent relatively steep gradients in both abiotic and biotic variables, they facilitate the ecological analysis of mechanisms contributing to biodiversity and flow of energy or materials across the landscape [156].

2.2.1. Terrestrial and Lotic Ecotones

In view of the ecological importance of ecotones, they can be divided into three broad classes.

Terrestrial ecotones may be pragmatically defined as discontinuities between systems. Consequently, the most obvious ecotones in terrestrial

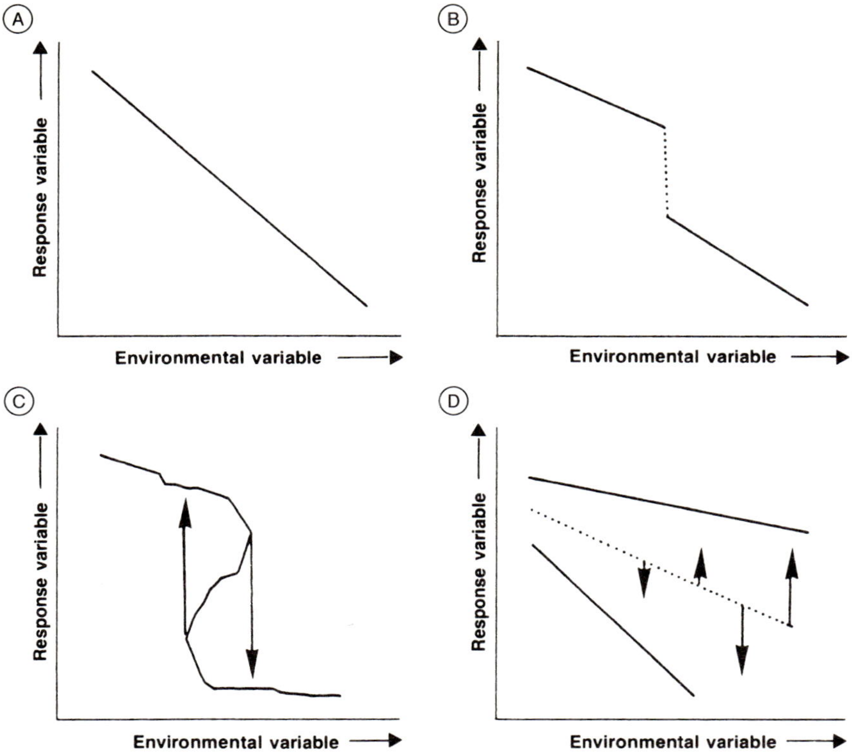

Figure 20. Ecosystem response to different types of environmental gradients (after [154a], modified)
A) Case in which the response of the ecosystem is smooth with respect to the environmental variable; B) Case in which the response is discontinuous; C) Case in which the response is folded and results in hysteresis; D) Case in which the system can have multiple responses

ecosystem complexes involve transitions between plants of communities dominated by different life forms. Large-scale examples are transitions from tundra to taiga along thermal gradients or transitions from prairie to forest along continental moisture gradients; a frequent small-scale type is the transition from field to woodlot in agricultural landscapes. This pragmatic view of terrestrial ecotones depends highly on the contrast between the different elements contained in the definition of the landscape, and the degree of contrast is necessarily a function of the variables used by the observer to categorize the landscape elements [138], [157]. Thus, a more precise and generally applicable method of defining ecotones is based on areally valid measurements of ecologically relevant parameters and subsequent multivariate cluster analysis. The results are operationally homogeneous spatial units (patches) separated by distinct boundaries whose location and geometry depend on the variables used to characterize the primary measuring points or elementary spatial units, respectively; the fusion level of clustering adopted; and finally the scale selected [158], [159].

Lotic ecotones can be defined, in an analogous way, as fluvial boundaries. This definition indicates that resource patches are separated by both longitudinal (i.e., upstream–downstream) and lateral (landward) ecotones that operate over highly variable temporal and spatial scales. This type of ecotone appears to be highly sensitive to landscape changes caused by physical and biotic disturbances. Examples of direct influences of disturbances are removal of riparian vegetation by floods or landslides, and edaphic or vegetative modifications by animals and, in particular, humans. Indirect influences, such as changes in concentration gradients of dissolved chemical compounds and pathways of chemical reactions, may result from changes in edaphic and vegetative properties following direct impacts [2], [35], [160], [161].

2.2.2. Lentic and Wetland Ecotones

Lentic ecotones. (i.e., boundaries between water bodies such as lakes, ponds, or reservoirs and adjacent terrestrial patches) also play an important role in coupling ecosystems. With their steep gradients and typically great heterogeneity, lentic ecotones can contribute considerably to mitigating unfavorable changes in aquatic ecosystems. Furthermore, they regulate the landscape mosaic by controlling energy and nutrient flow between adjacent patches in the ecosystems [156], [162].

Wetlands are defined as lands transitional between terrestrial and aquatic ecosystems where the water table is at or near the surface or the land is seasonally covered by shallow water. Therefore they support, at least periodically, hydrophytes, and their substrate is classified predominantly as undrained hydric soil [163]. A clear-cut ecological distinction separates *tidal* and *inland wetlands*, which may be subdivided further [36]. Wetlands have internal and external boundaries separating distinct vegetation patches [164], [165]. They can be categorized into various types according to the flows that dominate in the tidal and inland environments. In both tidal and inland wetlands, vertical and horizontal transfers occur across a series of surficial and lateral boundaries. Flows across *surficial boundaries* (e.g., graded sediment beds or soil horizons) include transfer from aerobic to anaerobic soils, from soils to surficial vegetation and then to litter and vice versa, and from open water to the atmosphere.

Transfers across *lateral boundaries* include material flows from the upland to the wetland (upland–wetland ecotones) or from the wetland into adjacent open water (wetland–open-water ecotones), from groundwater aquifers into soils, or across vegetation patches with each community dominated by different species (wetland–wetland ecotones).

Figures 21 and 22 illustrate a combined lentic–wetland–upland ecotone situation in the Bornhöved Lake District (Schleswig-Holstein). The figures show that lentic ecotones form unusually differentiated habitats [136]. Various microenvironments are formed within the shore zone of the water body, and similar microenvironments occur in certain sections of water bodies contrasting in size, depth, productivity, and catchment area [166]. Lentic ecotones have rich plant and animal communities characterized by a shifting proportion of terrestrial and aquatic organisms.

Allochthonous and autochthonous material is the source of organic matter in ecotonal shore zones: matter originating from primary production by macrophytes and algae and accumulated material of terrestrial origin supplied by the drainage basin. A fast decomposition rate in several ecotonal habitats makes this material available to higher trophic levels much more quickly than in most land and water habitats [136], [162].

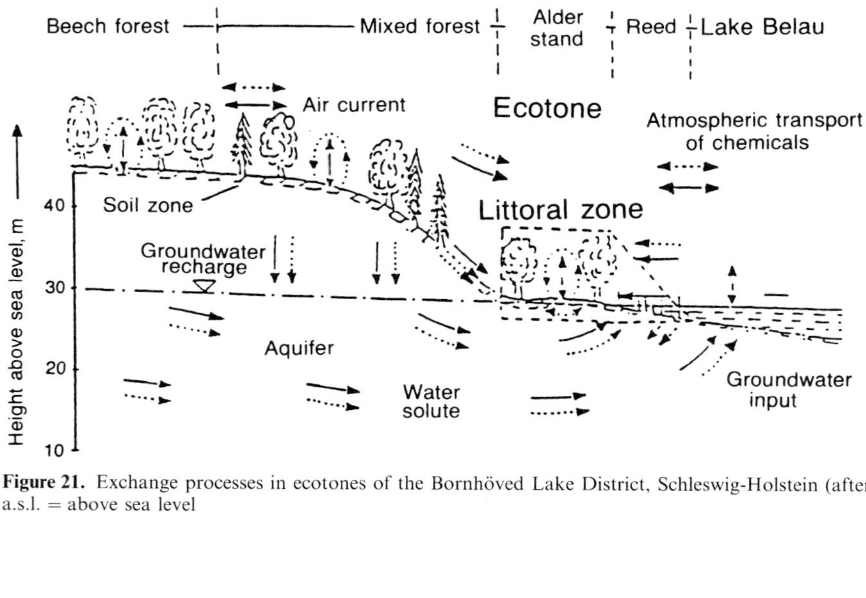

Figure 21. Exchange processes in ecotones of the Bornhöved Lake District, Schleswig-Holstein (after [156], modified)
a.s.l. = above sea level

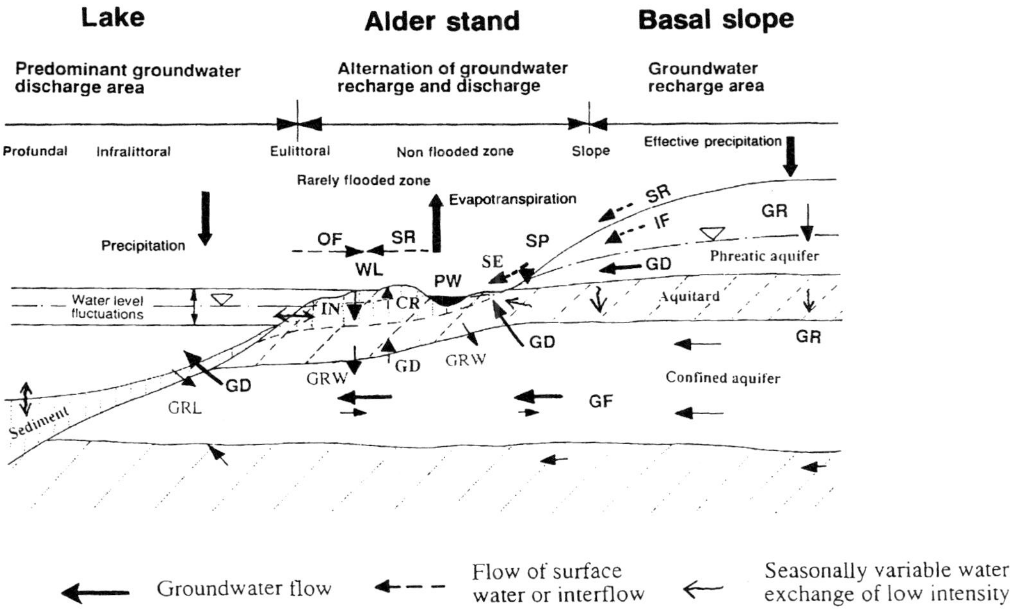

Figure 22. Water balance of ecotones adjacent to Lake Belau, Schleswig-Holstein (after [156], modified)
CR = Capillary rise; GD = Groundwater discharge; GF = Groundwater flow; GR = Groundwater recharge; GRL = Groundwater recharge by lake water; GRW = Groundwater recharge by wetland water; IF = Interflow; IN = Infiltration; OF = Overflow; PW = Ponding water; SE = Seepage on slope; SP = Spring at foot of slope; SR = Surface runoff; WL = Waterlogging

The majority of lentic ecotones in densely populated areas of the world is under visible human impact. Lake eutrophication, air pollution, sewage input, agricultural practices, and tourist activities often damage them. Disappearance of terrestrial and aquatic vegetation is the most common response to such unfavorable conditions.

2.3. Sensitivity of Ecosystems and Ecotones

In ecology, the sensitivity of biota or ecosystems to disturbances may be defined in terms of community stability. Questions of persistence and the probability of extinction are of particular importance in this connection; hence, these measures are also frequently defined as aspects of stability. Stability denotes the ability of a system to return to an equilibrium state after a limited disturbance, while persistence is an indication of resilience in the sense defined by Holling [167]: i.e., the capacity to absorb changes of state and driving variables and parameters. The more rapidly a system returns to equilibrium with the least fluctuation, the more stable it is, which means in terms of thermodynamics that stability is coupled with a (relative) minimum of entropy production [2], [168]. Further differentiations lead to various notions of community stability and nondemographic stability measures.

In the light of these definitions, a sensitive ecosystem is one that responds readily to a particular stress; an insensitive ecosystem may be oblivious to the stress. Sensitivity has also a temporal component: a system that responds more rapidly than another is considered more sensitive. Because of the complex character of system sensitivity to stress, various measurable aspects of sensitivity are recognized, which are summarized below:

Stability
how an (eco)system responds to perturbation and how readily it recovers

Resistance
rate at which (eco)system indicators change in response to a given stress—more resistant systems change more slowly

Recovery
how the system responds following removal of the stress. Recovery has two components: how rapidly and how effectively. (The temporal aspect is characterized as the system's resilience.)

Resilience
the inverse of the length of time required to return to near normal. According to WESTMAN [169], four components of resilience can be distinguished:

Elasticity: time required to restore the system to its initial steady state
Amplitude: zone of deformation from which the system will still return to its initial state
Hysteresis: extent to which the paths of degradation under chronic disturbance, and of recovery when disturbance ceases, are mirror images of each other
Malleability: degree to which the new steady state established following recovery differs from the original steady state

Most of the above definitions explicitly or implicitly assume that the system is in a steady state before a stress is applied. This is not necessarily true because ecosystems are generally continuously changing and developing. Therefore, the system need not return to a steady state after recovery, merely to a "near-normal" state. In some instances the ecosystem can switch with relatively minor disturbances into another steady state. This is called bifurcation: rapid transition at a critical point into a new steady state. Examples of bifurcation of aquatic ecosystems are, for example, the rapid change of lakes from a clean-water state with low amounts of harmless algae to a highly polluted state with highly noxious blue-green algae, or the switching of prairie lakes with healthy fish populations to a state with fish-kills.

2.3.1. Different Notions of Community Stability

A first distinction is between local and global stability. *Local stability* or stability in the vicinity of an equilibrium point describes the tendency of a community to return to its original state when subjected to a minor perturbation. *Global stability* describes this tendency when the community is subjected to a major disturbance. The graphic visualization of solutions to the equations of population dynamics for an m-species community may be represented on some m-dimensional surface, where each point marks a set of populations. Then, equilibrium situations are in principle characterized by those points at which the surface is flat; by equating the configuration with that of a landscape, in customary geographical notation this means on hilltops and in valley bottoms. The hilltop equilibria, however, are obviously labile (i.e., unable to survive the smallest displacement); only the valley bottoms are stable configurations (see Fig. 23 A). In looking beyond the realm of linearized stability in the immediate neighborhood of equilibrium points for which straightforward mathematical tools exist the situation becomes more complicated. Appropriate representation of the *global stability* of a system implies recourse to nonlinear equations of population biology (see Fig. 23 B). Again in geographical terms, such a global analysis aims at comprehending the stability of an entire landscape, not just the immediate vicinity of equilibrium points. This involves appropriate recognition of the fact that real-world environments are uncertain and stochastic, which means that the corresponding environmental parameters in the model equations exhibit random fluctuations [6], [170], [171].

This leads to a yet more general meaning of stability, termed *structural stability*, which refers to the qualitative effects on solutions of the model equations that are due to gradual variations in the model parameters. Thus, a system may be considered structurally stable if these solutions change in a continuous manner. Conversely, a system is structurally unstable if gradual changes in the system parameters (e.g., alterations in site factors of a community) produce qualitatively discontinuous effects [2], [37], [171].

2.3.2. Biodiversity and Stability of Model Communities

An important issue of theoretical ecology is how biodiversity relates to community structure and which community-level properties emerge from the disparate interactions of organisms, populations, and site qualities. Possible examples of such "emergent properties" include trophic and guild structures, stability, resilience, and successional stages. A generalized offshoot of the concept of emergent properties is the notion that biological systems are hierarchically organized, with new properties at each level of the

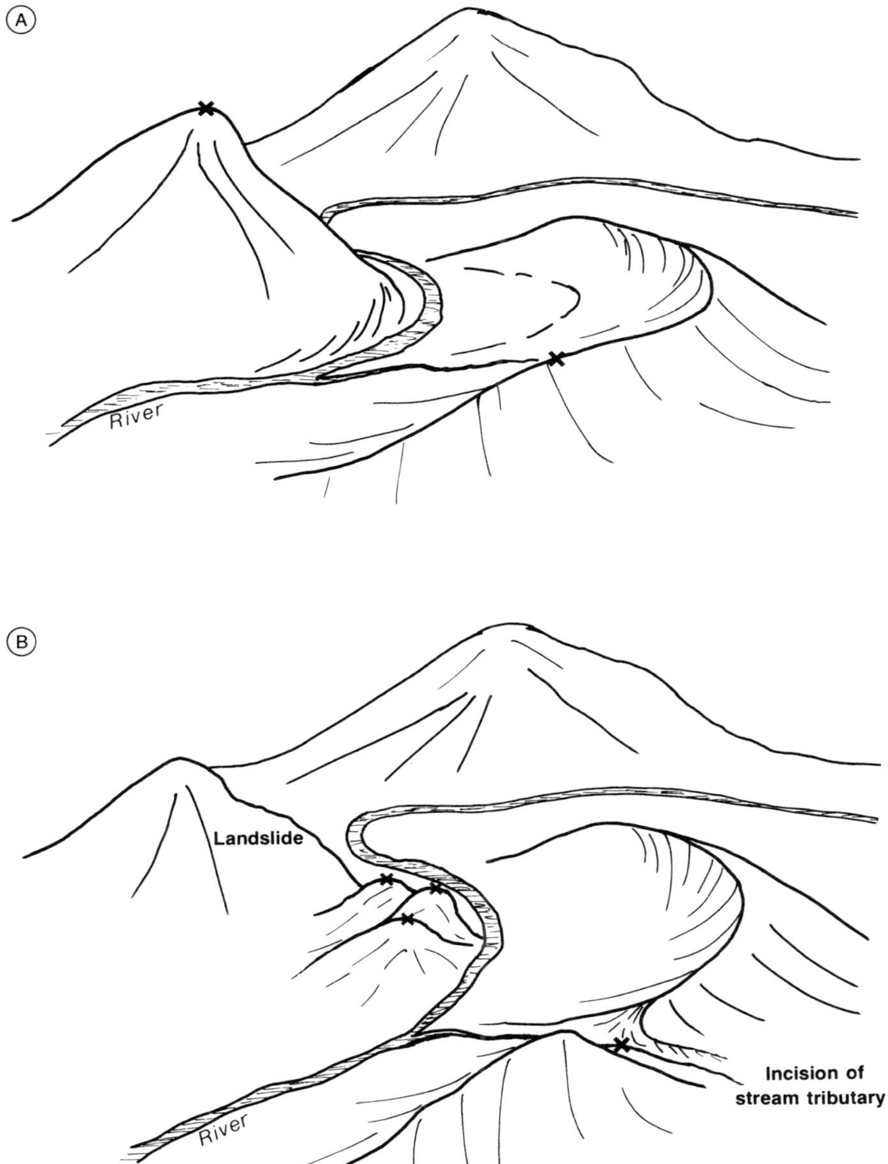

Figure 23. A) Visualization of system dynamics and stability conditions, equating phase–space characteristics with a landscape—labile equilibria (i.e., both low local and global stability); B) Illustration of global stability conditions by means of a landscape configuration, indicating different states of low local, but generally high stability

hierarchy [11], [38], [172]. According to this view, diversity is better understood if ecosystems or biota are decomposed hierarchically so that each process can be viewed as a stabilizing or disruptive factor at each level in a hierarchy of time and space scales.

During the 1950s and 1960s, the "conventional wisdom" in ecology [6] was that increased complexity within or diversity of a community lead to increased stability [39], [173]. MAY [174], however, came to contrasting conclusions by means of model food webs comprising a number of interacting species. The term β_{ij} was used to measure the effect of species j's density on species i's rate of increase. Thus β_{ij} would be zero when there was no effect, whereas both β_{ij} and β_{ji} would be negative for two competing species, and β_{ij} would be positive and β_{ji} negative for a predator (i) and its prey (j). Setting all self-regulatory terms (β_{ii}, β_{jj}) in his randomly connected networks at -1, MAY distributed all other β-values at random, including a certain number of zeros. Thus, the cybernetic webs serving as ecosystem models could be described by three parameters: S, the number of species; C, the connectance of the web, i.e., the fraction of all possible pairs of species interacting directly ($\beta_{ij} \neq 0$); and β, the average interaction strength, i.e., the average of nonzero β-values, regardless of sign.

Comparative analysis of these networks showed that they were likely to be stable (i.e., the populations represented would return to equilibrium after a small disturbance) only if

$$\beta(SC)^{1/2} < 1 \qquad (25)$$

In other words, MAY's model suggests, like others, that "too rich a web connectance or too large an average interaction strength leads to instability. The larger the number of species, the more pronounced the effect." This inference is clearly in contradiction to the above conventional wisdom, but it indicates that no general, unavoidable connection exists between complexity or diversity, respectively, and community stability. Another question is, how much is MAY's result an artifact arising from particular characteristics of the model and the interpretative techniques applied [6], [37].

The picture also alters if, instead of focusing on local stability and correspondingly minor disturbances, larger perturbations are considered or "species-deletion stability," which is particularly interesting from an ecotoxicological point of view. Under the assumption of the deletion of one species of a community owing to a persistent impact, the system is said to be species-deletion stable if all of the remaining species are retained at locally stable equilibria. In the simple six-species community of Figure 24 containing two top and two intermediate predators and two basal species (either plants or categories of dead organic matter), interaction strengths were defined at random, while connectance was varied systematically; thus, complexity could be equated with connectance. Again, stability generally de-

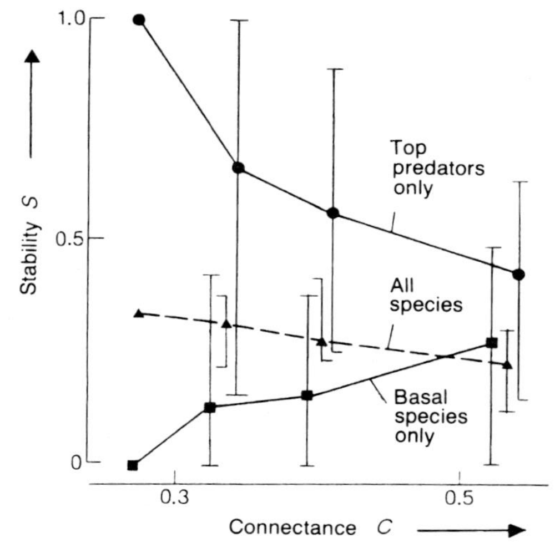

Figure 24. Species-deletion stability plotted against interspecific connectance (after [6])

creased with increasing diversity, but this trend was reversed when basal species or elements were eliminated. However, within the simulated small subset of stable communities, resilience actually increased with complexity.

Other theoretical approaches to the stability problem are due to MCMURTRIE [175], PIMM [10], and ULANOWICZ [40], who established criteria for the stability of ecosystems under certain restrictive conditions, such as zero-sum games. In particular, they have shown that ecosystems can become unstable against weak perturbations when the complexity increases beyond a certain value, although the transition is not always sharp [176].

BACHAS and HUBERMAN [177] related the complexity of hierarchical structures, as measured by their diversity [178], to dynamical behavior. The result suggests that there might be a quantifiable relation between the diversity of a hierarchical structure and its stability. Furthermore, the results of DOREIAN [179] and of BRIAND and COHEN [180] suggest that the effective dimensionality of the space in which the interactions occur, and the overlaps in food webs, contribute to the nature of the structural representation of the community. Hence, it may be concluded that their stability will likewise be affected, which is in agreement with the results of a comparative model analysis of unstructured cooperative interactions between arbitrary species on the one hand and ecosystems with pyramidal organization on the other [181].

In returning to the original problem set out by MAY [174], this author derived a condition of stability for more general systems as a function of species diversity and the strength of interactions. The stability of such systems is determined by the behavior of the largest eigenvalue of matrices governing the response of the system to small perturbations. As a result, MAY and coworkers showed in the case of nonhierarchical organizations how the removal of a zero-sum game condition can lead to further reduction of stability. Examining hierarchical ecologies, they demonstrated that these hierarchical structures are intrinsically more stable than unstructured ones [181].

The conflicting results among models developed so far must be considered in the light of model structure, particularly the number and composition of elements (i.e., species) simulated. In comparison to real communities with hundreds or thousands of species [136] and highly variable degrees of interaction, most models are extremely "impoverished" in species. Furthermore, the models quite often refer to randomly constructed communities, whereas real communities and ecosystems are far from being randomly constructed and normally display a complex hierarchical structure [2], [172]. Unstable communities are liable to collapse when they experience environmental conditions that reveal their instability, but the range and predictability of conditions may vary markedly from place to place. Under stable and predictable site conditions, a community will experience only limited fluctuations; thus, even a dynamically fragile one may still persist. By contrast, in a variable and largely unpredictable environment, only dynamically robust communities are likely to persist.

2.3.3. Stability-Oriented Biodiversity Analyses of Real Communities

Following MAY's approach, studies have been performed on the general importance of dynamic stability constraints by examining the relationship among S, C, and β in real communites. It follows from the inequality of Equation (25) that increases in S will lead to decreased stability unless compensatory decreases occur in C or β. Since data on interaction strengths for entire communities are unavailable, β is usually assumed to be constant. In such a case, communities with more species would retain stability only with a corresponding decline in average connectance, which means that the product SC should remain approximately constant.

REJMANEK and STARY [182] studied geographically distinct plant–aphid–parasitoid communities in Central Europe and found a decrease of C with an increase in S, with SC lying between 2 and 6. Evaluating the literature, BRIAND [183] analyzed 40 terrestrial, freshwater, and marine food webs and calculated a single value for connectance on the basis of both predator–prey and presumed competitive interactions (i.e., C_{max}). Once again, connectance decreased in an approximately hyperbolic way with species number, and SC was about 7.

MCNAUGHTON's field test [184] of MAY's inference, carried out in 17 grassland stands in Tanzania's Serengeti National Park, showed that "both average interaction strength and connectance declined as species richness of the grassland increased. The correlation was somewhat stronger for interaction strength than for connectance." Hence, if the grassland communities analyzed are representative biocenoses, species-poor stands are likely to be characterized by strong interactions among species, while species interacting with many others do so relatively weakly. Diffuse competition obviously increases with the number of species; in addition, MCNAUGHTON's findings suggest that communities have a blocklike or patchy organization consisting of species interacting among themselves but only little with species in other blocks.

MCNAUGHTON [184] also tested the prediction that complex communities are less likely to return to their state prior to perturbation. In the first experiment, the perturbation resulted from addition of plant nutrients; in the second, it involved the action of grazing buffalo. The results summarized in Table 7 indicate that each perturbation significantly reduced the diversity of the species-rich but not the species-poor community.

According to these analyses and studies on the species diversity of tropical rain forest ecosystems in particular, a climax community is more likely to have both low fluctuations in composition (i.e., high stability) and low resilience as the homogeneity of its environment in space and time increases. Figure 25 summarizes the diversity and abundance spectra of numerous neo- and paleotropical rain forest communities as related to soil properties [2]. The diagram clearly shows that in most cases, rain forest stands on highly

Table 7. Influence of nutrient addition on species richness, equitability ($H/\ln S$), and diversity in two fields; and influence of grazing by African buffalo on species diversity of two vegetation plots (after [184])

	Control plots	Experimental plots	Statistical significance
Nutrient addition			
Species richness per 0.5 m^2 plot			
Species-poor plot	20.8	22.5	n.s.*
Species-rich plot	31.0	30.8	n.s.*
Equitability			
Species-poor plot	0.660	0.615	n.s.*
Species-rich plot	0.793	0.740	$p < .05$
Diversity			
Species-poor plot	2.001	1.915	n.s.*
Species-rich plot	2.722	2.532	$p < .05$
Grazing			
Species diversity			
Species-poor plot	1.069	1.357	n.s.*
Species-rich plot	1.783	1.302	$p < .005$

* n.s. = not significant.

nutrient-depleted ferralsols, acrisols, and podzols of tropical lowlands attain diversity indices greater than 90 % of the theoretical maximum. However, a further differentiation according to local nutrient status and water budget is not possible with the data available. Nevertheless, stand 38 (Borneo) is of particular interest since it is representative of a forest that developed on a tropical bog; its exceptionally low index value allows the conjecture that its nutrient supply has fallen below a critical value, which is discussed in a broader thermodynamic context in Section 2.3.7.

In terms of MAY's stability criterion (Eq. 25), these findings would indicate a very low degree of connectance and average interaction strength between plants of the same species. With regard to nutrient uptake from the soil, this means a well-balanced low-grade intra- and interspecific competition involving a marked selection of K strategies (i.e., care for the offspring in the widest sense) and the formation of numerous ecological niches. Therefore, highly complex rain forest communities with their relatively constant environments appear to be more susceptible to outside, unnatural disturbances than the simpler, more robust communities of ectropical regions.

The example of Vietnam's inland forests is illuminating in this respect, since they still bear the scars of a ten-year herbicide spraying program by the U.S. Air Force during the Vietnam war. According to official U.S. figures, some 10.3 % of these forests, in addition to 36.1 % of mangrove forests, 3 % of cultivated land, and 5 % of other land, were sprayed. Thus, some 90 000 t of herbicides and antiplant chemicals was used between 1962 and 1971 [185]. Most of the spraying was with Agent Orange [i.e., a 1:1 mixture of (2,4-dichlorophenoxy)acetic acid (2,4-D) and (2,4,5-trichlorophenoxy)acetic acid (2,4,5-T)]; Agent White (2,4-D and picloram); and Agent Blue (cacodylic acid). Some 20 years after the spraying ended, little regrowth had occurred in areas sprayed three or four times. The fertility of the soils is rapidly reduced further if forests are replaced with grassland or bamboo, as has happened in a large part of the sprayed region. In particular, a loss of minerals and nitrogen has occurred, and in some cases a decrease in soil pH, which all inhibit diversified recolonization.

Defoliation has clearly affected the faunal assemblages, not only in the areas totally defoliated and subsequently converted into grassland but also in those regions less frequently sprayed where the plants in the upper reaches of the forest were damaged. In both areas, the number of animals and birds has dropped dramatically, and animals are also at considerable risk in some of the forest areas that have been isolated from the main body by herbicide application. Furthermore, the concern is that Agent Orange contained a high concentration of the highly toxic contaminant 2,3,7,8-tetrachlorodibenzo-*p*-dioxin, which is both teratogenic and carcinogenic in animals, and toxic to humans as well.

In comparison to tropical lowland rain forests where species diversity and nutrient content of the soil are negatively correlated, the situation in temperate regions is different. The composition of 132 plant communities from all parts of the continental United States (except Alaska) evaluated in relation to climate and soils shows that the amount of water available for evapotranspiration is of decisive importance for species diversity. This is further corroborated by the highly significant correlation between diversity and the temporal distribution of rainfall. Conversely, the analysis of 194 plant communities in northern Germany led to the conclusion that, most frequently, high diversity coupled with medium to low proportional abundances occurs on sites with high nutrient supply and medium to low soil moisture, whereas the combination of both high diversity and high abundance with high nutrient and medium to low moisture status

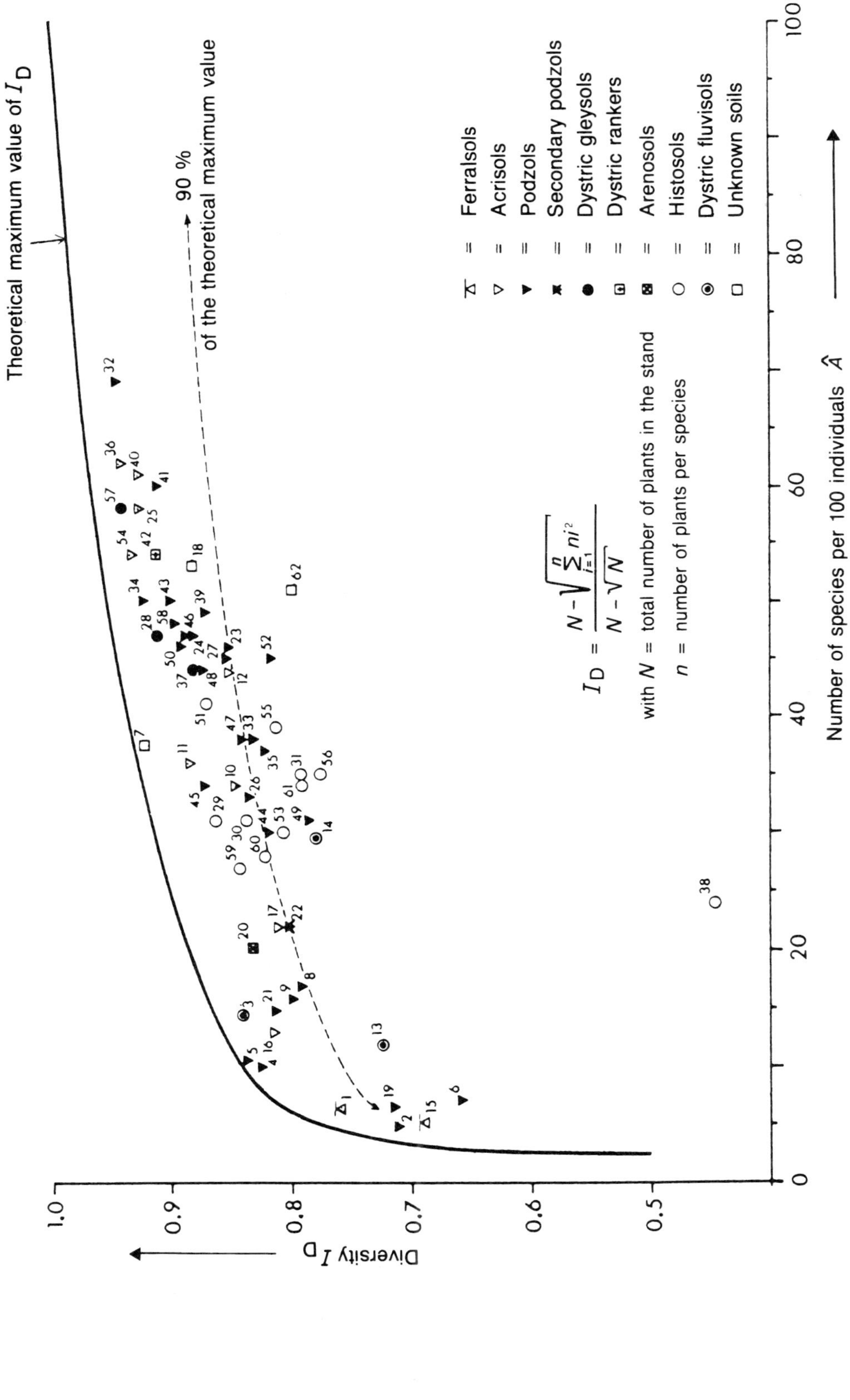

Figure 25. Diversity and abundance spectra of neo- and paleotropical rain forest communities

is comparatively rare. These inverse diversity–nutrient relationships in tropical and ectropical environments, on the one hand, and the fact that soil moisture controls diversity in the United States but not in northern Germany, on the other, lead to the final question of whether a unifying interpretation of these apparently incoherent results is possible. This is given in Section 2.3.7 by considering ecological communities as dissipative structures.

2.3.4. Nondemographic Measures of Stability

The introductory remarks on stability and resilience indicate that community and ecosystem stability can be viewed from perspectives other than demographic ones. The highly different aspects of functioning (e.g., primary productivity, nutrient cycles, energy and water balance) and the composite aspects of structure such as standing-crop biomass appear to be particularly important [6].

McNaughtons's [184] studies on Serengeti grassland communities showed that species-rich associations responded to perturbations with a significant drop in species diversity, while species-poor communities did not. In terms of *primary productivity*, however, grazing as a natural perturbation reduced the standing-crop biomass in the species-poor grassland much more than that of species-rich communities. Similarly, several studies have revealed a marked influence of *food web* structure on community resilience in response to perturbations of energy and nutrient supplies. O'Neill [186] considered a three-compartment system consisting of active (photoautotrophic) plant tissue, heterotrophs, and dead organic matter, and found that the rate of change in standing crop depends on the energy transfer among these compartments. With regard to the autotrophic compartments, it depends on the input variable (net primary productivity) and two output variables (fraction consumed by heterotrophs and fractions lost to soil by litterfall) in addition to a great variety of translocation processes [85]. The rate of change of the heterotrophic compartment depends on two inputs (consumption of living plant biomass and dead organic matter) and two outputs (respiratory heat loss and defecation). Finally, as Figure 26 shows, the energetic exchange rate of soil organic matter depends on two inputs (litterfall and defecation) and two outputs (physical transport into neighboring ecosystems and consumption by heterotrophs).

On the basis of this simple configuration, and inserting real data from communities representing a pond, a spring and tundra, tropical forest, a temperate deciduous forest, and a salt marsh, O'Neill [186] subjected the models of these communities to a standard perturbation that consisted of a 10% decrease in the initial standing-crop vegetation. The rates of recovery toward equilibrium thus monitored and plotted as a function of the *energy input* per unit standing crop of liv-

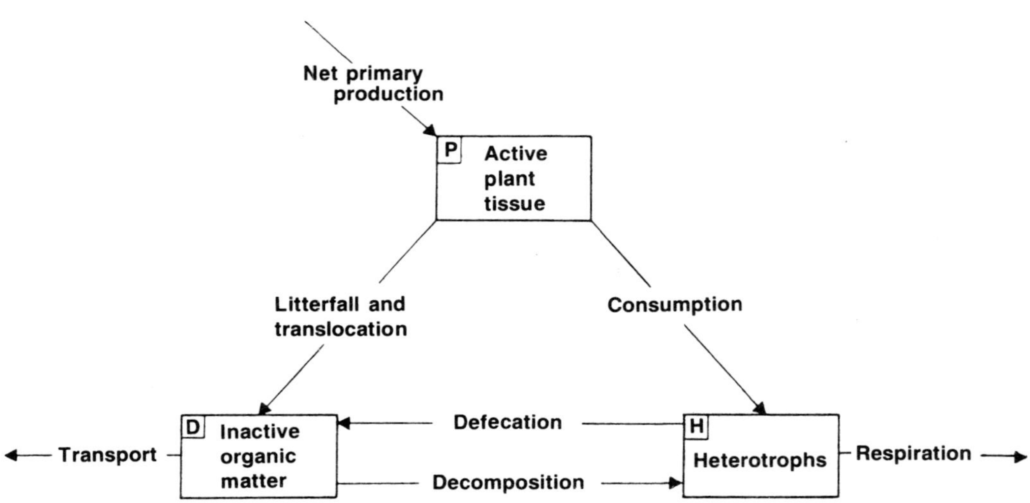

Figure 26. Three-compartment model of an ecosystem with typical energy transfers (after [186], modified)

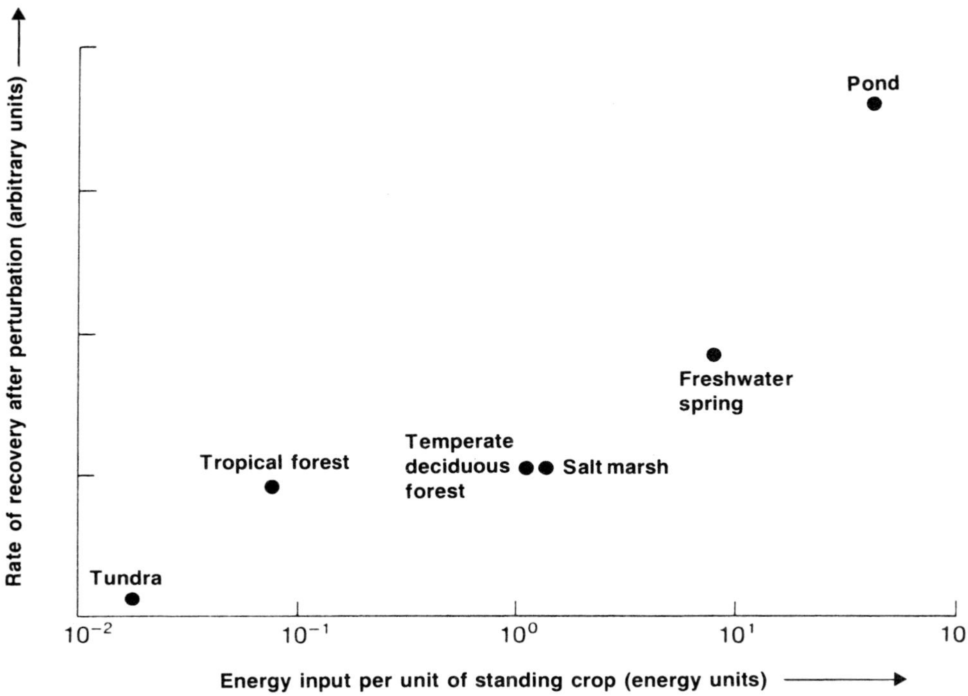

Figure 27. Recovery rate as an index of resilience after perturbation as a function of energy input per unit standing crop (after [186])

ing tissue are presented in Figure 27. It shows that the flux of energy through the different subsystems has an important influence on community resilience; the higher the flux is per unit standing crop, the more quickly will the effects of perturbation be compensated. This, in turn, seems to depend in part on the relative importance of heterotrophs in the system considered. The pond as the most resilient system has a heterotrophic biomass 5.4 times that of autotrophs, reflecting the high turnover rate of phytoplankton; the least resilient tundra has a heterotroph:autotroph ratio of only 0.004. De Angelis [187] has come to analogous conclusions in terms of residence times of various nutrients in different compartments of woodland communities. Thus, a unit of nitrogen exists in soil for an average of 109 years, is taken into forest biomass for 88 years, remains in the litter for up to 5 years, and passes through the detrivore compartment in a few days. In contrast to nitrogen, calcium is a less tightly cycled element whose residence times in soil and biomass amount to only 32 and 8 years, respectively.

More detailed long-term investigations in the forest and agroecosystems of the Bornhöved study area (Schleswig-Holstein, Germany) corroborate these findings. Here, the nitrogen deposition amounted to 19.9 and 18.2 kg $ha^{-1}a^{-1}$ in 1989 and 1990, respectively, with an ammonia fraction of 47 % (60 %) and 23 % (7 %) of organic nitrogen-compounds, while nitrate deposition yielded 6.1 and 5.9 kg nitrogen per hectare and year, respectively. The nitrogen balance of a field is summarized in Table 8.

The wet deposition of nitrates via canopy flow in a neighboring beech stand (*Asperulo–Fagetum*) of relatively uniform age structure (95 years) amounted to 30.3 kg nitrogen per hectare and year in 1989 and 37.5 kg in 1990, again with a major ammonia fraction. Comparably high were the deposition rates to the soils of a neighboring alder stand (*Carici elongatae–Alnetum*), where additional nitrogen fixation is due to the symbiotic activity of *Frankia*.

With regard to internal nitrogen fluxes the beech and alder stands differ considerably. In the first case, litterfall transported 78 kg nitrogen

Table 8. Nitrogen balance of a field in the Bornhöved Lake District

Process	Amount of nitrogen (measured, $kg\,ha^{-1}\,a^{-1}$)	
	1989	1990
Initial N content	51 [a]	38
Total N input	384	222
NO_3-N	30	14
NH_4-N	118	64
N_{org}	236	144.5
Mineralization	221 [b]	215 [b]
Nitrification		
Plant uptake	206	121
N × leakage	110	129
Other N losses [c]	7	7
Residual N_{min} content	39	4.5
Balance	+59	+70

[a] Deduced value. [b] Net mineralization. [c] Denitrification and volatilization of ammonia.

per hectare and year to the soil surface, where the chemical and microbial transformation processes induce seasonal changes in N_{min} content as illustrated by Figure 28. In the light of the site characteristics of this stand (Dystric Cambisol), the figures are indicative of rather limited nitrogen export from the community, which is further corroborated by measurements of the nitrogen fluxes in the related seepage water. In the distinctly more resilient alder stand, the nitrogen input due to litterfall amounted to 109.4 $kg\,ha^{-1}\,a^{-1}$ in 1989 and 104.5 $kg\,ha^{-1}\,a^{-1}$ in 1990, respectively. In comparison to the above beech stand, microbial degradation of the litter is very high in the *Alnetum*, and within 18 months almost all of the leaf tissue is degraded.

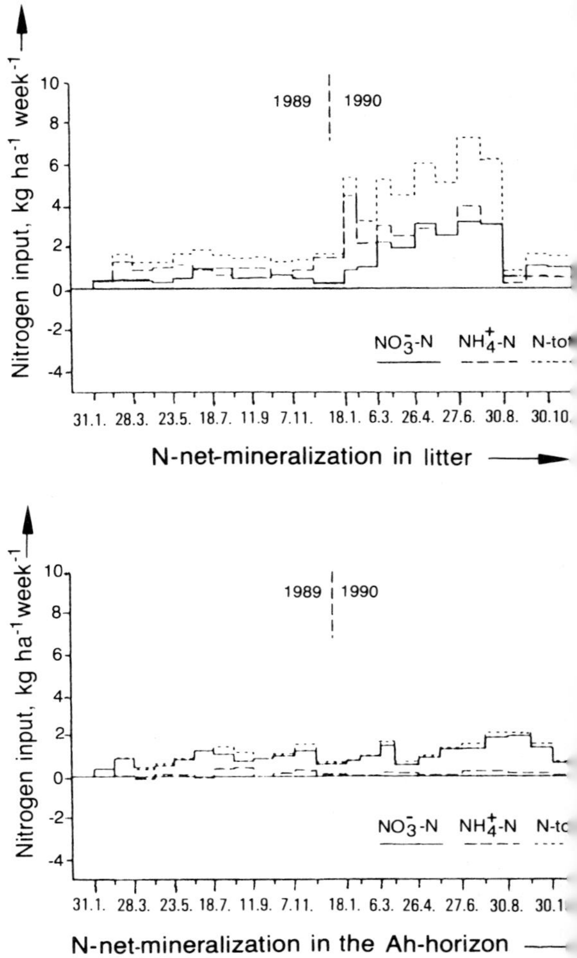

Figure 28. Seasonal variation of the N_{min} content of beech litter of a Schleswig-Holstein *Asperulo–Fagetum* stand in 1989 and 1990
Ah-horizon = part of the surface soil (→ Soil, Chap. 1)

2.3.5. Sensitivity of Aquatic Ecosystems

Aquatic ecosystems are natural collectors of materials from their surroundings, and water washes matter out from the land surface. Human activities increase this natural washout enormously by destroying the natural vegetation cover of the landscape due to agricultural, mining, construction, and forest-cutting activities. For organic matter of natural origin, with its contents of nutrients necessary for further growth of vegetation, a relatively closed cycle originally existed: the remains of organic matter consumed by humans and their domestic animals were returned to fields. In medieval times, increasing population concentrations in towns created local accumulations of decaying organic matter leading to the spread of infectious diseases. This problem consequently was solved by disruption of the feedback cycle. Organics with their content of elements essential for life were directed to streams, rivers, lakes, and ultimately the sea. The result was continuous deprivation of nutrients in soil and a need to subsidize them to maintain and enhance agricultural productivity. The open flow of nutrients from soil and deeper earth layers started to create unpleasant conditions by increasing aquatic production not only in streams, lakes, and ponds but also in the greatest lakes of the world and marginal seas such as the Adriatic. In addition, humans pollute water by applying various chemicals to the environment—both at home and in industry—and by extracting materials from the earth and transporting them over large distances, in part by means of waterways. Tanker catastrophes are becoming more and more common, and road transport is no less hazardous to waters.

Pollution sources are usually classified in two categories: point and nonpoint sources. Although *point sources* coming from a factory or town are technically much easier to handle, *nonpoint* sources are not only difficult to detect but still more difficult to eradicate. Typical nonpoint pollution sources originate from fertilizers and other chemicals spread on fields. Another source of water pollution is deposition of airborne pollutants, some of which reach water bodies in remote regions. The findings of elevated concentrations of DDT in Antarctica show the global scale of pollution problems.

The sensitivity of aquatic ecosystems to disturbances differs depending on the type and size of the respective ecosystem and the nature of the disturbance. In scientific terminology, all disturbing factors are stressors and the term "sensitivity to stress" is used. Stress is usually defined as a measurable alteration of the state of organisms as induced by an environmental change that renders the individual, population, or community more vulnerable to further environmental change.

2.3.5.1. Reactions of Aquatic Ecosystems to Stress

Aquatic ecosystems under stress undergo both changes of structure and changes of function. *Changes of structure* are manifest by changes in the composition of populations of different species in the ecosystem. *Changes of function* are represented by differences in the organic matter productivity of the respective ecosystem and in the rates of release and utilization of different gases and minerals. Thus, both a sensitivity of ecosystem structure and a sensitivity of ecosystem function occur. No uniformity of opinion exists as to whether the structure or function of ecosystems in general is more sensitive to various stressors. Several researchers feel that functional variables, especially those that are substrate limited, will always be less sensitive than structural measures because of functional redundance in the community. They suggest that any loss of functional capacity by one organism is immediately compensated by increased activity of another organism. Other scientists feel that functional capacity can be affected before compensatory mechanisms operate. This is especially likely when the compensatory mechanisms themselves are adversely affected by the stress or when they operate on a more lengthy time scale relative to the functional measure [188]. These authors have also summarized various functional reactions of ecosystems:

1) Community respiration increases
2) Productivity–respiration ratio becomes unbalanced
3) Productivity–biomass ratio increases as energy is diverted from growth and reproduction into acclimation and compensation
4) Importance of auxiliary energy increases (import becomes necessary)
5) Export of primary productivity increases
6) Nutrient turnover and nutrient loss increase
7) One-way transport increases and internal cycling decreases
8) Life span decreases; turnover of organisms increases
9) Trophic dynamics shift, food chains shorten, and functional diversity declines
10) Efficiency of resource use decreases
11) Condition declines

A well-documented example of greater sensitivity of functional variables of a terrestrial ecosystem is presented by VAN VORIS et al. [189], who measured structural and functional changes in small patches of a meadow subjected experimentally to cadmium stress. Functional changes such as spectral complexity and calcium export were affected at concentrations that did not affect structural measures such as the presence or absence of certain species.

In the direction of slowly and rapidly reacting ecosystems at opposite ends of the scale are pelagic lake and reservoir communities composed of very short-living (few hours to few days) species that react to any stress by their rapid appearance or disappearance, and forests with very long-living tree species that are incapable of rapid presence–absence types of reaction. Benthic communities of rivers and fish populations are intermediate: their life span is measured in months to years.

In aquatic ecosystems, many reports have shown greater relative sensitivity of structural than functional variables. For example, during his extensive comparative studies of Canadian Shield lakes, SCHINDLER [190] found no significant changes in decomposition or nutrient cycling in whole lakes treated with acid, but species composition of phytoplankton was among the earliest indicators of change. CROSSEY et al. [191] found in impaired rivers that measures of production and respiration were more variable than the macroinvertebrate composition. CRUMBY et al. [192] studied the biological reaction of Roaring River in north-central Tennessee to stress caused by construction around the river and by inadequate agricultural practices in the watershed. Species composition changes were reflected in a general decline in numbers of intolerant species and simultaneous increase of tolerant species.

However, many reports of the greater sensitivity of functional variables of aquatic ecosystems also exist. RODGERS et al. [193] found that process rate changes were more sensitive than biomass or chlorophyll concentration in detecting the effects of diverse chemicals on periphyton in artificial streams. When dealing with enrichment, functional measures are often a good warning indicator. UHLMANN et al. [194] and GNAUCK [195] summarized structural and functional changes in aquatic ecosystems and gave examples of experiments equivalent to those of VAN VORIS et al. [189]. Figure 29 compares the degree of variability (expressed by a relative index of instability) during a 50-d experiment with an experimental purification pond for several structural and a few functional variables. The greatest instability equivalent to the highest sensitivity is observed (with the exception of the alga *Scenedesmus obliquus*) for the biomass of individual species of organisms (index values between 0.85 and 1.45), while functional variables such as oxygen concentration, primary production, and turbidity and the summarizing biomass of functional groups of organisms such as zooplankton or phytoplankton or chlorophyll-a levels have an index close to or below 0.5. The least sensitive appeared to be the organic carbon elimination capacity of the pond, which characterizes, from the human point of view, its most important function.

Once the stress has ceased, two possibilities exist:

1) Reversible changes are induced
2) The changes prove irreversible

The deforestation in the coastal hills of Venezuela is a drastic example of irreversible changes of an extensive ecosystem. The soil structure, seed sources, and local physical environment have changed to such a degree that forests have not returned even after the areas were abandoned by humans. Perhaps the example merely reflects an exceedingly long recovery period, and the system will eventually recover. Yet for practical purposes, these ecosystem changes are permanent. A dried lake and a watershed with homogeneously distributed amounts of toxic substances are examples of irreversible changes of aquatic ecosystems.

If the disturbance is sufficiently regular and of near-natural origin, components of the ecosystem may adapt and eventually require disturbance to maintain a normal, resilient system (tidal pools require daily exchange with the sea, pastures require mowing or grazing, many forests are adapted to periodic fires). In aquatic ecosystems, this system of reaction is seen in periodic pools, which are inhabited by animals adapted to desiccation. Their characteristic fauna will disappear when water is permanently present because they are inferior in competition with longer-lived species not needing desiccation.

For chemical stress, the recovery of aquatic ecosystems depends mainly on the degree of accumulation of the respective chemical in the environment and the rate of flow. This conclusion can be corroborated by comparing the findings in lakes with those in rivers. The recovery of eutrophic or highly polluted lakes after the sources of organic pollution and phosphorus causing lake eutrophication have been reduced is very slow. Recovery takes as much as ten to twelve years. The delayed reaction is due to the enormous quantities of decomposable organic matter and phosphorus stored in the bottom mud. Oxygen at the bottom is consumed during the decomposition of organic matter and, under such conditions, phosphorus is continuously released. This phenomenon is called internal P load, which indicates that when the external load ceases, an internal one plays a major role in continuing eutrophication. At the other end of the spectrum are rivers with high flushing. YOUNT et al. [196] made an overview of the recovery of rivers after a stress and found that almost full recovery may occur within about two years.

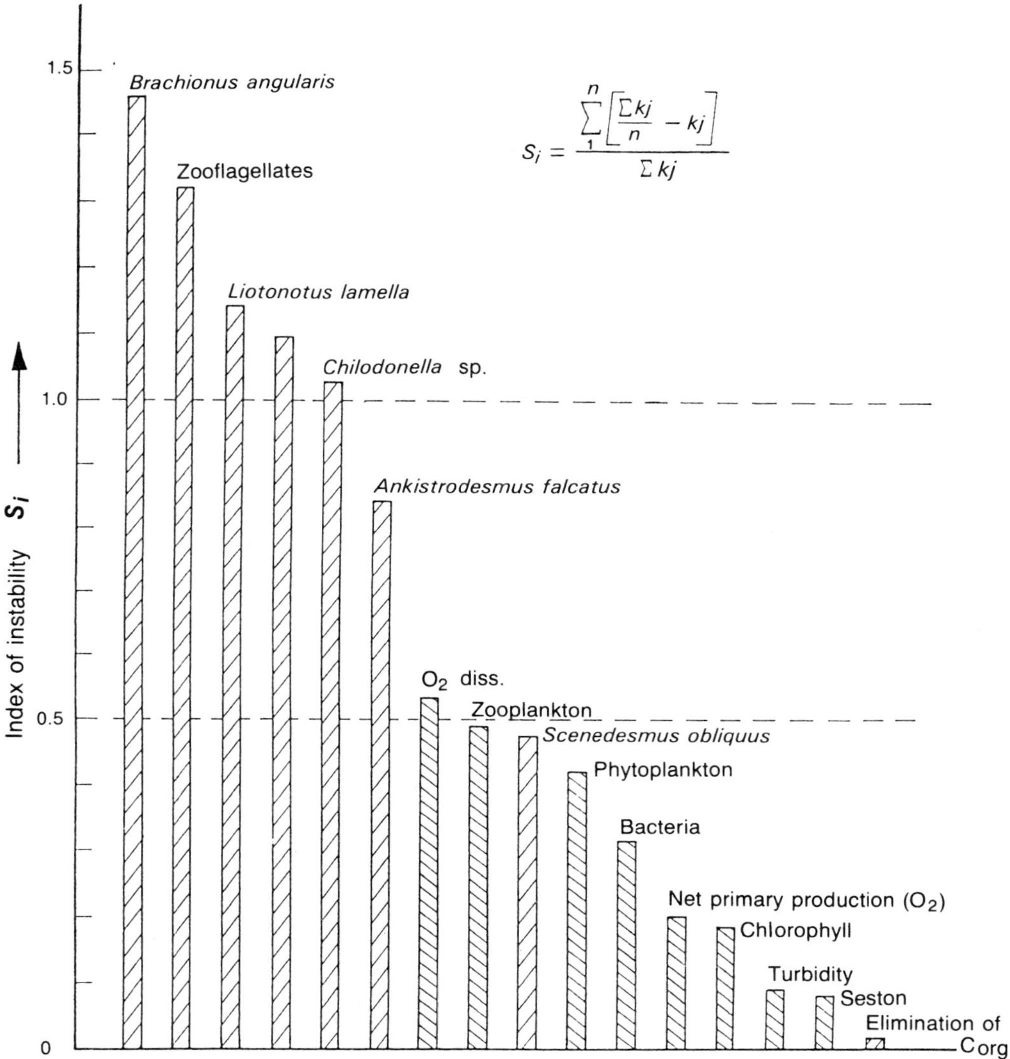

Figure 29. Comparison of the degree of variability, expressed in terms of an instability index, during a 50-d experiment with an artificial purification pond (after [194], modified)

2.3.5.2. Sensitivity to Different Stressors

The sensitivity of marine and freshwater fish worldwide to recent changes in the environment was determined by means of changed species diversity [41]. The numbers of extinct, endangered, or threatened species and species of special concern represent 21 – 69 % of the total fauna in different regions of the world, with the exception of only 9 % in Latin America. The highest numbers were found in California and South Africa; low figures in Iran, Australia, Sri Lanka, and also Arkansas. Major causes of this situation are the extraction of water for various human uses, physical alteration of fish habitats (mainly by channelization, construction of weirs and reservoirs, siltation from mass erosion and degradation of wetlands), and pollution (mainly municipal and industrial point-source pollution, agricultural pollution, and acid rain). The sensitivity of aquatic ecosystems to different specific stressors depends on the type of ecosystem and the agent. For chemical agents the degree of interference with the structure and metabolism of organisms is important. However, even such components as structural macroelements of organisms may be harmful, once their concentrations in the environment exceed certain levels. This is the case, for example, with plant nutrients. When present in large excess, they decrease life functions (e.g., plant growth; see Section 2.1.4.3). Sensitivity to toxic chemicals is discussed in Section 2.4. Sensitivity to acidification as an environmental process is described in Section 2.3.5.3 together with other signs of global change.

The sensitivity of aquatic ecosystems to *ionizing radiation* is not well known. However, some reactions parallel to those observed for terrestrial ecosystems are expected to be valid. The latter were studied in detail by WOODWELL [197]. He demonstrated that in a mixed forest–shrub–grassland exposed to different levels of ionizing radiation, the diversity of species is more sensitive than organic production. The sensitivity increases with the size of the organisms. The reason is that increased size puts greater demand on the photosynthetic mechanism for maintenance, leaving less for growth and repair. Sorting of the sensitivity was also observed on the basis of chromosome size: plants with large chromosomes were more sensitive.

Sensitivity to *communal organic pollution* is well known for both stream and lake ecosystems; this is the classical type of pollution studied by quantitative methods since the beginning of this century, the methods being continuously improved and new methods being continuously developed. From the pollution point of view, organic matter is not characterized as a chemical species but in a summarizing way, namely, the total of easily decomposable or refractory matter. The difference between the two categories is arbitrary: easily degradable matter is decomposed by the natural population of microorganisms within five days' incubation. In streams, systematic changes of the composition of benthos are observed with certain amounts of organic pollution. This is the basis for the bioindicator approach to detect the degree of organic pollution. However, geographic differences in the benthic fauna of different regions; dependence on stream characteristics such as size, rate of flow, and bottom type; and combination of pollution types all complicate the evaluation.

2.3.5.3. Sensitivity to "Global Change"

One of the effects anticipated by part of the scientific community to cause major changes in world climate and consequently local weather and food production within the next decades is "global change." Since the 1940s, compounds such as CO_2, chlorofluorocarbons (CCl_2F_2 and CCl_3F), methane (CH_4), and dinitrogen monoxide (N_2O) have displayed notable and deleterious increases or decreases in the case of tropospheric ozone (O_3). These gases (except O_3) are classified as "greenhouse gases" because they have a greenhouse effect on earth's atmosphere.

About a hundred years ago, CO_2 values of 280 ppm (mL/m^3) were measured in the atmosphere; now (in the 1990s), concentrations are about 25% higher. Carbon is both fixed and released by oceans, biota, and human activities. Approximately 30% of the global annual carbon fixation by plants occurs in the seas, which also form an integral part of the nitrogen flux of the earth. An estimate for 1980 gave the following figures for the fluxes of carbon among atmosphere, ocean, biosphere, and human activities [198]: 78×10^{12} kg C per year (flux ocean↔atmosphere), 65×10^{12} kg C (flux biosphere↔atmosphere), and 5×10^{12} kg C (flux due to human activities, mainly burning of fossil fuels and land clearance↔atmosphere). Although the contribution by human activities is far less than the natural fluxes, it shifts the total balance significantly.

In the United States, the sources of CO_2 from human activities were estimated for 1987 as follows [42]: 11% residential, 24% industry, 30% transportation, and 35% electric and power generation. Industry is therefore an important CO_2 source, even when the significant portions of electric power generation and transport associated with industry are neglected.

The climate has changed throughout the last few thousand years due to natural causes; the influence of human activities superimposed on these variations has also contributed significantly. Scientific documentation demonstrates that the overall rise in CO_2 has already had marked effects on the biosphere. Comparison of different recent plants with herbarium specimens collected many years ago disclosed that the leaf area and the nitrogen content of leaves in 15 species of trees, shrubs, and herbs decreased over the last 240 years. Also, a systematic increase in the growth rate of bristlecone pines and a decrease in stomatal density of the leaves of plants over the last 200 years was observed [199]. Although no aquatic plants were included in the studies, they are expected to be subject to similar effects. The reported changes may seem unimportant; however, this may be only the easily observable tip of the iceberg. More relevant parameters cannot be compared with those of earlier times when no scientific measurements existed. The detection of changes increases the credibility of predictions by demonstrating that biological changes have already taken place and are not merely scientists' anticipation.

Despite many uncertainties in predictions, a certain consensus exists among part of the scientific community that the increase in CO_2 and the other greenhouse gases due to human activities will warm the climate by the middle of the next century between 1 and 5 °C ("global warming"). Greater temperature increases are expected at high latitudes, especially in winter, with consequences for ice melting and sea-level rise. Consensus also exists that changes of this magnitude in temperature will be accompanied by substantial regional anomalies in temperature, rainfall, evaporation, cloudiness, windiness, drought, and so forth that could have substantial impact on agriculture, water supply, human and animal health, biological diversity, energy demand, and sea level [198], [200].

The changes expected in aquatic environments due to these increasing quantities of greenhouse gases are of both a physical and a chemical nature. Physical changes mainly involve global warming and an increased level of ultraviolet radiation at the earth's surface. Ultraviolet radiation in its present spectral composition is readily adsorbed in water, and only organisms living in the uppermost (few centimeters) water layers can be damaged. However, alterations in spectral composition of ultraviolet radiation expected as a consequence of global changes can markedly increase the layers affected.

One indirect effect of global warming on streams is expected in connection with increased variability in stream flow, associated with increased rates of sediment transport and sedimentation. Sediments are expected to fill in some pools and generally cover surfaces. However, streams may be flushed of material periodically. These conditions would favor invertebrates with shorter life cycles adapted to softer sediments, while larger invertebrates could be adversely affected by these conditions because they graze rock surfaces [43].

Indirect effects on aquatic ecosystems are also expected due to the high sensitivity of forest ecosystems to environmental variables. The sensitivity of forests to climate changes was estimated from changes observed during the geological past and from some geographical dependencies observed recently [201]. The results indicate that during global change, shifts will occur in the geographical ranges of tree species, along with changes in forest productivity. Because of the interrelation of soil nutrients, soil moisture, groundwater, and tree growth, changes in forest ecosystems will have major consequences for aquatic ecosystems, primarily rivers. River flow will be changed and, with it, leaching of minerals and organics from soil will occur to varying extents, with consequences for river inhabitants [44]. Additional indirect effects will be caused by changes in agricultural production.

The probable effect of the expected climate warming on fish of the Great Lakes Basin was estimated by Meisner et al. [202]. They concluded that no full extirpation of species is to be expected, but the populations of the most valued fish, salmonids and coregonids, will decrease, whereas the species less valued by fishermen (cyprinids, esocids, centarchids, and ictalurids) will probably increase. Thus, a decrease of fishery yields of preferred species is expected.

Other global changes in aquatic ecosystems are already observed locally in acidified regions. They result from direct and indirect chemical effects caused by increasing CO_2 levels. Doubling of atmospheric CO_2 levels, which is expected to occur at the beginning of the next century, is supposed to decrease the pH of rainwater by about 0.4 unit, which means a pH drop in some localities to 4.0. The lower pH will increase the solubility of carbonate minerals and increase the calcium, magnesium, bicarbonate, and sulfate content of rivers draining carbonate terrain. The greatest secondary effects on organisms are caused by the increased solubility of aluminum at low pH. Aluminum would be released, predominantly in bedrock terrain. The speciation of aluminum in river water is important in determining the toxicity of elevated aluminum levels as well as the acidity of soil solutions. Inorganically complexed or free aluminum is more toxic than organically complexed aluminum. Lakes with present pH levels of 6.5 are expected to become acidified to pH <5, with considerable effects on inhabiting organisms. Major impoverishment of lakes due to acidification has already occurred in northern Europe and Canada, and in mountainous lakes of more southern localities [45].

2.3.6. Sensitivity of Ecotones

Like communities and ecosystems, ecotones also differ in physical stability or resistance to changes (resilience) caused by disturbances. In contrast to lotic ecotones, adjacent terrestrial ecosystems and ecotones may be viewed as occupying distinctly more stable positions in a landscape where natural disturbances due to geomorphic processes are less frequent than the time for patch adjustment to them (see Sections 2.3.5 and 2.3.7).

In distinguishing reactions, recovery and persistence phases, and recurrence intervals as characteristic parameters of system response to physical disturbance [46], the recovery and disturbance recurrence intervals are of particular ecological importance. The ratio of the recovery to disturbance recurrence intervals R/D suggests differences in recovery or sensitivity characteristics of both ecotones and ecosystems. For unstable systems, the R/D ratios exceed 1. Typically, this is the case for predominantly transient systems such as many *lotic ecotones* [161]. Normally, deposition and erosion interact with riparian vegetation at varying intensities and frequencies so that communities are kept in the early stages of succession. A decrease in seasonal flooding of the riparian vegetation results in changes via several possible successional pathways, which may lead to the formation of purely terrestrial communities when the system is no longer influenced by flooding.

Examples of this type support the assertion of NAIMAN et al. [154a] that a paradox of modeling interfaces is predicting when and where ecotone parameters become unpredictable relative to adjacent homogeneous sites. Chaos theory may be useful in dealing with this uncertainty because ecotones are not simple averages of adjacent systems; the spatiotemporal pattern of environmental factors and the interactions between patches are as important in determining ecotone dynamics on different scales as are the characteristics of adjacent patches.

2.3.7. Stability and Resilience of Ecosystems in Evolution

On a very general level, the apparently unusual—because inverse—relationship between the nutrient status of soil and the diversity of related tropical rain forest communities described in Section 2.3.3 may be interpreted in terms of biological thermodynamics [168]. Biotic communities are open systems in exchange of energy and matter with their environment; their entropy production $\mathrm{d}S$ thus comprises the terms $\mathrm{d_i}S$ and $\mathrm{d_e}S$, where $\mathrm{d_i}S$ denotes the entropy production within the system while $\mathrm{d_e}S$ is a flux term describing entropy "export" into the environment:

$$\mathrm{d}S = \mathrm{d_i}S + \mathrm{d_e}S \tag{26}$$

The term $\mathrm{d_i}S$ can only be >0, but $\mathrm{d_e}S$ can also be negative. By identifying entropy with disorder, Equation (26) shows that an isolated system can evolve only toward greater disorder. For an open system, however, the "competition" between $\mathrm{d_e}S$ and $\mathrm{d_i}S$ permits the system, subject to certain boundary conditions, to adopt new states or structures. These are stationary if

$$\mathrm{d}S = 0 \tag{27}$$

or

$$\mathrm{d_e}S = -\,\mathrm{d_i}S < 0 \tag{28}$$

$\mathrm{d_i}S$ can be expressed in terms of thermodynamic "forces" X_i, and rates of irreversible phenomena J_i [164]; X_i may be gradients of temperature or concentration; the corresponding "rates" are then heat flux and chemical reaction rate. Hence,

$$\frac{\mathrm{d_i}S}{\mathrm{d}t} = \sum_{i=1}^{n} J_i X_i \tag{29}$$

Around equilibrium, the relationship between fluxes and forces is linear

$$J_i = \sum_{j=1}^{m} L_{ij} X_j \tag{30}$$

where L_i represents specific coefficients (e.g., coefficients of thermal conductivity or diffusion).

Provided the reservoirs of energy and matter in the environment of the open system are sufficiently large to remain essentially unchanged, the system can tend to a nonequilibrium stationary state far beyond the domain of linear thermodynamics. This state may be associated with dissipative structures [203] (i.e., structures resulting from a dissipation of energy rather than from conservative molecular forces). In considering phytocenoses from the viewpoint of stationary dissipative structures, the relationship of d_iS and d_eS as expressed in Equation (28) and the specific boundary conditions controlling entropy production and flux rates appear to be particularly important. A consequence of Equation (28) is that, thermodynamically speaking, stability or the capacity to maintain a nonequilibrium steady state is coupled with a (relative) minimum of total entropy production dS. Clearly this can be accomplished by either minimizing d_iS or maximizing d_eS or by a combination of both strategies.

Concentration processes involved in the normal metabolic activities of living systems play an important role in this connection, as can be seen from

$$\Delta G^\circ = R \cdot T \cdot \ln \frac{C_2}{C_1} \qquad (31)$$

where ΔG° is the difference in standard free energy, $R = 8.31\ \mathrm{J\,mol^{-1}\,K^{-1}}$, T = temperature in kelvin, and C_2, C_1 are higher and lower thermodynamic concentrations, respectively. Changes in concentration are a physical prerequisite for the production of a great many compounds, and an absolutely cogent one if substances are produced whose free energy is higher than that of the corresponding "raw materials."

On the nutrient-depleted soils of Amazonia (Fig. 25), for example, forest stands with their mosaic-like structure owing to the patchiness of rejuvenation processes (see Section 2.1.1) developed specific adaptations to facilitate these concentration processes. A dense root mat is formed over the soil in intimate contact with litter, with root tips growing upward into the fallen litter. Mycorrhizal associations, as the most common plant root–soil microorganism symbioses, benefit their host plants by effectively increasing the absorptive surface area of the root system, thus providing for increased uptake of nutrients. The dense fabric of predominantly fine roots also plays an important part in exchange and adsorption of nutrients from throughfall water. In addition, the structure of the foliage favors the use of nutrients during a long active life, the retransport of certain nutrients before leaf shedding, and a high polyphenol content and coriaceous nature, which both reduce herbivory. Another factor of great importance is the multilayered structure of the forest and the activities of epiphytes and microorganisms on much of the exposed surfaces, which together form highly efficient filtering systems scavenging nutrients from rainwater. Finally, biological nitrogen fixation in the root–humus–soil interface appears to play an important role in the nutrient budget [204]. Taking into account that even in the simplest cells the normal metabolic pathways imply several thousand complex chemical reactions that must be coordinated by means of an extremely sophisticated functional network, means that hierarchical order in both functional and spatiotemporal respects constitutes a further and most powerful negentropic factor. It characterizes every living system from the submicroscopic level to gigantic rain forest biomes such as the Amazonian Hylaea.

The effectiveness of these negentropic processes is further enhanced by most efficient entropy fluxes related to the transpiration and nocturnal respiration of plants. The molal entropy of H_2O increases from $63\ \mathrm{J\,mol^{-1}\,K^{-1}}$ (liquid) to $189\ \mathrm{J\,mol^{-1}\,K^{-1}}$ (gas) in the course of evaporation, and CO_2 has a molal entropy of $214\ \mathrm{J\,mol^{-1}\,K^{-1}}$. Consequently, the reverse process, the photosynthetic fixation of CO_2, is of comparable importance for the negentropy balance in the light of Equations (26) and (28).

As a consequence of these mechanisms, the results of the above comparative diversity analyses may be given a unified interpretation:

1) Species diversity is not a monotonous function of the nutrient status of soils. If soils form on basic or intermediate parent materials, pedogenic nutrient supply and species diversity may both increase during periods of $10^3 - 10^4$ (10^5) years, provided the water and energy factors are not limiting. In Figure 30, the period of ascending evolution is termed phase I. Shorter fluctuations of variable but generally decreasing amplitude are likely to be superimposed on this long-term trend. In phase II, which can last for several hundred thousand years, soils degrade in regard to nutrient supply, but species diversity keeps in-

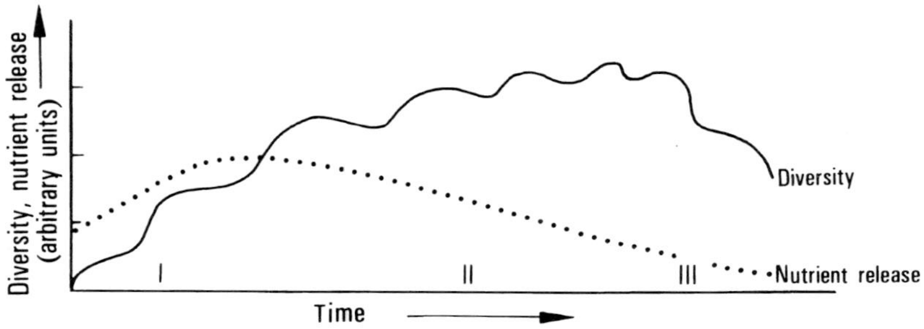

Figure 30. Generalized model of long-term ecosystem evolution in reaction to nutrient supply of soil

creasing for negentropic reasons (see the tropical rain forest stands in comparison to the ectroprical associations described in Section 2.3.3). Clearly such an evolution toward and including the following phase requires a sufficient degree of climatic and geomorphic stability (as has been the case in central Amazonia and considerable parts of Malesia), or else the continuity of pedogenesis will be interrupted by truncation processes that rejuvenate the soils (as happened in the major part of the Congo Basin). Phase III marks the eventual and comparatively acute decrease in diversity of phytocenoses once the nutrient status of soils has fallen below a critical level.

2) Spatial heterogeneity as a characteristic feature of the sites during phase I results in variability in the number of species and plant individuals. With this variability the associations can simultaneously retain genetic and behavioral types that can maintain their existence in low populations together with others capitalizing on opportunities for dramatic increase. The more homogeneous the environment in space and time (i.e., during phase II, and probably also phase III), the more likely is the system to have low fluctuations (i.e., high stability). Tropical rain forests represent climatically buffered and (largely) self-contained systems with relatively low (natural) variability on highly impoverished soils of the ferralsol, acrisol, and podzol classes. Temporary external disturbances are consequently likely to affect a proportionally greater number of species than in phase I populations, where the interspecific differences in ecological potency are usually distinctly higher. Hence, resilience of mature biocenoses is low during phase II. In contrast, biocenoses may be highly resilient in phase III of evolution, although unstable.

The greatly reduced resilience of phase II stands is particularly well illustrated by the influence of clearing activities on tropical forests. The soil, with its low retention capacity, is irrevocably impoverished after a brief period of fertilization due to sudden release of nutrients in the biomass. The brief period of fertility is used by shifting cultivators to produce crops, but even this system, developed by small native populations "in close contact with the environment" [205], can induce severe losses if forced to fit a market-oriented economy or to feed larger populations [206], [207].

2.3.8. Appraisal

The analysis of real communities and the comparative assessment of ecological models [2], [37] show that real communities are not those homogeneous and temporally invariant systems described by simple Lotka–Volterra equations and exemplified by most laboratory microcosms. Population dynamics of real-world ecosystems are spatially distributed in general, associated with a marked temporal variation, providing a multitude of ways in which the probability of coexistence is enhanced and biodiversity increased. Coexistence under stochastic, nonequilibrium conditions as described by patch-dynamics models at different scales can be just as strong and stable as that occurring under a deterministic, niche-differentiation model [208].

In a single patch, species extinction can occur as a result of competitive exclusion, overexploitation, and other destabilizing interspecific interactions or be due to environmental instability (e.g., unpredictable disturbances or changes in conditions). De Angelis and Waterhouse [209] and Remmert [210] have emphasized that the macroscale integration of unstable patches that are out of phase with each other can lead to persistent species-rich communities.

An important parallelism seems to exist between the properties of a community and the properties of its component populations, which will be subject to a (relatively) high degree of *K* selection in stable environments, while *r* selection (selection in the direction of high numbers of offspring) predominates in variable environments. The *K*-selected populations with their high competitive capacity, high inherent survivorship, and low reproductive rates are normally resistant to disturbances but, once perturbed, have little possibility of recovery (i.e., low resilience). The *r*-selected populations, by contrast, have less resistance but distinctly higher resilience. This inverse relationship between resistance and resilience is of particular importance in the ecotoxicological context (Section 3.2).

The above stability considerations mean that food chains should be shorter in "fluctuating" environments, since only the most resilient food webs would be expected to persist and short food chains are more resilient. An analysis of Briand's findings [183], however, shows that further critical tests of these logical ideas are required because the average maximal food chain lengths for 40 communities from both unstable and "constant" environments did not differ significantly. In conclusion, most communities are probably organized by a temporally and spatially varying mixture of "forces"—namely, competition, predation, and disturbances, with competition and predation being presumably less important in more disturbed environments [2], [211], [212]. Consequently, no single stability or sensitivity measure exists for a community. Stability varies with the aspect of the community or ecosystem under study and the nature of the disturbance. Without pertinent semantic specifications relating to defined temporal and spatial scales (duration and dimensionality of observations), indicator variables and the nature of perturbation statements about system stability would be meaningless in a general ecological and a more specific ecotoxicological context.

2.4. Exposure and Effect Criteria of Chemicals

Methodologies for estimating fluxes, transformations, and resultant environmental concentrations of chemicals depend on the knowledge of how rapidly the chemicals are discharged to a specific compartment in the environment and how rapidly they are transformed or removed by physical, chemical, and biological processes. Apart from the specific structure of the receiving compartment, particularly its physical and chemical characteristics, these processes are largely controlled by the inherent molecular properties of the chemicals released. These may be appropriately classified as sets of equilibrium and kinetic constants that define the exposure potential, and a complementary set of data that characterizes the effect potential, i.e., the chemical, biotic, and particularly, ecotoxicological effects of a substance.

In the following sections the essential exposure and effect criteria are described. Other chemical and physical properties such as flash points, flammability limits, and autoignition temperature are omitted because they are not of direct concern to the environmentalist. A comprehensive review of these and other dangerous properties of chemicals is given in [213]. With regard to the following quantitative specifications, however, note that chemicals are never 100 % pure and that the very nature and quantity of the impurities can have a significant impact on many environmental qualities. In particular, the following parameters are very sensitive to the presence of impurities: water solubility, odor characteristics and threshold values, biological oxygen demand, and toxicity [47].

2.4.1. Equilibrium Constants for Chemical Distribution

Density, Molar Mass, and Structure. The density of a substance, particularly in comparison to that of water or air, is of special environmental importance since it largely controls dispersion processes. Hydrocarbons are usually lighter than water. Figure 31 illustrates the relationships of molar mass and density.

Melting and Boiling Points. Normally, melting involves an increase in volume by about 10 %; exceptions are water, bismuth, and galli-

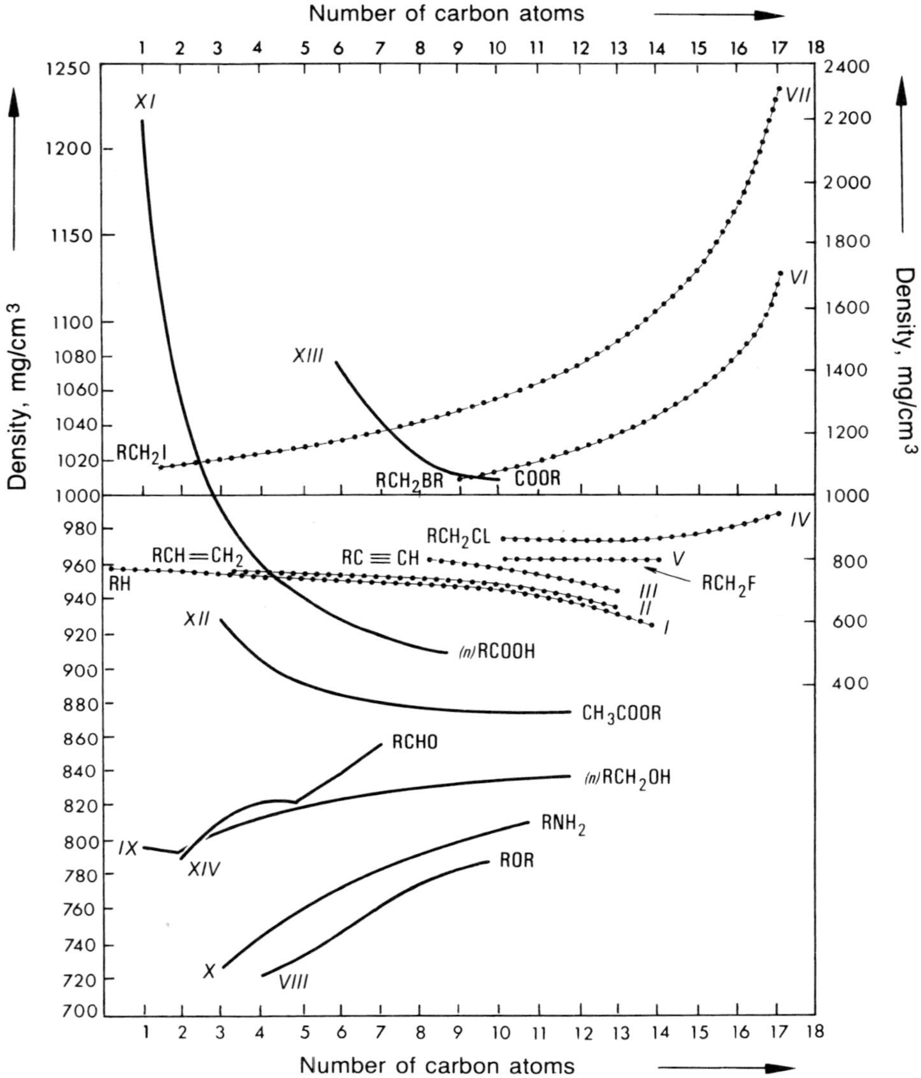

Figure 31. Relationships of molar mass and density (after [47], modified)
Dotted lines *I–VII* refer to right scale, lines *VIII–XIV* to left-hand scale

um. According to the sign of this change in volume, melting temperature rises or falls by 0.01–0.1 K/bar, which means that, under environmental conditions, pressure has little effect on melting point.

The melting and boiling points of the members of homologous series of hydrocarbons increase with increasing molar mass. When comparing organic compounds containing different halogens, however, the mass effect can be (partly) counterbalanced by polarization effects (e.g., in the case of CH_3F and CH_3Br). Substitution of another atom or group for hydrogen in an alkane results in an elevation of the boiling point. Thus, alkyl halides, alcohols, aldehydes, ketones, and acids boil at higher temperatures than ("elementary") hydrocarbons with the same carbon skeleton. The introduction of groups promoting association brings about a marked rise in boiling point, which is especially pronounced in alcohols and acids since hydrogen bonds can be formed [47].

Vapor Pressure, Vapor Density, and Volatility. The *vapor pressure* of a liquid or solid is the pressure exerted by the gas in equilibrium with the liquid or solid at a given temperature. The evaporative loss of a substance, volatilization, is a very important source of material for airborne transport and subsequent transformation and deposition processes. Vapor pressure data permit, in combination with solubility data, calculations of rates of evaporation of dissolved compounds from water by using Henry's law constant H, which is normally defined as

$$H = PS^{-1} \tag{32}$$

where P is the equilibrium vapor pressure of the pure chemical and S its saturation concentration or solubility at a fixed temperature.

Vapor density, compared to the density of air (=1), indicates whether a gas will be transported along the ground, possibly subjecting ecosystems to high exposure, or whether it will disperse rapidly. Vapor density is related to equilibrium vapor pressure through the fundamental state equation for a gas:

$$PV = nRT \tag{33}$$

If the mass of the substance and its molar mass are substituted for the number of moles n, this yields for the vapor density Vd [47]:

$$Vd = PM(RT)^{-1} \tag{34}$$

where P is the equilibrium vapor pressure (bar), $R = 8.31\ \mathrm{J\,mol^{-1}\,K^{-1}}$, M = molar mass (g/mol), and T = absolute temperature (K).

Volatilization of chemicals from water is an important transport process for a number of substances that have low water solubility and low polarity. Despite their very low vapor pressure, many compounds can volatilize rapidly because of their very high activity coefficients in solution, and by using the mathematical approach of MACKAY and PATERSON [214], [215] the rate constant for volatilization from water is given by

$$K_{\mathrm{vw}} = \frac{A}{V}\left(\frac{1}{K_{\mathrm{L}}} + \frac{RT}{HK_{\mathrm{G}}}\right)^{-1} \tag{35}$$

where A is the evaporating surface (cm²), V the liquid volume (cm³), H Henry's law constant (bar L/mol), K_{L} the liquid film mass-transfer coefficient (cm/h), K_{G} the gas film mass-transfer coefficient (cm/h), R the gas constant ($8.31\ \mathrm{J\,mol^{-1}\,K^{-1}}$), and T the temperature (K).

Volatilization processes from soil surfaces are complicated by variable contributions due to volatilization of the chemical from the water at the surface, evaporation of water itself, and the related capillary rise, which transports more water and dissolved chemical to the surface [216].

Water Solubility. Water solubility is an important parameter for the environmental assessment of all solid or liquid chemicals. In general, molecules with highly polar bonds will be relatively strongly attracted to one another; consequently, water and alcohol are potent solvents for ionic compounds such as salts, as well as polar nonionic molecules such as sugar. In contrast, nonpolar molecules such as tetrachloromethane, benzene, or gasoline are more weakly attracted to one another. They mix easily and are good solvents for nonpolar molecules, such as fats, grease, and paraffins, which are not soluble in water. Some solids are so firmly bonded that they are not significantly soluble in any common solvents except those with which they react chemically, whereby the nature of the molecule being dissolved is altered. Quartz is of this type: its solubility in pure water is of the order of 5 mg/L, but if the water contains a 0.01 M concentration of orthodiphenols, quartz is transformed into soluble esters and the solubility is increased to 40 mg/L [217].

The composition of natural waters varies considerably in space and time (see, e.g., [48]). Variables such as pH; water hardness; and concentration of cations, anions, and organic substances such as gelatin or fulvic and humic acids, all affect the solubility of chemicals in water.

Solubility of Mixtures. Mixtures of compounds behave differently from the individual compounds when brought in contact with water because each component partitions between the aqueous phase and the mixture. Compounds with a high water solubility tend to move into the aqueous phase, while the less soluble components have a greater tendency to remain in the mixture phase. Thus, the fractional composition of the water-soluble fraction differs from the original composition of the mixture, and the concentrations of the individual components of the water-soluble fraction are generally lower than their maximum solubilities [218].

***n*-Octanol–Water Partition Coefficient.** The *n*-octanol–water partition coefficient (K_{OW}) is defined as the ratio of the equilibrium concentration C of a dissolved substance in the two-phase system *n*-octanol and water (two solvents that are largely immiscible):

$$K_{OW} = \frac{C_{octanol}}{C_{water}} \tag{36}$$

Usually the decadic logarithm of K_{OW} is given ($\log K_{OW}$). Because K_{OW} is a dependent variable with regard to water solubility simultaneous use of both in structure–activity relationships or multivariate classification procedures will give spurious correlations. Experimental determination of both the aqueous solubility and the *n*-octanol–water partition coefficient led to the following regression [219]:

$$\log K_{OW} = 5.00 - 0.670 \log S \tag{37}$$

where S is the solubility in water (µmol/L). When solubility is expressed in mg/L, Equation (37) becomes

$$\log K_{OW} = 4.5 - 0.75 \log S \tag{38}$$

The regression covers many classes of both liquid and solid compounds with highly different polarities, from hydrocarbons and organic halides to aromatic acids, pesticides, and polychlorinated biphenyls (PCBs). The corresponding correlation coefficient is .97, which normally allows an estimation within one order of magnitude of the partition coefficient of a given chemical from its solubility in water. However, the scatter increases considerably for solubilities greater than 100 mg/L, as the example of L-tyrosine shows, where the experimental value of $\log K_{OW}$ is -2.26 while calculations using Equation (37) yield values of $+2.7$ or $+2.5$.

Furthermore, whether the above equations would also apply to chemicals with high dissociation constants, such as salts and strong acids or bases, is questionable since their activities in solution cannot be approximated by their concentration. Basically, the same restriction holds for aliphatic acids or bases whose partition coefficients can vary drastically with changes in pH [47], [220]. Nevertheless, the *n*-octanol–water partition coefficient has proved useful as a simple means of predicting biological uptake, lipophilic storage, and biomagnification factors (see Sections 2.4.2, 3.3.5 and Fig. 57), whereas extrapolations of the multifactorial soil adsorption processes are much more difficult.

2.4.2. Kinetic Constants for Environmental Processes

Every chemical reaction takes place at a definite rate depending on the specific conditions under which it occurs. These boundary conditions are the concentrations or pressure of the reacting substances, temperature, presence or absence of a catalytic agent, and radiation. Chemical kinetics deals with the rates and mechanisms of these reactions whose understanding is rendered difficult by the fact that both time and intermediate products are involved, compared to the subject of chemical equilibrium, which is concerned only with the initial and final states of reactants. Because of the complexity of these subjects, the following presentation is limited to determination of the essential rate constants for environmental processes from structure alone by using structure–activity relationships (see also → Toxicology, Chap. 2).

2.4.2.1. Hydrolysis

Hydrolysis of compounds usually results in the introduction of a hydroxyl function and is most commonly associated with the elimination of a leaving group X:

$$RX + H_2O \rightarrow ROH + HX \tag{39.1}$$

In water, the reaction is catalyzed mainly by H^+ and OH^- ions; whereas in moist soil, loosely complexed metal ions such as copper or calcium may also constitute important catalysts for the hydrolysis of several types of chemical structures. In addition, sorption (→ Soil) of the chemical may increase its reactivity toward hydronium or hydroxyl ions [221], [222].

Mabey and Mill [223] reviewed kinetic data for hydrolysis of a variety of organic compounds in aquatic systems in relation to the chemical characteristics of most freshwater systems. Quantitative structure–activity relationships, used as predictive test methods to develop the essential kinetic data, generally take the form of linear free-energy relationships such as the Hammett and Taft

$$\log K_x/K_o = \varrho\sigma \tag{39.2}$$

or Brønsted

$$\log K_x/K_o = m\mathrm{p}K_a + C \tag{39.3}$$

approaches, where ϱ is a constant for a given reaction under a given set of conditions, and σ is the Hammett constant, characteristic of a certain substituent; K_o and K_x are the rate constants for the unsubstituted and substituted structures, respectively, while the dissociation constant $\mathrm{p}K_a$ refers specifically to the acidity of the leaving group.

Highly variable hydrolysis reactions have been observed in soil and sediment. This appears to be the result of unusual pH relationships at the surface of soil constituents, possible occurrence of metal-ion-catalyzed processes, and general acid- and base-catalyzed reactions, possibly involving phenolic, amine, or sulfide groups in soils [222]. In some cases (e.g., in the presence of clay minerals), hydrolysis rates of organic chemicals are accelerated considerably compared to bulk solution, while adsorption to humic components frequently impedes hydrolysis. A detailed understanding of the mechanisms involved is limited, however, and structure–activity relationships appear to be available for only a few compounds.

2.4.2.2. Photolysis

Direct Photolysis. Natural direct photolysis results from absorption of photons from the solar spectrum above 290 nm by the chemical. The specific absorption rate ($K_{a,\lambda}$) is a measure of the overlap between the solar spectrum and the electronic absorption spectrum of the compound (i.e., an inherent property of a photoreactive chemical) [224]:

$$K_{a,\lambda} = 2.303\, E_0(\lambda)\, \varepsilon_\lambda \tag{40}$$

where $E_0(\lambda)$ is the scalar irradiance and ε_λ the molar absorptivity at wavelength λ.

Since pollutant concentrations are low under most environmental conditions, photolysis can be described by a first-order expression; i.e., the rate constant of direct photolysis k_d is directly proportional to pollutant concentration. From Equation (40), then

$$k_d = 2.303\, \Phi_d \int E_o(\lambda)\, \varepsilon\, d\lambda \tag{41}$$

where integration is over the range of $E_o(\lambda)$, which is absorbed by the compound, and Φ_d denotes the quantum yield, which may attain a maximum value of 1. However, rapid direct photolysis can occur even with compounds that react with very low quantum efficiencies [224].

Since more than one primary reaction may be involved in a photochemical reaction, the primary quantum yield may be defined with respect to a specific reaction. Moreover, the sum of the quantum yields of all primary reactions, including deactivation, must be unity. The overall quantum yield, which is the quantity most often determined, on the other hand, does not distinguish between the effects of primary and secondary reactions and it may exceed unity.

Modeling photochemical reactions in natural waters is still in its infancy. No proven general model applicable to every kind of natural water is presently available for estimating the rates of consumption or production of chemical species through light-induced reaction pathways. A lot of fundamental and essential information for constructing a precise and realistic model is still not available. For example, the chemical composition of natural waters and their temporal and spatial variabilities are still inadequately known. Thus, the reactants that may participate directly and indirectly in photochemical reactions in a competitive, synergistic, or catalytic way cannot yet be fully identified. The rate at which a particular photochemical reaction may proceed in this poorly defined medium, and to which extend it will proceed cannot be quantified extensively by empirical data. Besides, as stated previously, even in a well-defined medium, the detailed reaction pathways of photochemical reactions can still rarely be identified definitively.

Indirect photolysis may occur if a natural material can absorb solar photons and transfer part of the energy to a chemical. Thus, compounds that do not absorb light directly can still undergo mediated photolysis, and chemicals that photolyze directly may photolyze much more rapidly in natural waters containing organic substances in dissolved or particulate form. Despite these difficulties, general rate expressions for sensitized photolysis can be defined. The rate of light absorption is

$$I^s_{a,\lambda} = 2.3\, \varepsilon_{\lambda,s}\, E_o(\lambda)\, [S] \tag{42}$$

where $\varepsilon_{\lambda,s}$ is the molar absorptivity of the sensitizer and [S] is its concentration [224]. Thus, from Equation (42), the rate constants for indirect photolysis should increase with increasing [S] if $E_o(\lambda)$ is held constant. When the concentration of pollutant P is very dilute, the quantum yield is directly proportional to this concentration. By denoting C_λ as the proportionality constant, the total rate expression becomes

$$(\text{Rate})_\lambda = 2.3\, E_o(\lambda)\varepsilon_{\lambda,s} C_\lambda [\text{S}][\text{P}] \tag{43}$$

Provided the sensitizer concentration is constant, which is usually the case in nature, the rate expression follows first-order kinetics (→Air, Chap. 1).

2.4.2.3. Oxidation and Reduction

Direct and indirect phototransformation as described in the preceding section are possible removal pathways only for those chemicals that, either by their very nature or due to fixation onto light-absorbing carrier substances, can absorb radiation in the range of the solar spectrum. However, the main transformations that lead to the removal of chemicals from the atmosphere and water involve reactions with photochemically generated *oxidants* such as the hydroxy radical ($OH^{\cdot}$), ozone (O_3), hydroperoxy radical ($HO_2^{\cdot}$), organic peroxy radicals ($RO_2^{\cdot}$), singlet oxygen ($O_2\,\Delta_g$), and nitrate radical ($NO_3^{\cdot}$). Table 9 summarizes the reactivity of different classes of organic compounds toward some of the above oxidants.

The *reduction* of organic compounds requires an environment of low redox potential, i.e., eutrophic or deeper levels of aqueous systems and a soil depth at which oxygen is limited [225]. Reduction of organics can follow various mechanisms such as hydrogenation, halogenation, or coupled oxidation–reduction, many of which require catalysts. Like with many other chemical reductions, rates of reductive alteration of organics are temperature and pH sensitive [2]. Since the most common mechanism for reduction of organic molecules in the environment is via microbial attack, separating biologically induced from straight chemical reduction in situ is difficult. From the viewpoint of prospective hazard assessment, however, identifying those compounds that are potentially susceptible to reduction and the transformation products that might be deposited in the environment as a result is more important.

Table 9. Rates of oxidation in air and water (after [221])

Type of oxidant	Class or structure	Half-life of reactant
$HO^{\cdot}$ radical (in air)	*n*-, iso-, cycloalkanes	1–9 d
	alkenes	0.05–1 d
	halomethanes	0.2–100 a
	alcohols, ethers	1–3 d
	ketones	0.2–6 d
	aromatics	0.1–3 d
$RO_2^{\cdot}$ radical (in water)	alkanes, alkenes	220–2000 a
	alkyl derivatives	220–2000 a
	phenols, arylamines	1 d
	hydroperoxides	150 min
	polyaromatics	10 d
$O_2\,\Delta_g$ (in water)	aliphatic compounds	100 a
	substituted or cyclic alkenes	8–40 d
	alkyl sulfides	1 d
	dienes	19 h
	enamines	15 min
	furans	1 h

2.4.2.4. Sorption and Ion Exchange

Further processes of fundamental importance for the distribution of environmental chemicals in water and soil are sorption and ion exchange. Adsorption is the adhesion of ions or molecules to surfaces or to solid–liquid, liquid–gas, and liquid–liquid interfaces, causing an increase in the concentration of chemicals on the surface or interface over the concentration in solution. Adsorption occurs as a result of a variety of processes with a multitude of mechanisms when the concentration of chemicals lowers the free interfacial energy. The higher the free surface energy at an interface, the greater is the number of substances capable of lowering it.

Physical Adsorption. Molecules of a given material may be adsorbed from the gas or liquid phase to a solid by van der Waals forces. This process resembles a condensation that reaches equilibrium in a time limited chiefly by molecular diffusion rates. The weakness of van der Waals forces is reflected in the low adsorption enthalpy of physiosorptive (or unspecific) bonding; it amounts to hardly more than 40 kJ/mol [226]. Reversibility is a characteristic of physical adsorption that results from rapid equilibration; i.e., at a given temperature and pressure the same amount of material is adsorbed regardless of the direction from which equilibrium pressure and temperature have been approached. From the environmental point of view, however, the fact that physical adsorption of gases by highly porous media such as silica gel, charcoal, and soil often exhibits "hysteresis," a form of irreversibility due to condensation of liquid in pores is very important. Analogous phenomena occur with chemicals bonded at solid–liquid interfaces.

Chemical Adsorption. Molecules of a given substance may also be adsorbed from the liquid or gas phase by a solid through the activation of valence forces. Consequently, they may be considered as reacting with atoms or molecules on the surface of the solid in a way analogous to any other chemical reaction, and the products of this reaction are usually conveniently considered to be species quite different from the original reacting molecules of the atoms (chemisorption or specific bonding in terms of soil chemistry). Temperatures at which chemisorption occurs are related not to the boiling point of the adsorbate (material being adsorbed), as is the case with physical adsorption, but to the stability of the surface compound and its rate of formation. Therefore, rates of chemisorption vary enormously with material and temperature, as do those of ordinary chemical reactions, and may be very much slower than rates corresponding to diffusion of gases or solutes to the surface. Furthermore, the rates are likely to vary strongly with the amount of material adsorbed. Because of these rate characteristics, the amount of adsorbate at a given pressure or concentration and temperature—when observed over practical time periods—may depend strongly on the way these boundary conditions have been reached. Therefore, irreversibility is extremely common in chemisorption, which is characterized by enthalpy values >100 kJ/mol [226].

Ion Exchange. The presence and extent of ionization also have a great influence on the chemical behavior of a substance, particularly on solubility, sorption, toxicity, and other biological characteristics. Ion exchange as an important sorption mechanism for substances prone to ionization is viewed as an exchange with some other ion that initially occupies an adsorption site on the solid (→ Ion Exchangers, **A 14**) [225]. For example,

$$\mathrm{H^{+}}\overset{\mathrm{Ca^{2+}\,Mg^{2+}}}{\underset{\mathrm{K^{+}\,Na^{+}}}{\boxed{\qquad}}}\mathrm{Al^{3+}} + 10\,\mathrm{NH_4^+} \rightleftharpoons \mathrm{NH_4^+}\overset{4\,\mathrm{NH_4^+}}{\underset{2\,\mathrm{NH_4^+}}{\boxed{\qquad}}}3\,\mathrm{NH_4^+} + \mathrm{H^+} + \mathrm{K^+} + \mathrm{Na^+} + \mathrm{Ca^{2+}} + \mathrm{Mg^{2+}} + \mathrm{Al^{3+}} \qquad (44)$$

The ion-exchange property of a soil results mostly from the clay and silt fractions, organic matter, and hydrous oxides of iron. Soil particles have an amphoteric character but normally carry a net negative charge. Positive charges, as indicated by the amphoteric nature of the clay fraction, may originate from hydrous oxides of iron (ferrihydrites), aluminum, and manganese, and from exposed octahedral groups that react as bases by attracting protons from the surrounding soil solution. The negative charge increases and the positive charge decreases with rising pH as a result of increasing ionization of the acid groups and decreasing protonation of the basic groups. For decreasing pH, the corresponding change occurs in the opposite direction. The carboxyl groups of humus ionize under acid conditions, the phenolic hydroxyl groups mainly above pH 6. Because of the highly irregular shape of many clay minerals (and clay–humus complexes, in particular) and the nonuniform distribution of charges in the particles, the surface charge density is quite variable.

To describe ion-exchange processes, several formulas have been proposed in the course of exchange studies. The first group is essentially empirical and intended mainly to provide mathematical formulations best fitting the experimental data. Among this group, the Freundlich [49], Langmuir, and BET (i.e., Brunauer–Emmett–Teller) are particularly important [227]. The second group of formulas is related to the law of mass action.

Empirical Adsorption Formulas. In the *Freundlich equation*, the concentration of a substance in solution is related to that adsorbed as follows [49]:

$$\frac{x}{m} = k c^{\frac{1}{n}} \qquad (45)$$

where x is the amount of substance adsorbed; m the mass of the adsorbent; c the concentration of substance remaining in solution; and k, n are specific constants.

The Freundlich formula is of a parabolic type and therefore cannot give a maximum adsorption value as does the *Langmuir theory*. According to the latter, adsorption is considered proportional to the pressure of gas (or concentration of a solute) and to the amount of surface not yet covered with adsorbed molecules. The rate of desorption (or evaporation) of adsorbed molecules is considered proportional to the number of such molecules. At equilibrium, these two rates must be equal; thus,

$$\frac{n}{n_{\mathrm{m}}} = \frac{bp}{1 + bp} \qquad (46)$$

This equation, in which n is the number of molecules adsorbed, n_{m} the number of molecules required for a completely unimolecular surface film, p the concentration of a chemical in solution, and b a constant at a given temperature, is the *Langmuir adsorption equation* (*Langmuir isotherm*).

A third theory of adsorption widely used for representing the adsorption of gases or vapors on solids was advanced by Brunauer, Emmet, and Teller [227]. In contrast to Langmuir's theory, the *BET theory* makes allowance for adsorption of multimolecular layers as is commonly observed when nonporous solids adsorb vapors at pressures approaching saturated vapor pressure. The BET equation is therefore

$$\frac{n}{n_{\mathrm{m}}} = \frac{b\frac{p}{p_0}}{\left(1-\frac{p}{p_0}\right)\left[1+(b-1)\frac{p}{p_0}\right]} \quad (47)$$

where p_0 denotes the saturation vapor pressure, while the other quantities have the same meaning as in Equation (46). The equation is designed for physical adsorption and predicts a very sharp increase in the amount of material adsorbed as the saturated vapor pressure is approached, and this is found to be the case [50].

2.4.2.5. Biotransformation

Biotransformation is a more accurate and general term than biodegradation, because many chemicals are transformed to products of comparable molecular complexity instead of being degraded. Chemically speaking, biotransformation in water and soil systems includes hydrolysis, oxidative cleavage, and reduction, even in aerobic systems. Kinetics of biotransformation are normally based on the Monod model [51], which couples the rate of loss of the chemical to the growth of organisms involved. In the natural environment, however, growth is usually controlled by nutrients present in comparatively constant amounts, and the kinetic expression for biotransformation processes simplifies to a second- or pseudo-first-order reaction:

$$\mathrm{d}[\mathrm{C}]/\mathrm{d}t = k_{\mathrm{bt}}[\mathrm{C}][\mathrm{B}] = k'_{\mathrm{bt}}[\mathrm{C}] \quad (48)$$

where $\mathrm{d}[\mathrm{C}]/\mathrm{d}t$ is the rate of transformation of C; [B] is the number of organisms per unit volume; k_{bt} the second-order rate constant for biotransformation; and k'_{bt} the pseudo-first-order rate constant. The half-life of the chemical under transformation ($t_{1/2}$ at a given cell concentration) can then be calculated as

$$t_{1/2} = \frac{\ln 2}{k_{\mathrm{bt}}[\mathrm{B}]} \quad (49)$$

In deriving this equation, the microbial community in water, sludge, or soil is assumed to already be acclimated to the chemical, so no lag time is involved in production of the necessary level of biotransforming organisms or appropriate mutants or enzymes.

Because of the remarkable versatility of microorganisms in their catabolic activity, many substances are subject to microbial mineralization or cometabolism in water and soil. If, however, microorganisms have no enzymes active in a necessary catabolic sequence or if the substrate is protected from microbial attack because of special properties of the environment, the compound tends to persist. In some instances, persistence may be attributed to absence of the requisite enzymes or to the lack of permeability of microorganisms to the substrate [228]. Other mechanisms of recalcitrance may be inherent in the molecule itself; in some cases, biotransformable compounds are degraded quickly in one environment but are partially or totally refractory in another.

2.4.2.6. Chemical Structure and Biodegradation

Biodegradation is the most important transformation mechanism for organic compounds in nature. It can occur under conditions in which oxygen is present (aerobic) or absent (anaerobic). In aerobic processes, organic carbon is oxidized to CO_2, while in anaerobic processes it may ultimately be reduced to CH_4. In either case, the more important environmental variables affecting the rate and extent of biodegradation are temperature, pH, salinity, oxygen concentration, concentration of the chemical involved, ionic strength, composition of microbial associations and concentration of viable microorganisms, quantity and quality of nutrients (other than xenobiotic substances), trace metals, and vitamins.

In general, microbial degradation is impeded by ether bonds and by triple-substituted or condensed phenyl groups—particularly, if the latter have methoxyl, sulfone, nitro, or chlorine substituents in the *meta* position. Among the nitrophenols and chloroanilines, the *ortho*-isomers are relatively recalcitrant. Multiple substitution increases the recalcitrance of aromatic molecules considerably because their aerobic degradation is effected by relatively specific mono- and dioxygenases [228], [229]. Halogen substituents delay the electrophilic attack of these oxygenases considerably by reducing enzyme affinity. Therefore, multiple chlorine-substitued aromatics form relatively persistent compounds, the more so because the delayed cometabolic attack yields "dead-end" metabolites, which cannot be used for cell-building purposes or as an energy source.

Anaerobic Processes. As a consequence, reliable conclusions about biodegradation are not generally possible on the basis of chemical structure alone [47]; therefore, many laboratory methods have been developed for studying the aerobic biodegradative potentials of chemicals [230]–[233]. In comparison, tests for assessing anaerobic biodegradation have received less attention [84]. Nevertheless, anaerobic degradation is an important process for substances that are not aerobically biodegradable or whose physicochemical properties are such that their occurrence in an aerobic environment is restricted (e.g., chemicals that are strongly adsorbed or insoluble). For these substances, anaerobic biodegradation can be the major process responsible for breakdown in the environment.

Anaerobic biodegradation is complex and usually considered to occur in at least three consecutive main stages, the first of which involves hydrolysis of complex organic molecules such as carbohydrates, proteins, and lipids by the action of extracellular enzymes. In the subsequent stage (acidogenic step), the hydrolysis products are fermented to yield mainly short-chain fatty acids, alcohols, hydrogen, and carbon dioxide. Alcohols and acids are then converted in the acetogenic step. The last degradation step is effected by methanogenic bacteria (methanogens), which utilize acetate and hydrogen to form methane.

Anaerobic processes tend to be self-inhibitory because the hydrolysis of complex substrates such as fats, proteins, and carbohydrates yields volatile organic acids. By lowering the pH of the system, these may inhibit the growth of methanogens unless the system has sufficient buffer capacity [84]. Inhibition of gas production as a result of the toxicity of chemicals is also possible. Finally, some indications have been found that certain compounds (e.g., nitrilotriacetic acid) are degraded anaerobically only when aerobic organisms acclimatized to the compound are incorporated in the system [234].

When the organic compound liable to degradation or enzymatically induced metabolic transformations is utilized as a carbon source, the growth rate of the responsible microorganism depends on the concentration of the former. The rate of substrate utilization then becomes [52]

$$\begin{aligned} -\mathrm{d}C/\mathrm{d}t &= \mu X/Y = (\mu_\mathrm{m}/Y)\cdot CX/(K_\mathrm{s}+C) \\ &= k_\mathrm{b}\cdot CX/(K_\mathrm{s}+C) \end{aligned} \qquad (50)$$

where μ is the specific growth rate; X the biomass per unit volume; μ_m the maximum specific growth rate; K_s the concentration of the substrate to support half-maximum specific growth rate ($0.5\,\mu_\mathrm{m}$); $k_\mathrm{b} = \mu_\mathrm{m}/Y$, the biodegradation constant; and Y is the biomass produced from a unit amount of substrate consumed. The constants μ_m and K_s depend on temperature, pH, salinity, oxygen concentration, concentration of chemical, composition of microbial associations, and concentration of viable microorganisms, quantity and quality of nutrients (other than xenobiotic substances), trace metals, and vitamins.

2.4.2.7. Bioconcentration, Bioaccumulation, and Ecological Magnification

Bioconcentration is the concentration of a chemical in an organism due to direct uptake from the environment (e.g., ambient air or water), which does not include contaminated food. *Bioaccumulation* includes the latter pathway of concentration, while *biomagnification* considers only concentration processes via food uptake. *Ecological magnification* defines the increase in concentration of a substance in an ecosystem or a food web when passing from a lower trophic level to a higher one [53].

Bioconcentration or bioaccumulation is the result of both kinetic (diffusional transport and biodegradation) and equilibrium (partitioning) processes. It thus involves the following set of fundamental events [47]:

1) Partitioning of the xenobiotic between the environment and some surface of the organism
2) Diffusional transport of the molecules across cell membranes
3) Transport processes mediated by body fluids, such as exchange between blood vessels and serum lipoproteins
4) Concentration of the xenobiotic in various tissues depending on its affinity for specific biomolecules such as nerve lipids
5) Metabolism, cometabolism, or biodegradation of the xenobiotic

The relative importance of the above processes is likely to differ considerably as a result of both the physicochemical properties of the xenobiotic and the morphological and physiological structure of the target organisms. Therefore, an essential ecotoxicological requirement is that the determination of bioconcentration factors (*BCF*s) be complemented by an assessment of biomagnification values. Under real-world conditions, these would provide deeper insight into the complex bioaccumulation processes of xenobiotics as a result of transport and partitioning into higher levels of the food web, which may ultimately result in toxic concentrations [234].

On the basis of bioconcentration studies in different test organisms, correlations with both water solubility and *n*-octanol–water partition coefficients have been established. In some cases, the results proved excellent; in others, the correlations yielded only satisfactory structure–activity relationships (i.e., estimates of *BCF*s to within an order of magnitude for chemicals of $\log K_{ow}$ ranging from 1 to nearly 7).

3. Ecotoxicology

The production, use, and disposal of chemicals lead to their dispersion in the environment. The way in which substances are released depends on their physical and chemical properties (see Section 2.4.1), the production process (→ Production-Integrated Environmental Protection; → Plant and Process Safety), including extraction of natural resources (→ Raw Materials and Energy) and manufacturing; and the use pattern and means of disposal (→ Waste).

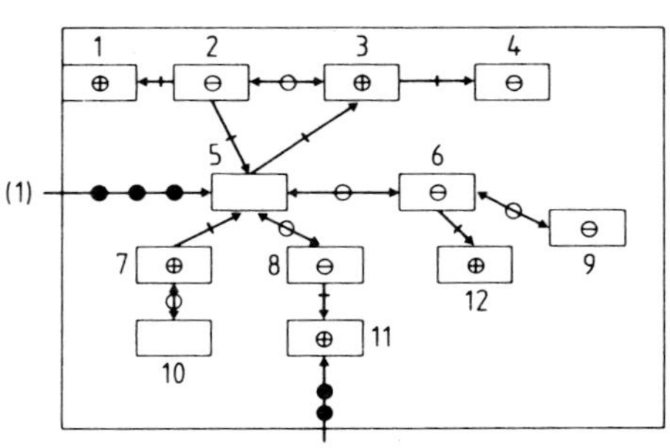

Figure 32. Component-specific influence of two xenobiotics on an ecosystem (after [158], modified)
Substance (1) affects a functionally central component, thus causing breakdown of the whole ecosystem. Substance (2) affects a peripheral component; even its deletion leaves the ecosystem practically unchanged.

Compartments of ecosystems have a limited capacity to assimilate chemicals by breakdown or dilution; if this capacity is exceeded, damage may ensue. Thus, the hazard posed by a single substance or a mixture of chemicals is a function of its inherent toxicity to organisms and the concentration attained. At the community and ecosystem levels, the sensitivity has to be considered specifically (Section 2.3), i.e., the functional position of the organisms affected with regard to the entire system. Figure 32 illustrates schematically the different influence of xenobiotics, affecting two elements of an ecosystem. In case (2), only a peripheral element (in the functional sense of the term) is damaged or deleted, which leaves the system relatively intact. In case (1), however, a component of central importance is eliminated, which leads to breakdown of the entire system.

Thus, the assessment of whether a substance presents a hazard to organisms in the environment is based on a comparison of its predicted environmental concentration (PEC) with the predicted no-effect concentration (PNEC) for organisms in the ecosystem affected. Whenever the PEC:PNEC is <1.0, a substance may be concluded not to present an environmental hazard under the boundary conditions considered. In the opposite case, hazard assessment should be complemented by risk assessment approaches, which define the probability that a hazard will occur. Figure 33 illustrates the temporal and spatial scales that must be considered when analyzing the impact of environmental chemicals on biota or ecosystems.

PEC values can be measured in the field. The identification of unequivocal spatial structures related to emission, transport, and sedimentation normally involves the construction of isopleth maps by means of metric interpolation. A pertinent statistical method that permits testing the validity of interpolation procedures is variogram analysis, normally applied in conjunction with kriging methods (i.e. methods in geostatistics for valid derivation of isopleths) [235], [236] (Section 3.1.4). Alternatively, the predicted environmental concentration can be estimated, if the manner and quantity in which a chemical enters the environment, and its subsequent distribution and transformation, are known (see Section 2.4.2; [2], [237]). This normally involves the application of models that are available at different levels of complexity, ranging from empirical models for practical purposes to rather abstract ones aiming at generalized insights via simulation.

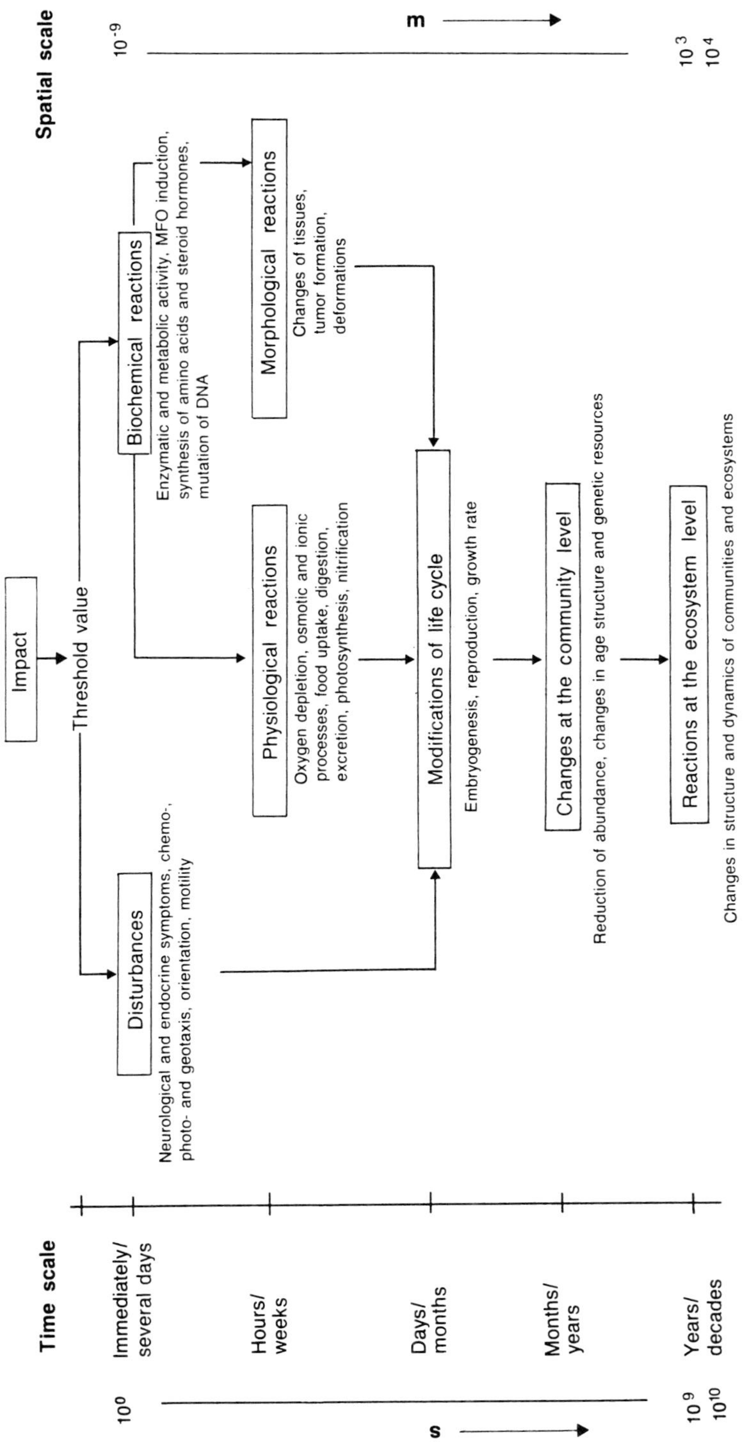

Figure 33. Temporal and spatial scales of the impact of chemicals on biota (after [53], modified)
MFO = monooxygenases

Effects assessment establishes a maximum concentration of a substance that does not cause adverse effects. The PNEC values may be derived from whatever ecotoxicological data are available, by using simple mono- and oligo-specific test systems, or micro- and mesocosms, and ideally from toxicity measurements at the ecosystem level. The lower the level of assessment approach, the more appropriate are assessment factors to compensate for restrictions in the data.

3.1. Exposure Assessment

Exposure assessment involves the identification of point or diffuse emission rates and the subsequent distribution tendencies, based on predictive models. In the first assessment step, the initial concentration ($PEC_{initial}$) in each compartment of concern must be estimated. If necessary, a revision in light of relevant new information may lead to a revised $PEC_{initial}$ and a PEC_{local} or a $PEC_{regional}$. The PEC_{local} is the estimated concentration of a substance when released to a particular environmental compartment, with further fate processes, (e.g., volatilization, sorption, biodegradation) taken into account. The $PEC_{regional}$ denotes the predicted concentrations of a substance in all relevant compartments, when the further distribution and fate of the chemical are taken into account.

3.1.1. Release Estimation

For realistic exposure assessments, basically the entire life cycle of a chemical should be analyzed and PEC values calculated for each relevant process. To standardize the release estimation for screening purposes [238], the EEC and OECD have defined three different categories of use pattern: main, industrial, and use. The main category classifies substances in four groups (see Table 10).

Table 10. Release fractions in the screening phase of hazard assessment of chemicals related to the main category (after [238])

Main category	Release fraction of production volume	Examples
Closed system	0.01	chemical intermediates
Enclosed in a matrix	0.1	plastic additives
Nondispersive	0.2	photochemicals
Wide dispersive use	1.0	solvents, plant protection products, detergents

In addition to the main categories, there are 15 different "industrial" categories and 55 different "use" categories [239]. This classification scheme allows a stepwise approach in release estimation. In the screening phase a rough estimate based on the main category may be sufficient, whereas in the confirmatory phase more refined release data (e.g., based on other categories) are needed. In the investigation phase, finally, detailed information on specific releases of the substance is required (→ Products, Chap. 3).

3.1.2. Elimination

A reduction in the concentration of chemicals in gaseous or aqueous discharges prior to their release to the environment is termed elimination. Elimination may be due to physical, chemical, or biochemical processes. The most significant elimination mechanisms are biochemical processes, especially aerobic biodegradation by bacteria or an alternation of aerobic and anaerobic phases (see Section 2.4.2) [2], [35]. Physical processes transfer a substance from one phase to another. In the case of volatile substances, the aeration process normally enhances their removal by stripping them from the solid or liquid phases to the atmosphere [47]. Substances from exhaust gas streams may be eliminated by scrubbing, e.g., by adsorption on a suitable material or passage through a suitable aqueous solution.

3.1.3. Dispersion

Pollution transport and dispersion are directly related to wind speed or flow velocity of rivers, etc. The higher the velocity, the greater is the volume of air passing a point source per unit time, and the lower is the resultant concentration per unit volume. Higher wind speed also means greater turbulence, which involves fluctuations of the local wind vector on various scales, and these eddies act to diffuse a plume. In addition to this "stretching" phenomenon [240], wind speed is important for the transport distance. In strong winds, pollutants may be transported long distances, but the resultant dilution over 10–50 km is usually such that most individual plumes have lost their identity and contribute to a more general (background) contamination of the lower atmosphere. Consequently, weaker regional wind systems may exhibit much greater potential for pollution since both horizontal transport and eddy diffusion are reduced.

This simple fact becomes all the more important because, under these conditions, local wind systems tend to develop that are difficult to predict with any accuracy [54]. In so far as they are closed smaller-scale circulations (e.g., land and sea breezes, mountain and valley winds, city winds), they often produce recirculation of pollutants coupled with a progressive increase in pollutant loading with time. The city thermal wind system operating mainly in the urban canopy layer is particularly dangerous; it is produced by microscale processes in the streets [241]–[243]. Here, flow converges on the city (or its variable thermal subcenters) from all directions, rises, diverges, and then moves outward to subside on the urban–rural fringe, where it rejoins the inflow.

The worst conditions for dispersion, however, occur in the case of temperature inversion and consequent formation of a stable boundary layer. With the warmer air layer acting as a lid on the mass of cooler air below, pollutants cannot escape from the cooler zone and thus increase in concentration in the confined space. This situation can be brought about by cooling (usually radiative) from below, by warming (usually adiabatic) from above, or by advection of warmer or cooler air.

Since the 1970s, modeling of dispersion processes in air has become more and more sophisticated owing to the progressive consideration of pertinent boundary layer effects and concentration or removal processes ([244]–[247], e.g.). This involves not only a considerable increase in computer time but also, and often more prohibitive, much more detailed input data. In the light of these circumstances, a question of practical importance is whether simpler models are not likely to yield comparable results on the condition that their data base can be made proportionally more comprehensive.

Box models, including dry and wet deposition and chemical transformations, are likely to provide calculated concentrations that are in reasonable or good agreement with measured values on a local scale as well as with transport data over a distance of 1000 km. An exemplary prototype whose fundamentals have often been used was developed by [248], describing SO_2 and NO_x concentrations in the industrial areas of North Rhine–Westphalia along the Rhine and Ruhr rivers.

Statistical models define immission concentrations by means of multiple regression relating immissions to emissions, meteorological parameters, topography, or other dispersion-controlling factors. Models of this type vary considerably with respect to both regression analytical approach and selection of data.

Observations of looping, coning, or (to a lesser degree) fanning plumes on an instantaneous basis reveal that the concentration is very peaked, with a marked maximum in the plume center. Averaged over growing time intervals, an increasingly smooth and wide plume envelope appears, which contains all of the short-period plume fluctuations. The wider spread results in a flatter concentration curve, with a correspondingly lower peak centered on the time mean axis of the plume (i.e., the mean wind direction). Analogous bell-shaped or Gaussian concentration profiles also characterize the vertical plane through the plume, although there may be a difference between the width of the horizontal and vertical profiles indicating that turbulence is greater in one or another direction.

Eulerian or Grid Model. The Eulerian or grid model is a mathematical representation of the pertinent transport processes and chemical reactions occurring in a grid of cells yielding the spatial and temporal variations of the ground cell concentrations of primary and secondary pollutants. It is characterized by fixed coordinate systems with respect to the ground and consists of a set of nonlinear coupled partial differential equations expressing the conservation of mass of each pollutant considered. Inputs to the model are numerical representations of the meteorological and emission conditions as a function of time of day, initial concentrations, and boundary conditions of inflow.

Lagrangian Model. In the trajectory or Lagrangian model, the time behavior of the concentration c of a pollutant is described by a differential equation of the type

$$\frac{\partial c}{\partial t} = \frac{\partial}{\partial z}\left(\underline{K} \cdot \frac{\partial c}{\partial z}\right) + S \qquad (51)$$

where c depends only on height above ground z and the time of travel t of an air parcel, while S refers to the net source–sink term representing the combined effects of emissions and chemical processes during transport, deposition, and flow across the upper boundary. $\underline{K}$ represents a vertical eddy diffusion coefficient whose values characteristically depend on time of day, height, and location. In comparison with the above Eulerian models, trajectory models have commendable mathematical advantages resulting from their moving coordinate systems with respect to a fictitious vertical air column moving horizontally with the advecting wind [249]. Considered as the present state of the art in operational long-range transport modeling, the Lagrangian approach has been used in numerous studies in the United States and Canada relating to transboundary air pollution.

3.1.4. Local and Regional Concentration Estimates

After emission, the environmental fate of a substance is governed by both transport and transformation processes that may be assessed by measurements or modeling approaches of which there are two basic types. The *local* or *monocompartment approach* is focused on the distribution of a substance within a particular environmental compartment, which leads to a PEC_{local}. The *regional* or *multicompartment approach* is directed toward the overall distribution, where transport occurs in all relevant environmental compartments (i.e., air, water, or some other medium) leading to $PEC_{regional}$ values.

The first approach is appropriate if the degradation rate is (distinctly) greater than the intercompartmental transfer rate; otherwise a multicompartment or multimedia approach is indicated (Section 3.3) [250]. Irrespective of the type of model to be used, the environmental parameters such as compartment volumes or compartment ratios must be defined; they are sensitive points for the final result. For substances that enter the environment from diffuse emission sources, a realistic PEC estimate cannot be calculated by means of the habitual point-source models. In such cases, the exposure assessment must be based on scale considerations and spatial modeling [251].

3.1.4.1. PEC Estimates in the Air Compartment

For the primary exposure assessment, models of advection and related dispersion processes may be used that compute concentrations of a substance at different receptor compartments. These require different minimum data sets consisting of discharge emission rates, effective height of source, and meteorological conditions, (stability classes of atmosphere, wind speed etc.). For regulatory purposes, the TA Luft model [252] is widely used (see also Section 3.1.3; →Air).

For subsequent revisions of the exposure assessment, information on the long-term fate of the chemical in the atmosphere is required. This involves the appropriate consideration of direct and indirect photodegradation and other chemical reactions [244], [253]. The reactivity of the hydroxy radical formed predominantly via photolysis of ozone by reaction of electronically excited oxygen atoms with water vapor is of fundamental importance in this connection.

A balance must be struck between simple models that do not represent the real-world situation and models that are too difficult to use. Thus, level 1 (equilibrium, conservative, steady state) and level 3 models (nonequilibrium, nonconservative, steady state) in the sense of the MACKAY et al. classification [250] are considered useful simple approaches. Level 2 models (equilibrium, nonconservative, and steady state) usually are of minor interest, whereas level 4 models (nonequilibrium, nonconservative, non-steady state) should be used only for estimating disappearance of chemicals from the environment [237].

An important drawback of these "unit world" models is that they compute only one concentration value for each compartment, whereas measurements normally indicate that actual concentrations in the environment may range over several orders of magnitude. Thus, they are suited only to provide an indication of concentrations in places far from the source (i.e., background levels). Improvements are possible by performing a sensitivity analysis, by using statistical approaches, or by adapting the unit world to the region of interest [254], [255]. An alternative method is the application of dispersion models, provided that appropriate input data, representative dispersion parameters, and sufficient meteorological information are available (see Sections 3.1.3, 3.3).

Table 11. Average concentration of selected heavy metals in Schleswig-Holstein soils in mg/kg (after [256])

Substrate	As		Cd		Co		Cr		Cu		Hg		Ni		Pb		Zn	
	A	C	A	C	A	C	A	C	A	C	A	C	A	C	A	C	A	C
S	2.61	**2.67**	**0.36**	**0.38**	1.96	**3.33**	2.55	**5.00**	**4.63**	**3.75**	0.082	**0.010**	8.97	**11.53**	**21.25**	**7.49**	**15.89**	**14.97**
Su2			0.46				5.67		7.70	5.61					23.91		25.95	16.41
Su3			0.31				18.83		13.26						34.72		43.26	
Su4																		
Su	7.13		0.37	0.42	3.30		11.45	16.03	**9.88**	4.75			11.02		29.00	6.29	**32.66**	15.61
Slu																		
Sl2			0.35				11.42		11.12				9.06		29.61		35.04	
Sl3	8.11		0.40	0.26	5.40		**22.73**	28.43	**9.21**	11.82			10.61	22.19	**30.67**	13.20	**43.91**	34.20
Sl4	7.21	**6.52**	0.51	**0.32**	6.73	**6.63**	**34.52**	**34.75**	**12.56**	**12.63**			**15.98**	19.65	**30.31**	**16.66**	**50.69**	**41.26**
Sl	**7.57**	**6.32**	0.43	**0.28**	**5.75**	**6.37**	**24.42**	**31.75**	**10.69**	**11.88**			**13.09**	20.46	**31.51**	**15.04**	**43.54**	**38.57**
Ls2																		
Ls3	7.05	**6.99**	0.70	**0.63**	7.84	**8.21**		**41.26**	13.83	**13.61**		0.022	19.61	**24.41**	31.99	**17.10**	63.59	**50.38**
Ls4				0.52				30.26		13.74				31.44		13.57		39.75
Ls	6.68	**6.68**	0.68	**0.60**	7.15	8.00	40.45	**39.07**	13.19	**13.63**	0.100	**0.022**	19.08	**25.44**	30.00	**16.49**	58.53	**48.38**
Lts		**6.45**		**1.15**		**11.62**				**15.35**		**0.023**		**31.09**		**15.58**		**52.92**
Lt2																		
Lt3																		
Ltu																		
Lt		**6.53**		**1.14**		**11.48**		59.71		**17.10**		**0.024**		**34.98**		**16.00**		**57.72**
Total	**6.78**	**5.46**	**0.45**	**0.55**	**5.34**	**7.24**	**21.04**	**30.20**	**10.05**	**10.70**	**0.085**	**0.019**	**13.63**	**22.90**	**29.04**	**13.34**	**39.64**	**36.75**

A = Average of A horizons (0–20 cm)
C = Average of C horizons
Total = Average heavy-metal content except marshland and peat
PNEC: As 20 mg, Cd 3 mg, Co 50 mg, Cr 100 mg, Cu 100 mg, Hg 2 mg, Ni 50 mg, Pb 100 mg, Zn 300 mg
0.00 for $n \geq 10$ and < 30
0.00 for $n \geq 30$

Su = Silty sand
Su2 = Low silt, Su3 = Medium silt, Su4 = High silt content

Slu = Loamy-silty sand
Sl = Loamy sand
Sl2 = Low loam, Sl3 = Medium loam, Sl4 = High loam content

S = Sand
Ls = Sandy loam
Ls2 = Low sand, Ls3 = Medium sand, Ls4 = High sand content

Lts = Clayey–sandy loam
Lt = Clayey loam
Lt2 = Low clay, Lt3 = Medium clay content
Ltu = Clayey–silty loam

3.1.4.2. PEC Estimates in Soil and Subsoil

Xenobiotics in soil and subsoil generally result from agricultural or industrial activities. The main problems in recent years were related to the application of manure and fertilizers, pesticides, sewage sludge, and to contaminated sites (see → Soil, Chap. 5, [2], [48]). Table 11 illustrates, on a comparative basis, heavy-metal concentrations in Schleswig-Holstein soils. In the light of the very high spatial variability of the soil cover, representative sampling is particularly important for concentration measurements. Regionalized variables should be applied to adequately characterize the chemical soil properties under consideration; consequently, variogram analysis in combination with kriging procedures is the appropriate method (see → Soil, Chap. 6; [158], [235]).

3.1.4.3. Variogram Analysis

Mathematical Concept. A regionalized variable can be considered the realization of a random function Y. Usually, this function is assumed to be stationary; this implies the following:

1) The expectation (m) of Y at any point of x is constant and independent of x.

$$E[Y(x)] = (m) \tag{52}$$

2) The covariance function of any pair of points $\boldsymbol{x}$ and $\boldsymbol{x} + \boldsymbol{h}$ depends exclusively on the vector $\boldsymbol{h}$ and is independent of $\boldsymbol{x}$

$$E[Y(\boldsymbol{x})\,Y(\boldsymbol{x} + \boldsymbol{h})] - m^2 = K(\boldsymbol{h}) \tag{53}$$

In many cases, only the increments of the functions are supposed to be stationary. The intrinsic hypothesis for a vector $\boldsymbol{h}$ concerning expected value and variance is

$$E[Y(\boldsymbol{x} + \boldsymbol{h}) - Y(\boldsymbol{x})] = 0 \tag{54}$$

$$Var[Y(\boldsymbol{x} + \boldsymbol{h}) - Y(\boldsymbol{x})] = 2\gamma(\boldsymbol{h}) \tag{55}$$

The so-called semivariogram (designated "variogram" further on), i.e., the function of the vector $\boldsymbol{h}$ is defined as

$$\gamma(\boldsymbol{h}) = 1/2\, Var[Y(\boldsymbol{x} + \boldsymbol{h})\, Y(\boldsymbol{x})] \tag{56}$$

From Equations (55) and (56),

$$\gamma(\boldsymbol{h}) = 1/2\, E[Y(\boldsymbol{x}_i + \boldsymbol{h}) - Y(\boldsymbol{x}_i)]^2 \tag{57}$$

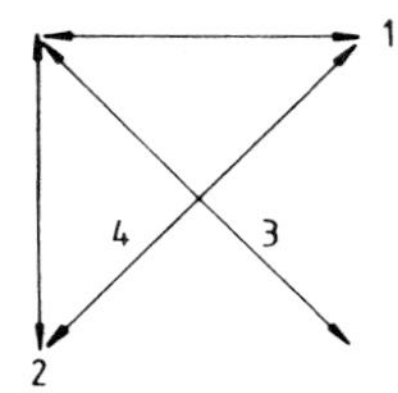

Figure 34. Principal axes of a variogram

In case of discontinuous data the expected value $(\boldsymbol{h})$ can be estimated by the formula

$$(\boldsymbol{h}) = \frac{1}{2n} \sum_{i=1}^{n} [Y(\boldsymbol{x}_i + \boldsymbol{h}) - Y(\boldsymbol{x}_i)]^2 \tag{58}$$

where n is the number of pairs of points.

The points are situated in either one-, two- or three-dimensional space. In two-dimensional space as dealt with here, the coordinates h_1 and h_2 determine the vector $\boldsymbol{h}$. Hence, the variance of measured values depends on the distance and the direction of the difference vector $\boldsymbol{h}$.

In practice, the variogram is usually computed for four main axes (Fig. 34) to account for possible directional effects. Distinctions of the range in different directions (anisotropy) then enable a more detailed interpretation.

A mean variogram that is independent of such directional effects is usually computed on the basis of the four directional variograms. A variogram is geometrically characterized by two boundary criteria called "sill" and "range" (Fig. 35). The latter denotes the maximum extent of influence, while the former is the analogous limiting value of influence as measured on the ordinate.

Interpretation of the Variogram. *Zone of Influence.* The foregoing indicates that a variogram does not necessarily have a maximum or a level of stabilization. Because of this, two basic types of variograms are distinguished (Fig. 36). In Figure 36 A the maximum range is reached when the correlation between $Y(\boldsymbol{x})$ and $Y(\boldsymbol{x} + \boldsymbol{h})$ becomes nil

$$\gamma(\boldsymbol{h}) = 1/2\, Var[Y(\boldsymbol{x} + \boldsymbol{h}) - Y(\boldsymbol{x})] \tag{59.1}$$

$$= 1/2\{Var[Y(\boldsymbol{x} + \boldsymbol{h})] + Var[Y(\boldsymbol{x})]\} \tag{59.2}$$

$$= \frac{2\sigma^2}{2} = \sigma^2 \tag{59.3}$$

This means that a mathematically meaningful interpolation is possible if and only if the

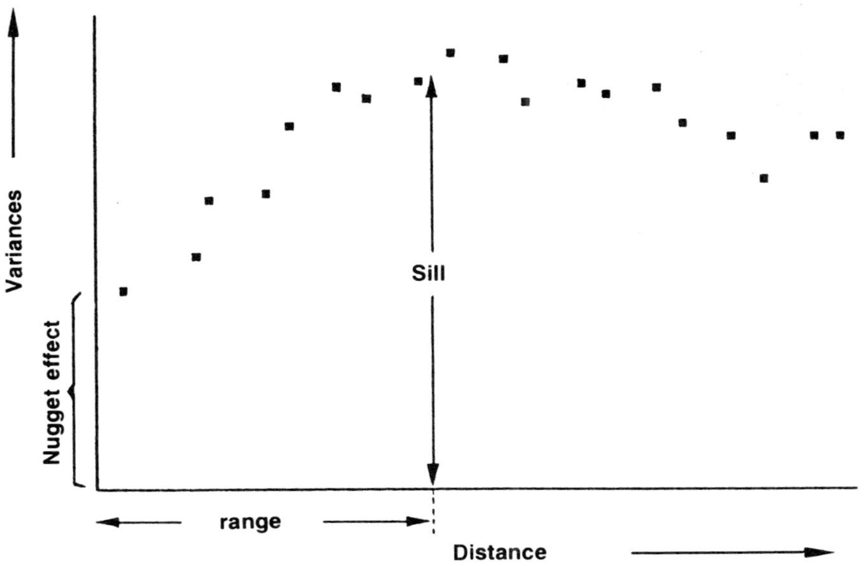

Figure 35. Sill and range of an experimental variogram

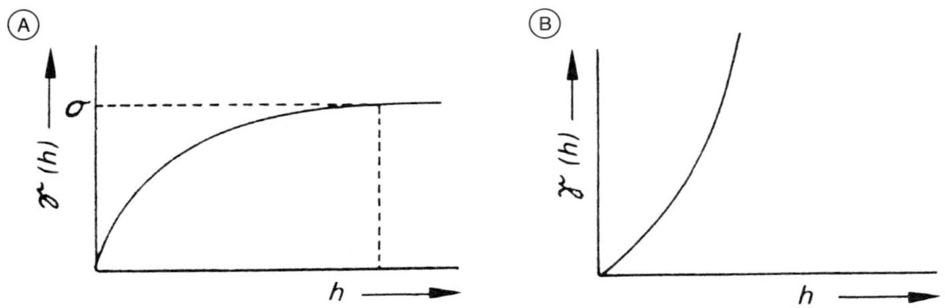

Figure 36. Bounded (A) and unbounded (B) variogram

maximum distance between neighboring points of measurement is shorter than the distance indicated by the range value of the variogram. In Figure 36 B the zone of influence extends beyond the area examined.

Mathematical Models of the Curve. To best fit the curve of the variogram under construction to the sequence of points obtained primarily from computation of the variance function, various mathematical models are used, the most important of which are described briefly (Fig. 37).

Power functions (Fig. 37 A):

$$\gamma(\boldsymbol{h}) = C|\boldsymbol{h}|^{\gamma} \quad \text{with } 0 < \gamma < 2 \tag{60}$$

$\gamma(\boldsymbol{h}) = C|\boldsymbol{h}|$ is the special case of a linear model.

Spherical model (Fig. 37 B):

$$\gamma(\boldsymbol{h}) = C\left[3/2\frac{\boldsymbol{h}}{a} - 1/2\frac{|\boldsymbol{h}|^3}{a^3}\right] \quad \text{for } |\boldsymbol{h}| \leq a \tag{61}$$

$$\gamma(\boldsymbol{h}) = c \quad \text{for } |\boldsymbol{h}| > a$$

Exponential model (Fig. 37 C):

$$\gamma(\boldsymbol{h}) = C[1 - e^{\frac{-|\boldsymbol{h}|}{a}}]; \quad a \cong 1/3 \text{ of range} \tag{62}$$

Gaussian model (Fig. 37 D):

$$\gamma(\boldsymbol{h}) = C[1 - e^{\frac{-|\boldsymbol{h}|^2}{a^2}}]; \quad a \cong 0.58 \times \text{range} \tag{63}$$

Often, for a more exact fit the models must be combined. The variance of the sample is approximately the same as the value of the sill as far as bounded variograms are concerned. The model

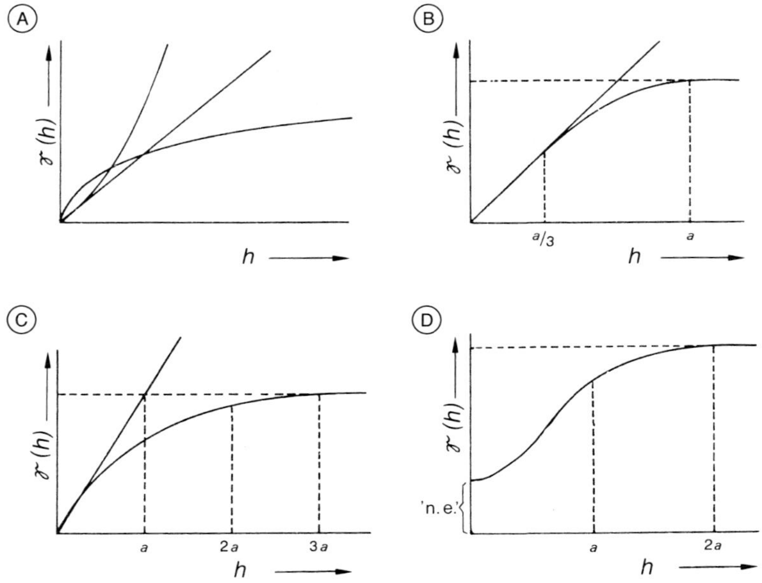

Figure 37. Mathematical models of a variogram
A) Power function model; B) Spherical model; C) Exponential model; D) Gaussian model with "nugget effect"

of the curve is important primarily for determination of the "nugget effect," a phenomenon that must be dealt with when the curve's behavior near the origin is considered.

Behavior of the Curve near the Origin.

1) A parabolic shape shows a high degree of continuity of the regionalized variable. It is differentiable.
2) A linear shape shows continuity "in average" according to MATHERON [235].
3) The curve does not intersect the abscissa at the origin. The nugget effect reveals great irregularity and can be caused fundamentally either by an extremely discontinuous distribution in the immediate neighborhood of the sample taken or by observational errors. Supplementary information is consequently needed.
4) A straight line parallel to the abscissa indicates that no correlation exists between any points $Y(\boldsymbol{x} + \boldsymbol{h})$ and $Y(\boldsymbol{x})$ whatever their distance might be. The sample does not show any spatial structure.

Anisotropies. Distributions characterized by different variabilities in different directions are reflected in the resulting variogram and called anisotropy.

1) Geometrical or affine anisotropy exists whenever elliptical zones of influence can be deduced in a two-dimensional space
2) In a three-dimensional space, variations might also appear in the vertical direction; this is the case of stratified anisotropy

3.1.4.4. Modeling Chemical Distribution in Soil and Subsoil

Assessing chemical concentrations in soil by means of models normally involves two basically complementary approaches. One type of model simulates chemical movement in the unsaturated zone, (i.e., from the soil surface down to the water table); the other describes chemical transport in the saturated zone or aquifer. BONAZOUNTAS [257], FRÄNZLE [2], and MATTHESS [48] provide state-of-the-art reviews of these models.

Figures 38 and 39, representing one of the most comprehensive hierarchical models developed so far [251], illustrate that the two most important factors influencing chemical behavior in the unsaturated soil zone are sorption and biotransformation. The essential boundary conditions and processes involved (Section 2.4) are indicated in the model that is free from limiting

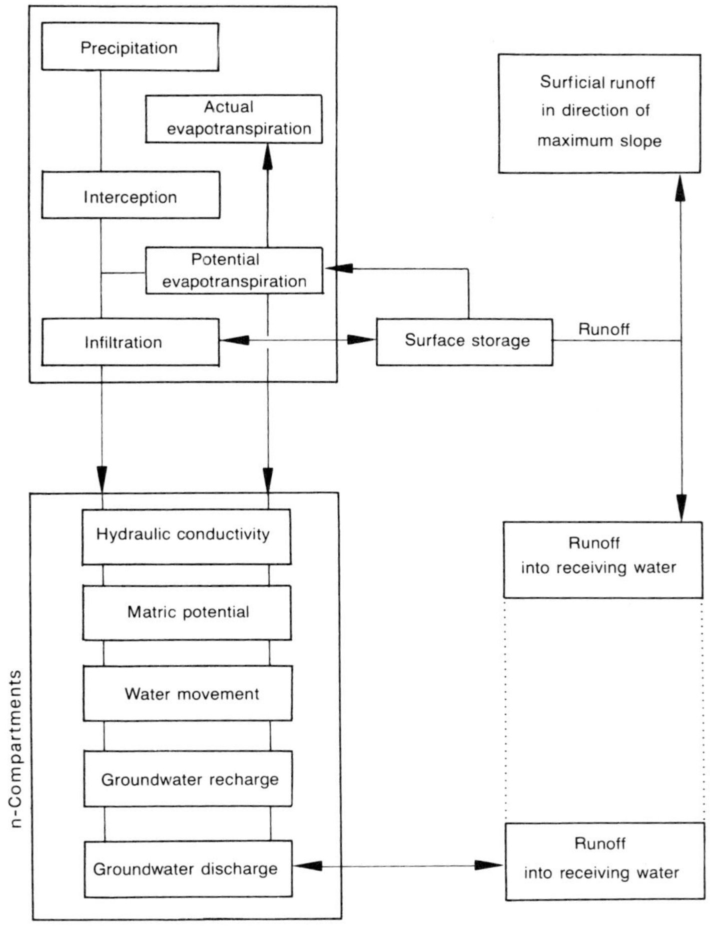

Figure 38. Soil moisture module of the integral WASMOD/STOMOD model (Wasserhaushaltsmodell/Stoffflußmodell, after [251], modified)

meso- and macro-scale assumptions such as homogeneity (i.e., homogeneous compartments and uniform distribution of chemicals) and stationarity (i.e., large-scale equilibria in extensive compartments). Extended versions of this model serve specific purposes, e.g., modeling of carbon and nitrogen fluxes through ecosystems (Fig. 40).

Small-Scale Assessment of Chemical Distribution. Fate and behavior of chemicals can be described approximately by means of table functions, see Table 12 [258]. In combination with geographic information systems, they permit the production of smaller-scale single-species sensitivity or distribution maps. Six parameters, defined in score form on a 0 (practically none) to 5 (very much, very high) scale, are employed to describe chemical reactions in soil:

1) Solubility in mg/L (20°C)
   0: <0.1, 1: 0.1–1, 2: 1–10, 3: 10–100, 4: 100–1000, 5: >1000
2) Adsorption, as defined by the Freundlich k values for organic matter (k_{oc}) and clay (k_c) at a soil–solution ratio of 1:2
   0: <0.5, 1: 0.5–50, 2: 50–150, 3: 150–500, 4: 500–5000, 5: >5000
   pH influence + or − (i.e., adsorption increases with increasing or decreasing pH)
3) Degradability, defined in terms of time required to produce an approximately 90% transformation at a soil temperature ranging from 11 to 16°C:
   1: >3 years, 2: 1–3 years, 3: 18 weeks to 1 year, 4: <18–6 weeks, 5: <6 weeks
4) Volatility, defined by the vapor pressure of the pure substance in (hPa) at 20–25°C:

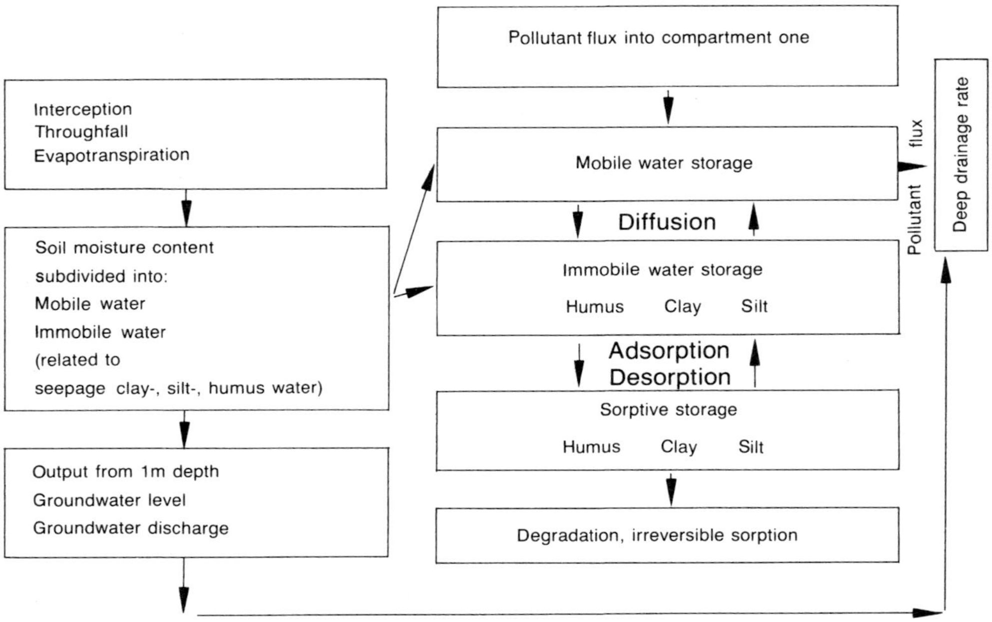

Figure 39. Solute module of WASMOD/STOMOD (Wasserhaushaltsmodell/Stofffflußmodell) (after [251], modified)

1: $<10^{-3}$, 2: 10^{-3}–100, 3: 100–500, 4: >500
or Henry constants
1: $<4\times10^{-6}$, 2: 4×10^{-4}, 3: 4×10^{-4}–0.04,
4: >0.04

In considering the influence of organic matter content on the adsorption process, the average scores of Table 12 must be defined more precisely according to Table 13. Further refinements relate to texture, pH value, temperature, etc. (Tables 14–16).

3.1.4.5. PEC Estimates in Aquatic Compartments

Establishing valid monitoring networks or sampling grids poses distinctly lesser problems in surface waters than in soils, but the inherent difficulties should not be underestimated. For instance, determination of the surficial nitrate concentration pattern in a dimictic–holomictic lake of about 1-km^2 size requires 40–50 stations [166]. For other chemical species, network densities must be higher still in the light of variogram-analytical evaluations.

In the light of these technical difficulties, modeling approaches play a considerable role [237]. Differentiation between large-scale dilution models operating at the regional level, (e.g., a whole river), and site-specific models requiring a more specific characterization of the projected site is indicated. Simple dilution models assume a homogeneous distribution of chemicals in a water body; i.e., for a given set of stream or river and sewage treatment (→ Water, Chap. 1) or discharge conditions, an average concentration is calculated. Site-specific models may yield concentration profiles of a chemical in the water body. Also concentrations of chemicals released into the river at many points can be estimated by dilution modeling. The simulation then comprises a relatively large region, typically a major catchment or an entire country. When a sufficient amount of data on dilution factors is available, chemical exposure can be expressed in the form of PEC frequencies in the rivers considered.

The data required to run a surface water model relate to the character of the chemical, the hydrological structure of the catchment, and the discharge pattern [237]. The chemical data include molar mass, melting and boiling points, water solubility, ionization constant, and kinetic parameters relating to sorption, volatilization, photolysis, hydrolysis, oxidation, and biotransformation processes (see Section 2.4). Environmental data vary according to the complexity of

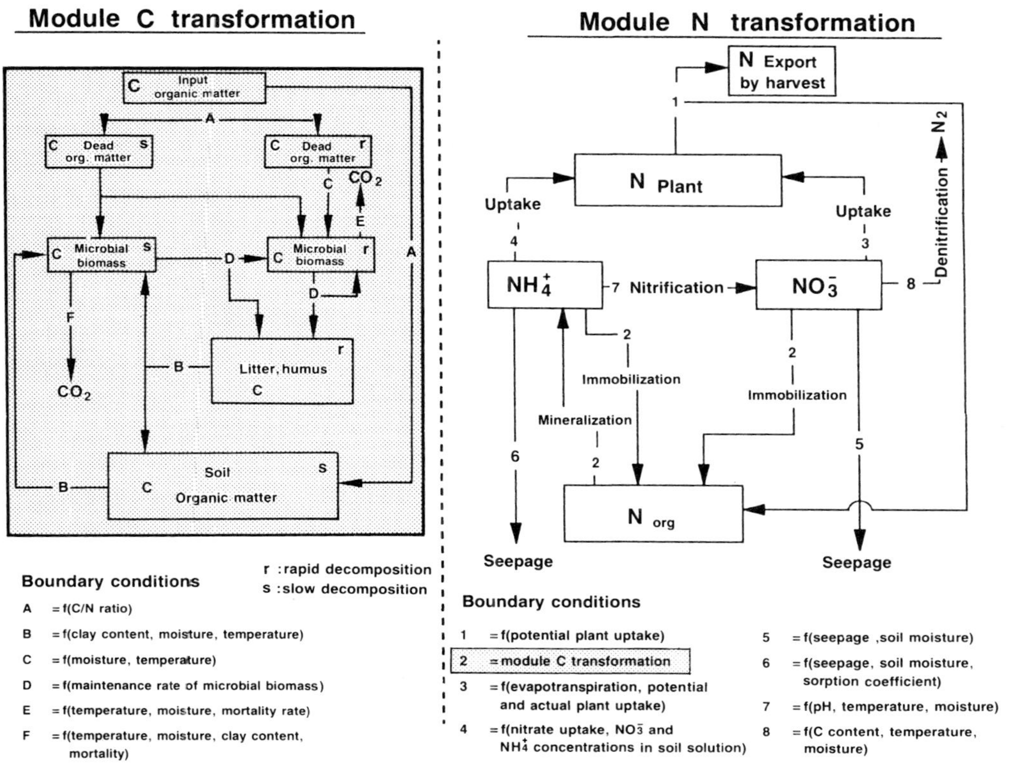

Figure 40. Carbon and nitrogen modules of the extended WASMOD/STOMOD (Wasserhaushaltsmodell/Stofffluß-modell) model (after REICHE, unpublished, modified)

the model, beginning with a stream dilution figure for a whole region and ending with more sophisticated approaches based on data on the number and discharge flows of sewage treatment plants and the flow regimes of rivers of a major catchment. Additional site-specific input data may include hydraulic geometry of channels, sediment load, terms of the catchment water balance, physical and chemical characteristics of the water body and its suspended and bottom sediments, etc. In addition, basic dilution models require data on the fraction of the chemicals released to the drain and its removal in different types of sewage treatment. Site-specific models require additional data on chemical loading for each simulated compartment. For spill models, volume and density of the spill are needed [259], [260].

For specific purposes, aquatic equilibrium modeling may become necessary to predict environmental concentrations more reliably. To this end, fast reactions of inorganic species dissolved in aqueous media can be classified under the following categories:

1) Reactions with solvent molecules, in particular dissociation
2) Substitution reactions with solvent or dissolved species
3) Redox reactions with dissolved gases or other ionic species

These reactions can lead to the formation of new complex ions with different ligands in the coordination sphere, ions with different oxidation state of the metal centers, or still other ions. The reactions are essential, since the medium in which the particular inorganic substance is dissolved largely determines the speciation of the particular metal ion. In addition, upon mixing of solutions of these metals with the environment, chemical modification of the species will occur initially by thermodynamically favored rapid re-

Table 12. Behavior of organic chemicals in soil (after [258], for explanation of figures see pp. 93–94)

Chemicals	Aqueous solubility	Adsorption to		pH influence	Degradation		Volatility	Concentrations*, mg/kg		
		Organic matter	Clay		Aerobic	Anaerobic			A	BC
Chlorinated alkanes and alkenes										
Dichloromethane	4	1	1	0	3–4	2	4	0.1	5	50
Trichloromethane	4	2	2	0	2–3	1–2	4	0.1	5	50
Tetrachloromethane	4	1–2	2	0	2–3	1–2	4	0.1	5	50
Dichloroethane	4	2–3	2	0	3	2	4	0.1	5	50
Monochloroethylene	4	1	1	0	4	4	3			
Trichloroethylene	4	2	1	0	3	2	4	0.1	5	50
Tetrachloroethylene	3	1–2	1	0	3	2	3	0.1	5	50
Carbocyclic compounds										
Benzene	4	1	1	0	3	2	4	0.01	0.5	5
1,2-Dichlorobenzene	3	3	2	0	2	2	3	0.05	1	10
1,2,4-Trichlorobenzene	2	3–4	1–2	0	2–3	2–4	3	0.05	1	10
Phenol	4	1–2	1–2	–	4	3	2	0.05	1	10
2,5-Dichlorophenol	4	3	2	–	3	4		0.01	0.5	5
2,4,5-Trichlorophenol	4	3	1	–	4	4		0.01	0.5	5
Pentachlorophenol	2	4	2	–	3–4	4	2	0.01	0.5	5
Aniline	4	1–2	1	+	4–5	4	2			
4-Chloroaniline	4	3	2–3	+	2	2	2			
Toluene	3	1	1	0	4	2–3	3	0.05	3	30
Toluidine	4	2–3	1	+	2	1				
Xylene	3	2–3	1–2	0	3		2	0.05	1	10

* A = reference level, can be regarded as unpolluted; B = indicative value for further investigation; C = clean-up necessary.

Table 13. Sorption capacity of soil related to organic matter content (after [258])

Organic matter content of soil, %	Relative sorption capacity (Table 12)				
	1	2	3	4	5
0.5–1	0	0	1	1.5	2
1–2	0	0.5	1.5	2	3
2–8	0.5	1	2	3	4
8–15	0.5	1.5	2.5	3.5	4.5
>15	1	2	3	4	5

* Average value of the uppermost 30 cm of soil (i.e., the Ah or Ap horizons).

Table 14. Sorption capacity of soil related to texture (after [258])

Texture classes*	Relative sorption capacity (Table 13)			
	1	2	3	4
1. S Ss'2*	0	0.5	1	1.5
2. Sc2 Sl2, 3 Ss' 3, 4 S's S'	0.5	1	1.5	2
3. Sls' Sl4 S'l2, 3 S'ls Ls Ls' Sc3 Cs4 S'C2, 3	0.5	1	2	3
4. S'l4 Cl Cs2, 3 C's Lcs Lc2, 3 Lcs' S'c4	1	1.5	2.5	3.5
5. C	1	2	3	4

* Medium texture of the uppermost 30 cm of soil (i.e., the Ah or Ap horizons). S = Sand, S' = Silt, L = Loam, C = Clay, s = Sandy, s' = Silty, l = Loamy, c = Clayey; 2 = low, 3 = medium, 4 = high content.

actions and subsequently by slower reactions [261]–[263]. The following slower reactions are important in aqueous media [257]:

1) Ligand substitution reactions of kinetically relatively inert ions
2) Electron-transfer reactions involving inner-sphere mechanisms for relatively inert ions and some outer-sphere reactions
3) Reactions with dissolved gaseous species or bacterially induced reduction
4) Precipitation of solids by formation of insoluble species through substitution reactions
5) Formation of metal hydroxides and sulfides by oxidation reactions

Table 15. pH control of relative sorption capacity of soil (after [258])

pH influence	pH ($CaCl_2$)			
	>6.5	6.5–5.5	5.5–4	<4
+	+0.5	0	−0.5	−1
−	−0.5	0	+0.5	+1

Table 16. Degradability of organic chemicals related to mean air temperature (after [258])

Degradability (Table 12)	Mean temperature of vegetation period, °C			Mean annual temperature, °C		
	21–16	16–11	11–6	12–9	9–6	6–3
1	1.5	1	0.5	0.5	0.5	0
2	2.5	2	1.5	1.5	1	0.5
3	3.5	3	2.5	2.5	2	1.5
4	4.5	4	3.5	3.5	3	2.5
4–5	5	4.5	4	4	3.5	3

These slower reactions impact the speciation and concentration in the aqueous phase and determine both type and extent of further interactions in a way comparable to fast reactions. Therefore, the time frame for some of the above reactions may indicate that characterizing the speciation in the original aqueous phase will also define the chemical behavior of the compound in the environment, whereas in other cases, understanding speciation implies exact knowledge of environmental interactions. To this end, a larger number of speciation models have been developed to address a series of specific situations. They are based on chemical thermodynamic reactions and principles, and also account for additional processes such as redox reactions [264], adsorption [265], complexation [266]–[268], ligand exchange [269], and others [270].

3.2. Effects Assessment

The assessment of possible effects of chemicals on the environment involves the determination of no-effect concentrations (PNECs) for aquatic species, edaphic species, and for species that may be affected by ecological magnification, i.e., by accumulation of substances through the food web. In addition, assessment procedures are indicated for substances that may have a negative impact on natural physicochemical processes in the atmosphere (for definitions and terminology of risk, see Appendix, Chap. 4).

Hitherto the predictive assessment of dose–effect relationships between chemicals and biotic elements in the environment has usually started with laboratory tests. Their results need extrapolation to the complex real-world situation, which proves rather difficult, as a rule, since the experiments are performed with defined amounts of chemicals in contact with a single species or restricted group of organisms. In contrast, the integral approach to assessing the effects of potential pollutants in the natural environment has to deal with the manifold interactions and complicating factors or boundary conditions, which are carefully avoided or controlled in the laboratory situation (see also → Toxicology, Chaps. 3–5).

The following are most prominent of these complicating factors in the natural environment [271]:

1) Uneven distribution of the pollutant in an ecosystem compartment because of gradients or currents, dilution, volatilization, sorption processes, etc. Bioaccumulation in successive elements of the food web may cause considerable variations in the individual load of different species.
2) Differences in specific sensitivity may lead to shifts in the abundance and diversity proportions of the biocenoses affected. Interspecific relationships, synergistic relations, and food web effects can be analyzed only in multispecies systems.
3) In a natural ecosystem, the effect of a pollutant on the biocenosis will be greatly influenced by weather, diurnal and seasonal variations, simultaneously occurring variable loads such as thermal and chemical loads, along with intermittent discharges.

To a certain extent, the influence of these complications can be accounted for by simulation techniques in the laboratory, but practical limitations necessarily exist. On the other hand, retrospective field experiments alone cannot be relied on since the doses are not known exactly, intentional pollution for experimental purposes is frequently prohibited, and biological observations are time-consuming and expensive. As a consequence, comparative tests should be carried out with selected model chemicals under both laboratory and natural conditions to determine the relative representativity of the test procedure. This validation of the tests should also pay appropriate attention to the important question of how far the test system or the experimental field plot is representative from a regional point of view [159].

In comparing this comprehensive, normative list of ecotoxicological models with the (still) predominant present-day practice, a fairly marked tendency exists to favor the bioindicator approach and aquarium or microcosm tests or technically comparable procedures [272], [273]. This has the appreciable advantage of relative simplicity and ensures a fairly high amount of reproducibility, but also poses the above problems as to rationally and reliably extrapolating to the complex real-world situation.

3.2.1. Test Systems for Aquatic Ecosystems

3.2.1.1. Level 0

Test procedures on the infraorganismic level (i.e., organells, cells, tissues) are omitted here and the hierarchical system suggested in the OECD Chemicals Testing Programme is followed [274]; consequently three successive levels of sophistication labeled 0, 1, and 2 are distinguished. The results of level 0 tests together with an increase in production of the chemical considered and its type of use pattern should indicate that subsequent tests on higher levels have to be carried out.

The following set of studies is considered relevant for the level 0 dossier:

1) Acute toxicity study with *Daphnia magna*
2) Acute toxicity study with fish
3) A static biodegradation test

According to the 7th Amendment to the Directive 67/548/EEC, Annex VII A, the minimal data basis (base set) for effects assessment purposes comprises acute tests for fish, *Daphnia*, and algae. In the OECD testing hierarchy, the latter forms part of level 1 enquiries.

Toxicity Test with *Daphnia magna*. The water flea *Daphnia magna* (Daphniidae, Cladocera) is a suitable test organism for static and semistatic tests and for flow-through systems. The results of these tests are relevant for initial hazard assessment because of the trophic importance of daphnids as primary consumers, their sensitivity to chemicals, and their relationship to various other ecologically important aquatic invertebrates. Also, automatic biomonitoring devices are feasible. The tests usually consist of two phases:

1) An acute phase that gives the 24-h EC_{50} value, the highest concentration that still causes no immobilization (EC_0), and the lowest concentration that causes 100% immobilization (EC_{100})
2) A reproduction phase yielding the EC_{50} and LC_{50} values at 24, 48, and 96 h; 7 and 14 d; and the end of the test (further information is provided by time span until the emergence of the first brood, the number of live and dead young, their condition, etc.)

The slope of the time–response curve depends largely on the speed and degree of absorption, inherent detoxification, and excretion mechanisms. While a steep slope usually indicates rapid absorption and toxic effects appearing within a short period of exposure, a shallow slope may be due either to low absorption or to toxic effects delayed beyond the period of observation.

The 14-d effect level in relation to the 24-h LC_{50} is, to a certain extent, indicative of the potential for chronic toxicity of a compound. Since the test duration is limited to two weeks, however, it may provide misleading results if the chemicals considered are taken up slowly or have delayed toxic activity. Yet the *Daphnia* test is in any case useful for determining whether further ecotoxicity studies are necessary to indicate which parts of an ecosystem or which species are most sensitive. A low effect concentration may trigger further tests while, conversely, a relatively high *Daphnia* effect concentration may guide any further testing away from invertebrates [275], [276].

Acute Toxicity Studies with Fish. Fish are of particular importance in aquatic environments, and their survival in waters bears witness to the absence of toxic substances or adverse conditions for the species in question. Consequently, toxicity tests with fish have been carried out in laboratories for about a century, but not infrequently the results obtained have differed widely because of lack of standardization of test organisms and boundary conditions.

Therefore, considerable numbers of studies have been undertaken since the 1970s to define the parameters involved in performing toxicity tests: taxonomy, age, and size of fish; composition of environments, in particular pH value, buffering capacity, mineral and organic content, temperature, and oxygenation; duration of contact with toxic substances; etc. The result has been the perfection of numerous techniques that can be applied in fresh water or seawater; they yield satisfactory results, depending on their objectives, although they cannot always be applied readily to other cases in view of major differences in the experimental conditions and the sensitivity of species selected.

Species frequently used in such studies are *Brachydanio rerio*, *Carassius auratus*, *Leuciscus idus*, *Phoxinus phoxinus*, *Rasbora heteromorpha* (Cyprinidae); *Jordanella floridae*, *Oryzias latipes* (Cyprinodontidae); *Poecilia reticulata* (Poeciliidae); and *Oncorhynchus mykiss* (Salmonidae). They comprise species that occur in aquatic environments into which the chemicals under study are released and so-called aquarium fish, which (frequently) are not representative of natural environments but are more suitable for screening tests.

The acute toxicity test on fish [274] is specifically designed to estimate the LC_{50} with observations after 24, 48, 72, and 96 h. However, the longer durations are reported to yield possibly misleading results because test organisms are not fed during the experiment [277]. The resulting data are used to prepare concentration–mortality curves with LC_{50} values, which also permit estimation of the maximum concentration causing no mortality and the minimum concentration causing 100% mortality. Where possible, LC_{50} values are also estimated for each of the recommended observation times, thus giving a more precise indication of the time–toxicity curve.

As for Daphniidae, comparison of the LC_{50} values with the potential environmental concentrations of a chemical will provide a first crude estimate of the potential hazard to fish. When the slope of the curve is steep and soon tends to be asymptotic to the time axis, longer exposure is not likely to result in a significant increase in mortality unless noxious effects that result from different mechanisms are delayed. Conversely, where the slope is shallow, cumulative toxicity and subsequent chronic effects may be indicated and an extension of the test period could result in a considerably lower LC_{50} estimate. For a more detailed review of the increasing range of types of acute toxicity tests on fish, see [47], [274], [277], [278]. In view of the numerous tests suggested, however, one should note that environmentally acute lethal toxicity is not the major problem. Acute lethal poisoning is easily detectable and should be readily controllable as well. Thus, subchronic and chronic exposures are likely to be the sources of problems. Consequently, obtaining information on the approximate magnitude of acutely lethal concentrations over an exposure period of up to 96 h might appear sufficient: "The greater the range of water qualities in which the acute LC_{50} is measured and the greater the variety of fish species used, the more likely is it that sufficiently cautious estimates can be arrived at. In this field of toxicity testing it is more important to get some measure of the variability existing, and ever perhaps try to account for it, than it is to mask any real differences which exist by conveniently standardizing tests. (Standard tests with fish appear, in fact, not to be concerned with the assessment of toxicity as such but rather with reproducibly bioassaying the concentration of a given chemical in some particular quality of water)" [279].

Tests for Biodegradation. Assessment of the environmental hazard of a particular chemical involves a prediction of interactions that may lead to degradation of the chemical or its eventual accumulation in particular environmental compartments. In the framework of the OECD Chemicals Testing Programme [274], the following test procedures *defining ready biodegradibility* have been suggested:

1) Modified AFNOR test
2) Modified Sturm test
3) Closed bottle test
4) Modified OECD screening test

These tests provide limited opportunities for biodegradation and acclimatization. Hence, a chemical giving a positive result may be assumed to biodegrade in the environment rapidly and, therefore, may be classified as "readily biodegradable."

Modified AFNOR Test. The modified AFNOR test requires a test material that must at least be soluble at the concentration tested [40 mg dissolved organic carbon per liter (DOC/L)] have negligible vapor pressure, not be inhibitory to bacteria, and not significantly adsorb on glass surfaces. The test substance is degraded by chemicoorganotrophic microorganisms added in low concentrations with the products used as the sole source of carbon and energy. The organic carbon remaining in solution after 3, 7, 14, 28 (and 42) d is measured, and the corresponding level of biodegradation calculated. A detailed description of the test, which is sensitive to an order of magnitude of ±10%, is given in AFNOR T-90-302 Test and the OECD Guideline 301 A [230].

Modified Sturm Test. The modified Sturm test has the same requirements as the preceding one, but uses the amount of CO_2 produced as indicator variable [280]. To this end, a chemically defined liquid medium, essentially free of other organic carbon sources, is spiked with the test substance and inoculated with a low concentration of sewage microorganisms. The CO_2 released is trapped as $BaCO_3$. By reference to suitable blank controls, the total amount of CO_2 produced by the test compound is determined and calculated as the percentage of total CO_2 (TCO_2) that the chemical tested could have theoretically produced in relation to its carbon content. In the absence of toxic effects on the inoculum, the reproducibility of the Sturm test is ±5%. Unless the test is adapted to handle ^{14}C-labeled compounds, test chemical concentrations cannot be lower than 5 mg/L [230].

Closed Bottle Test. The purpose of the closed bottle test is measurement of the biodegradability of organic compounds in an aerobic, aqueous medium at a test concentration of 2 (i.e., standard concentration) to 10 mg/L of active material. The degradation or biotransformation is given as the biochemical oxygen demand (BOD) within 28 d as a percentage of either the theoretical oxygen demand (TOD) or the chemical oxygen demand (COD) [230]. Most data elaborated with this test pertain to water-soluble compounds, but volatile compounds and those of low water solubility may also be tested, at least in principle. In the closed bottle test, a predetermined amount of test substance is dissolved in a mineral nutrient solution to provide the above concentration of active substance per liter. The solution is inoculated with a small number of microorganisms from a mixed population (derived from unfertilized garden soil and activated sludge) and kept in closed bottles in the dark in a constant temperature bath at 20 ± 1 °C. Degradation is followed by oxygen analyses over a 28-d period. A control with inoculum, but without test material, is run parallel for the determination of oxygen blanks. The reproducibility of the test is appropriate for screening purposes but not sufficient for a final decision regarding biodegradability. Although most experience has been gathered with water-soluble chemicals, the structure of the tests also permits, in principle at least, application to volatile and insoluble compounds.

Modified OECD Screening Test. The modified OECD screening test aims at measuring the ultimate biodegradability of water soluble, non-volatile organic compounds in an aerobic, aqueous medium with a starting test concentration of 5–40 mg DOC per liter. Degradation is calculated as the percentage DOC removal within 28 d with respect to the test material. The compound to be tested is dissolved in an inorganic medium (i.e., a mineral nutrient solution, fortified with a trace element and essential vitamin solution), providing the above concentration. The solution is inoculated with a small number of microorganisms from a mixed population and aerated at 20–25 °C in the dark. Degradation is determined by DOC analysis over the 28-d period, which is checked by means of a standard. A control with inoculum, but without test material or standard, serves to determine DOC blanks [230]. Data obtained from these and similar tests are applicable only to aerobic biodegradation. Since a more comprehensive evaluation of the possible fate of a chemical must also take anaerobic situations into account, other test procedures may become necessary.

Other Tests for Ready Biodegradability. At least some 20 additional methods have been suggested to evaluate biodegradation in surface waters or sewage treatment plants [47]. On the basis of comparative degradation tests, GERIKE [281] came to the conclusion that determination of DOC removal usually yields the most reliable and relevant results and could therefore constitute the major source of primary information on ready biodegradability.

Tests for Inherent Biodegradability. Substances that are not readily biodegradable in terms of the above or comparable tests may be examined with the aim of assessing their intrinsic potential for biodegradation (i.e., their liability to microbial breakdown under favorable conditions). At present, the modified Zahn–Wellens test and several methods for assessing the biodegradation potential of soil are available for determining inherent biodegradability [282].

The *Zahn–Wellens test* [283] has a number of advantages. As a batch die-away test, it employs activated sludge, mineral nutrients, and the test substance in solution as the only source of carbon. The mixture is agitated and aerated in a 1–4-L glass under diffuse illumination for up to 28 d. Degradation is monitored by determining the DOC (or COD) values in the (filtered) solution at regular intervals. The ratio of eliminated DOC (or COD) after each interval to the initial concentration is expressed as percentage biodegradation; plotted versus time it gives the biodegradation curve. This test provides a good indication of inherent biodegradability and permits acclimatization phenomena to be detected from the degradation – time relationships recorded. Its drawback lies in the high initial concentration of the test substance, which may exert toxic effects on the inoculum [283]. Like some others, the Zahn – Wellens test is suitable only for water-soluble, nonvolatile compounds.

The principle of the inherent biodegradability test in soil is described in OECD Test Guideline 304 A [233]. A small (50-g) sample of soil is treated with the ^{14}C-labeled test chemical in a biometer flask apparatus. The amount of $^{14}CO_2$ released from the test substance is absorbed in a 0.1 N KOH solution and determined by liquid scintillation counting. The test is one example of a class of indirect estimations of microbial degradation processes that generally rely on some unique characteristic of the compound in question, such as the release of $^{14}CO_2$ or of chloride from organics containing organically bound chlorine. Destruction of aromatic compounds in model systems can also be detected in the loss of ultraviolet light absorbance as the aromatic ring is cleaved [284].

Direct Measurement of Ecotoxic Substances. Among the direct measurement methods, gas chromatography provides a rapid, simple, and highly sensitive tool, especially in combination with mass spectrometry and in the form of multidimensional headspace capillary gas chromatography. The latter method increased the reliability of the qualitative and quantitative analysis of volatile compounds by the simultaneous chromatographic separation of samples in two capillary columns of different polarity. It is not affected by disturbances due to salts and compounds with higher molar mass and avoids effects otherwise exerted by the extractant used. By virtue of its high degree of mechanization, headspace injection can be automated easily.

Bioassays. In bioassays for the degradation of toxic compounds, a plant, insect, microorganism, or other suitable indicator organism that is sensitive to the treated sample is exposed at regular intervals after initial modification of the soil. The substance or its inhibitory derivatives are assumed to remain as long as the susceptible species or strain fails to grow or develops atypically, and the length of incubation of the soil required before the indicator organism starts developing in the test sample as well as in untreated soil is considered a reflection of the time at which the toxic component has dissipated [285].

3.2.1.2. Level 1

Subject to positive results from the acute toxicity tests of level 0, an increase in production of the compound, and a risk estimate associated with the use pattern intended, the following test procedures are foreseen for the level 1 dossier:

1) Chronic toxicity study with *Daphnia magna*
2) Subacute toxicity study with fish
3) Algal test
4) Test for species accumulation
5) Additional biodegradation test

Chronic Toxicity Study with *Daphnia Magna*. The chronic toxicity study with *Daphnia magna* is thought expedient if the LC_{50} appeared lower than 100 mg/L, and it should also include determination of the "no-toxic-effect levels" for both reproduction and lethality. For higher LC_{50} values, such a test appears justified only on the condition that biodegradation has not been proved by the relevant test carried out on level 0.

Subacute Toxicity Study with Fish. The subacute toxicity study with fish is a medium-term test with an exposure time of at least 14 d. As such, it is generally dynamic and based on the substantiation of lethal or sublethal effects on growth, behavior, or reproduction, including the determination of "threshold levels." If the LC_{50} for fish has been assessed at values lower than 1000 mg/L, the same species is to be used for the subacute toxicity study. For higher LC_{50} values, such a test appears necessary only if the previous biodegradation study yielded inconclusive results [286].

Algal Test. An algal test [274] enables a determination of the effects of a substance on the growth of unicellular green algae under laboratory conditions over a period of several cell generations. Since algae, as primary producers, are at the very base of aquatic food webs, their increase or decrease may influence higher trophic levels in natural aquatic ecosystems. In this respect, the possible toxic influences of a chemical on algal growth are highly relevant for the assessment of potential effects in the natural environment. The value of the test is, however, necessarily limited by the inherent difficulty of extrapolating laboratory conditions to the natural situation with its involvement of many other biotic and abiotic factors.

To obtain reproducible results, the algae (frequently *Chlorella pyrenoidosa*, *Scenedesmus quadricauda*, or *Scenedesmus subspicatus*) are exposed to the test substance at defined concentrations for 96 h. The algal growth, measured by the increase in biomass or cell number, is determined at least 24, 48, 72, and 96 h after the start of the test, and the corresponding concentrations per milliliter are calculated. The EC_{50} value is calculated graphically or by computation.

For assessment purposes, both the intrinsic growth rate (i.e., slope of the curve during logarithmic growth) of the control and experimental algal population and the total yield have proved useful. Even if the 96-h EC_{50} value does not seem to reflect high toxicity, this may be due to the fact that the test chemical has degraded considerably in the course of the test.

Test for Species Bioaccumulation. The potential of a chemical to bioaccumulate in aquatic organisms is determined by means of a sequential static test for species bioaccumulation, for which fish are preferably the animals of choice. Bioaccumulation is expressed by the bioconcentration of the chemical in the test organism relative to that in water under steady-state conditions [233]. In the first step of the test procedure, the time interval necessary to reach the steady-state concentration is determined by exposing one group of fish to one concentration of the test chemical without renewal of the test solution. In the second step, the dependence of bioaccumulation on various (i.e., at least three) concentrations of the test chemical is determined, and the result obtained is expressed in form of a sorption isotherm. In the third step of the procedure, the duration rate is measured. To this end, fish exposed to the maximum test compound concentrations in the preceding step are transferred to noncontaminated water from which the depurated compound is continually removed. The residual concentration in the fish is determined periodically until about 80 % depletion is reached or for a period of up to ten days, whichever comes first.

The test for species bioaccumulation was established with ^{14}C-labeled pesticides such as 2,4-D, atrazine, DDT, and others; consequently, such compounds can be appropriately used for standardization. The use of nonlabeled chemicals as standards requires highly sensitive (ppb-range) and compound-specific analytical methods. With respect to specificity, the test is suitable for nonionized organics that are not readily biodegradable, are relatively stable in the aquatic environment (i.e., half-life >4 d), are soluble in water at <2000 ppm ($\mu g/g$), and have an *n*-octanol–water partition coefficient (K_{ow}) >1000 and a volatility (i.e., water–air partition coefficient) >100.

With regard to direct bioaccumulation in individual species, it clearly correlates with the degree of lipophilicity, expressed in terms of the *n*-octanol–water partition coefficient (see Section 2.4.1). Exceptions normally relate to cases where K_{ow} overestimates the bioconcentration factor. The higher the K_{ow} of a chemical (i.e., the more lipophilic it is) the slower is it depurated (desorbed) from the organism. Since this process does not always follow first-order kinetics, however, apparent half-lives should not be extrapolated beyond the ranges measured.

In addition, reactions of chemicals in living organisms may result in derivatives that are more lipophilic than the parent compound or in binding (parts of) the chemical to special cell constituents. The latter applies, in particular, to inorganic or organometallic chemicals. Clearly, in either case, the K_{ow} value of the parent compound as such is not an appropriate indicator of the bioaccumulation potential. For organic compounds ionizing under physiological conditions (i.e., in the pH range 3–9), K_{ow} is predictive only if the chemical has no specific reactivity to cell constituents.

If the studies carried out on level 0 did not yield a satisfactory amount of information about biodegradation, another *static biodegradation test* can be chosen. If its result is negative again, a dynamic test in a flow-through system on the basis of lower concentrations, together with another inoculum, should then be used.

3.2.1.3. Level 2

The results of studies carried out on test level 1 are preferably to indicate which tests are to be made on level 2; this level is characterized by

1) An additional test for bioaccumulation
2) A chronic toxicity study with fish

3) An acute and a subacute toxicity test with another organism

Additional Bioaccumulation Studies. In the framework of supplemental bioaccumulation studies, the water solubility and *n*-octanol–water partition coefficient of the chemical in question merit particular attention (Section 2.4.1). Depending on the results of the previous accumulation test on level 1 and the physical and chemical properties of the compound, a flow-through test may be appropriate. It provides an opportunity for more adequate regulation of test conditions, including actual exposure concentrations, and is not limited in duration, thus allowing more comprehensive assessment of the significance of exposure time in evaluating bioaccumulation. The test substances are dissolved, if necessary with surfactant, in water. The solution is introduced continuously into an aquarium with constantly flowing water where the test fish are kept. The concentrations of chemicals in both water and test fish are analyzed, and the BCF is calculated. Since the variability of bioconcentration factors in several fish species is fairly high, it may be useful to achieve a higher degree of normalization by relating BCFs to the lipid content of the fish.

A technical advantage of the easily applicable test [287] is the possibility of automatic control. The dissolved oxygen concentration, pH, water temperature, etc. can be automatically and continuously measured and recorded.

Chronic Toxicity Study. A chronic toxicity study with fish is indicated if the "no-effect level" for reproduction is more than a factor 10 lower than the no-effect level for lethality of *Daphnia magna*. The same applies if the threshold level for fish as determined under level 1 is 10 times lower than the LC_{50} for fish. According to the relevant OECD Test Guideline [286], fish such as the aforementioned *Poecilia*, *Oryzias*, and *Jordanella* species should preferably be used as test organisms. On the basis of present limited experience, the tendency is to believe that the results of a toxicology study with a second species are not likely to change the estimated "no-toxic-effect" level to a great extent. Therefore, acute and subacute toxicity studies with another aquatic or terrestrial species, for instance the Japanese quail, should be performed only if the need is really indicated.

Representatives of the trophic levels described are listed below. Standard organisms are underlined. The assignment of an organism to a trophic level is based on the energy balance of the ecosystem considered and does not depend primarily on the species. Therefore, a population may represent more than one trophic level in dependence on the nutrition available for the species. In addition, earlier life stages often live on completely different nutrition compared to adults of the same species.

Primary Producers. Primary producers photo- or chemoautotrophically synthesize organic compounds using inorganic precursors. They include

1) Chlorophyll-containing species of vascular plants
2) Algae (e.g. green algae: *Selenastrum*, *Scenedesmus*, *Chlorella*; blue-green algae: *Microcystis*)
3) Purple sulfur bacteria, chlorobacteria
4) Chemoautotrophic bacteria (nitrifying bacteria, sulfur bacteria)

Primary Consumers. They live mainly on living or dead autotrophic organisms or on microorganisms. Representatives of this trophic level are especially plant-eating animals (i.e., species that are not carnivorous) of the following taxonomic groups:

1) Protozoa (e.g., *Uronema*, *Entosiphon*, *Tetrahymena*)
2) Annelids (e.g., *Tubifex*, *Enchytraeus*)
3) Crustacea (e.g., *Artemia*, *Daphnia sp.*, *Copepoda*, *Gammarus*, *Asellus*)
4) Mollusks (e.g., *Dreissensia*, *Mytilus*, *Ostrea*; several gastropods: *Patella*, *Viviparus*)
5) Insects (some insect larvae that are not carnivorous)
6) Nematodes (species that live in water)

Secondary Consumers. They live mainly on primary consumers. Among them are

1) Predatory insects and larvae of insects (e.g., *Chaoborus*)
2) Carnivore protozoa
3) Rotatoria
4) Coelenterates (e.g., *Hydra*)
5) Predatory copepods
6) Fish (Teleostei: e.g., *Cyprinus carpio*, *Brachydanio rerio*, *Poecilia reticulata*, *Oryzias latipes*, *Pimephales promelas*, *Lepomis macrochirus*, *Oncorhynchus mykiss* (former: *Salmo gairdneri*) *Leuciscus idus melanotus*, *Cyprinodon*, *Carassius*
7) Amphibians (e.g., *Rana*, *Xenopus*)

Decomposers. Organisms of this trophic level break down dead organic material to inorganic constituents. Representatives are

1) Bacteria (most species)
2) Fungi

Test organisms for benthic ecosystems

Biological group/test organism	Parameter tested
Benthic invertebrates	
Panagrellus redivivus	survival
Caenorhabditis elegans	survival
Tubifex tubifex	survival
Stylodrilus heringianus	survival, repellency, rate of transformation of the sediment, growth
Hyalella azteca	survival, growth, reproduction
Pontoporeia hyi (*Diporeia sp.*)	survival, repellency
Corbicula fluminea	survival, growth
Anodonata imbecilis	survival
Chironomus tentans	survival, growth, hatching
Chironomus riparius	survival, growth
Hexagenia limbata	survival, frequency of exuviation
Macrobenthic biocenoses	indices for biocenoses and populations
Macrophytes	
Hydrilla verticilata	length of shoots and roots, dehydrogenase activity, chlorophyll-a, peroxidases

3.2.1.4. PNEC Values and Assessment Factors

The function of hazard or risk characterization and assessment is the overall protection of the environment. With regard to aquatic ecosystems, two habitual assumptions are therefore made to extrapolate from the above single-species short-term toxicity data to ecosystem effects: (1) ecosystem sensitivity depends on the most sensitive species, and (2) protecting community structure protects community function. While the latter assumption is hardly questionable, the former needs further examination in light of the introductory statements to Chapter 3.

Table 17. Assessment factors for aquatic toxicity testing (after [288]–[290])

	Factor		
Toxicity data	EC	OECD	ECETOC*
One short-term toxicity value: algae, daphnia, or fish		1000	
Short-term toxicity: algae, daphnia, and fish	1000 [lowest $L(E)C_{50}$]	100	200
Long-term toxicity: fish or daphnia	100 (NOEC)		
Long-term toxicity: algae, daphnia, and fish	10 (lowest NOEC)	10	5
Long-term toxicity: two species in two taxonomic groups	50 (lowest NOEC)		
Field or mesocosm toxicity	case-by-case evaluation		1

* ECETOC = European Chemical Industry Ecoloy and Toxicology Centre.

For many existing and most new substances, the pool of data on ecosystem effects is very limited; i.e., only short-term toxicity data are available as a rule. This implies the application of assessment factors that permit one "to predict a level at or above which the balance of probabilities suggests an environmental effect will occur. It is not intended to be a level below which the chemical is considered safe. However, again, the probability is that there will be no effects" [288]. When defining assessment factors, a number of uncertainties must be taken into account in attempting to extrapolate from mono- or oligospecific laboratory test systems to multispecies ecosystems: (1) intra- and interspecies variations, (2) acute to chronic toxicity extrapolation, and (3) extrapolation of laboratory data to field impact. Additional uncertainty results from the different testing methods themselves or interlaboratory variation in toxicity data.

When reviewing data from laboratory test systems the assessment factors of Table 17 are proposed. Instead of using assessment factors on the lowest-effect data, statistical extrapolation methods may be applied to a (relatively large) set of long-term data as an alternative approach. The underlying assumptions are [291]:

1) The distribution of species sensitivities follows a theoretical distribution function
2) The groups of species tested in the laboratory are a random sample of this distribution

This implies the use of no-observed-effect concentrations (NOECs) for identical parameters (e.g., reproduction) of the species considered [292], especially if only limited data from different species are available [293]. Normally long-term toxicity data are log-transformed and fitted according to the distribution function, and a prescribed percentile (e.g., 95 % [291]) of this distribution is used as a criterion. This means that the NOEC may be exceeded for 5 % of the species in the community. The validity of some of the above assumptions may be questioned (see also [293], [294]). First of all, it must be taken into account that for most compounds few long-term NOECs are available so that most tests on distribution functions have low power. Second, the NOECs available are not a random sample. Emans et al. [295] compared NOECs derived from multiple-species experiments for aquatic systems with extrapolated values [296], [297] and came to following conclusions:

1) Validation is rendered difficult by the lack of reliable data from multiple-species tests and long-term monospecific tests
2) Results from both statistical extrapolation approaches do not differ significantly
3) Results of the statistical extrapolation with 5 % confidence are in good agreement with results from (semi-) field tests

3.2.2. Testing Strategies for Terrestrial Ecosystems

Soil constitutes a central compartment of terrestrial ecosystems. Chemicals introduced into soil can be adsorbed on soil constituents or transformed by soil organisms; furthermore, they may be taken up by plants, washed out by rain or irrigation water, or evaporated in gaseous form. The extent of adsorption is controlled by various soil properties, including composition and content of organic matter, type and content of clay minerals, ion-exchange capacity, and acidity; it is also closely connected to the physical and chemical properties of the substance in solution (see Section 2.4.2). In view of this complex situation, the empirical knowledge of the behavior of environmental chemicals is still fairly limited, and the possible danger of excessive fertilizer or manure application, pesticides, and heavy-metal contamination of soils and crops has been recognized only in the last decades, mostly in relation to the agriculture use of sewage sludge or in connection with atmospheric deposition of organics or heavy metals.

Given the fundamental importance of soil for biological conservation, fertility, element fluxes, biodegradation of organic materials, and retention and filtration of groundwater, broad agreement exists as to the necessity of specific, use-oriented risk assessment procedures. As for risk characterization in aquatic ecosystems, the predicted no-effect concentration in soil ($PNEC_{soil}$) must be compared with the predicted environmental concentration in soil and subsoil (PEC_{soil}) as described in Section 3.1.4.

3.2.2.1. PNEC Estimates for Soil and Subsoil

Terrestrial Tests and Assessment Factors. A substance discharged into soil can have noxious effects on both the soil's physicochemical properties and its biocenoses. If the pore structure of a soil is affected by the chemical, the soil might silt up and be compacted; thus, the percolation rate is reduced considerably, the water balance is impaired (see Section 2.1.2), and the air supply is impeded. Since soil acts as a pH buffer system and a reservoir for bound substances, impairment of the soil adsorption capacity by xenobiotics can infer exhaustion of the buffer capacity. Impairment of microbial activity implies a reduction of the catabolic and metabolic capacity of the soil, which may lead to retarded liberation of nutrients and accumulation of mainly organic substances; this implies further effects on soil-dwelling organisms [35]. A chemical represents a risk to the soil compartment of an ecosystem if its concentration exceeds the degradation potential of the soil or if it changes the soil properties—in particular, the filtration, buffering, and transformation functions—in such a way that adverse effects result for the environment [298]. Thus, data on geoaccumulation as obtained in the framework of exposure assessment may give a first indication of possible changes in soil properties. The $PNEC_{soil}$ can be derived from the results of toxicity tests on soil and subsoil organisms by the application of assessment factors. If no ecotoxicological data are available for such organisms, the equilibrium partitioning method may be applied as a first approximation. It assumes that the bioavailability and therefore the toxicity of chemicals are determined by their concentration in the pore water of soil. The applicability of the method has been tested less for soil- than for sediment-dwelling organisms; Van Gestel [299] has demonstrated its validity for

Table 18. Assessment factors for terrestrial ecosystems (after [289])*

Information available	Assessment factor
$L(E)C_{50}$ for preliminary stage toxicity tests (e.g., acute toxicity to plants and earthworms or microorganisms)	1000
NOEC for one additional long-term toxicity test	100
NOEC for additional long-term toxicity tests for two species of two trophic levels	50
NOEC for additional long-term toxicity tests for three species of three trophic levels	10
Field data/data of model ecosystems	case by case

* The $PNEC_{soil}$ is calculated on the basis of the lowest effect value measured.

earthworm toxicity for several chlorophenols, chlorobenzenes, and chloroanilines. Whether this also holds true for other compounds or species is questionable, however (e.g., lipophilic compounds or organisms that are exposed primarily via food).

For few existing chemicals, short-term terrestrial tests are available with microorganisms, earthworms, and higher plants; in addition, a long-term test is used with springtails. For new substances, two standardized short-term tests with higher plants and earthworms are required at level 1. If no acute tests are available, tests on primary producers, consumers, and decomposers should be applied for the preliminary stage of the concentration-effect assessment [300]–[303]. The Umweltbundesamt [289] suggests use of the same assessment factors for terrestrial as for aquatic ecosystems, depending on the type of investigations, (i.e., acute or long-term toxicity test) and the number of trophic levels studied (see Table 18).

Standardized long-term tests for earthworms [304] and for springtails [303] are available, both of which analyze effects on reproduction. The standardization of two long-term tests is close to completion: the test on staphylinids (Coleoptera) registers degree of parasitism, hatching rate, and reproduction [305]; the test on *Enchytraeus* (Annelida) can replace the reproduction test on earthworms [306]. Van Straalen and Van Gestel [307], Stavola [308], and Pedersen and Samsøe-Petersen [309] discuss test methods with various terrestrial species and different degrees of standardization.

Evaluation of Stress Tolerance of Soils as Ecosystem Compartments. To better comply with the exigencies of a provision-oriented environmental policy, the critical loads of soil with regard to nutrients, heavy metals, and xenobiotics require a generalized ecological evaluation, in addition to estimates of human exposure to contaminants in soil (see → Toxicology, Chap. 5; → Soil, Section 6.2). This implies the rational definition of soil standards that aim at sustaining appropriate physicochemical and biotic soil functions.

A recent approach of this type is based on the regionalized determination of geogenic and anthropogenic concentrations of potentially hazardous chemicals, and the assessment of different degrees of adverse chemical effects on soil and related biota [298] (see Table 19). Thus, the

Table 19. Complex ecological threshold values for soil quality characterization [298]

Natural background "clean soil" C_p	Range of tolerable disturbance no further chemical stress C_{eff}	Range of permanent heavy disturbances changes in land use, hazard assessment, and restoration indicated C_{dis}
No reduction of ecological soil functions		
Multifunctionality: optimum range of ecological soil functions	first low-level adverse effects on ecological soil functions	long-lasting or permanent disturbances of essential ecological functions
No detrimental fluxes of chemicals	no detrimental fluxes of chemicals	detrimental fluxes of hazardous chemicals into neighboring ecosystems
Optimum conservation of species	almost complete conservation of species (HC_5)	species deletion (HC_{50})
Site-specific quasi-natural climax communities	→	possibilities of land use reduced
Any form of ecologically indicated land use possible	gradually increasing reduction of soil functions	biotic and abiotic soil functions reduced

preventive threshold value C_p characterizes chemical concentration levels not affecting any essential ecological soil function. The C_{eff} value is indicative of first and low-level adverse effects and is defined by means of ecotoxicological test procedures and determination of the physico-chemical behavior of xenobiotics in soil. By contrast, the C_{dis} value indicates long-lasting or permanent disturbance of essential biotic and abiotic soil functions. Among the extrapolation methods for determining ecologically relevant effect concentrations [292], [310]–[315], the last mentioned appears most appropriate. It defines a hazard concentration (HC_p) that ensures the protection of a certain number of soil-living animals. Thus, the approach is based explicitly on resilience considerations (see Section 2.1), using NOECs as reference values. The hazard assessment involves four steps:

1) Determination of at least five NOECs
2) Standardization of NOECs values with regard to a standard soil from at least three taxonomic groups according to the following equation

$$\mathrm{NOEC}' = \mathrm{NOEC}(L,H)\frac{R(25,10)}{R(L,H)} \quad (64)$$

where NOEC′ is the NOEC of standard soil; $\mathrm{NOEC}(L,H)$ ist the experimental NOEC with given clay (L) and organic matter content (H); $R(25,10)$ is the reference value of standard soil with 25 % clay and 10 % organic matter; and $R(L,H)$ is the reference value on the basis of the clay and organic matter content of the test soil.
3) Determination of a safety factor T in relation to the different sensitivity of indicator species

$$T = \exp\left[\frac{3\,S_m d_m}{\pi^2}\ln\left(\frac{1-\delta_1}{\delta_2}\right)\right] \quad (65)$$

where S_m is the standard deviation of ln NOECs; d_m is a factor dependent on the number of test species; δ_1 is the proportion of nonprotected species, related to defined hazard concentration (HC_p); δ_2 is the probability of overestimating HC_p; and T is a safety factor for NOEC average.
4) Based on this safety factor the hazard concentration HC_p is defined as

$$HC_p = \frac{\exp(x_m)}{T} = \frac{\overline{\mathrm{NOEC}}}{T} \quad (66)$$

where m ist the number of test species; x_m the mean value of ln NOECs; and $\overline{\mathrm{NOEC}}$ the geometric mean value of NOECs.

The exemplary application of Equation (66) to cadmium and a test system of *Dendrobaena rubida, Lumbricus rubellus, Eisenia foetida, Helix aspera, Porcellio scaber, Platynothrus peltife*, and *Orchsella cincta* would yield an HC_5 (i.e., $\delta_1 = 0.05$) of 0.16 mg Cd per kilogram of soil. In comparison, the present A level of the Dutch "Leidraad Bodemsanering" is equivalent to 0.80 mg Cd per kilogram of standard soil.

As long as the C_{eff} value, defined by means of the above prodecure, is not exceeded, no essential reduction of ecological soil functions occurs and the long-term multifunctionality of soil is sustained. In contrast to this threshold, the C_{dis} value indicates a distinctly higher concentration of pollutants in soil that exerts long-lasting or irreversible effects on biotic and regulative soil functions. For assessment purposes, basically the same approach is foreseen as for the definition of C_{eff} (i.e., multispecies determination of HC_p with a p around 50). Unter these circumstances, neighboring ecosystems may also be affected due to export of primary pollutants or their metabolites. Thus, the C_{dis} value characterizes a threshold that necessitates protective or rehabilitation measures. Depending on both quality and intensity of land use, these may be quite different, including, e.g., limitations of crop production or reduction of fertilizer and pesticide application [298].

Complex Biotic Indicator Systems for PNEC Estimates. Toxic effects of chemicals in a biocenosis change according to the species composition of the community exposed to them. This is particularly important when a community is continually exposed to a concentration of a substance since its species are either least sensitive to this substance or adapted to it. Consequently, the effects of the substance on a selected laboratory test organism or a monospecific assemblage not previously exposed to that particular substance may be quite different. In consequence of these and other limitations of the aforementioned tests, artificial laboratory microecosystems (microcosms that simulate some of the characteristics of natural ecosystems) are more frequently used in toxicological studies [287], [316].

Model Ecosystem Approach. The model ecosystem approach uses physical models of natural ecosystems to simulate a selected subset of the relevant physical, chemical, and biotic characteristics of a potentially affected environment. These models may vary in complexity from a bacterial culture in a test tube to a net cage isolating a part of an ecosystem. Normally, the model ecosystem involves an aquarium or other small container in which soil, sediment, water, and several trophic levels of organisms are kept. Thus, METCALF and coworkers [317] studied the effects of chemicals in a terrestrial–aquatic interface to determine their potential to accumulate in food chain organisms and to degrade. Tests with the *Metcalf microcosm* have been standardized to permit comparison of different chemicals. Although this system does not allow for prediction of absolute concentrations of a substance expected or observed in a wide variety of actual environmental conditions, it has shown positive relationships among physical, chemical, and biotic properties, (e.g., water solubility and bioconcentration potential). However, it has not been used typically to determine rates of dominant processes, nor has it sufficiently incorporated many of the dynamic properties found in other aquatic environments.

FREHE's *greenhouse approach* [318] is an example of microcosm research related to agriculture. Each microcosm, measuring $207 \times 193 \times 200\ cm^3$, was fixed on a lysimeter containing a soil mixture (loess loam, sand, and peat) with an earthworm (*Lumbricus terrestris*) population; an association of pea (*Pisum sativum*), vetch (*Vica sativa*), oat (*Avena sativa*), and mustard (*Sinapis alba*) was grown in it. The greenhouses were provided with ventilation and sprinkling systems permitting the application of di-(2-ethylhexyl) phthalate for testing purposes in aerosol form.

Studies in such small-scale laboratory systems are valuable tools to demonstrate some principles of concern in ecotoxicology and have to be included in the general knowledge of the fate of chemicals in natural large-scale ecosystems. They are indispensable additions to single-species experiments in order to more reliably evaluate the influence of interspecific interrelationships on the toxic potential of compounds. An example is zinc toxicity with respect to marine diatoms and dinoflagellates, which, in the majority of cases, proved to be definitely more pronounced in five-species tests than in monospecific cultures [319]. Hence, an abbreviated model ecosystem approach may prove useful in environmental concentration–fate studies by providing data on the rate constants of some of the processes that govern the transport and transformation of chemicals. Thus limited-scope model ecosystems contribute to determine whether a mathematical model accurately or approximately predicts concentrations in the various environmental compartments included in the model ecosystem. Expanded types of microcosms may be envisaged to include a substantially higher amount of that variety of physical, chemical, and biotic properties that a chemical is likely to encounter in the real environment. In terms of structural diversity, these comprehensive, and consequently more expensive, *mesocosms* lead over to biocenotic experiments under open-air conditions, which constitute the highest assessement level possible.

Biocenotic Experiments. Individual *populations* display marked changes in abundance as a direct consequence of the exposure pattern and an indirect consequence of the induced variations of intra- and interspecific competition and resource availability. Thus, almost all collembolan species exposed to cypermethrin displayed a longer increase in population density after an initial reduction, followed by a secondary decrease, probably due to excessive exploitation of food resources or an increase of predators [320]. In numerous experiments, complete annihilation of species was the exception; recedent taxa were mostly affected. Yet in a few cases, populations with dominant species were almost completely suppressed (e.g., the predatory mite *Rhodacarellus silesiacus* and the collembolan species *Isotomina bipunctata* on experimental plots treated with aldicarb [321]). Also, nontarget organisms experienced changes in their abundance spectra that lasted for several years [322].

At the *community level*, chemicals nearly always induce changes in the abundance spectra, whereas changes in species composition occur less frequently, particularly as a result of persistent multifactorial stress [323], [324].

Ecosystem PNEC Assessment. Integral PNEC indicators at the *ecosystem level* are

1) Fluxes of macro- and micronutrients (K, Ca, Mg, N, P, S and Fe, Mn, Cu, Zn, Mo, B, Cl)
2) Duration of biochemical cycles
3) Energy flux rates and entropy production of phytocenoses
4) Changes in biodiversity and population dynamics

5) Structure of trophic networks or experimental food chains

Examples of the first groups of indicators are the influence of pentachlorophenol (PCP), 2,4-D, 2,4,5-T, and aldicarb on the nitrogen balance and litter degration of beech, pasture, and fallow ecosystems [325], [326]. In either case, degradation was impeded and the nitrogen content of the litter bags reduced. Shortly after application of the chemicals, the release of inorganic nitrogen compounds increased considerably at the expense of microflora and fauna. Furthermore, the retention capacity and the availability of many nutrients were diminished, and these effects lasted for more than four years depending on the chemical character and application rate of the compounds tested.

3.2.2.2. Complex Assessment and Modeling of Vegetation Damage

Pollutant gases such as sulfur dioxide and nitrogen oxides are frequently emitted in combination and normally with other pollutants such as soot and heavy-metal particles; in many places, photooxidants are emitted simultaneously. In these cases, the effects of combinations of pollutants are not simply the sum of the effects of individual components, but the relative concentrations of components or the specific sensitivity of the plants to one of them may determine the response. Thus, pollutants in combination may display antagonistic (reduced), additive (sum of individual pollutant effects), or synergistic (superadditive) effects.

In the past, most experiments on exposure to combinations of pollutants were carried out under laboratory conditions using higher concentrations of gases than are typical in the field. Newer fumigation techniques, for instance open-top chambers [327] or field fumigation systems [328], have now enabled the exposure of plants under field conditions with increasingly realistic concentration ranges. These experiments indicate that pollutant combination effects depend on pollutant concentrations; SO_2 and O_3 are often antagonistic at levels where the individual gases cause severe leaf injury, while superadditive effects predominate at lower concentrations (i.e., SO_2 <140 µg/m^3, O_3 <100 µg/m^3), particularly near the threshold dosage [329]. In addition, the injury symptoms for SO_2 and O_3 in combination are not a simple mixture; frequently, the symptoms of one pollutant occur more clearly or earlier than those of the other [330].

With regard to combinations of NO_2, SO_2, and O_3, effects were produced in many plants by lower concentrations of NO_2 and the other gases than if they acted alone, especially at the lower exposure levels [331], [332]. Synergistic effects of NO_2 in combination with SO_2 or O_3 apparently result from a decrease of the nitrate and nitrite reductase activities. While the latter tend to rise in response to NO_2 acting on its own, activation of these reducing enzymes is limited or even suppressed by the simultaneous presence of SO_2 or O_3 [333]. This adverse effect is further enhanced by the formation of free radicals. An inhibitory influence on the regulation of transpiration occurred at less than 20 µL/m^3 SO_2 + 20 µL/m^3 NO_2 [334]. The combination of NO_2 with O_3 is likely to endanger vegetation more than NO_2 + SO_2 in combination and should receive greater attention since these pollutants form essential components of photochemical smog. The threat to vegetation may increase still further if all three pollutants interact simultaneously [331], [332], [335], [336].

Combination effects, however, also depend on the temporal order in which the pollutants affect the plants. If they are first exposed to NO_2 and thereafter to SO_2, the usual SO_2 effects are reduced, while the opposite sequence causes superadditive effects because SO_2 affects nitrite reductase [333], [337]. Attention must also be drawn to the combination of gaseous SO_2 with acid precipitation; the effects of this combination are discussed in connection with the new types of forest damage.

Critical Levels. In conclusion, and with some reservations resulting from the above remarks, critical levels for short-term and long-term exposures may be defined. In view of the fact that most horticultural and agricultural plants do not show adverse effects at *sulfur dioxide concentrations* <30 µg/m^3, while particularly sensitive species of trees, moss, lichens, and bushy and grassland vegetation are adversely influenced by concentrations of ca. 20 µg/m^3, the critical level for SO_2 acting on its own is to be set at 20 µg/m^3 (≈ 0.007 mL/m^3) as the annual mean value. Experimental fumigation trials that indicate first adverse effects from 70 µg/m^3 SO_2 (≈ 0.025 ppm) for cultivated plants form the basis for setting the short-term value at this concentration [330].

Leaf necroses were the predominant effects of short-term, mainly single exposures of plants to >1 mL/m^3 *nitrogen dioxide*; lower concentrations caused alterations in photosynthesis and respiration as well as in organelle ultrastructures. The particular sensitivity of plants at night must also be taken into account, and the fact that high NO_2 concentrations are often associated with high NO levels. For long-term exposures, a seasonal differentiation of sensitivity is indicated. Thus, for summer, when the physiologically more active plants can rapidly metabolize NO_2, a half-year mean of 0.03 mL/m^3 appears tolerable. In winter, however, when plants are particularly sensitive, a half-year mean of 0.02 mL/m^3 must not be exceeded to protect both short-lived plants and long-lived arboreal flora.

Experimental investigations of *ozone damage* [336] permit definition of the following critical levels for the protection of vegetation against O_3 as a single pollutant. For longer-lasting exposures, 50 μg/m^3 (≈0.025 mL/m^3) O_3 appears indicated as a vegetation period average (composed of 7-h daylight mean values).

For *NO_2 in combination with O_3 and SO_2*, the World Health Organization has adopted an Air Quality Guideline of 30 μg/m^3 NO_2 as the arithmetic annual mean for areas in which SO_2 and O_3 may occur in annual and vegetation period means of up to 30 μg/m^3 SO_2 and 60 μg/m^3 O_3 [332]. However, nitrogen oxides do not only have direct effects, they may also affect plants indirectly via the soil, with additional nitrogen input greatly endangering nutrient-poor plant formations such as heath and moor communities. An amount of 10 μg NO_2 per cubic meter may suffice to damage the particularly sensitive ombrotrophic mires [338].

On the basis of exposure–effect relationships as determined for a variety of plants of widely differing sensitivity by various authors, Posthumus [339] deduced the following critical levels for *ammonia*:

100 μg/m^3 (≈0.14 mL/m^3)
for monthly mean concentration

600 μg/m^3 (≈0.86 mL/m^3)
for 24-h mean concentration

10 mg/m^3 (≈14.3 mL/m^3)
for 1-h mean concentration

Impacts on Forest Productivity. Since the 1970s the indirect effects of acid precipitation on forest trees and crops has been a major area of both scientific and economic concern. In Germany, forest decline attributed to flue gases, particularly SO_2, was described in the neighborhood of industrial plants more than a century ago. The situation did not change essentially until the late 1960s, when a new type of forest injury became evident that was no longer related to local emitters but occurred to an increasing extent in presumably clean air areas far away from conurbation and industrial centers [340]–[344]. In the mid-1970s, severe dieback of silver fir (*Abies alba*) was observed in southern Germany. In the early 1980s, other conifers, especially Scotch pine (*Pinus sylvestris*) and Norway spruce (*Picea abies*) showed similar symptoms over rapidly increasing areas, and deciduous species such as beech (*Fagus sylvatica*) and oak (*Quercus robur*) have recently been affected. According to the 1985 damage inventory, 52% (i.e., 3.8×10^6 ha) of the forest areas in the 58 growth districts of the Federal Republic of Germany were injured. Level 1 of damage (reduction of vitality) characterizes about two-thirds of this portion, while medium to high damage (levels 2–4) occurs on 19% of the forested area [345].

Comprehensive studies [346]–[349] show that the lack of nutrients, climatic factors [350], mistakes in forest management, infections, and pests may play an important role in the present complex forest diseases; but a rapidly growing number of findings indicate that the above anthropogenic air pollutants, and possibly others such as triethyl lead or nitrophenols [351], [352], and related soil acidification are the essential direct and indirect causes of damage [327], [353]–[356].

At the cellular level, pollutants may lead to a disturbance of photosynthesis and respiration and the following primary and secondary metabolisms. In some cases, impairment of photosynthesis occurs before visible damage can be observed [357]. At the same time, photooxidants and acid rain cause weathering of cuticular structures, destruction of membrane systems, and leaching of nutrients from leaves. Quantitative and qualitative changes in photosynthesis [358] lead to reduced primary production, increased respiration, and a changed assimilate allocation to the root–mycorrhiza system, stem, branches and leaves, and also to the reproduction and phytogenous defense system. Such stressed trees emit unsaturated hydrocarbons such as ethylene, isoprene, and terpenes, which react with secondary pollutants (e.g., ozone) and highly phytotoxic tertiary pollutants in the immediate neighborhood of the plants. Weakening by reduced assimilate allocation to the roots leads to reduced uptake of nutrients from soil that has already been disturbed by acidification (i.e., impoverished in potassium and magnesium).

MURACH [359] determined a calcium deficiency in the roots of declining spruce and fir, but the same levels of aluminum as in healthy trees; REHFUESS et al. [360] noted magnesium and possible calcium deficiencies by foliar analyses even on base-rich soils. Contrary to ULRICH's [361], [362] contention that a change in fine root biomass is paralleled by toxically high Al:Ca ratios, REHFUESS points out that the two phenomena are not necessarily synchronized, so that marked decreases in fine root biomass can precede the increase in soil solution aluminum.

Mineral deficiency due to leaching and soil acidification is frequently coupled with a relative surplus of nitrogen with respect to assimilated carbon as a consequence of enhanced deposition of nitrate and ammonia and the intensity of subsequent nitrification processes. KRIEBITZSCH [363], who studied nitrification in many types of acidic forest soils, divided them into groups A, B, C, and D. In group A, no nitrification occurred and ammonium was the only nitrogen source. In soil of groups B and C, partial nitrification occurred, while group D soils had complete nitrification. Former heathland soils in Germany now cultivated with various *Pinus* species and *Pseudotsuga* belong mainly to group A, where the nitrate level is low, whereas the ammonium concentration is high.

Quantitative and qualitative changes in the living biomass of forest trees will bring about changes in the food web of the forest ecosystem; the population of some consumer (e.g., insects) or decomposer organisms (e.g., fungi) may grow explosively while others are completely suppressed. The disturbance of soil biocenoses may then induce a reduced litter decomposition that leads to a further shortage of minimal nutrients. Trees subject to these stress factors are in general extremely sensitive to secondary damage by frost, drought, or insect infestation.

Visible *injury to crops* as a result of ambient acidic precipitation has not been reported from modern high-productivity agrarian ecosystems. Therefore, attempts to evaluate impacts of acidic deposition must be based on simulated precipitation in field experiments or controlled environments. Yet most of the crop cultivars studied in the field and the crop varieties studied in controlled environments exhibited no effect on growth or yield as a result of exposure to simulated rain usually up to ten times more acidic than ambient [364]. Likewise, out of five cultivars of soybeans studied, only one exhibited reduction in yield [365]. Genetic variations, possibly in combination with characteristics of the soil or simulated rain applied in these experiments, may account for the observed variability.

3.2.2.3. Statistical Analysis of Forest Dieback

Descriptive or static models of forest decline are analyses of the spatial correlation pattern of site qualities and damage levels [347], [366] that corroborate the importance of pedogenic nutrient supply and acidification processes for forest decline. Figure 41 illustrates, on the basis of comprehensive cross-tabulations of several hundred German stands, the extent of forest damage in terms of relevant site factors. The standardized damage level S (i.e., proportion of injured: unaffected forest area) correlates best with the pedogenic nutrient supply, frequency of fog situations, and elevation, while the correlation with other site factors is also significant on the 99.9% level although of little statistical relevance; $S = f(Sq)$ is a monotonous function, while $D = f(Nf)$ has a marked maximum. Thus, stands on members of the ranker, cambisol, and podzol soil groups with pH <4, marked nutrient deficiency, and concomitant reduction of buffering capacity down to the aluminum and iron ranges are most liable to dieback. Among airborne pollutants, SO_2 is particularly important. In combination with other pollutants, its effect on both vegetation and soil is enhanced by fog (contingency coefficient after PEARSON $C = 0.53$). The amount of sulfur deposition is controlled mainly by precipitation ($C = 0.80$). Comparison of the pH and sulfur deposition curves ($C = 0.89$) summarizes the major influence of gaseous and particulate pollutants on forests as a result of the above mechanisms. Their inherent complexity implies the recognition that critical levels for tolerable pollutant concentrations can be defined adequately only with regard to the entire set of concomitant stress factors affecting a plant community.

A more sophisticated analysis of the data of forest decline monitoring programs can be brought about by the CHAID approach (see

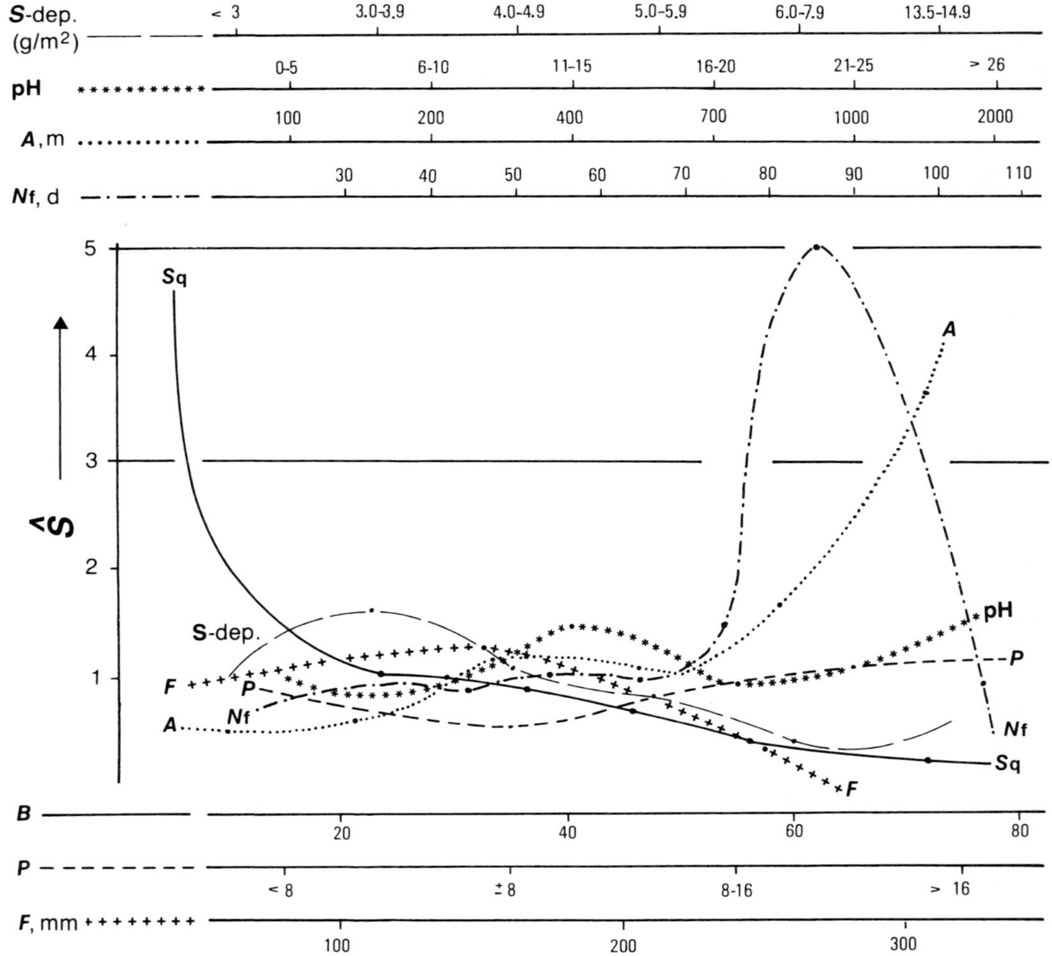

Figure 41. Forest decline as related to selected site factors
$\hat{S}$ = Ratio of damaged : undamaged forest area; *A* = Altitude (m); *Nf* = Number of foggy days; *Sq* = Soil qualitiy (nutrient supply); *P* = Porosity (percentage of pores > 50 μm); *F* = Plant available field capacity (mm); pH = Percentage frequency of acid rain (i.e., pH < 4.5); *S*-dep. = Sulfur deposition

Section 2.1.1.1) [103], [367]. CHAID divides the total data set successively into subsets that are, at each level of partitioning, characterized by the very variable that best explains the distribution and intensity of damage. Thus, the explicative value of different combinations of decline-inducing factors can be tested on high levels of statistical significance. A further advantage is the easy interpretability of the CHAID dendrograms.

Thus, a dendrogram of forest decline of about 500 stands in North Rhine-Westphalia clearly indicates the predominant role of the factors age, crown density, and composition of stand as boundary conditions of forest decline, while climatic factors such as annual precipitation and anthropogenic pollutant gases play a distinctly lesser role. These results are corroborated by comparable findings in other parts of Germany [103].

Dynamic Models of Forest Dieback. Simulation models of forest decline have to deal with two types of change — (1) change in complexity through increasing or decreasing differentiation, and (2) adjustment to extrasystemic changes. Negative and positive feedback processes can be correlated with systemic continuity or discontinuity, respectively. Continuity results when deviation-correcting mechanisms operate to ensure that structure and function are kept viable within the given parameters of the system. Quantization occurs when deviation is amplified to the point at which no deviation-correcting mechanism can prevent rupturing of the basic systemic framework (i.e., when resilience mechanisms are no longer able to contain and canalize the energies and thrust that have been generated). The process alters the relationships of the system to its environment, creating new spatiotemporal, structural, and functional boundaries — in short, the system is transformed to a new level of internal organization and environmental integration. Therefore, and specifically with regard to forest decline, models must employ system concepts to explain the circumstances in which quantization can occur from one developmental stage to another. Hence, they have to account for both systematic levels of biocenotic organization and cybernetic processes that demonstrate (1) systemic self-stabilization within a given organizational level and (2) systemic transformation resulting in a biocenotic quantum jump across an environmental frontier.

Models with Nonlinear Dose–Effect Relationships. Following models of reduced photosynthesis and leaf formation, and increased leaf aging and shedding under air pollution stress [368], [369], KOHLMAIER and PLÖCHL [370] developed a more comprehensive model concept for damage of Norwegian spruce. It includes the root system with reduced nutrient uptake and higher feeder root turnover, thus coupling the canopy and the soil root system. Reduced photosynthesis and changed assimilate allocation are simulated during the entire exposition to airborne pollutants, which may be a part of or the whole life span of a tree. Assimilate demand and supply have a hierachic structure with regard to, e.g., respiration, growth of foliage, roots, and living wood. The effect of an air pollutant on leaves and roots depends on both the specificity of instantaneous effects and the cumulative effects during the entire exposure time involving elimination and repair mechanisms of the plant. The complicated interaction pattern of these response mechanisms normally results in nonlinear dose–effect relationships with respect to concentration and accumulated pollutant and damage level of the plant. During one time step, the model first calculates the photosynthesis rate as dependent on external factors such as light intensity and air temperature, and internal parameters such as photosynthetic leaf efficiency; water uptake by roots is the limiting factor. Pollutants can enter the system by two different paths (i.e., leaves or needles and roots); in the first case, the photosynthetic efficiency of a particular needle age class is reduced; in the second, fine root turnover and consequently assimilate demand are increased. The momentary pollutant flux into the leaf is described in terms of gas exchange between the intercellular air and the ambient air by Fick's first law:

$$I_{\mathrm{in}}^{\mathrm{leaf}} = (c_{\mathrm{a}} - c_{\mathrm{i}})/r_{\mathrm{ai}} \qquad (67)$$

where the flux $I_{\mathrm{in}}^{\mathrm{leaf}}$ is proportional to the concentration difference $(c_{\mathrm{a}} - c_{\mathrm{i}})$ between ambient and intercellular air, and r_{ai} is a specific resistance factor for a particular pollutant. The corresponding flux per unit canopy or unit ground area is then obtained by multiplication with the leaf area index LAI. For the boundary condition $c_{\mathrm{a}} \gg c_{\mathrm{i}}$ Equation (67) reduces to

$$I_{\mathrm{in}} = (r_{\mathrm{ai}}/LAI)c_{\mathrm{a}} = u_{\mathrm{dep}}c_{\mathrm{a}} \qquad (68)$$

which is approximately equivalent to the dry deposition rate, with u_{dep} being the deposition velocity.

The pollutant load in the leaf or at its surface may be assumed to be eliminated either by physical processes (e.g., litterfall), by transport to other parts of the plant, or by metabolic processes. It appears more appropriate, however, to model the elimination process according to Michaelis–Menten kinetics of the type

$$\mathrm{d}\bar{c}_{\mathrm{i}}/\mathrm{d}t = u_{\mathrm{dep}}c_{\mathrm{a}} - k_{\mathrm{out}}\bar{c}_{\mathrm{i}}/[1 + (\bar{c}_{\mathrm{i}}/K_{\mathrm{c}})] \qquad (69)$$

where $\bar{c}_{\mathrm{i}}$ is the concentration per unit ground area and k_{out} the characteristic elimination constant, while the other symbols have the above meaning. The mode of interaction and the effect of a pollutant inside the plant are likely to be specific for the particular pollutant j. In the simplest case, the effect, above the specific no-effect level, will be proportional to the interior concentration $\bar{c}_i^j$:

$$eff^j(t) = a_j \bar{c}_i^j(t) \qquad (70)$$

where a_j is a specific effect constant relating to a particular process (e.g., photosynthesis). The effect variable $eff(t)$ may be composed of linear and higher-order terms of n interacting species:

$$eff(t) = \sum_{j=1}^{n} a_j \bar{c}_i^j(t) + \sum_{j=1}^{n} \sum_{k=1}^{n} a_{jk} \bar{c}_i^j \bar{c}_i^k \tag{71}$$

where synergetic effects are included in the second term with $j \neq k$.

According to BOSSEL et al. [369], the effect variable affects the leaf senescence function

$$ALT(t) = 0.1\,[1 + eff(t)] \tag{72}$$

with $0 \leq eff \leq 1.5$ and $0.1 \leq ALT \leq 0.25$, implying that the value 0.1 is the natural senescence rate without pollutant damage. The senescence rate influences the number of needle age classes:

$$NJ = 0.8/ALT(t) \tag{73}$$

and both determine the leaf's photosynthetic efficiency. Hence, the photosynthetic efficiency of a particular needle age class is

$$E(N) = 1 - ALT(t)[(N-1) + TJ] \tag{74}$$

where $0 < TJ < 1$ characterizes the season and $1 < N < J$ the age of a needle class.

The average photosynthetic efficiency is then expressed by

$$EFF = \sum E(N)\,L(N)/LEAF \tag{75}$$

where $L(N)$ is the leaf mass of a particular needle age class and $LEAF$ the total leaf mass. In unaffected systems, EFF ranges between 0.55 and 0.65 [370].

Acidification of soil and related leaching of mono- and bivalent cationic nutrients leads to higher feeder root turnover rates while a gradual loss of feeder root mass occurs. Following a titration curve of a buffered chemical system, this effect may be described by a logistic curve of the type

$$W_{to}(t) = f[W(t)] = W_{max}/[1 + F_k(e^{-KIt})] \tag{76}$$

where $W(t)$ is the accumulated pollutant at time t; W_{max} the maximum accumulation; F_k the correction factor ($W_{max} - 1$); K the buffer constant; I the constant input; and $W_{to}(t)$ the fine root turnover factor at time t [369].

This supply–demand model of photosynthetic efficiency and assimilate allocation appropriately describes the decline of a forest ecosystem under the influence of chronic air pollution stress. It clearly indicates that nonlinear effects may drastically accelerate the observed dieback phenomena, but more experimental data on both the biochemical and the physiological levels are required to better account for the combined effect of two or more pollutants [55], [371]–[373].

Time Series Analysis and Mapping. A common feature of the various methods of time series analysis of dynamic systems such as ecosystems is that they involve the observation of a complete set of state variables. This may constitute a major drawback in studies of empirical systems whose inherent dynamics are not known a priori. Under these circumstances, however, use can be made of the fact that in dynamic systems with a finite number of state variables, information on the momentary value of all state variables can be replaced by information on the recent history of a part of the variables. GROSSMANN et al. [374] therefore used time series analysis in combination with geographic information systems (GISs) to develop a scenario that describes forest damage in sequential form by means of maps. A geographic information system is a computer hardware and software system designed to collect, manage, analyze, and display spatially referenced data. Geographic information systems are emerging as the major spatial data-handling tools for solving complex natural resource planning problems, and the use of GIS technology has revolutionary implications for conducting research and presenting research results.

The GROSSMANN et al. approach POLLAPSE is a combination of dynamic models writing information into the data bank of a GIS, which thus produces a time series of maps of forest growth or damage [374]. These maps can be compared with the actual development so that deviations become readily discernible. The hypotheses underlying this type of regionalized modeling are

1) Photochemical oxidants cause foliar damage, and subsequent attack of airborne pollutants leads to leaching, eventually resulting in nutrient deficiency [375]
2) Ulrich's soil acidification hypothesis [361]
3) A combination of the two preceding assumptions

The predictive quality of these hypotheses was tested in the Pfaffenhofen area (Bavaria) on the basis of 462 forest stands, and the POL-

LAPSE methodology showed that the combined hypothesis (3) yielded the relatively best interpretation of forest damage phenomena observed in terms of both spatial pattern and degree of damage. The above methodology of time series analysis and mapping also offers excellent possibilities for the management of forests or other biotic systems as a base for sustained viability.

3.2.3. Impact of Xenobiotics on Aquatic Ecosystems

Xenobiotics released to aquatic environments have a great impact on biocenoses inhabiting streams, wetlands, and lakes. This adversely affects not only the respective aquatic ecosystems, but also humans who are in contact with this water. Aquatic organisms show how far-reaching consequences can be caused by effects that are difficult to detect on a short-term scale and therefore easily overlooked. Similar to effects on aquatic biocenoses, in most instances the effects of xenobiotics on humans cannot be recognized, not only because the necessary analyses cannot be made, but also because small, difficultly detectable effects can grow into major and irreparable population damage. The following describes how difficult it is to foresee the effect of a chemical on an aquatic ecosystem merely from (easy to make) short-term tests. This is a warning that care for human health and life means that society cannot wait until some clearly seen negative effect is proved beyond any doubt but must react on early warnings obtained by sensitive methods. For more detailed information, see [56]–[64].

3.2.3.1. Approaches and Methods of Studying the Impact of Xenobiotics on Aquatic Life

Aquatic life is organized into several hierarchic levels (Fig. 9 in Section 2.1.4):

1) Individual organisms
2) Monospecific populations
3) Communities of different species inhabiting the same environment, e.g., the pelagic communities freely swimming in water and benthic communities inhabiting the bottom of rivers and standing waters (ponds, lakes, reservoirs)
4) Ecosystems with their nonliving components interacting with organisms; within an ecosystem, organisms are organized into food

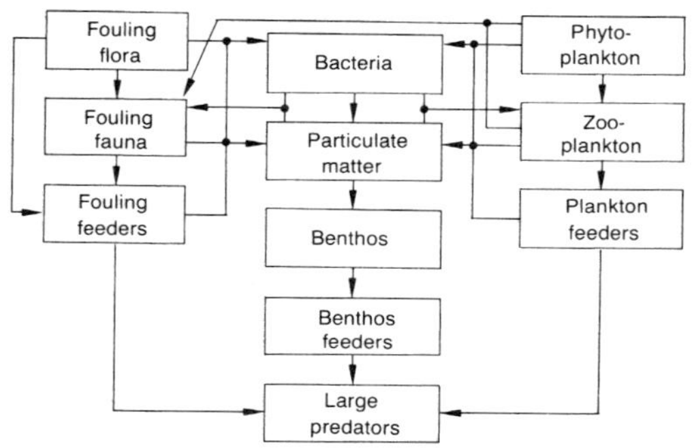

Figure 42. Food chains or food webs as the organization of organisms within an ecosystem according to food selection; arrows indicate directions of nutrient flow

chains (or more exactly, food webs) according to their selection of food items (Fig. 42)

Detection of the effects of xenobiotics is performed at all hierarchic levels of biological organization. In addition to this organizational hierarchy, a hierarchy of time scales of the effects can be distinguished, from very short ones measured in minutes up to very long ones measured in human generations [376], [377].

Methodology. The most common approaches to detect the impact of xenobiotics on aquatic life are listed below:

1) Classical toxicity tests on test individuals or cultivated populations of aquatic organisms to detect short-term lethal effects
2) Methods for detecting longer-term sublethal effects on test individuals and populations
3) Methods for detecting genotoxicity
4) Use of biomarkers, a relatively new approach for detecting stress effects of long-term exposure of individual organisms living in contaminated environments
5) Microcosm and mesocosm studies
6) Monitoring of long-term consequences of contaminants in nature by following the species diversity and loss of sensitive species in biocenoses and ecosystems

Other approaches are, e.g., predictions of toxicity based on molecular structure of the respective substances using quantitative structure–activity relationships (QSARs, see, e.g. [10], [219]), and tests for changes of morphologic or morphometric characteristics of organisms and detection of pathological individuals.

Classical Short-Term Toxicity Bioassays. Water organisms employed for these tests are usually some species of easily cultivated algae, water flies, and fish (see Section 3.2.1). Benthic, bottom-dwelling organisms commonly used are worms of the family Tubificidae living in the bottom mud of streams and lakes that are fairly polluted with organic matter of natural origin. In the simplest version, the tests consist of adding different concentrations of xenobiotics to jars (aquaria) containing the respective organisms and determining the time until half of the population dies (LC_{50} or LD_{50}). Recently, much care has been taken to standardize aquatic toxicity tests to maximize their major advantage — the rapid comparative evaluation of many xenobiotics [65].

Long-Term Sublethal Effects of Xenobiotics. In the aquatic environment, long-term sublethal effects are studied by measuring different reactions of test organisms. An example is the behavioral test with the freshwater bivalve mollusk *Dreissensia polymorpha* [378]. Its filtering rate, which is determined anemometrically in this test, represents a vital function, providing food and therefore vitality, growth, and reproduction of the organism. Detection limits of long-term sublethal tests are several orders of magnitude higher than those of short-term toxicity tests.

Genotoxicity. Compounds such as diethylstilbestrol that disrupt the endocrine systems of fish, wildlife, and humans were found to induce large-scale dysfunction at the human population level [66] by causing genetic and reproductive changes. Genotoxicity is studied by observations of the offspring of the affected population.

Determination of Biomarkers. Measurements of body fluids, cells, or tissues are carried out that indicate in biochemical or cellular terms the presence of contaminants or the magnitude of the host response [63]. Examples are, e.g., determination of marker enzyme activity in cells of target tissues, serum enzyme activity, appearance or loss of reaction products of an enzyme; major changes in the concentration of a cellular component such as glycogen, fat, mucus, or metallothionein; and measurements of damage to DNA.

Microcosm and Mesocosm Studies (see Section 3.2.2.1). Microcosms are laboratory model ecosystems consisting of mixed natural populations of organisms transferred from nature to a laboratory jar or aquarium, sometimes an arrangement that imitates a stream. Mesocosms are of the same origin, but of much larger volume and usually kept outdoors in their natural environment (e.g. [59], [379]–[381]). When xenobiotics are added, the changes in some common properties are measured (e.g., species composition, diversity, photosynthesis, or respiration).

Studies of Changes in Structure and Function of Communities and Ecosystems. Toxic chemicals can affect independently both the structure and the function of aquatic ecosystems. Low levels of perturbation may cause changes in community composition or structure without a change in function. Alternatively, a process, the rate of which is changed by stress (such as respiration or photosynthesis), may be reduced under sublethal stress, but the community structure is left intact. According to [382], the most predictable changes seem to be the disappearance of sensitive species from communities under environmental stress.

The integrated bioindicator approach currently gives the best results. It involves monitoring a suite of selected exposure and effects indicators at several levels of biological organization from the biomolecular to the community level [63], [383].

Critical Evaluation of the Above Approaches to Aquatic Ecosystems. For aquatic environments, *short-term lethal toxicity tests* have many advantages, as well as many disadvantages. Their major advantage is the ability to rapidly compare the relative toxicity of specific chemicals or specific mixtures. The limitations become particularly obvious when trying to use their results for higher levels of biological organization in the aquatic ecosystem. Thus, they do not account for the effects of

1) Chemical speciation in the environment
2) Kinetics and hysteresis in sorption of chemicals to sediments
3) Bioaccumulation through the food chain from the lowest level of algae and plants to the highest level of fish
4) Modes of toxic action on survival, growth, and reproduction of individual species populations

For aquatic sediments, an additional inadequacy is that sediment toxicity tests often rely on equilibration of chemicals in a sediment–water mixture or employ benthic organisms whose survival, growth, or reproductive success may depend not only on the presence of the toxicant but also on physical characteristics of the sediment.

The application of results of (laboratory) toxicity tests to natural populations is especially risky for the following reasons:

1) The way in which laboratory organisms are exposed to pollutants differs from exposure in the environment
2) Laboratory tests deal with single chemicals, and organisms are exposed to complex mixtures in the environment
3) Criteria for effects in the laboratory are not important functional endpoints in population and systems dynamics

The inadequacy of short-term tests for the aquatic environment may be shown by conclusions drawn from careful monthly bioassay results with the survival and reproduction of a water fly (*Ceriodaphnia*) and larvae of fathead minnow. These indicated that the effluent of an energy plant at Oak Ridge (Tennessee) causes no observed effect. However, community surveys demonstrated clear effects on diversity and species richness downstream of the power plant [63].

The use of *biomarkers* is more relevant to conditions in nature than acute toxicity methods. Biomarkers provide evidence of exposure to rapidly metabolized contaminants and in this way integrate pharmacodynamic and toxicological interactions. Correlations between biomarkers of exposure and ecological effects are expected to be better than those between population-monitoring parameters and indicators of exposure.

Observations in nature can be performed only in rare instances when the concentration of the xenobiotic in nature is increased for a prolonged period. These observations cannot be used for testing new materials. Therefore an integrated system of xenobiotic hazard evaluation has recently been developed (see Sections 3.2.1, 3.2.3.4).

3.2.3.2. Behavior of Xenobiotics in the Aquatic Environment and their Influence on Aquatic Organisms and Indirectly on Humans

A chemical of potentially xenobiotic character follows a complicated flow from production site to the environment, as shown in Figure 43. The situation in fresh water depends not only on the direct emission of the respective chemical to the water body but also on its emissions and fates in soil and groundwater (both associated with vegetation growth and decay), on the chemical and biological processes in the water body, and on the bioaccumulation in different organisms of a food chain. Environmental processes connected with local hydrometeorology, with precipitation and flow, and with temperature and light conditions modify the fate of xenobiotics. Figure 44 illustrates the behavior of hydrophobic xenobiotics in standing waters, particularly the exchange between the pelagic and benthic environment (i.e., between water and sediments). The participating major processes are shown: sedimentation, resuspension, and mud–water exchange. For hydrophilic xenobiotics, another important process is bioaccumulation (i.e., accumulation inside the organism's tissues and organs, mainly in fat and liver). In addition to these processes of an individual chemical compound, complex interactions occur between different chemical species in nature.

Many organic contaminants are readily taken up and *bioaccumulated* by aquatic organisms [61], [68] if the chemicals are dissolved in water. The bioavailability is much lower if the same compounds are sorbed to soil or sediment particles. Many toxic chemicals do not bioaccumulate but rather are metabolized. In many cases the metabolites are more toxic than the parent compounds present in the environment. Bioaccumulation, the process by which a substance is taken up by an aquatic organism from water and through food, is found to depend strongly on the size (weight) of the respective organisms (see, e.g., for pesticides [56], [386]). This is due to the dependence of most physiological processes on size (e.g., [59]). Biomagnification within aquatic food chains (i.e., the increasing concentration of toxic substances in higher organisms) was found important only for some chemicals (e.g., DDT [387]).

Humans are affected by the processes in water in different ways, as a result of their direct and indirect contact with water, but also through air, soil, and food. Large differences exist in the biological availability of contaminants associated with different environmental media, as well as individual and species-specific differences in pharmacodynamic disposition.

In most instances, one thinks of direct effects of the xenobiotic on organisms. However, recent ecological findings demonstrate that in nature the indirect effects are more important [388].

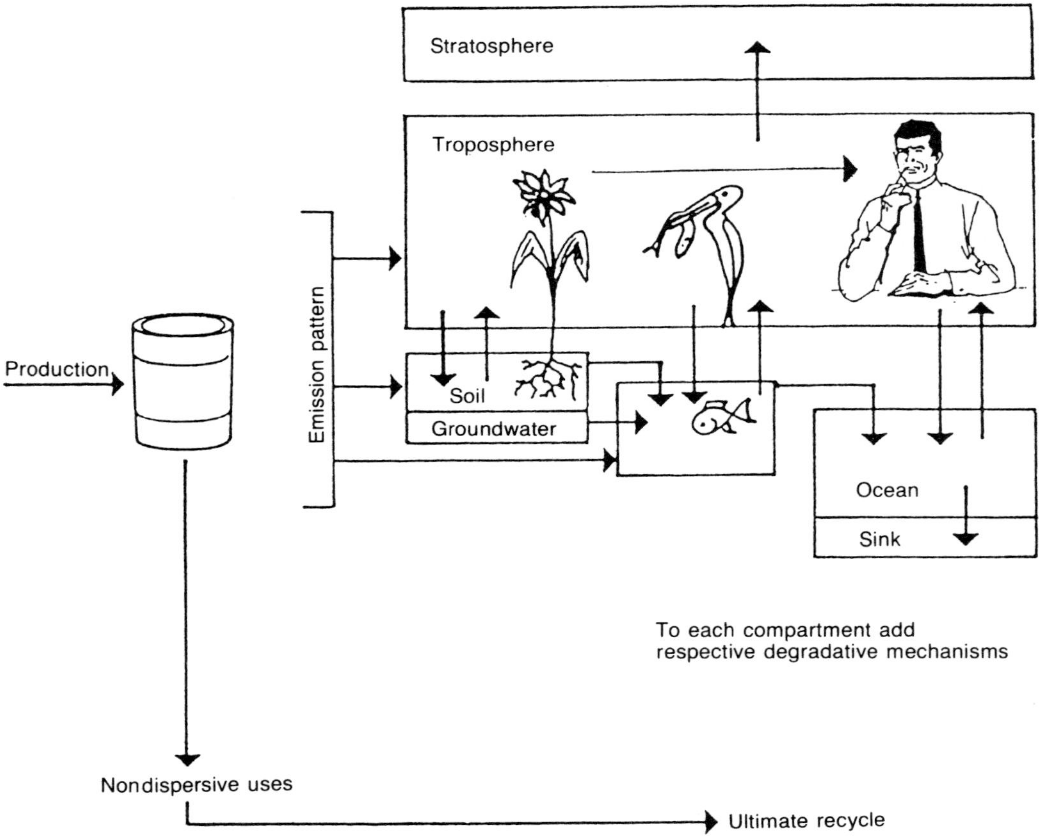

Figure 43. Schematic representation of the flow of a chemical from production to the environment (after [384])

They are caused, e.g., by changes in food composition when some organisms become rare or disappear because of toxic effects and by changes in competition and other possible relations between organisms. An easily understandable example is the eradication of an aquatic weed species. Indirectly, this will cause a complete change of abiotic conditions not only for other species that depend on the weed directly, such as invertebrates living on it, but also for fish species finding their shelter.

Adaptation. Organisms are capable of adapting to the effects of xenobiotics to a certain degree, and this can have both positive and negative effects on individual organisms (including humans) and on the total environment, as well as on environmental management. Adaptive acquisition of degradative abilities for some organic compounds or of resistance to heavy metals has been demonstrated for bacteria in laboratory ecosystems [389]. Such an adaptive response may be caused by different molecular and biochemical processes: (1) induction of specific enzymes in members of the community, resulting in an increase in the observed degradative capacity of the total community; (2) growth of a specific subpopulation of a microbial community able to take up and metabolize the substrate; and (3) selection of mutants that acquire altered enzymatic specificity or novel metabolic activities at the onset of exposure of the community to the introduced compounds.

The *negative effects of adaptation* may be exemplified by pests (e.g., insects or weeds) in the framework of eradication programs [67]. In West Africa, the widespread parasitic disease filariosis is caused by the *Filaria*. This parasite is transmitted to humans by blackflies. To eradicate the flies, the organophosphate Temephos was applied on a large scale in about 18 000 km of rivers

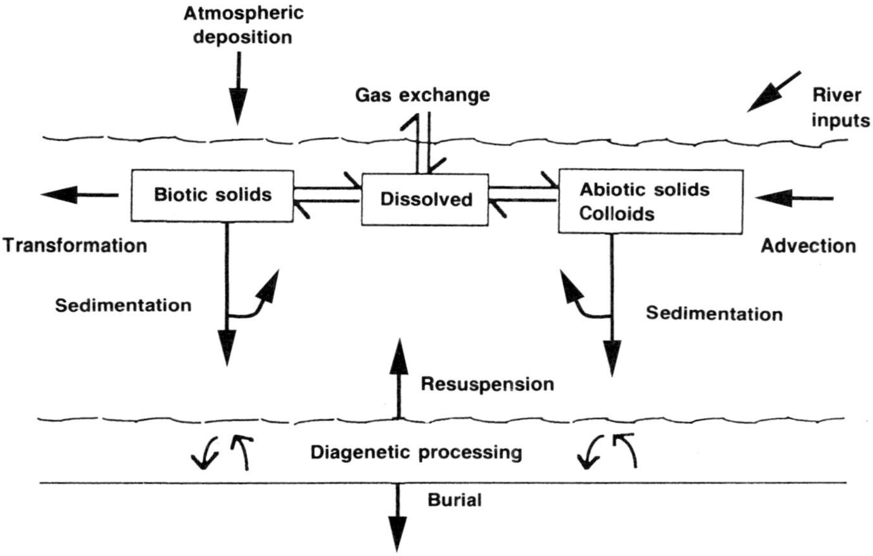

Figure 44. Behavior of hydrophobic xenobiotics in standing waters (after [385])

in the region during 1974–1980. However, resistance to this xenobiotic developed soon in one region and spread rapidly. Therefore, other means of eradication were sought that would act more persistently and not show the negative effects of the insecticide on other organisms in the rivers, since Temephos causes up to 100% mortality of many insect species living on the river bottom. No toxicity to fish was observed. However, the insecticide accumulated in fish tissues, which may have negative effects on humans that consume the fish.

In addition to the use of another more toxic organophosphate (chlorophoxime), a biological method of eradication was developed. This method demonstrated the general advantage of using biological agents. The bacterium *Bacillus thuringiensis* appeared to be very selective in killing only *Simulium* and few other species, and no adaptation of the fly populations was observed. Therefore, the application of a biological agent has the following advantages: (1) it is selective for the target goal; (2) no negative side effects are observed; (3) no rapid adaptation of pests takes place; and (4) the cost of application is lower.

Because of adaptation and other differences, the applicability of basic ecotoxicological principles derived in the temperate region to warm, arid, and cold climates is not without problems [67]. No simple straightforward relation exists for assessing regional differences of toxic effects due to differences in temperature and associated climatic factors. The general belief that metabolic rates are much higher in tropics has not been supported; nevertheless, an increased risk of toxic effects in tropics is to be expected.

3.2.3.3. Impact of Xenobiotics on Pelagic and Benthic Biocenoses

The data bank of the International Register of Potentially Toxic Chemicals (IRPTC) of UNESCO contains a record of some 8000 chemicals, and the number is growing rapidly. Therefore, only a few illustrative examples are described in the following, with an emphasis on related problems. The quantitative aspects of the processes and predictions of the impacts of xenobiotics are described in Section 3.3, dealing with ecotoxicological modeling.

An example of the dependence of toxicity not just on one element, but on the ratio of several, can be demonstrated by the case of arsenic. REUTHER [390] found for Swedish lakes that,

Range of cadmium concentrations of different waters in the Ruhr area

	Concentration, µg/L
Groundwater	0.05 – 0.15
Drinking water	0.05 – 0.2
Falling rain at Essen	0.2 – 1.1
Storm water flow from paved areas	1 – 10
Seepage water from sewage sludge deposits	<1
Seepage water from industrial sludge deposits	5 – 50

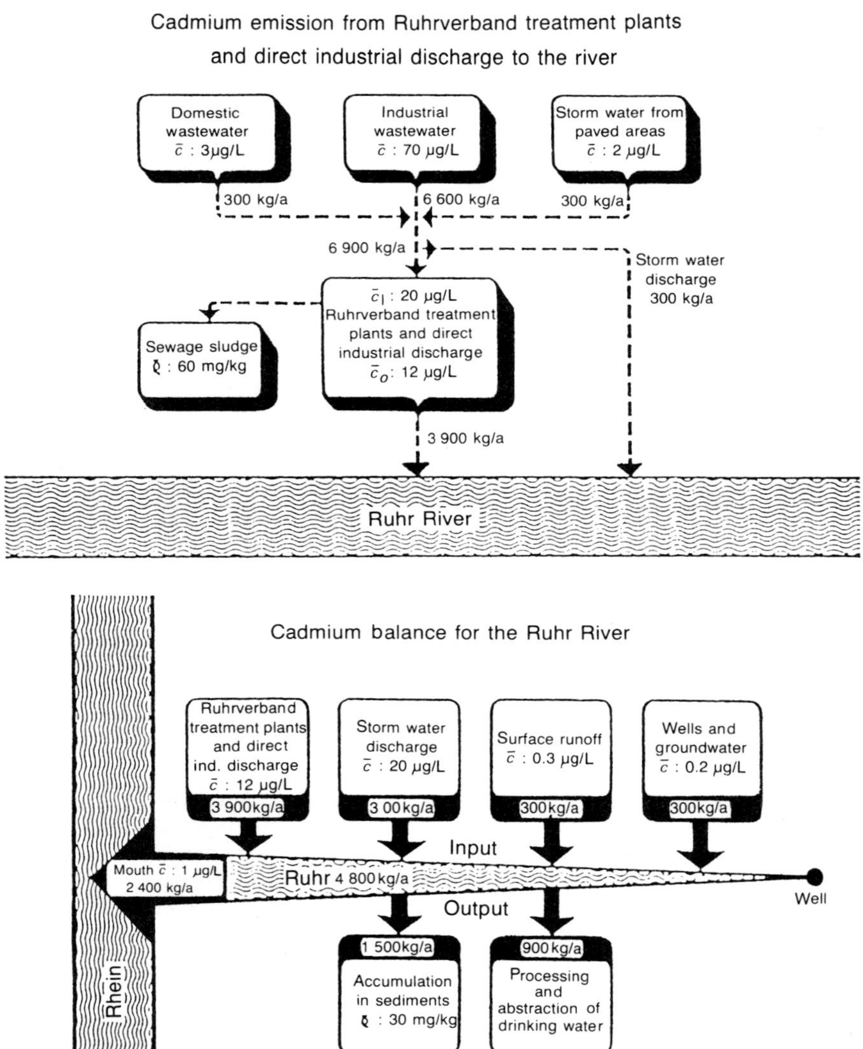

Figure 45. Origin, fate, concentrations, and annual input–output figures for cadmium in the river Ruhr (Germany) (after [392])

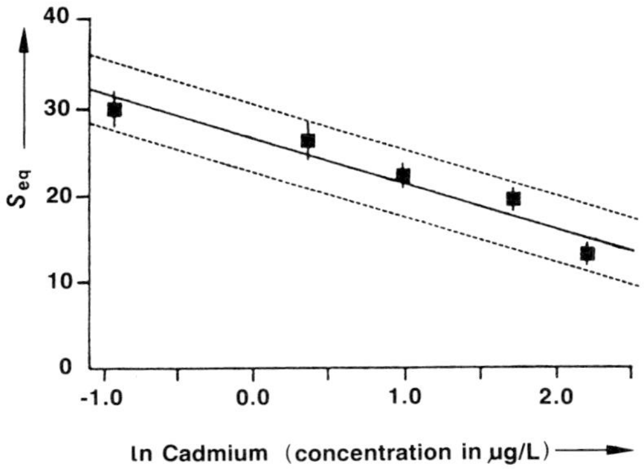

Figure 46. Microcosm analysis of the long-term effect of cadmium on the equilibrium number of species (S_{eq}) in aquatic biocenoses (after [379])

when added as arsenate, arsenic was readily accumulated in plankton at concentrations close to those found in nature. If the ratio P:As was low, arsenic proved toxic at much lower concentrations.

For the trace metals cadmium, lead, copper, and zinc, the effect of external variables on the amounts present in the environment was specific for each element [391]; especially important for Cd and Pb were sediment- and water-related abiotic variables. For the essential trace metals, the variance expressed by the effect of the common physical variables of the surrounding environment was very low: from 25 % for Zn up to 37 % for Cd and Cu. The quantities of the common xenobiotic cadmium in the aquatic environment can be demonstrated by the example of the river Ruhr in Germany (Fig. 45). A long-term effect of Cd on aquatic biocenoses was measured by using the microcosm method [379]. Figure 46 shows a steady decrease of the number of species in the microcosm with increasing cadmium concentration.

Some detergents greatly affect nature not only because of the amount of phosphates they contain (leading to eutrophication of water bodies), but also because of their toxicity. The detergent FIT was found to be highly toxic to frogs and their larvae [393].

Evidence exists of the sometimes close relation between the occurrence of some xenobiotics and the amounts produced in the respective region. Figure 47 shows how the quantities of PCBs, DDT, mirex, and chlorobenzene accumulated per year in layers of some U.S. lake sediments follow the sales or production rates of the respective years.

A systematic study of 50 chemical compounds demonstrated that their joint toxic effects can be considered simply as additive [394]. The authors proposed a quantitative structure–activity relationship (QSAR) technique for predicting the concentrations of the components of mixtures that would cause 50 % inhibition of the toxic effect by joint action. How far the effect of other compounds can act synergistically is not clear.

3.2.3.4. Ecological Risk Assessment of Xenobiotics

Ecological risk is a condition in which the normal functions of a population, ecosystem, or entire landscape are threatened by external forces or stress factors that, at present or in the future, may diminish the health, productivity, genetic structure, economic value, or aesthetic quality of the system [394]. In addition to xenobiotics, ecological risk may be caused by excessive inputs of nutrients; disturbance of normal energy flow by thermal pollution or global warming due to greenhouse gases; drastic changes in ecosystem structure from drying peatlands; and physical destruction of ecosystems by excessive grazing of rangelands, burning of forests, or covering with heavy loads of industrial dust.

Ecosystem risk assessment is defined as a set of procedures for measuring risk to the environment associated with the use of xenobiotics through an objective and probabilistic exercise based on empirical data and scientific judgment. The results of such a risk assessment can then be used to provide a consistent means of estimating the limiting concentrations of substances that will produce no unacceptable negative effects on ecosystems that may be exposed to the substance (see Section 3.2.5). The phases of making and reviewing an environmental risk assessment are presented in Figure 48. More about the method can be found in [68], [396], [397].

Concluding Remarks. The consequences of massive application of xenobiotics in Vietnam are drastic. Pesticide damage to nature involved the death of millions of trees and often their ultimate replacement by grasses, maintained in turn to this day by subsequent periodic fires; deep and lasting inroads into the mangrove habitat; widespread site debilitation via soil erosion and loss of nutrients in solution; decimation of terrestrial

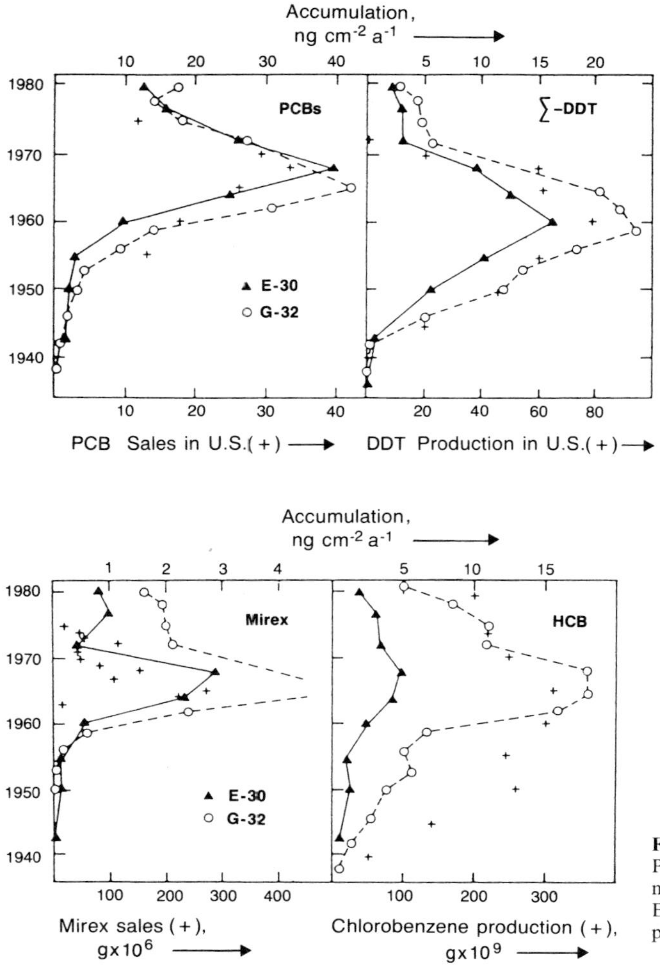

Figure 47. Annual accumulation rates of PCBs, DDT, mirex, and chlorobenzene in sediments of selected U.S. lakes (after [385]) E-30 and G-32 represent two different sampling stations (cores)

wildlife, primarily through destruction of their habitat; losses in freshwater fish, largely because of reduced availability of food species; and a possible contribution to the decline of offshore fishery. Humans are suffering both direct and indirect effects. In addition to a variety of health problems of exposed humans, the above-mentioned widespread damage to the inland and coastal forest ecosystems has had major indirect effects due to large-scale devastation of crops [67].

In the future, additional complications must be expected in the framework of xenobiotic application. These will be connected not only with the continuing adaptation of target organisms but also with changes in the total environment that are expected to be global. These include global warming accompanied by shifts in species composition, in food crops and production, and in conditions for human life. A mistake has been made in conceding to chemicals the right to be judged innocent until proven guilty. Therefore, a new strategy toward the application of xenobiotics is necessary, based on the following rules:

1) Minimize the number of newly invented chemicals in production and commercial use
2) Minimize the number of chemicals used in social and economic activities

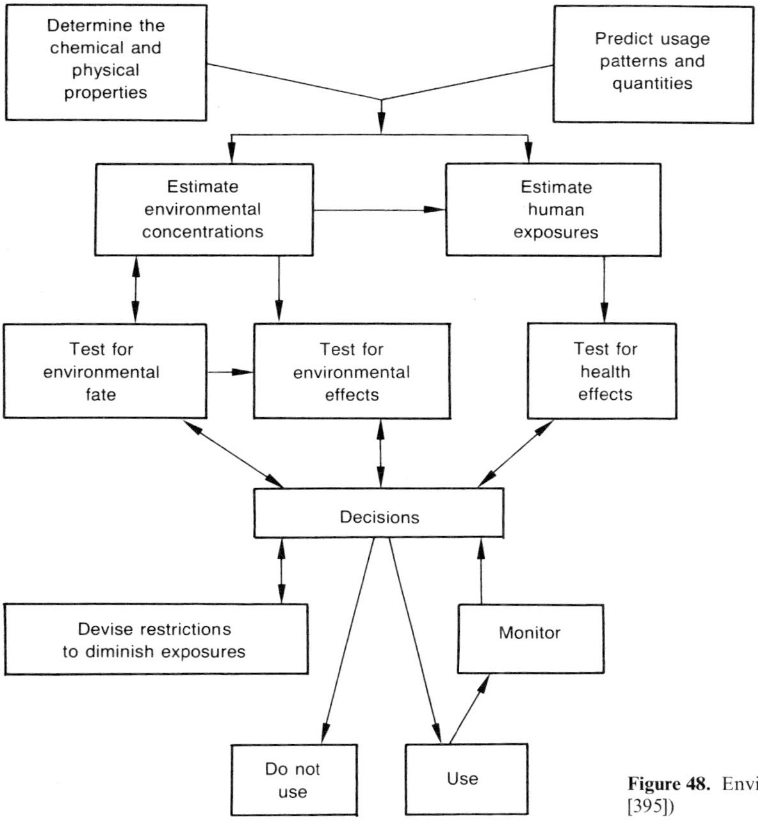

Figure 48. Environmental risk assessment (after [395])

3) Give priority to naturally occurring rather than synthetic chemicals
4) Give priority to the production of solid substances over liquids and liquids over gases
5) Minimize the spatial scope and time of their presence in the environment

3.2.4. Testing Approaches to Atmospheric Effects

Biotic Effects. For atmospheric hazard assessment, biotic and abiotic approaches are possible, although internationally accepted test guidelines to more precisely define the effects of airborne contaminants on organisms have not yet been developed. In most cases, only data on acute mammalian toxicity are available, which can be used for rough estimation of the risk a chemical poses for animals. However, atmospheric concentrations are normally not high enough to cause acute toxic effects. Therefore, data on long-term or chronic toxicity or on mutagenic and teratogenic effects would be more appropriate (→Air, Chap. 3).

Concerning the toxicity to terrestrial plants [2], no standardized test procedures are available for chemicals applied directly in gaseous or deposited form since a guideline for tests with herbaceous species has not yet been accepted. For some chemicals, enquiries into invertebrate toxicity—mostly insects (e.g., *Apis mellifera, Syrphus corollae*)—are available, which were conducted according to guidelines for the testing of plant protection agents.

Abiotic Effects. As regards abiotic effects, consideration of the following atmospheric processes is indicated [398]: global warming, stratospheric ozone depletion, tropospheric ozone formation, and acidification. A potential greenhouse gas has absorption bands in the 800–1200-cm^{-1} atmospheric window. Its *global warming potential* (GWP) is defined as the ratio of calculated warming for each mass unit of a gas emitted into the atmosphere relative to the calculated warming for a mass unit of the reference gas CFC 11 (CCl_3F). An approximate GWP value of a chemical S is given by

$$GWP = \frac{\tau_s M_{CFC11} S_s}{\tau_{CFC11} M_s S_{CFC11}} \tag{77}$$

where τ_s is the atmospheric lifetime (τ_{CFC11} = 60 a); M_s the molar mass (M_{CFC11} = 137 g/mol); and S_s the IR absorption strength in the interval 800–1200 cm^{-1} (S_{CFC11} = 2389 cm^{-2} bar).

In the above formula, the atmospheric lifetime is defined as

$$\tau = \frac{1}{k_d + k_w + k_c} \tag{78}$$

where k_d is the removal rate for dry deposition; k_w the removal rate for wet deposition; and k_c the pseudo-first-order chemical transformation rate.

A substance may have an impact on *stratospheric ozone* when its atmospheric lifetime is sufficiently long to allow for transport to the stratosphere and it contains chlorine or bromine substituents [289]. The ozone-depletion potential (ODP) is then defined as the ratio of calculated ozone change for a unit mass of gas emitted into the atmosphere relative to the depletion determined for an equal mass of the reference gas CFC 11 (ODP = 1). Thus, the ODP value can be approximated by

$$\text{ODP} = \frac{\tau_s M_{CFC11} n_{Cl} + \alpha n_{Br}}{\tau_{CFC11} M_s 3} \tag{79}$$

where τ_s is the atmospheric lifetime (τ_{CFC11} = 60 a); M_s the molar mass of gas considered (M_{CFC11} = 137 g/mol); n_{Cl}, n_{Br} the numbers of Cl and Br atoms, respectively, per molecule; and α a measure of the effectiveness of Br in ozone depletion with respect to Cl ($\alpha = 30$).

The generation of *tropospheric ozone* depends on

1) The reactivity of the substance and the degradation pathway
2) The meteorological situation, with highest O_3 concentrations expected at high temperatures, intense solar radiation, and low wind speeds
3) The concentration of air pollutants, where the concentration of nitrogen oxides has to exceed several μL/m^3

Less reactive species (eg., CH_4 or CO) are important for long-term O_3 concentrations, while highly reactive species such as alkylenes or aldehydes contribute significantly to ozone peaks.

In analogy to the above depletion potential a photochemical ozone formation potential (POCP index) has been defined, based on ethylene, which has a POCP value of 100. Comparative studies showed, however, considerable variability in the POCP values assigned to each organic compound, which is difficult to explain by recourse to the basic chemical characteristics alone [2]. Therefore, ozone formation is best derived from an approximate reactivity scale based on the rate constant for the (hydrocarbon + $OH^{\cdot}$) reaction and molar mass [289]:

$$OH^{\cdot}\ \text{scale} = \frac{k_s M_{ethylene}}{M_s k_{ethylene}} \cdot 100 \tag{80}$$

where k_s is the rate constant at T = 298 K for the reaction with OH radicals ($k_{ethylene}$ = 8.5 $\times 10^{-12} cm^3 mol^{-1} s^{-1}$) and M_s is the molar mass ($M_{ethylene}$ = 28 g/mol). For a presentation of pollutant impact on metals, organic and inorganic nonmetallic materials, see [2].

3.2.5. Appraisal

Hazard assessment of chemicals is based on stepwise determinations of predicted environmental concentrations and ecotoxicological no-effect concentrations followed by comparison of the two values. Figures 49 and 50 [238] indicate that whenever the PEC/PNEC ratio is <1.0, the chemical may be concluded not to present an environmental hazard under the conditions analyzed.

In addition to and as a substitute for specifically designed networks yielding spatially representative data, fate and exposure models are useful tools for estimating concentrations and are increasingly applied in many situations. In view of the considerable number of models available, careful selection is indicated to comply with specific problems. The fundamental problem is to define the set of questions a model is expected to answer and accuracy required at a given scale. Even if a suitable model has been chosen and

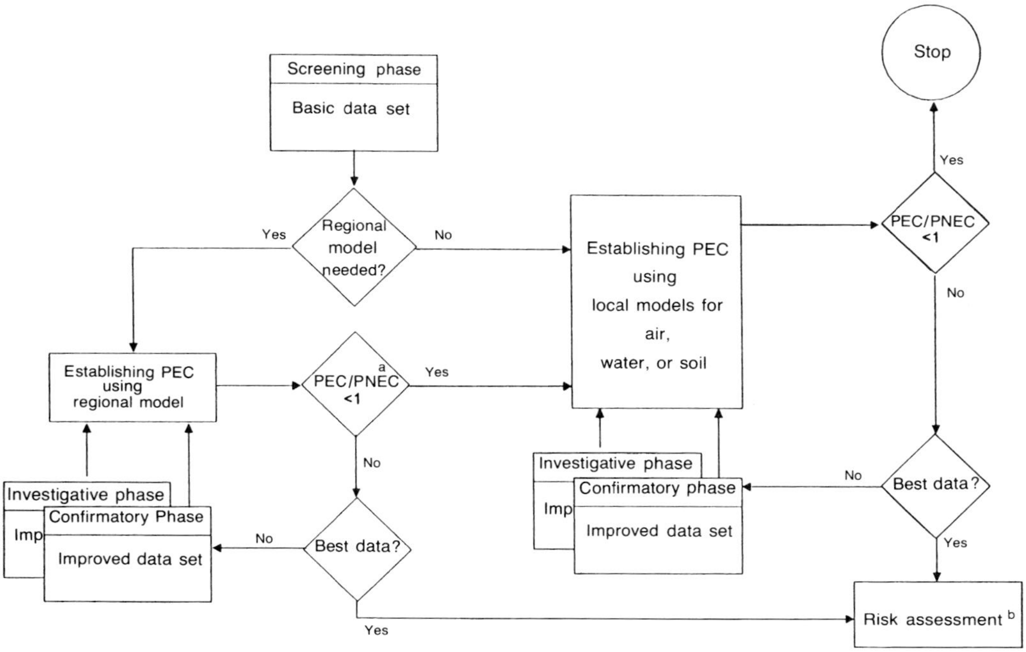

a Use PNEC from Figure 50

b Only if best data for PEC and PNEC were used

Figure 49. Exposure assessment scheme (after [238])

valid input data have been obtained, the results can still be misinterpreted if the underlying principles are not understood properly. Thus, both the assumptions and the limitations of different modeling approaches must be recognized (see Section 3.3).

The first field of model application is the assessment of the fate of chemicals at a specific location (i.e., predicting concentrations related to well-defined sources of emission). The second is the screening of chemicals whose purpose is the evaluation of potential exposure, typically at larger or even global scales. Therefore, consideration of an average environment appears justified in the latter case.

As in the case of monitoring networks [2], the networks providing input data for the calibration and validation of models also require a geostatistical control to allow for interpolation and spatial extrapolation. Only in the case of site-specific modeling is strict validation possible, while for screening models proper validation cannot be ensured, because no specific situation is available as a reference [237]. Validation of probabilistic models is possible, but this requires a substantial amount of valid environmental concentration measurements in different locations to allow geostatistical analysis of the primary data.

As summarized in Figure 49, the principle of effects assessment is to use toxicity data from short- and long-term experiments and appropriate assessment factors to calculate a PNEC value. In principle, this is accomplished separately for each environmental compartment, but practical difficulties may arise from the fact that most of the ecotoxicity data on existing chemicals were obtained for the aquatic compartment only.

Aquatic Hazard Assessment. According to the base set definition [288], [289], acute effect data for fish, *Daphnia*, and algae should be available. In addition, chronic toxicity investigations with standard or nonstandard organisms (see Section 3.2.1.3) may also be available. Multispecies tests, micro- and mesocosm studies, and field tests are comparatively rare for most chemicals [295]. The standard organisms used represent

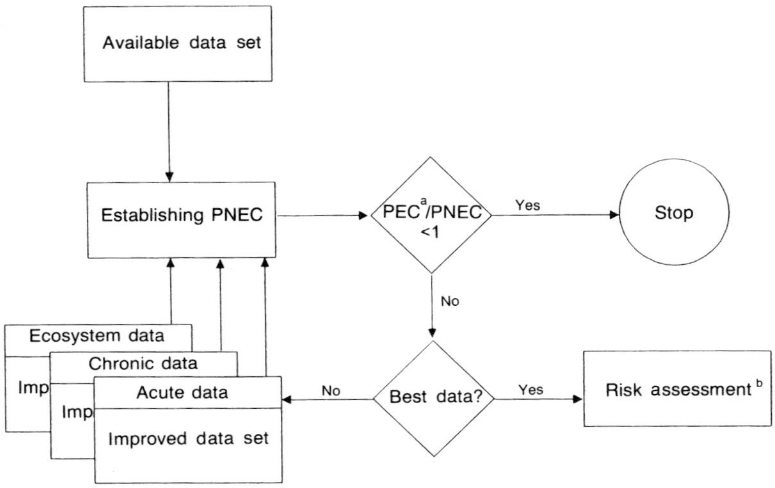

[a] Use PEC from Figure 49
[b] Only if best data for PEC and PNEC were used

Figure 50. Effects assessment scheme (after [238])

three trophic levels—primary producers and primary and secondary consumers—but in view of the great variability of aquatic phyto- and zoocenoses, their functional representativeness should be evaluated more precisely. In addition, a given population may represent more than one trophic level depending on the amount and spectrum of nutrition for the species; earlier life stages not infrequently live on completely different food than adults of the same species.

Therefore, ecotoxicological data collected in laboratory experiments can be difficult to translate into accurate predictions of effects in ecosystems. This applies in particular if the duration of a test is different from that of standard tests or if the test parameters (e.g., end points) are not comparable to those used in standardized test procedures (e.g., investigations of photosynthesis or behavior, studies conducted on the cellular or subcellular level).

In general, the use of laboratory test data will lead to an overestimate of chemicals' effects on complex aquatic ecosystems, which implies very conservative assessment factors, for the following reasons:

1) Bioavailability of chemicals is maximized in filtered-water laboratory systems, while in natural waters the higher concentrations of dissolved organic matter and suspended solids may reduce it substantially as a result of adsorption or complexation [290].
2) In laboratory tests, exposure concentrations are kept constant as far as possible, whereas in natural waters, exposure is likely to vary over time and degradation is greater. An overestimate is obvious when relatively long-term laboratory toxicity studies are the basis for extrapolation.
3) Laboratory experiments are normally conducted with a limited number of individuals or monospecific communites (e.g., *Daphnia magna*) and therefore exclude the various homoeotic mechanisms operating in complex natural multispecies communities.

The following features of laboratory experiments are likely to lead to an underestimate of toxic effects in ecosystems:

1) Few test species are used in laboratory test systems (see Section 3.2.1.3); therefore, they cannot be expected to encompass the most susceptible species in the field. In addition, the definition of sensitivity is dependent largely on the choice of the respective end point.
2) Information on the toxicity of substances normally relates to individual substances,

while in nature complex mixtures are often present. Their effects may be additive, but the extent to which toxicant concentrations lower than the NOEC can contribute to toxicity is not yet clear [399]–[401].

3) In the laboratory, efforts are normally made to remove stressors other than the substance tested. In the field, however, a variety of stressors (e.g., temperature, drought, or disease) may contribute to increase the susceptibility to a chemical well above that found in the laboratory.
4) The use of fixed assessment factors between acute EC_{50} values and chronic NOECs excludes the possibility that differences in toxic mechanisms exist between acute and chronic exposure.
5) The same applies to fixed assessment factors between chronic NOECs and ecosystem NOECs since the history of the ecosystem in a monitoring study is not always known precisely enough. Therefore, an observed "anomaly" in ecosystem behavior cannot be conclusively attributed to the presence of a substance at the time of monitoring.

Hazard Assessment for Sediments. In comparison to aquatic hazard assessment, analogous approaches for sediments have not yet reached the same level of standardization. Various techniques (e.g., equilibrium partitioning, determination of interstitial water quality, or spiked sediment toxicity) are under discussion [291]. As long as ecotoxicological effect data on sediment-dwelling organisms are lacking, the $PNEC_{sed}$ may be approximated by the PNEC for aquatic taxa and the sediment–water partitioning coefficient K according to the following formula:

$$PNEC_{sed} = K \cdot PNEC_{water} \qquad (81)$$

Terrestrial Hazard Assessment. Hitherto, hazard assessment schemes have been based predominantly on data obtained on aquatic organisms, while the amount of valid data on sediment and soil organisms, particularly of data obtained under realistic exposure conditions, is distinctly lower. Therefore, data obtained for aquatic test organisms should be used as an inital approach only to indicate whether terrestrial tests must be performed.

Aquatic organisms might display a similar susceptibility to chemicals compared to soil organisms exposed mainly to the pore water of the soil, but the transferability of aquatic effect values to pore water organisms requires further examination on the basis of comparative tests with chemicals of various groups. These have to make proper allowance for the fact that most chemicals relevant for soil either are resistant to degradation, show a strong adsorption to soil, or have a certain bioaccumulation potential. However, even readily biodegradable chemicals such as detergents may pose a problem for the terrestrial environment if applied in considerable amounts. Therefore, in many cases an adsorption potential may be more a supposition of biodegradibility than an exclusion criterion.

If no acute tests are available, one test with higher plants and one test with earthworms or microorganisms merit particular attention since they are involved in the transformation of organic matter and contribute considerably to the transformation potential of a terrestrial ecosystem.

Extrapolation of test results obtained for vertebrates (i.e., mice, rats, or birds) on the effects of a chemical in the field is difficult because of behavior- or stress-related individual or specific responses of the test organisms [254]. Furthermore, the assessment of critical environmental concentrations must be based on concentration estimates in the soil, which lead to effects via ingestion [402].

Fuzzy Set and Fuzzy Logic Applications in Ecological Research. A fuzzy set and a fuzzy logic approach [403] can be used to solve some problems of uncertainty in ecotoxicological research and practice, namely, the uncertainty of data and uncertainty of expert knowledge. Compared with the conventional methods of information processing, the application of the fuzzy set theory offers a better applicability of imprecise data and vague knowledge. This involves

1) The representation and processing of imprecise data in the form of fuzzy sets
2) The representation and processing of vague knowledge in the form of linguistic rules with imprecise terms defined as fuzzy sets

Two main applications of the fuzzy set theory in ecological and ecotoxicological research are data analysis and knowledge-based modeling. Conventional clustering methods definitely placed an object within only one cluster. With *fuzzy clustering*, an object need not be definitely placed within one cluster since the membership value of this object can be split up between different clusters. In comparison to conventional clustering methods, distribution of the membership value thus provides additional information from which the membership value can be interpreted as the degree of similarity between properties of a particular object and properties represented by particular clusters.

Kriging belongs to the most popular methods of spatial interpolation of data, but its application is sometimes restricted owing to an insufficient amount of data. The amount of available measurement data is often too low for conventional kriging methods, or exact data may be unavailable at some locations in the study area. In such a case, the data set can be completed by using additional imprecise data estimated subjectively by an expert. *Fuzzy kriging* is a modification of the conventional kriging procedure that utilizes exact (crisp) measurement data as well as imprecise estimates obtained from an expert. These imprecise data are defined as so-called fuzzy numbers (a fuzzy number is a fuzzy set whose membership function is convex and takes its maximal value 1 for only one parameter value). Fuzzy kriging can be used to estimate interpolated values in the form of fuzzy numbers reflecting the imprecision of the input data.

Fuzzy knowledge-based modeling can be particularly useful when no analytical model of the relations to be examined exists or data are insufficient for statistical analysis [404]. To facilitate the construction of fuzzy knowledge-based models of complex ecological or toxicological systems, a modeling support system based on fuzzy logic (FLECO) has been developed [405]–[407]. The FLECO procedure employs fuzzy logic to handle inexact reasoning and fuzzy sets to handle the uncertainty of data. The main tasks of this system are

1) To facilitate the creation of the fuzzy knowledge base
2) To simplify changes within the fuzzy knowledge base
3) To create and facilitate the connections between submodels
4) To offer a set of inference methods
5) To facilitate the simulation process

3.3. State of the Art of Modeling in Ecotoxicology

3.3.1. Characteristics of Ecotoxicological Models

An increasing interest in management of pollution by toxic substances has emerged during the 1980s, and this has caused an equally large interest in toxic substance modeling. Models for toxic substances attempt to model the fate and effect of these substances in ecosystems. These models are management models, deterministic models, more frequently compartment models than matrix models, holistic models, and lumped models (because the difficulty of determining parameters is greater than for ecological modeling in general). Toxic substance models are most often biogeochemical models because they attempt to describe the mass flows of the toxic substances considered, although models of population dynamics also exist, which include the influence of toxic substances on birthrate or mortality and therefore should be considered toxic substance models.

Toxic substance models differ from other ecological models by the following:

1) The need for parameters to cover all possible toxic substance models is great, and general estimation methods are therefore used widely. Section 3.3.5 is devoted to this question.
2) The safety margin should be high when, for instance, expressed as the ratio between the actual concentration and the concentration that produces undesired effects.
3) The possible inclusion of an effect component, which relates the output concentration of a toxic substance to its effect, is easy. However, finding a well-examined relationship to base it on is often a problem.
4) The need is for simple models because of items (1) and (2), and because of limited knowledge of process details, parameters, sublethal effects, antagonistic and synergistic effects.

Several questions must be clarified before a toxic substance model can be developed in accordance with the general procedure used for development of ecological models. The following information must be available:

1) The best possible knowledge about the processes in which the toxic substance considered is involved in the ecosystem (as far as possi-

ble, knowledge about the quantitative importance of the processes should be obtained)
2) Obtaining parameters from the literature and/or from experiments (in situ or in the laboratory)
3) Estimation of all parameters by the methods presented in Section 3.3.5
4) Comparison of the results from items (2) and (3) and an attempt to explain the discrepancies
5) Estimation of which processes and state variables should be included in the model (if the slightest doubt exists at this early stage, too many processes and state variables should be included rather than too few)
6) Use of sensitivity analysis to evaluate the significance of individual processes and state variables, which in many cases will lead to further simplification

Section 3.3.2 reviews the characteristic features of ecotoxicological models and tries to classify them. Section 3.3.3 gives an overview of some of the ecotoxicological models published during the last 10–15 years. The description of the chemical, physical, and biological processes in the models will, in general, be in accordance with equations generally applied for physico-chemical and biological processes. Some submodels of particular ecotoxicological interest are presented in Section 3.3.4. Section 3.3.5 deals with parameter estimation methods, which are of particular importance in ecotoxicological models. Section 3.3.6 is devoted to fugacity models in general, including their basic equations, and a case study in which the fugacity modeling approach is used is presented.

3.3.2. Classification of Ecotoxicological Models

Ecotoxicological models differ from ecological models in general by

1) Most often being more simple
2) More parameters are unknown, cannot be found in the literature
3) Wider use of parameter estimation methods
4) Possible inclusion of an effect component

Ecotoxicological models can be divided into six classes. The classification presented here is based on differences in the modeling structure. The decision on which model class to apply will be based on the ecotoxicological problem to be solved. The definitions of model classes are given below, along with the most appropriate use of each model type.

Class 1. Food Chain or Food Web Dynamic Models. Class 1 models consider the flow of toxic substances through the food chain or food web. Such models are relatively complex and contain many state variables. The models furthermore contain many parameters, which often have to be estimated by one of the methods presented in Section 3.3.5. These types of model are used when a great number of organisms are affected by the toxic substance, or the entire structure of the ecosystem is threatened by the presence of a toxic substance. Because of their complexity, these models have not been used widely. They are similar in many aspects to the more complex eutrophication models that consider the flow of nutrients through the food chain or even through the food web. In some instances, they are even constructed as submodels of a eutrophication model (see, for instance, [408]). Figure 51 shows

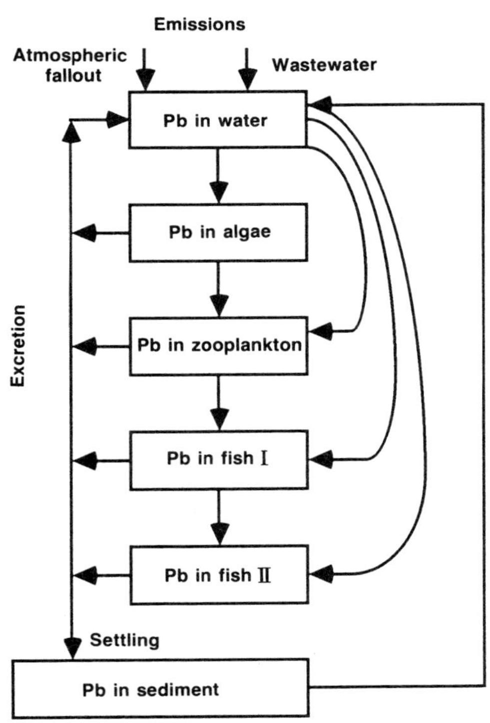

Figure 51. Conceptual diagram of lead accumulation in the food chain of an aquatic ecosystem

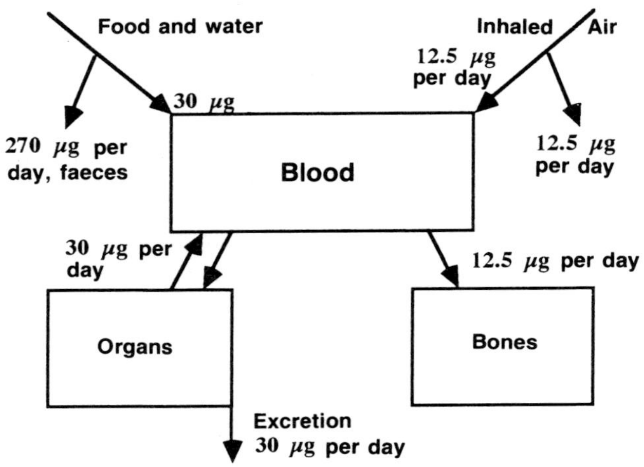

Figure 52. Static model of the daily lead uptake of an average Dane

a conceptual diagram of an ecotoxicological food chain model for lead. Lead from atmospheric fallout and wastewater enters an aquatic ecosystem, where it is concentrated through the food chain. Simplification is rarely possible for this model because the aim of the model is to describe and quantify the bioaccumulation and ecological magnification through the food chain.

Class 2. Static Models of the Mass Flow of Toxic Substances. If seasonal changes are minor, or of minor importance, a static model of the mass flows is often sufficient to describe the situation and even indicate the expected changes if the input of toxic substances is reduced or enlarged. This type of model is based on a mass balance as shown in the example in Figure 52. It often, but not necessarily, contains more trophic levels, and the modeler often is concerned with the flow of the toxic substance through the food chain. If some seasonal changes occur, this type, which in most cases is simpler than class 1, can still be used advantageously, for instance, if the modeler is concerned with the worst case of seasonal changes.

Class 3. Dynamic Model of a Toxic Substance in a Trophic Level. Often only the toxic substance concentration in one trophic level is of concern. This includes the zero trophic level, which is understood as the medium soil, water, or air. Figure 53 shows a model of copper contamination in an aquatic ecosystem as an example. The main concern is the copper concentration in the water, which may reach a toxic level for phytoplankton. Zooplankton and fish are much less sensitive to copper contamination, so the "alarm clock" rings first at the concentration level that is harmful to phytoplankton. However, only dissolved copper ions are toxic, and therefore the partition

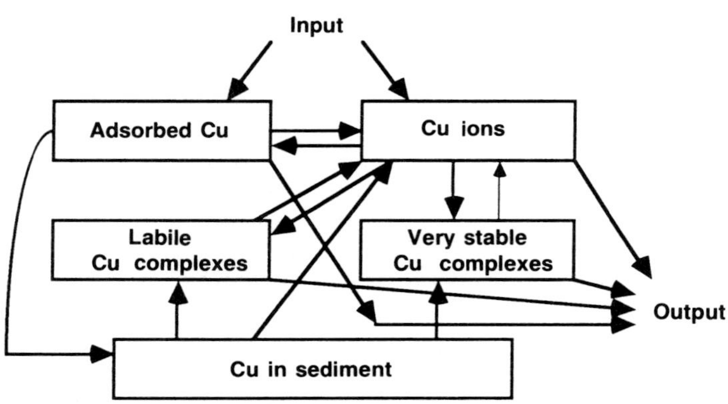

Figure 53. Conceptual diagram of a simple copper partition model

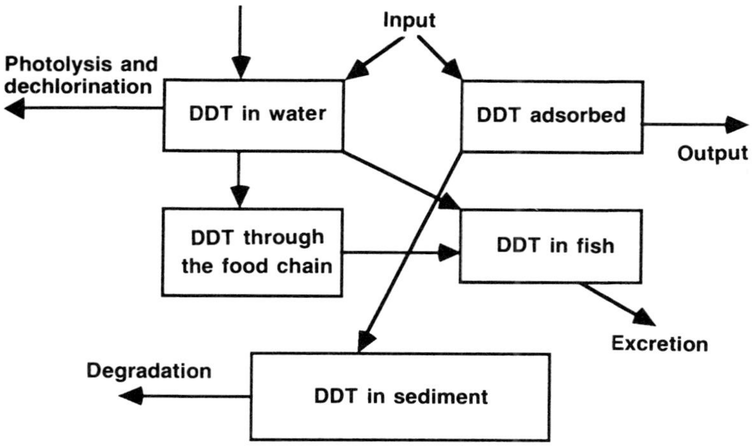

Figure 54. Conceptual diagram of a simple DDT partition model

of copper in ionic, complex-bound, and adsorbed forms must be modeled. The exchange between copper in the water phase and in sediment is also included because sediment can accumulate relatively large amounts of heavy metals. The amount released from sediment may be significant under certain circumstances, for instance, at low pH.

Figure 54 gives another example. Here, the main concern is the DDT concentration in fish, which may be so high that, in accordance with WHO standards, the fish are unfit for human consumption. The model can therefore be simplified by including not the entire food chain but only fish. Some physicochemical reactions in the water phase are, however, important and are included in the diagram.

The so-called HESP model [259] estimates exposure, or the total concentration of a toxic substance for individuals living at a contaminated site. A number of possible exposure routes are identified and subsequently quantified in the model. The model can be used to initiate a hazard assessment process, which can provide a sound basis for decisions regarding possible measures.

As seen from these examples, simplifications are often feasible when the problem is well defined. Simplifications take into account which component is most sensitive to toxic matter and which processes are most important for concentration changes.

Class 4. Ecotoxicological Models in Population Dynamics. Population models are biodemographic models and therefore have numbers of individuals or species as state variables. The simple population models consider only one population. Population growth is the difference between natality and mortality:

$$dN/dt = B \cdot N - M \cdot N = r \cdot N \quad (82)$$

where N is the number of individuals; B is the natality (i.e., the number of new individuals per unit of time and unit of population); M is the mortality (i.e., the number of organisms that died per unit of time and unit of population); and r is the increase in number of organisms per unit of time and unit of population (i.e., $B - M$). The concentration of a toxic substance in the environment or in the organisms influences natality and mortality, and if the relations between the concentration of a toxic substance and these population dynamic parameters are included in the model, it becomes an ecotoxicological model of population dynamics.

Population dynamic models may include two or more trophic levels, and ecotoxicological models will include the influence of the concentration of the toxic substance on natality, mortality, and interactions between these populations. In other words, an ecotoxicological model of population dynamics is a general model of population dynamics with the inclusion of the relation between toxic substance concentrations and some of the model parameters.

Class 5. Ecotoxicological Models with Effect Components. Although class 4 models include re-

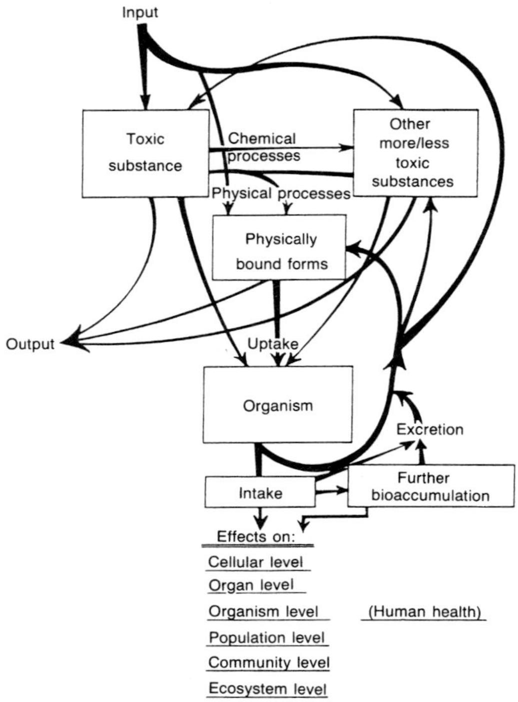

Figure 55. Conceptual diagram of ecotoxicological models with effect components on various levels of biological complexity

lations between concentrations of toxic substances and their effects, these are limited to population dynamic parameters. In comparison, class 5 models include more comprehensive relations between concentrations of toxic substances and effects. These models may include not only lethal or sublethal effects but also effects on biochemical reactions or on the enzyme system. As Figure 55 shows, the effects on various levels of the biological hierarchy may be considered from cells to ecosystems.

In many problems, more details about the effects may be required to answer the following relevant questions:

1) Does the toxic substance accumulate in the organism?
2) What will the long-term concentration in the organism be when uptake rate, excretion rate, and biochemical decomposition rate are considered?
3) What is the chronic effect of this concentration?
4) Does the toxic substance accumulate in one or more organs?
5) What is the transfer between various parts of the organism?
6) Will decomposition products eventually cause additional effects?

A detailed answer to all of these questions may require a model of the processes that take place in the organism and a translation of concentrations in various parts of the organism into effects. This implies, of course, that the intake = (uptake by the organism) × (efficiency of uptake) is known. Intake may be from either water or air, which can also be expressed by the use of concentration factors (the ratios between the concentration in the organism and that in air or water).

However, if all of the aforementioned processes are considered for just a few organisms, the model will become too complex, contain too many parameters to calibrate, and require more detailed knowledge than can be provided. Therefore, most models in this class do not consider too many details of the partition of toxic substances in organisms and their corresponding effects, but rather are limited to simple accumulation in the organisms and their effects. In most cases, accumulation is rather easy to model, and the following simple equation is often sufficiently accurate:

$$\mathrm{d}C/\mathrm{d}t = (e_\mathrm{f} \cdot C_\mathrm{f} \cdot F + e_\mathrm{m} \cdot C_\mathrm{m} \cdot V)/W - Ex \cdot C = (INT)/W - Ex \cdot C \quad (83)$$

where C is the concentration of the toxic substance in the organism; e_f and e_m are the efficiencies for the uptake from the food and medium, respectively (water or air); C_f and C_m are the concentration of the toxic substance in the food and medium, respectively; F is the amount of food uptake per day; V is the volume of water or air taken up per day; W is the body weight as either dry or wet matter; and Ex is the excretion coefficient (d^{-1}). As seen from the equation, INT covers the total intake of toxic substance per day.

This equation has a numerical solution:

$$C/C(\max) = \{INT \cdot [1 - \exp(Ex \cdot t)]\}/(W \cdot Ex) \quad (84)$$

where $C(\max)$ is the steady-state value of C:

$$C(\max) = INT/(W \cdot Ex) \quad (85)$$

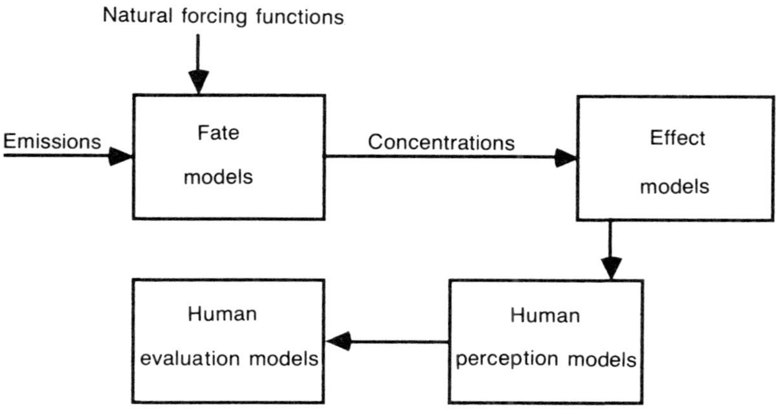

Figure 56. Components and general structure of comprehensive ecotoxicological modeling systems

i.e., the concentration of the toxic substance approximates a steady-state value after a certain time interval.

Class 6. Fate Models with or without a Risk Assessment Component. The last type of ecotoxicological models focuses on the fate of toxic substances: Where in the ecosystem will the toxic substance be found? In what concentration? These models are sometimes called multimedia models because they focus on the distribution of toxic substances in the four spheres. The complete solution of an ecotoxicological problem requires in principle four (sub)models, of which the fate model may be considered the first model in the chain; see Figure 56. The four components are [409] (1) a fate or exposure model; (2) an effect model, translating the concentration into an effect (see class 5); (3) a model for human perception processes; and (4) a model for human evaluation processes. The first two submodels are in principle "objective," predictive models, whereas the latter two are value oriented. The development of submodels (1) and (2) is based on physical, chemical, and biological processes. They are very similar to other environmental models and are based on mass transfer; mass balances; and physical, chemical, and biological processes.

The second submodel requires good knowledge of the effects of the toxic components. Submodels (3) and (4) differ from the generally applied environmental management models and are presented in some details below in the discussion of risk assessment.

Fugacity models (see Section 3.3.6) are a special type of fate model. They attempt to answer the following questions: In which of these six compartments (air, water, soil, sediment, suspended sediment, and biota) can the greatest problem be expected for chemicals emitted to the environment? What concentration will be expected in each compartment? What are the implications of these concentrations? The fugacity models are to a large extent based on physicochemical parameters and the use of these parameters to estimate other required parameters. The applicability of the fugacity model is therefore very dependent on the use of the estimation methods presented in Section 3.3.5.

The risk assessment component, associated with the fate model comprises human perception and evaluation processes (see Fig. 56). These submodels are explicitly value laden but must of course build on objective information about concentrations and effects. Factors that may be important to consider in this context are

1) Magnitude and time constant of exposure
2) Spatial and temporal distributions of concentrations
3) Environmental conditions determining the process rates and effects
4) Translation of concentrations into magnitude and duration of effects
5) Spatial and temporal distribution of effects
6) Reversibility of effects

The uncertainties relating to the information on which the model is based and the uncertainties related to the development of the model, are crucial in risk assessment. The uncertainty in risk assessment may be classified in one or more of the following five categories:

1) Good direct knowledge and statistical evidence on the important components (state variables, processes, and interrelations of the variables) of the model are available.
2) Good knowledge and statistical evidence on the important submodels are available, but the aggregations of the submodels are less certain.
3) No good knowledge of the model components for the considered system is available, but good data are available for the same processes from a similar system, and these data are considered applicable for use directly or with minor modifications, to the model development.
4) Some, but insufficient, knowledge ist available from other systems. Attempts are made to use these data without the necessary transferability. Attempts are made to eliminate gaps in knowledge by the use of additional experimental data as far as possible within the limited resources available for the project.
5) The model is to a large extent based on the subjective judgment of experts.

Acknowledgment of the uncertainty is of great importance and may be taken into consideration, either qualitatively or quantitatively. Another problem is of course: Where and how should the uncertainty be taken into account? Should the economy or the environment benefit from the uncertainty? If the results of the management strategy are to be ensured, the uncertainty must be used to reduce the emission standards.

Until 10–15 years ago, researchers had very little understanding of the processes by which people actually perceive the exposures and effects of toxic chemicals, but these processes are just as important for risk assessment as the exposure and effect processes. The following list of risk characteristics summarizes the concerns:

1) Is the risk voluntary or involuntary?
2) Are the levels known to the exposed people or only to science?
3) Is the risk novel, or old and familiar?
4) Is it common or dreaded (e.g., does it involve cancer)?
5) Does it involve death?
6) Are mishaps controllable?
7) Are future generations threatened?
8) Is the risk global, regional, or local?
9) Is the risk a function of time? How does it depend on time (e.g., increase or decrease)?
10) Can it be reduced easily?
11) Is the expected effect immediate or delayed?
12) Are many or a few people involved?
13) Are the effects observable immediately?

A factor analysis has been performed by Slovic et al. [410] that shows, among other results, a not surprising emphasis on people's perception of dreadful and unknown risks. Broadly speaking, two methods of selecting the risks exist. The first may be described as the *rational actor model*, involving people that look systematically at all the risks they face and make choices about which they will live with and at what levels. For decision making, this approach would use some single, consistent, objective functions and a set of decision rules. The second method may be called the *political–cultural model*. It involves interactions among culture, social institutions, and political processes for the identification of risks and the determination of those that people will live with and at what level.

Both methods are unrealistic, since they are both completely impractical in their pure form. Therefore, a strategy for risk abatement must be selected that is founded on a workable alternative based on the philosophy behind both methods.

Several risk management systems are available. Here, some recommendations should be given for the development of risk management systems:

1) Consider as many of the characteristics listed above as possible and include human perception of these characteristics in the model.
2) Do not focus too narrowly on certain types of risks; this may lead to suboptimal solutions. Attempt to approach the problem as broadmindedly as possible.
3) Choose strategies that are pluralistic and adaptive.
4) Benefit–cost analysis is an important element of the risk management model, but it is far from being the only important element, and the uncertainty in evaluation of benefit and cost should not be forgotten. The variant of this analysis applicable to environmental risk management may be formulated as follows:

Net social benefit
= social benefits of the project minus "environmental" costs of the project

5) Use multiattribute utility functions, but remember that people in general have trouble thinking about more than two to three (at most, four) attributes in each outcome.

Presentation of the six classes of models above clearly shows the advantages and limitations of ecotoxicological models. The simplifications used in classes 2, 3, and 6 (at least without risk assessment components) often offer great advantages. They are sufficiently accurate to give a very applicable picture (overview) of the concentrations of toxic substances in the environment, because of the use of large safety factors. Application of the estimation methods presented in Section 3.3.5 makes feasible the construction of such models, even though knowledge of the parameters is limited. The estimation methods obviously have a high uncertainty, but the great safety factor helps in accepting this uncertainty. On the other hand, knowledge about the effects of toxic substances is very limited, particularly at the organism and organ level. Therefore, models with effect components must not be expected to give more than a first rough picture of what is known today in this area.

Because of the character of ecotoxicological models, a few questions should be clarified before beginning the modeling procedure:

1) Obtain the best possible knowledge about the processes of the toxic substance under consideration
2) Attempt to get parameters of the processes in which the toxic substance is involved in the environment from the literature
3) Estimate all parameters using methods presented in Sections 3.3.4 and 3.3.5
4) Compare the results from items (2) and (3) and attempt to explain discrepancies, if present
5) Use sensitivity analysis widely to estimate which processes and state variables would be feasible and relevant to include in the model

3.3.3. An Overview: Application of Models in Ecotoxicology

Some toxic substance models are reviewed in Table 20 to give an idea of the types of models available today. Most models reflect the proposition that good knowledge of the problem and the ecosystem can be used to make reasonable simplifications. Model characteristics indicated in Table 20 are state variables or processes considered in the model. Note the number of modeled toxic substances and the processes taken into account. The model class, according to the classification 1 to 6 discussed in Section 3.3.2, is given in brackets after the toxic substance. Only a few class 4 models are included in Table 20. Ecological modeling has been approached from two sides: population dynamics and biogeochemical flow analysis. Since the second approach has been most involved in environmental issues, the toxic substance problem has also been approached from this angle. The few class 4 models, including the example given in Table 20, are population dynamic models, with some additional equations to account for the influence of toxic substances on natality and mortality. Provided these relations are available, constructing this type of model should be relatively easy.

The most difficult part of modeling the effect and distribution of toxic substances is obtaining relevant knowledge about the behavior of toxic substances in the environment and using this knowledge to make feasible simplifications. The modeler of ecotoxicological problems is particularly challenged to select the proper, balanced complexity, and many examples can be found of rather simple ecotoxicological models that are able to solve the basic problem.

It can be seen from the overview in Table 20 that most ecotoxicological models have been developed during the last decade. Before ca. 1975, toxic substances were rarely associated with environmental modeling, since the problems seemed straightforward. The many pollution problems associated with toxic substances were believed to be easily solved simply by eliminating the source of the toxic substance. During the 1970s, the environmental problems originating from toxic substances were acknowledged to be very complex because of the interaction of many sources and of many processes and components simultaneously. Several accidental releases of toxic substances into the environment have reinforced the need for models. Thus, various ecotoxicological models have been developed since the late 1970s. The list in Table 20 represents a comprehensive survey of available ecotoxicological models, but it should not be considered complete since it is not the result of an exhaustive literature review.

Table 20. Examples of toxic substance models

Toxic substance, model class	Model characteristics	Reference
Cadmium (1)	food chain similar to a eutrophication model	[411]
Mercury (6)	6 state variables: water, sediment, suspended matter, invertebrates, plant, fish	[412]
Vinyl chloride (3)	chemical processes in water	[413]
Methyl parathion (1)	chemical processes in water and microbial degradation and adsorption of benzothiophene, 2–4 trophic levels	[414]
Methyl mercury (4)	a single trophic level: food intake, excretion, metabolism, growth	[415]
Heavy metals (3)	concentration factor, excretion, bioaccumulation	[416]
Pesticides in fish, DDT and methoxychlor (5)	ingestion, concentration factor, adsorption on body, defecation, excretion, chemical decomposition, natural mortality	[417]
Zinc in algae (3)	concentration factor, secretion, hydrodynamic distribution	[418]
Copper in the sea (5)	complex formation, adsorption, sublethal effect of ionic copper	[419]
Lead (5)	hydrodynamics, precipitation, toxic effects of free ionic lead on algae, invertebrates, and fish	[420]
Radionuclides (3)	hydrodynamics, decay, uptake, and release by various aquatic surfaces	[69]
Radionuclides (2)	in grass, grains, vegetables, milk, eggs, beef, and poultry—state variables	[421]
SO_2, NO_x, and heavy metals on spruce forests (5)	threshold model for accumulation effect of pollutants (air and soil)	[368]
Toxic environmental chemicals in general tests (6)	hazard ranking and assessment from physicochemical and a limited number of laboratory data	[422]
Heavy metals (3)	adsorption, chemical reactions, ion exchange	[423]
Polycyclic aromatic hydrocarbons (3)	transport, degradation, bioaccumulation	[423]
Persistent toxic organic substances (3)	groundwater movement, transport and accumulation of pollutants in groundwater	[424]
Cadmium, PCB (2)	hydraulic overflow rate (settling), sediment interactions, steady-state food chain submodel	[408]
Hydrophobic organics (3)	gas exchange, sorption/desorption, hydrolysis compounds, photolysis, hydrodynamics	[425]
Mirex (3)	water–sediment exchange processes, adsorption, volatilization, bioaccumulation	[426]
Toxins (aromatic hydrocarbons, Cd) (3)	hydrodynamics, deposition, resuspension, volatilization, photooxidation, decomposition, adsorption, complex formation, (humic acid)	[427]
Heavy metals (2)	hydraulic submodel, adsorption	[428]
Oil slicks (3)	transport and spreading, influence of surface tension, gravity, and weathering processes	[429]
Acid rain (soil) (3)	aerodynamics, deposition	[70]
Acid rain	C, N, and S cycles and their influence on acidity	[430]
Persistent organic chemicals (6)	fate, exposure, and human uptake	[431]
Chemicals, general (6)	fate, exposure, ecotoxicity for surface water and soil	[432]
Toxicants, general (4)	effect on populations of toxicants	[433]
Chemical hazard (6)	basin-wide ecological fate	[434]
Pesticides (4)	effects on insect populations	[435]
Insecticides (2)	resistance	[436]
Mirex and lindane (6)	fate in Lake Ontario	[426]
Pesticides (3)	degradation in soil	[437]
Pesticides (3)	degradation in soil	[438]
Pesticides (3)	leaching to groundwater	[439]
Acid rain (5)	effects on forest soils	[70]
Acid rain (5)	cation depletion of soil	[440]
pH, calcium, and aluminum (4)	survival of fish populations	[441]
Photochemical smog and NO_x (6)	fate and risk	[442]
Nitrate (3)	leaching to groundwater	[443]
Oil spill (6)	fate	[444]
Toxicants (4)	effects on populations	[445]
Chromium (2)	distribution and accumulation in mussels	[71]
Toluene, benzene, Zn, Ni, Pb (3)	assess exposure levels of soil contaminants for humans	[446]

3.3.4. Ecotoxicological Processes

Ecotoxicological processes differ from other ecological processes, and a review of the relevant processes and their quantitative description is therefore necessary as basis for model development. A survey of the most crucial ecotoxicological processes and the most applied mathematical descriptions of these processes is given below.

Concentration Factor. The concentration factor relates the concentration of the xenobiotic in the medium to the concentration in organisms, sediment, or suspended matter. The concentration in organisms is a result of uptake by respiration, and the concentration in sediment or suspended matter is caused by adsorption and sedimentation processes. The concentration factor *CF* can often be described with sufficient accuracy as a constant independent of the concentration in the medium (water):

$$CF = C_S/C_W \qquad (86)$$

where C_S is the concentration in the organisms, suspended matter, or sediment and C_W is the concentration in water (or air). If the uptake includes the indirect route through food, the term bioconcentration factor *BCF* is often used ($BCF \geq CF$), which can be described by the same equation.

Table 21 gives equations relating water solubility and the *n*-octanol–water partition coefficient to *CF* and *BCF*. The *n*-octanol–water partition coefficient K_{ow} is defined in Section 2.4.1. An overview on relevant literature references dealing with the determination of the *n*-octanol–water partition coefficient of various chemicals in fish is given in Table 22. If the *n*-octanol–water partition coefficient is not known, it can be estimated from the water solubility of the corresponding substance (see Section 3.3.5 and Fig. 58).

Table 22. Determination of *n*-octanol–water partition coefficients in fish

Animal	Number of chemicals	References
Fish species	36	[447]
Mosquito fish	9	[448]
Mosquito fish	11	[449]
Trout	8	[450]
Fish species	26	[447]
Fathead minnow	59	[451]
Fathead minnow, bluegill	59	[452]
Fish species	13	[447]
Fish species	50	[447]
Mosquito fish	9	[448]
Fish species	36	[447]
Mosquito fish, whole	15	[449]

Biomagnification describes the process whereby pollutants are passed from one trophic level to the next and increasingly concentrated in organisms with higher trophic status (see Section 2.4.2.7). The magnification (i.e., the increasing concentration through the food chain) is caused by greater retention of toxic substances than of general food components by the organisms. The process may be described by the biological or ecological magnification factor, which indicates the magnification from one level in the food chain to the next.

Further important processes in ecological modeling are adsorption (see Section 2.4.2.4) and desorption.

Excretion of toxic substances by an organism can be shown to follow first-order kinetics to a good approximation, and the rate coefficient depends on the chemical compound and the size

Table 21. Regression equations for estimation of concentration, bioconcentration, and ecological magnification factors

Indicator	Relationship	Correlation coefficient	Range (indicator)
K_{ow}	$\log CF = -0.973 + 0.767 \log K_{ow}$	.76	2.0×10^{-2}–2.0×10^{6}
K_{ow}	$\log CF = 0.7504 + 1.1587 \log K_{ow}$	.98	7.0–1.6×10^{4}
K_{ow}	$\log CF = 0.7285 + 0.6335 \log K_{ow}$	.79	1.6–1.4×10^{4}
K_{ow}	$\log CF = 0.124 + 0.542 \log K_{ow}$	.95	4.4×100–4.2×10^{7}
K_{ow}	$\log CF = -1.495 + 0.935 \log K_{ow}$	.87	1.6×100–3.7×10^{6}
K_{ow}	$\log CF = -0.70 + 0.85 \log K_{ow}$	.95	1.0–1.0×10^{7}
K_{ow}	$\log CF = 0.124 + 0.542 \log K_{ow}$	.90	1.0×100–5.0×10^{7}
S, µg/L	$\log BCF = 3.9950 - 0.3891 \log S$	.92	1.2–3.7×10^{7}
S, µg/L	$\log BCF = 4.4806 - 0.4732 \log S$	.97	1.3–4.0×10^{7}
S, µmol/L	$\log BCF = 3.41 - 0.508 \log S$	.96	2.0×10^{-2}–5.0×10^{3}

of the organism considered (see [37]). This process is in most cases insignificant for the entire mass balance of the components considered but inclusion of the excretion process may be important for a model of a toxic substance on the organism level.

Biodegradation. It is rather difficult to estimate the reaction rate of biodegradation in ecotoxicology because a large number of factors influence the reaction and thereby also the concentration of microorganisms. The most important of these factors are: temperature, sorption, redox potential, ionic strength, adaptation of microbial populations, macronutrients and micronutrients available [453], [454].

Distribution of a Chemical Compound. Distribution of a chemical compound in the four spheres is of particular interest in fate models.

Distribution air–water

$$\frac{C^*}{To^*} \equiv F_L = \frac{H\,18}{1000\,R\,T} \text{(approximation)} \qquad (87)$$

or at (103.33 kPa) and 20 °C

$$F_L = 7.49 \cdot 10^{-4}\,H \qquad (88)$$

where H is Henry's constant (bar); C^* and To^* are equilibrium concentrations (mass/volume) in air and water, respectively.

Rate of evaporation (distribution liquid–vapor phase)

$$Ev = -D \cdot A \frac{\mathrm{d}To}{\mathrm{d}x} = b K_L \cdot A \cdot (To - To^*) \qquad (89)$$

where D is the molecular diffusion coefficient (area/time), A is the area, and b is a coefficient that adjusts the evaporation coefficient K_L (length/time) to substances other than oxygen ($b = 1.0$ for oxygen); b can be found in [72] for other compounds. Several empirical equations for calculation of K_L can be found in [73], [74].

$$b = \frac{K_L(\text{compound})}{K_L(\text{oxygen})} = \frac{D(\text{compound})}{D(\text{oxygen})} \qquad (90)$$

For lakes and streams,

$$To(t) = To(0) \cdot \mathrm{e}^{-b \cdot K_L \cdot a \cdot t} \qquad (91)$$

where

$$a = \frac{A}{V} = \frac{\text{area}}{\text{volume}} \qquad (92)$$

Distribution water–solid (sediment)

A distribution coefficient may be used at low concentrations (see also Langmuir's and Freundlich's adsorption isotherms):

$$F_S = \frac{C^*}{To^*} \qquad (93)$$

where C^* and To^* are the concentrations in the solid and water phases at equilibrium. Since the dimension of C^* is mass/mass (e.g., mg per gram dry matter) and To^* has the dimension of mass/volume (mg/L), F_S has the dimension volume/mass (L/g or L/kg).

F_S may be found from

$$F_S = F_{oc} \cdot f_{oc} \qquad (94)$$

where F_{oc} is the distribution coefficient for organic matter and f_{oc} is the fraction of organic carbon in the solid being considered.

The following relationship between F_{oc} and K_{ow} exists [455]:

$$\log F_{oc} = -0.006 + 0.937 \log K_{ow} \qquad (95)$$

Matter-Müller et al. [456] found the following relationship for activated sludge:

$$\log F_S = 0.39 + 0.67 \log K_{ow} \qquad (96)$$

where F_S has the units L/kg.

F_{oc} and F_S values for selected chemical compounds can be found in the literature.

Complex Formation. As for other chemical processes, the law of mass action is valid for complex formation reactions. For the process

$$\mathrm{Me}^{n+} + \mathrm{L}^{m-} = \mathrm{MeL}^{(n-m)+} \qquad (97)$$

(where Me denotes a metal and L a ligand), the law of mass action has the following form:

$$\frac{[\mathrm{MeL}^{(n-m)+}]}{[\mathrm{Me}^{n+}][\mathrm{L}^{m-}]} = K \qquad (98)$$

K is termed the stability constant if the process—as in this case—is the formation of a complex.

The formation of coordination complexes plays a major role in modeling the distribution as well as the effect of metal ions in aquatic ecosystems because of the following influences:

1) Increase in metal solubility

$$\mathrm{MeY(S)} + \mathrm{L} = \mathrm{MeL} + \mathrm{Y} \qquad (99)$$

where Me represents a metal that is bound in the sediment in the compound MeY; L denotes a ligand

2) Alteration of the distribution between oxidized and reduced forms of metals. The equilibrium constant for the process

$$Me^{n+} + me^- = Me^{(n-m)+} \quad (100)$$

differs from the one for

$$MeL^{n+} + me^- = MeL^{(n-m)+} \quad (101)$$

if the complexes, MeL^{n+} and $MeL^{(n-m)+}$, have different stability constants, which often is the case

3) Alleviation of toxicity due to alteration of the availability to aquatic life
4) Change in adsorption and ion exchange on sediment and suspended matter
5) Change in stability of the metal-containing colloids

The presence of ligands in water generally increases the transfer of metals from sediment, soil, and suspended matter and thereby increases the solubility. However, since complexes show lower toxicity than metal ions, the harm to aquatic life in most cases does not increase in proportion to the increased solubility.

The application of coordination chemistry to environmental problems is very complex because many ligands are present in aquatic ecosystems and all ligands compete simultaneously for formation of complexes with metal ions.

Photolysis. *The Stark–Einstein law* states that if a species absorbs radiation, one particle is excited for each quantum of radiation absorbed. The energy of the particle is limited by the energy of the photon absorbed. This energy is given by

$$E = h\nu \quad (102)$$

where E is the energy of the photon, h is Planck's constant, and ν is the frequency of the photon. Thus, depending on the nature of the incident light and the property of the chromophore, different excited states of the chromophore may be formed. Although not exclusively true, in general only light absorption that brings about electronic excitation of the chromophore will result in photochemical reactions. Excitation of the vibrations and rotations of the absorbing species in its ground electronic state is less likely to induce chemical changes.

The effectiveness of photochemical reactions in the consumption or production of a given chemical species is given by the *overall quantum yield* of that process (see Section 2.4.2.2). The overall quantum yield is the number of molecules of a given reactant consumed for each photon absorbed. It may be further subdivided into the primary quantum yield (molecules consumed by primary reactions) and the secondary quantum yield (molecules consumed by secondary reactions).

If the quantum yield is not strongly wavelength dependent, for an incident beam of light with a certain spectral characteristic, the complete rate law for direct photolysis is

$$-(\mathrm{d}[CA]/\mathrm{d}t) = k_a[CA] \quad (103)$$

where CA is the concentration of the reactant and k_a is the sum of the k_a values of all the wavelengths represented by the photons in the incident light. Models for estimating the rates of direct photolysis in natural waters under restrictive conditions have been attempted by several investigators [457]–[465].

Acid–base reactions are of a great environmental interest because almost all processes in the environment depend on pH. A few illustrative examples are included in the following list:

1) Ammonia is toxic to fish and the ammonium:ammonia ratio depends on pH.
2) Carbon dioxide is toxic to fish and the ratio of hydrogencarbonate to carbon dioxide depends on pH.
3) The fertility of fish and zooplankton eggs is highly pH dependent.
4) All biological processes have a pH optimum, which is usually in the range 6–8. This implies that algal growth, microbiological decomposition, nitrification, and denitrification are all influenced by pH.
5) The release of heavy-metal ions from soil and sediment increases very rapidly with decreasing pH. Heavy-metal hydroxides have a very low solubility product, which implies that most heavy-metal ions are precipitated at pH 7.5 or higher.

Acid rain illustrates the problems related to a low pH value, but too high a pH is also of environmental concern because of items (1) and (4) above.

Other Chemical Processes. In addition, hydrolysis and redox processes may also be important for the speciation and general presence of toxic substances. Many metal hydroxides may be dissolved by hydrolysis, and the release of metal ions from lake sediments is highly dependent on the redox potential at the interface between water and sediment.

Organic compounds can also undergo hydrolysis reactions. Many compounds react with water, with introduction of a hydroxyl group into the molecule. Most toxic organic compounds are oxidized by oxygen, and carboxyl groups may be formed. For all of these reactions, a chemical pathway should be found and the application of simple first-order reaction kinetics is often sufficient to describe the process in an ecotoxicological model.

3.3.5. Parameter Estimation

About 50 000 chemicals are produced on a large scale (1000 kg or more per year), but only about 4000 chemicals are produced at a quantity of $\geq 1 \times 10^6$ kg/a. Approximately one-third of these chemicals are well-defined organics. Their environment-relevant properties are listed below:

Physical properties
vapor pressure, water solubility, Henry's constant, adsorption coefficient, partition coefficient octanol–water, mass-transfer coefficient, density, viscosity, diffusion coefficient in air and water, boiling and melting point

Chemical properties
rate of hydrolysis, rate of oxidation or reduction, rate of photolysis, acid dissociation coefficient(s)

Biological properties
bioconcentration factor, rate of biodegradation in soil and water under aerobic or anaerobic conditions

Most of the biological parameters are needed for all species or at least a representative fraction of the species on earth. This implies that for construction of all possible ecotoxicological models, about 20 000 parameters are needed for 50 000 chemicals (i.e., a total of ca. 1×10^9 parameters). A review of the literature reveals that hardly 1 % of these parameters, or 10×10^6 values are known. Providing the 9.9×10^8 parameters lacking will be an enormous task—a task that will take several decades to accomplish. How can ecotoxicological models be constructed under these circumstances? The answer is to estimate the parameters, which is acceptable, because ecotoxicological models do not require a high accuracy, or rather, one is forced to interpret the model results by the use of large safety factors, which makes it less necessary to demand great accuracy of the modeling results.

Many estimation methods are available for ecotoxicological parameters, and too many pages would be required to present even an overview of these methods. Jørgensen [37] presents a rather simple method that can be applied easily and still give sufficient accuracy for development of most ecotoxicological models.

The *physical properties* of interest in ecotoxicology are vapor pressure P, water solubility S, partition coefficient K_{ow}, and Henry's constant H; for determination of the transfer between air and water, adsorption isotherms K_{ac} and diffusion coefficients D are also necessary. For organic compounds, all these properties can be estimated just on the basis of their chemical structure. Some estimation methods are based on estimation of the boiling point and critical properties, which are then used later to estimate the aforementioned physical properties. Experience shows that generally better estimates can be given if the boiling point is known and not found by estimation methods. Simultaneous use of two or more estimation methods will always give a better overall estimation (an average value is used) than if only one method is applied. All relevant equations for the described estimation system are given in [37]. The system itself is shown conceptually in Figure 57 while the correlation between the partition coefficient *n*-octanol–water and the water solubility is given in Figure 58.

Chemical properties (i.e., rate of hydrolysis, rate of oxidation or reduction, and acid dissociation coefficients) can also be estimated.

Obviously, *biological properties* are very important in ecotoxicological modeling, and since these properties are generally even less well known than physical and chemical properties, the estimation methods are even more important. Figure 59 shows how to estimate CF, BCF, and the ecological magnification EM by use of solubility data S and partition coefficients K_{ow}, see also the relation in Figure 60. From the struc-

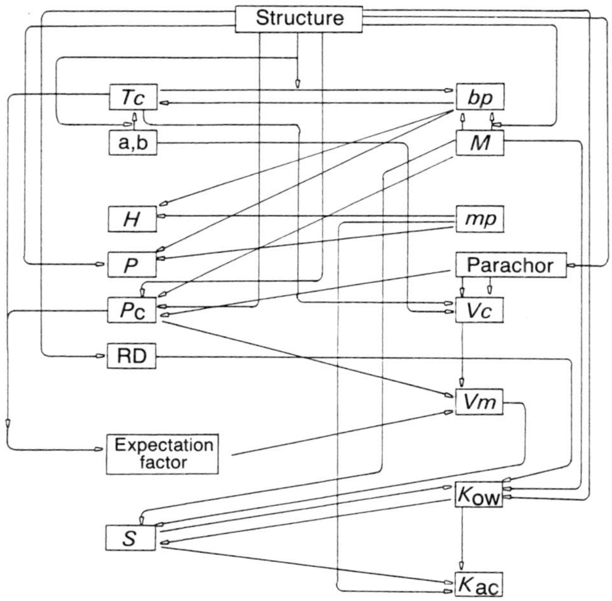

Figure 57. Conceptual diagram of an estimation system for ecotoxicological parameters (after [37])

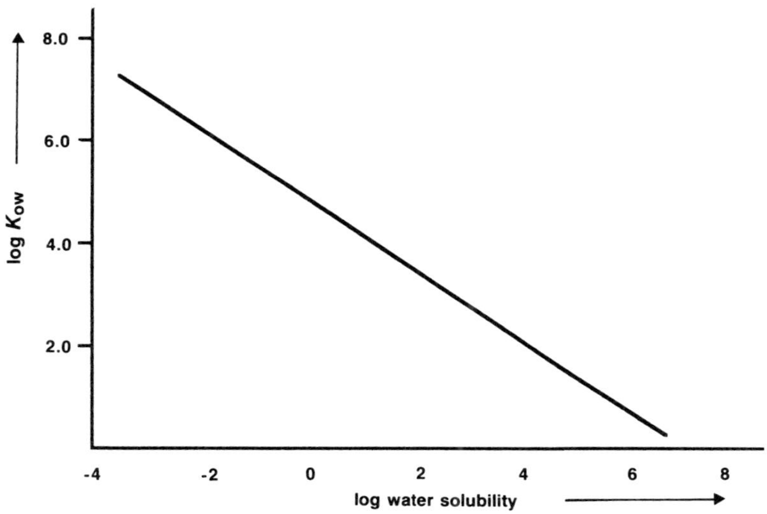

Figure 58. Linear regression between log K_{ow} and log S (water solubility in μmol/L), based on a range of organic compounds including aliphatic and aromatic hydrocarbons, aromatic acids, organochlorines, organophosphorus insecticides, and PCBs (after [466])

ture, furthermore, biodegradation rate coefficients and toxicity can be estimated. Some of the methods work via molar mass M and estimations of mp and bp (see Fig. 59).

The estimation methods used in practical ecotoxicological modeling have been tested widely on compounds with known properties and have been found to have a standard deviation of 10–60 % for most parameters, although the standard deviations are higher for toxicological parameters. This is an acceptable accuracy since large safety factors are recommended and applied in ecotoxicological modeling. Although the number of relevant parameters that can be found in the literature is limited, review of the literature is advantageous to find the parameters

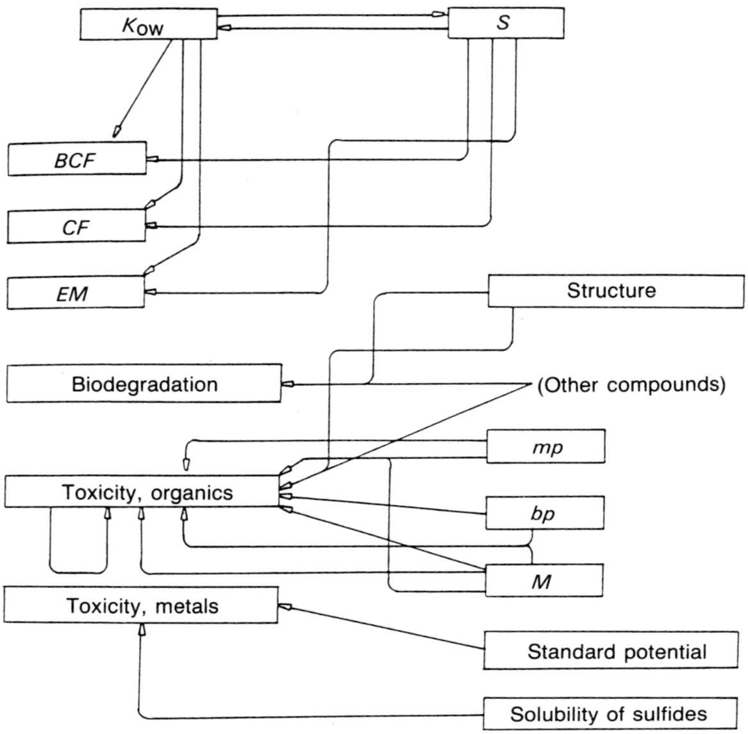

Figure 59. General structure–activity relationships of potentially toxic substances

that actually have been determined. Furthermore, a comparison of the parameters that have been estimated by various methods is helpful to get an idea of the accuracy of these methods for the compounds considered. The following references may be used as primary sources: [37], [47], [71], [72], [75].

3.3.6. Fugacity Models

The application of fugacity in environmental modeling has been discussed by Mackay and Paterson [214], [215]. Fugacity has the dimension of pressure and may be considered a measure of "escaping tendency" from the phase. When phases are at equilibrium, their fugacities are equal. Fugacity f is related to concentration by the following equation:

$$f = C/Z \qquad (104)$$

where C is the concentration (mol/m^3) and Z is the fugacity capacity (mol m^{-3} Pa^{-1}). Equilibrium involves

$$C_a/C_w = Z_a/Z_w = K_p \qquad (105)$$

where the indices a and w indicate air and water, respectively, as examples of the phases; K_p is a partition coefficient. Transfer rates between two phases by diffusion are expressed by the following equation (models per unit of area and time):

$$N = D \cdot \Delta f \qquad (106)$$

where N is the rate of transfer, D is the diffusion coefficient, and Δf is the difference in fugacity. D is the total resistance for the transfer consisting of the resistances of the two phases in series. Note that D may be determined as $K \cdot Z$, where K is the transfer coefficient and Z is the fugacity capacity defined in Equation (104).

The so-called unit world model consists of six compartments: air, water, soil, sediment, suspended sediment, and biota. This simplified model aims at identification of the partition of toxic substances emitted to the environment among these six compartments. The conceptual diagram of the unit world model is shown in Figure 61.

The fugacity model may be used at four levels. The *first level* calculates the equilibrium dis-

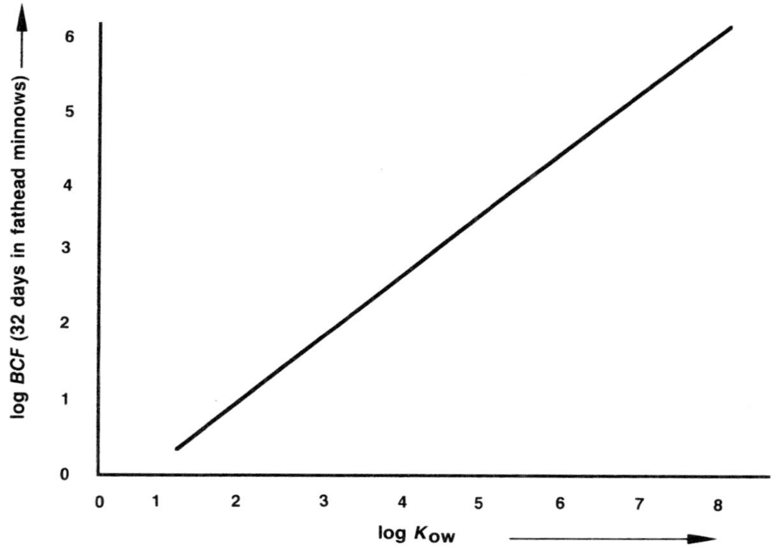

Figure 60. Relationship between log K_{ow} and log BCF, valid for fish ca. 25 cm in length

tribution of a chemical between phases. It assumes that each compartment is well mixed, and no reaction or advection into or out of the system occurs.

If M is the total amount of a chemical in moles, m_i and V_i are the amounts of the chemical and the volumes in each compartment, then:

$$M = \sum m_i = \sum C_i \cdot V_i = \sum f_i \cdot Z_i \cdot V_i = f \cdot \sum Z_i \cdot V_i \quad (107)$$

Since M is known, f may be calculated and the amount of substance in each compartment is given by

$$m_i = f \cdot V_i \cdot Z_i \quad (108)$$

The percentage distribution is found as m_i/M and the concentrations are found by

$$C_i = Z_i \cdot f \quad (109)$$

Level two considers equilibrium but includes reaction and advection as well. Reactions comprise photolysis, hydrolysis, biodegradation, oxidation, and so on. All the processes are assumed to be first-order reactions. If the rate constants are not known or cannot be estimated, they are set to zero. Since process rates are expressed in first-order form, they are additive and the total removal rate of a chemical from a compartment is

$$\sum V_i \cdot C_i \cdot \sum k_j = \sum V_i \cdot C_i \cdot k_i \quad (110)$$

where k_j and k_i denote the rate constants of conversion of substances j and i.

As in level one, a common fugacity f prevails, and if $f \cdot Z$ is substituted for C the total rate E is

$$E = f \cdot \sum V_i \cdot Z_i \cdot k_i \quad (111)$$

The average residence time due to reactions t_r may be found by use of the following equation:

$$t_r = M/E \quad (112)$$

and the overall rate constant K is E/M or $1/t_r$.

The *third level* is devoted to a steady-state, nonequilibrium situation, which implies that the fugacities are different in each phase. Equation (105) is used to account for the transfer. The D values may be calculated from quantities such as interface areas, mass-transfer coefficients (as indicated above, D is the product of the transfer coefficient and the fugacity capacity: $D = K \cdot Z$), release rate of chemicals into phases such as biota or sediment, and Z values, or by use of the estimation methods presented in Section 3.3.5.

Level four involves a dynamic version of level three where emissions, and thus concentrations, vary with time. This implies that differential equations must be applied for each compartment to calculate the change in concentrations with time, for instance:

$$V_i \cdot dC_i/dt = E_i - V_i \cdot C_i \cdot k_i - \sum D_{ij} \cdot \Delta f_{ij} \quad (113)$$

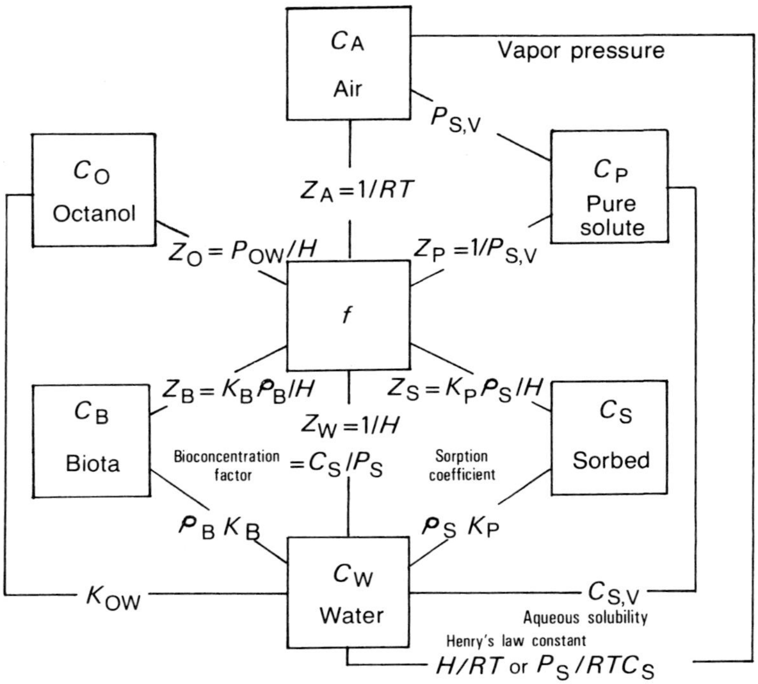

Figure 61. Conceptual diagram of fugacity models

This model level is similar in concept to the EXAMS model [250].

Levels one and two are sufficient in most cases, but if the environmental management problem requires the prediction of (1) the time required for a substance to accumulate to a certain concentration in a phase after emission has started, or (2) the length of time for the system to recover after the emission has ceased, the fourth level must be applied.

This approach has been used widely, and a typical example is given by MACKAY et al. [250]. It concerns the distribution of PCB between air and water in the Great Lakes. In this case, H was 49.1 bar and the air–water distribution coefficient ($= H/RT$) 0.02. The fugacity capacity for water ($= 1/H$) was found to be 0.0204 bar and the fugacity capacity for air ($= 1/RT$) 0.000404 bar. The distribution coefficient of PCB between water and suspended matter in the water was estimated to be 100000. Since the concentration of suspended matter in the Great Lakes has been found to be 2×10^{-6} on a volume basis (approximately 4 mg/L) the fraction dissolved was $1/(1 + 0.2) = 0.833$.

The PCB concentration in water of the Great Lakes was 2 ng/L, and in the air 2 ng/m^3. The fugacity can be calculated in water and air as C/Z, and the fugacity in water was found to be 17 times higher than in air, which implies that volatilization will occur. If the transfer coefficient in water is assumed to be 10^{-5} m/s and in air 10^{-3} m/s, the volatilization rate can be calculated from the traditional two-resistance model, using the relation $D = K \cdot Z$ to find the overall diffusion coefficient, D:

$$1/D = 1/10^{-5} \cdot 0.0204 \text{ bar} + 1/10^{-3} \cdot 0.000404 \text{ bar} \quad (114)$$

D is 1.36×10^{-7}; N is calculated by use of Equation (106):

$$N = D(f_w - f_a) = D(2.8 \times 10^{-7} \text{ Pa} - 1.53 \times 10^{-8} \text{ Pa}) \quad (115)$$

N is found to be 35.9×10^{-15} mol m^{-2} s^{-1}.

The transfer with precipitation is negligible compared to the volatilization rate, while the washout of particles and dry deposition are important processes. If these processes are considered, the net flux to the atmosphere becomes about 75% of the flux found above.

4. Appendix [290]

Risk Assessment Directive 93/67/EEC	Document XI/730/89 rev. 3 (Ispra)	Notes
Risk Assessment		
A generic term describing an administrative and technical process that entails some or all of the elements below	no equivalent term	much wider meaning of the term than at Ispra
Effects Assessment		
1) Hazard Identification Identification of the adverse effects a substance has an inherent capacity to cause; and, where possible or appropriate, assessment of a particular effect 2) Dose–response assessment Estimation of the relationship between dose (or level of exposure) and the incidence and severity of an effect	Hazard identification identification of a substance as a substance of concern: takes into account the inherent hazardous properties and exposure-related aspects	these two terms are synonymous they describe inherent properties
Exposure assessment	Hazard assessment (part only)	
Determination of the emissions, pathways, and rates of movement of a substance and its transformation or degradation in order to estimate the concentrations/doses to which human populations or environmental compartments are or may be exposed	= "environmental exposure"	this part of hazard assessment (Ispra) is synonymous with exposure assessment (draft directive) leads to PEC
	Hazard assessment (part only)	
	= "effect data with reference to the environmental compartment of concern"	this is aided by a knowledge of hazard identification (Ispra) leads to PNEC
Effects assessment plus exposure assessment	Hazard assessment	synonymous terms comparison of PEC and PNEC
Risk characterization	Risk assessment (see below) plus hazard assessment	synonymous terms, i.e., PEC:PNEC ratios against a probability scale
Estimation of the incidence and severity of the adverse effects likely to occur in a human population or environmental compartment due to actual or predicted exposure to a substance and may include		
Risk estimation	Risk assessment	synonymous terms
Quantification of the likelihood of the incidence and severity of the adverse effects	estimation of the probability that a substance causes adverse effects as a result of its presence in the environment at a given concentration	
Risk reduction	Risk management	synonymous terms, e.g., special use and disposal instructions, emission control measures, restrictions on types of use, total ban of use
Measures that would enable the risks for man or the environment in connection with the marketing of the substances to be lessened	the taking of measures appropriate at least to diminish significantly the presence of a substance in the environmental compartments of concern	

5. References

General References

[1] A. Leroi-Gourhan: *Le Fil du Temps*, Fayard, Paris 1983.

[2] O. Fränzle: *Contaminants in Terrestrial Environments*, Springer Verlag, Berlin 1993.

[3] L. Trepl: *Geschichte der Ökologie*, Athenäum, Frankfurt/Main 1987.

[4] E. Haeckel: *Generelle Morphologie der Organismen*, Reimer, Berlin 1866.

[5] C. J. Krebs: *Ecology*, Harper & Row, New York 1972.

[6] M. Begon, J. L. Harper, C. R. Townsend: *Ecology: Individuals, Populations and Communities*, Blackwell, Boston – Melbourne 1990.

[7] G. C. Butler (ed.): *Principles of Ecotoxicology*, J. Wiley, Chichester 1978.

[8] H. Parlar, D. Angerhöfer: *Chemische Ökotoxikologie*, Springer Verlag, Berlin 1991.

[9] H. Ellenberg: *Ökosystemforschung*, Springer Verlag, Heidelberg 1973.

[10] S. L. Pimm: *Food Webs*, Chapman & Hall, London 1982.

[11] R. V. O'Neill, A. R. Johnson, A. W. King "A Hierarchical Framework for the Analysis of Scale," *Landscape Ecology* **3** (1989) 193–205.

[12] K. Dierßen: *Einführung in die Pflanzensoziologie*, Wissenschaftliche Buchgesellschaft, Darmstadt 1990.

[13] W. Schröder, L. Vetter, O. Fränzle (eds.): *Neue statistische Verfahren und Modellbildung in der Geoökologie*, Vieweg, Wiesbaden 1994.

[14] H. G. Gauch: *Multivariate Analysis in Community Ecology*, Cambridge University Press, Cambridge 1982.

[15] W. Ziechmann: *Huminstoffe*, Verlag Chemie, Weinheim 1980.

[16] P. Schachtschabel et al.: *Lehrbuch der Bodenkunde*, Enke Verlag, Stuttgart 1989.

[17] K. Y. Kondratyev: *Radiation Processes in the Atmosphere*, WHO, Genever 1972.

[18] M. I. Budyko: "Solar Radiation and the Use of it by Plants". Agroclimatological Methods, *Proc. Reading Symp.*, UNESCO, Paris 1968, 39–53.

[19] H. Ellenberg, R. Meyer, J. Schauermann (eds.): *Ökosystemforschung. Ergebnisse des Solling-Projekts*, Enke Verlag, Stuttgart 1986.

[20] B. J. Nebel: *Environmental Science. The Way the World Works*, Prentice Hall, Englewood Cliffs, N.J., 1990.

[21] M. G. Barbour, J. H. Burk, W. D. Pitts: *Terrestrial Plant Ecology*, Benjamin/Cummings, Menlo Park, CA, 1980.

[22] BUA (Beratergremium für umweltrelevante Altstoffe der Gesellschaft Deutscher Chemiker): *Umweltrelevante Alte Stoffe III – Prioritätensetzung und eingestufte Stoffe der dritten Stoffliste*, VCH Verlagsgesellschaft, Weinheim 1992.

[23] O. Hutzinger: *The Handbook of Environmental Chemistry*, 5 vols., Springer Verlag, Berlin 1989–1991.

[24] T. Rosswall, R. R. Woodmanse, P. G. Risser (eds.): *Scales and Global Change*, J. Wiley, Chichester 1988.

[25] F. W. Schaller, G. W. Bailey (eds.): *Agricultural Management of Water Quality*, Iowa State University Press, Ames, IA, 1983.

[26] *FAO Yearbook: Fertilizer*, vol. 39. Statistical Analysis Service, Statistics Division, Food and Agriculture Organization, Rome 1989.

[27] D. K. Northington, Y. R. Goodin: *The Botanical World*, Times Mirror/Mosby College Publ., St. Louis, MO, 1984.

[28] M. Lieth, R. H. Whittaker: *Primary Productivity of the Biosphere*, Springer Verlag, Berlin 1975.

[29] L. R. Pomeroy, J. J. Alberts: *Concepts of Ecosystem Ecology. A Comparative View*, Springer Verlag, New York 1988.

[30] J. Cole, G. Lovett, S. Findlay (eds.): *Comparative Analyses of Ecosystems*, Springer Verlag, New York 1991.

[31] H. Walter, S. W. Breckle: *Ökologie der Erde*, vol. 1, "Ökologische Grundlagen in globaler Sicht," G. Fischer Verlag, Stuttgart 1991.

[32] D. Pimentel (ed.): *Handbook of Energy Utilization in Agriculture*, CRC Press, Boca Raton, FL, 1980.

[33] K. E. F. Watt: *Understanding the Environment*, Allyn and Bacon, Boston 1982.

[34] R. G. Wetzel: *Limnology*, Saunders College Publ., New York 1983.

[35] H.-P. Blume (ed.): *Handbuch des Bodenschutzes*, Ecomed, Landsberg/Lech 1992.

[36] W. J. Mitsch, J. G. Gosselink: *Wetlands*, Van Nostrand Reinhold, New York 1986.

[37] S. E. Jørgensen: *Modelling in Ecotoxicology*, Elsevier, Amsterdam 1990.

[38] T. H. F. Allen, T. B. Starr: *Hierarchy*, Univ. Chicago Press, Chicago 1982.

[39] C. Elton: *The Ecology of Invasion by Animals and Plants*, Methuen, London 1958.

[40] R. E. Ulanowicz. *Growth and Development: Ecosystems Phenomenology*, Springer Verlag, New York 1986.

[41] P. B. Moyle, R. A. Leidy: "Loss of Biodiversity in Aquatic Ecosystems: Evidence from Fish Faunas," in P. S. Fielder, K. J. Subodh (eds.): *From Conservation Biology: the Theory and Practice of Nature, Conservation, Preservation and Management*, Chapman and Hall, New York 1992, pp. 127–169.

[42] J. C. White (ed.): *Global Climate Change Linkages*, Elsevier, Amsterdam 1967.

[43] A. K. Ward: "Geological Mediation of Stream Flow and Sediment and Solute Loading to Stream Ecosystems due to Climate Change," in P. Firth, S. G. Fischer (eds.): *Global Change and Freshwater Ecosystems*, Springer Verlag, Berlin 1991.

[44] F. H. Borman, G. E. Likens: *Pattern and Process in a Forested Ecosystem*, Springer Verlag, New York 1979.

[45] C. E. W. Steinberg, R. F. Wright (eds.): *Acidification of Freshwater Ecosystems. Implications for the Future*, J. Wiley, Chichester 1994.

[46] R. J. Chorley, S. A. Schumm, D. E. Sudgen: *Geomorphology*, Methuen, London 1984.

[47] K. Verschueren: *Handbook of Environmental Data on Organic Chemicals*, Van Nostrand Reinhold, New York 1983.

[48] G. Mattheß: *Die Beschaffenheit des Grundwassers*, Borntraeger, Berlin – Stuttgart 1994.

[49] H. Freundlich: *Kapillarchemie*, Akad. Verlagsgesellschaft, Leipzig 1930.

[50] J. N. Israelachvili: *Intermolecular and Surface Forces. With Applications to Colloidal and Biological Systems*, Academic Press, London 1985.

[51] S. L. Faust, J. V. Hunter (eds.): *Kinetics of Biologically Mediated Oxidation of Organic Compounds in Aquatic Environments*, Marcel Dekker, New York 1978.

[52] J. W. Moore, S. Ramamoorthy: *Organic Chemicals in Natural Waters*, Springer Verlag, Berlin 1984.

[53] F. Korte (ed.): *Lehrbuch der ökologischen Chemie*, Thieme Verlag, Stuttgart 1987.
[54] R. Geiger: *Das Klima der bodennahen Luftschicht*, Vieweg & Sohn, Braunschweig 1961.
[55] B. Ulrich (ed.): *Internationaler Kongreß Waldschadensforschung: Wissensstand und Perspektiven*, Kernforschungszentrum Karlsruhe, Karlsruhe 1990.
[56] R. Dallinger, P. S. Rainbow: *Ecotoxicology of Metals in Invertebrates*, Lewis Publ., Boca Raton, FL, 1992.
[57] J. A. Foran, L. E. Fink: *Regulating Toxic Substances in Surface Waters*, Lewis Publ., Boca Raton, FL 1993.
[58] P. H. Howard: *Handbook of Environmental Fate and Exposure Data for Organic Chemicals*, vols. II–III, CRC Press, Boca Raton, FL, 1991.
[59] K. C. Jones: *Organic Contaminants in the Environment*, Elsevier Applied Science, New York 1991.
[60] R. A. Lewis: *Environmental Toxicology*, CRC Press, Boca Raton, FL, 1991.
[61] M. Mansour (ed.): *Fate and Prediction of Environmental Chemicals in Soils, Plants and Aquatic Systems*, Lewis Publ., Boca Raton, FL, 1993.
[62] S. Matsui, B. F. D. Barrett, J. Banerjee, (eds.): *Guidelines of Lake Management*, vol. 4, "Toxic Substances Management in Lakes and Reservoirs," International Lake Environment Committee Foundation, Otsu, Japan 1991.
[63] J. F. McCarthy, L. R. Shugart: *Biomarkers of Environmental Contamination*, CRC Press, Boca Raton, FL, 1990.
[64] R. G. Tardiff (ed.): "Methods to Assess the Impact of Pesticides on Non-target Organisms," SCOPE **49** (1992) IPCS Joint Symposia 16, SGOMSEC 7.
[65] A. M. V. M. Soares, P. Calow (eds.): *Standardization of Aquatic Toxicity Tests*, Lewis Publ., Boca Raton, FL, 1993.
[66] T. Colborn, C. Clement: *Chemically-induced Alterations in Sexual and Functional Development: The Wildlife/Human Connection. Advances in Modern Environmental Toxicology*, vol. XXI, Princeton Scientific Publ. Co., Princeton, N.J., 1993.
[67] P. Bourdeau, J. A. Haines, W. Klein, C. R. Krishna Murti (eds.): "Ecotoxicology and Climate with Special Reference to Hot and Cold Climates," SCOPE **38** (1989).
[68] S. M. Bartell: *Toxicological Risk Assessment in Aquatic Ecosystems*, CRC Press, Boca Raton, FL, 1991.
[69] M. J. Gromiec, E. F. Gloyna: "Radioactivity Transport in Water," Final Report no. 22 to U.S. Atomic Energy Commission, Contract AT (11-1)-490, 1973.
[70] P. Kauppi et al.: "A Model for Predicting the Acidification of Forest Soils: Application to Acid Deposition in Europe," IIASA Research Report 1984.
[71] S. E. Jørgensen, S. Nors-Nielsen, L. A. Jørgensen: *Handbook of Ecological Parameters and Ecotoxicology*, Elsevier, Amsterdam 1991.
[72] S. E. Jørgensen (ed.): *Modelling in Environmental Chemistry*, Elsevier, Amsterdam 1991.
[73] S. E. Jørgensen, M. Gromiec: *Mathematical Submodels of Water Quality Systems*, Elsevier, Amsterdam 1989.
[74] S. E. Jørgensen, I. Johnsen: "Principles of Environmental Science and Technology," *Stud. Environ. Sci.* **33** (1989).
[75] Y. Samiullah: *Prediction of the Environmental Fate of Chemicals*, Elsevier Applied Science, London 1990.
U. Schlottmann (ed.): *Prüfmethoden für Chemikalien*, S. Hirzel Verlag, Stuttgart 1994.
[76] BUA (Beratergremium für umweltrelevante Altstoffe der Gesellschaft Deutscher Chemiker): *Umweltrelevante Alte Stoffe – Auswahlkriterien und Stoffliste*, VCH Verlagsgesellschaft, Weinheim 1986.
[77] BUA (Beratergremium für umweltrelevante Altstoffe der Gesellschaft Deutscher Chemiker): *Umweltrelevante Alte Stoffe II – Auswahlkriterien und Zweite Stoffliste*, VCH Verlagsgesellschaft, Weinheim 1988.
[78] H. Bick: *Ökologie*, G. Fischer Verlag, Stuttgart 1989.
[79] M. Straškraba, A. Gnauck: *Freshwater Ecosystems. Modelling and Simulation*, Elsevier, Amsterdam 1985.
[80] J. Gilbert, M.-J. Dole-Olivier, P. Marmonier, P. Vervier in R. J. Naiman, H. Décamps (eds.): *The Ecology and Management of Aquatic-Terrestrial Ecotones*, UNESCO, Paris; Parthenon, Park Ridge, 1990, pp. 199–225.

Specific References

[81] L. G. Freeman, *Am. Anthr.* **38** (1973) 3–44.
[82] J. Seymour, H. Girardet: *Fern vom Garten Eden*, Fischer Verlag, Frankfurt/Main 1985.
[83] E. H. Blair, *Proc. of the Workshop on the Control of Existing Chemicals under the Patronage of the Organisation for Economic Co-operation and Development*, Umweltbundesamt, Berlin 1981, pp. 252–260.
[84] ECETOC: "Concentrations of Industrial Organic Chemicals measured in the Environment," Technical Report no. **29**, Brussels 1988.
[85] Leitungsgremium des Projektzentrums "Ökosystemforschung im Bereich der Bornhöveder Seenkette" (ed.): "Arbeitsbericht 1988–91," *EcoSys*, vol. 1, Kiel 1992.
[86] I. H. Koeman, *Ecotoxicol. Environ. Saf.* **6** (1982) 358–362.
[87] W. Takken, F. Balk, R. C. Jansen, J. H. Koeman, PANS **24** (1978) 455–466.
[88] Umweltbundesamt: Collection of Minimum Premarketing Sets of Data Including Environmental Residue Data on Existing Chemicals, Umweltbundesamt, Berlin 1982.
[89] A. G. Tansley, *Ecology* **16** (1935) 284–307.
[90] L. Finke: *Landschaftsökologie*, Westermann, Höller und Zwick, Braunschweig 1986.
[91] C. Elton: *Animal Ecology*, Macmillan, New York 1927.
[92] R. L. Lindeman, *Ecology* **23** (1942) 399–413.
[93] O. Fränzle in B. Markert (ed.): *Environmental Sampling for Trace Analysis*, VCH Verlagsgesellschaft, Weinheim 1994, pp. 305–320.
[94] G. G. Simpson: *The Major Features of Evolution*, Columbia University Press, New York 1953.
[95] R. P. McIntosh, *Ecology* **48** (1967) 392–404.
[96] C. E. Shannon, *Bell Syst. Tech. J.* **27** (1948) 379–423, 623–653.
[97] O. Fränzle, W. F. Killisch, *Kieler Geographische Schriften* **50** (1979) 211–245.
[98] L. Vetter: "Evaluierung und Entwicklung statistischer Verfahren zur Auswahl von repräsentativen Untersuchungsobjekten für ökotoxikologische Problemstellungen," Dissertation, Universität Kiel 1989.
[99] R. M. G. Jongman, C. F. R. Ter Braak, O. F. R. Van Tongeren: "Data Analysis in Community and Landscape Ecology," PUDOC, Wageningen 1987.
[100] K. R. Gabriel, *Biometrika* **58** (1971) 453–467.

[101] O. Fränzle, W. F. Killisch in W. Schröder, L. Vetter, O. Fränzle (eds.): *Neuere statistische Verfahren und Modellbildung in der Geoökologie*, Vieweg & Sohn, Braunschweig 1994, pp. 129–143, 255–281.
[102] O. Fränzle et al.: "Synoptische Darstellung möglicher Ursachen der Waldschäden: Untersuchungen zu Langzeitwirkungen von Kompensationskalkungen auf Buchen-, Fichten- und Kiefernforstökosysteme," Forschungsbericht 10803046/13 im Umweltforschungsplan des BMU, Kiel 1989.
[103] W. Schröder: "Ökosystemare und statistische Untersuchungen zu Waldschäden in Nordrhein-Westfalen: Methodenkritische Ansätze zur Operationalisierung einer wissenschaftstheoretisch begründeten Konzeption," Dissertation, Universität Kiel 1989.
[104] J. H. R. Gee, P. S. Giller (eds.): *Organization of Communities: Past and Present*, Blackwell, Oxford 1987.
[105] J. H. Connell, R. O. Slatyer, *Am. Nat.* **111** (1977) 1119–1144.
[106] G. Hemprich: "Landschaftsökologische Untersuchungen im Bereich des Belauer Sees und des Schmalensees," Diplomarbeit Geographie, Universität Kiel 1991.
[107] V. Volterra: Variations and Fluctuations of the Numbers of Individuals in Animal Species Living Together (1926) (reprinted in R. N. Chapman: *Animal Ecology*, McGraw Hill, New York 1931).
[108] J. H. Connell, *Am. Nat.* **122** (1983) 661–696.
[109] P. Grubb, *Biol. Rev. Cambridge Philos. Soc.* **52** (1977) 107–145.
[110] L. R. Lawlor, *Am. Nat.* **116** (1980) 394–408.
[111] G. Müller, E. Foerster, *Acker- Pflanzenbau* **140** (1974) 161–174.
[112] J.-F. Dhôte, F. Houllier in J.-D. Lebreton, B. Asselain (eds.): *Biométrie et Environnement*, Masson, Paris 1993, pp. 175–204.
[113] P. Yodzis in J. Diamond, T. J. Case (eds.): *Community Ecology*, Harper & Row, New York 1986, pp. 480–491.
[114] C. D. Canham, *Ecology* **69** (1988) 786–795.
[115] D. J. Levey, *Ecology* **69** (1988) 1076–1089.
[116] J. L. Monteith, *Symp. Soc. Exp. Biol.* **29** (1965) 205–234.
[117] C. T. de Wit: *Simulation of Assimilation, Respiration and Transpiration of Crops*, PUDOC, Wageningen 1978.
[118] C. W. Thornthwaite, *Geogr. Rev.* **38** (1948) 85–94.
[119] L. Turc, *Ann. Agrar.* **5** (1954) 491–596.
[120] B. L. Dzerdzeevskii in: *Climat et Microclimatologie. Actes du Colloque de Canberra*, (Arid Zone Research/Recherches sur la Zone Aride XI), UNESCO, Paris 1958, pp. 315–325.
[121] M. Fuchs, G. Stanhill, *Isr. J. Agric. Res.* **13** (1963) 63–78.
[122] R. O. Slatyer in: *The Use of Soil Water Balance Relationships in Agroclimatology* (Natural Resources Res. VII) UNESCO, Paris 1968, pp. 73–87.
[123] P. Benecke, R. R. van der Ploeg, *Forstarchiv* **49** (1978) 1–7, 26–32.
[124] J. v. Hoyningen-Huene, S. Nasdalack, *Landschaftsökol. Messen Auswerten* **1** (1985) 107–116.
[125] B. Sevruk: "Methodische Untersuchungen des systematischen Meßfehlers des Hellmann-Regenmessers im Sommerhalbjahr in der Schweiz," Dissertation, ETH Zürich 6789, 1981.
[126] H. Braden, *Mitt. Dtsch. Bodenkd. Ges.* **42** (1985) 294–298.
[127] F.-J. Löpmeier: *Verhandlungen 45. Dtsch. Geographentages*, Steiner Verlag, Stuttgart 1987, pp. 434–438.
[128] J. R. Philip in E. S. Hills (ed.): *Water Resources, Use and Management*, Melbourne University Press, Melbourne 1964, pp. 257–275.
[129] O. Schwarz in G. Einsele et al. (eds.): *Das landschaftsökologische Forschungsprojekt Naturpark Schönbuch*, VCH Verlagsgesellschaft, Weinheim 1986, pp. 161–179.
[130] R. E. Smith, J. Y. Parlange, *Water Resour. Res.* **14** (1978) 533–538.
[131] A. C. Imeson, *CATENA Suppl.* **4** (1983) 79–89.
[132] G. Einsele, G. Agster, M. Elgner in G. Einsele (ed.): *Das landschaftsökologische Forschungsprojekt Naturpark Schönbuch*, VCH Verlagsgesellschaft, Weinheim 1986, pp. 209–234.
[133] J. Drescher, R. Horn, M. de Boodt (eds.): "Impact of Water and External Forces on Soil Structure", *CATENA Suppl.* **11** (1988).
[134] E. Petzold, *Verh. Ges. Ökol.* **12** (1984) 487–498.
[135] M. Renger, O. Strebel, *Wasser Boden* **32** (1980) 362–366.
[136] Arbeitskreis Grundwasserneubildung, *Geol. Jahrb. Reihe C* **C 19** (1977) 3–98.
[137] O. Fränzle et al.: "Auswahl der Hauptforschungsräume für das Ökosystemforschungsprogramm der Bundesrepublik Deutschland," Forschungsbericht 10104043/02 im Umweltforschungsplan des BMU, Universität Kiel 1987.
[138] H. J. Heckmann, K. F. Schreiber, R. Thöle, *Mitt. Dtsch. Bodenkd. Ges.* **41** (1985) 353–356.
[139] H. Sager, Informationsbericht, part 1, Bayer. Landesamt für Wasserwirtschaft, München 1983, pp. 59–72.
[140] P. Greminger: "Physikalisch-ökologische Standortuntersuchung über den Wasserhaushalt im offenen Sickersystem Boden unter Vegetation am Hang," *Mitt. Eidgen. Anst. forstl. Versuchswesen* **60** (1985) no. 2.
[141] I. D. White, N. D. Mottershead, S. J. Harrison: *Environmental Systems*, Allen & Unwin, London 1984.
[142] M. Schroeder, *Dtsch. Gewässerkd. Mitt.* **20** (1976) 8–13.
[143] C. W. Thornthwaite, B. Holzman, *Tech. Bull. U.S. Dep. Agric.* **817** (1942).
[144] A. I. Dyer, *Q. J. R. Meteorol. Soc.* **87** (1961) 401–412.
[145] J. L. Monteith, G. Szeicz, *Q. J. R. Meteorol. Soc.* **86** (1960) 205–214.
[146] S. v. Stamm: "Untersuchungen zur Primärproduktion von *Corylus avellana* an einem Knickstandort in Schleswig-Holstein und Erstellung eines Produktionsmodells," *EcoSys*, suppl. vol. 4, Kiel 1992.
[147] S. Gäb, H. Parlar, F. Korte, *Chemosphere* **3** (1974) no. 5, 187–192.
[148] M. Straškraba, *Mitt. Int. Ver. Limnol.* **14** (1968) 212–230.
[149] J. Blake in K.-H. Zwirnmann (ed.): "Nonpoint Nitrate Pollution of Municipal Water Supply Resources: Issues of Analysis and Control," *IIASA Collab. Proc. Ser.* **CP-82-S4** (1982) 231–245.
[150] L. Procházková, P. Blažka, personal communication, 1993.
[151] L. R. Brown: *State of the World*, W. W. Norton & Co., New York 1985.

[152] H. Lieth, E. Box in R. Y. Mather (ed.): *Publications in Climatology*, vol. 25, C. W. Thornthwaite Associates, New Jersey 1972, pp. 37–46.

[153] S. McNaughton, M. Oesterheld, D. A. Frank, K. J. Williams in [30, pp. 120–139].

[154] I. Aselmann, H. Lieth, *Prog. Biometeorol.* **3** (1983) 101–119.

[154a] R. J. Naiman, H. Décamps, J. Pastor, C. A. Johnston, *J. North Am. Benthological Soc.* **7** (1988) 289–306.

[155] F. E. Clements: *Research Methods in Ecology*, University Publ. Comp., Lincoln 1905.

[156] W. Kluge, O. Fränzle, *Ges. Ökol.* **21** (1992) 401–407.

[157] D. L. Urban, R. V. O'Neill, H. H. Shugart, *BioScience* **37** (1987) 119–127.

[158] O. Fränzle in E. Bayer, H. Behret (eds.): "Bewertung des ökologischen Gefährdungspotentials von Chemikalien," *GDCh-Monogr.* **1** (1994) 45–90.

[159] O. Fränzle in B. Markert (ed.): *Sampling of Environmental Materials for Trace Analysis*, VCH Verlagsgesellschaft, Weinheim 1994, pp. 217–227.

[160] W. M. Stigliani: "Changes in Valued 'Capacities' of Soil and Sediments as Indicators of Nonlinear and Time-delayed Environmental Effects," IIASA Ecoscript **35** (1988), International Institute for Applied Systems Analysis, Laxenburg.

[161] R. C. Wissmar, F. J. Swanson in R. J. Naiman, H. Décamps (eds.): *The Ecology and Management of Aquatic-terrestrial Ecotones*, UNESCO, Paris; Parthenon, Park Ridge 1990, pp. 171–198.

[162] E. Pieczinska in R. J. Naiman, H. Décamps (eds.): *The Ecology and Management of Aquatic-terrestrial Ecotones*, UNESCO, Paris; Parthenon, Park Ridge 1990, pp. 103–140.

[163] U. Schleuß: "Böden und Bodenschaften einer norddeutschen Moränenlandschaft," *EcoSys*, suppl. vol. 2, Kiel 1992.

[164] M. M. Holland, D. F. Whigham, B. Gopal in R. J. Naiman, H. Décamps (eds.): *The Ecology and Management of Aquatic-terrestrial Ecotones*, UNESCO, Paris; Parthenon, Park Ridge 1990, pp. 171–198.

[165] D. Scholle: "Vegetationskundliche Untersuchungen im Raum Bornhöved und deren Auswertung mit Hilfe eines Geographischen Informationssystems (GIS)," Diplomarbeit, Universität Saarbrücken, 1991.

[166] G. Schernewski: "Raumzeitliche Prozesse und Strukturen im Wasserkörper des Belauer Sees," Diplomarbeit Geographie, in *EcoSys*, suppl. vol. 1, Kiel 1992.

[167] C. S. Holling in E. Jantsch, C. H. Waddington (eds.): *Evolution and Consciousness*, Addison-Wesley, Reading, Mass., 1976, pp. 73–92.

[168] I. Prigogine in E. Jantsch, C. H. Waddington (eds.): *Evolution and Consciousness*, Addison-Wesley, Reading, Mass., 1976, pp. 93–133.

[169] W. E. Westman, *BioScience* **28** (1989) 705–710.

[170] A. Gigon, Ber. Geobot. Inst. Eidg. Techn. Hochsch., Stift. Rübel Zürich **50** (1983) 149–177.

[171] E. van der Maarel, *Ber. Dtsch. Bot. Ges.* **89** (1976) 415–443.

[172] O. T. Solbrig, G. Nicolis: *Perspectives in Biological Complexity*, IUBS, Paris 1991.

[173] R. H. MacArthur, *Ecology* **36** (1955) 533–536.

[174] R. M. May, *Nature (London)* **238** (1972) 413–414.

[175] R. E. McMurtrie, *J. Theor. Biol.* **50** (1975) 1–11.

[176] J. E. Cohen, C. M. Newman, *Ann. Probab.* **12** (1984) 283–310.

[177] C. P. Bachas, B. A. Huberman, *J. Phys. A Math. Gen.* **20** (1987) 4995–5014.

[178] H. A. Ceccato, B. A. Huberman, *Physica Scr.* **37** (1988) 145–150.

[179] P. Doreian, *J. Soc. Biol. Struct.* **9** (1986) 115–139.

[180] F. Briand, J. E. Cohen, *Science (Washington D.C. 1883)* **238** (1987) 956–960.

[181] T. Hogg, B. A. Huberman, J. M. McGlade, *Proc. R. Soc. London B* **237** (1989) 43–51.

[182] M. Rejmanek, P. Stary, *Nature (London)* **280** (1979) 311–313.

[183] F. Briand, *Ecology* **64** (1983) 253–263.

[184] S. J. McNaughton, *Am. Nat.* **111** (1977) 515–525.

[185] A. Hay, *Nature (London)* **302** (1983) 208–209.

[186] R. V. O'Neill, *Ecology* **57** (1976) 1244–1253.

[187] D. L. de Angelis, *Ecology* **61** (1980) 764–771.

[188] J. Cairns Jr., B. R. Niederlehner, *Environ. Professional* **15** (1993) 116–124.

[189] P. van Voris, R. V. O'Neill, W. R. Emanuel, H. H. Shugart, *Ecology* **6** (1980) 1352–1360.

[190] D. W. Schindler, *Can. J. Fish. Aquat. Sci.* **44** (1987) Suppl 1, 6–25.

[191] M. J. Crossey, T. W. LaPoint, *Hydrobiologia* **162** (1988) 109–121.

[192] W. D. Crumby, M.-A. Webb, F. J. Bulow, H. J. Cathey, *Trans. Am. Fish. Soc.* **119** (1990) 885–893.

[193] J. H. Rodgers Jr., K. L. Dickson, J. Cairns, Jr., in R. G. Wetzel (ed.): "Methods and Measurements of Periphyton Communities," *ASTM*, Philadelphia, PA, 1980, pp. 142–167.

[194] D. Uhlmann, H. Mihan, A. Gnauck, *Acta Hydrochim. Hydrobiol.* **6** (1978) 421–444.

[195] A. Gnauck: "Strukturelle und funktionelle Änderungen in aquatischen Ökosystemen," Kongreß- und Tagungsberichte, Martin-Luther-Universität Halle-Wittenberg, 1982, pp. 335–344.

[196] J. D. Yount, G. J. Niemi: "Recovery of Lotic Communities and Ecosystems from Disturbance: Theory and Applications," *Environ. Manage. (N.Y.)* **14** (1990) no. 5, 132–137.

[197] G. M. Woodwell, *Science (Washington, D.C. 1883)* **156** (1967) no. 3774, 461–470.

[198] T. Rosswall, R. G. Woodmanse, P. G. Risser: *Scale and Global Change: Spatial and Temporal Variability in Biospheric and Geospheric Processes*, J. Wiley, Chichester 1988.

[199] J. Peñuelas in J. D. Ross, N. Prat (eds.): "Hommage to Ramón Margalef: or why there is such pleasure in studying nature," Universitat de Barcelona Publications, Barcelona 1991, 367–385.

[200] S. H. Schneider, N. J. Rosenberg in N. J. Rosenberg, W. E. Easterling, P. J. Crosson, J. Darmstadter (eds.): "Greenhouse Warming: Abatement and Adaptation," Proceedings of a Workshop held in Washington D.C., June 14–15, 1988, Resources for the Future, Washington 1989, pp. 7–34.

[201] J. S. Clark in R. L. Wyman (ed.): *Global Climate Change and Life on Earth*, Routledge, Chapman and Hall, New York 1991, pp. 65–98.

[202] J. D. Meisner et al., *J. Great Lakes Res.* **13** (1987) no. 3, 340–352.

[203] P. Glansdorff, I. Prigogine: *Thermodynamic Theory of Structure, Stability, and Fluctuations*, Wiley-Interscience, New York 1971.

[204] E. J. Fittkau: "Tropische Regenwälder – Ökologische Zusammenhänge," *Bensberger Protokolle* **66** (1991) 27–63 (Thomas-Morus-Akademie, Bensberg).

[205] R. Herrera, C. F. Jordan, E. Medina, H. Kluge, *Ambio* **10** (1981) 109–114.

[206] H. Sioli: *Amazonien. Grundlagen der Ökologie des größten tropischen Waldlandes,* Enke Verlag, Stuttgart 1980.

[207] C. F. Jordan (ed.): *An Amazonian Rain Forest: the Structure and Function of a Nutrient Stressed Ecosystem and the Impact of Slash and Burn Agriculture,* UNESCO, Paris, Parthenon, Carnforth 1989.

[208] D. L. de Angelis, *Ecology* **56** (1975) 238–243.

[209] D. L. de Angelis, J. C. Waterhouse, *Ecol. Monogr.* **57** (1987) 1–21.

[210] H. Remmert, *Verh. Ges. Ökol.* **16** (1987) 27–34.

[211] H. M. Wilbur, *Ecology* **68** (1987) 1437–1452.

[212] C. R. Townsend, *J. North Am. Benthological Soc.* **8** (1989) 36–50.

[213] N. I. Sax: *Dangerous Properties of Industrial Materials,* Van Nostrand Reinhold, New York 1975.

[214] D. Mackay, S. Paterson, *Environ. Sci. Technol.* **15** (1981) 1006.

[215] D. Mackay, S. Paterson, *Environ. Sci. Technol.* **16** (1982) 654A.

[216] W. F. Spencer, W. J. Farmer in R. Haque (ed.): *Dynamics, Exposure and Hazard Assessment of Toxic Chemicals,* Ann Arbor Sci., Ann Arbor 1980, pp. 143–161.

[217] O. Fränzle, *Z. Geomorphol. N.F.* **15** (1971) 212–235.

[218] M. C. Lee et al., *Water Res.* **13** (1979) 1249–1258.

[219] ECETOC: "Structure-activity Relationships in Toxicology and Ecotoxicology: an Assessment," *ECETOC Monogr.* **8** (1986).

[220] A. Leo, C. Hansch, D. Elkins, *Chem. Rev.* **71** (1971) 525.

[221] T. Mill in Umweltbundesamt (eds.): "Proc. Workshop Control of Existing Chemicals under the Patronage of the Organisation for Economic Cooperation and Development," June 10–12, 1981, Umweltbundesamt, Berlin 1981, pp. 207–227.

[222] G. Mattheß et al.: "Der Stofftransport im Grundwasser und die Wasserschutzgebietsrichtlinie W 101 – Statusbericht und Problemanalyse," *Ber. Umweltbundesamt Ger.* **7** (1985).

[223] W. Mabey, T. Mill, *J. Phys. Chem. Ref. Data* **7** (1978) 383–415.

[224] ECETOC: "Experimental Assessment of the Phototransformation of Chemicals in the Atmosphere," Technical Report no. 7, Brussels 1983.

[225] P. Schachtschabel, H.-P. Blume, G. Brümmer et al.: *Lehrbuch der Bodenkunde,* Enke Verlag, Stuttgart 1989.

[226] R. Brdicka: *Grundlagen der physikalischen Chemie,* Deutscher Verlag der Wissenschaften, Berlin 1971.

[227] G. Sandstede, *Z. Phys. Chem.* (*Munich*) **29** (1961) 120–133.

[228] M. Alexander in R. Haque (ed.): *Dynamics, Exposure and Hazard Assessment of Toxic Chemicals,* Ann Arbor Sci., Ann Arbor 1980, pp. 179–190.

[229] J. C. G. Ottow, *Landwirtsch. Forsch.* **35** (1982) 238–256.

[230] OECD Guidelines for Testing of Chemicals, no. 301 A–E, OECD, Paris 1981.

[231] OECD Guidelines for Testing of Chemicals, no. 302 A, OECD, Paris 1981.

[232] OECD Guidelines for Testing of Chemicals, no. 303 A, OECD, Paris 1981.

[233] OECD Guidelines for Testing of Chemicals, no. 304 A, OECD, Paris 1981.

[234] H. Bernhardt (ed.): *NTA: Studie über die aquatische Umweltverträglichtkeit von Nitrilotriacetat (NTA),* Richarz, Sankt Augustin 1984.

[235] A. G. Journel, C. J. Huijbregts: *Mining Geostatistics,* Academic Press, London 1978.

[236] G. Verly, M. David, A. G. Journel, A. Maréchal (eds.): "Geostatistics for Natural Resources Characterization," *NATO Adv. Sci. Inst. Ser.,* Reidel, Dordrecht 1984.

[237] ECETOC: "Estimating the Environmental Concentrations of Chemicals Using Fate and Exposure Models," Technical Report no. **50**, Brussels 1992.

[238] ECETOC: "Environmental Hazard Assessment of Substances," Technical Report no. **51**, Brussels 1993.

[239] HEDSET (1992): Harmonized Electronic Data Input Set for the Council Regulation on the Evaluation and Control of Existing Substances and the OECD, Existing Chemicals Programme (05/07/92).

[240] T. R. Oke: *Boundary Layer Climates,* Methuen, London 1978.

[241] T. R. Oke, *J. R. Meteorol. Soc.* **1982**, 1–24.

[242] D. O. Lee, *Prog. Phys. Geogr.* **8** (1984) 1–31.

[243] E. Jauregui, *Erdkunde* **41** (1987) 48–51.

[244] G. M. Hidy, C. S. Burton in G. M. Hidy et al. (ed.): *The Character and Origin of Smog Aerosols,* Wiley, New York 1980, pp. 385–433.

[245] T. Schneider, L. Grant (eds.): *Air Pollution by Nitrogen Oxides,* Elsevier Sci., Amsterdam 1982.

[246] B. Henderson-Sellers: *Modeling of Plume Rise and Dispersion – The University of Salford Model: U.S.P.R.,* Springer Verlag, Heidelberg 1987.

[247] J. Pankrath, *Staub Reinhalt. Luft* **47** (1987) 239–244.

[248] W. Klug in [255, pp. 243–248].

[249] A. Eliassen, Ø. Hov, I. S. A. Isaksen, J. Saltbones in [255, pp. 347–356].

[250] D. Mackay, S. Paterson, W. Y. Shiu, *Chemosphere* **24** (1992) 695–717.

[251] E. W. Reiche: "Entwicklung, Validierung und Anwendung eines Modellsystems zur Beschreibung und flächenhaften Bilanzierung der Wasser- und Stickstoffdynamik in Böden", *Kieler Geogr. Schr.* **79** (1991) 1–150.

[252] TA Luft, Aug. 28, 1974, GMBl I, pp. 426, 525, GMBl 1983, p. 98.

[253] BUA (Beratergremium für umweltrelevante Altstoffe): "OH-Radikale in der Troposphäre: Konzentration und Auswirkung," BUA-Stoffbericht 100, Hirzel Verlag, Stuttgart 1993.

[254] OECD: "Compendium of Environmental Exposure Assessment Methods for Chemicals," *Environ. Monogr.* **27** (1989).

[255] OECD, Workshop on the Application of Simple Models for Environmental Exposure Assessment, Berlin, Sep. 11–13, 1991.

[256] S. Wiegmann: "Die geogenen Schwermetallgehalte der Böden Schleswig-Holsteins," Diplomarbeit Geographie, Universität Kiel 1994.

[257] M. Bonazountas in H. Barth, P. L'Hermite (eds.): *Scientific Basis for Soil Protection in The European Community,* Elsevier, London 1987, pp. 487–566.

[258] DVWK (Deutscher Verband für Wasserwirtschaft und Kulturbau e.V.): *Abschätzen des Verhaltens organischer Chemikalien in Böden,* DVWK, Bonn 1989.

[259] ECETOC: "Hazard Assessment of Floating Chemicals after Accidental Spill at Sea," Technical Report no. **35**, Brussels 1990.
[260] D. G. Farmer, M. J. Rycroft (eds.): *Computer Modelling in the Environmental Sciences*, Clarendon Press, Oxford 1991.
[261] A. Dahmke: "Lösungskinetik von feldspatreichen Gesteinen und deren Bezug zu Verwitterung und Porenwasser-Chemie natürlicher Sander-Sedimente." *Berichte – Reports Geol. Paläont.-Inst. Universität Kiel* **20** (1988).
[262] A. Petersen: "Laboruntersuchungen zum Einfluß organischer Komplexbildner auf die Kinetik der Feldspatverwitterung," *Berichte – Reports Geol. Paläont.-Inst. Universität Kiel* **21** (1988).
[263] D. Boening: "Untersuchungen zur Modellierung des Stofftransportes in einem durch Deponiesickerwasser verunreinigten Grundwassergerinne," *Berichte – Reports Geol.-Paläont. Inst. Universität Kiel* **30** (1989).
[264] R. D. Lindberg, D. D. Runnels, *Science (Washington, D.C. 1883)* **425** (1984) 925.
[265] D. Langmuir in E. A. Jenne (ed.): "Chemical Modeling in Aqueous Systems," *ACS Symp. Ser.* **93** (1979).
[266] D. J. Kirkner, T. L. Theis, A. A. Jennings, *Adv. Water Resour.* **7** (1984) 120.
[267] P. G. Daniele, A. de Robertis, C. de Stefano, S. Sammarfano, *J. Chem. Soc. Dalton Trans.* **1985**, 2353–2361.
[268] R. Katlein: "Bestimmung von Komplexkapazitäten mit der Mangandioxid-Methode," Dissertation, Universität Marburg 1986.
[269] W. Stumm, R. Kummert, L. Sigg, *Croat. Chem. Acta* **48** (1980) 291–312.
[270] R. V. James, J. Rubin in E. A. Jenne (ed.): "Chemical Modeling in Aqueous Systems," *ACS Symp. Ser.* **93** (1979).
[271] H. J. Hueck, D. M. M. Adema, W. C. de Kock, J. Kuiper, *Ber. Umweltbundesamt Ger.* **10** (1978) 159–167.
[272] H. Biek, D. Neumann (eds.): "Bioindikatoren," *Decheniana Beih.* **26** (1982).
[273] L. Steubing, H. J. Jäger (eds.): *Monitoring of Air Pollutants by Plants. Methods and Problems*, Junk, The Hague 1982.
[274] OECD Guideline for Testing of Chemicals, no. 305 E, OECD, Paris 1981.
[275] W. Ernst, *Decheniana Beih.* **26** (1982) 55–66.
[276] E. A. Nusch, *Decheniana Beih.* **26** (1982) 87–98.
[277] B. Hamburger, *Decheniana Beih.* **26** (1982) 78–81.
[278] W. K. Besch, *Decheniana Beih.* **26** (1982) 67–77.
[279] C. C. Brown in G. C. Butler (ed.): *Principles of Ecotoxicology*, J. Wiley, Chichester 1978, pp. 115–148.
[280] P. Gerike, in [283, pp. 139–143].
[281] R. N. Sturm, *J. Am Oil. Chem. Soc.* **50** (1973) 159–167.
[282] R. Zahn, M. Wellens, *Chem. Ztg.* **98** (1974) 228–232.
[283] Umweltbundesamt: "Tests for the Ecological Effects of Chemicals," *Ber. Umweltbundesamt Ger.* **10** (1978).
[284] J. S. Whiteside, M. Alexander, *Weeds* **8** (1960) 204–213.
[285] A. Krotzky et al., *Z. Pflanzenernähr. Bodenkd.* **146** (1983) 634–642.
[286] OECD: Data Interpretation Guides for Initial Hazard Assessment of Chemicals, Provisional OECD, Paris 1984.
[287] P. Friesel, P. D. Hansen, R. Kühn, J. Trénel: "Überprüfung der Durchführbarkeit von Prüfungsvorschriften und der Aussagekraft der Stufe 1 und 2 des Chemikaliengesetzes," part VI, Umweltforschungsplan des Bundesministers des Innern, F + E-Vorhaben 10604001/08 (1984).
[288] European Communities (EC): Technical Guidance Documents in Support of the Risk Assessment Directive (933/67/EEC) for New Substances Notified in Accordance with the Requirements of Council Directive 67/548/EEC, Brussels 1993.
[289] Umweltbundesamt: "Technical Guidance on Environmental Risk Assessment of Existing Chemicals in Accordance with the Requirements of Council Regulation (EEC)" no. 793/93, draft, Berlin 1994.
[290] ECETOC: "Aquatic Toxicity Evaluation," Technical Report no. **56**, Brussels 1993.
[291] OECD: "Report of the OECD Workshop on the Extrapolation of Laboratory Aquatic Toxicity Data on the Real Environment," *Environ. Monogr.* **59** (1992).
[292] N. M. van Straalen, C. A. J. Denneman, *Ecotoxicol. Environ. Saf.* **18** (1989) 241–251.
[293] H. Lokke in M. H. Donkers, H. Eijsackers, F. Heimbach (eds.): "Ecotoxicology of Soil Organisms," *SETAC Special Publication Ser.*, Lewis, Michigan, 1994, pp. 425–441.
[294] T. L. Forbes, V. E. Forbes, *Funct. Ecology* **7** (1993) 249–254.
[295] H. J. B. Emans et al., *Environ. Toxicol. Chem.* **12** (1993) 2139–2154.
[296] T. Aldenberg, W. Slob, *Ecotoxicol. Environ. Saf.* **25** (1993) 48–63.
[297] C. Wagner, H. Lokke, *Water Res.* **25** (1991) 1237–1242.
[298] O. Fränzle et al.: "Grundlagen zur Bewertung der Belastung und Belastbarkeit von Böden als Teilen von Ökosystemen," Texte 59/93 des Umweltbundesamtes, Berlin 1993.
[299] C. A. M. van Gestel in H. Becker et al. (eds.): *Ecotoxicology of Earthworms*, Intercept, Andover 1992, pp. 45–54.
[300] OECD Guideline for Testing of Chemicals, no. 207, OECD, Paris 1984.
[301] OECD Guideline for Testing of Chemicals, no. 208, OECD, Paris 1984.
[302] NEN (Nederlandse Norm): Soil-determination of the Influence of Chemicals on Soil Nitrification, no. 5795, Nederlands Normalisatie Instituut, Delft 1988.
[303] BBA (Biologische Bundesanstalt für Land- und Forstwirtschaft): "Bestimmung der Reproduktionsleistung von *Folsomia candida* (WILLEM) in künstlichem Boden," (Draft), Braunschweig 1990.
[304] ISO (International Organisation for Standardisation): Soil Quality Effects of Pollutants on Earthworms (*Eisenia foetida*) – part 2: Method for the Determination of Effects on Reproduction. Draft Int. Standard 1993.
[305] L. Moreth, E. Naton: Richtlinie zur Prüfung der Nebenwirkung von Pflanzenschutzmitteln auf *Aleochara bilineata* Gyll. (Col., Staphylinidae). Unveröff. Richtlinienvorschlag. 1992.
[306] J. Römbke: "Entwicklung eines Reproduktionstests an Bodenorganismen – Enchytraen," UBA F + E-Vorhaben no. 10603051/01. Batelle-Institut, Frankfurt/Main 1991.

[307] N. M. van Straalen, C. A. M. van Gestel: "Ecotoxicological Terrestrial Test Methods Using Terrestrial Arthropods," Detailed Review Paper for the OECD Test Guideline Programme, Amsterdam 1992.
[308] A. Stavola: Detailed Review Paper on Terrestrial Ecotoxicology Test Guidelines, OECD Updating Programme, Periodical Review. OECD, Paris 1990.
[309] F. Pedersen, L. Samsøe-Petersen: Discussion Paper Regarding Guidance for Terrestrial Effects Assessment (Draft), Water Quality Institute, Horsholm, Denmark, 1993.
[310] H. Blanck, *Ecol. Bull.* **36** (1984) 107–119.
[311] C. A. J. Denneman, C. A. M. van Gestel: "Bodemverontreiniging en bodemecosystemen: voorstel voor C-(toetsings)waarden op basis van ecotoxicologische risico's," Rijksinstituut voor Volksgezondheid en Milieuhygiene, Bilthoven 1990.
[312] EPA (Environmental Protection Agency): "Pesticide Assessment Guidelines," Office of Pesticides and Toxic Substances, Washington D.C., 1982.
[313] S. A. L. M. Kooijman, *Water Res.* **21** (1987) no. 3, 269–276.
[314] W. Sloof, J. A. M. v. Oers, D. Zwart, *Environ. Toxicol. Chem.* **5** (1986) 841–852.
[315] C. E. Stephan et al.: "Guidelines for Deriving Numerical National Water Quality Criteria for the Protection of Aquatic Organisms and their Uses," U.S. Environmental Protection Agency, Washington 1985.
[316] K. Terytze: Qualitative and quantitative Aspekte der Bewertung von Böden. Manuskript 1994.
[317] R. L. Metcalf, *Environ. Qual. Saf.* **5** (1976) 141–151.
[318] C. Frehe, *Verh. Ges. Ökol.* **12** (1984) 507–510.
[319] H. Kayser, *Helgol. Wiss. Meeresunters.* **30** (1977) 682–696.
[320] A. M. Albert in B. Scheele, M. Verfondern (eds.): "Auffindung von Indikatoren zur prospektiven Bewertung der Belastbarkeit von Ökosystemen," Endberichte der geförderten Vorhaben, Teil 2, Jül.-Spez. 503, A1–A116, Jülich 1989.
[321] G. Weidemann, K. Mathes, H. Koehler in B. Scheele, M. Verfondern (eds.): "Auffindung von Indikatoren zur prospektiven Bewertung der Belastbarkeit von Ökosystemen," Jül-Spez. 439, 7–223, 1988.
[322] L. Beck, K. Dumpert, U. Franke, W. Schönbor in B. Scheele, M. Verfondern (eds.): "Auffindung von Indikatoren zur prospektiven Bewertung der Belastbarkeit von Ökosystemen," Jül-Spez. 439, 548–702, Jülich 1988.
[323] T. Basedow, H. Rzehak, W. Liedtke in B. Scheele, M. Verfondern (eds.): "Auffindung von Indikatoren zur prospektiven Bewertung der Belastbarkeit von Ökosystemen," Jül-Spez. 439, 223–369, Jülich 1988.
[324] G. Kneitz, W. J. Kloft in B. Scheele, M. Verfondern (eds.): "Auffindung von Indikatoren zur prospektiven Bewertung der Belastbarkeit von Ökosystemen," Jül.-Spez. 439, 702–907, Jülich 1988.
[325] G. Weigmann et al.: "Wirkung und Verbleib von PCP-Na in einem Brachland-Ökosystem," Abschlußbericht 037288, 037289, 037294 eines im Programm: "Auffindung von Indikatoren zur prospektiven Bewertung der Belastbarkeit von Ökosystemen" geförderten Verbundvorhabens, 1986.
[326] M. Runge et al. in B. Scheele, M. Verfondern (eds.): "Auffindung von Indikatoren zur prospektiven Bewertung der Belastbarkeit von Ökosystemen," Endberichte der geförderten Vorhaben, Teil 2, Jül.-Spez. 503, B1–B218, Jülich 1989.
[327] R. Guderian (ed.): *Air Pollution by Photochemical Oxidants*, Springer Verlag, Berlin – Heideberg – New York – Tokyo 1985.
[328] L. Steubing, A. Fangmeier, *Environ. Pollut. Ser. A* **44** (1987) 297–306.
[329] L. W. Kress, J. M. Skelly, *Plant Dis.* **66** (1982) 1149–1152.
[330] H.-J. Jäger, E. Schulze: "Critical Levels for Effects of SO_2," ECE Critical Levels Workshop, Bad Harzburg, March 14–18, 1988, Draft Report part II, pp. 15–50, UN Econ. Comm. Europe, Geneva 1988.
[331] R. Guderian: "Critical Levels for Effects of NO_x," ECE Critical Levels Workshop, Bad Harzburg, March 14–18, 1988. Draft Report part II, pp. 79–104, UN Econ. Comm. Europe, Geneva 1988.
[332] R. Guderian: "Critical Levels for Effects of Ozone (O_3)," ECE Critical Levels Workshop, Bad Harzburg, March 14–18, Draft Report part II, pp. 51–78, UN Econ. Comm. Europe, Geneva 1988.
[333] D. C. Robinson, A. R. Wellburn, *Environ. Pollut. Ser. A* **32** (1983) 109–120.
[334] E. A. Wright, P. W. Lucas, D. A. Cottam, T. A. Mansfield, *CEC Air Pollut. Res. Rep.* **4** (1986) 187–200.
[335] J. Mooi, *Forst. Holzwirt* **39** (1984) 438–444.
[336] R. Guderian, K. Küppers, R. Six, *VDI-Ber.* **560** (1985) 657–701.
[337] A. R. Wellburn, C. Higginson, D. Robinson, C. Walmsley, *New Phytol.* **88** (1981) 223–237.
[338] J. A. Lee, M. C. Press, S. J. Woodin in CEC: "Study on the Need for a NO_2 Long-term Limit Value for the Protection of Terrestrial and Aquatic Ecosystems," Final Report, EUR 10546 EN, 1985.
[339] A. C. Posthumus: "Critical Levels for Effects of NH_3 and NH_4^+," ECE Critical Levels Workshop, Bad Harzburg, March 1988, pp. 117–127, UN Econ. Comm. Europe, Geneva 1988.
[340] J. Bauch, P. Klein, A. Frühwald, H. Brill, *Eur. J. For. Pathol.* **6** (1979) 321–331.
[341] S. Athari, *Mitt. Forstl. Bundesversuchsanst. Wien* **39** (1981) 7–27.
[342] P. Nogler, *Allg. Forst Z.* **36** (1981) 709–711.
[343] S. Athari, H. Kramer, *Forst. Holzwirt.* **38** (1983) 204–206.
[344] G. K. Kenk, *Forst. Holzwirt* **39** (1984) 435–438.
[345] Bundesministerium für Ernährung, Landwirtschaft und Forsten: Waldschadenserhebung 1985, Bonn 1985.
[346] Bundesminister für Forschung und Technologie: "Umweltforschung zu Waldschäden," 2nd report, Bonn 1985.
[347] O. Fränzle, W. Schröder, L. Vetter: "Saure Niederschläge als Belastungsfaktoren: Synoptische Darstellung möglicher Ursachen des Waldsterbens," Umweltforschungsplan des Bundesministers des Innern, Forschungsbericht 10607046/13, Berlin 1985.
[348] Umweltbundesamt (eds.): "Wissenschaftliches Symposium zum Thema Waldschäden 'Neue Ursachenhypothesen'," *Schriftenr. "Texte" Umweltbundesamtes* **19** (1986).
[349] Umweltbundesamt (eds.): "IMA-Querschnittsseminar zur Waldschadensforschung 'Belastung und Schäden auf Ökosystemebene und ihre Folgen'," *Schriftenr. "Texte" Umweltbundesamtes* **18** (1986).

[350] H. H. Cramer in [348, pp. 308–312, 423–426].
[351] H. Faulstich in [348, pp. 22–27, 381–382].
[352] G. Rippen et al., *Environ. Technol. Lett.* **8** (1987) 475–482.
[353] B. Ulrich, *Umschau* **11** (1984) 348–355.
[354] B. Prinz, *Umschau* **18** (1984) 544–549.
[355] E. F. Elstner, W. Osswald, R. J. Youngman, *Experientia* **41** (1985) 591–597.
[356] A. Hüttermann, *Allg. Forst Jagdztg.* **4** (1978) 67–70.
[357] U. Arndt, M. Kaufmann, *Allg. Forst Jagdztg.* **156** (1985) 16–20.
[358] P. Benner, A. Wild, *J. Plant Physiol.* **129** (1987) 59–72.
[359] D. Murach, *Göttinger Bodenkd. Ber.* **77** (1984).
[360] K. E. Rehfuess, C. Bosch, E. Pfankuch: "Nutrient Imbalances in Coniferous Stands in Southern Germany," Int. Workshop on Growth Disturbances of Forest Trees, Oct. 1982, IUFRO/FFRJ-Jyvaskyla, Finland.
[361] B. Ulrich, *Forstwiss. Centralbl.* **100** (1981) 228–236.
[362] B. Ulrich, *Allg. Forst Z.* **38** (1983) nos. 26/27, 670–677.
[363] W. U. Kriebitzsch: "Stickstoffnachlieferung in sauren Waldböden Nordwest-Deutschlands," *Scr. Geobot.* (1978).
[364] R. Linthurst, *VDI Ber.* **500** (1983) 175–185.
[365] L. S. Evans et al., *Water Air Soil Pollut.* **16** (1981) 469–509.
[366] W. Schröder, O. Fränzle, L. Vetter, *Allg. Forst Z.* **41** (1986) no. 22, 543–544.
[367] W. Saager: "Multivariate Bestimmung waldschadensdisponierender Faktoren in Schleswig-Holstein auf der Grundlage der terrestrischen Waldschadensinventur," Diplomarbeit Geographie, Universität Kiel 1990.
[368] G. H. Kohlmaier et al., *Ecol. Modell.* **22** (1984) 45–65.
[369] H. Bossel, W. Metzler, H. Schäfer (eds.): *Dynamik des Waldsterbens,* Springer Verlag, Berlin – Heidelberg – New York 1985.
[370] G. H. Kohlmaier, M. Plöchl in [349, pp. 41–54].
[371] H. Bossel, *Ecol. Modell.* **34** (1986) 259–288.
[372] D. L. Godbold, A. Hüttermann, *Water Air Soil Pollut.* **31** (1986) 509–515.
[373] R. Schultz: "Vergleichende Betrachtung des Schwermetallhaushalts verschiedener Waldökosysteme Norddeutschlands," Berichte des Forschungszentrums Waldökosysteme/Waldsterben, Reihe A **32** (1987).
[374] W.-D. Grossmann, J. Schaller, M. Sittard, *Allg. Forst Z.* **39** (1984) no. 38, 837–843.
[375] B. Prinz, G. H. M. Krause, H. Stratmann, *LIS Ber.* **28** (1982).
[376] S. M. Adams et al., *Mar. Environ. Res.* **28** (1989) 459–464.
[377] S. Gerlach: *Marine Pollution: Diagnosis and Therapy,* Springer Verlag, Berlin 1981.
[378] A. Moubadan, J.-C. Pihan, *Hydroécol. Appl.* **5** (1993) 97–109.
[379] J. Cairns, Jr., B. R. Niederlehner in J. Cairns, Jr., B. R. Niederlehner, D. R. Orvos (eds.): *Advances in Modern Environmental Toxicology,* vol. XX: "Predicting Ecosystem Risk," Princeton Scientific Publ. Co., Princeton, New Jersey, 1992, pp. 327–343.
[380] P. H. Pritchard in R. A. Conway (ed.): *Environmental Risk Analysis for Chemicals,* Van Nostrand Reinhold, New York 1991, pp. 257–353.
[381] P. H. Pritchard, H. W. Bourquin in H. H. White (ed.): *Concepts in Marine Pollution Measurements,* University of Maryland, 1984, pp. 117–138.
[382] R. W. Howarth in J. Cole, G. Lowett, S. Findlay (eds.): *Comparative Analyses of Ecosystems,* Springer Verlag, New York 1991, pp. 169–195.
[383] J. Cairns, Jr., J. R. Pratt: "Multispecies Toxicity Testing Using Indigenous Organisms – A New, Cost-effective Approach to Ecosystem Protection," *Environmental Conference, TAPPI Proceedings,* TAPPI Press, Atlanta, Georgia, 1985, pp. 149–153.
[384] W. B. Neely: *Chemicals in the Environment,* Marcel Dekker, New York 1980.
[385] D. I. Swackhamer, S. J. Eisenreich in K. C. Jones (ed.): *Organic Contaminants in the Environment,* Elsevier Applied Sci., London 1991, pp. 33–86.
[386] S. Griesbach, R. H. Peters, S. Youakim, *Can. J. Fish. Aquat. Sci.* **39** (1982) 727–735.
[387] K. J. Macek, S. R. Petrocelli, B. H. Sleight in L. L. Marking, R. A. Kimerle (eds.): "Aquatic Toxicology," *Am. Soc. Test. Mater.* (1979) 251–268.
[388] B. C. Patten in M. Higashi, T. P. Burns (eds.): *Theoretical Studies in Ecosystems: The Network Perspective,* Cambridge University Press, Cambridge 1991, pp. 288–351.
[389] J. Roelof van der Meer, W. M. DeVos, S. Harayama, A. J. B. Zehnde, *Microbiol. Rev.* **56** (1992) 677–694.
[390] R. Reuther, *Sci. Total Environ.* **115** (1992) no. 3, 219–237.
[391] B. van Hattum, K. R. Timmermans, H. A. Govers, *Environ. Toxicol. Chem.* **10** (1991) 275–292.
[392] R. R. Imhoff, P. Koppe, E. A. Nusch in S. Matsui, B. F. D. Barrett, J. Banerjee (eds.): "Toxic Substances Management in Lakes and Reservoirs," *Guidelines of Lake Management,* vol. 4, International Lake Environment Committee, UNEP, Shiga, Japan, 1991, pp. 127–159.
[393] J. Plotner, R. Gunther, *Int. Rev. Gesamten Hydrobiol.* **72** (1987) no. 6, 759–771.
[394] N. Nirmalakhandan et al., *Water Res.* **28** (1994) 543–552.
[395] A. W. Maki, M. W. Slimak in W. Grodzinski, E. B. Cowling, A. I. Breymeyer (eds.): *Ecological Risks – Perspectives from Poland and the United States,* National Academy Press, Washington, D.C., 1990, pp. 77–87.
[396] J. Cairns, J. R. Pratt, *Water Sci. Technol.* **19** (1987) no. 11, 1–12.
[397] C. N. Haas, *Water Qual. Int.* **4** (1993) 30–32.
[398] F.-A. A. M. de Leeuw, *Chemosphere* **27** (1993) no. 8, 1313–1328.
[399] J. S. Alabaster, *Chem. Ind.* **1** (1981) 529–534.
[400] J. Hermens, E. Broekhuyzen, H. Canton, R. Wegman, *Aquat. Toxicol.* **6** (1985) 209–217.
[401] J. W. Deneer: "The Toxicity of Aquatic Pollutants: QSARs and Mixture Toxicity Studies, Chap. VI: The Joint Acute Toxicity to *Daphnia magna* of Industrial Organic Chemicals at Low Concentrations," Proefschrift 81–89, Rijsuniversiteit Utrecht, 1988.
[402] ECETOC: "Assessment of Non-occupational Exposure to Chemicals," Technical Report no. **58**, Brussels 1994.
[403] L. A. Zadeh: "Fuzzy Sets," *Inf. Control* **8** (1965) 338–353.
[404] A. Salski, *Ecol. Modell.* **63** (1992) 103–112.
[405] A. Salski, P. Kandzia, *Springer Informatik Fachber.* **296** (1991) 303–310.

[406] M. Schepers: "Ein Unterstützungssystem zur wissensbasierten Modellierung und Simulation von ökologischen Prozessen," Diplomarbeit, Inst. f. Informatik und Prakt. Mathematik, Universität Kiel 1991.
[407] M. T. Bui: "Weiterentwicklung des Unterstützungssystems zur wissensbasierten Modellierung unter dem Einsatz der Fuzzy-Set-Theorie," Diplomarbeit, Inst. f. Informatik und Prakt. Mathematik, Universität Kiel 1991.
[408] R. V. Thomann, *Ecol. Modell.* **22** (1984) 145–170.
[409] M. G. Morgan in J. V. Rodricks, R. G. Tardiff (eds.): "Assessment and Management of Chemical Risks," *ACS Symp. Ser.* **239** (1984) chap. 8.
[410] R. C. Schwing, W. A. Albers (eds.): *Societal Risk Assessment: How Safe is Safe Enough?* Plenum Press, New York 1982.
[411] R. V. Thomann et al., *Water Res.* **8** (1974) 841–851.
[412] D. R. Miller in G. C. Butler (ed.): "Principles of Ecotoxicology," *SCOPE* **12** (1979) 71–90.
[413] J. W. Gillet et al.: *A Conceptual Model for the Movement of Pesticides Through the Environment,* National Environmental Research Center, U.S. Environmental Protection Agency, Corvallis, OR 1974, Report EPA 660/3-74-024, pp. 79.
[414] R. R. Lassiter: *Principles and Constraints for Predicting Exposure to Environmental Pollutants,* U.S. Environmental Protection Agency, Corvallis, OR, 1978, Report EPA 118-127519.
[415] T. Fagerstrøm, B. Aasell, *Ambio* **2** (1973) 164–171.
[416] I. Aoyama, Yos. Inoue, Yor. Inoue, *Water Res.* **12** (1978) 837–842.
[417] D. K. Leung: *Modelling the Bioaccumulation of Pesticides in Fish,* Center for Ecological Modelling, Polytechnic Institute, Troy, NY, 1978, Report 5.
[418] K. L. Seip, *Ecol. Modell.* **6** (1978) 183–198.
[419] G. T. Orlob, D. Hrovat, F. Harrison, *Adv. Chem. Ser.* **189** (1980) 195–212.
[420] D. C. L. Lam, T. J. Simons in J. O. Nriago (ed.): *Metals Transfer and Ecological Mass Balances. Environmental Biochemistry,* vol. 2, Ann Arbor Sci., Ann Arbor 1976, pp. 537–549.
[421] T. B. Kirchner, F. W. Whicker, *Ecol. Modell.* **22** (1984) 21–44.
[422] F. Bro-Rasmussen, K. Christiansen, *Ecol. Modell.* **22** (1984) 67–85.
[423] S. M. Bartell, R. H. Gardner, R. V. O'Neill, *Ecol. Modell.* **22** (1984) 109–123.
[424] C. G. Uchrin, *Ecol. Modell.* **22** (1984) 135–144.
[425] R. P. Schwarzenbach, D. M. Imboden, *Ecol. Modell.* **22** (1984) 171–213.
[426] E. Halfon, *Ecol. Modell.* **22** (1984) 213–253.
[427] J. R. W. Harris et al., *Ecol. Modell.* **22** (1984) 253–285.
[428] N. Nyholm, T. K. Nielsen, K. Pedersen, *Ecol. Modell.* **22** (1984) 285–324.
[429] J. C. J. Nihoul, *Ecol. Modell.* **22** (1984) 325–341.
[430] P. A. Arp, *Ecol. Modell.* **19** (1983) 105–117.
[431] S. Paterson, D. Mackay, *Ecol. Modell.* **47** (1989) 85.
[432] M. Matthies et al., *Ecol. Modell.* **47** (1989) 115.
[433] J. T. de Luna, T. G. Hallam, *Ecol. Modell.* **35** (1987) 249.
[434] T. Morioka, S. Chikami, *Ecol. Modell.* **31** (1986) 267.
[435] G. B. Schaalje, R. L. Stinner, D. L. Johnson, *Ecol. Modell.* **47** (1989) 223.
[436] B. C. Longstaff, *Ecol. Modell.* **43** (1988) 303.
[437] D.-S. Liu, S.-M. Zhang, Z.-G. Li, *Ecol. Modell.* **41** (1988) 75.
[438] D.-S. Liu, S.-M. Zhang, *Ecol. Modell.* **37** (1987) 131.
[439] R. F. Carsel, L. A. Mulkey, M. N. Lorber, L. B. Baskin, *Ecol. Modell.* **30** (1985) 101.
[440] G. R. Gorban, E. Bosatta, *Ecol. Modell.* **40** (1988) 25.
[441] J. E. Breck, D. L. de Angelis, W. van Winkle, S. W. Christensen, *Ecol. Modell.* **41** (1988) 1.
[442] D. S. Wratt et al., *Ecol. Modell.* **64** (1992) 185–204.
[443] G. Wuttke, B. Thober, H. Lieth, *Ecol. Modell.* **57** (1991) 263–276.
[444] J. K. Cronk, W. J. Mitsch, R. M. Sykes, *Ecol. Modell.* **51** (1990) 161–192.
[445] T. C. Gard, *Ecol. Modell.* **51** (1990) 273–280.
[446] ECETOC: Hazard Assessment of Chemical Contaminants in Soil, Technical Report no. **40**, Brussels 1990.
[447] E. E. Kenaga, C. A. I. Goring: "Relationship Between Water Solubility, Soil Sorption, Octanol-water Partitioning and Bioconcentration of Chemicals in Biota," Special Technical Publication (STP) 707, ASTM, Philadelphia, PA, 1980.
[448] R. L. Metcalf, G. K. Sangha, I. P. Kopoor, *Environ. Sci. Technol.* **5** (1975) 709–713.
[449] P.-Y. Lu, R. L. Metcalf, *Environ. Health Perspectives* **10** (1975) 269–284.
[450] W. B. Neely, D. R. Branson, G. E. Blau, *Environ. Sci. Technol.* **8** (1974) 1113–1115.
[451] G. D. Veith, D. L. Defoe, B. V. Bergstedt, *J. Fish. Res. Board Can.* **36** (1979) 1040–1048.
[452] R. R. Lassiter: "Modelling Dynamics of Biological and Chemical Compartments of Aquatic Ecosystems," U.S. Environmental Protection Agency, Washington, D.C., 1975, EPA-660/3-75-012.
[453] R. J. Larson, R. L. Perry, *Water Res.* **15** (1981) 697.
[454] P. H. Pritchard, A. W. Bourquin, H. L. Fredrickson, T. Maziarz in A. W. Bourquin, R. H. Pritchard (eds.): "Proceedings of the Workshop: Microbial Degradation of Pollutants in Marine Environments," U.S. Environmental Protection Agency, Gulf Breeze, FL, 1979, p. 251.
[455] D. S. Brown, E. W. Flagg, *J. Environ. Qual.* **10** (1981) 382–386.
[456] C. Mater-Müller, W. Gujer, W. Giger, W. Stumm, *Water Technol. Sci.* **12** (1980) 299–314.
[457] O. C. Zafiriou, M. B. True, *Mar. Chem.* **8** (1979) 9–32.
[458] O. C. Zafiriou, M. McFarland, *J. Geophys. Res.* **86** (1981) 3173–3182.
[459] R. G. Zika, L. T. Gidel, D. D. Davis, *J. Geophys. Res.* **11** (1984) 353–356.
[460] C. van Baalen, J. E. Marler, *Nature* (*London*) **211** (1966) 951.
[461] R. M. Baxter, J. H. Carey, *Freshwater Biol.* **12** (1982) 285–292.
[462] W. R. Haag, J. Hoigne, E. Grassmann, A. M. Braun, *Chemosphere* **13** (1984) 631–640.
[463] P. G. Zepp, *Environ. Sci. Technol.* **12** (1978) 327–329.
[464] P. G. Zepp in J. Calkins (ed.): *The Role of Solar Ultraviolet Radiation in Marine Ecosystems,* Plenum Press, New York 1982, pp. 293–307.
[465] P. G. Zepp in O. Hutzinger (ed.): *The Handbook of Environmental Chemistry,* vol. 2, part B, Springer Verlag, Berlin 1982, pp. 19–40.
[466] C. T. Chiou, V. H. Freed, D. W. Schmedding, R. L. Kohnert, *Environ. Sci. Technol.* **11** (1977) 475–478.

Toxicology

Wolfgang Dekant, Spiridon Vamvakas, Institut für Toxikologie und Pharmakologie der Universität Würzburg, Würzburg, Federal Republic of Germany

The entire topic was coordinated by Wolfgang Dekant

3-527-20137-8/95/$ 5.00 + .50

In addition to the abbreviations defined in the front matter of this volume the following abbreviations are used:

Ah-R	arylhydrocarbon receptor
AP	apurinic/apyrimidinic site
APS	adenosine 5′-phosphosulfate
BHK	baby hamster kidney
BIBRA	British Industrial Biological Research Association
CoA	Coenzym A
DDT	1,1′-(2,2,2-trichloroethylidene)bis-(4-chlorobenzene)
DHHS	U.S. Department of Health and Human Services
DHP	delayed hypersensitive response
ECETOC	European Chemical Industry Ecology and Toxicology Centre
ED	effective dose
ELISA	enzyme-linked immunosorbent assay
FCA	Freund's complete adjuvant
FAD	flavine adenine dinucleotide
GABA	γ-aminobutyrate
GC/MS	gas chromatography/mass spectroscopy
GOT	glutamic acid oxalacetic transaminase
GSH	glutathione
GSSG	glutathione disulfide
GST	glutathione *S*-transferase
GTP	guanosine 5′-triophosphate
HGPRT	hypoxanthine – guanine phosphoribosyltransferase
IPCS	International Programme on Chemical Safety

LDH	lactate dehydrogenase
LOAEL	lowest-observed-adverse-effect level
LOEL	lowest-observed-effect level
MIF	migration inhibition factor
mRNA	messenger RNA
MTD	maximum tolerated dose
NADPH	nicotinamide dinucleotide phosphate (H)
NOEL	no-observed-effect-level
NTP	National Toxicology Program
PAPS	3′-phosphoadenosine-5′-phosphosulfate
PG	prostaglandin
rRNA	ribosomal RNA
SHE	Syrian hamster embryo
SMART	somatic mutation and recombination test
T, or TCDD	2,3,7,8-tetrachlorodibenzodioxin
TD	tumor dose
TK	thymidine kinase
tRNA	transfer RNA
UDP	uridine diphosphate
UDPG	uridine diphosphate glucose
UDPGA	uridine diphosphate glucuronic acid
UDS	unscheduled DNA synthesis

1. Introduction

1.1. Definition and Scope

Chemicals that are used or of potential use in commerce, the home, the environment, and medical practice may present variable types of harmful effects. The nature of these effects is determined by the physicochemical characteristics of the agent, its ability to interact with biological systems (hazard), and its potential to come in contact with biological systems (exposure).

Toxicology studies the interaction between chemicals and biological systems to determine the potential of chemicals to produce adverse effects in living organisms. Toxicology also investigates the nature, incidence, mechanisms of production, factors influencing their development, and reversibility of such effects. Adverse effects are defined as detrimental to the survival or the normal functioning of the individual. Inherent in this definition are the following key issues in toxicology:

1) Chemicals must come in close structural or functional contact with tissues or organs to cause injury.
2) All adverse effects depend on the amount of chemical in contact with the biological system (the dose) and the inherent toxicity of the chemical (hazard). When possible, the observed toxic effect should be related to the degree of exposure. The influence of different exposure doses on the magnitude and incidence of the toxic effect should be quantitated. Such dose–response relationships are of prime importance in confirming a causal relationship between chemical exposure and toxic effect (for details, see Section 1.7).

Research in toxicology is concerned mainly with determining the potential for adverse effects caused by chemicals, both natural and synthetic, to assess their hazard and the risk of human exposure and thus provide a basis for appropriate precautionary, protective, and restrictive measures. Toxicological investigations should permit evaluation of the following characteristics of toxicity:

1) The basic structural, functional, or biochemical injury produced
2) Dose–response relationships
3) The mechanisms of toxicity (fundamental biochemical alterations responsible for induction and maintenance of the toxic response) and reversibility of the toxic effect
4) Factors that modify response (e.g., route of exposure or species and sex of the individual)

For chemicals to which humans may potentially be exposed, a cricital analysis, based on the pattern of potential exposure or toxicity, may be necessary to determine the risk–benefit ratio for their use in specific circumstances and to devise protective and precautionary measures. Indeed, with drugs, pesticides, food additives, and cosmetic preparations, toxicology testing must be performed under government regulations before use.

1.2. Fields

Toxicology is a recognized scientific discipline encompassing both basic and applied issues. Although only generally accepted as a specific scientific field during this century, its principles have been appreciated for ages. The harmful or lethal effects of certain chemicals, present mainly in minerals and plants or transmitted by insects and other venomos animals, have been known since prehistoric times. In many countries, toxicology as a discipline has developed from pharmacology. Pharmacology and toxicology both study the effect of chemicals on living organisms and have often used identical methods. However, fundamental differences have developed. Years ago, only the dose dependence of the studied effects separated pharmacology and toxicology. Pharmacology focused on chemicals with beneficial effects (drugs) at lower doses, whereas toxicology studied the adverse health effects occurring with the same chemicals at high doses. Today, the main interest of research in toxicology has shifted to studies on the long-term effects of chemicals after low-dose exposure, effects such as cancer or other irreversible diseases; moreover, most chemicals of interest to toxicologists are not used as drugs.

The variety of potential adverse effects and the diversity of chemicals present in the human environment combine to make toxicology a very broad science. Toxicology uses basic knowledge from clinical and theoretical medicine and natural sciences such as biology and chemistry (Fig. 1.1). Because of this diversity, toxicologists usually specialize in certain areas.

Any attempt to define the scope of professional activities in toxicology must consider that the various subdisciplines are not mutually exclusive and frequently are heavily interdependent. Because of the overlap of mechanisms of toxicity, chemical classes, and observed toxic effects, clear divisions into subjects of equal importance are often not possible.

The professional activities of toxicologists can be divided into three main categories: descriptive, mechanistic, and regulatory. The *descriptive toxicologist* is concerned directly with toxicity testing. Descriptive toxicology still relies often on the tools of pathology and clinical chemistry; only since the 1970s more mechanism-based test systems have been included in toxicity testing [1]. The appropriate toxicity tests in experimental animals yield information that is extrapolated to evaluate the risk posed by exposure to specific chemicals. The concern may be limited to effects on humans (drugs, industrial chemicals in the workplace, or food additives) or may encompass animals, plants, and other factors that might disturb the balance of the ecosystem (industrial chemicals, pesticides, environmental pollutants).

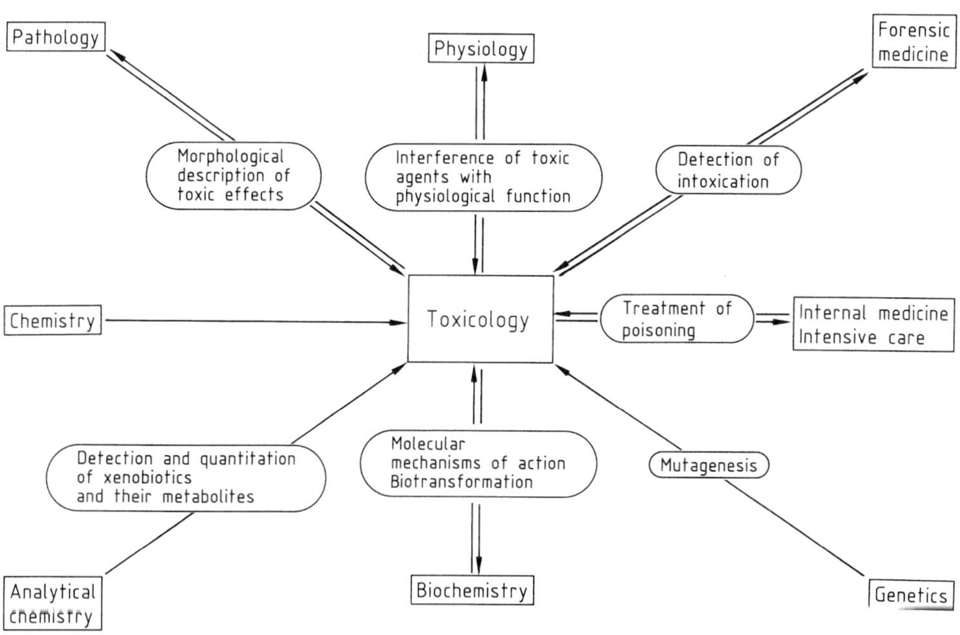

Figure 1.1. Scientific fields influencing the science of toxicology

The *mechanistic toxicologist* is concerned with elucidating the mechanisms by which chemicals exert their toxic effects on living organisms. Such studies may result in the development of sensitive predictive toxicity tests useful in obtaining information for risk assessment (see Chap. 4). Mechanistic studies may help in the development of chemicals that are safer to use or of more rational therapies for intoxications. In addition, understanding the mechanisms of toxic action also contributes to the knowledge of basic mechanisms in physiology, pharmacology, cell biology, and biochemistry. Indeed, toxic chemicals have been used with great success as mechanistic tools to elucidate mechanisms of physiological regulation. Mechanistic toxicologists are often active in universities; however, industry and government institutions are now undertaking more and more research in mechanistic toxicology.

Regulatory toxicologists have the responsibility of deciding, on the basis of data provided by descriptive and mechanistic toxicologists, if a drug or chemical poses a sufficiently low risk to be used for a stated purpose. Regulatory toxicologists are often active in government institutions and are involved in the establishment of standards for the amount of chemicals permitted in ambient air in the environment, in the workplace, or in drinking water. Other divisions of toxicology may be based on the classes of chemicals dealt with or the application of knowledge from toxicology to a specific field (Table 1.1)

Table 1.1. Areas of toxicology

Field	Tasks and objectives
Forensic toxicology	diagnoses poisoning by analytical procedures
Pesticide toxicology	studies the safety of pesticides, develops new pesticides
Occupational toxicology	assesses potential adverse effects of chemicals used in the workplace, recommends protective procedures
Drug toxicology	studies potential effects of drugs after high doses, elucidates mechanisms of side-effects
Regulatory toxicology	develops and interprets toxicity testing programs and is involved in controlling the use of chemicals
Environmental toxicology	studies the effects of chemicals on ecosystems and on humans after low-dose exposure from the environment

Forensic toxicology comprises both analytical chemistry and fundamental toxicologic principles. It is concerned with the legal aspects of the harmful effects of chemicals on humans. The expertise of the forensic toxicologist is invoked primarily to aid in establishing the cause of death and elucidating its circumstances in a postmortem investigation. The field of *clinical toxicology* recognizes and treats poisoning, both chronic and acute. Efforts are directed at treating patients poisoned by chemicals and at the development of new techniques to treat these intoxications. *Environmental toxicology* is a relatively new area that studies the effects of chemicals released by man on wildlife and the ecosystem and thus indirectly on human health.

Drug toxicology plays a major role in the preclinical safety assessment of chemicals intended for use as drugs. Drug toxicology also elucidates the mechanisms of side effects observed during clinical application. *Occupational toxicology* studies the acute and chronic toxicity of chemicals encountered in the occupational environment. Both acute and chronic occupational poisonings have exerted a major influence on the development of toxicology in general. Occupational toxicology also helps in the development of safety procedures to prevent intoxications at the workplace and assists in the definition of exposure limits. *Pesticide toxicology* is involved in the development of new pesticides and the safety of pesticide formulations. Pesticide toxicology also characterizes potential health risks to the general population caused by pesticide residues in food and drinking water.

1.3. History

Toxicology must rank as one of the oldest practical sciences, because humans, from the very beginning, needed to avoid the numerous toxic plants and animals in their environment. The presence of toxic agents in animals and plants was known to Egyptian and Greek civilization. The papyrus Ebers, an Egyptian papyrus dating from about 1500 BC and the surviving medical works of HIPPOCRATES, ARISTOTLE, and THEOPHRASTUS, published during 400–250 BC, all include some mention of poisons.

Greek and Roman civilizations intentionally used certain toxic chemicals and extracts for hunting, warfare, suicide, and murder. Up to the Middle Ages, toxicology was restricted to the use of toxic agents for murder. Poisoning was developed to an art in medieval Italy and has remained a problem ever since, and much of the earlier impetus in the development of toxicology was primarily forensic. Apparently, few advances were made in either medicine or toxicology between the time of GALEN (AD 131–200) and PARACELSUS (1493–1541). The latter laid the groundwork for the development of modern toxicology. He was clearly aware of the dose–response relationship. His statement, "All substances are poisons; there is none that is not a poison. The right dose differentiates a poison and a remedy," is properly regarded as a landmark in the development of the science of toxicology. His belief in the value of experimentation also represents a break with much earlier tradition. Important developments in the 18th century include the publication of RAMAZZINI's *Diseases of Workers* in the 1700s, which led to his recognition as the father of occupational medicine. The correlation between the occupation of chimney sweepers and scrotal cancer by PERCIVAL POTT in 1775 is also noteworthy.

ORFILA, a Spaniard working at the University of Paris, clearly identified toxicology as a separate science and wrote the first book devoted exclusively to it (1815). Scientists of the later 19th century who produced treatises on toxicology include CHRISTISON, KOBERT, and LEWIN. They increased knowledge of the chemistry of poisons, the treatment of poisoning, the analysis of both xenobiotics and toxicity, and modes of action and detoxification. A major impetus for toxicology in the 20th century was the use of chemicals for warfare. In World War I, a variety of poisonous chemicals were used on the battlefields of France. This provided stimulus for work on mechanisms of toxicity as well as medical countermeasures to poisoning. Since the 1960s, toxicology has entered a phase of rapid development and has changed from a science that was almost entirely descriptive to one in which the study of mechanisms has become the prime task. Among the many reasons for this are the development of new analytical methods since 1945, the emphasis on drug testing after the thalidomide tragedy, the emphasis on pesticide testing following publication of RACHEL CARSON's *Silent Spring*, and public concerns with environmental pollution and hazardous waste disposal.

1.4. Information Resources

Because of the complexity of toxicology as a science and the impact of toxicological investigations on legislation and commerce, a wide range of information on the toxic effects of chemicals is available. No single, exhaustive source of toxicological data exists; several sources are required to obtain comprehensive information on a particular chemical. Printed sources are often quicker and easier to use than computer data bases, but interactive online searching can rapidly gather important information from the huge number of sources present.

The information explosion in toxicology has resulted in two comprehensive volumes dedicated to toxicological information sources. These are:

P. Wexler: *Information Resources in Toxicology,* Elsevier Biomedical, Amsterdam – New York 1988. ISBN 0444012141.

J. K. Webster: *Toxic and Hazardous Materials, A Sourcebook and Guide to Information Sources* (*Bibliographical Indexes in Science and Technology No. 2*). Greenwood Press, New York – London 1987. ISBN 0313245754.

Textbooks. The easiest way to obtain information on general topics in toxicology and secondary references is the range of textbooks available on the market. Only a few selected books are listed below:

L. J. Casarett, J. Doull, C. D. Klaasen, M. O. Amdur: *Casarett and Doull's Toxicology; The Basic Science of Poisons,* 4th ed McGraw-Hill, New York 1991.

G. D. Clayton, F. E. Clayton (eds): *Patty's Industrial Hygiene and Toxicology,* Wiley, Chichester – New York 1985.

A. G. Gilman, L. S. Goodman, T. W. Rail, F. Murad: *Goodman and Gilman's, The Pharmacological Basis of Therapeutics,* 8th ed Pergamon Press, New York 1990.

T. J. Haley, W. D. Bernd (eds): *Toxicology,* Taylor and Francis, London 1988.

W. A. Hayes: *Principles and Methods of Toxicology,* Raven Press, New York 1989.

E. Hodgson, P. E. Levi (eds): *Textbook of Modern Toxicology,* Elsevier, New York 1987.

J. R. Mitchell, M. G. Horning (eds): *Drug Metabolism and Drug Toxicity,* Raven Press, New York 1984.

The huge volume by N. I. Sax and R. J. Lewis: *Dangerous Properties of Industrial Materials,* Van Nostrand Reinhold, New York, 1988 (7th ed.) contains basic toxicological data on a large selection of chemicals (almost 20000) and may serve as a useful guide to the literature for compounds not covered in other publications.

Monographs. The best summary information on toxicology is published in the form of series by governments and international organizations. Most of these series summarize the results of toxicity studies on specific chemicals. The selection of these chemicals is based mainly on the extent of their use in industry (e.g., trichloroethylene), their occurrence as environmental contaminants (mercury), or their extraordinary toxicity (e.g., 2,3,7,8-tetrachlorodibenzodioxin). The most important series include the following:

American Conference of Governmental Industrial Hygienists, Threshold Limit Values and Biological Exposure Indices (Cincinnati, OH). Published annually.

MAK-Begründungen, VCH Publishers, Weinheim, Federal Republic of Germany. This German series includes detailed information on the toxicity of chemicals on the German MAK list (ca. 150 reports are available; the series is continuously expanded).

British Industrial Biological Research Association (BIBRA): Over a period of six years the association has produced some 400 toxicity profiles.

The "Banbury Reports" issued by the Cold Spring Harbor Laboratory publish proceedings of conferences on toxicological and environmental subjects focusing mainly on aspects of risk assessment.

The Commission of the European Communities publishes the Reports of the Scientific Committee on Cosmetology and the Reports of the Scientific Committee for Food.

The Environmental Protection Agency (EPA) publishes a huge number of reports and toxicological profiles. They are indexed in "EPA Publications. A Quarterly Guide."

The European Chemical Industry Ecology and Toxicology Centre (ECETOC) issues "Monographs" (more than 20 have been published) and "Joint Assessments of Commodity Chemicals."

The monographs of the International Agency for Research on Cancer are definitive evaluations of carcinogenic hazards. The "Environmental Health Criteria" documents of the International Programme on Chemical Safety (IPCS) assess environmental and human health effects of exposure to chemicals, and biological or physical agents. A related "Health and Safety Guide" series give guidance on setting exposure limits for national chemical safety programs.

The National Institute for Occupational Safety and Health (NIOSH), has published 50 "Current Intelligence Bulletins" on health hazards of materials and processes at work.

The technical report series of the National Toxicology Program (NTP) reports results of their carcinogenicity bioassays, which include summaries of the toxicology of the chemicals studied. A status report indexes both studies that are under way and those that have been published. The program also issues an "Annual Review of Current DHHS [U.S. Department of Health and Human Services], DOE [U.S. Department of Energy] and EPA Research" related to toxicology.

Journals. Results of toxicological research are published in more than 100 journals. Those listed below mainly publish research closely related to toxicology, but articles of relevance may also be found in other biomedical journals:

Archives of Environmental Contamination and Toxicology
Archives of Toxicology
Biochemical Pharmacology
Chemical Research in Toxicology
CRC Critical Reviews in Toxicology
Clinical Toxicology
Drug and Chemical Toxicology
Environmental Toxicology and Chemistry
Food and Chemical Toxicology
Fundamental and Applied Toxicology
Journal of the American College of Toxicology
Journal of Analytical Toxicology
Journal of Applied Toxicology
Journal of Biochemical Toxicology
Journal of Toxicology and Environmental Health
Neurotoxicology and Teratology
Pharmacology and Toxicology
Practical In Vitro Toxicology
Regulatory Toxicology and Pharmacology
Reproductive Toxicology
Toxicology
Toxicology and Applied Pharmacology
Toxicology Forum
Toxicology and Industrial Health
Toxicology In Vitro
Toxicology Letters

Data Bases and Data Banks. Electronic sources, such as computer data bases or CD-ROM are a fast and convenient way to obtain references on the toxicity of chemicals. Since on-line searching of commercial data bases such as STN-International may be expensive, CD-ROM-based systems are increasingly being used. The major advantages are speed, the ability to refine searches and format the results, and non-text search options, such as chemical structure searching on Beilstein and Chemical Abstracts.

Useful information about actual research on the toxicology of chemicals may be obtained by searching Chemical Abstracts or Medline with the appropriate keywords. Specific data banks covering toxicology are the Registry of Toxic Effects of Chemical Substances, which gives summary data, statistics, and structures; Toxline (available in DIMDI) gives access to the literature.

1.5. Terminology of Toxic Effects

Toxic effects may be divided according to time scale (acute and delayed), general locus of action (local, systemic, organ specific), or basic mechanisms of toxicity (reversible versus irreversible). *Acute toxic effects* are those that occur after brief exposure to a chemical. Acute toxic effects usually develop rapidly after single or multiple administrations of a chemical; however, acute exposure may also produce *delayed toxicity*. For example, inhalation of a lethal dose of HCN causes death in less than a minute, whereas lethal doses of 2,3,7,8-tetrachlorodibenzodioxin will result in the death of experimental animals after more than two weeks. *Chronic effects* are those that appear after repetitive exposure to a substance; many compounds require several months of continuous exposure to produce adverse effects. Often, the chronic effects of chemicals are different from those seen after acute exposure (Table 1.2). For example, inhalation of chloroform for a short period of time may cause anesthesia; long-term inhalation of much lower chloroform concentrations causes liver damage. Carcinogenic effects of chemicals usually have a long latency period; tumors may be observed years (in rodents) or even decades (in humans) after exposure.

Toxic effects of chemicals may also be classified based on the type interaction between the chemical and the organism. Toxic effects may be caused by reversible and irreversible interactions (Table 1.3). When *reversible interactions* are responsible for toxic effects, the concentration of the chemical present at the site of action is the only determinant of toxic outcome. When the concentration of the xenobiotic is decreased by excretion or biotransformation, a parallel decrease of toxic effects is observed.

After complete excretion of the toxic agent, toxic effects are reduced to zero. A classic example of reversible toxic effects is carbon monoxide. Carbon monoxide binds to hemoglobin and, because of formation of the stable hemoglobin–carbon monoxide complex, oxygen binding is blocked. As a result of impaired oxygen transport in blood from the lung, tissue oxygen concentrations are reduced and cells sensitive to oxygen deprivation will die. The toxic effects of carbon monoxide are correlated directly with the

Table 1.2. Toxic effects of different chemicals categorized by time scale and general locus of action

Exposure	Site	Effect	Chemical
Acute	local	lung edema	chlorine gas
	systemic	liver damage	carbon tetrachloride
		narcosis	halothane
Subchronic	local	sensitization	toluene diisocyanate
	systemic	neurotoxicity	hexane
Chronic	local	bronchitis	sulfur dioxide
		nasal carcinoma	formaldehyde
	systemic	bladder carcinoma	4-aminobiphenyl
		kidney damage	cadmium

Table 1.3. Reversible and irreversible interactions of chemicals with cellular macromolecules as a basis for toxic response

Mechanism	Toxic response	Example
Irreversible inhibition of		
Esterase	neurotoxicity	tri-*o*-cresyl-phosphate
Covalent binding to DNA	cancer	dimethyl-nitrosamine
Reversible binding to		
Hemoglobin	oxygen deprivia-tion in tissues	carbon monoxide
Cholinesterase	neurotoxicity	carbamate pesticides

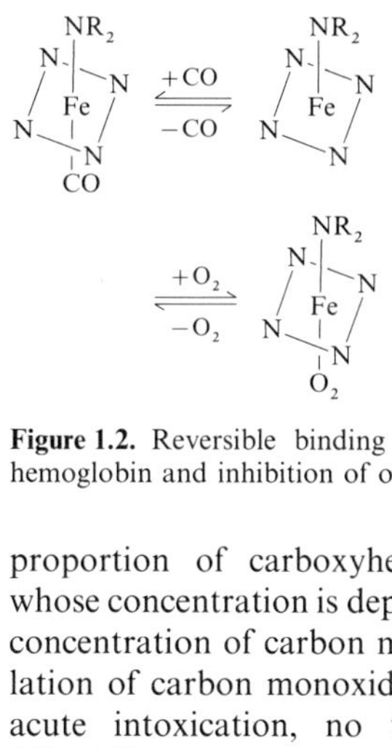

Figure 1.2. Reversible binding of carbon monoxide to hemoglobin and inhibition of oxygen transport

proportion of carboxyhemoglobin in blood, whose concentration is dependent on the inhaled concentration of carbon monoxide. After exhalation of carbon monoxide and survival of the acute intoxication, no toxic effect remains (Fig. 1.2).

Irreversible toxic effects are often caused by the covalent binding of toxic chemicals to biological macromolecules. Under extreme conditions, the modified macromolecule is not repaired; after excretion of the toxic agent, the effect persists. Further exposure to the toxic agent will produce additive effects; many chemical carcinogens are believed to act through irreversible changes of macromolecules (see Section 2.5.6).

Another distinction between types of effects can be made by the general locus of action. *Local toxicity* occurs at the site of first contact between the biological system and the toxic agent. Local effects to the skin, the respiratory tract, or the alimentary tract may be produced by skin contact with a corrosive agent, by inhalation of irritant gases, or by ingestion of tissue-damaging materials. This type of toxic response is usually restricted to tissues having direct contact with the agent. However, life-threatening intoxications may occur if vital organs such as the lung are damaged. For example, inhaled phosgene damages the alveoli of the lung and causes lung edema. The massive damage to the lung results in a substantial mortality observed after phosgene intoxication.

The opposite of local effects are *systemic effects*. These are characterized by absorption of the chemical and its distribution from the port of entry to a distant site at which toxic effects are produced. Except for highly reactive xenobiotics, which mainly act locally, most chemicals act systemically. Many chemicals that produce systemic toxicity cause damage only to certain organs, tissues, or cell types within organs. Selective damage to certain organs or tissues by systemically distributed chemicals is termed *organ-* or *tissue-specific toxicity* [2]; the organs damaged are referred to as target organs (Table 1.4). Major target organs for toxic effects are the central nervous system and the circulatory system, followed by the blood and hematopoietic system and visceral organs such as the liver or kidney. For some chemicals, both local and systemic effects can be demonstrated; moreover, chemicals producing marked local toxicity may also cause systemic effects as secondary responses to major disturbances in homeostasis of the organism.

1.6. Types of Toxic Effects

The spectrum of toxic effects of chemicals is broad; their magnitude and nature depend on many factors such as the physicochemical properties of the chemical and its toxicokinetics, conditions of exposure, and presence of adaptive or protective mechanisms. The latter factors include physiological mechanisms such as adaptive enzyme induction, DNA repair, and others. Toxic effects may be transient, reversible, or irreversible; some are deleterious and others are not. Toxic effects may take the form of tissue pathology, aberrant growth processes, or altered biochemical pathways. Some of the more frequently encountered types of injury constituting a toxic response are described in the following.

Table 1.4. Organ-specific toxic effects induced by chemicals that are distributed systemically in the organism

Chemical	Species	Target organ
Benzene	humans	bone marrow
Hexachlorobutadiene	rodents	damage to proximal tubules of the kidney
Paraquat	rodents, humans	lung
Tri-*o*-cresylphosphate	humans	nervous system
Cadmium	humans	kidney
1,2-Dibromo-3-chloropropane	humans, rodents	testes
Hexane	rodents, humans	nervous system
Anthracyclines	humans	heart

Immune-mediated hypersensitivity reactions by antigenic materials are toxic effects, often involved in skin and lung injury, that are caused by repeated contact with chemicals resulting in contact dermatitis and asthma. Inflammation is a frequently observed local response to the application of irritant chemicals, or it may be a component of systemic injury. This response may be acute with irritant or tissue-damaging materials or chronic with repetitive exposure of irritants. Necrosis (i.e., death of cells or tissues) may be the result of various pathological processes resulting from biochemical interactions of xenobiotics described in Chapter 3. The extent and patterns of necrosis even in the same organ may be different for different chemicals. Chemical tumorigenesis or carcinogenesis (induction of malignant tumors) is an effect often observed after chronic application of chemicals. Because of the long latency period and the poor prognosis for individuals diagnosed with cancer, studies for predicting the potential tumorigenicity of chemicals have developed into a major area of toxicological research. Developmental and reproductive toxicology are concerned with adverse effects on the ability to conceive and with adverse effects on the structural and functional integrity of the fetus. Chemicals may interfere with reproduction through direct effects on reproductive organs or indirectly by affecting their neural and endocrine control mechanisms. Developmental toxicity deals with adverse effects on the conceptus through all stages of pregnancy. Damage to the fetus may result in embryo reabsorption, fetal death, or abortion. Nonlethal fetotoxicity may be expressed as delayed maturation, decreased birth weight, and structural malformations. The most sensitive period for the induction of malformation is during organogenesis; neurobehavioral malformations may be induced during later stages of pregnancy.

1.7. Dose–Response: an Essential in Toxicology

In principle, a poison is a chemical that has an adverse effect on a living organism. This is, however, not a useful definition because toxic effects are related to dose. The definition of a poison thus also involves quantitative biological aspects. At sufficiently high doses, every chemical may be toxic. The importance of dose is seen clearly with molecular oxygen or dietary metals. Oxygen at a concentration of 21 % in the atmosphere is essential for life; in contrast, 100 % oxygen at atmospheric pressure causes massive lung injury in rodents, often resulting in death. Some metals such as iron, copper, and zinc are essential nutrients. When present in insufficient amounts in the human diet, specific disease patterns develop. At high doses these metals may cause fatal intoxications. Toxic compounds are not restricted to man-made chemicals but also include many naturally occurring substances. Indeed, the agent with the highest toxicity is a natural poison found in the bacterium *Clostridium botulinum* (LD_{50} 0.01 μg/kg).

Therefore, all toxic effects are products of the amount of chemical to which the organism is exposed and the inherent toxicity of the chemical; they also depend on the sensitivity of the biological system.

The word *dose* is most frequently used to characterize the total amount of material to which an organism is exposed; dose defines the amount of chemical given in relation to body weight. Dose is a more meaningful and comparative indicator of exposure than the term exposure itself. Dose usually implies the exposure dose, the total amount of chemical administered to an organism or incorporated into a test system. However, dose may not be directly proportional to the toxic effects because toxicity depends on the amount of chemical absorbed. Usually, dose correctly describes only the actual amount of chemical absorbed when the chemical is administered orally or by injection. Under these circumstances, the administered dose is identical to the absorbed dose; other routes of application such as dermal application or inhalation do not define the amount of agent absorbed.

Among different chemicals, a wide range of doses is needed to induce toxic effects or death. To characterize the acute toxicity of different chemicals, LD_{50} values are frequently used as a basis for comparison. The LD_{50} values (mg/kg) for a number of chemicals administered to rats are given below:

Ethanol	12 500
Sodium bicarbonate	4 220
Phenobarbital sodium	350
Paraquat	120
Aldrin	46
Sodium cyanide	6.4
Strychnine	5
1,2-Dibromoethane	0.4
Sodium fluoroacetate	0.2
2,3,7,8-Tetrachlorodibenzodioxin	0.01

Certain chemicals are very toxic and produce death after administration of microgram doses, whereas others are tolerated without serious toxicity in gram amounts. The data provided above clearly demonstrate that the toxicity of a specific chemical is related to dose. The dose dependence of the toxic effects of a specific chemical is termed the dose–response relationship. Before dose–response relationships can be appropriately used, several basic assumptions must be made. The first is that the response is due to the chemical given. The responses observed are usually assumed to be a result of the various doses of chemical administered. Under experimental conditions, the toxic response is usually correlated with the chemical given since both exposure and effect are well defined and can be quantified. However, a response is not always known to be the result of specific chemical exposure. For example, an epidemiologic study might result in the discovery of an "association" between a response (e.g., disease) and one or more variables, including estimated dose of a chemical. The true dose to which individuals have been exposed is often an estimate, and the specificity of the response for that chemical is doubtful. Further major assumptions necessary in establishing dose–response relationships are:

1) A molecular site (often termed receptor) exists with which the chemical interacts to produce the response. Receptors are macromolecular components of tissues with which a chemical interacts and produces its characteristic effect.
2) The production of a response and the degree of the response are related to the concentration of the agent at the receptor.
3) The concentration of the chemical at the receptor is related to the dose administered. Since, in most cases, the concentration of an administered chemical at the receptor cannot

be determined, the administered dose or the blood levels of the chemical are used as indicators for its concentration at the molecular site.

A further prerequisite for using the dose–response relationship is that the toxic response can be measured exactly. A great variety of criteria or end points of toxicity may be used. The ideal end point should be closely associated with the molecular events resulting from exposure to the toxin and should be readily determined. However, although many end points are quantitative and precise, they are often only indirect measures of toxicity. For example, changes in enzyme levels in blood can be indicative of tissue damage. Patterns of alterations may provide insight as to the organ or system that is the site of toxic effects. These measures usually are not directly related to the mechanism of toxic action.

The dose–response relationship combines the characteristic of exposure and the inherent toxicity of the chemical. Since toxic responses to a chemical are usually functions of both time and dose in typical dose–response relationships the maximum effect observed during the time of observation is plotted against the dose to yield time-independent curves. The time-independent dose–response relationship may be used to study dose–response for both reversible and irreversible toxic effects. However, in risk assessments that consider the induction of irreversible effects such as cancer, the time factor plays a major role and has important influences on the magnitude or likelihood of toxic response. Thus, for this type of mechanism of toxic action, dose–time–response relationships are better descriptors of toxic effects (see below).

The dose–response relationship is the most fundamental concept in toxicology. Indeed, an understanding of this relationship is essential for the study of toxic chemicals.

From a practical point of view, two different types of dose–response relationships exist. Dose–response relationships may be quantal (all or none responses; for example, death) or graded. The *graded* or *variable response* involves a continual change in effect with increasing dose (e.g., enzyme inhibition or changes in physiological function such as heart rate). Graded responses may be determined in an individual or in simple biochemical systems. For example, increasing concentrations of 2,3,7,8-tetrachlorodibenzodioxin (TCDD) added to cultured

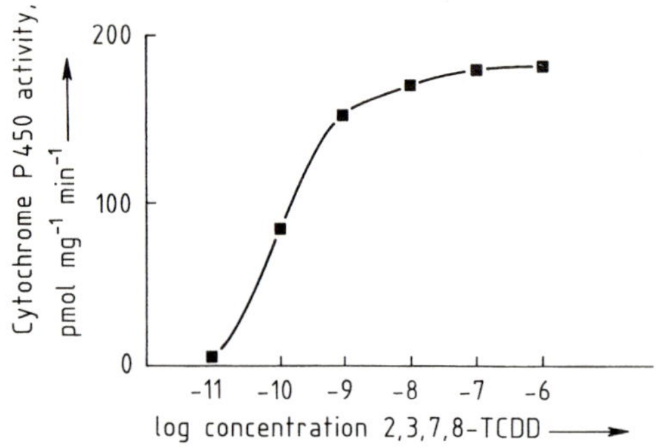

Figure 1.3. Dose-dependent induction of cytochrome P450 1A1 protein in cultured liver cells treated with 2,3,7,8-tetrachlorodibenzodioxin [3]

mammalian cells result in an increase in the concentration of a specific cytochrome P450 enzyme in the cells (for details of mechanisms, see Section 2.5.4.1). The increase is clearly dose related and spans a wide range (Fig. 1.3). An example of a graded toxic effect in an individual may be inflammation caused by skin contact with an irritant. Low doses cause slight irritation; as the amount increases, irritation turns to inflammation and the severity of inflammation increases.

In dose–response studies in a population, a specific end point is also identified, and the dose required to produce this end point is determined for each individual in the population. Both dose-dependent graded effects and quantal responses (death, induction of a tumor) may be investigated. With increasing amounts of a chemical given to a group of animals, the magnitude of the effect or the number of animals affected increases. For example, if an irritant chemical is applied to the skin, as the amount of the material increases the number of animals affected and the severity of inflammation increase.

Quantal responses such as death induced by a potentially lethal chemical will also be dose dependent. The dose dependence of a quantal effect in a population is based on individual differences in response to the toxic chemical. A specific amount of the potentially lethal xenobiotic given to a group of animals may not kill all of them, but as the amount given increases, the proportion of animals killed increases.

Although the distinctions between graded and quantal dose–response relationships are useful, the two types of responses are conceptually identical. The ordinate in both cases is simply labeled response, which may be the degree of response in an individual or the fraction of a population responding; the abscissa is the range of doses administered.

Graphics and Calculations. Even with a genetically homogeneous population of animals of the same species and strain, the proportion of animals showing the effect will increase with dose (Fig. 1.4 A). When the number of animals responding is plotted versus log of the dose, a typical sigmoid curve, with a log-normal distribution, which is symmetric about the midpoint, is obtained (Fig. 1.4 B).

When plotted on a log–linear scale, the normally distributed sigmoid curve obtained approaches a response of 0% as the dose is decreased and 100% as the dose is increased but theoretically never passes through zero or 100%. Small proportions of the population at the right- and left-hand sides of the curve represent hypo- and hypersusceptible members. The slope of the dose–response curve around the 50% value, the midpoint, gives an indication of the range of doses producing an effect. A steep dose–response curve indicates that the majority of the population will respond over a narrow dose range; a shallow dose–response curve indicates that a wide range of doses is required to affect the majority of the population. The curve depicted in Figure 1.4 B shows that the majority of individuals respond at about the midpoint of the curve. This point is a convenient description of the average response and is referred to as the median effective dose (ED_{50}). If mortality is the end point, then this dose is referred as median lethal dose (LD_{50}).

Death, a quantal response, is simple to quantify and is thus an end point incorporated in many acute toxicity studies. Lethal toxicity is usually calculated initially from specific mortality levels obtained after giving different doses of a chemical; the 50% mortality level is used most frequently since it represents the midpoint of the dose range at which the majority of deaths occur. This is the dose level that causes death of half of

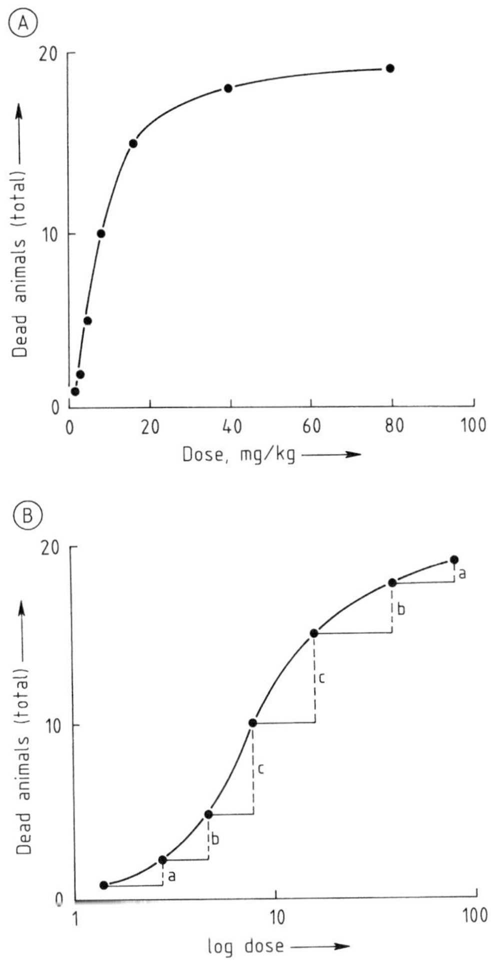

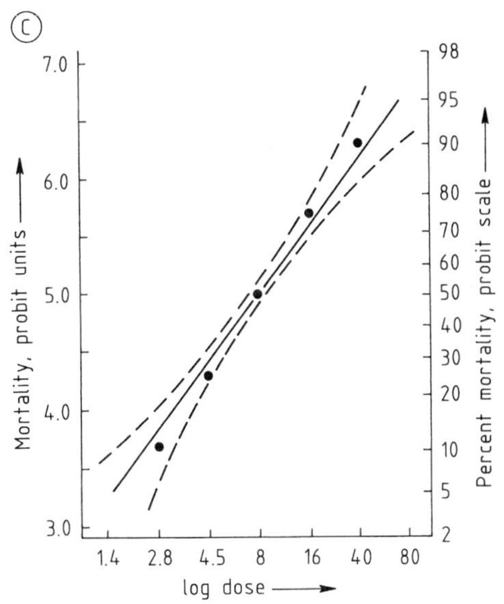

Figure 1.4. Typical dose response curves for a toxic effect
Plots are linear–linear (A); log–linear (B); and log–probit (C) for an identical set of data

the population dosed. The LD_{50} values are usually given in milligrams of chemical per kilogram of body weight (from the viewpoint of chemistry and for comparison of relative potencies of different chemicals, giving the LD_{50} in moles of chemical per kilogram body weight would be desirable). After inhalation, the reference is to LC_{50} (LC = lethal concentration), which, in contrast to LD_{50} values, depends on the time of exposure; thus, it is usually expressed as *X*-hour LC_{50} value. The LD_{50} or LC_{50} values usually represent the initial information on the toxicity of a chemical and must be regarded as a first, but not a quantitative, hazard indicator that may be useful for comparison of the acute toxicity of different chemicals [4].

Similar dose–effect curves can, however, be constructed for cancer, liver injury, and other types of toxic responses. To determine LD_{50} values and to obtain comparative information on dose–response curves, plotting log dose versus percent response is impractical since large numbers of animals are needed to obtain interpretable data. Moreover, other important information on the toxicity of a chemical (e.g., the LD_{05} and LD_{95}) cannot be determined accurately due to the slope of the sigmoid curve. Therefore, the dose–response curve is transformed to a log–probit (probit = probability units) plot. The data in Figure 1.4 C form a straight line when transformed into probit units. The EC_{50} or, if death is the end point, the LD_{50} is obtained by drawing a horizontal line from probit unit 5, which is the 50 % response point, to the dose–effect line. At the point of intersection, a vertical line is drawn, which intersects the abscissa at the LD_{50} point. Information on the lethal dose for 90 % or for 10 % of the population may also be derived by a similar procedure. The confidence limits are narrowest at the midpoint of the line (LD_{50}) and widest at both extremes (LD_{05} and LD_{95}) of the dose–response curve. In addition to permitting determination of a numerical value for the LD_{50} of a chemical with few groups of dosed animals, the slope of the dose–response curve for comparison between toxic effects of different chemicals is obtained by the probit transformation [5].

The LD_{50} by itself, however, is an insufficient index of lethal toxicity, particularly if comparisons between different chemicals are to be made. For this purpose, all available dose–response information including the slope of the dose–response line should be used. Figure 1.5 shows the dose–response curves for mortality of two chemicals. The LD_{50} for both chemicals is the same (10 mg/kg). However, the slopes of the dose–response curves are quite different. Chemical A exhibits a "flat" dose–response curve: a large change in dose is required before a significant change in response is observed. In contrast, chemical B exhibits a "steep" dose–response curve, and a relatively small change in dose causes a large change in response. The chemical with the steep slope may affect a much larger proportion of the population by incremental increases in dose than chemicals having a shallow slope; thus, acute overdosing may be a problem affecting the majority of a population for chemicals with steeper slopes. Chemicals with shallower slopes may represent a problem for the hyperreactive groups at the left-hand side of the dose–response curve. Effects may occur at significantly lower dose levels then for hyperreactive groups exposed to chemicals with a steep dose–response.

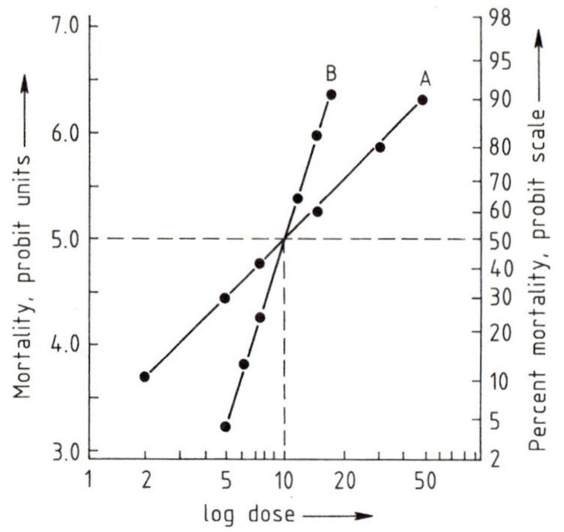

Figure 1.5. Comparison of dose–response relationships for two chemicals (log–probit plot)
Both chemicals have identical LD_{50} values, but different slopes of the dose–response curve

The LD_{50} values characterize the potential hazard of a chemical; the risk of an exposure is determined by the hazard multiplied by the exposure dose. Thus, even very toxic chemicals such as the poison of *Clostridium botulinum* pose only a low risk; intoxications with this compound are rare because exposure is low. Moreover, acute intoxications with other highly toxic agents such as mercury salts are rarely seen, despite detectable blood levels of mercury salts in the general population, since dose is also low. In contrast, compounds with low toxicity may pose a definite health risk when doses are high, for example, constituents of diet or chemicals formed during food preparation by heat treatment.

Therefore, to characterize the toxic risk of a chemical, besides information on the toxicity, information on the conditions of exposure is necessary. When using LD_{50} values for toxicity characterization, the limitations of LD_{50} values should be noted explicitly.

These limitations include

Methodological pitfalls influenced by
1) Strain of animal used
2) Species of animal used
3) Route of administration
4) Animal housing

Intrinsic factors limiting the use of LD_{50} values
1) Statistical method
2) No dose–response curve
3) Time to toxic effect not determined
4) No information on chronic toxicity

The most serious limitations on the use of LD_{50} values for hazard characterization are the lack of information on chronic effects of a chemical and the lack of dose–response information. Chemicals with low acute toxicity may have carcinogenic or teratogenic effects at doses that do not induce acute toxic responses. Other limitations include insufficient information on toxic effects other than lethality, on the cause of death, and on the time to toxic effect. Moreover, LD_{50} values are not constant, but are influenced by many factors and may differ by almost one order of magnitude when determined in different laboratories.

1.8. Dose–Response Relationships for Cumulative Effects

After chronic exposure to a chemical, a toxic response may be caused by doses that do not show any effects after single dosing. Chronic toxic responses are often based on accumulation of either the toxic effect or the chemical administered. Accumulation of the adminisitered chemical is observed when the rate of elimination of the chemical is lower than the rate of administration. Since the rate of elimination depends on plasma concentrations, after long-term application an equilibrium concentration of the chemical is reached in the blood. Chemicals may also be stored in fat (polychlorinated pesticides such as DDT) or bone (e.g., lead; see Section 2.4). Stored chemicals usually do not cause toxic effects because of their low concentrations at the site of toxic action (receptor). After continuous application, however, the capacities of the storage tissues may become saturated; xenobiotics may then be present in higher concentration in plasma and thus at the site of action, so toxic responses will result. Besides accumulation of the toxic agent, the toxic effect may also cumulate (Fig. 1.6).

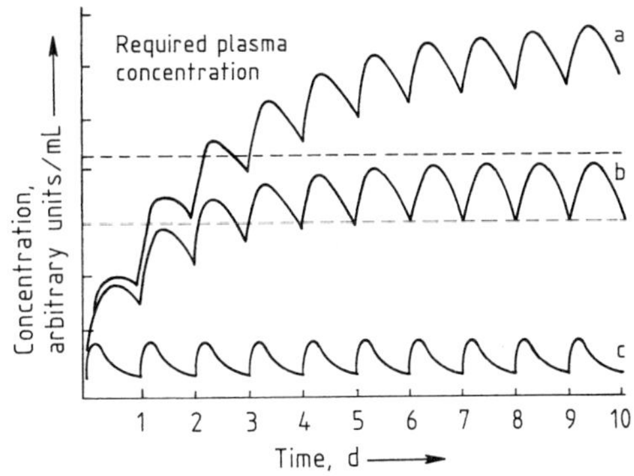

Figure 1.6. Accumulation of toxic chemicals based on their rate of excretion
a) The rate of excretion is equal to the rate of absorption, no accumulation occurs; b) Chemical accumulates due to a higher rate of uptake and inefficient excretion; the plasma concentrations are, however, not sufficient to exert toxic effects; c) The plasma concentrations reached after accumulation are sufficient to exert toxicity

For chemicals that bind irreversibly to macromolecules, the magnitude of toxic responses may be correlated with the total dose administered. In contrast to chemicals that act reversibly, the effect is not dependent on the frequency of dosing. Effect accumulation is often observed with carcinogens and ionizing radiation. In Figure 1.7, accumulation of effects is exemplified by the time and dose-dependent induction of tumors by 4-dimethylaminoazobenzene, a potent chemical carcinogen [6]. The TD_{50} values (50 % of the treated animals carry tumors) are used to characterize the potency. Identical tumor incidences were observed after high doses and a short exposure time or after low doses and long exposure; the tumor incidence was dependent only on the total dose administered.

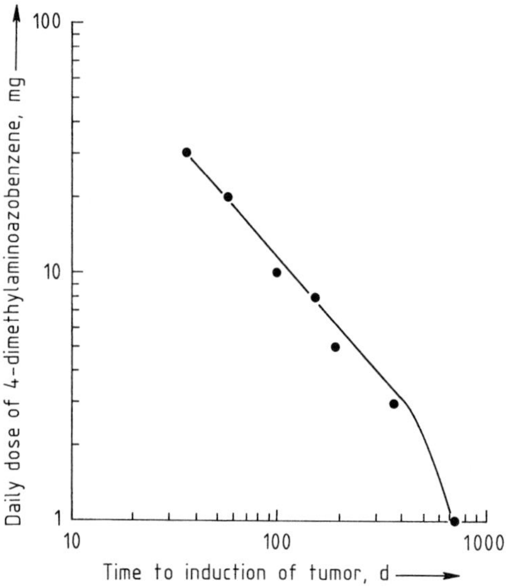

Figure 1.7. Time-dependent induction of tumors after different daily doses of 4-dimethylaminoazobenzene in rats [5]

The reversibility of toxic responses also depends on the capacity of an organ or tissue to repair injury. For example, after survival of the acute phase of intoxication, kidney damage by xenobiotics is often, without further consequence because of the high capacity of the kidney for cell proliferation and for repair of organ damage [7]. In contrast, injury to the central nervous system is largely irreversible since the differentiated cells of the nervous system cannot divide and dead cells cannot be replaced.

1.9. Factors Influencing Dose – Response

In animals and humans, the nature, severity, and incidence of toxic responses depend on a large number of exogenous and endogenous factors [8]. Important factors are the characteristics of exposure, the species and strain of animals used for the study, and interindividual variability in humans [9]. Toxic responses are caused by a series of complex interactions of a potentially toxic chemical with an organism. The type and magnitude of the toxic response are influenced by the concentration of the chemical at the receptor and by the type of interaction with the receptor. The concentration of a chemical at the site of action is influenced by the kinetics of uptake and elimination; since these are time-dependent phe-

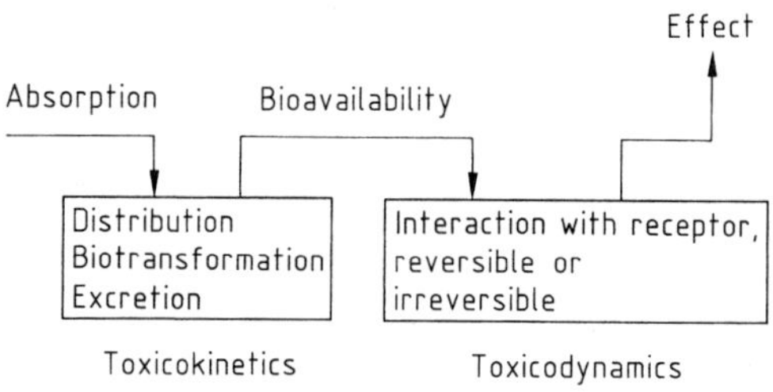

Figure 1.8. Toxicokinetics and toxicodynamics as factors influencing the toxic response

nomena, toxic responses are also time dependent. Thus, the toxic response can be separated into two phases, toxicokinetics and toxicodynamics (Fig. 1.8). *Toxicokinetics* describes the time dependence of uptake, distribution, biotransformation, and excretion of a toxic agent (a detailed description of toxicokinetics is given in Section 2.6). *Toxicodynamics* describes the interaction of the toxic agent with the receptor and thus specific interactions of the agent (see below and Section 2.6). Toxicokinetics may be influenced heavily by species, strain, sex, and exposure characteristics [10] – [12]. Differences in toxic response among species, route of exposure, and other factors often depend on the toxicokinetics. Since toxicodynamics (mechanism of action) is assumed to be identical between species, this provides the basis for a rational interspecies extrapolation of toxic effects when differences in toxicokinetics are defined.

1.9.1. Routes of Exposure

The primary tissue of system through which a xenobiotic comes in contact with the body, and from which it may be absorbed in order to exert systemic toxicity, is called the route of exposure. Frequent routes of environmental exposure are ingestion (peroral), inhalation, and skin contact. For investigational and therapeutic purposes, intramuscular, intravenous, and subcutaneous injections may also be routes of exposure.

The major routes by which a potentially toxic chemical can enter the body are — in descending order of effectiveness for systemic delivery — injection, inhalation, absorption from the intestinal tract, and cutaneous absorption. The relationship among route, exposure, biotransformation, and potential for toxicity may be complex and is also influenced by the magnitude and duration of dosing (Table 1.5).

Table 1.5. Toxicity of chemicals applied by different routes of exposure (data taken from [13])

Chemical	Species	Route of application	LD_{50}, mg/kg
DDT	rat	intravenous	68
	rat	oral	113
	rat	skin contact	1931
Atropine sulfate	rat	intravenous	41
	rat	oral	620
1-Chloro-2,4-dinitrobenzene	rat	oral	1070
	rat	intraperitoneal	280
	rabbit	skin contact	130
Dieldrin	rat	oral	46
	rat	intravenous	9
	rat	skin contact	10

The route of exposure has a major influence on toxicity because it affects the bioavailability of the toxic agent. The maximum tissue levels achieved, the time to reach maximum tissue levels, and thus the duration of the effect are determined by the rate of absorption and the extent of distribution within the system.

Direct injection into veins is usually restricted to therapeutic applications it is, however, important for the toxicology of intravenously used drugs in addicts. Chemicals applied by intravenous injection are distributed rapidly to well-perfused organs by blood and thus may result in the rapid induction of toxic effects. The rapid dilution of a chemical after intravenous injection by venous blood permits even the injection of locally acting or corrosive chemicals that are well tolerated. The likelihood of toxic effects from inhaled chemicals depends on a number of factors, especially the physical state and properties of the agent, concentration, and time and frequency of exposure. Major influences on the absorption and disposition of xenobiotics are exerted by species peculiarities since the anatomy of the respiratory tract and the physiology of respiration show major differences between rodents and humans. The water solubility of a gaseous xenobiotic strongly influences its penetration into the respiratory tract. As water solubility decreases and lipid solubility increases, penetration toward deeper regions of the lung, the bronchioli, and the alveoli is more effective. Water-soluble molecules, such as formaldehyde, are effectively scavenged by the upper respiratory tract and may have toxic effects on the eye and throat. In contrast, gases with low water solubility such as phosgene may penetrate through the bronchii and bronchioli to the alveoli. Damage to the alveolar surface may initiate a series of events finally resulting in lung edema. The degree to which inhaled gases, vapors, or particulates is absorbed and, hence, their potential to produce systemic toxicity depend on their diffusion rate through the alveolar membrane, their solubility in blood and tissue fluids, the rate of respiration, and blood flow through the capillaries.

Uptake through the alimentary tract represents an important route of exposure for xenobiotics accumulated in the food chain, for natural constituents of the diet, and for drugs. Absorption from the gastrointestinal tract depends on the lipophilicity of a chemical and the molecular mass of the xenobiotic. The presence of certain dietary constituents may influence the extent and rate of absorption. Chemicals absorbed from the gastrointestinal tracts are transported to the liver via the portal vein; hepatic metabolism ("hepatic first-pass effect") may efficiently reduce the concentration of the xenobiotic available in the systemic circulation after oral uptake. Compounds undergoing bioactivation in the liver usually exhibit greater toxicity when given orally than when absorbed across the respiratory tract, due to the high proportion of material passing through the liver. In contrast, chemicals that are toxic to extrahepatic, well-perfused organs such as the kidney often show a lower degree of toxicity when given orally.

Skin contact is an important route of exposure in the occupational and domestic environment. Local effects may include acute inflammation and corrosion, chronic inflammatory responses, immune-mediated reactions, and neoplasia. Percutaneous absorption may also be a significant route for the absorption of systemically toxic materials. Factors influencing the percutaneous absorption of substances include skin site, integrity of skin, temperature, formulation, and physicochemical characteristics, including charge, molecular mass, hydrophilicity, and lipophilicity.

1.9.2. Frequency of Exposure

The exposure of experimental animals may be categorized as acute, subacute, subchronic, or chronic. *Acute exposures* usually last less than 24 h; all aforementioned routes of exposure may be applied. With chemicals of low toxicity, repeated exposures may be used. Acute inhalation exposure is usually less than 24 h; frequently 4 to 8 h is selected as the time scale. Repeated exposure refers to application of the chemical for less than one month (*subacute*), one to three months (*subchronic*), and more than three months (*chronic*). Chronic exposures to detect specific toxic effects (carcinogenicity of a chemical) may span most of the lifetime of a rodent (up to two years). Repeated exposure may occur by any route; the least labor intensive route is oral, which involves mixing the chemical with the diet; only for specific chemicals or to simulate likely routes of exposure for humans, application in drinking water, by gastric intubation, or by inhalation is applied. These are more labor intensive and require skilled personnel or sophisticated techniques and thus are more costly.

The toxic effects observed after single exposure are often different from those seen after repeated exposure. For example, inhalation of high concentration of halothane causes anesthesia in animals and humans. In contrast, long-term application of halothane in lower doses causes liver damage in sensitive species. The frequency of exposure in chronic studies is important for the temporal characterization of exposure. Chemicals with a slow rate of excretion may accumulate if applied in short dosing intervals, and toxic effects may result (see Section 1.6). Also, a chemical that produces a severe effect when given in a single high dose may have no detectable effects when given in several smaller doses. Interspecies and strain differences in susceptibility to chemical-induced toxicity may be due to heterogeneity of populations, species-specific physiology (e.g., of the respiratory system), basal metabolic rate, size, and species-specific toxicokinetics and routes of metabolism or excretion (Table 1.6). In some cases, animal tests may give an underestimate, and in other cases an overestimate, of the potential toxicity to humans [14].

Table 1.6. Comparative LD_{50} values for four different chemicals in different animal species and estimated LD_{50} for humans

Chemical	Species	LD_{50}, mg/kg
Paraquat	rat	134
	mouse	77
	guinea pig	41
	human	32–48
Ethanol	rat	12 500
	mouse	8000
	guinea pig	5500
	human	3500–5000
Acetaminophen	rat	3763
	mouse	777
	guinea pig	2968
	human	42 800
Aspirin	rat	1683
	mouse	1769
	guinea pig	1102
	human	3492

1.9.3. Species Differences in Toxicokinetics

Species differences in toxic response are due to largely difference in toxicokinetics and biotransformation [15]. Distribution and elimination characteristics are quite variable among species. Both qualitative and quantitative differences in biotransformation may effect the sensitivity of a given species to a toxic response [16] (Table 1.7). For example, the elimination half-life of 2,3,7,8-tetrachlorodibenzodioxin in rats is 20 d and in humans up to 10 a [17]. An example of quantitative differences in the extent of biotransformation as a factor influencing toxic response may be species differences in the biotransformation of the inhalation anesthetic halothane. Rats, guinea pigs, and humans metabolize halothane to trifluoroacetic acid, a reaction catalyzed by a specific cytochrome P450 enzyme [18]–[20]. As a metabolic intermediate, trifluoroacetyl chloride is formed, which may react with lysine residues in proteins and with phosphatidylethanolamine in phospholipids (Fig. 1.9).

This interaction initiates a cascade of events that finally result in toxicity. The rate of metabolism of halothane in guinea pigs is much higher than in rats; thus, guinea pigs are sensitive

Table 1.7. Species and sex differences in the acute toxicity of 1,1-dichloroethylene after oral administration and inhalation in rats and mice (data from United Health Organization, Geneva, 1990)

Species	Dosing criteria	Estimated LD_{50}/LC_{50}
Rat, male	inhalation/4 h	7000–32 000 μg/L
Rat, female	inhalation/4 h	10 300 μg/L
Mouse, male	inhalation/4 h	115 μg/L
Mouse, female	inhalation/4 h	205 μg/L
Rat, male	gavage	1550 mg/kg
Rat, female	gavage	1500 mg/kg
Mouse, male	gavage	201–235 mg/kg
Mouse, female	gavage	171–221 mg/kg

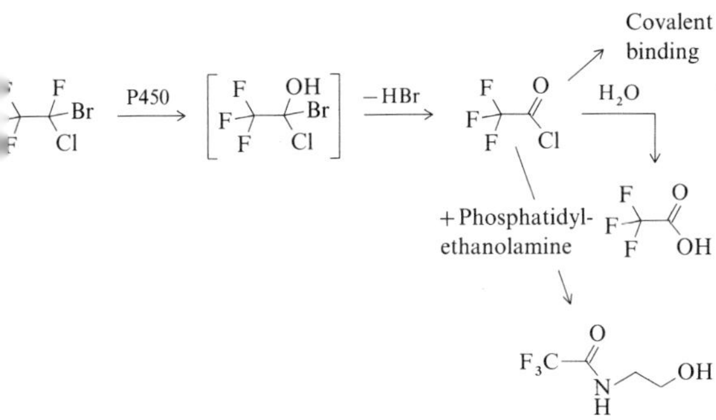

Figure 1.9. Halothane metabolism by cytochrome P 450 in rats, guinea pigs, and humans

to halothane-induced hepatotoxic effects and rats are resistant.

Qualitative differences in biotransformation are responsible for apparent differences in the sensitivity of rats and guinea pigs to the bladder carcinogenicity of 2-acetylamidofluorene. In rats, 2-acetylamidofluorene is metabolized by *N*-oxidation by certain cytochrome P 450 enzymes. The *N*-oxide is further converted to an electrophilic nitrenium ion, which interacts with DNA in the bladder; this biotransformation pathway explains the formation of bladder tumors in rats after long-term exposure to 2-acetylamidofluorene. In guinea pigs, 2-acetylamidofluorene is metabolized by oxidation at the aromatic ring, so nitrenium ions cannot be formed by this pathway; guinea pigs are thus resistant to the bladder carcinogenicity of 2-acetylamidofluorene (Fig. 1.10).

With some chemicals, the subject's age may significantly affect toxicity, likely because of age-related differences in toxicokinetics. Nutritional status may modify toxic response, probably by altering the concentration of cofactors needed for the biotransformation and detoxification of chemicals. Diet also markedly influences carcinogen-induced tumor incidence in animals [21] and may be a significant factor contributing to human cancer incidence [22].

The toxic response is influenced by the magnitude, number, and frequency of dosing. Thus, local or systemic toxicity produced by acute exposure may also occur by a cumulative process with repeated lower-dose exposures; also, additional toxicity may be seen in repeated-exposure situations. The relationships for cumulative toxicity by repetitive exposure compared to acute exposure may be complex, and the potential for

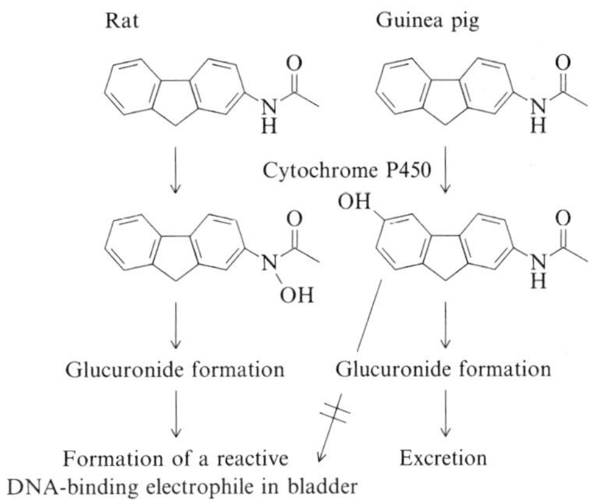

Figure 1.10. Biotransformation pathways of 2-acetylamidofluorene in rats and guinea pigs

cumulative toxicity from acute doses may not be quantitatively predictable. For repeated-exposure toxicity, the precise profiling of doses may significantly influence toxicity.

1.9.4. Miscellaneous Factors Influencing the Magnitude of Toxic Responses

A variety of other factors that depend on the conditions of the study may affect the nature and exhibition of toxicity. Examples are housing conditions, handling, volume of dosing, vehicle, etc. Variation in test conditions and procedures may result in significant interlaboratory variability in the results of otherwise standard procedures. For chemicals given orally or applied to the skin, toxicity may be modified by the presence of materials in formulations that facilitate or retard absorption of the chemicals. With respiratory exposure to aerosols, particle size significantly determines the depth of penetration and deposition in the respiratory tract and thus the site and extent of the toxic effects.

1.10. Exposure to Mixtures

In experimental animals, most data on the toxic effects of chemicals are collected after exposure to a single chemical; in contrast, humans are normally exposed to mixtures of chemicals at low doses. Moreover, prior, coincidental, and successive exposure of humans to chemicals is likely. Interactions between the toxic effects of different chemicals are difficult to predict; the effects of exposure to different chemicals may be independent, additive, potentiating (ethanol and carbon tetrachloride), antagonistic (interference with the action of another chemical, e.g., as seen with antidotes administered in case of intoxications), or synergistic. Ethanol exerts a potentiating effect on the hepatotoxicity of carbon tetrachloride. In rats pretreated with ethanol, the hepatotoxic effects of carbon tetrachloride are much more pronounced compared to control animals. This potentiation results from an increased capacity for bioactivation (see Section 2.5.4.) of carbon tetrachloride in pretreated rats due to increased concentrations of a cytochrome P450 enzyme in the liver [23]. Thus, important considerations for the assessement of potential toxic effects of mixtures of chemicals are toxicokinetic and toxicodynamic interactions. Toxicokinetic interactions of chemicals may influence absorption, distribution, and biotransformation, to both active and inactive metabolites. Mixtures of solvents often show a competitive inhibition of biotransformation. Usually, one of the components has high affinity for a specific enzyme involved in its biotransformation, whereas another component has only a low affinity for that particular enzyme. Thus, preferential biotransformation of the component with high affinity occurs. Different outcomes of enzyme inhibition are possible: if the toxic effects of the component whose metabolism is inhibited are dependent on bioactivation, lower rates of bioactivation will result in decreased toxicity; if the toxic effects are independent of biotransformation, the extent of toxicity will increase due to a slower rate of excretion. Toxic effects of mixtures may also result not from a major component but from trace impurities with high toxicity. For example, many long-term effects observed in animal studies on the toxicity of chlorophenols are believed to be due to 2,3,7,8-tetrachlorodibenzodioxin, which was present as a minor impurity in the samples of chlorophenols used for these studies.

2. Absorption, Distribution, Biotransformation, and Elimination of Xenobiotics

2.1. Disposition of Xenobiotics

The induction of systemic toxicity usually results from a complex interaction between absorbed parent chemical and biotransformation products formed in tissues; the distribution of both parent chemical and biotransformation products in body fluids and tissues; their binding and storage characteristics; and their excretion.

The biological effects initiated by a xenobiotic are not related simply to the inherent toxic properties of the xenobiotic; the initiation, intensity, and duration of response are a function of many factors intrinsic to both the biological system and the administered dose. Each factor influences the ultimate interaction of the xenobiotic with the active site (see Section 1.9). Only when the toxic chemical has reached the specific site and interacted with it can the inherent toxicity be realized. Toxicokinetics influences the concentration of the xenobiotic or its active metabolite at the receptor. In the dose—response concept outlined in Section 1.7, the toxic response is generally assumed to be proportional to the concentration of xenobiotic at the receptor. However, the same dose of a chemical administered by different routes may cause different toxic effects. Moreover, the same dose of two different chemicals may result in vastly different concentrations of the chemical or its biotransformation products in a particular target organ. This differential pattern is due to differences in the disposition of a xenobiotic (Fig. 2.1).

The disposition of a xenobiotic consists of absorption, distribution, biotransformation, and excretion, which are all interrelated. Volatile xenobiotics present in the inspired air can be taken up efficiently through the lung and distributed rapidly by blood. Chemicals present in the diet are absorbed from the intestinal tract and pass through the liver before reaching the systemic circulation for distribution. Xenobiotics in contact with skin may be absorbed slowly and delivered directly to the systemic circulation after absorption. The complicated interactions among the different processes of distribution strongly influence the concentration of a chemical at the receptor and thus the magnitude of toxic response. They may also be major determinants of organ-specific toxicity.

For example, in the absorption of a xenobiotic through the gastrointestinal tract, the chemical proceeds from the intestinal lumen to the epithelial cells. After intracellular transport, it passes through the basal membrane and lamina propria and enters the blood or lymph capillaries for transport to the site of action or storage. At that site, the xenobiotic is released from the capillaries into an interstitial area and, finally,

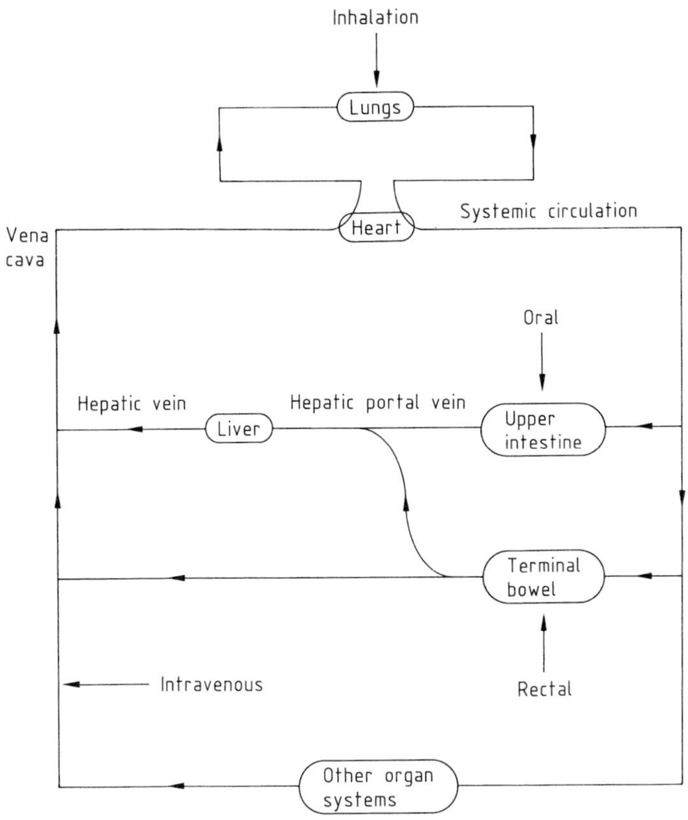

Figure 2.1. Possible fate of a xenobiotic in the organisms

through various membranes to its site of action, which may be a specific receptor, an enzyme, a membrane, or many other possible sites.

2.2. Body Membranes and Absorption

The skin, the lungs, and the cells lining the alimentary tract are major barriers to chemicals present in the environment. Except for caustic chemicals, which act at the site of first contact with the organism, xenobiotics must cross these barriers to exert toxic effects on one or several target organs. The process whereby a xenobiotic moves through these barriers and enters the circulation is termed absorption.

2.2.1. Membranes

Membranes are initially encountered whether a xenobiotic is absorbed through the dermal, oral, or inhalation route. These membranes may be associated with several layers of cells or a single cell. The absorption of a substance from the site of exposure may result from passive diffusion, facilitated diffusion, active transport, or the formation of transport vesicles (pinocytosis and phagocytosis). The process of absorption may be facilitated or retarded by a variety of factors; elevated temperature, for example, increases percutaneous absorption by cutanous vasodilation, and surface-active materials facilitate penetration. Each area of entry for xenobiotics into the organism may have specific peculiarities, but a unifying concept of biology is the basic similarity of all membranes in tissues, cells, and organelles.

All membranes are lipid bilayers with polar head groups (phosphatidylethanolamine, phosphatidylcholine). Polar groups predominate at the outer and inner surfaces of the membrane; the inner space of the membrane consists of perpendicularly arranged fatty acids [24]. The fatty acids do not have a rigid structure and are fluid under physiological conditions; the fluid character of the membrane is largely dominated by the fatty acid composition. The width of a biological membrane is approximately 7–9 nm. Figure 2.2 illustrates the concept of biological membrane

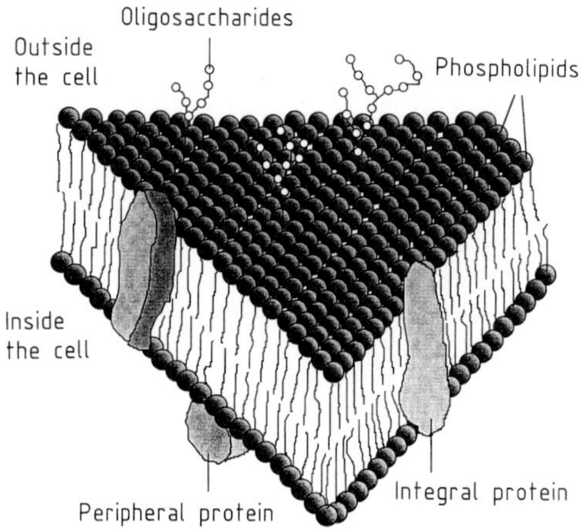

Figure 2.2. Simplified model of the structure of a biological membrane

(fluid-mosaic model). Proteins are intimately associated with the membrane and may be located on the surface of or inside the membrane structure, or may extend completely through the membrane. These proteins may also form aqueous pores. Hydrophobic forces are responsible for maintaining the structural integrity of both proteins and lipids within the membrane structure. The ratio of lipid to protein in different membranes may vary from 5:1 (e.g., myelin) to 1:5 (e.g., the inner membrane of mitochondria). Usually, pore diameters in membranes are small and permit only the passage of low molecular mass chemicals. However, some specialized membranes such as those found in the glomeruli of the kidney, where the pore size may be up to 4 nm, also permit the passage of compounds with molecular mass > 10 000.

The amphipathic nature of the membrane creates a barrier for ionized, highly polar compounds; however, changes in lipid composition, alterations in the shape and size of proteins, and physical features of bonding may change the permeability of membranes [25].

2.2.2. Penetration of Membranes by Chemicals

A chemical may pass through a membrane by two general processes: passive diffusion and active transport. Passive diffusion is defined by Fick's law and requires no energy (see below). Special transport processes involve the consumption of cellular energy to translocate the chemical across the membrane. Special transport may also act against a concentration gradient and result in the accumulation of a xenobiotic in a specific organ, cell type, or organelle.

Diffusion of Chemicals through Membranes. Many toxic chemicals pass through membranes by simple diffusion. Their rate of diffusion depends on their lipid solubility and is often correlated with their partition coefficient (ratio of solubility in organic solvents:solubility in water). Lipophilic chemicals may diffuse directly through the lipid domain of the membrane. However, a certain degree of water solubility seems to be required for the chemical to pass because many sparingly lipid-soluble chemicals penetrate easily. Once initial penetration has occurred, the molecule must necessarily traverse a more polar region to dissociate from the membrane. Compounds with extremely high partition coefficients thus tend to remain in membranes and to accumulate there rather than pass through them. Polar chemicals insoluble in the inner space of the membrane containing nonpolar fatty acids often may not penetrate membranes, although some low molecular mass polar chemicals may slowly penetrate the aqueous pores of the membranes.

The rates of movement of nonpolar xenobiotics through membranes can be predicted from Fick's law of diffusion. Polar compounds and electrolytes of small molecular mass are assumed to behave similarly. A first-order equation appears to be applicable to the majority of xenobiotics. The rate of diffusion of a xenobiotic is related to its concentration gradient across the membrane ($C_1 - C_2$); the surface area available for transfer A; the diameter of the membrane d; and the diffusion constant k. The latter is related to the size and structure of the molecule, the spatial configuration of the molecule, and the degree of ionization and lipid solubility of the xenobiotic.

$$\text{Rate of diffusion} = k\frac{A(C_1 - C_2)}{d}$$

Since the xenobiotic is removed rapidly from the membrane after absorption, C_2 can usually be ignored, and a log-linear plot of the amount of chemical that did not penetrate the membrane should be linear over time. When relatively comparable methods have been used, the half-live of penetration $t_{1/2}$ can be calculated. The rate constant of penetration k is:

$$k = \frac{0.693}{t_{1/2}}$$

When the half-lives of penetration after oral and dermal administration of several environmental contaminants were compared, rates were found to vary considerably. Rates of penetration through different routes in mammals show little or no correlation.

Ionization becomes particularly important when xenobiotics are introduced into the gastrointestinal tract, where a variety of pH conditions exist (see Section 2.2.4.2). Although many drugs are acids and bases and thus in a potentially ionizable form, most xenobiotics are neither acid nor base and thus are unaffected by pH. The amount of xenobiotic present in the ionized or nonionized form depends on the pK_a of the xenobiotic and the pH of the medium. When the pH of a soluton is equal to the pK_a of the dissolved compound, 50% of the acid or base exists in the ionized and 50% in the nonionized form. The degree of ionization at a specific pH is given by the Henderson–Hasselbalch equation

$$pK_a - pH = \log \frac{[\text{nonionized}]}{[\text{ionized}]}$$

$$pK_a - pH = \log \frac{[\text{ionized}]}{[\text{nonionized}]}$$

Since the nonionized, lipid-soluble form of a weak acid or base may penetrate membranes, weak organic acids diffuse most readily in an acidic environment and organic bases in a basic environment. Penetration occurs to a certain degree even when xenobiotics are not in the most lipid-soluble form, and a small amount of absorption can produce serious effects if a compound is very toxic.

2.2.3. Mechanisms of Transport of Xenobiotics through Membranes

Filtration. Passage of a solution across a porous membrane results in the retention of solutes larger than the pores (filtration). For example, filtration of solutes occurs in the kidney glomeruli, which have large pores and retain molecules with molecular mass > 10 000. Elsewhere in the body, filtration via pores may result in the passage of only relatively small molecules (molecular mass ca. 100); most larger molecules will be excluded. Thus, the uptake of xenobiotics via these pores is only a minor mechanism of penetration.

Special Transport Mechanisms. Special transport processes include active transport, facilitated transport, and endocytosis (Table 2.1). Often, the movement of chemicals across membranes is not due to simple diffusion or filtration. Even some very large or very polar molecules may readily pass through membranes.

Table 2.1. Special transport processes involved in the passage of xenobiotics through biological membranes

Type of transport	Carrier molecule required	Examples of substrates	Energy required	Against concentration gradient
Active transport	yes	organic acids in the kidney	yes	yes
Facilitated transport	yes	glucose	yes	no
Endocytosis	no	proteins	yes	?

Active transport systems have frequently been implicated in these phenomena. *Active transport* may be effected by systems that help to transport endogenous compounds across membranes. Such processes require energy and transport xenobiotics against electrochemical or concentration gradients. Active transport systems are saturable processes and exhibit a maximum rate of transport; they are usually specific for certain structural features of chemicals. A carrier molecule (likely a protein) associates with the chemical outside the cell and translocates it across the membrane for ultimate release inside the cell. This is particularly important for compounds that lack sufficient lipid solubility to move rapidly through the membrane by simple diffusion. Active transport plays a major role in the excretion of xenobiotics from the body, and major excretory organs such as the liver or the kidney have several transport systems that may accept organic acids, organic bases, or even metal ions as substrates.

In contrast to other special transport processes, some carrier-mediated processes do not require energy and are unable to move chemicals against a concentration gradient. These processes are termed *facilitated transport*. Facilitated transport is particulary benefical for compounds that lack sufficient lipid solubility for rapid diffusion through the membrane. Facilitated transport is more rapid than simple diffusion up to the point at which concentrations on both sides of the membrane are equal. For example, glucose is transported through a variety of membranes by facilitated transport. These transport mechanisms are not well understood.

Pinocytosis (liquids) and *phagocytosis* (solids) are specialized processes in which the cell membrane invaginates or flows around a xenobiotic, usually present in particulate form, enabling transfer across a membrane. Although of importance once the xenobiotic has gained entry into the organism, this mechanism apparently does not play a role in the initial absorption of a xenobiotic.

2.2.4. Absorption

Absorption is the process whereby xenobiotics cross body membranes and are translocated to the bloodstream. The primary sites of absorption of environmental contaminants are the gastrointestinal tract (gastointestinal absorption), the skin (dermal absorption), and the lung (respiratory absorption). Absorption of chemicals may also occur from other sites such as muscle, the subcutis, or the peritoneum after administration by special routes. In clinical medicine, many drugs are applied directly into the bloodstream to circumvent problems of absorption posed by the peculiarities of different routes.

2.2.4.1. Dermal Absorption

Human skin may come in contact with many potentially toxic chemicals. Skin is relatively impermeable to aqueous solutions and most xenobiotics present as ions. Therefore, it is a relatively good barrier separating the human body from its environment. However, skin is permeable in varying degrees to many xenobiotics, and some chemicals may be absorbed through the skin in sufficient amount to cause a toxic response [26]. For example, a large number of agricultural workers have experienced acute poisoning from exposure to parathion (dermal $LD_{50} \approx 20$ mg/kg) during application or from exposure to vegetation previously treated with this pesticide.

The human skin is a complex, multilayered tissue comprising ca. 18 000 cm^2 of surface in an average human male. Chemicals to be absorbed must pass through several cell layers before entering the small blood and lymph capillaries in the dermis. Transport in blood and lymph will then distribute absorbed chemicals in the body. Human skin consists of three distinct layers (Fig. 2.3) and a number of associated ap-

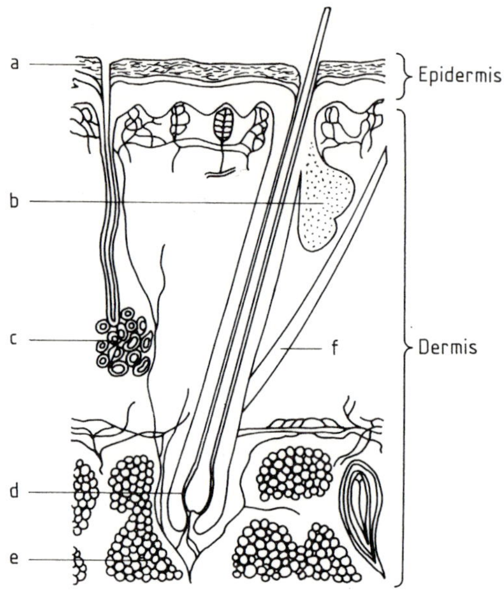

Figure 2.3. Cross section of human skin
a) Stratum corneum; b) Sebaceous gland; c) Sweat gland; d) Hair follicle; e) Fat; f) Muscle

pendages (sweat and sebaceous glands, hair follicles).

The *epidermis* is a multilayered tissue varying in thickness from 0.15 (eyelids) to 0.8 (palms) mm. This tissue is the greatest deterrent to the absorption of xenobiotics. The epithelial tissues of the skin develop and grow divergently from other tissues. Proliferative layers of the basal cells (stratum germinativum) differentiate and gradually replace cells above them as surface cells deteriorate and are sloughed from the epidermis. The cells in this layer produce fibrous, insoluble keratin that fills the cells and a sulfur-rich amorphous protein that comprises the cell matrix and a thickened cell membrane. This cell layer, the stratum corneum (Fig. 2.3 a), provides the primary barrier to the penetration of foreign compounds. It consists of several layers of flattened, stratified, highly keratinized cells. These cells are approximately 25–40 µm wide and have lost their nuclei. Although highly water retarding, the dead, keratinized cells of the stratum corneum are highly water absorbent (hydrophilic), a property that keeps the skin supple and soft. A natural oil covering the skin, the sebum, appears to maintain the water-holding capacity of the epidermis but has no appreciable role in retarding the penetration of xenobiotics. The rate-determining barrier in the chemical absorption of xenobiotics is the stratum corneum.

The *dermis and subcutaneous tissue* offer little resistance to penetration, and once a substance has penetrated the epidermis these tissues are rapidly traversed. The dermis is a highly vascular area, providing ready access to blood and lymph for distribution once the epithelial barrier has been passed. The extent of absorption of a chemical through the skin may be influenced by temperature and relative humidity [27]. The blood supply in the dermis is under complex, interacting neural and humoral influences, whose temperature-regulating function can have an effect on distribution by altering blood supply to this area.

The appendages of the skin are found in the dermis and extend through the epidermis. The primary appendages are the sweat glands (Fig. 2.3: eccrine and apocrine) (c), hair (d) and sebaceous glands (b). These structures extend to the outer surface; they therefore may play a role in the penetration of xenobiotics; however, because they represent only 0.1–1% of the total surface of the skin, their contribution to overall dermal absorption is usually minor.

Percutaneous absorption could occur via several routes, but the majority of nonionized, lipid-soluble xenobiotics appear to move by passive diffusion directly through the cells of the stratum corneum. Arguments for the importance of transepidermal absorption are that epidermal damage or removal of the stratum corneum increases permeability, that epidermal penetration rate equals whole-skin penetration, that epidermal penetration is markedly slower than dermal, and that the epidermal surface area is 100–1000 times the surface area of the skin appendages. Penetration of very small or polar molecules is apparently more favorable through the appendages or other diffusion shunts, but chemicals of this type are representative of only a small fraction of toxic xenobiotics. Polar substances, in addition to movement through shunts, may diffuse through the outer surface of the protein filaments of the hydrated stratum corneum, while nonpolar molecules dissolve in and diffuse through the noraqueous lipid matrix between protein filaments.

Human stratum corneum displays significant differences in structure from one region of the body to the other, which affect the rate of absorption. Penetration at certain body regions thus varies according to the polarity and size of the molecule, but for most nonionized xenobiotics the rate of penetration is generally in the following order:

scrotal > forehead > axilla = scalp > back
= abdomen > palmar and plantar.

The palmar and plantar regions are highly diffusible, but their much greater thickness (100–400 times that of other regions) results in an overall lag time in diffusion.

The condition of the skin greatly influences the absorption of xenobiotics. Damage to or removal of the stratum corneum causes a dramatic increase in the permeability of the epidermis to xenobiotics. Caustic and corrosive chemicals such as acids or alkali and burns will greatly enhance dermal absorption and thus influence the toxicity of a xenobiotic applied to the skin. Soaps and detergents are among the damaging substances routinely applied to skin. Whereas organic solvents must be applied in high concentrations to damage skin, 1% aqueous solutions of detergents increase the rate of penetration of solutes through human epidermis dramatically. Alteration of the stratum corneum by organic solvents may also be the cause of increased penetration.

Organic solvents can be divided into damaging and nondamaging categories. Damaging solvents include methanol, acetone, ether, hexane, and some solvent mixtures. These solvents and mixtures are able to extract lipids and proteolipids from tissues and are thus expected to alter permeability. Although the mechanical strength of the stratum corneum is unaltered, delipidization produces a more porous, nonselective surface. Solvents such as higher alcohols, esters, and olive oil, do not appear to damage skin appreciably. On the contrary, the penetration rate of solutes dissolved in them is often reduced. Surprisingly, lipid-soluble xenobiotics may be markedly resistant to washing even a short time after application. For example, 15 min after application, a substantial portion of parathion cannot be removed from contaminated skin by soap and water.

When comparisons across species are made, human skin appears to be more impermeable, or at least as impermeable, as the skin of the cat, dog, rat, mouse, or guinea pig. The skin of pigs and guinea pigs, in particular, serve as a useful approximation to human skin, but only after a comparison has been made for each specific chemical.

Temperature, surface area of applied dose, simultaneous application of another xenobiotic, relative humidity, occlusion, age, and hyperthermia are among a number of chemical, physical, and physiological factors that may alter skin penetration.

2.2.4.2. Gastrointestinal Absorption

The oral route of entry into the body is especially important for accidental or intentional (suicide) ingestion of poisonous materials. Food additives, food toxins, environmental xenobiotics accumulated in the food chain, and airborne particles excluded from passage to the alveoli are also introduced into the digestive system. The penetration of orally administered xenobiotics is confined primarily to the stomach and intestine [28].

The gastrointestinal tract may be viewed as a tube traversing the body. It consists of the mouth, esophagus, stomach, small and large intestines, colon and rectum (Fig. 2.4). The digestive tract is lined by a single layer of columnar cells, usually protected by mucus, that do not present a barrier to penetration. The circulatory system is closely associated with the intestinal tract (30–50 μm from membrane to vasculature), and once xenobiotics have crossed the epithelium of the intestinal tract, entry into capillaries is rapid. Venous blood flow from the stomach and small intestine rapidly removes absorbed xenobiotics and introduces them to the hepatic portal vein, resulting in transport to the liver. Absorption of chemicals may occur along the entire gastrointestinal tract; however, most xenobiotics are absorbed in the stomach and the small intestine. A major factor favoring absorption in the intestine is the presence of microvilli that increase the surface area to an estimated 100 m^2 in the small intestine (see Fig. 2.4 B). Thus, the absorption of xenobiotics is generally considered to be greatest in this area of the gastrointestinal tract. Although the gastrointestinal tract has some special transport processes for the

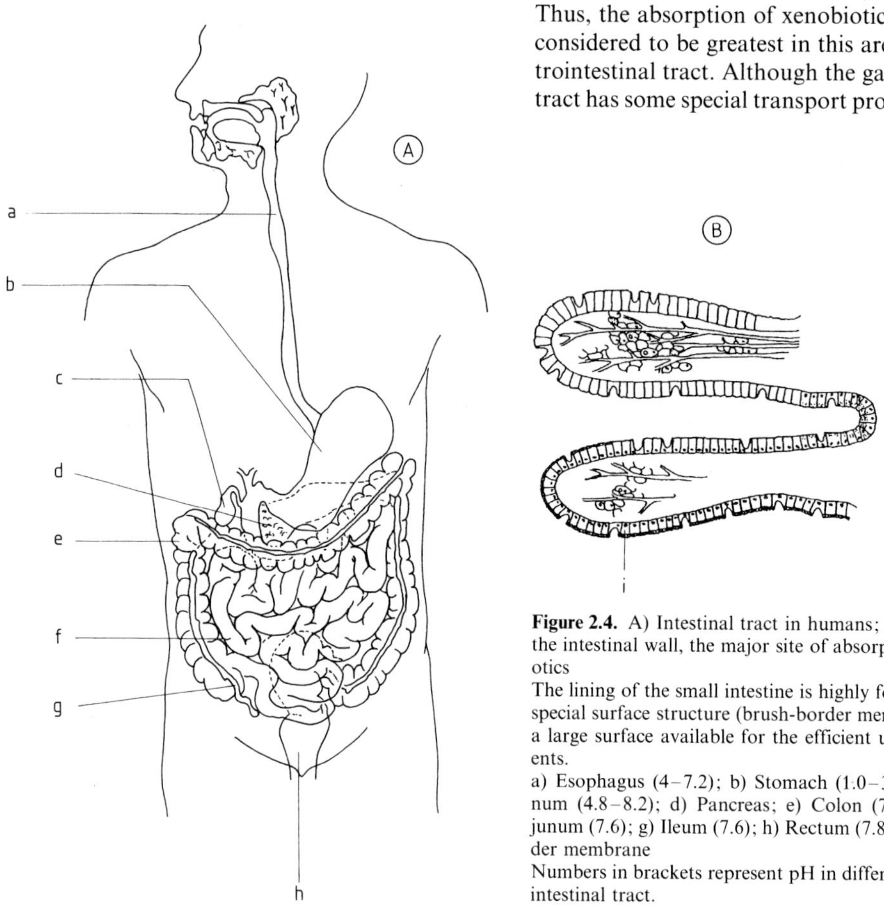

Figure 2.4. A) Intestinal tract in humans; B) Anatomy of the intestinal wall, the major site of absorption of xenobiotics
The lining of the small intestine is highly folded and has a special surface structure (brush-border membrane) to give a large surface available for the efficient uptake of nutrients.
a) Esophagus (4–7.2); b) Stomach (1.0–3.0); c) Duodenum (4.8–8.2); d) Pancreas; e) Colon (7.9–8.0); f) Jejunum (7.6); g) Ileum (7.6); h) Rectum (7.8); i) Brush-border membrane
Numbers in brackets represent pH in different parts of the intestinal tract.

absorption of nutrients and electrolytes, most xenobiotics seem to enter the body from the gastrointestinal tract by simple diffusion. Exceptions are some heavy metals such as thallium and lead, which mimic the essential metals iron and calcium, respectively. They are thus absorbed by active transport systems developed for the uptake of these nutrients.

The gastrointestinal tract has areas of highly variable pH, which can markedly change the permeability characteristics of ionic compounds. For example, passive diffusion is greatly limited except for nonionized, lipid-soluble chemicals. Although variable according to secretory activity, the pH of the stomach is ca. 1–3 and that of the intestine approximately 7. The measured pH of intestinal contents may not be the same as the pH of the epithelium at the site of absorption, which explains the entrance of compounds whose pK_a would suggest less favored absorption. The variations in pH in different sections of the intestinal tract may influence the absorption of acids and bases. Since most xenobiotics are absorbed by diffusion, only the nonionized, membrane-permeable form may be absorbed. Weak organic acids are present mainly in the nonionized lipid-soluble form in the stomach and in the ionized form in the intestine. Therefore, organic acids are expected to be absorbed more readily from the stomach than from the intestine. In contrast, weak organic bases are ionized in the stomach but present in the lipid-soluble form (nonionized) in the intestine. Absorption of such compounds should therefore occur predominantly in the intestine rather than the stomach.

However, other factors determining the rate of membrane penetration, such as surface area available for diffusion, blood flow (influencing concentration gradients), and law of mass action, also influence the site of absorption of acids or bases from the gastrointestinal tract. For example, although benzoic acid is present to only 1 % in the lipid-soluble, nonionized form in the small intestine, the large surface area of the small intestine and the rapid removal of absorbed benzoic acid with blood result in its efficient absorption.

Many other factors contribute to gastrointestinal absorption. A xenobiotic must, of course, be dissolved before absorption can take place. Factors such as particle size, presence of organic solvent or emulsifiers, and rate of dissolution thus also affect absorption. In addition, the presence of microorganisms and hydrolysis-promoting pH offer opportunities for the biotransformation of many xenobiotics. Other factors affecting gastrointestinal absorption include binding to gut contents, intestinal motility, rate of emptying, temperature of food, effects of dietary constituents, health status of the individual, and gastrointestinal secretion.

2.2.4.3. Absorption of Xenobiotics by the Respiratory System

The respiratory system is an organ in direct contact with evironmental air as an unavoidable part of living. A number of xenobiotics are in gaseous (carbon monoxide, nitrogen oxides), vapor (benzene, carbon tetrachloride), and aerosol (lead from automobile exhaust, silica, asbestos) forms and are potential candidates for entry via the respiratory system. Indeed, the most important cause of death from acute intoxication, carbon monoxide, and the most frequent occupational disease, silicosis, result from the absorption or deposition of airborne xenobiotics in the lung.

The respiratory tract may be considered as having three major regions: the nasopharyngeal, the tracheobronchial, and the pulmonary (Fig. 2.5). The nasopharynx begins in the mouth and extents down to the level of the larynx. The trachea (a), bronchii (b), and bronchioli (c) serve as conducting airways between the nasopharynx and the alveoli (d), the site of gas exchange between inhaled air and blood. The human respiratory system is a complex organ containing more than 40 different cell types. These cell types contribute to the pulmonary architecture and function over various zones of the lung, although to some extent, individual cell types can be found in several zones. The tracheobronchial system consists of airways lined with bronchial epithelium with associated submucosal glands and several different tissues with specific function and the lung vasculature.

The absorption of xenobiotics via the respiratory route is favored because of a short path of diffusion, large surface area (50–100 m^2), and large concentration gradients. At the alveoli (site of gas exchange), the membranes are very thin (1–2 μm) and are intimately associated with the vascular system. This enables a rapid exchange of gases, ca. 5 ms in the case of carbon dioxide and 200 ms for oxygen. In the resting adult,

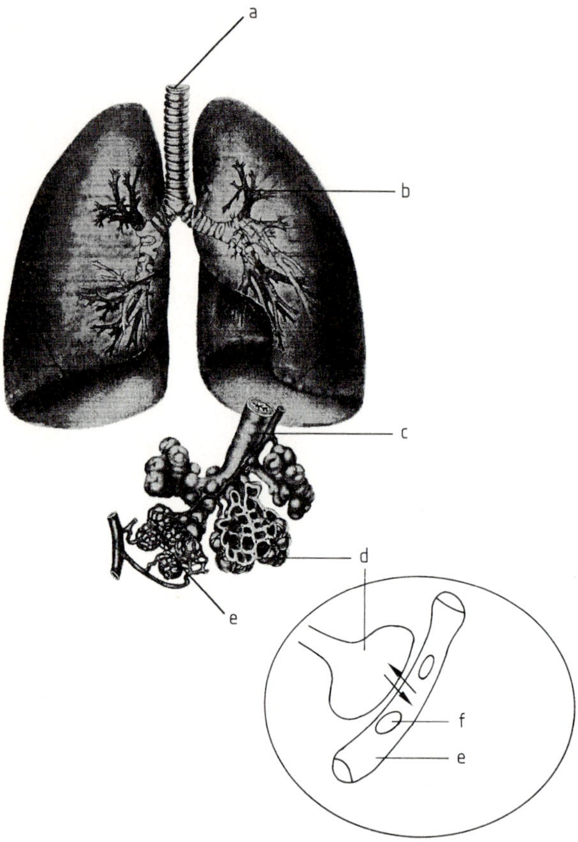

Figure 2.5. Anatomy of the human respiratory system
a) Trachea; b) Bronchii; c) Bronchioli; d) Alveoli; e) Capillary; f) Erythrocyte

280 mL/min of oxygen are taken up into the lungs and 230 mL/min of carbon dioxide are exhaled. A thin film of fluid lining the alveolar walls aids in the initial absorption of xenobiotics from the alveolar air. Simple diffusion accounts for the somewhat complex series of events in the lung regarding gas absorption. The sequences of respiration involving several interrelated air volumes define both the capacity of the lung and the factors important to particle deposition and retention. Among the elements important in total lung capacity is the residual volume, i.e., the amount of air retained by the lung despite maximal expiratory effort. Largely because of their slow release from this volume, gaseous xenobiotics in the expired air are not cleared immediately, and many expirations may be necessary to rid the lung air of residual xenobiotic. The rate of entry of vapor-phase xenobiotics is controlled by the alveolar ventilation rate, and a xenobiotic present in alveolar air may come in contact with the alveoli in an interrupted fashion about 20 times per minute. The diffusion coefficient of the gas in the fluid of pulmonary membranes is another important consideration, but doses are more appropriately discussed in terms of the partial pressure of the xenobiotic in the inspired air. Upon inhalation of a constant tension of a gaseous xenobiotic, the arterial plasma tension of the gas approaches the tension of gas in the expired air. The rate of entry is then determined by blood solubility of the xenobiotic and blood flow. In case of a high blood–gas partition coefficient, a larger amount must be dissolved in the blood to raise the partial pressure. Chemicals in inspired air with a high blood–gas partition coefficient require a longer period to approach the same tension in the blood than less soluble gases.

Aerosols and Particulates. The entry of aerosols and particulates is affected by many factors. A coal miner inhales 6000 g of coal dust particles

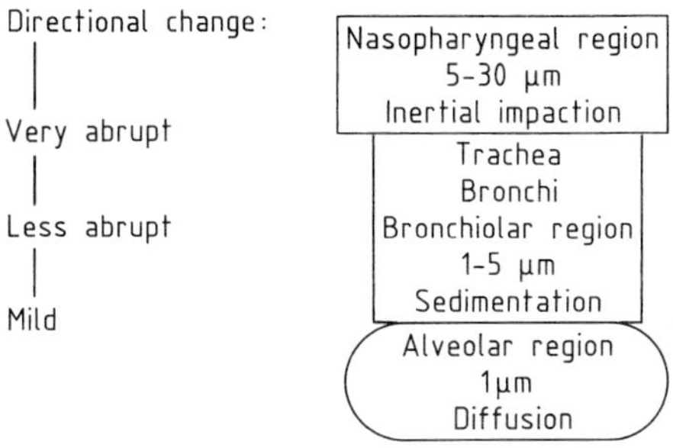

Figure 2.6. Effect of size on the disposition and sedimentation of particulates in the respiratory tract
The site of particle sedimentation is determined largely by particle size; only very fine particles are deposited in the alveoli; larger particles do not reach the lung but are deposited in the nasopharynx.

during his occupational lifetime, and only 100 g are found postmortem; therefore, effective protective mechanisms are operative. The parameters of air velocity and directional changes in air flow favor impaction of particles in the upper respiratory system. Particle characteristics such as size, chemistry of the inhaled material, sedimentation, and electrical charge are important to retention, absorption, or expulsion of airborne particles. In addition to the other lung characteristics mentioned, a mucous blanket propelled by ciliary action clears the respiratory tract of particles by directing them to the gastrointestinal system (via the glottis) or to the mouth for expectoration. This system is responsible for 80% of lung particulate clearance. The deposition of various particle sizes in different respiratory regions is summarized in Figure 2.6, which shows that particle size is important for disposition and particles larger than 2 μm do not reach the alveoli [29].

The direct penetration of airborne xenobiotics at alveolar surfaces or in the upper respiratory tract is not the only action of toxicological importance. Both vapor and particulates can accumulate in upper respiratory passages to produce irritant effects. Irritant gases may be deposited in the respiratory tract depending on their water solubility and may cause localized damage characterized by edema, swelling, mucus production, and increased vascular permeabilitiy. If major airways are obstructed by these processes or important anatomical structures of the respiratory tract such as the alveoli are damaged, life-threatening or deadly intoxications may be caused by the inhalation of irritant gases.

Despite the effectiveness of ciliary movement and phagocytosis, the cumulative effects of silica, asbestos, or coal dust ultimately cause important chronic fibroses even though direct absorption is of minor importance. Thus, phagocytosis prevents acute damage but may contribute to chronic toxicity. Little evidence exists for active transport in the respiratory system, although pinocytosis may be important for penetration. The lung is an area of extensive metabolic activity; enzymes present in the lung may catalyze both activation and detoxification of xenobiotics (Section 2.4).

2.3. Distribution of Xenobiotics by Body Fluids

After entering the blood by absorption or intravenous administration, xenobiotics are available for distribution throughout the body. The initial rate of distribution to organs and tissues is determined by the blood flow to that organ and the rate of diffusion of the chemical into the specific organ or tissue. Uptake of xenobiotics into organs or tissues may occur either by passive diffusion or by special transport processes. Within tissues, binding, storage, or biotransformation can occur. After reaching equilibrium, the distribution of a chemical into organs and tissues is determined largely by affinity; blood flow determines distribution only during the initial phase shortly after uptake.

Body fluids are distributed among three distinct compartments: vascular, interstitial, and intracellular water. Plasma water and interstitial water represent the extracellular water. Plasma water plays an important role in the distribution of xenobiotics. Human plasma accounts for about 4% of the total body weight and 53% of the total volume of blood. By comparison, interstitial tissue fluids account for 13% of body weight and intracellular fluids for 41%. The concentration of a xenobiotic in blood after exposure will depend largely on its apparent volume of distribution. If the xenobiotic is distributed only in plasma, a high concentration will be achieved within the vascular tissue. In contrast, the concentration will be markedly lower if the same quantity of xenobiotic were distributed in a larger pool including interstitial water or intracellular water.

Factors that affect distribution, apart from binding to blood macromolecules, include route

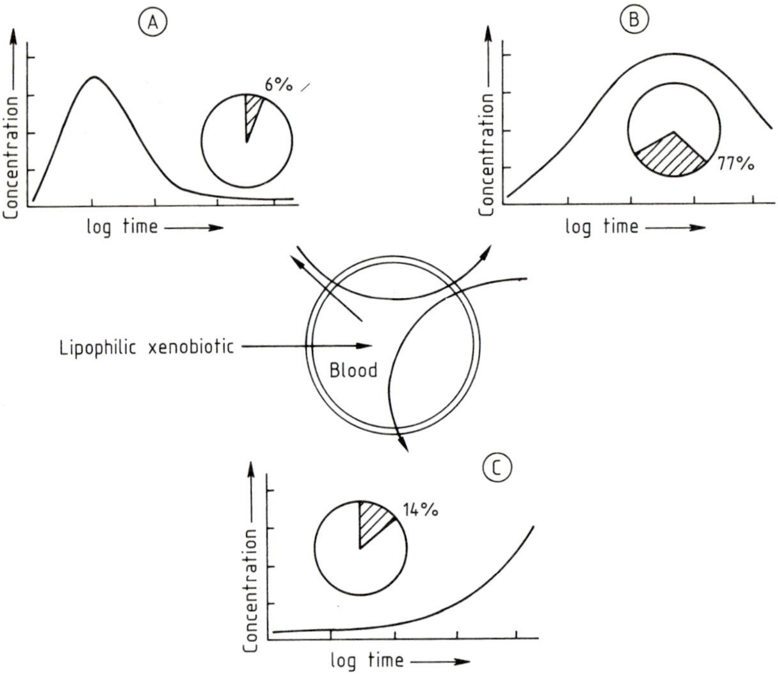

Figure 2.7. Uptake and redistribution with blood of lipophilic xenobiotics
Lipophilic xenobiotics in the blood are first distributed to well-perfused organs (A); after some time, they are redistributed to organs with lower blood flow representing a larger fraction of the body weight (B, C)

of administration, rate of biotransformation, polarity of the parent xenobiotic or biotransformation products, and rate of excretion by the liver or kidneys. Gastrointestinal absorption and intraperitoneal administration provide immediate passage of a compound to the liver, whereas dermal or respiratory routes provide at least one passage through the systemic circulation prior to reaching the liver. Most xenobiotics are metabolized to products that are more polar and thus more readily excreted than the parent molecules. Therefore, the rate of metabolism is a critical determinant in the distribution of a compound because those compounds that are readily metabolized are usually readily excreted and thus are proportionally less prone to accumulate in certain tissues. The same principle applies to polarity since very polar xenobiotics will be excreted readily. Chemicals may circulate either free or bound to plasma protein or blood cells; the degree of binding and factors influencing the equilibrium with the free form may influence the availability for biotransformation, storage, or excretion [30].

Patterns of xenobiotic distribution reflect certain physiological properties of the organism and physicochemical properties of the xenobiotics. An initial phase of distribution may be distinguished that reflects cardiac output and blood flow to organs. Heart, liver, kidney, brain, and other well-perfused organs receive most of a lipophilic xenobiotic within the first few minutes after absorption. Delivery to the smooth muscles, most viscera, and skin is slower; consequently, the time to reach a steady-state concentration of a xenobiotic in these organs may be several hours. A second phase of xenobiotic distribution may therefore be distinguished; it is limited by blood flow to an organ or tissue and involves a far larger fraction of the body mass than does the first phase of distribution (Fig. 2.7).

Only a limited number of xenobiotics are sufficiently soluble in blood to account for simple solution as a route of distribution; many xenobiotics are distributed in association with plasma proteins. The binding of drugs to plasma proteins is of key importance in transport. Many organic and inorganic compounds of low molecular mass appear to bind to lipoproteins, albumins, and other proteins in plasma and are transported as protein conjugates. This binding is reversible. Cellular components may also be responsible for the transport of xenobiotics, but such transport represents a minor route. The transport of xenobiotics by lymph is usually of little quantitative importance since intestinal blood flow is 500–700 times greater than intestinal lymph flow.

Many studies of plasma protein binding of drugs have demonstrated that binding to serum albumin is particularly important for these chemicals [31], [32]. Few studies have been performed on the reversible binding of toxic xenobiotics, but available evidence suggests a significant role for lipoproteins in plasma. These plasma proteins may bind xenobiotics as well as some physiological constituents of the body. Examples of plasma proteins, that may bind xenobiotics are albumin, α- and β-lipoproteins, and several metal-binding proteins such as transferrin. Lipoproteins are important for the transport of lipid-soluble endogenous chemicals such as vitamins, steroid hormones, and cholesterol, but they may also bind lipophilic xenobiotics. If a xenobiotic is bound to a protein, it may not reach the site of action; thus protein binding may influence the toxic effects exerted by a xenobiotic. The extent of binding to plasma proteins varies considerably among xenobiotics. Some will not be bound; with other xenobiotics, more than 90% of the dose administered may be present bound to plasma proteins. These ligand–protein interactions are reversible and provide a remarkably efficient means of transporting xenobiotics to various tissues. The xenobiotic–protein interaction may be described simply according to the law of mass action:

$$[\mathrm{T}]_{\mathrm{F}} + [\text{free sites}] \underset{k_2}{\overset{k_1}{\rightleftharpoons}} [\mathrm{T}]_{\mathrm{B}}$$

where $[\mathrm{T}]_{\mathrm{F}}$ and $[\mathrm{T}]_{\mathrm{B}}$ are free and bonded xenobiotic molecules, repectively, and k_1 and k_2 are the rate constants for association and dissociation; k_2, which governs the rate of binding to the protein, dictates the rate of xenobiotic release at a site of action or storage. The ratio k_1/k_2 is identical with the dissociation constant, K_{diss}. Among a group of binding sites on proteins, those with the lowest K_{diss} values for a given xenobiotic will bind it most tightly.

In contrast to the covalent binding to proteins observed with many xenobiotics or their electrophilic metabolites (see Section 2.5.6), the interaction of xenobiotics with plasma proteins is most often noncovalent and reversible. Noncovalent binding is of primary importance with respect to distribution because of the opportunities to dissociate after transport. Binding of xenobiotics to plasma proteins may be due to several types of interactions summarized below:

Ionic Binding. The degree of ionic binding varies with the chemical nature of each compound and the net charge. Dissociation of ionic bonds usually occurs readily, but some transition metals exhibit high association constants (low K_{diss} values) and exchange is slow. Ionic interactions may also contribute to the binding of alkaloids with ionizable nitrogen groups and other ionizable xenobiotics.

Hydrogen Bonding. As a rule, only the most electronegative atoms form stable hydrogen bonds. Protein side chains containing hydroxyl, amino, carboxyl, imidazole, and carbamoyl groups may form hydrogen bonds, as can the nitrogen and oxygen atoms of peptide bonds.

Van der Waals Forces. The binding force of van der Waals bonds depends critically on the proximity of interacting atoms and diminishes rapidly with distance. However, when these forces are summed over a large number of interacting atoms that "fit" together spatially, they can play a significant role in determining the specificity of xenobiotic–protein interactions.

Hydrophobic Interactions. When two nonpolar groups come together they exclude the water between them, and this mutual repulsion of water results in a hydrophobic interaction. Hydrophobic interactions may also be considered as a special case of van der Waals forces. Minimization of the thermodynamically unfavorable contact of a polar group with water molecules provides the major stabilizing effect in hydrophobic interactions.

Consequences of the binding to plasma proteins are reduced availability of the free xenobiotic in the cells and delayed excretion. The plasma protein-bound xenobiotic cannot cross capillary walls due to its high molecular mass. The fraction of dose bound is thus not available for delivery to the extravascular space or for filtration by the kidney. It is generally accepted that the fraction of xenobiotic that is bound may not exert toxic effects. Plasma proteins that can bind endogenous chemicals and xenobiotics are listed below, together with examples of bonded xenobiotics:

α-Lipoproteins	vitamins A, K, D steroid hormones dieldrin
Albumin	salicylate tetracyclines phenols vitamin C

The binding of xenobiotics to plasma proteins is, however, most often reversible; the amount of combined xenobiotic is in equilibrium with the amount of free xenobiotic; thus, binding usually slows excretion or delivery to cellular sites of action. Toxic consequences of the reversible binding of a xenobiotic to plasma proteins may arise after saturation of the binding capacities of plasma proteins and by displacement of the bound xenobiotic by another chemical with higher affinity, which increases the free fraction of the formerly bound xenobiotic. This will result in an increased equilibrium concentration of the xenobiotic in plasma and the target organ, with potentially harmful consequences.

2.4. Storage of Xenobiotics in Organs and Tissues

Absorbed xenobiotics may be concentrated in specific organs or tissues. The concentration of a xenobiotic in a tissue may cause toxic effects to that particular tissue; some xenobiotics actually attain their highest concentration at the site of toxic action. However, other xenobiotics may be concentrated in tissues without harmful consequences. Some tissues have a high capacity to accommodate certain xenobiotics and may release them only slowly. The compartment or tissue in which a chemical is concentrated can also be considered a storage depot for this xenobiotic. If a chemical is stored in a depot and thus removed from the site of action (e.g., polychlorinated biphenyls in fat or lead in bone), no manifestation of toxicity may be observed immediately, although a potential for adverse effects exists. For example, lead stored in bone does not cause a toxic response but has the potential for mobilization and thus for migration into soft tissues; toxic effects may appear after mobilization. Since the xenobiotic in storage depots is in equilibrium with the free xenobiotic in plasma, mobilization is constant, which results in constant exposure of the target organ to low concentrations of the xenobiotic. Some storage depots for specific chemicals in mammalian organisms are listed in the following material:

Lead	bone
Fluoride	bone, teeth
Cadmium	kidney
Iron	transferrin (a protein in blood)
Polychlorinated pesticides such as DDT	fat
Arsenic	skin

Liver, kidney, fat, bone, and plasma proteins may serve as storage depots for absorbed xenobiotics. Both liver and kidney have a high storage capacity for xenobiotics and are major storage sites for a multitude of chemicals. Accumulation of circulating xenobiotics from the blood by active transport systems and binding to certain tissue constituents are major mechanisms involved in renal and hepatic storage.

Several proteins rich in thiol groups present in liver and kidney have a high affinity for xenobiotics [33]. Ligandin, a binding protein, which also has enzymatic activity (ligandin is a glutathione *S*-transferase, see Section 2.5.5.6.) and thus participates in xenobiotic biotransformation, binds organic acids, some azo dyes, and corticosteroids. Metallothionein, a cysteine-rich protein present in liver and kidney, serves as a binding and storage protein for several metals including cadmium and zinc. Its biosynthesis is increased after exposure to metals; this may result in storage of a considerable percentage of a cumulative metal dose as metallothionein complex in liver and kidney.

Highly lipophilic chemicals rapidly penetrate membranes and are thus efficiently taken up by tissues. Because of their lipophilicity and in case of inefficient biotransformation, they are stored in the most lipophilic environment in the organism, fat. Most xenobiotics seem to accumulate by physical dissolution in neutral fats, which may constitute between 20 and 50% of the body weight in human males. Large amounts of lipophilic xenobiotics may therefore be present in fat; for xenobiotics not undergoing biotransformation (e.g., 2,3,7,8-tetrachlorodibenzodioxin), determination of their concentration in body fat is a good measure of exposure.

Usually, xenobiotics stored in fat do not induce toxic responses because the xenobiotic is not readily available at the target site for toxic action. However, during rapid mobilization of fat due to disease or starvation, a sudden increase in the plasma concentration and, thus the toxic effects in target organs may occur. For example, signs of organochlorine pesticide intoxication have been observed after starvation in animals pretreated with persistent organochlorine pesticides.

Specific xenobiotics have a high affinity for bone and may accumulate in the bone matrix. For example, 90 % of the lead and a major part of the strontium present in the body after chronic exposure are stored in the skeleton [34]. Lead and strontium accumulate in bone due to their similarity to calcium; inorganic fluoride, which is also a "bone-seeker," accumulates in bone due to similarities in size and charge to the hydroxyl ion. The storage of xenobiotics in bone may or may not be responsible for toxic effects. Lead stored in the skeleton is not toxic to bone; in contrast, both stored fluoride and stored strontium cause toxic effects to bone (fluorosis, osteosarcoma). Xenobiotics stored in bone are also in equilibrium with the unbound xenobiotic circulating in plasma and may thus be released.

Certain plasma proteins have a high affinity for xenobiotics; binding of a chemical to plasma proteins may constitute both a transport and a storage form. Globulins such as transferrin (involved in iron transport) and ceruloplasmin (copper) and α- and β-lipoproteins (lipophilic xenobiotics and endogenous chemicals) may be involved in binding.

Storage in tissue may greatly alter the rate of excretion of a xenobiotic. Only xenobiotics present in plasma are available for distribution, biotransformation, and excretion. However, excretion or biotransformation changes the plasma concentration of a xenobiotic, and some of the stored chemical is released into plasma from the site of storage. Because of this mechanism, the rate of excretion of a xenobiotic stored in tissues may be very small.

2.5. Biotransformation

Most xenobiotics entering the body are lipophilic. This property enables them to penetrate lipid membranes, to be transported by lipoproteins with blood, and to be rapidly absorbed by the target organ. However, the efficient excretory mechanisms of the organism require the solubility of the xenobiotic in aqueous media; thus, a certain degree of hydrophilicity is required for efficient excretion. Lipophilic substances can only be excreted efficiently by exhalation, this is, however, restricted to volatile xenobiotics. In the absence of efficient means for excretion of nonvolatile chemicals, constant or even intermittent single exposure to a lipophilic chemical could result in accumulation of the xenobiotic in the organism.

Therefore, animal organisms have developed a number of biochemical processes that convert lipophilic to hydrophilic chemicals and thus assist in their excretion. These enzymatic processes are termed biotransformation; the enzymes catalyzing these reactions are referred to as enzymes of biotransformation. The enzymes of biotransformation differ from most other enzymes active in organisms by having a broad substrate specificity and by catalyzing reactions at comparatively low rates. The low rates of biotransformation reactions are often compensated by high concentrations of biotransformation enzymes. For example, approximately 5 wt% of the protein in rat liver consist of cytochromes P450, major enzymes of biotransformation.

Biotransformation is most often the sum of several processes by which the structure of a chemical is changed during passage through the organism. The metabolites formed from the parent chemical are usually more water soluble; the increased water solubility lowers the ability of the metabolites to partition into membranes, restricts renal and intestinal reabsorption, and thus facilitates excretion with urine or bile.

2.5.1. Phase-I and Phase-II Reactions

Xenobiotic metabolism is catalyzed by many different enzymes. For solely operational purposes, the enzymes of biotransformation are separated into two phases. In *phase-I reactions*, which involve oxidation, reduction, and hydrolysis, a polar group is added to the xenobiotic or is exposed by the enzymes of biotransformation. *Phase-II reactions* are biosynthetic and link the metabolite formed by phase-I reactions to a polar endogenous molecule to produce a conjugate. Various endogenous molecules with high polarity and thus even higher solubility are utilized for conjugation; the conjugates formed are often ionized at physiological pH and thus highly water soluble. Moreover, the moieties used for conjugation are often recognized by specific active transport processes, which assist in their translocation across plasma membranes and thus further enhance the rate of excretion.

The fate of a particular chemical and the participation of various phase-I and phase-II enzymes of biotransformation are determined by its chemical structure; biotransformation is usually complex and often integrated. Many chemicals containing functional groups undergo conjugation without prior phase-I biotransformation, whereas others are oxidized or reduced prior to conjugation. In contrast, chemicals lacking functional groups may also undergo phase-II biotransformations without being subjected to a prior phase-I reaction (examples are 1,2-dibromoethane and perchloroethylene; see Section 2.5.4.).

2.5.2. Localization of the Enzymes of Biotransformation

The enzymes of biotransformation are localized mainly in the liver. A significant fraction of the blood from the splanchnic area, which also contains xenobiotics absorbed from the intestine, enters the liver. Therefore, the liver has developed the capacity to enzymatically modify most of these chemicals before storage, release, or excretion. However, most other tissues also contain enzymes that can catalyze biotransformation reactions. The contribution of extrahepatic organs to the biotransformation of a chemical depends on many factors, including chemical structure, dose, and route of administration. Biotransformation of a chemical in extrahepatic tissue may have toxic effects on this specific tissue and thus have important toxicological consequences.

Inside cells, phase-I enzymes are present mainly in the endoplasmic reticulum, a myriad of lipoprotein membranes extending from the mitochondria and the nucleus to the plasma membranes of the cell. The microsomal fraction obtained by ultracentrifugation of tissue homogenates, which is enriched in vesicles from the endoplasmic reticulum, is often used to study the enzymatic biotransformation of xenobiotics in vitro.

The presence of phase-I enzymes within membranes has important implications since lipophilic chemicals will distribute preferentially into lipid membranes; thus, high concentrations of lipophilic xenobiotics are present at this site of biotransformation. In contrast to phase-I enzymes, phase-II enzymes are often soluble, nonmembrane associated, and present in the cytoplasm of the cell. They are found in the supernatant (cytosol) obtained by ultracentrifugation of homogenized tissues. The subcellular localizations of enzymes responsible for biotransformation are listed below:

Phase-I enzymes	
Cytochrome P 450	microsomal
Flavin-dependent monooxygenase	microsomal
Prostaglandin synthase	microsomal
Epoxide hydrolase	microsomal/cytosolic
Phase-II enzymes	
UDP-glucuronyltransferases	microsomal
Sulfotransferases	cytosolic
N-acetyltransferases	cytosolic
Glutathione *S*-transferase	cytosolic/microsomal

2.5.3. Role of Biotransformation in Detoxification and Bioactivation

The general purpose of biotransformation reactions is detoxification, because xenobiotics should be transformed to metabolites, that are more readily excreted. However, depending on the structure of the chemical and the enzyme catalyzing the biotransformation reaction, metabolites with a higher potential for toxicity than the parent compound are often formed. This process is termed bioactivation and is the basis for the toxicity and carcinogenicity of many xenobiotics with a low chemical reactivity (see Section 2.5.6.). Interaction of the toxic metabolite initiates events that may ultimately result in cell death, cancer, teratogenicity, organ failure, or other manifestations of toxicity. Formation of reactive and more toxic metabolites is more frequently associated with phase-I reactions; however, phase-II reactions may also be involved in bioactivation, as well as combinations of phase-I and phase-II reactions.

2.5.4. Phase-I Enzymes and their Reactions

Phase-I reactions are catalyzed by microsomal monooxygenases and peroxidases, cytosolic and mitochondrial oxidases, reductases, and hydrolytic enzymes. In all of these reactions, functional groups are added or exposed that can be conjugated later.

2.5.4.1. Microsomal Monooxygenases: Cytochrome P450

Microsomal monooxygenases are the cytochrome P450 enzymes and the mixed-function amine oxidase of flavin-dependent monooxygenase. Both enzyme systems add a hydroxyl moiety to the xenobiotic. Cytochrome P450, a carbon monoxide binding hemoprotein in microsomes, is the most important enzyme system involved in phase-I reactions. The name cytochrome P450 is a generic term applied to a group of hemoproteins defined by a unique spectral property observed when reduced cytochrome P450 (containing Fe^{2+}) is treated with carbon monoxide. The complex formed has a maximum absorption at 450 nm, which results from the presence of an axial thiolate ligand to the heme iron. This spectral characteristic is present only when the protein is intact and catalytically functional. Denatured cytochrome P450, like other heme proteins, shows an absorbance maximum at 420 nm.

Cytochrome P450 enzymes represent a coupled enzyme system composed of the heme-containing cytochrome P450 and the nicotinamide, adenine dinucleotide phosphate (H) (NADPH)-containing cytochrome P450 reductase [35]. This flavoprotein has a preference for NADPH as its cofactor and transfers either one or two electrons from NADPH to cytochrome P450. Cytochrome P450 and the reductase are embedded in the phospholipid matrix of the endoplasmic reticulum. The phospholipid matrix is crucial for enzymatic activity since it facilitates the interaction between both enzymes.

In vertebrates, the highest concentrations of cytochrome P450 are found in the liver; however, cytochrome P450 enzymes are also present in the lung [36]–[40], kidney, testes, skin, and gastrointestinal tract [41]. The presence of several forms of cytochrome P450 with different substrate specificity and different rates of biotransformation for certain xenobiotics was indicated by studies in the 1970s. In the early 1980s, several different cytochrome P450 enzymes from rodents were purified to apparent homogeneity. Moreover, in the 1980s, a large number of cytochrome P450 enzymes have been purified from human organs. All of these cytochrome P450 enzymes share the heme, but they differ both in the composition and thus the structure of the polypeptide chain and in the reactions they catalyze [42].

Individual enzymes are regulated in their expression by a variety of factors such as treatment with xenobiotics, species, organ, sex, and diet. Because of the multitude of enzymes present, the term "superfamily" of cytochrome P450 is frequently used. In mammals, two general classes of cytochrome P450 exist: six families involved in steroid metabolism and bile acid biosythesis, and four families containing numerous individual cytochromes P450 mainly responsible for xenobiotic biotransformation.

Reactions Catalyzed by Cytochrome P450. Cytochrome P450 enzymes are monooxygenases; thus, these enzymes utilize one of the atoms of molecular oxygen and incorporate it into the xenobiotic in the following stoichiometry (RH is the substrate):

$$RH + O_2 + NADPH + H^+ \longrightarrow ROH + H_2O + NADP^+$$

The second oxygen atom is reduced to water with consumption of NADPH as reducing cofactor. The likely mechanisms of electron transfer and xenobiotic oxidation is shown in Scheme 1.

In the first step of the catalytic cycle, the xenobiotic combines with the oxidized form of cytochrome P450 (Fe^{3+}) followed by a one-electron reduction by NADPH–cytochrome P450 reductase to form a reduced (Fe^{2+}) cytochrome P450–substrate complex. This complex then combines with molecular oxygen, and another electron from NADPH is accepted. In a series of further steps, which are not completely understood, an oxygen atom from the intermediate is transferred to the substrate, while the other oxygen atom is reduced to water. In the last step of the catalytic cycle, the oxidized substrate dissociates and regenerates the oxidized form of cytochrome P450. Examples of oxidation reactions catalyzed by cytochromes P450 are shown in Table 2.2.

Cytochromes P450 may catalyze the hydroxylation of carbon–hydrogen bonds to transform hydrocarbons to the corresponding alcohols. In larger aliphatic chains, the $(\omega-1)$ position is often a favored point of attack. Oxidative *N*-, *O*- or *S*-dealkylations and oxidative dehalogenations are similar in mechanism to aliphatic hydroxylation, but give different end products due to secondary reactions of the intermediate products formed. Olefins are also oxidized by cytochrome P450, and with some substrates, epoxides are formed as products. The reaction, however, does

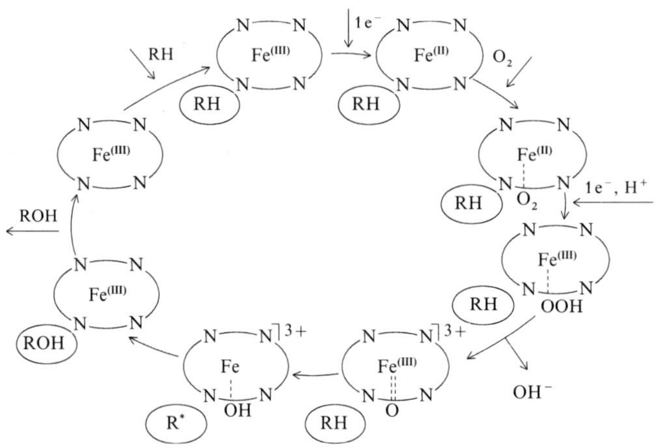

Scheme 1. Mechanisms of electron transfer and xenobiotic oxidation

Table 2.2. Oxidations catalyzed by cytochrome P450

Type of reaction	Examples
Aliphatic hydroxylation	$R-CH_2-CH_2-CH_3 \xrightarrow[P450]{O_2} R-CH_2-CHOH-CH_3$
N-Dealkylation	$RR^1N-CH_3 \xrightarrow{P450} [RR^1N-CH_2-OH] \longrightarrow RR^1NH + H_2C{=}O$
O-Dealkylation	$O_2N-C_6H_4-OCH_3 \longrightarrow [-O-CH_2-OH] \xrightarrow{P450} O_2N-C_6H_4-OH + H_2C{=}O$
Epoxidation	$H_2C{=}CH_2 \xrightarrow{P450}$ 

not proceed in a concerted fashion, but involves discrete ionic intermediates; these may also rearrange to products other than epoxides as shown for chloroolefins in Figure 2.8.

Sulfur or nitrogen is oxidized by the addition of oxygen at the lone electron pair on the sulfur or nitrogen atom. The products formed (sulfoxides or hydroxylamines) may be stable like many sulfoxides, may be further oxidized by

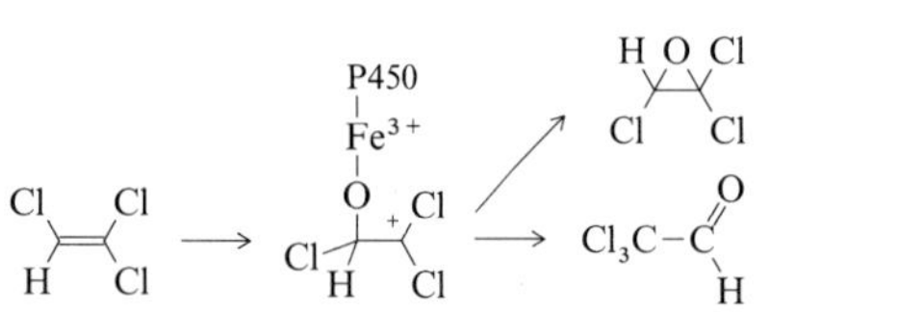

Figure 2.8. Mechanism of oxidation and rearrangement of trichloroethylene to chloral and trichlorooxirane, respectively

other enzymes in the organism (e.g., hydroxylamines), or may decompose to sulfur and the corresponding oxo compound.

The aforementioned reactions may be catalyzed by most cytochromes P450 involved in xenobiotic biotransformation; the type of reaction catalyzed seems to be influenced more by steric factors regarding the substrate binding site of individual cytochromes P450 than by electronic factors. As shown in Table 2.3, which gives an overview of the human cytochrome P450s involved in the biotransformation of xenobiotics and drugs, steric factors are likely major determinants of the substrate specificity of cytochrome P450 enzymes.

In addition to promoting oxidative metabolism, cytochrome P450 may also catalyze *reductive* biotransformation reactions [43]. These reactions are favored under reduced oxy-

Table 2.3. Human cytochromes P450 identified as major catalysts in the biotransformation of specific xenobiotics that seem to play major roles in the oxidation of substrates listed

Cytochrome P450			
1A1	1A2	2E1	3A4
Benzo[*a*]pyrene	phenacetin	vinyl chloride	aflatoxin B_1
Other polycyclic hydrocarbons	1-aminofluorene	trichloroethylene	17-β-estradiol
	2-amino-3-methylimidazo-[4,5-*f*]quinoline	halothane benzene	6-aminochrysene sterigmatocystine
	2-naphthylamine	dimethylnitrosamine acetaminophen	nifedipine ethinylestradiol

gen pressure or occur with xenobiotics lacking oxidizable C–H bonds or olefinic moieties. In these cases, the xenobiotic instead of oxygen accepts one or two electrons from the NADPH–cytochrome P450 reductase or from cytochrome P450. Reductive biotransformation catalyzed by cytochrome P450 has been demonstrated with some azo dyes and several aromatic nitro compounds (see below).

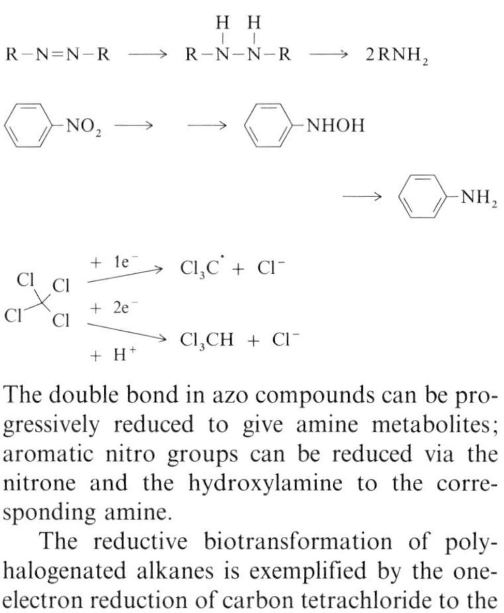

The double bond in azo compounds can be progressively reduced to give amine metabolites; aromatic nitro groups can be reduced via the nitrone and the hydroxylamine to the corresponding amine.

The reductive biotransformation of polyhalogenated alkanes is exemplified by the one-electron reduction of carbon tetrachloride to the trichloromethyl radical and chloride; reductive biotransformation of carbon tetrachloride by a two-electron reduction results in the formation of chloroform.

2.5.4.2. Microsomal Monooxygenases: Flavin-Dependent Monooxygenases

Tertiary amines and sulfur-containing drugs are metabolized to *N*-oxides or sulfoxides by a microsomal monooxygenase that does not depend on cytochrome P450 [44], [45]. This enzyme is a flavoprotein present in the endoplasmic reticulum. It is capable of oxidizing nucleophilic nitrogen and sulfur atoms in xenobiotics. However, this enzyme has a catalytic mechanism different from other heme- or flavin-containing enzymes. Like other monooxygenases, flavin-containing monooxygenases require molecular oxygen and NADPH as cofactors for oxygenation. In contrast to the other monooxygenases, flavin-dependent monooxygenases do not contain heme or iron, and substrate binding is not required for generation of the enzyme-bound oxygenating intermediate (Fig. 2.9). The active, oxygenating form of the enzyme is present in the cell, and any soft oxidizable nucleophile that can gain access to this oxygenating form will be oxidized. Precise fit of substrate to enzyme is not necessary. This property seems to be largely responsible for the broad substrate specificity of flavin-dependent monooxygenases. Flavin-dependent monooxygenases catalyze the oxidation of xenobiotics with few, if any, common struc-

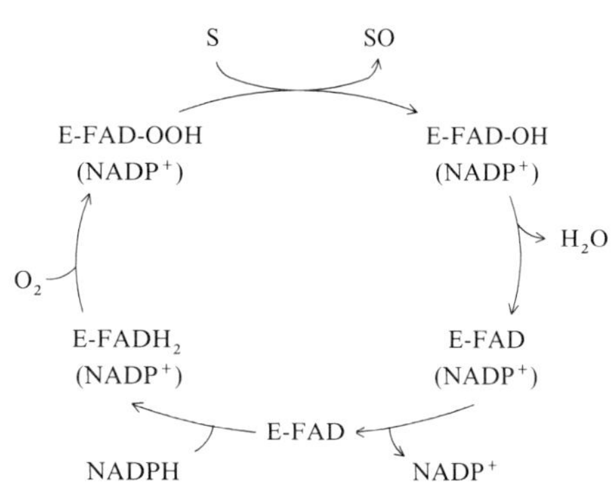

Figure 2.9. Mechanism of xenobiotic oxidation by the flavin-dependent monooxygenase
FAD = flavineadenine dinucleotide

Table 2.4. Oxidations catalyzed by flavin-dependent monooxygenases

Type	Examples
Amine oxidation	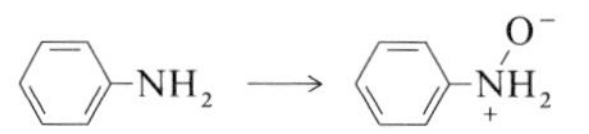
Hydroxylamine oxidation	$RN(OH)-CH_3 \longrightarrow RN^{+}(O^{-})=CH_2$
Thioamide oxidation	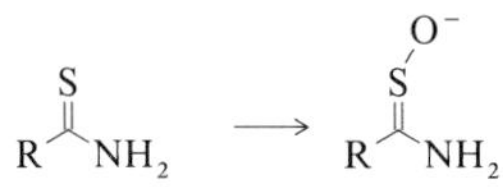
Thiol oxidation	2R–SH ⟶ R–S–S–R
Disulfide oxidation	R–S–S–R ⟶ 2R–SO_2

tural features at maximum velocity and catalyze the oxidation of a variety of xenobiotics (Table 2.4).

Many essential xenobiotics also carry functionalities that can be oxidized by flavin-dependent monooxygenases. However, these enzymes apparently discriminate between physiologically essential and xenobiotic soft nucleophiles and seem to exclude the former.

As with cytochrome P450, species- and tissue-specific forms of flavin-dependent monooxygenase have been described. Recent studies have shown that several isoforms of flavin-dependent monooxygenase exist; all isoforms show differences in their distribution in species and in organs within species [46].

The regulation of expression of flavin-dependent monooxygenase seems to be complex. Xenobiotics, which were shown to increase the concentration of cytochrome P450 in mammals, did not influence flavin-dependent monooxygenase concentrations. Recent evidence suggests that soft nitrogen and sulfur nucleophiles in the diet may act as inducers of flavin-dependent monooxygenases.

2.5.4.3. Peroxidative Biotransformation: Prostaglandin Synthase

During the biosythesis of prostaglandins, a polyunsaturated fatty-acid, arachidonic acid, is oxidized to yield prostaglandin G_2 (PGG_2), a hydroperoxy endoperoxide (→ Prostaglandins, **A 22**, pp. 263–264). Prostaglandin G_2 is further transformed to prostaglandin H_2 (PGH_2). Both the formation of prostaglandin G_2 and its further transformation to prostaglandin H_2 are catalyzed by the same enzyme, prostaglandin synthase [47]–[49]. This enzyme is a glycoprotein with a molecular mass of ca. 70 000 and contains one heme per subunit. The enzyme is found in high concentrations in germinal vesicles and renal medulla, and also in several other tissues such as skin and adrenals [50]. The cyclooxygenase and peroxidase activities of prostaglandin synthase generate enzyme- and substrate-derived free-radical intermediates (Fig. 2.10). Biotransformation of xenobiotics may be associated with both the cyclooxygenase and the peroxidase activity of prostaglandin synthase (Fig. 2.10). During the cyclooxygenase-catalyzed conversion of arachidonic acid to prostaglandin G_2, peroxy radicals are formed as intermediates. These lipid peroxy radicals represent a source of reactive oxygen metabolites (see below) and can, in turn, biotransform xenobiotics. Oxidation by the cyclooxygenase activity of prostaglandin synthase is important in the oxidation of diols derived from carcinogenic polycyclic aromatic hydrocarbons and transforms these diols to diol epoxides (Fig. 2.10, top).

During reduction of the peroxide prostaglandin G_2 to prostaglandin H_2, the peroxidase undergoes a two-electron oxidation. To return to the ground state, the enzyme requires two consecutive one-electron reductions, which are performed by abstracting electrons from available donors. In addition to endogenous substrates, xenobiotics may act as electron donors and thus be oxidized to radicals (Fig. 2.10 bottom). This process is termed cooxidation of xenobiotics. Classes of xenobiotics undergoing cooxidation during prostaglandin syntheses are aromatic amines, phenols, hydroquinones, and aminophenols. The role of prostaglandin synthase in the biotransformation of xenobiotics is somewhat unclear since many of the end products of xenobiotic oxidation catalyzed by prostaglandin synthase-mediated cooxidation are identical to those formed by cytochrome P450. Therefore, prostaglandin synthase is assumed to contribute to the oxidative biotransformation of xenobiotics in tissues low in monooxygenase activity [51].

In addition to prostaglandin synthase, other peroxidases may also participate in the oxidation of xenobiotics [52]–[54]. For example, mammary gland epithelium contains lactoperoxidase and leukocytes contain myeloperoxidase. The

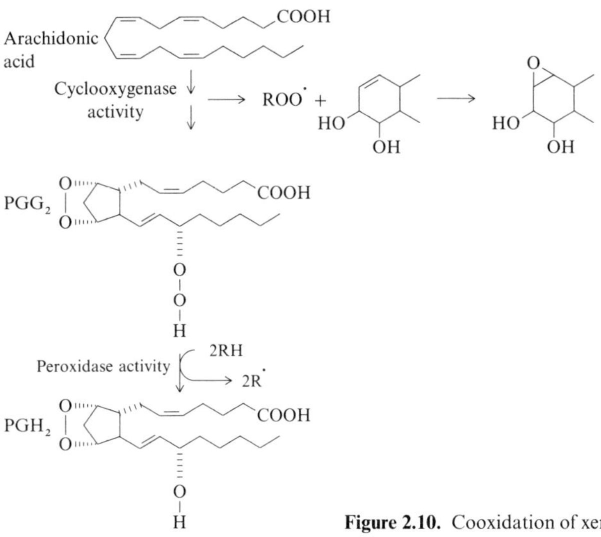

Figure 2.10. Cooxidation of xenobiotics during the biosynthesis of prostaglandins

general reaction catalyzed by this enzyme involves the reduction of hydroperoxide coupled to oxidation of the substrate:

$$ROOH + XH \longrightarrow ROH + XOH$$

The availability of peroxides in tissues likely controls the extent of peroxidative biotransformation; however, the availability of hydrogen peroxide is usually low because it is efficient scavenged by catalase and glutathione peroxidase. Therefore, peroxidative metabolism occurs mainly in tissues that can maintain an oxidizing environment.

2.5.4.4. Nonmicrosomal Oxidations

Several enzymes located in mitochondria or the cytoplasm of the cell may also catalyze the oxidation of xenobiotics. In contrast to cytochromes P450 with their broad substrate specificity, most of the nonmicrosomal oxidases have a more narrow substrate specificity and accept only xenobiotics bearing specific functional groups as substrates. *Alcohol dehydrogenases* catalyze the oxidation of alcohols to aldehydes or ketones:

$$RCH_2OH + NAD^+ \longrightarrow RCHO + NADH + H^+$$

The enzyme is found mainly in the soluble fraction of liver, but also in other organs such as the kidney and lung, and is responsible for the oxidation of ethanol. The expression of alcohol dehydrogenase is under genetic control, giving rise to a number of variants with differing activities. Usually, the oxidation of alcohols to aldehydes is reversible, since the reduction of aldehydes is also catalyzed efficiently by aldehyde reductases. However, in vivo, the reaction proceeds in the direction of alcohol consumption because aliphatic and aromatic aldehydes are further oxidized to acids:

$$RCHO + NAD^+ \longrightarrow RCOOH + NADH + H^+$$

The reaction may be catalyzed by *aldehyde dehydrogenase*, which has a broad substrate specificity, and several isozymes of aldehyde dehydrogenase are found in liver cytosol, mitochondria, and microsomes with characteristic substrate specificities. Other enzymes in the soluble fraction of liver that may oxidize aldehydes are the flavoproteins *aldehyde oxidase* and *xanthine oxidase*.

Monoamine oxidases are a family of flavoproteins present in many tissues including liver, kidney, brain, and intestine. These mitochondrial enzymes have a broad and overlapping substrate specificity and oxidize a variety of amines.

$$RCH_2NH_2 + O_2 + H_2O \longrightarrow RCHO + NH_3 + H_2O$$

The monoamine oxidase found in the central nervous system is concerned primarily with neurotransmitter turnover.

2.5.4.5. Hydrolytic Enzymes in Phase-I Biotransformation Reactions

Many tissues contain enzymes with *carboxylesterase* and *amidase* activity. These enzymes are located in both the microsomal and the soluble fraction and hydrolyze ester and amide linkages in xenobiotics.

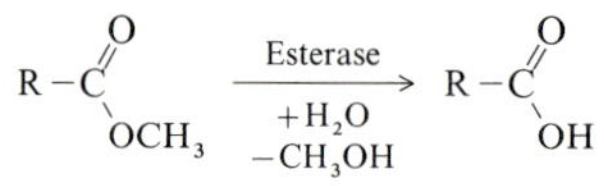

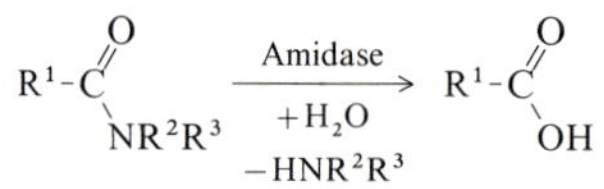

Although esterases and amidases were thougth to be different enzymes, all purified esterases have been demonstrated to have amidase activity; similarly, all amidases have esterase activity. In general, esters are cleaved more rapidly than amides. The expression of many esterases is under genetic control.

Epoxide hydrolase is an important enzyme that cleaves aliphatic and aromatic epoxides. The enzyme hydrates arene oxides and aliphatic epoxides to the corresponding *trans*-1,2-diols [55]. Water is required as cofactor; the catalytic mechanism of epoxide hydrolases involves ester formation of the oxirane with a carboxylic acid function at the active site of the enzyme and hydrolysis of this ester by water; no metals or other cofactors are required.

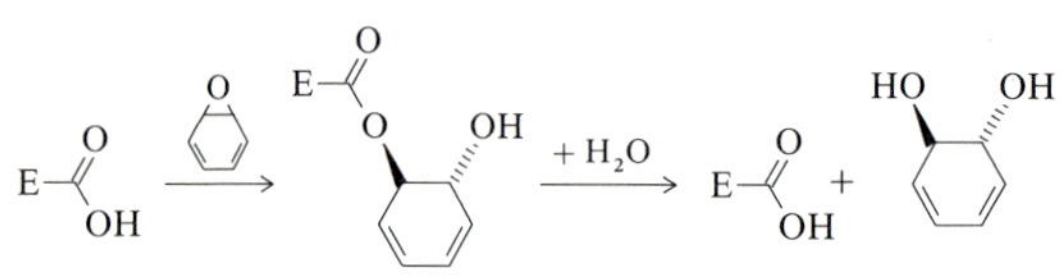

Microsomal epoxide hydrolases are thought to be present in close proximity to the microsomal cytochromes P450; in most cases, reaction of the epoxide with the less reactive diol is considered to represent an important detoxification reaction for metabolically formed oxiranes. Epoxide hydrolases are found in many tissues such as liver, kidney, testes, and intestine. Their distribution is heterogeneous between different cell types in a specific organ; in addition, several forms of microsomal epoxide hydrolase with broad substrate specificity have been found in different animal species. Moreover, in animals, in addition to the membrane-bound epoxide hydrolase, a soluble epoxide hydrolase is present in the cytoplasm of several tissues.

2.5.5. Phase-II Enzymes of Biotransformation and their Reactions

Products of phase-I biotransformation reactions having functional groups such as hydroxyl, amino, or carboxyl often undergo a conjugation reaction with an endogenous substrate. The endogenous substrates may include sugar derivatives, sulfate, amino acids, and small peptides (glutathione). The conjugation products are usually more polar and thus more readily excreted than their parent compounds.

In contrast to phase-I biotransformation reactions, phase-II reactions are biosynthetic and require energy. Energy is usually consumed to generate a high-energy cofactor or an activated intermediate and then utilized as cosubstrate. Thus, depletion of the cofactor or general interference with cellular energy status may interfere with the ability of cells to conduct phase-II biotransformation reactions.

2.5.5.1. UDP-Glucuronyltransferases

Glucuronidation represents one of the main phase-II biotransformation reactions in the conversion of both endogenous and exogenous compounds to water-soluble products [56], [57]. The glucuronides formed are excreted with bile or urine. The formation of an activated glucuronide (uridine diphosphate glucuronic acid, UDPGA), is required for glucuronide formation. UDPGA is formed in a sequential reaction from uridine and glucose 1-phosphate

$$\text{Uridine triphosphate} + \text{Glucose 1-phosphate} \xrightarrow[\text{Pyrophosphorylase}]{\text{UDPG}} \text{Uridine diphosphate glucose} + \text{Pyrophosphate}$$

$$\text{Uridine diphosphate glucose} + 2\,\text{NAD}^{+} + \text{H}_2\text{O} \xrightarrow{\text{Dehydrogenase}} \text{UDPGA} + 2\,\text{NADH}_2$$

where UDPG denotes uridine diphosphate glucose.

The enzymes that carry out the coupling of the xenobiotic with UDPGA are termed UDP-glucuronyltransferases. They couple D-glucuronic acid with a wide variety of xenobiotics carrying functional groups to give β-D-glucuronides [58]. These glucuronides are highly polar and ionized at physiological pH; thus, they are rapidly excreted. The membrane-bound UDP-glu-

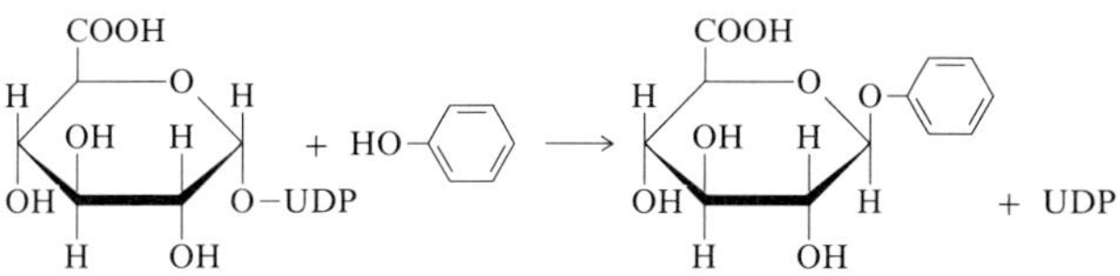

UDP-Glucuronic acid Phenyl-β-glucuronide

Figure 2.11. Conjugation of phenol to phenyl glucuronide catalyzed by UDP-glucuronyltransferase (UDP = uridine diphosphate)

curonyltransferases are found in highest concentration in the liver, but also in most other tissues studied. The reaction catalyzed involves a nucleophilic displacement (SN_2) of the functional group of the substrate with Walden inversion (Fig. 2.11).

Like cytochrome P450 UDP-glucuronyltransferases represent a family of enzymes. At least ten individual forms are known. The various forms respond differently to inducers and have preferences for certain classes of chemicals. UDP-glucuronyltransferases catalyze the conjugation of many functional groups of xenobiotics with glucuronic acids. Some typical examples are shown in Table 2.5.

Glucuronids formed in the liver are excreted with urine or bile. Aglycons with molecular masses > 300 are transformed to glucuronides surpassing the molecular mass threshold for biliary excretion (see Section 2.5.8.2) and thus excreted with bile into the intestine. Glucuronides can be cleaved there by β-glucuronidase present in intestinal microflora to the respective aglycon, which can be reabsorbed from the intestine and translocated back to the liver with blood. The resulting cycle is called enterohepatic circulation; compounds undergoing enterohepatic circulation are excreted only slowly and usually have a longer half-life in the body.

2.5.5.2. Sulfate Conjugation

The formation of water-soluble sulfate esters is observed with many xenobiotics carrying functional groups such as alcohols, phenols, and arylamines. These reactions are catalyzed by sulfotransferases, a large group of soluble enzymes found in many tissues [58], [59]. Sulfotransferases catalyze the transfer of a sulfate group from the "active sulfate" 3′-phosphoadenosine 5′ = phosphosulfate (PAPS) to hydroxyl groups and amines. The resulting products are referred as sulfate esters or sulfamates.

The sulfate esters are ionized at physiological pH and may therefore be excreted rapidly in

Table 2.5. Typical conjugation reactions catalyzed by UDP-glucuronyltransferases

Type of reaction	Examples
O-Glucuronide formation	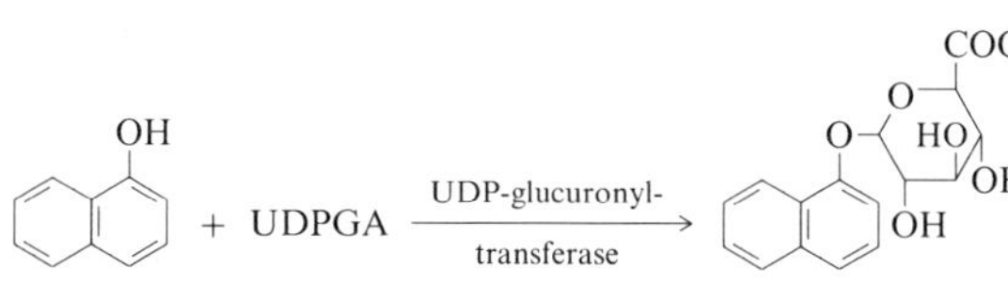 1-Naphthol
N-Glucuronide formation	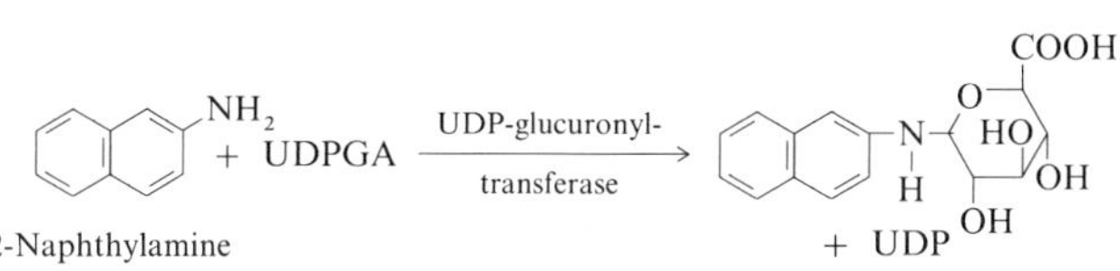 2-Naphthylamine
S-Glucuronide formation	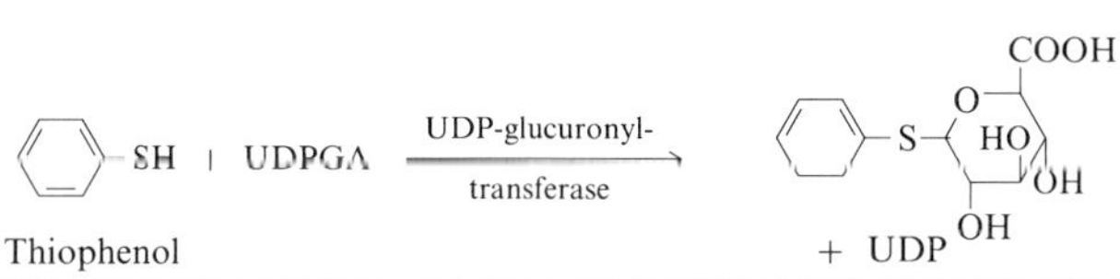 Thiophenol

urine; thus, sulfate conjugation is an effective mechanism to enhance the rate of excretion for many xenobiotics. Sulfotransferases are also a family of enzymes, and at least four different classes of sulfotransferases are involved in biotransformation reactions. These classes are aryl sulfotransferases, hydroxysteroid sulfotransferases, estrone sulfotransferases, and bile salt sulfotransferases; each class again has been divided in to several distinct forms differing in substrate specificity, pH optimum, and immunological properties.

Sulfate conjugation also requires a sulfate donor, 3′-phosphoadenosine 5′-phosphosulfate. PAPS is likely synthetized in the cytosol of most mammalian cells by a two-step reaction consuming ATP and utilizing inorganic sulfate coming from the catabolism of cysteine or from diet. In the first step of this sequence, the sulfation of ATP to adenosine 5′-phosphosulfate (APS) is catalyzed by a ATP sulfurylase. Adenosine 5′-phosphosulfate is further transformed to 3′-phosphoadenosine 5′-phosphosulfate by an adenosine 5′-phosphosulfate kinase. The equilibrium concentration of PAPS in mammalian cells may be low, however, due to the tight coupling of two enzymes in the formation of 3′-phosphoadenosine 5′-phosphosulfate, biosynthesis may proceed rapidly.

$$SO_4^{2-} + ATP \xrightarrow{\text{Sulfurylase}} APS + \text{Pyrophosphate}$$

$$APS + ATP \xrightarrow{\text{APS-phosphokinase}} PAPS + ADP$$

Despite the rapid synthesis, the sulfation of xenobiotics may be limited by reduced availability of 3′-phosphoadenosine-5′-phosphosulfate. The availability of the cofactor of the synthesis reaction, sulfate, may be limited by consumption due to sulfation of a xenobiotic or the limited availability of free cysteine for transformation to sulfate. Therefore, the capacity of sulfation for certain xenobiotics is dependent on dose. After administration of low doses, the compound may be excreted as sulfate; after high doses, the capacity of sulfate conjugation may be saturated and other biotransformation reactions such as glucuronide formation may become more important for biotransformation.

2.5.5.3. Methyltransferases

Many alcohols, phenols, amines, and thiols present as both endogenous and exogenous compounds may be methylated by several *N*-, *O*-, and *S*-methyltransferases. The most common donor for the methyl group is *S*-adenosyl methionine, which is formed from methionine and ATP. Often, these reactions do not increase the water solubility of a xenobiotic; they are, however, considered phase-II reactions since they mask potentially toxic functional groups and thus may serve as a detoxification reaction.

A large variety of different enzymes catalyze methylation of xenobiotics. The more important enzymes involved in methylation reactions are usually found in many tissues and are present in the soluble tissue fraction. Some of the enzymes have a high specificity for certain endogenous compounds such a histamine or norepinephrine; others, such as catechol-*O*-methyltransferase, metabolize both endogenous catechols and certain xenobiotics having aromatic rings with catechol functionalities.

2.5.5.4. *N*-Acetyltransferases

Aromatic amines, hydrazines, sulfonamides, and certain aliphatic amines are biotransformed to amines in a reaction catalyzed by *N*-acetyltransferases. Enzymes that catalyze the acetylation of amines are designated as acetyl-CoA: amine *N*-acetyltransferases. These enzymes utilize acetyl coenzyme A as a cofactor [60]. The acetylation reaction of arylamines occurs in discrete steps. In the first step, the acetyl group from acetyl coenzyme A is transferred to the *N*-acetyltransferase, which then acetylates the arylamine, thus regenerating the enzyme and forming the amide (Fig. 2.12).

N-acetyltransferases are found in a number of different forms in cytosol of many tissues. In many species, the expression of *N*-acetyltransferases is under genetic control; polymorphism for the expression of *N*-acetyltransferase have been found in several animal species and in humans. The transfer of an acetyl group to amines is reversible; deacetylation of amides occurs in

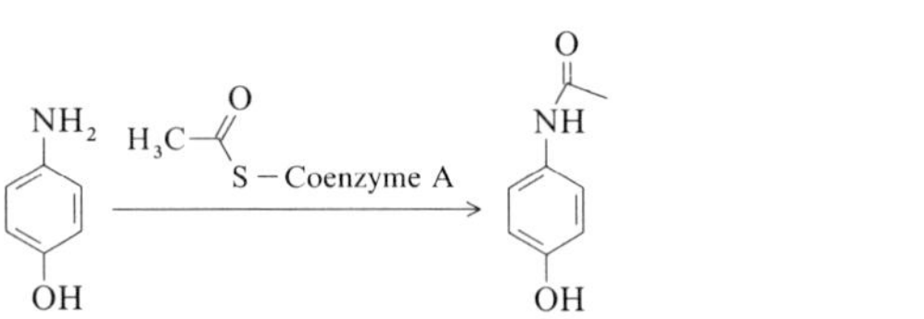

Figure 2.12. Mechanism of the *N*-acetylation of amines by *N*-acetyl transferase

many species. Large differences exist between strains, species, and individuals in the extent of expression of amidases (see Section 2.5.4.5).

2.5.5.5. Amino Acid Conjugation

Exogenous carboxylic acid are conjugated with a variety of amino acids to form amides. Substrates for conjugation are mainly carboxylic acids containing aromatic rings. Glutamate and glycine appear to be the most common amino acids involved in these conjugation reactions in mammals; in other species such as reptiles and birds, reaction with ornithine occurs. The reaction proceeds in two steps and is catalyzed by two different enzymes. In the first step, the carboxylic acid is activated to form a coenzyme A derivative in a reaction involving coenzyme A and ATP. The enzymes that catalyze this reaction are called ATP-dependent acid:coenzyme A ligases and are present in mitochondria. They appear to be identical to the intermediate chain length fatty acid:coenzyme A synthetase. The thus formed coenzyme A thioester then transfers its acyl group to the amino group of the acceptor amino acid. This reaction is catalyzed by an *N*-acetyltransferase (Fig. 2.13). The enzymes catalyzing both steps in amino acid conjugation exist in several forms with different substrate specifities.

2.5.5.6. Glutathione Conjugation of Xenobiotics and Mercapturic Acid Excretion

The conjugation of xenobiotics or their metabolites with the tripeptide glutathione is an important reaction. Glutathione is composed

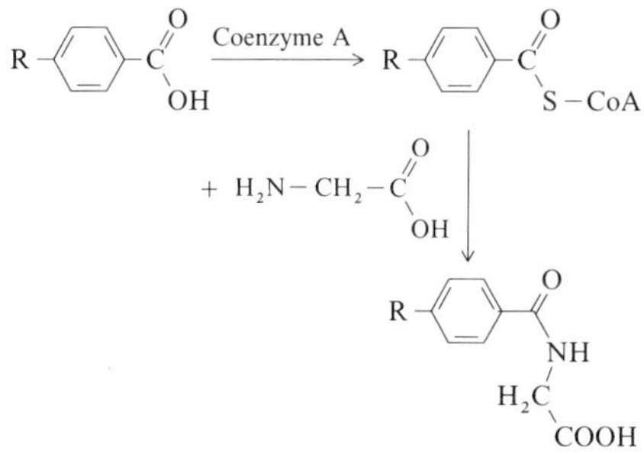

Figure 2.13. Amino acid conjugation of a xenobiotic carboxylic acid by ATP-dependent acid-coenzyme A ligases followed by *N*-acetyltransferase

of the amino acids cysteine, glutamic acid, and glycine (γ-glutamylcysteinylglycine) and is present in many cells in high concentration (up to 10 mmol/L in liver cells) [61]. Since glutathione conjugation captures reactive electrophiles and transforms them to stable, often nontoxic thioethers, the formation of glutathione conjugates protects cells from the harmful effects of these electrophiles and thus serves as a major detoxification reaction (see Section 2.5.6.5). Glutathione conjugation is catalyzed by a family of enzymes termed glutathione *S*-transferases, which are present in highest concentration in the liver, but are also found in kidney, testes, and lung in high activity [62]. Glutathione *S*-transferases are both membrane-bound and soluble enzymes; with most substrates, the activity of the soluble glutathione *S*-transferase is higher than that of the microsomal glutatione *S*-transferase [63].

Cytosolic glutathione *S*-transferases exist in many different iso-forms, each species being a dimer of different subunit composition [62], [64]. The glutathione *S*-transferase gene exists in at least six different families. In contrast to the multiple forms of soluble glutathione *S*-transferases thus possible, only one form of the membrane-bound enzyme is known. The glutathione *S*-transferases catalyze the reaction of the sulfhydryl group of glutathione with chemicals containing electrophilic carbon atoms (Table 2.6).

Thioethers are formed by reaction of the thiolate anion of glutathione with an electrophile. For enzyme catalysis both an electrophilic center in the xenobiotic and a low rate of reaction between xenobiotic and glutathione assistance by glutathione *S*-transferases are required. The enzyme accelerates the rate of this spontaneous reaction by several orders of magnitude. Glutathione thioethers formed in the organism are not excreted but are processed further to excretable mercapturic acids. Mercapturic acids are thioethers derived from *N*-acetyl-L-cysteine. Mercapturic acid formation is initiated by conjugation of the xenobiotic or an electrophilic metabolite with glutathione (see below).

This is followed by splitting of the glutamate moiety by γ-glutamyltranspeptidase, an enzyme specifically recognizing γ-glutamyl peptides and found in high concentration in the kidney and other excretory organs. Dipeptidases catalyze the elimination of glycine from the intermediary cysteinylglycine *S*-conjugate to give the cysteine

Table 2.6. Substrates for mammalian glutathione *S*-transferases

Type of reaction	Examples
Aryltransferase	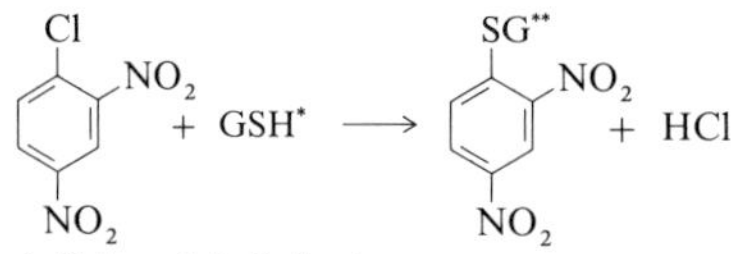 1-Chloro-2,4-dinitrobenzene
Arylalkyl-transferase	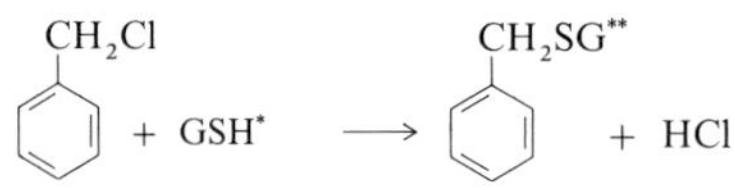
Alkylene-transferase	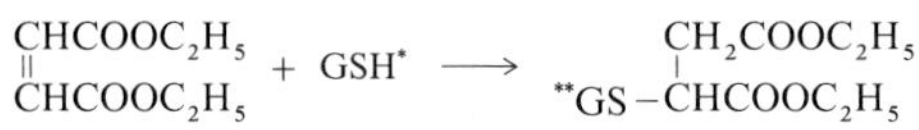
Epoxide-transferase	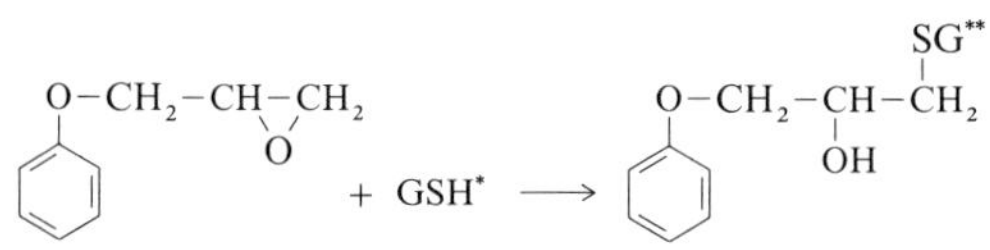

* GSH = Glutathione. ** GS = Glutathionyl residue.

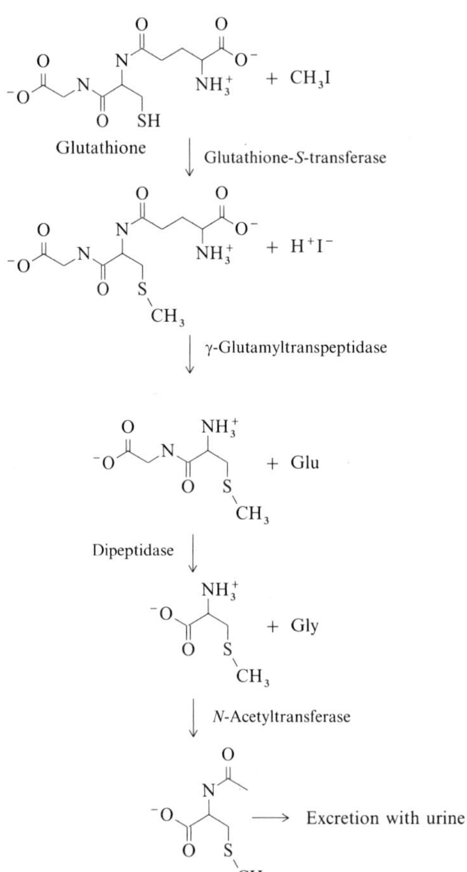

Mercapturic acid

S-conjugate, which, in the final step of mercapturic acid formation, is *N*-acetylated by a cysteine conjugate-specific *N*-acetyltransferase with acetyl coenzyme A as cofactor. The mercapturic acids formed are readily excreted into urine by active transport mechanisms in the kidney [65].

Glutathione conjugation is one of the most important detoxification reactions for reactive intermediates formed in organisms. Usually, metabolically formed intermediates are efficiently detoxified; however, under specific circumstances, glutathione conjugation may be overwhelmed by high concentrations of electrophiles, resulting in covalent binding of intermediates to cellular macromolecules causing disruption of important cellular functions cell death and necrosis (see Section 2.5.6.5).

2.5.6. Bioactivation of Xenobiotics

Many xenobiotics with low chemical reactivity (e.g., the solvent carbon tetrachloride, the environmental contaminant hexachlorobutadiene, and the heat exchanger fluid tri-*o*-cresyl phosphate) cause toxic effects. These toxic effects are initiated by the covalent binding to macromolecules of metabolites formed in the organisms by the enzymes of biotransformation—a

Table 2.7. Basic mechanisms involved in the bioactivation of xenobiotics based on chemical reactivity of intermediates formed

Mechanism	Structure and reactivity of the intermediate	Examples
Biotransformation to stable but toxic metabolites	different structures, selective interaction of formed metabolite with specific acceptors, or disruption of specific biochemical pathways	dichloromethane, acetonitrile, parathion
Biotransformation to electrophiles	reactive electrophiles	dimethylnitrosamine, acetaminophen, bromobenzene
Biotransformation to free radicals	radicals	carbon tetrachloride
Formation of reactive oxygen metabolites	radicals	paraquat, aromatic nitro compounds

process termed bioactivation. With many chemicals, reactive metabolites formed during bioactivation may be detoxified efficiently; thus, toxic effects occur only when the balance between the production of reactive metabolites and their detoxification is disrupted. For example, toxic effects may be observed with a certain chemical only when the formation of reactive intermediates is enhanced or when the capacity for detoxification is diminished.

The mechanismus of bioactivation of xenobiotics may be classified into four categories describing the basic types of reactive intermediates formed and their potential reactivity (Table 2.7) [66], [67].

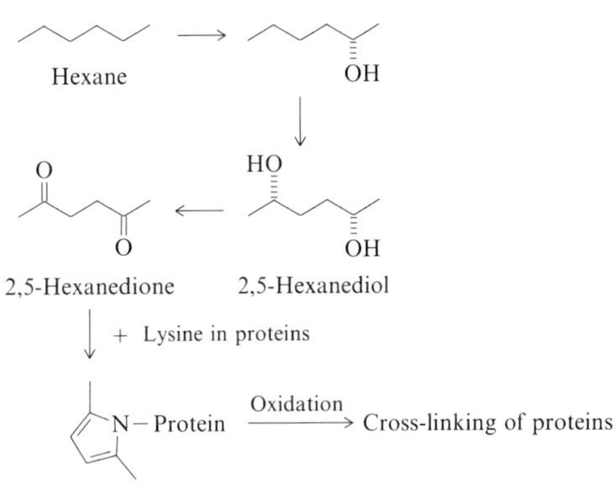

Figure 2.14. Bioactivation of hexane by cytochrome P450 to 2,5-hexanedione
Hexanedione reacts with lysine groups in proteins to form pyrroles; oxidation of two neighboring pyrrole residues causes the cross-linking of proteins.

2.5.6.1. Formation of Stable but Toxic Metabolites

The formation of stable but toxic metabolites is limited to a few selected chemicals because only few xenobiotic metabolites are both stable and toxic, examples are the bioactivation mechanisms of the solvents hexane and dichloromethane. Hexane produces a characteristic neuropathy and peripheral nerve injury after chronic exposure. The same typical manifestations of toxicity are also observed when the hexane metabolites 2-hexanone and 2,5-hexanedione are given to animals. The mechanism of hexane neuropathy thus involves oxidation of hexane by cytochromes P450 at both ends of the carbon chain ($\omega-1$ hydroxylation) and further oxidation of the thus introduced alcoholic function. The 2,5 hexanedione formed reacts with lysine residues in axonal proteins by Schiff base formation followed by cyclization to give pyrroles [68]. Oxidation of the pyrrole residues then causes cross-linking between two hexane-modified proteins; the resulting changes in the three-dimensional structures of proteins perturb axonal transport and function and cause damage of nerve cells (Fig. 2.14).

Carboxyhemoglobin formation is found after human exposure to dichloromethane. Dihalomethanes are oxidized by cytochrome P450, likely by P450 2E1, to carbon monoxide, which, because of its high affinity for iron(II)-containing porphyrins, binds to hemoglobin and interferes with oxygen transport by blood. Other examples of the formation of stable but toxic metabolites include the oxidation of acetonitrile to cyanide

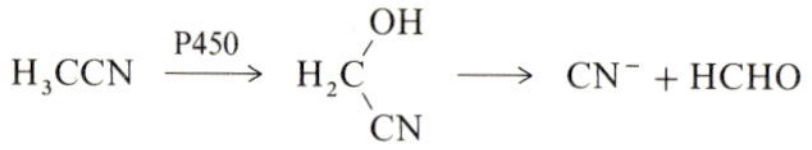

and the oxidative desulfuration of parathion.

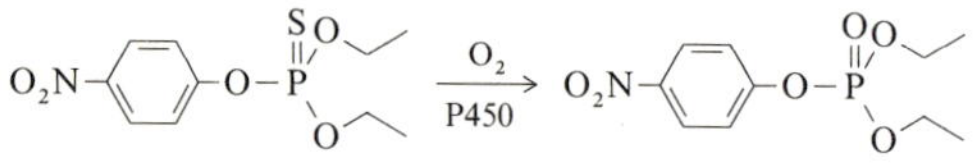

Parathion Paraoxon

Parathion is a potent insecticide acting as an inhibitor of cholinesterase, but itself only a weak cholinesterase inhibitor. Biotransformation of parathion by oxidative desulfuration to the potent cholinesterase inhibitor paraoxon is responsible for the high insecticidal potency.

2.5.6.2. Biotransformation to Reactive Electrophiles

Biotransformation to reactive electrophiles is the most common pathway of bioactivation. The cytotoxicity and carcinogenicity of many chemicals are associated with the formation of electrophiles and the ensuing alkylation or acylation of tissue constituents such as protein, lipid, or DNA. Reactive intermediates include such chemically diverse functionalities as epoxides, quinones, acyl halides, carbocations, and nitrenium ions. The metabolic formation of electrophiles may be catalyzed by many different enzymes, although the majority of examples elucidated to date involve cytochrome P 450-mediated oxidations.

Cytochrome P 450 catalyzes the transformation of olefins to reactive and electrophilic epoxides. For example, the carcinogenicity of the industrial intermediate vinyl chloride (Fig. 2.15) and the fungal toxin aflatoxin B_1 (Fig. 2.16) are dependent on their transformation to electrophilic epoxides [69], [70].

Carbocations are formed during the cytochrome P 450-mediated oxidation of *N,N*-dialkylnitrosamines. For example, the mutagen and potent carcinogen *N,N*-dimethylnitrosamine is hydroxylated by cytochrome P 450 followed by elimination of formaldehyde. Monomethylnitrosamine thus formed is unstable and rearranges to release an electrophilic carbocation.

Acyl halides are formed by the oxidation of carbon atoms containing at least two halogen atoms. The initially formed products are unstable α-halohydrins, which loose hydrogen chloride and thus give reactive acyl halides. An example for the formation of acyl halides as reactive intermediates is the cytochrome P 450-mediated oxidation of chloroform to phosgene [71].

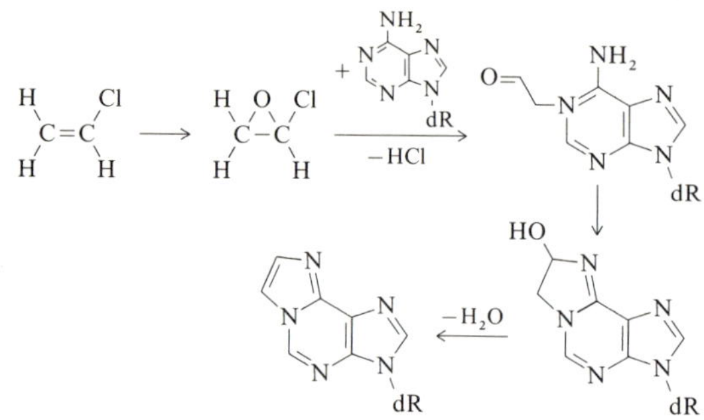

1,N^6-Ethenodeoxyadenosine

Figure 2.15. Bioactivation of vinyl chloride to chlorooxirane and reaction of the epoxide with critical macromolecules in the cell
dR = Deoxyribose

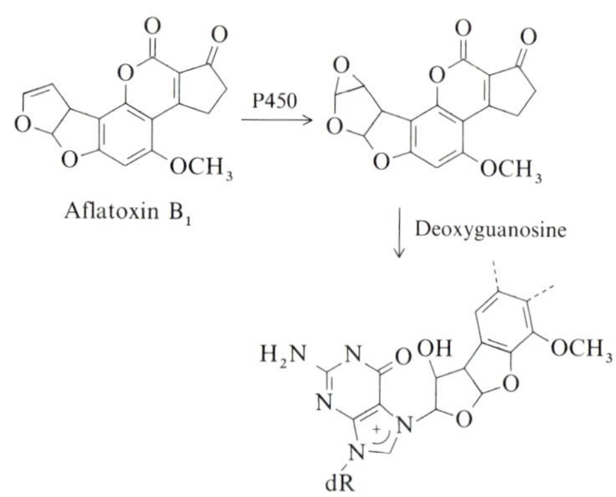

2,3-Dihydro-2-(deoxyguan-7-yl)-3-hydroxyaflatoxin B_1

Figure 2.16. Bioactivation of aflatoxin B_1 to an electrophilic oxirane which results in the formation of DNA adducts and is believed to initiate tumor induction by aflatoxin B_1 in the liver

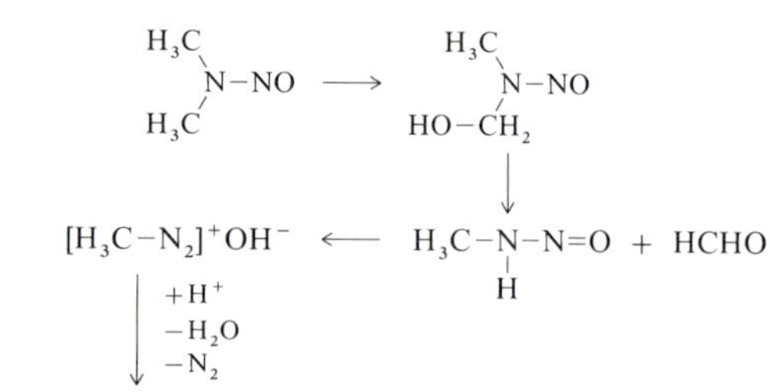

Scheme 2.

Besides cytochromes P450, other monooxygenases, such as flavin-dependent monooxygenase, or enzymes of phase-II biotransformation, such as UDP-glucuronyltransferases, sulfotransferases, or even the glutathione *S*-transferases (GST), may catalyze the bioactivation of xenobiotics [72]. For example, *N*-acetylamidofluorene is oxidized to *N*-hydroxyacetylamidofluorene by cytochrome P450. However, this metabolite is not electrophilic and requires further biotransformation via sulfate conjugation to the highly reactive *O*-sulfate ester. This sulfate ester fragments to a reactive intermediate (a nitrenium ion) that covalently binds to tissue constituents such as DNA.

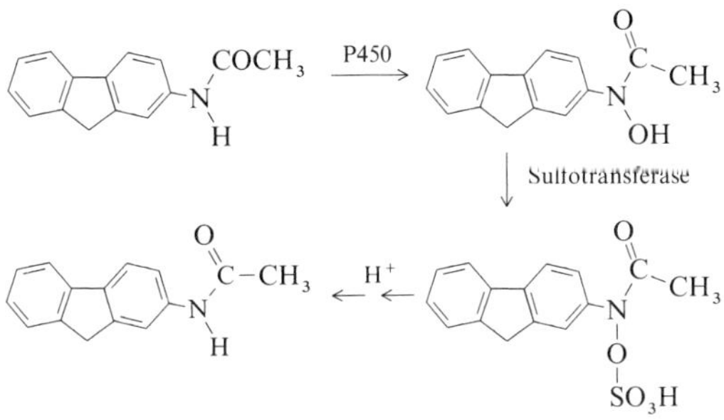

Some glutathione *S*-conjugates, which are biosynthetized to detoxify electrophiles, are toxic and mutagenic [73], [74]. 1,2-Dibromoethane is metabolized by glutathione conjugation to *S*-(2-bromoethyl)glutathione. Intramolecular displacement of the bromine on the adjacent carbon atom gives a highly strained, electrophilic episulfonium ion.

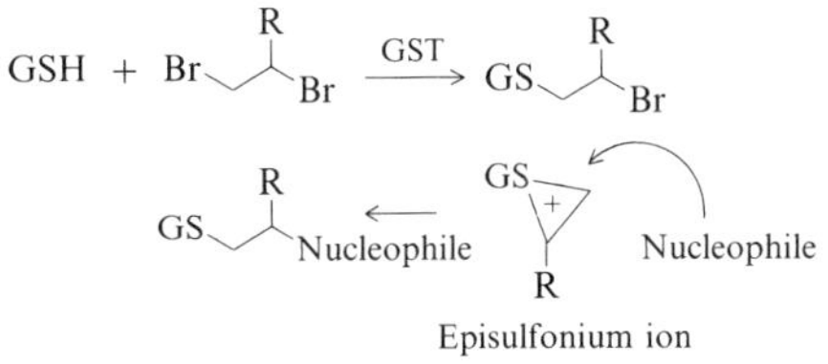

Other toxic glutathione *S*-conjugates require processing by the enzymes involved in mercapturic acid formation to give electrophiles. A minor pathway in perchlorethylene biotransformation results in biosynthesis of *S*-(1,2,2-trichlorovinyl)glutathione in the liver [75]. This glutathione *S*-conjugate is translocated to the kidneys and cleaved by γ-glutamyltranspeptidase and dipeptidases to *S*-(1,2,2-trichlorovinyl) L-cysteine, which is a substrate for renal cysteine conjugate β-lyase. Cysteine conjugate β-lyase cleaves it to pyruvate, ammonia, and dichlorothioketene whose binding to cellular macromolecules is likely responsible for the renal toxicity of perchloroethylene.

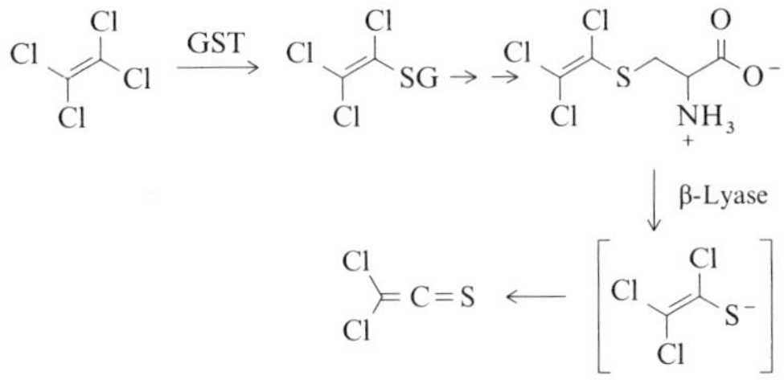

Because of the high concentrations of *S*-(1,2,2-trichlorovinyl)-L-cysteine obtained by active transport to the kidney, covalent binding of the dichlorothioketene formed via this pathway occurs only in the kidney, despite the presence of cysteine conjugate β-lyase in many other organs.

2.5.6.3. Biotransformation of Xenobiotics to Radicals

Free radicals are chemical species that may be formed by a one-electron oxidation to give a radical cation, by a one-electron reduction to give a radical anion, or by homolytic fission of a σ-bond to give a neutral radical.

Free radicals are highly reactive and, when formed in biological systems, can react with a variety of tissue molecules. Radicals may abstract hydrogen atoms or undergo oxidation–reduction reactions, dimerizations, and disproportionation reactions. Radicals may also participate in a radical chain mechanism, which is initiated by a reaction causing a free radical and propagated by a sequence of reactions giving further radicals as products.

The toxic and tumorigenic solvent carbon tetrachloride is the outstanding example of a bioactivation reaction to a free radical. Carbon tetrachloride is biotransformed by a one-electron reduction to yield the trichloromethyl radical and chloride.

$$Cl_4C + e^- \longrightarrow Cl_3C^{\cdot} + Cl^-$$

The trichloromethyl radical may abstract hydrogen atoms from tissue macromolecules to

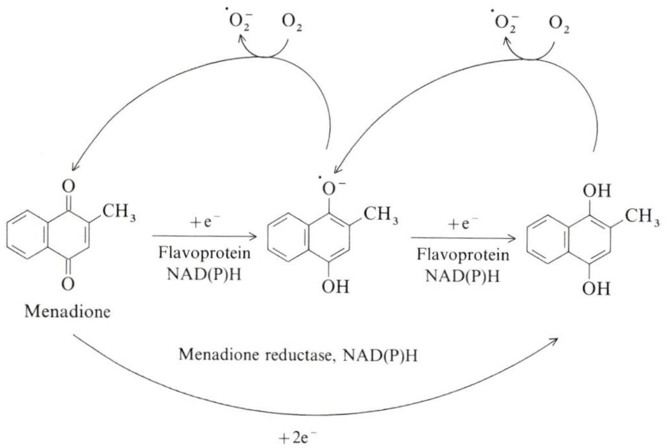

Figure 2.17. Biotransformation of menadione and induction of oxidative stress by reduction of triplet oxygen to the superoxide radical anion

give chloroform, a proven metabolite of carbon tetrachloride, or may dimerize to give hexachloroethane, another metabolite of carbon tetrachloride. Toxic effects of the formation of radicals during biotransformation reactions are lipid peroxidation and oxidative modification of proteins (see Section 2.5.6.4). Formation of radicals has been implicated in the bioactivation of many xenobiotics. Radicals may be formed by the NADPH-dependent cytochrome P450 reductase, by nitroreductases or by one-electron oxidations catalyzed by peroxidases such as prostaglandin synthase. Formation of free radicals from tissue constituents also plays an important role in the toxic effects of ionizing radiation [76]–[79].

2.5.6.4. Formation of Reactive Oxygen Metabolites by Xenobiotics

Xenobiotic-induced formation of reduced oxygen metabolites such as the superoxide (hyperoxide) radical anion $O_2^{-}\cdot$, hydrogen peroxide, and the hydroxyl radical is a mechanism for producing cell damage, the so-called oxidative stress [80]–[83]. The biotransformation of certain xenobiotics that can undergo redox cycling or enzyme-catalyzed oxidation–reduction reactions may be associated with the production of reduced oxygen metabolites. 2-Methylnaphthoquinone (menadione) has been used intensively to study the formation and cellular reactions of reduced oxygen metabolites. Menadione and other quinones are involved in enzymatic redox cycles; these one-electron oxidation reactions are associated with the formation of the superoxide radical anion by a one-electron reduction of triplet oxygen (Fig. 2.17). In aqueous solution, superoxide is not extremely reactive; the dismutation or further reduction of superoxide may, however, give rise to hydrogen peroxide. Hydrogen peroxide is also a poor oxidant in biological systems, but sufficiently stable to cross biological membranes. The toxicity of hydrogen peroxide is attributed to formation of the hydroxyl radical by the Fenton reaction catalyzed by metal ions such a Fe^{2+}:

$$M^n + H_2O_2 \longrightarrow M^{(n+1)} + HO^{\cdot} + HO^-$$

where M denotes a transition metal. The highly reactive hydroxyl radical may then initiate cellular damage by radical-based mechanisms. Besides menadione, oxidative stress may also be initiated by other xenobiotics such as the bispyridinium herbicide paraquat. Moreover, the formation of reduced oxygen metabolites plays an important role in host defense against infectious agents and in the initiation an propagation of certain diseases such as arteriosclerosis and polyarthritis.

Since oxygen radicals are also formed in low concentrations during cellular respiration, efficient mechanisms for their detoxification exist (see Section 2.5.6.5). Oxidative stress is thus observed only when the equilibrium between oxidants and reductants is disturbed and detoxification mechanisms are overwhelmed.

2.5.6.5. Detoxification and Interactions of Reactive Metabolites with Cellular Macromolecules

Reactive intermediates formed inside cells can react with low and high molecular mass cellular constituents. These interactions can result in the formation of less reactive chemicals and thus in detoxification, or they may perturb important cellular functions resulting in acute and/or chronic toxic effects such as necrosis or cancer. Generally, the interaction with low molecular mass constituents in the cell results in detoxification, whereas the irreversible interaction with cellular macromolecules results in adverse effects [65], [84]–[86].

Detoxification of reactive intermediates may be due to hydrolysis, glutathione conjugation, or interactions with cellular antioxidants. The reaction of electrophilic xenobiotics with the nucleophile water, present in high concentrations in all cells, is the simplest form of detoxification. Many of the products thus formed are of low reactivity and may be excreted rapidly. For example, acyl halides formed by the oxidation of olefins such as perchloroethylene are hydrolyzed rapidly to halogenated carboxylic acids; only minor amounts of the intermediary acyl halide reacts with proteins and lipids.

Glutathione-dependent detoxifications represent an important detoxification mechanism for metabolically formed electrophiles, free radicals, and reduced oxygen metabolites [87]–[90]. Electrophiles react with the nucleophilc sulfur atom of glutathione; this reaction may be spontaneous or enzyme catalyzed. Spontaneous reactions are observed at appreciable rates only with soft electrophiles (glutathione is a soft nucleophile); the conjugation of hard electrophiles with glutatione requires enzymatic catalysis; usually, the rates of conjugation catalyzed by glutathione *S*-transferase differ between hard and soft electrophiles, hard electrophiles being conjugated less efficiently. For example, the hard electrophile aflatoxin B_1 8,9-oxide does not spontaneously react with glutathione; only in the presence of a certain glutathione *S*-transferase enzyme, a glutathione *S*-conjugate of aflatoxin B_1 8,9-oxide is formed.

Glutathione also plays a major role in the detoxification of reactive oxygen metabolites and radicals. Selenium-dependent glutathione peroxidases are important enzymes catalyzing the detoxification of hydrogen peroxide. In this reaction, 2 mol of glutathione (GSH) are oxidized to glutathione disulfide (GSSG).

$$H_2O_2 + 2\,GSH \underset{\text{Glutathione reductase}}{\overset{\text{Glutathione peroxidase}}{\rightleftarrows}} GSSG + H_2O$$

Glutathione can be recycled by the reduction of glutathione disulfide by glutathione reductase.

The copper and zinc-dependent cytosolic and the manganese-dependent mitochondrial superoxide dismutase detoxify superoxide radical anions. Hydrogen peroxide formed by dismutation of superoxide is transformed to water and oxygen by catalase and thus detoxified:

$$2\,H_2O_2 \xrightarrow{\text{Catalase}} 2\,H_2O + \tfrac{1}{2}\,O_2$$

Several cellular *antioxidants* also play a role in the detoxification of radicals. α-Tocopherol is an important lipophilic antioxidant, whose presence in lipid membranes prevents damage to lipid constituents (e.g., unsaturated fatty acids) by radicals. The hydroxyl radical, the superoxide radical anion, and peroxy radicals react with α-tocopherol to yield water, hydrogen peroxide, and hydroperoxides, which may be detoxified further by catalase and glutathione peroxidase. α-Tocopherol is transformed during these reactions to give a stable radical of comparatively low reactivity. Ascorbic acid is an important antioxidant present in the cytoplasm of the cell that may also participate in the detoxification of radicals.

2.5.6.6. Interaction of Reactive Intermediates with Cellular Macromolecules

Although a substantial body of information is available on the biotransformation of xenobiotics to reactive metabolites and the chemical nature of those metabolites, considerably less is known about how reactive intermediates interact with cellular constituents and how those interactions cause cell injury and death. The reactions of toxic metabolites may result in the formation of covalent bonds between the molecule and a cellular target molecule, or they may alter the target molecule without formation of a covalent bond, usually by oxidation or reduction [91].

Table 2.8. Metabolically formed electrophiles and their prime targets for covalent binding in cells

Soft →				Hard
		Nucleophile		
SH of cysteine or glutathione	sulfur of methionine	primary or secondary nitrogen atoms in peptides (lysine, arginine, or histidine)	amino groups of purines and pyrimidines in RNA and DNA	oxygen in purines and pyrimidines in DNA and RNA
		Electrophile		
α,β-Unsaturated carbonyls, quinones	epoxides, alkyl sulfates, alkyl halides	nitrenium ions	benzylic carbocations	aliphatic and aromatic carbocations

Electrophilic metabolites may react with different nucleophilic sites in cells. Nucleophilic sites in cellular macromolecules are thiols and amino groups present in proteins, amino groups present in lipids, and oxygen and nitrogen atoms present in purine and nitrogen atoms present in purine and pyrimidine bases in DNA. The formation of a covalent bond may permanently alter the structure and/or activity of the modified macromolecule and thus result in a toxic response. The complexity of the reaction of electrophilic metabolites with the various nucleophilic sites in cells may be interpreted by the concept of hard and soft electrophiles/nucleophiles (hard and soft acids/bases) [92]. Thus, hard electrophiles formed during a biotransformation reaction (e.g., carbocations formed from *N,N*-dialkylnitrosamines) react predominantly with hard nucleophiles such as the oxgen and nitrogen atoms in DNA; soft electrophiles such as α,β-unsaturated carbonyl compounds (e.g., acrolein or benzoquinone) react predominantly with soft-tissue nucleophiles such as the sulfhydryl groups of cysteine in proteins (Table 2.8).

Covalent interactions of xenobiotics with proteins occur with several nucleophilic nitrogen atoms; both alkylation and acylation reactions of amino acids have been reported as the consequence of formation of reactive intermediates in cells. Besides the sulfur atom of cysteine, nitrogen atoms in the amino acids lysine, histidine, and valine are frequent targets for electrophilic metabolites. Consequences of the modifications may be inactivation of enzymes important for cellular function, changes in the tertiary structure of proteins, or changes in gene expression. Alkylation of the sulfhydryl-dependent enzymes of mitochondrial respiration is thought to play an important role in the initiation of mitochondrial dysfunction and thus cell damage.

Some modified proteins may also serve as immunogens. Hypersensitivity reactions, formation of immune complexes, and delayed hypersensitives may be the consequences of protein adduct formation. Indeed, many drug- and chemical-related hypersensitivity reactions observed in clinical medicine are based on the formation of covalent protein adducts and their recognition as "foreign" by the immune system [93].

Oxidative stress produces mixed disulfides of proteins with low molecular mass thiols such as glutathione, thus altering protein structure and function. In addition, oxidants and radicals promote the oxidation of amino acids in proteins, which may increase the susceptibility of these proteins to proteolysis [94], [95]. Increased protein oxidation has been implicated in cellular aging and in the mechanismus of toxicity of several redox active transition metals.

Radicals formed during the biotransformation of xenobiotics may abstract hydrogen atoms from cellular components [76]. The abstraction of hydrogen atoms from polyunsaturated fatty acids of lipids results in lipid peroxidation. The fatty acid radicals thus formed may react with molecular oxygen to give peroxy radicals, which react further to hydroperoxides. The initiated radical chain reactions cause the cleavage of carbon–carbon bonds in the fatty acids to short fragments such as α,β-unsaturated carbonyl compounds [96], [97] (see Section 3.3.1).

The disruption of membranes and the formation of toxic hydroperoxides and α,β-unsaturated carbonyl compounds may cause disruptions in cellular calcium homeostasis and thus cause biochemical changes ultimately leading to cell

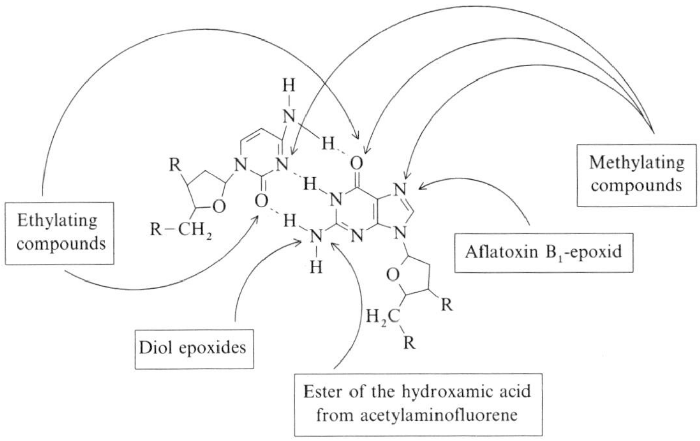

Figure 2.18. Regioselectivity of DNA alkylation by different electrophiles

death [98], [99]. The reaction of electrophilic metabolites with DNA constituents results in the formation of altered purine and pyrimidine bases or other DNA damage such as DNA strand breaks or the loss of single bases from the double helix. Many of these modifications are "premutagenic lesions." After gene expression, these lesions may be translated into mutations [100]. Mutations in certain genes are considered the basis for the evolution of neoplastic cells and cancer, and thus play a major role in chemical carcinogenesis. Other types of DNA damage may result in the activation of genes important for cellular differentiation or other regulatory functions. Electrophilic intermediates alkylate the nitrogen and oxygen atoms of the purine and pyrimidines bases in DNA; deoxyguanosine is often preferentially alkylated. The site of alkylation of a certain base in DNA again depends on the electrophilicity of the alkylation agent; hard electrophiles react preferentially with the oxygen atoms of guanosine; soft electrophiles alkylate the exocyclic amino groups (Fig. 2.18).

The pattern of base alkylation is different when DNA is modified in biological systems or when isolated DNA, nucleosides, and nucleotides are reacted with the xenobiotic or its metabolite(s). Regioselectivity is further modified by solvents, buffer salts, and concentration of reactants. Certain modification of deoxyguanosine result in the labilization of the glycosidic bond; loss of the deoxyguanosine derivative causes an "apurinic site" (AP) in DNA (Fig. 2.19). Other chemical modifications may result the lability of the five-membered ring and ring opening after reaction with water (Fig. 2.20).

Radicals formed as reactive intermediates may also cause DNA damage. Besides DNA strand breaks, which have been observed frequently, the reaction of oxygen-derived radicals may also result in the oxidation of purine and pyrimidine nucleotides. As a result of the development of sensitive techniques for the characterization of oxidative modifications in DNA, a number of modified bases have been indentified.

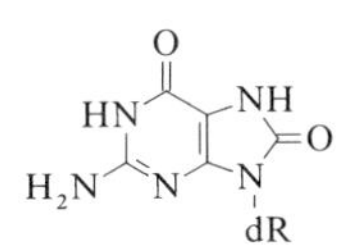

8-Hydroxydeoxyguanosine
(7,8-Dihydro-8-oxodeoxyguanosine)

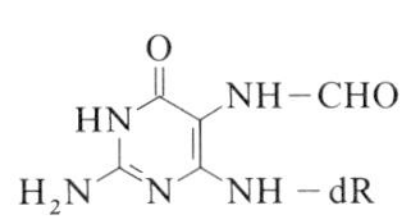

Formamidopyrimidine-deoxyguanosine

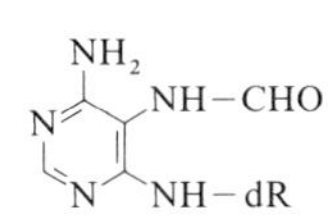

Formamidopyrimidine-deoxyadenosine

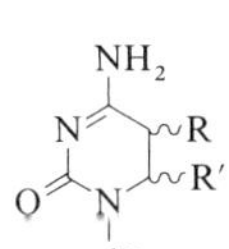

R,R' = H, OH, OOH
dR = Deoxyribose

5,6-Dihydrodeoxycytidine

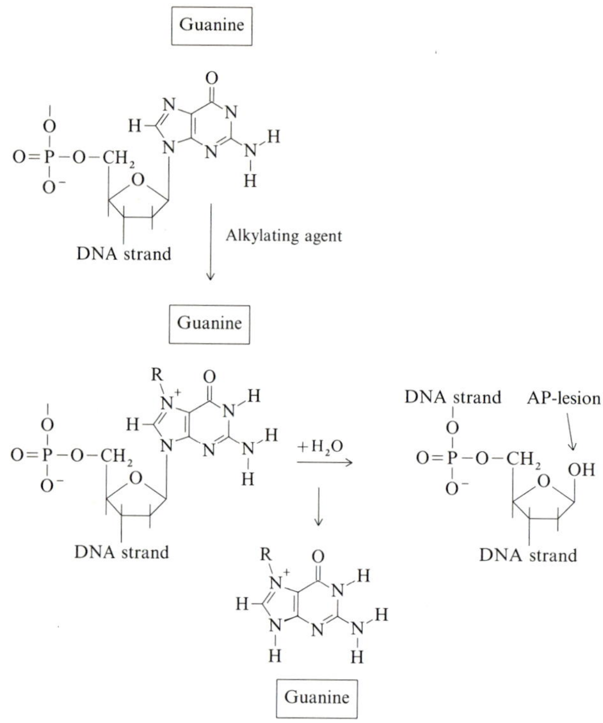

Figure 2.19. Labilization of the bond of guanosine to the DNA backbone by alkylation of the N-7 position resulting in the loss of the modified guanosine from DNA and the formation of an apurinic site (AP)

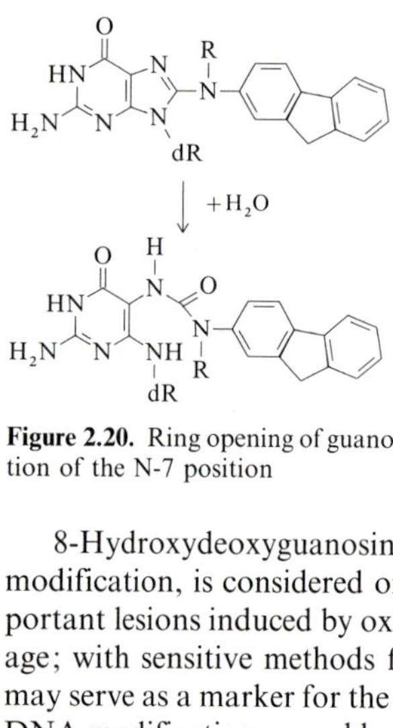

Figure 2.20. Ring opening of guanosine in DNA by alkylation of the N-7 position

8-Hydroxydeoxyguanosine, a premutagenic modification, is considered one of the more important lesions induced by oxidative DNA damage; with sensitive methods for quantitation, it may serve as a marker for the extent of oxidative DNA modification caused by a xenobiotic or by other processes.

2.5.7. Factors Modifying Biotransformation and Bioactivation

The biotransformation of xenobiotics may be modified by a variety of factors both intrinsic and extrinsic to the normal functioning of the organism. The changes in the extent of biotransformation may profoundly affect the toxicity of a specific chemical. When biotransformation results in detoxification and rapid excretion, increased rates of biotransformation will decrease toxicity. On the other hand, the toxicity of a chemical bioactiviated to reactive intermediates will increase by enhancing biotransformation. A variety of factors influence the extent of biotransformation; many of the effects listed in Sections 2.5.7.1 and 2.5.7.2 were decribed primarily in experimental animals. However, observations in humans (e.g., after drug treatment) indicate that similar effects, albeit not at the same magnitude or duration, must occur in humans.

2.5.7.1. Host Factors Affecting Biotransformation

Enzyme Induction. The activity of the enzymes of biotransformation can be enhanced by pretreatment with a range of structurally different chemicals. These chemicals can be drugs, pesticides, natural products, environmental contaminants, and even ethanol. The enhanced enzyme activities and the increased enzyme concentrations may results from an increased de

Table 2.9. Inducers of the enzymes of biotransformation and enzymes whose cellular concentrations are increased by pretreatment

Inducing agent	Induced enzymes
2,3,7,8-Tetrachlorodibenzodioxin	cytochrome P450, UDP-glucuronyltransferase
Ethanol	cytochrome P450
Phenobarbital	cytochrome P450, epoxide hydrolase, UDP-glucuronyltransferases
trans-Stilbenoxide	epoxide hydrolase
3-Methylcholanthrene	cytochrome P450, UDP-glucuronyltransferases

novo synthesis of the protein, reduced degradation, or other often-unknown factors. The increase in concentration of an enzyme of biotransformation in the organism, a certain organ, or cell type is termed "enzyme induction" [101], [102]. Several hundred different chemicals have been demonstrated to increase the biotransformation of other xenobiotics and to act as enzyme inducers. The majority of these studies have focused on the induction of the microsomal monooxygenases, mainly cytochrome P450 enzymes; however, other membrane-bound enzymes such as UDP-glucuronyltransferases may also be induced. Glutathione *S*-transferases are the only cytosolic enzymes of biotransformation whose activities can be increased by the administration of inducers to experimental animals (Table 2.9). The onset, magnitude, and duration of increases in the concentration of biotransformation enzymes after administration of an inducer and the associated biochemical and morphological effects depend on the chemical nature of the inducing agent, dose, and time of administration. For example, the times required for maximum induction of specific cytochrome P450 enzymes by the classical inducers phenobarbital and 3-methylcholantrene are different. Moreover, besides increases in the activity of hepatic monooxygenases, phenobarbital administration results in marked hepatic hypertrophy and proliferation of the smooth endoplasmic reticulum; these effects are absent in animals treated with 3-methylcholanthrene. Induction by parenteral application of 3-methylcholanthrene results in maximum enzyme concentrations in the liver within 48 h, whereas maximal induction by parenteral application of hypnotic doses of phenobarbital requires up to 5 d.

Enzyme induction is reversible. After withdrawal of the inducing agent, enzyme activities

Table 2.10. Mechanisms of cytochrome P450 enzyme induction by different xenobiotics

Cytochrome P450	Inducing agent	Mechanism of induction
1A1	2,3,7,8-tetrachlorodibenzodioxin	increased gene transcription
1A2	3-methylcholanthrene	stabilization of messenger RNA
2B1, 2B2	phenobarbital	increased gene transcription
2E1	ethanol, acetone	protein stabilization
3A1	dexamethasone	increased gene transcription, independent of glucocorticoid receptor
3A1	triacetyloleandomycin	protein stabilization
4A1	clofibrate	increased gene transcription, receptor mediated

return to basal levels within a characteristic time. Again, this time span is dependent on the chemical nature of the inducing agent. For example, cessation of phenobarbital treatment will result in a decline of enzyme activities to basal levels within one to two weeks. The mechanisms of enzyme induction are complicated and only partially understood. Current knowledge suggests that different chemicals influence the activities of the enzymes of biotransformation by different mechanisms; even the effect of a specific chemical on different enzymes may be due to separate mechanisms (Table 2.10).

Only a few of the mechanisms are well understood. A modulation of gene expression seems to be the basis for many inductive effects. For example, induction of cytochrome P450 1A1 is prevented by inhibitors of protein synthesis. In addition, recent studies using the potent inducing agent 2,3,7,8-tetrachlorodibenzodioxin have identified a high affinity binding protein (the arylhydrocarbon receptor, Ah receptor) with the properties of a receptor for 2,3,7,8-tetrachlorodibenzodioxin in rat liver cytosol [103], [104]. Binding of 2,3,7,8-tetrachlorodibenzodioxin to that protein results, after further interaction with other proteins, in the translocation of the formed complex from the cytosol to the nucleus. This translocation is followed by interaction with

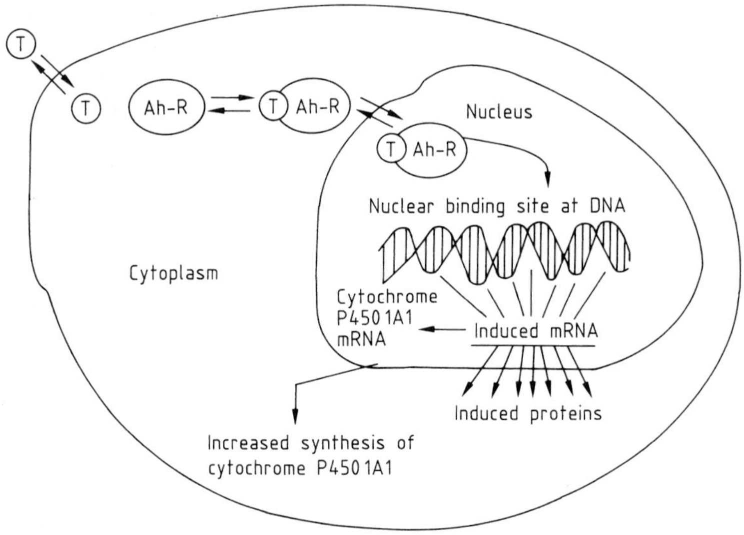

Figure 2.21. The Ah receptor (Ah-R) and mechanism of enzyme induction by 2,3,7,8-tetrachlorodibenzodioxin (T)

specific recognition sites on the genome, transcription, and translation of the specific gene for cytochrome P450 1A1 and for other enzymes of biotransformation such as UDP-glucoronyltransferases (Fig. 2.21).

In contrast, the mechanisms of enzyme induction by phenobarbital and, for example, ethanol are not defined. A specific receptor for phenobarbital has not been demonstrated; however, experiments in 1993 suggest an involvement of the glucocorticoid receptor in phenobarbital-mediated enzyme induction. Ethanol and other inducers, in contrast, seem to stabilize the cytochrome P450 2E1 protein against degradation by an unknown mechanism.

Enzyme Inhibition. The decrease in activity of specific enzymes of biotransformation is termed inhibition of biotransformation. As noted above, inhibition of biotransformation may increase or decrease the toxicity of a xenobiotic. For example, the inhibition of cytochrome P450 by 2-(diethylamino)ethyl-2,2-diphenylpentanoate (SKF-525A) causes an increase in hexobarbital sleeping time, but a decrease in the hepatoxicity of carbon tetrachloride.

Several mechanisms responsible for inhibition of xenobiotic-metabolizing enzymes are operative [105], [106]. Besides inhibition of protein synthesis and thus synthesis of the enzyme, xenobiotics may irreversibly bind to the active site of the enzyme (suicide inhibition) [107]. After activation of the xenobiotic by the enzyme, the intermediate formed reacts with constituents of the enzyme at or near the active site, thus blocking further catalytic activity. This effect has been demonstrated with several xenobiotics that are inhibitors of cytchrome P450, such as alkenes and compounds containing allylic and acetylenic derivatives. For example, the reactive metabolite formed by the cytochrome P450-catalyzed oxidation of ethylene—ethylene oxide—may alkylate the pyrrole nitrogens in the heme moiety, thus resulting in heme destruction.

During exposures to mixtures, chemicals with high affinity for certain enzymes of biotransformation will be preferentially metabolized, thus, the biotransformation of other constituents of the mixture will be reduced or even totally inhibited. This reduction in biotransformation may also alter the toxicity of a chemical present in a mixture compared to that of the pure chemical. For example, coadministration of ethanol markedly reduces the toxic effects of methanol (metabolic acidosis, reversible or even permanent blindness). These toxic effects are caused by the oxidation of methanol to formic acid as toxic metabolite; formic acid accumulates in the organism and causes damage to the optical nerve. Methanol oxidation to formaldehyde is competitively blocked by the administration of ethanol; under these circumstances, the toxic metabolite formic acid cannot be formed;

methanol is excreted from the organism unchanged in urine and by exhalation.

Inhibition of some enzymes of biotransformation may also be due to effects on the tissue levels of necessary cofactors. The availability of glutathione for conjugation is reduced by blocking glutathione biosynthesis; diethyl maleate and some other chemicals deplete intracellular glutathione concentrations by reacting with glutathione to a glutathione *S*-conjugate. Pretreatment of experimental animals with these chemicals, followed by the application of a xenobiotic that requires glutathione for detoxification, will result in an increased toxic response. Moreover, large doses of nontoxic chemicals metabolized by sulfotransferases may deplete the cofactor for sulfate conjugation and thus alter the disposition, and probably the toxicity, of other xenobiotics undergoing sulfate conjugation.

Genetic Differences in the Expression of Xenobiotic-Metabolizing Enzymes. The ability of different animal species to metabolize xenobiotics is related to evolutionary development and therefore to different genetic constitution; thus, major species differences in the extent and pathways of biotransformation exist. These variations can be divided into qualitative and quantitative differences. Qualitative differences involve metabolic pathways and are related to species defects or peculiar reactions of a species. For example, guinea pigs do not have the enzymatic capacity to catalyze the last step in mercapturic acid formation, the *N*-acetylation of cysteine *S*-conjugates, and therefore excrete cysteine *S*-conjugates as end products of this pathway. Certain species such as cats cannot form glucuronides from xenobiotics.

Quantitative variations are often due to species differences in gene and enzyme regulation. For example, interindividual differences have been described in humans in the biotransformation of many drugs. The *N*-acetylation of the tuberculostaticum isoniazid has a genetic basis. Some individuals are homozygous for a recessive gene, which may result in the absence of isoniazid *N*-acetyltransferase; they are "slow acetylators." In normal homozygotus or heterozygotus "rapid acetylators," isoniazid is transformed rapidly to the *N*-acetyl derivative. The *N*-acetyltransferase polymorphisms are correlated with different responses to isoniazid-induced toxicities; 80% of the Japanese and Eskimos are "rapid acetalators," whereas only 40 to 60% in some European populations belong to this category [108]. Polymorphisms in the expression of cytochrome P450 1A1 and a specific glutathione *S*-transferase in human lung have been implicated in an increased rate of lung cancer in individuals expressing high cytochrome P450 1A1 who are deficient in this glutathione *S*-transferase. Cytochrome P450 1A1 bioactivates aromatic hydrocarbons present in cigarette smoke to yield electrophiles; glutathione *S*-transferase detoxifies these metabolites.

Influence of Sex on Biotransformation Reactions. Sex differences in the extent and pathways of biotransformation may be based on sex-depentend expression of certain enzymes of biotransformation. For example, adult male rats metabolize many xenobiotics at higher rates than females; both phase-I and phase-II biotransformation reactions seem to be influenced by sex-dependent factors. With cytochrome P450, at least three different hepatic enzymes have been demonstrated to be under the control of sex hormones (Table 2.11).

In experimental animals, sex differences in the expression of biotransformation enzymes usually become apparent at puberty and are maintained throughout adult life. Despite the relatively large sex-dependent variations seen in animal studies, sex seems not to have a profound influence on the biotransformation of chemicals in humans.

Dietary Constituents and the Enzymes of Biotransformation. Nutritional factors influencing biotransformation may be mineral or vitamin deficiencies, protein content, starvation, and natural substances in the diet. Mineral deficiencies (calcium, copper, zinc) reduce the activities of cytochromes P450. On the other hand, an excess of dietary iron has been observed to in-

Table 2.11. Sex hormone-dependent hepatic cytochrome P450 in the rat

Enzyme	Sex specificity	Remarks
P450 2D	female	expressed at a hormone-independent basal rate, stimulated by estrogen, suppressed by androgen
P450 2C	male	
P450 3A	male	neonatally imprinted by androgen

crease monooxygenase activity. Dietary cobalt, calcium, and manganese may increase the hepatic levels of glutathione and thus influence glutathione *S*-conjugate formation. Several vitamins are directly or indirectly involved in the regulation of cytchrome P450. For example, diets deficient in vitamins C and E reduce the activity of monooxygenases, whereas deficiencies in other vitamins increase monooxygenases. Moreover, several vitamins serve as important cellular antioxidants and influence the energy and redox state of the cell, thus also affecting biotransformation reactions.

Low-protein diets generally reduce the activity of cytochrome P450 and certain phase-II biotransformation reactions. Thus, nutrient status may also modify the toxicity of xenobiotics. For example, *N,N*-dimethylnitrosamine is a potent hepatocarcinogen in rats kept on a high-protein diet, but almost without effect in rats kept on a low-protein diet. Food deprivation reduces the hepatic concentration of glutathione by as much as 50 % due to reduction of glutathione biosynthesis. Thus, xenobiotics detoxified by glutathione conjugation are more toxic in starved than in fed animals. Fasting has also been shown to increase the levels of cytochrome P450 2E1, but decrease the levels of cytochrome P450 2C11. Several natural ingredients in the diet of laboratory animals such as indoles, diallyl disulfide, and psoralens may increase the activities of cytochromes P450. However, some of these compounds may selectively increase some cytochrome P450 enzymes, but inhibit others. For example, diallyl disulfide, a constituent of garlic, has been shown to induce cytochrome P450 2B in rats, but inhibits cytochrome P450 2E1. Vegetable ingredients present in broccoli are potent inducers of phase-II enzymes of biotransformation and thus increase the capacity of the organism to detoxify reactive intermediates. These natural ingredients are thought to play a major role in the anticarcinogenicity of diets rich in those vegetables.

2.5.7.2. Chemical-Related Factors that Influence Biotransformation

Xenobiotic-related factors influencing biotransformation are its physicochemical properties (i.e., the chemical structure of the xenobiotic including the presence of functional groups) and dose. The major determinant of the rate of

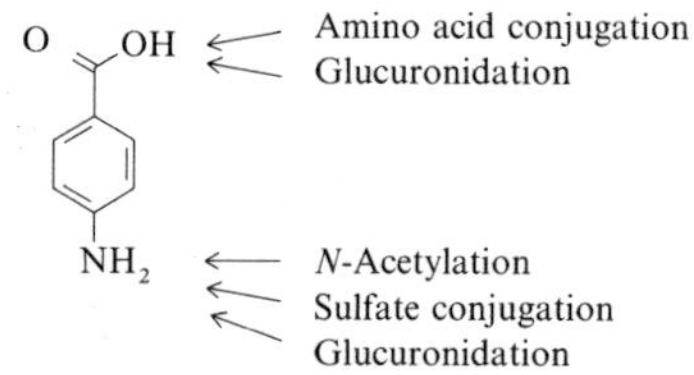

Figure 2.22. Possible biotransformation pathways for *p*-aminobenzoic acid

biotransformation is the concentration of the substrate at the active site of the enzyme. This concentration is determined by structure, lipophilicity, and dose. Lipophilic xenobiotics readily cross cell membranes and are rapidly absorbed and distributed in the organism. Moreover, lipophilic xenobiotics show a higher partitioning into lipid membranes. These factors contribute to higher concentrations of lipophilic xenobiotics at the active center, especially of membrane-bound enzymes. The presence of functional groups also influences rates and routes of biotransformation. Certain functional groups may compete for the same substrate for conjugation; also specific functional groups may undergo different reactions as indicated in Figure 2.22 for *p*-aminobenzoic acid.

The presence of other specific functional groups may have a major effect on biotransformation and its regioselectivity. For example, the presence of trifluoromethyl groups in alkanes renders the adjacent methylene carbon almost inert to hydroxylation and strongly influences the regioselectivity of enzymatic hydroxylation on aromatic rings.

Dose is one of the most important factors determining rate and route of biotransformation in more complex molecules. Certain enzymes of biotransformation have a high affinity but low capacity for a specific chemical, whereas other enzymes have a high capacity but low affinity. As dose increases, high-affinity, low capacity enzymes will become staturated and low-affinity, high-capacity pathway(s) will biotransform a larger percentage of the dose [109].

2.5.8. Elimination of Xenobiotics and their Metabolites

The evolution of complex forms of life necessitated the development of specialized mechanisms to eliminate waste products formed from endogenous compounds and to prevent the accumulation of toxic xenobiotics present in the diet and taken up with food. Excretion of wastes by earlier forms of aquatic life was largely passive and involved the loss of large volumes of water und nutrients. For land-living animals, conservation of water, minerals, and nutrients was necessary for survival. Therefore, complex mechanisms for the elimination of both endogenous chemicals and xenobiotics evolved. A wide variety of xenobiotics may be handled by the evolved excretory mechanisms and may thus be efficiently eliminated from the body.

Xenobiotics may be excreted as parent compound, metabolites or conjugates formed during phase-II biotransformation reactions. A major route of excretion of xenobiotics is via the kidney, and in some cases the urinary elimination of a parent compound or metabolite can be used to determine the absorbed dose. The kidney is the only organ that functions almost exclusively as an organ of elimination. The cells of the liver have more varied functions than those of the kidney; however, the liver is also important in the excretion of chemicals not effectively eliminated by the kidneys. Chemicals may be eliminated from the liver into bile and thus be finally excreted with feces. Active transport mechanisms (i.e., transport against a concentration gradient) play a major role in renal and hepatic excretion of xenobiotics. In contrast, in most other organs that serve as excretory systems for xenobiotics, passive excretion mechanism are operative. For example, volatile chemicals and metabolites may be eliminated in expired air; this route is quantitatively significant for some solvents and inhalation anesthetics. Specific xenobiotics may also be excreted by sweat, salvia, and milk.

2.5.8.1. Renal Excretion

The function of the kidneys accounts for the elimination of most of the byproducts of normal metabolism and most of the polar xenobiotics and metabolites of lipophilic xenobiotics that humans and experimental animals are exposed to [110]. The kidney is a complex structure that consists of a number of different cell types [111]–[113]. Essentially the kidney filters the blood. All components present in blood with a molecular mass of less than 50 000 (depending on structure and charge) enter the tubular system; there, important nutrients and most of the filtered water

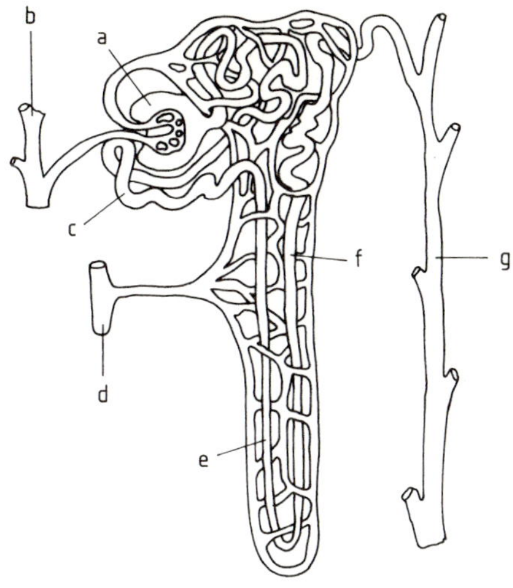

Figure 2.23. Structure of a nephron
a) Glomerulus; b) Renal artery; c) Proximal tubule; d) Renal vein; e) Loop of Henle; f) Distal tubule; g) Collecting duct

are recovered. Only a small fraction of the primary filtrate is excreted as urine (1–2 L/d). The human kidney consists of approximately 2×10^6 nephrons, which are the functional units that perform the filtration of blood and the recovery of essential nutrients. The structure and components of a single nephron are shown in Figure 2.23.

Glomerular Filtration. Renal excretion is the product of three complex and interactive processes, glomerular filtration, tubular reabsorption, and tubular secretion. Glomerular filtration is the passive filtering of the plasma as a result of its passage through glomerular pores (7–10 nm in diameter) under hydrostatic pressure generated by the heart. The average rate of glomerular filtration in adults is 125 mL/min or almost 200 L/d. Glomerular filtration shows little specificity other than molecular size. Free solutes in the plasma that pass through the glomerular pores will all appear in the filtrate. Only protein-bound low molecular mass xenobiotics will not appear in the filtrate and remain in blood. Glomerular filtration is influenced by factors that affect the hydrostatic pressure or integrity of the glomerulus; thus, these factors may result in elevated plasma concentrations of excretory products formed from endogenous chemicals and from xenobiotics.

Tubular Reabsorption. The daily volume of the glomerular filtrate exceeds that of total body water by a factor of four and contains many necessary nutrient such as glucose, amino acids, and salt; therefore, most of the glomerular filtrate must be recovered. Thus, the second major process occurring in the kidney is tubular reabsorption. A number of discrete mechanisms, both active and passive and with varying degrees of specificity, are involved in tubular reabsorption. Many of these mechanisms are located in the cells of the proximal segments of the tubules. These cells account for the reabsorption of 65–90 % of the glomerular filtrate. Glucose, certain cations, low molecular mass proteins, amino acids, and organic acids are actively reabsorbed. Water and chloride are passively reabsorbed as a result of the osmotic and electrochemical gradients generated by the active transport of sodium and potassium. The osmolarity of the fluid in the collecting duct is regulated in the loops of Henle; most of the remaining water and ions are reabsorbed in the distal tubules and collecting ducts. The rate of reabsorption in these segments of the proximal tubule is regulated to maintain the osmolar concentration of the blood. Most xenobiotics are also reabsorbed after glomerular filtration by passive diffusion during passage through the nephron. Passive tubular reabsorption of lipophilic xenobiotics is therefore greater than the reabsorption of polar xenobiotics or endogenous wastes.

Tubular Secretion. Xenobiotics present in blood may also be excreted by the kidney by tubular secretion, which transports xenobiotics from the peritubular fluid (blood) to the lumen (urine) in the tubule. Tubular secretion is often selective; active transport mechanisms account for the secretion of many organic acids, including glucuronides and sulfates, and strong organic bases. Secretion of weak bases and some weak acids may also occur by a passive mechanism that utilizes pH differences between peritubular fluid and urine. At the pH of the tubular lumen, these compounds become ionized and do not diffuse back across the cell wall.

Factors Affecting Renal Excretion of Xenobiotics. Xenobiotics are excreted by the same mechanisms that eliminate endogenous wastes; very *polar xenobiotics* in plasma water are removed primarily by glomerular filtration and excreted in the urine with minimal involvement of tubular reabsorption or secretion. The rate of renal elimination of most of these xenobiotics depends largely on the rate of glomerular filtration. Since *lipophilic compounds* cross cell membranes more readily, they distribute into a much larger tissue volume than polar compounds, which are more likely to be restricted to the vascular volume. However, lipophilic xenobiotics metabolized to more polar compounds are usually returned rapidly to the circulation and readily excreted. Therefore, the rate of metabolism of a xenobiotic may also play an important role in its rate of excretion.

2.5.8.2. Hepatic Excretion

Nutrients and xenobiotics are delivered from the gastrointestinal tract to the liver by the portal vein to be biotransformed there. The liver is thus located between the intestinal tract and the general circulation, and thus ideally located for the biotransformation of nutrients and xenobiotics taken up by this route. Besides participating in the biotransformation of xenobiotics, the liver can be a major excretory organ and contributes to the excretion of many xenobiotics by eliminating them with bile into the gut and thus into feces [114]–[116].

The bulk of the liver consists of cells arranged in plates two cells thick. These plates are arranged radially around the terminal branches of the hepatic veins and are exposed to blood from the portal vein and hepatic artery flowing through interconnecting spaces referred to as hepatic sinusoids. The sinusoidal walls are freely permeable even to relatively large particles; special transport may play a role only in the uptake of certain anions from the blood. The epithelial hepatic cell is the smallest unit of the liver and accounts for most of the varied functions of this organ, including storage, secretion, biotransformation, and excretion (Fig. 2.24).

Bile formation is thought to be a product of the active transport of certain ionized compounds and the passive transport of other solutes and water, which follow a concentration or electrochemical gradient. *Active secretion* of anions and cations appears to be controlled by different mechanisms, but the compounds actively excreted are usually amphipathic molecules (i.e., they have both polar and nonpolar moieties). Bile acids are the classic example of endogenous amphipathic molecules. Conjugates of lipophilic

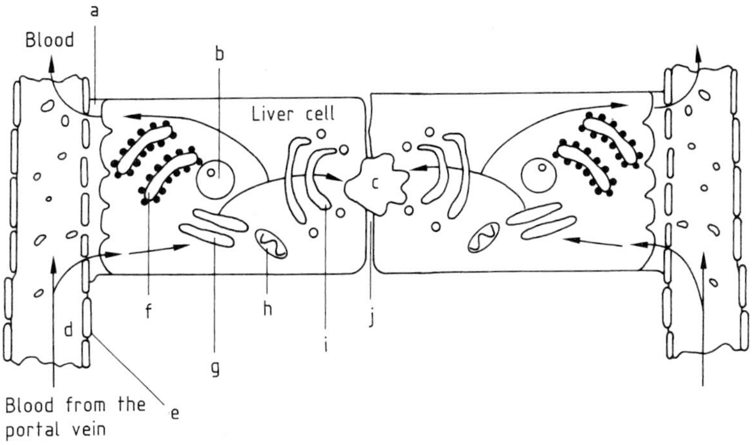

Figure 2.24. Structure of a liver cell (hepatocyte) and possible pathways for the uptake and elimination of xenobiotics and their metabolites
a) Space of Disse; b) Nucleus; c) Bile; d) Liver capillary; e) Sinusoidal wall; f) Rough endoplasmic reticulum; g) Smooth endoplasmic reticulum; h) Mitochondria; i) Golgy complex; j) Bile duct

xenobiotics are also examples of amphipathic molecules, and many of these conjugates are ionized, which facilitates excretion from the hepatocyte by active transport mechanisms.

As a result of both active and passive secretion, xenobiotics excreted into bile may be classified in different groups. Solutes found in bile may be divided according to their concentration ratio in bile versus blood. For example, the excretion of Na^+, K^+, Cl^-, and glucose, for which the bile:blood ratio is close to unity, is thought to be passive. Conjugates of xenobiotics and bile acids have a bile:blood ratio > 10 and are thought to be actively excreted from liver into bile. Other compounds such as proteins, inulin, sucrose, and phosphates do not cross the canalicular membrane of hepatocytes and have a bile:blood ratio $\ll 1$. Bile secreted by the liver cells into the bile canaliculi flows into the narrowest branches of the bile duct, the cholangioles, then into the hepatic duct, which carries the bile to the gallbladder and finally into the intestine. Lipophilic xenobiotics may appear in bile at low concentrations prior to metabolism. However, a compound absorbed from the stomach or intestine after ingestion is likely reabsorbed from the intestine if it is secreted in bile before it has been structurally modified by biotransformation. Moreover, conjugates excreted into bile such as glucuronides can be hydrolyzed by enzymes present in the bacteria of the intestinal microflora and the aglycone may be reabsorbed. Most xenobiotics reabsorbed from the intestine are returned to the liver. The process of excretion into bile, reabsorption from the intestine, and return to the liver is termed enterohepatic circulation. Enterohepatic circulation serves as an efficient physiological recovery mechanism for bile acids and certain hormones. When xenobiotics are trapped in this cycle, their rate of excretion from the body may be significantly reduced and their toxicity may be significantly increased.

Effect of Molecular Mass on Hepatic Excretion of Xenobiotics. From the liver cell, xenobiotics may be excreted into bile or returned to blood. The molecular mass of nonvolatile organic xenobiotics or their metabolites determines the primary route by which they will be excreted from the hepatocyte. In the rat, xenobiotics, their metabolites, and conjugates formed in phase-II reactions with molecular masses > 500 are most often excreted into bile; xenobiotics and their metabolites with molecular masses < 350 return from the hepatocyte to blood and are thus delivered to the general circulation and to the kidney for excretion. In humans, the critical molecular mass threshold for biliary excretion is approximately 500.

2.5.8.3. Xenobiotic Elimination by the Lungs

Any xenobiotic present in blood with sufficient volatility will pass from the blood across the alveolar membrane into the air space of the lung and may be exhaled. The rate of elimination from blood of a volatile xenobiotic depends on the solubility of the xenobiotic in blood, the rate of respiration, and blood flow to the lung. Xenobiotics such as ether, which are highly soluble in blood, are eliminated rapidly by exhalation; their elimination may be enhanced by a forced increase in the rate of respiration (hyperventilation). In contrast, volatile xenobiotics with low solubility in blood are only slowly cleared from the lung and thus from the body by exhalation; their rate of exhalation may not be markedly influenced by hyperventilation. The proportionality among xenobiotic volatility, blood solubility, and the concentration of a volatile xenobiotic in the blood is utilized to determine blood alcohol content and thus estimate sobriety by breath analysis.

2.6. Toxicokinetics

The toxic response to chemical exposure depends particularly on the magnitude, duration, frequency, and route of exposure. These will determine the amount of material to which an organism is exposed (the exposure dose) and, hence, the amount of material that can be absorbed (the absorbed dose). The latter determines the amount of material available for distribution and formation of toxic metabolites and, hence, the likelihood of inducing a toxic effect. Opposing absorption and metabolite accumulation is elimination. All of these factors define the disposition of a xenobiotic. The modeling and mathematical description of the time course of the disposition of a potentially toxic xenobiotic in the organism are termed toxicokinetics [117]. Most of the methodologies and principles applied in toxicokinetics were first used to model the kinetics of chemicals used as drugs (pharmacokinetics).

The objective of both toxicokinetics and pharmacokinetics is to quantitate the dynamic time course of xenobiotic absorption, distribution, biotransformation, and elimination processes in living organisms. Both the entire process of disposition and individual steps such as elimination may be characterized.

2.6.1. Pharmacokinetic Models [118], [119]

A common way to describe the kinetics of drugs is to represent the body as a number of interconnected compartments, which may or may not have an anatomical or physiological reality. These compartments represent all those tissues, organs, and fluids in the body that are kinetically indistinguishable from each other. A compartment might be represented by a cluster of cells within an organ, an organ, the blood, or the entire body taken together. Recently, so-called physiologically based pharmacokinetic and toxicokinetic models have been developed. These models permit the modeling of the time course of the concentration of a chemical in a tissue based on physiological considerations and may be particulary useful for the interspecies extrapolations necessary in risk assessment.

2.6.1.1. One-Compartment Model

The simplest toxicokinetic model depicts the body as a single homogeneous unit within which a xenobiotic is uniformly distributed at all times. The toxicokinetics of a xenobiotic may be analyzed by a one-compartment model if the plasma concentrations determined after a single dose decrease exponentially (the plot of the logarithm of the plasma concentration versus time yields a straight line). In this model, elimination of a chemical from the body occurs by first-order processes, i.e.,

$$\frac{\mathrm{d}X}{\mathrm{d}t} = -k_e X$$

where X is the amount in the body at time t, and k_e is the rate constant for first-order elimination. In compartmental analysis, k_e is often referred to as an apparent first-order rate constant, to emphasize that the underlying processes may in reality only approximate first-order kinetics. For example, the xenobiotic might actually be eliminated by active biliary secretion for which zero-order saturation kinetics would be observed under suitable conditions. If the xenobiotics is not present in the organism and a known amount X_0 is rapidly administered, the total amount of the xenobiotic initially in the body will be approximately

$$X_0 = \text{total dose}$$

The amount of the xenobiotic present in the organism at a certain time after administration is then:

$$X(t) = X_0 \times e^{-kt}$$

The first-order rate constant of elimination can be determined from the slope of the plot of the log plasma concentration versus time and can be applied to estimate the half-life of elimination for the xenobiotic

$$t_{1/2} = \frac{0.693}{k_e}$$

Within seven half-lives, a xenobiotic is almost completely eliminated (99.2%), although complete elimination will never theoretically be achieved. Important characteristics of the first-order elimination of a xenobiotic according to the one-compartment model are:

1) The half-life of the xenobiotic is independent of dose; the semilogarithmic plot of the plasma concentration of the xenobiotic versus time yields a straight line
2) The concentration of the xenobiotic in plasma decreases by a constant fraction per unit of time

Xenobiotics responsible for toxic effects are usually not injected into blood, so intake is not instantaneous compared to distribution and elimination, and a lag period is observed before maximal concentration in plasma. If intake also approximates a first-order absorption process, then the rates of absorption and elimination will determine the time course of the plasma concentration (Fig. 2.25).

The rate of change in the concentration of a xenobiotic in a one-compartment model with first-order absorption and elimination may be described by

$$\frac{dX}{dt} = -k_e X_e + k_a X_a$$

where k_a is the rate constant of absorption, X_a is the amount of xenobiotic present at the site of absorption, k_e is the rate constant of elimination, and X_e is the amount of xenobiotic eliminated.

The rate constant of elimination can be determined experimentally by treating the elimination phase as if it occurred after the intravenous injection described above. This is done by extrapolation of the terminal straight line of the semilog plot of the ordinate at a putative X_0. By determining the plasma concentration of the xenobiotic at time points shortly after administration, the rate constant for absorption may be determined by plotting the difference between the early plasma concentrations and the extrapolated portion of the elimination curve.

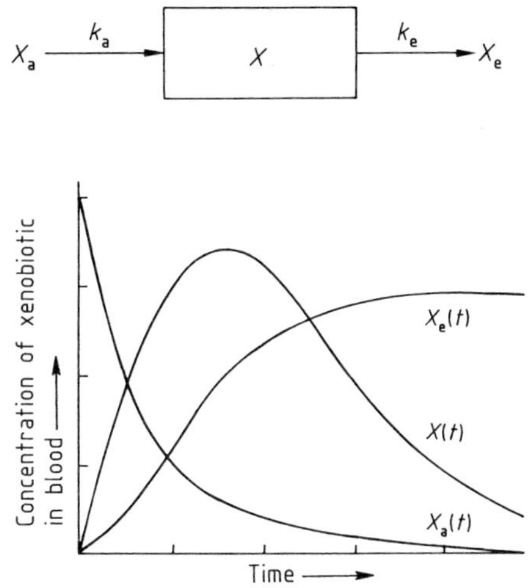

Figure 2.25. One-compartment model with first-order absorption and first-order elimination

2.6.1.2. Two-Compartment Model

If a xenobiotic does not distribute and equilibrate throughout the body rapidly, a two-compartment model provides a better description of the kinetics of disposition. In this case, the semilogarithmic plot of plasma concentration versus time does not yield a straight line. In two-compartment models, the main or central compartment is assumed to represent the blood and highly perfused organs and tissues such as the liver, heart, and kidneys that are in rapid distribution equilibrium with the blood, whereas the second or peripheral compartment corresponds to poorly perfused tissues such as muscles and fat. This model is depicted in Figure 2.26 and is called a two-compartment open pharmacokinetic model.

The time course of the plasma concentration of the xenobiotic can be decribed by two overlaying monoexponential terms of the type

$$X = A e^{-kt} + B e^{-k't}$$

where A and B are proportionality constants and k and k' are rate constants. Thus, the pharmacokinetics of a two-compartment model describe a biexponential decay in the amount of xenobi-

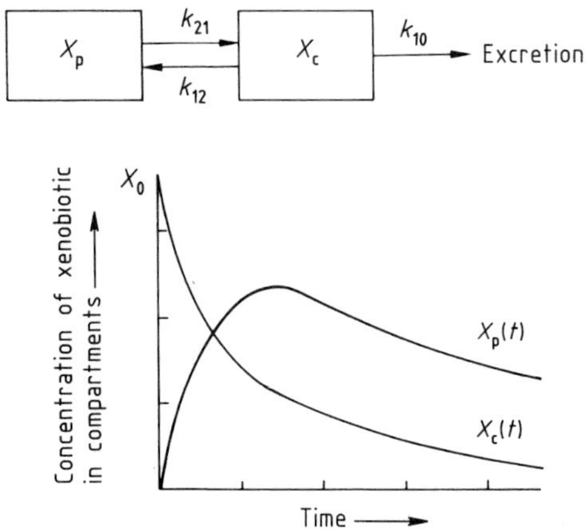

Figure 2.26. Two-compartment model for the distribution of a xenobiotic
X_c = concentration in the central compartment; X_p = concentration in the peripheral compartment; X_0 = initial dose; k_{21}, k_{12}, k_{10} = rate constants

otic in the body. Analogous to a one-compartment model, a distribution equilibrium is assumed to exist within each compartment that can be expressed in terms of blood concentration. According to this equation, a plot of the logarithm of blood concentration versus time will yield a biphasic decay from which the constant A, B, k, and k' can be estimated either graphically or by nonlinear regression analysis. Once these experimental constants have been determined, the pharmacokinetic parameters can be calculated. The numerical values of these parameters can aid in assessing the relative importance of tissue distribution and elimination to the disposition of the xenobiotic.

2.6.2. Physiologically Based Pharmacokinetic Models

Although compartmental analysis models are convenient to use and provide useful descriptions of the overall time course of xenobiotic disposition, the practical limitations of curve fitting to experimental data generally restrict such models to one or two compartments. Alternatively, a compartmental model can often be developed from physiological principles, thereby circumventing the inherent limitations of curve-fitting analysis. A physiological pharmacokinetic or toxicokinetic model is a mathematical description of the disposition of a xenobiotic in the organism or in part of it (e.g., a specific organ). Such a model is constructed by using physiological and biochemical parameters such as blood flow rates, tissue and organ sizes, binding, and biotransformation rates [120]–[122]. These are generally more complex and require the specification of many parameters; thus, physiological compartmental models are still simplified representations of biological systems. In addition, the determination of physical and biochemical parameters is often both difficult and inaccurate. However, the physiological framework provides several advantages: the physical definition of compartments and transfer rates facilitates the incorporation of existing knowledge about the quantitative behavior of biological systems into the model; physiological changes with time during chronic exposure to a xenobiotic, such as those due to physical growth or induction of biotransformation and changes in excretion rates, can be introduced (Fig. 2.27).

Most important, however, a reasonable basis exists for extrapolating the kinetics of a xenobiotic to predict its disposition following various types and patterns of exposure and, most important, to extrapolate from experimental data obtained in one species of experimental animal to other animal species and humans.

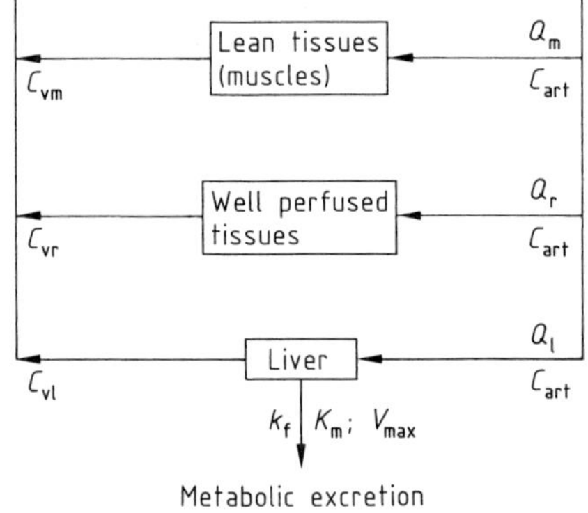

Figure 2.27. Schematic representation of a physiologically based toxicokinetic model

Physiologically based models use the concepts of mass balance and flow-limited transport under the following definitions: blood serves to distribute a xenobiotic from the site of absorption to other parts of the body. In the normal sequence of events, a chemical entering the bloodstream is often distributed rapidly within the blood, and its blood concentration can be considered essentially uniform.

The chemical enters and leaves the compartment with the blood flow and diffuses or is transported from blood to tissue and back. The chemical may also undergo a variety of physical interactions such as binding to macromolecules; the result is a partitioning between tissue and blood, depending on the chemical's particular affinity for each medium. Diffusion or transport directly between adjacent compartments, enzymatic biotransformations, and excretion may also occur. The net result of all of these changes is expressed by a mass balance differential equation, which is simply a mathematical statement of the conservation of mass. Frequently, these mass balance equations can be greatly simplified because many of the terms required may not apply to a particular compartment.

3. Mechanisms of Acute and Chronic Toxicity and of Chemical Carcinogenesis

3.1. Biochemical Basis of Toxicology

Since the 1960s, toxicology has changed from a descriptive discipline that observed and classified the harmful effects of chemicals in animals, by relying on the tools of pathology to a discipline able to explain the mechanisms of the basic changes in cell function responsible for toxic effects (mechanistic toxicology). This progress resulted from widespread application of techniques and concepts from a range of basic sciences, most notably biochemistry and cell biology. Mechanistic explanations of toxic phenomena aid in the prevention of toxicity by chemical or biological means and provide a rational basis for the use of animal data to assess the anticipated consequences of human exposure to a particular chemical.

Many toxic compounds are chemically stable and produce their characteristic effects by interference with biochemical or physiological homeostatic mechanisms. Many adverse events are the consequence of disturbance of normal physiology and do not result in cell death. A xenobiotic may, for example, activate plasma membrane receptors and induce physiological signal transduction pathways. Induction of a normal cascade at the wrong time or to the wrong extent (too much) may cause undesired or harmful effects. Toxic compounds interfering with homeostatic mechanisms do not necessarily cause cell death; however, the induced changes may have a harmful impact both on the altered cells and on the involved tissue or the entire organ. Therefore, the biochemical and molecular sites of action of a xenobiotic need to be known.

Cytotoxicity, resulting in cell death, is often the consequence of exposure to a harmful xenobiotic, but the number of cells that must be killed before the function of a tissue or organism is noticeably impaired is highly variable. Some cell types, such as the epithelia of the kidney and the liver, have the ability to regenerate in response to damage, while others such as neurons cannot. Furthermore, some organs, such as the liver, lung, and kidney, have a substantial functional reserve capacity in excess of normal requirements, and normal function can be maintained even in the presence of extended necrosis.

In addition to cell death, disturbances in the regulation of cell division induced by toxic xenobiotics may have harmful long-term consequences for the organism affected. Nonlethal alterations in the genome of somatic cells have been knwon to result in mutations and possibly lead to malignant transformation and tumor formation [123]. During the last 15 years, evidence has shown that compounds not directly interacting with the genomic DNA can also produce cancer by so-called epigenetic mechanisms [124]. These may involve a proliferative response of epithelial cells to cytotoxicity, which has been suggested to occur with high-dose carcinogens such as allyl isothiocyanate and chloroform, or a more direct action enhancing the rate of cell division in the absence of cytotoxicity, which is seen with 2,3,7,8-tetrachlorodibenzodioxin. Increased cell replication, whatever the cause, is accompanied by increased chance of unrepaired DNA lesions that may be fixed as mutations. Hyperplasia has long been suspected to precede neoplasia, but the inevitability of such a progression has never been established on a pathological basis alone. Even in the absence of foreign compounds, DNA is damaged to a considerable extent by reactive oxygen species formed during different biochemical processes in the cell. Non-genotoxic carcinogens may thus be inferred to act by enhancing the likelihood of this normal DNA damage being fixed as a mutation, leading to cancer [125].

Some sites or chemical reaction mechanisms of toxic action are described in this chapter. The following major mechanisms are neither comprehensive nor mutually exclusive.

3.2. Receptor – Ligand Interactions

3.2.1. Basic Interactions

Some xenobiotics can interact with physiological receptors because of structural similarities to endogenous compounds. The toxic effects of these chemicals are related to their ability to interfere with normal receptor – ligand interactions either as agonists or as antagonists of the physiological ligand. Receptors are macromolecular components (most often proteins) of tissues that interact with specific endogenous ligands or structurally related xenobiotics to induce a cascade of biochemical events and produce characteristic biological effects. The binding between a receptor R and a ligand L is usually reversible and can be described by an equilibrium reaction.

$$R + L \underset{k_2}{\overset{k_1}{\rightleftharpoons}} RL$$

The dissociation constant K_d that describes this relationship is

$$K_d = \frac{k_1}{k_2} = \frac{[L]\,[R]}{[LR]}$$

where [L], [R], and [LR] are the concentrations of ligand, unbound receptor, and ligand-bound receptor, respectively. The affinity of the ligand for a certain receptor is proportional to $1/K_d$. The ligand may be an endogenous substance that interacts with the receptor to produce a normal physiological response, or it may be a xenobiotic that either elicits (agonist) or blocks (antagonist) the response.

Receptor – ligand interactions are generally highly stereospecific. Usually, small changes in the chemical structure of the ligand can drastically influence its capability to bind to the receptor and thus reduce or completely abolish the effect elicited by ligand – receptor interaction. In other cases, changes in the chemical structure have an important impact on the response without altering the binding of the xenobiotic to the relevant receptor. Differences in the extent of activity are observed not only with structurally distinct chemicals but also with enantiomers of chiral compounds. Synthetic chemicals with stereo centers generally contain both enantiomers, often in a 1:1 ratio (racemic mixture); yet in many instances, only one of the enantiomers is biologically active. In pharmacologic or toxicologic studies, the inactive or weakly active enantiomer should be viewed as an "impurity," which may confound the interpretation of results obtained with the racemic mixture. Because the enzymes responsible for biotransformation of xenobiotics also contain active sites with specific steric requirements, stereoisomers may be differentially biotransformed. The selective biotransformation of only one enantiomer may markedly alter the potency and efficacy of one enantiomer compared to the other. Stereoselective differences in action among enantiomers should not be surprising because these chemicals are mirror images, much as the right hand is of the left, and receptors have a physical orientation that can be compared to a glove. Although the left-hand and right-hand gloves look remarkably similar, they do not fit both hands equally well. However, these theoretical considerations do not imply that stereoisomers always elicit different responses. Since the chiral center of a stereoisomer is not necessarily involved with the active site of the receptor, chirality does not always result in differences in toxic response among enantiomers.

The acute toxic effects of many xenobiotics are related directly to their ability to interfere with normal receptor – ligand interactions. This is most clearly the case with neurotoxins, acting within and outside the central nervous system (CNS). The proper function of the nervous system is highly dependent on a diverse array of receptor – ligand interactions. For example, the belladonna alkaloids atropine and scopolamine bind to and block the cholinergic receptors, the binding sites of the physiological agonist acetylcholine (Fig. 3.1). The interactions of acetylcholine and atropine are an excellent example for demonstrating that interference of a xenobiotic with physiological regulatory mechanisms may result in both desired (pharmacological) and undesired (toxic) effects.

Since atropine itself cannot elicit the physiological responses mediated by acetylcholine binding to this receptor, toxic effects may result.

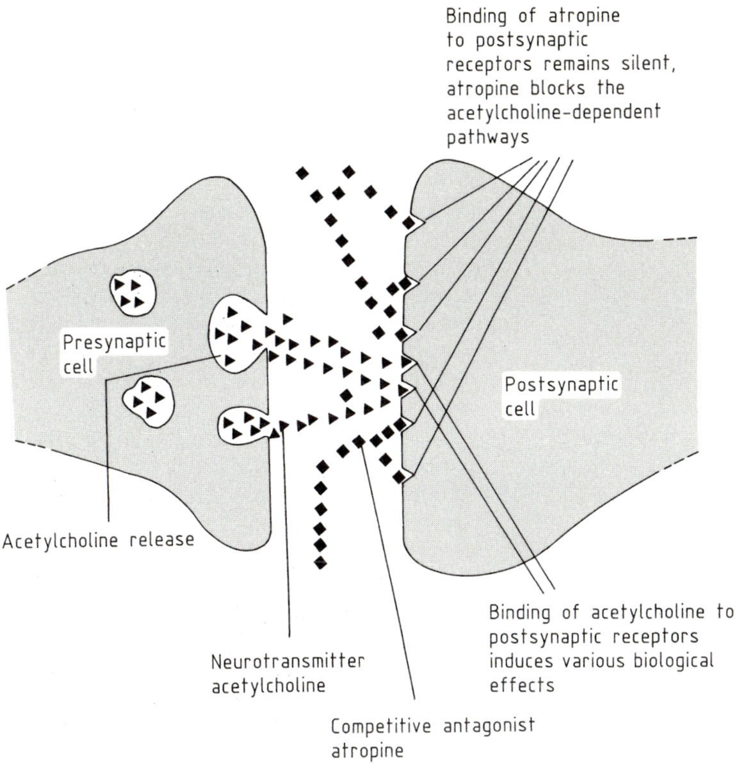

Figure 3.1. Induction of physiological response(s) by binding of the agonist acetylcholine to postsynaptic receptors; in contrast, binding of the antagonist atropine to the same receptors does not induce biological effects

In clinical therapy, atropine is often used as an antispasmodic to reduce the hypermotility state of the gastrointestinal tract or the urinary bladder. Its anticholinergic effects are also utilized in cardiovascular pharmacology to decrease pathologically elevated heart rates. At the same time, however, blockage of acetylcholine-mediated physiological cascades may cause dry mouth, blurred vision, and constipation. In addition, in the central nervous system, toxic effects of atropine include restlessness, confusion, hallucinations, and delirium.

Blockage of the receptors for the neutral amino acid glycine by strychnine represents a further example of important and, in this particular case, life-threatening agonist–antagonist interactions. Binding of glycine to its receptor increases the permeability of the plasma membrane to chloride ions, resulting in hyperpolarization and reduced activity of nerve cells (i.e., glycine exerts an inhibitory effect on the nervous system). Blockage of glycine receptors by the antagonist strychnine, which does not elicit a response, results in hyperactivity and severe convulsions.

In addition to acute toxicity, chronic influences exerted by xenobiotics on steroid hormone homeostasis may result in adaptive and proliferative responses in certain tissues. The long-term consequences may involve impairment or loss of physiological function. Moreover, binding to a specific receptor may cause changes in gene expression and contribute to tumor formation. For example, binding to the Ah receptor mediates many of the effects observed with the highly toxic and carcinogenic 2,3,7,8-tetrachlorodibenzodioxin (TCDD). A cytosolic binding protein (cytosolic receptor) has also been identified for peroxisome proliferators, such as clofibrate and diethylhexylphthalate [126]. This protein has DNA-binding domains and appears to activate gene transcription in the nucleus in a manner analogous to the steroid hormones. The increased cell proliferation produced in target organs such as the liver may be responsible for the

Table 3.1. Chemicals causing toxic response by interaction with receptors (Ah = aryl hydrocarbon)

Type of receptor	Chemical
Muscarinic receptor	atropine
Glycine receptor	strychnine
Ah receptor	TCDD
Peroxisome proliferator receptor	peroxisome proliferators (clofibrate, diethylhexylphthalate)

carcinogenicity of many peroxisome proliferators. Characteristic examples of receptor–xenobiotic interactions with harmful acute or chronic consequences are summarized in Table 3.1.

3.2.2. Interference with Excitable Membrane Functions

The maintenance and stability of excitable membranes is essential to normal physiology. Excitable membranes are critical to the function of nerves and muscles because of their ability to generate and propagate action potentials. Action potentials are elicited by the exchange of ions between the intra- and extracellular compartments; hence, they depend on the normal activity of ion channels and membrane ion pumps.

Xenobiotics may perturb excitable membrane functions in many different ways. They can block sodium channels, resulting in toxic effects such as paralysis. For example, tetrodotoxin from the puffer fish irreversibly blocks the sodium channel along the nerve axon, thus preventing the inward sodium current of the action potential while leaving the outward potassium current unaffected. The marine toxin saxitoxin, which is structurally quite different from tetrodotoxin, also produces its paralyzing effects by blocking sodium channels in excitable membranes in essentially the same manner. The insecticide 1,1′-(2,2,2-trichloroethylidene)bis(4-chlorobenzene) (DDT) produces its neurotoxic action by interfering with the closing of sodium channels, thus impairing the physiological repolarization of excitable membranes. This results in hyperactivity of the nervous system and repetitive discharges of neurons. The most striking symptoms in poisoned insects or mammals are persistent tremor and convulsive seizures. Other neurotoxins such as cyclodiene insecticides and picrotoxin interact with the γ-aminobutyrate (GABA) receptor in neurons. Blockage of the GABA receptor by these antagonists impairs the transport of chloride ions into the cell, resulting in partial depolarization of the membranes in the absence of adequate signals. The clinical symptoms of the pathophysiological changes can be summarized as a state of uncontrolled excitation. In addition, with cyclodiene insecticides this hyperexcitability is augmented by inhibition of the $Ca^{2+}-Mg^{2+}$-ATPase, which results in accumulation of intracellular free calcium and increased release of neurotransmitters from storage vesicles (Fig. 3.2). The partially depolarized cells can be excited by weak, normally inactive stimuli.

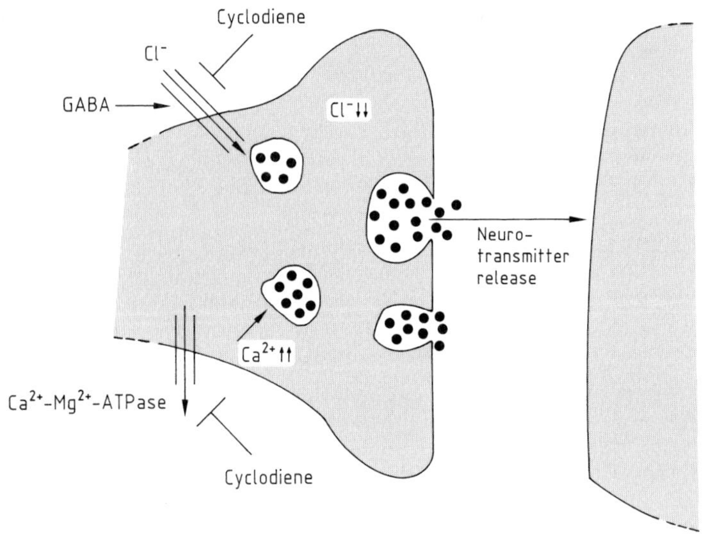

Figure 3.2. Sites of action of chlorinated cyclodiene insecticides on the plasma membrane of nerve cells

Organic solvents have a depressant effect on the CNS, resulting in anesthesia. This effect is produced via nonspecific alterations in membrane fluidity due to the high lipid solubility of these solvents, not via interference with specific ion channels. This depressant effect is not based on certain structural features or specific interactions; it is also observed after inhalation of the inert, but highly lipophilic, gas xenon.

3.2.3. Interference of Xenobiotics with Oxygen Transport, Cellular Oxygen Utilization, and Energy Production

Continuous, sufficient production of energy in the form of ATP in specialized cellular organelles (mitochondria) is essential for cell survival and function. The process of cellular energy production requires oxygen, cofactors, and a wide array of specialized enzymes acting in concert. Many chemicals produce their toxic effects by interfering with the transport and cellular utilization of oxygen or with the oxidation of carbohydrates, which is coupled to the synthesis of ATP by oxidative phosphorylation.

Interference with cellular energy production may occur at different sites in the uptake, transport, and cellular utilization of oxygen. *Oxygen transport* from the lung to tissues may be reduced or blocked by xenobiotics, which compete with oxygen for the binding site in the transport protein hemoglobin or chemically modify this binding site. Carbon monoxide competitively blocks the binding of oxygen to hemoglobin. Chemical oxidation of the iron in hemoglobin also impairs oxygen transport. Some xenobiotics oxidize the Fe^{2+} atom at the oxygen binding site in hemoglobin to Fe^{3+}. Hemoglobin containing Fe^{3+} (methemoglobin) is not able to reversibly bind oxygen. Thus, xenobiotics causing methemoglobinemia (e.g., nitrites, aromatic amines) effectively block oxygen transport.

Cellular utilization of oxygen in the tissues is blocked by cyanide, hydrogen sulfide, and azide because of their affinity for cytochrome oxidase. Cyanide exerts its toxic effects by interrupting electron transport in the mitochondrial cytochrome electron transport chain. In addition, interference with the enzyme systems producing ATP is a well-defined mechanism of toxicity. The ultimate formation of ATP in the cell by the oxidation of carbohydrates may also be blocked at other sites. For example, rotenone and antimycin A interfere with specific enzymes in the electron transport chain, nitrophenols uncouple oxidative phosphorylation, and sodium fluoroacetate inhibits the citric acid cycle, also called the Krebs, or tricarboxylic acid cycle. Nitrophenols interfere with the production of high-energy phosphates in mitochondria. They prevent the phosphorylation of ADP to ATP, but electron flow and oxygen consumption continue. In addition to inducing loss of cell function due to depletion of ATP, this type of uncoupler also causes a marked elevation of body temperature arising from excess heat production. Blockage of the tricarboxylic acid cycle affects organs that rely heavily on the availability of a continuous energy support and results in cardiac and nervous system toxicity. A classic example of a mechanism inhibiting the Krebs cycle is termed "lethal synthesis" and is exemplified by fluoroacetate, which is incorporated into the Krebs cycle as fluoroacetyl-coenzyme A (CoA). Fluoroacetyl-coenzyme A inhibits the aconitase-catalyzed conversion of citrate to isocitrate and interrupts the Krebs cycle (Fig. 3.3). The Krebs cycle is the major degradative pathway for the generation of ATP and also provides intermediates for biosynthesis. For example, the majority of carbon atoms in porphyrins stem from succinyl-CoA, and many of the amino acids are derived from α-ketoglutarate and oxalacetate. Hence, interruption of the Krebs cycle by the incorporation of fluoroacetate may be lethal for the cell. The consequences of ATP depletion include impairment of membrane integrity, ion pumps, and protein synthesis. Depending on the extent, energy depletion will inevitably lead to loss of cell function and cell death.

3.3. Binding of Xenobiotics to Biomolecules

Many toxic chemicals exert their effects by the covalent linkage of reactive metabolites to essential macromolecules of the cell. These include proteins and lipids (found, e.g., as membrane structural elements) and nucleic acids (see Sections 3.3.1–3.3.3). The linkage generally involves binding of electrophilic metabolites to nucleophilic sites such as thiol, amino, and hydroxyl groups in the side chains of proteins and is essentially irreversible, depending only on the turnover of the macromolecule in question for repair of the lesion. In addition, other toxins may

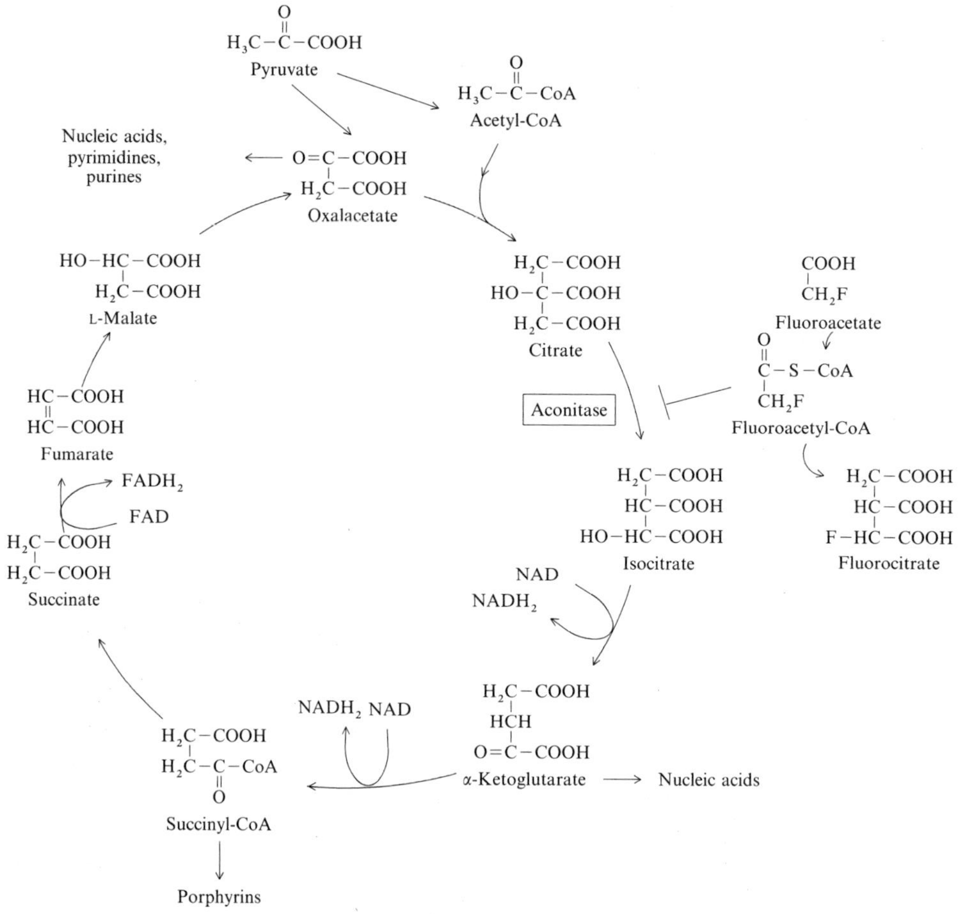

Figure 3.3. Lethal synthesis
Incorporation of fluoroacetate in the Krebs cycle; formation of fluorocitrate and inhibition of aconitase results in blocking of the tricarboxylic acid cycle and interruption of cellular energy production

bind to proteins and impair their normal function without being converted first to reactive intermediates.

3.3.1. Binding of Xenobiotics or their Metabolites to Cellular Proteins

Many toxic substances exert their effects via binding to the active sites of enzymes or other proteins that are critical to cellular function. For example, hydrogen cyanide binds with high affinity to the Fe^{3+} ion in cytochrome oxidase, thus blocking the last step in the mitochondrial electron transport chain, which is important for cellular energy production. This single and highly specific site of action is responsible for rapid induction of the often fatal toxic effects of cyanide. Carbon monoxide binds to iron-(II) in hemoglobin, whose physiological function is to bind, transfer, and deliver oxygen to tissues. Since carbon monoxide has a 210-fold higher affinity for this binding site than oxygen, very low concentrations of carbon monoxide in the atmosphere are sufficient to displace the physiological ligand and produce severe toxic effects. So, 0.1 % carbon monoxide in the atmosphere can occupy roughly 50 % of the available hemoglobin binding sites, because, taking into consideration the higher affinity, 0.1 % carbon monoxide correspond to 20 % oxygen.

Another example for the importance of protein binding in toxicity is the binding of metal ions (e.g., lead, mercury, cadmium, and arsenic) to thiol groups of proteins. This binding contributes, by yet largely unknown mechanisms, to the toxicity of these metals. Porphyria induced by lead, mercury, and other metals is based in part on the inhibition of specific enzymes of heme biosynthesis and results in the accumulation of specific intermediates in heme synthesis. The binding of cadmium to the protein metallothionein, which is rich in thiol groups, results in active concentration of the metal–metallothionein complex in the proximal tubules of the kidney. Catabolism of the metal complex in the lysosomes of proximal tubule cells, with concomitant liberation of the toxic metal, results in nephrotoxicity (functional impairment of and severe structural damage to the kidneys), a common effect of cadmium exposure in humans and experimental animals [127].

Binding to active sites in enzymes may result in the inhibition of biochemical pathways vital to the cell. The induction of toxic effects may be due to accumulation of a specific enzyme substrate or to insufficient amounts of substrate available for normal physiological function. For example, many organophosphate pesticides inhibit cholinesterases by covalently binding to the active site, the amino acid serine of cholinesterase. Since the phosphorylated cholinesterase is stable, the covalent binding results in the inhibition of enzymatic activity (Fig. 3.4). Because of the impaired cleavage, acetylcholine accumulates at cholinergic synapses and neuromuscular junctions. Organophosphates thus produce the typical signs of acetylcholine poisoning, such as increased salivation and lacrimation, abdominal cramps and diarrhea, cough, bronchonconstriction and breathing impairment, mental confusion, headaches, tremor, and coma.

Protein modifications may also be the basis for immunosuppressive effects and chemical-induced allergy. For many chemicals that cause necrosis, binding to cellular macromolecules is essential for the expression of toxic effects. However, in most cases, the temporal sequence of events that are triggered by the covalent binding and the cause–effect relationships between these events are not fully understood.

3.3.2. Interaction of Xenobiotics or their Metabolites with Lipid Constituents

Lipid peroxidation in biological membranes by free radicals initiates a series of events that finally cause cellular dysfunction and cell death. The formation of radicals during peroxidation is a self-propagating process; the reaction may be started by organic radicals formed during biotransformation or by oxygen radicals formed by disruption of cellular energy metabolism. The initiation of lipid peroxidation via interaction of free radicals with polyunsaturated fatty acids to form lipid peroxy radicals—which then produce lipid hydroperoxides and other lipid peroxy radicals—has been proposed as one critical step leading to cell injury and death. Peroxidative damage to plasma membrane lipids may cause impairment of membrane integrity and, finally, rupture of the plasma membrane. In addition, breakdown of the membranes of subcellular organelles, such as those of the mitochondria, the endoplasmic reticulum, and lysosomes, can also contribute to induction of cell death. The end products of the breakdown of membranes, mainly unsaturated aldehydes, can also produce toxicity in distal tissues. Induction of lipid peroxida-

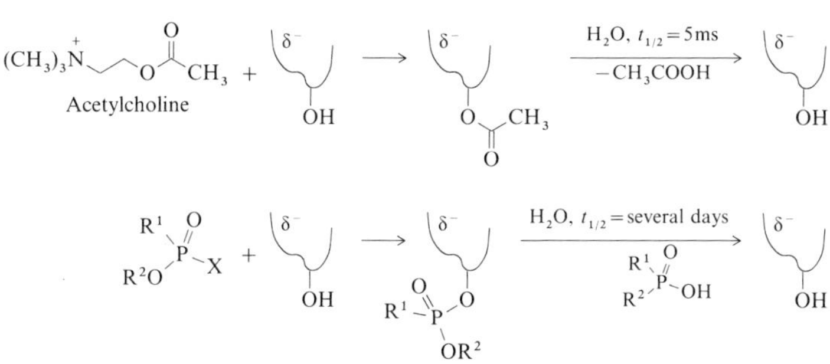

Figure 3.4. Mechanisms of cholinesterase inhibition by organophosphate pesticides
Compare the $t_{1/2}$ of the enzyme complex with the physiological substrate acetylcholine (a few milliseconds) to the $t_{1/2}$ of the enzyme complex with organophosphate pesticides (several days).

tion is involved in the toxicity of many chemicals, which are transformed to free radicals. A well-studied example is carbon tetrachloride, which is converted by cytochrome P450 to the trichloromethyl radical ($^{\bullet}CCl_3$) and the trichloromethylperoxy radical ($CCl_3O_2^{\bullet}$) (see Section 2.5.4.1). Radicals and reactive oxygen species may interact with the major soluble cellular thiols, glutathione, and thiol-containing proteins.

An important function of reduced glutathione is to protect sulfhydryl groups of proteins by keeping them in the reduced state. Depletion of intracellular glutathione stores appears to be a prerequisite for the onset of significant oxidative stress. Many critical enzymes in the cell depend on reduced thiol groups to maintain their activity. Hence, concentrations of reactive oxygen species in excess of that necessary to deplete intracellular gluthathione can induce oxidation of protein thiols to form disulfide linkages, thereby impairing enzymatic activity. Although direct covalent interaction of electrophilic chemicals with protein thiols may contribute to enzyme inhibition, reversible oxidation of the thiol group as a result of oxidative stress appears to be more important. The Ca^{2+}-transporting membrane systems constitute one group of thiol-containing enzymes whose impairment as a result of oxidative inactivation may be critical for cell injury and death.

3.3.3. Interactions of Xenobiotics or their Metabolites with Nucleic Acids

Electrophilic compounds, usually formed by the oxidative biotransformation of xenobiotics, may also interact with various nucleophilic sites in DNA, principally O-6, N-7, N-2, and C-2 of guanine. Moreover, other types of DNA damage may occur as a consequence of covalent binding of electrophiles or by interaction of DNA with reactive oxygen metabolites (Table 3.2).

These interactions may alter gene expression (see Sections 3.5 and 3.7). The changes in gene expression may be quantitative (i.e., formation of the wrong amount of a protein at a not appropriate period of cellular life) or qualitative (i.e., formation of a protein with altered or impaired properties). In both cases, the changes may lead to the death of the cell. However, the interaction of xenobiotics with nucleic acids is more important in generating somatic mutations, which can be the initiating event for a process ultimately leading to malignant transformation of cells and tumor formation. Alkylation of the O-6 position of guanine is apparently important for the mutagenic and carcinogenic effects of nitrosamines and other chemicals that readily form methyl carbonium ions. In addition, other sites such as the N-7, N-2, and C-2 positions of guanine may also play an important role in DNA adduct formation with other electrophilic chemicals. Ribonucleic acid also contains nucleophilic sites. Thus, critical intracellular functions of RNA (e.g., protein synthesis) may be perturbed by covalent interaction of electrophilic chemicals with RNA.

Table 3.2. Types of DNA damage induced by xenobiotics

Type of DNA-damage	Examples
Modifications of purines and pyrimidines	alkylation by electrophiles, oxidation to 8-hydroxyguanosine and other oxidized purines, UV-induced formation of thymidine dimers
DNA strand breaks	oxygen radicals, activation of calcium-dependent endonucleases
AP lesions (apurinic, apyrimidinic sites)	loss of DNA bases through labilization after covalent binding of xenobiotics to specific sites in purine and pyrimidines
DNA cross-links	bifunctional electrophiles react with nucleophilic sites in both DNA strands (interstrand cross-link) or with two bases on one strand (intrastrand cross-link)

3.4. Perturbation of Calcium Homeostasis by Xenobiotics or their Metabolites

Intracellular calcium concentrations are rigorously maintained at ca. 10^{-7} mol/L, against an extracellular concentration of $> 10^{-3}$ mol/L. Three principal buffer systems are important in maintaining this steep gradient in virtually all cells: the plasma membrane, the mitochondria, and the endoplasmic reticulum. All systems require ATP to transport calcium, directly or indi-

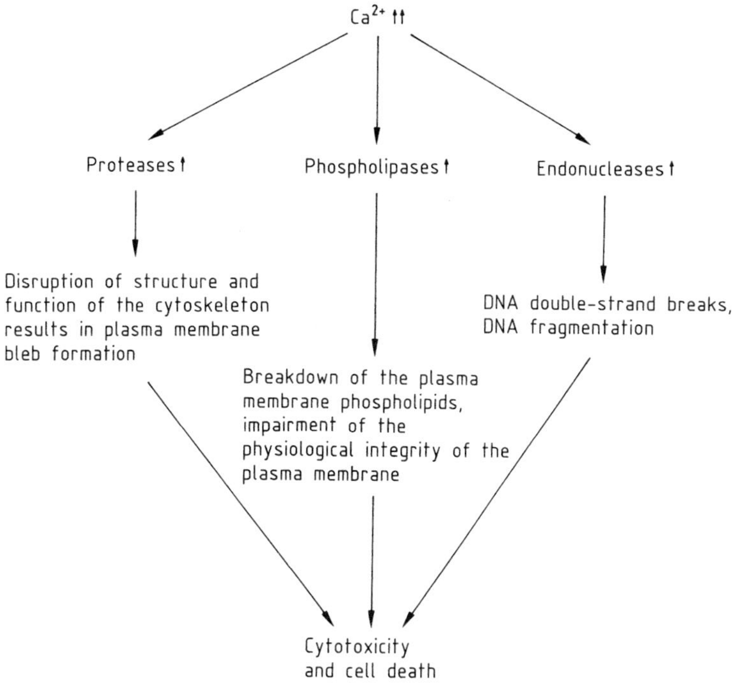

Figure 3.5. Increased concentrations of intracellular calcium may activate calcium-dependent degradative enzymes, resulting in the destruction of proteins and lipids and in DNA fragmentation, thus causing cell death

rectly. Exposure of freshly isolated or cultured cells to numerous toxins such as *tert*-butylhydroperoxide, quinones, paracetamol, and carbon tetrachloride induces a rapid and sustained rise of the cytosolic calcium concentrations that correlates well with the subsequent loss of cell viability [128], [129]. Prevention of this rise by omitting calcium in the extracellular buffer or by addition of calcium chelators may prevent cell death. A central event in calcium-induced toxicity is the activation of calcium-dependent degradative enzymes such as proteases, phospholipases, and endonucleases (Fig. 3.5).

Cytoskeletal integrity is important for a number of cellular functions including motility, shape, secretion, and division. Activation of calcium-dependent proteases results in disruption of cytoskeletal components and formation of plasma membrane blebs (protrusions). In addition, activation of calcium-dependent phospholipases may induce membrane phospholipid breakdown, and calcium-dependent endonucleases may lead to DNA fragmentation by the formation of DNA double-strand breaks. These toxic biochemical cascades may act in concert to cause cell death [98].

3.5. Nonlethal Genetic Alterations in Somatic Cells and Carcinogenesis

The induction of cancer by chemicals is a complex multistep process involving interactions between environmental and endogeneous factors. Tumors are formed as a result of aberrant tissue growth due to loss of control mechanisms of cell division. A model traditionally used for an operational description of carcinogenesis is the initiation–promotion model. The first stage—initiation—requires a genotoxic event such as binding of an electrophilic xenobiotic to DNA, causing a premutagenic lesion in a single cell [130], [131]. After DNA replication, the premutagenic lesion may be transformed into a heritable mutation (i.e., by base-pair substitution; (see Section 3.7). In the stage of promotion, several, primarily nongenotoxic mechanisms facilitate (promote) the preferential proliferation of the initiated cell, finally resulting in tumor formation. Initiators are usually genotoxic agents, whereas promoters act generally by interfering with extranuclear sites and processes; most promoters increase cell growth and cell proliferation.

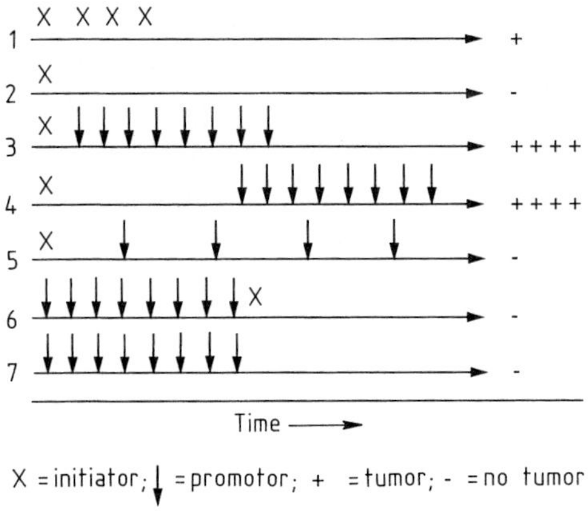

Figure 3.6. Schematic presentation of the initiation–promotion model of chemical carcinogenesis in the mouse skin Promotors reduce the latency period and increase the number of tumors only when applied after the initiator (3, 4). The application frequency required for promoting effects to occur depends on the initiating carcinogen and the target tissue. If the application frequency of the promotor is too low, the promoting effects are not observed (i.e., no tumors are formed or the tumor yield is not increased, 5). This is also the case, when the promotor is applied prior to the initiator (6) or alone without initiating agent (7). Tumors can also be produced by the initiating agent alone if it is applied at sufficient concentration or above a certain application frequency (1). In this case, the initiator is termed a complete carcinogen.

In contrast to *initiators*, which are believed to result in irreversible changes in the cellular genome without threshold exposure levels, *promoters* show some reversibility, and threshold exposure levels for promoters may be discussed. This stepwise nature of carcinogenesis was first shown in mouse skin, and this model has now been well characterized (Fig. 3.6). In a typical experiment, an initiating chemical such as dimethylbenzanthracene is applied to mouse skin at a low dose so that very few tumors, if any, are produced in the animals lifetime. After an interval of one week to one year, the treated (initiated) skin is exposed to multiple applications of a promoter, such as the phorbol esters found in croton oil. Tumors begin to appear within 5 to 6 weeks after application of the promoter, and all mice carry tumors by 10 to 12 weeks after the start of application (Fig. 3.6).

The initiation–promotion experiments on mouse skin and similar experiments in rodent liver have led to the following general rules of carcinogenesis:

1) The initiator must be given first; no tumors or very few tumors result if the promoter is administered first.
2) The initiator, if given once at a subcarcinogenic dose, does not produce tumors during the time of observation; however, repeated doses of the initiator may cause tumors even in the absence of promoter (the initiator is a complete carcinogen in this case).
3) The action of the initiator is irreversible; tumors result in nearly the same yield if the interval between initiation and promotion is extended from one week to one year.
4) The initiator is an electrophile or is bioactivated to an electrophile, which binds covalently to DNA causing a mutation after DNA replication.
5) In contrast, no evidence exists of covalent binding of promoters or their metabolites to DNA.
6) The action of the promoter is reversible at an early stage and usually requires repeated exposure; thus, a threshold level of exposure to promoters probably exists. However, threshold levels cannot be reliably defined, as long as the biochemical mechanism involved in tumor promotion is not known precisely. In contrast to the established target of the initiation process (DNA), the molecular mechanisms of tumor promotion are still largely unknown.

In addition to the promotion of skin carcinogenesis by phorbol esters, promoters for tumors in other organs are known or suspected [132], [133]. Bile acids are promoters of colon carcinogenesis in experimental animals. In humans, a strong association exists between high intake of dietary fat and cancer of the colon; since ingestion of fat increases the amount of bile acids in the colon, the increased incidence of colon cancer may be due to the promoting effect of bile acids on intestinal epithelia. In rat bladder, saccharine and cyclamate are promoters of tumors initiated by a single dose of dimethylnitrosourea; tryptophan is a promoter of urinary bladder tumors in dogs treated with an initiating dose of 4-aminobiphenyl or 2-naphthylamine. Hormones are also known modifiers of chemical carcinogenesis. An oral or intravenous injection of dimethylbenzanthracene produces mammary tumors in susceptible female mice. Prolactin increases and accelerates tumor development, whereas ovariectomy results in reduced tumor yield.

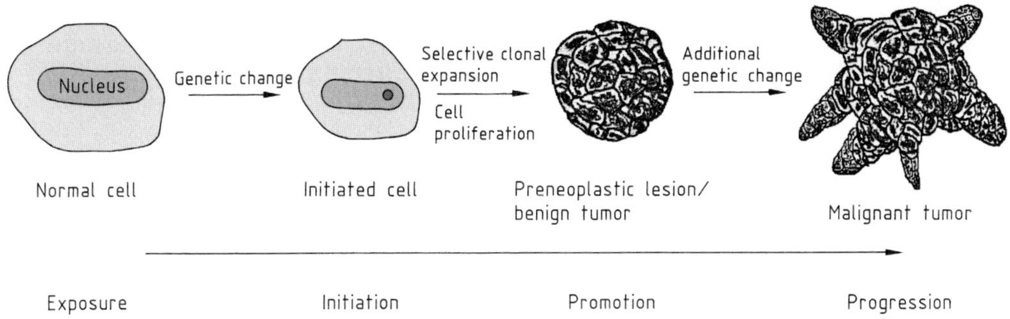

Figure 3.7. Multistep model of tumor formation
Genotoxic events resulting in heritable mutations cause the formation of an initiated cell (initiation). In the following stage of promotion, nongenetic (epigenetic) events contribute to preferential proliferation of the initiated cell. In the third stage of progression, additional genetic events increase the malignancy of the tumor tissue (i.e., its growth becomes increasingly destructive and metastases are formed in other organs).

In addition to these two stages, a third stage termed progression has been established as an integral part of the carcinogenic process in recent years. In this last stage, additional mutational events increase the malignancy of the tumor; i.e., the tumor grows in an invasive manner, destroying the surrounding tissues, and metastasizes to other organs (Fig. 3.7).

3.6. DNA Structure and Function

An introduction to the structure and replication of DNA is given in → Nucleic Acids, **A 18**, pp. 2–5; 7–11.

3.6.1. Transcription

(→ Nucleic Acids, **A 18**, pp. 9–10)

The linkage between DNA in the nucleus and proteins in the cytoplasm is not direct (Fig. 3.8). Information contained in the DNA molecule is transferred to the protein-synthesizing machinery of the cell via messenger RNA (mRNA) (→ Nucleic Acids, **A 18**, p. 6), which is synthesized complementary to the relevant DNA sequence by RNA polymerase. The mRNA molecules act as transport vehicles for information contained in the genes being expressed.

In eukaryotic cells, the initial mRNA copy contains homologues of both the intron and the

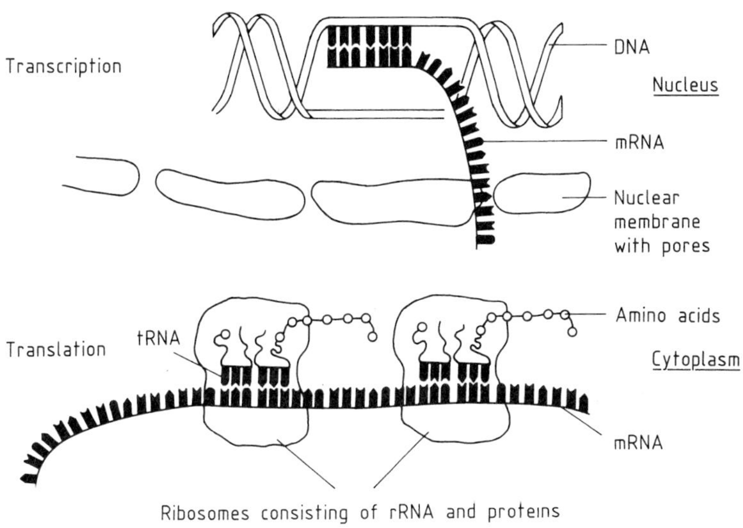

Figure 3.8. Scheme of the important steps in gene transcription and protein synthesis

exon regions. The intron regions are removed, and the exon regions are spliced together to form the active mRNA molecules, which are then transported through the pores of the nuclear membrane to the cytoplasm.

3.6.2. Translation

(→ Nucleic Acids, **A18**, p. 10)

The next process involves the translation of mRNA molecules into polypeptides. This procedure requires many enzymes and two further types of RNA: transfer RNA (tRNA) and ribosomal RNA (rRNA). A specific tRNA exists for each amino acid. The tRNA molecules are involved in the transport and coupling of amino acids into the resulting polypeptide (Fig. 3.8). Each tRNA molecule has two binding sites: one for the specific amino acid, the other containing a triplet of bases (the anticodon) that is complementary to the appropriate codon on the mRNA.

The rRNA is complexed with protein to form subcellular globular organelles called ribosomes. Ribosomes can be regarded as the reading heads, which allow the linear array of mRNA codons to base-pair with an anticodon of an appropriate incoming tRNA–amino acid complex.

3.6.3. Regulation of Gene Expression

All cells possess the same genetic information; however, different types of cells exhibit distinct gene transcription patterns. These differences in gene expression are critical to the morphological and biochemical properties of the many thousands of cell types in the human and animal body. Hence, mechanisms are required that regulate gene expression (i.e., determine which genes are expressed and to what extent and which genes are not expressed in a certain cell type at a particular time). The mechanisms involved in regulation of gene transcription are not entirely understood. The transcription of structural genes is regulated by a special set of codons, in particular, promoter sequences, the initial binding sites for RNA polymerase before transcription begins. Different promoter sequences have different affinities for RNA polymerases. Additional regulatory genes called operators regulate the activity of several genes or gene groups (called operons). The activity of the operator itself is further controlled by a repressor protein, which stops transcription of the whole operon by binding to the operator sequence. Because of these regulatory mechanisms, cells are able to express only the genes required at a given moment for their specialized function. This not only helps to conserve cellular energy but also is decisive for correct cellular differentiation, tissue pattern formation and function, and maintenance of the physiological integrity of the entire organism.

3.6.4. DNA Repair

All living cells possess several efficient DNA repair processes. DNA repair is crucial in protecting cells from spontaneous and exogenous lethal and mutating effects such as heat-induced DNA hydrolysis, UV light, ionizing radiation, DNA-reactive chemicals, free radicals, and reactive oxygen species. Among the various existing DNA repair mechanisms, the most comprehensively studied mechanism in eukaryotes is the excision repair pathway. This mechanism includes a group of enzymes acting cooperatively to recognize DNA lesions, remove them, and correctly replace the damaged sections of DNA.

The excision repair pathway is regarded as error free and does not lead to mutations. However, this pathway may become saturated after excessive DNA damage. In this case, the cell may be forced to activate other repair mechanisms that are not error free. Several of these mechanisms, such as error-prone repair, have been well characterized in bacteria, but their counterparts, if any, in mammalian cells have not been identified.

3.7. Molecular Mechanisms of Malignant Transformation and Tumor Formation

3.7.1. Mutations

Mutations are hereditary changes in genetic information, resulting from spontaneous or xenobiotic-induced DNA damage. The term mutation can be applied to point mutations, which are qualitative changes involving one or a few bases within one gene, as well as to larger changes involving parts of the chromosome detectable by light microscopy or even whole chromosomes and thus many thousands of genes (Table 3.3).

Table 3.3. Types of mutations

Gene mutations	base-pair substitutions, deletions, insertions, gene rearrangements, gene amplifications
Chromosomal mutations	
Structural	breaks, translocations
Numerical	loss or gain of an entire chromosome

Point mutations can occur when one base is substituted for another (base substitution) or when base pairs are deleted or inserted (deletions/insertions). Substitution of another purine for a purine base or of another pyrimidine for a pyrimidine base is called a *transition*, while substitution of purine for pyrimidine or pyrimidine for purine is called a *transversion*. Very slight alterations in the chemical structure of the DNA bases may be sufficient for a base pair substitution to occur. Guanine for example, normally pairs with cytosine, while O^6-methylguanine (a frequent DNA modification seen with methylating agents such as dimethylnitrosamine) pairs with thymine (Fig. 3.9).

These changes in certain codons may cause insertion of the wrong amino acid into a relevant polypeptide. In this case, the changes are called *missense mutations*. Such proteins may have dramatically altered properties if the new amino acid is close to the active center of an enzyme or affects the three-dimensional structure of an enzyme or a structural protein. Hence, the alterations may result in marked changes in the differentiation and proliferative characteristics of the affected cells. A base substitution can also result in the formation of a new inappropriate stop (or *nonsense*) codon. The result of nonsense mutations is the formation of a shorter and, most likely, inactive protein. Owing to the redundancy of the genetic code, about a quarter of all possible base substitutions will not result in amino acid replacement and will be silent mutations.

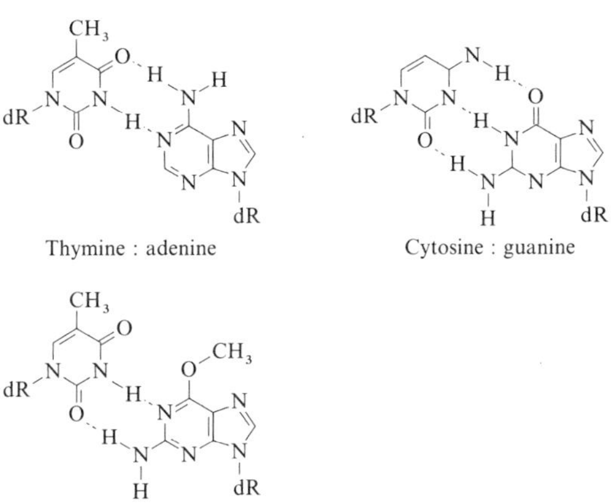

Figure 3.9. Formation of a base substitution
Guanine normally pairs with cytosine and adenine with thymine (upper part); in contrast, O^6-methylguanine (a frequent DNA modification induced by methylating agents such as dimethylnitrosamine) pairs with thymine (lower part), resulting in a hereditary change of the genetic information
dR = Deoxyribose

Bases can be also deleted or added to a gene. Because each gene has a precisely defined length, these changes, if they involve a number of bases that is not a multiple of three, result in a change in the reading frame of the DNA sequence and are known as *frameshift mutations*. Such mutations often have a dramatic effect on the polypeptide coded by the affected gene, because most amino acids will differ from the point of the insertion or deletion of bases in the DNA strand onward.

Some forms of unrepaired alkylated bases are lethal, due to interference with DNA replication. Others, such as O^6-methylguanine lead to mutations, if unrepaired. These differences indicate that not all DNA adducts are of equivalent importance. In fact, some adducts appear not to interfere with normal DNA functions or to be rapidly repaired, others are mutagenic, and yet others are lethal. The most vulnerable base is guanine, which can form adducts at several of its atoms (e.g., N-7, C-8, O-6, and exocyclic N-2; see Section 2.5.6.6).

Intrastrand and Interstrand Cross-Links. Xenobiotics with bifunctional alkylating properties can also form links between adjacent bases on the same strand (intrastrand cross-links) or cross-links between bases on different strands (interstrand). The induction of a frameshift mutation does not necessarily require formation of covalent adducts. Some compounds with a planar structure, such as particular polycyclic aromatic hydrocarbons, can intercalate between strands of the DNA double strand. The intercalated molecules may interfere with DNA repair or replication and cause insertions and/or deletions of base pairs. The precise molecular event is still unclear, although several mechanisms have been proposed. Hotspots for frameshift muta-

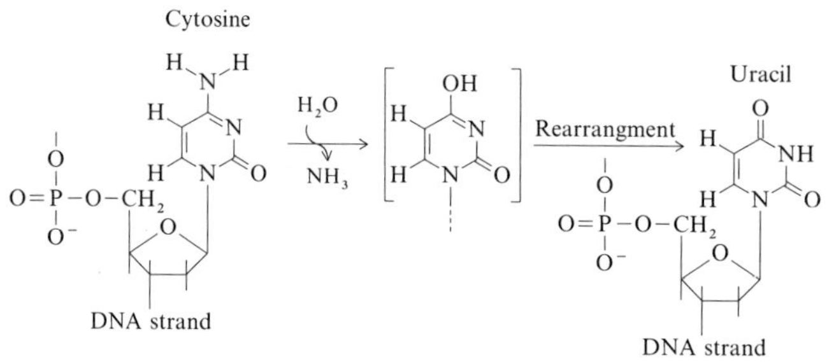

Figure 3.10. Deamination of cytosine to uracil

tions often involve sections of the DNA strand where there is a run of the same base (e.g., the addition of a guanine to a run of six guanine residues).

DNA strand breaks result from hydrolysis of the sugar–phosphate bond of a nucleotide. In a double strand, both single- and double-strand breaks may occur. DNA strand breaks are often induced by hydroxyl radicals, which are formed at high rates both spontaneously during normal cell life and in the presence of exogenous chemicals. AP lesions in the DNA strand result from spontaneous hydrolysis of the glucosidic bond and loss of the DNA base (see Fig. 2.18). Similar to DNA strand breaks, AP lesions are common spontaneous events; however, hydrolysis can be increased dramatically by various types of DNA adducts, such as the N^7 substitution of purines, a common target of alkylating chemicals (see Section 2.5.6.6). Another common spontaneous event is the deamination of cytosine to uracil (Fig. 3.10), approximately 100 deaminations take place in each cell every day.

3.7.2. Causal Link between Mutation and Cancer (see also, → Carcinogenic Agents; → Mutagenic Agents, treated in the A series)

The change from cells undergoing normal, controlled cell division and differentiation to cells that are transformed, dividing without control, and undifferentiated or abnormally differentiated does not occur as a single step. Malignant transformation is a multistage process. Evidence for the involvement of multiple stages comes from in vitro studies, animal models, and epidemiological observations. In humans, the latent period between exposure to a chemical carcinogen and the appearance of a tumor in the target tissue is approximately 10–25 a. Modern molecular biology techniques enable thorough investigation of the genome of malignant cells compared to the genome of their normal counterparts. These studies clearly show that a single mutation is not sufficient to induce malignant transformation. The number of genetic changes varies between two and seven in different tumor types. Also, several types of mutations are usually formed in a malignant transformed cell (i.e., base-pair substitutions, gene rearrangements, chromosomal breaks, and deletions) [134], [135].

3.7.3. Proto-Oncogenes and Tumor-Suppressor Genes as Genetic Targets [136], [137]

Normal control of cell division and differentiation is now generally accepted to be based on the interplay of two sets of genes, the proto-oncogenes and the tumor-suppressor genes. Abnormal activation of proto-oncogenes or inactivation of tumor-suppressor genes eventually leads to malignant transformation.

3.7.3.1. Proto-Oncogenes

Oncogenes were originally discovered in the genome of transforming retroviruses and were therefore named v-oncogenes [136], [138]. Subsequent studies showed that these viral genes were originally derived from the mammalian genome. In the normal cell, these proto-oncogenes have important functions in signal transduction pathways (Table 3.4). Proto-oncogenes are expressed in the course of physiological growth processes, such as organ formation during embryogenesis in the maternal organism, regeneration of damaged tissues, and stimulation of cell division by growth factors in the adolescent and adult organism. Activation of cellular proto-

Table 3.4. Examples of proto-oncogenes: physiological function and location of the derived proteins

Proto-oncogene	Function of protein	Location of protein
Proto-oncogenes as plasma membrane receptors		
erb R	epidermal growth factor	plasma membrane
ros	insulin receptor	plasma membrane
Proto-oncogenes with functions in intracellular signal transduction		
src	tyrosine-specific protein kinase	plasma membrane associated
abl	tyrosine-specific protein kinase	cytosol
fes	tyrosine-specific protein kinase	cytosol
mos	serine–threonine kinase	cytosol
raf	serine–threonine kinase	cytosol
ras	guanosine triphosphate-(GTP-)binding protein with GTPase activity	plasma membrane associated
Proto-oncogenes with functions in the cell nucleus		
myc	DNA-binding protein	cell nucleus
fos	DNA-binding protein	cell nucleus
myb	DNA-binding protein	cell nucleus

Table 3.5. Mechanisms of proto-oncogene activation and examples of activated proto-oncogenes (termed c-oncogenes) found in human or animal tumors

Oncogene	Type of mutation	Examples of tumors
c-*myc*	chromosomal translocation	murine plasmocytoma
c-*abl*		human Burkitt's lymphoma
		chronic myeloic leukemia
N-*myc*	gene amplification	neuroblastoma
c-*myc*		small-cell lung carcinoma
c-Ki-*ras*		adrenocortical tumor of mice
c-Ha-*ras*	point mutation	colon carcinoma
c-Ki-*ras*		lung carcinoma
c-N-*ras*		acute myeloic leukemia
		several chemically induced rodent tumors (i.e., mammary and skin tumors)

oncogenes into c-oncogenes in spontaneously and chemically induced tumors results in qualitatively (altered protein) or quantitatively (too much, at the wrong time) altered gene expression. As shown in Table 3.5, oncogenes can be activated by several types of genetic damage such as point mutations, gene amplifications, and chromosomal translocations.

A substantial number of human tumors (10–15 %) contain activated *ras* oncogenes. In all cases so far examined, activation relies on a point mutation in codon 12, 13, 59, or 61 of one of the *ras* proto-oncogenes. The p21 protein coded by *ras* genes exerts GTPase (GTP hydrolase) activity and is membrane bound. The physiological function of p21 is not precisely known; the protein is bound to the inner surface of the plasma membrane; therefore, it may have a function in transducing signals from growth factor receptors across the cytoplasm to the nucleus. The point mutations decribed above change the properties of the p21 protein and often result in reduced GTPase activity. Since this activity is also responsible for inactivation of the p21 protein itself, reduction of GTPase activity by point mutation results in prolongation of the *ras* signal transduction pathway.

Point mutations resulting in activation of *ras* proto-oncogenes have also been described in several chemically induced rodent tumors. In the majority of mammary tumors induced by nitrosomethylurea, the *ras* oncogene is mutated at codon 12. Similar activation of H-*ras* at codon 61 is detected in mammary and skin tumors induced by 7,12-dimethylbenzanthracene. Nitrosomethylurea is an alkylating agent that induces the formation of O^6-methylguanine in DNA, which is consistent with the type of point mutation (G→A transition) observed at the 12th codon of H-*ras*. The mutation at codon 61 in dimethylbenzanthracene-induced tumor cells (mainly A→T transversion) is consistent with the formation of adenine adducts, resulting from dimethylbenzanthracene metabolites binding to adenine residues.

The number of proto-oncogenes that must be activated to convert a normal cell into a malignant cell is unknown at present. Several oncogenes may, in fact, act in concert to achieve the neoplastic state (i.e., one oncogene associated with the plasma membrane or cytoplasm, the other acting at the nuclear level). However, the causal link between proto-oncogene activation and tumor formation has not been established.

3.7.3.2. Tumor-Suppressor Genes

Oncogenes act in a dominant manner; i.e., activation of one copy of the gene can result in perturbation of normal cell proliferation or differentiation. Tumor-suppressor genes code for proteins that negatively regulate cell proliferation (i.e., prevent inadequate increases of all division) and inhibit neoplastic transformation [139]. In these control functions, only one active gene is necessary for suppression; thus, both copies must be lost or inactivated before the cell is allowed to grow in an uncontrolled manner.

Retinoblastoma is a rare childhood cancer of the eye (frequency 1 per 20 000 births), and about 40 % of retinoblastoma cases are hereditary. The hereditary form of the disease evolves early in life; it often involves both eyes, and patients have also a high risk of tumors at other sites such as bone. In contrast, the sporadic form of this tumor type usually appears later in life and is restricted to one eye, and the risk of tumors in other organs is not increased. The normal function of the gene involved, termed the retinoblastoma (rb) gene, appears to be differentiation of the visual apparatus. In some cases in the hereditary form of retinoblastoma, peripheral lymphocytes and other cell types have lost parts of chromosome 13, resulting from a germ line clastogenic event (chromosome breakage). In sporadic cases, only the cells of the tumor have an abnormal pair of chromosomes 13. For tumor formation, both copies of the gene must be inactivated. In the hereditary form, the first mutation is inherited and the chromosome 13 deletion is found in all cells; the second mutation occurs randomly. Because cells in both retinas already carry the first mutation, the risk of tumor formation is equal and high for both eyes. In the sporadic form, two mutations must occur within the same cell. Because either mutation alone is unlikely, the probability is low that both events will occur simultaneously in the same cell and in practically all cases retinoblastomas are restricted to one eye [140].

In addition to these heritable tumors, mutated tumor-suppressor genes are found in many human cancers that are not hereditary. In fact, in most human cancers, the most frequently mutated genes are the tumor-suppressor genes. Compared to more than 50 oncogenes already identified, tumor-suppressor genes have not been so thoroughly studied. At present, eight tumor-suppressor genes have been cloned, and seven additional genes of this group have been localized in the genome. Similar to the function of the proto-oncogenes (Table 3.4), the tumor-suppressor genes cloned so far act at different sites in the signal transduction pathways from the plasma membrane to the nuclear genome, in some cases as negative regulators of gene transcription. Since the end of the 1980s, inactivation of the p53 tumor-suppressor gene has been recognized as an almost universal step in the formation of human cancers. Mutations of this tumor-suppressor gene are estimated to be present in roughly 50 % of the 65×10^6 new cancer cases diagnosed worldwide each year. The protective mechanism of the p53 protein ("guardian of the genome") [141] has been elucidated, and it may open new approaches to cancer therapy. In cells with DNA damage, the transcription of p53 is induced and the protein prevents cell proliferation until DNA repair has occurred. If the cell is unable to repair the DNA damage, the p53 induces programmed cell death (called apoptosis) and eliminates the genetically altered cell. Cells in which the p53 pathway is inactivated by mutation of the gene proliferate despite existing DNA damage so that mutations and malignant clones can arise among the survivors of the genetically altered cells. In contrast to the other tumor-suppressor genes, in which both copies must mutate for inactivation of their function, the altered p53 protein of the mutated copy can bind and thus inactivate the product of the second, normal p53 gene. Mutation of one copy of p53 can thus result in inactivation of the entire p53 pathway and contribute to tumor formation.

3.7.4. Genotoxic versus Nongenotoxic Mechanisms of Carcinogenesis

Oncogene activation and tumor-suppressor gene inactivation induced by mutations provide strong evidence for the involvement of genotoxic mechanisms in tumor formation. However, it has been recognized for many years, that cancers can arise without direct or indirect interaction between a chemical and cellular DNA (i.e., in the absence of direct mutations). The distinction between nongenotoxic and genotoxic carcinogens has recently been defined more sharply, following the identification of a comparatively large number of nongenotoxic carcinogens by the U.S. National Toxicology Program [142]. These carcinogens include a wide range of chemicals acting via a variety of mechanisms, such as disruption of normal hormonal homeostasis in hormone responsive tissues, peroxisome proliferation, and proliferation of urothelial cells of the ureter and urinary bladder following damage via kidney stones.

Genotoxic carcinogens tend to induce tumors in several tissues of both males and females in rats and mice. In contrast, nongenotoxic carcinogens usually induce tumors only at high doses, in one tissue, of one sex, or in one species. The experimental evidence available so far does not support the existence of real thresholds for DNA-reactive carcinogens, although very low concentrations may exist at which practically no clinically manifest tumors are observed in the animal (or human) life span. However, these concentrations cannot be considered thresholds. In the case of carcinogens that operate via other biologic effects, the carcinogenic activity would parallel dose–response relationships of the relevant biologic effects, a very important aspect for human risk assessment. Treatment regimens or exposure scenarios that do not elicit biological effects would not promote tumor formation. However, two major problems exist: (1) very few dose–response studies have been performed with nongenotoxic carcinogens, and (2) in most cases, the biochemical mechanisms responsible for the tumor-promoting action are not understood.

Usually, nongenotoxic carcinogens are divided into two major categories. The first includes compounds that induce cytotoxicity and regenerative cell proliferation, for example, 2,2,4-trimethylpentane and other branched hydrocarbons in the proximal tubules of the kidney. The second group of nongenotoxic carcinogens induces cell proliferation in the absence of cytotoxicity (they are directly mitogenic); relevant examples of this group are carcinogenic hormones or peroxisome proliferators such as di(2-ethylhexyl)phthalate and clofibrate. Induction of cell proliferation is involved in both categories of nongenotoxic carcinogens and may contribute to malignant transformation by increasing the number of spontaneous genetic errors, because DNA replication does not occur with 100% fidelity. Furthermore, in rapidly proliferating cells, DNA damage has a higher chance of being converted to heritable mutations. However, in all these cases, cell proliferation is the final result of a yet-unidentified molecular mechanism.

Genotoxicity and nongenotoxic mechanisms are not mutually exclusive. Rather, they cooperate in tumor formation, as can be seen with many genotoxic carcinogens. In most cases, genotoxic carcinogens induce tumors only after applications of high doses, concomitantly causing cytotoxicity, cell death, and regenerative proliferation. Hence, tumor is the final outcome of a complex, multistep interplay between genotoxic and extranuclear events.

3.8. Mechanisms of Chemically Induced Reproductive and Developmental Toxicity

Reproductive toxicity covers any detrimental effect on the male or female reproductive system due to exposure to toxic chemicals. Developmental toxicity refers to detrimental effects produced by exposures of developing organisms during the embryonic, fetal and neonatal stages. The main phases of reproduction are listed below:

Germ cell production
Spermatogenesis (man)
Oogenesis (woman)

Preimplantation phase
Fertilization
Formation of the blastocyst
Implantation

Embryonic phase
Organogenesis (in humans the first 12 weeks, in rat the first 2 weeks of gestation)

Fetal phase
Functional maturation and growth of the organs, in humans week 12 to 30

Peri- and postnatal phase
Last week of gestation, birth and first period after birth

Such effects can be either irreversible or reversible. Embryolethal effects are incompatible with survival and result in resorption or spontaneous abortion. Irreversible effects that are compatible with survival may cause structural or functional abnormalities in the offspring; these are called teratogenic. Embryotoxic chemicals may also cause overall growth retardation or delayed growth of certain organs.

3.8.1. Embryotoxicity, Teratogenesis, and Transplacental Carcinogenesis

For an agent to be classified as a developmental toxicant, it must be harmful to the developing organism at exposure levels that do not induce severe toxicity (e.g., substantial reduction in weight gain, persistent emesis, or convulsions) in the mother. Adverse effects on the developing organism under severe maternal toxicity may be secondary to perturbations in the maternal system. For practical purposes, however, test compounds can be initially administered at maternally toxic doses to determine the threshold level for adverse effects on offspring. At these exposure levels, conclusions can be qualified to indicate that adverse effects on the conceptus were obtained at maternally toxic exposure levels and may not be indicative of selective developmental toxicity.

The susceptibility of the developing organism to xenobiotic insults varies dramatically within the narrow time span of the major developmental stages (the preimplantation, embryonic, fetal, and perinatal periods) because developing organisms undergo rapid and complex changes in this short period. The major morphogenic events occurring during preimplantation development are the formation of a compact mass of cells (the morula) and of the blastocyst. The latter already exhibits a certain degree of cellular differentiation. As shown in Table 3.6, considerable similarity exists in the timing of preimplantation development across several mammalian species, regardless of the total length of gestation [143]. At the time of blastocyst formation, cell division and metabolic capacity increase dramatically. During the preimplantation period, biochemical changes under progesterone and estrogen control render the endometrium sensitive to blastocyst implantation. One important sign of blastocyst implantation is a prostaglandin-dependent increase in endometrial vascular permeability.

Table 3.6. Timing (days after fertilization) of key developmental events in rats, monkeys, and humans

	Rats	Monkeys	Humans
Blastocyst formation	3–4	5–7	5–8
Implantation	6	9	7
Organogenesis (all organs)	6–15	20–45	21–56
Length of entire gestation	22	165	267
Organogenesis (examples)			
Neural plate	9	19–20	18–20
Upper limb buds	11	27–29	28
Lower limb buds	12	29–31	31–32
Forepaw rays	14	35–37	37
Testes differentiation	14–15	37–39	43–48
Heart septation	16	36	46–47
Palate closure	16–17	45–47	56–58

Alterations in the hormonal milieu or direct excretion of specific xenobiotics into the uterine epithelia during this period can impair implantation and cause embryolethality. Limited amounts of data suggest that the embryo before implantation is susceptible to lethality but rarely to induction of structural aberrations (teratogenicity) with chemical insults. Sublethal exposures of preimplantation embryos have not yet been adequately explored.

After implantation, organogenesis occurs, which is characterized by the division, migration, and association of cells into primitive organs. The most characteristic susceptibility of the embryo to xenobiotics during the organogenesis period is the induction of structural birth defects (terata). Within the organogenesis period (embryonic period), individual organ systems possess highly specific periods of vulnerability to teratogenic insult (Fig. 3.11). Administration of

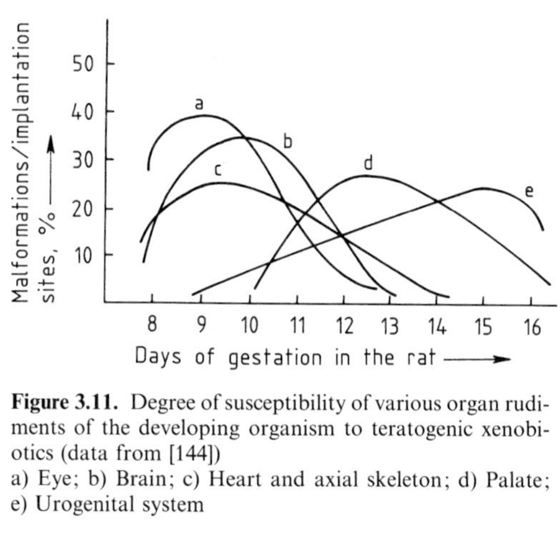

Figure 3.11. Degree of susceptibility of various organ rudiments of the developing organism to teratogenic xenobiotics (data from [144])
a) Eye; b) Brain; c) Heart and axial skeleton; d) Palate; e) Urogenital system

a teratogen on day 9 of rat gestation results in a high level of brain and eye defects, while on day 15, structural abnormalities of the kidney and urinary bladder predominate. As shown in Figure 3.11, critical periods for the susceptibility of different organs overlap so that exposure to teratogens usually results in a spectrum of malformations, differing in severity, in a number of organ systems [145].

The critical phase for inducing anomalies in individual organ systems may be as short as one day or may extend throughout organogenesis. Urogenital defects, for example, can result from drug treatment from days 9 to 16 of gestation in the rat. However, structural defects of the urogenital system depend on the day(s) of exposure to the teratogenic xenobiotic. Development of the urogenital system is multiphasic, and individual stages may have different sensitivities to chemical insult. The mechanisms governing embryonic differentiation are not well understood, but they are certainly involved in the intrinsic susceptibility of individual organs to teratogenic insult.

Functional maturation and growth are the major processes occurring after organogenesis, during the fetal and perinatal periods. Insult at these late developmental stages leads to growth retardation or to more specific functional (but not structural) disorders and transplacental carcinogenesis. The fetal and perinatal period of life is highly susceptible to carcinogenesis because of the high cellular replication rates, presence of xenobiotic biotransforming enzymes in the fetus, and immaturity of the immune system in the developing organism. Several childhood tumors occur so early after birth that prenatal origin is considered likely. Among these are acute lymphocytic leukemia, Wilms' tumor (nephroblastoma), and neuroblastoma. Studies with direct-acting transplacental carcinogens such as ethylnitrosourea indicate that susceptibility to carcinogens begins after completion of the organogenesis period in rodents. Tumors in offspring occurred primarily when ethylnitrosourea was given during the fetal period, whereas birth defects and embryolethality predominated with exposures in the embryonic phase [146], [147]. However, this does not imply that teratogenesis and carcinogenesis are mutually exclusive processes. Teratogenesis and carcinogenesis can be regarded as graded responses of the embryo to injury, with teratogenesis representing the grosser response involving major tissue necrosis in early, relatively undifferentiated embryos, combined carcinogenicity–teratogenicity damage occurring in older embryos, and finally carcinogenicity alone in the fetus.

3.8.2. Patterns of Dose–Response in Teratogenesis, Embryotoxicity, and Embryolethality

The major toxic effects from prenatal exposure observed at the time of birth are embryolethality, malformations, and growth retardation. The relationship among embryolethality, malformations, and growth retardation is quite complex and depends on the type of agent, the time of exposure, and the dose. Some developmental toxins may cause malformations of the entire litter at exposure levels that do not cause embryolethality (Fig. 3.12). If the dose is increased, embryolethality can occur, often in combination with severe maternal toxicity. Malformed fetuses are often retarded in growth, and the curve for growth retardation is often parallel to, and slightly displaced to the right of the curve for teratogenicity. Such a pattern of response is indicative of agents with high teratogenic potency.

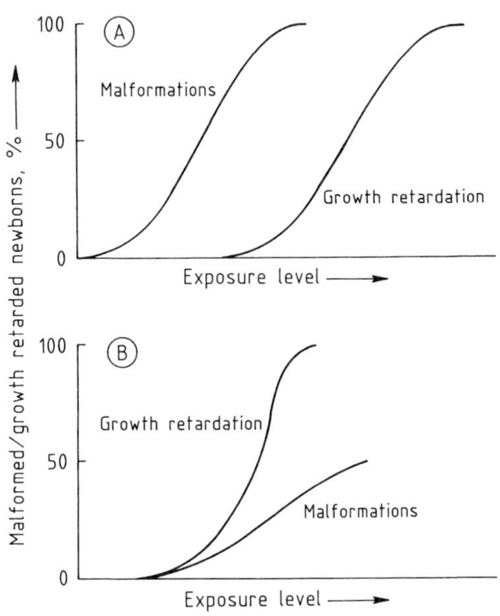

Figure 3.12. Possible dose–response relationship of teratogens.
A) Teratogens interfering with specific events in differentation; B) Teratogens acting via general cytotoxicity and induction of cell necrosis

A more common dose-response pattern involves embryolethality, malformations, and growth retardation of surviving fetuses. Exposure to these chemicals results in a combination of resorbed, malformed, growth-retarded, and "normal" fetuses. Depending on the teratogenic potency of the agent, lower doses may cause predominantly malformations. As the dosage increases, however, embryolethality predominates until the entire litter is resorbed. Growth retardation can precede both these outcomes or parallel the teratogenicity curve.

A third dose–response pattern consists of growth retardation and embryolethality without structural abnormalities. Growth retardation of surviving fetuses usually precedes significant embryolethality. Agents producing this pattern of response would be considered embryotoxic but not teratogenic and are also toxic to the maternal organism. In contrast, potent and specific teratogens often show only weak toxicity to the maternal organism.

The best-known example is the hypnotic sedative thalidomide, which was rarely associated with severe undesired effects in humans but induced malformations in thousands of children whose mothers had taken the drug as a sleeping aid at the recommended therapeutic doses during pregnancy. The existence of these three general patterns of response indicates that for some agents, embryolethality and teratogenicity are different degrees of manifestations of the same primary insult. For other agents, a qualitative difference in response exists, and the primary insult leads to embryotoxicity and embryolethality alone.

For practical purposes (for exact experimental procedures, see Section 4.7.2) a relatively small number of pregnant rodents (approximately eight per group) are exposed on days 6 through 15 of gestation to the test agent at doses up to those causing limiting maternal toxicity or severe embryotoxicity (death, severe growth retardation). The purpose of this dose-range finding study is to obtain a qualitative yes or no signal about the potential developmental toxicity of the agent and information on doses causing severe impairment of the maternal organism. For the main evaluation of developmental toxicity, the highest dose should cause measurable but slight maternal toxicity (i.e., significant depression of weight gain) or embryotoxicity (i.e., significant depression of fetal body weight, increased embryolethality, and/or structural malformations), and the low dose should cause no observable effects.

3.8.3. Mechanisms of Action in Teratogenesis

3.8.3.1. Cytotoxic Teratogens (see also, → Teratogenic Agents, treated in the A series)

The majority of well-known developmental toxins produce both structural malformations and embryotoxicity. This response pattern is typical for xenobiotics that are cytotoxic to replicating cells by impairing DNA transcription, DNA translation, or cell division. Examples include alkylating agents, often employed in cancer chemotherapy, and many pharmaceuticals or chemicals that are biotransformed to electrophilic metabolites. One important reason for the great susceptibility of the embryo to these agents is the extremely high rate of cell division during the period of organogenesis. In rats within days 8 to 11 of gestation, the DNA content of the embryo increases approximately 1000-fold [148]. Hence, the fact that many inhibitors of cellular proliferation are embryotoxic and teratogenic is not surprising.

A characteristic, early effect in target organs destined to become malformed is the onset of excessive cell death. For malformations to develop, the increased necrosis must occur at a selective site and within a critical period. Low doses of cytotoxic agents administered relatively early in the critical period may produce levels of cell death that can be replaced through compensatory hyperplasia of surviving cells, resulting in the formation of growth-retarded but morphologically normal fetuses. Higher doses administered later during the critical period may cause substantial depletion of cells at a selective site of the developing organ, leaving insufficient time for replacement prior to the occurrence of critical morphogenetic events. The resulting hypoplasia of the organ rudiments, as well as the reduced proliferative rate of surviving cells, are important events related to the induction of malformations. High levels of exposure may damage so many cells and organ systems that survival is not possible and embryolethality results.

A single application of a cytotoxic agent during organogenesis may result in all three outcomes. Some litters may be totally resorbed, others may contain only growth-retarded fetuses, and others may have a mixture of malformed or growth-retarded fetuses and resorption sites. The variation in response has been ascribed both to the differential delivery of chemicals to individual embryos depending on their position in the uterus and to variations in the developmental age of embryos within the same litter [148]. The fact that so many different parameters may influence the final outcome clarifies the difficulties in predicting the consequences of chemical exposure during pregnancy.

N-methyl-*N'*-nitro-*N*-nitrosoguanidine is an alkylating agent that induces replication-dependent mutations and inhibition of DNA synthesis. Administration of *N*-methyl-*N'*-nitro-*N*-nitrosoguanidine to rats on days 7 to 12 of gestation results in malformations of the brain, palate, vertebral column, ribs, and limbs. Teratogenicity predominates under these exposure conditions, but embryolethality is also elevated at all dose levels [148]. With *N*-methyl-*N'*-nitro-*N*-nitrosoguanidine, hindlimbs are more frequently malformed than forelimbs, and limbs on the left side are malformed more frequently than those on the right. At all time points examined, limbs on the left had higher levels of cell death than limbs on the right, and hindlimbs had higher levels than forelimbs. Thus, with *N*-methyl-*N'*-nitro-*N*-nitrosoguanidine, the level of cell death is quantitatively correlated with the frequency of limb malformations. Apparently, the limb buds in the four limb types have different intrinsic susceptibility to *N*-methyl-*N'*-nitro-*N*-nitrosoguanidine, possibly based on the percentage of cells undergoing proliferation in each limb type. However, the mechanistic background of this marked site specifity cannot be defined completely with present knowledge.

Sites of high proliferative activity in the embryo are susceptible to cell death after exposure to cytotoxic teratogens, and the frequency and morphologic pattern of the malformations are related to the localization and extent of necrosis in embryonic organ rudiments. Depression of DNA synthesis or DNA damage does not appear to be as important. The embryo can tolerate substantial inhibition of DNA synthesis that, unless accompanied by excessive necrosis, does not lead to malformations. Likewise, chromosomal aberrations have been observed in embryonic cells after exposure to a variety of alkylating agents during the organogenesis period. Cells bearing chromosomal aberrations after transplacental exposure appear to be rapidly eliminated, usually within 24 h, indicating that in most cases chromosomal aberrations contribute to cell death but not to heritable mutations in surviving cells [149]. Even stable chromosomal aberrations do not persist after several cell divisions following prenatal exposure to ethylnitrosourea [150]. The rapid rate of cell division may mean insufficient time to repair DNA damage. In summary, cell death is a more likely outcome of DNA damage in embryonic tissues than are heritable mutations.

Some degree of restorative growth (increased mitotic activity) occurs in embryonic tissues after cytotoxic insult. In the case of *N*-methyl-*N'*-nitro-*N*-nitrosoguanidine, mitotic activity remains elevated long after the necrotic episode has subsided, especially in the left hindlimbs, which have the highest level of cell death and malformation.

Restorative growth is a rather uncoordinated process that may have dual consequences resulting in either repair or exacerbation of the cytotoxic insult. Each cell type at the damaged site undergoes restorative growth as quickly as possible, independent of adjacent cells. This uncoordinated growth exacerbates the mistiming of inductive interactions between tissues that are initially pertubed by differential cell death and may increase the rate of malformations.

3.8.3.2. Teratogens Affecting Specific Events in Differentiation

Teratogens interfering with specific events in differentiation disrupt development by highly specific mechanisms not involving excessive necrosis and can cause highly specific malformations in the absence of embryolethality (Fig. 3.12). These specific teratogens usually induce a distinct subset of all possible malformations at a given time of exposure. No general mechanism of action underlies the effects of these agents; each agent appears to operate differently. In most cases, the mechanistic background of specific teratogenesis in the absence of gross cytotoxicity is not understood.

Glucocorticoids (→ Hormones, **A 13**, pp. 134–138) are a classic example of selective teratogens. Physiologic levels of glucocorticoids are required for normal growth and differentiation of embryonic tissue, whereas higher pharmacologic doses administered at certain times during gestation induce malformations in laboratory animals, most often of the palate and less frequently of the limbs. The organ-specific teratogenicity of glucocorticoids correlates with the concentration of steroid receptors, which is higher in the craniofacial region than in the rest of the embryonic body [151]. Different strains of inbred mice exhibit different degrees of susceptibility to glucocorticoid-induced cleft palate. The concentrations of glucocorticoid receptors in the target cells of the embryo correlate with the strain susceptibility (i.e., higher receptor levels are found in responsive strains compared to nonresponsive strains). Glucocorticoids cause selective growth of the target cells and alteration in production of extracellular matrix in the absence of massive cell death. Glucocorticoid induction of cleft palate alone in the absence of embryolethality, extensive necrosis, and overall growth retardation of the embryo is a good example of a teratogen operating through a specific mechanism of action involving receptor-mediated events.

Another example of a teratogen that produces a distinct malformation syndrome by a specific mechanism of action is the drug *acetazolamide*, currently utilized in the pharmacotherapy of glaucoma. Acetazolamide induces an unusual, distinct type of limb malformation. At teratogenic doses of acetazolamide, embryolethality is not significantly increased and growth retardation is relatively minor. The principal pharmacological activity of acetazolamide is inhibition of the enzyme carbonic anhydrase. This enzyme catalyzes the hydration of CO_2 leading to H^+ and HCO_3^- and is an important factor in the control of blood pH. Inhibition of carbonic anhydrase induces severe metabolic acidosis. In addition to acetazolamide, several structurally diverse carbonic anhydrase inhibitors also produce the same pattern of malformations; therefore, inhibition of the enzyme is widely assumed to be involved in the teratogenic effect. This hypothesis is further supported by the fact that exposure of mice at the critical period to a high CO_2 environment produces a pronounced respiratory acidosis and the same limb defects as acetazolamide. A strong correlation exists between pH and cell proliferation (i.e., reduced pH results in reduced proliferation in a number of tissues). The absence of cell death in acetazolamide-exposed limbs suggests that the malformations result from reduced cell proliferation [152], [153]. Further evidence in support of the pH hypothesis can be obtained from the fact that many human teratogens such as valproic acid, isotretinoin, bimethadione, phenytoin, thalidomide, warfarin, and aminopterin are weak acids and none is a weak base.

3.8.3.3. Nonspecific Developmental Toxins

Nonspecific developmental toxins induce growth retardation and embryolethality but are not teratogenic. The mitochondrial protein synthesis inhibitors chloramphenicol and thiamphenicol are known examples of selective embryotoxic compounds. They induce dose-dependent inhibition of mitochondrial respiration, ATP content, and cytochrome oxidase activity in embryonic tissues, and the mitochondrial toxicity is quantitatively correlated with growth retardation and death of embryos [148]. No basis exists for target organ susceptibility when such fundamental cellular processes are perturbed, and consequently all tissues appear to be affected. The early signs of perturbation are overall growth retardation progressing to lethality of the entire litter above a critical threshold. This mechanism is incompatible with teratogenicity in which some tissues are damaged and others are spared, permitting the survival of abnormal embryos.

3.8.3.4. Developmental Toxicity due to Perturbations in the Maternal Organism and the Placenta

The best examples for indirect-acting developmental toxins are those that lead to maternal nutritional deficiencies. Generalized malnutrition, caloric restriction, and protein deficiency during pregnancy lead to severe growth retardation, thyroid deficiencies, and retardation of maturation of the central nervous system of the fetus that are not reversible with sufficient or even augmented food supply to the neonate after birth. Deprivation of specific nutrients in the maternal diet, such as vitamin A, zinc, and folic acid, induces not only growth retardation and embryolethality but also malformation. In addition, agents that reduce the transport of nutrients from the maternal body to the embryo through interference with placental function can have an indirect embryotoxic effect. Demonstration that developmental toxins of this group do not penetrate embryonic tissues is the usual approach taken to prove the "indirect" mode of action. This is often difficult because few analytic techniques are sensitive enough to detect low levels of xenobiotics in the embryo.

Alteration of uteroplacental blood flow is an important mechanism of action of indirect-acting developmental toxins. Uteroplacental blood flow is reduced in women with hypertension, a condition that is associated with the birth of growth-retarded infants. Clamping of the uterine vessels during the organogenesis period for up to an hour produced a spectrum of malformations specific for the developmental age at the time of obstruction, but negligible lethality or growth retardation. In contrast, partial but permanent reduction in uteroplacental blood by uterine artery ligation is generally associated with growth retardation. Drugs inducing vasoconstriction in the uterus and placenta such as serotonin, epinephrine, and ergotamine cause malformations, embryolethality, and growth retardation [148].

3.8.4. Distribution and Biotransformation of Xenobiotics during Pregnancy in the Maternal Organism

In terms of xenobiotic disposition and biotransformation, the maternal, placental, and fetal compartments comprise independent, yet interacting systems that undergo profound changes throughout the course of pregnancy. Important changes in the disposition of xenobiotics during pregnancy are increased intestinal absorption and reduced plasma protein binding, both of which may contribute to high concentrations of xenobiotics in the embryo. In the first six months of human pregnancy, fat-soluble xenobiotics are stored in the increased fat content of the maternal body; however, in the last trimester these fat stores tend to be mobilized, which may result in increased exposure to the stored xenobiotic of both mother and fetus. In addition, a decrease in hepatic biotransformation is observed during pregnancy. The above alterations, which tend to favor retention of xenobiotics, are in part counterbalanced by increased rates of renal plasma flow and glomerular filtration.

3.8.5. Transport and Biotransformation of Xenobiotics in the Placenta and the Developing Organism

Generally, xenobiotics that are lipophilic, nonionized at physiologic pH, and have molecular masses < 500 readily cross the human placenta. Although the human placenta contains most of the important phase-I and phase-II enzymes of xenobiotic biotransformation, the activities of these enzymes are negligible compared to the liver. Hence, no significant metabolism is expected to occur in the placenta unless the mother has been exposed to enzyme inducers. Important for the human situation is the fact that smoking appears to increase significantly the activity of aryl hydrocarbon hydroxylase (AHH; cytochrome P450 IA 1) in the human placenta [154]. Xenobiotic biotransformation does not occur or is negligible in the embryo or fetus. Also, the majority of experimental evidence available so far does not indicate inducibility of xenobiotic metabolism in the fetus until late in gestation.

4. Methods in Toxicology

4.1. Toxicological Studies: General Aspects

The aim of toxicology is the assessment and management of potential hazards from exposure of humans and the general environment, including animals and plants, to chemicals. To achieve this objective, detailed knowledge of the inherent hazard of a xenobiotic (for definition, see Section 5.2.1), i.e., its acute and chronic toxicity, its no-observed-effect-level (NOEL), as well as its teratogenic, mutagenic, and carcinogenic effects, is required. This information cannot be obtained from a single experiment. A battery of in vivo and in vitro toxicity tests must be utilized. As required by law in most industrialized countries, all toxicity testing must be performed under rules of good laboratory practice with exact documentation of all relevant conditions and results. Adequate planning of toxicity tests to obtain optimal information from the experiments may greatly improve the basis for risk assessment of a chemical and may reduce the number of animals needed and the expenses associated with toxicity studies. Therefore, all test batteries should be part of an integrated approach to toxicity studies and include not only methods to determine the toxic effects of a chemical and their dose dependence but also toxicokinetics, biotransformation, and mechanisms of action. This chapter provides an overview of currently used methods for the assessment of a chemical's toxic profile.

For evaluation of the toxic effects of a new chemical in laboratory animals, two types of studies are carried out: acute toxicity and repeated-dosing studies.

Acute Toxicity. Following administration of a single dose of the test substance or of multiple doses given over a period of up to 24 h, potentially adverse effects are usually monitored during the following 14 d. Acute toxicity studies in animals aim to assess the human risk from single exposure to high doses, for example, in industrial accidents, after drug overdoses, or after suicide attempts.

Repeated-Dosing Studies: Subacute, Subchronic, and Chronic Toxicity. The purpose of repeated daily doses of a chemical for part of the animal's life span is to study subchronic and chronic effects. Studies on *subacute toxicity* are carried out for two to four weeks, while studies on *subchronic toxicity* usually last for a period of three months. These studies are helpful in assessing the human risk resulting from frequent exposure to household or workplace chemicals and from intake of chemicals used for therapeutic purposes. Studies to determine *chronic toxic effects* are carried out for at least six months; studies aiming to investigate the carcinogenic effects of a test compound are carried out over the animal's entire lifetime. Lifetime exposure of humans may occur to widespread environmental pollutants, food additives, or residues of agricultural chemicals in food.

In addition to the acute and repeated-dose toxicity studies, the *reproductive* and *developmental toxicity* as well as the *genotoxicity* of a new chemical must be investigated in separate experiments.

Many thousands of new and potentially toxic compounds are synthesized every year. Money, resources, and manpower would be wasted if the entire battery of toxicity tests were automatically performed with every new chemical. Therefore, toxicity testing is undertaken on the basis of a decision point approach in several stages as shown in Figure 4.1. At the end of every stage, the decision must be made whether development will be continued or whether, on the basis of toxicity data available so far, the potential human risk of exposure to this chemical is unacceptable. If the latter is true, the development and consequently the toxicity testing are stopped.

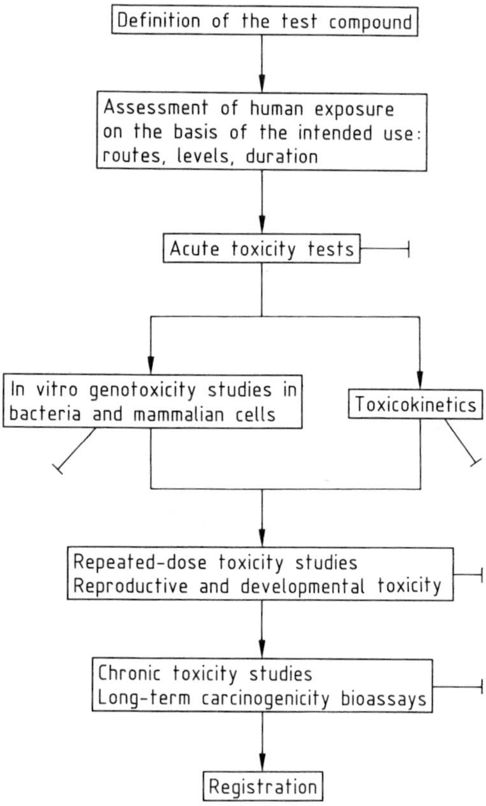

Figure 4.1. Evaluation of the toxicity profile of a new chemical compound on the basis of a decision point approach
At every step of toxicity testing, further development may be interrupted if, on the basis of data collected up to that point, the human risk is considered unacceptable.

Animal Husbandry. The use of standardized conditions for the housing of animals play a major role in the planning, evaluation, and interpretation of toxicity tests. Animals must be kept in a controlled environment (i.e., constant temperature of 22 ± 3 °C, sufficient ventilation, relative humidity between 30 and 70 %, and a 12-h light–dark cycle). Diet composition and quality of drinking water must also be standardized and controlled throughout the experiment. Only healthy young adult animals should be enrolled in the studies, and the animals should be allowed to acclimatize to experimental conditions for at least one week prior to the first dosing. After the acclimatization period, animals with poor health or body weights varying by more than 20 % from the group's mean body weight are either excluded from the studies or randomized to ensure a homogeneous population in the different control and treatment groups. The basic guidelines—choice of species, number of animals, dosing regimens, duration and frequency of observation, assessment of specific body functions—for acute, subchronic, and chronic oral toxicity tests—are summarized in Table 4.1.

4.2. Acute Toxicity

4.2.1. Testing for Acute Toxicity by the Oral Route: LD_{50} Test and Fixed-Dose Method

The objectives of acute toxicity tests are

1) To assess the intrinsic toxicity of the test compound

Table 4.1. Basic guidelines for acute, subchronic, and chronic oral toxicity tests

	Acute oral	Subchronic oral	Chronic oral
Animals	rats preferred	rodent and nonrodent species	rodent and nonrodent species
Sex	males and females equally distributed per dose level		
Age	young adult, weight variation within 20% of mean	rodents, 6 weeks; dogs 4–6 months old	
Number of animals	at least 10 (5 per sex)	at least 20 for rodents (10 per sex)	50 per sex group for rodents
Number of treatment groups	3; mortality rates between 10 and 90% should be produced	3; mortality should not exceed 10% in high-dose group	3; low dose should reflect expected human exposure and high dose must produce not more than 10% mortality
Untreated control	not necessary	yes	yes
Vehicle control	yes, if vehicle of unknown toxicity is used	yes	yes
Dosing	gavage; single dose, same dose of vehicle; if necessary use divided doses over 24 h	diet, gavage, drinking water	diet, gavage, drinking water
Duration of study	at least 14-d observation period	90 d	6–24 months in rats
Body weight determination	before dosing, weekly thereafter, and at death	weekly and at termination	weekly for first 13 weeks; every 2 weeks thereafter and at termination
Necropsy	all animals	all test animals; organ weights of liver, kidney, heart, lungs, brain, gonads, adrenals, and spleen	
Histopathology	examination of organs showing evidence of gross pathological change	all tissues high-dose and control groups; liver, kidney, heart, lungs, target organs, and any gross lesion in mid- and low-dose groups	all tissues of animals
Frequency of cage-side observations	frequently during day of dosing; once each morning and late afternoon thereafter	daily	daily
Observations, assessments	nature, onset, severity, and duration of any effect observed		
		ophthalmoscopy: pretest and at termination in control and high-dose groups	
		hematology/clinical chemistry: pretest, dosing midpoint, termination	pretest and at 3, 6, 12, 18, 24 months
		urine analysis: dosing midpoint, termination	pretest and at 3, 6, 12, 18, 24 months

2) To identify target organs of toxicity affected by the xenobiotic
3) To provide information concerning the dose selection and treatment regimens for repeated-dose studies
4) To provide information for human risk assessment after a single high-dose exposure to the chemical
5) To provide essential data for the classification, labeling, and transportation of the chemical (regulatory view report)

LD_{50} Test. Determination of the mean lethal dose (LD_{50}) is still often considered the first step in evaluation of the acute oral or inhalation toxicity of a new chemical; with present knowledge and recent experiences, formal determination of LD_{50} should no longer be considered necessary. However, since the test is still widely used, it is described briefly.

In the LD_{50} test, groups of animals are treated with graduated doses of the test compound and by using mathematical models the dose that causes death in 50% or more of the population is determined. Internationally accepted guidelines recommend the use of at least three dose groups with five males and five females for each dose or the use of three dose groups with five animals of one sex and one dose with five animals of the other sex [155]. The LD_{50} is then determined from data obtained as described in Section 1.7. Chemicals with LD_{50} values ≤ 25 mg/kg are considered very toxic; between 25 and 200 mg/kg, toxic; and between 200 and 2000 mg/kg, harmful. Some potentials and limitations of the LD_{50} test are listed below:

Potential

1) Useful as a first approximation of hazards in the workplace
2) Basis for the design of subchronic studies
3) Properly conducted test may give useful information on other relevant toxicity parameters
4) Rapid completion

Limitations and Problems

1) Lethality only criterion applied, other toxic effects not considered
2) Animal welfare is major point of concern because large number of animals are required to obtain statistically acceptable values
3) Large variations in LD_{50} in different laboratories with identical chemicals, many influencing factors
4) Species and strain differences cause difficulties in extrapolation
5) No information on chronic toxicity obtained (chronic effects are more important for regulating exposure)

The scientific significance of the LD_{50} test has been questioned repeatedly in the last years not only because the lethal dose is not relevant for human risk assessment but also on the basis of the variability of the test results and, last but not least, animal welfare reasons. Comparative assessment of LD_{50} values in 60 laboratories under controlled conditions resulted for example in LD_{50} values differing considerably (up to a factor of 14). Therefore, numerous alternatives to the LD_{50} have been proposed for the evaluation of acute toxicity that rely on signs of toxicity rather than on mortality. One of these procedures, the fixed-dose method, has recently gained acceptance by the OECD and the EU.

Fixed-Dose Method. The fixed-dose method relies on the observations of clear signs of toxicity at one of a series of fixed-dose levels (i.e., 5, 50, 500, 2000 mg of the chemical per kilogram of body weight). The dose levels at which signs of toxicity, but no deaths, are detected are used to classify the test compounds according to their toxic potential (Table 4.2).

As can be seen from Table 4.2, the fixed-dose method allows a classification identical to that previously obtained in the LD_{50} test. Moreover, comparative investigations utilizing both the LD_{50} and the fixed-dose procedures revealed that in the majority of test compounds (80–90%), the toxicity class assigned by determining the LD_{50} was identical to the fixed-dose method [155], [156].

In the fixed-dose method, at least ten animals (five per sex) are used for each dose investigated. The initial dose chosen (5, 50, 500, or 2000 mg per kilogram of body weight) is one judged likely to produce evident toxic effects, but not mortality. When such a judgment cannot be made because of lack of information on the potential toxic effects of the xenobiotic, an initial "sighting" study should be carried out. If clear signs of toxicity do not occur at the chosen dose during the two-week observation period, the dose is increased to the next level. Careful clinical exam-

Table 4.2. Classification of toxicity of a xenobiotic with the fixed-dose method

Dose (oral), mg/kg	Results	Classification
5	less than 90% survival	very toxic
	90% or more survival but evident toxicity	toxic
	90% or more survival, no evident toxicity	retest at 50 mg/kg
50	less than 90% survival	toxic, retest at 5 mg/kg
	90% or more survival, but evident toxicity	harmful
	90% or more survival, no evident toxicity	retest at 500 mg/kg
500	less than 90% survival or evident toxicity and no death	harmful, retest at 50 mg/kg
	no evident toxicity	retest at 2000 mg/kg
2000	less than 90% survival	harmful
	90% or more survival, with or without evident toxicity	unclassified, does not represent a significant acute toxic risk if swallowed, no further testing necessary

ination of the animals is performed at least twice on the day of administration and once daily thereafter for the next two weeks. Animals obviously in pain or showing severe signs of distress or toxicity will be killed humanely. Cage-side examinations include skin and fur, eyes and mucous membranes, respiratory system, blood pressure, somatomotor activity, and behavior (for procedures, see Section 4.11). Particular attention is directed to observation of tremors, convulsions, hypersalivation, diarrhea, and coma as indices of neurotoxicity. Food consumption and weight development are also monitored constantly. At the end of the observation period, all animals in the study are killed and subjected to gross autopsy. Organs showing macroscopic evidence of gross pathology are further subjected to histopathological examination.

The fixed-dose method offers several important advantages compared to the traditional LD_{50} test:

1) Available evidence suggests that the fixed-dose method produces more consistent results without substantial interlaboratory variation.
2) It provides information on the type, time of onset, duration, and consequences of toxic effects. This information is more relevant for assessing the risks of human exposure to the chemical than the mean lethal doses of the LD_{50} test.
3) It requires lower numbers of animals than the LD_{50} test (roughly 50%) and subjects the animals to less pain and distress.
4) It enables the classification of chemicals according to regulatory requirements.

Presently, the fixed-dose method has been validated only for assessing acute oral toxicity, but it is also applicable to acute toxicity testing by the dermal or inhalation route.

4.2.2. Testing for Acute Skin Toxicity

Knowledge of the dermal toxicity of a new chemical is one of the prerequisites for assessment of the risks associated with human exposure to the compound because skin contact may represent a very important route of exposure in the occupational setting and in the home. Testing for dermal toxicity is usually performed in rabbits. Three types of application of the test chemical are employed: the unocclusive, semiocclusive, and occlusive. The test compound is applied uniformly to the back or to a band around the trunk (clipped free of hair); ca. 10% of the body surface of the animal should be covered. Solid substances are pulverized and moistened to a paste with physiological saline solution or another appropriate solvent whose effects have been fully evaluated prior to the skin test. For occlusive or semiocclusive testing, the application site is covered with a plastic sheet (or other impervious material) or with a porous gauze dressing, respectively. For unocclusive exposure, the application site should be as close to the head as possible to prevent ingestion of the chemical by the animal by licking the site of application. The duration of exposure varies between 4 and 24 h. If no chemical-related toxic effect on the skin or systemic toxicity is observed after doses up to 2 g per kilogram of body weight, testing at higher doses is unnecessary. At the end of the exposure

Table 4.3. Evaluation of skin reactions according to the Draize scoring system

Erythema	Score
No erythema	0
Very slight erythema (barely perceptible)	1
Well-defined erythema	2
Moderate to severe erythema	3
Severe erythema (beet redness)	4
No edema	0
Very slight edema (barely perceptible)	1
Slight edema (edges of area well defined by definite raising)	2
Moderate edema (area raised ca. 1 mm)	3
Severe edema (raised more than 1 mm and extended beyond area of exposure)	4

Table 4.4. In vitro test systems for detection of dermal irritation potential

System	End point
Mouse skin organ culture	leakage of LDH* and GOT**
Human epidermal keratinocytes	release of labeled arachidonic acid, cytotoxicity
Cultured BHK21/C13 cells	growth inhibition, cell detachment
SKINTEX–protein mixture	protein coagulation

* LDH = Lactate dehydrogenase.
** GOT = Glutamic acid oxalacetic transaminase.

period, the compound is removed with cotton wool soaked in an appropriate solvent and the skin irritation is scored according to the Draize scoring system as shown in Table 4.3 [157]. In addition, any adverse systemic effects caused by percutaneous absorption of the test compound are monitored.

However, substantial differences exist in skin anatomy between humans and experimental animals. In general, the penetration of chemicals through human skin is similar to that of pig, miniature swine, and squirrel monkey and clearly slower than through the skin of rat and rabbit. For example, administration of the insecticides lindane and parathion to rabbit skin results in an absorption of 51.2 and 99.5% of the dose, respectively; the corresponding absorption rates for human skin are 9.3 and 9.7%.

Since the 1980s, *in vitro studies* using human skin samples have been increasingly conducted to estimate percutaneous absorption of chemicals. The following experimental design is commonly applied: A piece of excised human skin is attached to a diffusion apparatus that has a top chamber for the test compound, an O-ring to hold the skin in place, and a bottom chamber to collect samples for analysis. The flow of a chemical across the skin can be calculated with models based on chemical, thermodynamics, by taking into consideration the octanol–water partition coefficients, the saturation concentration in aqueous solution, and the molecular mass of the test compound [158]. However, the method has not been evaluated sufficiently for routine application.

Furthermore, an increasing number of *toxicokinetic models* for estimating the extent of percutaneous absorption of chemicals has appeared in the literature since the 1980s. Among them, a physiologically based toxicokinetic model was recently developed to describe the percutaneous absorption of volatile and lipid-soluble organic contaminants in dilute aqueous solution [159]. This toxicokinetic model considers both physiological parameters such as volumes of body compartments and blood flow rate and the properties of the test compound. Presently, these models do not play a role in regular toxicity testing and are therefore not discussed here.

Extensive progress has been also made in the 1990s in developing in vitro systems for evaluating the dermal irritation potential of chemicals. An overview of the systems that have been evaluated so far for a range of compounds by comparison of their predictive accuracy with animal test results is presented in Table 4.4 (or a review see [160]).

4.2.3. Testing for Acute Toxicity by Inhalation

Studying the toxicity of a chemical by inhalation exposure requires considerable technological input. Therefore, inhalation exposure is usually not tested if this pathway of absorption is not expected to occur because the test chemical is nonvolatile or the physicochemical properties of solids do not allow the generation of respirable particles. Particles with diameters > 100 µm are unlikely to be inhaled because they settle too rapidly. Particles with diameters of 10–50 µm are likely retained in the nose and upper part of the respiratory tract, while particles with diameters of < 7 µm can reach the alveoli of the human lung. When performing toxicity studies with

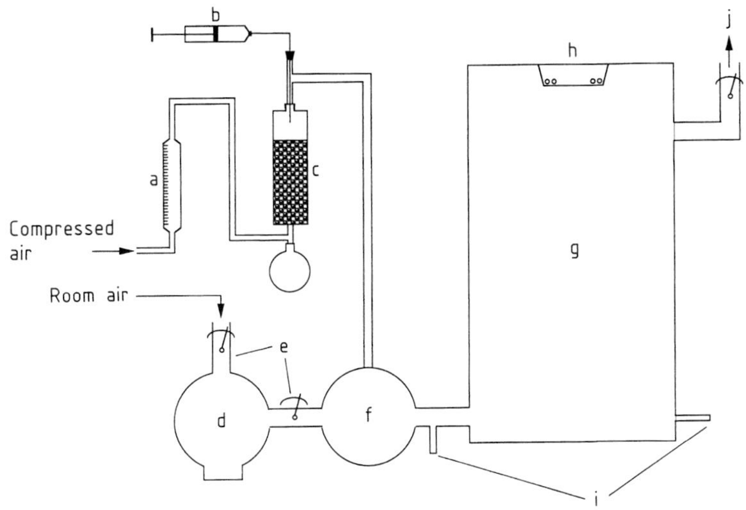

Figure 4.2. Example of an open inhalation chamber for exposure to volatile liquids
a) Meter; b) Syringe coupled to infuser; c) Heated beads of silica or glass; d) Compressor; e) Metering valves; f) Mixing chamber; g) Exposure chamber; h) Ventilation; i) Sampling valves for determination of atmospheric concentration; j) Exhaust

inhalation exposure, the differences in respiratory physiology between humans and small laboratory rodents must be considered. In contrast to humans, the rat is an obligate nose-breather with a complex nasal turbinate structure, which filters many small particles. Therefore, the upper size limit for particles reaching the alveolar region in rats is in the range of 3–4 μm in diameter.

The duration of exposure in acute inhalation studies is usually 4–6 h and may be either whole-body or head(nose)-only exposure in specific exposure chambers. A number of important considerations should be taken into account in planning and evaluating inhalation studies. The frequently made assumption that on the basis of their physical form, gases and vapors will be absorbed uniformly throughout the respiratory tract is incorrect. Many gases or vapors, such as ammonia, formaldehyde, and sulfur dioxide, have a high water solubility and are absorbed rapidly by the humid epithelial surface of the upper respiratory tract. Therefore, toxic effects observed after inhalation of this type of chemical will generally be confined to these regions, especially the nasal passages. In contrast, chemicals with low water solubility such as nitrogen dioxide, phosgene, and ozone will penetrate readily to the lower pulmonary regions, even at relatively low concentrations in the respiratory air. In mixed atmospheres, where both gas or vapor and particulates are present, the vapor or gas may be adsorbed to the particulate fraction so that the deposition pattern of the vapor or gas is governed by the size of the particulate fraction and not by the water solubility of the vapor or gas.

Inhalation exposure of experimental animals may be studied in a dynamic or static mode. In *dynamic systems* (Fig. 4.2), the test atmosphere is renewed continuously to ensure atmospheric stability and identical concentrations of the chemical in the gas phase. The liquid xenobiotic is added continuously to a stream of air on a heated bed of glass beads. The vaporized xenobiotic is then mixed with room air and introduced into the exposure chamber. The concentration of the xenobiotic in the atmosphere of the chamber is determined by taking air samples and quantifying xenobiotic content by gas chromatography. This mode of inhalation exposure is more complicated and requires larger amounts of the test chemical and suitable systems to ensure complete mixing of the continuously applied test compounds with the stream of air flowing through the exposure system.

Static systems are sealed and the volatile xenobiotic is injected into the system to generate an atmosphere with a specific concentration, which is cycled through the system. The concentration of the chemical in the gas phase decreases during exposure due to uptake and biotransformation of the test compound by the experimental animals. Carbon dioxide exhaled by the ani-

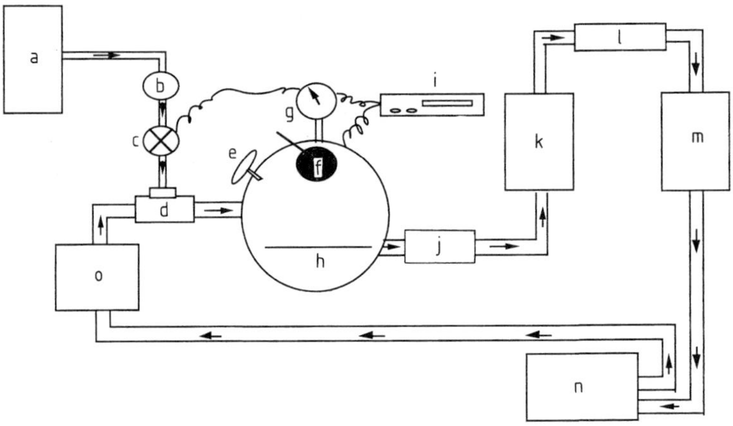

Figure 4.3. Closed exposure system
a) Oxygen cylinder; b) Metering valve; c) Solenoid valve; d) Mixing chamber; e) Thermometer; f) Oxygen sensor; g) Pressure gauge; h) Exposure chamber; i) Oxygen monitor; j) Injection port; k) Condenser; l) Flow meter; m) Carbon dioxide absorber; n) Gas chromatograph; o) Pump

mal is trapped, and oxygen concentrations are monitored and adjusted automatically. Samples of the atmosphere can be analyzed automatically by gas chromatography to determine the concentration of xenobiotic over time (Fig. 4.3). Static systems are used predominantly in acute toxicity studies and in research laboratories because they are relatively inexpensive and consume only a small fraction of the chemical compared to the amounts need to generate dynamic exposure. By analogy to the LD_{50} values obtained in acute oral studies, the mean lethal air concentrations LC_{50} are assessed in acute inhalation studies, usually at a 4-h exposure time.

4.3. Repeated-Dose Toxicity Studies: Subacute, Subchronic, and Chronic Studies

Repeated-dose toxicity studies assess the effects resulting from the accumulation of a compound or its toxic effects in the organism and, unlike acute studies, can also reveal toxic effects that appear after a latency period. A classic example is the delayed neuropathy caused by many organophosphorus insecticides and cresyl phosphates, which is manifested several weeks after the first administration of the test compound. In contrast to the marked clinical symptoms observed in the course of delayed neuropathy, these compounds rarely cause any acute symptoms immediately after the first administration. Hence, false-negative results are obtained in acute toxicity testing.

Testing for *subacute toxicity* is usually performed over a time period of two to four weeks with three doses as an aid to select the doses for subchronic studies. Studies to determine *subchronic effects* are usually performed in rats and dogs over 10 % of the animals life span (3 months in rats, 12 months in dogs). At the start of the study, rodents should be 6–8 weeks and dogs 4–6 months old. The animal numbers enrolled are 10–20 rats and 6–8 dogs per sex per dose group. The main goal of the subchronic study is to assess a no-observed-effect level and to determine the organ with the greatest susceptibility to the toxic effects of the test compound. Ideally, the lowest dose of the chemical should not induce toxic effects, the intermediate dose should be slightly toxic, and the high dose should induce clear signs of toxicity without causing death in more than 10% of the animals.

In both subchronic and chronic toxicity studies (see below), the test compound is often incorporated into the diet or added to the drinking water. Food consumption varies from weaning to maturity, with younger animals consuming more food on a body weight basis. Therefore, the changes in body weight and food consumption must be predicted on a weekly basis and concentration of the test compound in the diet adjusted to ensure constant dosing throughout the study. Compounds not stable in the diet or water or not accepted by the animals may also be applied by gavage. Application by gavage directly into the stomach ensures constant dosing; however, gavage studies require skilled personnel, and gavage-related trauma may reduce survival in all groups. Practically all oral dosing regimens are performed on a seven-day-per-week basis, while a five-day-per-week scheme is generally used in skin, gavage, and inhalation studies.

Studies on the *chronic toxicity* of chemicals are usually performed for at least 6 months in rodents and 12 months in dogs. Dose levels are usually based on the results of studies of acute and subchronic toxicity. The highest dose applied should be toxic, i.e., suppress body weight up to 10 % (maximum tolerated dose, MTD; see Section 4.8). The two other dose levels are usually one-fourth and one-eighth of the MTD. Xenobiotics showing no adverse effects in the short-term studies are usually tested at doses that are 100–200 times higher than the anticipated human exposure.

In both subchronic and chronic studies, cage-side examination as well as clinical chemistry are performed routinely during the test period. After termination of the study, much emphasis is put on the histopathological evaluation of treatment-induced adverse effects.

Cage-side observations during the study

1) Body weight, food and water consumption
2) Skin and fur, eyes, mucous membranes
3) Respiration and blood circulation
4) Motor activity and behavioral pattern

Clinical chemistry during the study

1) *Blood:* erythrocyte, leukocyte, and differential leukocyte counts; hemoglobin concentration; hematocrit, platelet, and reticulocyte counts; electrolytes; inorganic phosphorus and alkaline phosphatase; glucose, protein, albumin, creatinine, urea, lipids, enzymes
2) *Urine:* volume and coloration/turbidity, osmolality and pH, glucose and protein, urine enzymes and cytology

Toxicologic pathology after termination of the study and in animals dying during the study

1) Organ weights and macroscopic evaluation
2) Histopathological examination of brain, liver, kidney, spleen, testes, and every organ with macroscopic changes

4.4. Ophthalmic Toxicity

The majority of injuries to the eyes from direct contact with a chemical occur with substances that are handled in an uncontrolled manner (e.g., by children in the home); this type of injury is easily prevented in the occupational setting by simple protective measures. However, many irritant gases and vapors may also produce ophthalmic toxicity; these toxic effects are of practical importance in occupational medicine.

The conventional in vivo eye irritation test in the rabbit was formalized by Draize some 50 years ago and still remains the only fully validated method to assess ophthalmic toxicity [157]. Increasing criticism based primarily on the discomfort and deliberate injury caused to animals led to the development of several in vivo and in vitro alternative methods, which are described at the end of this chapter.

Eye Irritation Test according to Draize. In the conventional in vivo test, the chemical (0.1 mL of liquid or 100 mg of solid chemical) is instilled into one eye of each of the test rabbits, the contralateral eye serving as control. Eyes are then examined periodically (usually after 1, 24, 48, and 72 h and 7 d), and the ocular lesions are scored as shown below, essentially according to Draize et al. [157].

Cornea

A) Opacity	
No opacity	0
Scattered or diffuse area, iris details clearly visible	1
Easily discernible translucent areas, iris details slightly obscured	2
Opalescent areas, iris details not visible, pupil size barely discernible	3
Opaque, iris invisible	4
B) Area of cornea involved	
One-quarter or less, but not zero	1
Greater than one-quarter, but less than half	2
Greater than half, but less than three-quarters	3
Greater than three-quarters, up to whole area	4

Corneal score = (A) × (B) × 5 (maximum total score = 80)

Iris

A) Normal	0
Folds above normal, congestion, swelling, circumcorneal injection (any or all), iris still reacting to light	1
No reaction to light, hemorrhage, gross destruction	2

Iris score = (A) × 5 (maximum total score = 10)

Conjunctivae

A) Vessels normal	0
Vessels definitely injected, above normal	1
Diffuse, deep crimson red, individual vessels not readily discernible	2
Diffuse beefy red	3
B) No chemosis (swelling)	0
Any swelling above normal (includes nictitating membrane)	2
Obvious swelling with partial eversion of lids	2
Swelling with lids about half closed	3
Swelling with lids about half to completely closed	4
C) No discharge	0
Any amount of discharge different from normal	1
Discharge with moistening of lids and hairs adjacent to lids	2
Discharge with considerable moistening around eyes	3

Conjunctival score = [(A) + (B) + (C)] × 2 (maximum total score = 20)

Total maximum score (cornea + iris + conjunctiva) = 110

Evidence is increasing that a volume of 0.01 mL of liquid xenobiotic is as sensitive as the conventionally used 0.1 mL and probably more appropriate for comparison with human exposure situations. Ocular toxicity testing for exposure to gases, vapors, and aerosols is carried out in appropriate exposure chambers.

Other Test Methods. Since the 1980s, a number of in vivo and in vitro alternatives to the conventional eye irritation test have been suggested. The in vivo alternatives aim at reducing the discomfort of animals by employing lower doses of the test material and increasing the sensitivity of the test by using noninvasive objective measurements. Among them, the assessment of corneal thickness and of intravascular pressure seem to be sensitive parameters for identifying mild to moderately irritant chemicals. Despite the large number of suggested in vitro tests, no single in vitro test has currently proved effective in prediciting eye irritation response. Therefore, in vitro tests cannot yet replace the rabbit eye test. However, in vitro assays are useful as screens for product development to reduce the number of tests performed in animals later.

Among the different *in vitro test systems* for ophthalmic toxicity, the chorioallantoic membrane of the hen's egg is probably the best system presently developed [161]. This isolated nonocular preparation is highly vascular and exhibits a high sensitivity to ocular irritants. On day 10 of incubation, the test compound is applied onto the chorioallantoic membrane, and irritant effects such as hyperemia, hemorrhage, and coagulation are scored after 0.5, 2, and 5 min. Because of the sensitivity of the skin preparation, the correlation of the data obtained in this system with those obtained with the Draize irritation test is best when tenfold dilutions of the test compounds are applied.

Tests in enucleated eyes and isolated corneal preparations are promising possibilities for the future, but they are at earlier stages of development than the test with the chorioallantoic membrane of the hen's egg. They require further development and validation. Finally, a variety of cytotoxicity end points cause by application of a test compound (i.e., crystal violet staining, neutral red uptake, and fluorescein leakage) have been determined in many cell lines in culture as possible parameters for predicting eye irritation by a chemical. Cell lines include corneal epithelial cells, lung fibroblasts, Chinese hamster ovary cells, canine renal cells, and human fibroblasts. Obviously, although such tests cannot replace the eye irritation, they can become part of a test battery, that, as a whole, may prove effective in predicting eye irritation response.

Since the beginning of the 1990s, much attention has been paid to the *Eutex system*, a method for predicting in vivo ocular irritation via a target biomacromolecular approach [162]. The test was developed on the basis of the biochemical properties of the cornea. The normal cornea is transparent because the diameter of collagen fibrils and the interfibrillar distance is less than half the wavelength of light, thus minimizing light scattering. Alteration of this arrangement by irritant substances results in light scattering and corneal cloudiness or opacity. The Eutex test system is based on the aggregation of a synthetic matrix of plant proteins that mimic the corneal opacification observed in vivo. The system is calibrated by spectrophotometric measurement of the aggregation of the protein reagent caused by well-characterized eye irritants. In addition to the standard procedure for soluble and miscible materials, several protocols are available such as the membrane partition assay for water-insoluble, opaque, or pigmented materials and the alkaline membrane assay for chemicals with $pH > 7.8$. In a recent evaluation of 142 products with known ocular effects in the rabbit, the Eutex system showed sensitivity and specificity values of 94 and 89 %, respectively, as well as good intra- and interlaboratory reproducibility. Further advantages of the test are the short time required to obtain results (approximately 24 h) and its cost effectiveness because it does not require animals, animal housing, and care. Hence, the Eutex system may become a valuable tool for screening eye irritation in the future.

4.5. Sensitization Testing

Chemicals that have the potential to elicit allergic reactions are introduced continuously into the human environment. Therefore, allergic skin reactions are becoming an increasingly important problem, especially in the workplace. Allergic contact dermatitis is one of the most common occupational diseases and may become debilitating unless the causative agent is identified and exposure ceases. Although irritant dermatitis is generally produced by direct interaction of the chemical with skin constituents, allergic dermatitis is the result of a systemic immune reaction that, in turn, induces effects in the skin. An important characteristic of allergic reactions that must be taken into account when testing allergenic potential is that allergic responses usually have a biphasic course. The *induction period* between initial contact with the causative agent and the development of skin sensitivity may be as short as two days for strong sensitizers such as poison ivy extract or several years for a weak sensitizer such as chromate; for most chemical compounds with allergenic potential the induction period usually takes from 10 to 21 d. After this initial development of sensitivity to a certain allergenic chemical, the time between reexposure to this agent and the occurrence of clinical allergic symptoms is generally 12–48 h; in animal testing, this period is called the *challenge phase*.

On the basis of induction phase–challenge phase biphasic testing, several sensitization tests have been developed (see Table 4.5), [163], [164]. Although they differ by route and frequency of treatment, all utilize the guinea pig as test species. In general, for the induction phase the chemical is administered to the shaved skin intradermally, epicutaneosly, or both, several times over a period of two to four weeks. Adjuvants, i.e., Freund's complete adjuvants (FCA; a mixture of heat-killed *Mycobacterium tuberculosis*, paraffin oil, and mannide monooleate), are often included to increase the immunological response. During the challenge phase, a nonirritating concentration of the chemical is applied. The concentration of the test chemical and the application route (epi- or intradermal) often differ between the two phases. Sensitization is assessed by examining the skin reactions (edema, erythema) after the challenge phase and comparing them with any skin reactions observed immediately after the induction phase; the latter reactions are considered to result from direct irritating (toxic) properties of the test chemical. Hence, the difference between the symptoms observed after the induction phase and after the challenge phase is attributed to the allergenic effects of the chemical.

Table 4.5. Currently used guinea pig sensitization tests

Test	Induction: route/number of applications	Challenge: route/number of applications
Draize	intradermal/10	intradermal/1
Open epicutaneous	epidermal open/20	epidermal open/1
Buehler	epidermal occlusive/3	epidermal occlusive/1
FCA*	intradermal in FCA*/3	epidermal open/1
Split adjuvants	epidermal occlusive/4 + FCA* intradermal/1	epidermal occlusive/1
Optimization	intradermal + FCA*/10	intradermal/1 epidermal occlusive/1
Maximization	intradermal + FCA*/1 epidermal occlusive/1	epidermal occlusive/1

* FCA = Freund's complete adjuvant.

Guinea Pig Maximization Test. The guinea pig maximization test is the most widely used and is considered very sensitive. The first part of the induction phase includes simultaneous injection of FCA alone, the test compound in saline solution, and the test compound in FCA into three different areas in close proximity to each other. The second part of the induction phase seven days later employs epicutaneous application of the chemical on a filter paper, which is occluded and left in place for 48 h. The challenge phase is conducted epicutaneously for 24 h, two weeks after the induction phase. The maximization test is very sensitive and may produce false-positive results. The original procedure (injection of the test compound) does not allow testing of final product formulations. Therefore, a modified procedure has been developed. In the first week, FCA is injected four times and the test product formulation is administered epidermally as in the second induction week.

Mouse Ear Swelling Test. In addition to the established guinea pig tests, the mouse ear swelling test is in a rather advanced stage of development. The induction phase is conducted by abdominal injections of the chemical and FCA, and the challenge dose is applied to the left ear of the mouse, the right ear serving as the control [164]. The mouse ear swelling test is clearly less expensive and may be at least as sensitive as the guinea pig tests; however, it has not yet been accepted by regulatory authorities as an alternative.

In addition, several in vitro approaches are under development that aim to assess sensitizing potential. These in vitro methods use lymphocytes or other types of cultured cells and determine as end points either biochemical indices of allergic responses such as the production of migration inhibition factor (MIF) or cellular events such as cell migration. None of these tests has undergone sufficient comparative evaluation with a battery of compounds with established sensitizing activity in vivo. Because of the complexity of the pathogenetic mechanisms of chemical allergy, a satisfactory and reliable in vitro replacement system is rather unlikely to become available in the near future.

4.6. Phototoxicity and Photosensitization Testing

The biologically active spectrum of light consists of the ultraviolet (UV) (220–400 nm) and visible light (400–760 nm) ranges. The UV spectrum is further divided into UVA (315–400 nm), UVB (280–315 nm), and UVC (220–280 nm); the latter is absorbed in the stratosphere and does not reach the earth surface. The primary source of toxic effects on the skin is UVB light, although UVA light may also play a critical role in some reactions.

Xenobiotics localized within the skin may be activated by UVB light (perhaps by visible light, as well) and induce phototoxicity or photosensitization (photoallergy). *Photoallergy* is similar to allergic contact dermatitis both mechanistically and clinically, the only difference being that the chemical must react with light to become allergenic. Photoallergic reactions are not necessarily dose dependent and show great interindividual variability.

By analogy, *phototoxicity* may be compared to irritant dermatitis. Many phototoxic reactions may be caused by the formation of free radicals followed by lipid peroxidation and localized inflammation. In addition to phototoxicity and photoallergy, light-induced activation of chemicals may cause depigmentation or induction of an endogenous photosensitizer or of a disease characterized by photosensitization such as lupus erythematosus or pellagra. However, these reactions are comparatively rare. The tests to identify photoallergenic and phototoxic chemicals are performed analogously to the tests for skin sensitization and irritation, respectively.

Photoallergenic potential is evaluated by repeated application of the test compound to the skin of guinea pigs. This area is exposed to UV light after every application; and the UV light treatment should cause a very slight erythema. Several days after this induction phase, the challenge phase is conducted by treatment with a low dose of the test compound together with UV light. *Phototoxic reactions* are usually observed after the first exposure to the test compound together with UV light; in addition to guinea pigs, mice and rabbits are also utilized in tests for phototoxicity.

Numerous in vitro tests for phototoxicity have been suggested since the 1970s. These include tests in *Candida albicans* and bacteria that do not reliably predict phototoxicity for the majority of the compounds. A more recent approach utilizes three cultured cell types: a human epidermal cell line derived from epidermal carcinoma, a primary cell line of human epidermal keratinocytes derived from cosmetic surgery samples, and the 3T3 mouse fibroblast cell line. The cells are treated with a nontoxic concentration of the test compound for several hours, followed by exposure to UV light for varying lengths of time. Cytotoxicity is assessed by the extent of reduction of 3-(4,5-dimethylthiazol-2-yl)-2,5-diphenyltetrazolium bromide, an established assay for mitochondrial toxicity. Preliminary evaluation of 30 compounds with known phototoxic potential has shown a sufficiently high degree of correlation; however, more validation is needed. The development of in vitro tests for photosensitization faces the same problems and limitations described above for in vitro assays to assess sensitization potential.

4.7. Reproductive and Developmental Toxicity Tests

Reproductive and developmental toxicity is a very broad term including adverse effects on any of the following:

1) Male or female sexual structure and function (fertility)
2) Development of the new organism through the period of major organ formation, organogenesis (embryotoxicity and teratogenicity)
3) Development of the new organism during the peri- and postnatal periods

The field of reproductive toxicology has become increasingly important in the last 50 years with the recognition that viral and bacterial infections and xenobiotic chemicals can produce severe and irreversible defects in the offspring. The first reports of malformations due to rubella virus infections, ionizing radiation, sex hormones, dietary deficiencies, and chemicals appeared in the 1930s and 1940s, but the potential impact of reproductive toxicity on public health was recognized more than two decades later. In 1960, a large increase in newborns with specific limb malformations, which are rarely seen otherwise, was recorded in Germany and other parts of the world. One year later, the sedative/hypnotic thalidomide was recognized as the causative agent. Of the mothers with characteristically malformed children 84% had definitely taken the drug during pregnancy, 8% may have taken it, and only 8% had a negative history of thalidomide consumption. The thalidomide epidemic resulted in 10000 malformed children and subsided after the drug was withdrawn from the market at the end of 1961. One important consequence of the thalidomide disaster was that most countries in the world introduced requirements for the testing of potential new drugs for reproductive toxicity before they could be released for general use. All test batteries required include detailed tests for reproductive toxicity. In 1966, the FDA published guidelines for a three-segment study to evaluate adverse effects on fertility in the parental organisms and the development of the new organism. This three-segment design is still used with minor modifications throughout most of the world. In addition, multigeneration studies monitoring the cumulative effects of continuous application of the test chemical during two or three generations are very useful to assess human risk from environmental pollutants (i.e., pesticide residues).

4.7.1. Segment I: Fertility and General Reproductive Performance

Experiments are usually conducted in rats (20 animals of each sex per dose) with three doses of the test chemical, most often administered with diet. The treatment must not cause general systemic toxicity in the parental organism; therefore, in dose selection the dose levels are chosen according to observations in studies of subacute and subchronic toxicity, which are usually performed before testing for reproductive toxicity. Young adult male rats are treated for 60–80 d prior to mating to cover a whole period of spermatogenesis. Female rats are pretreated for 14 d to cover three estrous cycles. Treatment of both sexes is continued during the mating period and that of females throughout pregnancy. Half of the females are sacrified just before term, and the numbers of resorbed and dead fetuses as well as structural abnormalities in the developed fetuses are assessed. In the United States, pregnancy is interrupted midterm in half of the females. Treatment of the remaining females is continued through parturition and lactation until weaning of the newborns, usually 21 d after birth. The young animals (F_1 generation) are reared without receiving the test compound until sexual maturity, when their fertility is assessed. During the rearing period, the development of the young animals is monitored with cage-side carefully clinical observations. If an adverse effect on fertility, pregnancy, or development of the offspring is observed, it is necessary to evaluate whether the effect is due to toxicity to the male or female reproductive system or both. This information can be obtained by separate mating of treated males with untreated females and vice versa.

Furthermore, evaluation of toxic effects on the male and female reproductive systems with specific test systems may be required. The effect of the test compound on male reproductive performance may be evaluated by monitoring mating behavior (e.g., frequency of copulation). Structural and functional impairment of the male reproductive organs is assessed by conducting gross pathology and histology of the testes and sperm analysis (viability, motility, and morphology). Histological examination of the ovaries plays an important role in assessment of toxic effects on the female reproductive system.

4.7.2. Segment II: Embryotoxicity and Teratogenicity

Segment II studies assess adverse effects during the period of organogenesis. Xenobiotics interfering with the developing organism during this extremely sensitive period may cause severe and irreversible structural malformations. These studies are carried out in two species, usually rats (20 per dose) and rabbits (10 per dose); in most cases, two dose levels and an untreated control group are included. Pregnant animals are tested during the period of organogenesis: days 6 to 15 for rats and 6 to 18 for rabbits. The fetuses are delivered by cesarean section one day prior to the estimated time of delivery: day 21 for rats and day 31 for rabbits. The main reason for avoiding natural delivery is to prevent loss of deformed or dead fetuses by cannibalism, which happens in rodents and rabbits. The uterus of the maternal animal is excised, weighed, and examined for implantation sites and resorbed fetuses. The pups are weighed, and one-half of each litter is usually examined for skeletal defects and the remaining one-half for soft-tissue defects.

4.7.3. Segment III: Peri- and Postnatal Toxicity

For segment III studies, treatment of pregnant rats with three dose levels (10–12 animals per dose) begins on day 16 of gestation and is carried on through delivery and lactation, until weaning of the offsprings, normally on day 21 postpartum. Treatment during parturition and lactation is also performed in segment I studies; however, segment III studies may offer additional information since higher doses can be used than in segment I. The peri- and postnatal segment evaluates effects on birth weight and survival as well as development of the offspring in the postnatal period. However, extrapolation of results of segment III studies from the rat to the human situation should be performed with care and consideration of the specific circumstances in each species. In contrast to the human situation, where in addition to the mother the social environment takes care of the newborns, young rats depend completely on the functional integrity of the maternal organism. Furthermore, the organism of newborn rats is significantly less mature by the time of delivery than that of newborn humans. Hence, treatments that affect the maternal organism simply by causing sedation or fatigue may significantly impair the development of newborn rats in the first 21 days of life. Such effects should not be automatically interpreted as relevant for postnatal toxicity in humans.

4.7.4. Multigeneration Studies

Multigeneration studies assess the cumulative effects on reproduction and development of continuous application of a test chemical for two or three generations. This application mode is relevant for long-term exposure to chemicals in the environment, such as pesticide residues in food or contamination of drinking water with agricultural chemicals or nonbiodegradable solvents.

Although exact guidelines concerning the number of animals enrolled per group and number of the offspring used for further breeding may differ among regulatory authorities, the basic principles of multigeneration tests are essentially identical. Three dose levels are usually given to groups of 25 female and 25 male rats shortly after weaning at days 30 to 40 of age. In the multigeneration study, these rats are referred to as the F_0 generation. The F_0 generation is treated throughout breeding, which occurs at about 140 days of age, and the female animals during pregnancy and lactation, as well. Hence, the offspring (F_1 generation) have been exposed to the test compound in utero, via maternal milk, and thereafter in the diet. In many protocols, the F_1 generation is standardized to include certain numbers of animals (e.g., eight animals per litter). In analogy to the F_0 generation, the F_1 generation is bred at about 140 days of age to produce the F_2 generation. In some of the parents (F_0 and F_1 generations), gross necropsy and histopathology are performed with greatest emphasis on the reproductive organs. In addition, necropsy and histopathology are studied in all animals that die during the study.

The percentage of F_0 and F_1 females that become pregnant, the number of pregnancies carried out to full term, the litter size, and the number of resorptions, stillborn, and live births are recorded. Viability counts and pup weights are recorded at birth and at days 4, 7, 14, 21, and 28 of age. With these data, the following parameters are calculated for assessment of the long-term reproductive toxicity of the test compound:

$$\text{Fertility index (\%)} = \frac{\text{Number of pregnancies}}{\text{Number of matings}} \times 100$$

$$\text{Gestation index (\%)} = \frac{\text{Number of litters}}{\text{Number of bred females}} \times 100$$

Birth index (%)

$$= \frac{\text{Number of pregnancies resulting in live offspring}}{\text{Number of pregnancies}} \times 100$$

Viability index (%)

$$= \frac{\text{Number of animals alive at day 4 after birth}}{\text{Number of newborns}} \times 100$$

Lactation index (%)

$$= \frac{\text{Number of animals alive at day 28 after birth}}{\text{Number of animals alive at day 4 after birth}} \times 100$$

4.7.5. Role of Maternal Toxicity in Teratogenesis

If an agent with selective developmental toxicity is administered throughout the organogenesis period (days 6 to 15 in the rat), identifying the most sensitive target organs becomes difficult. In addition, teratogenic effects may be masked by embryolethality with repeated dosing during the organogenesis period. Developmental toxicity in the form of increased resorption and decreased fetal body weight is generally accepted to occur at maternally toxic dose levels. The role of maternal toxicity in causing congenital malformations, however, is not clear. Doses causing maternal toxicity, as indicated by reduced maternal body weight, clinical signs of toxicity, or death, commonly cause reduction in fetal body weight, increased resorption, and rarely, fetal deaths. Three patterns of association between maternal toxicity and malformations can be observed: (1) for some compounds, maternal toxicity is not associated with malformations; (2) for others, maternal toxicity is associated with a diverse pattern of malformations, often including cleft palate; and (3) the maternal toxicity of still others is associated with a characteristic pattern of malformations.

Compounds in the second category are the most difficult to classify in terms of teratogenic potential. Cleft palate is the principal malformation resulting from food and water deprivation during pregnancy in mice; however, cleft palate is also a malformation specifically induced in mice by a number of teratogens, most notably the glucocorticoids, without apparent maternal toxicity. Complete determinations of food and water consumption, maternal body weights, and impairment of the maternal organism are necessary to distinguish between cleft palate caused by the teratogenic effect of a chemical on the embryo and that resulting from systemic maternal toxicity, which secondarily affects embryonic development. The association of maternal toxicity with major malformations, such as exencephaly and open eyes, is not generally accepted, although most investigations agree that maternal toxicity can cause minor structural abnormalities such as variants in the ribs.

4.7.6. Assessment of Human Risk on the Basis of Animal Data

The timing of exposure and patterns of dose-response from animal studies have important implications for extrapolating animal data to humans. A wide spectrum of end points can be produced, even under the controlled conditions of timing and exposure achieved in animal studies. In some cases, the spectrum includes a continuum of responses, with depressed birth weight or functional impairment occurring at low doses, birth defects at intermediates doses, and lethality at high doses. Less commonly, birth defects alone or lethality alone is produced. Therefore, in estimating human risk, all exposure-specific adverse outcomes must be taken into consideration, not just birth defects.

A similar response pattern is observed in humans exposed to developmental toxicants. The spectrum of responses is determined by the time and duration of exposure, magnitude of exposure, and interindividual differences in sensitivity. Hence, manifestations of developmental toxicity cannot be expected to be identical across species; that is, an animal model cannot be expected to forecast exactly the human response to a given exposure. For instance, an agent that induces cleft palate in the mouse may elevate the frequency of spontaneous abortion or cause growth retardation in humans. However, any manifestation of exposure-related developmental toxicity in animal studies can be regarded as indicative of a spectrum of response in humans.

Epidemiological data suggest that the majority of human embryos with chromosomal or morphologic abnormalities are lost through early miscarriage and relatively few survive to term. Consequently, determination of malformations or growth retardation at the time of birth alone (malformations, stillbirths, low birth weight) is likely to result in a substantial underestimate of the true risk, insofar as the onset of embryolethality would be missed.

The sensitivity (ability to detect a true positive response in humans) and specificity (ability to detect a true negative response in humans) of laboratory animal studies have been evaluated [165]. Of 38 compounds having demonstrated or suspected teratogenic activity in humans, all except one (tobramycin, which causes otological deficits in humans) tested positive in at least one animal species. Approximately 80 % of the compounds were positive in multiple species. The positive responses elicited were 85 % in the mouse, 80 % in the rat, 60 % in the rabbit, 45 % in the hamster, and 30 % in the monkey. These findings indicate that the usual laboratory animal species are highly sensitive in detecting human teratogens.

In contrast to the high sensitivity, laboratory animals show low specificity for predicting human teratogenesis; of 165 test compounds with no evidence of teratogenic activity in humans, 65 (41 %) were positive in more than one animal species. The high percentage of false-positive results (i.e., compounds with no evidence of human teratogenicity that induce malformations in animals) may in part be accounted for by the high doses usually applied in animal tests. Also, human studies determine effects from the time of birth onward, which may result in underestimation of the true risk.

4.7.7. In Vitro Tests for Developmental Toxicity

Models to elucidate the mechanism of embryogenesis have been under development for several decades, and because of this, developmental toxicology is one of the fields in which alternatives to animal experimentation are available. However, because of the complicated, multistep nature of the development of a new life, none of the in vitro systems presently available can replace animal tests. In vitro tests serve, rather, for screening purposes (i.e., to preclude the extensive, traditional whole-animal test protocol for compounds with marked toxicity on reproduction and development). The existing alternative test systems fall into six groups: lower organisms, cell culture systems, organ culture systems, whole embryo cultures, embryos, and others. Since none of the in vitro methods is sufficiently validated with a set of compounds for which the human or animal effects are known and the field is much too extensive to be reviewed comprehensively here, an overview of the existing systems is given in Table 4.6. For two comprehensive reviews, see [166], [167].

Finally, in addition to in vivo and in vitro test systems, mathematical models that correlate structural features with potential teratogenic activity have also been utilized to predict developmental toxicity. Structure–activity relationships for teratogens have been reviewed [168].

4.8. Bioassays to Determine the Carcinogenicity of Chemicals in Rodents

Despite the many available short-term in vivo and in vitro tests to determine the genotoxic and carcinogenic potential of chemicals and the vast amount of literature on the subject, the lifelong carcinogenicity bioassay remains the main instrument for reliable evaluation of the carcinogenic properties of a xenobiotic. The principal guidelines for performing bioassays were established some 25 a ago by the U.S. National Cancer Institute and have essentially been adopted with slight alterations by all regulatory authorities (Table 4.7).

Many factors may interfere with the analysis and interpretation of data from animal carcinogenicity studies. A variety of statistical techniques have been developed to adjust for confounding factors and to estimate confidence intervals and significance of results. Significance tests are used to assess neoplastic response in treated groups compared to control groups or historical controls (cancer incidence in the identical strain and species observed in control groups for other cancer bioassays under identical housing conditions in the same facility). To estimate the absolute cancer risk posed by a specific chemical, background or spontaneous cancer incidences (induction of neoplasms not related to administration of the test chemical) must be well defined. In general, high background incidences of cancer such as liver cancer observed in specific

Table 4.6. In vitro test systems for developmental toxicity

Group	Test system/organisms	End points monitored
Lower organisms and small animals	sea urchins drosophila trout, medaka (fish species) plania brine shrimp animal virus	growth
Cell culture	pregnant mouse and chick lens epithelial cell	protein synthesis
	avian neural crest	cell differentiation
	neuroblastoma	cell differentiation
Organ culture	frog limb	regeneration
	mouse embryo limb bud	morphological and biochemical differentiation, toxicity
	metanephric kidney organ culture from day 11 mouse embryos	morphological and biochemical differentiation
Whole embryo cultures	chick embryo	embryotoxicity, malformations
	frog embryo teratogenesis assay	lethality, no observed-effect-level, development stage, attained growth, motility, pigmentation, gross anatomical malformations
	rat embryo culture (postimplantation embryo)	viability, growth and macromolecular content, gross structural and histological abnormalities

Table 4.7. Basic procedures of rodent carcinogenicity bioassays

Species	rat (Fischer 344, Sprague–Dawley, Wistar) mouse (B6C3F1, CD)
Age of animals at the beginning	4 to 6 weeks (shortly after weaning)
Number of animals	50 per sex per dose for carcinogenicity 10–20 for additional studies during the course of experiment
Dose	at least three doses and vehicle control maximum tolerated dose intermediate dose nontoxic dose
Duration	24 months
Application	gavage, in feed, drinking water, inhalation (only if absolutely necessary)
Toxicologic pathology	all animals: gross necropsy weight of all important organs histopathology of all tissues (ca. 40) and all tumors and preneoplastic lesions by two independent pathologists

strains of mice require larger numbers of animals in the treatment groups to detect increases in cancer incidence induced by administration of the test chemical and to obtain statistically significant results. The demonstration of a dose–response curve for cancer incidence in groups of animals treated with different doses of the carcinogen will increase the confidence in positive results of an animal cancer bioassay. The same holds true for identical results observed with cancer as an end point in an independent study.

However, despite the importance of animal cancer bioassays for characterizing chemical carcinogens, this approach has been criticized recently (see also Section 5.2.4.4). Because of the influence of the rodent carcinogenicity assay on the development of new chemicals and pharmaceutical, the pros and cons of this type of assay are discussed in depth.

Practically all chemicals identified as human carcinogens produce tumors in the rodent bioassay. Hence, the test has a very good predictive value and every chemical exerting carcinogenicity in rodents should be handled as a potential human carcinogen. When considering the thousands of new chemicals developed each year, the main disadvantages of the lifelong assay are its high costs and long duration. A two-year gavage study in only one species costs more than $\$ 1 \times 10^6$ and requires a total of three to four years or longer for completion. For optimized evaluation of carcinogenic properties, animal species should be used that are closest to humans in terms of biotransformation and the toxicokinetics of test compounds. However, for practical and financial reasons, only rodents can be used. Long-term carcinogenic bioassays in dogs or primates, for example, require seven to ten years for completion and are much more costly.

Application of the maximum tolerated dose in rodent bioassays has been the subject of much controversy [169]–[175]. The experimental design limitations of in vivo studies require, however, the application of high doses. If, for example, at a specific dose a chemical causes a 0.5% increase in human cancer incidence, this would result in several hundred thousand additional cancer cases in a country such as the Federal Republic of Germany each year and would definitely pose an unacceptable risk. However, identification of this 0.5% increase in cancer incidence with statistical confidence in the rodent bioassay would require a minimum of 1000 animals, provided the spontaneous tumor incidence is zero. Therefore, general agreement appears to exist that the use of the MTD, although not an optimum solution, is necessary for risk characterization. According to the U.S. National Cancer Institute, the MTD is defined as "the highest dose that can be predicted not to alter the animals' normal longevity from effects other than carcinogenicity." In practical terms, the MTD is the dose that, in the subchronic three-month toxicity study does not cause more than a 10% weight decrement in treated compared to control groups and does not produce mortality, clinical signs of toxicity, or pathological lesions other than those that may be related to a neoplastic response that would be predicted to shorten an animal's life span. Ideally, this is exactly what should also happen in the 24-month carcinogenicity study i.e., a slight impairment of weight gain in the group receiving the MTD. However, due to cumulation of toxic effects and alterations in toxicokinetics of the xenobiotics during the study (e.g., induction of toxification or detoxification pathways by application of the xenobiotic in high doses), the MTD dose group often shows reduced survival rates in the lifelong bioassay. This may invalidate the study (make it inadequate for evaluation of carcinogenic potential). Indeed, it is not a rare event in carcinogenicity studies. The opposite effect may also occur; because of toxicokinetic differences between the three-month and 24-month studies, the MTD chosen may turn out to be too low in the long-term bioassay. Despite all these problems, because of the absence of a satisfactory alternative solution, use of the MTD is currently the only method to compensate for the fact that in relation to the human population exposed to potential carcinogens, the numbers of rodents used in the carcinogenicity bioassay are extremely low. The legitimate argument against the MTD is that any chemical given at a sufficiently high dose level will induce adverse effects. This understanding, which is beyond dispute in toxicology, has been tentatively generalized by several scientists in recent years as the idea that carcinogenic effects obtained in the MTD may result exclusively from target organ toxicity. The increased cell proliferation may contribute to tumor formation by increasing the rate of spontaneous mutations because DNA replication does not occur with 100% fidelity. Furthermore, during increased cell turnover, the time available to repair DNA damage is reduced so an increased number of damaged DNA sites may be converted into heritable mutations. Although this may be the mechanism underlying the carcinogenic effects of some nongenotoxic chemicals, it cannot be generalized to every tumor observed in the MTD. Toxicity and cell proliferation do not necessarily result in tumor formation. Table 4.8 summarizes the important differences between rodent bioassays and human exposure to carcinogens.

In addition to these problems, which are inherent in the bioassay procedure, the evaluation of toxicologic pathology has repeatedly become

Table 4.8. Some important differences between currently conducted carcinogenicity tests in rodents and human exposure to potential carcinogens

Rodent carcinogenicity test	Human exposure
High doses	(usually) low doses
Continuous exposure	(often) infrequent or not regular exposure
Single compound, no interactions	simultaneous exposure to several carcinogenic chemicals, interactions probable
Homogeneous population	heterogeneous population

an issue of debate because differences in evaluations among pathologists are frequent. This does not necessarily indicate the incompetence of one of the pathologists. The different evaluations may be the result of difference in terminology. Also, evaluations are sometimes conducted years apart and during this time understanding of the pathogenesis of lesions may have changed. This is shown by the fact that even the same pathologist may not come to the same conclusion when reevaluating tissue slices several years after the first examination.

Because of the uncertainities of rodent bioassays and the extremely high financial and personnel expenses, a multitude of short-term tests has been developed in recent years. These tests are aiming to predict the carcinogenic potential of chemicals. Most of these in vitro tests are based on damage to genetic material (genotoxicity) by the chemical or its metabolites. Genotoxicity is without doubt the field in toxicology with the best-established and validated in vivo and in vitro short-term tests.

4.9. In Vitro and In Vivo Short-Term Tests for Genotoxicity

Genetic toxicology is a comparatively new field of research that has grown rapidly since the 1960s. It was established as a formal discipline some 25 years ago when the Environmental Mutagen Society in the United States and the European Environmental Mutagen Society were founded. During the following years, evidence increasingly accumulated that the majority of carcinogens are also mutagenic and a large number of short-term in vitro and in vivo tests were developed. Most of these tests are more or less well validated and aim to assess genotoxic properties of chemicals. Today, the majority of potential carcinogens are first identified as mutagens or chromosome-damaging agents in short-term tests and subsequently as carcinogens in the rodent carcinogenicity bioassay. The numerous in vivo and in vitro assays can be categorized into two major groups:

1) Short-term tests detecting gene mutations
2) Short-term tests detecting structural and/or numerical chromosomal aberrations

In Vitro Test Systems for Mutagenicity. The in vitro tests aiming to detect gene mutations can be categorized into two groups, the microbial and mammalian cell assays. An important step in the history of modern genetic toxicology was the development of genetically precisely defined strains of bacteria carrying mutations in particular genes coding for enzymes involved in the biosynthesis of amino acids.

4.9.1. Microbial Tests for Mutagenicity

4.9.1.1. Ames Test for Bacterial Mutagenicity

The most common method of detecting mutations in microorganisms is selecting for reversions in strains that have a specific nutritional (i.e., amino acid) requirement differing from wild-type members of the species; the tester strains are "auxotroph" for this particular nutrient. The *Salmonella typhimurium* mutant strains developed by BRUCE AMES are not able to synthesize histidine because each strain carriers one of a number of mutations in the operon (group of genes) coding for histidine biosynthesis [176], [177]. The result of this mutation is that the tester strains cannot grow and form colonies in histidine-free medium. The mutation may revert to the wild-type sequence or a functionally equal sequence either spontaneously (a rare event) or by exposure of the tester strains to genotoxic compounds. Like wild-type bacteria, the revertant colonies are capable of synthesizing histidine and can form colonies in histidine-free medium. For the common tester strains, the DNA sequence at the site of the original mutation in the relevant histidine gene has been determined. According to the type of mutation leading to the inability to synthesize histidine, the strains can be categorized into two groups, the base-pair substitution and the frameshift strains. The difference between these two categories can be illustrated with the following sentence, in which every letter represents a DNA base, every word a triplet coding for an amino acid, and the whole sentence a gene coding for a protein (i.e., enzyme of histidine biosynthesis). The correct sentence represents the gene in the wild-type strain:

THE NUN SAW OUR CAT EAT THE RAT	(original sentence "wild-type")
Base-pair substitution:	
THE **S**UN SAW OUR CAT EAT THE RAT	missense mutation coding for a wrong amino acid

THE NSN SAW OUR CAT EAT THE RAT	nonsense mutation resulting in interruption of gene transcription
Frameshift mutation:	
THE NUN SAS WOU RCA TEA TTH ERA T	+ 1-frameshift mutation through base insertion

This example illustrates that a base substitution can revert because of another base substitution, and a frameshift mutation can revert another frameshift mutation. Hence, the Ames test provides information not only on the genotoxic potential of the test compound but also on the nature of the DNA damage.

Genetic Makeup of the *Salmonella typhimurium* Strains Used in the Ames Test. *Histidine Mutations of the Tester Strains.* The base-pair substitution strains can be categorized into two families:

1) The *Salmonella typhimurium* strains TA100 and TA1535 carry the sequence CCC (for leucine) instead of the wild-type sequence CTC (for proline). These missense mutations may be efficiently reverted by mutagens with alkylating properties.
2) The second group of base-pair substitution strains carry a nonsense mutation (TAA instead of CAA). These strains (TA2638, TA102) may detect mutations induced by radicals or oxidizing agents such as hydrogen peroxide and reactive oxygen metabolites.

The commonly used frameshift strains TA98 and TA1538 carry a +1-frameshift mutation near a GCGCGCGC sequence. This strain may be used to detect frameshift mutagens such as polycyclic aromatic hydroxycarbons, certain aromatic amines, and certain aromatic nitro compounds.

In addition to the mutation in one of the genes of histidine biosynthesis, the Ames strains carry other genetic alterations that increase their sensitivity to detect mutagens (an overview of the genetic makeup of *Salmonella typhimurium* tester strains is provided in Table 4.9).

rfa Mutations. The *rfa* mutation results in a defective lipopolysaccharide membrane, thus increasing the permeability of the cell wall to bulky hydrophobic chemicals. In addition, strains with defective cell walls are not pathogenic to experimental animals and humans.

uvrB Deletion. Wild-type bacteria possess several effective DNA repair systems, that operate practically error free and can repair DNA damage without allowing mutations to occur. To overcome the problem of DNA repair and thus the decreased sensitivity of the test system, Ames constructed a series of strains with a deletion of the *uvrB* gene, which codes for a subunit of an important enzyme (the cor-endonuclease-I) of the error-free excision repair system. This change increases the sensitivity of the tester strains to mutagens by several orders of magnitude.

Table 4.9. Genetic makeup of commonly used *Salmonella typhimurium* tester strains

Strain	Histidine mutation		Additional genetic alteration
	Location	Route	
TA100	his G46	CCC instead of CTC base substitution results in proline instead of leucine	*uvrB* *rfa* *pkM*101
TA1535	his G46	CCC instead of CTC base substitution results in proline instead of leucine	*uvrB* *rfa*
TA2638	his G428 his G8476	ATT nonsense mutation results in interruption of transcription	*rfa* *pkM*101
TA102	his G428	multiple copies of plasmid pAQ1 carry the revertible nonsense mutation ATT	*rfa* *pkM*101
TA98	his D3052	-1-frameshift mutation with a GCGCGCGC sequence	*uvrB* *rfa* *pkM*101
TA1538	his D3052	-1-frameshift mutation with a GCGCGCGC sequence	*uvrB* *rfa*

The Plasmid pkm 101. Wild-type *Salmonella typhimurium* strains do not process an "error-prone" DNA repair, which is found for example in *Escherichia coli* and some other members of the Enterobacteriaceae. In contrast to the aforementioned error-free excision repair pathway, the error-prone system operates with low fidelity, introducing new mutations into the genome while repairing a damaged DNA site. To overcome this deficiency and increase sensitivity to mutagens, the gene for error-prone repair system has been introduced to some of the *Salmonella typhimurium* tester strains with the plasmid pkm 101. This plasmid also carries the genetic information for ampicillin resistance, an important property to monitor the physiologic integrity of tester strains in the course of the experiments.

The Plasmid pAQ 1. The *Salmonella typhimurium* strain TA 102 carries the revertible histidine mutation on multiple copies (approximately 30) of the plasmid pAQ 1 and not on the chromosome. Reversion of one of these copies returns the bacteria's capability to synthesize histidine. This increase in sensitivity is partly offset because strain TA 102 has an intact excision repair system (it does not have the *uvrB* deletion). This strain has been constructed because DNA interstrand cross-linking agents such as the quinone mitomycin C or the combination of psoralens und UV light require an intact excision repair system to generate mutations. DNA interstrand cross-links must first be removed from one strand along with a small number of adjacent bases by excision repair. The gap left behind is repaired, and the remaining broken cross-link attached to the other strand is a premutagenic lesion that may give rise to a mutation by error-prone repair.

Problem of Bioactivation in the Ames Test. The majority of mutagenic xenobiotics require bioactivation (toxification) to reactive electrophiles to induce DNA damage. The main disadvantage of bacterial mutagenicity assays is that the tester strains do not possess many of the enzymes that bioactivate xenobiotics in mammals. For example, cytochrome P450 activity is not detectable with most substrates in *Salmonella typhimurium*. Therefore, the biotransformation occurring in the intact animal must be simulated by supplementing the test system with the enzymes of bioactivation and the necessary cofactors. For this purpose, numerous exogenous metabolic systems have been used in the last decades. They can be grouped into cell-free and cell-based systems, and among the cell-free systems the 9000 *g* supernatant (called S-9 fraction; the S-9 fraction with the cofactors necessary for enzyme activity is called S-9 mix) from centrifugation of rat liver homogenates is the most widely used and best validated.

Properties of Standard S-9 Mix. The S-9 fraction is usually prepared from the liver of male adult rats. The standard S-9 mix is capable of performing phase-I biotransformation reactions (e.g., oxidation, reduction) but is deficient in most phase-II conjugation reactions. The latter are often involved in detoxification reactions, while phase-I enzymes usually result in toxification of xenobiotics to more electrophilic metabolites. Hence, the S-9 mix may efficiently simulate the bioactivation of xenobiotics in the liver, but not their detoxification. This may be regarded as an advantage since it usually increases the sensitivity of the system. On the other hand, the discrepancy between phase-I and phase-II enzymatic activities may be a source of false-positive results because the genotoxicity observed in vitro may not reflect the in vivo situation, and detoxification may predominate in the intact animal. The deficiency in conjugation reactions may be partly overcome by adding appropriate cofactors for phase-II conjugation reactions (i.e., glutathione). The S-9 fraction from other organs may also be prepared and used for the bioactivation of organ-specific carcinogens (i.e., S-9 fraction of the renal cortex for compounds inducing renal cell tumors). Alternatively, cell-based systems may be used to bioactivate xenobiotics in bacterial mutagenicity assays. Freshly isolated hepatocytes or hepatocytes in primary culture retain the activity of the phase-I and phase-II enzymes of the intact liver and do not require the addition of cofactors for enzyme activity. However, besides greater technical difficulties in obtaining hepatocytes of good quality compared with S-9 fraction, additional problems arise when using intact cells because the electrophiles formed may have a very short half-life and may be trapped and react with macromolecules before diffusing out of the hepatocytes, resulting in false-negative results.

A recent elegant approach to overcome the deficiencies of the bacteria in biotransformation reactions is to clone genes of mammalian biotransformation enzymes into plasmids and introduce these into tester strains.

Although this approach is very new (the first report appeared in the 1990s) and has not been widely used thus far, it seems to be a promising method for the future. Specific potentials and limitations of genotoxicity test are listed below.

Potentials

High reliability when testing known carcinogens and non-carcinogens
Extensive data base available
High sensitivity
Information about the mechanism of mutations may be obtained by comparing results in different strains
Rapid and inexpensive
May be used as a bioassay to detect mutagenic components in complex mixtures

Limitations

Some important carcinogens are not active due to specific mechanisms of action (metals, particles, asbestos)
Data are only qualitative
Sample must be sterile
Problems in testing bactericidal chemicals

Experimental Procedure for the Ames Test. Presently three different protocols are widely used in the Ames test: the plate incorporation assay, the preincubation assay, and the fluctuation test.

Plate Incorporation Assay. The basic procedure of the plate incorporation assay is illustrated in Figure 4.4. Briefly, 2.0-mL aliquots of soft agar overlay medium (top agar, 0.6% agar and 0.5% sodium chloride in distilled water) containing a trace of histidine and excess biotin and maintained in the liquid state at 45 °C, 100 μL of the tester strain, 20–100 μL of the test compound, and when needed, 500 mL S-9 mix (or another bioactivation system) are added. After mixing, the solution is poured on dried Vogel–Bonner minimal medium plates. The plates including untreated, solvent, and positive (UV light or established mutagen) controls are incubated for 2 d at 37 °C. At least ten concentrations of the test compound are usually tested with two plates per concentration.

Preincubation Assay. Some mutagens, particularly those metabolized to short-lived reactive electrophiles, may not be detected in the standard plate incorporation assay due to reaction of the electrophile with constituents of the medium. This type of xenobiotic may be detected more efficiently by using the preincubation assay protocol, in which the bacteria are preincubated with the test compound and S-9 mix in suspension for 30–120 min in the absence of top agar. After the end of this preincubation period, 2.0 mL of soft agar is added to each tube and the reaction mixture is poured into Vogel–Bonner plates, which are incubated as described above.

Fluctuation Test. In the fluctuation test, the number of mutants in a series of small independent replicate cultures is detected. Overnight bacterial cultures are incubated with the test compound and the appropriate metabolizing system (when necessary) in the presence of a trace of histidine. As in the other protocols, this trace of histidine allows a few replication cycles, which are necessary for expression of the mutations following the initial DNA damage. The above mixture is then divided into a large number (usually 50–100) of test tubes or microtiter plates. When the histidine trace is consumed, only revertant cells can grow. Test tubes or wells of the microtiter plates containing revertants become turbid, and the media turn acidic as a result of acid release during growth. This pH drop to 5.2–6.8 can be demonstrated by the color change of a pH indicator such as bromoethyl blue.

Scoring for Colonies in the Plate Incorporation and Preincubation Assay. The limited growth of nonrevertant colonies due to the trace of histidine added to enable mutational expression of DNA damage results in a slight background lawn of growth. Therefore, before scoring the plates for revertant colonies, the background lawn should be examined either macroscopically or under the low magnification of a light microscope. At toxic concentrations of the test chemical, the plates appear clear. Also, at toxic concentrations of the test chemical, tiny colonies (minicolonies) may be formed. This happens when, at bactericidal concentrations of the test chemical, very few bacteria survive. The histidine trace usually added to the system may then be sufficient for formation of these minicolonies (pseudorevertants). Revertant colonies resulting from gene mutations are clearly larger and can be counted either by hand or with an automatic colony counter.

Evaluation of Results. A typical mutagenic dose–response curve is illustrated in Figure 4.5. From the linear part of the dose–response curve, the mutagenic potency of the test compound can be calculated and is usually expressed in number of revertants per nanomole of test compound. At higher, toxic concentrations of the test chemical, the curve may flatten. At concentrations that are very toxic to the bacteria, the number of revertant colonies may start to decline after reaching

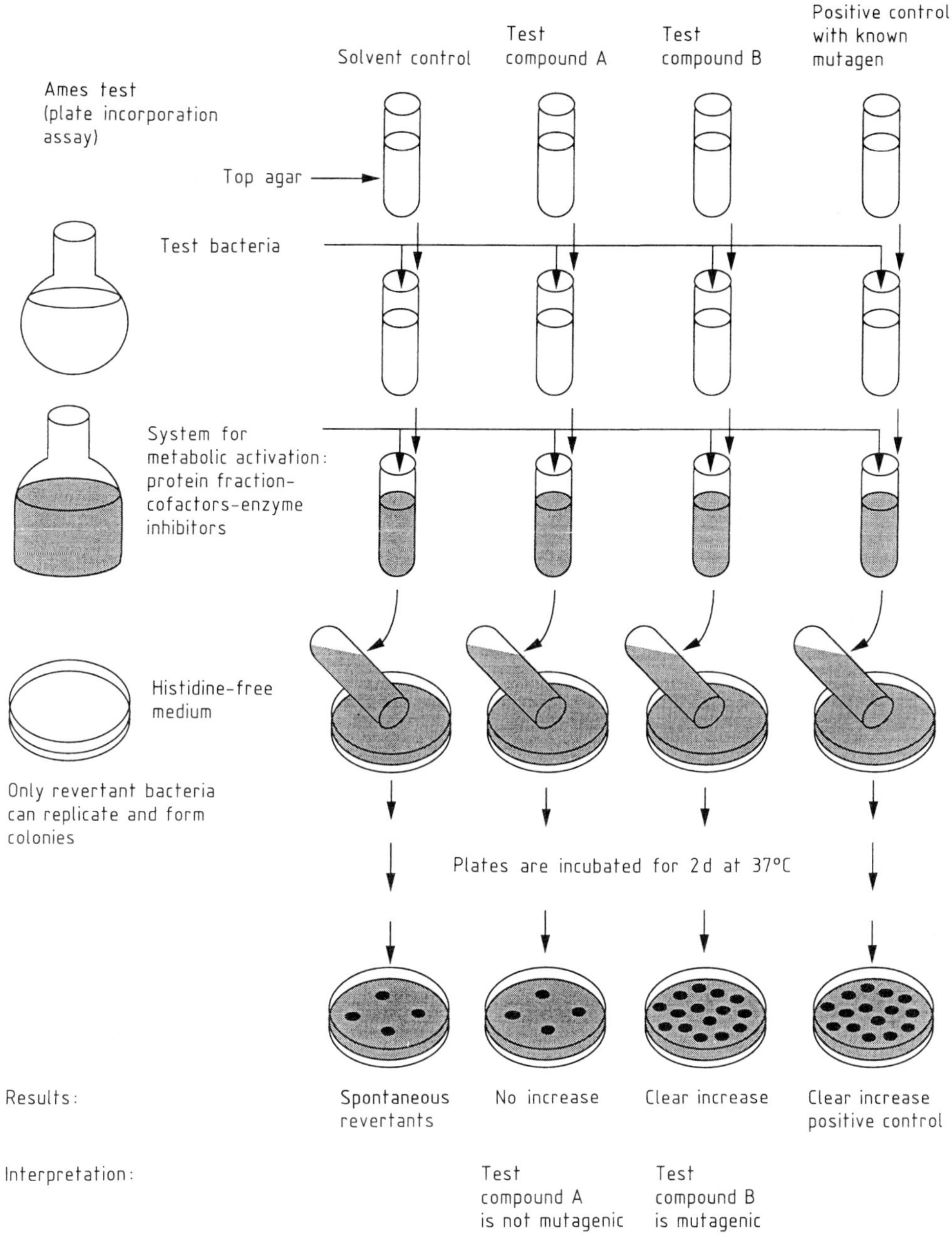

Figure 4.4. Basic procedure for the Ames test/plate incorporation assay
50 to 200 plates are usually prepared and evaluated per test. At least ten concentrations of the test compound are examined with two plates per concentration level.

a maximum. The significance of the results is analyzed by the usual statistical methods.

Role of the Ames Test in Evaluating Carcinogenic Properties. The Ames test is a system to detect mutagenicity that has gained great practical importance as a short-term predictive test for the potential carcinogenicity of a xenobiotic. Hence, the Ames test may predict the carcinogenic properties of genotoxic but not of

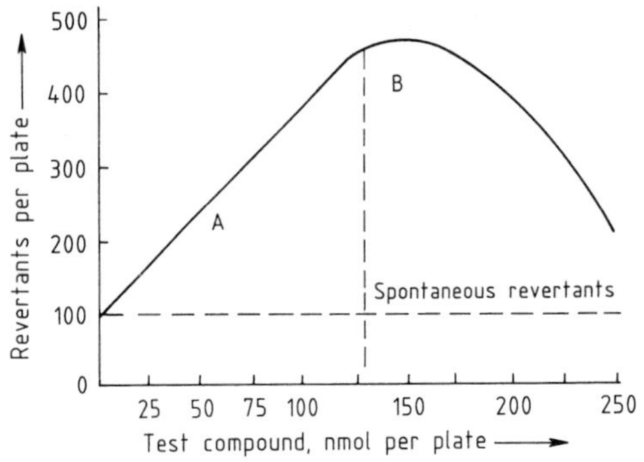

Figure 4.5. Typical dose–response curve in the Ames test
A) At nontoxic mutagen concentrations the number of revertants increases in a dose-dependent manner; B) Higher, toxic concentrations inhibit growth, which is reflected in flattening of the curve followed by decreasing numbers of obtained revertants

nongenotoxic carcinogens. The very high correlations between mutagenicity and carcinogenicity found in previous reports—values of 90–95% were repeatedly estimated—are certainly not correct. One reason for this overestimation may be that in previous decades, mainly structurally alerting chemicals (xenobiotics carrying functional groups present in other carcinogens and known or suspected to be metabolized to electrophiles) were tested in the long-term rodent carcinogenicity test and most of these structurally alerting carcinogens are mutagenic. In contrast, in recent years, more important environmental pollutants and hazardous occupational chemicals are increasingly being tested for carcinogenicity, independent of their chemical structure. As a result, an increasing number of carcinogenic compounds has been found, that do not react directly with DNA to cause mutations. These xenobiotics are thought to act via additional and (unfortunately) only partly defined so-called epigenetic (not involving genetic changes) mechanisms. Nongenotoxic carcinogens cannot be detected by the Ames test. In contrast to the genotoxic carcinogens that are usually active in two species or at multiple sites, nongenotoxic carcinogens are often active in one species and at a single site (i.e., branched hydrocarbons in the male rat kidney or peroxisome proliferators in the mouse liver). This is confirmed by comparing the carcinogenicity results with bacterial mutagenicity. While more than 70% of the "two species/multiple sites" carcinogens are positive in the Ames test, only 40% of the "one species/single site" carcinogens exert bacterial mutagenicity. Therefore, a reasonable evaluation of the results obtained in bacterial mutagenicity systems can be conducted only by taking into consideration addition information on the chemistry, toxicokinetics, biotransformation, and biological effects of the test compounds.

Presently, the Ames test is an important part of a screening battery for possible carcinogenic properties in the course of development of new pharmaceuticals and chemicals. Depending on the intended use of the chemical, development may be interrupted if a compound is clearly positive in the Ames test and in an additional short-term test, often the in vivo micronucleus test (see Section 4.9.5.3). With this "decision point approach," the number of unnecessary lifelong bioassays in rodents is significantly reduced and large amounts of money and time may be saved.

4.9.1.2. Mutagenicity Tests in *Escherichia coli*

The *E. coli* tester strain WP2 and additional tester strains subsequently developed originate from the wild-type *E. coli* B strain; all tester strains are tryptophan (trp) auxotrophic. For example, the *E. coli* WP2 trp E group possesses a terminating mutation at an A–T base pair of the trp E gene, which codes for anthranilate synthetase, an enzyme of the tryptophan biosynthetic pathway. Hence, the *E. coli* mutagenicity assays, like the Ames test, detect reversion of the deficiency to synthesize tryptophan. The *E. coli* tester strains possess the *umuDC*$^+$ genes, coding for the error-prone repair system, and their membrane structure permits the passage of many large molecules. The experimental procedures and the inherent advantages and limitations are practically identical to the *Salmonella typhimurium* test. However, the *E. coli* test systems are not as well validated as the Ames test, which renders the selection of appropriate strains for investigation of the mutagenicity of a specific xenobiotic and the interpretation of the obtained results more difficult.

4.9.1.3. Fungal Mutagenicity Tests

The most widely used species in fungal mutagenicity tests is *Saccharomyces cercevisiae*; the test detects reversion of the isoleucine auxotrophy induced by exposure to mutagens in analogy to the Ames test.

4.9.2. Eukaryotic Tests for Mutagenicity

Although prokaryotic systems are fast, inexpensive, and versatile, eukaryotic test systems are necessary because of the important differences between prokaryotic and eukaryotic organisms in the structure and function of the genome and in the biotransformation and transport of xenobiotics. The eukaryotic systems for detecting mutations include tests in the fruit fly *Drosophila melanogaster*, in many different types of cultured mammalian cells, and in the intact animal in vivo.

4.9.2.1. Mutation Tests in *Drosophila Melanogaster*

Drosophila melanogaster has a short generation time of 10 d and a similar cellular and chromosomal structure and function to mammalian cells. In addition, the genome of the fly is well characterized and *Drosophila* species can perform many of the phase-I and phase-II reactions occurring in mammals. The *Drosophila* sex-linked recessive lethal test was a popular screening system in the 1970s and was used to screen many hundreds of compounds. The popularity of the test has, however, decreased drastically in recent years because the test did not perform well in international collaborative trials that investigated the utility of mutagenicity assays as predictive tools. Another *Drosophila* test, devised in the 1980s, the somatic mutation and recombination test (SMART) seems promising, but it has not been sufficiently validated. Since both tests are not widely used for screening at present, the experimental procedures are not described here (for more information, see [160]). Some potentials and limitations of the mutation tests in *Drosophila melanogaster* are listed below:

Potentials

1) Large number of test organisms can be raised
2) Metabolic activation endogenous
3) Genotoxic effects evaluated in germ cells
4) Large data base on mutagens

Limitations

1) Some known carcinogens such as polycyclic hydrocarbons respond poorly
2) Problems with toxic compounds (insecticides)

4.9.2.2. In Vitro Mutagenicity Tests in Mammalian Cells

Many short-time mutagenicity tests in mammalian cell lines has been developed. These systems have the advantage of determining genotoxic effects in relevant cell types and may permit the study of mechanisms of carcinogenicity. The potentials and limitations of these in vitro tests follow.

Potentials

1) Full range of mutational responses may be observed
2) Low costs when compared to in vivo assays
3) Respond to particulates and other types of compound not detected in the Ames test
4) More relevant end point for mammals

Limitations

1) Most cell lines used are transformed cells and do not represent "normal" cells
2) Only specific loci from the entire genome are monitored
3) Most cell lines used are deficient in biotransformation enzymes

In contrast to popular bacterial assays that detect reversion mutations, the commonly used mammalian mutagenicity assays are based on the detection of forward mutations. A large number of cells is treated with the test compound. After a certain period of time, the cells are exposed to a toxic agent that is lethal to all cells not carrying mutations (i.e., only mutated cells can survive). Cultured mammalian cells are normally diploid and have two copies of each gene. Hence, recessive mutations may be missed if a normal copy is present on the homologous chromosome because the probability that a mutagenic compound alters genes is very low. Therefore, mutations are assessed in genes on the X chromosome in male cells, where only one copy of the gene is present. Alternatively, heterozygous cells are used, where one copy of the gene is already inactivated.

Mammalian cells possess genes that allow the cell to salvage nucleotides from the surrounding medium. These genes, although not essential for cell survival, conserve cellular energy because the cell need not synthesize these molecules from simple precursors by energy-consuming pathways. If the medium is supplied with altered nucleotides, the normal (not mutated) cell will incorporate them into the DNA and this will result in cell death. However, if this salvage property is lost due to mutation, the mutated cells are not able to incorporate the toxic nucleotides from the surrounding medium and will survive and form colonies that can be detected as a parameter of genotoxicity. Although a gene may be inactivated by a mutation, the mRNA and the corresponding enzyme produced prior to the mutational event may be present for some time after exposure to the mutagenic test compound. Therefore, the cells have to be left for some time before challenging with the toxic nucleotide; this period is called the expression time.

Basically, two genes of the salvage pathway are utilized in mutagenicity assays: the hypoxanthine–guanine phosphoribosyltransferase (HGPRT) gene and the thymidine kinase (TK) gene. The HGPRT gene is located on the X chromosome in humans and in the Chinese hamster, from which many useful cell lines have been developed. The genetic changes detected in the HGPRT gene as the target site are mainly point mutations because chromosomal deletions usually extend to flanking regions of the target gene, which may contain essential genes. Since only one copy of these essential genes exists in male cells, this results in cell death so that the cells in which the HGPRT gene is inactivated by deletion are usually lost for the test. This is not the case with the TK gene, because it is autosomal and heterozygous cells (TK^+/TK^-) are used in the assay. Here, both point mutations and changes at the chromosomal level can be detected. The HGPRT gene is the target gene in the Chinese hamster CHO and V79 cell lines, while mouse lymphoma L5178Y $TK^{+/-}$ cells detect mutations at the TK locus. The mechanistic background of these tests is shown in Figure 4.6.

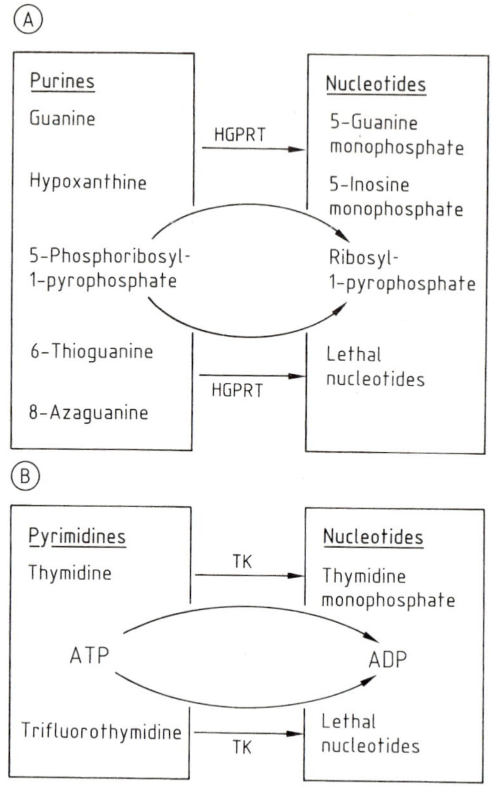

Figure 4.6. Assessment of mutagenicity in mammalian cells
A) The V79/HGPRT (hypoxanthine–guanine phosphoribosyltransferase) assay; B) The L5178Y $TK^{+/-}$ (thymidine kinase) mouse lymphoma assay
Mutant cells do not form the lethal nucleotides and survive in the presence of 6-thioguanine or trifluorothymidine.

The use of Chinese hamster cell lines for mutagenicity screening is limited due to their low sensitivity. Chinese hamster cell lines grow on monolayers, and owing to metabolic cooperation, only a relatively small cell population can be used. In contrast, L5178Y $TK^{-/-}$ cells grow in suspension, and the system is not impaired by problems of metabolic cooperation because intercellular bridges do not occur. As described above, a wide variety of genetic events, including gene mutations, recombinations, and mitotic nondisjunction, can result in the formation of the $TK^{-/-}$ genotype from the heterozygote $TK^{+/-}$. Two protocols have been devised for mutation assays with mouse lymphoma L5178Y cells. The plate test in soft agar and the fluctuation test in suspension. For dose finding, a preliminary cytotoxicity assay is usually conducted when testing new compounds. In this assay, the cloning efficiency (for the plate test) or the relative suspension growth (for the fluctuation test) is determined. The highest test concentration for the mutation assay is usually the concentration that reduces cloning efficiency or suspension growth to approximately 10–20% of the control values. In addition, a moderate concentration causing survival reduction to 20–70% and a low concentration (> 70% survival) are also used. After treatment with the test compound (3 to 7 h; in the presence or absence of an exogenous metabolic system, i.e., S-9 mix), the cells are incubated for 2 d for mutation expression before exposure to trifluorothymidine for 10–12 d. In the presence of trifluorothymidine, only mutant cells can survive. Thus, the relative growth of the cells at the end of the experiment can be used as a parameter for mutagenicity.

4.9.3. In Vivo Mammalian Mutation Tests

In contrast to the well-established and widely used micronucleus tests in rodents (see Section 4.9.5.3), the existing in vivo tests to detect gene mutations are not used for routine genetic toxicology testing; therefore, only brief descriptions are given here (for exact experimental procedure, see [160]).

4.9.3.1. Mouse Somatic Spot Test

The mouse somatic spot test detects the inactivation of genes at a set of heterozygous loci controlling hair pigmentation. Inactivation can be induced by a wide spectrum of genetic alterations ranging from gene mutations to major chromosomal changes.

4.9.3.2. Mouse Specific Locus Test

The mouse specific locus test consists of treating parental mice homozygote for a marker gene controlling coat pigmentation, intensity, and pattern or size of the external ear. Treated mice are mated with a tester stock that is homozygous recessive at the marker loci. The resulting F_1 generation is normally heterozygous at the marker genes and therefore expresses the wild-type phenotype. In the case of mutations at any of these genes, the F_1 offspring express the recessive phenotype. The mouse specific locus test has a comparatively low sensitivity, thus requiring that many thousands of newborns be scored. However, it remains an important method for investigating heritable mutations.

4.9.3.3. Dominant Lethal Test

The dominant lethal test assesses embryonic death resulting from genetic changes in parental germ cells. The procedure is based on observing the viability of uterine implantations. Implantations that die at an early stage form a deciduoma or mole. The genetic event resulting in early death is predominantly chromosomal damage in the parental germ cells resulting in a dominant lethal mutation. Although the assay was popular in the 1960s and 1970s as a screen for germ cell damage, it is not widely used at present because it is considered relatively insensitive.

4.9.3.4. Future Development of In Vivo Tests to Detect Point Mutations

In contrast to the well-established in vitro systems detecting point mutations (i.e., in bacteria), no validated in vivo test exists for that purpose. With the new DNA technologies, several promising approaches may result in the development of a satisfactory in vivo model for the detection of point mutations in the future. Three of the most promising current developments are described briefly below.

Use of Tissue-Specific Genetic Polymorphisms. Mutations in intestinal crypt cells can be studied. Changes in the genes coding for glycoproteins/glycolipids in the epithelial membranes of the small intestine are assessed by loss of the ability to bind the lectin *Dolichos biflorus* agglutinin. Due to the large number of scored cells (the whole intestine is scored for mutant clones), the test requires only a small number of animals. The test seems promising in evaluating the activity of mutagens present in diet.

Restriction Site Mutation Analysis. Restriction enzymes are endonucleases that recognize specific DNA sequences and induce DNA strand breaks at certain sites in these sequences. The restriction site mutation analysis detects DNA damage by making these recognition sites insensitive to the cutting action of the restriction enzyme. Experimental procedures currently being developed use the polymerase chain reaction for amplification of DNA sequences from treated and control cells. The polymerase chain reaction technique allows the production of a large number of copies of any known DNA sequence for which primers are available. The amplified DNA sequence can then be identified as a characteristic band on a polyacrylamide gel. These primer sites are lost after successful nuclease attack; therefore, control DNA will not be amplified by polymerase chain reaction. In contrast, DNA damage leading to resistance to the cutting action of the endonucleases will allow amplification by polymerase chain reaction to occur and the amplified product will be observed on the gel [178].

Transgenic Animal Models. Microinjection of foreign DNA into fertilized mouse eggs and integration of this DNA into the mouse genome results in the formation of transgenic organisms (i.e., all cells of the adult animal contain a copy of the introduced "foreign" DNA). Currently, two transgenic animal models are being developed in mice for assessing mutations. In the first, a lac Z (β-galactosidase) gene from *E. coli* carried by a λ-bacteriophage vector was microinjected into eggs of (BALB/CXDBA/2) $CD2F_1$ mice. The transgenic mouse produced contains 80 copies of the lac Z gene per haploid chromosome set. In the second transgenic strain, a repressor (lac I) gene for the galactosidase gene was integrated via a λ shuttle vector into the genome of a C57B1/6 mouse. In both cases, after single or multiple treatment with the test compound the animals are allowed at least seven days for expression of the mutation. The λ vector is then rescued from the isolated DNA of the chosen organ or tissues and packed into a λ phage, which is used to infect an *E. coli* strain. These bacteria are incubated with the chromogenic agent X-gal, which turns blue upon biotransformation by galactosidase. If the lac Z gene has been inactivated by mutation, galactosidase activity is lost and no color is produced (i.e., the clear plaques are indicative of mutation). In the second model, if the lac repressor gene has been inactivated by mutation, the galactosidase gene is transcribed and the enzyme produced metabolizes the chromogenic agent, which turns blue. Hence, in this case, the blue plaques are the mutants. Early studies with model mutagens gave promising results [179], but much more validation is required to evaluate the sensitivity and specificity of the system.

4.9.4. Tests Providing Indirect Evidence for DNA Damage

4.9.4.1. Unscheduled DNA Synthesis (UDS) Assays

Two major DNA repair mechanisms have been identified in mammalian cells. The first mechanism involves direct reversal of DNA damage, such as cleavage of pyrimidine dimers that have been induced by UV light or removal of potentially mutagenic methyl groups from the O^6 position of guanine by O^6-methylguanine-DNA methyltransferases. The second mechanism recognizes damaged DNA bases, removes them after nicking the DNA backbone, fills the gap formed with the correct bases using the opposite strand as a template, and finally seals the second strand break.

The most widely used method to study DNA repair is the "unscheduled" DNA synthesis assay. The principle of the UDS assay consists in detection of the incorporation of radioactive DNA bases or chemical analogues of nucleotides such as bromodeoxyuridine into DNA in the course of excision repair after damaged bases have been removed. This unscheduled DNA synthesis induced by DNA damage must be distinguished from the semiconservative DNA synthesis during DNA replication. DNA replication can occur in the same cells that undergo repair or, alternatively, in other cells of the test population and usually contributes to an incorporation of radiolabeled base or nucleotide analogue in concentrations several orders of magnitude higher than repair. Three approaches are currently available to overcome this problem and to detect selectively DNA repair-dependent incorporation.

1) Use of quiescent cell populations, that do not replicate
2) Suppression of DNA replication by incubation in serum-reduced and arginine-depleted medium and pretreatment with the ribonucleotide reductase inhibitor hydroxyurea
3) Evaluation of radioactivity by autoradiography of single cells, which allows the distinction between heavily labeled replicating cells and lightly labeled DNA-repairing cells

The UDS assay can be conducted in vitro and in vivo.

In Vitro Assay in Cultured Mammalian Cells. After inhibition of DNA replication by serum reduction, arginine depletion, and treatment with hydroxyurea (if replicating cell lines are used), the cells are exposed to the test compound in the presence of radiolabeled thymidine. Among the various cell types used, primary cultures of rat hepatocytes are often preferred to permanent cell lines because their biotransformation capabilities are closer to those of liver cells in vivo. In addition, primary cultures of hepatocytes do not divide, so no supression of cell proliferation is necessary.

In Vivo Assay. Radioactivity incorporation can be measured by scintillation counting of the isolated DNA or by autoradiography of whole cells. For determination of UDS in vivo, the animals are treated with the test compound and radiolabeled thymidine (most often applied by implanted minipumps). Radioactivity in DNA is determined after isolation of the DNA.

Combined In Vivo – In Vitro Approach. In this case, animals are treated in vivo followed by cell isolation from the appropriate organs and labeling of the cell with the radioactive nucleotide in vitro. Assessment of radioactivity incorporation after in vivo treatment is usually conducted by autoradiography.

UDS assays are not as popular as they were in the 1970s and 1980s. Several factors have contributed to this development. Detection of excision repair demonstrates that the test compound has caused repairable damage to DNA, but does not tell whether this has any significance for the mutagenic potential of the chemical. This is very important because measurements of UDS are averages over all cells of the study population, whereas mutation is a rare event in an individual cell. In addition, extranuclear events such as mitochondrial toxicity resulting in oxidative stress and increased production of reactive oxygen species or increased calcium concentrations in the nucleus and activation of endonucleases may also result in indirect DNA damage and induction of UDS.

4.9.4.2. Sister-Chromatid-Exchange Test

Sister-chromatid exchanges are reciprocal exchanges between sister chromatids thought to occur at homologous loci. Although detection of sister-chromatid exchanges can also be conducted after in vivo treatment, the test is widely applied in established cell lines in vitro (Chinese hamster cell lines, human fibroblast cell lines) and in freshly isolated human lymphocytes. Any cell type that is replicating or can be stimulated to divide can be used. The test is carried out essentially according to the following basic procedure. Cells in exponential growth are usually exposed simultaneously to 5-bromo-2′-deoxyuridine and the test compound for a period equivalent to approximately two cell cycles (incubation time depends on the cell type used). In the final 1 – 2 h of incubation, a spindle inhibitor (colchicine or demecolcine) is added to arrest the cells in metaphase. For detection of sister-chromatid exchanges, the two chromatids are differentially stained with a fluorescent dye such as Hoechst 33258 plus Giemsa. With this method, sister-chromatid exchanges appear as color change on one chromatid (Fig. 4.7).

The sister-chromatid exchange test is one of the quickset, easiest, and most sensitive methods for detecting genetic damage. A large number of compounds have been evaluated with this method, which is usually part of the test battery for genotoxicity screening.

The problem is that the precise genetic event detected by this test is basically not known. Sister-chromatid exchanges are obviously not related to chromosomal aberrations since they often fail to detect potent clastogens. They relate more

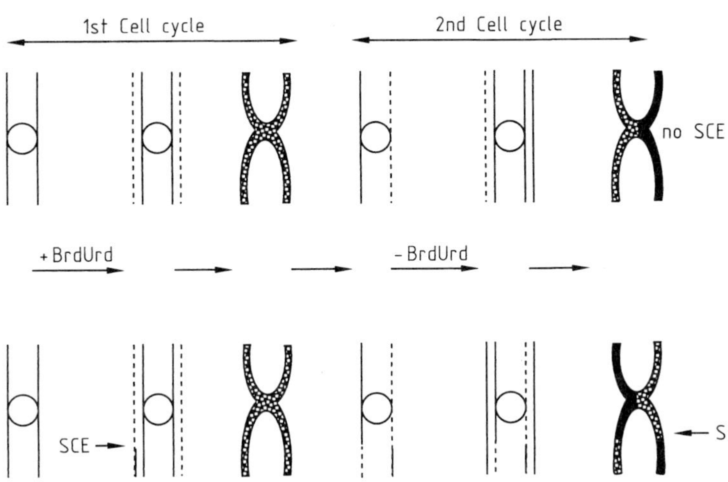

Figure 4.7. Sister chromatid exchange (SCE)
The exchange is visualized by treating cells with 5-bromo-2′-deoxyuridine (BrdUrd) during two DNA replication cycles. The distribution of BrdUrd between chromatids is then determined by staining with a fluorescent dye and UV irradation.

to gene mutations, although there are examples of compounds that clearly induce sister-chromatid exchanges in the absence of mutations as well as the converse. Despite the uncertainty resulting from the unknown mechanism, the sister-chromatid-exchange test is very useful because it detects important groups of potentially carcinogenic compounds such as alkylating agents and nucleotide analogues, as well as compounds inducing DNA single-strand breaks or acting through DNA binding.

4.9.5. Tests for Chromosome Aberrations (Cytogenetic Assays)

4.9.5.1. Cytogenetic Damage and Its Consequences

In addition to gene mutations, most tumor cells investigated thus far exhibit structural and/or numerical chromosomal aberrations. Chromosomal aberrations such as deletions and translocations may result in activation of proto-oncogenes or inactivation of tumor-suppressor genes. In certain cancers of the lymphatic system (human Burkitt's lymphoma and mouse plasmocytoma), the normal proto-oncogene *c-myc*, which is involved in the regulation of cell proliferation, is translocated to the immunoglobulin locus. Immunoglobulin genes show very high transcription rates during development of B lymphocytes. The translocated *c-myc* is subjected to the same control mechanisms, resulting in inappropriately elevated expression of the gene and increased cell proliferation. In addition to these quantitative changes, chromosome aberrations may result in alterations of structure and function of cellular proteins. In many cases of human chronic myeloid leukemia, a piece of chromosome 9 carrying the proto-oncogene *c-abl* undergoes reciprocal translocation with a piece of chromosome 22. This results in joining of the proto-oncogene *c-abl* with a gene on chromosome 22 and production of a fusion protein encoded by both DNA sequences.

During cell division, chromosome segregation depends on the functional integrity of the proteins of the spindle apparatus, which correctly divides the replicated genome into the two new nuclei. Impairment of this process (i.e., by chemical damage to the spindle apparatus) may result in nondisjunction, which means that both copies of a particular chromosome move into one daughter cell while the other cell receives none.

Approximately 10% of the tumor types investigated so far are monosomic or trisomic for a specific chromosome. Aneuploidy is also an important cause of severe birth defects. Trisomy or monosomy of large chromosomes usually leads to early death of the embryo; trisomy of smaller chromosomes may allow survival of the embryo, but the newborn shows severe anatomical abnormalities and major physiological impairment. The most common syndrome of this group is Down's syndrome, which results from trisomy of chromosome 21.

4.9.5.2. In Vitro Cytogenetic Assays

In the in vitro cytogenetic assays, proliferating cultured cells are treated with the test compound, and chromosomal aberrations are investigated after one or more cell cycles. Although cytogenetic assays can, in principle, be carried out in any cell type, Chinese hamster cell lines and human peripheral blood lymphocytes from healthy donors are generally used. In contrast to the spontaneously dividing Chinese hamster cell lines, peripheral blood lymphocyte cultures must be stimulated to divide by a mitogenic agent, such as phytohemagglutinin. Treatment is performed about 44 h after phytohemagglutinin stimulation, when the cells are proliferating. At least three doses of the test compound should be investigated, with the highest being in the low cytotoxic range. The usual recommendation is to treat the cells for ca. 1.5 normal cycles times, which is 15 h for Chinese hamster cells and 12–14 h for human peripheral lymphocytes. Many authors recommend longer treatment periods because some chemicals induce a mitotic delay at clastogenic doses (i.e., they increase the duration of the cell cycle). At the end of treatment, cells are induced to accumulate in the metaphase stage of mitosis with a spindle inhibitor such as colchicine or demecolcine, and at least 200 cells per treatment group are scored under the microscope for chromosomal aberrations. Chromosome breaks, fragments, and exchanges are considered specific structural aberrations, whereas gaps are usually excluded from the quantitation because the mechanism underlying their formation is not understood. Chromosome pulverization (complete fragmentation in small pieces) may also be regarded not as a parameter of clastogenicity, but rather as an indicator of severe cytotoxicity and cell death induced by the test chemical. Chemicals that are clastogenic in vitro at noncytotoxic concentrations are likely to be clastogenic in vivo. When bias resulting from lack of or insufficient biotransformation in the in vitro situation can be excluded, negative results in the in vitro assay provide a strong indication for absence of in vivo clastogenesis.

4.9.5.3. In Vivo Cytogenetic Assays

Induction of chromosome aberrations can be detected in intact animals either by examination of metaphases as described above for cultured cells or by the formation of micronuclei. Rats and Chinese hamsters are usually employed in metaphase analysis, while mice are commonly preferred in the widely used micronucleus test in bone marrow cells.

Rodent Micronucleus Test. The rodent micronucleus assay is an important part of the test battery for genotoxicity and is usually included in the toxicity evaluation of both pharmaceuticals and chemicals. The assessment of micronuclei is commonly conducted in polychromatic erythrocytes of bone marrow after treatment of young mice for six to eight weeks. Induction of numerical chromosome aberrations or chromosome breaks in the immature erythroblast of the bone marrow results in the formation of a micronucleus after cell division because the chromosome fragments are not participating in the chromosome segregation accomplished by the spindle apparatus. In the course of the physiological maturation of the red blood cells, the main nucleus is extruded from the erythroblast, while the micronucleus remains in the cell and can be detected in the young polychromatic erythrocyte after staining (Fig. 4.8). The lack of nucleus in these bone marrow cells facilitates scoring for micronuclei under the microscope; however, peripheral blood cells and liver cells have also been used.

Usual treatment protocols include application of a single dose of the test compound and harvesting of bone marrow cells after 24, 48, and 72 h or, alternatively, multiple applications (two or three doses 24 h apart) followed by a single harvest another 24 h after the last dose. The bone marrow is aspirated with a needle from the proximal end of the femur shafts and flushed with serum to obtain a homogeneous cell suspension,

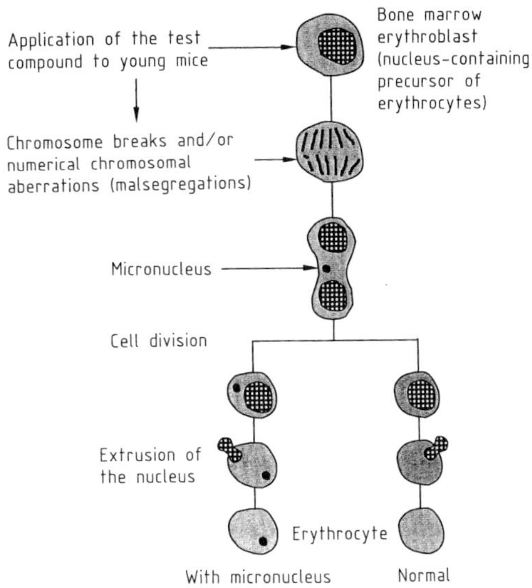

Figure 4.8. Mechanism of micronucleus formation in bone marrow erythroblasts

and after staining, the number of micronucleated immature polychromatic erythrocytes (not the number of micronuclei) is scored under a microscope. The immature erythrocytes stain bluish and can be distinguished clearly from the orange-red mature erythrocytes. Only tests with positive and negative control values within the historical control value of the laboratory are considered adequate. The minimum number of cells scored depends on the spontaneous frequency of micronuclei; 2000 cells should be scored if the control incidence is $< 0.2\,\%$ [180]. The ratio of polychromatic immature erythrocytes:monochromatic mature erythrocytes indicates the integrity of proliferation and differentiation processes in the bone marrow and is used as a measure of general toxicity. If the test compound has been administered at the maximum tolerated dose (as determined in the subacute toxicity test) and no increase in micronucleated erythrocytes can be detected, it is classified as negative for clastogenic activity in vivo. Clear positive results in the in vivo assay indicate clastogenic potential also in humans.

Rodent Bone Marrow Metaphase Analysis. Metaphase analysis is usually carried out in rats at 6, 24, and 48 h after administration of a single dose or alternatively at 6 and 24 h after administration of multiple doses at 2 to 4 h prior to sampling. The animals are treated with spindle inhibitor such as colchicine and demecolcine to induce accumulation of cells in the metaphase. Metaphase analysis is much more time-consuming than scoring for micronuclei; therefore, only 50 cells per animal (500 per experiment with five male and five female animals) are usually analyzed [180]. In contrast to the rodent micronucleus test, the bone marrow metaphase analysis is not automatically included in the general screening scheme for genotoxicity; it is performed only when specific questions arise or for research purposes.

4.9.6. Malignant Transformation of Mammalian Cells in Culture

The term transformation in the present context means that a cultured cell line has taken on one or more morphological and/or malignant changes to give tumor cells. Among the various changes observed by exposure of cell lines to carcinogenic compounds, the following are two clearly established parameters indicating morphological transformation:

1) Inducation of anchorage independence (i.e., cells become capable of forming colonies in soft agar)
2) Loss of contact inhibition resulting in a tendency to grow in a piled-up criss-crossed pattern and to form foci

The term malignant transformation indicates that a cell line is capable of producing an invasive tumor in a suitable host. Hence, not every morphologically transformed cell line has necessarily undergone malignant transformation. On the other hand, human cancer cells are by definition malignantly transformed because they may metastasize under appropriate conditions.

The vast majority of cell transformation studies have been carried out with Syrian hamster cells [i.e., Syrian hamster embryo (SHE) cells and baby hamster kidney (BHK) cells] and with human fibroblasts.

SHE cells, isolated from 13-day-old embryos, can be induced to form piled-up criss-crossed colonies (foci) by treatment with carcinogens. Cells derived from these foci can produce tumors in athymic mice after extended subculturing (35–70 population doublings). Hence, the induction of malignantly transformed tumorigenic cells is a laborious, time-consuming procedure with several inherent problems (e.g., contamination of the plates in the course of the multiple passaging steps over several months). Therefore, the tendency developed to establish morphological transformation that can be induced in a comparatively short time as a short-term test for carcinogens. However, anchorage-independent cells showing criss-crossed growth pattern are usually not carcinogenic in animals. In addition, interlaboratory comparative elaboration studies indicated that scoring may be arduous and subjective, and in many cases the lack of dose-response and difficulties in obtaining consistently reproducible results in repeated assays may occur in the same laboratory [181].

Malignant transformation of cultured cells is an important tool to investigate the molecular mechanisms (i.e., mutations in oncogenes and tumor-suppressor genes) likely involved in carcinogenesis. However, the use of cell transformation assays for screening purposes and for human risk assessment is very limited.

4.9.7. In Vivo Carcinogenicity Studies of Limited Duration

In vivo tests of limited duration provide evidence for carcinogenicity in the whole animal in a short period (i.e., 52 weeks or less) compared to the lifelong bioassay. The experimental procedure usually involves administration of several doses of a known initiating carcinogen for a specific target organ to investigate promoting effects of the subsequently administered test compound. Alternatively, a known potent promotor for a specific tumor can be used to facilitate detection of initiating properties of the test compound. Also, many limited bioassays currently established or being developed aim to predict carcinogenicity by detecting cell lesions that consistently precede the appearance of the relevant tumor (i.e., induction of altered foci in the liver). In vivo carcinogenicity studies of limited duration are not automatically included in the usual test battery for carcinogenicity and are not accepted by regulatory authorities as a replacement for the conventional lifelong carcinogenicity study. The decision to carry out limited bioassays is made individually according to the available and required information on the test compound. Therefore, the present chapter includes only a brief description of the limited bioassays currently available.

4.9.7.1. Induction of Altered Foci in the Rodent Liver

In the course of rat liver carcinogenesis, cell foci exhibiting specific biochemical alterations precede the formation of liver tumors. These foci have, for example, abnormal concentrations of γ-glutamyl transpeptidase, glucose 6-phosphatase, adenosine triphosphatase, and the placental form of glutathione *S*-transferase. Unlike normal liver parenchyma, they do not accumulate iron following iron loading. These parameters allow reliable and objective histochemical identification of preneoplastic foci as soon as three weeks after the initiation of treatment with known carcinogens. The yield is usually highest by 12 to 16 weeks of exposure; therefore, the recommended approach is exposure for 12 weeks to the test compound with subcutaneous injection of iron during the last 2 weeks to produce the iron load [182].

4.9.7.2. Induction of Lung Tumors in Specific Sensitive Strains of Mice

Certain strains of mice (i.e., the A/Heston) exhibit a high spontaneous incidence of lung tumors and are extremely sensitive to pulmonary carcinogens such as polycyclic aromatic hydrocarbons, certain nitrosamines, aflatoxin B_1 ethyl carbamate, hydrazines, and certain alkylating agents. The sensitivity of the test is highest 30 to 35 weeks after exposure initiation. Extension of treatment for a longer period is not recommended, because after 35 weeks the spontaneous lung tumor incidence increases rapidly in control animals as well, resulting in decreased sensitivity of the test.

4.9.7.3. Induction of Skin Tumors in Specific Sensitive Strains of Mice

The carcinogenic activities of several chemicals and crude mixtures can be readily revealed by their continuous application to the skin of a highly sensitive mouse strain, the Sencar mouse [183]. Tars from coal, petroleum, or tobacco, as well as the pure polycyclic aromatic hydrocarbons and their congeners contained in such mixtures, give clearly positive results in this limited bioassay; in contrast, some arylamines and other established carcinogens do not elicit a positive response in this mouse skin test. These discrepancies are due mainly to differences in toxification or detoxification reactions between the systemic and the local applications. The mouse skin contains specific cytochrome P450 enzymes and peroxidases that can bioactivate polycyclic aromatic hydrocarbons, hence the differences in response. Therefore, this test cannot be considered a reliable predictor of the potential carcinogenicity of a xenobiotic or of human risk.

4.9.8. Methods to Assess Primary DNA Damage

4.9.8.1. Alkaline Elution Techniques

Many biochemical and analytical procedures are available to detect and quantitate the damage to DNA after contact with xenobiotics. Some of these procedures are very sensitive and thus permit detection and quantitation of DNA damage or DNA modifications after application of relevant doses of carcinogens to animals and after occupational or environmental exposure of humans.

Alkaline DNA filter elution was developed based on the observation that the rate at which large DNA single strands pass through a membrane filter under alkaline denaturation conditions depends on the length of the strands. A broad spectrum of DNA damage types—both direct and indirect—caused by effects of the toxicant on cellular function, can be detected with this technique. The indirect effects include DNA single-strand breaks, DNA–protein cross-links, and interstrand cross-links (DNA–DNA).

The basic procedure (Fig. 4.9) operates essentially as follows. After treatment of cells in vitro or isolation of cells from an organ or a tissue of an animal treated in vivo, cell samples are placed onto membrane filters and lysed with a detergent-containing solution. This lysis solution is allowed to flow through the filter, thus removing most of the cellular protein and RNA, the intact DNA of cellular chromatin being retained on the filter. An elution solvent with a pH generally > 12.0 is then pumped slowly through the filter to disrupt the hydrogen bonds between DNA strands. Treatment-induced DNA single-strand breaks produce DNA fragments with re-

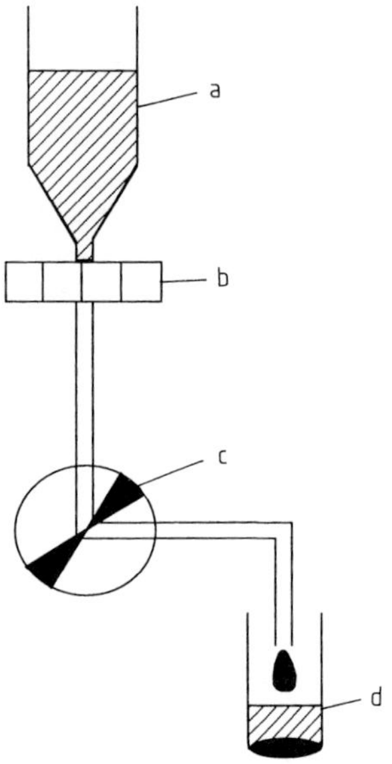

Figure 4.9. Schematic of apparatus for the filter elution of DNA to detect treatment-induced DNA strand breaks, DNA–DNA cross-links, and DNA–protein cross-links
a) Reservoir; b) Filter; c) Pump; d) Autosampler

duced molecular masses, thus increasing the rate at which DNA passes through the filters, whereas DNA–DNA or DNA–protein cross-links decrease the rate compared to untreated controls. The amount of DNA eluted can be quantified either by using radiolabeled cell populations (by pretreating the cells with ^{3}H-thymidine) or by fluorogenic DNA-reactive compounds. The major limitations of the test are essentially the same as described in Section 4.9.4 for DNA repair. DNA strand breaks do not necessarily indicate mutagenicity or interaction of a xenobiotic with DNA; they may also result from extranuclear damage such as increased production of reactive oxygen species due to impairment of mitochondrial functions. Hence, although a vast amount of data exists on the ability of mutagens and carcinogens to induce single-strand breaks, the test has only limited value as a screening method for mutagenicity.

4.9.8.2. Methods to Detect and Quantify DNA Modifications

The increased sensitivity of analytical methods has resulted in the development of several sensitive and selective methods to detect and quantify modifications of DNA bases induced by xenobiotics. These methods often rely on sophisticated chromatographic techniques to separate unchanged DNA constituents from modified DNA bases.

High-performance liquid chromatography (HPLC) with fluorescence detection may be used to detect and quantify DNA adducts formed by highly fluorescent xenobiotics, such as aflatoxin B_1 and polycyclic aromatic hydrocarbons. In addition, some DNA adducts of xenobiotics are highly fluorescent, such as alkylations of the N^7 atom of guanosine [184] and cyclic derivatives formed from vinyl chloride [185]. HPLC with electrochemical detection may detect oxidative DNA modifications, such as 8-hydroxydeoxyguanosine with high sensitivity and has been used to study DNA damage by cellular aging [186].

Gas chromatography/mass spectrometry (GC/MS) techniques are also becoming more widely used in the detection and quantitation of DNA modifications. DNA adducts formed by methylating and ethylating agents may be detected, after derivatization to volatile derivatives, by GC/MS with selected ion monitoring [187]. Because of its high sensitivity (attomole range), GC/MS coupled with chemical ionization and negative-ion detection (e.g., after electrophore labeling with pentafluorobenzyl bromide) may have wide application in DNA adduct monitoring [188], [189].

A widely used method to detect DNA adducts of bulky organic substituents, which cannot be transformed to volatile derivatives, is the ^{32}P-postlabeling technique [190]. The method does not require the use of radiolabeled xenobiotic and sophisticated or expensive mass spectrometry. Due to the availability of ^{32}P-adenosine triphosphate with high specific activity and the possibility of concentrating DNA adducts, this procedure may detect adduct frequences as low as one adduct in 10^{10} nucleotides. The method involves isolation of DNA from an animal or cells treated with the xenobiotic, enzymatic hydrolysis of the DNA to the 3′-nucleoside monophosphates, and enzymatic phosphorylating of the 3′,5′-nucleotide diphosphates with ^{32}P-adenosine triphosphate as phosphate donor. The obtained mixture of ^{32}P-labeled nucleosides and modified nucleotides is then separated by multidimensional thin layer chromatography. Modified nucleotides are detected by putting the thin-layer plates onto radiation-sensitive film; quantitation may be performed by liquid scintillation spectrometry

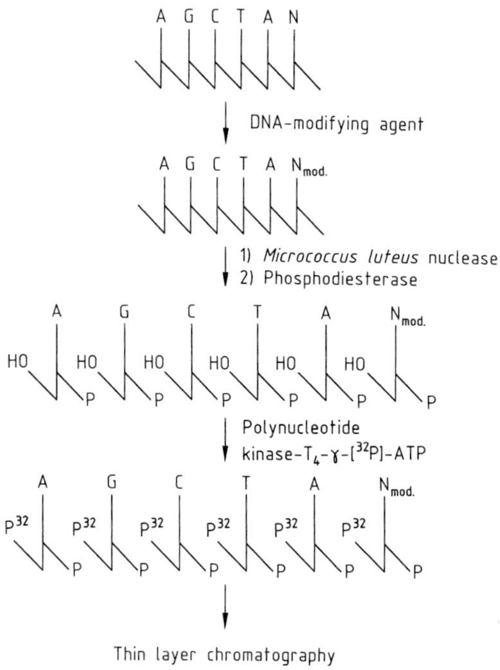

Figure 4.10. The ^{32}P-postlabeling procedure for detection of DNA damage ($N_{mod.}$ = modified deoxynucleotide)

after cutting the adduct spots from thin-layer plates (Fig. 4.10).

Several procedures to detect DNA adducts based on immunological methods have also been developed. These procedures have the advantage of being rapid and simple; however, they are usually selective only for a specific type of adduct and require a time-consuming procedure to generate the antibody. In addition, quantitation of adducts is complicated.

4.9.9. Interpretation of Results Obtained in Short-Term Tests

Although no single rule applies in every case and each test compound should be evaluated individually, some general recommendations may be useful for a reasonable interpretation of results from short-term tests. Positive results in one in vitro test for mutagenicity (i.e., the Ames test) and one in vivo test (i.e., the rodent bone marrow micronucleus assay) indicate potential carcinogenicity. The test compound is highly suspect for potential carcinogenicity if clear-cut, dose-dependent evidence for genotoxicity in more than one in vivo and one in vitro test has been obtained, especially if the compound or its metabolites are structurally alerting for DNA reactivity.

This is supported by recent evaluation of the activity of known human carcinogens in the *Salmonella* mutagenicity test and the rodent bone marrow micronucleus test [191]. As shown in Table 4.10, most of the human carcinogens

Table 4.10. Mutagenicity in *Salmonella typhimurium* and clastogenicity in rodent bone marrow of known human carcinogens [191]

Human carcinogens*	*Salmonella* mutagenicity	Rodent bone marrow chromosomal aberrations or micronuclei
Organic compounds		
Aflatoxins	+	+
4-Aminobiphenyl	+	+
Analgesics containing phenacetin	+	+
Azathioprine	+	+
Benzene		+
Benzidine	+	+
Betel quid and tobacco	+	+
Bis(chloromethyl) ether	+	(+)
Chlorambucil	+	+
Chlornaphazine	+	+
Cyclophosphamide	+	+
Melphalan	+	+
Mustard gas	+	ND(+)
Myleran	+	+
2-Naphthylamine	+	+
Tobacco, smokeless	+	+
Tobacco, smoke	+	ND(+)
Treosulphan	+	+
Vinyl chloride	+	+
Soot, tars, and oils		
Coal tar pitch	+	ND
Coal tar	+	ND
Mineral oil (untreated and mildly treated)	+	ND
Soot	+	ND(+)
Metals		
Arsenic compounds		[illegible]
Chromium compounds (hexavalent)	+	+
Nickel and nickel compounds		ND

+ = positive response; (+) = predicted positive response; ND = not tested so far.

* The carcinogenic hormones (essentially natural and synthetic estrogens) and fibers (asbestos, erionite, and talc containing asbestiform fiber) are not included; the members of these groups tested so far gave negative or inconclusive results in the *Salmonella* mutagenicity and rodent bone marrow clastogenicity assays.

identified so far are positive in these popular tests. Apart from hormones, no nongenotoxic organic chemical has been shown to cause cancer in humans thus far. In contrast, since the late 1980s an increasing number of organic chemicals has emerged that are clearly carcinogenic in rodents but are not mutagenic in *Salmonella* or clastogenic in rodent bone marrow. In addition, most of these nongenotoxic rodent carcinogens (and their metabolites) do not exhibit structural alerts for DNA reactivity. The reason for the differences between the epidemiological data in humans and the chronic rodent bioassays are not known. However, these considerations also suggest that although positive results in the short-term assays for genotoxicity indicate carcinogenic potential and may give reason to interrupt further development of a compound, negative results cannot be taken as a guarantee of the absence of carcinogenic activity in vivo, but rather as a green light to proceed to the lifelong rodent bioassay.

4.10. Evaluation of Toxic Effects on the Immune System

Since the 1970s, experimental and epidemiological evidence has accumulated that the mammalian immune system may be altered (i.e., suppressed or induced) by a wide range of environmental chemicals and drugs. Well-known examples, are the polychlorinated and polybrominated biphenyls; dibenzo-*p*-dioxins; benzene; isocyanates; metals such as chromium, lead, and nickel; and organometalics such as di-*n*-octyltin chloride and tri-*n*-butyltin oxide.

The current status of the methods available for evaluation of toxic effects on the immune system in animals and in in vitro models is not satisfactory. The available methods are not sufficiently validated, and additional tests with higher sensitivity, specificity, and reproducibility are required. Although the immune system is considered as one target organ in terms of systemic toxicology, it consists of several different organ systems. A very high number of different cell types present in practically all tissues and compartments of the human or animal body participate in the immune response. Because of its complexity, the immune system is described here only very briefly to help understand current practical approaches to evaluation of the adverse effects on immune function induced by chemicals (see also, → Immunotherapy and Vaccines, **A14**, pp. 50–55).

The overall immune system can be categorized into two major subsystems, the humoral and the cell-mediated subsystems. The effects on the *humoral system* are mediated by B lymphocytes, which produce antibodies that react with antigenic (usually foreign) material (i.e., antibodies attack bacteria and viruses before they can enter the host cell). The *cell-mediated system* involves primarily the mobilization of phagocytic leucocytes (macrophages) to ingest foreign organisms such as bacteria and the activation of T lymphocytes.

The two systems do not function independently in different situations; rather, they interact by complex feedback mechanisms that are only partly understood. One of the main properties of the immune system is the rapid production of a large number of cells capable of reacting with a specific antigen when this antigen is presented (again) to the organism. This important property is based on the presence of a variety of memory cells that were specifically adapted to the antigen at the time of initial contact.

Immunocompetent cells are required for host resistance; thus, exposure to immunotoxicants can result in increased susceptibility to any type of toxicity and disease including cancer. Despite its generally recognized importance, immunotoxicity testing is not included in the routine toxicity evaluation of chemicals under development and official guidelines are not yet available. However, since immunotoxicology will probably play a very important role in the near future, a practical approach for assessing effects on the immune system in the course of conventional toxicity studies in the rat is presented below, according to guidelines of the National Institute of Public Health and Environmental Hygiene in the Netherlands and an international workshop on immunotoxicology held in 1990 [192], [193].

Evaluation of adverse effects on the immune system can be carried out in two tiers. The first tier evaluates immune-related parameters (hematology, blood chemistry, organ weights, and histopathology) that can be included without major difficulties in standard testing protocols of repeated-dose toxicity studies (Table 4.11).

Compounds showing some immunotoxic properties in this first tier and chemicals suspected of having immunotoxic effects based on information from prior studies or structure–activity relationships are further evaluated by functional

Table 4.11. Immunological parameters that can be included in repeated-dose toxicity studies in rats for compounds for which no prior immunotoxic potential has been identified (tier I)

Parameters assessed	Tests
Hematology and blood chemistry	differential white blood cell counts bone marrow cellularity albumin:globulin ratio serum immunoglobulin classes
Organ weights	thymus, spleen, lymph nodes
Histopathology	thymus, spleen, lymph nodes, bone marrow, Peyer's patches (if oral administration) bronchus-associated lymphoid tissue (in case of pulmonary administration)

Table 4.12. Selected functional assays to assess immunotoxicity of compounds for which some immunotoxic properties have been implied in tier I evaluation (see Table 4.11) or in prior studies (tier II)

Parameter assessed	Tests
Cell-mediated immunity	mixed leukocyte response to antigenic determinants; induction of cell proliferation by the T-lymphocyte mitogen concanavalin A; T-lymphocyte cytotoxicity; delayed hypersensitivity response (DHR) to keyhole limpet hemocyanin (in contrast to the first three assays, that are carried out in cultured cells in vitro, the DHR is induced in vivo)
Antibody-mediated immunity	antibody plaque-forming cell response; serum antibody titer after exposure to specific antigens determined by enzyme-linked immunosorbent assay (ELISA)
Natural and induced host resistance	natural killer cell cytotoxicity; macrophage-mediated phagocytosis and intracellular killing; assessment of macrophage and T-lymphocyte competence by inoculation of *Listeria monocytogenes*; host resistance to melanoma cells

assays that assess the competence of immune cells (Table 4.12). For a detailed description of experimental procedures, see [194].

Perspectives for the Future. Since the 1980s, molecular immunopharmacology and immunotoxicology have evolved as very active areas of research. Many immunoregulatory cytokines has been cloned, thus enabling molecular analysis of gene expression at the level of DNA, RNA, or protein. Although the methods are not yet standardized or sufficiently validated to be included in regular toxicity testing, they may play an important role in the near future, especially the quantitative assessment of immune-related mRNA species in tissues (spleen, thymus, lymph nodes) of in vivo exposed animals.

4.11. Toxicologic Evaluation of the Nervous System

The identification and characterization of neurotoxic properties of new chemical substances and pharmaceuticals is one of the essential goals of every toxicity screening program. Neurotoxicity has traditionally been associated with structural pathological modification of nervous system constituents (i.e., neurons, glial cells, or endothelial cells). Hence, the most important tool for assessing neurotoxic effects in the past has been histopathological examination of the nervous system of animals acutely or chronically exposed to xenobiotics. Over the last 10 or 20 years, however, several reasons have spurred the development and use of functional tests in neurotoxicity screening. First, detailed histopathological analysis of the nervous system is very time-consuming and requires experienced neuropathologists. Second (and probably more important), many chemical compounds disturb the nervous system without causing identifiable structural lesions. This clear need for functional tests to assess neurotoxicity in preclinical safety studies has led to the development of a functional observational battery and an automated test to detect locomotor activity in rats based on methods that have been used by neuropharmacologists for many decades to evaluate psychoactive neurologic and autonomic pharmaceuticals.

4.11.1. Functional Observational Battery

The functional observational battery can be carried out in both acute and repeated-dose toxicity studies. The sequence of tests is usually arranged to progress from the least to the most interactive with the animal. Assessment begins with home cage observations followed by measurement made while handling the animals and assessment of activities in the open field. Assessment of reflexes, as well as physiologic and neuromuscular parameters, is carried out at the end of the examination. Specific parameters assessed at these different stages are summarized in Table 4.13, and more detailed descriptions of some important tests are given below (for recent reviews, see [195]–[197]).

Catalepsy can be measured by placing the rat on four corks (35 mm high, 40-mm diameter, 100-mm distance between fore- and hindfeet, 60-mm distance between right and left feet). Duration of immobility at this position is measured for up to 60 s. For assessment of catalepsy, a number of experimental modifications exist (i.e., placing the rats on a horizontal bar, 12 cm above the ground). However, the four-cork procedure allows the best distinction between cataleptic and heavily sedated animals.

Table 4.13. General parts of the functional observational battery to assess neurotoxicity and specific examples of parameters recorded

A	home cage observations	posture palpebral closure (eyelids wide opten to completely shut) convulsions (clonic, tonic) biting
B	observations while handling the animal	ease of removal from cage and of handling in hand, lacrimation, salivation, piloerection, fur appearance
C	open-field activity (observations over 3-min period)	time to first step number of rears (supported and unsupported) number of urine pools and defacations mobility and gait tremors, ataxic gait, convulsions
D	reflexes	approach response (e.g., to a pencil) touch response click response tail response pupil and eye blink response forelimb and hindlimb extension righting reflex: hold rat in supine position, drop from approximately 30 cm and note ease of landing
E	physiologic observations	catalepsy body temperature body weight
F	neuromuscular observations	rotarod performance grip strength hindlimb extension strength hindlimb foot splay

Grip Strength. The rat is allowed to grip a triangular ring with its forepaws and is pulled back along a platform until its grip is broken. As the pulling back continues, the animal's hindpaws reach a T-shaped rearlimb grip bar, which it is allowed to grasp and then forced to release by continued pulling. Special devices are used to measure the maximum strain required to break forelimb and hindlimb grip.

Rotarod Performance (Rotating Rod Test). This test requires preliminary training of the animals to walk on a rotating rod (7-cm diameter, rotation rate 5–12 rpm). During the training period, containers underneath the rod are filled with water to prevent the rat from jumping off. For testing, the time that each rat remains on the rotarod is measured up to 2 or 3 min. Because of its objectivity and reproducibility, this test is one of the most popular for evaluation of adverse effects on the nervous system.

Hindlimb Foot Splay. The hindfeet are painted, animals are dropped from a horizontal position 30 cm above a table onto a paper, and the distance between the center of the ink spots is measured.

4.11.2. Locomotor Activity

Locomotor activity is not automatically part of the screening battery for neurotoxicity; the decision to carry out the test is made on an individual basis according to preliminary observations. Locomotor activity is assessed in special test chambers using a photocell detection procedure. Animal movement inside the chamber interrupts the photobeams and is translated into activity counts. Data on animal activity are usually recorded over three 5-min intervals.

A major problem in neurotoxicity screening is that the general behavior of rats is not stable and uniform, even when they are kept under strictly controlled conditions and not exposed to neurotoxic compounds. Therefore, many authors suggest carrying out the tests without knowledge of the treatment protocols. Futhermore, since many observations are subjective, experienced and well-trained scientists and technicians must be used in the evaluation.

5. Evaluation of Toxic Effects

One of the major environmental and occupational issues of concern to both scientists and administrators is the control of potential health hazards to humans due to the production, use, and disposal of chemicals. The concern arises from the increasing numbers of chemicals in production and use and the increasing numbers of chemicals demonstrated to exert toxic effects in one or several of the sensitive toxicity testing systems available. This situation has afforded growing legislative control of the production and application of chemicals to ensure adequate protection of human health. Control measures based on the recognition of potential adverse health effects may limit the presence of hazardous chemicals in the environment or regulate the use of hazardous chemicals, thus reducing the potential health risks to humans (Table 5.1).

The assessment of potential human health risks resulting from the exposure to chemicals provides the basis for appropriate regulatory and control measures. The health risk assessment determines whether a xenobiotic may cause adverse health effects, at what level and frequency of exposure, and the probability that adverse health effects will occur. The term "risk assessment" is increasingly used in the context of potentially toxic chemicals. Scientific risk assessment considers the available data on the toxicology of a specific chemical when judging which agents potentially pose a significant risk to the human population. Toxicology focuses on the identification and quantitation of potential hazards by using animal studies as surrogates for humans. Permissible exposure levels for humans are derived from the results of the animal studies by using margins of safety or defining "acceptable" incidences of adverse health effects in exposed humans [198].

Table 5.1. Possible measures to reduce human exposure to hazardous chemicals

Application or exposure to chemical in question	Measures to reduce exposure
Industrial chemicals	reduction or cessation of application; protective measures in the workplace; alternative chemicals with lower hazard
Pharmaceuticals	cost–benefit analysis
Alcohol, smoking, drugs of abuse	education
Environmental chemicals	quantitation of exposure, strategies for avoidance or reduction of environmental pollution

Health risk assessment and its use in regulatory decisions have recently generated intense controversy. The debate over risk assessment is politically and emotionally charged, and creates an adverse atmosphere heightened by the extraordinary sums of money at stake. Industry complains that the costs of complying with possible overregulation based on inappropriate risk assessments may be excessive; moreover, lawsuits on potential environmentally caused diseases, especially in the United States, involve huge sums of money. On the other hand, environmentalists claim that risk assessment practices and policies do not adequately protect human health; moreover, health care costs for the treatment of environmentally caused diseases may also be very high. These considerations have led to an intensive rethinking of the health risk assessment process and have increased the awareness that in many cases, the scientific foundation for risk assessment is weak. This rethinking led to the conclusion that resolution of the controversies by the development of effective prevention strategies and rational priority setting may be achieved only by strengthening the scientific background and available data by research and by developing better methods to estimate risks due to chemical exposures [141], [198], [199].

In discussions on the health effects of potentially toxic chemicals, the terms hazard and risk are often used with an identical meaning; however, they are clearly different. The term hazard defines the intrinsic toxicity of a chemical and is not identical to risk. Risk is the estimated or measured probability of injury or death resulting from exposure to a specific chemical. Risk may be described either in semiquantitative terms such as high or low risk or in quantitative terms such as one person experiencing an adverse effect per 10000 persons exposed. Risk may also be described in absolute terms (probability of adverse effects due to a specific chemical exposure) or in comparative terms by comparing the probability of adverse effects between a population exposed to an agent and an unexposed population.

The health risks due to the contact with potentially toxic chemicals depend on the conditions of exposure, because not only the intrinsic toxicity but also the dose of a chemical determine the magnitude of the adverse effect. As noted in Chapter 1, the magnitude of toxic effects is the product of the intrinsic toxicity of a chemical multiplied by the dose taken up by exposed animals or humans; thus, all toxic effects are dose dependent, and even very toxic chemicals may not cause toxic effects when the dose is low. If the dose is zero the toxic effect and the risk of adverse health effects will be zero. On the other hand, chemicals with low intrinsic toxicity may induce toxic effects when the dose is high and may thus pose significant risk. In toxicological terms, risk is therefore the product of the intrinsic toxicity of a chemical and the exposure characteristics.

5.1. Acceptable Risks, Comparison of Risks, and Establishing Acceptable Levels of Risk

In earlier phases of risk assessment, the basic belief was that few chemicals are toxic and all of these toxic chemicals are derived from synthetic processes. To achieve a zero risk, chemical exposure must be reduced below a threshold level, under which it causes absolutely no risk. However, where such a threshold cannot be demonstrated, one must assume that a finite risk may occur at any exposure level, consequently, absolute control of risk is possible only if the source of exposure is eliminated altogether. These considerations resulted in the zero-risk concept. The Delaney Clause of the Federal Food and Drug Act in the United States is an example of a zero-risk approach in the regulation of food additives. This law states that no xenobiotic whose carcinogenic potency in animals has been demonstrated may be used as a food additive.

However, the more widespread testing of chemicals for toxicity, the increased sensitivity of analytical instruments to detect chemicals in the environment and the workplace, and developments in the science of toxicology put the basic assumption of the zero-risk concept—that only synthetic chemicals are toxic—in question. These developments led to the recognition that zero risk was unachievable and, perhaps, unnecessary for the regulation of chemicals. The observation was based mainly on a few facts: (1) all chemicals, both of synthetic and natural origin, are toxic under specific exposure conditions; (2) most of the hazardous chemicals routinely encountered by humans are of natural rather than synthetic origin; (3) most of the exposure to hazardous synthetic chemicals cannot be avoided entirely or be eliminated from the environment without changing profoundly the way of life in many countries; and (4) in the case of cancer risk assessment, DNA damage and mutations, assumed to be of major significance in the process of carcinogenesis, occur spontaneously, albeit at a low rate. Examples of endogenously occurring DNA damage are hydrolytic deamination, depurination, oxidative modification, and endogenously formed DNA adducts. Well-known examples of hazardous synthetic chemicals are benzene, which is present in the environment as a result of its emission from motor vehicles, cigarette smoking, and other sources, or 2,3,7,8-tetrachlorodibenzodioxin formed in forest fires. The effect of naturally occurring chemicals and chemical exposure due to life-style factors is best exemplified by the estimated contribution of different factors to the incidence of avoidable cancers in man (Fig. 5.1).

According to the large epidemiological study of Doll and Peto, natural chemicals in diet and chemicals inhaled by cigarette smoking are the major causative agents of human cancer. Occupational and environmental exposure to synthetic chemicals constitutes only a minor causative factor.

Because of these facts, the acceptable risk concept was developed as an alternative. The acceptable risk concept realizes that it is not possible to eliminate all potential health risks associated with chemical exposure due to the life style. According to the concept of acceptable risk, safety—the reciprocal of risk—is no longer an absolute term but is redefined as a condition of certain, but very low and thus acceptable, risk. This conceptual change improves the ability to deal with potentially very low risks identified by the increased sensitivity of analytical instrumentation and with increasingly sensitive scientific

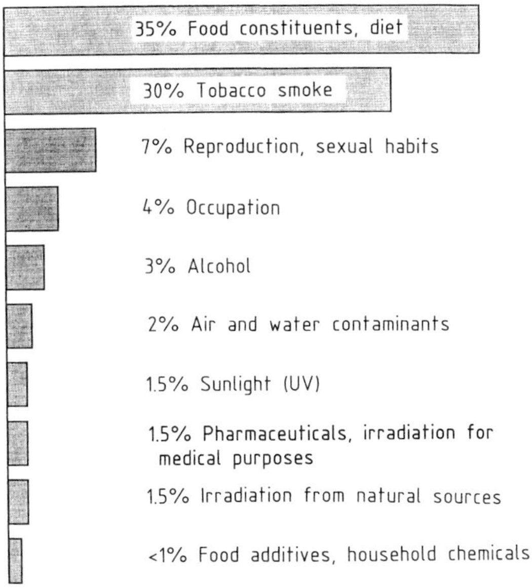

Figure 5.1. Contribution of chemical exposure and life-style factors to the incidence of avoidable cancer in human [200]

methods to detect potential adverse effects of xenobiotics. The concept of acceptable risk also permits the definition of limits for the exposure to toxic chemicals that can be considered to have a negligible impact on the incidence of adverse effects in an exposed population. Risk assessment is therefore unavoidable and must implicity or explicity involve a balance of risk and benefit [141], [198], [199], [201], [202]. Some of the main factors considered in establishing acceptable risk levels for exposure to a chemical are outlined below.

Beneficial aspects

1) Economic growth
2) Employment
3) Increased standard of living
4) Increased quality of life
5) Taxes generated

Detrimental aspects

1) Decreased quality of life
2) Health effects
3) Lawsuits
4) Loss of environmental resources
5) Loss of work
6) Medical expenses

5.2. The Risk Assessment Process

Several individual elements make up the risk assessment process. In the first step, the potential adverse health effects of a chemical, a mixture of chemicals, or a specific technical process are evaluated with toxicity tests (for details, see Chap. 4) and, if the chemical is already in widespread use and humans are exposed, by considering the data from epidemiological studies. The second step in risk assessment determines the dose-response for the observed adverse effects. In parallel, the exposure of humans to this xenobiotic is quantitated by analytical procedures or, if the chemical is not yet in widespread use, by the estimation of likely exposure scenarios. In the third step, the results obtained in the toxicity studies are extrapolated. This involves an extrapolation both from adverse effects seen in experimental animals to humans and, often, from the effects seen after high doses in animals to the much lower doses usually encountered by humans. The last step of risk assessment, risk characterization, involves the combination of steps one to three to judge the existence and magnitude of the public health problem and characterize the uncertainities inherent in the risk assessment process.

5.2.1. Hazard Identification Techniques

Hazard identification is the step in which the adverse effects of the xenobiotic are determined. Evaluation of both acute and chronic toxicity is performed by using animals as experimental models for humans. The use of animals as surrogate for humans is based on the following assumptions: (1) xenobiotics with a likely adverse effect in humans will manifest some degree of toxicity in other living systems when the dose is sufficiently high; (2) if a sufficient number of animal species are dosed with the xenobiotic, at least one should exhibit a similar pattern of biotransformation and toxicokinetics to that seen in humans; and (3) if a sufficient number of different animal species are dosed with the xenobiotic, at least one is likely to exhibit the toxic responses and clinical symptoms occurring in humans.

Typical end points in studies aimed at hazard identification in intact animals include mortality, reproductive and developmental effects, target organ toxicity, and cancer. Hazard identification studies at present also include the determination of a range of biochemical end points related to specific toxic effects such as toxicokinetics, routes and extent of biotransformation, structure of reactive intermediates, and binding of reactive intermediates to cellular macromolecules.

In addition, many short-term tests for specific toxic effects such as DNA damage, mutagenicity, or clastogenicity are increasingly included in hazard identification procedures.

The acute toxicity of xenobiotics is evaluated by a number of procedures from which the LD_{50} may be calculated. With the recently recommended fixed-dose method, information on target organs affected and types of toxic effects may also be obtained (see Section 4.2.1). Repeated-dose toxicity studies last between two weeks (subacute toxicity studies) and 6 to 24 months (chronic toxicity studies), the life span of the animals. The studies include postmortem examination, histopathology, clinical chemistry, and hematology at termination and at specified time points during the study. From the chronic toxicity studies, the lowest-observed-effect level (LOEL), also referred to as the lowest-observed-adverse-effect level (LOAEL), as well as the NOEL, are obtained for toxic effects other than cancer. The NOEL is the highest dose administered that does not induce an observable toxic effect. The NOEL may not be identical to the no-effect level when insensitive methods are applied or the wrong end point is selected. Moreover, the number obtained for the NOEL depends on the number of animals used in the study and the spacing of the doses applied.

Short-Term Tests for Specific Toxic Effects In Vivo and In Vitro. A variety of short-term tests has been developed for the detection and quantification of toxic effects. Most of the more established and well-evaluated test systems are designed to evaluate the genotoxic activities of xenobiotics and employ well-defined genetic changes (DNA damage, gene mutations, chromosome defects, cell transformation) in prokaryotes, lower eukaryotes, and mammalian cells. Knowledge that a genotoxic chemical is active in vivo in the target organ of carcinogenesis enhances confidence that the genotoxicity of the chemical is important in the process of cancer induction by that chemical. Before any short-term test can be used with confidence to assess potential toxic effects, its validity should be thorougly evaluated and its major drawbacks should be explicitly noted. Moreover, it should be kept in mind that most available short-term tests are well designed to give qualitative information and their major use for risk assessment purposes is therefore the confirmation or exclusion of a specific toxic response. The magnitude of toxic response in intact animals or humans depends on both toxicokinetics and toxicodynamics. The toxicokinetic phase of the toxic response is not considered in most of the in vitro short-term tests. Therefore, short-term tests in vitro may be used for hazard identification only in combination with studies on adverse effects in animals and studies on the toxicokinetics of a xenobiotic.

Biotransformation and Toxicokinetics. Studies on the extent of biotransformation including structural identification of the metabolites formed from the xenobiotic both in intact animals and in appropriate in vitro systems such as organ homogenates or fractions with enzymatic activity also contribute to hazard identification [203]. The structures of metabolites formed and the presence and extent of covalent binding of metabolites to macromolecules such as protein and DNA indicate the formation of electrophilic metabolites and thus a potential hazard.

For example, the structure of excreted mercapturic acids may give information on the structure and reactivity of the electrophilic metabolite and sites of cellular interactions [204]. Information on the rate of absorption and elimination may indicate a possible accumulation of the xenobiotic in humans, with the consequence of potential adverse effects. These studies should be performed on at least two animal species in vivo. For in vitro studies, human tissue samples should be included to confirm that biotransformation reactions are identical to those, observed in animals occur in humans. Different mechanisms of toxicity may operate at different dose levels; in these instances, toxicokinetic data may help in understanding dose-dependent mechanistic differences. The toxicokinetics of a xenobiotic in humans may be extrapolated by physiologically based pharmacokinetic models from the results obtained in experimental animals. Information on potential pathways of biotransformation may also be made by computerized structure analysis of the xenobiotics with specific computer programs [142], [205]. These are designed to predict routes and rates of biotransformation based on the presence of functional groups in the molecule; at present, however, the available programs are far from perfect and should be used only in conjunction with expertise and as a basis for experimental planning.

Structure–Activity Relationships and Chemical Structure Analysis. Predictive data on the potential of xenobiotics (mainly organic chemicals) to induce adverse effects may be derived from relationships between chemical structure (physicochemical properties, presence of functional groups, atomic configuration) and biological activity, termed structure–activity relationships.

Two approaches to hazard prediction utilize structure–activity relationships. The first is essentially a qualitative approach and involves comparison of the structure of the xenobiotic with that of other compounds already known to cause specific toxic effects. The comparison with known structures and the knowledge of biotransformation reactions and mechanisms of toxicity permit the identification of "toxophores" in the structure of xenobiotics. Toxophores are functional groups present in the molecule that are likely converted to toxic metabolites or possess chemical reactivity related to mechanisms of toxicity. Examples of toxophores are olefinic moieties, which may be oxidized to epoxides, or terminal carbon atoms in aliphatic hydrocarbons carrying two halogen atoms, which may be oxidized to acyl halides. The scope and usefulness of structure–activity relationships depend on the availability of a sufficiently large data base on the toxic effects of chemicals with common subgroups and structures [142], [205]. The main limitation of this approach is the qualitative nature of the resulting estimate of potential toxicity.

The second approach using structure–activity relationships is termed *quantitative structure–activity relationships* and relies on the computerized analysis of data bases on toxic effects of chemicals. A basic feature of the applied techniques is the use of pattern recognition schemes or substituent weighting factors coupled with regression analysis. Some useful predictions using these techniques have been made; however, because of the complex nature of toxic effects and the multitude of factors governing the toxic response, the applicability of quantitative structure–activity relationships to predict toxicity profiles is still severely limited.

Clinical and Epidemiological Studies. In principle, the best evidence for the toxic effects of a chemical in humans is derived from clinical and epidemiological studies. These studies assess effects in the species of interest for risk assessment and at relevant concentrations. Potentially confounding extrapolations from high to low dose and from animals to humans are not required. However, several weaknesses limit the applicability of these studies to hazard identification. In most cases, reliable exposure data are lacking, consequently, dose–response relationships cannot be established. In addition, the sensitivity of epidemiological studies to detect health problems is comparatively low. Unless the toxic effect of a particular xenobiotic is very unusual in control (unexposed) groups, it may pass unnoticed in a normal survey. Examples are the identification of asbestos exposure as a cause of mesothelioma or vinyl chloride as a cause of hemangiosarcoma, very rare forms of cancer in humans not exposed to asbestos or vinyl chloride, respectively. Moreover, the results of epidemiological studies, especially on cancer risk, may reflect the risk associated with exposure to chemicals decades ago because of the long latency period. Other limitations are confounding variables such as smoking and concomitant exposure to other xenobiotics, which often impair the interpretation of carcinogenicity data in humans. Thus, evidence based on epidemiological observations has identified only a limited number of chemicals as human carcinogens; many of the identified compounds are used in cancer chemotherapy and have the intrinsic property of genotoxicity (Table 5.2).

Table 5.2. Examples of established human carcinogens based on epidemiological observations

Chemical or agent	Site of tumor formation
Aflatoxin	liver
Alcoholic drinks	mouth, esophagus
4-Aminobiphenyl	bladder
Benzidine	bladder
2-Naphthylamine	bladder
Arsenic	skin, lung
Asbestos	lung, pleura, peritoneum
Azathioprine	reticuloendothelial system
Benzene	bone marrow
Bis(chloromethyl) ether	lung
Cadmium	prostate
Chlorambucil	bone marrow
Chlornaphazine	bladder
Chromium	lung
Cyclophosphamide	bladder
Bis(2-chloroethyl) sulfide	larynx, lung
Nickel compounds	nasal cavity, lung
Estrogens	endometrium, vagina
Phenacetin	kidney and lower urinary tract
Polycyclic aromatic hydrocarbons	skin, scrotum, lung
Steroid hormones	liver
Tobacco	mouth, pharynx, larynx, esophagus, lung, bladder
Vinyl chloride	liver

5.2.2. Determination of Exposure

The quantitation of exposure, both in individuals and in populations, is a prerequisite for the quantitation of risk. Reliable data on exposure are needed to assess the adverse effects of the xenobiotic and to recognize specific risk factors such as occupation, life style, or social status. The dimensions of exposure include intensity, frequency, route, and duration of exposure; in addition, the nature, size, and makeup of the exposed population should be characterized. The assessment of exposure is a difficult and complex task that is often neglected. Typically, estimations and field measurements are required. The estimation of human exposure to a particular xenobiotic involves an initial estimation of the possible sources of the chemical and the possibilities of exposure. A good inventory of sources may provide important information on critical pathways of exposure, populations at particular risk, and levels of exposure.

In many cases, the duration and level of exposure, especially after chronic contact, may be estimated only from ambient levels of the xenobiotic in the environment, and estimations may thus be crude; because of the large number of potentially exposed people, only in special situations (e.g., occupational exposure, after disasters) will exposure data, including determination of the internal dose, be available.

Specific procedures to detect exposure to a certain xenobiotic and procedures to estimate exposure by determining biological effects may be used. These include

1) Direct measurement of the chemical in environmental samples such as water, air, and soil
2) Measurement of the chemical, its metabolites, or products of the interaction of the chemical or its biotransformation products with cellular macromolecules (protein or DNA) in body fluids and tissues (biomonitoring)

Biological end points in exposure assessment may be

1) Assessment of biochemical indicators for specific adverse effects known to be caused by the xenobiotic (e.g., inhibition of specific enzymes such as cholinesterase activity in persons exposed to organophosphate pesticides).
2) Observation of pathological evidence of exposure, such as cytogenetic changes in lymphocytes from workers exposed to chromosome-damaging chemicals in the workplace. The use of biological end points for exposure assessment, however, lacks the resolving power to discriminate between endogenous changes and the effects of xenobiotics; therefore, specific chemical indices of exposure should be favored for quantitation of exposure [187], [206].

Due to the time-consuming, costly and labor-intensive procedures required, data on the exposure to xenobiotics are usually limited. Difficulties in identifying concomitant exposures, interactions with other xenobiotics or activities, special risk groups such as the very old or very young and pregnant women, and patterns of exposure result in a high level of uncertainty in the exposure assessment in human populations.

5.2.3. Dose–Response Relationships

Establishment of the dose-response for adverse effects in animals is the decisive step in risk assessment. This step quantitates the relationship between received or administered dose and biological response, and may be performed on an individual or population basis. Dose–response assessment includes exposure intensity and duration as well as factors modifying toxic response such as sex, age, health status, and route of administration. Since the dose–response assessment can be performed only rarely in humans, extrapolation from data obtained in animals to humans is usually required. Moreover, many animal experiments, particularly carcinogenicity bioassays, are performed that give high doses to increase the sensitivity of the assay. Therefore, besides species extrapolation, an extrapolation from effects seen after the high doses in animals to the low doses usually encountered by humans is necessary. This extrapolation step from high, sometimes toxic, doses in animals to low doses in humans is controversial.

Extrapolations from high to low dose are performed differently depending on the type of toxic response elicited by the xenobiotic. As explained in Chapter 1, toxic effects of a chemical may be caused by both reversible and irreversible interactions of the xenobiotic or its metabolite(s) with macromolecules in the organism. Many acute toxic responses such as carbon monoxide poisoning are based on reversible interactions and are associated with thresholds. Threshold doses are doses below which the probability of a response is zero. The biological basis for thresholds is well founded and may be demonstrated on a mechanistic basis. On the other hand, many chronic toxic responses, particulary chemical carcinogenesis, are often considered nonthreshold effects. Since a negative can never be proven, the absence of thresholds cannot be demonstrated by experiments and is based on consideration of the mechanisms of chemical carcinogenesis. Different approaches have thus been developed in establishing acceptable levels of exposure to threshold and nonthreshold responses (see below).

5.2.4. Risk Characterization

Hazard identification, exposure assessment, and dose–response assessment merge into risk characterization. Risk characterization estimates the incidence of expected adverse health effects in exposed populations. As noted in Section 5.2.1, risk characterization and the establishment of acceptable exposure levels are handled differently for carcinogenic and noncarcinogenic xenobiotics. For chemicals that cause adverse effects by mechanisms with thresholds, the "safety factor" approach was developed. For nonthreshold responses such as cancer, both quantitative and qualitative risk assessment procedures are used.

5.2.4.1. The Safety Factor Methodology

When the safety assessment of a chemical is based on animal toxicity testing, a different evaluation of the data is required compared to safety evaluations relying primarily on human observations. Since animal experiments to determine health hazards of chemicals were widely used in the 1940s and human health risks had to be extrapolated from these data, the idea of a safety factor, which was to be applied to the results from animal studies, was developed. The safety factor approach was first proposed in the 1950s by Arnold Lehman of the FDA as a protective measure for human health and is intended to compensate for uncertainties in the extrapolations of animal data to humans. This concept was applied to the determination of acceptable daily intakes (ADIs) by the World Health Organization from the 1960s onward. The ADI was defined as "the daily intake of a chemical which, during an entire lifetime, appears to be without appreciable risk on the basis of all known facts at that time."

The safety factor—or as it is sometimes called, the uncertainty factor—has been introduced to consider both interspecies and interindividual differences in response to potential toxic effects of the chemical under consideration. The major purpose of the safety factor is the protection of human health by establishing safe exposure levels; the exposure levels defined do not mean that exposure above these levels will result in adverse effects. Despite several limitations and criticisms, the safety factor approach has been used for many years in Western Europe and the United States, and has proven useful and reliable. It was thus adopted internationally as the standard procedure for assessing the ADI.

This risk assessment approach is based on the establishment of a NOEL in animal studies on chronic toxicity and then defining permissible human exposures by the application of a safety factor. In this context, the NOEL is defined as the lowest dose of the xenobiotic in an animal experiment that produced no detectable effect in the most sensitive animal species treated. Once a NOEL has been determined, a safety factor for human exposure is introduced and, often, a permissible level of exposure of one-hundreth of the NOEL in animals is defined for humans. The 100-fold safety factor is justified on the basis of a 10-fold difference to reflect an interspecies difference in susceptibility and a 10-fold difference to reflect possible interindividual variations in susceptibility in humans [201].

$$\frac{\text{No-observed-effect level from the study}}{\text{Safety factor}} = \text{Acceptable daily intake}$$

The safety factor approach assumes that toxic effects exerted by the chemical exhibit a dose–response curve with a threshold, that the results on the toxicity studies in animals are relevant to the human situation, and that the extrapolation of dose is reliable.

As noted above, safety factors of 100 are frequently employed. However, when reliable data for adverse effects of the chemicals in humans are available, a safety factor of only 10 may be used. On the other hand, a safety factor of more than 100 is appropriate when no or only limited data on the toxicity of the chemical in animals are available. Note that the safety factor is not based on scientific evidence. Additional data on the mechanisms of toxic effects in animals and on the toxicokinetics and biotransformation of the xenobiotic in animals and humans are required and add scientific credibility to the safety factors proposed. Therefore, the choice of a safety factor should consider, in addition to the NOEL, a series of additional qualitative parameters shown below:

Evaluation of animal toxicity studies

1) Number of studies and effects observed
2) Type of toxic effects
3) Time course for toxic effects
4) Tumorigenicity

Evaluation of biochemical end points

1) Biotransformation and toxicokinetics
2) Mechanism of action, covalent binding to macromolecules
3) Short-term test for genotoxicity and other nonthreshold effects

Evaluation of species differences

1) Interspecies variations in biotransformation and toxicokinetics
2) Influence on anatomical and physiological differences of toxic effects between species

Application of the most appropriate safety factor requires careful analysis of the data available on the toxic effects for every chemical on a case-by-case basis. Consideration should be given to the quality and completeness of the data and the number of spread of the dose levels used. Besides the derived NOEL, all further information obtained in long-term animal toxicity studies is valuable for characterizing toxic effects and determining safety factors. The type of toxic effect and the shape of the dose–response curve should also be taken into account in determining the safety factor.

Also important are considerations of the ability to extrapolate from toxic effects seen in animals to human exposure scenarios and the results of human epidemiology studies, when available. Because of the complexity of information required, no universally accepted guidelines can be developed for determination of the precise magnitude of safety factors; therefore, expert judgment on an individual basis is a major factor contributing to the size of safety factors.

The safety factor approach has several problems. In experiments to determine the NOEL in animals, the experimental group size and the spacing of dose levels are major determinants of the numerical value of the NOEL obtained; the smaller the size of the individual dose groups and the larger the spacing of doses, the less likely is it that an effect will be observed. This phenomenon has the effect of rewarding poor experimental design because small group sizes in experiments will tend to produce higher NOELs. Moreover, nonthreshold effects may not be detected due to the low number of animals enrolled or the dose levels applied. In addition, the slope of the dose-response curve is often not considered or may not be determined with sufficient accuracy, and chemicals with steep and shallow dose–response curves are treated alike. Advantages and disadvantages of safety factors in the risk assessment process are listed in the following material:

Advantages

1) Simple application
2) Ease of understanding
3) Flexibility of use
4) Use of expert judgment

Disadvantages

1) Uncertainties of threshold values and size of safety factor
2) No risk comparison possible
3) Slope of dose–response curve not adequately considered
4) Experimental NOELs are dependent on group size in the animal toxicity testing and end point selected

5.2.4.2. Risk Estimation Techniques for Nonthreshold Effects

The procedures outlined below are most often applied to the low-dose risk estimation of human or animal carcinogens. Since dose–response data are not available for effects in animals at doses relevant to human exposure, extrapolations are required for determining the potential human cancer risk. The methodology employs mathematical modeling to characterize the relationship between exposure and response or to place an upper bound on the dose–response relationship. Dose–response data, available from specific study situations (mainly animal bioassays using high doses of the chemical and, sometimes, heavily exposed population groups), are extrapolated to the often much lower exposures of the general population in order to calculate the possible risk. Therefore, cancer risk assessment generally involves extrapolating risk from the relatively high exposure levels employed in animal studies, or occupational studies where cancer responses can be measured, to risks at the relatively low exposure levels that are of environmental concern. However, since the majority of carcinogenicity experiments use only two or three doses, it is impossible to assess the shape of the dose-response with a reasonable degree of precision. Risk assessment must therefore rely on some arbitrary assumptions about the shape of the dose-response relationship at low doses. Risk estimates thus obtained are not true or actual risk but values obtained by extrapolation well below the range of experimental observations. Advantages and disadvantages of quantitative risk assessment are summarized below:

Advantages

1) Gives numerical values on risk that may be used for setting exposure limits
2) Permits the comparison of risks due to different chemicals
3) Provides a reasonable basis for setting exposure limits by identifying compounds with high risk

Disadvantages

1) Extrapolation of data obtained at high doses to low doses, relevant for human exposure using mathematical models, which are not based on cancer biology and pathophysiology
2) Mechanistic and kinetic data are not used for the risk estimation process
3) Expensive and time-consuming lifelong bioassay required

The risk extrapolation techniques used are based on several conservative default assumptions, some unsupported by any direct empirical evidence. Conservatism is introduced to ensure maximum protection for those exposed to presumed chemical carcinogens. These assumptions were adopted to achieve some consistency in the application of risk-estimating techniques. The major assumptions are (1) carcinogenic risks are estimated from the data obtained in the most sensitive animal model only using positive responses (data from bioassays that do not show a treatment-related increase in cancer incidence are ignored); (2) linear, nonthreshold dose-risk models are applied; (3) statistical upper confidence limits are used rather than best estimates; and (4) a linear dose-response curve is assumed at low doses.

5.2.4.3. Mathematical Models Used in High- to Low-Dose Risk Extrapolation

Mathematical models for quantifying human cancer risk from exposure to carcinogens were first developed in the 1950s. These models were based on the one-hit or the multistage model of chemical carcinogenesis. A probit model was proposed in 1961 for assessing low-dose risk by extrapolation. An improved probit model was introduced in the 1970s by the FDA for computing the level of carcinogenic chemicals permissible in food. This procedure included the suggestion that the dose causing a very low risk (one additional cancer in 1 000 000 exposed people) to be considered as a "virtually safe dose." Introduction of this mathematical model into the risk assessment process stimulated the development of a variety of other mathematical models for carcinogen risk assessment. The basis of all the above methods is to apply a mathematical model to the tumor incidence observed in a long-term animal bioassay.

A major problem in dose–response and risk extrapolation is the determination of an appropriate mathematical model to predict effects at hypothetical low levels of exposure. Several models have been developed, and most of them contain analytical functions that appear to fit the experimental dose range quite well and also contain a dose–response functionality. Most models differ in the functions used to estimate response in the very low dose range (i.e., as dose approaches zero). The mathematical models in common use in carcinogen risk assessment are the following:

1) The linear model
2) Statistical or distribution models: log–probit, Mantel–Bryan, logit, Weibull
3) Mechanistic models: one-hit (linear), multihit, multistage (Armitage–Doll), linearized multistage, Moolgavkar
4) Other models: statisticopharmacokinetic, time-to-tumor

Linear Model. Linear extrapolation involves the intersection of a straight line between the origin (zero dose) and the upper confidence limit of the response at the single, lowest experimental dose. This model is based on the assumption that the increase in tumor incidence by the applied xenobiotic augments an already ongoing process.

Distribution Models. These are based on mathematical functions of presumed population characteristics (i.e., on the assumption that every member of a population has a critical dosage or threshold below which the individual will not respond to the exposure in question). The *probit* *model* assumes that log dose–responses have a normal distribution. This model serves as the basis for the Mantel–Bryan risk extrapolation procedure. Other distribution models on which carcinogenicity dose–response models have been based include the *logit* and *Weibull models*.

Mechanistic Models. Such models are based on the currently presumed mechanisms of chemical carcinogenesis. Each model reflects the assumption that a tumor originates from a single cell. The concept underlying the one-hit model is that a tumor can be induced by exposure of DNA to a single molecule of a carcinogen. This model is essentially equivalent to assuming that the dose–response is linear in the low-dose region and as a consequence tends to produce very low calculated "virtually safe doses" compared with the other currently applied models. The *multi-hit model* is a generalized version of the one-hit model, which assumes that more than one hit is required at the cellular level to initiate carcinogenesis.

The biological justification for the *multistage* (*Armitage–Doll*) model is that cancer is assumed to be a multistage process that can be approximated by a series of multiplicative linear functions. The assumption is made that the effect of a chemical carcinogen occurs in multiple steps and that the effect of each step is additive. The dose–response predicted by this model is approximately linear at low doses, resulting in estimates of potential risk that are similar to those of the one-hit model.

The *Moolgavkar–Venson–Knudson model* attempts a more comprehensive consideration of the biologic processes of cancer formation than the other mechanistic models and may provide a more accurate estimate of human risk by reducing some of the interspecies uncertainties. It is based on a two-stage growth model, considers the birth and death of cells and the effect of cell proliferation on the number of available cells for malignant transformation, and assumes that two specific, irreversible, and rate-limiting mutational events are necessary for cancer development. This model may quantitatively consider major phenomena influencing cancer formation, such as cell proliferation, initiation/promotion, genetic predisposition, and synergism/antagonism. Moreover, model parameters may be obtained experimentally. At present, the major limitation of the application of this model is the lack of availability of many of the important biological parameters.

Other Models. The carcinogenicity of many chemicals is based on their biotransformation to reactive metabolites. The *statisticopharmacokinetic model* arises from a consideration of competing metabolic activation and deactivation (e.g., detoxification and DNA repair systems) reactions and estimates the "effective dose" (i.e., the level of reactive metabolites formed and interacting with critical macromolecules, rather than the administered dose).

A modification of the probit model relates it to the time at which a tumor is detected. For risk assessment, this *time-to-tumor model* uses the time to observance (latency) in addition to the proportion of animals bearing tumors at each dose.

5.2.4.4. Interpretation of Data from Chronic Animal Bioassays

The results of experimental animal studies may provide a variety of data in addition to a simple indication of the presence or absence of cancer. This information may be useful for the evaluation of potential human cancer risk and includes the number of neoplasms per animal, the number of different types of neoplasms observed, and the number of species affected. The organ or target tissue in which the carcinogenic response occur is also important because some rodents have extremely high and variable spontaneous incidence of certain tumor types. Where a chemical increases the number or accelerates the formation of neoplasms that occur spontaneously in high and variable incidence, the response carries less weight than the appearance of tumors with very low spontaneous rates. The time to development of tumor will also give an indication of potency. Further considerations of the interpretation of the obtained data are given in Section 4.8.

5.2.4.5. Problems and Uncertainties in Risk Assessment

The practice of using tumor incidences obtained in long-term animal experiments with high doses (MTD, maximum tolerated dose; see Section 4.8) for human cancer risk assessment and several of the conservative default assumptions used in the extrapolation processes have come under major criticism in the 1990s. Some of the assumptions made when using experimental

Table 5.3. Uncertainties in quantitative risk assessment and the application of scientifically based methods for reduction of these uncertainties

Uncertainty	Reduction by
Extrapolation between species	physiologically based toxicokinetic modeling, comparative studies on toxicokinetics
Extrapolation from high to low dose	toxicokinetic modeling, quantitative determination of dose–response for biochemical effects of xenobiotics (e.g., DNA adducts, genetic changes), mechanisms of carcinogenesis
Extrapolation from controlled experimental conditions to variable human situations	none
Conservatism	elucidation of mechanism of actions
Toxicity and increased cell proliferation	dose–response for cell proliferation and cytotoxic effects
Dosimetry	toxicokinetic modeling and physiologically based toxicokinetic modeling
Mixtures	research on mechanisms of interactions
Bioassay-inherent factors in risk assessment	expert judgment

animal data for human risk assessment are not directly testable experimentally. Nevertheless, these assumptions are widely used because risk assessment would be difficult or impossible without them [201].

However, recent results on the mechanisms of carcinogenesis have revealed a complex biological process with many variables; a comprehensive consideration of all these parameters by a single, generally applicable mathematical model is not possible. Moreover, observations such as nonlinear toxicokinetics and delineated mechanisms of action for nongenotoxic carcinogens have indicated that a more scientifically based approach to carcinogen risk assessment, likely in a case-by-case examination by expert panels, may be more appropriate for defining actual risk than mathematical modeling. The major points of controversy in risk extrapolation from animal experiments and possible solutions to reduce uncertainties are described in Table 5.3.

Extrapolation between Species. In the absence of information on the possible mechanism of carcinogenesis by a particular xenobiotic, the tumor incidences observed in an animal experiment are often assumed to be useful in assessing human risk. However, the differences in the physiology and anatomy of laboratory animals and humans are well recognized. For example, the life span of the laboratory rodent is approximately two years whereas that of humans is approximately seventy years. Cancer appears to develop in rodents over a time scale that is proportional to the life span, and it is generally assumed that this will apply to all chemicals being assessed. The assumption that humans will respond in a similar fashion to laboratory animals is frequently shown to be inappropriate. Mechanistic studies demonstrate that qualitative and quantitative aspects of toxicokinetics and biotransformation, DNA repair and tissue susceptibility, and immune and other defense mechanisms may explain observed differences in the response of laboratory species and humans to exposure to carcinogens [207]. Expert judgment may be required to assess the nature of the end point or the mechanisms of carcinogenic action of the chemical in the experimental animal and to decide whether they are relevant to humans.

Extrapolation from High to Low Dose. The use of very high doses in animal cancer bioassays and the required extrapolation from high to low dose are major points of criticism. Most cancer bioassays are performed in relatively small groups of rodents (between 50 and 100) with the administration of high doses. The highest dose selected is usually the MTD in order to have maximum sensitivity and to ensure that a carcinogenic response is not overlooked simply because the dose was too low. The pros and cons of cancer bioassays with high doses in animals are given below:

Information obtained

1) Definite identification of compounds with carcinogenic activity in the species
2) Information about relative potency of different chemicals
3) Information about the carcinogenic activity of the test chemical, when administered alone
4) Characterization of tumor types, target organs, presence or absence of dose-response, which permit comparison of different chemicals and help in establishing structure–activity relationships to improve predictive capabilities
5) Information about the lack of carcinogenicity of many chemicals to assist priority setting in public health

Limits

1) No direct information about effects at doses lower than dose studied
2) No information on the mechanism of carcinogenicity
3) No information on the effects of the test compound, when administered together with other chemicals (synergistic/antagonistic effects)
4) Use of high doses that may cause unspecific toxic effects contributing to carcinogenicity
5) Acute toxic effects that may prohibit long-term administration of specific chemicals in sufficiently high doses to cause tumors

Human exposure to carcinogens usually occurs at doses several orders of magnitude lower than those used experimentally. Clearly, the possible shape of the dose–response relationship is vitally important in establishing the likely effects at doses substantially below those in the observable range. Mathematical models used for extrapolation only give the upper limits of risk; the real risk may be somewhere between zero and the number calculated. Since the dose–response cannot be determined experimentally, the application of different mathematical models results in markedly different curve slopes in the low-dose range (Fig. 5.2), which influence risk assessment and the establishment of ADI values.

Only one study to determine the actual shape of the dose–response curve at low doses with group sizes large enough to constitute statistical significance was performed, the so-called ED_{01} study using the potent carcinogen 2-acetylaminofluorene as a model compound administered in the diet to mice. Urinary bladder and liver neoplasms were found to be related to

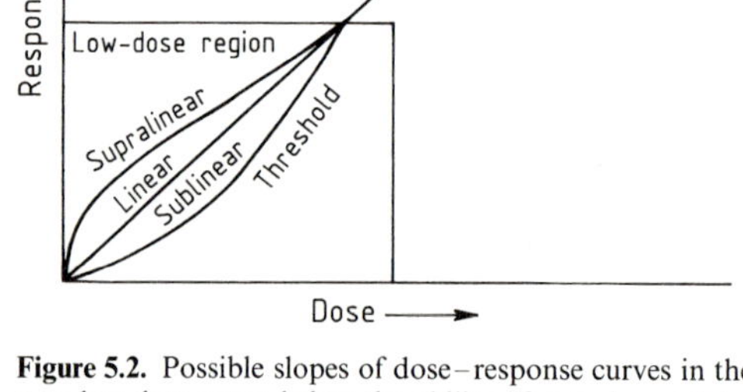

Figure 5.2. Possible slopes of dose–response curves in the very low dose range below the ability of experimental determination in cancer bioassays
● = Experimental data point

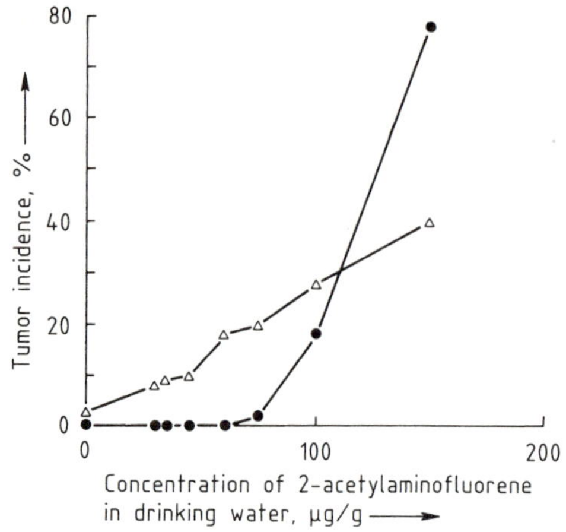

Figure 5.3. Dose–response data for the induction of liver (Δ) and urinary bladder (●) tumors induced by 2-acetylaminofluorene in mice after 24-month administration in diet

2-acetylaminofluorene administration. The tumor incidences obtained showed that even within one species and one specific carcinogen, major differences occur in the shape of the dose–response curve at low doses, depending on the target organ affected. The zero incidence of bladder tumors at low concentrations suggested the presence of a threshold for this end point, whereas the dose–response data for liver tumors were nearly linear at all concentrations and did not indicate a threshold (Fig. 5.3).

Extrapolation from Controlled Experimental Conditions to Variable Human Situations. In most animals studies, all critical factors are carefully controlled to guarantee the reproducibility of experimental observations. This is in contrast to the sometimes very large differences and time-dependent changes in the environmental exposure and other circumstances of the human population, which cannot be simulated in animal experiments. The complexity of these circumstances (i.e., changing life-style factors or workplaces) would necessitate an assessment of each human individually. Thus, risk assessment normally extrapolates directly from the animal data unless good evidence exists to suggest that an important confounding factor has been introduced. Besides the variable exposure situations, the human populations exposed to chemicals often differ in age, sex, and ethnic background and will certainly have a more heterogeneous genetic makeup than experimental animals. For the majority of carcinogenic risk assessments, individual variability can be taken into account only by the application of safety factors to compensate for these uncertainties.

Dosimetry. The problems of estimating environmental or occupational exposure to potential carcinogens have been described above. Equally difficult problems exist in extrapolating the doses applied in animal experiments to those that might be encountered by humans due to environmental or occupational exposure and in distinguishing between administered dose and effective dose.

Doses may be extrapolated from animal experiments to humans by

1) Expression of dose as a function of body weight (mg/kg or mmol/kg).
2) Expression of dose as a concentration in food or water (usually parts per million, i.e., μg/g or μL/L).
3) Expression of dose as a concentration in inhaled air (usually parts per million, i.e., μL/L).
4) Correction of dose for surface area. This is achieved in a complicated process by first raising the body weight of the laboratory animal to the power of $^2/_3$ or $^3/_4$. The correction converts doses that are expressed as concentrations in air or food to milligrams per kilogram of body weight in the laboratory species. The dose is then corrected for body surface area and converted back to concentrations in air or food by using appropriate conversion factors for humans. Corrections using surface area are based on the observation that metabolic rate is proportional to body surface area. For acute toxicity, this method may have some merit, but the rate of biotransformation may differently affect the potency of a carcinogen, depending on the role of biotransformation in detoxification or activation. Using correction factors for surface area provides data suggesting that humans may be more susceptible to potential carcinogens than laboratory animals compared to the data obtained by extrapolation on the basis of body weight.
5) Extrapolation assuming that the tissue dose is the primary determinant of carcinogenic response. This approach requires a study of the pathways of biotransformation and the kinetics of the pathway that generates the ultimate electrophile. In vitro studies may be necessary to obtain appropriate human data.

Further problems occur if the experiment is carried out by a protocol that is completely dissimilar from the human experience, for example, if animal exposure is carried out for a lifetime and human exposure is for a shorter period or if the animal experiment uses an exposure route that is not relevant for humans. As a general rule for chemical carcinogens, where the target tissue dose is presumably the most important determinant of the carcinogenic potency, the total body burden must be computed from the various routes of exposure to assess overall dose.

Mixtures. Cancer bioassays in animals are most often performed with a single chemical of definite high purity; exposure of experimental animals to other chemicals or other confounding factors such as vector-based disease is carefully avoided. In contrast, humans are continuously exposed to mixtures of chemicals and other agents. Thus, even when animals or humans are exposed to two or more chemicals, further complications of data interpretation are introduced. In general, the response of the animal and the human will depend on whether the chemicals' activities are interactive or not and whether the effects of the chemicals are similar or not. If the chemicals are not interactive and the response is similar, an additive effect may be assumed to occur. If they are not interactive, the overall response may be less than additive. In addition, a variety of metabolic or infectious diseases in the course of a human's life may have major impact on the development of a certain tumor due to exposure to the chemical under question. These real-life factors cannot be simulated in the experimental situation [208].

Bioassay-Inherent Factors in Risk Assessment. A variety of biological factors exist that influence the assessment of carcinogenic hazard. Among these are the quality of the experiment, including the quality of the pathology, environmental control of animal facilities, estimation and standardization of chemical administration, purity of the chemical admistered, and many other factors that together make up compliance with good laboratory practice. A further problem may be encountered if differences occur in outcome from different experiments. In the majority of cases, some aspect of an exceptional experiment may explain why the response was different (e.g., use of a different strain of animal). Since only positive data are used and negative data are ignored for carcinogenic risk assessment purpose to provide a conservative estimate of risk, this factor may also contribute to an overestimation of risk.

Conservatism and the Mechanisms of Chemical Carcinogenesis. To provide maximum protection, a conservative approach to risk assessment is preferred. This approach assumes that a single molecule of a carcinogen can interact with DNA and produce cancer; therefore, there can be no threshold for chemical carcinogenesis. Moreover, selection of the most sensitive response, irrespective of mechanism of action, and the assumption that the dose–response is linear at low doses may overestimate risk. This overestimation may be due to the unknown degree of conservatism at each step and to the amplification of previous bias in the assumption by the next step. The magnitude of overstatement of risk by these conservative assumptions is unknown, but is claimed to amount to several orders of magnitude.

Role of Toxicity and Increased Cell Proliferation. Critics point out that the high doses used in rodent carcinogenicity studies may have nonspecific effects such as an increase in the rate of cell proliferation due to cytotoxic effects [209]. These effects can be unique to high doses, and mitogenesis may itself be mutagenic in many ways, either by errors in replication or by conversion of endogenous DNA damage (e.g., by oxidative processes) and exogenous DNA damage to mutations before repair can occur. Moreover, sustained increases in the rate of cell proliferation may also yield secondary mutational events and could be important in the promotional phase of carcinogenicity by increasing the clonal expansion of initiated cells and thus increasing the chance that multiple critical mutational events will occur [210]. Thus, cell division will increase the chance of tumor formation. In this case, the tumor incidence of the same chemical applied at much lower doses is likely to be much lower than a linear model would predict and may be zero [173], [211], [212]. Therefore, adding methods for determining cell division to animal cancer bioassays and applying the results obtained to estimate low-dose risks more adequately would be important. Cell proliferation has been implicated as a major contributor to the carcinogenicity of several chemicals such as phenobarbital, 1,4-dichlorobenzene, *d*-limonene, and peroxisome proliferators.

Selection of Mathematical Model. The various models used in risk extrapolation fit the experimentally observed data well; however, they predict widely differing potential risks at low doses. Depending on the data used in the calculation, the predicted risks may differ by several orders of magnitude. These differences are inherent in the application of mathematical models since only two or three doses are used in the experimental dose groups.

5.3. Future Contributions of Scientifically Based Procedures to Risk Assessment and Qualitative Risk Assessment for Carcinogens

Since quantitative cancer risk assessment can neither claim to be a scientific basis for development of historical models nor attempt to incorporate the large amount of scientific data on the mechanism of carcinogenesis into the assessment of risk, qualitative approaches considering all relevant data may be the best available solution for risk assessment. This approach, referred to a "weight-of-evidence determination of risk" is increasingly emphasized by regulatory authorities. The weight-of-evidence approach includes critical evaluation of the animal bioassay and all other available information on adverse effects of the chemical together with biotransformation, toxicokinetics, and expert judgment [213].

The criticisms outlined above have demonstrated the need for further refinement of the extrapolation procedures by toxicokinetics and studies on the mechanisms of tumor formation. The role of animal experiments in predicting the potential risk of human carcinogen exposure will likely remain an important step in the risk assessment process, despite the obvious limitations of this approach [214], [215]. However, the data obtained with animals must be interpreted with caution and in light of other data, both quantitative and qualitative, on the adverse effects of the compound (Table 5.4).

Application of Toxicokinetic Models in Risk Assessment. Information about the way in which a chemical is absorbed, biotransformed, and excreted may be critical in extrapolating the rele-

Table 5.4. Descriptive dimensions proposed as a framework to facilitate the use of mechanistic data in evaluation of carcinogenic risk to humans

Data set	Example of information required
Evidence of genotoxicity	DNA adduct formation, mutagenicity, bioactivation
Evidence of effects on the expression of genes relevant to the process of carcinogenesis	alterations of the structure or quantity of product of a protooncogene or suppressor gene
Evidence for effects on cell behavior	mitogenesis, cell proliferation, hyperplasia
Evidence of time and dose–response relationships and interactions	initiation, promotion, progression

vance of the results obtained in experimental animals to humans [120], [121]. For the purposes of risk extrapolation, the administered dose is generally assumed to be proportional to the effective dose. However, many chemicals are known to be carcinogenic only after they have been activated to reactive electrophiles by enzymatic reactions. The amount of reactive metabolite formed might not be directly related to dose because saturable enzymatic processes are involved in both bioactivation and detoxification. The extent of bioactivation and detoxification may therefore be highly dose dependent, and the relationship between administered dose and target dose may not be proportional at all applied doses. For example, when high-affinity, low-capacity enzymes catalyze the detoxification of a xenobiotic and low-affinity, high-capacity enzymes catalyze its bioactivation, the amount of reactive metabolite formed is likely highly dependent on the administered dose (i.e., a tenfold higher administered dose may result in a 100-fold higher effective target dose). Moreover, cosubstrates required by the enzymes catalyzing bioactivation or detoxification may be depleted. After the reactive metabolite is formed, it is often deactivated by a second enzyme such as epoxide hydrolase or glutathione *S*-transferase. These enzymes can also be saturated. The reactive metabolites that are not destroyed by these detoxification pathways may bind to DNA. The metabolites bound to DNA can be removed by saturable DNA repair systems. These effects are referred to as nonlinear toxicokinetics. Nonlinear kinetics are also seen with chemicals inhibiting or inducing drug-metabolizing enzymes. Application of a carcinogen at high doses may result in the induction of enzymes catalyzing its bioactivation and may leave detoxifying enzymes unaffected. Thus, the expected steady-state levels of reactive intermediates present in the cell and capable of binding to DNA are expected to be disproportionately higher after application of high doses, which induce biotransformation enzymes, than after low dose, which leave the levels of biotransformation enzymes unaffected. Moreover, the metabolism of a xenobiotic may be changed as a consequence of the effects of its long-term administration [216].

To estimate the dose-dependent relationship between the dose administered and the effective dose, toxicokinetic models incorporating saturable processes have been applied. The examples in Figure 5.4 show the theoretically derived relationships between administered dose and effective dose for the same chemical in animals by varying the kinetic parameters for bioactivation, detoxification, and DNA repair [217]: saturation of enzymatic bioactivation (A); saturation of detoxification and activation (B); and saturation of activation, detoxification, and DNA repair (C).

The "hockey stick" shape of the dose-response curve may remain unnoticed in animal experiments with high dose. The nonlinear correlation between effective dose and administered dose in the low, relevant dose region for human exposure may result in a decrease in the potential risk of exposure when compared to risk estimation based on experiments with high doses. Effective doses may be determined by measuring the amount of DNA adducts formed after administration of a carcinogen based on the assumption that for genotoxic agents the carcinogenic response is related to the extent of DNA adduct formation in target tissues. Unfortunately, nonlinear dose–response curves for effective doses have not yet been observed experimentally; all attempts to determine the dose-dependent formation of DNA adducts of potent carcinogens have shown linear relationships between administered dose and effect [218], [219]. However, the increased sensitivity of analytical instruments currently available offers the opportunity to study the dose–effect curve for a wide range of carcinogens transformed to intermediates with different electrophilic reactivity in the low-dose range by quantifying the dose-dependent concentrations of DNA lesions. Quantitation of

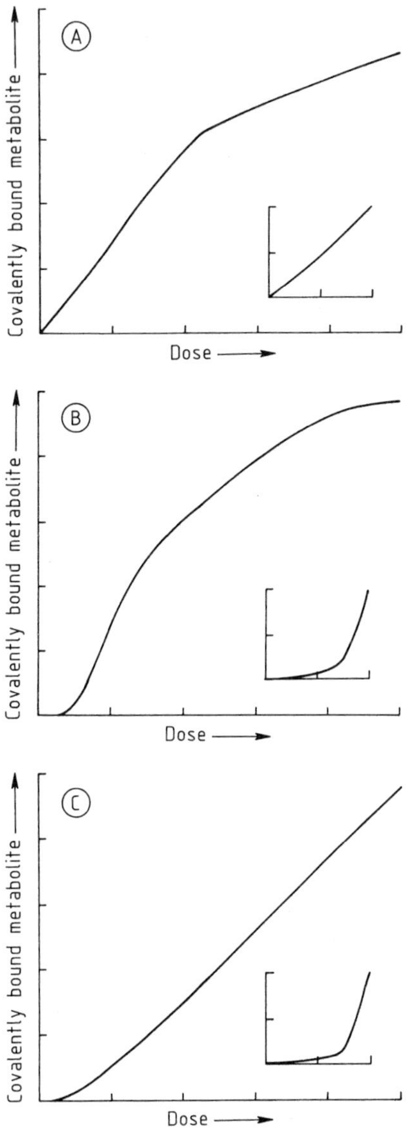

Figure 5.4. Possible relationships between administered dose and effective dose for the same data set including correction for nonlinear toxicokinetics
A) Saturation of enzymatic bioactivation; B) Saturation of detoxification and activation; C) Saturation of activation, detoxification, and DNA repair
The shapes of the curves in the low-dose range are shown in detail in the small sections.

xenobiotic–hemoglobin adducts and chromosomal abnormalities in lymphocytes (e.g., sister-chromatid exchange frequency) offers other examples of biological markers that may prove useful in the definition of dose–response curves. The application of these procedures offers the advantage of obtaining quantitative information on a nonstochastic effect thought to be relevant in the mechanisms of carcinogenesis and thus in risk assessment.

As noted in Section 5.2.4.5, dosimetry may be a major cause of uncertainities in risk extrapolation from animals to humans. In addition to saturable enzymatic reactions, the effective dose of a carcinogenic chemical may also be influenced by species differences in absorption, distribution, and elimination, which may also be influenced by the dose administered.

To account for such factors, physiologically based pharmacokinetic models are increasingly used in the process of risk assessment. The principal purpose of the application of physiologically based pharmacokinetic models is to predict the concentration of carcinogen at the target site and describe the relationship between administered dose and target dose over a range of concentrations. By application of these models, a more accurate dose extrapolation is possible over broad ranges and may also incorporate nonlinearities in bioactivation and detoxification. Since the principles employed in the development of these models apply across species, the definition of the relationship between administered dose and effective dose and the important rate processes that cause a deviation from linearity permits a more accurate species extrapolation. The application of known physiological parameters may also enable the prediction of target organ concentrations in humans when direct measurements are not possible and may permit the comparison of different routes of application for effective dose. Physiologically based pharmacokinetic models are increasingly used to support quantitative risk assessments. Due to the risk estimates based on effective dose rather than external dose levels and the consideration of potential nonlinear relationships, an overestimation of risk obtained by linear extrapolations may be overcome and result in more scientifically founded risk assessments.

Mechanisms of Carcinogenicity and the Risk Assessment Process. Research on the molecular effects of particular agents has increased understanding of the mechanisms of carcinogenicity. A range of biological processes has been implicated in carcinogenesis. Some of these mechanisms may be common for most carcinogens; others may be restricted to particular classes of chemicals or specific circumstances. Despite the

Table 5.5. Examples of nongenotoxic carcinogens and their presumed mechanisms of action

Class of nongenotoxic carcinogen	Typical examples	Presumed mechanism of action
Peroxisome proliferators	diethylhexylphthalate	receptor-mediated increase in gene transcription, oxidative stress
α_{2u}-Globulin-binding agents	unleaded gasoline, limonene, decalin	regenerative cell proliferation due to cytotoxicity
Phenobarbital		mitogenic activity
Chlorinated dioxins	2,3,7,8-tetrachlorodibenzodioxin	receptor-mediated increase in gene transcription
Hormones	17-α-ethynylestradiol	receptor-mediated increase in gene transcription

incomplete knowledge of carcinogenesis and a limited understanding of many processes, mechanism-based decisions are increasingly introduced in risk assessment [220].

The emerging field of molecular toxicology may hold promise for the future of human risk assessment, and the application of methods developed for molecular biology may offer more accurate ways of assessing human risk. The most promising applications of mechanistic results for risk assessment are nongenotoxic carcinogens [221]. Many xenobiotics are carcinogenic in animals but do not cause any detectable mutagenicity in in vitro studies. In contrast to other chemical carinogens that affect multiple organs or both sexes of rat and mouse, many nongenotoxic carcinogens affect only a single organ in a single sex of a single species [124]. Numerous examples exist of xenobiotics or treatment regimens that produce tumors in animal experiments unlikely to involve the interaction of the xenobiotic or its metabolite with DNA. Examples include the induction of subcutaneous sarcomas by repeated injection of glucose or saline solutions, the induction of skin cancer after chronic skin damage, and the induction of bladder cancer by the implantation of solid materials [222]. Recent experimental evidence also suggests that a number of xenobiotics that do not damage DNA, may induce cancer in specific organs after systemic administration. Examples include chemicals binding to the circulating protein α_{2u}-globulin [223], peroxisome proliferators, and several structurally unrelated compounds interacting with receptors [224] (Table 5.5).

Definition of the terms genotoxic and nongenotoxic is presented by BUTTERWORTH [124]: "A genotoxic agent is one for which a primary biological activity of the chemical or a metabolite is alteration of the information encoded in the DNA. These can be point mutations, insertions, deletions or changes in chromosome structure or number. Chemicals exhibiting such activity can usually be identified by assays that measure reactivity with the DNA, induction of mutations, induction of DNA repair or cytogenetic effects. Nongenotoxic chemicals are those that lack genotoxicity as a primary biological activity. While these agents may yield genotoxic events as a secondary result of other induced toxicity, such as forced cellular growth, their primary action does not involve reactivity with the DNA."

Xenobiotics such as unleaded gasoline, limonene, and 1,4-dichlorobenzene induce renal cancer in male rats, but not in female rats or in either sex of mice [225]. Metabolites of these xenobiotics bind to the circulating protein α_{2u}-globulin, whose synthesis in the rat is under the control of androgens. The modified α_{2u}-globulin is concentrated in the kidney proximal tubules and is not degradable by the processes responsible for degradation of α_{2u}-globulin. Thus, accumulation of the modified protein causes cytotoxicity and regenerative cell proliferation, which has been implicated as the cause of tumor formation in the kidney.

Tumors in organs whose function is regulated by the endocrine system are often observed after hormonal therapies in humans and hormone treatment in animals. Because of the absence of genotoxicity of most hormones, nongenotoxic processes involving receptor-mediated transcription and activation or repression of specific genes have been implicated in hormonal carcinogenesis [226].

For both examples, threshold mechanisms may be postulated. Kidney cancer induced by α_{2u}-globulin-binding agents requires cytotoxicity and cell death. Nontoxic low concentrations of these xenobiotics may not be tumorigenic. For receptor-mediated processes a disproportional relationship between receptor occupancy and hormonally mediated cancer is also considered likely [227]–[229]. Hence, any assessment of the carcinogenic risk of these chemicals must take into account the mechanisms by which they produce their effects.

The use of molecular toxicology in risk assessment is also advocated by the International Agency for Research on Cancer [220]. A recent monograph on the identification of cancer risks by mechanistic investigations suggests that "when available data on mechanisms are thought to be relevant to evaluation of the carcinogenic risk of an agent to humans, they should be used in making the overall evaluation, together with the combined evidence for animal and/or human carcinogenicity." The consensus report further states that no definite guidelines for the inclusion of mechanistic data in the evaluation of carcinogens can be elaborated. However, a range of options are available: "First, information concerning mechanisms of action may confirm a particular level of carcinogen classification as indicated on the basis of epidemiological and/or animal carcinogenicity data. Second, for a particular agent, strong evidence for a mechanism of action that is relevant to carcinogenicity in humans could justify 'upgrading' its overall evaluation. Third, an overall evaluation of human cancer hazard on the basis of animal carcinogenicity data could be downgraded by strong evidence that the mechanism responsible for tumor growth in experimental animals is not relevant to humans. In keeping with the goal of public health, priority must be given to the demonstration that the mechanism is irrelevant to humans."

6. References

[1] D. Henschler, *Trends Pharmacol. Sci. Fest. Suppl* (1985) 26–28.
[2] W. Dekant, H.-G. Neumann: *Tissue Specific Toxicity: Biochemical Mechanisms*, Academic Press, London 1992.
[3] Zacharewski et al., *Arch. Biochem. Biophys.* **272** (1989) 344–355.
[4] D. O. Chanter, R. Heywood, *Toxicol. Lett.* **10** (1982) 303–307.
[5] Y. Alarie, *EHP Environ. Health Perspect.* **42** (1981) 9–13.
[6] H. Druckrey, D. Schmähl, W. Dischler, A. Schildbach, *Naturwissenschaften* **49** (1962) 217–228.
[7] P. H. Bach, C. P. Ketley, I. Ahmed, M. Dixit, *Food Chem. Toxicol.* **24** (1986) 775–779.
[8] P. D. Anderson, L. J. Weber, *Toxicol. Appl. Pharmacol.* **33** (1975) 471–483.
[9] W. W. Weber, D. W. Hein, *Pharmacol. Rev.* **37** (1985) 25–79.
[10] T. Green, M. S. Prout, *Toxicol. Appl. Pharmacol.* **79** (1985) 401–411.
[11] C. L. Alden, *Toxicol. Pathol.* **13** (1985) 135–140.
[12] L. R. Pohl, J. W. George, H. Satoh, *Drug Metab. Dispos.* **12** (1984) 304–308.
[13] N. I. Sax, R. J. Lewis: *Dangerous Properties of Industrial Materials*, 7th ed., Van Nostrand Reinhold, New York 1988.
[14] E. J. Calabrese, *J. Pharm. Sci.* **75** (1986) 1041–1067.
[15] R. T. Williams, *Biochem. Soc. Trans.* **2** (1974) 359–367.
[16] H. B. Hucker, *Annu. Rev. Pharmacol.* **10** (1970) 99–116.
[17] M. VandenBerg, J. DeJongh, H. Poiger, J. R. Olson, *CRC Crit. Rev. Toxicol.* **24** (1994) 1–74.
[18] R. C. Lind, A. J. Gandolfi, P. d. I. M. Hall, *Anesthesiology* **70** (1989) 649–653.
[19] R. A. van Dyke, J. A. Gandolfi, *Drug Metab. Dispos.* **2** (1974) 469–476.
[20] I. G. Sipes et al., *J. Pharmacol. Exp. Ther.* **214** (1980) 716–720.
[21] A. Turturro, R. Hart, *Exp. Gerontol.* **27** (1992) 583–592.
[22] W. C. Willet, *Science (Washington, D.C. 1883)* **264** (1994) 532–537.
[23] K. O. Lindros, Y. A. Cai, K. E. Penttila, *Hepatology (N.Y.)* **12** (1990) 1092–1097.
[24] S. J. Singer, G. L. Nicolson, *Science (Washington, D.C. 1883)* **175** (1972) 720–731.
[25] M. Eisenberg, S. M. McLaughlin, *Residue Rev.* **31** (1976) 1–8.
[26] R. C. Scott et al., *Br. J. Dermatol.* **115** (1986) 47–48.
[27] P. H. Dugard, *Food Chem. Toxicol.* **24** (1986) 749–753.
[28] T. R. Bates, M. Gibaldi in J. Swarbick (ed.): *Current Concepts in the Pharmaceutical Sciences: Biopharmaceutics*, vol. 1, Lea and Febiger, Philadelphia 1970.
[29] A. R. Brody, *EHP Environ. Health Perspect.* **100** (1993) 21–30.
[30] J.-P. Tillement, E. Lindenlaub: *Protein Binding and Drug Transport*, F. K. Schattauer Verlag, Stuttgart 1986.
[31] M. S. Rose, *Br. Med. Bull.* **25** (1967) 227–234.
[32] U. Westphal: *Steroid-protein Interactions*, Springer Verlag, New York 1971.
[33] M. P. Waalkes, P. L. Goering, *Chem. Res. Toxicol.* **3** (1990) 281–288.
[34] R. A. Goyer, *EHP Environ. Health Perspect.* **100** (1993) 177–187.
[35] F. P. Guengerich, T. L. Macdonald, *Acc. Chem. Res.* **17** (1984) 9–16.
[36] G. S. Yost, A. R. Buckpitt, R. A. Roth, T. L. McLemore, *Toxicol. Appl. Pharmacol.* **101** (1989) 179–195.
[37] F. M. Guengerich, *Annu. Rev. Pharmacol. Toxicol.* **29** (1989) 241–264.
[38] F. P. Guengerich, *J. Biol. Chem.* **266** (1991) 10019–10022.
[39] F. P. Guengerich, *Life Sci.* **50** (1992) 1471–1478.
[40] F. P. Guengerich, *FASEB J.* **6** (1992) 745–748.
[41] L. S. Kaminsky, M. J. Fasco, *CRC Crit. Rev. Toxicol.* **21** (1992) 407–422.
[42] P. Guengerich, *FASEB J.* **6** (1992) 667–668.
[43] D. R. Koop, *FASEB J.* **6** (1992) 724–730.
[44] D. M. Ziegler, *Drug Metab. Rev.* **19** (1988) 1–32.

[45] R. N. Hines et al., *Toxicol. Appl. Pharmacol.* **125** (1994) 1–6.
[46] R. E. Tynes, R. M. Philpot, *Mol. Pharmacol.* **31** (1991) 569–574.
[47] T. E. Eling, D. C. Thompson, G. L. Foureman, M. F. Hughes, *Annu. Rev. Pharmacol. Toxicol.* **30** (1990) 1–45.
[48] B. J. Smith, J. F. Curtis, T. E. Eling, *Chem. Biol. Interact.* **79** (1991) 245–264.
[49] T. Eling et al., *Drug Metab. Rev.* **14** (1983) 1023–1053.
[50] W. L. Smith, L. J. Marnett, *Biochim. Biophys. Acta.* **1083** (1991) 1–17.
[51] B. H. Bach, J. W. Bridges, *CRC Crit. Rev. Toxicol.* **15** (1985) 217–329.
[52] R. J. Duescher, A. A. Elfarra, *J. Biol. Chem.* **267** (1992) 19859–19865.
[53] P. D. Josephy, T. E. Eling, R. P. Mason, *Mol. Pharmacol.* **23** (1983) 461–466.
[54] A. J. Kettle, C. C. Winterbourn, *J. Biol. Chem.* **267** (1992) 8319–8324.
[55] H. Thomas, F. Oesch, *ISI Atlas Sci.: Biochemistry* **1** (1988) 287–291.
[56] B. Burchell, W. H. Coughtrie, *Pharmacol. Ther.* **43** (1989) 261–289.
[57] T. R. Tephly, B. Burchell, *Trends Pharmacol. Sci.* **11** (1990) 276–279.
[58] K. W. Bock, B. S. Bock-Hennig, G. Fischer, L. W. D. Ullrich in H. Greim et al. (eds.): *Biochemical Basis of Chemical Carcinogenesis*, vol. 1, Raven Press, New York 1984, pp. 13–22.
[59] G. J. Mulder: *Sulfation of Drugs and Related Compounds*, CRC Press, Boca Raton, Fla., 1981.
[60] D. A. P. Evans, *Pharmacol. Ther.* **42** (1989) 157–234.
[61] I. M. Arias, W. B. Jakoby: *Glutathione: Metabolism and Function*, Raven Press, New York 1976.
[62] B. Ketterer, D. J. Meyer, A. G. Clark in H. Sies, B. Ketterer (eds.): *Glutathione Conjugation: Mechanisms and Biological Significance*, Academic Press, London 1988, pp. 73–135.
[63] R. Morgenstern et al., *Biochem. Pharmacol.* **33** (1984) 3609–3614.
[64] J. L. Cmarik et al., *Cancer Res.* **50** (1990) 2747–2752.
[65] J. Caldwell, W. B. Jakoby: *Biological Basis of Detoxification*, Academic Press, New York 1983.
[66] M. W. Anders: *Bioactivation of Foreign Compounds*, Academic Press, Orlando 1985.
[67] M. W. Anders in I. M. Arias et al. (eds.): *The Liver: Biology and Pathology*, 2nd ed., Raven Press, New York 1988, pp. 389–400.
[68] G. M. B. StClair et al., *Chem. Res. Toxicol.* **1** (1988) 179–185.
[69] F. P. Guengerich, D. C. Liebler, *CRC Crit. Rev. Toxicol.* **14** (1985) 259–307.
[70] D. C. Liebler, F. P. Guengerich, *Biochemistry* **22** (1983) 5482–5489.
[71] L. R. Pohl et al., *Drug Metab. Dispos.* **9** (1981) 334–339.
[72] M. W. Anders, W. Dekant: *Conjugation-dependent Carcinogenicity and Toxicity of Foreign Compounds*, Academic Press, San Diego 1994.
[73] T. J. Monks et al., *Toxicol. Appl. Pharmacol.* **106** (1990) 1–19.
[74] M. W. Anders, W. Dekant, S. Vamvakas, *Xenobiotica* **22** (1992) 1135–1145,
[75] W. Dekant, S. Vamvakas, M. W. Anders in W. Dekant, H. G. Neumann (eds.): *Tissue Specific Toxicity: Biochemical Mechanisms*, Academic Press, London 1992, pp. 163–194.
[76] S. D. Aust et al., *Toxicol. Appl. Pharmacol.* **120** (1993) 168–178.
[77] B. D. Goldstein, G. Witz, *Free Radicals Res. Commun.* **11** (1990) 3–10.
[78] B. D. Goldstein, B. Czerniecki, G. Witz, *EHP Environ. Health Perspect.* **81** (1989) 55–57.
[79] J. P. Kehrer et al., *Toxicol. Appl. Pharmacol.* **95** (1988) 349–362.
[80] P. A. Cerutti, B. F. Trump, *Cancer Cells* **3** (1991) 1–7.
[81] P. A. Cerutti, *Eur. J. Clin. Invest.* **21** (1991) 1–5.
[82] R. H. Burdon, V. Gill, C. Rice-Evans, *Free Radicals Res. Commun.* **11** (1990) 65–76.
[83] R. W. Pero, G. C. Roush, M. M. Markowitz, D. G. Miller, *Cancer Detect. Prev.* **14** (1990) 551–561.
[84] W. B. Jakoby, D. M. Ziegler, *J. Biol. Chem.* **265** (1990) 20715.
[85] H. Sies, A. Wendel: *Functions of Glutathione in Liver and Kidney*, Springer Verlag, New York 1978.
[86] D. M. Ziegler in W. B. Jacoby (ed.): *Enzymic Basis of Detoxification*, vol. I, Academic Press, New York 1980, pp. 201–227.
[87] E. Boyland, L. F. Chasseaud, *Adv. Enzymol. Relat. Areas Mol. Biol.* **32** (1969) 173.
[88] T. A. Fjellstedt, R. H. Allen, B. K. Duncan, W. B. Jakoby, *J. Biol. Chem.* **248** (1973) 3702–3707.
[89] R. C. James, R. D. Harbison, *Biochem. Pharmacol.* **31** (1982) 1829–1835.
[90] L. I. McLellan, C. R. Wolf, J. D. Hayes, *Biochem. J.* **258** (1989) 87–93.
[91] J. A. Hinson, D. W. Roberts, *Annu. Rev. Pharmacol. Toxicol.* **32** (1992) 471–510.
[92] S. D. Nelson, P. G. Pearson, *Annu. Rev. Pharmacol. Toxicol.* **30** (1990) 169–195.
[93] L. R. Pohl, *Chem. Res. Toxicol.* **6** (1993) 786–793.
[94] E. R. Stadtman, C. N. Oliver, *J. Biol. Chem.* **266** (1991) 2005–2008.
[95] E. R. Stadtman, *Science* (*Washington, D.C. 1883*) **257** (1992) 1220–1224.
[96] H. Esterbauer, H. Zollner, R. J. Schaur, *ISI Atlas Sci.: Biochemistry* **1** (1988) 311–317.
[97] H. Kappus, *Biochem. Pharmacol.* **35** (1986) 1–6.
[98] J. L. Farber, *Chem. Res. Toxicol.* **3** (1990) 503–508.
[99] J. P. Kehrer, D. P. Jones, J. J. Lemasters, J. L. Farber, H. Jaeschke, *Toxicol. Appl. Pharmacol.* **106** (1990) 165–178.
[100] K. Hemminki, *Carcinogenesis* (*London*) **14** (1993) 2007–2012.
[101] K. W. Bock, H.-P. Lipp, B. S. Bock-Hennig, *Xenobiotica* **20** (1990) 1101–1111.
[102] D. R. Koop, D. J. Tierney, *Bioassays.* **12** (1990) 429–435.
[103] L. Wu, J. P. J. Whitlock, *Proc. Natl. Acad. Sci. U.S.A.* **89** (1992) 4811–4815.
[104] J. P. J. Whitlock, *Chem. Res. Toxicol.* **6** (1993) 754–763.
[105] B. Testa, *Xenobiotica* **20** (1990) 1129–1137.
[106] B. Testa, P. Jenner, *Drug Metab. Rev.* **12** (1981) 1–117.
[107] J. R. Halpert, F. P. Guengerich, J. R. Bend, M. A. Correia, *Toxicol. Appl. Pharmacol.* **125** (1994) 163–175.
[108] M. Blum et al., *DNA Cell Biol.* **9** (1990) 193–203.
[109] M. E. Andersen, *CRC Crit. Rev. Toxicol.* **11** (1981) 105–150.

[110] M. W. Anders et al., *Renal Disposition and Nephrotoxicity of Xenobiotics*, Academic Press, San Diego 1993.
[111] M. W. Anders, *Kidney Int.* **18** (1980) 636–647.
[112] R. M. Brenner, F. C. J. Rector: *The Kidney*, W. B. Saunders Co., Philadelphia 1981.
[113] E. C. Foulkes, *Proc. Soc. Exp. Biol. Med.* **195** (1990) 1.
[114] B. A. Saville, M. R. Gray, Y. K. Tam, *Drug Metab. Rev.* **24** (1992) 49–88.
[115] C. D. Klaassen, J. B. Watkins, *Pharmacol. Rev.* **36** (1984) 1–67.
[116] C. D. Klaassen, *Drug Metab. Rev.* **5** (1976) 165–196.
[117] M. Gibaldi, D. Perrier: *Pharmacokinetics*, Marcel Dekker, New York 1982.
[118] J. G. Filser, *Arch. Toxicol.* **66** (1992) 1–10.
[119] D. A. Smith, *Drug Metab. Rev.* **23** (1991) 355–373.
[120] M. E. Andersen et al., *Toxicol. Appl. Pharmacol.* **87** (1987) 185–205.
[121] M. E. Andersen, D. Krewski, J. R. Withey, *Cancer Lett. (Shannon, Irel.)* **69** (1993) 1–14.
[122] R. B. Conolly, M. E. Andersen, *Annu. Rev. Pharmacol. Toxicol.* **31** (1991) 503–523.
[123] S. M. Cohen, L. M. Ellwein, *Science (Washington, D.C. 1883)* **246** (1990) 1007–1011.
[124] B. Butterworth, *Mutat. Res.* **239** (1990) 117–132.
[125] Q. Chen, B. N. Ames, *Toxicologist* **13** (1993) 110.
[126] S. Green, *Biochem. Pharmacol.* **43** (1992) 393–401.
[127] J. Abel, D. Höhr, H.-J. Schurek, *Arch. Toxicol.* **60** (1987) 370–375.
[128] S. Orrenius, P. Nicotera, *Arch. Toxicol. Suppl.* **11** (1987) 11–19.
[129] S. Orrenius, D. J. McConkey, G. Bellomo, P. Nicotera, *Trends Pharmacol. Sci.* **10** (1989) 281–285.
[130] E. C. Miller, J. A. Miller, *Pharmacol. Rev.* **18** (1966) 805–838.
[131] E. C. Miller, J. A. Miller, *Cancer* **47** (1981) 2327–2345.
[132] P. A. Cerutti, *Science (Washington, D.C. 1883)* **227** (1985) 375–381.
[133] International Agency for Research on Cancer: *Models, Mechanisms and Etiology of Tumour Promotion*, IARC Sci. Publ., Lyon 1984.
[134] C. C. Harris et al., *EHP Environ. Health Perspect.* **75** (1987) 109–119.
[135] P. Lawley, *Mutat. Res.* **213** (1989) 3–26.
[136] J. M. Bishop, *Science (Washington, D.C. 1883)* **235** (1987) 305–311.
[137] J. M. Bishop, *Cell* **64** (1991) 235–248.
[138] J. M. Bishop, *Cell* **42** (1985) 23–38.
[139] J. A. Boyd, J. C. Barrett, *Mol. Carcinogen.* **3** (1991) 325.
[140] A. G. Knudson, *Cancer Res.* **45** (1985) 1437–1443.
[141] C. C. Harris, *Science (Washington, D.C. 1883)* **262** (1993) 1980–1981.
[142] J. Ashby, R. W. Tennant, *Mutat. Res.* **257** (1991) 229–306.
[143] R. L. Brinster in T. H. Shepard, J. R. Miller, M. Marois (eds.): *Methods for Detection of Environmental Agents that Produce Congenital Defects*, Elsevier New York, 1975, pp. 113–124.
[144] J. G. Wilson: *Environment and Birth Defects*, Academic Press, New York 1973.
[145] J. M. Manson, L. D. Wise in M. O. Amdur, J. Doull, C. D. Klaassen (eds.): *Casarett and Doull's Toxicology. The Basic Science of Poisons*, Pergamon Press, New York 1991, pp. 226–254.
[146] R. P. Bolande in J. G. Wilson, F. C. Fraser (eds.): *Handbook of Teratology*, vol. 2, Plenum Press, New York 1977, pp. 293–328.
[147] S. Ivankovic: "Perinatal Carcinogenesis, DHEW Publication no. (NIH) 7a–1063", *Nat. Cancer Inst. Monogr.* **51** (1979) 103–116.
[148] D. Neubert, H. J. Barrach, H. J. Merker, *Curr. Top. Pathol.* **69** (1980) 242–324.
[149] I.-D. Adler, *Teratog. Carcinog. Mutagen.* **3** (1983) 321–343.
[150] I. Theiss, A. Basler, G. Röhrborn, *Teratog. Carcinog. Mutagen.* **3** (1983) 219–230.
[151] R. M. Pratt, D. S. Salomon in M. R. Juchau (ed.): *The Biochemical Basis of Chemical Teratogenesis*, Elsevier, New York 1981, pp. 179–1799.
[152] L. B. Holmes, R. L. Trelstad, *Teratology* **20** (1979) 289–296.
[153] W. J. Scott, C. A. Duggan, C. M. Schreiner, M. D. Collins, *Toxicol. Appl. Pharmacol.* **103** (1990) 238–254.
[154] H. L. Gurtoo et al., *Int. J. Cancer* **31** (1983) 29–37.
[155] E. Schlede, U. Mischke, R. Roll, D. Kayser, *Arch. Toxicol.* **66** (1992) 455–470.
[156] M. J. Van den Heuvel, A. D. Dayan, R. O. Shillaker, *Hum. Toxicol.* **6** (1987) 279–291.
[157] J. H. Draize, G. Woodard, H. O. Calvery, *J. Pharmacol. Exp. Ther.* **82** (1944) 377–390.
[158] B. Berner, E. R. Cooper in A. F. Kydonieu, B. Berner (eds.): *Transdermal Delivery of Drugs*, CRC Press, Boca Raton 1987, pp. 107–130.
[159] J. A. Shatkin, H. S. Brown, *Environ. Res.* **56** (1991) 90–108.
[160] B. Ballantyne, T. Marrs, P. Turner: *General and Applied Toxicology*, Macmillan Press, New York 1993.
[161] C. A. Reinhardt, M. Aeschbacher, M. Bracker, J. Spengler in A. M. Goldberg (ed.): *In Vitro Toxicology—Approaches to Validation. Alternative Methods in Toxicology*, vol. 5, Mary Ann Liebert, New York 1987, pp. 463–470.
[162] J.-F. Regnier, C. Imbert, J.-C. Bautounet, *Altern. Lab. Anim.* **22** (1994) 39–50.
[163] P. A. Botham et al., *Food Chem. Toxicol.* **29** (1991) 275–286.
[164] G. M. Henningen in D. W. Hobson (ed.): *Dermal and Ocular Toxicology. Fundamentals and Methods*, CRC Press, Boca Raton 1991, pp. 153–192.
[165] V. H. Frankos, *Fundam. Appl. Toxicol.* **5** (1985) 615–622.
[166] E. M. Faustman, *Mutat. Res.* **205** (1988) 355–384.
[167] G. P. Daston, R. A. D'Amato in M. I. Mehlman (ed.): *Benchmarks: Alternative methods in Toxicology*, Princeton Sci. Publ., Princeton, N.Y., 1989, pp. 77–109.
[168] V. K. Gombar et al., *Quant. Struct. Act. Relat. Pharmacol. Chem. Biol.* **10** (1990) 306–332.
[169] D. B. Clayson, F. Iverson, R. Mueller, *Teratog. Carcinog. Mutagen.* **11** (1991) 279–296.
[170] J. K. Haseman, A. Lockhart, *Fundam. Appl. Toxicol.* **22** (1994) 382–391.
[171] D. G. Hoel et al., *Carcinogenesis (London)* **9** (1988) 2045–2052.
[172] B. N. Ames, R. Magaw, L. S. Gold, *Science (Washington, D.C. 1883)* **236** (1987) 271–280.
[173] B. N. Ames, L. S. Gold in T. Slaga, B. Butterworth (eds.): *Chemically Induced Cell Proliferation: Implications for Risk Assessment*, Wiley-Liss, New York 1991, pp. 1–20.

[174] L. S. Gold et al., *Science (Washington, D.C. 1883)* **258** (1992) 261–265.

[175] D. J. Waxmann, L. Azaroff, *Biochem. J.* **281** (1992) 577–592.

[176] B. N. Ames, J. McCann, E. Yamasaki, *Mutat. Res.* **31** (1975) 347–364.

[177] D. Maron, B. Ames, *Mutat. Res.* **113** (1983) 173–215.

[178] J. M. Parry, M. Shamsher, O. O. F. Skibinski, *Mutagenesis* **5** (1990) 209–212.

[179] S. W. Kohler et al., *Proc. Natl. Acad. Sci. U.S.A.* **88** (1991) 7958–7962.

[180] M. Richold et al. in D. J. Kirkland (ed.): *UKEMS Sub-Committee on Guidelines for Mutagenicity Testing*, report part 1: Revised Basic Mutagenicity Tests, UKEMS Recommended Procedures, Cambridge University Press, Cambridge 1990, pp. 115–141.

[181] J. J. McCormick, V. M. Mahler, *Environ. Molec. Mutag.* **14** (1989) 105–113.

[182] G. M. Williams, J. H. Weisburger, *Mutat. Res.* **205** (1988) 79–90.

[183] M. Robinson, *EHP Environ. Health Perspect.* **68** (1986) 3–151.

[184] N. Ozawa, F. P. Guengerich, *Proc. Natl. Acad. Sci. U.S.A.* **80** (1983) 5266–5270.

[185] H. Bartsch et al., *Drug Metab. Rev.* **26** (1994) 349–371.

[186] C. Richter, J.-W. Park, B. N. Ames, *Proc. Natl. Acad. Sci. U.S.A.* **85** (1988) 6465–6467.

[187] P. B. Farmer, H.-G. Neumann, D. Henschler, *Arch. Toxicol.* **60** (1987) 251–260.

[188] J. A. Swenberg et al., *Carcinogenesis (London)* **13** (1992) 727–729.

[189] I. A. Blair, *Chem. Res. Toxicol.* **6** (1993) 741–747.

[190] K. Randerath, E. Randerath, *Drug Metab. Rev.* **26** (1994) 67–85.

[191] M. D. Shelby, E. Zeiger, *Mutat. Res.* **234** (1990) 257–261.

[192] J. G. Vos, *Arch. Toxicol.* **4** (1980) 95–108.

[193] A. D. Dayau et al., *Immunotoxicity of Metals and Immunotoxicology*, Plenum Press, New York 1990.

[194] K. L. J. White in K. Miller, J. L. Turk, S. Nicklin (eds.): *Principles and Practice of Immunotoxicology*, Blackwell Scientific, Oxford 1992, pp. 304–323.

[195] V. C. Moser, J. P. McCormick, J. P. Creason, R. C. MacPhail, *Fundam. Appl. Toxicol.* **11** (1988) 189–206.

[196] V. C. Moser, *J. Am. Coll. Toxicol.* **8** (1989) 85–93.

[197] C. C. Haggerty, *J. Am. Coll. Toxicol.* **8** (1989) 53–69.

[198] D. M. Maloney, *EHP Environ. Health Perspect.* **101** (1993) 396–401.

[199] J. V. Rodricks, *EHP Environ. Health Perspect.* **102** (1994) 258–264.

[200] R. Doll, R. Peto, "The Causes of Cancer: Qualitative Estimates of Available Risks of Cancer in the United States Today," *J. Natl. Cancer Inst.* **66** (1981) 1190–1308.

[201] F. R. Johannsen, *CRC Crit. Rev. Toxicol.* **20** (1990) 341–367.

[202] A. M. Jarabek, W. H. Farland, *Toxicol. Ind. Health* **6** (1990) 199.

[203] D. W. Nebert, *Mutat. Res.* **247** (1991) 267–281.

[204] R. T. H. van Weie, R. G. J. M. van Dijck, N. P. E. Vermeulen, N. J. van Sittert, *CRC Crit. Rev. Toxicol.* **22** (1992) 271–306.

[205] R. W. Tennant, J. Ashby, *Mutat. Res.* **257** (1991) 209–227.

[206] A. C. Beach, R. C. Gupta, *Carcinogenesis (London)* **13** (1992) 1053–1074.

[207] C. C. Travis, *Toxicology* **47** (1987) 3–13.

[208] H. Vainio, M. Sorsa, A. J. McMichael: *Complex Mixtures and Cancer Risk*, International Agency for Research on Cancer (WHO), Lyon 1990.

[209] J. A. Swenberg, R. R. Maronpot in (ed.): *Chemically Induced Cell Proliferation: Implications for Risk Assessment*, Willey-Liss, New York 1991, pp. 245–251.

[210] S. M. Cohen, L. B. Ellwein, *Cancer Res.* **51** (1991) 6493–6505.

[211] B. Ames, L. S. Gold, *Mutat. Res.* **250** (1991) 3–16.

[212] B. N. Ames, L. S. Gold, *Science (Washington, D.C. 1883)* **249** (1990) 970–971.

[213] G. M. Gray, J. T. Cohen, J. D. Graham, *EHP Environ. Health Perspect.* **101** (1993) 203–208.

[214] J. C. Barrett, *EHP Environ. Health Perspect.* **100** (1993) 9–20.

[215] J. A. Swenberg, D. G. Hoel, P. N. Magee, *Cancer Res.* **51** (1991) 6409–6414.

[216] T. Green, *Xenobiotica* **20** (1990) 1233–1240.

[217] D. G. Hoel, N. L. Kaplan, M. W. Anderson, *Science (Washington, D.C. 1883)* **219** (1983) 1032–1037.

[218] J. A. Swenberg, N. Fedtke, T. R. Fenneli, V. E. Walher in D. B. Clayson, I. C. Munro, P. Shubik, J. A. Swenberg (eds.): *Progress in Predictive Toxicology*, Elsevier Pub. (Biochemical Division), New York 1990, pp. 161–184.

[219] W. K. Lutz, *Carcinogenesis (London)* **11** (1990) 1243–1247.

[220] H. Vainio, P. N. Magee, D. B. McGregor, A. J. McMichael: "Mechanisms of Carcinogenesis in Risk Identification," *IARC Sci. Publ.* **116** (1992).

[221] C. Shaw, H. B. Jones, *Trends Pharmacol. Sci.* **15** (1994) 89–93.

[222] P. Grasso, M. Sharrat, A. J. Cohen, *Annu. Rev. Pharmacol. Toxicol.* **31** (1991) 253–287.

[223] J. A. Swenberg, *EHP Environ. Health Perspect.* **101** (1993) 39–44.

[224] L. Poellinger, M. Göttlicher, J.-A. Gustafsson, *Trends Pharmacol. Sci.* **13** (1992) 241–245.

[225] S. J. Borghoff, B. G. Short, J. A. Swenberg, *Annu. Rev. Pharmacol. Toxicol.* **30** (1990) 349–367.

[226] R. M. McClain, *Toxicol. Lett.* **64/65** (1992) 397–408.

[227] A. M. Tritscher et al., *Cancer Res.* **52** (1992) 3436–3442.

[228] C. Portier et al., *Fundam. Appl. Toxicol.* **20** (1993) 48–56.

[229] J. P. V. Heuvel et al., *Cancer Res.* **54** (1994) 62–68.

Legal Aspects

JEAN-MARIE DEVOS, CEFIC, Brussels, Belgium (Chaps. 1, 3)

LUDWIG KRÄMER, Head of Legal Services, DG XI, European Commission, Brussels, Belgium (Chap. 2, Chap. 5 in part)

GABRIELLE H. WILLIAMSON, LeBoeuf, Lamb, Greene & MacRae, Brussels, Belgium (Chap. 4)

PASCALE KROMAREK, Elf Aquitaine, Direction Environment, Paris-La-Defense, France (Chap. 5)

TURNER T. SMITH Jr., Hunton & Williams, Washington, D.C., United States (Chap. 6)

MUNETO ITODA, JCIA, Tokyo, Japan (Chap. 7)

The entire topic was coordinated by JEAN-MARIE DEVOS

3-527-20137-8/95/$5.00 + .50

1. Industrial Chemistry and Law

1.1. Law, Society, Chemistry, and Ethics

Law is the normative instrument that society uses to determine in a peaceful way the rules governing the relations between people (private law); people and authorities (public law); and states (international public law). It reflects to a large extent the values and ethics of a given society at a given time (→ Introduction/Ethics). Normative activities have – at an early stage – addressed the question of humans and their interface with industrial and trade activities either to promote such activities, to control or restrict them, or in some cases, even to prohibit them.

Although the requirements of an organized society, as reflected in law, are supposed to be integrated, complementary, and consistent, they might actually conflict or at least appear contradictory. For example, the legal system might aim at economic objectives – for example, the free movement of products in an economic area – as reflected in one of the four basic freedoms of the European Communities Treaties. On the other hand, society and lawmakers might impose severe restrictions on the movement of products because of their impact on organized collectivities and the environment (see Section 4.3).

Law might actually impose restrictions and administrative burdens on research, production, transport, and distribution activities, either because of the "dangerous" nature of such activities or because they involve products that are, to various degrees, considered dangerous or simply perceived as negative or hostile to people or the environment. The end of the twentieth century has seen the emergence of the "environmental society". With the growing satisfaction of basic needs in the developed world and the collapse of traditional ideologies, growing awareness of environmental problems has become a major factor influencing sociology, politics, law, and economics. In parallel to the new "environmental" awareness, and in apparent contrast, industry and trade activities are more and more characterized by increased competition and "globalization." After the success of the General Agreement on Tariffs and Trade (GATT) round in Uruguay and the creation of a World Trade Organization, the dominant economic concepts and trade realities are clearly hostile to trade barriers and restrictions. The resolution of potential and actual conflicts between these two approaches should lie in a coherent and humanistic vision, based on human values, on responsibility, and on confidence in sustainable progress and science.

The main objective of this chapter is to focus on a few key trends which affect, or will influence, the socio-legal framework in which industry has to operate. As shown in the various chapters of the keyword Legal Aspects, innovation, production, distribution, and marketing strategies must integrate environmental and other ethical values which determine their legal framework.

At the same time, industry's response should be based on sound and credible science, on economic sense and on appropriate communication.

In other words, modern *legal systems* tend to address production, distribution activities and products from "cradle to grave" and respond to a growing social concern about perceived risks generated by "dangerous" activities and products. This is particularly relevant in the area of chemicals which are subject to comprehensive sets of rules in the various regions of the world (see Chaps. 3, 6, 7). Perceived risks may be real and have to be seriously addressed. They may also result from fears generated by some media or pressure groups. In contrast, *science* has never been more advanced but, in a way, has become a victim of its own success as technology allows society to better measure even minor risk factors (or simply traces) which are then deemed to be unacceptable. [See for example the progress of detection limits of modern analytics using ultrasensitive equipment and the debate on the revision of the Drinking Water Directive [1.1]. The key issues are the allowance of greater flexibility for nonhealth parameters and reliance on toxicological data in setting maximum admissible concentrations (MAC) for health-related parameters. Progress in toxicological, analytical, and epidemiological knowledge and methods may imply that MAC should evolve, in line with WHO guidelines.]

The importance of innovation as a driving force of modern economy and society was defined by SCHUMPETER at the beginning of the 20th century as the entire process from the birth of an idea through to its widespread application in society. Innovation implies changed technologies, economic and social evolutions, and adaptability. The driving forces of innovation and changes are rooted in human curiosity and needs and are influenced by their evolving hierarchy.

Social psychology based on Maslow's approach, distinguishes five hierarchical levels of needs

1) Physical survival and physiological needs
2) Safety
3) Social belonging
4) Respect and recognition
5) Self-fulfilment

According to classical marketing sciences, innovation was responding to "technology pushes" on the one hand and almost unlimited "market demands" on the other hand. Until the lower needs are satisfaction of human needs as described in Maslow's ladder in industrialized countries and with the achievement of high standards of living, perceived priorities have evolved. Public awareness has become more and more attentive, and critical of "side effects" of innovation.

In other words, the better and the more social needs are fulfilled, the more acute is the perception and concern of the society about the instruments developed to respond to such needs and about their effects. This phenomenon has had a direct and massive impact on the lawmaking process and has led to a vast number of regulations, particularly in the health, safety, and environment area.

The Rio United Nations Conference on Environment and Development (UNCED) Declaration is probably the widest international effort to address environmental concerns on a global basis, although reflecting different Regional sensitivities and priorities. The growing environmental awareness has generated many "law–regulatory"-driven technologies and market developments (e.g., the compulsory introduction of catalytic converters for automobiles) and has led to the phased withdrawal of products considered as harmful to the environment (as for example, the phasing out of CFCs, initiated by the 1987 Montreal protocol on substances that deplete the ozone layer). Market and competition effects are far-reaching.

In the face of these developments and in parallel to the international, European and national lawmaking process, industry has gradually developed comprehensive "pro-active" programs. Such programs are meant to meet the public's and the authorities' concerns and to offer a sound managerial and operational response from industry. Such a response may take the form of general policy declarations or more operational programs. [See for example the "Business Charter for Sustainable Development" (November 1990) of the International Chamber of Commerce. The Charter lists sixteen principles for Environmental Management.] Typically, one leading feature of such policies is the improvement of information flows betwen industry, the public and the authorities.

On a more sectorial basis, the "Responsible Care" program developed by the chemical industry on a worldwide and a Regional basis reflects this sector's commitment towards achieving continuous and measurable progress in the health, safety, and environment areas. Association members of the International Council of Chemical Associations, such as the European Chemical Industry Council (CEFIC), the Japanese Chemical Industry Association (JCIA), and the U.S. Chemical Manufacturers Association (MA) have identified Responsible Care as a priority and a contribution to sustainable development.

1.2. Policy and Legal Principles Affecting Industrial Chemistry

Recent years have seen the emergence of two major policy principles that influence and will continue to impact with special significance the lawmaking process affecting the exercise of activities related to chemicals.

1.2.1. Precautionary Principle

The "precautionary principle" has been used more and more frequently and quoted in public statements adopted in several international forums and according to some opinions has become a principle of (customary) international law [1.2]–[1.5]. Examples of international forums include:

1) Bergen Conference on Action for a Common Future, Norway, May 8–16, 1990. The Bergen Conference was organized by the government of Norway in cooperation with the UN Economic Commission for Europe as part of the preparation of the UN Rio Conference on Environment and Development. The Bergen Conference established that "in order to achieve sustainable developments, policies must be based on the Precautionary Principle. Environmental measures attack the causes of environmental degradation."

2) Principle 15 of Rio Declaration on Environment and Development, Rio de Janeiro, Brazil, June 13, 1992. "In order to protect the environment, the precautionary approach shall be widely applied by States according to their capabilities. When there are threats of serious or irreversible damage, lack of full scientific certainty shall not be used as a reason for postponing cost-effective measures to prevent environment degradation."

The precautionary principle is part of the wider concept of sustainable development and has been integrated in the European Communities' law since its introduction in the Maastricht Treaty [1.6]. Article 130r, § 2, of the Treaty explicitly refers to the precautionary principle: "Community policy on the environment shall aim at a high level of protection .... It shall be based on the 'Precautionary Principle' and on the principles that preventive action should be taken, that environmental damage should as a priority be rectified at source and that the polluter should pay ...."

The new European Constitution is a major example of the fast recognition in a legally binding instrument of the two major principles for future environmental policies.

The precautionary principle institutionalizes caution when (sufficient) evidence exists that an activity or substance is likely to cause unacceptable harm to the environment. Its actual meaning, interpretation, and definition, however, are much debated because the implications of the principle could be far reaching, if not devastating. Definitions vary widely from encouraging a cautious and responsible attitude, based on scientific evidence, not to generate or discharge in an uncontrolled way elements hostile to humans and the environment, to a much more radical "anti-industrialization" attitude, which is leading to the prohibition of substances or activities even in the absence of conclusive scientific evidence. According to the precautionary principle, as defined by part of the doctrine and in several marine pollution conventions, a substance of activity posing a threat to humans or to the environment should therefore be prohibited (i.e., banned or restricted) even if there is no conclusive scientific proof of a causal link between that particular substance, or activity, and humans or environmental damage.

This rather radical approach was first used in the Ministerial Declaration of the Second North Sea Conference [1.7]. "The Parties agree to ... accept the principle of safeguarding the marine ecosystem of the North Sea by reducing polluting emissions of substances that are persistent, toxic and liable to bioaccumulate at source by the use of the best available technology and other appropriate measures."

This applies especially when there is reason to assume that certain damage or harmful effects on the living resources of the sea are likely to be caused by such substances, even where there is no scientific evidence to prove a causal link between emissions and effects ("the principle of precautionary action").

The overall goal was to achieve a substantial reduction in the total quantity of persistent, toxic and bioaccumulative substances reaching the North Sea. Actual policies of Member States are aiming at a target of 50 %. The national interpretations of the London Declaration have been far from homogeneous, most of them requiring some certainty of (scientific) evidence regarding environmentally damaging effects (i.e., France) or calling for actual evidence before taking action.

Those variations clearly reflect the embarrassment of public authorities who are confronted by complex situations in which multiple causation factors make scientific assessment and conclusions difficult or uncertain. [This embarrassment was reflected in the Third International Conference on the Protection of the North Sea, The Hague, March 7–8, 1990 (Third North Sea Conference). The final declaration confirmed the continuous application of the precautionary principle but limited it to those substances that are persistent, toxic, and liable to bioaccumulate.] A radical interpretation of the precautionary principle as reflected in the London Declaration is seriously objectionable – from both legal and scientific standpoints – because it conceptually opposes and rejects scientific evidence and causality as instruments for sound policy actions.

Such an interpretation necessarily leads, at the extreme, to an impasse and might be translated into a "paralysis principle." Indeed, a requirement based on the existence of "reasons to believe" that damage or harmful effects are "likely" to be caused becomes meaningless if opposed to or separated from scientific evidence. For example, a mere trace of a substance in the environment would equal pollution of that environment leading to possible prohibitons. The progress of analytical methods today allows the detection of minute quantities of a given substance (e.g., a piece of sugar in a lake). The progress of scientific and analytical methods without parallel progress of impact and risk assessment techniques could actually multiply purely "emotionally"-based policies. The exclusion of scientific evidence as a determining factor for shaping environmental policy might actually lead to the opposite of a precautionary and reasonable attitude. At the extreme, it might lead to decision-making processes entirely dominated by "beliefs" and "opinions" dominated by ideology without due regard for an objective and scientifically sound analysis of facts.

The radical interpretation of the precautionary principle seems to be disregarded by provisions of Article 130 r of the Maastricht Treaty. While endorsing the concept of the precautionary principle, the Treaty does not define it and obliges the Community to take account of available scientific and technical data [Article 130 r (§ 3)].

Realistically, complex situations and interactive systems might lead to a degree of uncertainty where scientific evidence is incomplete or even appears contradictory. The lack of full scientific evidence might encourage policy makers to take measures "where there are threats of serious or irreversible damage." Such measures must, however, be based on sufficient scientific evidence.

It remains to be seen if and how a moderate and reasoned interpretation will finally emerge in the international and national spheres, where scientific evidence and causation will actually serve and precondition the applicability of the precautionary principle.

[A dramatic and illustrative evolution has been the apparent change of policy of the U.S. administration, which, under the Reagan and Bush presidencies, carefully avoided endorsing the principle, while the EPA, under President Clinton, developed a strategy that would prohibit, reduce, or substitute the use of chlorine and chlorinated compounds on the basis of such a principle.

Clean Water Initiative Submitted to Congress, 1 February 1994. While aiming at a strategy for prohibiting, reducing, and substituting the use of chlorine and chlorinated compounds, the administration proposed a time schedule allowing federal agencies and "other experts" to comprehensively assess the use the environmental and health impacts of chlorine and chlorinated compounds, and availability and relative efficiency and safety of substitutes. Within 30 months, a plan "for our appropriate actions" should be developed. The National Academy of Sciences is expected to complete a study for the Congress on the current knowledge of chemicals "that exhibit endocrine immune and nervous system health effects in human and wild life." The highly political and emotion-driven nature of this issue is demonstrated by the more moderate approach of the administration, which resulted from fierce reactions from many circles, including industry.]

The sole answer for policy makers is to develop more, not less, science and technology assessment and to assist policy and decision makers to take the most appropriate decisions.

1.2.2. The Polluter-Pays Principle

Historically, an economic and preventive principle, the "polluter-pays principle" is gradually influencing public and administrative law, or civil law.

The polluter-pays principle was initially adopted by the OECD in its Recommendations [1.8]. The 1991 OECD Recommendation [1.9] extended the polluter-pays principle to the "cost of damage" generated by pollution.

The Parliamentary Assembly of the Council of Europe summarized the extension of the polluter-pays principle to compensation of damage as follows:

"States, industry and all persons shall be liable to pay for the environmentally harmful consequences of their actions and development programmes ... The principle of the polluter's liability, or the "Polluter Pays Principle" as it is commonly known, shall be strictly applicable." [Recommendation 1130 (1990)]

The polluter-pays principle has been introduced into the European Communities Constitution via Article 130 r of the Single European Act [1.10] and later in the Treaty on the European Union. The extension of the polluter-pays principle as an environmental law principle has received priority consideration by the OECD. In its preventive form, the principle implies that those whose emissions could create a risk to the environment should bear the full cost of reducing or eliminating such emissions so as to prevent this risk from materializing.

The principle is thus also a means of internalizing environmental costs [1.11]. However, it is not yet clear what impact it will have in specific areas such as civil liability for damage, although it may be argued that it is based on a similar concept, i.e., the author of a damage, to persons or to property, has to compensate for such damage.

One of the important, and complex, questions raised by the polluter-pays principle within civil liability mechanisms is linked to the very question of "what is a damage?" since there is a tendency to extend the notion of damage to "damage to the environment", or "ecological damage". These are extremely difficult to define and to quantify. (For example, how is the degradation of the ozone layer to be evaluated in terms of damage and to which cause can it be rationally attributed? Who is the polluter? The "global climatic change" issue raises a number of complex questions of this nature.)

Other difficulties are linked to the establishment of causation, when several sources of pollution exist, and to the establishment of evidence. Should the plaintiff bring the full evidence that damage was caused by a given attributable incident?

Or is "prima facie" evidence sufficient? Should the burden of the proof be reversed with the risk of jeopardizing the very concept of civil liability? These, and other questions, have been debated in the context of the "Council of Europe" Convention and are still much argued within the European Union.

Efforts have also been made to make civil liability rules not only a legal instrument for compensation of injury to people and damage to property but also to use them as a (limited) tool for environment remediation. It is sometimes argued that strict civil liability could play a role in the prevention of environmental degradation and therefore be used as an instrument of environmental policy.

1.3. Economic Integration and Environmental Protection

The interface between economic integration and environmental protection is a major emerging issue at the end of the 20th century. Both on the world trade scene and within the European Union, law and policymakers have to reconcile potential conflicts between the logic of economic integration implying and enforcing similar rules and policies to facilitate trade and competition and, on the other hand, the logic of national/regional prohibitions and restrictions on products and activities. Global issues call for global solutions. Achieving cohesion is far from obvious, as illustrated by the present debates concerning the tensions between the Internal Market of the Community and environmental policies, but also in the debate within the World Trade Organization.

International trade of products which might have an impact on the environment is increasingly subject to severe restriction and/or to authorization and information procedures. For example, the case of waste movement restrictions and free trade is illustrative of such tensions. The EC Regulation on Control of Waste of February 1, 1993 [1.12] is intended to implement the Basel Convention on the "Control of Transboundary Movements of Hazardous Waste and their Disposal" (adopted and opened for signature on March 22, 1989). This Regulation completes and tightens the monitoring system already applicable in the Community. It covers five categories of transport: transfer between Member States, transfer within Member States, export to non-EC countries, imports from outside the EC and waste transfers. The role of the OECD, in seeking a workable solution especially for waste for recovery should be stressed.

The Regulation provides for a prior authorization system, recognizes a broad right for Member States to object to shipment and implements the self-sufficiency and the proximity principles which both aim at encouraging local waste disposal. Paradoxically, a piece of Community legislation, supposed to have direct effects, is opposing one of the fundamental freedoms of the EC Treaty, i.e., free circulation of goods. This has raised much criticism within the Community and the legal debate continues on the (impossible?) reconciliation of free movement of goods principles and restrictions of transboundary shipments of waste and recyclable materials (secondary raw materials) (see Chap. 4). Rather than concentrating its policy on the proximity principle, which may actually create inefficiencies and additional unnecessary costs, the Community should develop sound and uniform criteria for operating waste facilities and movements of waste.

Another example of restrictions to trade motivated by environmental concerns is the prior informed consent (PIC) procedure. The PIC procedure is based on the principle that international shipment of a chemical that is banned or severely restricted in order to protect human health and the environment should not proceed without the agreement, or contrary to the decision of, the designated national authority in the importing country [1.13]–[1.17]. These developments raise the question of their interface and compatibility with world trade liberalization agreements under the GATT. There is no doubt that the whole "trade and environment" debate will constitute a major global policy issue of the end of this century. For a comprehensive review of the current state of affairs, see [1.18].

1.4. Other General Trends of Health, Safety and Environmental Laws Affecting Industrial Chemistry

While legislations dealing with environmental aspects of industrial activities have to be situated within general public and civil law, the evolution of health, safety and, especially, environmental (HSE) laws has been characterized by the following elements:

1) A more autonomous, systematic system of law is gradually developing both in the national, European, and international spheres. The fast development of a European Community comprehensive environmental law is based on policy programs which aim at fully integrating environmental policies in other Community policies and in setting long-term objectives and targets [1.19], [1.20]. The primacy of Community law over national legislation, where sufficient, clear, and precise provisions exist, and the direct effect into national legal systems have far-reaching implications.
2) The gradual emergence of a "right to know" both in the European Union and in the United States is a good example of the impact of environmental law concepts on public and administrative law. This right to know gives citizens broad access to information obtained by public authorities in the environmental area, subject to some limitations [1.21].
3) The impact of environmental issues on competition policy. In a growing number of cases, "environmental benefits" are claimed in industrial and marketing strategies both vis-à-vis consumers and vis-à-vis public authorities. This development raises many questions about ethics and (fair) competition. The eco-label concept is an illustration of this evolution [1.22].

Other aspects are linked to state aids given for environmental purposes which might distort competition in a given industrial sector.

The European Commission has issued guidelines describing its policy on state aid for environmental protection. Such guidelines would authorize up to 30% state subsidy for (voluntary) investment related to the environment, while investment aimed at compliance with environmental legislation could be supported up to 15% of the cost [1.23].

1.5. Scope of this Keyword

The keyword Legal Aspects does not aim at providing a comprehensive or exhaustive review of European, national, or international rules affecting chemicals. For a recent publication relating to international aspects of chemical control see [1.24]. The mere coverage of, for example, European Community legislation would require several volumes of Ullmann's. Many specific aspects have already been addressed in other keywords dealt with in volumes B7/B8 or in relevant sections of them. The main objective of the keyword Legal Aspects is to make better understood the interface between the legal system in general and chemical-related activities and to present in this respect the major trends characterizing the evolution of law on the eve of the year 2000.

2. European Community Legislation Affecting Chemicals

Definition of Terms Used in this Chapter. The European Union (EU) was set up by the Treaty of Maastricht, which came into force in November 1993. Part of the Treaty on the EU are the Treaties on the European Communities (EC), the EURATOM Treaty, and the Coal and Steel Treaty.

The EU does not have legal personality, but the EC has. Therefore, any legal instrument is adopted not by the EU but by the EC. Whenever references to legal texts are made, the term EC is correct, EU may only be used in a political context.

Prior to the coming into force of the Maastricht Treaty the correct name for the EC was European Economic Community (EEC). Thus, all legal texts adopted before November 1993 have in their title the reference EEC.

2.1. The European Community Framework

The European Community has, under different Treaties, the objective of creating an integrated economic community in which industrial frontiers are no longer economic frontiers, and which includes the free circulation of goods and services, capital, and labor as well as a high level of health, safety and environmental protection. The mandate to establish and monitor the Internal Market for goods and services, capital, and labor is in the Treaties given to Community institutions: Council and Commission, the European Parliament, the Economic and Social Committee, and the European Court of Justice (ECJ). Member States have the objective of facilitating achievement of the Community's tasks, and abstaining from any measures that could jeopardize the attainment of Community objectives [2.1]. In contrast to that there is the joint responsibility between the European Community and its Member States acting in isolation to ensure an appropriate protection of the environment [2.2]. This principle of subsidiarity, enshrined since autumn 1993 in the general provisions of the EC Treaty, has existed since the 1987 amendment of the environmental chapter of the EEC Treaty. However, despite six years of practical experience, no single piece of legislation exists in which subsidiarity questions played any role during legislative procedure. Proposals for legislation made by the EC Commission have been adopted or rejected without express recurrence to this principle.

Chemical legislation in the European Community must find its way through this framework of responsibilities. In very broad terms, past legislation on product standards was generated mainly at Community level and thus had a considerable degree of uniformity. Legislation banning products or restricting their use was more often adopted at a national level; with the completion of the internal Community market, a marked trend exists towards integrated rules for bans and restrictions, but progress is slow. With regard to production-orientated legislation, Community legislation in the past has been unsystematic and limited to framework rules. The tendency has developed recently to arrive at a more systematic approach, although the responsibility for rule making remains shared between Member States and the Community, the Community being limited to umbrella legislation.

2.2. Product-Oriented Legislation

Community legislation on chemical substances and preparations as well as on specific chemical products – food additives, pharmaceuticals, cosmetics, fertilizers – started early in the 1960s, shortly after the creation of the EEC. This legislation was based almost entirely on Article 100 of the EEC Treaty, which reads:

"The Council shall, acting unanimously on a proposal from the Commission, issue Directives for the approximation of such provisions laid down by law, Regulations or administrative action in Member States which directly affect the establishment or functioning of the common market...".

Total or Optional Harmonization? The approximation of national legislation aimed at diminishing the differences between national rules that created, or were capable of creating, barriers to the free circulation of goods. Since in all the above-mentioned sectors of chemical products, questions of health and safety considerably influenced rule setting, the EEC did not attempt the legislative technique called "optional harmonization" but rather used a "total harmonization" approach.

Under *optional harmonization*, Member States fix standards for national production at their discretion and at a level they consider appropriate. However, they are obliged to admit on their market those products from other Member States that comply with Community standards. This kind of approach was, for instance, largely used by the EEC in the area of technical legislation.

Under the *total harmonization* approach, which was used in the area of chemical products, national legislation is replaced by Community legislation so that, finally, one rule exists within the Community. This approach is obviously more ambitious, more delicate, and more time consuming. Since questions of health and safety are central to all Member States' national policy, progress in replacing the different national rules by one single Community rule has been slow and even slower in the areas of food and pharmaceuticals than in other areas.

In the area of chemical substances and preparations, national legislation in the 1960s was not very sophisticated. Therefore, attempts to create uniform rules for chemicals were relatively successful. Community legislation on chemical substances [2.3], solvents [2.4], pesticides [2.5], and paints and varnishes [2.6] was adopted soon after the transition period for setting up the EEC. This legislation set uniform rules for the classification, labeling, and packaging of substances and preparations, and ensured that chemicals conforming to these requirements were allowed to be circulated freely within the Community. Since this legislation in the 1960s and 1970s had to be adopted unanimously, Member States were able to block the adoption of Community legislation that did not conform to their national rules on health and safety, until their concern was met.

A Dynamic Cooperative System under the Control of the Commission. The possibility of blocking legislation served to protect the interests of Member States in terms of health and safety, before such legislation was adopted. However, this need could also arise after the adoption of Community legislation, for instancce, when an accident in a Member State demonstrated that the Community rule was not strict enough or where new scientific evidence was found. Since an amendment of Community legislation was slow and required an initiative from the EEC Commission, this mechanism was not considered able to adequately cover all cases. Therefore, the Community legislation on chemicals contained a safety clause: where a Member State found that a chemical, although complying fully with Community rules, presented a risk to health and safety, the Member State was allowed to take provisional measures to counteract that risk. These measures could even go so far as to remove the chemical from the market. The Member State that took such a measure had to inform the Commission and other Member States of its action and give a reason for its decision. Then, a Community procedure was started to decide whether the national measure should be extended to the entire Community. Until such a decision, the national safeguard measure was maintained. By way of illustration, the text of such a safeguard clause is given here [2.7]:

"Where a Member State has detailed evidence that a substance, although satisfying the requirements of this Directive, constitutes a hazard for man or the environment by reason of its classification, packaging and labelling, it may provisionally prohibit the sale of that substance or subject it to special conditions in its territory. It shall immediately inform the Commission and the other Member States of such action and give reasons for its decision. The Commission shall consult the Member States concerned within six weeks, then give its view without delay and take the appropriate measures. If the Commission considers that technical adaptations to this Directive are necessary, such adaptations shall be adopted, either by the Commission or by the Council in accordance with the procedure laid down in Article 21; in such a case, the Member State which has adopted safeguard measures may maintain them until the adaptations enter into force."

In the mid-1970s, two elements were added to this situation – attempts to introduce Community-wide bans and restrictions on chemicals, and rising environmental concerns. Clearly, the feeling was that legislation on chemicals that tried to open up national frontiers and set uniform rules was insufficient since national prohibitions and restrictions for use continued to exist, and with growing concerns for health, safety, and the environment, new restrictions were being introduced at the national level. To ensure uniform standards, the EC felt the need to adopt Community-wide restrictions or bans. This legislation [2.8] was adopted on an ad hoc basis: whenever nationally – or indeed in international bodies – the need was felt to ban or restrict the use of a chemical product, the European Economic Community tried where appropriate, to reach Community-wide standards. The difficulty of reaching a consensus on such restrictions under the unanimity rule had, as a consequence, the adoption of a relatively limited number of such Community rules. Several attempts to accelerate the decision-making procedure at the Community level were unsuccessful, probably mainly because Member States were reluctant to see responsibility for health and safety questions shift more and more to the Community. Thus, between 1976 and 1992 some 30 substances were banned or severely restricted in use, covering only a part of the national bans or restrictions.

The second element was upcoming environmental concern in legislation. This concern gradually led some Member States to legislate in the area of chemicals, in order to protect, preserve, or improve the quality of the environment. The EEC reacted by including environmental considerations into its existing legislation on chemicals. The evolution of Directive 67/548/EEC, adopted in 1967 at Community level, is characteristic in this regard: in the mid-1970s, France informed the Commission of its intention to introduce new chemical legislation that took into a account environmental considerations. The Commission was of the opinion that such legislation would be capable of creating new obstacles to trade within the Community. Therefore, it suggested an amendment to Directive 67/548/EEC that finally led to the adoption of Directive 79/831/EEC, which, for the first time in such precise terms, talked of the need "to protect man and the environment" [2.9]. Since environmental policy is based, among other factors, on the principle that preventive action should be taken [2.10], that Directive introduced the principle that before new chemical substances were placed on the market, information had to be provided to competent authorities, together with test results on the substances, in order to enable the authorities to evaluate the foreseeable risk for humans and the environment. This evaluation is made Community-wide in concert between national and Community administrations. Conditions for marketing of new chemical substances based on health, safety, or environmental considerations are fixed for the entire Community.

A similar approach regarding notification prior to placing a substance on the market was used when the Community adopted legislation for genetically modified organisms on the market (deliberate release) [2.11]. Again, evaluation of the risk of such deliberate release is made Community-wide prior to the release.

However, this approach is by no means generalized; it is not applied to all chemicals, i.e., chemical preparations can be placed freely on the market without prior provision of test results and other information to competent authorities.

In 1987, the EC Treaty was amended. Specific rules were introduced to ensure achievement of the Internal Market by the end of 1992. For that purpose, a new Article 100a provided that measures with the objective of achieving the Internal Market could be adopted by majority decision of Member States. At the same time, a new Article 100a (4) was included in the Treaty, which reads:

"If, after the adoption of a harmonization measure by the Council acting by a qualified majority, a Member State deems it necessary to apply national provisions on grounds of major needs referred to in Article 36, or relating to protection of the environment or the working environment, it shall notify the Commission of these provisions. The Commission shall confirm the provisions involved after having verified that they are not a means of arbitrary discrimination or a disguised restriction on trade between Member States ...".

Furthermore, Community environmental policy was considerably strengthened, by the introduction of an environmental chapter, Articles 130r to 130t, into the Treaty. Environmental measures were, in principle, to be adopted unanimously (Art. 130s). Article 130t stipulated, "The protective measures adopted in common pursuant to Article 130s shall not prevent any Member State from maintaining or introducing more stringent measures compatible with this Treaty."

The borderline between the provisions of Article 100a and Article 130s seems, with regard to chemical products, to be rather clear: since chemical preparations are products and should – under the concept of an Internal Market, where national frontiers are no longer economical frontiers – circulate freely within the Community, the legal foundation for rules on their classification, labeling, or packaging – including their ban or restrictions on use – should be based on Article 100a. Article 130s would be reserved to production-related legislation that did not affect the establishment or the functioning of the Internal Market.

The majority of Community chemical legislation did indeed follow this approach. However, not all legislation respected this pattern. When the Council adopted legislation on chlorofluorocarbons (CFCs) and other ozone-depleting substances, this legislation was based on Article 130s [2.12]; the same approach was followed on waste legislation, although, from the point of view of the environment, there exists little difference between poisoning by chemicals or by waste [2.13]. Thus, because of national initiatives remaining possible under Article 130t, uniformity of rules within the Community was not achieved.

Even where new legislation on chemicals was based on Article 100a, the establishment of uniform rules was questioned. Indeed, certain interpretations of the above-mentioned clause of Article 100a (4) are equivalent to the wording of Article 130t. In other words, this interpretation would allow Member States, after the adoption of Community standards, to introduce new, diverging rules at the national level, not merely to continue to apply national legislation that existed prior to the adoption of Community rules. Whether Article 100a really has the same meaning as Article 130t or should be interpreted more narrowly, remains to be decided by the European Court of Justice, which, until now, has not had occasion to pronounce itself on this question.

Pending this question and more generally, Member States remain responsible, together with Community institutions, for the protection of health, safety, and the environment. They have thus not lost the responsibility for banning or restricting the use of chemical products, as long as no comprehensive legislation setting up Community-wide bans and restrictions exists. Article 100a, seen in more general context, enables the Community to draft uniform rules for chemicals that are based on a high level of protection of health, safety, and the environment. To the extent that the Community manages to elaborate, adopt, and permanently adapt such rules to technical and scientific progress, recurrence to national bans and restrictions will not be necessary; thus Article 100 a (4) would lose practical importance.

The Maastricht Treaty on European Union has not changed the principles on rule making at the Community level. Thus, the trends described above are likely to remain applicable in the coming years.

2.3. Legislation on Chemical Installations

As stated above, Community legislation on chemical installations has not been very systematic in the past. The establishment and functioning of a common market were not in the past perceived to require the elaboration of Community-wide standards for industrial installations. Thus, rules affecting chemical installations were set up only with the introduction of an environmental policy at the Community level and under environmental protection auspices. Industrial installations that discharge dangerous substances into waters or into groundwater or emit pollutants into the air need specific authorization for such emission [2.14]. Such permitting rules were, however, far from covering all installations or all pollutants.

An industrial accident in Seveso, Italy, provoked Community legislation on the prevention of major accident hazards of certain industrial activities [2.15]. This so-called Seveso Directive, adopted in 1982 and amended several times, requested installations that process or stock chemicals to take measures to prevent industrial accidents, to inform the adjacent population of potential risks, and to elaborate on-site and off-site emergency plans. This legislation is described in detail in Chapter 5.

In 1985, the EEC adopted a Directive that required an environmental impact assessment for a number of public or private projects, before planning consent was given [2.16]. In 1993, the EEC adopted a Regulation on the voluntary participation in a system of eco-auditing [2.17]. This Regulation aims at the improvement of eco-management by industrial installations. The Regulation will take effect in spring 1995. Also, in 1993, the Commission proposed a legislative measure at the Community level on integrated pollution control, which aims at a generalized, integrated permitting system for industrial installations, in order to improve, rationalize, and harmonize authorizations for industry and adequately protect the environment [2.18].

These Community rules on chemical installations were either umbrella rules or left Member States relatively large discretion in complementary rule making at the national level. This reflects the fact that while chemical products should circulate freely within the EC, the chemical installations that produced these products are not movable. Therefore, where legislation on installations was adopted prior to the 1987 amendment of the EC Treaty, it was based on Articles 100 and 235 of the EC Treaty at the same time. After the single Act 1987 took effect, Community legislation on installations was based on Article 130 s; the one exception until now – legislation harmonizing national programs on disposal of waste from the titanium dioxide industry, which was finally based on Article 100 a of the EEC Treaty [2.19] – seems isolated. In general, legislation on industrial installations will for the foreseeable future probably be based on Article 130 s of the EEC Treaty. This is likely also to apply to accessory environmental legislation, such as emission of pollutants into the air or soil.

The consequence of the use of Article 130 s is that Member States, under Article 130 t, will remain entitled to maintain or introduce more protective legislation with regard to chemical installations. Different national legislation, however, might lead to different production costs and gradually influence the competitive situation of industrial installations. This, in turn, might have an influence on the free circulation of goods produced in such installations. Where Community rules on installations have a significant impact on the cost of production, good reasons exist to consider whether the establishment of an Internal Market does not require equivalent standards over the entire Community territory. This would suggest that for such rules, Article 100 a be chosen as a legal basis, since standards that are laid down on the basis of Article 130 s will – because of the existence of Article 130 t – never be capable of achieving uniformity. For example, Directive 82/501/EEC on the prevention of major accidents, which affects mainly industrial installations that produce or process chemical products, might well request such considerable investments for accident prevention that rules within different Member States would distort competition for the chemical products that are placed on the market. According to the content of such a revised Directive, good reason might thus exist to base such a Directive on Article 100a and not on Article 130s of the Treaty. Indeed, the precise content of the Directive will determine whether one can argue that the objective of such an accident prevention Directive was principally the protection of humans and the environment and whether the establishment of local competitive conditions among installations was only an accessory objective.

3. Towards the Adoption of New Civil Liability Rules: A New Challenge for the Chemical Industry

"Toutes les pertes et tous les dommages qui peuvent arriver par le fait de quelque personne, soit imprudence, légèreté, ignorance de ce que l'on doit savoir, ou autres fautes semblables, si légères qu'elles puissent être, doivent être réparées par celui dont l'imprudence ou autre faute y a donné lieu. Car c'est un tort qu'il a fait, quand même il n'aurait pas eu l'intention de nuire." Extract from "Lois Civiles, II" by Jean Domat (1625–1696). The original French text can be translated as follows: "All losses and all damage caused by the deeds of any person, through negligence, thoughtlessness, ignorance of what one should know or similar shortcomings, however trivial they may be, must be compensated by the person whose fault or negligence has created the damage. Because he generated a tort, even though this was unintentional."

3.1. Introduction

Significance of Civil Liability and its Evolution. A significant number of research, production, and processing activities are carried out by a host of individuals and companies. Once manufactured, the most diverse products – from the simplest to the most technologically complex – are put on the market and are subject, among other things, to transport, packaging, labeling, and storage operations. All of these activities may cause damage either to persons, to goods, or to the environment. Thus, one can easily imagine the importance of "civil liability law" or "third-party liability" and its economic impact on industry. "Manufacturer's liability" must be considered in the framework of general legal systems of third-party liability, or "responsabilité civile ou quasi-délictuelle" in French civil code systems.

Basics of Civil Liability. Traditionally speaking, civil liability for damage means an obligation, under certain conditions, to compensate for damage caused to persons, property, or proprietary rights. The concept of environmental damage is relatively new and refers only to the nature of the damage. It basically addresses damage to the unowned environment, for example, disappearance of a fish species in a river. The plaintiff has to prove the existence of the following factors: (1) that he/she has suffered damage to his/her person or property; (2) that the damage was caused by an offense of negligence by the defending party (i.e., was based on fault); and (3) that a causal link exists between the damage and the incident.

These fundamental principles of civil law are similar in most European countries but are subject to evolution as described below. Liability instruments of direct interest to industry have been drafted in relation to

1) Defective products
2) Transport of (hazardous) goods
3) Industrial activities

Chapter 3 addresses liability, and compensation questions, in the context of industrial and marketing or trade activities. In the following material, the major trends affecting the evolution of civil liability are outlined, and various major international legal instruments recently adopted or still under discussion in this area are reviewed. Some of the major issues at stake and their perspective on the eve of the year 2000 are outlined in this chapter.

3.2. General Trends and Debates Affecting Modern Civil Liability Systems

3.2.1. New Trends

When can a Person or Company Be Held Liable? In most European countries, existing civil liability laws are based on fault ("continental law systems") or on the "negligence concept" ("common law systems"). However, in some countries new laws with a "strict" liability regime as a basis for liability have been enacted or are under consideration. The strict liability regime usually applies to damage caused by activities that are deemed especially dangerous, such as handling hazardous waste, dangerous substances, or nuclear energy. In other cases, the case law (theoretically based on the fault concept), has led to a similar result. Under common law, the "duty of care" has been considerably extended and case law developed in the direction of strict liability. Similarly, fault-based liability is interpreted by courts and jurisdictions in such a stringent manner that the difference between these regimes has shrunk. Even without such specific law, the case law of most countries has developed a strict liability regime for damage caused by dangerous activities, such as handling explosives or operating dangerous installations.

At European level, the Directive on defective products (see Section 3.3) is based on strict liability of producers for defective products put on the market. A similar trend toward strict liability has characterized international conventions in the context of land and maritime transport of dangerous goods [the so-called International Institute for the Unification of Private Law (UNIDROIT) convention, and the present "Hazardous Noxious Substances" (HNS) draft convention of the International Maritime Organization].

Caution is needed and "labels" can be misleading because a fault-based liability regime can actually be more severe for potentially liable parties than a well-defined strict liability system. Enormous conceptual and practical differences also exist between the so-called strict liability systems.

For example, little common ground exists between the "operator's liability" approach, which is the basis of the Council of Europe Convention (see Section 3.7), and the absolute "generator's liability," which was once proposed in the context of a draft protocol on liability for waste to the Basel Convention [3.1]. Both texts are supposed to be based on strict liability. The operator's liability approach focuses on actual responsibility of the person in control of certain activities and substances; the generator's liability approach merely indicates who should bear the damage consequences of products (here waste) whether this person is, or is not, in control. The generator can be held liable even for acts committed by third parties, professional or not.

Arguments in Defense. Certain arguments may be used to defend oneself. One cannot be held liable if one proves the existence of "force majeure" (i.e., the damage was caused by an act of war or hostilities or by a natural phenomenon of an exceptional, inevitable, and irresistible character). A similar defense is, for instance, a claim that the damage was caused by an intentional act by a third party or that the person who suffered the damage had by his own fault contributed to the damage.

Additionally, different laws contain defenses of a special nature typical of the area concerned. The state-of-the-art defense is one example of regulatory exclusion of liability. According to the Directive on defective products, producers are not held liable if they prove that the state of scientific and technical knowledge at the time they put the product into circulation was not such as to enable the existence of a defect to be discovered.

In the context of liability related to industrial activities, compliance with permits is taken into consideration although it does not lead to an absolute defense. Emissions within "tolerable levels under local circumstances" (Council of Europe Convention) also offer possible defenses depending on legislation or case laws.

The Emergence of the Polluter-Pays Principle in Civil Liability Law (see Section 1.2.2). The polluter-pays principle was introduced in the European Communities Constitution via Article 130 r of the Single European Act and later in the Treaty on the Economic Union. Scientific literature on the polluter-pays principle is considerable but in what concrete and practical manner it will be applied is still unclear because interpretations of questions such as "Who is the polluter?" still differ.

3.2.2. Recognition of Ecological Damage in International and Community Law of Civil Liability

One of the most significant aspects of the present debates and current work in the international sphere is the extension of the concept of damage (death, personal injury, damage to property) to ecological damage. This same ecological damage is defined in various ways, depending on the instruments analyzed. Some common features can, however, be identified: (1) environmental damage concerns the community, not the individual; (2) environmental damage is difficult to quantify. For example, what is the value of a variety of flowers that has disappeared because of pollution? U.S. jurisdictions and legal professions have made considerable efforts to quantify the ecological damage. Another interesting source is the case law developed in connection with major maritime pollution cases (e.g., the Amoco Cadiz decision).

The European Commission Green Paper on Remedying Environmental Damage of 1993 (see Section 3.2) envisages (1) the harmonization of civil liability rules for environmental damage and (2) the possibility of remedying environmental damage not met by the application of civil liability principles. The Green Paper, being a discussion document only, does not propose any definition of environmental damage but refers to definitions contained in the Council of Europe Convention and in the amended proposal for a Council Directive for damage caused by waste. The fundamental importance of and difficulties related to the definition are emphasized.

The 1991 amended proposal for a Council Directive on civil liability for damage caused by waste (see Section 3.4) defines "impairment of the environment" as any significant physical, chemical or biological deterioration of the environment in so far as this is not considered to be damage within the meaning of "damage to property." "Damage" as such, in turn, is traditionally defined to mean death or physical injury and damage to property. The compensation obligation under the Directive would apply for both damage and impairment of the environment. The Council of Europe Convention refers to "loss or damage by impairment of the environment ... [which] shall be limited to the costs of measures of reinstatement actually undertaken or to be undertaken." This definition includes the definition of the "CRTD Geneva Convention" of October 10, 1989 (see Section 3.5).

The word environment is itself defined very broadly in the Council of Europe Convention; it covers natural abiotic and biotic resources, such as air, water, soil, fauna, and flora, and the interaction among the same factors, assets making up the cultural heritage, and the characteristic aspects of the landscape. These last two elements of the definition illustrate the erosion of the traditional concept of damage, because it is now a matter of covering damage to collective property by civil liability mechanisms. Such damage is particularly difficult to quantify. The Commission, unlike the Council of Europe, has avoided giving a definition of the word environment as such in the mentioned proposal for the Directive on liability for waste.

3.2.3. International Harmonization in the Context of Liability for Dangerous Activities

Opinion is growing that civil liability rules should be addressed at the international level not only in relation to products and goods, but also in respect of industrial activities. International rules are said to be necessary because pollution is no respecter of national boundaries and Member States have been establishing different Regulations in this field that could distort competition and hamper the smooth operation of a single market. Protection of the environment is also one of the Community aims, equivalent to creation of the single market, as enshrined in the Treaty of Rome Articles 130 r–t.

This approach was much debated in the early 1990s. Although harmonization of civil laws is conceptually attractive from the point of view of the Internal Market, it is effectively desirable for industry only if it would simplify existing systems and if new international instruments do not simply add another layer of possible legal actions. Harmonization efforts are also seriously challenged by the existence of different procedural rules linked to national judicial systems.

The two most important efforts to introduce harmonization of liability for dangerous activities have been developed within the Council of Europe and the EEC. The Commission Green Paper on Remedying Environmental Damage envisages harmonization of (strict) civil liability rules for environmental damage, as well as the possibility of remedying environmental damage not met by the application of civil liability principles. The latter means creating compensation funds to be used when no liable party can be identified or in case of the insolvency of the liable party. Civil liability is also considered the domain of Community legislation because it is a legal principle closely linked to general European enviromental policy principles, such as the prevention of damage and the polluter pays.

Another important international document is the already-mentioned Council of Europe Convention on civil liability for damage resulting from activities dangerous to the environment. The convention was influenced by major pollution of the Rhine river by Sandoz that affected several countries. It is based on the belief that issues relating to civil liability and compensation for damage resulting from dangerous activities need to be addressed at the international level.

3.2.4. Actions and Requests by Organizations

The right to sue is normally given only to the party with a legal (i.e., direct and certain) interest in recovering compensation for injuries or for damage suffered to property. Where damage occurs to property that is not owned (i.e., environmental damage), no injured party with the right to bring legal action can be identified. At the European level the tendency is to grant such a right – although limited and under the control and at the discretion of the courts – to environmental associations. This right to protect the "public interest" normally rests with the public authorities that operate under public law and democratic institutions.

For example, according to Article 4(3) of the amended proposal for a Council Directive on civil liability for damage caused by waste, common interest groups or associations, which have as their objective the protection of nature and the environment, are given the right either to seek remedy or to join in legal proceedings that have already been brought. Also the Council of Europe Convention gives certain restricted rights to associations seeking to protect the environment; namely, the group may request of courts or jurisdictions (1) that an unlawful, dangerous activity, which poses a grave threat of damage to the environment be prohibited; (2) that the operator be ordered to take measures to prevent an incident or, after an incident, to prevent damage; or (3) that the operator be ordered to take measures of reinstatement. However, any party to the convention may reserve the right not to apply this provision.

3.2.5. Financial Security Schemes

Several proposals suggest compulsory insurance or other financial security schemes to cover the liability under various pieces of legislation. Article 11 of the amended proposal for a Council Directive on civil liability for damage caused by waste provides that the liability of the producer, who in the course of a commercial or industrial activity produces waste, and of the eliminator of waste, be covered by insurance or other financial security. The amended proposal for a Council Directive on the landfill of waste, in turn, contains an obligation on the operator of the landfill to have a financial guarantee or any other equivalent, whose purpose is to cover the estimated costs of the closure procedures and after-care operations of the landfill (Art. 17).

3.2.6. Essential Requirements

The trends described above represent a real evolution in fundamental civil liability rules. However, some of them are still quite flexible, and others imply more theoretical than practical changes compared to the existing situation. An example of this is the discussion of whether liability should be fault based or strict.

Nevertheless, some essential requirements exist for any legislation on civil liability, one of the most significant of which is maintaining the establishment of causation. It is of utmost importance to the entire system that the causal link between the incident and the damage suffered be shown by the plaintiff. Likewise, in the context of damage caused by industrial activities, causation must be proved to exist between the operator in charge of the activity and the incident that led to the damage.

3.3. Liability for Defective Products and European Community Law

3.3.1. Background and Legal Basis

The Directive on defective products was approved by the Council on July 25, 1985 [3.2]. European Community Member States had to implement the Directive by July 1988. In early 1994, most Member States had effectively introduced the Directive into their internal law. Only France and Spain had not yet enacted the Directive, which led the Commission to open legal proceedings in the European Court of Justice for the noncompliance of these two countries with their obligations under the EC Treaty.

The declared objective of approximation of the laws has been partially attained since the text allows Member States to introduce some variations to three specific provisions. Article 100 of the Treaty of Rome provides the legal basis for the Directive and lays down the following conditions:

1) Unanimous council decision, based on a proposal from the Commission
2) Compulsory consultation of the assembly (European Parliament) and the economic and social committee if there is a need to

amend national laws. Both institutions were consulted and expressed their opinion.

3) Grounds for the Directive: According to indent one, the approximation of the laws of the Member States concerning the liability of producers for damage caused by the defectiveness of products is necessary because existing differences may distort competition, affect the free movement of goods within the Common Market, and lead to differing degrees of consumer protection.
4) Social–political justification: The preamble assumes that "liability without fault on the part of the producer is the sole means of adequately solving the problem, peculiar to our age of increasing technicality, of a fair apportionment of the risks inherent in modern technological productions."

The industry has challenged strict liability and expressed preference for liability regimes based on fault or negligence in accordance with traditional principles of the law of torts. This battle was lost for the chemical industry, however, when even Italy, once a strong opponent of strict liability, accepted this regime in the political compromise that led to the adoption of the Directive in 1985. Was this a real source of concern to industry and should strict liability a priori be perceived as hostile by industry? The answer very much depends on three interrelated factors:

1) The content of actual provisions of the Directive
2) The content of national laws implementing the Directive
3) Case law, including that of the EC Court of Justice, that could play an important role in clarifying ambiguities

As stressed in Chapter 5, more important than the "labels" or names used to qualify liability regimes are their specific features. Other elements are also important: the practice of the courts, the judicial organization and its cost, and the general attitude of the public toward the liability and compensation system. In this connection, the Directive, although far from perfect from the point of view of legal certainty, was probably a reasonable compromise acceptable to governments, industry, and consumers (although some would not share this view, both in consumer circles and in industry).

As shown below, the question of recognition of the state-of-the-art defense was an important aspect of the "package" finally adopted and illustrates quite well the conceptual tensions and arguments that prevailed during the European Commission and the Council of Ministers discussions.

3.3.2. Nature of the Liability Regime

Principle: No-fault Liability. The Directive provides for the following system: the injured party has to prove the damage, the defect of a product, and the causal relationship between the defect and the damage (Art. 4). Thus, negligence and fault are no longer relevant as sources of producer's liability. Suppression of the "fault" standard should ease the task of the victim. However, it should be stressed that the case law of most countries had already introduced a softening of the victim's position, for example, in reversing the burden of proof or in extending considerably the concept of fault or "negligence" of the producer. According to common law concepts, the duty of care is now very strict indeed, with the important exception of the state-of-the-art defense.

Lines of Defense and Exemptions from Liability (Art. 7 of the Directive). The producer must prove that his/her is one of the cases of exemption provided for, namely,

1) That he/she did not put the product into circulation
2) That the defect did not exist when the product was put into circulation or that it came into being afterwards
3) That the product was neither manufactured for sale or any other form of distribution, nor manufactured or distributed by him/her in the course of his/her business
4) That the defect is due to compliance of the product with mandatory Regulations issued by the public authorities
5) State-of-the-art defense ("development risk", see Section 3.3.3)
6) In the case of a manufacturer of a component, that the defect is attributable to the design of the product in which the component has been fitted or to instructions given by the manufacturer of the product

This last point is very important and was stressed by the chemical industry. All sectors of the chemical industry should be fully aware of this defense. It will be more and more important to have clear agreements between purveyors and buyers about the purity and characteristics of component products.

3.3.3. The Three Options of the Directive

A fundamental question that had to be addressed by the authors of the Directive was the so-called harmonization question. Should the EEC Directive provide for strict harmonization ("maximum Directive") or should it create a "minimum system" allowing the Member States to establish different and additional standards, depending on their concepts of consumer protection? Real harmonization being one of the raisons d'être of this text was preferable. The EEC Member States delegations recognized that a maximum Directive would be the best solution from the Treaty of Rome point of view, but at the level of details, compromises became necessary and the maximum approach turned into a limited approximation of laws.

State-of-the-Art Defense. A highly important question is, should the manufacturer of a defective product be held liable if the state of scientific and technological development did not enable him to detect the existence of a defect at the time the product was put in circulation. The inclusion of development risks in the scope of liability was predicted to interfere with technological innovation. Until now, in the vast majority of Western legal systems, no party has had to accept the risk of the unknown and unforeseeable. As CEFIC wrote in a position paper addressed to the EEC Council:

"While it is perfectly understandable that a system of liability should oblige producers to take all possible precautions against any possible injury to the consumer, it seems totally unreasonable to require that these same producers be liable for the unknown, and for risks over which they have no control whatsoever. Injury caused by a defect in a product cannot be a source of liability if the defect was undetectable and unforeseen at the time of its being put into circulation because of the state of the art."

After long and difficult debates, the EC Council finally introduced in the Directive what is known as the state-of-the-art defense. Accordingly, the producer has the possibility of proving that the state of scientific and technical knowledge when the product was put into circulation was not such as to enable the existence of the defect to be discovered. This defense may therefore be introduced in the legislation of Member States and was effectively introduced in most national laws of implementation.

However, the Community regime allows individual Member States to maintain the liability of the producer for "development risk" if it already existed in their internal laws when the Directive was approved (i.e., July 1985). Only Germany was affected by this provision, because its Pharmaceutical Act did not allow producers to use the state-of-the-art defense.

The Community regime also allows Member States to introduce derogations to that principle in new legislation, although this possibility is tempered by a clause of Community "standstill" and the obligation to follow a specific procedure under the supervision of the Commission [Art. 15 (1) of the Directive]. The importance of recognizing a state-of-the art defense for sectors such as the chemical and related industries is clear if its impact on research and development is considered, which in certain sensitive areas might otherwise be discouraged. Evidence already exists, especially in the United States, showing that the liability regime based on absolute liability and disregarding important factors such as the state of technology and the plaintiff's behavior has had adverse effects on new products.

In the United States, the impact of certain court cases based on "absolute producer liability" on insurance costs and the (non)availability of coverage has been tremendous. In the Federal Republic of Germany, the Pharmaceutical Law has not – thus far – generated similar experiences. However, nobody can predict what the consequences would be for the German pharmaceutical industry of a serious "developmental risk" case similar to the thalidomide affair. The compromise achieved in the Directive on the state-of-the-art defense seems to preserve a fair balance and efficiency.

Derogation in Respect of Financial Limitation of Liability. The Directive makes no provision for a Community system of financial ceilings to liability. It does, however, allow Member States to introduce on an individual basis a liability "ceiling" for personal injury, which may then not be less than 70×10^6 ECU (European Currency Units). The German and Belgian laws have, for example, introduced such ceilings.

The Commission has been instructed to submit a report to the Council ten years after the date of notification of the Directive (1995).

Acting in the light of this report, the Council is to decide whether or not to cancel the option of introducing ceilings. As with the state-of-the-art issue, such a provision will mirror the Council's political embarrassment at the time of its adoption. The Council will, however, be free to decide to maintain the state-of-the-art and the ceilings option if a political majority so wishes.

Derogation in Respect of Agricultural Products. The Directive normally excludes agricultural products from its scope. However, Member States are authorized to include them within their laws and to apply the regime of the Directive.

3.3.4. Other important Aspects of the Directive

The Directive covers not only personal injury but damage to property (with a "lower threshold" of 500 ECU in the second stage). This provision has led to some divergent interpretations.

Definition of Producer. "Producer" is extensively defined in the Directive as

1) The manufacturer of a finished product
2) The producer of any raw material
3) The manufacturer of a component part
4) Any person who, by putting his name, trademark, or other distinguishing feature on the product, presents himself as its producer

Clearly, the concern of the Directive's drafters has been to avoid the possibility of an injured person finding himself in a situation for which no one is responsible. Hence, care is taken in defining the producer. So true is this that the Directive provides for treating the importer into the Community as a producer within the meaning of the Directive and, should it prove impossible to identify the actual producer, treating the supplier as the producer.

Product Defectiveness. For the purposes of the Directive, a product is defective not according to objective or material criteria but with reference to the "safety which a person is entitled to expect." Thus, § 6 of the recital states that defectiveness does not mean the product's lack of fitness for its intended use but the lack of safety that the public at large is entitled to expect. Such a reference creates difficulties of interpretation. In reality, it will be for the judge to assess. However, the Directive provides that all circumstances be taken into account in assessing the safety the product should provide.

Some of the criteria to be considered in assessing the safety of a product are: (1) the presentation of the product; (2) the use to which it could reasonably be expected to be put; and (3) the time when the product was put in circulation. Safety is to be assessed by taking into account the expected use of the product, its marketing, the information given and user instructions, and similar factors related to the product.

Conclusion. The new Community regime on liability for defective products contains both pros and cons and seems generally balanced. Despite imperfections and weaknesses, it genuinely contributes to the progressive establishment of similar legislations in Europe: one major weakness is that it leaves open the question of diverging procedural rules in the judicial systems of the Member States.

Another important question is the possible impact of other more recent Community legislation, i.e., the Product Safety Directive and the proposed Directive on Liability of Suppliers of Services on the Product Liability System.

The Product Safety Directive was adopted on June 29, 1992 and had to be implemented by the Member States in June 1994. The Directive was challenged on grounds of the Maastricht Treaty "subsidiarity" principle by the Federal Republic of Germany. A very wide and strict "general safety requirement" is introduced, although the Directive does not apply to sectors that are already subject to specific legislation. The Directive introduces a general obligation to market "safe products" only, which is not the same concept as a (non)defective product of the Product Liability Directive. Other provisions address the following issues:

1) Monitoring of products
2) Provision of information to consumers concerning possible risks involved in use of the products
3) Adoption of measures to ensure information of consumers
4) Ensuring uniform management of tests, checks, warnings, prohibitions, and market withdrawals

In practice, this legislation should be seen in relation to the already existing comprehensive laws on safety and should contribute to the highest safety standard for products marketed in the European Union.

The proposed Directive on the Liability of Suppliers of Services would make the supplier liable for damage to the health and physical integrity of persons or to property caused by a fault committed by the supplier. However, the proposal "revises the burden of proof" since the supplier would have to establish the absence of fault. The prospects of the Directive being adopted were not very good in 1994 because most Member States were reluctant to accept Community legislation in this area.

The practical impact of the new Community regime, particularly on insurance costs, has so far not been dramatic, but it is still uncertain and will depend on the content of national laws and their application by national courts under the control of the European Court of Justice.

3.4. Community-Proposed Directive on Liability for Waste

Implementation of the polluter-pays principle provides for consideration of the scope of civil liability to better define responsibility in the environmental field. So as to answer this requirement, also expressed in the previous Directive on the control of transfrontier movements of waste (84/631/EEC) and several other Community documents, the Commission in 1989 submitted a proposal for a Directive on civil liability for damage caused by waste. This proposal was amended two years later, in 1991 [3.3].

The proposal raised a number of arguments, one of them being that the Community Directive on civil liability should be compatible with the UNIDROIT Convention and, on a broader level, with the approach developed in the Council of Europe Convention. With the content of the amended proposal, this compatibility did not seem ensured. For example, contrary to the international conventions mentioned, the proposal channelled liability onto the waste producer regardless of whether or not the producer was in control of waste. To create a system that is workable and practicable, all other equivalent instruments must be studied and considered. Although the proposal was put aside pending the results of comprehensive discussions on the basis of the Green Paper on Remedying Environmental Damage, the main aspects of the draft Directive are described.

3.4.1. The "Channelling" of Liability

Article 3 of the proposal from 1991 poses the principle of the liability of the waste producer for damage and impairment of the environment caused by waste, irrespective of fault on the part of the producer. Simultaneously, the proposal is intended to implement the polluter-pays principle. However, this principle – the legal ambit of which remains ill-defined – does not necessarily imply a strict liability regime. A liability regime based on the polluter's fault would be compatible with this principle if the regime ensures that the polluter actually suffers the consequences of the pollution. The Commission's choice of a strict liability regime channelled onto the producer is explained by the concern to use civil liability as an instrument to eliminate or reduce waste. Such an objective, however, has no direct relation to the function of civil liability, which is to ensure compensation of damage arising from wrong use or handling of products whether they are waste or not.

A positive feature of the proposal is Article 2.2.C., providing a transfer of liability whenever the producer delivers waste to a licensed or approved treatment or disposal facility. Applying the same logic, it is difficult to see how a similar transfer of liability would not occur upon delivery to a licensed carrier.

3.4.2. Definition of the Producer

The Commission proposal refers to any person who, in the course of a commercial or industrial activity, produces waste and/or anyone who carries out preprocessing, mixing, or other operations resulting in a change in the nature or composition of this waste. Article 2(2), however, states that the following are to be deemed the producer of the waste:

> "The person who imports the waste into the Community;
>
> The person who had actual control of waste when the incident giving rise to the damage to or impairment of the environment occurred if the producer is not identified;
>
> The person responsible for the installation, establishment or undertaking where the waste was lawfully transferred to such installation, establishment or undertaking licensed pursuant to Directive 75/442/EEC on waste (amended thereafter), 75/439/EEC on waste oils or 78/319/EEC on hazardous wastes (amended thereafter)."

The method chosen in the draft leads to a complex definition based on a system of references to previous Directives, all of them providing for treatment, storage, and disposal operations subject to licensing. This leads to a contradiction in the underlying philosophy when it "channels" liability onto the producer while acknowledging the responsibility of the operator or the person having "actual control" of the waste, but only in cases where the producer has not been identified.

3.4.3. The Concept of Waste

The Commission proposal refers to the definition of the original waste Directive 75/442/EEC, which was amended in 1991. Waste is defined under Article 1(a) of the amended Directive (91/156/EEC) as any substance or object in the categories set out in Annex 1 that the holder discards or intends or is required to discard. The concept of waste is rather broad by definition but excludes, for instance, gaseous effluents emitted to the atmosphere, radioactive waste, waste arising from mining operations, and decommissioned explosives. Given the fact that the definition is also fundamentally based on a subjective approach – the intent of the holder is decisive – judging whether or not a certain material is waste is far from simple. The categories set out in the mentioned annex do not make this easier, giving only rough examples of what should be considered waste.

Problems related to the definition of waste would complicate implementation of the Directive proposal. Different interpretations of waste in various Member States might also distort competition. Problems related to the definition have been under discussion for years, but a satisfying solution has not yet been found. Finally, one more question is still to be raised: Why is a harmonization of civil liability laws required especially for damage caused by waste? Would that not create an uneven situation when other possible sources of damage are left untouched?

3.4.4. Insurance and Compensation Fund

Article 11 of the draft Directive states the obligation of the producer, who produces waste in the course of a commercial or industrial activity, and of the eliminator of waste to be covered by insurance or other financial security. The article also provides for the Commission to study the feasibility of establishing a European fund for compensating damage caused by waste when the person liable cannot be identified or is incapable of providing full compensation.

3.5. Civil Liability for Damage Caused during the Carriage of Dangerous Goods by Land: The Convention of October 10, 1989

The International Institute for the Unification of Private Law and the United Nations Economic Commission for Europe have actively cooperated in framing an international instrument on liability and compensation for damage caused during the carriage of dangerous goods by road, rail, and inland waterways. Prepared within the Institute of Rome, the draft was subsequently forwarded to the UN Economic Commission for Europe. CEFIC, along with other international organizations representing transport and insurance interests, participated as an observer in the work of UNIDROIT and the United Nations. The draft convention was finalized at the extraordinary session of the Inland Transport Committee of the UN Economic Commission for Europe, held in Geneva from October 2–10, 1989.

The main thrust of the Convention rightly establishes the linkage between civil liability and operational responsibility. This last, dated October 10, 1989, has been open for signature by the states since February 1990. Five ratifications are required for it to come into force. Among the countries that expressed at that time their active support are West European countries (especially The Netherlands, Denmark, the Federal Republic of Germany, and France with some reservations, Finland, and Austria) and also Central and Eastern European countries, and the United States.

The salient features of the Convention are as follows:

1) A "strict" liability regime is in effect.
2) For the purposes of the Convention, carriage comprises the period from the beginning of loading of goods to the completion of unloading operations.
3) Liability is channelled onto the carrier (save in the case of loading or unloading when such operations are carried out under the control of another person).
4) Reference is made to the ADR list for the definition of substances to which the Convention will apply (including hazardous wastes).
5) The principle of compulsory insurance (or any system of equivalent financial security) exists.
6) Total financial liability of the carrier (road, rail) is limited to a ceiling of 30×10^6 ECU per vehicle (18×10^6 ECU for personal injury; 12×10^6 ECU for damage to property and the environment). In the event of the "personal injury" ceiling being used up, the amounts of unmet claims for personal injury

could compete with claims for damage to property within the "damage to property" ceiling. The liability ceiling specified for inland waterway transport is 15×10^6 ECU (8×10^6 ECU for personal injury; 7×10^6 ECU for other damage).

7) The carrier (road, waterway) is defined as being the person having use of the vehicle. The carrier is presumed to be the person on behalf of whom the vehicle was registered. This presumption may be reversed.
8) The rail carrier is defined as being the person(s) operating the railway on which the incident giving rise to damage occurred.
9) Grounds for exemption from liability are based on:
   Lack of information by the shipper or any other person as to the nature of the goods
   The fact that a third party acted or failed to act with intent to cause damage
   Deliberate fault of the injured party
   Occurrence of certain exceptional events: act of war, civil war, or natural phenomena of an exceptional and irresistible nature
10) The definition of "damage" is extended to cover ecological damage in addition to the more traditional categories of "death or personal injury" and "damage to property." Ecological damage is defined as "any loss or damage by contamination of the environment caused by dangerous substances, provided that the compensation for impairment of the environment ... is limited to costs of reasonable measures of reinstatement actually undertaken and to be undertaken."

The states will, however, be able to introduce some reservations if they so wish (particularly concerning defenses and the limiting of liability). From the economic and practical standpoints, it does not look as though the Convention will result in upheaval in land or rail transport. The Conseil des Bureaux (international insurers' body designed to facilitate the settlement of damage caused by automobile traffic – "green card system") has stated that the "green card" system will adapt fairly easily to the introduction of the Convention into the laws of Member States. As far as rail is concerned, the vast majority of railway companies have their own systems of insurance coverage. The Convention is, by contrast, a novelty for waterway transport, for which no compulsory insurance regime has existed up to now.

The authors of the draft have forwarded a resolution to the International Maritime Organization for a similar convention to be adopted for maritime transport, thereby complementing the work carried out in regard to land transport.

3.6. Liability in Connection with the Carriage of Dangerous Substances by Sea (HNS)

The adoption of an international draft convention on civil liability has for some years been included in the agenda of the IMO Legal Committee. An initial draft based on a "two-tier" system of liability – both complicated and costly – failed to achieve adoption in 1984. Despite this setback, the International Maritime Organization still sees the issue as a priority. The IMO Legal Committee is discussing several alternatives:

Alternative 1:
Sole liability of the shipowner coupled with a general raising of amounts of the total limitation stipulated in the 1976 Convention on the Limitation of Liability for Maritime Claims (CLLMC).

Alternative 2:
Sole liability of the shipowner coupled with an additional specific amount provided for in the CLLMC Convention for damage caused by HNS substances.

Alternative 3:
Shipowner liability limited in accordance with the provisions of the CLLMC and complemented by compulsory insurance, the cost of which would be borne by the shipper.

Alternative 4:
Shipowner liability limited in accordance with the provisions of the CLLMC and complemented by a fund financed by the shipper through either a pre- or a post-event financing mechanism.

Other options are possible, including the adoption, envisaged at the beginning of 1994, of a maritime carrier liability regime complemented by setting up an international compensation scheme, to come into operation beyond the limits provided by the shipowners' liability coverage or when the person responsible is not solvent. The system would be based on financial contributions calculated as a function of the type of cargoes and their hazard. Various options have been discussed, including the possibility of requesting importers, not shippers, to contribute. In any event, users of transport services would be asked to contribute.

The issue of adoption by the IMO of an international convention on liability and compensation is expected to be addressed in a 1996 diplomatic conference and will remain an important priority in years to come.

3.7. The Council of Europe Convention on Compensation for Damage Caused by Dangerous Activities

3.7.1. Background

In the 1986 Oslo Conference of European Ministers of Justice, the decision was taken to consider options to extend the reach of civil law to promoting environmental protection. Accordingly, the European Council of Ministers Committee set up a committee of experts in 1987 to propose measures for compensation for damage caused to the environment.

After its sixth meeting, the committee of experts instructed the Secretariat General and the Legal Affairs Directorate of the Council of Europe to prepare an interim activity report. This report was to be submitted to the European Committee on Legal Cooperation (CDCJ), a senior Council of Europe body in the legal field, and to Member States of the Council. The report enclosed preliminary draft rules on compensation for damage resulting from dangerous activities. The content of the Convention was adopted by the European Council of Ministers on March 24, 1993. The Convention was opened for signature on June 21, 1993 in Lugano and was signed by seven Council of Europe Member States (Cyprus, Greece, Finland, Italy, Liechtenstein, Luxembourg, and The Netherlands). Other countries were expected to follow suit, and debate within the European Union concerning the Commission's Green Paper has directly influenced this process.

The main aim of the Convention is to ensure prompt and effective compensation of damage to people, property, and the environment during the exercise of dangerous activities carried out in a professional capacity in installations or on sites. As stated in Article 1 of the Convention, it aims at providing adequate compensation for damage resulting from activities that are dangerous to the environment and also provides means of prevention and reinstatement.

3.7.2. Key Features of the Council of Europe Convention

3.7.2.1. Liability Regime

The Convention provides for a regime involving strict (no-fault) liability of "the operator of a dangerous activity." The operator's responsibility is not based on the concept of fault, but on the causal link existing between the damage caused and the incident occurring at the time or during the period when the operator was in charge of a dangerous activity. This general approach is in line with the one persistently advocated by the author and CEFIC, which have always wanted to see a strict link maintained between civil liability for damage and operational responsibility. The explanatory report to the Convention rightly points out major reasons for making the person in charge of the activity from which the damage resulted bear the cost of damage; the activity of this person is the source of the damage, and this person is best placed to prevent the damage and limit its extent.

3.7.2.2. Concept of Operator

The operator, as already discussed above, is defined as being "the person who exercises the control of a dangerous activity." Essential in determining who is in charge of a certain activity is to consider who has the power to decide on the way in which the activity is carried out.

3.7.2.3. Incident

Incident is defined as any sudden occurrence or any series of occurrences having the same origin that cause damage or create a grave and imminent threat of damage.

3.7.2.4. Concept of Dangerous Activity

Dangerous activity is defined as being "one or more of the following activities provided that it is performed professionally, including activities conducted by public authorities:

1) The production, handling, storage, use, or discharge of one or more dangerous substances or any operation of a similar nature dealing with such substances;
2) The production, culturing, handling, storage, use, destruction, disposal, release, or any other operation dealing with one or more:
   - Genetically modified organisms which as a result of the properties of the organism, the genetic modification and the conditions under which the operation is exercised, pose a significant risk for man, the environment or property;
   - Microorganisms which as a result of their properties and the conditions under which the operation is exercised pose a significant risk for man, the environment or property, such as those microorganisms which are pathogenic or which produce toxins;
3) The operation of an installation or site for the incineration, treatment, handling, or recycling of waste, such as those installations or sites specified in Annex II, provided that the quantities involved pose a significant risk for man, the environment or property;
4) The operation of a site for the permanent deposit of waste.

Dangerous substances, in turn, are defined as follows:

1) Substances or preparations which have properties which constitute a significant risk for man, the environment, or property. A substance or preparation which is explosive, oxidizing, extremely flammable, highly flammable, very toxic, toxic, harmful, corrosive, irritant, sensitizing, carcinogenic, mutagenic, toxic for reproduction, or dangerous for the environment within the meaning of Annex I, Part A to the Convention shall in any event be deemed to constitute such a risk;
2) Substances specified in Annex I, Part B to the Convention. Without prejudice to the application of sub-paragraph a above, Annex I, Part B may restrict the specification of dangerous substances to certain quantities or concentrations, certain risks or certain situations."

The operators of waste installations and sites for permanent waste deposit are explicitly subject to the new system. However, specific transitional provisions and liability rules are to be applied in the case of permanent waste deposit.

3.7.2.5. Concept of Damage

The concept of damage is defined as follows:

1) Loss of life or personal injury
2) Loss of or damage to property other than to the installation itself or property held under the control of the operator, at the site of the dangerous activity
3) Loss or damage by impairment of the environment in so far as this is not considered damage within the meaning of the preceding subparagraphs, and provided that compensation for impairment of the environment, other than for loss of profit from such impairment, is limited to the costs of measures of reinstatement actually undertaken or to be undertaken
4) The costs of preventive measures and any loss or damage caused by preventive measures, to the extent that the loss or damage referred to in subparagraphs (1) to (3) arises out of or results from the hazardous properties of the dangerous substances, genetically modified organisms, or waste.

The Convention covers legal categories traditionally included under the concept of "damage" in civil law (i.e., loss of life or personal injury). Additionally, the text introduces the concept of loss or damage by impairment of the environment, provided compensation is limited to the costs of measures of reinstatement actually undertaken or to be undertaken. Since "the environment" is defined in a remarkably broad way, the Convention potentially goes beyond the precedents established in the UNIDROIT Convention or in the EC draft Directive on liability for waste.

3.7.2.6. Environment

As already mentioned, the word "environment," in the confines of the Convention, includes the following factors:

1) Natural resources both abiotic and biotic, such as air, water, soil, fauna, and flora and the interaction among these factors
2) Property that forms part of the cultural heritage
3) Characteristic aspects of the landscape

The definition contained in the Convention is not exhaustive. On the contrary, the explanatory report to the Convention explicitly states that damage to any other aspect of the environment may also give rise to liability under the Convention. Even without this note, the definition is considered broad. Point (1) covers the factors traditionally understood to constitute "the environment." Points (2) and (3) refer to more immaterial values.

A legal definition of the environment (or damage to the environment) will influence the process of determining the type and scope of the necessary remedial action. The Commission Green Paper on Remedying Environmental Damage refers to this definition of the environment and also notes the problems related to the matter.

3.7.2.7. Right of Action of Private Groups

Article 18 of the Convention to some extent authorizes actions by private associations that according to their statutes, aim at the protection of the environment, while it restricts the scope of such actions before the courts:

1) Requests for prohibition of any unlawful dangerous activity posing a grave threat of damage to the environment
2) Requests for an order to the operator to prevent an incident or damage
3) Requests for an order after an incident to prevent damage or reinstate the environment

3.7.2.8. Causal Link and Administration of Proof

According to Article 10, while maintaining the requirement of a causal relationship between the incident giving rise to damage and the activity of the operator, the court "shall take due account of the increased danger of causing such damage inherent in the dangerous activitiy."

3.7.2.9. Financial Security Scheme

According to Article 12 of the Convention, each Member State shall ensure that where appropriate, taking due account of the risks of the activity, operators conducting a dangerous activity on its territory be required to participate in a financial security scheme or to have and maintain a financial guarantee up to a certain limit, of such type and terms as specified by internal law to cover the liability under this Convention.

The article leaves a lot of freedom for national legislators, because it provides for a financial security scheme for operators covered by the Convention where appropriate, of the type and to such a limit as wanted, respectively. This solution is to be welcomed as leaving Member States free in a matter that does not require strictly prescribed international rules.

3.7.2.10. Access to Information

A system of access to information held by authorities or operators is also contained in the Convention. The same system explicitly recognizes the right to protect the confidentiality of data. The system is based broadly on the provisions of Community law governing the subject.

3.8. The 1993 European Commission Green Paper on Remedying Environmental Damage

In addition to international organizations, new civil liability regimes have been enacted by various EC Member States. In its Green Paper on Remedying Environmental Damage of 1993, the European Commission assumed that different approaches adopted might lead to distortion of competition and hamper the smooth operation of the single market. For this reason, the Commission suggested that a Community system of civil liability in relation to activities dangerous to the environment was necessary. Civil liability is also considered to belong to the domain of the Community legislation, since it presents many interfaces with EC environmental policy (see Section 3.2.1.1).

The purpose of the Green Paper, adopted in March 1993, is to inspire discussion on whether and how requirements to remedy environmental damage might be introduced appropriately and effectively within the Community to recover the costs of such restoration [3.4]. As its name envisages, the communication is a discussion document that does not formally propose a certain regime on the matter. Instead, it presents alternatives for the legislator and summarizes different arguments for and against different solutions. Nonetheless, the document outlines the main principles as follows:

1) The Commission aims at harmonization of civil liability systems for environmental damage in cases when a liable party can be identified, and it expresses a clear preference for strict (objective) liability, e.g., as defined in the Council of Europe Convention
2) Civil liability is not seen as a means for compensating damage when no liable party can be found; thus, the Green Paper looks for another kind of compensation mechanism, such as a collective compensation fund

Therefore, the Commission aims at a strict liability regime combined with a nonliability-related system (i.e., the cost of repair could be shared among several economic sectors).

The Green Paper suggests that such a joint compensation mechanism is needed in the event of the following:

1) Damage is unbounded or latent
2) Cumulative acts or incidents have occurred
3) Liable parties cannot be identified
4) No basis exists for liability
5) No causal link is determinable
6) No party has a legal interest to bring action

Questions to be answered are on whom the burden to contribute to such funds should rest and whether such funds are necessary in addition to existing mechanisms. The Green Paper does not provide the answer but asks whether the polluter-pays principle should be respected to the greatest possible degree so that contributions are requested from "the economic sectors most closely linked to the type of damage needing restoration." However, a number of difficulties exist in interpretation of the principle and in identification of those sectors. For example, should the "sector" that is today most closely linked to the damage pay for damage caused by operators in the past? In addition, the identification of sectors that are most closely linked to the damage raises serious principle and technical objections. It can be argued that postpollution, unrelated to faulty or illegal pollution, is the result of a given stage of development of society. Therefore, collective and public law answers and compensation are more suitable for such "historic" pollution.

Apart from the U.S. Superfund [Comprehensive Environmental Response Compensation Liability Act 1980 (CERCLA)] and the funds established by the oil industry [Brussels Convention on the Establishment of an International FUND for Compensation for Oil Pollution Damage (1971) and parallel financial structures, such as TOVALOP (Tankers Owners Voluntary Agreement Concerning Liability for Oil Pollution), CHRISTAL (Contract Regarding Interim Supplement to Tanker Liability for Oil Pollution), and OPOL (The Offshore Pollution Liability Agreement)], such compensation schemes are still unusual and of minor significance. In the Netherlands, a fund on damage from air pollution was created in 1972. In France, a fund for noise, which compensates persons living around Paris airports, was created in 1973. The U.S. Superfund, as is widely agreed, has proved inefficient and unfair and therefore should not serve as a model. Solutions for the oil industry, in turn, serve only for purposes characteristic of that industry (i.e., those resulting from major transport accidents).

The Green Paper considers a number of issues related to civil liability and aims at stimulating discussions among all interested parties. Therefore, it has a major role in opening discussions on the fundamental principles of Community environmental policy and also on the Community's role and need for harmonization in this area. The Council of Europe Convention recently adopted will, without doubt, have a great influence on the discussions since although it is binding on only a few states so far, it is the first existing international legal instrument covering civil liability in the context of dangerous activities for environmental damage as well as damage to persons and property.

The communication does not suggest any definition of environmental damage but merely refers to definitions contained in the Council of Europe Convention and in the amended proposal for a Council Directive for damage caused by waste. However, the fundamental importance of and difficulties related to the definition are emphasized.

3.9. Economic Aspects and Questions Bearing on the Insurance of the Industrial Risk

In the last few years, industry has encountered growing difficulties in renewing its "civil liability operation" insurance policies, in terms of both the cost of premiums and the extent of risks covered. Insurance companies, as well as the reinsurance market, have been very hesitant, chiefly for the following reasons:

1) The impact of the U.S. liability crisis, characterized by the number and magnitude of claims for damages
2) Fear of a similar situation developing in Europe, particularly in view of the introduction of new liability laws and increasingly stringent technical Regulations
3) The potential liability of businesses involved in actions understood to be based, among other things, on damage to the environment whose cause is uncertain
4) Anticipated development at the national and international levels of new laws that would deal specifically with liability for pollution

This situation has prompted industry to study new additional methods of risk financing. In the United States, chemical and insurance companies alike have created insurance pools to increase available capacity.

In Europe, industry has found it increasingly difficult to find appropriate coverage, especially for "gradual pollution." Studies are therefore being undertaken to assess the "feasibility" of and arrangements for setting up a European Industrial Insurance Pool designed to provide added capacity and better conditions. Coverage of major industrial accident risks and of the risk of environmental pollution is reportedly being envisaged in particular. However, the actual adoption and operation of a pool depend largely on the development of the traditional insurance market. In many cases, conditions surrounding the renewal of policies can be said to have improved in recent years. One reason for this favorable trend could be that the envisaged creation of an insurance pool under direct industry control has served to make the insurance market more buoyant. The concept of a pool has rightly been perceived as mirroring the grave concern of industry and as offering an alternative to insurance mechanisms that no longer appear to be meeting industry's demand adequately.

The insurance crisis has not disappeared, however. Its acuteness has merely been mitigated a little. Major incidents or legislative initiatives unreasonably extending the scope of industrial civil liability could adversely affect the world insurance and reinsurance market, especially in the area of pollution.

3.10. Conclusions

Seeking harmonization of civil liability laws for all European Community member countries was the basis of the "product liability" Directive. Harmonization, or at least approximation of laws regarding liability for dangerous activities, was offered as a working hypothesis in the Green Paper published by the European Commission in 1993.

The issue could, however, be addressed in an even wider context, given that other economic or geographic areas might be affected by the transfrontier effects of pollution, as illustrated by major industrial accidents outside EC territory. With regard to the European Community, a harmonization of national laws is to be supported, if it eliminates trade distortions and simplifies existing provisions. A harmonization of this kind could, moreover, be achieved in a broader framework than that of the Community. However, real harmonization – a simplification of laws and a reduction of potential conflicts – is a most difficult and challenging task for lawmakers. No genuine harmonization can be achieved in this area without parallel progress in harmonization of the procedural rules governing civil liability. Therefore, this is the beginning, not the end, of a complex process.

The U.S. liability and insurance crisis has demonstrated that the dilution of basic legal requirements related, for example, to causation and evidence can lead to complete economic and legal failure. The question of need, justification, and possible creation of nonliability-related compensation mechanisms will have to be addressed carefully in the coming years, especially in relation to compensation for "historic" gradual nonattributable pollution damage and old dumping sites. Public authorities, the chemical industry, the insurance world, and the public at large have a joint interest in finding consensus-based solutions in a most important societal issue as the 20th century draws to a close.

4. Waste Shipments in the European Union: The Interface between Environmental Regulation and the Internal Market

4.1. Introduction

The last few years have seen an increase in the interaction between international trade policies and environmental protection. The first international convention on this subject was the 1933

Convention Relative to the Preservation of Fauna and Flora in Their Natural State. However, until the late 1980s, environmental regulators were developing their policies in large part separately from economic or trade considerations. This situation is changing rapidly, and today almost every international, intergovernmental organization has a program on trade and the environment. Moreover, environmental considerations are addressed as integral components of trade policy decisions, and trade or economic implications of environmental regulations are, conversely, being seen as key components of environmental regulatory decisions as well.

What is responsible for this dramatic change? First has been an increased recognition of the need to focus on transnational pollution issues such as the question of carbon dioxide emissions, chlorofluorocarbons (CFCs), and industrial accidents, for example, the Sandoz Rhine river pollution case. These transnational issues are increasingly addressed by multilateral environmental agreements, which also contain trade provisions. One example is the Montreal Protocol on CFCs of the United Nations Environment Programme (UNEP, 1987). Another example is the Basel Convention on Control of Transboundary Movements of Hazardous Wastes (UNEP, 1989).

In addition, focus has increased on the impact on competitiveness caused by some countries' having strict environmental standards or by attempts by such countries to erect barriers against imports produced in countries having less stringent environmental standards. This has led international, intergovernmental organizations, such as the General Agreement on Tariffs and Trade (GATT), to address the danger of countries taking unilateral actions under the guise of protection of environment, health, or safety, with the effect (and perhaps even the intention) of discriminating against imports. Included here would be such unilateral actions as discriminatory product standards, testing/certification requirements, and environmental subsidies.

A series of GATT disputes have been related directly to environmental regulation brought under the GATT nondiscrimination and national treatment provisions. In addition, the GATT, now the World Trade Organization (WTO), has instituted a Trade and Environment Program to facilitate the enforcement of "free" trade principles while simultaneously encouraging sustainable development.

Related subjects involve the effect of environmental policies of the developed or "Northern" countries on developing or "Southern" countries. Complicated issues of market access, technology transfer, commodity pricing of natural resources, debt levels, paternalism, and protectionism must be addressed here.

Regrettably, writings on the subject of free trade and environmental regulation tend to represent one polarized view or the other, either environmental protection or free trade. On the *free trade side*, the position is that environmental objectives do not justify violating the multilateral trading system rules of national and nondiscriminatory treatment. Moreover, the free traders believe that the sovereignty of countries in setting their own environmental priorities and requirements should be respected; that environmental policy should not impede the free flow of goods, services, persons, and capital; and that international environmental issues, while perhaps a component of trade policy, should be subservient to trade policy in the event of a dispute.

In contrast, *environmentalists* generally believe that economic expansion leads to environmental degradation; that harmonized standards reduce environmental protection; that unilateral trade-restrictive actions are valid or even desirable to promote domestic or global environmental objectives; and that environmental subsidies and other incentives or disincentives should be encouraged. The environmentalist perspective is not concerned with what it perceives as "outdated" issues of nondiscriminatory trade. Protection of the environment justifies almost any measures necessary.

4.2. European Community/Union Measures

In the European Community/Union, the interrelationship between trade and environmental regulation is best seen in the development of the Internal Market.

4.2.1. The Internal Market, "1992"

The Treaty of Rome was enacted in 1957 to establish the European Economic Community and to guarantee, among members of the Community, four basic freedoms: the free movement of goods, persons, services, and capital between Member States.

The concept of 1992, "the Internal Market," was one of the best public relations coups in recent political history. Under the Treaty of Rome and other early documents of the European Economic Community, a single market and the elimination of physical, technical, and fiscal barriers (explanation see below) between Member States of the Community had already been anticipated. However, discouraged by the slow progress toward such a single market, the Community developed the 1992 program to accelerate its progress.

The first major document in the 1992 process was the White Paper on Completing the Internal Market [4.1]. It was published in 1985 and included a variety of proposals for legislation and timetables in which the 300-plus measures necessary to eliminate barriers to the free movement of persons, goods, services, and capital in the Community had to be enacted by the end of 1992. The barriers the Community was trying to eliminate among its Member States were physical, technical, and fiscal ones.

Physical barriers are such things as customs services, passports, and transport formalities. *Technical barriers* include health and safety requirements, standards, and product packaging. *Fiscal barriers* are among the most difficult to eliminate. Included here are harmonization of general tax rates and value-added tax (VAT) rates. Of the 300-plus measures necessary to implement the Internal Market, between one-third and one-half were environmental, health, and safety initiatives.

4.2.2. The Single European Act

(see Section 2.1)

The Single European Act took effect on July 1, 1987 (Single European Act, signed at Luxembourg on February 17, 1986, and at The Hague on February 28, 1986). It contains numerous provisions amending the Treaty of Rome, many of which were necessary for the 1992 process to be successful. The most significant provisions affecting trade and waste shipments are the following:

1) Inclusion of a new Article 100 A in the Treaty of Rome providing that harmonization measures can generally be adopted by the Council by a qualified majority vote, rather than by unanimity as before
2) Inclusion for the first time in the Treaty of Rome of provisions specifically relating to environmental protection
3) Revision of the consultation procedure with the Parliament, giving Parliament a stronger role in the policy-making process

New Article 100 A of the Treaty of Rome (Article added by Article 18 of the Single European Act) provides that measures can be adopted by the Council by qualified majority, if their objective is the approximation of the provisions of national laws in order to further the establishment and functioning of the Internal Market, i.e., harmonization measures (see Section 2.1).

In the past, Council decisions had to be reached on a unanimous basis. As result, many Council actions, especially controversial ones, were deadlocked because one or another Member State could not agree with the rest. To facilitate action on measures necessary to implement the Internal Market, a new Article 100 A was included to permit the Council to act on Internal Market measures by qualified majority vote. (The votes of the Council are weighted. The total number of votes among the 12 Member States is 76; a qualified majority consists of 54 votes.) Exemptions from qualified majority voting exist here for measures related to fiscal provisions; worker rights issues; as well as health, safety, environmental, or consumer protection issues. Here, unanimity still had to prevail under the Single European Act.

Enactment of Community legislation by unanimous Council votes is usually more complicated and time consuming than reaching qualified majority decisions. One of the early steps in the EC legislative process under the Single European Act, therefore, always involves "jockeying for position" between the European Commission and Member States to determine the legislative basis of the measure in question. A recent dispute on the legal basis of a piece of Community legislation concerning titanium dioxide went as far as the European Court of Justice [4.2]. In this case, the Court of Justice ruled on the legal basis of Community legislation that was aimed both at the achievement of the Internal Market and at environmental protection. The Court upheld the legal basis of the legislation as an Internal Market, rather than environmental, measure and annulled the Directive. The Court ruled that the choice of legal basis should be governed by the aim and content of the legislation in question.

As stated above, the Single European Act also specifically provides that protection of the environment is a valid area of Community legislation. In practice, the Community began legislating in the environmental area in the late 1960s. However, the original Treaty of Rome is primarily a trade agreement and does not specifically provide authority for such legislation. Under the Single European Act, the Community must now consider environmental implications in adopting its other legislation, as well [4.3]:

> "Action by the Community relating to the environment shall be based on the principles that preventive action should be taken, that environmental damage should as a priority be rectified at source, and that the polluter should pay. Environmental protection requirements shall be a component of the Community's other policies."

The Member States were rather suspicious of this amended Treaty language, since environmental, health, and safety regulation had traditionally been within their own purviews. Therefore, the Treaty language now also includes two other provisions: (1) the issue of subsidiarity (see Section 2.1), and (2) unanimous voting on environmental matters. "Subsidiarity," a term that has received much attention in the context of the Maastricht Treaty, in this environmental context, means merely that environmental legislation or enforcement should be undertaken at the lowest feasible regulatory level.

New Article 130 T of the Treaty of Rome [4.3] specifies that Member States will not be prevented from adopting more stringent protective environmental measures than those enacted at the Community level, provided these measures are compatible with the Treaty of Rome (see Section 2.2). This is very useful for Member States, such as Denmark, the Netherlands, and Germany, that periodically adopt more stringent legislation. At the Community level, there has also been a trend to "harmonize up" (i.e., to harmonize toward more stringent environmental regulation). Invariably, the Community has usually taken existing legislation from one of the more stringently regulated Member States as a model for its own Community-wide legislation.

A number of procedural changes also occurred as a result of the Single European Act, the majority of which increased the power of the European Parliament. The general procedure of parliamentary advisory review for Community actions is one of the Commission proposing legislation and the Council acting on it. Prior to such Council action, Parliament and ECOSOC review the proposal and issue advisory opinions, which the Commission and/or Council can accept or reject.

The Single European Act introduced another procedural step in certain situations, the so-called Cooperation Procedure. Under this procedure, two readings in the Parliament and two Council reviews (under new Article 149, as replaced by Article 7 of the Single European Act) take place.

4.2.3. The Maastricht Treaty – Environmental Provisions [4.4]

Mention should also be made of the environmental provisions of the Maastricht Treaty. The Maastricht Treaty was agreed to by the Heads of State and Government of the European Community Member States in December 1991, to revise the Treaty of Rome to provide for such changes as gradual political integration, a common currency, and a Single European Bank.

The Maastricht Treaty is notable for its further integration of environmental considerations into all other policies of the European Community; the text goes even further than that of the Single European Act. (Since the entry into force of the Maastricht Treaty changed the name "European Community" to "European Union," henceforth, reference is to either the European Union or the EU, even though the measures described may have been initiated in EC times.) It also contains language on the interrelationship between transportation, energy, and agricultural policies and environmental protection. Moreover, the Treaty deals with the need to consider trade or competition and economic development issues within the framework of environmental considerations. In fact, protection of the environment is explicitly contained in the introductory section of the Maastricht Treaty (i.e., impact on the environment should be considered in the development of all other EU policies and the EU has competence to regulate in this area). The concept of "sustainable development" is also found in this Treaty.

The Treaty also eliminates the need for unanimous votes for environmental measures and adds a new "Cooperation Procedure" which expands the role of the European Parliament. In addition, the question of the meaning of "subsidiarity" is supposedly clarified. However, confusion about the exact definition of this term, and when it is to be applied, is still present under this new Treaty.

Finally, the Maastricht Treaty provides for the creation of a "Cohesion Fund" to assist the least financially sound of the European Member States, Ireland, Spain, Portugal, and Greece (including to help them improve environmental protection and enforcement measures).

4.2.4. Other Applicable Provisions of the Treaty of Rome

Notification and Standstill Procedure on Technical Standards and Regulations. Significant European Court of Justice case law exists on "mutual recognition"; i.e., goods lawfully produced and marketed in one EU Member State must be accepted in all other Member States (except where refusal is justified by Articles 30 and 36). Directive 83/189/EEC [4.5] specifies notification procedures for technical standards and regulations in the EU. The procedure was developed to discourage national, unilateral actions in this area. Member States must inform the Commission on each draft technical standard or regulation. The Commission then informs the other Member States. The Commission and the other Member States have the option of consenting to the draft technical standard or regulation text. The acting Member States must take their positions into account. Waiting periods for adoption of the measure(s) are also specified.

Article 30. Article 30 of the Treaty of Rome provides that "quantitative restrictions on imports and all measures having equivalent effect shall be prohibited." Quantitative restrictions have never constituted a serious obstacle to intra-EU trade. Therefore, the emphasis in practice has been on measures having an "equivalent effect." All legislative rules and administrative provisions, as well as administrative practices that form a barrier to importation or exportation, are governed. Included here are provisions and practices that render importation or exportation more expensive or difficult compared to sales of home products on the domestic market. The *Cassis de Dijon* decision [4.6] established the basic principles to be applied in judging the legitimacy of measures adopted by EU Member States that hinder the free movement of goods among them. In the absence of harmonized EU law, Member States are free to establish rules affecting the free movement of goods.

However, goods lawfully produced or marketed in one Member State must be accepted in other Member States, except where refusal is based on the "rule of reason" or on exceptions provided in Article 36. The rule-of-reason exceptions are applicable only to noneconomic matters, i.e., (1) environmental protection, (2) consumer protection, (3) prevention of unfair commercial practices or (4) effectiveness of fiscal supervision – provided, however, that the national measure in question is not unreasonable, disproportionate, or protectionist.

Exceptions provided in Article 36 include national restraints on trade, if necessary to protect (1) public morality, public policy, or public security; (2) health and life of persons, animals, or plants; (3) national treasures of artistic, historic, or archaeological value; and (4) industrial or commercial property.

However, not all trade restrictions based on the "rule of reason" or on Article 36 exemptions are acceptable. A restriction may not, for example, constitute a means of arbitrary discrimination or disguised restriction on trade between Member States. These concepts are referred to as the principles of "nondiscrimination" and "proportionality;" the restriction must apply equally to nationally produced goods and goods imported from other Member States and must be necessary to achieve legitimate, proportional objectives.

For environmental cases, there should also be a balance between environmental and free trade purposes. A regulatory measure must be necessary to achieve a stated goal, and no less restrictive measures can exist to achieve this goal. A case in point is the Danish Bottles case [4.7], which demonstrates the complexities of justifying the compatibility of environmental measures with rules on the free movement of goods. In question was a Danish law providing that mineral water, soft drinks, and beer could be sold only in refillable containers. A deposit and return system to ensure refilling of a large portion of containers was also included. The result of this law was that beverages lawfully sold in other Member States could not be sold in the same containers in Denmark, a clear violation of Article 30 of the Treaty of Rome.

The European Commission took Denmark to the European Court of Justice to contest this legislation. The Court found that the Danish system of mandatory deposit on containers for mineral water, soft drinks, and beer, although placing a greater burden on non-Danish producers, did not infringe Article 30. The Court held that environmental protection is an objective that may justify restraints on trade. The Danish deposit system appeared necessary to attain the desired objective, and no other less restrictive means existed.

A more recent example of one State's unilateral trade action, ostensibly to protect the environment, is the German Packaging Waste Regulation [4.8]. Under it, a complicated take-back and recycling system for packaging and packaging waste (set up and funded by industry because of the constraints of the Regulation) imposes, in practice, significantly disproportionate problems on importers into Germany than on the domestic industry.

A number of complaints have been filed with the European Commission alleging Germany's violation of Article 30 of the Treaty of Rome. One benefit of the German Regulation is that it has resulted in the EC's accelerating the development of its own Directive on Packaging and Packaging Waste [4.9] to harmonize all EU Member State rules on this subject. Germany, however, has steadfastly opposed the text of this EC draft Directive and refused to amend its own Regulation. As a result, the European Commission decided, on July 6, 1994, to initiate legal proceedings against Germany because its unilateral packaging Regulation creates an unjustified barrier to trade between EU Member States. In all likelihood, this is merely a political ploy to get Germany to drop its opposition to the EC's draft Packaging and Packaging Waste Directive and to bring the German regulatory requirements for waste packaging into line with the EC Directive's requirements (i.e., reduce the German recycling and reuse targets and build up domestic German recycling capacity more rapidly).

4.3. Waste Shipments: Conflict between European Union Free Movement of Goods Principles and Environmental Regulation

4.3.1. The New Regulation 259/93: Introduction

Recent regulatory initiatives on waste shipments provide perhaps the most dramatic example of the conflict between the European Union's principles of free movement of goods and environmental regulation. On May 6, 1994, the European Union Regulation "On the Supervision and Control of Shipments of Waste within, into and out of the European Community" [4.10] became effective (hereinafter Regulation 259/93). Regulation 259/93 imposes a complex system of shipment notification and consent on all shipments of waste, not just hazardous waste, that cross borders in the European Union, as well as all waste imports into and waste exports from the EU. Although the notification and consent procedures vary depending on the type of shipment, its destination, and its purpose; waste shipments for recovery or recycling operations are covered, as is waste for disposal.

Regulation 259/93 must be viewed in the context of other key EC waste legislation, most notably, Directive 75/442 EEC on Waste [4.11], as amended by Directive 91/156/EEC [4.12]; and Directive 91/689/EEC on Hazardous Waste [4.13], replacing Directive 78/319 on Dangerous Waste [4.14]. For these pieces of legislation, see → Waste, Chap. 2. International commitments of the EU on this subject (e.g., Basel Convention) are also interrelated.

On the specific topic of transfrontier movements of waste, the EEC had Regulations as early as 1984, with Council Directive 84/631/EEC, as amended [4.15], "On the Supervision and Control within the European Community of the Transfrontier Shipments of Hazardous Waste." Directive 84/631 required Member States to take measures to ensure the supervision and control of transfrontier shipments of hazardous waste within, into, and out of the Community to protect human health and the environment.

Under Directive 84/631, all cross-border movements of hazardous waste were subject to a notification requirement to be effected by a uniform Consignment Note and typically made to the Competent Authority of the Member State of destination. Procedures for acknowledgment of and objections to the shipment, if any, were provided. However, given the narrow definition of "hazardous waste," Directive 84/631 applied to only a limited number of transfrontier shipments of waste. According to 1992 figures of the Organization for Economic Cooperation and Development, ca. 200×10^6 t of waste cross OECD borders each year (140×10^6 t cross EU borders). Of these 200×10^6 t, < 1% was estimated by the OECD to involve so-called hazardous waste.

In addition, since this 1984 legislation was a Directive, EC Member States tended to implement it inconsistently. To fill this regulatory gap, address implementation inconsistencies, and adapt the rules regulating transfrontier movements of hazardous wastes to the requirements of the March 22, 1989 Basel Convention on Transboundary Movements of Hazardous Waste and Their Disposal and the Lomé Convention of December 15, 1989 (the Fourth ACP–EEC Convention), the European Council adopted the Regulation "On the Supervision and Control of Shipments of Waste Within, Into and Out of the Community" (Regulation 259/93). Since this legislation is in the form of a Regulation, as opposed to a Directive, it is directly applicable and binding on EU Member States.

Regulation 259/93 was also drafted to ensure consistency with the OECD, March 1992 Council Decision Concerning the Control of Transfrontier Movements of Waste Destined for Recovery Operations, specifically its "Green," "Amber," and "Red" Lists of Waste. These lists are categorized according to hazard ("Green" being least hazardous, with "Red" most hazardous) and appear at Annexes II through IV of Regulation 259/93. Regulation 259/93 was also deemed necessary to ensure continued and improved environmental protection in EU waste management, in view of completion of the Internal Market and its resultant removal of border controls.

Finally, the issues of waste management in general in the EU, and waste shipments in particular, are highly emotional and political ones. Regulation 259/93 is at least partially the result of successful political pressure by environmental groups, some Member State regulatory authorities, and "the public" for stricter and harmonized European Union legislation on this subject.

The overriding purpose of the Regulation is, of course, to ensure prior notification and consent for waste shipments to Competent Authorities, Member States, enabling them to be informed of the particular type, movement, and disposal or recovery of waste, so that they can take all necessary measures for the protection of health and the environment, including the possibility of raising reasoned objections to the shipments (e.g., de facto "PIC" or prior informed consent).

Regulation 259/93 provides for a Consignment Note notification system for the covered waste shipments to the Competent Authorities of destination, dispatch, and transit and to the consignee (see Sections 4.3.2.1 and 4.3.2.2). The content of the notification and the subsequent regulatory avenues available to the Competent Authorities to "control" these shipments depend on the nature and composition of the waste (i.e., its hazard); the shipment's purpose (disposal or recovery); and its destination (to or through other countries in the EU, to non-EU OECD Decision signatories, to other parties to the Basel Convention, to countries having equivalent bilateral or multilateral agreements with EU Member States – the possibilities are numerous).

The bases on which Member States may object to a waste shipment for disposal include the principles of proximity (waste should be disposed of near the point of generation – a policy encouraging local waste disposal); self-sufficiency (to ensure that each Member State can dispose of its waste in its own territory); and priority for recovery. Differing positions have been taken by commentators as to whether the first two of these principles can or should be extended to waste shipments for recovery under the Regulation. Political pressures by local authorities, NIMBY("not-in-my-backyard")-ism by the neighboring public if such an interpretation is upheld, and resulting negative competitive impacts on industry in the EU can be anticipated if this interpretation is sustained.

4.3.2. Specific Provisions of Regulation 259/93

Exclusions from the scope of Regulation 259/93 are:

1) Offloading to shore of waste generated by normal operation of ships and offshore platforms, including wastewater and residues, provided that such waste is subject to a specific binding international instrument
2) Shipments of civil aviation waste
3) Shipments of radioactive waste
4) Shipments of waste already covered by other relevant legislation
5) Shipments of waste into the EU subject to the environmental protocol of the Antarctic Treaty
6) Shipments of waste contained on Annex II (the OECD Green List of Waste) destined only for recovery, subject to certain conditions

The main text of the Regulation is divided into sections, depending on whether the shipments are between EU Member States or involve imports or exports and, in each section, whether disposal or recovery is involved.

4.3.2.1. Shipments of Waste between EU Member States

Shipments of Waste for Disposal. If a notifier intends to ship waste for disposal from one EU Member State to another or pass it in transit through one or several other Member States, the Competent Authority of destination must be notified, and a copy of the notification must be sent to the Competent Authority of dispatch and of transit and to the consignee. Notification must also cover any intermediate stage of the shipment from the place of dispatch to its final destination. The notification is to be effected by means of a Consignment Note, which is to be issued by the Competent Authority of dispatch. In making the notification, the notifier must complete the Consignment Note and must supply any additional information and/or documentation requested by the Competent Authorities.

Moreover, the notifier must enter into a contractual agreement with the consignee for disposal of the waste. Specific provisions are included as acknowledgments of the Consignment Note and authorization of the shipments. The shipment may be effected only after the notifier has received authorization from the Competent Authority of destination.

To implement the principles of (1) proximity, (2) priority for recovery, and (3) self-sufficiency at the EU and national levels, Member States may take appropriate measures generally or partially to prohibit shipments of waste for disposal.

Shipments of Waste for Recovery. The notification requirements of Regulation 259/93 with respect to shipments of waste between Member States for recovery are generally consistent with the requirements described above for shipments between Member States for purposes of disposal, with certain key deviations. First, the content of the Consignment Note differs. Moreover, tacit consent for a shipment is deemed to exist if 30 days have passed following notification, and no objection has been lodged.

In addition, Competent Authorities with jurisdiction over specific recovery facilities may decide, notwithstanding the notification and consent requirements of the Regulation, that they will not raise objections concerning shipments of certain types of waste to specific recovery facilities. Such decisions may be limited for a specific time period and may be revoked at any time. The Commission must be notified of any such special arrangements; it will inform the Competent Authorities of the other EU Member States and the OECD. However, Member States of transit or dispatch may object.

Shipments of Waste within one Member State. The notification and consent provisions described above do not apply to shipments of waste within one Member State. Here, the individual Member State in question has the role of establishing an appropriate system for the supervision and control of shipments of waste, "coherent with" the system established by Regulation 259/93. The Member States must inform the Commission of these individual systems. However, shipments between Regions are covered if the Region has jurisdiction over waste management (see also Section 4.3.3.1).

4.3.2.2. Exports of Waste from the EU

Exports for Disposal. Exports of waste from the EU for disposal are prohibited under Regulation 259/93, except to European Free Trade Area (EFTA) countries, which are also parties to the Basel Convention. Exports of waste for disposal to an EFTA country may also be banned if the EFTA country of destination prohibits imports of such waste, if it has not given its written consent to the specific import in question, or if the Competent Authority of dispatch in the EU has reason to believe that the waste will not be managed in accordance with environmentally sound methods in the EFTA country of destination.

The notification and consent requirements with respect to exports of waste from the EU for disposal generally track those described above for intra-Union shipments for disposal, with certain deviations. For example, the notification and consent time periods are significantly expanded here. The Regulation also contains language as to obligations of EU Customs offices of departure.

Exports of Waste for Recovery. Exports from the EU of waste destined for recovery are prohibited under Regulation 259/93, except for those exports to countries to which the OECD Decision applies and other countries that are parties to the Basel Convention (or with which the EU or the EU and its Member States have concluded bilateral, multilateral, or regional agreements/arrangements compatible with the Basel Convention and EU legislation). The Regulation also provides specific requirements that such agreements or arrangements must meet.

In addition to the procedures above, exports of waste from the EU for recovery to countries to which the OECD Decision, Basel Convention, and/or bilateral or multilateral agreements or arrangements apply may also be banned if the country in question prohibits all imports of such waste, if it has not given its consent to the specific import in question, or if the European Union Competent Authority of dispatch has reason to believe that the waste will not be managed in accordance with environmentally sound methods in the country of destination.

Generally, waste specified on Annex II of Regulation 259/93 (OECD Green List of Waste) for recovery is not subject to the notification and consent requirements of the Regulation, again with certain exceptions. First, the Commission is required to inform every country to which the OECD Decision does not apply of the waste categories contained in Annex II and request written confirmation that such waste is not subject to control in the country of destination and that the latter will accept such waste to be shipped without recourse to the control procedures. Annex II waste must be exported for recovery operations within a facility that is locally authorized and operating in the importing country. Moreover, such exports will be subject to surveillance.

As to exports of Annex III and IV waste for recovery (OECD Amber and Red Lists, respectively), the notification and consent procedures discussed earlier with respect to intra-Union shipments generally apply.

Finally, the Regulation provides that all exports of waste to ACP (African Caribbean, Pacific) States will be prohibited. This prohibition does not prevent an EU Member State to which an ACP State has decided to export waste for processing from returning that processed waste to the ACP State of origin.

4.3.2.3. Imports of Waste into the EU

Imports for Disposal. Imports of waste into the EU for disposal will be prohibited, except those from EFTA countries that are Parties to the Basel Convention, other countries Party to the Basel Convention, or countries that have concluded appropriate bilateral or multilateral agreements or arrangements with the EU as a whole or with individual Member States. Such agreements or arrangements must be notified to the European Commission, which will review them.

As in the case in the intra-EU and export shipment situations discussed above, notification of imports of waste for disposal will be subject to the Consignment Note procedure. Notification must be made to the Competent Authority of destination, and the Consignment Note will be issued by that Competent Authority. Again, specific deadlines for notification, objections, and decisions are provided.

Imports for Recovery. Imports of waste into the EU for recovery are prohibited under Regulation 259/93, with the exception of imports from countries to which the OECD Decision applies and, again, countries Party to the Basel Convention or with which the EU/Member States have concluded appropriate bilateral, multilateral, or regional agreements or arrangements.

The notification and control procedures of Regulation 259/93 also apply to waste imports into the EU for recovery from countries and through countries to which the OECD Decision applies for waste listed on Annexes III and IV of the Regulation or for waste that has not yet been assigned to an Annex.

4.3.2.4. Transit of Waste from Outside and Through the EU for Disposal or Recovery Outside the European Union

Here again, the Consignment Note and control procedures apply. The notification is to be effected by means of the Consignment Note sent to the last Competent Authority of transit within the EU, with copies to the consignee, the other Competent Authorities concerned, and the Customs offices of entry into and departure from the EU.

The shipment may be admitted into the European Union only if the notifier has received the written consent of the last Competent Authority of transit, which authority will also issue the Consignment Note. Specific time deadlines for notifications and decisions are also provided. Transits of waste listed on Annexes III and IV for recovery through one or more EU Member States from a country, and transit of waste for recovery to a country to which the OECD Decision applies require notification of all Competent Authorities of transit in the Member States concerned. Such shipments may be effected only in the absence of objections.

4.3.2.5. Common Provisions

Regulation 259/93 contains certain common provisions applicable to all shipments, whether they are for disposal or recovery, or involve waste imports, intra-EU shipments, or exports from the European Union. These cover general notification possibilities, illegal traffic, financial guarantees and insurance, inability to facilitate recovering disposal and return, regulatory authority designation, documents retention, etc.

4.3.3. The Trade–Free Movement and Environmental Protection Conflicts in Complying with EC Regulation 259/93

4.3.3.1. Definitions

One of the most difficult issues to anticipate in compliance with this new EC Regulation is a definitional one. Even though much progress has been made in the definitions of "waste" and "hazardous waste" in the OECD and in recent EC Directives on the subject, clear EC definitions have still not been developed, and differing EU Member State definitions continue to exist. Therefore, regulatory inconsistencies on this subject can be expected to remain.

One case in point is the German distinction between something having commercial value (Wirtschaftsgut) and waste, and the German treatment of secondary raw materials. However, the new German Federal Comprehensive Waste Reform Law (Kreislaufwirtschaftsgesetz) intends to change German waste definitions to comply with EU definitions. This law was finally passed by the Bundesrat before its summer 1994 recess and is expected to come into effect, at the latest, by 1996.

The EC Waste and Hazardous Waste Directives attempt to define waste in their texts. The definition of waste in the 1991 Waste Directive (91/156/EEC) is as follows:

"Waste" shall mean any substance or object in the categories set out in Annex I which the holder discards or intends or is required to discard" (Article 1).

The "hazardous" waste definition under the 1991 Hazardous Waste Directive (91/689/EEC) is much more complicated and includes various references to Annexes, origin, composition, properties, and a final umbrella category of "any other waste" considered by a Member State to be hazardous, in effect negating the usefulness of the harmonized European-level definition.

In addition, both the 1991 Waste and the Hazardous Waste Directives mandate the compilation of lists or catalogues of waste under their domain. The EU has recently published its Waste Catalogue [4.16] under the 1975 and 1991 Waste Directives [4.17], and the so-called Waste Catalogue II or List of Hazardous Wastes under the 1991 Hazardous Waste Directive was published on December 31, 1994 [4.18].

The process of developing both European Commission lists has been highly controversial. Officials at the Commission have taken varying positions as to whether these should be listings of products and substances or whether they should, instead, be criteria-based. Moreover, a number of key Member States, most notably Germany and France, are still lobbying to have the new European List of Hazardous Waste be similar to those in place in their countries (i.e., a listing of substances or products that have been found to be hazardous; therefore, waste containing any of these products or substances in whatever quantities would, likewise, be categorized as hazardous).

As was the case with Waste Catalogue I, the scope and content of the EU Hazardous Waste List (Waste Catalogue II) are of major importance to a number of industries and companies operating in Europe. The content of this new List will affect such commercial questions as whether and where a material can be recycled or remediated, whether it can be shipped across borders, and what other regulatory constraints and liabilities will apply vis-à-vis manufacturers, generators, and transporters. Sales price and contractual and marketing questions of customers are intricately involved here as well.

The new List of Hazardous Waste Pursuant to Article 1 (4) of Directive 91/689/EEC on Hazardous Waste (January, 1994), while still product related, deals only with broad product categories and is keyed to the thresholds and concentrations developed for hazardous properties under EU chemicals legislation [e.g., the 0.1% concentration for category 1 and 2 carcinogens in the 1988 Preparations Directive (88/379/EEC)]. The inclusion of these thresholds and concentrations has been the subject of some controversy.

Wallonian Waste Decree. As illustrated in the German "Wirtschaftsgut" example above, another definitional issue of concern in complying with Regulation 259/93 is whether a material is a "product" or "good," hence coming under the Treaty of Rome's free movement of goods principles, or whether it is something else entirely. Under the new EC Regulation 259/93 and such recent legal interpretations as the July 1992 ECJ Judgment on Wallonian Waste Imports [4.19], waste should be granted special treatment for environmental protection purposes, tantamount to at least a partial exemption from the European Union's free movement of goods and Internal Market principles. This case was particularly interesting in that the Wallonian Decree at issue not only impeded the flow of waste across EU Member State borders, but also impeded intra-Belgian trade on a regional basis. While the ECJ held that the challenged Wallonian Decree did not conform to the 1984 Hazardous Waste Movements Directive (the precursor to Regulation 259/93), it found that the Decree's banning of the disposal of wastes in Wallonia that were generated outside Wallonia was not "discriminatory" within the meaning of the Treaty of Rome. This decision, was based on the "particularité" of wastes, in that Article 130S of the Treaty of Rome requires that, as a priority, "environmental damage" should be rectified at the source. The Court also referred to the Basel Convention's principle of waste self-sufficiency.

The ECJ appears to have erroneously interpreted the Treaty of Rome provisions here. The Wallonian Decree clearly violates not only the 1984 Hazardous Waste Shipments Directive, but also the free movement of goods provisions of the Treaty of Rome. Relying on Article 130 provisions is also questionable since a Directive that deals with shipments of materials and border transfers within the EU should more appropriately be considered an Internal Market measure and, therefore, judged under Article 100A of the Treaty of Rome, not Article 130S.

In addition, because the Wallonian Judgment did not address all of the issues on this subject raised by Regulation 259/93, a definitive legal interpretation is needed on the free trade versus environmental protection argument for waste shipments in the EU.

Article 130S vs. Article 100A. The debate continues – the ECJ has just rejected the European Parliament's request to annul Regulation 259/93 because it was improperly based on Article 130S (environment) of the Treaty of Rome. The Parliament said that this Regulation was essentially an Internal Market (trade) measure, hence based on Article 100A of the Treaty of Rome and subject not only to the Treaty's free movement of goods provisions, but also to Article 113 of the Treaty dealing with external trade (It is not just happenstance that an Article 100A determination for the new Regulation would give the Parliament greater powers as to the form and content of the Regulation.) However, the ECJ rejected the Parliament's arguments and upheld the Article 130S basis of the Regulation. Waste, it seems, is indeed special and, at least for now, will not be called a "good" in the EU.

The European Court of First Instance has also just rejected as inadmissible a request by two French waste management companies and one Luxembourg one that it annul Article 4 of Regulation 259/93, which permits individual Member States to ban imports of waste, even nonhazardous waste, from other EU countries. (At issue was a French Decree banning imports into France of household waste for disposal in landfill.) The Court of First Instance did not address the substance of the plaintiffs' complaint. Instead, the case was dismissed on the procedural ground of lack of "standing" of the plaintiffs, since they were not "directly concerned." In this ruling, the President of the Court said that Article 4 of the 1993 Council Regulation was not aimed at companies, "except in their objective quality as economic operators in the waste management and transport sector, the same as for any other operators in an identical situation." The Court's President also rejected the request for damages and interest. Case T-475/93, SA Buralux, SA Satrod, SA Ourry vs. Council of the European Union, judgment published in O.J.E.C. C188, July 9, 1994.

Disposal and Recovery. Another definitional concern with Regulation 259/93 is seen in the question of how to define "disposal" and "recovery." Member States of the European Union have, to date, taken inconsistent positions as to whether operations to be carried out are either of the two. This is obviously a critical point for purposes of defining the type of notification and the content of information to include in the Consignment Note under Regulation 259/93. Defining an operation as "recovery" or "recycling/further use," as opposed to "disposal," will also have a critical effect on how the proximity principle would apply. This issue is particularly difficult for operations involving energy generation and incineration.

Perhaps the most difficult practical problem resulting because of these vague definitions is in the area of waste for recovery. The impact on byproducts, coproducts, secondary raw materials, and product or container take-back programs must be clarified.

The practical, commercial difficulties posed by all of the above definitional issues for EU companies in attempts to adhere to Regulation 259/93 are discussed in Sections 4.3.3.2–4.3.3.4. These problems are magnified even more in the international arena.

In addition to the dilemmas caused by inconsistencies between the EU-level definitions and those in some EU Member States, OECD and Basel Convention strictures, obviously, also have to be adhered to. It is already evident that definitional divergencies between these two organizations and the EU may cause problems in the future.

In addition, many affected companies are multinational enterprises, with manufacturing, waste-generating plants in and outside the EU. Company attempts to impose one comprehensive waste management scheme throughout all of such companies' international activities will be stymied. A case in point is seen in the differences in the U.S. and EU definitions of "waste," "hazardous waste," "disposal," and "recovery," and the shipment restrictions imposed. For example, unlike the ECJ, the U.S. Supreme Court has determined that waste is an article of commerce (i.e., a "good") meriting constitutional protection against discriminatory state and local Regulations that impede its free movement and that nothing special or distinctive exists in waste that should permit discriminatory restriction of its flow across state lines to apply in the United States.

A second example is seen in the fact that the United States attempts to define waste in a way that encourages beneficial recycling (exemption of certain kinds of materials destined for certain recycling). The EU definition follows a less desirable, inconsistent approach: It covers all secondary materials as "waste" as well, regardless of whether disposed or recycled, but then differs in the regulatory treatment of conventional waste and recyclable waste.

Some parties, frustrated by these definitional differences and the length and complexity of the debate believe that "waste prevention" should be the focus in the future, instead of waste shipment restrictions, especially when the words "waste" and "hazardous waste" cannot even be defined properly.

4.3.3.2. Liabilities

An attempt must also be made to address liability issues under Regulation 259/93 at the EU level to ensure some degree of consistency in determining liable parties and in enforcement. From past experience, this issue cannot be left solely to the Member States. Adoption of EU legislation on Civil Liability for Damage Caused by Waste is a first step (see Section 3.4), but continued Member State inconsistencies on this subject will further confuse companies in compliance.

4.3.3.3. Administrative Issues

European industry, while favoring harmonized regulation of waste shipments throughout the EU, has raised some of the issues noted in the previous sections. However, it appears to be focusing currently on the failure of the EU and its Member States to provide the administrative structures and procedures to allow industry to comply with Regulation 259/93 by its effective date of May 6, 1994, or shortly thereafter.

Among the most active EU-level industry associations on this subject have been UNICE, Eurométaux (which has particular concerns about disruption of shipments of scrap metals as secondary raw materials), the European Chemical Industry Council (CEFIC), and the European Federation of Waste Management (FEAD). According to a UNICE discussion paper on the implementation of Regulation 259/93,

> "... the adminstrative backup necessary for the proper implementation of the Regulation is widely lacking in most of the Member States. It is both paradoxical and disastrous for a Community Regulation, directly applicable according to Community law to be paralyzed in this way ...."

The concern about the lack of administrative systems in the Member States to implement the Regulation and the enormous burden to be placed on Competent Authorities in the Member States in homogeneous or integrated implementation is well justified. As noted in Sections 4.3.2.1–4.3.2.4, the notification and approval system under Regulation 259/93 is very complex and inconsistent, depending on the types of shipments, their purposes and destinations, and quite bureaucratic and paper-intensive. It is questionable whether the Competent Authorities will be able to deal with the large number of notifications to be expected in the time anticipated by the Regulation. Given the large degree of discretion still left to the Member States in implementation, it is also questionable whether a truly harmonized system will evolve. Therefore, it cannot be expected that a significant degree of increased environmental protection will ensue as a result of Regulation 259/93. (Of course, on the other hand, trade, will be severely hampered.)

Moreover, given the fact that Regions having responsibility for waste management will also be in a position of appointing Competent Authorities, an additional administrative question is raised. It is rare that the proper Regional infrastructure exists in each of the Regions throughout the European Union which may appoint such Competent Authorities.

If this administrative situation is not solved quickly, it could lead at worst to failure of the entire system or, at a minimum, to the need to store a large number of wastes prior to disposal or recovery, pending clearance. Such storage may result in more environmental, health, and safety problems than the new requirements in the Regulation will solve.

Other administrative concerns include the fact that on the eve of implementation of the Regulation, the standard Consignment Note to be drafted by the Commission was not yet available; the list of competent authorities to be designated by Member States was also not available (indeed, most of the Member States had not even submitted their Competent Authority designations to the Commission); the list of customs offices called for had not yet been published; and the list of wastes on the OECD Green List that will be subject to specific controls in certain Member States was not available, nor was information on which third countries have bilateral agreements with which Member States or the EU authorizing such trade.

Thus, according to one industry observer, "chaos" will result as companies attempt to comply with the terms of the Regulation, with clearly inadequate and inconsistent administrative structures and procedures in place.

The Commission Staff, while admitting certain transitional administrative problems, does not appear to believe that all of these concerns are fully justified, and international environmental organizations believe that the administrative and compliance concerns of industry are excessive. Reports that illegal traffic in toxic waste, particularly in a West-to-East direction in Europe and to Asian countries, is increasing have led to growing environmental group pressure to strengthen EC legislation even further on this subject. Greenpeace, in particular, is reportedly concerned with increased shipments to Asian countries, as markets in Africa and in Central and Latin America become increasingly difficult to enter, and with reports of mislabeling waste shipments as bound for recycling or recovery facilities, when, in fact, they are being sent to developing countries for disposal.

4.3.3.4. Trade and Competition Issues

Since broad disparities exist among EU Member States on the administrative structures and procedures to implement Regulation 259/93 and definitional and liability disputes continue, competitive differences will arise state-to-state in the EU. In addition, possible uses and abuses of the proximity and self-sufficiency principles by Competent Authorities could give rise not only to political problems, but also to major anticompetitive ones as well.

Although disposal of waste at sites as close as possible to the point of generation is generally preferable, in some circumstances moving waste over greater distances to ensure its proper disposal may be desirable. Moreover, many companies in Europe have one or a limited number of waste disposal or recycling sites to ensure the proper disposal or recovery of their company's waste throughout the EU. Sending waste to these sites to be disposed of will involve movement, possibly over considerable distances, from one site to another and perhaps across national boarders in the EU. This growing tendency for large, usually multinational, companies to be as self-sufficient as possible in the integrated management of their own waste should be encouraged rather than penalized by possibly inappropriate application of the proximity principle.

How the growing number of contractual arrangements involving manufacturers of products taking back contaminated products or packaging from customers (under a life-cycle philosophy) will be merged with the requirements of Regulation 259/93 is also unclear. Therefore, under this new Regulation, industry could well face disruption of its waste disposal and recovery programs, delays in operations, and the need to store waste at significant cost and potential environmental risk, pending administrative resolution of these questions.

In addition, some would argue that it is unwise for the EU, given the current economic situation in particular, to adopt more stringent legislation on this subject than its major trading partners have and than is required by its international commitments.

The Parties to the Basel Convention agreed on March 25, 1994, to adopt the EU Council Common Position to amend the Basel Convention to impose additional restrictions on exports of hazardous waste to developing countries (potentially non-OECD states), for disposal immediately and for recycling or recovery as of December 31, 1997. As a result, at the Second Conference of the Parties to the Basel Convention (Basel COP 2) recently held in Geneva, the Basel signatory countries agreed, in Decision II/12, to ban trade of hazardous waste between OECD and non-OECD countries. The U.S. Government and U.S. industry oppose this action and estimate that such a ban on trading waste for recycling between OECD countries and non-OECD (developing) countries could cost the United States about $ 2.2×10^9 a year in commodity trade.

A major indusry sector affected by this ban would be the scrap metal industry. In addition, many non-OECD countries are desperate for raw materials to fuel their economies and may, in response, redefine what is considered a Basel waste in order to circumvent the Treaty.

In the European Union, Greenpeace has pressured the European Commission immediately to bring it into conformance with these Basel Convention commitments. The Commission, on the other hand, does not believe it is necessary to modify the Regulation and certainly not at this time.

Such a ban on hazardous waste shipments between OECD and non-OECD countries, especially for recycling or recovery, would arguably not be compatible with the GATT. In addition, if the hazardous waste definition is overly broad, valuable goods not usually considered waste – for example, chemical byproducts, secondary raw materials, and used electronic goods that constitute legitimate trade between the EU and non-OECD countries – would be blocked. Again, all of these issues and the differing ways in which the EU and its trading partners handle them will likely result in competitive disadvantages for European industry vis-à-vis its non-EU competitors.

4.4. Conclusion

With its Regulation 259/93, the European Community has undertaken an ambitious regulatory program in a much needed area of harmonized EC legislation, with laudable environmental protection goals. However, the scope and complexity of the Regulation, without adequate administrative implementation procedures in place at the EC level and in the Member States, will give rise to significant legal, trade and competition, and administrative compliance problems for European industry. Moreover, the undesirable inconsistencies on this subject among EU Member States are likely to continue to exist in the application and enforcement of this Regulation. In addition, despite ECJ determinations to the contrary, Regulation 259/93 is primarily a free movement of goods/Internal Market measure, not an environmental one. Wastes, especially waste for recovery, should be treated as "goods" under the free movement provisions of the Treaty of Rome. Failure to do so for intra-Community movements of waste violates the letter and spirit of the Treaty of Rome, as amended – without providing additional harmonized environmental protection.

Finally, inconsistent ratifications and implementation of Basel Convention and OECD commitments by the EU and its trading partners will further complicate the competitive problems facing European industry. Recent Basel Convention amendments on this subject, especially with respect to shipments of waste for recovery, arguably violate GATT rules. It will be interesting to see whether, and how, this issue will be resolved in the GATT's–WTO's new Trade and Environment Programme or perhaps in a new GATT "Green Round."

5. The Seveso Directive: An Example of the Interface between Community Law and National Implementation Rules

5.1. Historical Background and Content

Council Directive 82/501/EEC of June 24, 1982 on the major accident hazards of certain industrial activities owes its existence to a number of industrial accidents. Subsequent to accidents in Flixborough in 1974 (United Kingdom), Beek (The Netherlands) in 1975, and in particular at an ICMESA plant in Seveso (Italy) in July 1976, the Commission undertook to draft legislation for the prevention of major industrial accidents, having been encouraged by the intentions of the United Kingdom, The Netherlands, France, and Italy to adopt national legislation in order to improve accident prevention.

The Commission's proposal was submitted to the Council in July 1979, about three years after the Seveso accident. The proposal was largely inspired by measures that existed for nuclear energy installations to avoid nuclear accidents. While the general lines of the proposed Directive were agreed to relatively quickly by the Council, adoption of the Directive was held up because France, in particular, was afraid that information on risks and safety measures to be given to other Member States might create difficulties for the siting and activity of its nuclear installations. Under pressure from the European Parliament, which had from its beginnings been in favor of the adoption of the Directive, and the Benelux countries, a compromise was finally found that enabled the Directive to be adopted in June 1982.

At the end of 1984, a leak of methyl isocyanate from a Union Carbide plant in Bhopal, India caused the death of more than 2000 persons, and some 20000 persons were injured. Following that accident, the Commission proposed an amendment to Directive 82/501/EEC, which was adopted in early 1987 (Directive 87/216/EEC). The Parliament, in its opinion on this proposal, had suggested a wide-ranging review of the Directive. However, the accident of Sandoz in Basel in October 1986 and the public concern raised by it necessitated immediate action. The Commission submitted in spring 1988 a proposal for a further amendment that considerably enlarged the existing provisions for accident prevention related to the storage of dangerous substances. That proposal was adopted within six months, a record time, as Directive 88/610/EEC.

The announced general revision of Directive 82/501/EEC was inserted into the Commission's legislative program 1993 but has, as of April 1995, not yet been adopted. Directive 82/501/EEC aims at preventing major accidents and limiting the consequences thereof. Approximating the legislation of the Member States is a means of attaining this objective.

Industrial activity is defined in Article 1 of this Directive. Industrial activity means any operation in an industrial installation listed in Annex I to the Directive that involves one or more dangerous substances and is capable of presenting major accident hazards, as well as storage of dangerous substances under conditions laid down by Annex II of the Directive.

Major accidents are also defined in Article 1: "an occurrence such as a major emission, fire or explosion resulting from uncontrolled developments in the course of an industrial activity, leading to a serious danger to man, immediate or delayed, inside or outside the establishment, and/or to the environment, and involving one or more dangerous substances."

The Commission thought it necessary to publish guidelines for the interpretation of this concept. Directive 82/501/EEC provides for a number of obligations:

1) Manufacturers who process or store certain chemicals have a general duty to prevent major accidents and limit their consequences for humans and the environment (Art. 3 and 4). Should a major accident occur, manufacturers have a duty to report it (Art. 10). The Competent Authorities of Member States have a general duty to organize inspections or other methods of control [Art. 7(2)].
2) Manufacturers, furthermore, have specific duties that apply either when any of 180 dangerous substances are used in quantities above specified threshold levels or when any of 28 dangerous substances are stored. In such cases, the manufacturer must produce and update periodically a safety report (Art. 5) and an on-site emergency plan.
3) The Competent Authorities must ensure that an off-site emergency plan is produced (Art. 5 and 7). They must further ensure that "persons liable to be affected by a major accident" are informed of safety measures and of the correct behavior to adopt in case of accident.

5.2. Implementation Requirements

Member States had to adapt their national legislation to the requirements of the Directive by January 8, 1984. This national legislation and the above mentioned amendments of the Directive phased in gradually. The following shows how three Member States – France, Germany, and Greece – have complied with the requirements under Directive 82/501/EEC.

Situation in France. On August 16, 1982 (i.e., only seven weeks after the adoption of the Directive), France issued an internal circular, by which it drew the attention of the Commissaires de la Republique (the Préfets) to the existence of the Directive and requested that specific care be given to the information requirements contained in its Article 8. It listed the installations that came under the Directive and were, in France, the subject of the legislation on "classified installations" (loi sur les installations classées) of July 19, 1976 and of several decrees issued under that legislation, in particular a decree of September 21, 1977.

The legislation of 1976 provides that all types of installations, listed in a specific register, that might present a risk must obtain an administrative license for operation. The corresponding list includes more than 300 activities and thus goes far beyond the number of installations that are covered by Directive 82/501/EEC. To fully adapt its existing legislation, in 1987, France adopted legislation on the organization of civil protection, forest protection against fire, and the prevention of major accidents. This legislation requested the establishment of national, regional, or departmental offices for organizing help in cases of accidents and natural or man-made catastrophes. Emergency plans among others were to be prepared by the departmental prefects around certain industrial installations.

These are the essential pieces of French legislation to which workers' safety provisions must be added in order to cover the area of the Directive. What is striking is the fact that the legislation on industrial installations is, to a large extent, umbrella legislation. The major part of the obligations that the Directive places on the individual manufacturer is to be determined by individual administrative decision. Some 500000 French plants are registered and subject either to a system of declaration or, for some 55000, to a system of authorization procedure under the legislation of 1976. Most of the elements of Directive 82/501/EEC are found in the 1976 legislation, although this legislation does not seem to cover all elements of Articles 3 and 4.

The requirement for safety reports (risk analysis) was enlarged after adoption of the Directive to cover existing installations. These safety reports under Article 5 of Directive 82/501/EEC refer to some 320 types of installations. On-site emergency plans were, prior to adoption of the Directive, mandatory only for oil industry installations; since the first circulars of 1982 and 1983, they have also been required for installations that come under Article 5 of the Directive. Off-site plans have been requested gradually since 1985, subsequent to legislation adopted that year.

The French system is a clear case of existing means being amended in order to comply better with a Directive that postdates the basic national legislation. Furthermore, the French system gives considerable authority to the departmental, regional, and national administrations, which operate via individual administrative decisions.

Situation in Germany. Germany effected Directive 82/501/EEC and its subsequent amendments by a great number of provisions, under essentially four different categories:

1) Legislation on workers' safety
2) Legislation on chemical substances and preparations
3) Legislation on catastrophes
4) Environmental legislation

Overall, more than 40 pieces of legislation, adopted at the federal or regional level, apply to the subject matter covered by the Directive.

The most relevant provisions to Directive 82/501/EEC are contained in the so-called "Störfallverordnung," the 12th Regulation on the implementation of the Federal Act on Quality Standards (Bundesimmissionsschutzgesetz). For political reasons, Germany thought it necessary to have its national legislation in place when Community legislation on this subject was adopted. This led the German government to submit a proposal for the 12th Regulation on February 20, 1980, while the Commission proposal for Directive 82/501/EEC dates from July 16, 1979. In 1980, the 12th Regulation was adopted in Germany, while the Community Directive was politically agreed to, though not yet formally adopted. The same coincidence occurred after the Sandoz accident in 1986, when both the German government and the Commission submitted proposals. This time the Community acted more quickly; Directive 88/610/EEC was adopted in 1988, while the German amendment was finally adopted in 1991.

The multiplicity of legal texts makes it difficult to identify discrepancies between the 12th Regulation and Directive 82/501/EEC, without being confronted with the objection that other legal provisions explicitly or implicitly cover the aspect in question. The text of the German 12th Regulation is more precise than that of the French legislation; it gives greater legal certainty to industrial installations and less discretion to the administration.

The manufacturers' general obligations are listed specifically, as well as the obligation to establish an on-site emergency plan and a safety analysis. The obligation to inform persons outside the plant has been placed on manufacturers. Off-site measures are to be agreed to with the competent regional or federal authorities. Generally, the threshold levels in German legislation are more stringent than those in the Directive. Also, the list of substances that make the provisions applicable is longer than that of Directive 82/501/EEC.

Situation in Greece. Greece is a Member State that, at the time of its accession to the Community in 1981, had relatively little legislation on the environment. Accident prevention measures for installations were a rather new item, and the Greek administration was not really accustomed to monitor manufacturers' behavior with regard to safety. These rather sketchy lines might help explain why national measures to transpose the amendment of Directive 82/501/EEC took several years. A similar attitude is to be observed regarding the amendment made by Directive 88/610/EEC: only in 1993 after the Commission had filed an application to the European Court of Justice for noncommunication of national implementation measures, Greece transformed the requirements of Directive 88/610/EEC into national law.

The national legislation reproduces the requirements of Directive 82/501/EEC and its amendments correctly. However, the responsibility of monitoring the legislation is given to no less than four bodies [environmental and industrial ministries, prefects at local–regional level, and a Coordinating Interministerial Committee (CIC)]. These rather dispersed responsibilities might have contributed to certain doubts being raised about effective application of the legislation. Indeed, several accidents suggest that effective application, safety analysis, and on-site or off-site emergency plans have not been made for all installations that come under the Directive, nor is regular inspection always made. One reason for this situation is that the different central and local–regional administrations did not receive any supplementary financial or human resources, when the new monitoring tasks were assigned to them in 1988. Thus, the administration does not act as counterpart interlocutor to manufacturers. Since the requirement of informing the public of measures to be taken in case of accident has been placed on prefects, a manufacturer does not incur too serious risks regarding compliance with the Directive's requirements.

5.3. Conclusions

If one looks horizontally at Directive 82/501/EEC and its implementation, a number of observations can be made:

1) The Directive owes its existence and its tightening-up to industrial accidents and public concern; these accidents made possible the adoption of preventive provisions that had not been acceptable at the national or Community level prior to the accidents.
2) Of the three Member States considered, France undoubtedly gives the greatest discretion to the administration and Germany the greatest legal certainty to manufacturers.
3) Effective application of the requirements of the Directive requires flexible cooperation between manufacturers and administration regarding inspection, monitoring, safety analysis, and risk assessment. The administration itself needs to be trained and experienced to be able to contribute to accident prevention without unduly hampering manufacturers' activity. The absence of a trained, effective, and active administration that effectively monitors the practical application seems to be the greatest handicap in Greece.
4) The Directive is not entirely clear on the question of who has to inform the public on safety measures and on behavior in case of accident (Art. 8) – the administration or the manufacturer. This sometimes leads to practical difficulties. Furthermore, if this kind of information is too detailed, it might provoke negative reactions among the population concerned.

5) The cost of prevention might be significant, while the benefits – the absence of major industrial accidents – are very difficult to estimate. Both France and Germany had, prior to adoption of the Directive, taken a number of steps to prevent major accidents, although these measures focused more on human health than on environmental protection.
6) It is not clear under the Directive whose task it is to set up the off-site emergency plan, to ensure its consistency with the on-site emergency plan, and to manage off-site emergency operations. Obviously, the manufacturer is not capable of providing for all details of the off-site plan such as blocking roads or evacuating the population. Therefore, active cooperation by the administration is necessary, not only in case of accident, but also to learn about cooperative mechanisms and gain experience for any accidental event.
7) French administration has a tendency to ask for measures to prevent accidents in the worst possible scenario, while German administration normally is satisfied with the degree of accident prevention that is reached by using the best available technology.
8) German administration seems to attach limited importance to the question of siting of industries, whereas this aspect plays an increasing role in the French administration's approach to accident prevention.
9) Safety reports play an essential role among all measures required by the Directive. They are considered a means to obtain more precise information on installations, to identify their accident potential, and to prepare preventive measures more adequately. However, which role the administration has to play in this context is not made clear in the Directive. The differences among Member States seem to be important. The administration in Greece seems to be easily satisfied with the safety reports submitted. In France, the authorities examine the safety reports and sometimes require other preventive measures, but never "accept" the report, thus leading to a kind of uncertainty by manufacturers. It might be useful to systematically check, as German administration does, the reports' completeness, correctness, and their effective monitoring and updating.

The hope is that further amendments of the Directive will not be the result of new industrial accidents.

6. U.S. and European Environmental Legislation

6.1. Introduction

This Chapter discusses the nature and role of environmental Regulation, particularly as it applies to the chemical industry. It makes no attempt to summarize the vast and rapidly evolving body of current U.S. and EC environmental legislation. Other sources can be found for this (see, e.g., [6.1], [6.2]). The environmental legal provisions applicable to all phases of the chemical industry are now so extensive, detailed, and complex that they cannot be treated exhaustively under a single keyword of Ullmann's.

Rather, the attempt here is to step back from the system and compare briefly the style and context found in the world's two major centers of regulatory activity, the United States, and Europe, to note some of the key features of the legislation in both systems; and to comment briefly on what may lie ahead.

6.2. The Environmental Regulatory System [6.3]

The United States and Western Europe have had significant environmental laws on the books since the early 1970s. The United States has erected an elaborate edifice of Regulations. At the EC level and in many European countries (Germany being a noteworthy exception), however, a great deal has not happened until recently. Over the past several years, the European Community and its Member States have been generating a flow of environmental initiatives of ever-increasing importance. Now developing in Europe is an extensive framework of environmental regulatory and liability law (see Chaps. 3, 4) that presents business operating in Europe with many of the same kinds of risks found in the United States. The ways these laws are developed in the European Community and in the United States differ fundamentally.

6.3. Methods of Lawmaking in the United States and the European Community

United States. In the United States, the law and lawyers have been central to the development of environmental policy, and highly structured efforts at the federal level have led the way. U.S. Regulation relies heavily on generic, federal administrative rule making to achieve specificity and uniformity, and to gather the technical, scientific, and economic data that form the premise for public decision making. Administrative rule making proceedings provide public, formal, and structured opportunities for citizens and industry to participate in the development of Regulation through comments, public hearings, and sometimes judicial challenges to rules. The "political" aspects of the U.S. environmental regulatory process take the form of lobbying directed at the U.S. Congress, as well as less formal efforts to lobby the administrative agencies that establish, implement, and enforce Regulations.

In short, in the United States, contending interests use legal processes in the judicial arena and before administrative agencies to influence the formulation and execution of policy. This is particularly true of environmental groups who, with some frequency, use citizen suit provisions in federal environmental legislation as a means of influencing environmental policy.

The result of this dynamic, contentious law and rule making process is a highly articulated, but often inflexible, regulatory system, i.e., a regulatory system that is at times too rigid, complex, and confusing.

EC Regulation (see also Chaps. 3–5). Regulation in Europe, on the other hand, has traditionally been regarded as "technical" or "political," rather than "legal" in nature, although this attitude appears to be changing. European industry, as a whole, has not in the past looked to legal counsel to assist it in handling environmental matters, but has relied primarily on its technical staff, who often operate through national and European-wide trade associations. The EC regulatory process, furthermore, is largely nonadversarial and is the result of a slow, nonpublic, lawmaking structure designed to achieve political consensus among the Member States. A close industry–government relationship exists. Industry operates through contacts with national authorities; through "expert" advisory committees at the Commission of the European Community; and through lobbying of the Commission, Parliament, and COREPER (the permanent representation of Member States, which assists the Council on day to day matters). Proposed Directives are generally made publicly available and are commented on by Parliamentary committees, but until recently, this was not required at the early stages. This situation has been changed, at least in theory, in 1992–1993 with the adoption of Council and Commission Resolutions calling for early consultation of interested bodies, including industry associations. No structured open administrative process for promulgating Directives or Regulations exists, however, such as the U.S. notice and comment rule making process.

Implementation of EC Directives is left to the authorities of individual Member States. The degree to which Directives are effectively implemented varies widely from country to country. While some Member States do not give full effect to Directives until long after adoption, other states adopt even stricter measures than those existing at the Community level. For example, Denmark, The Netherlands, and Germany have well-developed national environmental policies. Ireland, Greece, Portugal, Spain, and to some extent, Italy, are in earlier stages of development. Belgium, France, and the United Kingdom are somewhere in the middle, although they have had environmental policies for many years. British law is now developing especially rapidly. As a result of the wide disparity in Member State national environmental laws, chemical companies operating in Europe must be familiar with both Community and national environmental legislation in order to minimize the risk of unexpected environmental Regulation or liability.

6.4. U.S. and European Environmental Laws

Environmental law, to date, has progressed chiefly in two broad streams – one regulating the manufacturing process and the other regulating the products themselves (→ Products). A third stream of natural resource law regulates extraction of the oil, gas, coal, and various minerals that provide the raw material for the chemical manufacturing process. (A number of minor subjects in environmental law deal, for example, with the special protection of certain types of resources, such as wetlands, endangered species, wild and scenic rivers, archaeological sites, and historic preservation. They are not dealt with here since they are less important to and less specific to the chemical industry.) This third broad stream of natural resource Regulation will be treated in → Raw Materials and Energy. The focus in this section is on chemical process and chemical product Regulation.

Chemical process Regulation encompasses chiefly pollution control Regulation – the classic areas of air emissions, water effluents, and solid and hazardous wastes. It centers on the chemical manufacturing plant, although it typically also applies to all related commercial or industrial facilities (e.g., storage depots, means of transport, and disposal facilities). Generally, the basic tools are construction or operating permits that impose relevant operating constraints, monitoring and reporting requirements, etc. Beyond these standard provisions, however, environmental impact assessment laws, accident planning laws, and public information laws have also been developed, sometimes implemented in permits or the permit process and sometimes separately.

6.4.1. Environmental Impact Assessment Laws

Both the United States and the EC have environmental impact assessment (EIA) laws [6.4]. These apply chiefly during the licensing of new chemical or other manufacturing facilities or the major expansion of existing plants. They require a systematic, detailed, and written analysis of the likely environmental impacts of the proposed project. This analysis is designed to force both chemical industry management and government regulators to consider environmental factors in their project decision making. In both the United States and the EC, however, the EIA laws themselves mandate no particular result. Rather, they are procedural laws premised on the assumption that a better decision-making process leads to better decisions. U.S. law goes beyond EC law in many ways. It requires consideration of all reasonable alternatives to the project, and has more detailed guidance and "teeth" than its EC counterpart.

6.4.2. Accident Planning and Public Information Laws

The United States and the EC likewise have accident planning and public information laws of great significance for the chemical industry. The EC developed broad-gauged legislation on this subject earlier than the United States, and most European countries have a more systematic approach to risk assessment and analysis built into the new (and modified) plant licensing process than does the United States to date. The EC law (the Seveso Directive, see Chap. 5) is designed to prevent and limit the consequences of major accidents resulting from industrial operations involving dangerous substances [6.5].

The corresponding U.S. law, the Emergency Planning and Community Right-to-Know Act, requires (1) emergency planning, (2) emergency release notification, (3) community right-to-know reporting, and (4) toxic chemical release inventory reporting [6.6]. While the EC does not yet have an analogue to the U.S. toxic chemical release inventory requirements, one is on the drawing board. For many years, the United States has also had separate oil spill legislation requiring the well-known Spill Prevention, Control and Countermeasure (SPCC) plans [6.7] and has much better developed legislation on aboveground and belowground storage tank Regulation than either the EC or most European countries (Germany being an exception in this regard).

6.4.3. Air and Water Pollution Legislation

The air and water pollution legislation in both the United States and the EC is extensive [6.8]. In both cases, the legislation uses "best-technology"-type substantive emission and effluent standards (emission limits in German terms) as well as case-specific emission and effluent limits to be set at levels necessary to attain air and water quality requirements (the latter called immission standards in German terms).

U.S. water pollution legislation has for years built its control program around a comprehensive permit system, and U.S. air pollution legislation is now adopting the same technique. The EC legislation has had no comprehensive permit system requirements to date, although the Commission is now developing a multimedia permit system modeled after the British Integrated Pollution Control system. The U.S. air and water legislation has regulated new plants more stringently than existing sources through technology-based new source performance standards for industry categories. The legislation has evolved industry-specific technology-based limits for existing sources on the water side and is now moving to do likewise on the air side, and has developed complex antidegradation requirements (in effect, special limits on degradation of an air or water resource that meets existing ambient requirements). The EC legislation has tended, in practice, chiefly to apply technology-based limits for new sources in a few large industrial categories, and the pervasive air and water quality-based Regulation of existing sources called for by EC legislation has never been put in place. Indeed, not even the air and water quality monitoring systems necessary to implement such ambient-based Regulation have been developed. The Commission is still struggling to come up with a politically and administratively feasible vehicle for setting air and water technology-based limits for the industry, such as are pervasive in the United States and in Germany. Antidegradation requirements, to the extent theoretically present under EC law, have been largely ignored.

In both the United States and the EC, efforts are now being made to move beyond Regulation of the few parameters dealt with in both the air and the water context in the early years and to tackle the difficult and complex job of effectively regulating a multitude of toxic pollutants.

6.4.4. Solid and Hazardous Waste Regulation

U.S. Regulations. In the field of solid and hazardous waste Regulation, the United States has for years had a much more effective regulatory system in place for hazardous waste than the EC (although the German waste regulatory system is comprehensive, except for a loophole in the case of "recyclable" wastes, now to be closed by the new German Eco-Cycle Waste Law of 1994) [6.9]. The U.S. cradle-to-grave regulatory system uses registration requirements, permits, and transport manifests to regulate virtually all aspects of the generation, storage, transport, treatment, and disposal of a broad range of hazardous wastes (including many destined for recycling), under an extremely complex set of rules and both written and unwritten regulatory practices. All types of treatment, storage, and disposal facilities must meet specific design and operating performance standards, including statutory minimum technology and groundwater monitoring requirements and a prohibition on the land disposal of untreated hazardous wastes. The U.S. system has complex rules for mixtures of wastes; for wastes derived from other wastes; and for distinguishing wastes from products, coproducts, byproducts, chemical intermediates, and secondary raw materials. It regulates some wastes destined for recycle or reuse and does not regulate others. Despite all this hazardous waste Regulation, however, the U.S. system has virtually no comprehensive or systematic Regulation of solid waste at the national level, this area being left almost entirely to state and local governments.

EC Regulations (see Section 4.3). The EC waste management framework is more coherent, but less effective, than that in the United States. The European system is built on a basic framework Directive that deals with all forms of waste and is supplemented by a more specific Directive dealing with hazardous waste [6.10]. The entire system is now in flux (see Section 4.3).

Further, the European waste regulatory system has depended, from the outset, on Regulation of the *shipment of waste* (see Section 4.3.1). Both the United States and the EC countries have Regulations governing how hazardous materials must be transported, but EC waste shipment Regulation governs whether hazardous waste can be shipped across national borders at all, and under what conditions. As early as 1984, the Directive on Transfrontier Shipments of Hazardous Wastes regulated hazardous waste shipments across Member State borders and into and out of the EC by creating prior notification, manifesting, and packaging requirements [6.11]. More recently, the community has adopted a new Regulation on Shipment of Waste that implements the Basel Convention in Europe (the Basel Convention governs shipment of hazardous waste on an international scale) and that imposes more stringent requirements (including prior authorization in some cases) on the shipment of all forms of waste, both hazardous and normal, across Member State borders and into or out of the EC (see Section 4.3.1). Shipments of waste for recovery operations would, depending on how they were listed in three categories (Red, Amber, and Green; see Section 4.3.1) on lists prepared by the "Committee Procedure," be (1) largely excluded from Regulation (green), (2) subject to a prior notification regime (amber), or (3) subject to a prior authorization regime (red; waste not listed would be subject to the authorization system). This Regulation could balkanize the EC on waste matters, with local authorities lodging objections motivated by protectionism of local disposal or recovery facilities and NIMBY-ism. Chemical facilities may be faced with disruption of their waste disposal and recovery arrangements, and those dependent on off-site waste disposal or recovery (and having only limited on-site waste storage capacity) may find their operations at risk of interruption.

To implement the EC's Waste and Hazardous Waste Directives as recently amended, the Commission has drawn up a comprehensive list of solid and hazardous wastes (the European Waste Catalogue). A separate list has been adopted by the Council, which will be critical to EC waste legislation because it will determine the threshold question, for particular wastes, of whether they must be regulated as "solid" or as "hazardous" (see Section 4.3.3). The Commission is also having great difficulty establishing and implementing the definition of "waste" found in relevant Directives. All of the difficulties described in Section 4.3.4 are likely to prolong the period before the EC achieves effective implementation of waste and hazardous waste Regulation.

U.S. Superfund and EC Packaging Legislation. Two idiosyncratic areas of waste Regulation remain to be discussed – one in the United States and one in the EC. In the United States *Superfund legislation* uses government-funded cleanup backed up by a civil liability scheme to deal with soil and groundwater contamination [6.12]. Originally designed for old dump sites, this broad-reaching legislation extends in fact to all instances of such contamination, both old and new and both on-site at chemical manufacturing facilities and off-site. The government-funded cleanup process has been slow and has been required by Congress to use extremely stringent cleanup levels. The liability regime imposes retroactive, joint and several, strict liability regarding site cleanup liability and responsibility for natural resources. This liability is imposed on virtually everyone connected with the contaminated site since its contamination, including lenders in some cases. The primary Superfund program has been widely criticized and drives broad secondary efforts of private remediation at sites where the potentially responsible parties hope to avoid becoming entangled in the main cleanup and liability program. The European Commission has been considering legislation to deal with soil and groundwater contamination in Europe, and the Council of Europe has adopted a convention on this subject (see Chap. 3).

The second piece of idiosyncratic legislation is the effort by the EC and a number of Member States to regulate one form of solid waste, *packaging*, through product take-back (or advance collection, sorting, recycle, or disposal fees) and thus to develop a new paradigm for regulating all forms of solid waste (see Section 4.3.4). The Commission, reacting to the so-called Töpfer Ordinance on packaging waste in Germany and to similar efforts in France and differently structured efforts in other Member States, has proposed a Directive on Packaging and Packaging Waste, which is aimed at eliminating and preventing trade barriers arising from varying national packaging waste initiatives. The essence of this legislation is the ability to impose product charges to force those in the packaging chain, from manufacturer to consumer, to begin to internalize some elements of the costs of collection, sorting, recycle, or disposal of solid waste. The German system achieves the same end by imposing on the commercial parts of the packaging chain a legal "product take-back" obligation. By forcing internalization of some of the costs of the end-of-life disposition of products, the scheme aims at correcting market conditions so as to favor recycling, while letting the market rather than direct Regulation sort out how much recycling is warranted and in which cases. Particularly at the EC level, however, extensive direct regulatory measures, which may well not be environmentally wise or economically efficient, are also being considered.

6.4.5. Regulations on Chemical Products

EC Regulations. Environmental legislation regarding chemical products was one of the first areas pursued by the European Community, although initially its motivations had more to do with the Common Market than with environmental protection. The first major piece of chemical legislation was adopted in 1967, and established a common Community system for the classification, packaging and labeling of dangerous chemical substances [6.13]. This basic Directive was modified in 1979 by the so-called Sixth Amendment, which established a premarketing clearance requirement [6.14]. Under this amendment, companies marketing new "dangerous substances" are required to submit the new chemical substance to a prior notification procedure that involves testing the substance for toxicity and eco-toxicity and preparing technical dossiers to provide information on the substance's characteristics (see Section 2.2). In 1988, the Community adopted similar requirements for dangerous preparations (i.e., mixtures of dangerous substances) [6.15].

The Commission has also recently adopted material safety data sheet (MSDS) rules and has extended its Regulation to include the marketing and use of certain dangerous substances and preparations, as well as enacting Directives in the field of biotechnology designed to control the handling and release of genetically modified organisms. The Commission is also moving to impose on manufacturers and importers of existing chemical substances some significant data reporting obligations for assessment of their risk characteristics.

U.S. Regulations. In the United States, chemical product Regulation on a comprehensive basis began with the Toxic Substances Control Act in 1976 [6.16]. That Act places on manufacturers the responsibility of providing data on the health and environmental effect of chemical substances and mixtures and gives the Environmental Protection Agency comprehensive authority to regulate the manufacture, use, distribution in commerce, and disposal of chemical substances. The central regulatory mechanism is the requirement for premanufacture review of new chemical substances prior to their commercial production and introduction to the marketplace. The Environmental Protection Agency also has authority to require testing of chemicals, although no requirement for a "base level" of tests exists as a precondition to premanufacture review as in the Sixth Amendment to the EC Regulation.

Regulation of environmental harms from chemicals in the workplace takes place in the United States under the Occupational Safety Health Act. Under that act, the Hazard Communication Standard requires that employees be informed about the potential chemical exposures involved in their work and be trained to deal with them [6.17]. Chemical manufacturers and importers are required to evaluate the hazards of the chemicals they produce or import, and to use MSDS to transmit this information to their own employees and to employers to whom they ship the chemicals. Further, chemicals shipped must contain labels on the containers with similar information.

Beyond these core areas, both the U.S. and the European systems regulate use of polychlorinated biphenyls, asbestos, and agricultural chemicals. Further, hazardous chemicals are regulated to varying degrees and in varying ways in consumer products, foods, drugs, and other end uses such as cosmetics and medical devices as well.

6.5. Some Contrasts in U.S. and EC Environmental Legislation [6.3]

For obvious constitutional and political reasons, the scope, specificity, and rigor of EC legislation do not equal those of U.S. federal legislation.

6.5.1. Regulatory Gaps

Important regulatory gaps exist in the EC (e.g., on wetlands, underground storage tanks, and until recently, municipal sewage), along with weaknesses (e.g., Member State ability to determine, on a relatively subjective basis, the applicability of certain EC Directives on fish protection and bathing water). Nonetheless, in some areas, EC provisions or proposals go beyond present U.S. practices or requirements (e.g., eco-labeling, eco-auditing, packaging waste, and economic and fiscal measures such as a carbon or fuel tax).

6.5.2. Implementation of Legislation

Implementation of the EC Directives is spotty and frequently ineffective. Directives, particularly the early ones, tend to be general policy statements, with little of the detailed standards and permit requirements of their U.S. counterparts. For example, the European Community as yet has no specific standards for surface impoundments (e.g., pits, ponds, lagoons), underground tanks or pipes, waste piles, and various forms of waste treatment. Where Member State implementation takes place, standards and permit conditions are often negotiated on a case-by-case basis, and systematic monitoring and reporting obligations are not always imposed.

United States. In the United States, the implementation of legislation at the state level is reasonably well assured by U.S. federal legislation through a series of devices lacking in EC legislation. First, a well-staffed, trained, and funded agency (the EPA) can be found at the federal level in the United States. Also available is a great deal of public, comparable monitoring data and other information, as well as effective freedom-of-information legislation at the federal and state levels. Second, the EPA may as a formal matter withdraw a state's right to implement many of the federal regulatory programs if it judges the proposed implementation to be inadequate. The EPA also normally has direct prior approval authority over state-implemented Regulations, standards, and individual permits before those measures can go into effect under federal law (although they can and do sometimes go into effect under state law without regard to EPA approval at the federal level). In the absence of an EPA-approved state implementation program or Regulations, standards, or permits, the EPA is frequently required to promulgate directly applicable federal provisions. Should it fail to supervise state implementation where it has a mandatory duty to do so, it may be sued in federal court by citizens and environmental groups and forced to act. Lastly, the EPA has frequently threatened to withhold federal funding on which the state programs depend if state implementation is deemed inadequate.

6.5.3. Enforcement of Legislation

The enforcement of EC legislation is perhaps the weakest element of the EC regulatory framework. All the European Community can really do is insist that Member States enact legislation (i.e., transpose EC legislation into national law). The European Court of Justice has new power to impose penalties on Member States for failing to implement EC environmental law, but it remains to be seen how effective this tool will prove. An enforceable obligation on the part of Member States to undertake effective, concrete implementation through more specific regulatory standards and requirements, comprehensive permit or approval systems, and an adequately staffed and trained bureaucracy with sufficient funding and legal authority is generally lacking in many countries. Furthermore, the European Environment Agency, which is only just getting started has only a data collection and dissemination function. It does not, for the time being, have enforcement powers.

Enforcement in the United States is vigorous and takes a number of forms. Nearly all of the major U.S. environmental regulatory statutes require some form of publicly available self-monitoring and reporting. The system thereby automatically highlights lapses in compliance and the need for enforcement. When the EPA deems state enforcement inadequate, it normally has the authority at any time to override the state's failure to enforce and to issue an administrative order or, in some cases, to assess civil penalties on an administrative basis. In either case, the EPA may enforce its action in federal court or take the violator directly into federal court for injunctive relief and civil or criminal penalties. Should the EPA fail to do so, citizens and environmental groups can themselves normally take the violators into federal court.

Further Development in the EC. The EC institutions currently have none of these tools at their disposal. At the level of enforcement of specific standards, requirements, or permit conditions against an individual regulated facility, the Community plays virtually no role. Such enforcement is left up to the Member States. Enforcement is difficult, even when attempted, since self-monitoring and reporting obligations have not always been imposed either by Regulation or by permit conditions. Government failure to prosecute generally cannot effectively be challenged except through publicity and the political process.

The Commission is now engaged in a series of moves to surmount the difficulties it has encountered in trying to develop an effective "command-and-control" EC regulatory system. It is considering the use of fiscal and economic measures instead of administratively driven control requirements. It is trying to make greater use of directly applicable Regulations in the place of Directives that must be implemented by Member States; and in any event, EC legislation, overall, is becoming more specific, thereby providing the detail necessary for ensuring the consistent implementation that has been absent in earlier legislation. The Commission is drafting an Integrated Pollution Control Directive, modeled on British legislation, that would impose comprehensive permitting requirements covering all environmental media, i.e., air, water, soil, groundwater (but, initially, applying only to a restricted list of larger chemical, metallurgical, and paper plants). This Directive would impose substantive and management requirements as permit conditions and would implement on a case-by-case basis the existing media-specific air and water quality legislative requirements. The Commission has also adopted an eco-audit Regulation that provides a (presently "voluntary") comprehensive, universal system that is a partial, periodic substitute for the self-monitoring and reporting provisions that are lacking in existing Directives. These eco-audits would have to be verified by independent outside auditors, with public disclosure of certain information. The likely result of these eco-audits will be public and political pressure for more effective standard setting, implementation, and enforcement.

At a fundamental level, for the first time, the Treaty on Political Union (the Maastricht Treaty), would provide the Court of Justice with limited powers to sanction with penalties those Member States failing to implement EC law. Also, the Court of Justice has been developing the following legal principles to increase the significance of EC law: (1) obligations on national bureaucrats at all levels to apply EC Regulations, even where the national legislature has failed to implement that EC Regulation adequately (i.e., "direct applicability"); (2) "direct effect" of provisions in EC Directives that are sufficiently precise and unconditional, even where the national legislatures have failed to implement adequately; and (3) state responsibility for harm to individuals arising from the government's failure to implement EC legislation adequately. The cumulative effect of these developments will be to increase the importance of EC environmental law and to render more likely its effective, if not consistent, implementation and application throughout the European Community.

6.6. Outlook [6.3]

While the basic framework of U.S. environmental law has been fairly settled (the U.S. Clean Air Act and new initiatives by the first Republican-controlled Congress in 40 years being exceptions), environmental law in Europe is changing rapidly. The past and present roles of European environmental law, of course, indicate that the EC environmental proposals may not necessarily have more than limited importance for business. As is well known, a gap has existed between the rhetoric and the reality of European environmental law. The European Community and many Member States have a good deal of environmental "law" on the books, but much of the EC legislation requires little real action. Moreover, an inability at the EC level and, in some cases, a reluctance at the Member State level to implement and enforce existing environmental law, have been evident. Further, a remarkable lack of comparable data to document many European environmental problems, as well as a lack of data on government actions and enforcement in regard to them, are noticeable.

The politics of the environment appear to be changing, however, As a result of such crises as Seveso, Chernobyl, and Sandoz, and of the palpable, long-neglected environmental problems facing Western and Eastern Europe, public attitudes toward environmental risks in the European Community are evolving quickly. Environmental groups, although perhaps not yet as sophisticated and well funded as their U.S. counterparts, are becoming increasingly well organized and active. Adoption of the EC Directive on Freedom of Information on the Environment and creation of the European Environment Agency, by improving the quantity and quality of publicly available environmental information, will in all likelihood result in heightened public awareness, increased effectiveness of environmental groups, and consequently, increased political pressure for consistent and effective environmental legislation and enforcement.

7. Environment, Safety, and Chemical Safety in Japan

7.1. Current Situation Concerning Environmental Protection

Since the early 1970s, Japan has made a strong effort to tackle the problem of environmental preservation. Various measures, such as reinforcement of pollution control Regulations and amplification of pollution control facilities have been implemented.

As a result, some serious environmental conditions have been alleviated and the situation has generally improved since the 1980s. Moreover, because the social demand for environmental preservation, particularly in metropolitan areas, tends to be more and more diversified and sophisticated, the Environment Agency is promoting various measures in response to this situation.

In addition, finding solutions to the recent environmental problems that are developing into international and global issues and global climate problems are considered important political subjects by the Japanese government as in other countries. In November 1993 the basic law for environmental pollution control was put in force in response to the result of the UNCED held in June 1992.

7.1.1. Air Pollution Control

Regulatory standards to protect human health (environmental standards) have been set for SO_2, NO_2, CO, photochemical oxidants, and airborne particles based on the Basic Law for Environmental Pollution Control. For Regulation of exhaust gas emission accompanying industrial activities, emission control Regulations are in force for SO_x, NO_x, cadmium and its compounds, Cl_2 and HCl, F_2, HF and SiF_4, lead and its compounds, and exhaust dust. The Regulations are based on the Air Pollution Prevention Law. With regard to SO_x and NO_x, Regulations on total amounts of emissions are also valid within the designated area in the country. In addition to these nationwide uniform standards, special emission standards must be met by newly constructed or expanded facilities in the specified area. In certain areas, additional emission standards have been set by local governments (in municipalities designated by government ordinance).

To meet these standards, industries have taken measures such as desulfurization of exhaust gas [176×10^6 m^3/h (STP) at 1810 facility units], denitration of exhaust gas [142×10^6 m^3/h (STP) at 379 units], development of combustion processes that emit lower amounts of NO_x, absorption of HCl and HF, electrostatic dust separation, etc.

The regulatory system for total NO_x emission control for each industrial installation was introduced into three areas with a high population density, and the system was implemented in Yokohama and in Kawasaki in April 1982, and in Osaka and Tokyo in November 1982.

However, NO_2 concentration in the environment shows an increasing trend recently, and reinforcing the Regulations on NO_2 mobile emission sources such as automobiles is considered necessary.

In June 1989, the Air Pollution Prevention Law was amended. Fibrous materials such as asbestos that are hazardous to human health have been treated as specific dangerous forms of dusts in this amendment.

7.1.2. Control of Water Pollution

Wastewater Regulation standards were implemented in 1970 based on the Water Pollution Prevention Law. Together with additional standards implemented by the respective local governments (in municipalities designated by government ordinance), water pollution has been controlled with regard to the following items:

1) **Substances Posing a Health Hazard (Toxic Substances).** Concentration limits are set for cadmium and its compounds, cyanides, organic phosphorus compounds, lead and its compounds, chromium(VI) compounds, arsenic and its compounds, mercury, alkylmercury and other mercury compounds, polychlorinated biphenyls, trichloroethylene, and tetrachloroethylene.
2) **Items Affecting Living Environment.** Limits have been set on pH range, BOD, COD, suspended solids, concentration of oily substances, phenols, copper, zinc, soluble iron, soluble manganese, chromium, fluorine, and a number of *Escherichia coli* colonies.

Industries have promoted various measures such as wastewater treatment and process rationalization. As a result, the environmental standards on toxic substances have been satisfied by 99.9%, and a general trend of improvement is observed for household effluents. However, in closed water areas such as rivers, inner bays, lakes, and swamps, pollution was still proceeding and eutrophication was advancing. Since the cause of such situations was the insufficent installation of sewage systems for treatment of household wastewater, the Water Pollution Prevention Law was revised.

The Law on Special Measures for Preservation of Water Quality of Lakes and Swamps was enacted in July 1984, and an environmental standard and wastewater discharge control standard for nitrogen and phosphorus concentration were introduced.

The Regulation on the total quantities of contaminants in water was implemented in 1978 for three closed sea areas, namely, Tokyo Bay, Ilse Bay, and Seto Inland Sea, and all-around measures for reduction of the total COD were implemented. In January 1991 the third Regulation for the reduction of total quantities of contaminants was prepared. The countermeasures for water quality control of the closed sea will become more important in the future because tremendous growth of algae resulting from eutrophication will seriously damage fishery by causing red tide.

To counteract groundwater pollution, trichloroethylene and tetrachloroethylene were classified as hazardous substances under the Ordinance for Enforcement of the Water Pollution Prevention Law in April 1989; in June, upon amendment of the Water Pollution Prevention Law, percolation of specific water containing hazardous materials into the groundwater was prohibited. In this connection, the ordinances for reinforcement of the Law related to the Disposal and Cleaning of Wastes were also revised.

7.1.3. Revision of Compensation Systems for Patients with Pollution-Related Diseases

The compensation system for patients suffering from pollution-related diseases was implemented in September 1974, and compensation has been paid, at the cost of polluting parties, to the approved patients in Class no. 1 (directed to the four designated diseases caused by air pollution) and Class no. 2 (directed to diseases caused by water pollution or soil contamination). For Class no. 1, the Central Pollution Study Committee submitted a report in October, 1986 proposing a system that reflects a substantial improvement in environmental pollution.

Based on the above report, a revision bill of the Law on Compensation of Pollution-Related Patients was approved by Parliament. The contents of the revision are (1) all-out cancellation of class I area designation due to the improvement of the environmental situation; (2) continued compensation to approved patients; and (3) promotion of comprehensive environmental hygiene measures (establishment of a fund). As a result, a long-awaited rational revision of this system was implemented in March 1988. By the end of 1993 the number of approved patients was 84471 and expenses totaled 93.4×10^9 Yen.

7.1.4. Measures for Waste Disposal

Since 1970, the Law on Waste Treatment and Cleaning has been in force, with the two groupings of "Domestic Wastes" and "Industrial Wastes." Section 3 of the law stipulates that an entrepreneur shall be responsible for, e.g., self-accountability, appropriate treatment, discharge reduction, not making products difficult to treat, etc. With respect to industrial waste, the law provides for standards on collection, transportation, and disposal and Regulations concerning the industrial waste treatment industry, industrial wastes treatment facilities, and hazardous industrial wastes.

As far as domestic wastes are concerned, awareness is growing of the need for waste reduction, including recycling of waste. This is due to the fact that the quantities of wastes discharged have been increasing in recent years, existing incineration and final treatment facilities have reached their maximum capacity, and need exists to utilize limited resources more effectively. The Law Concerning the Utilization of Recycled Resources was published in April 1991. The bill for revision of the Law on Waste Treatment and Cleaning was deliberated by Parliament in September 1991.

7.1.5. Response to Global Environmental Problems

The Law Concerning Protection of Ozone Layers by Regulation of Specified Substances was enforced in Japan in May 1988. The law specifies measures that are to be taken to regulate the manufacture of specific substances, to restrict their discharge, and to rationalize the use of these substances.

In June 1990, the Montreal Protocol was revised to strengthen Regulations by placing a complete ban on the use of certain fluorocarbons by the year 2000 and adding to the list substances such as carbon tetrachloride and 1,1,1-trichloroethane. The related industries in Japan are making efforts to meet these requirements by developing alternative substances for use in place of existing ones.

In August 1990, IPCC of United Nations Environment Program (UNEP)/WHO (the intergovernmental panel regarding climate fluctuation) issued a report about the forecast influences and countermeasures of global warming, which had been studied for three years since 1988. In Japan, corresponding studies have been conducted under the leadership of relevant government authorities. The issue of global warming is also of utmost importance to industries.

It is estimated that ca. 244×10^6 t/a of CO_2 (in terms of C) is discharged in Japan, accounting for 4.3% of the amount discharged worldwide. Of this amount, about 4.9% originates from the chemical industry. Since the chemical industry has curtailed energy consumption by about 40% since the 1980s, little room is available for further energy saving. Thus, reduction of CO_2 discharge appears to be a difficult problem.

7.1.6. Basic Policy of the Japanese Chemical Industry Association about Environment and Safety

The Executive Committee of the International Council of Chemical Associations (ICCA) has set up the ICCA Principles of Responsible Care with the aim of promoting a program for the improvement of voluntary environmental and safety measures.

The Japan Chemical Industry Association concluded that the chemical industry clearly realizes its social responsibility and continues its endeavors to undertake voluntary measures in order to achieve sound growth of the chemical industry at large, at a time when humans are faced with new international problems such as global-scale environmental issues, waste disposal, and the safety of chemical substances. Japan's chemical industry has already made great progress in environmental pollution countermeasures and is at a very advanced level compared with to its U.S. and European counterparts.

7.2. Current Situation Concerning Safety and Fire Prevention Measures

As a result of a series of explosions and fires that occurred in 1973–1974 in chemical and oil refinery complexes in Japan, the chemical and related industries have implemented safety and fire prevention measures. Due to such efforts, large-scale explosion accidents in the plants and works have decreased drastically since 1978.

However, the number of accidents is gradually leveling off, and further safety and fire prevention measures should be taken and promoted.

7.2.1. Revision of Fire Service Law

Ignitable and flammable materials are classified as Hazardous Materials based on the Fire Service Law.

On May 24, 1988, the law for partial revision of the Fire Service Law (Law no. 55 of 1988) was promulgated, and in this relation, the revision of the government and ministerial ordinances. The essential part of this revision is the "review of the scope of hazardous materials" to define the hazardous material more clearly and introduce methods to test whether a substance is hazardous or not. The tests should be performed and judged by the enterprises themselves, following and considering these laws and Regulations.

Another important point is a review of the scope of the term "dangerous articles."

1) The term dangerous article is clearly defined, and judgment of a system by testing is introduced. The test methods and criteria are established by taking into account the definition and international test methods.
2) A designated quantity is determined by government ordinance by taking into account the danger of the articles concerned.
3) Quasi-hazardous materials and special combustible materials are regulated under the category of combustible materials.

The provisions relating to the revision of the scope of dangerous materials in this law were enforced as of May 23, 1990.

7.2.2. Safety Measures for the Transport of Dangerous Goods

For the air and marine shipments involved frequently in international transport, international Regulations have been introduced so that Japanese domestic legislation is consistent with international transport standards.

7.2.2.1. Marine Transport Standards for Hazardous Substances

Breakbulk Transport. The Ministry of Transport provided Rules for the Marine Transport and Storage of Dangerous Goods with respect to the breakbulk transport. The rules were determined with reference to the International Marine Dangerous Goods Code (IMDG Code) set by the International Maritime Organization. The notice of revision incorporating the 25th revision of the IMDG code was enforced in January 1991. With regard to container testing, the test system and UN label indication system have been internationally enforced since January 1991, and the use of qualified containers having UN labels is obligatory for the transport of dangerous goods.

Transportation in Bulk. With regard to the marine transport in bulk of hazardous liquids by chemical tankers and liquid gas tankers, the Rules for the Massive Transport and Storage of Dangerous Substances were amended as of July 1986, and the International Bulk Chemical Code (IBC) and International Gas Carrier Code (IGC) of the IMO were officially adopted.

The amendment of IBC Codes was adopted in March 1989. In response to this amendment, the Ocean Pollution Prevention Law, the Regulations on the Marine Transport and Storage of Dangerous Materials, etc. were revised and enforced in October 1990. In particular, as for the Ocean Pollution Prevention Law, it is not necessary to make prior assessment for pollution classification of goods mixed with pollutants already assessed or that contain pollutants already classified by the International Marine Organization.

7.2.2.2. Standards for Air Transport of Hazardous Cargoes

With respect to air transport, standards for the transport of hazardous cargoes (notification) in Japan are determined on the basis of the Aviaiton Act. These standards are in accordance with the Regulations on hazardous cargoes (International Civil Aviation Convention Annex no. 18) and the technical guidelines of the International Civil Air Organization (ICAO). These standards (notice) were revised on December 20, 1990, and enforced on January 1, 1991. Testing of containers and the use of containers with UN labels have been obligatory since 1991 as in the case of marine transport.

7.2.2.3. Standards for Transportation Containers for Poisonous and Toxic Substances

The Control Act on Poisonous and Toxic Substances stipulates the handling of poisonous and toxic substances.

The Pharmaceutical and Supply Bureau of the Ministry of Health and Welfare issued the standard on transportation containers for poisonous and toxic substances (no. 3; a standard for small-size transportation containers having an inner volume of 450 L or less). The standard went into effect on October 1, 1991.

This standard is in line with the United Nations Hazardous Substances Experts Committee Recommendation-no.6 Edition from the standpoint of international harmonization. Also on October 1, notice was given of the establishment and revision of standards concerning each of the poisonous and toxic substances with respect to emergency measures in transportation accidents as well as disposal.

7.2.2.4. Safety Measures in the Use of Gases Under High Pressure

Safety Regulations with respect to high-pressure gas are in accordance with the High-Pressure Gas Control Act and governmental ordinances associated therewith.

The Industrial Location and Environmental Protection Bureau of the Ministry of International Trade and Industry has been holding meetings of the High-Pressure Gas Safety Policy Committee since September 1989 to study the best possible way of organizing high-pressure gas safety administration in the future. In June 1991, a report was published, proposing the following:

1) To promote self-security (i.e., self-protection measures and attitude for security)
2) To strengthen safety measures in response to diversification of applications
3) To review import Regulations from the standpoint of internationalization
4) To promote international cooperation in the field of safety
5) To strengthen execution systems by making full use of private-sector inspection organizations
6) To enhance systems supporting voluntary security activities

Since July 1991, the Ministry of International Trade and Industry has been holding meetings of the High-Pressure Gas and Explosives Safety Review Committee to study the future administration of high-pressure gas safety, including revision of the High-Pressure Gas Control Act.

7.3. Present Situation Regarding Safety Measures for Chemical Substances

In Japan, two laws exist for maintaining the safety of chemical substances under the notification system for new chemical substances prior to the manufacture thereof: (1) The Law Concerning Examination and Regulation of Manufacture, etc. of Chemical Substances, which is designed to regulate substances that are hazardous to human health because of contamination of the environment and (2) the system of Investigation of Toxicity of Chemical Substances under the Industrial Safety and Health Law, which is to regulate the concentration of carcinogenic substances in the workplace.

7.3.1. The Law Concerning Examination and Regulation of Manufacture, etc. of Chemical Substances

The Law Concerning Examination and Regulation of Manufacture, etc. of Chemical Substances had been, until its revision in April 1987, implemented effectively for the 14 years since its enforcement in 1973 to regulate PCBs and similar substances.

However, as Regulations became stricter in Europe and the United States, the need arose to coordinate Regulations on chemical substances in various countries and to expand Regulations on harmful chemical substances that are slowly degraded in the environment. The Japanese government then enacted a Law for the Partial Revision of the Law Concerning Examination and Regulation of Manufacture, etc. of Chemical Substances in May 1986 and implemented it in April 1987. The gist of the revised law is as follows:

1) The system of prior examination of new chemical substances was reinforced by demanding that toxicity tests, namely, mutagenicity tests (Ames test, chromosonal aberrations test) and repeated dose tests are to be conducted for 28 days, in addition to the biodegradability and accumulation tests now in force. The tests must be performed, at the time the government is notified of the chemical substances, prior to their manufacture and import.
2) Substances that are difficultly degradable and are suspected to have chronic toxicity are to be termed "Designated Chemical Substances" and must be put under surveillance with respect to their production volume and application after manufacture and import.
3) When such "Designated Chemical Substances" are suspected to be contaminating the environment and be hazardous to human health, the manufacturer or importer will be advised to perform a study on hazards of such substances.
4) If such studies reveal that the substances are hazardous, they are termed "Class II Specified Chemical Substances," and their production volume and handling methods must be regulated.
5) "Specific Chemical Substances" with respect to the old Law Concerning Examination and Regulation of Manufacture, etc. of Chemical Substances are now termed "Class I Specified Chemical Substances," and the same control measures, such as ban on their manufacture in principle and on their use in open systems, are used as in the past.
6) As a result of tests conducted by the government on existing chemical substances (which correspond to a prior examination for new substances), some of them may be termed "Designated Chemical Substances."

7.3.2. Recent Situation Regarding the Enforcement of Law Concerning Examinations and Regulations of Manufacture, etc. of Chemical Substances ("Chemical Substances Law")

Under the old Chemical Substances Law (until March 1987) the reported number of new chemical substances admitted for manufacture and import was 3864; under the new Chemical Substances Law (April 1987–December 1993), which applies more stringent Regulations, it is 1492.

The designated chemical substances as of the end of September 1993 amounted to 97, including chlorohydrocarbons and organotin compounds. Among them are 135 substances (excluding the 23 substances that were regrouped to the "Class II Specified Chemical Substances") whose manufacture and import quantities must be reported. The 23 specified chemical substances inlcude chlorohydrocarbons (trichloroethylene, tetrachloroethylene, carbon tetrachloride) and organotin compounds (e.g., triphenyl- and tributyltin compounds). Furthermore, safety measures for each substance such as the preparation of proper use manuals and labeling of containers are being taken by individual industry organizations.

As far as existing chemical substances are concerned, the government continues to conduct systematic safety monitoring. The Ministry of International Trade and Industry checked the safety of 970 substances from 1974 through 1993. The result was that 358 substances decomposed in solvents to nonhazardous compounds, 594 substances showed low and 18 substances high accumulation of hazardous compounds upon decomposition. Of these, the nine substances or classes of substances listed below are designated Class I Specified Chemical Substances. In addition, the Ministry of Health and Welfare conducts toxicity testing of each substance.

Polychlorinated biphenyls
Polychlorinated napththalenes
Hexachlorobenzene
Aldrin
Dieldrin
Endrin
DDT
Chlordane
Bis(tributyltin)oxide

More careful handling and discharge, not only of those substances that are regulated specifically but of all chemical substances in open systems, is desired.

7.3.3. International Cooperation on Chemical Substance Regulations

In Japan, the Good Laboratory Practice (GLP) system was introduced in 1986, and the coordination of testing methods with those of OECD and the mutual acceptance of test data were promoted.

Moreover, the OECD started to promote the work of international cooperation for investigating the safety of existing chemical substances. Specifically, the preparation of data bases (EXICHEM) on the testing status in member countries and joint investigation of chemical products of high production volume are being put in force. Concerning the joint investigation of chemical products with a high production volume, 53 substances were allocated in April 1990 to each country as Phase 1 in accordance with the investigation priority substances (P 1) list. Japan took charge of nine substances and collected data on them. Based on the result of this investigation, tests in 1991 were conducted with respect to characteristics on which no data were available or the data were indefinite. In addition, investigation of about 100 substances (Phases 2 and 3) was initiated toward the end of 1991. In this program Japan is to evaluate data on 24 substances. Furthermore, until the year 2000 about 5000 chemical substances of high production volume including the above figures will be surveyed. Japan will evaluate one-fourth of those substances.

8. References

References for Chapter 1

[1.1] Council Directive 80/778/EEC Relating to the Quality of Water Intended for Human Consumption, OJL 229, August 30, 1980, p. 11.J.
[1.2] J. Cameron, J. Aubouchay: "Precautionary Principle: a Fundamental Principle of Laws and Policy for the Protection of the Global Environment," Boston College International and Comparative Law Review, 1991.
[1.3] E. Rehbinder, "The Precautionary Principle in an International Perspective," in Miljørettens grundspøgsmål – bidrag til en nordisk forskeruddannelse – G-E-C Gad, København 1994.
[1.4] M. Rémond-Gouilloud, "Le risque de l'incertain: la responsabilité face aux avancées de la science" – La Vie des Sciences, *Comptes rendus, série générale* vol. 10 (1993) no. 4, 341–357 – Archives de l'Académie des Sciences, Paris.
[1.5] Oslo and Paris Conventions, Paris, 21–22 Sep. 1992, vol. 14 (1991) no. 1, p. 16.
[1.6] Treaty on European Union, Maastricht, 7 Feb. 1992.
[1.7] "London Declaration," *2nd Int. Conf. on the Protection of the North Sea*, 24–25 Nov. 1987.
[1.8] [C(72)128], OECD 1972; [C(74)223, OECD 1974.
[1.9] [C(90)177 (final)], OECD 1991.
[1.10] O.J.E.C., L 169 1–29, June 26, 1987.
[1.11] H. Smets: "Le principe pollueur payeur, un principe de l'environment," *Rev. Gen. Droit International Public* **2** (1993) April/June.
[1.12] Regulation 93/259/EEC on the Supervision and Control of Shipments within, into and out of the Community.
[1.13] International Information Exchange and the PIC Procedure (UNEP/FAO).

[1.14] UNEP (1989): London Guidelines for the Exchange of Information on Chemicals in International Trade (1987, amended 1989), UNEP, Nairobi.
[1.15] FAO (1989): International Code of Conduct on the Distribution and Use of Pesticides (1985, amended 1989, FAO, Rome).
[1.16] FAO/UNEP (1991): Guidance for Governments: Operation of the Prior Informed Consent Procedure for Banned or Severely Restricted Chemicals in International Trade, FAO/UNEP, Rome/Geneva.
[1.17] Council Regulation (EEC) No 2455/92 or July 23, 1992 (amended on January 11, 1994) concerning the Export and Import of Certain Dangerous Chemicals.
[1.18] Rege, "GATT Law and Environment-Related Issues Affecting the Trade of Developing Countries," in Journal of World Trade Law (June 1994).
[1.19] "Towards Sustainability," European Commission Fifth Environmental Programme [*COM* (92) 93 final]
[1.20] Resolution of the Council and the representatives of the Governments of the Member States on the fifth Community Policy and Action Programme for the Environment and Sustainable Development, adopted on March 18, 1992, OJC 138, May 17, 1993.
[1.21] Council Directive 90/313/EEC On the Freedom of Access to Information on the Environment of June 7, 1990 and Communication to the Council, Parliament and the Economic and Social Committee 93/C 156/05 on Public Access to the Institutions' Documents.
[1.22] Council Regulation 880/92/EEC on a Community Award Scheme for an Eco-Label (OJL 99, April 11, 1992)
[1.23] OJC 72/3, March 10, 1994.
[1.24] R. Lönngren: *International Approach to Chemical Control: A Historic Overview,* Kemi, The National Chemical Inspectorate, Stockholm 1992.

References for Chapter 2

[2.1] EEC Treaty, Article 5.
[2.2] EEC Treaty (1987–1993), Article 130r (4). Maastricht Treaty on European Union, Article 3b.
[2.3] Directive 67/548/EEC Relating to the Classification, Packaging and Labelling of Dangerous Substances, O.J. 1967, no. 196, p. 1 with later amendments.
[2.4] Directive 73/173/EEC relating to the Classification, Packaging and Labelling of Dangerous Preparations (Solvents), *O.J.* 1973, no. L189, p. 7, with later amendments.
[2.5] Directive 78/631/EEC Relating to the Classification, Packaging and Labelling of Dangerous Preparations (Pesticides), *O.J.* 1978, no. L206, p. 13, with later amendments.
[2.6] Directive 77/728/EEC Relating to the Classification, Packaging and Labelling of Paints, Varnishes, Printing Inks, Adhesives and Similar Products, *O.J.* 1977, no. L303, p. 23 with later amendments.
[2.7] Example taken from Directive 67/548/EEC (note 3) in its version of the 6th amendment, Directive 79/831/EEC, *O.J.* 1979, no. L259, p. 10.
[2.8] Directive 76/769/EEC on the Approximation of the Laws, Regulations, and Administrative Provisions of the Member States Relating to Restrictions on the Marketing and Use of Certain Dangerous Substances and Preparations, *O.J.* 1976, p. 201 with later amendments.
[2.9] Directive 79/831 (note 7), recital 1.
[2.10] EEC Treaty, Article 130r (2).
[2.11] Directive 90/220/EEC on the Deliberate Release into the Environment of Genetically Modified Organisms, *O.J.* 1990, no. L117, p. 15.
[2.12] Regulation 594/91/EEC on Substances that Deplete the Ozone Layer, *O.J.* 1991, no. L67, p. 1; now replaced by Regulation 3093/94, *O.J.* 1994 no. L333, p. 1.
[2.13] Directive 91/156/EEC on Waste, amending Directive 75/442/EEC, O.J. 1991, no. L78, p. 32; Directive 91/689/EEC on Hazardous Waste, *O.J.* 1991, no. L377, p. 20.
[2.14] Directive 76/464/EEC on Pollution caused by Certain Dangerous Substances Discharged into the Aquatic Environment of the Community, O.J. 1976, no. L129, p. 23; 80/68/EEC on the Protection of Groundwater Against Pollution Caused by Certain Dangerous Substances, *O.J.* 1980, no. L20, p. 43; 84/360/EEC on the Combating of Air Pollution from Industrial Plants; *O.J.* 1984, no. L188, p. 20.
[2.15] Directive 82/501/EEC on the Major Accident Hazards of Certain Industrial Activities, *O.J.* 1982, no. L230, p. 1 with later amendments.
[2.16] Directive 85/337/EEC on the Assessment of the Effects of Certain Public and Private Projects on the Environment, *O.J.* 1985, no. L175, p. 40.
[2.17] Regulation 1836/93 allowing Voluntary Participation by Companies in the Industrial Sector in a Community Eco-management and Audit Scheme, *O.J.* 1993, no. L168, p. 1.
[2.18] Commission Proposal for a Council Directive on Integrated Pollution Prevention and Control, O.J. 1993, no. C311, p. 6.
[2.19] Directive 92/112/EEC, *O.J.* 1992, no. L409, p. 11.

References for Chapter 3

[3.1] UNEP, "Liability and Compensation for Damage resulting from the transboundary Movement of Hazardous Wastes and their Disposal." Basel Convention, Montevideo, Dec. 1992, draft articles.
[3.2] Directive on Civil Liability caused by Defective Products, *O.J.E.C.* no. L210, August 7, 1985.
[3.3] Amended proposal for a Directive on Civil Liability for Damage caused by Waste, *O.J.E.C.*, no. L192/6 (July 23, 1991).
[3.4] Communication to Council and Parliament: Green Paper on Remedying Environmental Damage, *COM* (93), 47, March 1993.

References for Chapter 4

Sections of this chapter have been taken, in amended form, from two previous publications by G. H. Williamson

G. H. Williamson: "Environmental Protection versus Free Trade: A European Community Perspective," in: *Private Investments Abroad,* Southwestern Legal Foundation, Dallas, Texas, 1993, Chap. 3.
G. H. Williamson: "The European Union's New Regulation on Waste Shipments: Its Content and Practical Implementation Concerns," *Environment Watch, Western Europe*, May 6, 1994.

[4.1] "Completing the Internal Market: White Paper from the Commission to the European Council," Commission of the European Communities, June 1985.
[4.2] Commission vs. Council, Judgment of June 11, 1991, on Dir. 89/428 of the Council.
[4.3] Paragraph 2, new Article 130 R, added to Part Three of the Treaty by Article 25 of the Single European Act.
[4.4] Treaty on European Union, signed at Maastricht on Feb. 7, 1992 (*O.J.E.C.* C191, July 25, 1992).
[4.5] *O.J.E.C.* 1983, L 109/8, amended by Directive 88/182/EEC, *O.J.E.C.* 1988, L 81/75.
[4.6] Rewe-Zentral AG vs. Bundesmonopolverwaltung für Branntwein (Case 120/78), Judgment of Feb. 20, 1979. ECR (1979) p. 649.
[4.7] Commission vs. Denmark (Case 302/86), Judgment of Sep. 20, 1988, ECR (1988) p. 4607; 1 CMLR 408.
[4.8] Verordnung über die Vermeidung von Verpackungsabfällen, Bundesgesetzblatt 1991, part I, p. 1234.
[4.9] Common Position No. 13/94 with a View to Adopting a European Parliament and Council Directive on Packaging and Packaging Waste, 94/C/137/08, *O.J.E.C.* C 137/65, May 19, 1994.
[4.10] Regulation 259/92, *O.J.E.C.* L 30/1, Feb. 2, 1993.
[4.11] Directive 75/442/EEC, *O.J.E.C.* L 194/39, July 25, 1975.
[4.12] Directive 91/156/EEC, *O.J.E.C.* L 78/32, March 26, 1991.
[4.13] Directive 91/689/EEC, *O.J.E.C.* L 377/21, Dec. 31, 1991, as amended.
[4.14] Directive 78/319/EEC, *O.J.E.C.* L 84/43, March 31, 1978.
[4.15] Directive 84/631/EEC, *O.J.E.C.* L 326/31, Dec. 13, 1984.
[4.16] Commission Decision 94/3/EC, *O.J.E.C.* L 5/15, Jan. 7, 1994.
[4.17] 75/442/EEC and 91/156/EEC.
[4.18] 91/689/EEC; List published at *O.J.E.C.* L 356/14 seq., Dec. 31, 1994.
[4.19] Commission vs. Belgium, E.C.J. Case no. C-2/90, decided July 9, 1992.

References for Chapter 6

[6.1] S. M. Novick et al. (eds.): *Law of Environmental Protection*, Environmental Law Institute, Washington, D.C. 1987.
[6.2] T. T. Smith, Jr., R. D. Hunter (eds.): *European Community Deskbook*, Environmental Law Institute, Washington, D.C. 1992.
[6.3] T. T. Smith, Jr., et al.: "Understanding European Environmental Regulation," *Conference Board Monograph*, report no. 1026, New York 1993.
[6.4] 42 U.S.C. §§ 4321-70C; Directive 85/337, *O.J.E.C.* L 175/40, July 5, 1985.
[6.5] Directive 82/501, *O.J.E.C.* L 230/1, Aug. 5, 1982.
[6.6] 42 U.S.C. $ 11001 (a)–(c).
[6.7] 33 U.S.C. § 1321.
[6.8] The Clean Air Act, 42 U.S.C. §§ 7401 ff, as amended; the Clean Water Act, 33 U.S.C. § 1251–1376. The EC Air and Water Legislation is Summarized and Cited in Ref. [6.2].
[6.9] The Resource Conservation and Recovery Act, 42 U.S.C. §§ 6901 ff. The EC Waste Legislation is Summarized in Ref. [6.2], pp. 24–28.
[6.10] Directive 75/442, *O.J.E.C.* L 194/26 (July 25, 1975), as amended; Directive 78/319, *O.J.E.C.* L 78/32 (Mar. 31, 1978), as amended.
[6.11] Directive 84/631, *O.J.E.C.* L 326/31 (Dec. 13, 1984), as amended.
[6.12] 42 U.S.C. §§ 9601 et seq.
[6.13] Directive 67/548, *O.J.E.C.* L 196/1 (Aug. 16, 1967), as amended.
[6.14] Directive 79/831, *O.J.E.C.* L 259 (Oct. 15, 1979).
[6.15] T. T. Smith, Jr., R. D. Hunter (eds.): *European Community* in [6.2, pp. 28–31].
[6.16] 15 U.S.C. §§ 2601–29.
[6.17] 29 C.R.R. § 1910.1200, issued under 5 U.S.C. 553.
[6.18] T. T. Smith, Jr., et al.: in [6.3, p. 7–10].
[6.19] T. T. Smith, Jr., et al.: in [6.3, p. 10].

Environmental Management in the Chemical Industry

LOUIS JOURDAN, European Chemical Industry Council (CEFIC), Brussels, Belgium

3-527-20137-8/95/$5.00 + .50

1. Introduction

1.1. Historical Development

A few decades ago, the term "environmental management," not to mention "environmental management system," would have raised eyebrows. Despite the fact that some advanced companies were already behaving quite properly as far as environmental protection was concerned—at least relative to the prevailing standards—the term had not yet been coined, and companies did not have the type of environmental organization considered necessary today.

Nevertheless, environmental considerations had been taken into account, consciously or unconsciously, in the management of chemical industry for a long time: one might quote, for example, a French imperial decree from 1810 controlling "dangerous, noxious and infectious establishments" which could be considered as a precursor to the EC "Seveso" Directive. Closer to the present, the Solvay process for the manufacture of soda ash, which in the second half of the nineteenth century replaced the Leblanc process, brought substantial environmental improvements.

In fact, safety considerations were integrated into industrial management long before the need to protect the environment was recognized. Following dramatic accidents and severe health damage that were common in the early stages of the industrial revolution, governments, supported by a few enlightened industrialists, established the first of many standards for process, product, and human safety. The efforts of certain politicians to protect workers against the effects of white lead, and of Chancellor Bismarck to lay the foundations for accident prevention and social security, are well known examples. Some major companies, such as E.I. du Pont de Nemours, have built their reputations on the basis of very strict safety standards.

But it was not before the 1960s or 1970s that the impact of chemical products and chemical industry on the environment were widely publicized – first in 1962 with Rachel Carson's "Silent Spring," then with the UN Conference in Stockholm in 1972. Since that time, local and national governments as well as intergovernmental organizations, in particular the Organization for Economic Cooperation and Development (OECD) and various United Nations Agencies, have established a number of legislative actions, conventions, or recommendations that have necessarily been integrated into the management of chemical companies.

More recently the concept of "Sustainable Development" was introduced in the 1987 report "Our Common Future" from the UN Commission for Environment and Development, led by G. H. Brundtland [1]. The report recognized the threats imposed on the environment by both increasing population and ongoing industrial development, but it also identified challenges that must be met if such development is to be sustained without compromising the needs of future generations. Industry has an indispensable role to play if these objectives are to be achieved, in particular with respect to energy consumption, food production, and chemicals.

The Brundtland report paved the way to organization of the UN Conference on Environment and Development (UNCED), held in Rio de Janeiro in June 1992 [2]. Whatever can be said about the meager immediate results of the Conference, its products, in particular the "Agenda 21," will have to be taken into consideration by industrial managers. Two chapters of Agenda 21 are of direct relevance to chemical industry, and a United Nations Commission for Sustainable Development (UNCSD) has been set up to implement the Conference's decisions.

The integration of environmental considerations into industrial management is influenced most obviously by the level of development of the country in question: even within the European Community, after more than twenty years of legislative actions to harmonize ("approximate") national legislation, some differences still exist between the behavior of certain companies in different Member States. When one looks at developing countries, the differences are often even greater. Political regimes also have an influence on environmental awareness and requirements: the state of the environment left by the centrally planned economy regimes of the former communist countries clearly shows that free-market economy, democracy, and a high state of development all contribute to superior environmental protection in industrialized or industrializing countries.

1.2. Factors Influencing Companies to Improve Their Environmental Performance

While companies undoubtedly must satisfy their first "bottom line" (profitability), as well as a second crucial one—socio-economic responsibility—they now must satisfy a third one as well, namely environmental responsibility. Several factors are pushing industrial managers to meet this third challenge and to improve the environmental performance of their companies. Some of these are listed below, in an order that might vary depending on a particular manager's personal commitment, the company's culture, the local legal and economic conditions, etc.

Legal Requirements and Regulations. In all industrialized countries, environmental regulations are now being established at an ever increasing pace (Fig. 1), and they are setting ever tighter standards. Although enforcement requirements may differ between countries, no reputable company can afford not to comply with at least national and local regulations.

Environmental Liability. Fears of prosecution triggered by civil or even criminal liability for environmental damage are a strong incentive toward excellence in environmental performance. Some well-known "environmental scandals" have been settled, either between the parties involved or by the courts, at levels that could endanger the financial survival of a company. A matter of especially great concern is "past pollution;" i.e., the ecological damage created by an industrial operation that was complying fully with the legal requirements of some past period. Even if not yet common, criminal liability for environmental damage may soon become a reality, as is already the case for safety failures. Several pieces of legislation (Council of Europe, European Community) based on the concept of "strict" liability currently being adopted, will make the matter even more important (see Chap. 4 for more detailed information).

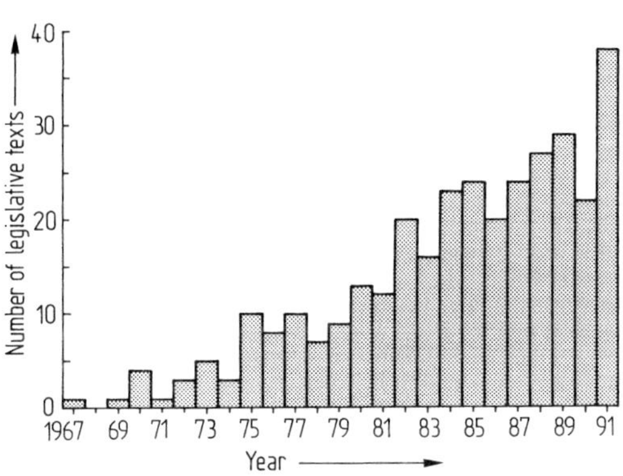

Figure 1. Evolution of environmental legislation in the European Community (EC)

Image. As the public, under the influence of the media and many environmental non-governmental organizations (NGOs), becomes increasingly aware of the need to protect the environment, companies will need to build public images as "good environmental citizens" if they hope to maintain their licenses to operate. There have already been cases of companies boycotted by consumer organizations in response to some environmental affair. "Eco-labels" or "Green Awards" have become strong elements in marketing strategies, and environmental performance criteria are becoming subject to more and more consideration in relations between suppliers and consumers.

Employees. Employees and unions have traditionally been particularly concerned by labor relations, and especially wages, working conditions, and safety, but they are becoming increasingly interested as well in the environmental performance and reputation of the companies with which they are associated. Some companies have established "Environment Committees" at the shopfloor level paralleling the long-established Safety Committees.

There are also clear indications that a good environmental image helps a company recruit high-quality personnel, especially at the management and executive level.

Finance. It is now recognized by the financial community–shareholders, bankers, insurers – that good environmental performance is an asset, even if in the short term it entails some capital expenditures and increased operating costs. There are more and more cases of corporate annual general meetings of shareholders where questions on environmental performance are debated. Moreover, insurance companies now adjust their premiums according to the environmental reputations of their clients. Some organizations have proposed to their customers – mostly in the financial community – an "eco-rating" of major industrial corporations, similar to financial ratings.

Professional Associations. Industrial associations at the national or international level have no real power over their member companies, but they are nevertheless an effective platform for establishing common principles and guidelines that set a high-level yardstick of environmental performance for all their members.

Ethical Considerations. Some industry leaders are already well-known for their personal commitments to environmental protection, and they manage their companies accordingly. Many other company executives are beginning to transfer to their management activities their personal ethics with respect to the environment.

1.3. Scope of This Contribution

When speaking about environmental management it has become current practice to discuss health-protection and safety-management aspects together with purely ecological matters. This contribution devoted to "environmental management" will therefore include various aspects of health, safety, and environmental (HSE) management. It must be recognized, however, that in some companies these elements are not integrated, and that health protection (H), safety (S), or health protection and safety (H + S), are managed separately from environmental protection (E). Conversely, some companies include quality under the same management structure. Because of their specificity, safety and environmental aspects of transport and distribution are integrated into an HSE management structure by only a few companies.

2. Environmental Management in Chemical Companies

2.1. Preamble

As noted in the previous chapter, the concept of environmental management is relatively new; it has developed without established standards, in a "learning-by-doing" process. It is therefore no wonder that the environmental management structures adopted by individual companies differ significantly according to company size, geographical location, and coverage; the legislative framework of the country(ies) in which they operate; their culture; etc.

What is described below is the typical structure, organization, and content of an environmental management system adopted by a typical large or multinational company with headquarters in one of the industrialized countries. Companies of smaller size or operating under different conditions should adapt it accordingly.

2.2. Environmental Management Systems

Basically, the functioning of a company relies on two systems:

1) An *operational system*, which achieves the basic purposes of the company (e.g., the manufacture of a certain range of products under certain financial and social conditions)
2) A *management system*, which acts on the operational system to help it function and improve

Management systems applied to the various aspects of management are an important key to the long-term success of any company. Fundamentally, such a system necessitates defining a policy, establishing standards and objectives, planning for achievement, and allocating whatever resources are required, as well as documenting, measuring, auditing, and reviewing for improvement. An environmental management system (EMS) is built along this "loop," as represented by Figure 2. Its main elements are described below.

2.3. Environmental Policy

Although it must be recognized that establishing and publicizing a policy for HSE matters is not, or at least not yet, a part of the culture of some companies, a clearly stated environmental policy is undoubtedly the foundation of any environmental management system.

An *environmental policy* can be defined as a public statement of the intentions, principles of action, and objectives that govern the management of a company with respect to the environment. Although the details and extent of such a

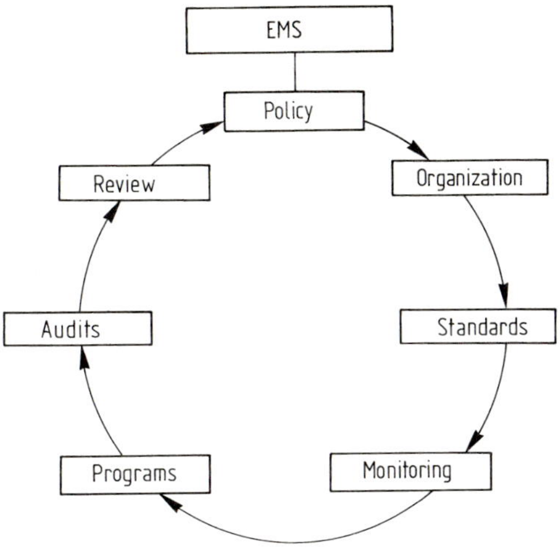

Figure 2. The environmental management system (EMS) "loop"

policy may vary from company to company, the policy itself is generally expressed in a concise statement. Typical elements include:

1) A commitment to conduct the company's activities in a manner that safeguards the environment, the health, and the safety of its personnel, local communities, consumers, and the general public
2) Compliance with environmental legislation, or even an intention to go beyond legislative compliance when appropriate
3) Communication of environmental information to concerned stakeholders within society
4) Information for customers on the nature and safe use of the company's products
5) Environmental education and training of employees and contractors
6) Support for environmental research
7) Access to environmentally sound technologies

A corporate environmental policy is a document approved and endorsed by the highest level of management; the President, Chief Executive Officer (CEO), or Executive Committee. The policy is also made available to the public and communicated to all interested parties, in particular the company's employees, shareholders, and major customers.

2.4. Organization and Structure

To implement a company's environmental policy the necessary human resources must be allocated and organized in such a way that they work with the greatest possible effectiveness, in particular in their relationship with the operating system. Organization implies a structure that identifies areas of responsibility and authority for key personnel.

2.4.1. Typical Organization

Again, the personnel structure of an environmental management system may differ quite considerably from company to company, but a typical organization is that described by Figure 3 (note, however, that in some companies the functions H + S and E are separated). The basic principles are as follows:

The *Board of Directors* is responsible for defining the company's environmental policy. Some companies have established an "HSE Advisory Committee" to assist the Board in this function; members of the Advisory Committee are members of the Board or persons external to the company selected for their interest, competence, or representativity in HSE matters.

The *Managing Director* (President, CEO) is responsible for the adoption and implementation of the company's environmental policy. His personal involvement should be made clearly visible across the company (in the famous words of DuPont's E. Woolard: "as Chief Officer of this company, I am also its Chief Environmentalist" [3]).

A *senior manager* (Vice-President, Director) reporting directly to the President, heads the company's central HSE department. This centralized function is not directly responsible for

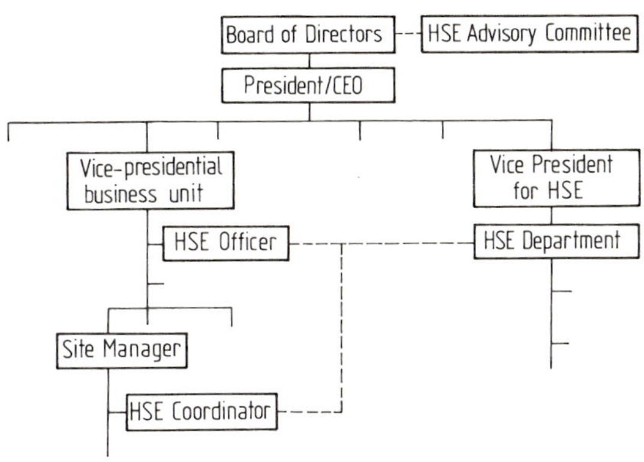

Figure 3. Structure of a typical health, safety, and environmental (HSE) organization

policy implementation, but ensures the proper working of the various elements of the EMS. It includes specialized activities such as toxicology/ecotoxicology, occupational health, emissions, waste management, environmental audits, product safety, regulatory affairs, etc. Most of the members of the central HSE departments act in an advisory capacity to support the "line" management.

Business units (product divisions, manufacturing companies, geographical areas) are responsible for policy implementation. Here again, HSE directors or coordinators at the level of the division or site function in an advisory capacity.

2.4.2. Staffing of HSE Functions

The number of employees holding a full-time HSE position varies widely, depending on such factors as process and product hazards, local regulations, etc. An approximate range is from 2 to 25 HSE staff members per 1000 employees, with an average around ten.

Allocation of professionals between a central corporate HSE department and the HSE departments in various divisions and sites is again a matter to be decided by the company: Some have large centralized departments, with up to 150 experts competent in all aspects of the corporation; others prefer to decentralize to the greatest extent possible at the level of divisions or sites, maintaining only a light core group with mainly coordinating functions.

Whether the HSE functions should be staffed by generalists or by professionals educated in HSE matters is much debated, especially since the educational system now provides explicit technical or university degrees in HSE disciplines. Equally important are concerns about the possibility of career development for specialists in HSE. Many companies regard an HSE position as a necessary step toward general management; others ensure that HSE specialists have career development opportunities similar to those offered to generalists.

2.4.3. Employee Motivation and Training

As noted previously, employees in business units and operational sites are responsible for the implementation of a company environmental policy. This requires that staff at all levels be clearly aware of their accountability, as well as properly trained and motivated.

Most companies hold employees accountable through evaluation of their individual environmental performance, or at least safety performance, which is more easily measurable. Plant and site managers are often required to set HSE objectives for their specific areas of responsibility, with possible monetary bonuses linked to achieving the goals. Several companies have prize programs with monetary awards or other forms of recognition for environmental improvements.

Training in HSE matters is a key element in the effectiveness of an EMS. Training is especially necessary when a generalist is appointed to an HSE position. Several companies include HSE elements in their training programs for managers at various levels. Training of all employees in HSE matters is more and more considered to be a major priority, even if for the moment only a few companies have implemented such comprehensive programs.

The case of line middle-managers deserves special consideration, because these persons sometimes perceive environmental demands as coming from people outside the plant with little knowledge about what is happening in practice. It is therefore important that senior management demonstrate its commitment to environmental quality, for example, by facilitating a two-way communication channel on HSE matters, allocating the necessary human and financial resources, and properly rewarding positive achievements. Also, programs that include a strong motivation element, such as Responsible Care (see Section 5.4), are helpful in convincing line managers, as well as staff people like sales managers, of their environmental duties.

Another difficult case is that of contractors working on behalf of a company for construction or maintenance work, waste disposal, transport, etc. It is normally part of a company's policy to insist that contractors work under the same HSE standards as the company's own personnel, but accountability, communication lines, training possibilities, etc. are obviously different. The problem requires special attention on the part of all those involved.

2.4.4. Research and Development

Most major chemical companies have R & D programs devoted exclusively to HSE matters and representing some 10–25% of all their

R & D expenditures. Environmental R & D topics are linked most obviously to a company's products and processes, and to its HSE priorities and programs.

2.5. Standards, Regulations, Monitoring, and Environmental Programs

A fundamental element of an EMS is a comprehensive register of applicable legislation and company standards. It is against these governmental or company standards that environmental performance will be measured, permitting the necessary programs for correction or improvement to be developed.

2.5.1. Register of Standards

The number and complexity of environmental statutes at the local, national, and international level is constantly increasing. It is the role of a central business unit or individual site HSE functions to determine those that are applicable in the various sectors of the company. It may also happen that, because of gaps in the applicable legislation or difficulties in adapting it to the specialities of a certain activity, a given company is forced to establish its own standards, which would of course go beyond the legislated criteria. In general such standards are established at the highest level of the company, and are applicable to all its sectors.

To ensure proper implementation by line personnel, legislation and company standards may require more detailed guidelines or guidance notes. Such guidelines, generally drafted by professionals in the central HSE department on the basis of their own experience and outside information, provide site managers with key principles that are to be incorporated into local procedures.

Maintaining the currency of all documentation of applicable laws and standards, guidelines, and local procedures, and communicating modifications and additions to the affected personnel, is a difficult task; a written procedure with clear allocation of responsibilities may be necessary.

2.5.2. Measuring Environmental Performance

"You can't manage what you can't measure." Therefore, to measure the effectiveness of an environmental management system, real data are required on the precise effects of the company's activities on the environment and individual people. It is thus necessary to conduct a planned, regular sampling and monitoring program at the various company sites as well as outside the company (e.g., regarding the effects of certain products).

It is generally accepted that the parameters to be monitored should include:

- Point sources and fugitive emission to the atmosphere
- Point sources and fugitive discharges to water or sewers
- Solid and other wastes, particularly hazardous wastes
- Contamination of land
- Use of water, fuels, and energy
- Discharge of thermal energy, noise, odor, and dust
- Effects on specific parts of the environment and ecosystems
- On-site accidents and near-misses
- Personnel injuries
- Transport accidents
- Complaints from community residents
- Complaints from customers related to HSE matters

All these data should be obtained by prescribed or accepted (standardized) measurement techniques and collected in registers established at various levels in the company. Some companies collect only absolute figures, while others combine certain data with appropriate weighting factors into an index or several indices. Such indices may be useful management tools for fixing objectives and demonstrating improvements.

Results of all measurements are communicated to the various interested parties, from plant operators to senior management, at the appropriate frequency and in the relevant form.

2.5.3. Environmental Programs

On the basis of global, long-term targets, set at the highest achievable level (e.g., 50% reduction in waste generation within the next 10 years), together with data from the register of environmental effects, individual environmental programs are established for the various sites and units of a company. Programs usually cover a one-year period within a multi-year global program.

Proposals for individual programs are checked for feasibility, relevance, and priority, usually by the central HSE department. A proposal for a corporate environmental program is then submitted to senior management, which decides on the proposals to be accepted and allocates the necessary human and financial resources. The corresponding investments are included in the global investment program of the company.

It is not easy to determine the amount of capital investment currently devoted to HSE by chemical companies, in part because there is no established methodology for computing these expenses. Furthermore, while it may be reasonably easy to determine the cost of end-of-pipe equipment, how is one to isolate the environmental component in the cost of a "cleaner," integrated process? Estimates of capital expenditure related to HSE average 10–20% of total capital investment for most companies. Operating expenses for HSE are in a similar range, and these are expected to increase, at least as long as processes intrinsically safe and "clean" are not generalized.

2.6. Environmental Audits

All companies have long been accustomed to having their financial management systems audited by accredited experts. A few decades ago the auditing of safety practices became a current practice for many companies. Only a few are so far familiar with the concept of "environmental auditing," but a trend in this direction is spreading rapidly across industry.

2.6.1. Definition

The International Chamber of Commerce (ICC) has proposed the following definition for an environmental audit [4]: "A management tool comprising a systematic, documented, periodic, and objective evaluation of how well environmental organization, management and equipment are performing with the aim of helping to safeguard the environment by:

1) Facilitating management control of environmental practices
2) Assessing compliance with company policies, which would include meeting regulatory requirements"

Such a definition might appear somewhat narrow when one considers that many corporate policies tend to cover unregulated risks. Indeed, the United States Environmental Protection Agency (EPA) has included such unregulated risks in its definition of environmental auditing.

Environmental audits are therefore a necessary component of a company's environmental management system, permitting a determination of whether or not the system is implemented effectively and is suitable for fulfilling the company's environmental policy.

Given the many activities of a company that have a bearing on the environment, environmental audits may serve various purposes:

Compliance audits verify that an industrial activity meets its legal obligations (permits, emission limits, etc.). They do not differ in principle from inspections by government inspectors.

Health and safety audits are still carried out separately in many companies, even though the tendency now is to integrate them into global environmental audits.

Site audits are the most common type of environmental audits. They are designed to check individual site activities against those elements of a company's environmental policy and management system applicable to the site in question. For large sites, audits may be limited to a single plant or a series of units.

Specific issues may also be the subject of an environmental audit at the corporate or site level, such as waste management, transport and distribution activities, etc.

2.6.2. The Audit Plan

To ensure proper verification of a global environmental management system, auditing of the various facets of a company's activities should be carefully planned.

The central HSE department is generally responsible for establishing an appropriate *audit plan*, which should deal with the following points:

1) Setting priorities
2) Establishing specific areas and activities to be audited and the frequency for auditing each particular site as a function of the nature and risk of its activities as well as conclusions reached in the previous audit; experience shows that a frequency of one to five years covers most situations encountered in the chemical industry
3) Objectives of the various audits
4) Personnel requirements

5) A protocol for conducting the audits
6) Procedures for reporting audit findings

2.6.3. Environmental Auditors

It is generally accepted, even in legislation (see Section 4.4), that environmental audits should be performed either by personnel from within the company (but wherever possible independent of the unit being audited) or people external to the company. A growing number of organizations and consulting companies are offering their expertise for environmental audits, which may be useful when internal expertise or capacity is lacking. Because of the rapid development of this market, great care should be exercised in the selection of an external consultant. It is likely that in the future a system of accreditation of environmental auditors will be developed; for the moment certification of auditors on the basis of the ISO 10011 standard for quality-system audits may provide a certain degree of assurance.

Larger companies would certainly prefer to carry out environmental audits using their own teams of auditors, supplied either by the central HSE department or drawn from various line activities within the company. A typical team includes three to four persons under the recognized authority of a leader. At least two members of the team should be experts in environmental matters, and one or more should be familiar with the particular type of activitiy to be audited. All members should have the necessary skills to obtain, from observation and interviews, clear, unbiased information on the operation in question without being regarded as "policemen" by the local employees. To facilitate contacts and technical arrangements an employee of the unit subject to audit may be nominated to act as a liaison with the auditing team.

The time required for an environmental audit varies considerably according to the size of the site, the nature of the operation, the cooperation of local personnel, etc. It may range from a few days to several weeks.

2.6.4. Audit Methodology and Process

The detailed actions required to carry out an environmental audit are highly dependent on the type and nature of the activities to be audited, but the following features are common to most audits:

1) Identification of elements of the management system, including instruction manuals, written procedures, communication, and reporting schemes
2) Evaluation of the control system
3) Assessment of strengths and weaknesses of the management system, especially deviations from written procedures and failures in the reporting lines
4) Examination of records of data on controlled or accidental emissions, energy use, complaints, etc.
5) Evaluation of training and motivation of personnel

Auditors collect the necessary information through direct observation and interviews of personnel at the plant, drawn from all levels. Cross-checking of information is a common practice, but care should be exercised to avoid friction with and between employees. Contacts may be facilitated by starting the audit activities with a brief meeting with plant personnel, where the audit objectives, plan, and methods are explained. The full cooperation of local management is in any case required.

Immediately at the conclusion of an audit the leader of the auditing team should present the major findings to the local (plant or site) management. Already at this stage a number of immediate corrections can often be achieved.

2.6.5. Audit Report and Action Program

The final report is drafted by the team leader, and usually checked by the site manager before being sent to senior management (President and/or Board).

A typical report includes three parts:

1) An executive summary of major findings
2) A section on detailed observations, the part most useful to local management for improving the operating procedures
3) A proposal for corrective actions, including capital investment, major equipment modification, or changes to managerial procedures when appropriate

The corrective action program is discussed at senior management level. When approved, in conjunction with an appropriate time scale, it is integrated into the global environmental program of the company.

2.7. Environmental Management Reviews

The suitability and effectiveness of an environmental management system must be assessed at regular intervals. This is accomplished at regular management meetings involving senior managers of the company together with the head of the central HSE department. Such meetings are intended to review the latest data on environmental performance, trends, deviations from predetermined objectives and targets, how effectively the environmental program has been implemented, etc.

When necessary, a review meeting may lead to decisions to modify or adapt the environmental objectives, the program, or even the policy if this seems justified.

With this last element the environmental management system "loop" is closed.

3. Environmental Communication

3.1. Environmental Communication: A New Aspect of Management

Traditionally, industry and particularly the chemical industry have not been models of good communication with society; indeed, secretiveness and confidentiality were deeply entrenched in the cultures of many reputable companies. Starting with the disclosure of financial information, however, enterprises have increasingly adopted a more open attitude.

A similar evolution has occurred in the environmental domain: a few decades ago, no company would have considered going beyond minimal "instructions for use" on the labels of their products. Then came recognition of the "duty-to-inform," mostly in the form of information provided to the local community in the vicinity of an industrial site regarding actions to be taken in case of an accidental emergency at the plant. Finally, the public "right-to-know" was recognized by industry, first in North America, then in Europe, and ultimately even in Japan, where public-opinion pressures are less severe.

This is not to say that there is now full transparency on HSE matters between industry and society. It is worth noting in this context that environmental organizations talk in terms of "corporate disclosure," whereas industry prefers to talk in terms of "corporate communication and reporting." Some companies still believe that the public is interested only in being told about risks to which they are exposed, not absolute figures regarding emissions of dangerous substances. Industrial property rights are often used as an argument to limit disclosure.

Nevertheless, environmental communication is now featuring in the environmental policy of a growing number of companies, and it is developing rapidly across the world chemical industry.

A number of factors explain this trend:

1) The disclosure of environmental information has become compulsory under some existing national legislation (see Section 4.4). In the future more legal requirements are expected in many countries.
2) Within the broad concept of "sustainable development" (→ Ethics and Industrial Chemistry, Section 1.3.1.) environmental reporting has been identified as a major challenge for industry. As mentioned in the Agenda 21 document [5], "companies are encouraged to report annually on their environmental records, as well as on their use of energy and natural resources."
3) The credibility of industry, in particular of the chemical industry, is poor in the eyes of the public. Providing data on environmental performance (showing not only positive achievements but also difficulties and failures) can help restore industry's image and build credibility.
4) Customers have begun to integrate environmental considerations, including the environmental performance of suppliers, into their purchasing decisions.
5) Shareholders, financial institutions, and the insurance branch require information to help them assess the real present and future financial risks to which companies they are dealing with are exposed.
6) Employees and local communities need information about their working and living environments.
7) Environmental NGOs are likely to campaign more aggressively against corporations that give the impression they have something to hide about their environmental performance.

In fact, more and more companies consider environmental reporting to be a definite competitive advantage and a major factor in their "license to operate."

3.2. Audiences and Vehicles

Many different stakeholders in society have an interest in environmental information supplied by industry. The key ones include:

Local and national governmental agencies
Local communities
The general public
Media (general and specialized)
Employees
Shareholders
Banks, financial institutions, and insurers
Domestic and trade customers
Schools, colleges, universities, and business schools
Environmental NGOs

Satisfying the needs of these various audiences may require specific vehicles for environmental communication. Some examples of appropriate communication forms are:

Labels. The chemical industry has for several decades been using regulatory or standardized labels, in particular for hazardous products. Specification sheets, such as the standardized "Material Safety Data Sheets" (MSDS), provide supplementary information on the hazards and risks of chemicals for various user categories.

Product advertising increasingly includes some environmental information. To limit the use of ill-founded or misleading statements, certain industrial organizations have drafted special guidelines on the matter [6]. The eco-labeling schemes operative in the EC and other countries represent a form of regulated environmental advertisement.

Media, including the general press and TV, are becoming increasingly interested in the environmental performance of enterprises. However, they are attracted more by bad news than good news, and the publicity afforded by the media to unfortunate accidents or incidents may seriously endanger a company's reputation.

Company newsletters and *magazines* offer a privileged form of support to environmental information directed at employees or shareholders.

"Open house" days and *plant visits* provide the best opportunity to convey environmental information about a site to local communities (including local media and environmental organizations), to employees, and to schools and other educational bodies.

"Hot lines", *toll-free telephone numbers*, and *information centers* are being made available by some companies to their customers and the public.

3.3. The Environmental Report

The concept of an environmental report, especially an annual environmental report, is becoming more and more popular among corporations in North America and Europe, and also Japan. Many companies consider the annual environmental report as a key element in their environmental communication strategy.

Because this concept is still in its infancy (the first full free-standing environmental reports were published in the late 1980s or early 1990s), there is as yet no standard or commonly accepted format for such reports. What follows is therefore simply a review of current trends, which may evolve significantly in the coming years.

3.3.1. Audiences

The annual environmental report is certainly the communication vehicle best suited to the needs of a number of audiences, in particular the employees, local communities, shareholders and financial institutions, environmental organizations, etc.

This underscores a major difficulty in the concept of the annual environmental report: finding the right style and the right balance between technical information, financial data, and easy-to-understand words, graphs, and figures. From the experience gained so far it is apparent that some reports would benefit from an improvement in style and presentation.

3.3.2. Corporate or Site Reports

Perhaps for fear of local adverse reactions, some corporations publish only a report covering the company as a whole, with consolidated data. However, if the report is to meet the needs of local communities and local audiences it is obvious that data pertaining to the site in question must also be published.

This fact is accepted by a growing number of companies. Some publish a single report with a section for the entire corporation together with various sections dealing with individual sites; others publish a corporate report supplemented by free-standing individual reports for their (main) sites.

3.3.3. Contents of the Report

Some reports are still mainly textual, providing only a few numeric figures for principal indicators, generally consolidated at the corporate level. Certainly reports should include a significant textual part, presenting, for example, the company's environmental policy, its environmental program and organization, the major achievements, objectives, and targets. However, a growing need is perceived to publish numerical data for both corporate and site reports as a response to the now recognized "right-to-know" of the public. Indeed, publishing such numerical data within a report offers a definite advantage over provision of the "dry" figures required under certain legislation: the figures are put in the perspective of a real business (size of the site, products manufactured, etc.), thus avoiding—at least partially—easy misinterpretation by certain groups.

It is mentioned in Section 2.5.2 that some companies establish "environmental indices" from measured data on emission. Certainly such indices are useful for demonstrating the trends in environmental performance, but using them as a substitute for the publication of absolute figures can generate harsh criticism from some concerned stakeholders.

Another issue of concern is whether reports of all companies (at least in the same industrial sector) should include data covering the same parameters for a company or a site. This is certainly reasonable in terms of environmental relevance. Furthermore, one cannot deny the right on the part of many stakeholders (including regulatory authorities) to undertake certain comparisons between companies and sites, and this consideration would also militate in favor of a certain comparability of data. At least a "core list" of common parameters should be included in all reports.

3.3.4. Links to Environmental Management Systems

Environmental communication is not *stricto sensu* an element of the EMS. However, it is becoming more and more accepted that the results of environmental audits—at least those not containing confidential information, the disclosure of which could endanger competitiveness—must be made public. The EC Regulation on Eco-Audits (see Section 4.3) requires the publication of a "public statement" after every audit. Site environmental reports would be a perfectly appropriate support vehicle for these statements.

3.3.5. Trends

There are many indications that corporate environmental reports, published at regular intervals, will eventually become obligatory for all major companies, in particular in the chemical industry.

The quality and quantity of information disclosed in the reports is likely to evolve. A research study [7] has identified five stages in the way environmental reports develop to meet the stakeholders' information needs:

1) Green "glossies," newsletters, videos; short statements in annual reports
2) One-off environmental reports, often linked to a first formal policy statement
3) Annual reporting, linked to an environmental management system but providing more text than figures
4) Provision of full environmental performance data on an annual basis, available on diskette or on-line, with an environmental report cited in the annual report
5) Sustainable-development reporting linking environmental, economic, and social aspects of corporate performance and supported by indicators of sustainability

Presently, most examples of environmental reports meet the definition of stage 3; only a few have reached stage 4. To attain stage 5 companies will need to develop and adopt methodologies for full environmental cost pricing and life-cycle assessment of their activities and products.

3.4. Stakeholder's Reactions

The argument most commonly used by those reluctant to embark upon environmental communication is a fear of adverse reactions from various groups within society. Experience shows, however, that with the necessary skills and precautions, the release of environmental information and in particular of annual environmental reports generates only occasional negative reactions, and that the benefits definitely outweigh the difficulties.

For example, local communities and employees always welcome environmental information, especially that related to specific sites. Shareholders and financial analysts have begun to take an interest in environmental reports, but are likely to demand deeper assessments. The media have reacted positively to the release of environmental reports, but may be critical of reports that are too "textual." Environmental organizations are for the moment granting the benefit of doubt to corporations strongly involved in environmental reporting.

Certainly the credibility of information released by companies, in particular by those with a poor public image, may become an issue, and it will take some time for companies to demonstrate that the information they publish is candid and accurate. This has prompted some companies to request validation of their information, in particular of their environmental reports, by independent third parties, such as accredited inspection organizations. For the moment, few reports have received such external validation, but several companies envisage having future reports validated by reputable, undisputed third-party organizations.

3.5. Who Communicates?

Many companies still have a policy of relegating the responsibility for environmental communication to their professional communicators, backed as far as technical information is concerned by staff people in the HSE department. Professional communicators will always be needed to coordinate a company's information policy and define its rules and practices, but there is a recent trend toward inviting line and staff managers to participate in environmental communication. For example, open house days obviously involve many employees, who must be prepared to answer questions about a company's environmental performance. Site managers or superintendents may also be interviewed by journalists or media reporters, and employees are the company's best ambassadors in the local community.

Involving various levels of HSE managers in communication becomes all the more important when a communication crisis bursts following an accident in a plant, for example, or at any event attracting the attention of the media. All companies and sites should establish crisis management plans, duly tested and identifying all the major players, especially with respect to communications. There have been too many examples of accidents or isolated incidents that have ruined several years of effort by reputable companies to build trust with the public, simply because communications have been poorly managed during the event.

Many companies now include communication as an element in their training programs, in particular in the training of their HSE specialists.

4. The Impact of Legislation on Environmental Management

Complying with regulatory standards is clearly an important element of any environmental management system. The purpose of this section is not to describe all legislative texts applicable to industrial activities, but rather to identify major trends in environmental legislation influencing the management of chemical companies.

4.1. Operating Permits

Any industrial operation is subject to the granting of a permit by the competent authority. The dossier submitted with an application should include all information required to evaluate the environmental impact of the planned operation. It is then up to the authority to establish conditions that may be attached to the permit. These conditions might in some cases mandate a revision of certain processes or procedures as they were initially envisaged by the applicant. The permit process therefore implies a dialogue between the authority and the applicant, and it also implies the release of information to members of the public who might be affected by the planned operation. Even if the power of decision remains with the authority, the result of such consultation might be quite important for some "sensitive" operations, such as waste disposal sites. The whole permit process can extend over several years, which underscores the extent to which environmental questions can impact on industrial management decisions.

Certain conditions attached to such a permit are of a managerial nature (e.g., emergency plans, reporting lines, etc.). Most of them, however, establish emission limit values (ELVs) for a list of relevant substances or indicators (waste, energy, noise, etc.) that the planned operation is expected to meet at all times.

In theory, these ELVs should result from a risk assessment process, starting with a definition of environmental quality objectives (EQOs) for the area concerned. Defining an EQO is a political decision by the authority: the fact that the area in question is of a residential nature, for example, or that rivers should continue to sustain aquatic life, and so forth. EQOs lead to the determination of environmental quality standards (EQSs) for every environmental compartment, such as the maximum concentration of SO_2 permitted in the atmosphere or heavy-metal limits in water, the maximum level of noise, etc., which in turn make it possible to achieve the EQOs. EQOs are fixed at the local level, or more often internationally, on the basis of scientific assessments. Finally, a model is used to calculate various ELVs of the planned operation to ensure that the EQOs are met, taking into account the existing levels of various parameters in the area concerned. This process, although complicated and possibly leading to discrepancies between the conditions of similar operations in different areas, is the only one that can assure that the environmental risk of an industrial activity is limited to an acceptable level while at the same time being optimally cost-effective.

Nevertheless, another process is becoming increasingly popular for establishing ELVs: the use of *"best available technique"* (BAT). A BAT may be defined as that set of technical and managerial processes and procedures available at a given time which provides for the most cost-effective manufacture of a certain product with minimal impact on the environment. The use of BAT is attractive to the authorities because it simplifies the process of establishing ELVs. It may also attract industrialists, because it ensures a "level playing field" for all companies in a particular business. However, it does raise certain questions, such as how long a BAT remains a BAT before becoming outdated by a new development, the possibility that some developers might enjoy a monopolistic situation, or the best way of ensuring that EQOs are met in heavily industrialized areas.

Whatever the process chosen for determining the ELVs, it is necessary that an integrated approach be adopted so that the impact on all compartments of the environment is minimized: taking care that atmospheric pollution, for example, is not transferred to the aquatic environment. A series of modern legislative texts follows this trend, including the EC Directive on Integrated Pollution Prevention and Control (IPPC).

4.2. Process Safety Legislation

Safety, because of its probabilistic aspects, cannot be regulated simply by standards. Assessing the accidental risk of an activity with respect to both equipment and management practices is therefore the key to safe operation (→ Plant and Process Safety, **B8**).

In principle, regulation is impossible because, at least in the chemical industry, every case is a special one. However, legislation can require an operator to carry out a risk assessment and present the results to the authorities. Here again, a dialogue between the competent authority and the applicant will determine the level of acceptable risk and various conditions the activity should satisfy to meet this acceptable risk.

EC Directive 82/501/EEC, the "Seveso" Directive, now being revised, is typical of legislation that fixes not a standard, but a management practice (i.e., the risk assessment procedure) in order to achieve its objective.

4.3. The EC Eco-Management and Audit Regulation

The EC Council has recently adopted a Regulation (EEC/1836/93) introducing a "Voluntary Community Scheme of Eco-Management and Audit." The initial intent was to facilitate development of environmental auditing practices within industry, but it was soon recognized that environmental audits could not be isolated from environmental management systems, and the title and content of the regulation were modified accordingly.

An enterprise deciding to take part in the scheme is committed to take the following actions:

1) Carry out an initial review of its environmental performance
2) Establish an EMS
3) Conduct periodic environmental audits of its various sites
4) Produce a "public statement" after each audit
5) Have the public statement "verified" by an independent and accredited third party

When the competent national authorities have satisfied themselves that a particular enterprise has complied with the requirements of the regulation they then grant what is known as a "statement," which the company is free to use at its discretion.

The regulation includes annexes specifying the elements of an EMS, the content and methodology of environmental audits, the role of verifiers, and the requirements for accreditation.

To what extent industry will be attracted by this scheme, at least as long as it remains voluntary, is yet to be seen. There are indications that external pressures, such as image, requests from trade consumers, and ratings by the financial community, may prompt companies to join the scheme.

It may also happen that when the regulation comes up for revision (in principle in 1998) the scheme could be made mandatory, at least for some categories of industrial activities. This would thus be a typical example of an intrusion by legislation into the environmental management of private enterprise.

4.4. Mandatory Environmental Communication

Operating permits require an operator to monitor emissions and all other parameters subject to ELVs, and to report the measured results to the competent authority. In many countries, including the EC Member States, legislation of the "Freedom of Information" type make any information (especially environmental information) collected by governmental authorities available to the public, either upon request or directly at the initiative of the authorities themselves. In addition, many statutes related to accidental risk, such as the EC Directive on Major Hazards (the "Seveso" Directive) require active informing of any population likely to be affected by an accident from a chemical site [8]. Through this mechanism a form of mandatory environmental communication already exists in many countries.

Some countries have gone further in requiring industrial activities to measure the emissions of a specific list of substances and then report them to a "register" freely available to the public, sometimes on-line or on diskettes. In the United States, for example, the 1986 Emergency Planning and Community Right-to-Know Act (EPCRA) introduced under Section 313 of Title III the concept of the Toxics Release Inventory (TRI) [9]. Under this legislation, companies employing more than 10 full-time employees must provide the EPA with annual emission data for some 300 toxic chemicals manufactured, processed, or used in excess of certain threshold amounts. Data on emissions released and waste generated by individual sites are made available to the public through printed reports and electronic databases.

Certainly the introduction of the TRI has met the expectations of many stakeholders; it has also been a powerful incentive to industry in the United States to monitor and reduce its emissions and waste generation. It should be recognized, however, that making raw data available out of context opens the door to considerable misinterpretation, for example, identifying the largest companies or sites as "major polluters."

Certain other countries, in Europe and elsewhere, also maintain registers of the TRI type, albeit ones that are less comprehensive and less readily available. The EC has been considering introduction of a "Polluting Emission Register," possibly modeled along the lines of the TRI.

5. Voluntary Action by Industry

Environmental legislation, however comprehensive and detailed it may be in many industrialized countries, cannot cover all the cases encountered in day-to-day industrial practice. Moreover, legislation is rigid, and it cannot be adapted to follow the pace of technological and managerial development. It is also not the most cost-effective approach, because it requires that resources be committed to inspection and enforcement. Finally, the pressure of legislation is not perceived as a real motivation on the part of the workforce.

This has induced industry, led by a few large, advanced companies, to develop voluntary environmental activities and programs. These steps have been further encouraged by industrial associations and supported by certain enlightened governments. Some of the resulting acts can be seen as full substitutes for legislation. More often they complement existing legislation in facilitating its implementation and bridging gaps that legislation cannot cover.

Voluntary corporate action can take several forms. These are described briefly below, before focusing on the worldwide chemical industry initiative known as Responsible Care.

5.1. Internal Statements

Several elements of the environmental management systems discussed in Chapter 3 constitute voluntary actions, provided they are not made mandatory, for example, through an extension of the EC Eco-Management and Audit Regulation (see Section 4.3). These include in particular:

1) The HSE policy (see Section 2.3); this policy, endorsed at the highest level of corporate management, states the principles and objectives that govern management as far as HSE matters are concerned
2) The environmental program (see Section 2.5.2), which defines actions to be taken in a certain area to reach an objective within the framework of a set timetable

Both overall environmental policy and specific environmental programs are internal management tools, even if many companies publicize them (or at least a few of their environmental programs, avoiding the disclosure of competitively sensitive information) in their annual environmental reports. Experience shows that the drafting of a set of environmental programs contributes greatly to the development of environmental awareness within the workforce.

5.2. Voluntary Agreements

At the other extreme of the spectrum are "voluntary agreements" signed between a company and an authority for the purpose of achieving a target, with the choice of means left to the company.

The term "voluntary agreement" may not appear strictly appropriate: in one sense, entering into an agreement is always a voluntary action; on the other hand, the alternative to not signing such an agreement is, in most cases, a law. Other terms are sometimes used as well, in particular "covenant."

The components of a voluntary agreement are:

1) The *aim*; e.g., an emissions reduction target, a labeling scheme, the phaseout of a certain product, etc.
2) The *parties to the agreement*:
   a) On the governmental side, a local authority (e.g., for reduction of discharges to water from a site), a national government (for a global plan involving an industrial sector), or the EC Commission (although monitoring the implementation may be left to national authorities).
   b) For industry, the most straightforward case is when the agreement is signed by a company, but when the agreement covers an entire industrial sector at the national or regional level, the responsible party is usually the trade association concerned. This immediately raises the problem of the extent of the authority an association may be able to wield over its members.
3) The *penalties* in case of failure to achieve the aims of the agreement. Some agreements include financial penalties; this may lead to serious difficulties if the agreement has been signed by a trade association. More often the penalty is simply that a statute is subsequently enacted, which is more burdensome and less cost-effective for both parties.
4) A *monitoring system* and an *administration* to facilitate implementation of the agreement. Experience has shown that this can be done through light, cost-effective organizations.

The following examples are representative of voluntary agreements within the chemical sector:

1) The "33/50 Program" signed between the U. S. Environmental Protection Agency and a number of chemical companies for the voluntary reduction of their emissions of 17 of the most polluting substances by 33 % in 1992 and by 50 % in 1995 relative to 1988 levels
2) EC Commission Recommendation 90/437/EEC of June 1990 formalizing a voluntary program of reduction in the use of chlorofluorocarbons by the foam plastics industry, represented by its European associations
3) EC Commission Recommendation 89/542/EEC introducing a voluntary system of detergent labeling under the control of the European industry federation of detergent manufacturers
4) The various "covenants" signed between the Dutch government and Dutch industry associations for implementing the National Environment Policy Plan (NEPP) of 1989, which established a long-term policy and a program of reduction of emissions, negotiated within the various sectors

5.3. Public Commitments

Half-way between internal statements and binding agreements are the various forms of public commitments by which a company, through its President or CEO, commits itself to adopting an environmental behavior expressed in a set of principles.

There are indeed certain similarities between this form of voluntary action and the voluntary agreements discussed above:

1) The principles define the aims of the commitment. In general they are drafted in the form of a set of 10–15 short guidelines. For effective, practical implementation these must be supplemented by Codes of Practice or Guidance Notes.
2) The principles are drawn up by an organization such as the national or international trade association of the industrial sector concerned, and then endorsed publicly by the senior management of individual member companies.
3) The other party is the public, in general represented by environmental non-governmental organizations (NGOs) that check company environmental performances against the content of the corresponding guiding principles. Companies may suffer a severe penalty when failure to implement the principles is publicized by a credible NGO.

The best known example of such a commitment is the "Business Charter for Sustainable Development" launched by the International Chamber of Commerce (ICC) in April 1991, and now endorsed by over 1100 major corporations around the world [10]. The Charter is a set of 16 principles covering various aspects of the integration of environmental considerations into management practice, such as: environmental policy, continuous improvement of environmental performance, employee education, research, technology transfer, impact assessment, product and services stewardship, sound design and operation, contractors and suppliers, emergency preparedness, dialogue with stakeholders, and reporting.

The ICC Business Charter for Sustainable Development is recognized as a major driving force for continuous improvement in the environmental performance of enterprises through voluntary public commitment. Similar sets of guiding principles have been established by a number of national and international organizations. Some are less comprehensive than the ICC Business Charter, especially with respect to the requirements on research, product stewardship, and reporting; others stress specific aspects, such as emergency preparedness or overseas operations. Two of these Codes are perhaps worthy of special mention.

One is the set of CERES principles (formerly the "Valdez" principles), drawn up in 1989 by the Coalition for Environmentally Responsible Economies (CERES) after the Exxon Valdez oil spill [11]. Signatory companies, based mostly in the United States, are particularly committed to environmental reporting (see Chap. 3) and compensation for environmental damage.

A second is the Global Environment Charter, launched in 1991 by Keidanren, the Japanese Federation of Economic Organizations [12]. The Keidanren Charter emphasizes the role of corporations in technology transfer, including transfer to developing countries, but does not require public reporting.

5.4. Responsible Care

Several chemical companies are signatories to the ICC Business Charter or other sets of principles like those mentioned above. But most also take part in "Responsible Care," the chemical industry's own initiative that originated in 1984 in Canada and is now spreading throughout the world [13].

5.4.1. Responsible Care; What Is It?

Responsible Care relies on two fundamental features:

1) A commitment by company leaders to seek continuous improvement in the environmental performance of a company's activities, including employee involvement and relations with customers and the society's stakeholders. This commitment is expressed by the signature of the company President or CEO on a set of Guiding Principles.
2) A program established and operated by national chemical industry associations to assist their member companies in implementation of the Guiding Principles and verification of the true achievements resulting from their commitments.

5.4.2. The Guiding Principles

There is not a single set of Responsible Care Guiding Principles: these are instead established on a national basis by the corresponding chemical industry associations with consideration directed to local circumstances. However, the Principles everywhere are remarkably similar in content and format; indeed, they are all modeled along the lines of other well-known principles, such as the ICC Business Charter for Sustainable Development or the CEFIC Guidelines for Protection of the Environment.

5.4.3. National Programs

Similarly, the details involved in facilitating and monitoring specific programs are left to national chemical industry associations, who must adapt them to local considerations such as the national legislative framework, the state of industrial development, the companies' cultures, etc. However, to ensure consistency in the initiative in the various countries, all national programs are expected to include seven fundamental features:

1) A formal *commitment* on behalf of each company to a set of guiding principles (see above. In the majority of cases the commitment is signed by the company CEO, but in some countries it is fulfilled by endorsement by the Board of the national association).
2) A series of *codes, guidance notes*, and *checklists* to assist companies in implementing the commitment. In some countries (e.g., the United States) a set of detailed codes has been drafted to cover the following areas: control of emissions, process safety, occupational health and safety, transport and distribution, community awareness and emergency response, and product stewardship. Elsewhere, especially in Europe, relevant associations had already developed a number of guidance documents on environmental management long before the introduction of Responsible Care. Producing and updating such codes and guidance documents is a continuing process.
3) The progressive development of *indicators* against which improvements in performance can be measured. In general, these indicators include sets of emission data for a number of selected substances and data on waste generation and recycling, as well as data on energy consumption and accidents.
4) An ongoing process of *communication* on HSE matters with interested parties outside the industry. The issue of environmental communication, discussed in Chapter 3, is therefore an integral part of Responsible Care. Obviously the recommendations provided by the association to its member companies regarding the communications process influence the selection of indicators of performance.
5) Provision of *fora* in which companies can share views and exchange experiences on implementation of their commitments. Some associations provide extensive service to their member companies in the organization of seminars, workshops, local cells, etc. This is particularly valuable for increasing the motivation of people who in general do not respond easily to environmental concerns, such as sales managers or line middle-managers.
6) Adoption of a *title* and *logo* that clearly identify a national program as being consistent with and part of the concept of Responsible Care. The standard logo (Fig. 4) has been registered in many countries, and is used throughout the world by the chemical industry. Many associations, even in non-English speaking countries, have retained the registered term "Responsible Care;" others have adopted a different expression that communi-

Figure 4. The Responsible Care logo

cates in translation the spirit of the term Responsible Care.

7) Consideration of the best ways to encourage all member companies to *commit to* and *participate in* Responsible Care. Many associations have made or will make company commitments to Responsible Care a condition of membership. This implies that if a company does not fulfill its commitment, the association has the right to expell it. Before taking this extreme action, however, strong peer pressure will be exerted by other members on the noncomplying company.

5.4.4. The Scope of Responsible Care; Product Stewardship

Responsible Care is sometimes understood as covering nothing more than the manufacturing activities of a company. Fundamentally, however, the Guiding Principles of Responsible Care apply to all aspects of industrial operation, from basic research to product development, manufacture, distribution, and use. They are thus more nearly consistent with a "cradle-to-grave" approach.

In this respect, product stewardship is a vital component of Responsible Care. Product stewardship may be defined as "the responsible and ethical management of the health, safety, and environmental aspects of a product throughout its life cycle, from its design to manufacture and distribution, including resource consumption, transport, and final disposal options."

Responsible Care therefore aims at developing an "environmental culture" among all employees of a company, whatever their position and function. Researchers are motivated to assess the HSE effects of a new product at an early stage of development, and to find substitutes if the risk appears unacceptable. Engineers develop processes with a concern for minimal consumption of resources and the lowest possible impact on the environment. Production people operate their plants safely, and with full respect for environmental standards. People in charge of transport and distribution seek the safest means of transport. Sales managers advise their customers on the safe use of products, including problems raised after normal use to avoid creating waste problems.

5.4.5. The Challenges of Responsible Care

The successful development of the Responsible Care initiative by chemical industry in many countries throughout the world entails two major challenges:

1) Finding ways to maintain the integrity of the initiative as national associations adapt their programs to local conditions and cultures—which may vary significantly among countries like the United States, Japan, and the nations of South America, Europe, or Africa. In addition, Responsible Care is a chemical industry initiative, but other industrial sectors more or less closely related to the chemical industry would benefit from similar programs adapted to their specific needs.

   The International Council of Chemical Associations (ICCA) has established a procedure for checking the consistency of programs developed by national associations against the seven fundamental features, both at the time of introduction of the program and periodically thereafter. The ICCA has also established rules for "partnership" with other industrial sectors that deal with chemicals (transport and distribution companies, waste disposal contractors, etc.) and are also interested in Responsible Care.

2) Finding ways to make Responsible Care acceptable and credible to the public, regulatory authorities, and environmental organizations. Responsible Care is not simply a public-relations exercise: it is based strictly on improved performance—not on words, but on actions. Credibility therefore will be gained by demonstrable achievements, which will need to be communicated effectively. Many national programs include "public advisory panels" consisting of representatives of the various stakeholders; others have developed alternative channels of communication, but all companies are encouraged to adopt an open attitude toward information.

Honest dialogue between industry and the public and between plants and local communities is a crucial element in bringing about the industry-wide cultural change required by Responsible Care, and in building public confidence and trust, a key to enabling the chemical industry to retain its "license to operate."

6. Standardization and Certification of Environmental Management Systems

As discussed in Chapter 2, companies first establish policies for health, safety, and environmental matters, and then set up environmental management systems to ensure that objectives defined in the policies are effectively achieved.

An EMS cannot really be defined by law, even if the EC Eco-Management and Audit Regulation (see Section 4.3) does describe the basic elements of an EMS and encourages companies to establish such EMSs. Conversely, leaving the companies without adequate guidance on how to establish an EMS may result in serious discrepancies in the structure and organization of EMSs as adopted by individual firms, with the accompanying risk of some stakeholders coming to question the credibility of the EMSs of certain companies.

This dilemma has led to the concept of standardization of environmental management systems. In addition to helping companies achieve their objectives, appropriate standards can, through certification, offer assurance to others that the stated objectives will be achieved.

6.1. Standards for Environmental Management Systems

6.1.1. Generic or Specific Standards

Although there is a long history associated with the drafting of standards for product specifications, testing methods, etc., the concept of standards for management systems is quite new: it indeed first appeared in the mid 1980s with the issuance of the series of standards for quality management systems ISO 9000 (and its national or regional equivalents, EN 29000, BS 5750).

Whereas the ISO 9000 standards were restricted in their initial scope to that which was necessary to achieve product quality, it was soon recognized that the same requirements could also have a generic interpretation, and that, for example, the ISO 9001 and ISO 9004 standards could be adapted to cover environmental management systems.

Nevertheless, a few standardization bodies undertook the drafting of standards specific to EMS, such as BS 7750 (see below), issued in 1992. A few similar national standards have since been issued elsewhere, and the International Organization for Standardization (ISO) is taking the same route.

A certain number of standardization bodies share with some industrialists the view that there should be further development of the concept of a generic management-system standard. The standard would include appendices covering specific requirements for product quality, environmental management, etc. The vision is that this structure would eventually replace the emerging sets of separate management-systems standards, but it probably will not come to pass for several years.

6.1.2. National EMS Standards

In March 1992 the British Standards Institute (BSI) issued the first standard specifically designed for environmental management systems, under the reference BS 7750. In April 1993 the Association Française de Normalisation (AFNOR) published an experimental standard (X30-200) which, although somewhat different in its wording, follows the same structure as the British Standard. Other national initiatives can be expected if a widely recognized standard is not drafted soon at the international level.

The BS 7750 standard defines a series of relevant terms, and then specifies the following requirements related to EMSs:

The need to establish an EMS
Drafting an environmental policy
Organization and personnel
Environmental effects evaluation and register
Environmental objectives and targets
Environmental management program
Environmental management manual and documentation
Operational control
Environmental management records
Environmental management audits
Environmental management reviews

From an analysis of the text of the standard and its annexes it is clearly apparent that, first, the standard covers only the "environment" and not the health and safety aspects that are normally encompassed in any EMS; furthermore, it is very similar to the annex dealing with the requirements concerning EMS in the EC Eco-Management and Audit Regulation 1836/36.

In the course of 1992–1993 a certain number of companies conducted a "pilot test" of the standard. The standard was submitted for review at the end of this exercise, and may be altered if necessary.

6.1.3. ISO Standards

Under the impulse of S. SCHMIDHEINY from the Business Council for Sustainable Development, the International Organization for Standardization (ISO) established in 1991 a Strategic Advisory Group on the Environment (SAGE). The ISO-SAGE in turn created six subgroups, one of them in charge of making recommendations on a standard for EMS. In June 1993 the work of ISO-SAGE was taken over by a new Technical Committee (TC): ISO/TC 207, "Environmental Management."

A subcommittee of TC 207 is in charge of producing a standard for EMS, together with a specification that will include general guidance regarding its use. The standard is expected by 1995. There are indications that here, also, the standard will cover only the "E" part of an HSE management system.

Another subcommittee of TC 207 is dealing with the issue of environmental audits, in particular as an element of EMS.

6.1.4. Comité Européen de Normalisation

CEN, the European Committee for Standardization, has recently set up a Technical Board "Environment" and a TC "Environmental Management Tools" for the specific purpose of implementing the mandate they received from the EC Commission to draft a standard compatible with the EC Eco-Management and Audit Regulation. CEN, however, seems to have left to the ISO the responsibility of drafting the actual standard, which they will then convert into an EN standard under the procedures outlined in the "Vienna agreement."

6.1.5. The ISO 9000 Standards as a Standard for EMS

There is a growing perception that a separate standard for EMS might not be necessary, and that the standard for quality management systems (the ISO 9000 series) can be extended to cover the health, safety, and environmental aspects of the EMS. This has indeed been partially recognized by the BSI, which included an annex in BS 7750 establishing links to BS 5750, and by ISO, which intends to include in their standard for EMS a set of general guidelines based on ISO 9004.

The Chemical Industry Association (CIA) in the United Kingdom went even further in 1992 by issuing guidelines for applying ISO 9001 to the EMSs of chemical companies [14]. These guidelines translate the requirements of ISO 9001 in the context of HSE management systems, including terms specific to the ISO 9000 series, such as design control, purchasing, servicing, etc.

Here, also, a pilot exercise carried out by a number of private companies has demonstrated the feasibility of the approach. Indeed, the approach is equivalent to that of BS 7750—but with the advantage of covering all aspects of H, S, and E management.

The South African Standards Board issued in 1993 a similar guidance document, but it was restricted to the environment.

This route clearly demonstrates the validity of the concept of a generic management-system standard, complete with specific guidance to address such individual elements as the EMS.

6.2. Certification of Environmental Management Standards

Standards for management systems, in addition to their "educational" aspects, also allow for "certification" (or "registration") of the activity covered by the standard by an independent, accredited third party, thus giving confidence to the company and outside stakeholders that the objectives of the management system will be achieved.

6.2.1. Certification of an EMS

Certification is a process in which a certification body (often a commercial firm) assesses a particular system against the relevant standard and then issues a certificate of compliance. This is followed up with a series of regular surveillance visits (usually once or twice a year). Certification bodies may themselves be accredited for this function by a governmental or similar agency, lending additional credibility to their work and resulting in an "accredited certification."

In the case of certification of an EMS, specific accreditation procedures have yet to be developed in Europe and elsewhere. Several organizations are working on guidelines and criteria for the accreditation of EMS certification bodies. These will be derived from the criteria used for accreditation of ISO 9000 certification bodies because of the similarities between the standards for quality management systems and those for EMS.

Incidentally, another reason for interest in the use of an ISO 9000 standard extended to EMS is that a single accredited body could then certify both the quality management system and the EMS of a company, with significant savings for the industry as a whole. If the standard could be recognized internationally, as in the case of an ISO standard, this would facilitate the mutual recognition of certification and accreditation systems, a problem that is in urgent need of solution.

It can be expected that the existence of a certificate awarded by an international or national accredited body will become increasingly important to regulators and stakeholders as a form of assurance regarding the professionalism of the company and its EMS. This will assist many companies in meeting the key objective of being able to demonstrate and promote their management of environmental issues in a tangible and credible way. Lack of certification will in time increasingly expose a company to multiple audits by customers seeking to satisfy themselves with regard to the company's environmental probity.

6.2.2. Certification Against the EC Regulation

One article in the EC Eco-Management and Audit Regulation stipulates that "Companies implementing national, European or international standards for environmental management systems and certified, according to appropriate certification procedures, as complying with those standards, shall be considered as meeting the corresponding requirements of this Regulation ..." (certification, however, will not exempt the "environmental statement" from being validated by a certified verifier).

The introduction of this article was of crucial importance as a way of linking the Regulation and voluntary participation on the part of companies to certification against the standards for EMS; because the public statement is an integral part of the scheme, it will quite normally be "verified" by the certification body as a standard part of the certification process.

7. International Aspects of Environmental Management

Most chemical companies, at least above a certain size, operate in several countries. In some cases all the countries may be at a similar state of development, but both industrialized and developing countries may also be involved. Especially in the latter case, special consideration must be given to environmental management to avoid the problems that too often have tarnished the images of "multinational" or "transnational" companies.

7.1. Environmental Requirements in Different Countries

Many reasons, mostly of a commercial nature, may prompt a company to establish an operation in another country, one of which might be a less stringent environmental legislative atmosphere in the host country. There have been a few cases in which "relocation" of a company's activity from its home country to another was perfectly justified because of the sudden introduction of exceptionally stringent environmental legislation not considered sensible or appropriate by most countries, even the most developed ones. Unfortunately, however, there have been more cases in which companies have established operations in countries with little environmental legislation simply to benefit from standards less demanding and costly than those reasonable standards in effect in their home country. Such "double-standard" practices, including in the environmental domain, have contributed significantly to the bad reputation associated with "multinationals" by some environmental organizations and the governments of some developing countries.

These practices are no longer considered to be ethical by reputable corporations that, in their foreign operations, comply not only with local legislation but also with their own company standards, which are generally equivalent to or exceed the requirements of their home country.

This is a basic principle expressed by many companies in their environmental policies.

It must be recognized, however, that companies of foreign parentage following this course of action put themselves in a position of competitive disadvantage vis-à-vis local companies satisfied with complying, at lower cost, only with national environmental requirements. Foreign companies can sometimes overcome this difficulty by "educating" both national companies (e.g., through the intermediacy of national chemical industry associations) and local governments in modern aspects of environmental protection. The role of transnational corporations in raising the environmental awareness and behavior of developing countries is now clearly recognized [1, Chap. 3. III].

7.2. Technological Cooperation

Technological cooperation between countries at different levels of development has been identified as a key component of sustainable development.

Technological cooperation between companies in industrialized and developing countries can take several forms:

Technical Assistance. A company or an organization sends experts to a company requesting expertise and assistance, perhaps for an audit of environmental performance, a trouble-shooting mission, an education or training session, etc. The mission might be accomplished on a commercial basis, but more often it is carried out under the aegis of a voluntary program operated by an industrial association, a government, or an international organization.

Transfer of Technology. A company makes available, within the terms of a commercial contract, a process or a technique to which it holds title, generally as the result of a research and development program on its own premises. Such a transfer of technology should obviously include technological cooperation before, during, and after actual start-up of the operation.

Foreign Investment. A company may elect to set up its own operation in a foreign country, either in the form of a fully owned subsidiary or as a partnership with some other company at an appropriate level of equity. The industrial operation might be built from a green site, or the foreign investor might acquire an existing facility. In general, any such foreign investment implies a transfer of technology from the home company to the foreign one.

Several industrial associations have produced guidelines for an environmentally sound transfer of technology [15].

8. References

[1] The World Commission on Environment and Development: *Our Common Future*, Oxford University Press, Oxford 1987.

[2] S. P. Johnson: *The Earth Summit: The United Nations Conference on Environment and Development (UNCED)*, Graham and Trotman Ltd., London 1993.

[3] S. Schmidheiny (Business Council for Sustainable Development): *Changing Course*, MIT Press, Cambridge 1992.

[4] International Chamber of Commerce: *Environmental Auditing*, Publication 486, ICC Publishing, Paris 1989.

[5] S. P. Johnson: *The Earth Summit*, Graham and Trotman, London 1993, Agenda 21.

[6] Confederation of Finnish Industries: *Guidelines for Environmental Arguments in Marketing*, Helsinki 1990.

[7] Deloitte Touche Tohmatsu International, International Institute for Sustainable Development, Sustainability: *Coming Clean, Corporate Environment Reporting*, Deloitte Touche Tohmatsu, London 1993.

[8] M. S. Baram, D. G. Partan: *Corporate Disclosure of Environmental Risks, US and European Law*, Butterworth Legal Publ., Salem 1992.

[9] U.S. Environmental Protection Agency: *Emergency Planning and Community Right-to-Know Fact Sheet*, EPA, Washington D.C., 1988.

[10] International Chamber of Commerce: *Business Charter for Sustainable Development*, Publication 210/356A, ICC Publishing, Paris 1991.
J. O. Willums, U. Goluke: *From Ideas to Action*, Publication 504, ICC Publishing, Paris 1992.

[11] The Coalition for Responsible Economies: *The CERES Principles*, CERES, Boston 1989.

[12] Keidanren: *The Global Environmental Charter*, Keidanren, Tokyo 1991.

[13] European Chemical Industry Council: *Responsible Care*, CEFIC, Brussels 1993.

[14] Chemical Industries Association: *Responsible Care Management Systems, Guidelines for Certification to ISO 9001*, CIA, London 1992.

[15] *CEFIC Guidelines on Transfer of Technology (Safety, Health and Environment Aspects)*, CEFIC, Brussels 1991.

Noise and Vibration

Wolfgang Daiminger, Klaus R. Fritz, Edwin Schorer, Burkhard Stüber, Müller-BBM Schalltechnisches Beratungsbüro, Planegg bei München, Federal Republic of Germany

The entire topic was coordinated by Burkhard Stüber

1. Introduction

Noise abatement is an essential consideration with respect to chemical and petrochemical plants for the protection of people living in the vicinity of such a plant as well as employees working at the facility. Reduction of workplace noise within a plant also increases the intelligibility of public announcements or warning signals, at the same time that it facilitates communication among personnel on the plant floor and in the control rooms. Moreover, extraordinary noise events are more conspicuous in a quiet environment, so measures that limit background noise contribute to improved safety conditions. Effort to counter excess vibration of plant equipment in general has the further advantage that it can prevent unnecessary damage and consequent disruption of productive activities.

2. Definitions

2.1. Sound and Vibration Levels

Audible sound refers to mechanical vibration with components falling in the approximate frequency range 16 Hz–16 kHz. Sound in gases and liquids is always related to vibrational motion of the particles constituting the fluid. Important *field quantities* with respect to sound are the time variation in the associated alternating pressure $p(t)$ (which is superimposed over some static

3-527-20137-8/95/$5.00 + .50

pressure) and the time variation of the velocity $v(t)$ with which the affected particles move back and forth.

The root mean square (rms) value p of the alternating pressure $p(t)$ is commonly expressed in terms of a logarithmic function, the *sound pressure level* L, which is measured in decibels (dB). L is defined by the relationship

$$L = 20 \log (p/p_0) \text{ dB} \tag{1}$$

The reference value p_0 (for airborne sound) has been internationally fixed at $p_0 = 2 \times 10^{-5}$ N/m², a value that corresponds to the human auditory threshold at medium frequencies. A compelling reason for selecting a logarithmic scale is the fact that the alternating pressures of interest from a technical standpoint cover a range of eight orders of magnitude.

Other field quantities used to describe sound energy or sound-energy transfer can be derived through multiplication. Thus, the *intensity* I

$$I(t) = \overline{p(t) \cdot v(t)} \tag{2}$$

specifies the power (i.e., energy per unit time) that will penetrate a standard surface perpendicular to the velocity vector. The overbar in Equation 2 indicates a time-average value. The *sound power* $W(t)$ can be derived by forming the surface integral:

$$W(t) = \oint_S \overline{p(t) \cdot v(t)} \cdot \cos\varphi \, \mathrm{d}S \tag{3}$$

where φ is the angle between the velocity vector $v(t)$ and a line normal to the surface element dS.

The rms value W of the sound power $W(t)$ is also expressed in terms of a logarithmic measure (again with the unit decibels) known as the *sound-power level* L_W:

$$L_W = 10 \log (W/W_0) \text{ dB} \tag{4}$$

For airborne sound sources, the value $W_0 = 10^{-12}$ W is normally used as a reference point.

Mechanical vibrations in solids can be characterized by three motion quantities and their variations as a function of time. These are the time variations of the displacement $s(t)$, velocity $v(t)$, and acceleration $a(t)$. The three quantities can be interconverted by integration or differentiation:

$$a(t) = \frac{\mathrm{d}v(t)}{\mathrm{d}t} \quad \text{and} \quad v(t) = \frac{\mathrm{d}s(t)}{\mathrm{d}t} \tag{5}$$

All three measurable variables cover large ranges, so—just as in the case of sound pressure—rms values are often reported rather than the quantities themselves, in logarithmic form (once again in decibels) as the corresponding levels L. These levels are defined by the relationships

$$\begin{aligned} L_s &= 20 \log (s/s_0) \text{ dB} \\ L_v &= 20 \log (v/v_0) \text{ dB} \\ L_a &= 20 \log (a/a_0) \text{ dB} \end{aligned} \tag{6}$$

The following have been established as reference values for computing velocity and acceleration levels (according to international standards; see ISO 1683 [1]):

$$\begin{aligned} v_0 &= 10^{-9} \text{ m/s} \\ a_0 &= 10^{-6} \text{ m/s}^2 \end{aligned} \tag{7}$$

However, in the field of acoustics it is usually recommended that a somewhat different set of reference values be utilized instead:

$$\begin{aligned} s_0 &= 8 \times 10^{-12} \text{ m} \\ v_0 &= 5 \times 10^{-8} \text{ m/s} \\ a_0 &= \pi \times 10^{-4} \text{ m/s}^2 \end{aligned} \tag{8}$$

Using these reference values, the three structure-borne sound levels L_s, L_v, and L_a all share a common numerical value at a frequency of 1000 Hz, and the average sound-pressure level directly in front of a vibrating structure corresponds to the average structure-borne sound level.

2.2. Level Calculations

Studies in the field of sound and vibration usually involve individual sound signals that are not coherent relative to one another (i.e., they do not depend on each other). The power quantities in such a case are found to be additive according to the following addition rule:

$$L_{\text{total}} = 10 \log \left(\sum_{i=1}^{n} 10^{0.1 \cdot L_i} \right) \text{dB} \tag{9}$$

where L_{total} = total level
L_i = ith individual level
n = number of individual levels to be added

This relationship can of course also be reversed. If, for example, level L_1 is to be subtracted from level L_2, the following calculation leads to the level L_{Diff} representing the differential value:

$$L_{Diff} = 10 \log (10^{0.1 \cdot L_2} - 10^{0.1 \cdot L_1}) \text{ dB} \quad (10)$$

Just as in addition, any averaging of incoherent signals must also be carried out using power quantities. The following expression can be used for obtaining an energetic mean value:

$$L_m = 10 \log \left(\frac{1}{n} \sum_{i=1}^{n} 10^{0.1 \cdot L_i} \right) \text{ dB} \quad (11)$$

where L_m = energetic mean level (level of the average value)
L_i = ith individual level,
n = number of individual levels to be averaged

The above arithmetic rules apply to all the various level quantities, including the sound-power level, acceleration level, etc.

The average level of a time-dependent (variable) sound-pressure level $L(t)$, expressed over an average time T, is called the (energy) *equivalent continuous level* L_{eq}, defined according to the following expression:

$$L_{eq} = 10 \log \left(\frac{1}{T} \int 10^{0.1 \cdot L(t)} \mathrm{d}t \right) \text{ dB} \quad (12)$$

The significance of L_{eq} is that it permits a time-dependent noise function to be replaced by an equivalent constant noise with the same energy.

If a time-dependent variable noise consists of n time periods with individual durations T_i and corresponding sound-pressure levels L_i, where the latter are constant over the course of the stated durations, then the following expression is valid:

$$L_{eq} = 10 \log \left(\frac{1}{T} \sum_{i=1}^{n} T_i 10^{0.1 \cdot L_i} \right) \text{ dB} \quad (13)$$

where $T = \sum_{i=1}^{n} T_i$

The equivalent continuous level L_{eq} is the function most often used to assess the effect of time-dependent sound on human beings.

2.3. Spectra

In addition to considering its variation as a function of time (i.e., sound pressure vs. time), a particular vibration event can also be described in terms of its spectrum, in which case the vibrational quantity is expressed as a function of frequency f. Measuring instruments with filters—so-called fast Fourier transform (FFT) analyzers—are generally used for this purpose, since they make it possible to establish levels within discrete frequency ranges. Most important for practical applications is filtering in octave bands, where the transmission band selected has a range of one octave. The relationship between the lower and upper frequency limits for such a transmission band is 1:2. The octave-band central frequencies are uniformly determined; sound-pressure levels L_{oct} in octave bands are generally provided for the following central frequencies f_n:

$$f_n = 31.25 \times 2^{n-1} \text{ with } n = 1-9 \quad (14)$$

In addition, it is also common to use third-octave bands or narrow-band spectra with constant bandwidths.

3. The Impact of Noise and Vibration

3.1. Sound Perception and Noise Effects on Humans

Audible sound covers a frequency range of 3 decades and a sound-pressure range of ca. 7 decades; in a double-logarithmic presentation of sound pressure versus frequency (Fig. 1), the *audible range* is represented by the area between the threshold in a quiet situation ("threshold in quiet," the lowest audible sound level for a pure tone) and the threshold of pain.

Vibrations at very low frequencies ($f < 16$ Hz, infrasound) and very high frequencies ($f > 20$ kHz, ultrasound) cannot be perceived by the human ear. In Figure 1 the region encompassed by speech and music is hatched, and a limit of hearing-damage risk has been indicated for a noise exposure of eight hours per day.

At a constant sound level, sound with a low or very high frequency will not elicit the same loudness sensation as that in the medium-frequency range. This is illustrated by curves of equal loudness (Fig. 2). Figure 2 shows that a 100-Hz tone at $L = 50$ dB, for example, is perceived as being of the same loudness as a 1000-Hz tone at $L = 40$ dB.

In order to simulate this hearing characteristic in the measurement of sound levels, a *weighting filter network* (*A-filter*) with defined attenuation at low and high frequencies is used to approximate a loudness-equivalent measuring

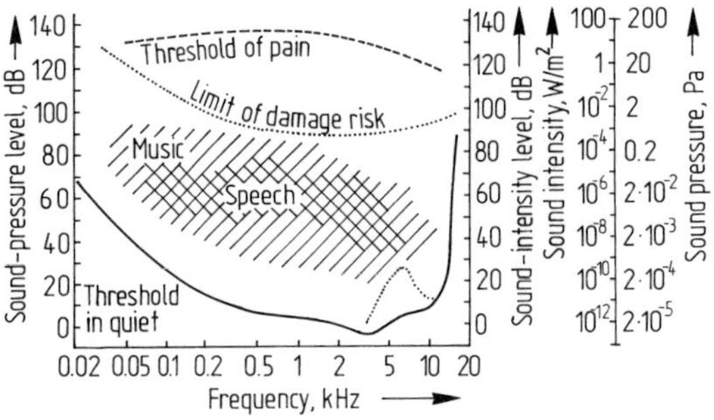

Figure 1. Hearing area; i.e., the area between the threshold in quiet and the threshold of pain; the area encompassed by music and speech and the limit of damage risk are also indicated.
The ordinate scale is expressed not only in terms of sound-pressure level but also with respect to sound intensity and sound pressure. The dotted part of the threshold in quiet relates to subjects who frequently listen to very loud music (based on ZWICKER and FASTL [2]).

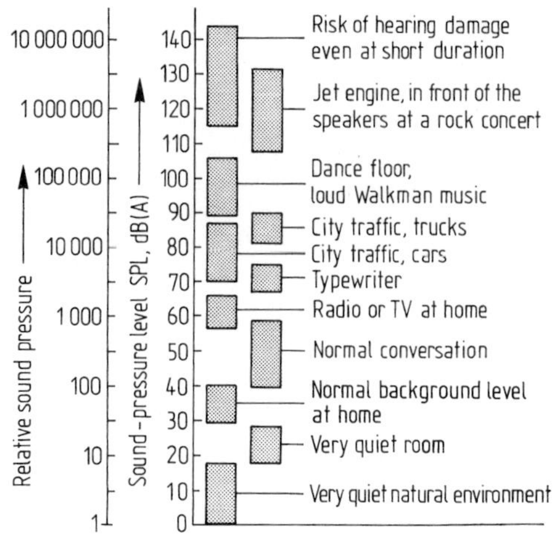

Figure 3. Typical sound-pressure levels for everyday noise sources

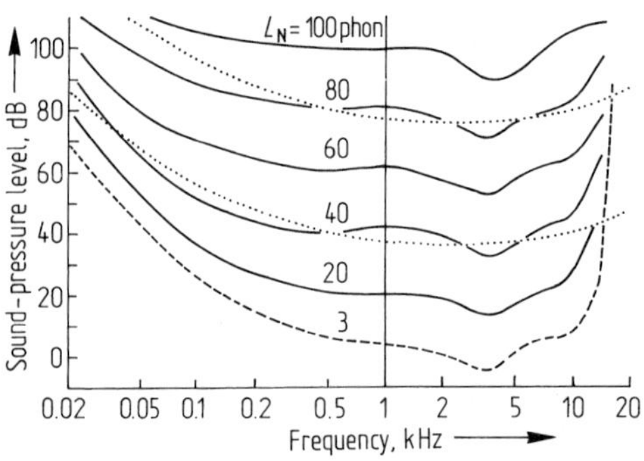

Figure 2. Equal-loudness contours for pure tones in a free sound field.
The parameter is expressed in terms of loudness level, L_N. The 3-phon curve also represents the threshold in quiet. A-Weighting curves have been provided for comparison (dotted) at two levels (based on ZWICKER and FASTL [2]).

result. This A-weighting curve stems from the early days of sound measurement; examples of weighting curves are provided in Figure 2.

The A-weighting curve is constant at all levels, while the loudness sensation (i.e., the shape of the equal-loudness contours) varies with level. From Figure 2 it is apparent that an A-weighted sound-pressure level approximates the loudness sensation better in the medium-level range than at high sound levels. In spite of this, A-weighted sound-pressure levels, indicated by dB(A), are generally used for the measurement and assessment of sound events. The A-weighting curve is internationally standardized (IEC 651 [3]). Additional weighting curves are defined for special applications (e.g., B-, C-, D-weighting). Currently accessible, more accurate measurement methods for ascertaining hearing-adjusted loudness have not yet gained general acceptance (e.g., ISO 532 [4]). Figure 3 illustrates a number of typical everyday sound-pressure levels.

Noise can impair or destroy human hearing, and it can also cause physical or mental stress (increased pulse rate and/or blood pressure, nervousness, sleep disorders). Furthermore, noise can increase the risk of an accident if warning signals go undetected because of excessive background noise. These effects are particularly important to consider in the assessment of workplace noise.

There is no evidence to indicate that noise with a sound-pressure level below 85 dB(A) can directly impair the ear. In terms of continuously present noise and its average A-weighted sound-pressure level $L_{A,m}$, biological effects are now commonly accepted as follows:

disturbance of a night's rest	$L_{A,m} > 40$ dB(A)
disturbance of concentration	$L_{A,m} > 45$ dB(A)
disturbance of communication	$L_{A,m} > 50$ dB(A)
increased risk of cardiovascular disease	$L_{A,m} > 65$ dB(A)

These noise effects are relevant factors in any assessment of noise in residential neighborhoods located near industrial zones or roads.

The hearing-impairment effect of noise depends heavily on the sound-pressure level of the noise itself as well as on the corresponding exposure time. Irreparable damage can be caused even by single transient sound events (shots, explosions) with peak values exceeding ca. 140 dB(A). Long-duration noise exceeding 85 dB(A) (e.g., at the workplace) leads to short-term impairment of hearing capacity (*Temporary Threshold Shift,* TTS), from which the ear will recover during periods of silence. Based on empirical data, ISO 1999 [5] provides a mathematical model for calculating damage-probability distributions for populations exposed to noise. The calculation model uses the following input quantities:

1) The A-weighted equivalent continuous noise level over an 8-h period (noise-rating level) expressed in dB(A)
2) Exposure time in years
3) Age in years
4) Sex

A rule of thumb for calculating a maximum permissible continuous noise level at the workplace, one that will not lead to permanent hearing impairment even after years of exposure, is expressed by the following formula:

$$L_{A,m} \leq 85 + 10 \log (24/T_n) \text{ dB(A)} \qquad (15)$$

where T_n = daily noise-exposure time in hours

According to this model, no irreparable hearing damage is to be expected at levels up to 90 dB(A) with a working time of eight hours per day.

If the ear is exposed to noise levels higher than those recommended, sensory cells (hair cells) located in the inner ear will be destroyed, leading to permanent hearing impairment (*Permanent Threshold Shift,* PTS).

Wearing ear protectors at continuous noise levels > 85 dB(A) can prevent or reduce temporary shifts of the hearing threshold, and the danger of permanent damage is also considerably reduced.

3.2. Effects of Vibration on Humans and Equipment

Vibration becomes distinctly noticeable above a certain magnitude. The threshold for vibration perception depends on the frequency. In the range 5–100 Hz, the perception of vibration is nearly proportional to velocity. Vibrations in this frequency range will be felt if their velocity exceeds 0.1 mm/s. Vibrations with velocities of ca. 0.4–1.6 mm/s are clearly recognizable, and those up to ca. 6.3 mm/s are strongly felt. Vibrations with frequencies < 5 Hz must be more intense before they will be perceived.

Increasing vibration impact results in an observable impairment of well-being and fitness; higher values endanger the health and may even lead—depending upon the exposure time—to permanent injury. The effect produced by vibration depends on the point at which the vibrations penetrate the human body. Sitting or standing individuals may suffer alteration of the skeleton (spine) and internal organs. Vibrations passing through the hands and arms can cause peripheral circulatory problems, nervous disease, and alterations of the bones, joints, and muscles of the hand–arm system (white-finger syndrome).

The vibration of plant equipment may also lead to physical damage or work disruption if it exceeds a certain magnitude (e.g., vibrational cracks at ducts, mechanical breakdown, structural damage, etc.).

4. Legal Aspects

In all highly industrialized countries, noise is recognized as a dominating incident factor in the human environment. Impairments caused by the effects of other vibration are of minor importance, but they do occur. Measures designed to reduce noise and vibration in industry and trade are likely to prevail only if appropriate laws are enacted, because achieving the desired end entails extra costs, and it may sometimes complicate plant operation and maintenance. Regulations aimed at protecting individuals from noise and vibration are directed toward immission levels in the vicinity of industrial plants as well as immission at the workplace. It is only natural that such rules have developed differently in various countries, excluding the possibility of our providing a detailed examination in the context of this article of the many specific regulations.

4.1. Occupational Noise

Regulations directed toward noise within the plant (e.g., noise at the workplace) are essential to protect workers from hearing impairment. As indicated in Section 3.1, most countries have established 85 dB(A) as the limit for permissible noise at the workplace, and this value should not be exceeded if it can be avoided by technical means. Ear protection should be worn at any workplace with a noise-rating level ≥ 85 dB(A). Moreover, workers exposed to such high noise levels should have their hearing regularly checked by a specialist in industrial medicine.

4.2. Occupational Vibration

Rules related to the human effects of vibration at the workplace are necessary in order to protect the staff from injury and restrictions with respect to their performance and fitness, particularly at workplaces associated with plant safety (e.g., control rooms, etc.). ISO 2631 [6] defines limits for vibration effects based on the criteria "reduced comfort boundary," "fatigue-decreased proficiency boundary," and "exposure limits" for whole-body exposure as a function of frequency and daily exposure time (see Fig. 4). ISO 5349 [7] deals with criteria for assessing vibrational effects on the hand–arm system.

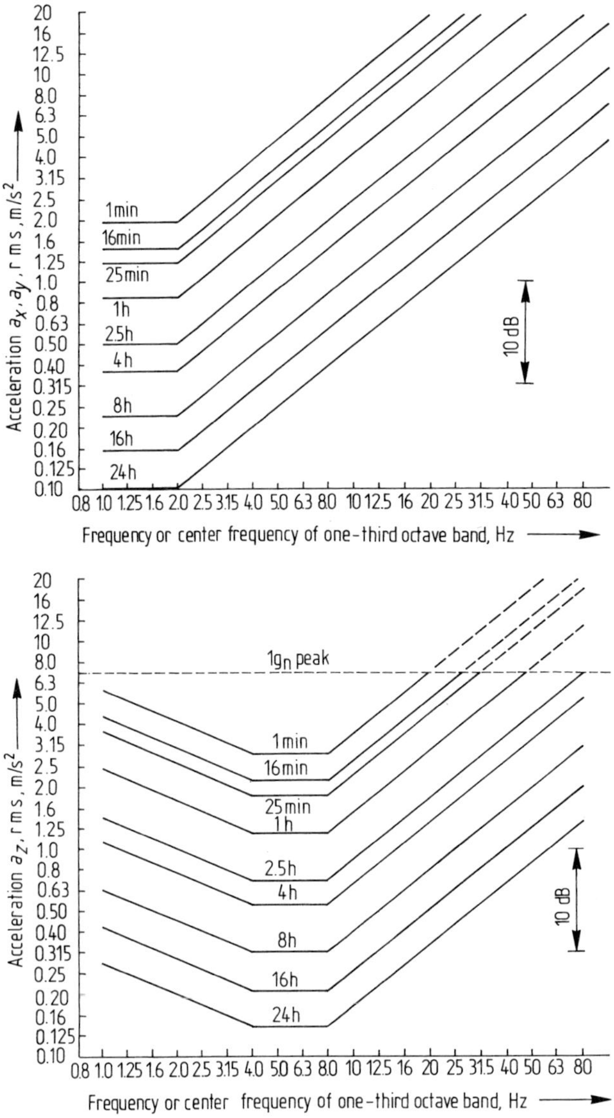

Figure 4. Transverse (A) and longitudinal (B) acceleration limits as a function of frequency and exposure time. "Fatigue-decreased proficiency boundaries" (based on ISO 2631-1 [6]); to obtain "exposure limits," multiply the acceleration values by 2 (6 dB higher); for "reduced-comfort boundaries" divide acceleration values by 3.15 (10 dB lower)

4.3. Environmental Noise and Vibration

Regulations with respect to noise in the general vicinity of industrial plants are concerned not with potential hearing damage but rather with protection from disturbance and noise pollution. Although the annoyance caused by industrial noise and vibration from industrial plants is relatively minor compared to road and rail-traffic noise, industrial noise sources tend to be over-represented from the standpoint of complaints.

ISO 1996 [8], consisting of three parts, provides a set of international regulations for noise protection in residential neighborhoods located near industrial areas. The basic quantity used for describing sound immission is the A-weighted energy-equivalent continuous-level L_{Aeq}. The noise-rating level L_{Ar} that has been proposed assumes the probable need for additional charges K_1 related to tonality and K_2 for impulsive noise components as follows:

$$L_{\mathrm{Ar}} = L_{\mathrm{Aeq}} + K_1 + K_2 \,\mathrm{dB(A)} \quad (16)$$

ISO 1996 lists additional measurement quantities that permit a detailed description of sound immission even for fluctuating or single-sound events. National or local authorities must establish appropriate noise-limiting values that should not be exceeded by the noise-rating levels in affected neighborhoods. The magnitude of the limiting value, the extent of additional charges for tonality and impulsive noise, and the evaluation procedure itself vary not only from country to country but sometimes even within different states or regions of the same country. In general, nighttime noise limits are 10–15 dB lower than those for the daytime; some countries also define intermediate values for the early morning and evening hours.

Noise-limiting values also depend on the type of structures present in the immission area. In some countries, the lowest limit is 35 dB(A) at night for residential areas. The highest allowed value is typically 70 dB(A) for both day and night for industrial zones in which apartments are permitted only in exceptional cases. It can generally be expected that residential use without major disturbance will be possible if a value of 45 dB(A) is not exceeded at night. Compliance with respect to the limits is usually checked by taking measurements 0.5 m in front of an open bedroom or living room window.

Like the limiting values for sound immission, limits for vibrational immission also differ from country to country. The vibrational immissions of chemical plants operating up to current technical standards are considerably below the threshold for vibration perception, so problems with vibrational immission are quite unlikely in the vicinity of such plants.

5. Vibration

5.1. Measurement

Generally speaking, all three of the motion quantities described in Section 2.1 can be used for the measurement of vibration.

The displacement of a vibrating structure has an easily measurable magnitude only at very low frequencies. For this reason measurements of displacement itself are rarely performed. It is much easier to measure the related motion quantities (velocity and acceleration) and then convert these into the quantity of interest.

Geophones are used for detecting vibrational velocity. These are moving-coil sensors with characteristic frequencies tuned to the lower end of the transmission range. Such characteristic frequencies often fall within the relevant measuring range, so appropriate correction networks are required. Because of the way geophones are constructed, their industrial range of application is rather narrow.

Accelerometers are piezoelectric transducers with characteristic frequencies tuned to the upper end of their transmission range. These are universally applicable for measuring vibrations, and many designs and models are available for use in a wide variety of situations. Accelerometers are almost always utilized for modern industrial vibration measurements.

Measurement techniques are prescribed in such individual regulations as ISO 2631 [6] (effects on man) and ISO 2372 [9] (machine vibrations). Modern digital instruments are used for registering and evaluating the signals, a process that may include frequency analysis, use of weighting filters, etc. Continuously registering instrumentation is recommended for monitoring vibrations from critical equipment in which standstill might lead to expensive downtime or safety-relevant incidents.

5.2. Principal Sources

Vibration is produced by all types of machinery involving rotating or reciprocating masses, including pumps, compressors, turbines, motors, centrifuges, and mills.

ISO 2372 [9] provides severity criteria for assessing machine vibration as a function of the nature and size of the device. Equipment with strong vibrational emissions must be isolated from areas subject to protection by specially installed foundations or resilient mountings. This is especially true for machines associated with extensive vibration, such as rocking conveyors, oscillating grates, etc. Figure 5 illustrates severity criteria for various types of machine vibration.

Pipelines may be subject to high vibration levels due to pulsating flow (e.g., piping associated with piston or screw compressors). Such vibration may even be amplified by system resonance caused by an unfavorable arrangement of pipe supports. Tear-offs and, in the case of continuous stress, fatigue fracture of the pipes themselves can be the result. Furthermore, vibrating pipes may cause vibrations in other parts of a plant. Figure 6 illustrates criteria applicable to the evaluation of piping vibration.

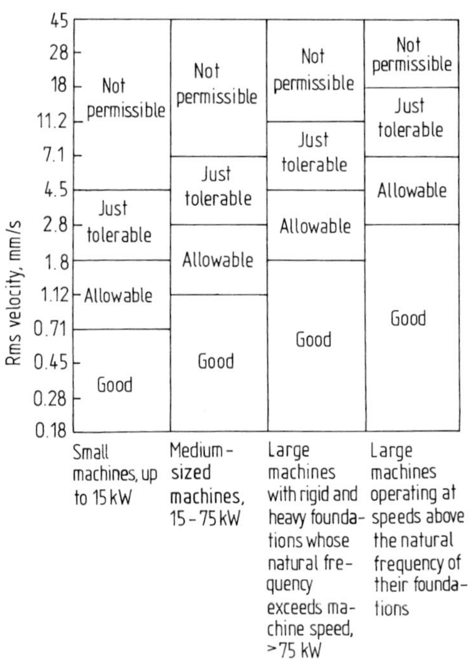

Figure 5. Vibration-severity criteria for the range 10–1000 Hz (based on ISO 2372 [9])

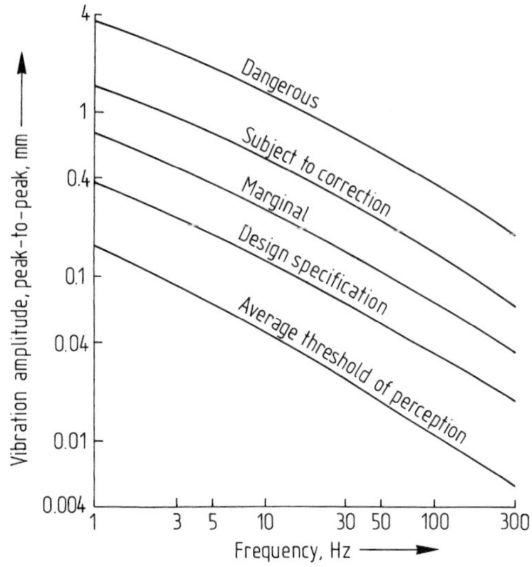

Figure 6. Criteria for piping vibration levels

6. Sound Emission

The sound emission from a sound source can be described fundamentally in terms of a corresponding sound-pressure level L. Since this measure depends upon distance from the sound source, some indication must also be provided of the measuring distance. However, it is always preferable to work with a single numerical value for the description of a sound emission, namely the sound-power level L_W (see Section 2.1), which is generally supplemented with a spectrum (e.g., in octave bands).

6.1. Determination of Sound-Power Levels

There are various methods available for laboratory or field measurements intended to establish sound-power levels. The most important procedures applicable to industrial plants are described below.

Measurement in Hemianechoic Space. The hemianechoic space method can be applied to sound sources or sound-source groups characterized by free sound propagation. The corresponding temporally and locally averaged sound-pressure level L_m is measured on a surface with a total area S that encompasses the sound source and terminates at a set of reflecting boundary surfaces (e.g., the floor). The measuring surface is in most cases a cuboid with surfaces located (preferably) 1 m from the contour of the sound source. Based on the assumption that sound waves pass perpendicularly through the measuring surface, the sound-power level L_W is then:

$$L_W = L_m + 10 \log (S/S_0) \text{ dB} \quad (17)$$

where $S_0 = 1 \text{ m}^2$

For more details (exact position of the measuring points and correction for surroundings, noise level, etc.), ISO 3744 [10] should be consulted.

Method Applicable to Multi-Source Industrial Plants. This method, described in ISO/DIS 8297 [11], can be applied to large industrial plants with numerous sound sources provided the major extension of the plant (up to ca. 300 m) is horizontal. Sound-pressure levels are determined on a single measuring path that encloses the entire plant. The required height of the microphone depends on the size of the plant and the distance between the plant contour and the measuring path; a typical value is 5–15 m. A unidirectional microphone can be used to reduce the influence of background noise (for more details see ISO/DIS 8297 [11]). This measuring approach is easy to apply (measuring time ca. 4 h for a plant occupying 40 000 m^2), and it establishes both a sound-power level and an octave-band spectrum for all noise radiated by the plant.

Vibration Measurement. The airborne noise power radiated by a vibrating surface can be determined by measuring the corresponding vibration. This technique is most often applied when a direct measurement of airborne sound is impractical because of excessive background noise. A vibration-measuring instrument is used to establish a temporally and locally determined velocity level L_{vm} at the surface of the radiating structure (surface S). The radiation behavior of a structure can be described either by the radiation efficiency σ or the logarithmic measure 10 log σ. The radiation efficiency depends on the ratio of the radiated airborne wavelength to the size of the radiator, the geometric form of the radiator, and the phase distribution of structure-borne sound over the radiator surface. The sound-power level L_W is obtained from the quantities L_{vm}, S, and σ as follows:

$$L_W = L_{vm} + 10 \log (S/S_0) + 10 \log \sigma \text{ dB} \quad (18)$$

where $S_0 = 1 \text{ m}^2$

For more details see ISO/TR 7849 [12]. This procedure is generally not as accurate as the direct measurement of airborne sound because of an often insufficient knowledge of the radiation efficiency. A value of $\sigma = 1$ provides an upper limit for the sound-power level.

Measurement of Sound Intensity. In contrast to indirect methods for the determination of sound intensity and sound power by sound-pressure measurements, sound intensity [see Eq. (2)] can also be measured directly with the aid of a two-microphone method and digital two-channel FFT analyzers for data processing. Details regarding the measurement of sound power by the intensity method are available in ISO 9614 [13]. Part 1 of this standard describes the method for obtaining data at discrete points on the measuring surface. Part 2 (in preparation) describes a scanning method that is often easier to apply. With the scanning method, a measuring probe is led continuously along a set of adjacent measuring paths that cover the entire measuring surface. Intensity measurements are seldom applicable to industrial plants.

6.2. Flow-Induced Noise

In refineries and chemical or petrochemical plants much of the noise generated is due to the flow of gases or liquids. For this reason it is appropriate at this point to provide a brief introduction to sound generation by flow mechanisms.

Part of the flow energy in any flow field is converted into sound energy. At subsonic flow speeds, flow noise is generated by sources of three types: monopole, dipole, and quadrupole.

A *monopole* source exists if an unsteady volume flow is responsible for the noise generation. Typical examples result from periodic afflux of fluid into a pipe or collapsing cavitation bubbles. Periodic afflux into a pipe is responsible for the low-frequency noise emitted by piston and screw compressors. If the local pressure in a flowing liquid falls below a certain critical value (approximately equal to the vapor pressure of the liquid), then the presence of appropriate nuclei can cause gas- and vapor-filled bubbles or cavities to develop and grow. This phenomenon, which is typical of all liquids, is called (hydrodynamic) *cavitation*. Such bubbles can collapse violently if the external pressure increases above the critical pressure, giving rise to very high local pressure peaks (sometimes exceeding 10^5 bar with strong cavitation, sufficient to cause material damage) and also to high noise levels. Cavitation usually generates a rattling, broad-band type of noise. Noise generation by cavitation is important with respect to valves and pumps.

In the case of a *dipole* source the volume flow is locally constant, but alternating forces are present as well, such as at the surface of a solid object encountering a stream generated by vortex shedding or interacting with a turbulent incoming flow or a turbulent efflux. This type of noise generation is particularly important in the case of fans and valves.

If the alternating forces also tend to compensate each other, then the corresponding noise sources are of the *quadrupole* type, as observed with free turbulence. Such quadrupole sources are sometimes important in conjunction with valves or vents to the open air.

Theoretical considerations have shown that with respect to the sound power W of the three different types of elementary sources, velocity dependencies can be anticipated as listed below provided the frequencies typical for the noise behavior and the characteristic velocity U (e.g., the flow velocity) are proportional to each other and the sources are also "compact."

Source	*Sound power*
Volume flow (monopole)	$W \sim \varrho c^3 M^{2 \ldots 4}$
Alternating force (dipole)	$W \sim \varrho c^3 M^{4 \ldots 6}$
Free turbulence (quadrupole)	$W \sim \varrho c^3 M^{6 \ldots 8}$

where ϱ = density of the flow medium
c = sound velocity in the flow medium
U = characteristic velocity
$M = U/c$ (the Mach number)

This table shows that in a sound field composed of different types of sources that are active simultaneously, the sources of higher order become more conspicuous with increasing Mach number compared with those of lower order. This is the reason why quadrupole sources are of no significance in liquids, where the Mach number is small.

6.3. Major Sources of Noise

The main individual sources of noise in chemical and petrochemical plants are:

- Burners (process furnaces, steam boilers, and flares)
- Fans
- Compressors
- Pumps
- Electric motors
- Steam turbines
- Gears
- Valves
- Exhausts to the open air
- Granule conveyors and silos
- Airborne splash noise from cooling towers
- Coal mills
- Belt conveyors

Reference values for the A-weighted sound-power levels of noise generated and radiated to the outside by these sound sources as a function of installed power are illustrated in Figure 7 based on state-of-the-art noise-control measures.

6.3.1. Burners (→ **14.** Combustion, **B3**)

The noise-radiation level of a process furnace, steam boiler, or flare is determined by the nature of the burner. Burner noise is generated by turbulent combustion of gas and oil, turbulent efflux of gaseous fuel through the burner nozzles, and the efflux of steam added to improve combustion. Important quantities associ-

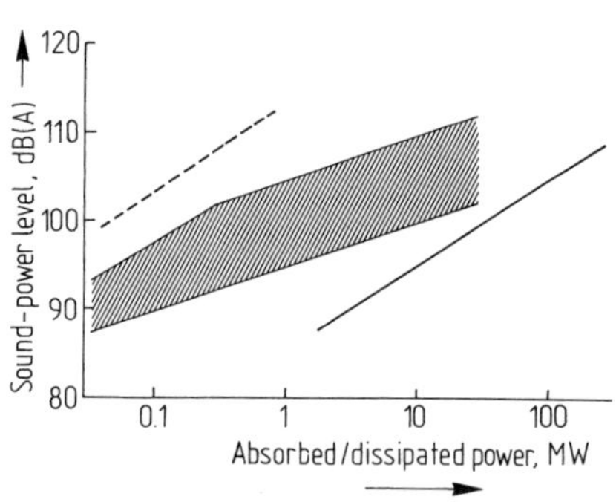

Figure 7. A-weighted sound-power levels of important noise sources in chemical and petrochemical plants based on state-of-the-art noise abatement; solid line = process furnaces, steam boilers; broken line = induced-draft cooling towers; hatched area = fans, compressors, pumps, steam turbines, gears, valves, coal mills

ated with combustion noise include heat release and the extent of turbulence in the flame.

In the case of *furnaces* and *boilers*, the burner nozzles are located inside a room. The resulting noise is of a broadband nature with peaks attributable to characteristic frequencies of the combustion room and the gas or air supply of the burner. Noise radiation takes place via air intakes from the surroundings (natural-draft burners), air flow into the air ducts and then into the open air (forced-draft burners), the furnace walls, and flue-gas ducts into the stack and then into the open air. Stack walls themselves are generally of minor importance.

The following noise-control measures are used to reduce burner-noise radiation:

1) Installation of silencers in the air inlets of natural-draft burners
2) Acoustically optimized furnace-wall construction
3) Acoustical treatment of the air ducts in forced-draft burners

Newly designed furnaces and boilers outfitted with currently available noise-control facilities are capable of meeting A-weighted burner-noise sound-power levels of

$$L_{WA} = 83 + 10 \log N \text{ dB(A)} \quad (19)$$

where N = heat release in MW (0.1...10 MW)

Flares are used for burning off small masses of gas resulting from leakage and changes in operating conditions, burn-off during controlled start-up and shut-down procedures, and burn-off occasioned by emergency shut-down of a process plant. Only elevated flares can be used for the emergency elimination of large masses of gas. Elevated flares are designed for gas-mass flows up to ca. 500 t/h. Ground flares are often used for smaller mass flows because combustion is more satisfactory with a ground flare than with an elevated flare. This applies not only to environmental stress occasioned by unwanted light radiation but also with respect to the emission of residual hydrocarbons in the flue gas. Furthermore, a ground flare is less noisy with the same gas-mass flow relative to the corresponding elevated flare [by about 10–20 dB(A)]. Ground flares are designed for gas masses up to ca. 70 t/h.

With a ground flare, noise is radiated via the air-intake openings and the flare mouth. It is customary to install silencers at the air-inlet side, which can reduce the A-weighted sound-power level to

$$L_{WA} = 102 + 15 \log q \text{ dB(A)} \quad (20)$$

where q = gas-mass flow in t/h (3...70 t/h)

6.3.2. Fans

Process plants most often utilize centrifugal and axial fans. The noise from a fan consists of a broadband component accompanied by specific tones. Noise sources of the dipole type are responsible for the broadband noise; these are caused by vortex shedding from the moving blades and turbulent flow acting upon solid construction parts. The specific tonal components are produced mainly by forces resulting from interactions between the moving blades and guide vanes in the case of axial fans and between blades and cut-off with centrifugal fans, both of which lead to periodically fluctuating streams flowing toward solid components.

The noise is radiated primarily via free openings on the suction or pressure sides of the fan, by ducts connected to the fan, and by the fan casing.

The following secondary noise-control measures are useful for reducing this radiated noise:

1) Installation of silencers at the suction and pressure sides of the fan, particularly for fans with free openings on either the suction or the pressure side
2) Acoustical enclosures, or wrapping of the fan casing
3) Acoustical wrapping of connected ducts

Such noise-control measures permit the achievement of an A-weighted sound-power level of

$$L_{WA} = 75 + 10 \log N_a \text{ dB(A)} \quad (21)$$

where N_a = rated electric power in kW (1...3000 kW)

Air coolers and induced-draft cooling towers are equipped with single-stage axial fans. In the case of low-noise fans (broad blades, large diameter, tip speed < 35 m/s), the corresponding fan noise can be brought down to an A-weighted sound-power level of

$$L_{WA} = 78 + 10 \log N_a \text{ dB(A)} \quad (22)$$

where N_a = rated electric power in kW (0.1... 500 kW)

6.3.3. Compressors (→ High-Pressure Technology, **B4**, pp. 611–616; → 19. Refrigeration Technology, **B3**, pp. **19**-10–**19**-14)

Noise radiation from a compressor depends on the type of construction, the power, and the rate of revolution. The most common compressor types in chemical and petrochemical plants are piston compressors, turbocompressors (axial and centrifugal compressors), and Roots and screw compressors.

Piston Compressors. A piston compressor conveys a fixed mass of gas with each piston stroke. This periodic conveyance is in turn related to periodic pressure variations on the suction and pressure sides of the compressor. The noise spectrum that results is responsible for the noise level in the low-frequency range. High-frequency noise is produced by relief within the pressure valves, where flow rates often reach sonic speeds.

Sound radiation of the noise takes place via the compressor casing and the connected suction and pressure pipes, including pulsation dampers.

Turbocompressors. In axial-flow compressors, compression of the gas is achieved by rotating blade rings and rows of stationary or adjustable guide vanes. The corresponding noise spectrum consists of specific tones superimposed over broadband noise. The broadband noise is mainly caused by (a) alternating forces on the rotating blades due to vortex shedding at the back edge and by turbulent flow down stream from struts and intake nozzles, as well as (b) alternating forces on guide blades in the turbulent flow down stream from the rotor blades. Individual tones in the spectrum result mainly from periodic alternating forces due to interaction of the rotor blades with the guide vanes.

The noise generated by such a compressor is radiated via the connected suction and pressure piping (major portion), the compressor casing, and interstage coolers and separators.

Roots and Screw Compressors. Compressor noise in this case is due mainly to the periodic conveyance of gas and the flow and exhaust noise accompanying pressure-compensation phenomena at the pressure outlet.

Noise generated by such a compressor is radiated via the connected piping (major portion), the compressor casing, and the inlet (for air compressors). A sound-power spectrum for noise introduced into the connected piping can be determined quite accurately provided a time-dependent volume flow can be derived from the compressor's construction data.

Important secondary noise-control measures applicable to compressors include installation of pulsation absorbers and silencers in the suction and pressure pipes as well as acoustical wrapping of the connected pipes. These noise-control measures can result in the following A-weighted sound-power levels in the case of piston and turbocompressor noise:

$$\text{Piston compressor } L_{WA} = 86 + 5 \log N \, \text{dB(A)} \qquad (23)$$

where N = required power in kw (50...1000 kW)

$$\text{Axial turbocompressor } L_{WA} = 105 + 5 \log N \, \text{dB(A)} \qquad (24)$$

where N = required power in MW (0.3...20 MW)

If stringent acoustical requirements are in force, installation of a compressor within a building requires provisions for a suitable acoustical enclosure.

With Roots and screw compressors, auxiliary sound-absorbing enclosures are common, leading to an A-weighted sound-power level of

$$L_{WA} = 72 + 10 \log N \, \text{dB(A)} \qquad (25)$$

where N = required power in kW (10...300 kW)

6.3.4. Pumps (→ High-Pressure Technology, **B4**, pp. 600–611)

Pumps can be divided into two main categories: hydrostatic and hydrodynamic. *Hydrostatic* pumps are used for the generation of very high pressures. One characteristic of these pumps is a mechanical separation between the suction and pressure sides. Furthermore, depending on the extent of this separation, the discharge pressure is almost independent of the counterpressure or lift of the pump. The most important types of hydrostatic pumps are the vane cell pump, the gear pump, the axial piston pump, and the spindle pump.

In contrast to hydrostatic pumps, *hydrodynamic* pumps (rotary pumps) provide no mechanical separation between the inlet and discharge ports. The pressure difference between the two ports must be maintained by hydrodynamic forces. Rotor blades introduce kinetic energy into the axially flowing liquid, energy that is transformed mainly into pressure energy. Fluid conveyance is continuous.

The major cause of noise radiation from pumps is alternating pressure within the pumped liquid. With respect to hydrostatic pumps, three different mechanisms can be responsible for noise generation:

1) Periodic alternation of the discharge
2) Impulse-like pressure-compensation processes at the point of intersection of liquid volumes that differ in pressure
3) Cavitation and gas exhaust during the suction process

Hydrodynamic pumps are associated with noise-generation mechanisms similar to those of fans. Furthermore, cavitation may arise under certain operating conditions in conjunction with the suction process at the rotating blades.

The major noise radiation from a pump occurs via connected pipes and the pump casing. An important secondary noise-control measure is acoustical wrapping of the pipes, leading to an A-weighted sound-power level of

$$L_{WA} = 70 + 10 \log N \, \mathrm{dB(A)} \qquad (26)$$

where N = required power in kW (1...2000 kW)

If the acoustical requirements are rather strict, pumps must be installed within sound-absorbing enclosures or in separate buildings.

6.3.5. Electric Motors

Three-phase line-voltage electric motors are commonly employed for rated powers up to ca. 400 kW. Three-phase high-voltage electric motors (6000 V) are used for rated powers > 160 kW. The noise generated by electric motors is due mainly to associated blowers (in the case of air-cooled motors), magnetostriction, and the motor bearings. The extent of noise radiation from a modern electric motor depends on the motor's power and rate of revolution as well as the extent of built-in noise-control measures. In the case of 3000-rpm motors, level differences > 10 dB(A) may distinguish "normal" from "low-noise" types. Enclosures are necessary for large motors when the acoustical requirements are stringent.

For air-cooled line-voltage motors the following A-weighted sound-power level can be achieved assuming the use of low-noise blowers:

$$L_{WA} = 75 + 7 \log N \, \mathrm{dB(A)} \qquad (27)$$

where N = nominal power in kW (3...400 kW)

For high-voltage motors with integrated noise control or a suitable enclosure, an appropriate target for the A-weighted sound-power level is

$$L_{WA} = 70 + 7 \log N \, \mathrm{dB(A)} \qquad (28)$$

where N = nominal power in kW (160...3000 kW).

6.3.6. Steam Turbines

The noise generated by a steam turbine is due mainly to flow noise from within the turbine itself. This noise is probably caused by the same mechanisms operative in turbocompressors. Condensing turbines generate additional noise in the condensers.

Noise generated by a steam turbine is radiated via the casing and the steam-feed pipe as well as the steam-exhaust pipe and any condensers that may be present. The following secondary noise-control measures are effective for reducing noise radiation:

1) A sound-absorbing enclosure for the turbine casing
2) Combined heat and noise wrapping of the steam-feed pipe
3) Wrapping of the steam-exhaust pipe (and condensers) or installation of a silencer in the steam-exhaust pipe adjacent to the steam turbine

These measures can lead to an A-weighted turbine sound-power level of

$$L_{WA} = 95 + 7 \log N \, \mathrm{dB(A)} \qquad (29)$$

where N = rated power in MW (0.3...20 MW)

6.3.7. Gears

Many investigations of gear noise and its origin have been described in the literature. The evidence suggests that there is a close relationship between dynamic load on the gear teeth (and therefore the service life) and resulting noise. In some cases it is possible to derive information regarding gear quality and gear defects from the nature of the radiated gear noise.

Geometric deviations and elastic deformation of the teeth are major sources of gear-noise generation. Uneven spots on the surface of the teeth arise during manufacture, and these cannot be avoided. As the teeth engage, the uneven spots are "played" like a phonograph record. Although the shape deviations are very small (ca. 10 μm), they can cause significant noise, which is generally of a broadband nature.

Flank deformations of the teeth are often due to defects in the driving gear of the tooth-cutting machine, and therefore display spatial periodicity related to tooth frequency and the rate of revolution of the gear-cutting machine. Flank defects generally cause frequency maxima that are related not to the meshing frequency of the gear in question but rather to the gear-cutting machine used for its manufacture.

A special case of tooth defect is *pitch defect*, caused by unequal spacing between the teeth. The noise produced by pitch defect displays peak levels at the harmonics of the rate of revolution. Under loaded condition, the teeth of a gear wheel are deformed slightly as a result of their elastic properties. This is an important factor with respect to noise generation, because such deformation changes as a function of time.

The most important factors associated with the generation of gear noise are geometrical deviations, alterations in tooth stiffness, specific tooth power, circumferential speed, alignment of the shafts, shaft bearings, and stiffness of the casing. These parameters interact in a complicated way that depends to some extent on design data (tooth geometry, wheel diameter, tooth width, etc.).

In order to reduce sound radiation, gears are often provided with acoustical wrappings and mounted resiliently. These measures permit the following A-weighted sound-power level to be met:

$$L_{WA} = 74 + 7 \log N \text{ dB(A)} \qquad (30)$$

where N = rated power in kW (1 ... 20000 kW)

6.3.8. Valves (→ High-Pressure Technology, B4, pp. 621–622)

Gas flow through control valves generates noise mainly as a result of the flow field downstream from the narrowest cross section, especially via vortex shedding and turbulent-flow impact on solid construction elements (with alternating intensity, thereby constituting a dipole source), free radiation in the mixing zone (a quadrupole source), and compression shock when sonic speeds are reached or exceeded within the valve.

Noise control is generally unnecessary with valves for liquids so long as there is no cavitation or evaporation (flashing).

Noise generated by control valves is radiated mainly via the connected piping, not the valve casing itself. Noise-reduction features have been successfully incorporated into gas control valves in recent years, but process plants still typically require extensive additional secondary valve-noise control measures. These include sound-absorptive wrapping of the pipes (and sometimes the valve casing) and/or the installation of silencers between a valve and the connected pipes. These noise-control measures can lead to an A-weighted sound-power level for control-valve noise of

$$L_{WA} = 70 + 10 \log N_a \text{ dB(A)} \qquad (31)$$

where N_a = power dissipated in the valve in kW (0.1 ... 3000 kW)

6.3.9. Exhaust Noise

Chemical and petrochemical plants contain numerous valves for venting gaseous materials to the atmosphere. The corresponding noise is often characterized by very high sound power. Many of these are emergency valves engaged only in the case of an unforeseen incident, but during start-up or shut-down of a plant it often becomes necessary to blow out superfluous steam or air via so-called start-up control valves. Pressure relief and noise generation with respect to such installations is confined mainly to the valve and the vent opening.

The major sound-transmission paths are blow-out openings, pipes, and valve casings. Important noise-control measures include installation of valves with low-noise design characteristics, acoustical wrapping of valve casings and connected piping, installation of silencers (which generally consist of separate pressure-relief and absorption units) at the blow-out openings, as well as addition of sound-absorbing cladding to the silencer casing. These measures make it possible to achieve an A-weighted sound-power level of

$$L_{WA} = 95 + 10 \log q \text{ dB(A)} \qquad (32)$$

where q = mass flow in t/h

6.3.10. Granule-Conveyor Piping (→ 8. Solids Handling, **B2**, pp. **8**-19–**8**-24)

Pneumatic conveyance is widely used for transporting plastic granules between polymerization plants and silo facilities. The extent of sound radiation from a pneumatic conveyor line depends on the piping parameters (diameter, wall thickness, material), transport parameters (product throughput, transport speed), and parameters associated with the material conveyed (density, granule size, hardness).

Noise is produced by impact between the conveyed material and the pipe walls. Maximum excitation of the pipe occurs at bends because of the large impulse interchange that accompanies a turn.

The frequency spectrum of noise produced by granule conveyance is broadband in nature, consisting mainly of high-frequency components. Common noise-control measures include use of special steels instead of aluminum for the pipes (especially at bends), together with acoustical wrapping of both pipes and bends. If these measures are taken, the sound-power level of a granule-conveyor pipe can be reduced to less than

$$L_{WA} = 75 + 10 \log l \, \mathrm{dB(A)} \qquad (33)$$

where l = pipe length in m

6.3.11. Water-Splash Noise from Cooling Towers

Apart from air coolers, most cooling towers in process plants are of a modular, rectangular design for the purpose of water recooling. The towers generally operate according to the principle of countercurrent or crosscurrent air flow, and are equipped with axial fans (see Section 6.3.2). Cooling of descending water takes place by evaporation and convective heat transfer (→ 17. Cooling Tower Technology, **B3**).

Fan drive is accomplished with an electric motor connected via a gear box. The fan hub itself is flanged directly to the output shaft of the gear box. In addition to noise produced by the fans, gears, and drive motors, much of the emitted noise is caused by water splash in the basin. Radiation of these water noises occurs via air intakes and the diffuser opening. A reasonable estimate of the A-weighted sound-power level of water noise is provided by the expression

$$L_{WA} = 69 + 10 \log q \, \mathrm{dB(A)} \qquad (34)$$

where q = water throughput in m^3/h

In general, secondary noise-control measures are not necessary with the cooling towers found in chemical or petrochemical plants.

6.3.12. Coal Mills (→ 5. Size Reduction, **B2**)

Coal mills are an important source of sound in coal processing plants (power stations and coal gasification or liquification facilities). The following types of mill are used for crushing and pulverizing coal, depending on the quality of the crude coal and the technical requirements: ball triturators; bowl mills, cylinder crushers, and roller mills; and impact rotor mills or air-jet impact pulverizers.

The milling itself occurs according to the following principles:

In a ball triturator, steel balls are lifted by friction in a rotary drum at the interior steel cladding, causing coal to be ground by the impact of backfalling balls [A-weighted sound-power level: 114...129 dB(A)].

In bowl, cylinder, and roller mills the coal is crushed between rolling bodies and an appropriate grinding bowl [A-weighted sound-power level: 105...118 dB(A)].

In principle, an impact rotor mill resembles a centrifugal fan with blades designed to act as baffle plates. Crushing of the coal occurs as a result of impact with the rotating baffle plates and the armored casing [A-weighted sound-power level: 105...111 dB(A)].

Noise radiated by coal mills is derived from the following sound sources:

1) Impact noise produced as grinding balls strike the interior armoring of the milling drums (ball triturators)
2) Impact noise from coal striking a rotating impact rotor and the armored casing (impact rotor mills and air-jet impact pulverizers)
3) Crushing and rolling noise (bowl mills, cylinder mills, and roller mills)
4) Impact noise produced at the crude coal inlet
5) Driving motors and gears
6) Fan noise (impact rotor mills)

In order to reduce noise transmission, coal mills are installed in their own separate buildings. Provided careful attention has been paid to acoustical planning in the building concept, it is possible to achieve a coal-mill A-weighted sound-power level of

$$L_{WA} = 85 + 5 \log N \text{ dB(A)} \tag{35}$$

where N = driving power in kW (10...2000 kW)

6.3.13. Belt Conveyors (→ 8. Solids Handling, B2, pp. 8-8–8-9)

Belt conveyors are usually used for transporting coal and slag in coal-processing plants. Such a conveyor consists of an endless belt that passes over two drums, the driving pulley and the tail pulley. Between the two pulleys the belt is supported by support rollers. One or more driving stations may be located at the head of the conveyor.

In defining the sound radiation due to a belt conveyor it is necessary to differentiate between noise from the driving station and noise due to motion of the band. The noise produced by the driving station is a function of the corresponding gears and driving motors. Sound radiation from motion of the band is due mainly to rolling noise generated by belt contact with the support rollers, together with bearing noise from the support rollers themselves. The belt speed is an important parameter with respect to overall noise level.

With a belt conveyor system an A-weighted sound-power level of

$$L_{WA} = 84 + 5 \log q \text{ dB(A)} \tag{36}$$

where q = output in t/h

can be achieved if the following measures are taken:

1) Limiting the band speed to a maximum of 3 m/s
2) Use of low-noise bearings for the support rollers
3) Regular maintenance of the support rollers (i.e., replacement of defective bearings and support rollers)
4) Provision of an acoustical enclosure for the driving station

6.4. Sound-Power Levels Associated with an Entire Plant

Large-scale industrial plants typically exhibit the following noise characteristics:

1) Many individual sound sources spread over a large surface area
2) Relatively constant noise emission as a result of 24-h operation
3) Constant broadband background noise that provides little specific information since it results from the interaction of many discrete sound sources

For large-scale industrial plants, especially refineries and petrochemical complexes, the area-specific A-weighted sound-power level (also called the specific sound-power level) is an important parameter for estimating community noise levels, one that is being increasingly accepted, particularly with respect to land-utilization and building codes.

In order to avoid misunderstanding it is essential that uniform definitions be adopted. We therefore suggest the following definitions of specific sound-power levels $L_{W''A}$:

For a single plant within a larger complex:

$$L_{W''A} = L_{WA} - 10 \log \left(\frac{S_a}{S_0}\right) \tag{37}$$

where L_{WA} = A-weighted sound-power level of the plant in question
S_a = "Plant area"; i.e., the area occupied by an individual process plant or an auxiliary plant, including access areas within the plant but not the surrounding areas (access areas include open areas, roads, and main pipelines within the plant)
S_0 = 1 m^2

For an overall industrial complex:

$$L_{W''A} = 10 \log \left(\sum_{\nu=1}^{n} 10^{0.1 \cdot L_{WA\nu}} \right) - 10 \log \left(\frac{S_C}{S_0}\right) \tag{38}$$

where n = number of individual plants within the factory
$L_{WA\nu}$ = fraction of A-weighted sound-power level attributable to the νth individual plant within the complex
S_C = "Complex area"; i.e., the total area occupied by the complex, including storage tanks, administrative buildings, traffic areas, and car parks. Large open areas (fallow land) within the limits of the complex are not included.

On average, the following area ratios are typical:

$$\Sigma S_a/S_C \approx 0.05 \text{ for refineries} \tag{39}$$
$$\approx 0.15 \text{ for petrochemical complexes}$$

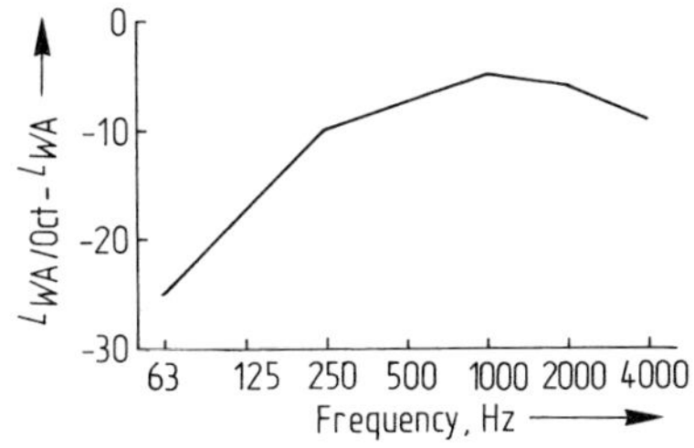

Figure 8. Typical A-weighted sound-power spectrum in octave bands for refineries as well as chemical and petrochemical complexes
L_{WA} = A-weighted overall sound-power level; $L_{WA/oct}$ = A-weighted sound-power level in octave bands

For individual plants, production facilities, and auxiliary plants subject to currently available noise-abatement measures, specific sound-power levels fall within the following limits:

$$68 \leq L_{W''A} \leq 82\ \text{dB(A)} \qquad (40)$$

The dispersion reflected in the specific A-weighted sound-power levels $L_{W''A}$ for individual plants is relatively large. Therefore, acoustic predictions applicable to individual plants in relation to their area-specific sound-power levels are possible only for plants with similar operational conditions (including throughput, installed power, etc.).

In the course of developing zoning and land-utilization plans, and for the determination of minimum distances applicable to refineries and chemical or petrochemical complexes, an A-weighted sound-power level ($L_{W''A}$) of 62 dB(A) represents a reasonable estimate for assignment to the overall premises.

Figure 8 provides a standardized, typical octave sound-power spectrum of 0 dB(A) for refineries and chemical or petrochemical complexes.

7. Sound Immission

The sound immission associated with an industrial plant in its surroundings depends not only on sound emission within the plant and the distance from the point of immission but also on various topographical factors and, especially at distances of more than 200 m, factors related to the weather. Direct measurement of the noise-immission level due to an individual plant is often impossible as a result of inevitable background noise; in this case, immission calculations must be based on emission data collected within the plant itself.

7.1. Measurement of Sound Levels in the Vicinity of an Industrial Complex

The sound-pressure levels generated by a distant sound source vary considerably; e.g., by more than 20 dB(A) at a distance of 1000 m. This is due largely to changeable weather conditions, which are in turn associated with highly variable sound propagation. The highest sound-power levels are generally recorded when the wind is blowing directly from the plant to the receiver (i.e., under downwind conditions). The dispersion of measured values under downwind conditions is relatively small, so an *average* sound-pressure level under downwind conditions is the appropriate quantity for describing noise immission due to a plant. A downwind situation is defined as one in which the wind direction deviates no more than 60° from the line joining the sound source and the point of measurement.

Noise immission in the vicinity of a continuously operating industrial plant is generally measured at night because of the correspondingly lower traffic background noise levels. In general, all measurements should be conducted under downwind conditions, and an energetic mean value should be used based on at least three measurements. Addition of appropriate charges for tonality and impulsive noise components leads to a corresponding noise-rating level (see Section 4.3). Such an assessment of noise immission should normally be based on the weather situation that actually prevails in the vicinity of the plant. If a downwind situation is to be expected at least 25 % of the year, then this weather situation, which supports sound propagation, is the one that must be taken as prevailing. If a downwind situation is relatively rare, a reduction of up to 5 dB(A) is permissible in establishing the long-term noise-rating level as a function of wind direction and speed, the frequency of temperature inversions, and the distance between the plant and the immission point. It is important to verify, however, whether relevant governmental regulations include provisions for such long-term assessments.

7.2. Outdoor Sound Propagation

Sound immission to be anticipated at a greater distance from the sound source can be determined through a sound-propagation calculation based on sound emission at the source. This calculation method is useful for an acoustic prognosis with respect to a plant still in the planning stage, as well as for an existing facility if background noise at the immission point prevents direct measurements.

ISO/DIS 9613-2 [14] suggests an international calculation approach that could be described as a semiempirical method. The complex physical phenomenon of sound propagation is divided into a set of individual steps described in empirical ways, based partly on physical considerations. Real sound sources are represented by point sound sources characterized by sound-power levels L_W. The sound-pressure level L at the immission point of interest under downwind conditions is then calculated according to the following equation:

$$L = L_W - A \qquad (41)$$

where A is the sum of a series of attenuation terms, the most important of which are described below.

1) A_{div}: Attenuation due to geometrical divergence

Sound from a point sound source propagates within a hemispherical space. This leads to a geometrically based level reduction:

$$A_{\mathrm{div}} = 10 \log \left[4\pi \times \left(\frac{d}{d_0} \right)^2 \right] = 20 \log \left(\frac{d}{d_0} \right) + 11 \text{ dB} \qquad (42)$$

where d = the distance (sound source–immission point) in m, and $d_0 = 1$ m.

This means that a level reduction of 6 dB can be anticipated with a doubling of the distance.

2) A_{atm}: Attenuation due to atmospheric absorption

During airborne sound propagation, part of the sound energy is transformed into heat. This airborne absorption depends on the ambient temperature and the relative humidity of the air as well as on the wavelength (or frequency) of the sound. It is represented by the absorption coefficient α. The following expression characterizes the attenuation due to atmospheric absorption:

$$A_{\mathrm{atm}} = \alpha \times d \text{ dB} \qquad (43)$$

The absorption coefficient increases with increasing frequency (e.g., at 500 Hz, $\alpha \approx 0.21$ dB/100 m).

3) A_{ground}: Attenuation due to ground effects

At larger distances, sound-level reductions are observed as a result of interference between the direct ray and sound rays reflected by the ground. Sound scattering and absorption in the surface layer of air is another important factor. The two effects are difficult to differentiate, and are expressed by the damping attenuation A_{ground}, which in turn depends on the height of the sound source, the height of the receiver, and the distance by which the two are separated. A_{ground} increases from 0 dB to 5 dB with increasing distance.

4) A_{screen}: Attenuation due to screening

A "sound shadow" is always observed behind a physical barrier (e.g., a building, a row of houses, a wall, or a dam). Thus, the corresponding sound-pressure level is reduced relative to the value predicted for free sound propagation. A_{screen} can be calculated on the basis of geometric quantities (height of the source, screen height, receiver height, distance between source and screen and between screen and receiver).

Apart from these damping attenuations, other influences also lead to level reductions. A level *increase* can be caused by the presence of reflecting barriers, an effect that can be calculated by the use of mirror sound sources.

The calculations are carried out in accordance with ISO/DIS 9613-2, generally in octave bands. It is common to use PC programs in performing calculations of this type. The following formula is useful for an approximate immission calculation:

$$L_{\mathrm{A}} = L_{W\mathrm{A}} - 27 \log d + 4 \text{ dB} \qquad (44)$$

where L_{A} = A-weighted sound-pressure level under downwind conditions
$L_{W\mathrm{A}}$ = A-weighted sound-power level of the sound source
d = distance (sound source – immission point) in m

This relationship is applicable to free sound propagation over an even surface in the absence of screening, and for the average octave spectrum shown in Figure 8.

7.3. Typical Sound Levels at the Workplace and Near an Industrial Complex

It is necessary today to take modern acoustics into account during the technical planning phase of plant development. This is the only way of meeting acoustical requirements established by the responsible authorities (see Section 4) for sound immission limits both at the workplace and in the immediate neighborhood.

In new chemical plants equipped with state-of-the-art noise-control measures, sound-pressure levels exceeding 85 dB(A) under normal operating conditions are found only inside machinery structures and accessible enclosures. Outside the plant, sound-pressure levels should not exceed 80 dB(A).

With modern noise-control measures, the minimum distance between a plant boundary and an adjacent residential neighborhood required to meet an immission-level target of 45 dB(A) is ca. 500 m for a complex occupying 1 km^2, and ca. 350 m for a plant size of 0.3 km^2. These distances are sufficiently small to permit the resolution of acoustic problems related to protection of the neighborhood for most refineries and chemical or petrochemical plants (→ Oil, Oil Refining, **A 18**, p. 89; → Iron, **A 14**, pp. 581–582).

8. References

[1] ISO 1683 (E), Acoustics—Preferred Reference Quantities for Acoustic Levels, 1983.
[2] E. Zwicker, H. Fastl: *Psychoacoustics—Facts and Models*, Springer Verlag, Berlin – Heidelberg 1990.
[3] International Electrotechnical Commission, IEC Standard, Publication 651, Sound Level Meters, 1979.
[4] ISO 532 (E), Acoustics—Method for Calculating Loudness Level, 1975.
[5] ISO 1999 (E), Acoustics—Determination of Occupational Noise Exposure and Estimation of Noise-Induced Hearing Impairment, 1990.
[6] ISO 2631-1 (E), Evaluation of Human Exposure to Whole-Body Vibration, part 1: General Requirements, 1985; ISO 2631-2 (E), Evaluation of Human Exposure to Whole-Body Vibration, part 2: Continuous and Shock-Induced Vibration in Buildings (1 to 80 Hz), 1989; ISO 2631-3 (E), Evaluation of Human Exposure to Whole-Body Vibration, part 3: Evaluation of Exposure to Whole-Body *z*-Axis Vertical Vibration in the Frequency Range 0.1 to 0.63 Hz, 1985.
[7] ISO 5349 (E), Mechanical Vibration—Guidelines for the Measurement and the Assessment of Human Exposure to Hand-Transmitted Vibration, 1986.
[8] ISO 1996-1 (E), Acoustics—Description and Measurement of Environmental Noise, part 1: Basic Quantities and Procedures, 1982; ISO 1996-2 (E), Acoustics—Description and Measurement of Environmental Noise, part 2: Acquisition of Data Pertinent to Land Use, 1987; ISO 1996-2 (E), Acoustics—Description and Measurement of Environmental Noise, part 3: Application to Noise Limits, 1987.
[9] ISO 2372 (E), Mechanical Vibration of Machines with Operating Speeds from 10 to 200 rev/s—Basis for Specifying Evaluation Standards, 1974.
[10] ISO 3744 (E), Acoustics—Determination of Sound Power Levels of Noise Sources—Engineering Methods for Free-Field Conditions over a Reflecting Plane, 1981.
[11] ISO/DIS 8297 Acoustics—Determination of Sound Power Levels of Multi-Source Industrial Plants for the Evaluation of Sound Pressure Levels in the Environment—Engineering Method, 1988.
[12] ISO/TR 7849 (E), Acoustics—Estimation of Airborne Noise Emitted by Machinery Using Vibration Measurement, 1987.
[13] ISO 9614-1 (E), Determination of Sound Power Levels of Noise Sources Using Sound Intensity, part 1: Measurement at Discrete Points, 1993: part 2: Measurement at Planned Points (incomplete).
[14] ISO 9613-2 (E), Acoustics—Attenuation of Sound during Propagation Outdoors, part 2: A General Method of Calculation, 1992.

Air

JOHN WOLF, Kommission Reinhaltung der Luft im VDI und DIN, Düsseldorf, Federal Republic of Germany (Chap. 1)

GERHARD MANIER, Institut für Meteorologie der Technischen Hochschule, Darmstadt, Federal Republic of Germany (Chap. 2)

THOMAS EIKMANN, Universität Gießen, Hygiene-Institut, Gießen, Federal Republic of Germany (Sections 3.1–3.3)

GERHARD SCHOLL, Bad Oeynhausen, Federal Republic of Germany (Section 3.4)

GERHARD FELDHAUS, Bonn, Federal Republic of Germany, KLAUS GREFEN, Kommission Reinhaltung der Luft im VDI und DIN, Düsseldorf, Federal Republic of Germany (Chap. 4)

PETER BRUCKMANN, Ministerium für Umwelt, Raumordnung und Landwirtschaft, Düsseldorf, Federal Republic of Germany, HANS-ULRICH PFEFFER, Landesumweltamt Nordrhein-Westfalen, Essen, Federal Republic of Germany (Chap. 5)

KLAUS LÜTZKE, RWTÜV Anlagentechnik GmbH, Essen, Federal Republic of Germany (Chap. 6)

EBERHARD SCHMIDT, Universität Karlsruhe, Institut für Mechanische Verfahrenstechnik und Mechanik, Karlsruhe, Federal Republic of Germany (Section 7.1)

TORSTEN SCHMIDT, Engelhard Process Chemicals GmbH, Hannover, Federal Republic of Germany (Section 7.2)

THEO PILHOFER, QVF Glastechnik GmbH, Wiesbaden, Federal Republic of Germany (Section 7.3)

HELMUT KRILL, Lurgi Energie und Umwelt GmbH, HARALD MENIG, Fachhochschule Frankfurt am Main, Frankfurt/Main, Federal Republic of Germany (Section 7.4)

STEPHAN SCHIRZ, Fachhochschule Münster, Fachbereich Versorgungstechnik, Steinfurt, Federal Republic of Germany (Section 7.5)

HELMUT POLLACK, Gelsenkirchen, Federal Republic of Germany (Section 7.6)

WALTER SIMMLER, Bayer AG (retired), Leverkusen, Federal Republic of Germany (Chap. 8)

The entire topic was coordinated by JOHN WOLF

3-527-20137-8/95/$5.00 + .50

1. Introduction: Emissions due to the Chemical Industry

The protection of the earth's atmosphere is one of the greatest environmental-political challenges worldwide. International agreements are needed to reduce man-made emissions and to introduce effective countermeasures. Precautionary measures on a national scale are certainly necessary, but alone are not sufficient to prevent the global changes to the earth's atmosphere that could result from the greenhouse effect or the ozone hole. Using the chemical industry as an example, the following chapters describe measures taken in Germany, and discuss the proportion of the total emissions for which the chemical industry is responsible.

The air—the natural mixture of gases that forms the earth's atmosphere—contains 78 % nitrogen and 21 % oxygen at sea level, excluding water vapor. The trace gases present in air are extremely important. Without them, the temperature at the earth's surface would be ca. −18 °C.

Concentrations of these trace gases have been measured continuously only in recent times. The longest series of records is for carbon dioxide (CO_2). The continuous increase in CO_2 concentration since the beginning of industrialization over 200 years ago is shown in Figure 1.1. The growth rates for nonmethane hydrocarbons (NMHC), nitrogen oxides (NO_x), carbon monoxide (CO), methane (CH_4), dinitrogen oxide (N_2O), and ozone (O_3) are given in Table 1.1. As a result of the tendency of these concen-

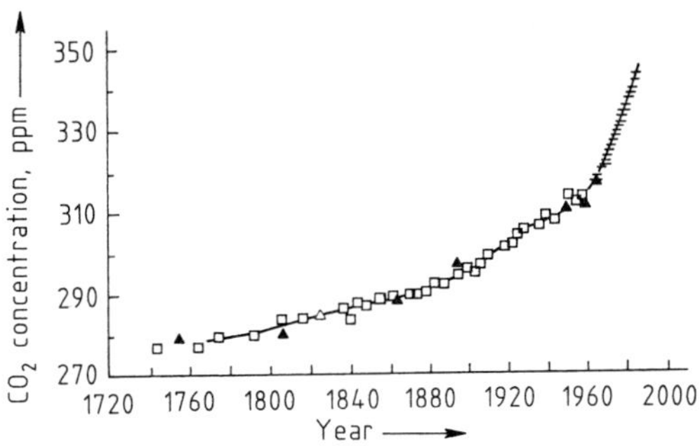

Figure 1.1. Increase in the CO_2 content of the atmosphere

trations to increase, some scientists have predicted a temperature increase of 3–5 °C, based on mathematical models. It might be thought that such a small temperature rise would not make any action necessary, as temperature changes have often been observed in the earth's history. The most recent change was a reduction of 6 °C following a temperate period, which led to an ice age. Today we are in a warm period, and a further temperature increase of 3–5 °C could change the precipitation distribution, shift climatic and vegetation zones, and hence contribute to an increase in climate-induced damage, degradative changes to the soil, and deterioration in the world food supply. The main contributors to this greenhouse effect are carbon dioxide (ca. 50 %), followed by methane (19 %) and the chlorofluorohydrocarbons (17 %). Other contributors include the ozone of the troposphere (ca. 8 %), nitrogen oxides (ca. 4 %), and stratospheric water vapor (ca. 2 %). The increase in the methane concentration is often directly related to the food requirements of the increasing world population (e.g., a byproduct of animal husbandry). The increase in CO_2 concentration correlates with the growth of energy production and traffic, associated with industrialization. The aim of German legislation is to reduce the energy-related emission of carbon dioxide by 25 % by the year 2005. The volume emitted during the year 1987 is the reference figure: emission from the pre-unification Federal Republic in 1987 was 726×10^6 t; total emission from the reunified Federal Republic in 1990 was 1031×10^6 t [1.1], [1.2]. The chemical industry will contribute to the reduction in CO_2 emissions by utilizing industrial waste heat, by increasing the efficiency of its fuel-fired equipment (and other energy-saving measures), and by rationalizing energy utilization by matching supply to demand.

Table 1.1 Mean global concentrations and rates of increase of the most important trace gases

Compound	Concentration, ppb	Rate of increase, %/a
CO_2	350×10^3	0,5
C_2H_6	1	0,8*
NO_x	very variable	1
CO	100	0.8*
CH_4	1640	1
N_2O	310	0.3
O_3	30	1*

* Increase observed in the northern hemisphere only.

Whereas the environmental consequences of increased CO_2 emission cannot at present be estimated exactly, information is available on the effect of atmospheric impurities which have accumulated in the past as a result of the limited technology then available for reducing emissions (see Chap. 3). The environmental effects first became evident as damage to forests caused by acid rain [1.3].

A considerable proportion of the acid in acid rain is derived from nitrogen oxides (NO_x), most of which is produced by combustion processes in factories and internal combustion engines by oxidation of nitrogen in the fuel and combustion air. A relatively small proportion is produced in the production of nitric acid.

The emission of nitrogen oxides in the Federal Republic of Germany in 1990 was ca. 2.6×10^6 t in the 11 old Länder and ca. 550×10^3 t in the 5 new Länder [1.3], [1.4]. Oxidation of the sulfur in fuel during combustion processes gives rise to sulfur dioxide emissions, e.g., from sulfuric acid production, from natural gas desulfurization plants (Claus plants), and from oil refineries. The emission of sulfur dioxide in 1990 was ca. 10^6 t (old Federal Republic). The new Länder emitted 4.8×10^6 t SO_2. The corresponding emissions of carbon monoxide in 1990 were 9×10^6 t (3.8×10^6 t), and of airborne particulate 480×10^3 t (1.8×10^6 t) [1.5].

Emissions of volatile organic compounds (VOCs) were 2.5×10^6 t (1.2×10^6 t). About half of this resulted from incomplete combustion in internal combustion engines, mainly in road and rail vehicles. Approximately 20 % of the VOCs are emitted during storage and handling, during the filling of vehicles, and by evaporation from vehicles.

The annual total emission of 56×10^3 t of benzene is mainly (>90 %) due to motor vehicles. The benzene content of gasoline for Otto engines is currently 1.7–2.5 vol %. The main use of benzene is in industrial chemistry for the synthesis of aromatic intermediates and cycloaliphatic compounds. Approximately 80% of the benzene is used in the production of ethylbenzene and isopropylbenzene (cumene), and as a precursor for styrene, phenol, and cyclohexane, which is further processed to give intermediates for polyamide production.

Approximately 30% of VOC emission results from the use of solvents, and nearly half of this is emitted in the course of painting. Next in order of quantity come degreasing in the metal industry, printing, and adhesives. A high proportion of these emissions are of halogenated hydrocarbons (HHCs), especially the chlorinated hydrocarbons (CHCs) such as perchloroethylene, trichloroethylene, dichloromethane, and 1,1,1,-trichloroethane. In 1991, a total of 259×10^3 t were produced, of which 202×10^3 t were exported. The main uses of CHCs are metal degreasing, solvents for adhesives, plastics processing, solvents in chemical syntheses, paint removal, and textile cleaning.

The use of CHC solvents decreased from 180×10^3 t in 1986 to 53×10^3 t in 1992 following the introduction of recycling systems. Increased use of recycled materials will further reduce consumption to 40×10^3 t by 1995. In Western Europe, total consumption of CHC solvents in 1990 was 600×10^3 t (510×10^3 t new material, 90×10^3 recycled material). In the EU, 75% of all CHC solvents used are lost to the atmosphere, while the corresponding figure for Germany is 53% owing to legal restrictions on emissions [1.6]. The possibilities of finding substitutes for these applications are limited, so that continued efforts are being directed to the improvement of waste-gas purification techniques.

Production of the following compounds was discontinued as early as the 1980s, mainly in order to reduce emissions of dioxin: trichlorophenol (TCP), 2,4,5-trichlorophenoxyacetic acid (2,4,5-T), pentachlorophenol (PCP), and polychlorinated biphenyls (PCB).

Other man-made greenhouse gases and compounds that contribute to the destruction of ozone in the stratosphere include the chlorofluorohydrocarbons (CFCs) and halons, both of which are products of the chemical industry. Fully halogenated CFCs are stable in the troposphere. However, they can also reach the stratosphere where they are photolytically cleaved by UV radiation. This produces free chlorine atoms which can break down the ozone in a catalytic chain reaction.

A detailed treatment of the photochemical reactions of these and other trace gases in the atmosphere can be found in Chapter 7. As a result of the Montreal Protocol of 1987 and the follow-up London Conference of 1990, the production of carbon tetrachloride, fully halogenated CFCs, and halons is to cease worldwide by the year 2000. Production has been discontinued in Germany in 1995, and in the EU in 1997. Production of hydrogen-containing CFCs, CH_3CCl_3, and CHF_2Cl will cease in the year 2005 [1.9]. In the meantime, emissions of CFCs in Germany will be much reduced as a result of programs of recycling solvents and cooling agents recovered from refrigerators and air conditioners. CFC mixtures which cannot be processed in this way will be thermally decomposed in an environmentally friendly manner. In most areas, suitable replacements for CFCs are available [1.6], [1.10]. CFCs have not been used since 1988 in hair sprays, deodorants, and other domestic aerosol spray dispensers. CFC-free medicaments are being developed to replace the 900 t of CFCs which are currently used in medical aerosol sprays (e.g., asthma sprays). Since mid-1994 household appliances in Germany have been produced without CFCs; propane–butane mixtures or fluorohydrocarbons (FHCs) are now used as refrigerants. The latter compounds (e.g., R 134 a, CH_2FCF_3) are more short-lived in the atmosphere because of their hydrogen content, and have lower ozone depletion potential (ODP). Pentane is a suitable substitute for producing foamed plastics.

Halons were mainly used in fire extinguishers and industrial firefighting equipment. They have now been replaced by CO_2, water, foam, and powder. In 1990, 80683 t of CFCs were produced in the Federal Republic of Germany. The production of halons was discontinued in 1991.

Guidelines for the limitation of VOC emissions are being prepared in the European Union. These are intended to lead to a 30% reduction in VOC emissions in all areas by the year 1998, compared to the year 1990. Also, solvent management plans will be set up. The above-mentioned EU Guidelines will considerably accelerate the conversion to low-solvent water-based paints or even solvent-free powder paints in the paint industry (annual production of paints in 1992: 1.6×10^6 t). Meanwhile, the major German automobile manufacturers now have plants available for painting motor vehicles with water-based paints [1.11]. As halogenated hydrocarbons are not used in these plants, VOC emissions have been considerably reduced. Initial tests show that medium-sized and small automobile painting and repair workshops, numbering around 12000, can also use water-based paints. In any case, it should be possible to reduce the emission of organic solvents by up to 70%. The printing and paper industries, which emit ca. 60×10^3 t VOCs (mainly hydrocarbons) could also make a considerable contribution to the planned reduction in emissions by changing to cleaning agents of vegetable origin, (e.g., in offset printing) and by finding substitutes for the solvent-containing paints and adhesives used in finishing processes.

In these and many other cases, further reductions in the emission of hydrocarbon solvents will be achieved by the use of substitute materials or by minimizing the solvent content of products rather than by improving waste-gas cleaning technology, which can only be considered as secondary or end-of-the-pipe technology.

Approximately 25×10^3 t VOCs were emitted in 1990 by chemical plants [1.12], mainly during the production of plastics, rubber, pharmaceuticals, dyes, paints, solvents, and adhesives.

The emissions due to the German chemical industry from 1979 to 1990 are listed in Table 1.2 [1.12].

These data refer to the 11 Länder of the "old" Federal Republic. Comparable data from the Democratic Republic for this period are not available. The following data on emissions have been published by the new Länder for the year of reunification (1990): sulfur dioxide 460×10^3 t, nitrogen oxides 35×10^3 t, carbon monoxide 25×10^3 t, airborne particulates 150×10^3 t, and VOCs 90×10^3 t. Further reductions will take place following the shutdown of obsolete plants and the improvement of other plants.

Unfortunately, it has not been possible to obtain comparable data for the emissions due to the chemical industries of other European countries.

Table 1.2 shows that there has been a drastic reduction in the pollutants emitted into the atmosphere by the chemical industry (pre-unification Federal Republic), which on average accounts for less than 3% of total air pollution. This success is due in large measure to the changes brought about by the TA Luft 1986 (see Chap. 3) [1.7], which has led to an extensive reorganization of the regulations for the control of emissions, and to the updating of the waste gas cleaning technology to correspond to the current state of the art. The TA Luft states that all old plants must be provided with state-of-the-art waste gas cleaning equipment within eight years. Similar regulations (Großfeuerungsanlagenverordnung, 13. BImSchV), were implemented in 1986 for power stations (flue-gas purification, especially by desulfurization and removal of nitrogen compounds). These regulations applied to a total of ca. 50000 installations.

The costs of environmental protection were correspondingly large, accounting for an average of 15% of the total investment in chemical plants. Of these costs, ca. 40% are for air purification [1.8].

Since ca. 1970, the German chemical industry has increased its production by ca. 200%, while at the same time emissions of air pollutants have been reduced by over two-thirds. This shows that industrial growth and the conservation of the environment need not be in conflict with each other. The following chapters consider which techniques can be used (Chap. 7.6) and the legal measures necessary to achieve the desired standard of environmental protection.

2. Physics and Chemistry of the Atmosphere

2.1. Chemistry of the Atmosphere

2.1.1. Formation of the Primeval Atmosphere

The primeval atmosphere was formed by the evolution of gas from the molten planet earth. As oxygen was chemically combined, the gases of

Table 1.2 Emissions from the German chemical industry in 10^3 t (1979–1990)

	1979	1981	1983	1984	1985	1986	1987	1988	1989	1990
SO_2	178,0	158,5	131,0	125,0	111,0	104,0	101,0	91,0	71,0	56,0
NO_x	114,0	105,0	91,5	89,5	88,0	85,0	80,0	69,0	63,5	52,0
Dust	27,5	18,0	15,5	14,5	13,0	11,0	9,0	7,5	6,5	6,5
CO	92,0	60,5	51,0	43,0	44,5	47,0	47,5	56,0	54,5	57,5
VOCs	76,0	57,5	41,5	34,0	31,5	30,5	20,5	21,0	27,0	25,5

this first atmosphere must have been largely in reduced form. The atmosphere consisted mainly of methane, with some ammonia, hydrogen, and water. In the molten planet, a gigantic blast furnace process started, in which nickel and iron oxides were reduced to nickel and iron. These components eventually formed the earth's core. Oxygen was liberated and increased the degree of oxidation of the earth's mantle and of the outgassed substances. The atmosphere then contained water vapor, carbon dioxide, and nitrogen. It is possible that this atmosphere had the same composition as the gases produced today by volcanic action, which have the following composition:

H_2O	80%
CO_2	10%
H_2S	7%
H_2	0.5%
CO	0.5%
CH_4 and NH_3	traces

2.1.2. Formation of the Present Atmosphere

The present atmosphere was formed along with the hydrosphere, lithosphere, and biosphere. This can be envisaged as follows:

The earth cooled by radiating energy into space. The upper layers of the atmosphere became cool enough to be supersaturated with water vapor, and clouds formed. Droplets that were formed in the clouds led to raindrops. The rain did not yet reach the earth's surface, but vaporized in the lower atmosphere, which was still very hot. The water vapor so formed was carried back to high altitude by convection currents, where it again condensed to form clouds. Rain formed, and again fell back into the lower levels of the atmosphere and evaporated.

This water cycle was associated with a very large transfer of energy, as the heat needed to evaporate the water in the lower levels of the atmosphere was later liberated in the upper levels by condensation to form clouds. This is known as latent heat flow, as the heat is latent in the water vapor transported upwards. This latent heat flow led to further radiative cooling. After ca. 100000 years, the rain reached the earth's surface, and a period of rain began which lasted ca. 40000 years. Initially, the rain evaporated immediately at the earth's surface. In the course of time, the earth's surface cooled more and more, so that liquid water collected in the deeper regions, eventually forming the hydrosphere. The rain washed carbon dioxide out of the atmosphere, forming the carbonate sediments of the lithosphere in the seas.

A precondition for the formation of the biosphere was the presence of oxygen in the atmosphere.

The oxygen was formed by plants according to the following equation:

$$6\,CO_2 + 6\,H_2O + \text{light} + \text{chlorophyll} \rightarrow C_6H_{12}O_6 + 6\,O_2$$

This process is known as assimilation. At the same time, plant respiration and decay of the plant mass convert oxygen into carbon dioxide. A continuous increase in the oxygen content of the atmosphere can only happen if part of the plant mass does not decay, but is preserved with exclusion of air (i.e., forms sediments). It has been calculated that 1×10^{22} g organic carbon—the amount of carbon preserved with exclusion of air—corresponds to 2.7×10^{22} g oxygen. This is 20 times the amount present in the atmosphere today. The difference was consumed in oxidizing the earth's crust and the primeval atmosphere.

The chronology of these events was probably as follows:

The first appearance of cells in an oxygen-free environment (energy gain by fermentation processes) occurred 4×10^9 years ago.

The first oxygen was produced in water $(2-3) \times 10^9$ years ago at a depth great enough to avoid damage to the organisms by UV radiation from the sun, but with adequate availability of light. Part of this oxygen was released from the ocean into the atmosphere and formed the ozone layer. Part of the UV radiation was absorbed by this ozone layer, and life formation moved closer to the surface of the water, the oxygen content of the atmosphere increased, and the ozone layer grew. The ozone layer protected the earth from the UV radiation of the sun to an ever-increasing extent, until eventually life became possible on land. This process was complete ca. 600×10^6 years ago.

2.1.3. Composition of the Present Atmosphere

It is convenient to consider the composition of the atmosphere in three parts:

1) Dry air
2) Water in all three states of aggregation
3) Atmospheric aerosol

Table 2.1. Composition of the dry atmosphere [2.1]

Component	Concentration	Probable residence time in the atmosphere
Principal components		
N_2	78.09 %	2×10^7 a
O_2	20.95 %	
Ar	0.93 %	
CO_2	0.033 %	5–10 a
Other components (constant concentration)		
Ne	18 ppm	
He	5 ppm	3×10^6 a
Kr	1 ppm	
Xe	0.09 ppm	
CH_4	1.5 ppm	3 a
CO	0.1 ppm	0.35 a
H_2	0.5 ppm	
N_2O	0.25 ppm	<200 a
Other components (variable concentration)		
O_3	≤10 ppm (stratosphere) 5–50 ppb (pure air) ≤500 ppb (polluted air near ground)	
H_2S	0.2 ppb (over land)	10 d
SO_2	0.2 ppb (over land)	5 d
NH_3	6 ppb (over land)	1–4 d
NO_2	1 ppb (over land) 100 ppb (polluted air near ground)	2–8 d
CH_2O	0–10 ppb	

Composition of the Dry Atmosphere (Table 2.1). Nitrogen and oxygen together account for >99 % of the atmosphere. If the noble gas argon is included, this figure becomes 99.96 %.

The carbon dioxide is present at a much lower concentration which varies somewhat, especially close to the surface. If the carbon dioxide is added to the above three gases, the total then accounts for 99.997 % of the dry atmosphere. The small remainder of 0.003 % (30 ppm) is nevertheless extremely important, and includes both highly aggressive gases and gases upon whose presence the very existence of life earth depends, such as ozone. The concentrations of some of these gases can fluctuate considerably.

Variation in Composition with Altitude. If there is a concentration gradient within a given volume of a gas, Brownian motion and diffusion of the gas eventually lead to a uniform spatial distribution. In addition, the gases in the atmosphere are subject to gravitational acceleration towards the earth's center. In the equilibrium state, this force is compensated by a pressure gradient force acting upwards. This force results from the decrease in air pressure with altitude. The partial pressure of a light gas decreases more slowly with altitude than that of a dense gas, a phenomenon known as segregation.

For example, the light gas helium (relative molecular mass 4 g/mol) is present at the earth's surface at a partial pressure of 5.24×10^{-3} hPa. The other air components (mean relative molecular mass 29 g/mol) have a total partial pressure at the earth's surface of 1000 hPa. At an altitude of 113 km, the partial pressures of the helium and of the other air components might be expected to be identical, but this is not the case. Measurements have shown that the composition remains constant in the lower 80–100 km, but above 100 km the concentration of the helium decreases less rapidly than that of the heavier constituents of the atmosphere. Segregation in the lower atmospheric region is prevented by other mechanisms, mainly turbulent mixing of the atmosphere by rising and falling air streams. This region, in which the composition of the atmosphere is constant, is named the homosphere, and extends to an altitude of ca. 100 km. In the overlying atmospheric layers, the concentration of the light gases decreases more slowly with altitude than those of the heavy gases, and segregation takes place. The topmost layer of the atmosphere consists almost entirely of hydrogen and helium. The atmosphere above 100 km is known as the heterosphere.

The Ozone Layer. At an altitude of 20–30 km, there is a layer in which the concentration of ozone is very high. There are several reactions that lead to the formation and destruction of atomic oxygen and ozone [2.2].

$$O_2 + h\nu \rightarrow 2\,O \quad (1)$$

$$O_3 + h\nu \rightarrow O_2 + O \quad (2)$$

These are the primary photochemical reactions. Equation (1) represents cleavage of the oxygen molecule into two oxygen atoms by UV radiation. In Equation (2), the action of radiation up to the IR range of the solar spectrum on an ozone molecule leads to formation of an oxygen molecule and an oxygen atom. Two further, secondary reactions are involved in the formation and destruction of ozone.

$$O + O_2 + M \rightarrow O_3 + M \quad (3)$$

$$O + O_3 \rightarrow 2\,O_2 \quad (4)$$

In Equation (3), a neutral collision partner M must be present to remove the excess energy and momentum liberated by ozone formation.

The probability of a particular reaction taking place depends on the available radiation and the density of the atmosphere. Both of these parameters vary with altitude, the density decreasing, and the radiation increasing with increasing altitude. As these two parameters have opposite effects, there is an altitude at which the ozone concentration is a maximum. This altitude can be calculated, and is ca. 20 km at a latitude of 10° N. At higher latitudes, the maximum is displaced upwards and the concentration decreases. Measurements have shown that at lower latitudes the calculated concentrations are considerably higher than the measured concentrations. At higher latitudes, the measured concentrations are greater and occur at lower altitudes than calculations predict. These differences between measured and calculated values can only be explained by transport of ozone from lower to higher latitudes. Human activities (e.g., CFC production, air traffic) increase the concentration of nitrogen and chlorine radicals (NO_x and ClO_x) in the stratosphere. Both substances lead to ozone destruction, but because the individual reaction cycles are interlinked and extremely complex, the prediction of future changes in the ozone layer is difficult [2.2].

Uppermost Layers of the Atmosphere. Above 500 km, the atomic forms of the atmospheric gases dominate, and above 1000 km, the light atoms helium and hydrogen are virtually the only species present. Moreover, above 500 km, the mean free path between collisions becomes so large that the particles follow ballistic trajectories. This is known as the spray region, and the atmospheric region is known as the exosphere. When the individual atoms follow ballistic trajectories, they can overcome the attraction of the earth and escape into space. Only hydrogen can reach the escape velocity of 12 km/s to an appreciable extent, so that the amount of hydrogen in the atmosphere is very small. Hydrogen is produced by the dissociation of water vapor. The probability of escape of an oxygen atom is very low, so that it can be assumed that all the oxygen formed on the earth is still present.

2.1.4. Trace Gas Cycles

Sulfur Cycle. The following sulfur compounds are present in the atmosphere: H_2S, SO_2, SO_3, and SO_4^{2-}.

All these compounds are chemically reactive. There is a link to the water (hydrological) cycle. Their residence times are short, so that their concentrations can show very large spatial and time-dependent variations.

The most important natural source is the bacterial degradation of organic substances in coastal regions and marshy areas, forming hydrogen sulfide (ca. 32×10^{12} g/a, as sulfur). High wind speeds and the breaking of ocean waves lead to the formation of small water droplets which then evaporate, leaving behind salt (NaCl) as the main component. Sulfates (44×10^{12} g/a) are also formed. Volcanic activity causes the transport of 3×10^{12} g/a S into the atmosphere, mainly as H_2S and SO_2. The most important man-made source of sulfur compounds is the burning of coal and mineral oil, which forms sulfur dioxide (65×10^{12} g/a).

The most important sink for the sulfur dioxide and sulfates is wet deposition (106×10^{12} g/a) by the atmospheric processes of washout and rainout. Gases and sulfate can dissolve in raindrops and be transported back to the earth's surface, a process known as washout. The condensation nuclei from which cloud droplets are formed, and which can then become raindrops, contain gases and sulfates. This process is known as rainout.

In addition to wet deposition, dry deposition of sulfates and sulfur dioxide on plants and on the surface of the earth or water (3.8×10^{13} g/a) can occur. The total turnover of sulfur, is 144×10^{12} g/a.

Nitrogen Cycle. The nitrogen molecule is the main constituent of the atmosphere. Nitrogen is chemically unreactive, and its contribution to the nitrogen compounds cycle is of minor importance, due to the enormous size of the reservoir.

Dinitrogen oxide, N_2O, has a long residence time in the atmosphere. It is formed by the reduction of nitrates and nitrites by bacteria in the soil. The compound probably disappears by photochemical dissociation:

$$N_2O + h\nu \rightarrow N_2 + O$$

There is no link to the other nitrogen cycles. N_2O takes part in catalytic ozone degradation reactions.

Ammonia and its compounds are chemically reactive. There is a link to the water cycle. Because the residence time of these gases in the atmosphere is short, the concentration can show large spatial and time-dependent variations.

The main source is the degradation of organic substances in the soil. Figures for the amount (as nitrogen) that enters the atmosphere fluctuate considerably [$(113-244)\times 10^{12}$ g/a]. Dry and wet deposition carry NH_3 back to the earth's surface. Reactions with water vapor, sulfur dioxide, and carbon dioxide can form ammonium sulfate or ammonium bicarbonate as aerosol particles. A small sink for ammonia is provided by the reaction with the radical $^{\bullet}OH$, which forms NO_x and water. This represents the only link between the ammonia cycle and the NO_x cycle.

Nitrogen oxides, NO_x are chemically reactive. There is a link to the water cycle. The residence time in the atmosphere is short, and the concentrations can therefore show very large spatial and time-dependent variations. The most important substances are nitrogen monoxide (NO) and nitrogen dioxide (NO_2). Other compounds include dinitrogen trioxide (N_2O_3), nitrous acid (HNO_2), dinitrogen pentoxide (N_2O_5), nitric acid (HNO_3), nitrites (NO_2^-) and nitrates (NO_3^-). The nitrates have the highest oxidation state, and are therefore the stable final state after the reaction of NO via NO_2 to form NO_3^-. Each oxidation step occurs by reaction with ozone (O_3).

The main sources of nitrogen oxides are bacterial nitrate degradation in the soil ($2.1-8.9\times 10^{13}$ g/a) and the combustion of oil in industry and by motor vehicles (19×10^{12} g/a). Further sources are the link to the ammonia cycle and the formation of nitrogen oxides by lightning. Organic nitrogen compounds are very important in the soil and in the ocean, but they can be harmful constituents of the atmosphere.

Carbon Cycle. Four atmospheric carbon cycles exist, involving the substances CO_2, CO, CH_4, and other carbon compounds.

Carbon dioxide is one of the most important constituents of the atmosphere. It has a long residence time. The main sources are the combustion of coal, the respiration of living organism, and the degradation of organic substances. The main sink is photosynthesis, a further sink being the dissolution of carbon dioxide in the oceans and formation of calcium carbonate sediments. The carbon dioxide content of the atmosphere increases by ca. 0.75 ppm annually.

The greenhouse effect of the atmosphere (solar radiation passes through, thermal radiation is retained) depends on the concentrations of water vapor and carbon dioxide. Therefore, an increase in the carbon dioxide content can lead to a change in the greenhouse effect, and hence to climatic change. However, because these processes are extremely complex, predictions are difficult [2.3].

Carbon monoxide can be oxidized to carbon dioxide, but this reaction does not proceed in the atmosphere, so that CO has a relatively long residence time. Sources of carbon monoxide include incomplete combustion and generation by microorganisms on the surface of the ocean. Bacteria in soil consume carbon monoxide. It also reacts with $^{\bullet}OH$ radicals. Carbon monoxide concentrations can show very large spatial and time-dependent variations, especially in towns.

Methane is a relatively common gas and has low chemical reactivity. It is produced and consumed as part of the biological cycle. Its concentration is presently increasing by ca. 1.5%/a. Reaction with oxygen in the stratosphere leads to the formation of the methyl radical and $^{\bullet}OH$, which also contributes to catalytic ozone breakdown. Methane also contributes to the greenhouse effect.

Water Cycle. The total mass of water on the earth is 1.36×10^{21} kg, corresponding to 2.66×10^{6} kg water per m^2, or a layer of water 2660 m thick. Of this water, 97% is contained in the oceans, and 2.3% in ice. Subterranean water (groundwater) accounts for 0.7%, and lakes and rivers 0.01%. The atmosphere contains 0.001% or 26.6 kg/m^2. This amount of water corresponds to a depth of 2.66 cm. As the average annual precipitation is 857 mm, and only 26.6 mm is present in the atmosphere, the equivalent of the total atmospheric water content must be deposited on the earth's surface in 10 d to give the above mean annual precipitation figure, i.e., the cycle time of the water in the atmosphere is 10 d. The water cycle causes cleansing of the atmosphere, as all substances are eventually converted to water-soluble products and return to the earth's surface with the rain. Evaporation produces only water vapor.

2.1.5. The Aerosol

The atmosphere contains particles as well as gaseous components. These particles are much larger than single molecules, but are still so small that their settling rate is generally negligable. Their size varies between 5×10^{-3} and 20 µm, and their number density between 10^2 and

10^6 cm^{-3}. Particles with a radius of 0.1 μm are especially abundant. Particles larger than 10 μm are extremely rare, as their settling rate is so great that they become deposited on the earth's surface close to their source. Very small particles also do not occur, because they agglomerate rapidly to form larger particles.

Sources and Chemical Composition of the Aerosol. There are four mechanisms by which aerosols are produced:

1) Combustion (industrial combustion, forest fires, traffic), forming salts, carbonates, and lead.
2) Reactions in the gas phase (sulfur and nitrogen cycle) to form sulfates and nitrates.
3) Whirling up of solid material, whereby silicates and salts of calcium, potassium, and sodium enter the atmosphere.
4) Whirling up of liquid. At high wind speeds, air bubbles form on the ocean. When these bubbles burst, many small water droplets are formed. These then evaporate, leaving salt as an aerosol.

Sinks for the Aerosol. Large particles leave the atmosphere owing to their high rate of settling, becoming deposited preferentially on vegetation. Larger particles are also washed out of the atmosphere by rain.

The residence time of large particles (> 10 μm) amounts to minutes or hours. Particles of 2–10 μm have a residence time of hours or days, and those of 0.1 μm a residence time of several days. Particles smaller than 0.05 μm agglomerate to form larger particles and have a short residence time (< 1 h).

Effect of Altitude on Aerosol Concentration. In the lowest 5 km, the concentration of the aerosol decreases much more rapidly than that of the other atmospheric constituents. Whereas the altitude of the homogeneous atmosphere is 8 km, it is found that a hazy atmosphere is homogeneous to an altitude of 3 km. (The altitude of a homogeneous atmosphere is that for which the density of the air or of the aerosol, independent of the altitude, is equal to the density at the earth's surface.) Above an altitude of 5 km, the aerosol behaves like air, i.e., the densities decrease at the same rate. Between 20 and 30 km, the aerosol concentration remains constant. Aerosol particles are produced in this region.

2.2. Energetics of the Atmosphere

2.2.1. Solar Radiation

The sun is the sole source of energy for all processes on the earth. At the upper limit of the atmosphere, the radiation flux density is 1376 W/m^2. This radiation is attenuated in the atmosphere by scattering and absorption caused by air molecules, certain gases, the aerosol, and clouds.

Attenuation by the gases is proportional to λ^{-4}. With increasing path length through the atmosphere, the energy maximum shifts from the blue to the red region. One consequence of this is the red appearance of the sun at sunrise and sunset.

The aerosol also attenuates the shorter wavelengths more strongly than the longer ($\sim \lambda^{-1}$). Different particle size distributions of the aerosol can lead to very different directional distributions of the scattered radiation. Forward scattering dominates, in contrast to scattering by air molecules.

The absorption of solar radiation by gases leads to excitation. Owing to the high density of the air, the collision number is so large that the energy of excitation can be transferred as kinetic energy to other air molecules. This leads to an increase in the temperature of the gas. Thus, unlike scattering, absorption leads to the transformation of radiant energy into thermal energy.

The nitrogen molecules and the oxygen atoms or molecules absorb UV radiation (< 200 nm) even at high altitudes. Ozone absorbs at wavelengths of < 300 nm, so that radiation in this range does not reach the earth's surface. Ozone absorbs only weakly in the visible region. Water vapor absorbs in the infrared.

Even a thin cloud layer prevents the sun from being visible, i.e., direct solar radiation is zero.

Part of the direct solar radiation is absorbed at the earth's surface. The energy yield increases with the altitude of the sun and also with increasing darkness of the earth's surface. Forests and water surfaces are very dark, while snow has the brightest surface.

Diffuse solar radiation is produced by scattering of direct solar radiation by air molecules and the aerosol. The sum of the direct and diffuse solar radiation is the global radiation. In the following, some values of the global radiation F_G for a cloudless sky at various zenith angles ζ

(measurements were carried out in Frankfurt/Main, Germany) are given:

ζ,	30°	40°	50°	60°	70°	80°
F_G, W/m^2	963	830	670	481	293	119

If the sky is completely covered with clouds, the following percentages of the global radiation of a cloudless sky are obtained:

Fog	15%
Rain clouds	19%
Low clouds	25–33%
Medium altitude clouds	40–50%
High ice clouds	80%

2.2.2. Infrared Radiation

All bodies whose temperature is above absolute zero emit radiation. In the atmosphere, these includes the following emitters: the earth's surface, clouds, water vapor, and carbon dioxide.

The earth's surface and clouds radiate like black bodies, i.e., the radiation flux density F depends on the surface temperature T and the longwave emissivity ε of the material and is given by:

$$F = \varepsilon \times 5.67 \times 10^{-8} \times T^4$$

For most surfaces, ε lies between 0.92 and 0.98.

Unlike solid or liquid bodies, gases do not radiate at all in certain wavelength ranges, and radiate like a black body in other ranges.

2.2.3. Radiation Equilibrium

It can be assumed that the planet earth is in thermal equilibrium. Thus, the absorbed shortwave radiation from the sun is equal to the long wave radiation emitted from the earth. In the absence of an atmosphere, this would occur if the earth's surface temperature were −21 °C. But the average temperature of the earth's surface is +14.3 °C, and a temperature of −21 °C is found at an altitude of ca. 6 km. This altitude corresponds to the average altitude of the upper surface of the clouds. The atmosphere and the clouds provide effective protection from radiation. This is also known as the greenhouse effect, whereby shortwave radiation is allowed to reach the earth, while water vapor, carbon dioxide, aerosol, and clouds only allow the partial transmission of long-wave radiation.

2.2.4. Energy Balance

The radiation balance of the earth's surface is positive, and that of the atmosphere negative. The energy balances of the earth and the atmosphere are evened out by the turbulent flows of sensible and latent heat.

A turbulent transport process, whether of temperature, water vapor, or pollutants can be pictured as follows: The air flow is split up into several densely packed flow filaments. In each flow filament, the mean flow rate, temperature, water vapor content, and concentration of pollutants is constant. There are differences between the individual flow filaments. In a turbulent flow, there is an intensive exchange of air parcels between the individual flow filaments. These air parcels transfer the property of one flow filament into another. If there is a temperature difference between the two flow filaments, this exchange of air parcels implies a transfer of heat. The transfer is always in such a direction that temperature differences are evened out. The same considerations also apply to water vapor (latent heat flow) or any other substance.

If we consider the effect of geographical latitude on the energy balance, there is a surplus at the lower latitudes, while, at high latitudes where the sun does not shine for a large part of the year or the angle of incidence of the solar radiation is low even in summer, the energy balance is negative. As the temperature does not increase continuously in the lower latitudes or decrease continuously in the polar regions, there must be a transfer of energy from the lower latitudes to the higher. This transfer exhibits a maximum in the geographical latitude at which the energy balance is zero. This is the case at a latitude of ca. 40° N or 40° S. At this latitude, the energy transferred must be 4.2×10^{15} W. This transfer takes place as a result of normal weather processes, in high latitudes by the interchange of low- and high-pressure areas, and in low latitudes by the circulation of trade winds.

2.3. Dynamics of the Atmosphere

2.3.1. Forces

Five forces act in the atmosphere and determine wind flows.

The gravitational force causes all objects, including the air, to accelerate towards the center of the earth.

The pressure gradient force operates wherever pressure differences exist. It accelerates the air from high- to low-pressure regions. In the vertical direction, it opposes the gravitational force.

The Coriolis force is an apparent force that must be introduced because the air movements are observed on a rotating earth. The Coriolis force diverts flows to the right in the northern hemisphere, and is proportional to the flow rate.

The centrifugal force acts in an outwards direction on curved flow paths (e.g., low-pressure areas or tornados). It is proportional to the square of the flow rate, and is inversely proportional to the radius of the flow path.

The frictional force is always present if there is a spatial change in flow rate, as is always the case in the vicinity of the earth's surface. The frictional force is in the opposite direction to the direction of the flow and its magnitude increases with increasing flow rate and with decreasing stability of the atmosphere (see also Section 2.3.3.)

2.3.2. Motions

There is very often an equilibrium between two or three forces in the atmosphere, with the other forces being of minor importance. (This is true even though continuously observable changes in the flow pattern can only be explained by nonequilibrium conditions.) In the free atmosphere, pressure gradient forces and Coriolis forces dominate. In the equilibrium condition, flow takes place parallel to the isobars (lines of equal atmospheric pressure). In the northern hemisphere the winds in low-pressure regions rotate anticlockwise, and in high-pressure areas the rotation is clockwise.

Close to the earth's surface, the frictional force is also important.

Two different layers close to the earth's surface can be distinguished. In the lower (Prandtl) layer, which extends to a height of ca. 100 m, the wind velocity increases with altitude, but the wind direction remains constant. In the overlying layer, planetary boundary layer, which can reach an height of 1500 m, the wind velocity increases further and the wind direction changes.

The variation of wind speed with height in the Prandtl layer over a land surface with constant roughness can be expressed by the following equation:

$$\bar{u}(z) = u_*/\varkappa\,[\ln(z/z_0)] - \psi(z/L)$$

$\bar{u}$ = mean wind speed, z = altitude coordinate, u_* = friction velocity, $\varkappa = 0.4$, z_0 = roughness height, ψ = stability function, L = Monin–Obukhov length.

Disregarding the stability function, the flow rate varies with the logarithm of the altitude. At the roughness height ($z = z_0$), the velocity is zero. z_0 depends on the height of the roughness elements. It varies from a few millimeters over ice and water surfaces to several meters over large towns and forests. The friction velocity u_* represents the effect of friction. u_* is ca. $0.1 \times \bar{u}$ (10 m). The stability function represents the change with height as a function of the stability of the layer (see Section 2.3.3). For a stable layer, the function is <0, and for an unstable layer >0.

The flow direction in the Prandtl layer deviates from the direction of the ground flow at the top of the planetary boundary layer. There is a shift to the left, which increases with increasing roughness and stability (the extent of shift being 10–40°). At the top of the planetary boundary layer, the wind vector changes into the equilibrium flow resulting from the pressure gradient force and the Coriolis force. That is, with increasing height in the planetary boundary layer, the wind rotates towards the right and the velocity increases.

Deviations from the ideal conditions (flat surface and constant roughness) can lead to considerable changes in the wind field.

For very large ground roughness values (e.g., in large towns or forests) the wind speed does not become zero at height z_0, but at greater heights. A zero point shift, ($z = z + d_0$) must be introduced. The zero point shift d_0 is somewhat less than the mean height of the obstructions.

The flow is very strongly deviated by topographical obstructions. The deviation depends on the orientation of the topographical structures relative to the ground flow (flow at the top of the planetary boundary layer). In valleys, the flow is channeled in the direction of the axis of the valley, independent of the direction of the ground flow. The channeling effect becomes greater as the depth of the valley increases, but even very wide valleys such as the Rhine valley at Ludwigshafen, Germany (width ca. 30 km, height of sides ca. 300 m) cause channeling of the flow. The wind speed at the base of the valley is lower for larger differences between the ground flow direction and the valley direction. The air preferably flows around individual high and steep mountains, but goes over the top of wide high

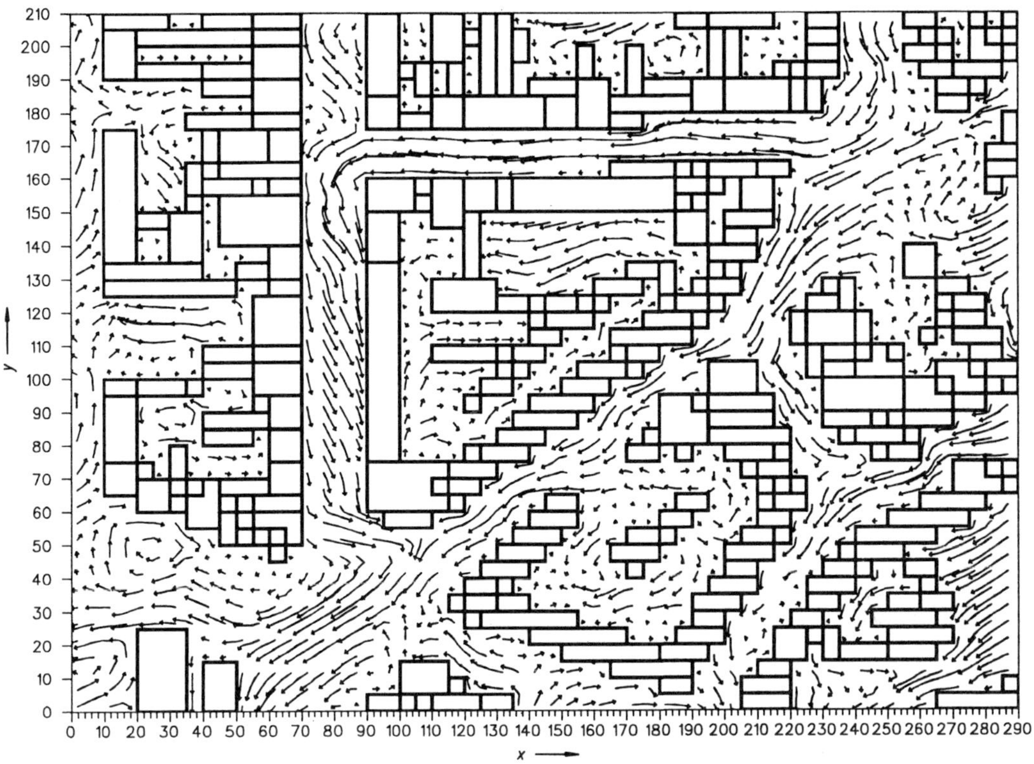

Figure 2.1. Calculated wind field 1.5 m height in a city area for undisturbed flow at 60°, showing 10-s trajectories

ridges. In basins, the flow rate is significantly lower, and in open locations on mountains it is higher than over flat terrain.

Single buildings modify the flow considerably. On the side facing the wind, a ground eddy forms with a horizontal axis and a countercurrent flow near ground level. A disturbed region also forms on the downwind side, i.e., an eddy with a horizontal axis and a countercurrent flow at ground level. Further downwind, the flow changes back to the undisturbed state. The distance to which this disturbance extends in the flow direction can be many times the height of the building. Undisturbed flow is reestablished at a distance equal to 100 times the height of the building [2.4].

In streets, an eddy is produced whose intensity increases as the ground flow component perpendicular to the street increases.

The flow field in cities and industrial areas is extremely complex owing to the complexity of the configurations of the buildings.

Flow fields of this kind cannot be adequately understood by measurements in the field. Wind tunnel tests [2.4] or mathematical models [2.5] must be used. An example of the mathematical modeling of the wind field in a city area is shown in Figure 2.1.

2.3.3. Stability

Thermodynamics plays a part in many atmospheric processes. This is very important in the suppression or production of vertical movement. The most spectacular manifestation of this is a thunderstorm, in which intense vertical movements of the air are a characteristic feature. On clear nights, the thin clusters of very long, meandering smoke trails that can be observed owe their existence to a much diminished vertical movement.

In each case, the vertical temperature gradient is decisive. Normally, the decrease in temperature with altitude is 0.7 °C/100 m. On cloudless nights, reversal of the temperature gradient occurs such that the temperature can increase by several degrees per 100 m (inversion). On sunny days and during the summer months the temperature decrease can be >1 °C/100 m.

These temperature changes in the atmosphere can be compared to the temperature changes experienced by a parcel of air that is ideally thermally isolated from its surroundings, and either rises or falls like a body floating in a liquid. The temperature change of the air parcel can be deduced from the First Law of Thermodynamics, which gives a temperature decrease of 1 °C for a 100 m ascent, and an equal temperature increase if the air parcel sinks by 100 m. In the normal case, the rising air parcel is colder and hence heavier than its surroundings. As it is heavier, it accelerates downwards, back to the starting level. The air parcel that has sunk is warmer than its surroundings and hence lighter, so that it accelerates upwards, again towards the starting level. Thus, vertical movements involving organized flows (e.g., upward moving or "anabatic" winds) or turbulence are damped, i.e., the layer is stable. The layer is especially stable if temperature increases with altitude (inversion).

If the temperature decrease in the atmosphere is >1 °C/100 m the temperature of the air parcel above the starting level is higher than that of its surroundings. It is therefore lighter and becomes accelerated upwards, away from the starting level. This also holds when an organized or accidental displacement in a downwards direction occurs. Thus, vertical movements are amplified, and the layer is unstable.

The effect of stability or instability on the turbulence of the flow is especially relevant to the dilution of harmful substances in the atmosphere (see Section 2.4). With turbulent fluctuations, momentum is also transferred, and this transfer determines the friction of the flow. Hence, the friction is greater for unstable layers, and the wind profile has a different form (see Section 2.3.2).

Measures of stability are, e.g., the Richardson number, the Monin–Obukhov stability length, and stability classes.

The Richardson number Ri is defined as [2.6]:

$$Ri = \frac{g}{\bar{\Theta}} \frac{\partial\bar{\Theta}/\partial z}{/(\partial\bar{u}/\partial z)^2}$$

where g = gravitational acceleration, $\bar{\Theta}$ = potential temperature.

It is a dimensionless number that represents the ratio of buoyancy to frictional forces.

The Monin–Obukhov stability length L is defined as [2.7]:

$$L = -\frac{\bar{T}}{g} \frac{u_*^3}{\overline{w'T'}\varkappa}$$

where $\bar{T}$ is the temperature of the air in Kelvin, w' the turbulent fluctuation of the vertical velocity, T' the turbulent fluctuation in temperature, and $\overline{w'T'}$ the turbulent heat transfer.

This number also represents the ratio of the frictional forces to the buoyancy. Its dimension is length.

The Richardson number and the Monin–Obukhov stability length are negative for unstable layers and positive for stable layers.

The formerly much used stability classes (e.g., those due to Klug–Manier [2.8]) should no longer be employed, as these do not correspond to the present state of the art.

2.4. Air-Pollution Control

Pollutants are always present in the atmosphere, and may be of natural origin (see Section 2.1.3). The only effective method of reducing additional pollution of human origin is to reduce emissions.

The subject of this section is not the reduction of emissions, but what happens to pollutants after they have been released into the atmosphere, and what can be done to prevent threshold values from being exceeded. The observance of threshold values can be monitored by measuring pollution concentrations. In general, point measurements are taken, so that a result is only applicable to that location. Threshold values are often statistical quantities which must be determined from a number of single measurements. To include all variations of the influential factors (source parameters and meteorological conditions), measurements must be carried out over long periods (one year or more). Only then can it be decided whether threshold values are being exceeded or observed. The measurements provide only a present time and past for each measuring location. To obtain spatial information, and to enable the effects of preventive measures to be forecast (new sources, reduction of emissions, relocation of sources), the extent of the pollution must be calculated.

To obtain statistical information, the concentrations for time series, with the corresponding variations of the input quantities, must be calculated. The calculation is highly time consuming. Frequency distributions of the emissions and the meteorological input quantities are therefore used. The use of several one-dimensional frequency distributions is only possible if the individual parameters are independent of each other. For the most important meteorological input quantities, i.e., wind direction, wind velocity, and stability, this is certainly not the case, so that multidimensional frequency distributions are required.

2.4.1. Methods of Calculating the Dispersion of Pollution

There are four methods of calculating pollution concentration:

1) Empirical formulas
2) Gauss formulas
3) Solution of the diffusion equation
4) Particle models

2.4.1.1. Empirical Formulas

Empirical formulas give the relationship between measured emissions, immissions, meteorological influences, and geometrical parameters. The simplest formula expresses the fact that the concentration is proportional to the source intensity, is inversely proportional to the wind velocity, and decreases in proportion to a negative index of the distance between the source and the point under consideration. Using this method, cause and effect can be investigated and, within limits, concentration distributions at ground level can be obtained. The method is not transferable, and the formulas must be recalculated if there are changes in the emissions. It is not suitable for forecasting the effects of preventive measures.

2.4.1.2. Gauss Formulas

Diffusion experiments using continuous sources at various heights led to the Gauss formula:

$$\bar{C}(x,y,z) = Q/[2\pi\sigma_y(x)\sigma_z(x)\bar{u}] \cdot \exp\left(-\frac{y^2}{2\sigma_y^2(x)}\right)\exp\left(-\frac{z^2}{2\sigma_z^2(x)}\right)$$

where $\bar{C}$ is the concentration (time average), x, y, z are space coordinates (x in the wind direction), Q is source intensity, σ_y horizontal spread of concentration perpendicular to the wind direction, σ_z the spread of concentration in the vertical direction, and $\bar{u}$ the mean wind velocity.

Experiments show that the concentration distribution perpendicular to the mean direction of propagation can be represented by Gauss functions [2.9]. The experiments measure the spreads and show that these depend on the distance from the source, the height, and the stability of the atmosphere.

There are numerous Gauss formulas, all derived from an original formula, which applies to a spreading spontaneous emission that drifts with the wind:

$$C(x,y,z,t) = Q/[(2\pi)^{3/2}\sigma_x(t)\sigma_y(t)\sigma_z(t)] \cdot \exp\left(-\frac{x^2}{2\sigma_x^2(t)}-\frac{y^2}{2\sigma_y^2(t)}-\frac{z^2}{2\sigma_z^2(t)}\right)$$

t = traveling time of the cloud, $\sigma_{x,y,z}$ = diffusion in the x, y, z directions.

Other formulas can include the effects of influence such as:

1) Height h of the source
2) Reflection at the earth's surface, at an inversion, or at a wall
3) Chemical conversions
4) Settling rate of the pollutant
5) Time-dependent emissions
6) Point, line, two-dimensional, and volume sources

Gauss formulas are still extensively used. They are easily understood and simple to use, and the computation is not complicated. However, the conditions under which the Gauss formula can be used are so restrictive (constant meteorological conditions over space and time, flat surface of uniform roughness) that the results calculated can only be used as a guide, and the formula is only suitable for rough calculations. Moreover, better methods are available for calculating concentrations.

2.4.1.3. Diffusion Equation

The balance equation for a given substance (diffusion equation) is:

$$\frac{\partial\bar{C}}{\partial t} = -\bar{u}\frac{\partial\bar{C}}{\partial x} - \bar{v}\frac{\partial\bar{C}}{\partial y} - \bar{w}\frac{\partial\bar{C}}{\partial z} + \frac{\partial}{\partial x}\left(K_x\frac{\partial\bar{C}}{\partial x}\right) + \frac{\partial}{\partial y}\left(K_y\frac{\partial\bar{C}}{\partial y}\right) + \frac{\partial}{\partial z}\left(K_z\frac{\partial\bar{C}}{\partial z}\right) + P$$

where time averages are indicated by a bar, K_x, K_y, K_z are exchange coefficients and P is the source term.

The change in concentration with time at a point (x, y, z) (left-hand side of equation) is determined by:

1) Mass transport to or from the test point by an average wind $(\bar{u}, \bar{v}, \bar{w})$ at a nonzero concentration gradient (1st to 3rd summand on the right-hand side of the equation).
2) Turbulent transport (4th to 6th summand on the right-hand side of the equation). Mass is transported from the region with an excess to the region with a deficit.
3) Sources and sinks for masses of material.

Analytical solutions to the diffusion equation exist (the Gauss formula is a solution of the diffusion equation given certain preconditions for the relationship between diffusion and distance from the source). However, they cannot be used in practice if meteorological fields vary over space and time, and topographical structures and buildings are to be included. The diffusion equation can then only be solved numerically.

Computation Grid. To obtain a numerical solution of the differential equation, a three-dimensional grid is superimposed on the region in which the concentration calculation is to be carried out. Depending on the mathematical model, the region being studied, and its size, the grid can be rectangular or can follow the topography. The spatial solution depends on the size of the region, the size of the structures to be analyzed, and the speed and storage capacity of the computer. The grid element size lies between 1m and some kilometers. Regions studied can vary from a few meters (e.g., a back yard) to many kilometers (e.g., transport over a large area).

Finite Difference Equation. To obtain a numerical solution, the differential equation must be converted into a difference equation. This gives the following:

$$\begin{aligned}
\frac{\Delta C_{i,j,k,t}}{\Delta t} = \frac{1}{\Delta x}\Big\{ & u_{i-\frac{1}{2},j,k} C_{i-\frac{1}{2},j,k,t} \\
& - u_{i+\frac{1}{2},j,k} C_{i+\frac{1}{2},j,k,t} + K X_{i+\frac{1}{2},j,k} \\
& \cdot \frac{C_{i+1,j,k,t} - C_{i,j,k,t}}{x_{i+1,j,k} - x_{i,j,k}} \\
& - K X_{i-\frac{1}{2},j,k} \frac{C_{i,j,k,t} - C_{i-1,j,k,t}}{x_{i,j,k} - x_{i-1,j,k}}\Big\} \\
+ \frac{1}{\Delta y}\Big\{ & v_{i,j-\frac{1}{2},k} C_{i,j-\frac{1}{2},k,t} \\
& - v_{i,j+\frac{1}{2},k} C_{i,j+\frac{1}{2},k,t} + K Y_{i,j+\frac{1}{2},k} \\
& \cdot \frac{C_{i,j+1,k,t} - C_{i,j,k,t}}{y_{i,j+1,k} - y_{i,j,k}} \\
& - K Y_{i,j-\frac{1}{2},k} \frac{C_{i,j,k,t} - C_{i,j-1,k,t}}{y_{i,j,k} - y_{i,j-1,k}}\Big\} \\
+ \frac{1}{\Delta z_k}\Big\{ & w_{i,j,k-\frac{1}{2}} C_{i,j,k-\frac{1}{2},t} \\
& - w_{i,j,k+\frac{1}{2}} C_{i,j,k+\frac{1}{2},t} + K Z_{i,j,k+\frac{1}{2}} \\
& \cdot \frac{C_{i,j,k+1,t} - C_{i,j,k,t}}{z_{i,j,k+1} - z_{i,j,k}} \\
& - K Z_{i,j,k-\frac{1}{2}} \frac{C_{i,j,k,t} - C_{i,j,k-1,t}}{z_{i,j,k} - z_{i,j,k-1}}\Big\} \\
+ & \frac{P_{i,j,k,t}}{\Delta x \, \Delta y \, \Delta z_k}
\end{aligned}$$

where i, j, k are the indices of the grid element for the coordinate directions x, y, z and $\Delta C_{i,j,k,t}$ is the concentration change in the grid element with indices i, j, k, in time Δt.

To determine all the summands on the right hand side of the equation, the following are required:

1) The three-dimensional wind field u, v, w
2) The three-dimensional field of the exchange coefficients K_x, K_y, and K_z
3) The three-dimensional concentration field at time $t = 0$
4) The three-dimensional field of the production and dissipation terms $P_{i,j,k,t}$

Also, the boundary conditions and the time interval Δt must be fixed.

The wind field largely determines the concentration field. Over a flat surface of uniform roughness, the wind field is horizontally homogeneous, and only changes with height (see Section 2.3). If the land is topographically organized, the land use diverse, and buildings present, the wind field becomes highly structured and cannot be determined from measurements. A mathematical model must be used to determine the three-dimensional wind field.

Only one equation is necessary to determine the concentration. Wind velocity is a vector, and three components u, v, and w must be calculated to define it (i.e., three differential equations must be solved). Other factors are the pressure-gradient force (which depends on the air pressure), the bouyancy forces determined from the temperature field, and the air density. Hence, the number of variables increases to six, and a coupled system of six equations (4 prognostic and 2 diagnostic) must be solved. The method is used in a number of models. The computation is very sophisticated, and only a limited number of structures can be solved, even using a large computer.

If the extensive information available from wind tunnel experiments and field measurements on wind flows over obstacles is utilized, a wind distribution roughly corresponding to reality can be arrived at. To use this in the propagation calculation, this wind field must simply be made to be free of divergence, which can be realized by the iterative solution of a single differential equation. The calculation time is considerably shorter, and very large regions with many obstacles can be used [2.10].

Whichever of the two methods is used, a three-dimensional wind field is obtained for the diffusion calculation.

Turbulence Field. The variance in the Gauss formulas corresponds to the exchange coefficients in the diffusion equation, i.e., they represent the dilution of pollutants by the turbulence of the atmosphere. Measurements of the exchange coefficients for the transport of pollutants in the atmosphere are currently not available. The exchange coefficients cannot be calculated from the variances, as the Gauss formulas used are not solutions of the diffusion equation, and the environmental conditions during the measurements were horizontally homogeneous.

There are theoretical approaches to finding the exchange coefficients for the vertical transport momentum. This method is valid for flat land of uniform roughness, and is established by measurements of the exchange coefficients. It is assumed that the theoretical relationships for any oriented surfaces can be used, and that the exchange coefficients for pollutants can be determined from the exchange coefficients for momentum. The three-dimensional field of exchange coefficients is then obtained from the wind field and the distance to the solid boundary.

Concentration Field. In general, it is assumed that the concentration field at time $t = 0$ is zero. However, this assumption is not obligatory.

Pollution sources have various spatial structures and behavior patterns with time, depending on their nature. An accident can often be regarded as a point source, and is sometimes an instantaneous emission of material. Emissions from vehicles are linear and vary with time. A sewage treatment plant is a two-dimensional source with an almost constant emission. The spatial structure of the source is fitted to the grid. Thus each discrete source becomes a volume source with at least the size of a grid element. The temporal behavior of the source corresponds to the emissions in the time interval Δt.

Pollution Sinks. Pollutants can disappear by chemical reactions, deposition on the ground, and washout by precipitation; either another substance is formed or the original substance is transferred to the soil or groundwater. The quantity disappearing per unit of volume and time interval is always a certain percentage of the concentration.

Boundary Conditions. Solid boundaries (the earth's surface, walls, and roofs) are impermeable, but material can be deposited there. The side boundaries are permeable, and the upper boundary can be permeable or impermeable (in the case of an inversion). These conditions can be represented by appropriate limits to the flow components and exchange coefficients at the boundaries.

Time Interval. A mathematically stable solution is only obtained if the time interval is sufficiently short. The time interval becomes shorter as the grid element becomes smaller and the wind velocity and exchange coefficient increase. It can vary between 0.5 s and 1 min.

Results. The result of the calculation is the variation with time of the structure of the pollutant cloud. An example is shown in Figure 2.2, which shows a section at a height of 2.5 m through a calculated pollutant gas cloud 400 s after the onset of emission.

Figure 2.2. Pollutant cloud from a continuous source, 400 s after start of emission Horizontal cross-section at 2.5 m height, undisturbed flow at 135°

Accuracy. There are a number of sources of error that affect the accuracy of the calculations:

1) Inaccurate emission data
2) Errors in the numerical determination of the wind field
3) Errors in the determination of the exchange coefficients
4) Numerical errors due to the method of computation

As it is not possible to estimate the individual errors, the reliability of the computation must be found by comparing measured with computed figures. Investigations using wind tunnel experiments and field measurements [2.10] show that the accuracy can be given by the factor 2 or 1/2.

2.4.1.4. Stationary Solution of the Diffusion Equation

Continuous sources of constant emission must often be considered. In these cases, only the stationary concentration distribution is of interest. An equation for the determination of this concentration distribution is obtained if the change with time $\partial C/\partial t$ is set at zero in the differential equation. In the stationary case, advective and turbulent flows counterbalance production and dissipation. In this case, too, an analytical solution for normal practical use is not possible, and the differential equation must be converted into a difference equation. The stationary solution is obtained by iteration of the initial field with zero concentration and constant sources and sinks.

Figure 2.3 shows results for the propagation of car exhaust emissions.

2.4.1.5. Particle Models

An alternative method for determining concentration fields is by following the paths of a large number of individual particles in a given wind and turbulence field (Lagrange particle model). The use of particle models avoids numerical errors originating from the computing method.

2.4.2. Use of the Diffusion Calculation

TA-Luft. For plants requiring a license, a diffusion calculation must be carried out in certain circumstances according to the guidelines of the TA-Luft [2.11]. A Gauss formula that gives the concentration distribution for a contin-

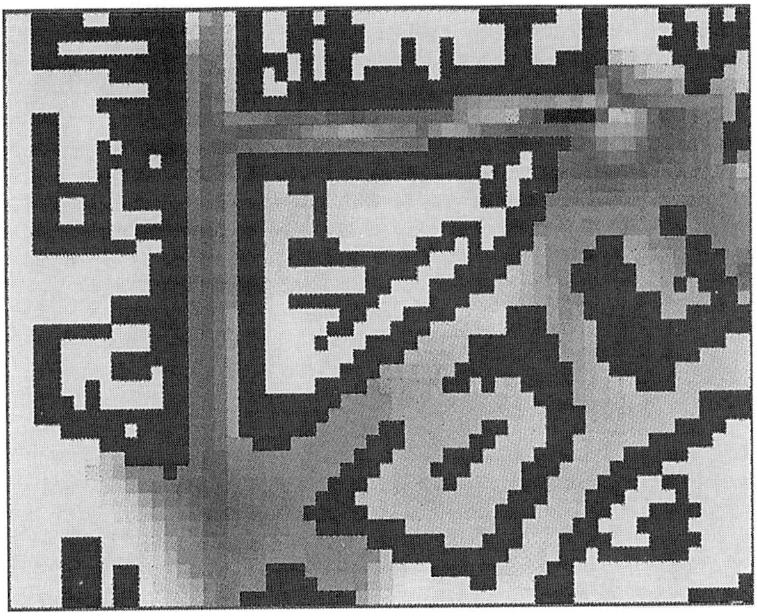

Figure 2.3. Diffusion of car exhaust emissions
Cross-section at height 1.5 m, undisturbed flow at 320°, wind speed at height 10 m 1.7 m/s; traffic only in through streets, 500 vehicles per hour (1000 h^{-1} in upper part of longitudinal street)

uous source is used. The concentration at ground level is determined by the intensity of the source, its height (including the effects of the buoyancy of a hot gas if applicable), wind direction, wind speed, and diffusion class. The computation method was developed before 1985, and no longer corresponds to the state of the art. This is especially true for topographically organized land or if land use changes. Some important influences cannot be taken into account, e.g., inversions, chemical changes, and variations in meteorological parameters with location and time. An EC Regulation may possibly supersede TA-Luft.

Safety analysis is compulsory for many plants, especially in the chemical industry [2.12]. A propagation calculation may be required in the course of such a safety analysis.

Two different questions can arise:

In the case of a flammable gas, a worst-case distance and an average distance from the source at which the gas is still capable of ignition (lower ignition limit) must be stated. With toxic gases, the region in which there is the possibility of a hazard must be established. Here, too, the worst case and the average case are required.

The substances released can be light, density-neutral, or heavy gases, or can be particulate. A gas is assigned to one of these three groups according to its density at the place of release relative to the density of the air. The heavy gases therefore also include cryogenic gases that under normal conditions are not heavier than air and would be grouped with the light gases, and high molecular mass gases at high temperatures are regarded as light.

The spread of *heavy gases* cannot be calculated by the method described above, as in this case propagation is controlled by the force of gravity, and the sudden change in density at the interface with the surrounding air prevents mixing. A heavy gas spreads in a flat layer close to the ground, like a liquid. Obstacles such as fences, walls, houses, and dams can prevent propagation. In streets and on sloping country, channeling takes place, extending propagation.

Analytical or numerical models to determine ignition limits cannot be used, as the effect of obstructions cannot be included. Therefore, VDI guidelines recommend the use of the results of systematic wind tunnel investigations [2.13]. Dimensional analyses enable the results of small-scale experiments to be transferred to the full scale.

For a continuous source of heavy gas, three characteristic quantities are obtained:

$L_{cc} = (V_0^2/g_e)^{1/5}$ (length scale)
$T_{cc} = (V_0/g_e^3)^{1/5}$ (time scale)
$U_{cc} = (V_0/g_e^2)^{1/5}$ (velocity scale)

where V_0 is the source volume flow in m^3/s in the wind tunnel or in reality, and g_e is the acceleration due to gravity, normalized for the relative density difference.

Wind tunnel experiments show that, in the worst case, the dilution depends only on the normalized distance x/L_{cc}. As the lower limit of flammability is defined by a particular dilution, a constant value is obtained for the normalized distance. If this is multiplied by the characteristic length, which can be determined from the source data, the distance of the lower limit of flammability on the full scale can be found, and it can then be seen whether a safety hazard exists. The VDI guidelines list the results of a large number of experiments with various shapes of obstacles.

In the case of *gases of neutral density*, a modified Gauss model is used in the VDI guidelines [2.14]. Each emission is divided up into a number of individual clouds emitted sequentially, so that emissions that change with time can also be considered. The worst case is given by:

1) The test point lying on the downwind side of the source
2) A stable temperature stratification
3) A barrier layer at a height of 20 m, or at the height of the upper edge of the highest building

As an average case, the most common wind velocity and diffusion class is used, with no barrier layers.

The variation with time of the concentration at a given test point is calculated. From this, depending on the substance, the relevant parameters, e.g., average value or dose, can be calculated, or the highest value can be found. Point sources, line sources, flat sources, and volume sources can be treated.

Heavy gases are treated like density-neutral gases if the concentration of the heavy gases is $<1\%$ and the gas can have toxic effects. In this case, the propagation model for heavy gases gives the input values for the diffusion calculation for density-neutral gases.

Light gases have either a relative molecular mass lower than that of the gas mixture air, or the temperature of the gas is significantly higher than that of the surrounding air. Light gases experience a buoyancy. They are accelerated upwards, and the gas only finds its way back to the earth's surface at considerable distances from the source by turbulent diffusion. The maximum concentration is shifted to greater distances. In the guidelines [2.14], this is accommodated by a displacement of the source in an upwards direction. The height of the source is obtained from the heat emission, and is calculated as described in the TA-Luft Guidelines [2.13]. The worst case propagation situation is given by the combination of wind velocity and diffusion class that gives the highest concentration at the test point.

Odor. Factory farming, sewage works, composting plants, waste dumps, and chemical plants can be sources of unpleasant odors. Immissions are directly perceived by people, and can constitute a nuisance. The most important factors determining the degree of the nuisance are frequency, duration, intensity, and nature of the odor, and the opinions of the people affected. Gauss models or solutions of the diffusion equation are not in themselves suitable for calculating odor nuisance, as they give average values for stationary diffusion conditions. The averaging times in the diffusion experiments used to produce the diffusion figures were 30–60 min, and the average figures calculated by the Gauss model are therefore also averaged over this time period. In the human nose, the test volume is a few cubic centimeters, and the averaging time is one to a few seconds. Variations in concentration due to atmospheric turbulence lead to large fluctuations in the intensity of the odor.

A VDI guidline [2.15] enables the frequency of occurrence of the odor of a particular material above a given concentration of the odoriferous substance in the vicinity of single or multiple sources to be determined. This is based on a model that gives average concentrations. It is assumed that these average values are derived from a frequency distribution of concentrations of short averaging time. The frequency distribution can be represented by an exponential function. As well as the probability of detecting an odor, the probability that an odor sensation in excess of the nuisance level will enter the respiratory tract is also obtained.

Accidental Release (see also Plant and Process Safety, **B8**). The main problem with accidental releases of flammable or toxic gases is detecting the accident. Only if an accident is immediately recognized can the appropriate measures be taken to prevent hazards to personnel. The region in which threshold values have been exceeded and the expected duration can be predicted by means of a diffusion calculation. Preconditions for this include:

1) Immediate establishment of the location of the accident and of the temporal behavior of the pollutant source
2) Computational methods significantly more rapid than real time (by a factor of >10) for calculating the future concentration of the pollutant
3) Criteria for the noxiousness of immission concentrations resulting from the accident

The methods currently used to detect accidental discharges are inadequate, especially where the transport of hazardous goods is involved. In the future, it will be possible to use remote monitoring systems which, when fully developed, will

1) Immediately detect an accident
2) Identify the pollutant
3) Give the spatial distribution of the pollutant concentration
4) Give the changes in the pollutant source with time

Diffusion calculations can already be carried out by personal computers appreciably more rapidly than real time, especially as the accuracy of the diffusion calculation need not be better than the accuracy with which the source term can be determined.

Current threshold values for toxic substances are only partially suitable for assessing short-term accidental immissions.

After accidental release a reconstruction of events is often required, or the amount of pollutant material deposited on the ground may have to be determined. An accidental-release calculation can enable the propagation process to be reconstructed, and the emissions can be assessed from the measured immission concentrations. The amount of material deposited and the extent of the region affected can also be determined.

2.4.3. Environmental Compatibility and Simulation of the Environment

If the earth is to remain habitable, all human activities must be investigated in terms of whether their consequences are environmentally tolerable. Also, all man-made objects are exposed to the effects of the environment, so that they become contaminated and are eventually destroyed. Consequently, repair, replacement, and even disposal may be necessary, so causing further pollution of the environment.

Attempts can be made to solve the first problem by means of environmental compatibility studies, experiments, and tests [2.16]. For this, all the effects on the soil, water, and the atmosphere must be taken into account. The effects on the atmosphere can be divided into effects on air quality and effects on the climate.

Environmental simulations aim to provide information on the durability of manufactured objects. Here, too, the effects of air pollution and climatic influences are investigated [2.17].

Environmental Compatibility. The question of the effects of air pollutants is closely linked with the diffusion calculation. For given emissions, immission concentrations and amounts of material deposited or absorbed can be found. From this, the effects on humans, animals, plants, monuments, and natural features can be derived. Synergistic effects can never be excluded, but assessment of these is difficult. Climatic effects (air temperature, air humidity, rain, sunlight, and wind) on the human population are difficult to quantify. Normally, humans, animals, and plants have a wide tolerance range within which they can live without harm. The lack of specific receptors for the individual components of the climate and the variable susceptibility of people (e.g. due to age and state of health) means that different effects can produce the same stress.

Combinations of meteorological parameters can be regarded as separate entities, and subjective descriptive terms can be applied to them, such as harmful, stress-inducing, causing nuisance, neutral, beneficial, or useful.

The effects of corrective measures on, e.g., air temperature and humidity, ventilation by the wind, and sunlight can be predicted by mathematical models.

The combined effects of climate and pollution are sometimes very important, but it is not at present possible to quantify these.

Environmental Simulation. Simulation of the environment aims to predict the functionality and lifetime of products. In environmental simulation experiments, the essential characteristics of the environment must be modelled, and a time-compression technique is necessary. Synergistic effects must be included.

In laboratory experiments, only one parameter is usually investigated, and the time compression can be achieved by increasing the dose. Limits are imposed by the necessity for a linear dose–effect relationship. In field experiments, the magnitude of the causal influences are often not known, and only the effects can be analyzed. Also, forecasts of future events must be made from minimal changes and short exposure times. Numerical environmental simulation is only in the early stages of development. The degree of time compression depends on solely computer capacity. However, there is in particular a lack of essential information about processes occuring at the atmosphere–object interface.

3. Effects of Air Pollutants

3.1. Effects on Humans

Air pollution has always been an undesirable byproduct of human activities, and has presumably had harmful effects on human health ever since the cave dwellers lit their first fire. The problem increased with the growth of human settlements, the practise of crafts and trades, and the extensive development of large industrial concentrations starting in the mid-1800s. Smog episodes, e.g., in 1930 in the Maas Valley (Belgium), in 1948 in Donora, Pennsylvania, and in 1952 in London attracted the attention of a wide section of the public to the problems of health damage by air pollutants and the consequent increase in the death rate of especially vulnerable sections of the community.

Continuing structural change in post-industrial society, including a decrease in classical industrial operations (e.g., the coal and steel industries) and a simultaneous increase in the importance of the service industries, has led to considerable changes in the release of emissions, with consequent reductions in the extent to which the population is exposed to air pollutants.

Land can no longer be divided up into highly polluted industrial regions and so-called clean air regions according to the amount of pollution, mainly because of the increasing impact of traffic and the reduction in the release of harmful substances from stationary emitters (e.g., power stations, domestic heating, and coking plants). Today, areas of a high traffic density in rural regions can sometimes have higher immission concentrations of pollutants than areas of low traffic density in large cities.

The increasing importance of hitherto unknown emitions has led to increasing effects of these other pollutants on the health of the population. As well as the substances typical of motor vehicles (mainly benzene, nitrogen dioxide, and ozone), persistent chlorinated organic compounds play an ever-increasing role in the exposure of the human population to pollution (including secondary accumulation in foodstuffs).

Furthermore, there has been a significant increase in indoor air pollution in recent years. Changes in living conditions (longer periods spent in enclosed spaces), the use of new types of construction materials, and changes in behavior patterns (e.g., inadequate ventilation) have increasingly led to health damage (sick building syndrome).

As a result of this, and also of the general population's continuously increasing awareness of the environment, a tendency towards individual risk assessment has become noticeable. The statistical observation of health hazards is carried out by (epidemiological) methods different from those used to investigate individual risks, so that it has been necessary to develop new techniques of observation and investigation. Examples of this are the provision of out-patient departments specializing in environmental medicine for the clinical examination of patients and the rapid development of environmental medicine as a branch of general medicine.

3.2. General Effects

The effects of air pollutants on humans can range from simple nuisance (e.g., an odor) to serious health damage (e.g., periods of smog in London in 1952 caused over 4000 deaths). A large number of substances cause air pollution, and their concentration in the air varies greatly with time and location, depending on weather conditions (transmission effect) and the various types of emitters involved. Today, their concentrations are in general very low (e.g., compared with those at the workplace), so that an assessment of health damage can only be carried out by considering combination effects in the low dose range. Competing effects from the private actions of individuals (e.g., smoking, medication) and exposure by other routes (e.g., water and food) must also be considered. Especially when long-term (chronic) effects are being investigated, epidemiological studies on representative population groups are carried out, (e.g., investigations of health effects in the "clean air" zone of North Rhine-Wesphalia, Germany).

The respiratory tract, with its large inner surface area, is the main site of damage by air pollutants. However, there can also be effects on the heart and circulatory system, blood formation, the kidneys, the immune and nervous systems, and the skin. Air pollutants relevant to environmental medicine include sulfur dioxide, nitrogen dioxide, ozone, carbon monoxide, dusts, heavy metals, and a wide range of hydrocarbons and chlorinated hydrocarbons. Of special significance are air pollutants that are regarded as potential carcinogens. Important substances of this type include arsenic, benzene, cadmium, diesel motor emissions (DME), polycyclic aromatic hydrocarbons (PAH), and 2,3,7,8-TCDD (dioxin).

3.2.1. Specific Pollutants

It is considered that an effect on humans by an environmental pollutant has been caused if reversible or irreversible changes to normal physiological processes are caused in the human organism. The scheme of continuous biochemical activities is changed by the substance in question, and this can be detected and observed. In the case of substances with a reversible effect, it is possible to specify a dose or exposure concentration below which no effect is produced or can be observed. The presence of the substance at concentrations below this threshold can therefore be regarded as harmless to health.

In the case of substances that have an irreversible, carcinogenic, mutagenic, or allergenic effect, it is assumed that an effect threshold does not exist. Since a risk is present even at low concentrations, special efforts must be made to reduce or limit such pollutants.

The specific characteristic effects of a substance are mainly a function of a combination of dose, exposure time, the nature of the effects, and the fundamental mechanism of these effects. In the context of environmental pollutants, "effect" means any change which is brought about by a substance after acute or chronic exposure. Damage to health means reversible or irreversible undesired changes caused by a substance or factor. Toxicity is the ability of a substance or factor to cause such damage, depending on the applied dose and exposure time. Risk is defined as the probability that a given damage occurs in that part of a population exposed to a harmful agent.

The effects of environmental pollutants are assessed by using the results of animal experiments, in vitro experiments on biological material, human case studies, and epidemiological investigations. The derivation of standard values for the protection of human health is mainly based on case studies of humans.

The toxicological basis for such values is the lowest concentration which leads to detectable harmful effects on humans and/or animals (LOAEL = lowest observed adverse effect level) and/or the highest dose at which no known adverse effects have so far been detected in humans or animals (NOAEL = no observed adverse effect level). In accordance with the WHO definition, the LOAEL is the lowest concentration of a substance at which definite detrimental changes to morphology, functionality, growth, development, and life expectancy of the organism concerned is to be expected.

When standard values are derived from the LOAEL, safety and protection factors must be used, including the following:

1) A safety margin in the estimation of a no-effect level
2) Interindividual differences (toxicokinetic and toxicodynamic differences)
3) Interspecies differences (toxicokinetic and toxicodynamic differences)
4) The simultaneous effect of several substances, or their combined effects
5) Variation in the sensitivity of persons (special risk groups)
6) Severity of the effects
7) Exposure by other routes (e.g., food contamination)

3.2.2. Persons Needing Special Protection

Because of their general sensitivity to the pollutants, children, pregnant women, old and ill persons, and persons with a genetic predisposition must receive special consideration when assessing the effects of (air) pollutants. For example, persons with chronic bronchitis or bronchial hypersensitivity are especially susceptible to substances which attack the respiratory tract or lungs. This is exacerbated in children, because they have higher volume per minute respiration rates per unit body weight than adults, and are more sensitive to infections.

Persons sensitive to (air) pollutants also include allergic persons for whom recent investigations have shown that the frequency and severity of allergic reactions can be intensified by certain nonallergenic foreign substances in the air. Thus, the percentage of children with hay fever is significantly higher in areas of high traffic density than in traffic-free areas, even though the pollutants themselves do not have an allergenic effect [3.1].

Setting guideline values for the protection of especially sensitive persons (e.g., those with congenital immune deficiency or other special idiosyncrasies) is difficult. Thus it is currently not possible to specify a dose of an allergen that will have no effect on an allergic person. For these persons, the standard values for pollutants derived for a normal population cannot guarantee freedom from complaints. Here, further measures should be taken in addition to those recommended in the standard regulations.

3.2.3. Carcinogenicity, Mutagenicity, and Teratogenicity

Carcinogenicity is the ability of a substance or physical factor to cause cancer in humans or animals (→Carcinogenic Agents, treated in the A series). Cancer-risk factors can lead to the uncontrolled and unorganized formation of neoplasms, which include both benign and malignant tumors. Carcinogenic factors can include ionizing and UV radiation, numerous chemical substances, and oncogenic viruses. Also, immune deficiencies, hormonal influences, and inherited predispositions affect the formation of cancer.

The development of malignant tumors takes place in three phases, according to the multistage model:

1) Initiation: in general, the sudden appearance of a persistent cell modification that is transmitted to the descendants of the cell. It gives the affected cell the ability to form a tumor when acted upon by a promotor.
2) Promotion: a lengthy process (weeks to years) in which the action of promotors (cancer-risk factors) cause cells modified by initiation to multiply. Promotors by themselves are not sufficient for the production of malignant cells, but increase the risk of progression to a malignant tumor.
3) Progression: increase in autonomy and malignity (manifestation of a tumor).

Processes that intensify the initiating action of a substance are known as cocarcinogenic processes. Syncarcinogenicity means the cooperation of several carcinogenic substances, leading to the additive or superadditive intensification of the action of the individual substances. Anticarcinogenic processes are those in which cancer formation is inhibited or prevented by the effects of chemical factors.

Humans are exposed to a large number of carcinogenic substances via various environmental media (e.g., water, soil, air, food), and physical and biological factors can also play an important role. The risk of cancer can be obscured by general environmental factors, individual behavior patterns (e.g., smoking, diet, and hobbies), and/or by stress at the workplace. Risk estimates in the United States indicate that the effect of environmental factors in the narrower sense (exposure to pollutants in water, soil, and air) on the overall death rate due to cancer is quite small (ca. 5%).

Causative factors in cancer death rates in the United States [3.2], [3.3], given as percentages of all cancer deaths, are as follows

General anthropogenic pollution (e.g., indoor air, air, drinking water, soil, abandoned polluted areas, pesticides)	2%
Geophysical factors (e.g., exposure to sunlight, pollution of indoor air by radon, general radiation)	3%
Workplace	4%
Industrial products	<1%
Tobacco consumption	30%
Alcohol	3%
Food	35%
Food additives	<1%
Reproductive and sexual behavior	7%
Medication and medical procedures	1%
Infections	10?
Unknown	?

Although there is clearly a risk of cancer even when the above-mentioned effects are excluded any additional burden from the environment should be minimized. An additional cancer risk of 5% would mean an additional 22000 cancer deaths per year in the United States.

The carcinogenicity of a substance is usually expressed as the unit risk (risk coefficient), which indicates the risk of cancer from lifelong exposure, e.g., to 1 µg of a carcinogenic air pollutant per cubic meter of respired air. Generally, the "upper-bound risk value" is given; i.e., there is a probability of 95% that the actual risk is less than the unit risk. The virtually safe dose (VSD) is the daily dose based on a lifetime's exposure (70 years) that leads to an acceptable theoretical additional risk over and above the background risk.

Mutagenicity is the ability of a substance or physical agent to cause permanent structural changes in the genetic information of cells in vivo and in vitro (→Mutagenic Agents, treated in the A series). The changes can involve one or more genes, one or more chromosomes, or sections of chromosomes. Mutations in gametes are transmitted to their descendants. Mutations in body cells remain in the organism in which they were induced.

Teratogenicity is the ability of certain chemical substances, microorganisms, or ionizing radiation to cause nontransmissible malformations, mainly in mammalian embryos (→Teratogenic Agents). According to German chemical legislation on hazard data for substances and preparations, a substance with teratogenic effects has the ability to harm prenatal human life, causing irreversible disturbance to fetal development in the mother's body or irreversible impairment of the post-natal development of the child. In theory, all substances can have a harmful effet in all phases of reproduction. Reproductive toxicology deals with the effects of compounds on the sperm and ova, on conception and implantation, on organ formation, and on fetal maturity. It is not yet known precisely how the formation of defects in embryos occurs. It is thought that teratogenic effects arise through gene damage due to chromosome aberrations, disturbances to mitosis, or inhibited or faulty synthesis of nucleic acids.

3.3. Environmentally Relevant Air Pollutants

3.3.1. Sulfur Dioxide

Sulfur dioxide (SO_2) is an irritating gas which can be retained to 90% in the nasal passages on inhalation owing to its high solubility in water. It can reach the lower respiratory passages if its concentration is very high, if it is adsorbed on dust particles, or if breathing is mainly through the mouth. Dust particles can then catalyze the production of sulfur trioxide (SO_3). SO_2 is absorbed from the respiratory tract into the blood. After biological conversion to sulfate, it is mainly excreted in the urine. Only a few minutes after exposure, the distribution of sulfates by the blood into the tissues and organs and their excretion in the urine can be detected. Radioactively labelled sulfur compounds can be found in the lungs up to one week after exposure.

No systemic toxicity or damage to organs due to SO_2, including mutagenic or carcinogenic effects, have so far been observed. The effects of sulfur dioxide and of the acids formed from it vary in severity, depending on the state of health of the exposed person. Persons already suffering from lung and bronchial diseases must be regarded as high-risk groups. Furthermore, short- and long-term exposure, and the effects of physical stress under certain conditions of exposure require special assessment.

Numerous studies have shown that changes in lung function only occur in healthy subjects at very high SO_2 concentrations, untypical of the environment. However, in the case of respiratory systems already affected by chronically obstructive diseases and asthma, sensitivity to SO_2 can be many times higher. SO_2 concentrations of $>10000\ \mu g/m^3$ can lead to restriction of ciliary motion and slowing or cessation of mucociliary transport, with consequent impairment of lung clearance. Acute effects are bronchoconstriction, bronchitis, and tracheitis. Physical exertion, with increased respiration rate and volume, can significantly lower the threshold concentration of 10000 µg SO_2/m^3 [3.4], [3.5].

Young asthma sufferers exposed to SO_2 by inhalation through a mouthpiece, with simultaneous physical exertion, show noticeable effects on lung function at the low concentration of 100 µg/m^3. However, with asthmatic adults, a concentration of at least 350 µg SO_2/m^3 is necessary. Simultaneous exposure to 300 µg SO_2/m^3, 100 µg NO_2/m^3, and 50 µg ozone/m^3, leads to an increase in bronchial sensitivity in the provocation test with histamine, carbachol, acetylcholine, etc. [3.6].

In chronic exposure to SO_2, the main effect is an increased incidence of bronchitis and other symptoms of infection. Furthermore, investigations of the effects of exposure to SO_2 at typical environmental concentrations using spirometric tests showed that lung function was affected. An investigation of 200 schoolchildren in an industrial area showed that exposure to increasing SO_2 levels (up to 281 μg SO_2/m^3) and particle concentrations led to impairment of the enforced vital capacity and the 1 s capacity. Also, ca. 25 % of the children studied showed a ca. four-fold sensitivity to pollution [3.7].

Reduction of the enforced lung function was observed in adult patients with chronic obstructive respiratory ailments on a 5-d exposure to 190 μg SO_2/m^3 (peak figure 720 μg SO_2/m^3) and 44 μg dust particles/m^3. Also there was a significant increase in the observable effects at low air temperatures (< 0 °C). Impairment of lung function (oscillatory resistance) in normal population samples was observed on increasing the 24-h mean value of 60 μg SO_2/m^3 to 256 μg SO_2/m^3 with simultaneous exposure to airborne particles in concentrations of 69 μg/m^3 and 179 μg/m^3 respectively (24-h mean values) [3.8].

In Japan, using normal population samples, an increase in the incidence of bronchitis in men from 5.5 % to 6.7 % was observed, and in women from 1.1 % to 3.9 % when the sulfur dioxide level increased from 60 μg SO_2/m^3 to 140 μg SO_2/m^3. The corresponding concentrations of airborne particles were 100 μg/m^3 and 200 μg/m^3, respectively. The smoking behavior of the sample groups investigated were standardized. Investigations on the normal population in Germany showed that when the sulfur dioxide level increased by ca. 50 μg/m^3, a 10 % increase in the incidence of bronchitis occurred [3.9].

In children, an increase in the long-term chronic exposure to 181 – 275 μg SO_2/m^3 and 230 – 310 μg airborne dust/m^3 led to increased incidence of diseases of the lower respiratory tract compared with children from a control region with lower air pollution. Also, in other studies of children from variously polluted regions, a correlation between the frequency of diseases of the lower respiratory tract and the degree of pollution was found [3.10].

A series of epidemiological investigations into the effect of SO_2 on health were carried out on chronically exposed population groups in polluted areas during smog episodes in which the concentrations of pollutants (mainly SO_2 and dust) were extremely high for brief periods. Here, it was shown that exposure over several days led to an increase in the death rate of older people with chronic obstructive lung diseases or heart diseases when the mean 24-h concentration of SO_2 reached ca. 700 μg/m^3 and the airborne dust concentrations simultaneously rose to 500 μg/m^3 [3.10].

During the smog periods in North Rhine-Wesphalia in January 1985, a significant increase (8 %) in the death rate occurred in the polluted area. Hospital admissions for respiratory and cardiovascular diseases increased by 17 %, and a marked correlation with age was noticeable, the increase being greater in older people. The highest 24-h mean values in the polluted area were 830 μg SO_2/m^3, and, in a control area, 320 μg SO_2/m^3. The corresponding airborne dust concentrations were 600 μg/m^3 and 190 μg/m^3 [3.11].

When setting threshold values for SO_2 in the air, it should be noted that in many cases SO_2 immission values were used as indicators of air quality when classifying effects from epidemiological investigations, whereas other pollutants that were also present as immissions in high concentrations were disregarded. However, especially in humans, a combination effect with particulate air pollutants present at the same time must be assumed.

In Germany, the TA-Luft Regulations for sulfur dioxide specify an arithmetic annual mean value of 140 μg/m^3 (IW 1) and a 98 % value of 400 μg/m^3 "for protection against health hazards". The maximum immission concentrations (MIK values) specified by the VDI Air Pollution Commission are 1000 μg/m^3 for the 30-min value and 300 μg/m^3 for the 24-h mean value. The World Health Organization (WHO) has recommended the following guidelines: 500 μg/m^3 as the 10-min maximum and 250 μg/m^3 as the 1-h maximum. When SO_2 is present with airborne dust (black smoke), the specified 24-h mean value is 125 μg/m^3 for both SO_2 and airborne dust, and the 1-a mean value is 50 μg/m^3 [3.12] – [3.14].

3.3.2. Nitrogen Oxides

The nitrogen oxides NO and NO_2 are important air pollutants. Health hazards are almost exclusively associated with NO_2 and its reaction products, while NO acts as a reservoir from which nitrogen dioxide is produced.

Nitrogen dioxide is a strongly irritant gas. On inhalation, the mucous membranes of the respiratory tract are attacked, the gas reacting immediately with the moisture present there and on alveolar surfaces. It probably does not reach the lung capillaries, although lung function is affected. On inhalation, 80–90% of NO_2 is absorbed in the respiratory tract. Absorption possibly occurs by formation of nitrous and nitric acids or their salts, as indicated by the detection of nitrite and nitrate in the urine and blood after NO_2 inhalation.

The odor threshold lies in the range 200–440 μg/m^3. Under controlled conditions, effects on the lung function of healthy experimental subjects are detectable above a concentration of 3800 μg NO_2/m^3. In the case of asthma sufferers under physical exertion, effects on lung function occur above 560 μg/m^3. The combined effects of NO_2, SO_2, and ozone at low concentrations increase the sensitivity of the respiratory tract to acetylcholine [3.4].

In the case of bronchitic patients, a fairly rapid increase in the respiratory tract resistance takes place in the concentration range 3000–3760 μg NO_2/m^3. Moreover, at higher concentrations, a decrease in the alveolar–arterial oxygen pressure difference, increased sensitivity to acetylcholine, and an increase in the respiratory tract resistance caused by carbachol is observed. The results of epidemiological studies on the relationship between NO_2 exposure and health effects should not be used for determining threshold limit values for the protection of personnel because other pollutants are also usually present [3.15].

The effects observed in animal experiments are of only limited applicability to humans, as combination effects with other pollutants are generally not considered. Short-term and single exposures of rats in the low concentration range (380 μg NO_2/m^3) led to inhibition of prostaglandin metabolism. At concentrations of ten times this figure, metabolic impairment was of the same order of magnitude, although the recovery time was significantly extended (< 70 h). The observed decrease in prostaglandin breakdown can lead to changes in the respiratory tract and disturbance of the blood flow through the lungs [3.16].

Long-term exposure of experimental animals at 190–1900 μg NO_2/m^3 led to thickening of the alveolar capillary membrane, loss of the ciliary epithelium, and collagen formation in untypical areas of the lungs. In rats, the inhalation of 900 μg NO_2/m^3 for 4 h led to the degranulation of fat cells, followed by excretion of histamine [3.17].

In a series of studies, increased sensitivity towards viral and bacterial infections after chronic exposure to NO_2 has been demonstrated. Prolonged inhalaton of 940 μg NO_2/m^3 by mice leads to increased sensitivity towards *Klebsiella pneumoniae*, the magnitude of this effect appearing to depend more on the concentration than on the total dose. On the basis of the results of animal experiments, it can be assumed that humans will also experience weakening of the immune system on exposure to NO_2 or combinations thereof with other air pollutants.

In Germany, the TA-Luft Regulations specify the following levels of nitrogen dioxide "for protection against health hazards": 80 μg/m^3 as the annual arithmetic mean (IW 1) and 200 μg/m^3 (IW 2) as the 98% value. The maximum immission concentrations (MIK values) laid down by the VDI commission on air pollution are 200 μg/m^3 as the 30-min mean value and 100 μg/m^3 as the 24-h mean value. The following guidelines are recommended by the World Health Organization (WHO): 400 μg/m^3 as the 1-h mean value and 150 μg/m^3 as the 24-h mean value [3.12], [3.14], [3.18] (see also Chap. 4).

3.3.3. Ozone

Ozone (O_3) is a highly reactive gas that can change or decompose enzymes, coenzymes, and proteins by oxidizing thiol groups, and can oxidize fatty acids to toxic fatty acid peroxides. Protein and lipid membranes are the main sites of attack by ozone.

Ozone is mainly absorbed via the respiratory system, where rapid reaction with the tissues occurs. In many animals, acute exposure to high concentrations of the gas (440–1000 μg/m^3) causes rapid, shallow breathing, increased resistance of the respiratory system, reduced vital capacity, and bronchial hyperreactivity. Younger animals show higher sensitivity to these concentrations of ozone. An 8 °C increase in the air temperature doubles the toxic effects. Recent investigations have shown that concentration is more significant than exposure time.

Numerous animal experiments have demonstrated that chronic exposure to concentrations between 200 and 1000 μg O_3/m^3 has a number of harmful effects, including inflammation of the bronchiole region leading to lung emphysema with fibrosis; reduction of the inhalation and exhalation capacity per second; and decreased lung compliance (elasticity). Inhalation capacity per second is the most sensitive indicator of lung function, and a significant and detectable reduction is apparent at 240 μg O_3/m^3. When rats were exposed for 93 d to as little as 120 μg O_3/m^3, an increase in the ketosteroids in the urine and a decrease in the cholinesterase activity occurred. At 200 μg O_3/m^3, biochemical changes in the lung tissue take place, and above 1000 μg/m^3, a decrease in the surfactant concentration can be detected [3.4], [3.19], [3.20].

Both short- and long-term exposure to ozone cause increased mortality rates in animal experiments (60–160 μg/m^3). This has been attributed to decreased antibacterial activity of the lung tissue [3.19], [3.20].

The odor threshold of ozone for humans is 20–40 μg/m^3. Lower concentrations lead to odor acclimatization. Subjective effects such as lachrymation (caused by other substances present with the ozone), irritation of the respiratory system, coughing, headache, and respiratory discomfort are observed above a concentration of ca. 200 μg/m^3. In school children and adults, changes in the lung function parameters (e.g., reduction of enforced exhaled air volume and increase in the resistance of the respiratory system) appear at concentrations between 160 and 300 μg/m^3. In clinical exposure experiments with volunteer subjects, detectable changes to lung function parameters were detected after 6.6-h exposure to 160 μg O_3/m^3 or 1–3 h exposure to 240 μg/m^3 combined with physical exertion [3.19], [3.20].

Deterioration in physical fitness was observed at 240–740 μg O_3/m^3. With maximum physical exertion (competitive sport), a decrease in performance was detected at considerably lower ozone concentrations. Inflammatory reactions of the tissues of the respiratory tract appear above 160 μg/m^3 after 6.6-h exposure with intermittent physical activity. Also, an increase in the incidence of asthma attacks was noted at 240–300 μg/m^3 [3.19], [3.20].

At ozone concentrations of >1000 μg/m^3, marked changes to lung function occur, with a significant reduction in the diffusion capacity and one-second capacity. The effects of ozone on lung function show considerable individual differences. Approximately 20–30 % of the population show pronounced reactions. Also, there is an increase in the number of cells of the immune system, the fragility of the erythrocytes increases, and visual impairment occurs (e.g., reduction in visual acuity and changes in the field of vision).

In general, children show similar changes in lung function to those exhibited by adults after exposure to ozone. Older persons generally react less often than young adults, but their symptoms are often more severe. In epidemiological investigations, a correlation between typical environmental concentrations of ozone and temporary changes to lung function are reported. When respiratory diseases are already present, a deterioration in health can occur. Investigations on children show a connection between levels of the daily maxima of the ozone immission concentrations and impairment of lung function. Such changes were observed above 120 μg/m^3. This exposure concentration is significantly below those used in experimental investigations. The difference is probably due to the simultaneous presence of other pollutants. Symptoms such as irritation of the eyes, nose, and throat; chest pain, coughing; tiredness; and nausea were observed in children subjected to ozone concentrations of >200 μg/m^3. In asthmatic patients, as the severity of the disease increased, a correlation between the frequency of attacks and the ozone concentration was observed. People with obstructive lung diseases and smokers generally do not react to ozone more sensitively than healthy subjects. However, patients with limited lung function show more severe reactions compared to persons not so affected [3.20].

Carcinogenic effects of ozone have not so far been proved. However, ozone can have effects resembling those caused by radioactivity, and even low ozone concentrations (200 μg/m^3) in cell cultures can cause chromosome rupture resembling the effects of X rays. Therefore, although there is no indication of increased cancer rates in occupational groups exposed to ozone, a carcinogenic effect of ozone cannot definitely be excluded.

High-risk population groups include all persons frequently exposed to prolonged strenuous physical activity in the open air on hot summer days. These are mainly healthy children and adults who take part in games or sport, and persons involved in strenuous work in the open air.

In Germany, the VDI Air Pollution Commission has set a 30-min mean value of maximum immission concentrations (MIK value) of 120 μg/m^3. Exposures of up to 400 μg/m^3 for a period of up to 30 min should not occur more frequently than once per week. The environmental ministers' conference has specified a maximum figure of 180 μg/m³. If this is exceeded, a recommendation should be made to the public along these lines: "Persons who have found from experience that they are especially sensitive to ozone (or other pollutants) should not take prolonged strenuous exercise in the open during the afternoon." The WHO recommends a 1-h figure for ozone of 150–200 μg/m³. To prevent chronic damage to health, an 8-h figure of 100–200 μg/m³ is recommended.

3.3.4. Carbon Monoxide

Carbon monoxide (CO) is a colorless, odorless, and tasteless gas, with low solubility in water. It is formed by the incomplete combustion of carbon or carbon-containing compounds.

Carbon monoxide affects humans by combining with hemoglobin (Hb) to form carboxyhemoglobin (COHb). Carbon monoxide blocks the divalent iron center of Hb by displacing oxygen without change of oxidation state. Carbon monoxide forms a bond with hemoglobin that is ca. 240 times as strong as that formed by oxygen. Therefore, even at comparatively very low concentrations it blocks the oxygen-binding site, and prevents oxygen transport by the blood. The bond between CO and hemoglobin is reversible, and when an intact organism breathes CO-free air, all the CO is eventually released and replaced by O_2, and Hb functionality is fully restored. If pure oxygen is breathed, the regeneration process is much faster. As CO is formed endogenously in normal metabolism, unexposed persons have 0.8% COHb in their blood. On average, smokers have ca. 7% COHb, and this rises to 15% COHb for heavy smokers [3.4].

All the symptoms of carbon monoxide intoxication are due to oxygen deficiency in the tissues and the accumulation of carbon dioxide during normal metabolism. Unlike reduced hemoglobin, COHb cannot carry away the carbon dioxide. The most important effects are on the central nervous system and the cardiovascular system. The symptoms of carbon monoxide intoxication depend primarily on the COHb concentration in the blood, which is in turn affected by a number of factors including the CO concentration in the respired air, the respiration-time-volume, the duration of the exposure, the oxygen demand of the tissues, and the Hb concentration.

The symptoms of carbon monoxide intoxication with increasing COHb content of the blood [3.22] are as follows:

5–10%	Slight, measurable visual impairment (threshold of fusion frequency)
10–20%	Slight headache, lassitude, unwellness, shortness of breath with exertion, palpitations
20–30%	Vertigo, partial loss of consciousness, limpness and paralysis of limbs
30–40%	Pink skin color, loss of consciousness, shallow breathing, circulatory collapse
40–60%	Deep unconsciousness, paralysis, Cheyne-Stokes respiration, decrease in body temperature
60–70%	Death within 10 min to 1 h
>70%	Death within a few minutes

Within limits, the healthy organism can compensate for oxygen deficiency caused by CO, in the long term by an increase in hemoglobin, and in the short term by increasing the flow of blood through the organs at risk. Neurophysiological investigations and experimental behavior studies show a significant decrease in the ability to concentrate in healthy persons with 2.5% COHb. This effect can be important in regions of high exposure concentrations (e.g., traffic jams in road tunnels).

In persons with cardiovascular diseases (heart muscle diseases) and sclerotic changes in the blood vessels of the heart and brain, harmful effects appear at relatively low concentrations (2.8–3%). For instance, in 1000 metallurgical workers exposed to CO for at least 4.2 years, a linear correlation between CO dose and angina pectoris symptoms was established. These industrial medical findings were supported by results of experimental investigations. For example, angina pectoris patients, after a 90-min journey in an open vehicle through Los Angeles, showed a significant reduction in the time required for the appearance of typical chest pain compared with the time required after a journey in which synthetic air was inhaled. The COHb level in the patients after the journey in the open vehicle was 5.08%, and 0.65% after breathing synthetic air [3.22], [3.23].

According to calculations, a 24-h exposure to air containing 10 mg/m³ CO produces a maximum of 1.5% COHb, a 30-min exposure to 30 mg/m³ CO produces 2.3% COHb, and a 30-min exposure to 50 mg/m³ CO gives a maximum of 2.8% COHb [3.22], [3.24].

In Germany, the TA-Luft Regulations for carbon monoxide "for protection against health hazards" specify an arithmetical annual mean of

10 mg/m^3 (IW 1), and 98 % figure of 30 mg/m^3 (IW 2). The maximum immission concentrations (MIK values) specified by the VDI commission for air pollution are 50 mg/m^3 for the 30-min figure and 10 mg/m^3 for both the 24-h and annual mean figures. The WHO recommends the following figures: 15 min: 100 mg/m^3, 30 min: 60 mg/m^3, 1 h: 30 mg/m^3, and 8 h: 10 mg/m^3 [3.12], [3.14], [3.25].

3.3.5. Dusts

Dusts are defined as dispersions of solid materials in gases. The particles can be of any shape, structure, and density, and can be up to ca. 100 μm in size. They are formed either by entrainment or mechanical processes. Like smoke and fog, dusts are types of aerosol.

The health hazard due to dusts depends on chemical composition, concentration, exposure time, and especially particle size. This last parameter distinguishes dusts from gases and vapors. Dusts are principally absorbed by respiration. The transport and deposition of a dust in the respiratory tract are mainly determined by the behavior of particles in flowing gases. The characteristic parameter is the aerodynamic diameter d_{ae} of a particle. The aerodynamic diameter of a particle of a given shape and density is given by the diameter of a sphere of unit density (1 g/cm^3) which has the same sedimentation rate as that of the particle in still air or air moving with laminar flow. Fine particles of size <10 μm are regarded as airborne dust [3.26].

Dust fractions are classified according to their particle size as follows: coarse dust: diameter >50 μm, medium-size dust: 10–50 μm, fine dust: 0.5–10 μm, and very fine dust: <0.5 μm. So-called respirable dust (airborne dust), which consists of particles of size <10 μm, is of relevance to health.

Particulate pollution (dust) is a mixture of many chemical substances that can have a wide range of effects both separately and in combination. They can be of natural or anthropogenic origin. The main constituents of atmospheric dust are sulfates, carbonates, nitrates, halides, silicates, salts and oxides of calcium, iron, potassium, sodium, aluminum, magnesium, and organic substances. The components with toxic potential include heavy metals, arsenic, polycyclic aromatic hydrocarbons (PAH), and chlorinated hydrocarbons.

Depending on their chemical and physical properties and the duration of their effect, dusts can cause morphological damage and changes to the function of the lungs or to their susceptibility to infection. The particular part of the respiratory tract affected by dusts depends mainly on their particle size. Particles of >10 μm diameter only reach the nose and throat area, and do not enter the middle to lower regions. They can also be carried into the digestive tract by the lung clearance mechanism. The tracheobronchial dust is the dust fraction deposited in the area of the mucociliary clearance apparatus of the tracheobronchial system, and this too can be carried into the digestive tract. Alveolar dust is the dust fraction deposited in the alveolae and in the region of the bronchioles without mucociliary clearance. Via the tracheobronchial system, it can appear both in the digestive tract and in the interstitial tissue of the lungs.

The effectiveness of dust transport mainly depends on the presence of an intact bronchial mucous membrane. The transport rate can be adversely affected by medication or toxic substances. Effects of gaseous substances such as SO_2 and NO_2 on ciliary activity and mucociliary clearance have also been observed.

Dusts can contribute to the acute effects of smog and can also cause chronic effects. Acute dust pollution always gives rise to health effects that can be unambiguously detected in experimental and epidemiological investigations. The most important first acute effects of dust are impairment of the clearance mechanism of the respiratory system and irritation of the bronchial mucous membrane. The lowest dust concentrations, derived from periods of smog in London and New York, at which an effect on death rate due to disturbances of respiratory function can be detected are in the region of 0.15 mg/m^3. The first changes to the lung function of children were detected at an airborne dust concentration of 0.18 mg/m^3 in the middle of the day [3.4], [3.27].

Longer exposure leads to cytotoxic effects and tissue damage. The first effects are observed at an annual average concentration of airborne dust of >0.09–0.15 mg/m^3. More recent industrial medical investigations have shown that dust at the workplace can promote chronic bronchitis, although cigarette smoke has a significantly greater influence. At today's low dust concentrations in the outside atmosphere and also indoors, it may prove very difficult to detect long-term effects on the incidence of chronic bronchitis by epidemiological methods against the background of active and passive smoking.

In Germany, the TA-Luft Regulations for airborne dust (irrespective of its constituents) specify an arithmetic annual average of 0.15 mg/m^3 (IW 1) and a 98% figure of 0.30 mg/m^3 (IW 2) "for protection against health hazards". For "protection against significant detriment and nuisance", the following figures for dust deposits (nonhazardous dusts) are prescribed: 0.35 $g m^{-2} d^{-1}$ (IW 1) and 0.65 $g m^{-2} d^{-1}$ (IW 2). The VDI commission for air pollution recommends the following maximum immission concentrations (MIK values): 1-a average: 0.100 mg/m^3, 24-h average: 0.250 mg/m^3, 3-h average: 0.500 mg/m^3. The following standard figures for simultaneous exposure to sulfur dioxide and airborne dust are laid down by the WHO: total suspended particulates (24-h figure): 0.120 mg/m^3, thoracic particles (deposition characteristics 50% <10 μm, 24-h figure): 0.070 mg/m^3 [3.12], [3.14], [3.26].

3.3.6. Lead

Lead is one of the most extensively investigated pollutants both in the environment and the workplace. There have been many investigations into the occurrence of lead in the various environmental media (air, water, soil, and food) and into the toxicity of lead to humans and animals. These are documented in the literature in a large number of comprehensive reports.

In the general population of Germany, the most important source of lead pollution is food. The daily intake has been estimated by the registry and advisory center for environmental chemicals of the German Health Office (ZEBS). The figure for men is 147.2 μg/d (1.03 mg/week), and for women 104.22 μg/d (0.73 mg/week). Of this, ca. 5–10% is absorbed in the digestive tract by adults, while small children have an higher absorption rate amounting to ca. 50% [3.28].

Of the total lead absorbed by adults (in air, food, and drinking water), the fraction absorbed by inhalation is ca. 17–25%. Because of the higher lead absorption rate in the digestive systems, this figure is much lower for children (2–3%).

Lead absorption by the human organism depends on the solubility of the lead compounds and the composition of the food. Deficiencies of calcium and vitamin D increase the absorption. The amount of inhaled lead-containing dust taken up by the organism is determined by the deposition, elimination, and absorption of the inhaled dust particles. It is assumed that 90% of inhaled lead is absorbed.

The absorbed lead first enters the blood, and is distributed among the various organs and tissues. Approximately 90% of the lead in the blood is bound to components of the erythrocytes. Lead inhibits various enzymes and thus affects severals stages of hemoglobin synthesis. Several precursors of hemoglobin appear in the blood in increasing amounts, and are excreted in the urine (e.g., δ-aminolevulinic acid and coproporphinogen-III). Effects of lead on the erythrocyte membrane and reduction of the lifetimes of erythrocytes have also been described.

Lead is mainly accumulated in the bones. It can pass through the placenta and into the fetal blood circulatory system. Lead can also pass through the blood–brain barrier to a small extent, especially in children, in whom this barrier is not fully formed. The excretion of lead takes place via the feces and kidneys, a smaller fraction being eliminated via the hair and nails.

Acute intoxication is rare, as most lead compounds are sparingly soluble. Symptoms such as vomiting, colic, constipation, and acute kidney failure occur. Sometimes, lead encephalopathy is diagnosed, which leads to vomiting, apathy, vertigo, coordination disturbances, hyperactivity, and other effects.

The best known effects of chronic lead intoxication include weakness, loss of appetite, sleeplessness, nervousness, constipation, headache, muscular and joint pains, shivering, encephalopathy, and colic. Objective signs include pallor, weight loss, anemia, increased excretion of δ-aminolevulinic acid in the urine, increased blood lead level, decreased level of δ-aminolevulinic acid dehydrase in the blood, and changes in peripheral nerve conduction velocity.

Many investigations into damage to the central and peripheral nervous system caused by exposure to lead in the course of employment or from the environment have been carried out. There is a clear correlation between increased exposure to lead in the prenatal period and/or in early childhood and neuropsychological effects such as hyperactivity, fine motor disturbances, and reduction in intelligence and learning ability.

In the typical exposure situation of the general population, the following observed effects can be correlated in quantitative terms with the lead concentration in whole blood (internal lead burden). At lead concentrations of >10 μg/100 mL in the blood of children and >15 μg/100 mL in adults, the first effects that appear are changes in the biosynthesis of hemoglobin by inhibition of certain enzymes. At blood concentrations of >30 μg/100 mL, there is a statistically significant increase in arterial blood pressure. In children with blood lead concentrations of >20 μg/100 mL, statistically significant impairment of individual functions of the CNS can be detected in neuropsychological tests. Impairment of fetal and infant development becomes statistically significant at concentrations of >10 to 15 μg/100 mL [3.29], [3.30].

Lead concentrations in the blood of the general population in Germany lie currently between 6 and 7 μg/100 mL. Lead concentrations of >15 μg/100 mL are observed in only <1% of the population. The WHO requires that 98% of the normal population of adults and children in Europe should have a lead concentration in the blood of <20 μg/100 mL [3.31].

There is no evidence of carcinogenic or cocarcinogenic effects of lead on humans, although the appearance of cancer in animals is observed at high doses. Mutagenic effects (chromosome aberrations) can occur with lead intoxication. Apart from its toxic effect on the fetus, there is no evidence that lead has teratogenic properties. In Germany, the TA-Luft Regulations for lead and inorganic lead compounds as components of airborne dusts "for protection against health hazards" specify an arithmetic annual average of 2.0 μg/m^3 (IW 1). For "protection against significant detriment and nuisance", the figure 0.25 mg m^{-2} d^{-1} is specified as the IW 1 value for lead and inorganic lead compounds. The VDI commission has prescribed the following maximum immission concentrations (MIK values): 1-a average: 1.5 μg/m^3, 24-h average: 3.0 μg/m^3. The WHO recommends a standard range of 0.5–1.0 μg/m^3 [3.12], [3.14], [3.32].

3.3.7. Cadmium

The heavy metal cadmium is a nonessential trace element. It is absorbed orally and by inhalation, skin absorption being practically insignificant. Absorption of cadmium by the German population is mainly through food. According to information provided by the registry and evaluation center for environmental chemicals in the German Health Office (ZEBS) for the year 1984, the average cadmium intake via food by men was 34600 ng/d (0.242 mg/week) and by women 26900 ng/d (0.189 mg/week). The amount taken in via respired air is small, and is estimated at ca. 40–50 ng/day. Smoking 20 cigarettes per day results in a cadmium absorption of ca. 360–2200 μg/a. In food cadmium is mainly chemically bound to proteins [3.28].

The effects of cadmium depend on the quantity of cadmium absorbed into the blood. The absorption rate by the stomach and intestines is usually ca. 5%, but this can be higher in cases of iron-deficiency syndromes. In contrast, cadmium entering the alveolar region of the lungs by inhalation is almost completely absorbed.

Cadmium has a marked tendency to accumulate in the human organism, mainly in the kidneys and liver. Thus, cadmium concentrations in the liver and kidneys increase with increasing age, and also in the blood and the urine. The most important factor is cigarette smoking. Cadmium concentrations in the blood of smokers are 3–5 times as high as those for nonsmokers, and cadmium concentrations in the renal cortex of smokers are ca. twice those for nonsmokers.

The health effects of increased chronic cadmium exposure in the workplace and the environment consist mainly of disturbances to kidney function, increase in arterial blood pressure, and the production of bronchial carcinomas.

The toxic effect of cadmium on the kidneys is mainly a result of the increase in the cadmium concentration in the kidney tissue. An early symptom is the increased excretion of low molecular mass proteins in the urine due to a disturbance to reabsorption in the proximal tubular section (tubular proteinuria). The diagnostic criterion is the increased excretion of β_2-microglobulin and retinol-binding protein. As well as tubular proteinuria, cadmium-induced glomerular proteinurias also occur, characterized by the excretion of high-molecular proteins. This symptom is very probably due to the deposition of immune complexes in the blood capillaries of the glomeruli [3.4].

If cadmium levels in the renal cortex exceed 200 μg/g, the first indications of a tubular proteinuria can appear in susceptible persons. At concentrations of >250 μg Cd/g tissue, the appearance of a tubular proteinuria can be expected in 50% of exposed persons.

Animal experiments have shown that the cardiovascular system is also sensitive to cadmium. Increase in blood pressure and myopathic heart damage can occur without any sign of damage to the kidneys. There are no corresponding results for humans at present. Lung and liver damage, anemia, and changes to the mineral balance of the bones (itai-itai disease) are also known.

There is no evidence of teratogenic effects of cadmium on humans, probably due to the low permeability of the placenta to cadmium. Information on the mutagenic effects of industrial or environmental exposure is contradictory, although genotoxic changes to human and animal cells can be caused by cadmium salts.

It is today considered certain that a number of cadmium compounds show carcinogenic potential after inhalation. The German MAK Commission classifies cadmium, cadmium chloride, cadmium oxide, cadmium sulfate, cadmium sulfide, and other compounds that could enter the human organism (in the form of respirable dusts/aerosols) as materials with a definite carcinogenic effect (Group II A 2). Orally ingested cadmium, however, has no carcinogenic effect, according to current knowledge [3.33].

The German TA-Luft Regulations for cadmium and inorganic cadmium compounds as constituents of airborne dust specify an arithmetic annual average of 0.04 µg Cd/m^3 (IW 1) "for protection against health hazards". For "protection against serious detriment and nuisance", the figure for cadmium and inorganic cadmium compounds is $5\ \mu g\,m^{-2}d^{-1}$ (IW 1). The WHO recommends that in urban and industrial areas, cadmium concentrations of 0.01 to 0.02 $\mu g/m^3$ can be tolerated. However, any increase in existing concentrations should be avoided so as to minimize the entry of cadmium into the food chain via deposition of dust. The Länder committee for immission protection (LAI) gives a "unit risk" for cadmium of 1.2×10^{-2} and an orientation figure of 1.7 ng/m^3. This figure represents the upper limit of background concentration in rural areas in Germany, and is regarded as a standard for future immission reductions in urban areas [3.12], [3.14], [3.34].

3.3.8. Arsenic

Arsenic is a metalloid which is ubiquitous in the earth's crust in various forms. Toxicologically, inorganic arsenic compounds, organic arsenic compounds, and arsenic hydride (arsine) must be considered separately.

Arsenic is mainly ingested orally by humans in food and drinking water, but is also inhaled. The extent of absorption depends on the type of arsenic compound. Arsenic can also be absorbed through the skin. Organic arsenic compounds are ingested mainly during the consumption of fish, mussels, and crustaceans, the main compounds concerned being arsenobetaine, arsenocholine, and trimethylarsonium lactate and its derivatives.

Arsenic is rapidly transported within the body by the blood. It becomes distributed in human tissue in varying concentrations, being more concentrated in the skin, hair and nails, the lungs, the bones, and the brain. Inorganic arsenic compounds in the human body are largely methylated to dimethylarsinic acid, and to a lesser extent to methylarsonic acid. These methylated compounds are excreted more rapidly in the urine, and are less toxic. The excretion of arsenic is mainly performed by the kidneys, ca. 75 % of the arsenic in the urine being present as dimethylarsinic acid and a lower percentage as monomethylarsonic acid and inorganic arsenic.

The toxic effect of arsenic is mainly due to its ability to form covalent bonds with thiol groups. Various enzymes, such as monooxidase, urease, glucose oxidase, choline oxidase, can be inhibited by arsenic. Acute effects, chiefly on the gastrointestinal tract, are only likely with high doses. A dose of 70–180 mg arsenic trioxide is fatal for humans. The main symptoms of acute intoxication are difficulties in swallowing, abdominal pain, nausea, vomiting, chest pains, and watery diarrhea. Inorganic arsenic compounds, especially the trivalent oxides and chlorides, cause irritation of lungs and skin.

Chronic arsenic intoxication leads to disturbances to the peripheral and central nervous system, and lesions to skin and mucous membranes. Skin lesions include irregular pigmentation of the skin (melanism and leucoderma), hyperkeratosis, the appearance of diagonally striped white fingernails (Mees line), and warts. Damage to the mucous membranes mainly takes the form of conjunctivitis, nasopharyngeal catarrh, perforations of the nasal septum, bronchitis, diarrhea and constipation, and stomach cramps.

Effects on the PNS take the form of paraesthesia, motor disfunction, and a decrease in nerve conduction velocity. Lassitude and apathy, and more rarely encephalopathy indicate damage to the CNS. The optic nerves and the organ of balance can also be affected. The liver can become inflamed, and morphological changes such as cirrhosis and carcinoma can occur. Other effects related to exposure to arsenic have been observed, including damage to the heart (ECG changes), disturbance of red corpuscle production, (anemia and agranulocytosis), allergic contact dermatitis, and peripheral vascular disorders, leading to gangrene (black foot disease).

Indications of long-term occupational or environmental exposure include hematological, dermatological, and neurological symptoms. The critical threshold dose for a daily intake of arsenic is stated by various authors to be 400 μg/d. Inhalation of arsenic can lead to increased incidence of cancer of lung and skin, liver carcinoma, and tumors of the brain and nervous system. Oral ingestion mainly leads to skin carcinoma [3.4].

Arsenic has a teratogenic effect, as it can penetrate the placenta of both animals and humans.

The carcinogenic effect of inorganic arsenic on the human skin, lungs, and occasionally the liver has been demonstrated by epidemiological studies. It is explained by the inhibition of repair processes. Possibly, the phosphate groups of nucleotides are replaced by arsenate, as the two ions are of similar structure. Organic arsenic compounds do not appear to have a carcinogenic effect. The available data to some extent show dose–effect correlations for lung cancer (bronchial carcinoma) due to inhalation and for skin cancer due to oral ingestion.

Immission limit values are not specified for arsenic either in the TA-Luft Regulations or in the Air Quality Guidelines of the WHO. The Länder Committee for Immission Protection (LAI) gives a "unit risk" for arsenic of 4×10^{-3} and an orientation value of 5 ng/m^3. The latter figure represents the upper limit for the background concentration in rural areas of Germany, and can be used as a measure for future immission reductions in densely populated areas [3.34].

Inorganic arsenic compounds are essentially negative in point mutation tests, but have been demonstrated to be effective inducers of chromosome aberrations and sister chromatid exchanges. Chromosome changes were also observed in the lymphocytes of therapeutically or occupationally exposed persons.

3.3.9. Benzene

Benzene is the simplest aromatic hydrocarbon and is used as a solvent for a wide range of substances. It occurs in nature only in very low concentrations, e.g., as a constituent of crude oil (max. 0.4 g/L). It can also be formed by the incomplete combustion of organic compounds. In Germany, the main source of the benzene to which the population is exposed in the outdoor air is traffic.

The amount of benzene absorbed by the population of densely populated areas can be 200 μg/day, and can reach 600 μg/day in regions close to benzene emitters. Benzene absorption can also occur through smoking (20 cigarettes per day corresponds to a maximum of 600 μg), and can sometimes be due to food sold in gas stations (up to 5 μg/d) [3.35].

Prolonged inhalation of benzene leads to ca. 50% absorption, oral ingestion probably to 100%, and skin contact to <0.2%. It is distributed throughout the organism, and becomes concentrated in fat-containing organs. Retained benzene is eliminated unchanged to a very variable extent by respiration (depending on the exposure concentration and individual circumstances) or is oxidatively metabolized and excreted with the urine.

The carcinogenic effect results from the metabolization of benzene to benzene epoxide by the monooxigenases in the endoplasmatic reticulum. Benzene epoxide can be converted nonenzymatically to phenol, but can also be metabolized by enzymatic action to catechol. Other metabolites include hydroquinone and hydroxyhydroquinone, benzyl glycol, and *trans-trans*-muconic acid. Also, a reaction of benzene epoxide with glutathion mediated by glutathion-*S*-transferase can occur. This thioether, after further metabolization, is then excreted as phenylmercaptic acid in the urine. Benzene epoxide can also react nonenzymatically with nucleic acids (to nucleic acid adducts, e.g., *N*-7-phenylguanine). In both humans and animals, the principal metabolite of benzene is phenol, which is mainly excreted in the urine in conjugated form. According to the latest opinion, the ultimate toxic and carcinogenic metabolites of benzene are the quinones and hemiquinones [3.36].

Like most solvents, benzene has a depressant effect on the CNS. Local irritant effects on the skin and mucous membranes can occur. The most important effect of benzene is damage to the blood-forming system, especially on long-term exposure. The classical symptoms of acute and chronic benzene intoxication are generally only observed at very high or medium concentrations and usually after prolonged exposure. The immission concentrations normally present in the environment are insufficient to cause these symptoms. It is unanimously accepted that chronic exposure to benzene can lead to a very rapidly accelerating depression of all bone marrow functions, which then gives rise to changes in the peripheral blood picture. The various bone marrow functions can be damaged either simultaneously, in succession, or in isolation.

There is some disagreement over the question whether damage to bone marrow functions caused by chronic exposure to benzene is a precondition for the appearance of leukemia. The more recent view is that, in addition to the frequently described acute myeloid leukemia in epidemiological studies, there are also indications of an increased occurrence of chronic myeloid and chronic lymphatic leukemias, and a dose-dependent increase in lymphatic and hematopoietic neoplasms.

In vitro and in vivo tests have given no indication of point mutations due to benzene. However, numerical and structural chromosome aberrations were observed in exposed humans and in animal experiments. Carcinogenic effects in rodents (leukemias, lymphomas, and an increase in the various types of tumor and tumor locations) could be found in long-term inhalation experiments. However, high concentrations were required.

Immission limit values are not prescribed either by the TA-Luft Regulations or the Air Quality Guidelines of the WHO. Epidemiological studies gave unit risk estimates of 6×10^{-6} to 20×10^{-6}, with the median figure of 9.2×10^{-6}. The LAI gives 9×10^{-6} as the unit risk, and an orientation figure of 2.5 μg/m³. This figure represents the upper limit of the background concentration in rural areas of Germany, and is the standard for future immission reductions in densely populated areas [3.34].

3.3.10. Polycyclic Aromatic Hydrocarbons

Polycyclic aromatic hydrocarbons (PAH) are formed pyrolytically in all incomplete combustion processes and in the low-temperature carbonization of organic materials. The relative concentrations of the many types of PAH (PAH profiles) mainly depend on the extent of heating. When different organic materials are pyrolyzed, the profiles obtained at a given temperature are very similar. Thus, as a general rule for the assessment of the carcinogenic hazard due to PAH in the open air, it is assumed that, as in the ideal case, the PAH profiles are identical, and that the concentration of an individual PAH can be used as a measure of the carcinogenic potential of the whole group. Therefore, the assessment of PAH concentrations in the air is usually based on the parameter benz[*a*]pyrene (BaP), as the most experience is with this substance. However, the BaP concentration in cigarette smoke or diesel engine emissions is not suitable for estimating the carcinogenic potential of the PAH in these media.

Compounds containing at least four rings are particularly relevant to the environment, as a number of these have carcinogenic effects. PAH have been detected in many vegetable and animal foodstuffs. Particularly high BaP concentrations are found in roasted or grilled products. Various vegetables such as kale, lettuce, carrots, spinach, and some types of grain sometimes contain high concentrations of PAH.

The absorption of PAH takes place mainly by the oral route (in food), while absorption by inhalation is almost exclusively associated with particles. The daily amount of BaP ingested in food in Germany is thought to be 1000–10 000 ng per capita. The daily intake by inhalation is estimated to be 10–200 ng per capita. Ingestion via drinking water is negligibly small, and cutaneous absorption in general plays no role. The amount of BaP ingested orally is certainly greater than that absorbed by inhalation in terms of quantity, but the intake of particles by inhalation is of considerably greater significance because of the formation of bronchial carcinomas [3.34].

The PAH compounds adhering to the particles are partly dissolved in the lungs and metabolized there. The metabolites can act on the cells, but can also be taken into the blood and detoxified in the liver, or can be excreted unchanged, especially in the bile, via the gut. Only a very small proportion of the PAH compounds is excreted with the urine in either a metabolized or an unchanged state.

PAH compounds have an inducing effect on microsomal enzymes. Little is known of acute or chronic toxic effects of the most important PAH compounds. Clinical signs of intoxication first appear after comparatively high doses. For example, newborn mice after a subcutaneous injection of 30 000 ng dibenz[*a, h*]anthracene (=15 mg/kg body weight) showed no acute effects. This shows that the carcinogenic effect predominates over other effects [3.37].

The mutagenic effects of some PAH compounds have been detected in many test systems. There is a good correlation between the results of in vitro mutagenicity tests and the carcinogenic effect. BaP has a toxic effect on rat and mouse embryos, and a reduction in the fertility of mice has been reported.

The carcinogenic effects (mainly skin and lung carcinomas in persons exposed to chimney soot and coal tar) have long been known, and the mechanism of the carcinogenic action of BaP has been well researched. In general, the most widely accepted view is that it is not the substance itself that is active, but the reactive epoxides formed by microsomal monooxigenases. The epoxides are converted by epoxidases to *trans*-dihydrodiols, which are the final carcinogenic metabolites. In laboratory animals, the carcinogenic effect is independent of the mode of application, and systemic tumors were also observed in other tissues.

Immission limit values are not specified for PAH compounds or BaP either in the TA-Luft Regulations or the WHO Air Quality Guidelines. From the risk estimates based on epidemiological studies, a unit risk of 7×10^{-2} can be derived for BaP. The LAI specifies an orientation value of 1.3 ng/m^3 for BaP. This figure represents the upper limit of the background concentration in rural areas of Germany, and is a standard for future immission reductions in densely populated areas [3.34].

3.3.11. Diesel Motor Emissions

Diesel motor emissions (DME, diesel engine exhaust) consist of a large number (>1000) of gaseous and particulate substances. DME particles, due to their small diameter (median diameter: 0.2–0.3 μm), mostly reach the alveoli, but are then to the extent of ca. 50% expelled in the exhaled air. The mass of diesel soot particles absorbed by inhalation by the general population can only be estimated at present. Exposure in the open air in rural areas is up to 1.5 μg/m^3, in densely populated areas 5–10 μg/m^3, and in regions close to the emitters 15–25 μg/m^3 [3.34].

Apart from the particles, which carry adsorbed organic substances, the main toxic components of DME are nitrogen oxides and aldehydes. The acute toxicity of DME is insignificant. In animal experiments, the inhalation of DME with high particle concentrations led to a reduction in lung clearance. In inhalation experiments, exposure of rats to DME caused benign and malignant lung tumors after exposures that were ca. 75–100 times as high as those in densely populated areas in the middle of the week.

The carcinogenic effect is clearly linked to the particle phase. The very low level of polycyclic aromatic hydrocarbons (PAH) in DME can account for at most 1% of the carcinogenic effect.

Particles and particle extracts from DME exhibit a mutagenic effect in several in vitro and in vivo test systems. Although this demonstrated a genotoxic mechanism, other mechanisms are also under discussion.

Epidemiological investigations on persons heavily exposed to DME in the course of their employment have so far provided "limited evidence" for carcinogenic effects of DME on humans, because insufficient attention was paid to extraneous effects.

DME is classified as a substance definitely known to cause cancer (MAK III A2), and is classified with the Group II cancer-producing substances ("serious hazard") in the hazardous substance regulations. Based on inhalation experiments on rats, a unit risk of 7×10^{-5} has been derived [3.34].

3.3.12. Dioxins and Furans

Polychlorinated dibenzodioxins (PCDD) and polychlorinated dibenzofurans (PCDF) are tricyclic, almost planar, aromatic compounds with similar physical, chemical, and biological properties. Depending on the degree of chlorination a number of isomeric compounds (congeners) can exist. The theoretical possible number of congeners is 75 for PCDD, and 135 for PCDF. None of the congeners are produced industrially, except as analytical standards. PCDD and PCDF are formed only as undesirable byproducts of chemical and thermal processes, and enter the environment from many sources. Their persistence, buildup in the food chain, ubiquitous occurrence, and the marked toxicity of some congeners of both groups can constitute a health hazard for humans. They are present as impurities at very low concentrations in many chemicals (e.g., pentachlorophenol) [3.4].

Adequate data are unavailable for most of the PCDD and PCDF congeners of toxicological significance. The most biologically active compound of the PCDD group is 2,3,7,8-tetrachlorodibenzo-*p*-dioxin (TCDD or "Seveso" dioxin). This does not occur in isolation, but can be detected in PCDD/PCDF mixtures in a wide variety of compositions. The toxic potential of PCDD/PCDF mixtures is estimated as TCDD equivalent figures (TEq), derived partially from subchronic toxicity studies and from certain biochemical properties. TEq figures represent the relative toxicity of a mixture in terms of the toxic effect of 2,3,7,8-TCDD. The scientific basis for the TCDD equivalence factors (TEF) is that there is a common effect mechanism, and there is therefore a strict correlation between the binding of the various PCDD/PCDF compounds to the cytoplasmic arylhydrocarbon (Ah) receptors and the biological effects resulting from this. The following properties of PCDD and PCDF based on animal experiments were taken into account [3.38]:

1) The binding affinity to the Ah receptors
2) The ability to induce various enzyme systems
3) The toxic effects on certain cell cultures (e.g., keratinization)
4) The acute and chronic toxicity

The strongest counterargument is that the isomers so compared with each other differ strongly in their toxicological behavior, and this cannot be adequately represented by the TEF, and TEq figures are not suitable for estimating the potential cancer risk of exposure to PCDD and PCDF. Any statements made about the possible risk to the population from other effects must be made with considerable reservations, especially concerning long-term exposure to very low concentrations.

In the environment PCDD and PCDF are almost exclusively associated with fine dust, and hence can be absorbed by inhalation. However, the amount absorbed by inhalation is less than 5% of the total. An inhaled volume of 20 m^3/d corresponds to an inhaled dose of 0.4–6 pg TEq per person per day. With 50% deposition in the lungs and 100% absorption, this represents 0.2–3 pg TEq absorbed per person per day [3.38].

The highest proportion of the daily intake of PCDD and PCDF by the general population comes from foodstuffs of animal origin. The average daily intake by an adult is estimated by BGA/UBA (1993) to be ca. 2 pg TEq/kg body weight. Bearing in mind variations in eating habits, an average intake by 4–9 year old children of 2–3 pg TEq/kg body weight can be derived. For one year old children, an average daily intake of ca. 4 pg TEq/kg body weight is calculated. Gastrointestinal absorption from foodstuffs is almost 100%. Intake with drinking water is small [3.39].

2,3,7,8-TCDD is rapidly distributed in the organism, and causes systemic damage to various organs; it can cross the placental barrier. 2,3,7,8-TCDD affects lipid metabolism, vitamin A balance, and thyroid gland function. Some of these disturbances are due to binding to the Ah receptors and the resulting induction of various enzyme systems. Various glucuronide and sulfate conjugates have been shown to be metabolites. The metabolism takes place through P450-dependent enzymes. There is a high percentage of nonmetabolized 2,3,7,8-TCDD. The PCDD compounds are mainly eliminated in the feces. Dioxins accumulate in the fatty tissue and are mobilized in the milk of nursing mothers. Animal experiments and human experience have shown that excretion decreases with increasing degree of chlorination. The mechanism of excretion in humans is unknown. The half-life of PCDD in humans is 5–10 years, but only a few weeks in rats [3.40].

Acute and subchronic intoxication by TCDD-containing PCDD mixtures can be recognized from a large number of organ manifestations. The skin and liver are particularly affected by disturbance of the porphyrin and fat metabolism, and by the induction of mixed function oxidases. The PNS, CNS and immune system can also be damaged. Direct contact can cause lesions of the skin. TCDD has strong acute toxicity compared with PCB. Acute exposure is first followed by irritation of the eyes and the upper respiratory tract, sometimes accompanied by vertigo, nausea, and vomiting. In later stages, reddening and swelling of the affected parts of the skin can occur. Typical skin lesions known as "chloracne" develop. The most affected organ is the liver. Swelling and liver parenchyma damage occurs, leading to an increase in the transaminases in the blood. Neurological damage following TCDD exposure seems to occur more often than has generally been assumed up to now [3.38], [3.40].

Animal experiments indicate fetotoxic and teratogenic effects, and impairment of reproduction by TCDD is also known. Following maternotoxic doses, teratogenic effects are observed in mice, but fetotoxic effects in rats and monkeys. At present, there is no indication of teratogenic effects in primates. Animal experiments also indicate immunotoxic effects such as thymic atrophy and suppression of cell-mediated humoral immune responses. TCDD does not appear to be mutagenic in the Ames test. Absence of mutagenicity is also shown in many other in vivo and in vitro tests, the binding of TCDD to DNA being very weak. In exposed industrial workers, no indication of chromosomal damage has been found either.

The carcinogenic effect of TCDD following oral and parenteral application in animal experiments has been thoroughly investigated and neoplasms at multiple sites have been recorded. Carcinogenic effects of other PCDD or PCDF compounds are presently unknown.

Persons exposed in Vietnam to TCDD-containing herbicides or to TCDD in Seveso were not found to have any increase in chromosome damage. Comprehensive epidemiological studies in the United States, Sweden, Canada, and New Zealand did not provide any solid evidence for carcinogenic effects from the use of herbicides containing TCDD or PCDD as impurities. However, an increased incidence of stomach carcinomas was reported in exposed workers. More recent investigations of this group revealed a slight increase in the standardized mortality rate, but no statistically significant increase in the incidence of tumors. Case control studies of workers who had worked with TCDD-containing herbicides for some years revealed an increased incidence of rare soft tissue sarcomas (hemangiosarcomas). The apparent absence of a mutagenic effect, the lack of DNA adduct formation, and the tumor-promoting activity in two-stage models in liver and skin indicate that TCDD must be regarded as a nongenotoxic carcinogen. This allows the conclusion that a threshold of carcinogenicity exists, below which promotion is not maintained [3.38]–[3.41].

In view of the lack of any indication of a direct genotoxic effect, and assuming a threshold concentration for the promoting effect, a temporarily tolerable daily uptake has been derived by national and international authorities. A figure of 1 pg TEq/kg body weight has been proposed for Germany [3.38].

If, however, contrary to the present state of knowledge, it is assumed as a precautionary measure that there is no effect threshold, linear extrapolation from animal experiments with oral application of 2,3,7,8-TCDD yields a putative unit risk of 1.4 for inhalative exposure. Combination effects with other PCDD compounds were not included.

Immission limit values for PCDD/PCDF have neither been specified in the TA-Luft Regulations nor in the WHO Air Quality Guidelines. The LAI gives an orientation figure of 16 fg/m^3 for 1,2,3,7,8-TCDD. This figure represents the upper limit of the background concentration in rural areas of the Federal Republic of Germany, and is the standard for future immission reductions in densely populated areas [3.34].

3.4. Effects on Plants [3.42]–[3.45]

The effects of air pollutants on plants are of interest both from the point of view of their economic exploitation and because of their great ecological significance.

Before they act on plants, air pollutants are subject to the effects of a large number of fortuitous and nonfortuitous factors (weather, etc.). The plants themselves have defence mechanisms that can influence the progress and severity of the action of a pollutant. They react towards air pollutants with a specific resistance behavior of genetic origin, which is also affected by the plants' state of development at the time and by the external growth conditions. Since plants are open systems, they are highly sensitive towards foreign substances present in the air.

3.4.1. Relationship between Immission and Effect

A complex structure of relationships exists between immission and effect. The reaction of the plants depends primarily on the presence of one or more immission components. The type and extent of the effect are essentially determined by the genetic structure, the growth stage of the plant at the time of exposure, and the modifying influence of various external factors. Depending on the severity and circumstances of the pollution, the stage of growth, and environmental conditions such as soil, weather or climate, and biotic effects, individual plants can suffer acute, chronic, or insidious damage. Plant communities react with changes in their composition and structure, up to complete destruction. The plants' economical and ecological viability can be affected, and eventually gene erosion by extinction of the species can occur.

Acute damage is mainly caused by the short-term effects of immissions at higher concentrations. This is manifested first on plant leaves by loss of turgescence and subsequent drying of the affected parts of the tissue. In general, a few days later, ivory colored, brown, red-brown, or blackish leaf necroses occur. Repeated effects of this kind weaken the vitality of plants. Severe, acute damage can lead to their complete destruction.

Chronic damage occurs after prolonged effects of low concentrations of immissions. Typical external signs include yellowing, or reddish-brown, turgescent marks on the leaves. In the course of continuing exposure, the symptoms of damage become more severe, causing decreased vitality, restriction of growth, and finally the death of the plant.

In additon to acute and chronic damage, erosive damage is also observed. This is produced by the action of acid fog or water-soluble deposits of reactive dusts. In these cases, the high ion concentrations that can be formed on moist leaf surfaces are very important.

Indirect damage can occur when air pollutants accumulate in soil penetrated by roots and change its properties as growth substrate (e.g., pH value). Damage can also be caused by absorption of accumulated substances through the roots.

Finally, immissions can cause a general reduction in the resistance of the plants towards plant diseases (e.g., from fungi and insects) and frost.

The solubility, reactivity, and state (gaseous or particulate) of the substance are the main properties that determine whether a pollutant can affect the plant surface on contact. The spectrum of activity is broad. Not all species and varieties of plants react in the same way to a harmful immission component. Also, some harmful substances produce specific reactions in certain plants. This peculiarity in the spectrum of effects on vegetation has been recognized from early times, and used to detect various air pollutants [3.46].

Effect Criteria. The large number of possible effects of air pollutants can be characterized by the corresponding effect criteria. These range from accumulation of harmful substances in the plants, disturbance of the physiological and biochemical processes in the cell, and submicroscopic changes to the cell organelles, to morphological changes to organs and impairment of the entire organism. On the level of a plant community, effects on structure and function, possibly including complete destruction, must be mentioned. Other effect criteria include reduced vitality, reduced yields, loss of quality, increased susceptibility to abiotic and biotic influences, and death of the entire plant. An overview of the criteria for the detection of the effects of immissions on plants is given in Table 3.1 [3.47].

Proof of Effects. In addition to the measurement of emissions and immissions, the use of plants as information systems for the protection of the environment from air pollution is indispensable. To obtain a complete survey, it is essential that every effect should be determined on both higher and lower plants. However, in practice, this requirement can only be met to a certain extent. As a rule, relevant effects on the most representative plants possible are chosen [3.48].

Table 3.1. Criteria for determining the effects of immissions on plants [3.47]

Criterion	Level of investigation
Species spectrum Abundance of species	ecosystem biocenosis population
Growth form Necroses Chloroses Leaf fall Biomass Accumulation of materials	population organism organ
Photosynthesis Gas metabolism Water balance Nutrient status	organism organ
Cell damage Chloroplast damage Tannin losses Tissue infiltration	tissue cells
Enzyme activity Metabolism (protein, aminoacids, carbohydrates, pigment status, energy balance, buffering capacity, mineral balance)	organism tissue cells organelles macromolecules

Representative plants are those for which the effects are broadly comparable with those on other plants, or for which an effect can be predicted from another effect on one and the same plant. Since the many factors cause a wide variety of harmful effects, the causal relationship between air pollution and effects often cannot be established with certainty in field investigations. For the qualitative and quantitative determination of immission effects, standardized methods whose results enable reliable conclusions to be drawn have now been in use for some years [3.49]–[3.52].

Dose–Effect Relationship. Greenhouse and open land experiments under defined boundary conditions and observations in open land enable the relationship between immission concentration and effect to be described. As absorption of the pollutants usually precedes their effect, an attempt is also made to determine the relationship between these two parameters rather than a relationship between immission concentration and effect. The amount of pollutant taken up from the immission is defined as the immission dose (milligrams pollutant taken up per kilograms acceptor substance or milligrams pollutant taken up per squared meter acceptor surface) and as the immission rate (milligrams pollutant taken up per kilograms acceptor substance per unit time, or milligrams pollutant taken up per squared meter acceptor surface per unit time) (see VDI 2309, Sheet 1). The use of the terms immission dose and immission rate is subject to the limitation that the analytically determined concentration in an acceptor is always a function of the immission flow (see VDI Guideline 2450, Sheet 1), as well as being the result of absorption by other routes, and processes of growth, destruction, excretion, and redistribution in the organism.

Immission dose, immission rate, and immission concentration are known as immission values. The relationships determined between one of these values and its effects on plants is the basis for the establishment of maximum immission values by the VDI commission for air pollution. The aim and significance of these values, which give the decision maker the essential information for fixing limit values, are described in the VDI Guideline 2310, Sheet 1 [3.53]. Concentration/dose–effect relationships determined in gas exposure experiments on test plants can only be translated to open land relationships to a limited extent, as in climatic chambers, greenhouses, or other closed experimental arrangements, the changing immission concentrations of an open location and the varying meteorological growth factors such as light, temperature, air humidity, and wind can only be approximated. Under such "artificial" conditions, plants can react towards air pollutants more sensitively than in open land conditions [3.54]. Observations under natural open land conditions or in open land experiments also enable only approximate conclusions to be drawn about the relationship between exposure to a harmful substance and its effect. Furthermore, a generally valid derivation of the relationship between immission concentrations and effect cannot be obtained with certainty from observations in open land or from the results of open land experiments. This is because in the open air very variable air pollution occurs, several immission components normally being present. This leads to effects that are not comparable to specific damage by a single pollutant.

Open top chambers occupy a position somewhere between open land and climatic cabinets [3.55]. These large chambers, which are open at the top, are placed in the open air. The dose–effect relationship is determined experimentally by introducing air from outside to the test plants. In practice, the immission values are deduced by careful investigation of the results of one or all three methods. Responsibility for determining immission values lies with various working groups of the German Association of Engineers (VDI). The maximum immission values obtained are not limit values, but standard figures at which the risk of damage to vegetation can be limited.

3.4.2. Effect of Individual Pollutants

3.4.2.1. Sulfur Dioxide

Many publications deal with the effect of sulfur dioxide on plants. Important knowledge was obtained in the 1930s, especially in the United States and Canada. A bibliography by the VDI for 1974–1978 lists almost 250 references [3.56].

Uptake and Plant Reaction. Sulfur is an essential nutrient for plants. It is taken up via the roots from the soil in the form of SO_4^{2-} ions. However, plants can also use the SO_2 from air pollutants to make up for sulfur deficiency.

To affect the leaf [3.57], SO_2 must penetrate via the stomata into the leaf tissue and cells. This leads to the formation of sulfite and hydrogensulfite ions as dissolved products. A significant part of the absorbed SO_2 is oxidized to sulfate. In general, it is known that damage can consist of reduced activity of various enzymes or increased peroxidase activity and breakdown of chlorophyll. However, these reactions are not specific to the effects of SO_2, and can also result from the action of ozone or phytopathogenic fungi. Reactions specific to SO_2 include the breakdown of tryptophan or indolylacetic acid and the rupture of S–S bonds in proteins.

Photosynthesis reacts to the effects of SO_2 with high sensitivity. However, any retardation of this process can be compensated to a certain extent by an increase in the concentration of CO_2.

Detoxification reactions include neutralization of the acid sulfite and oxidation to sulfate and the formation of organic sulfur compounds. This explains why the concentration or, more accurately, the SO_2 uptake per unit time determines the extent of damage. The physiological activity of a plant is therefore a decisive parameter for the risk of acute damage. This is at its greatest in the spring and summer months, and especially in the morning hours. Chronic damage, mainly nonspecific in character, is caused by the gradual accumulation of sulfur until the damage threshold is exceeded.

Plant species vary in their sensitivity towards SO_2. In general, plants are classified as very sensitive, sensitive, and less sensitive. However, even the most resistant plant species are not totally resistant to SO_2. The resistance behavior of various economically important cultivated plants is given in Table 3.2 [3.58].

Table 3.2. Resistance of plants to SO_2

Very sensitive plants	Sensitive plants	Less sensitive plants
Deciduous trees and bushes		
Walnut (*Juglans regia*)	Lime (*Tilia spec.*)	Plane (*Platanus spec.*)
Red currant (*Ribes rubrum*)	Red beech (*Fagus silvatica*)	Oak (*Quercus spec.*)
Gooseberry (*Ribes uva-crispa*)	Hornbeam (*Carpinus betulus*)	Poplar (*Populus spec.*)
	Apple (*Malus domestica*)	Maple (*Acer spec.*)
	Hazelnut (*Carylus avellana*)	Alder (*Alnus spec.*)
		Lilac (*Syringa vulgaris*)
		Woad (*Salix spec.*)
		Robinia (*Robinia pseudoacacia*)
		Birch (*Betula spec.*)
		Cherry, plum (*Prunus spec.*)
		Vine (*Vitis vinifera*)
		Rhododendron (*Rhododendron spec.*)
Conifers		
Fir (*Abies spec.*)	Pine (*Pinus silvestris*)	Black pine (*Pinus nigra austriaca*)
Spruce (*Picea spec.*)	Weymouth pine (*Pinus strobus*)	Arbor vitae (*Thuja spec.*)
Douglas fir (*Pseudotsuga menziesii*)	Larch (*Larix sepc.*)	Yew (*Taxus baccata*)
		Juniper (*Juniperus spec.*)
		Japanese Cypress (*Chamaecyparis spec.*)
Agricultural and horticultural plants		
Clover (*Trifolium spec.*)	Barley (*Hordeum vulgare*)	Potato (*Solanum tuberosum*)
Lucerne (*Medicago sativa*)	Oats (*Avena sativa*)	Tomato (*Lycopersicon esculentum*)
Lupin (*Lupinus spec.*)	Rye (*Secale cereale*)	Maize (*Zea mays*)
Vetch (*Vicia sativa*)	Wheat (*Triticum sativum*)	Turnips (*Beta vulgaris ssp.*)
Horse bean (*Vicia faba*)	Lettuce (*Lactuca sativa*)	Cabbage (*Brassica oleracea ssp.*)
Spinach (*Spinacia oleracea*)	Bean (*Phaseolus spec.*)	Carrot (*Daucus carota*)
Pea (*Pisum sativum*)	Rape (*Brassica napus*)	Strawberry (*Fragaria chiloensis*)
		Garlic (*Allium spec.*)
Ornamental plants		
Begonia (*Begonia spec.*)	Dahlia (*Dahlia variabilis*)	Rose (*Rosa spec.*)
Vetch (*Lathyrus odorata*)	Geranium (*Pelargonium zonale*)	Gladiolus (*Gladiolus spec.*)
	Petunia (*Petunia hybrida*)	Tulip (*Tulipa gesneriana*)
	Carnation (*Dianthus caryophyllus*)	Iris (*Iris spec.*)
	Fuchsia (*Fuchsia hybrida*)	Narcissus (*Narcissus pseudonarcissus*)
		Chrysanthemum (*Chrysanthemum leucanthemum*)
		Heather (*Calluna vulgaris*)

This classification is carried out by using as effect criteria visible damage and impairment of physiological vitality, adversely affecting the economic exploitation of the plant. Results of open land observations and experiments in combination with continuous and noncontinuous SO_2 measurements in the air yield the essential information for the establishment of maximum immission parameters for the protection of vegetation [3.58]. In risk assessment from epidemiological data on the effects of SO_2, the presence of other pollutants is not taken into account. The composition of the pollutants, their mixing ratio in the air, and also hitherto unrecognized factors can give rise to synergistic or additive effects on plants. Various findings concerning the combined effects of SO_2 with NO_2, HCl, HF, ozone, and heavy metals are reported in the literature.

Relationship Between Concentration/Dose and Effect. Investigations into exposure to SO_2 under experimental conditions have shown that most plants show considerably greater compatibility than in open land conditions [3.58]. In the last decade, short-term high SO_2 concentrations have in general been limited to accidental releases, so that there is now more interest in the harmful effects of SO_2 on long-term exposure to low concentrations. Investigations into tolerance thresholds carried out in various countries give figures of 0.04 mg/m^3 for very sensitive plants, $>$0.08 mg/m^3 for sensitive plants, and up to 0.12 mg/m^3 for less sensitive plants.

There are many publications on the subject of combined effects, especially of SO_2 and ozone. It is often difficult to link visible symptoms of damage with the pollutant causing them. The nature of the effect is also dependent on the degree of exposure to the pollutant. A combination of SO_2 and ozone in a concentration at which each individual gas would cause serious leaf damage often has an antagonistic effect. At lower concentrations, synergistic effects predominate [3.59]. The effect of combining SO_2 with other gases, according to available knowledge, can be antagonistic, additive, or synergistic. The effect of a combination of substances depends on the ratio of the concentrations of the two components and their absolute concentrations [3.56].

Threshold values for SO_2 for the protection of vegetation have not yet been established. The VDI Guideline 2310, Sheet 2, of Aug. 1978, in which the maximum immission values for SO_2 for the protection of vegetation are specified has not so far been finalized.

3.4.2.2. Hydrogen Fluoride

As early as the 1800s, hydrogen fluoride was known to be a pollutant harmful to plants. It has the strongest phytotoxic action of all gases. The gaseous compounds SiF_4 and H_2SiF_6 are as harmful as HF, while particulate fluorine compounds are less so. Among the many publications on the effect of fluoride immissions on plants. There is a comprehensive review of the subject in VDI Guideline 2310, Sheet 3 [3.60].

Plant Uptake and Plant Reaction. Plants can take up fluoride both from the air and from the soil. The fluoride absorbed from the soil is carried to the parts of the plant above ground by transpiration. Transport via the xylem and plasmatic transport causes fluoride to be deposited preferentially in the tips or edges of the leaves.

Absorption of gaseous fluorine compounds from the air takes place mainly through the stomata of the leaves. Particulate immissions containing water-soluble fluorides can cause local erosive damage, mostly of the epidermis, on plant surfaces covered with dew. The absorption of gaseous fluorine compounds through the stomata is influenced by the width of the openings, which depends on, e.g., light, temperature, atmospheric humidity, and the state of development of the plant. Damage to the plants takes place by direct action, which is concentration-dependent, and indirectly by accumulation in the leaf, especially in the tips and edges. In these parts of the tissue, the amount of fluoride deposited is often many times higher than in the other parts of the plant. Because of the marked concentration effects, plants are also damaged by low concentrations if they are exposed to the pollutant for prolonged periods. Increased fluorine contents are usually a sign of the effects of fluorine compounds, and are therefore an important diagnostic criterion.

Variable sensitivity towards fluoride immissions can occur with different plant species, varieties, and origins. However, absolute resistance does not exist. The species-specific resistance differences are utilized to build up resistant series and resistant groups. For practical protection against immissions, it is convenient to divide plant species into the groups "very sensitive", "sensitive", and "less sensitive" [3.60]. The very sensitive species include spruce, vines, apricots, onions, gladioli, and other bulbs, and corms. The group of sensitive plants includes pines, firs, larch, beech, walnut, cherry, lucerne, grain, fodder barley, strawberries, and caryophyllaceous plants. The less sensitive plants include yew, black pines, maple, oak, birch, pear, potatoes, tomatoes, turnips and beet, geraniums, and ornamental compositae. However, the various varieties and subspecies within a group (e.g., the gladioli) can differ greatly in sensitivity among themselves.

Relationship between Concentration/Dose and Effect. Extensive documentation of the relationship between HF concentrations in the air and the effects on higher plants, as a basis for hazard predictions, is provided in the VDI Guideline 2310, Sheet 3. This contains results of experiments on exposure to gases, and data from open land observations. The relationship between immission and effect does not generally obey the quantitative law of erosion, which states that equal concentrations/doses cause equal effects. A demonstration of the difference in reaction between individual species was provided by gladioli and ponderosa pines that were exposed for 2 weeks in the vicinity of an aluminum works [3.61]. At an average HF concentration of 0.49 ppb, the gladioli showed a damage index (ratio of necrotized leaf tips to total length × 100) of 25.4, compared with 32.7 for the pines exposed to a concentration of 2.36 ppb (i.e., 5 times as high). The damage to the pines increased progressively with the concentration. The difference in reaction may essentially be caused by differences in the secondary deposition of the absorbed fluorine at the leaf edges and tips, and also by an effect on the diffusion resistance of the leaves.

The combination of fluorine compounds with other gaseous air pollutants can lead to increased, decreased, or unchanged fluorine accumulation in plants. At present, there is no consistent model that can account for the combined effects of HF with other gaseous components. However, it is possible to make qualitative predictions for certain boundary conditions [3.60].

Maximum Immission Values. The necessary preconditions for the derivation of maximum immission values for hydrogen fluoride in combination with other pollutants have not been fulfilled. Hence, the VDI Guideline 2310, Sheet 3, of December 1989, only gives results of gas exposure experiments and open land investigations in which the amount of harmful substance was defined by the concentration and duration of its effect. Based on protection of the species and not of the individual, the maximum immission values for hydrogen fluoride are as follows (in µg/m^3):

Very sensitive plants	1.0[a]	0.3[b]	0.2[c]
Sensitive plants	2.0	0.6	0.4
Less sensitive plants	7.5	2.5	1.2

[a] Exposure time: 1 d
[b] Exposure time: 1 months
[c] Exposure time: 7 months (vegetation period)

3.4.2.3. Hydrogen Chloride and Chlorine

The harmful effects of hydrogen chloride on plants are comparable to those of SO_2. In gas exposure experiments, it was found that the lower limit for damage to sensitive plants after prolonged exposure lies between 0.1 and 0.3 mg/m^3.

The results of HCl gas exposure experiments cannot be applied directly to open land conditions, as the atmosphere often contains large quantities of neutral chlorides (e.g., salt) in the form of aerosols as well as HCl. These cannot be distinguished from HCl by measurement techniques. Salt aerosols, e.g., in coastal regions, can cause plant damage, but their effect is small compared with that of HCl. Laboratory gas exposure experiments have shown that salt aerosols have less than 10% of the harmful properties of HCl aerosols or HCl gas [3.62].

The VDI Guideline 2310, Sheet 4, gives the following limit values (mg/m^3) based on gas exposure experiments lasting 24 h and 1 month respectively:

Very sensitive plants	0.80	0.10
Sensitive plants	1.20	0.15

Very sensitive plants include vines, spruce, horse beans, and berberis, while sensitive plants include red beech, pine, oats, and primulas, and the less sensitive plants include oak, blue spruce, kale, and fuchsia.

Chlorine is as harmful as hydrogen chloride, but is only of relatively small significance for vegetation because of its very rare occurrence.

Several species of pine showed needle necroses after exposure to 3 mg Cl_2/m^3 for 3 h [3.62]. Very sensitive herbaceous plants, including lucerne, tobacco, and radish were damaged after exposure to as little as 0.3 mg/m^3 for 2 h.

3.4.2.4. Nitrogen Dioxide and Ammonia

Nitrogen dioxide is the most harmful of the oxides of nitrogen, but it is less phytotoxic than other acidic gases. Nevertheless, prolonged exposure of sensitive plants to >0.4 mg NO_2/m^3 leads to growth retardation and signs of chlorosis. In gas exposures of several hours, concentrations of 10–20 times this figure are necessary to cause any damage.

The VDI Guideline 2310, Sheet 5, for sensitive plants such as maple, spruce, wheat, and fuchsia, gives a limit value of 0.35 mg/m^3 as the mean figure for the vegetation period, and 6 mg/m^3 for a single exposure of 30 min. For very sensitive plants such as apple, birch, larch, oats, and rose, slight damage is possible at these concentrations. Less sensitive plants include beech, yew, brassicas, and gladioli.

Ammonia is absorbed through the stomata of the leaves, and can be used for protein synthesis in the cells provided no cell damage is caused at the concentration present. Damage caused by ammonia differs from that caused by acid gases in that the leaves and needles first turn brownish-red, later dark brown, and finally black. Relatively little is known about the dose–effect relationship. Gas exposure experiments by the LAI showed that the first signs of damage to cress, a sensitive plant, were visible after exposure to 1 mg NH_3/m^3 for 2–3 days [3.63].

3.4.2.5. Hydrogen Sulfide

Hydrogen sulfide was formerly thought to be a gas that causes relatively slight plant damage, plants in general being regarded as much more resistant to H_2S than to SO_2 [3.42, p. 99]. Following more recent investigations, this conclusion must be amended. Exposure of spinach and peas to 1.2 mg H_2S/m^3 led to an increase in transpiration which caused drooping of the leaves, often followed by the formation of necroses [3.65]. Also, the efficiency of photosynthesis can be reduced, leading to a reduced formation of biomass.

H_2S has been reported to be even more harmful than SO_2 [3.66]. In a gas exposure test lasting 14 d, damage occurred even at concentrations of 0.45 mg H_2S/m^3 in various garden plants (radishes, dwarf beans, and cress) comparable to that produced by SO_2 at 0.70 mg/m^3. Only red clover was more sensitive to SO_2 than to H_2S. However, it can be assumed that H_2S damage to vegetation only occurs on accidental release from industrial installations.

3.4.2.6. Ethylene

Ethylene is produced in the plant itself. It acts as a ripening hormone, and at the same time automatically stimulates increased ethylene production. Anthropogenic ethylene concentrations are also effective in extremely low concentrations. This differentiates it fundamentally from the other harmful gases. Necroses hardly ever occur. Instead, inhibition of the lengthwise and surface growth of leaves occurs, abnormal ripening such as the curling of younger leaves and the downwards curling of the leaf stalks (epinastia), dying off of end buds, premature aging, yellowing, and leaf fall.

The following results are selected from the many gas exposure investigations: 0.1 mg C_2H_4/m^3 caused a 30% reduction in the dry substance of dwarf beans and radishes after 14-d exposure. The blooms of ornamental plants lost their color at the same concentration after the second day. The first reactions can occur at a concentration of little more than 0.01 mg/m^3. As well as the above plant species, peach, clover, tomatoes, peas, potatoes, and roses are sensitive to ethylene. However, relatively insensitive plants include the monocotyledonous varieties, with the exception of orchids and the conifers, which are very sensitive.

The flat dose–effect curves are striking. An effect appears immediately at very low concentrations, but increases only very slowly with increasing concentration. This makes it very difficult to establish immission threshold values [3.63]. A comprehensive study of the effect of ethylene and other hydrocarbons on higher land plants was carried out by the Deutsche Gesellschaft für Mineralölwissenschaft und Kohlechemie (German Association for Mineral Oil Science and Carbon Chemistry) [3.64].

3.4.2.7. Ozone and Other Photooxidants

Ozone has a strong phytotoxic effect on plants. It is a component of photochemical smog, being an intermediate product of photooxidation which eventually leads to the production of the strongly phytotoxic peroxyacetyl nitrate as well as other smog components. The photochemical oxidants are taken up exclusively by direct absorption from the air. In the absorption of ozone, it has been found that the stomatic resistance is a decisive factor, and determines the severity of the effect of the gas in each individual case [3.67]. The action of ozone leads to a change in the permeability of the cell wall, which leads to an increased loss of the dissolved substances in the cell. The efficiency of photosynthesis is reduced by ozone to a measurable extent, often before the appearance of visible damage. Plants react to ozone with varying sensitivity.

It is estimated that in the United States ca. 90% of plant damage is due to photooxidants. The direct economic damage is mainly due to the effects on growth, yield, and quality. Plant products used as foods and feeds and ornamental plants show reduced yields, changes in the composition of the useful plant substances, and lower perceptible quality. Because photooxidants are always in gaseous form, damage to ornamental plants is also widespread in the United States [3.70].

Considerable damage is also reported in The Netherlands. The occurrence of damaging ozone concentrations in Germany was detected in many urban areas with the aid of highly sensitive indicator plants (e.g., the tobacco species Bel-W-3 or the small nettle). Also, the contribution of ozone to the so-called new forest damage is generally acknowledged. The phenomenon of extensive forest damage observed in the mid-1970s is linked to air pollution, e.g., by ozone, and is supraregional. A large range of effects appear over a narrow concentration range. Higher concentrations are found especially in the areas of Central Europe remote from sources of pollution and at higher altitudes. Moreover, ozone pollution has considerably increased in recent years [3.71].

The concentration – effect relationship is the subject of many investigations. In VDI Guideline 2310, Sheet 6 [3.59], air quality criteria for the protection of vegetation were derived from results from open land, climatic chambers, and greenhouses. According to this, plants react with varying sensitivity to the action of O_3. The more sensitive plants include apricot, walnut, Weymouth pine, lucerne, oats, lettuce, and tomato. The less sensitive plants include red beech, red oak, Norway maple, mustard, turnips, strawberries, celery, and cabbage lettuce.

Combination effects with other harmful substances are known in which the effects of ozone are intensified by the presence of, e.g., SO_2 and NO_2. However, in the present state of knowledge, it is not yet possible to quantify the effect of SO_2 and NO_2 on the overall effect.

The maximum immission concentrations for the protection of vegetation (see VDI Guideline 2310, Sheet 6) apply to O_3 as a single component acting for a brief period, i.e., 0.5 – 8 h. In general, very sensitive, sensitive, and less sensitive plants are protected up to concentrations of 320 μg/m³, 480 μg/m³, and 800 μg/m³ respectively for an exposure time of 0.5 h. The concentration decreases with increasing exposure time, and, for a period of 8 h, is 70 μg/m³, 160 μg/m³, and 320 μg/m³, respectively. There is no figure for prolonged exposure, as the necessary data are not available. The WHO specifies a figure of 60 μg/m³ for the protection of vegetation for long periods, as average value for the vegetation period [3.72].

3.4.2.8. Dust

The term dust refers to particulate air pollutants of very heterogeneous composition. Their phytotoxic effect is considerably less than that of harmful gaseous pollutants. The direct effect of dusts depends very much on their solubility in water. Indirect effects on roots depend on the condition of the soil. Dust deposits on plant surfaces do not usually have a serious effect. However, damage to vegetation can be caused by salt, gypsum, cement, and chalk dusts (erosion and incrustation). Heavy deposits, e.g., of fly ash, on leaf surfaces can impair the photosynthesis efficiency of the assimilation organs. As well as physical methods of determining the constituents of sedimentary (i.e., deposited) dust and airborne dust, chemical methods are also very important. Dust deposits containing the heavy metals lead, cadmium, and zinc are of particular significance. Zinc is an essential trace element for plants. Plants with an inadequate supply of zinc show deficiency symptoms, but the metal is toxic in excess. Lead and cadmium are not essential for plant life, and have a much greater toxic effect on plant tissue.

These harmful substances enter the plant directly from polluted air, and also indirectly via the roots from soil in which they have accumulated. The main subject of interest is not the slight phytotoxic effects, which are only observed at higher heavy metal concentrations.

More important is the fact that plants used for both human and animal food can become contaminated. This contamination can pass down the food chain and endanger the health of humans and animals [3.73], [3.74].

3.4.2.9. Acid Rain

It had already been established by the mid-1800s that smoke from industrial installations and domestic heating can affect the chemical composition of rain, and that sulfuric acid was one of the pollutants concerned. After 1850, rainwater analyses were carried out in Germany and other countries to determine the effect of air pollution on the penetration of substances into the soil. These investigations were carried out because it was thought that chronic pollution of the soil could also affect plants by this route.

From 1950 onwards, following worldwide discussions on acid rain, regional measurement networks were installed in large parts of Europe and North America [3.74]. The main measurements made were of the sulfate, nitrate, and chloride levels in the rain, though other substances, e.g., ammonium compounds, were also determined. The highest acid content was found in the industrialized areas of Central Europe, where a pH of 4 was recorded. Rain from areas without anthropogenic pollution had an average pH of 5.6–4.6.

Absorption and Plant Reaction. The action of acid rain on plants has mainly been investigated under laboratory conditions. Plants show a considerably greater reaction under laboratory conditions than has been found in the smaller number of open land observations. The absorption of acid rain is influenced by morphological and structural properties of leaves and needles. Under comparable exposure conditions, broad-leaved plants are usually more sensitive than narrow-leaved plants. Damage to the leaf organs takes the form of fossules (pitting), dent-like deformations, and punctiform necroses. The effect of acid rain apparently is to produce more rapid weathering of the cuticle. Chlorophyll degradation is also possible, and the rate of washout of the plant's own nutrients is increased by proton–cation exchange. However, the negative effects of the added hydrogen ions can sometimes be more than compensated for by plant nutrients in the acid rain. The many possible effects of acid rain have been reported in comprehensive studies carried out by VDI [3.74]. Investigations in the United States have shown that treatment with simulated acid rain led to reductions in the yield of maize and soya beans, these being the most economically important crops in the United States.

Acid rain can cause indirect effects on vegetation via the soil. Changes to the chemical and physical soil condition are particularly important, as borne out by the intensity of the discussion of this complex question. In particular, toxic metal ions that can damage the root system may be liberated. The most important aspect of the action of acid rain on the soil is the increase in the plant availability of toxic aluminum and heavy metal ions due to pH reduction.

A considerable increase in forest damage took place in the 1970s in Germany, and acid rain was immediately thought to be the cause. The first damage was observed to fir trees in so-called pure air regions of Southern Germany. Since 1980, widespread damage to spruce trees has also been recorded. There has also been increasing damage to deciduous trees, in which the symptoms, chronological development, and distribution of the damage were different from those for fir and spruce trees. At first, many different conclusions were drawn by experts on the possible causes of the forest damage. These ranged from the one-sided assumption that the effects were due to exclusively natural factors to the view that they were due only to anthropogenic factors. In the course of the continuing discussion and consideration of the results of new investigations, it appeared more and more probable that forest damage was linked to air pollution whose effects could be changed by natural and/or silvicultural influences. In the most recent state of the debate on the causes of forest damage, three possible effect mechanisms are favored [3.74]:

1) Nutrients are washed out of contaminated soil by acid rain, and toxic metal ions (e.g., aluminum) are released. These damage the plant roots, whose ability to absorb nutrients and water is reduced.
2) The plants are directly damaged in the crown region by acid rain.
3) There is a direct effect of gaseous air pollutants, leading to a certain susceptibility of the whole organism to the subsequent effects of acid rain.

The causes of forest damage are still under discussion. There are other possible factors, e.g., combination effects of various air pollutants, toxic substances as yet unknown, and various stress factors.

4. Laws and Technical Regulations

4.1. Introduction

In Germany, the maintenance of air purity and plant safety have their legal basis in the Federal Antipollution Law or *Bundes-Immissionsschutzgesetz* (BImSchG). This is the most important legislation for the protection of the environment from the point of view of both environmental and industrial politics. It is a demanding law of great importance in environmental politics, aiming to prevent air pollution, to provide protection from noise and vibration, to prevent environmental damage and other hazards, and to promote safety in production plant. Its significance for industrial politics is that almost all industrial and business enterprises require a license before starting operation, and must remain under continuous environmental supervision by the state during their operation. The BImSchG is therefore effectively a plant licensing law into which extensive environmental protection is integrated.

The law lays down the basic requirements which the plant must satisfy. In doing so, it makes considerable use of so-called undefined legal terms such as harmful effects on the environment, pollution, emissions, hazards, plant, state of the art, precautions, and compatibility. These legal terms are characterized by the fact that their meaning is only partially defined by the law. They therefore admit scope for discretion on the part of the authorities and courts, and thus affect legal security. However, they are unavoidable in law concerning technology in order to be able to keep abreast of scientific and technological developments. "The setting of a given legal safety standard by making rigid rules is more likely to limit both technological development and the preservation of basic rights (if it allows these at all) than to enhance them. It would constitute a backward step at the cost of safety" [decision of the German Federal Constitutional Court (Bundesverfassungsgericht) of Aug. 8th, 1978: the so-called Kalkar Ruling]. The use of undefined legal terms is thus asserted to be fundamentally and constitutionally permittable.

However, undefined legal terms constantly require further definition to give consistent legal applicability. For this, there are several possibilities:

1) Regulations enacted on the basis of BImSchG. The most important of these are those regulations issued by the Federal Authorities for the enforcement of the Federal Antipollution Law (BImSchG). Others are issued by the Länder (e.g., smog regulations). Regulations are legally binding on persons to whom they apply.
2) General administrative regulations. These define the undefined legal terms, with legally binding effect on the executive authorities, or they restrict the discretion of these authorities. The most important of these regulations are the technical instructions for the maintenance of air purity (TA Luft) and protection against noise (TA Lärm).
3) Standards set by groups or committees of nongovernmental technical legal authorities. These can be National Standards (DIN, VDI), EU Standards (CEN, CENELEC), or International Standards (ISO). These standards are only binding if government regulations refer to them. Otherwise, they generally have the status of generalized expert opinion.

Further development in the areas of air pollution prevention and plant safety will be increasingly determined by the following:

1) National legislation will be increasingly controlled by laws made by the EU.
2) In the national, supranational, and international areas, regulations relating to individual environmental pollutants are becoming increasingly incorporated into general antipollution regulations. Examples of the latter include environmental criminal law, the law dealing with tests for environmental compatibility, and the law of liability for environmental matters.

4.2. Federal Antipollution Law

The law relating to the protection of the environment from the harmful effects of air pollutants, noise, vibrations, and similar effects (Bundes-Immissionsschutzgesetz, BImSchG) has been in force since April 1st, 1974. It has been much modified since then, mainly by the 2nd amending law of Oct. 4th, 1985, the 3rd amending law of May 11th, 1990, and the law of April 22nd, 1993 relating to capital investment facilitation and residential building land, which has been in force since May 1st, 1993.

The BImSchG brought about a comprehensive legal reorganization of the law on pollution. In particular, it superseded the regulations of the §§ 16ff. Trade and Industry Code (Gewerbeordnung) and a large number of regulations made by the Länder; included nonlicensable installations; and created new regulations (e.g., for planning, traffic, and products).

The BImSchG makes regulations for the building and operation of installations, but not for effects on the environment that depend simply on human behavior, which are covered by law at the Länder level. As far as air purity and noise reduction are concerned, BImSchG applies to installations in all areas (e.g., industry, trade, crafts, agriculture, and forestry) and includes nationalized and private enterprises (§ 2). Antipollution Law in the wider sense applies only to profit-making enterprises. It basically does not apply to airports and nuclear power stations.

The law has three aims (§ 1):

1) The most important aim is protection against harmful effects on the environment.
2) For licensable installations, protection is extended to other hazards such as fire and explosion and risk of disturbances which may be associated with the production of chemical products in particular.
3) The long-term aim is prevention. Even if there are no harmful environmental effects, the aim is to improve the protection of human living space, especially by judicious utilization of modern technology (state of the art).

Pollution prevention law is divided into four main areas:

1) Installation-related pollution prevention
2) Product-related pollution prevention
3) Region-related pollution prevention
4) Traffic-related pollution prevention

Installation-related pollution prevention mainly relates to stationary installations, i.e., factory premises. A distinction is made between licensable installations (§§ 4ff.) and nonlicensable installations (§§ 22ff.).

Licensable installations are those whose nature or size render them especially likely to bring about harmful effects on the environment or to cause other hazards. They include stationary waste disposal plants in which waste materials are stored or treated. Licensable installations are specified in detail by regulations (4. BImSchV).

Licensable installations may only be erected, operated, or significantly modified if a licence has been granted (§§ 4, 15).

Licensable installations incur four fundamental duties (§ 5):

Duty to Protect. The construction and operation of the installation must not lead to any possibility of harmful effects on the environment or to any other hazards, or to serious detriment or nuisance. This duty extends also to the prevention of pollution. In general the limits of harmful pollutants are more exactly defined mainly in general administrative regulations, especially by the immission values of TA Luft and TA Lärm. These specify at what level of regular exposure there is a harmful effect on the environment. The immission figures are always determined by well-defined measurement and assessment methods. More precise definition is usually provided by nongovernmental regulations, such as DIN standards and VDI guidelines. Duty to protect is more precisely specified for installations where there is a hazard from potential disturbances. Here, methods of preventing such disturbances or of limiting their effects are prescribed in the accident regulations (12. BImSchV).

Duty to Take Precautions. When licensable installations are constructed and operated, precautions must be taken against harmful effects on the environment, in particular by state-of-the-art measures to limit emissions. These are usually assessed by comparison with emission limits as prescribed principally in the TA Luft Regulations, Power Station Emission Regulations (13. BImSchV), and Waste Incineration Plant Regulations (17. BImSchV).

Duty to Avoid, Reuse, or Dispose of Residues. Licensable installations are among the most significant producers of waste. The goal of waste prevention is best served by preventing the generation of residues, or, failing that, by the lawful and nonhazardous utilization of any residues generated. Residues are substances that arise during production and processing that are not desired by the proprietor of the business, who is only absolved from the duty of preventing or reusing these residues if this is technically impossible or unreasonable. The proprietor is then obliged to dispose of the residues as waste materials without impairment of the well-being of the general public.

Duty to Utilize Waste Heat. In 1985, the duty to utilize waste heat produced in an installation was introduced as the fourth fundamental duty. The prerequisite is that such utilization is technically possible and feasible for the type and location of the installation. In 1990, the duty to utilize waste heat was widened to include supplying it to a willing third party. This fundamental duty only applies to installations specified by order of the Federal Government.

The license is always granted by following a formal procedure, i.e., with participation of the public. The procedure is laid down in § 10 and 9th BImSchV, and is initiated by a written application to which the documents required for assessment should be attached. These include descriptions of installations and business enterprises, data on emissions, and proposed measures for preventing harmful effects to the environment and other hazards, including a safety analysis in the case of certain installations. The licensing authority receives the statement from the authority whose field of duties is affected by the proposal. If the application documents are completely in order, the proposal is published, and the application and documents must be exhibited for public inspection for one month. Any person then has the right to raise objections to the proposal up to two weeks after the end of this period, i.e., a total of about six weeks. If objections are raised, the licensing authority must discuss these with the objectors and the applicant. In those types of installation specified in column 2 of the Appendix to 4th BImSchV, a simplified license application procedure without the participation of the public is used (§ 19). Discussion is concluded without notification or display of the documents.

In order to speed up the licensing procedure, preliminary decisions can be made for individual licensing prerequisites and for the location of the installation (§ 9). Large installations can be licensed in stages by means of partial licenses (§ 8). If an installation is to be significantly altered, the licensing authority, before the required license is granted, can permit commencement of construction of the installation, if at that point in time it can be assumed that a license will be granted, if there is a public interest for environmental reasons in the earlier building of the installation, and the proposer of the project accepts the risk of subsequent refusal of the license (§ 15a).

If the basic duties of § 5 are fulfilled and the installation satisfies other public legal requirements and concerns of industrial safety the licensing authority must grant the license (§ 6), and the applicant has a legal right in this respect. In accordance with other regulations, the license for the installation includes most of the other licenses (e.g., planning permission) and permits (concentration effect of § 13).

The license authorizes the construction and operation of the proposed installation. Subsequent restrictions to the licensed business are only allowed to a limited extent, especially by subsequent regulations under § 17, but they may not be excessive.

Installations which do not require a license under § 4 are in the category of nonlicensable installations (§§ 22ff.), e.g., small and medium-sized commercial enterprises, domestic heating, and construction machinery. Less stringent basic duties apply to these (§ 22). For some types of installation, the duties are detailed in official regulations, e.g., 1st BImSchV (small furnaces), 2nd BImSchV (installations that contain halogenated hydrocarbons), 7th BImSchV (wood processing plants).

For both licensable and nonlicensable installations, supervisory procedures may be prescribed, usually according to § 52, or, with special reference to the determination of emissions and pollutants, according to §§ 26ff. Since Sep. 1st, 1990, anyone who operates a licensable installation must provide a statement of emissions, and this must be updated every 2 years (§ 27, 11. BImSchV). In the statement of emissions, figures must be provided for the harmful substances emitted from the installation during the previous calendar year.

Following the 3rd amending law of 1990, BImSchG has been upgraded to the status of a plant safety law. The authorities are empowered to order certain technical safety checks and safety reviews (§ 29a). The law also provides for the appointment of a technical committee for plant safety (§ 31a), an accident committee (§ 51a), and accident representatives (§ 58a ff.).

Regulations can demand the prevention of harmful effects on the environment from air pollutants, plant noise or vibration, combustible materials, fuels, and other materials and products (§§ 32ff.). Relevant regulations include: 3rd BImSchV (sulfur content of light heating oil and diesel fuel), and 19th BImSchV (chlorine and bromine compounds used as fuel additives).

A section of the BImSchG is dedicated to the prevention of pollution from traffic. The design and operation of motor vehicles can be regulated in accordance with § 38 with respect to air purity and noise prevention. In still weather, traffic restrictions can be ordered by the Land authorities; the smog regulations issued by the Länder provide the details. Since Sep. 1st, 1990, the street traffic authorities can order traffic restrictions, whether in still weather or not, to prevent harmful air pollution, especially in the inner cities (§ 40 Section 2). A complete system of noise prevention for the design of streets, railways, and roads is provided by §§ 41–43 in conjunction with § 50. Some details are covered by the Traffic Noise Prevention Order (16. BImSchV).

The prevention of pollution in individual regions is controlled by §§ 44 ff. and § 49. These regulations control regional or local problems of pollution prevention. For areas in which especially hazardous air pollutants occur or are likely to occur, a continuous comprehensive control of air polluting substances is prescribed (§ 44). All significant sources of air pollution are included in an emission register, i.e., industrial and other business enterprises, domestic heating, and motor vehicles (§ 46). The emission register is based on the statements of emissions made in accordance with § 27. If investigations show significantly high concentrations or a continuous increase in air pollution over long periods of time, the setting up of air antipollution plans must be considered for the whole area of investigation or for parts thereof (§ 47). Under certain conditions, the Land authorities must prepare a cleanup plan. In other cases, according to their judgment, they may prepare a precautionary plan. Noise reduction plans may be prepared under certain circumstances in accordance with § 47 a. For areas in special need of protection (highly polluted areas or areas receiving special protection), specially stringent controls may be prescribed (§ 49).

As well as stipulating requirements for environmental technology, the law also lists a number of duties in connection with management structure. Proprietors of businesses that have environmental implications resulting either from emissions or from the products manufactured must appoint an antipollution representative (§ 53). He is the company expert in questions of pollution. His most important task is, on his own initiative, to promote the introduction of environmentally friendly processes and products, and to support the managers of the business therein. Control of the working environment within the company is also entrusted to him (§ 54). He has a special position within the company, having the right to make direct representations to management in important questions. He may not be dismissed as a result of his actions (§§ 56, 57, 58). For especially hazardous installations, the appointment of an accident representative is prescribed (§ 58 a). His rights and duties essentially correspond to those of the antipollution representative (§ 58 b–58 d). The law on waste materials prescribes the appointment of a waste materials representative, also modeled on the antipollution representative, and similarly the water resources policy act prescribes a water pollution representative. These various functions are often assigned to a single member of staff, who then has the function of an environmental representative.

Since Sep. 1st, 1990, proprietors of licensable installations have been obliged to make reports of their management structure to the competent authorities (§ 52 a). In corporations whose authorized representation includes several members, and in unincorporated firms with several legally appointed staff representatives, the authority must be informed who within the management is responsible for pollution prevention. All proprietors of licensable installations must also inform the authority of their method of ensuring that regulations and orders intended to prevent harmful effects on the environment and other hazards will be observed in their organization. In this connection, the management structure must be made known to the authority. The law does not demand a more detailed management structure, but more details emerge, for example, from the stipulations concerning management representatives, accident regulations, and general rules of management structure. Further elements are contained in the EU Regulation of June 29th, 1993 concerning the voluntary participation of business undertakings in a cooperative system for environmental management, investigation of the environmental aspects of a company (ABl. Nr. L 168/1), and the system of quality assurance standards (DIN/ISO 9000–9004).

4.3. Federal Statutory Antipollution Regulations Relating to Air Pollution and Plant Safety

4.3.1. First BImSchV

The first statutory regulation for the implementation of the Federal Antipollution Law (regulation concerning small furnaces – 1. BImSchV) of July 15th, 1988, mainly based on § 23 BImSchG, controls the construction, structure, and operation of furnaces that do not require a license under BImSchG (§ 1), i.e., installations with an output of up to 1 MW using solid fuel, 5 MW using light heating oil, or 10 MW using gaseous fuel. Small furnaces are important from the point of view of pollution prevention because of their abundance and their relatively high contribution to pollution (low-altitude gas discharge).

Main contents: Limit values for particulates, CO, and oil derivatives (§§ 4ff.); limitation of NO_x emissions from oil- and gas-fired furnaces by means of state-of-the-art combustion technology (§ 9); limitation of waste-gas emissions for oil- and gas-fired furnaces (§ 11); restriction to certain fuels (§ 3); inspections (§ 14); annual measurements during supervision by the regional stack inspector (§§ 15ff.); discharge conditions for waste gases (§ 18). Existing furnaces must be modified after a transition period of 3–7 years to comply with requirements for new installations (§ 23).

4.3.2. Second BImSchV

The second statutory regulation for the implementation of the Federal Antipollution Law (regulation to limit emissions of volatile halogenated hydrocarbons – 2. BImSchV) of Dec. 10th, 1990, based on § 23 BImSchG, applies to nonlicensable installations in which volatile halogenated hydrocarbons are used as solvents, especially surface treatment plants, dry cleaning plants, textile finishing plants, and solvent extraction plants. The regulation incorporates new knowledge about the potential hazards of volatile halogenated hydrocarbons and the ability of volatile chlorofluorocarbons (CFCs) to destroy ozone. This regulation implements The Montreal Protocol of Sep. 17th, 1987 concerning CFCs.

Main contents: The only permitted volatile halogenated hydrocarbons are tetrachloroethylene, trichloroethylene, and dichloromethane, and these may be used only in certain installations (§ 2). Current authorization for CFCs expired on Dec. 31st, 1994. Technical requirements, handling methods, and emission limit values are prescribed to limit the emission of permitted solvents. Existing installations must be renovated by Dec. 31st, 1994 at the latest (§ 6). This regulation also contains instructions for measurement and supervision (§§ 10ff.).

Disposal of solvents from these or similar installations is regulated in the regulations for used solvent disposal (Oct. 23rd, 1989). Other restrictions and prohibitions relating to CFCs are contained in the EU Regulation for CFCs and Halons (Oct. 14th, 1988) and the CFC–Halon Prohibition Order (May 6th, 1991), which is based on the Chemicals Act.

4.3.3. Third BImSchV

The third statutory regulation for the implementation of the Federal Antipollution Law (regulation concerning the sulfur content of light heating oil and diesel fuel – 3. BImSchV) of Jan. 15th, 1975, based on § 34 BImSchG.

Emissions from domestic and small commercial furnaces and diesel vehicles have a major influence on SO_2 emissions in densely populated areas because of their low emission altitude. To reduce these emissions, this regulation limits the sulfur content to 0.20 wt% in light heating oil and diesel fuel (following the amending regulation of Dec. 14th, 1987). The EU Guideline 75/716/EEC of March 30th, 1987 prescribes the lowest possible limit value. Earlier limit values were: May 1st, 1975: 0.55 wt%; May 1st, 1976: 0.50 wt%; Jan. 1st, 1979: 0.30 wt%. Since Oct. 1st, 1996, a new limit value of 0.05 wt% applies for diesel fuel.

4.3.4. Fourth BImSchV

The fourth statutory regulation for the implementation of the Federal Antipollution Law (regulation concerning licensable installations – 4. BImSchV) in the draft of July 24th, 1985, based on § 4, Section 1, Clause 3, BImSchG specifies the range of licensable installations. Only those types of installations named in the Appendix require a licence, usually on the grounds of maintaining air purity. They include almost all industrial installations, e.g., those involving chemicals (see Nos. 4.1–4.10). The obligation to be licensed often depends on the capacity or size of an installation and extends to adjacent equipment which is associated with it geographically and operationally and which can have an important bearing on the emission or immission behavior or the safety of the installation, such as the storage of raw materials, finished products, and residues, and equipment for transport, waste materials, packaging, and shipping.

For installations named in Section 1, the license is granted in a formal procedure (§ 10 BImSchG) with participation of the public, and for installations named in Section 2 by a simplified procedure (§ 19 BImSchG), i.e., without participation of the public.

4.3.5. Seventh BImSchV

The seventh statutory regulation for the implementation of the Federal Antipollution Law (regulation to limit the emission of wood dust – 7. BImSchV) of Dec. 18th, 1975, based on § 23 BImSchG applies to nonlicensable installations for the working or processing of wood or wood-containing materials if wood dust or shavings are emitted from these installations.

Main contents: Wood dust and shavings must be stored in bunkers, silos, or other closed spaces. These must be emptied in such a manner that the emission of wood dust or shavings is prevented as far as possible (§ 3). Emission limit values are specified for wood dust and shavings. If the exhaust air contains particulates, a limit of 50 mg/m^3 in the exhaust air is imposed (§ 4).

4.3.6. Eleventh BImSchV

The eleventh statutory regulation for the implementation of the Federal Antipollution Law (regulation concerning the declaration of emissions – 11. BImSchV) of Dec. 12th, 1991, based on § 27, Section 4 BImSchV, concretizes § 27 BImSchG, which obliges proprietors of all licensable installations to provide a statement of emissions to the competent authority, i.e., information on the type, amount, and distribution over space and time of atmospheric pollutants emitted annually from the installation. The emissions statement is the basis for the emissions register, and for planning and clean-up procedures. Installations with low emissions may be excepted (§ 1). The time period for the statement is the even-numbered calendar year (§ 3). In general a complete statement of emissions must be supplied, with quantities of these emissions. In certain cases, a simplified statement may be provided, for which operational data and a list of materials handled are sufficient (§ 4). Since Jan. 1st, 1990, a supplementary statement must be provided every other year (formerly every year) (§ 5). The authority may publish the figures from the emission statement, unless trade and operational secrets can be deduced therefrom (§ 27, Section 3 BImSchG).

4.3.7. Twelfth BImSchV

The twelfth statutory regulation for the implementation of the Federal Antipollution Law (accident regulation – 12. BImSchV) of Sep. 20th, 1991, mainly based on § 7 BImSchG, basically applies to all licensable installations in which materials referred to in Appendices II and III or other materials that are highly toxic, toxic, able to support combustion, explosive, combustible, highly flammable, or flammable (Appendix IV) are present or can be formed in an accident. It does not apply if negligible quantities are involved (§ 1). The regulation is intended to prevent accidents and to limit their effects as far as possible. Accidents are disturbances to normal operation in which materials as specified in Appendices II, III or IV lead to a serious hazard from large-scale emissions, fires, or explosions, i.e., danger to life, serious effects on health, or environmental hazards affecting the well-being of the population (§ 2).

Main contents: Basic safety duties (§ 3), detailed safety duties (§§ 4–6), safety analysis (§ 7), duty to report accidents and near accidents (§ 11), information to the public (§ 11 a), duty of notification (§ 12). Safety duties requiring qualified personnel (e.g., safety analysis) need only be carried out for installations named in Appendix I (§ 1, Section 2); these are mainly chemical installations.

4.3.8. Thirteenth BImSchV

The thirteenth statutory regulation for the implementation of the Federal Antipollution Law (power station emission regulation – 13. BImSchV) of June 22nd, 1983, mainly based on § 7 BImSchG, applies to the construction and design of large furnaces, i.e., installations with a combustion heat output of ≥ 50 MW, or ≥ 100 MW for gaseous fuels (§ 1).

Main contents: Emission limit values for particulates, CO, NO_x, SO_2, and halogen compounds (§§ 3 ff.). A continuous monitoring method is desirable for measurement and assessment (§§ 21 ff.). For NO_x, emission limit values were not finalized, as the technical development was still in a state of flux. Here, so-called dynamic clauses were introduced, requiring the use of the existing state of art. These have since been specified in detail to enable them to be carried out by management. The part with most relevance to air purity consists of instructions for cleaning up old installations (§§ 17 ff.). The regulation gave the proprietors until July 1st, 1984 to decide whether they wished to reequip their installations or shut them down by April 1st, 1993. If they opted for reequipment, this had to be completed by July 1st, 1988. Most furnaces have now been reequipped. Compared with 1980 figures, emissions of SO_2 in 1993 from large furnaces in the "old" Länder of the Federal Republic have been reduced by ca. 80 %, of NO_x by ca. 70 %, and of particulates by ca. 75 %. In the area of power generation, $> 21 \times 10^9$ DM were invested for these measures, of which 14.3×10^9 DM were for desulfurization of flue gas and 6.8×10^9 DM for reduction of NO_x emissions.

4.3.9. Seventeenth BImSchV

The seventeenth statutory regulation for the implementation of the Federal Antipollution Law (regulation concerning incineration plant for combustible waste and similar materials – 17. BImSchV) of Nov. 23rd, 1990, based on § 5 Section 2, and § 7 BImSchG, applies to all licensable installations in which solid or liquid waste or similar solid or liquid combustible materials are burnt, including installations whose main purpose is different (e.g., cement furnaces).

Waste incineration plants are of great significance for the environment because of their emissions of gaseous inorganic chlorine and fluorine compounds, and also because of some constituents of emitted particulates, especially heavy metals and polyhalogenated aromatic compounds. The emission limit values laid down earlier in TA Luft 1986 were made much more stringent, and the requirements of the EU Guidelines of June 1989 with respect to existing and new incineration plants for domestic refuse were incorporated in national law.

Main contents: Emission limit value for dioxins and furans 0.1 ng TEq/m³ (i.e., reduction in dioxin emissions from waste incineration plants by > 99 %); emission limit values for other air pollutants (§ 5); constructional and operational requirements, e.g., dwell time and minimum temperature (§ 4); and the annual provision of information to the public (§ 18). For existing installations there will be a series of intermediate deadlines until March 1st, 1994 or Dec. 1st, 1996 (§ 17).

4.3.10. Nineteenth BImSchV

The nineteenth statutory regulation for the implementation of the Federal Antipollution Law (regulation concerning chlorine and bromine compounds – 19. BImSchV) of Jan. 17th, 1992, based on § 34 BImSchG, seeks to reduce pollution of the environment by dioxins and furans by prohibiting the addition of chlorine and bromine compounds (scavengers) to motor vehicle fuel. The marketing of chlorine and bromine compounds as fuel additives and the marketing of fuels that contain such additives is forbidden (§ 2).

4.3.11. Twentieth BImSchV

The twentieth statutory regulation for the implementation of the Federal Antipollution Law (regulation to limit hydrocarbon emissions during the transfer and storage of gasoline – 20. BImSchV) of Oct. 7th, 1992, based on § 23 BImSchG and § 24 Trade Regulation.

The object of the 20th BImSchV is to reduce the emission of hydrocarbons during the supply of gasoline. These substances may be directly hazardous to health (e.g., the carcinogen benzene). However, hydrocarbons are also important precursors for the formation of ozone and other photooxidants. The regulation applies to nonlicensable installations to which deliveries of gasoline are made, or from which gasoline is taken, e.g., filling stations, small supply tanks, and road and rail tankers (§§ 1, 2).

Main contents: In principle, the gas displacement method must be used in filling operations from storage tanks and during deliveries to filling stations. If this is technically infeasible, the gasoline vapors must be passed to a gas-cleaning device (§ 3). The filling of vehicles at filling stations is regulated by 21. BImSchV.

4.3.12. Twenty-First BImSchV

The twenty-first statutory regulation for the implementation of the Federal Antipollution Law (regulation to limit hydrocarbon emissions during the filling of motor vehicles – 21. BImSchV) of Oct. 7th, 1992, based on § 23 BImSchG and § 24 Trade Regulation.

The object of 21st BImSchV is to reduce the release of gasoline vapors during the filling of motor vehicles at filling stations. Customers, filling station personnel, and persons who live close to filling stations are exposed to gasoline vapors, which contain the carcinogen benzene.

The regulation applies to nonlicensable filling stations where motor vehicles are filled with gasoline (§ 1). In the filling operation, a gas recycling system must be used which collects the gasoline vapors from the tank of the vehicle and feeds them back to the storage tank of the filling station (§ 3). The regulation applies with immediate effect to new filling stations. Existing filling stations must in principle be provided with a gas recycling system within not more than 5 years (§ 9).

4.3.13. Twenty-Second BImSchV

The twenty-second statutory regulation for the implementation of the Federal Antipollution Law (regulation concerning levels of pollutants – 22. BImSchV) of Oct. 26th, 1993, based on § 48 BImSchG, incorporates three EU Guidelines into Federal Law: of July 15th, 1980 (80/779/EEC amended by June 21st, 1989 89/427/EEC) on limit values for SO_2 and suspended particulates, of Dec. 3rd, 1982 (82/884/EEC) on the limit value for the lead content of air, and of March 7th, 1985 (85/203/EEC) on air quality standards for NO_x. The regulation became necessary after the European Court of Justice in May 30th, 1991 judged that the Federal Republic of Germany had not duly incorporated these guidelines. The limit values named in the guidelines are established as pollution limit values by the regulation (§ 1). The following were also established: the main reference time periods (§ 2), criteria for the location of measuring points (§ 3), and the methods of measurement and analysis (§ 4). The actions necessary if the limit values are exceeded are specified in § 6 (air purification plans, official orders). TA Luft is unaffected according to § 5.

4.3.14. Other Federal Regulations

The Law on Testing for Environmental Compatibility of Feb. 12th, 1990 is based on the EU Guideline of June 27th, 1985 on testing for environmental compatibility for certain public and private projects (85/337/EEC).

Testing for environmental compatibility must be carried out for the most important licensable installations, nuclear installations, projects involving water, mining installations, trade installations, and other installations (Appendix to § 3 of the law).

The proposer of the project determines and describes the foreseeable effects of the project on the population and the environment. This involves a process of communication and consultation with the relevant authorities and the general public. The competent authority then presents and appraises all the information on the project in a comprehensive report. The findings concerning licensing of the project are considered using the prescribed official procedure, and must be made available to the public. For licensable installations, the test for environmental compatibility is carried out as part of the licensing procedure. Details of this are given in 9th BImSchV.

The Law of Liability for the Environment of Dec. 10th, 1990 establishes liability for damage to the environment caused by installations specified in Appendix I (§ 1). These include the most important (according to the BImSchG) licensable installations. Liability is not limited to accidents or disturbances, but also includes normal operation (§ 5, § 6 Section 2). Liability is considerably increased by the presumption of cause in § 6, i.e., if an installation is liable to cause the damage produced, it is presumed that the damage was in fact caused by the installation. The authorities can set aside the presumption of cause on receipt of proof that the installation was operated in accordance with regulations.

Environmental Sanctions. According to the 18th Criminal Law amendment (law for combating environmental crime) of March 28th, 1980, modified as of June 27th, 1994 a special section on criminal offences against the environment was introduced into the penal code. The penal code of §§ 324–330 d includes the following punishable offences: § 325 (air pollutants hazardous due to the improper operation of installations), § 327 (illicit operation of installations), § 329 (imperilment by air pollutants of areas needing protection), § 330 (serious hazard to the environment), § 330 a (serious hazard due to the release of toxic substances).

4.3.15. Smog Regulations of the Länder

By the powers given to them in §§ 49 Section 2 and 40 Section 1 of the BImSchG, the individual Länder (with the exception of Mecklenburg-Vorpommern) have issued smog regulations which control the designation of smog zones, and specify measures to be taken to restrict traffic and reduce emissions in the case of a smog alert. The hazard levels (warning, first-stage alert, second-stage alert) correspond with the following measures:

1) Temporary restriction of the operation of plant (including complete shutdown)
2) Restriction of the use of certain fuels (e.g., high-sulfur coals and heating oils) in industrial and business enterprises and communal and private households
3) Prohibition of traffic for certain periods, certain vehicles, and certain zones

The smog regulations of the Länder follow the so-called model regulation developed by the Länder antipollution commission and therefore have essentially the same structure, although each Land has its own special features. All smog regulations, as well as defining and prescribing the above warning stages, contain certain smog event criteria. These criteria are based on both the actual meteorological conditions and the air pollution situation. Whereas the definition of still weather conditions can differ from Land to Land (e.g., for North Rhine Westphalia the definition is an average wind speed of < 1.5 m/s at ground level over a period of 12 h, but for Bremen 4 m/s), in all smog regulations of the Länder uniform criteria apply to the decision as to when a given concentration of harmful substances corresponds to one of the alert stages. However, the number and distribution of the measuring stations in the network of pollution monitoring stations in the smog zones of the Länder are different, so that ultimately the methods of deciding on the various alert stages in the individual Länder also differ. For example, different rules apply to the issuing of a smog alert if the concentration figures found exceed the limit values (for SO_2 in combination with suspended particulates, for SO_2 alone, for NO_2, or for CO) at more than 50 % of the measuring stations, at all measuring stations, or at, e.g., two neighboring measuring stations.

A smog alert is issued if the concentration limit values for only one of the above-mentioned harmful substances or for SO_2 in combination with suspended particulates are exceeded (e.g., the 24-h average and the last 3-h average of the sum of the concentration of the SO_2 and twice the concentration of suspended particles, or if the 3-h average for SO_2, NO_2, or CO is exceeded).

4.4. General Administrative Regulations for the Prevention of Air Pollution

The Federal Government is empowered according to § 48 BImSchG, after hearing the parties concerned and with the approval of the Bundesrat, to issue general administrative regulations. These regulations, which in principle are binding on the administrative authority only, serve to implement BImSchG and the statutory regulations made under this law.

4.4.1. TA Luft

The first general administrative regulation (Allgemeine Verwaltungsvorschrift – 1. AVwV) of the Federal Antipollution Law (Technical Directions for the Prevention of Air Pollution, or Technische Anleitung zur Reinhaltung der Luft, TA Luft) is an important instrument for the implementation of the Federal Antipollution Law. As described in Part 1 of the four-part TA Luft, this is to be applied in particular

1) When considering applications for a license for the construction and operation of an industrial installation, for any significant change in its location, design, or operation, or for a cleanup of such an installation
2) When considering applications for a preliminary decision
3) For supplementary regulations
4) When ordering the determination of the nature and extent of the emissions from an installation and of the extent of pollution in the zone of influence of the installation

The first TA Luft regulation, issued in 1964, has since been amended several times. The chronological development was as follows:

Sep. 8th, 1964	General administrative regulation for licensable installations according to §§ 16 industrial code (TA Luft 64).
Aug. 28th, 1974	First general administrative regulation (TA Luft 74) of the BImSchG. This was brought into operation in March 15th, 1974.
March 4th, 1983	Amendment of TA Luft. This was largely restricted to a revision of the regulations for licensing procedures and for the assessment of pollution for "defense against hazards".
March 1st, 1986	Amendment of TA Luft following amendment of the BImSchG (Oct. 4th, 1985) and the earlier (July 24th, 1985) amendment of 4th BImSchV for licensable installations, with adjustment of the installation catalog to conform to technical developments. The "principle of prevention" was increasingly taken into account, and the cleanup concept for old installations was first established. Approximately 50000 licensable industrial and business installations are affected in the Federal Republic of Germany.

The four parts of TA Luft are as follows:

1) Scope
2) General instructions for the prevention of air pollution
3) Requirements for the limitation and determination of emissions
4) Requirements for old installations (cleanup)

Part 2 of TA Luft (see above for further details of Part 1) specifies in more detail the mainly undefined legal terms included in the expression "harmful effects on the environment due to atmospheric pollution". Explanations are given of, e.g., the assessment of health hazards, significant detriment, significant pollution, assessment in those cases where limit values have not been established, and abnormal occurrences. In No. 2.5, in explanation of § 5 of 1 BImSchG, the term "harmful effects on the environment" is defined for important pollutants by means of limit values. A distinction is made between pollution levels at which there are health hazards and levels at which significant detriment and pollution occur. In the latter case, an abnormal occurrence investigation (No. 2.2.1.3 TA Luft) is required if a pollution limit value is exceeded, and this must also be carried out for substances for which no pollution limit values are specified in TA Luft but where there is "reasonable suspicion of harmful effects on the environment". Other fundamental explanations in Part 2 of TA Luft concern, e.g., the term "precautions", and give information about methods of limiting carcinogenic substances, the consideration of applications for the granting of an interim action, a partial license, modifications to a license, and the discharge of waste gases. In No. 2.6, "determination of pollution parameters", important aspects of pollution measurement (measurement planning, sampling, and analysis) are described, and guidelines can be obtained from the VDI handbook of air purity prepared by the KRdL (Kommission Reinhaltung der Luft, or Commission for the Prevention of Air Pollution) formed by the VDI (Verein Deutscher Ingenieure, or Association of German Engineers) and DIN.

In Part 3 "limitation and determination of emissions", emission limit values corresponding to state-of-the-art technology for licensable installations are specified, and emission minimization requirements are laid down. Large furnaces are not included, as these are covered by 13th BImSchV. Methods for the measurement and control of emissions are specified in Part 3. Emission measurement techniques are described in more detail in the VDI Guidelines listed in Appendix G. Finally, No. 3.1.1 of TA Luft refers to continuing detailing of the state of the technology as defined by the BImSchG, and to the guidelines for process and gas cleaning in the VDI handbook on the prevention of air pollution. These are listed in Appendix F.

A comprehensive plan for cleaning up old installations is set up in TA Luft 86, Part 4. Old installations must be renovated within a given time limit (between "immediately" and 10 years), depending on the type, quantity, and hazardous nature of the emissions and on any special features of the installation. A distinction is made between "supplementary regulations" with immediate effect for the prevention of harmful effects on the environment under certain conditions (if pollution limit values are exceeded or carcinogenic substances are emitted) and precautionary supplementary regulations. As for new installations, the severity of the precautionary requirements (time limits, technical measures) depends on the potential hazard of the emitted substance. Measures to minimize emissions are in all cases fundamentally preferable to improving the condition of discharges.

Unlike laws and regulations, TA Luft is an instruction to the authorities, and does not directly affect individual citizens. However, as its provisions affect antipollution licensing procedures, they have an indirect effect on individual citizens or proprietors of installations with respect to the principle of equality. TA Luft sets measuring standards that are uniform nationally and therefore ensure uniform treatment of proprietors and uniform legal security. Any deviation from its provisions is possible only in exceptional circumstances, e.g., if these provisions no longer correspond to an indisputable technological development.

4.4.2. Fourth BImSchVwV: Determination of Air Pollution in Test Areas

The general administrative regulation (AVwV) of Oct. 26th, 1993 is based on § 44 Section 1 BImSchG and certain guidelines by the European Parliament. It contains definitions of terms and instructions on the object of measurement, measurement areas (test areas as defined by § 44 Section 3 BImSchG), number and position of measuring points, measuring methods, measuring equipment, and the assessment of results. In order to inform the public about the pollution situation, the data must be provided in a prescribed form suitable for publication. In Appendices A–E of this AVwV, directions are given on EU Guidelines to be considered, the choice of measuring points and instructions to be followed, and the DIN Standards and VDI Guidelines that provide measuring techniques and the determination of process parameters (see also Section 4.4.1). The data must be recorded in a standard form, i.e., following the pattern given in Appendix E, and must be transmitted in the form of a report to the EU Commission. Furthermore, the data must be transferred to the German Parliament in accordance with § 61 BImSchG and, in accordance with the highest Land Authorities, to the Federal Environment Office.

The EU (or EEC) Guidelines cited in Appendix C1 and the corresponding Federal Guidelines mentioned in C2 for the monitoring of pollution are of special importance:

1) Guidelines for the construction and testing of the suitability of measuring equipment for the continuous monitoring of pollution (Gemeinsames Ministerialblatt (GMBl), 1981, p. 355)
2) Guidelines for selecting the location and method of construction of automatic measuring stations in telemetric measurement networks (GMBl, 1983, p. 78)
3) Guidelines for establishing reference methods, choosing methods of measuring equivalence, and using calibration procedures (GMBl, 1988, p. 191)

The AVwV will make it possible to establish a uniform federal practice and increasingly a European practice in the determination and monitoring of pollution for the assessment and improvement of the state of air pollution prevention.

4.4.3. Fifth BImSchVwV: Register of Emissions in Test Areas

The 5th BImSchVwV of April 24th, 1992 (amended on March 24th, 1993) contains legal requirements which the competent authorities must take into account regarding the establishment and regular updating of emission registers. These registers must be established for the test areas created in accordance with § 44 Section 3 of the BImSchG, in particular for the preparation of air antipollution plans (§ 47 BImSchG). Groups of emitters particularly affected include licensable furnaces (as specified in 1. BImSchV), and other furnaces. This AVwV also applies to "other non-licensable installations" with emissions of volatile halogenated hydrocarbons (as specified in 2. BImSchV) and wood dust (as specified in 7. BImSchV).

They also apply to nonlicensable installations where organic solvents are used if 2nd BImSchV does not apply, e.g., plants for recycling residues, painting, printing, smoke-curing or fumigation, also roasting plants, filling stations, stationary internal combustion engines, surface-treatment plants, and chemical cleaning plants. Other groups to be considered as emitters include livestock, installations of the Federal Railroads, Federal Postal Administration, and Federal Armed Forces, road vehicles, railroad locomotives, shipping, aircraft, and agricultural and military traffic "if inclusion of the emissions is necessary for preparation of the emission register".

The following substances or groups of substances produced by the emitters listed above are to be considered: dust, lead and inorganic lead compounds, sulfur dioxide, nitrogen oxides (expressed as NO_2), carbon monoxide, and 11 other substances or groups of substances including asbestos, diesel exhaust particles, ammonia, and especially those substances specified in No. 2.3 of TA Luft (carcinogenic substances).

The measurements are carried out either continuously or as individual measurements.

The setting up and presentation of the emissions register (style and scope) is carried out in accordance with established principles to make the data consistent and readily assessible, and to enable and facilitate the updating of the emission register. Thus, the style, scope, and system of presentation are regulated for the individual groups of emitters. The conclusions of the emissions statement order are also important here. The use of 5th BImSchVwV should provide the authorities and other participants with effective instruments for developing measures for reducing emissions and for calculating the spread of pollution.

4.5. Technical Regulations for the Prevention of Air Pollution

The term "technical rules" includes not only technical rules (TR) referred to explicitly under that name, but also standards, guidelines, directions, instructions, worksheets, leaflets, and recommendations produced by various private institutions. However, a technical rule in the wider sense is not only to be regarded as the responsibility of the private maker of rules, but can also require Governmental regulation (e.g., TA Luft, see Section 7.3.4.1.). In this section, the issuing of technical rules for the prevention of air pollution by the private institutions DIN, VDI, CEN (Comité Européen de Normalisation), and ISO is described in more detail in the national, European, and international context. These private institutions make an important contribution to the prevention of pollution within the State by converting so-called anticipated expert reports into the format of technical rules. All areas of technology except electrical technology are included. In the DIN catalog of technical rules in the DITR (Deutsches Informationszentrum für Technische Regeln, or German Information Center for Technical Rules), more than 130 private organizations are listed as rule makers. Over 200 books of technical rules are recorded. In 1994, ca. 80000 documents were entered as "technical rules" in the DITR databank in Berlin. Every year, ca. 6000 technical rules are added, and ca. 3000 are withdrawn and hence invalid. The DITR also operates and makes available the multinational and multilingual standards databank PERINORM. This contains the bibliographical information for all valid national standards and proposed standards for Germany, the United Kingdom, France, Austria, Switzerland, and The Netherlands, all European and International Standards, all other valid technical rules in Germany and France, and information on technical rules originating from Government and other authorities, including EU Guidelines. In total, PERINORM 1994 contains over 120000 entries. PERINORM data can be obtained directly from the DITR, or by using the computer disks available from DITR. The DITR is thus an important instrument for providing reviews and information on technical rules, including those relevant to environment protection.

4.5.1 National Regulations

In mid-1990 in Germany, the then existing organizations (the VDI Commission for the Prevention of Air Pollution and the DIN Standards Committee for the Prevention of Air Pollution) merged to form the *Kommission Reinhaltung der Luft (KRdL) im VDI und DIN* (Commission for the Prevention of Air Pollution in the VDI and DIN). The task of this committee is the preparation of VDI Guidelines, DIN Standards, DIN Provisional Standards, DIN–EN Standards, and DIN–ISO Standards. The office of the KRdL took over the secretariat of the ISO/TC (Technical Committee) 146 "Air Quality" (see also Section 4.5.3.) of the DIN Standards Committee for the Prevention of Air Pollution. Furthermore, on the initiative of the KRdL, the CEN/TC 264 "Air Quality" was established in March 1991 (see Section 4.5.2.). The secretariat of the KRdL is also responsible for this technical committee. In the establishment of technical rules for the prevention of air pollution, the principles, rules of procedure, and priorities of the DIN (in accordance with DIN 820) and also those of CEN and ISO apply to the procedural guidelines created for the KRdL. In the preparation of VDI Guidelines, the procedural guidelines of the VDI (in accordance with VDI 1000) must be followed.

The KRdL comprises over 200 working groups made up of ca. 1700 specialists in economics, science, and administration working together in an honorary capacity. The KRdL is subdivided into several specialist groups:

Group I: formation and prevention of emissions
Group II: environmental meteorology
Group III: the effects of dust and gases
Group IV: measurement techniques
Group V: methods of off-gas purification and dust technology

The work of the KRdL is supported financially by the Federal Government, whose interest in the work of the Commission is apparent from the statement of the case for the budget, according to which "the Federal Ministry for Environment, Nature Conservation and Nuclear Safety is supported by the KRdL in the performance of tasks in the area of prevention of air pollution as defined by § 1 of the Federal Antipollution Law such that the KRdL, in voluntary self-responsibility and together with the participating authorities, science, and industry, determines the current state of the art in science and technology and records this in guidelines and standards whose contents can become part of the function of the executive and the legislature".

The KRdL is in charge of ca. 380 VDI Guidelines, which form the six-volume VDI/DIN Handbook "Prevention of Air Pollution". The actualization and upkeep of and assistance with ca. 30 DIN/ISO Standards also fall within its sphere of responsibility. The first European Standards (EN) for the prevention of air pollution are currently in course of preparation (see Section 3.5.2.).

4.5.2. European Regulations

The protection of the environment is becoming more and more a political topic of the European Union (EU). The 5th Antipollution Action Program of the EU came into force on Jan. 1st, 1993, and will continue until the year 2000. The EU is politically active in the area of technical regulations. In its resolution (85/C 136-01) of May 7th, 1985, which referres to a "New Concept" in the field of technical harmonization and standardization, the European Parliament resolved to limit the harmonization of official regulations of the EU to basic necessities, and commisioned the existing private legislative bodies CEN (Comité Européen de Normalisation) and CENELEC (Comité Européen de Normalisation Electrotechnique) with the preparation of European Standards (EN) for the regulation of technical/scientific matters taking into account the current state of the art of science and technology. The European Standards Committee, CEN, includes the national standards organizations of the 18 countries of the EU and EFTA.

DIN officially represents the Federal Republic of Germany as a member of CEN (office in Brussels). A European Standard (EN) is developed with general consent, and is adopted by qualified majority vote. It must then, irrespective of the way in which the individual member country has voted, be included unchanged in all national standards, and becomes a DIN – EN Standard in the Federal Republic of Germany. National standards that conflict with it must be withdrawn in all CEN member countries. During the development phase of an EN under the umbrella of CEN, the so-called Standstill Agreement applies to the national standards may be written or published on the subject of this EN.

To avoid trade restrictions that could arise through national technical rules, CEN, in cooperation with EU and EFTA, provides an information service for standardization work carried out in the European countries (Guideline 83/EEC – Information Service on Standards and Technical Regulations). This openness about national standardization plans enables possible conflicts between member countries to be largely avoided, and allows coordinated standardization programs to be set up. Moreover, following an agreement on cooperation made in 1984 between the EU, EFTA, and CEN/CENELEC, the European Standardization Committees of the EU and EFTA are mandated to carry out special standardization work. In applying the above mentioned "New Concept" to the field of technical harmonization and standardization, the results of this standardization work form the necessary supplementation for the completion of the EU Guidelines and their incorporation in national law. The European standardization work on air quality is an important example of this.

In March 1990, the Commission for the Prevention of Air Pollution (KRdL) in VDI and DIN proposed to CEN a technical committee "Air Quality". In October 1990, this application was approved by the CEN member organizations. The office of the KRdL was empowered to take over the tasks of the secretariat. The inaugural meeting of CEN/TC 264 "Air Quality" took place in March 1991 in Bonn.

The scope of CEN/TC 264 "Air Quality" was defined at the inaugural meeting in Bonn as follows: "Standardization of methods for the characterization of air quality with respect to emissions, pollution, air within buildings, gases in the soil and emitted by it, and waste dumping, especially:

– the establishment of uniform measuring procedures for air pollutants (e.g., particles, gases, odors, and microorganisms) and methods for determining the effectiveness of systems for cleaning waste gas,

but excluding:

– the establishment of limit values for pollutants
– work areas and clean rooms
– radioactive substances."

The subject of "air within buildings" is expressly mentioned in the definition of the scope of CEN/TC 264. It is probable that this topic will become even more important in connection with the EC building materials guideline. At the present time (1994), the EU Guideline on the incineration of hazardous waste is of special relevance to the European standardization work for the prevention of air pollution. The object of this EU Guideline is "to preserve the environment and improve its quality, and to contribute to the protection of human health". In contrast to the former EC Guidelines on domestic waste, this EU Guideline sets stringent requirements which must be adhered to by all EU Member States. These stringent requirements prescribe specific techniques of both pollution limitation and measurement to achieve the low limit values. The guideline thus sets new standards in the field of EU Regulations for the prevention of air pollution.

The EU Guideline follows the "New Concept" (see above). Thus, for example, technical details of measurement are not discussed in the guideline itself. Instead, for information on measurement techniques, reference is made to European Standards (EN), or, if these are not available, to ISO Standards, or even national standards.

The measurement techniques used in existing national standards were often designed for considerably higher emission concentrations than those to be expected when European pollution reaches the low levels predicted for the future in the above EU draft guideline. There is therefore a need to review procedural parameters or even to develop completely new reference methods. Thus the CEN/TC 264 Technical Committee is required in many cases to provide improvements to measurement techniques or to carry out comparative testing of existing methods. For this, the standardization work for verification of procedural parameters will be accompanied by extensive European measurement programs (interlaboratory testing). For example, in association with the standardization work, elaborate comparative tests of various methods of measuring dioxin levels at domestic waste incinerators in Germany and Austria have been carried out. Six institutes from EU and EFTA countries contributed to these investigations. The measuring methods for threshold/guideline value monitoring of 0.1 ng/I-TEq m^3 for the dioxins or 10 mg/m^3 for HCl (24-h average 5 mg/m^3) must be validated. These two projects are mandated and financially supported by the EU (100 % support for the measurements, 50 % support for preparing the standards). This financial support ensures or at least makes it probable that the work will be carried out in the time set by the EU (1996).

4.5.3. International Regulations

The ISO (International Organization for Standardization) founded in 1947 is a worldwide association of national standards organizations. Only one national standards organization of any country can be a member of ISO at any one time. Over 70 % of the national member organizations of the ISO are linked with the government, or become so linked by incorporation into the law of the member country. The national member of ISO for Germany is DIN. The results of the work of ISO are published as International Standards (ISO Standards). Standardization for electrical technology is covered by the IEC (International Electrotechnical Commission).

The work of ISO worldwide is carried out by ca. 20 000 experts in over 200 technical committees (TCs). Each TC has a secretariat which is based at one of the national member organizations. Important environmental protection secretariats of ISO are assigned to DIN.

Up to 1994, ca. 8000 ISO Standards have been published. An ISO Standard is the result of the cooperative work of experts in the individual national member organizations, and is carried out according to fixed rules. It can be valid both as an ISO Standard in its own right and after conversion into a national standard (e.g., a DIN–ISO Standard).

Every year, ca. 10000 working documents worldwide are sent for further processing in the national working committees for agreement and to obtain the national votes. Meetings in the international area are called as necessary. ISO Standards are reviewed every five years for their validity or in case of scientific or technological developments. The ISO/TC 146 "Air Quality" was established in 1971. The secretariat was taken over by the office of the Standards Committee for the Prevention of Air Pollution (N Luft) of DIN. As the "KRdL of the VDI and DIN" was formed in 1990 (see Section 5.1.), the office of the KRdL in the VDI and DIN in Düsseldorf took over the secretariat of the ISO/TC 146.

The scope of the ISO/TC 146 "Air Quality" is as follows: "Standardization in the area of air quality including definition of terms, air sampling, measurement techniques, and reporting of characteristic values for air. Limit values for air quality are excluded." The ISO/TC 146 comprises ca. 30 working groups from which four subcommittees (SCs) are formed. The meeting of the ISO/TC 146 in the autumn of 1993 in Philadelphia set up two new subcommittees for "Indoor Air" and "Meteorology."

To prevent trade restrictions in third party countries, close cooperation and the performance of standardization work by harmonization of the standardization programs of CEN and ISO are extremely important. The cooperative work between ISO and CEN is therefore regulated by the Vienna Agreement of July 1991. Also, the cooperative work of the national standardization organizations with CEN is carried out according to fixed rules and on the basis of formal agreements. New European standardization projects can be proposed by any CEN/CENELEC member, by any European organization, and by the Commission of the European Union.

5. Air Pollution Monitoring

Introduction. Air pollution monitoring is the systematic measurement of impurities in the atmosphere at locations where harmful and/or adverse effects on "acceptors" such as humans, animals, plants, or materials may occur (immission measurements). Atmospheric impurities are trace substances that are not natural components of the atmosphere [5.1] or substances whose concentration in the atmosphere deviates significantly from the natural level as a result of human activities [5.2].

Assessment is made by comparing measured levels of substances (e.g., concentrations, m/V; amounts deposited $m A^{-1} t^{-1}$, or doses, m (harmful substance/m (body weight) $\times t$ or $m V^{-1} t^{-1}$) with limit values based on their effects. These limit values are specified to protect acceptors from harmful environmental effects and to maintain natural living conditions [5.3].

However, attempts to provide the fullest possible monitoring of air pollution in a given area encounter the following main difficulties:

1) In densely populated and industrial areas, the number of hazardous substances that can be emitted into the atmosphere and are listed in the emissions registers can run into hundreds. These substances have very different chemical, physical, and toxicological properties, and their concentrations can extend over ca. 12 orders of magnitude (from mg/m^3 to fg/m^3). This precludes any simple, quick measurement and assessment.
2) The dynamics of the atmosphere and the variation in the emissions over space and time lead to large variations in the concentration of atmospheric impurities over space and time. The measurement of all possible air quality conditions in an area (statistically, the "population") is not practicable, so that measurements of pollution always have the character of spot checks.

The above difficulties have led to the development of a wide range of specific measuring techniques and make detailed measurement planning necessary.

In this chapter, both measuring methods and measurement planning are described for the most important fields of air pollution monitoring. Related topics such as the monitoring of air in buildings and the measurement of ground air are not discussed. Reviews can be found in [5.4], [5.5].

5.1. Measurement Planning

5.1.1. Types of Air Pollution Monitoring

The starting point of every measurement plan is the precise definition of the questions to be addressed by the results. Examples of important pollution measurement tasks are:

1) Measurements for specified areas
a) Pollution level measurement over a large area: general air pollution monitoring [5.6]
b) Representative measurement of expected pollution in the area affected by a planned factory (licensing procedures, TA Luft [5.7])
c) Real-time measurements of acute episodes of high atmospheric pollution (winter smog [5.8], high ozone concentrations [5.9])
d) Measurements in highly polluted areas to estimate maximum exposure of acceptors (e.g., EU Guidelines NO_2 [5.10] and SO_2/suspended particulate matter (SPM) [5.11], and measurements in street canyons in accordance with § 40 Section 2 BImSchG [5.12])
e) Monitoring of the effects of pollution-control measures, temporal trends [5.13]
f) Orientational (screening) measurements [5.14]
g) Scientific investigations, e.g., of the transport of atmospheric pollutants, chemical reactions, calibration of diffusion modeling
2) Measurements in and around industrial plants
a) Measurements following complaints
b) Investigation of emitters, cause analysis [5.15], [5.16]
c) Measurements in cases of fire and accidents [5.17]
d) Monitoring the effectiveness of pollution abatement measures [5.18]
e) Monitoring of installations producing diffuse emissions [5.16]

Each of these tasks requires its own dedicated measurement plan.

The most important area of activity is the assessment of air quality in given areas by comparing measured levels of pollution with standards based on the effects of such pollution. This applies in particular to countries that have imposed limit values on pollution levels [5.3]. In their antipollution policies, these countries usually employ the dual strategy of imposing limits on both emissions and levels of environmental pollution. Limit values for air quality are intended to reduce both harmful effects on the environment and hazards to health (smog), and can also have preventive effects (e.g., guide values from the EU directives for NO_2 [5.10] and SO_2/SPM [5.11]). If these limit values are exceeded, emission reduction must follow. They are therefore an element of pollution control.

5.1.2. Planning of Pollution Measurements

The measurement of environmental pollution should be planned in such a way that the results will answer all the questions for which they are designed at minimum cost (see Section 5.1.1). The measurement plan therefore includes an assessment of technical, financial, and personnel aspects, and also of the relevance of the data obtained to the information required. The important parameters that must be fixed in any measurement plan are listed in Table 5.1, which also represents a typical example of air pollution measurement in the area surrounding a planned industrial plant in accordance with the requirements of German licensing procedures [5.7].

5.1.2.1. Measurement Networks (Measurements Over a Large Area)

The example given in Table 5.1 is typical of an important procedure for measuring pollution over a large area with the aid of grid system [5.19]. This type of grid is used if exposures in an investigation area are to be determined as representatively as possible. The grid arrangement of measurement points ensures a representative selection of locations in the investigation area. If the structure of the measurement network is independent of the structure of the investigation area (e.g., location of residential and industrial areas), the number of measurement points is weighted by the random structure of the network to correspond with areas of different levels of pollution. The density of measurement points on the network determines the resolution of the system; hence structures smaller than the network unit area (1 km^2 in the example of Table 5.1) cannot be detected (e.g., points of high pollution in streets). The frequency of sampling at the measurement points determines the temporal resolution. If continuous measurement is performed (see Section 5.3.1.2), the temporal resolution can be practically perfect, except for periods when the equipment is out of commission.

The effect of data loss in sequential measurements can be determined by statistical methods [5.19], [5.20].

When using random measurements, not only must the size of the statistical sample be adequate [5.21], but also measurement times must be randomly selected to ensure accurate representation of the time dependence. As atmospheric pol-

Table 5.1. Parameters of pollution measurements to be established for a measurement plan

Parameter	Example [5]
Inquiry	preliminary (screening) pollution measurements in the licensing procedure, representative random sample measurements
Measurement area	circular area around pollution source with a radius of 30 times the effective stack height (simplified)
Assessment criteria (place and time dependent) needed to evaluate measured data used to obtain characteristic values	limit values IW 1 and IW 2 of TA Luft; arithmetic annual mean (I 1) or 98 percentile (I 2) based on 1 km^2 (assessed surface)
Arrangement, choice and density of measurement sites	uniform grid of 1 km^2 mesh, and "random" choice of measurement sites
Measurement period	one year, or at least six months
Measurement height	1.5–4 m above ground level
Measurement frequency	52 (104) measurements per unit assessment area for gaseous pollutants, depending on the pollution load
Duration of individual measurement	$^1/_2$ h for gaseous pollutants, 24 h for airborne dust, one month for dust deposition
Measurement times	random choice
Measurement objectives	emitted atmospheric pollutants from the planned installation
Measurement methods	national standard measurement methods (VDI Guidelines)
Required reliability of measurements	high
Quality requirements, quality assurance, calibration, maintenance	VDI Guidelines
Collection, validation, archiving, and assessment of measured data	calculation of characteristic values I1V and I2V for each assessment area
Costs	depends on measurement area and objectives

lution often shows marked variation during the day and correlation with the seasons, measurements only at certain times of the day or certain times of the year would not give representative results.

The number of samples needed for random sample measurements, in accordance with the boundary conditions listed in Table 5.1, has been investigated statistically for real sampled populations. It was found that the information content of the group (statistical sample) of random samples increased only slightly if the number of measured values was >100 [5.21].

The basic concept of grid measurement can also be used on area-representative measurement networks for general pollution monitoring over large areas [5.20]. Depending on the size of the area, larger distances between the lines of the network are chosen, with measurement points 4–16 km apart. Several German air pollution networks are designed in this way for densely populated areas (e.g., the Rhine-Ruhr area [5.19], Berlin, and Hamburg). These also serve as smog alert systems.

Another method of designing area-representative measurement networks uses a representative selection of measurement points in areas of different use and air quality (e.g., residential areas, city centers, industrial zones, green areas, and points of high traffic density). The results give a realistic picture of the exposure if the relative lengths of time spent by vulnerable objects (e.g., people) in the different areas (microenvironments) are estimated, and the measurements weighted accordingly. This system [5.22] is often used for studying the exposure of humans to certain substances, including the interiors of buildings, and possibly other environmental media.

The required number of measuring stations, and hence the density of measurement points, depends on the structure of the area under investigation (e.g., position and size of emitters, and areas of human habitation, business activity, and free areas). These factors determine the immission structure and hence the required local resolution. Several methods have been developed to optimize the position and number of measurement points. For the measurement networks in Paris [5.14] and Madrid, the immission structure for nitrogen dioxide was determined approximately at many measurement points (146 points in Paris) with a simple screening procedure (diffusion tubes, see Section 5.3.7) to obtain the basic information for planning the measurement grid.

Statistical methods [5.19], [5.23] were used to determine the standard error expected when figures determined at the measurement points were used to estimate corresponding figures at points in the surrounding area. By assuming "tolerable" margins of error, it can be shown whether existing measurement grids fulfil the required conditions or how they can be optimized.

The measurement procedures described above are intended to give a representation of the ambient air quality or exposure in given areas. They are less suitable for determining exposure at places where objects to be protected are likely to be exposed to the highest levels of pollution (e.g., residents in street canyons with high traffic densities) or for determining whether a limit value is being complied with at every point in an area of a country.

This type of question presupposes a definition of limit values that refers to a single point in space and time [e.g., EU directive limits, measurements at locations with high traffic density according to § 40 Section 2 of the Antipollution Law (BImSchG) [5.12]]. The measurement plan therefore concentrates on points of maximum exposure. Based on knowledge already available (e.g., structure of the area, emissions register with diffusion modeling, screening measurements), the most important areas of pollution are found, also taking into consideration the duration of exposure of the exposed objects, and are used as the measurement points [5.24]. A problem of this approach is that the determination of the most serious pollution points is generally not entirely unarbitrary, and the measurements can only be used to a limited extent to assess "real" exposure because of the worst-case assumption. However, the system has the advantage that if the correct choice of measurement points is made, the measurement cost is comparatively low.

5.1.2.2. Evaluation of Data; Period and Frequency of Measurements

The choice of measurement period and frequency and the evaluation of the results depend mainly on the definition of the assessment criteria (limit values) used. The assessment and measurement systems must complement each other. If the criteria are related to a one-year period, the measurement period must be the same. If, however, alerts or short-term abatement measures are based on periods of 1–3 h (e.g., as in the German Smog Regulations [5.8] or the EC Directive for atmospheric pollution by ozone [5.9]), the temporal resolution of the measurements must be smaller than the assessment interval, which makes continuous measurement necessary. Only for these decisions, which must practically be made in real time, are continuous measurements essential. In the determination of annual representative values, random samples are sufficient, even though in practice continuous-measurement grids have become established in many cases.

For regulatory purposes (e.g., comparison with limit values), it is generally sufficient to derive two characteristic figures from the statistical sample of measured values; the average pollution (arithmetic mean or median) and the peak pollution (e.g., a 98 percentile). A 98 percentile figure means that 98% of the measurements in the statistical sample lie below this figure and 2% above it. It is statistically more robust than the highest or second highest measurement, which are often used in the United States to characterize peak pollution figures.

Attempts to calculate the characteristic values from the statistical sample of measurements by assuming a statistical distribution (e.g., normal or log normal distribution) has not proved fruitful. Especially where effects of single sources are concerned, there can be very skewed distributions of values, and even distributions with several peaks. Estimation of quantiles that do not depend on assumed distributions is therefore preferable [5.25].

The question whether and how the uncertainty of measurement of pollution (e.g., expressed as a confidence interval with a given confidence level) should be taken into account when comparing measurements with limit values is the subject of controversy. A draft ISO Guideline contains recommendations on this subject [5.26]. In contrast to the measurement of emissions, the results of immission measurements are generally compared directly with the limit values to determine whether these have been complied with or exceeded without explicit reference to confidence intervals.

5.1.2.3. Planning Measurements on Industrial Plants

Careful measurement planning is needed for individual questions of pollution monitoring, such as measurements specifically relating to cause analysis in connection with complaints, proof of effectiveness of measures taken to reduce emissions from diffuse sources [5.18], or supervision of large petrochemical installations with many potential diffuse emissions [4.16]. Such planning can greatly reduce measurement costs compared with schematic use of measurement grids. The assessment of the validity of complaints or of the results of abatement measures for single sources often requires one or two measurement points only, these being located at the place specified by the complainant or the point of maximum pollution when determining peak and average figures due to a particular source. These are also evaluated as a function of wind direction. For cause analysis and the identification of sources of pollution, simultaneous measurements are made upwind and downwind; mobile measurements are taken downwind and assessed as a function of wind direction. In the case of multicomponent measurements, the methods of fingerprint or cluster analysis are used [5.15], [5.27], [5.28]. Especially when sources emit several substances simultaneously, with a characteristic emission pattern (e.g., ratios of different hydrocarbons or heavy metals to one other), analysis of the pollution pattern, e.g., as a function of the wind direction, can often enable the main cause or causes to be determined, and in favorable cases even the approximate relative immission contributions [5.15].

5.2. Quality Assurance

Immission measurements can be very costly with respect to capital investment, management, and personnel. Based on the results, very far-reaching decisions can often be made, with considerable economic or ecological consequences. Therefore, quality assurance forms an inseparable part of the measurement system [5.4]. Quality assurance can be divided into two different categories, which are described in more detail in Sections 5.2.1 and 5.2.2 [5.29].

5.2.1. Procedure-Based Methods

Every comprehensive immission-measurement procedure consists of several steps: sampling, sample preparation and cleanup, isolation of the pollutant, detection, data collection, and data evaluation. In a few cases, especially in the continuous measurement of inorganic gases, certain steps of the sequence can be omitted (e.g., separation of the pollutants).

In most types of pollution measurement, extensively documented, comprehensive procedures, covering all steps from sampling to data evaluation and standardized in DIN–ISO Standards, CEN Standards, or VDI Guidelines [5.30], are recommended.

An important part of the documentation of measuring procedures is the specification of characteristic values such as

1) Detection limit: the smallest concentration or amount of a substance that can be distinguished from zero with a given probability (e.g., 95 %)
2) Trueness (accuracy): the degree of agreement between a measured value and the "true" value or an accepted reference value
3) Precision: the degree of agreement between values measured under given conditions by the repeated application of a measuring procedure

There are various definitions of these and other statistical concepts. Details can be obtained from [5.31]–[5.34].

The comprehensive VDI Guidelines have hitherto been of the greatest importance in the standardization of pollution measurement in Germany [5.35]. However, as the EU becomes more and more firmly established, the importance of European and international standardization regulations will grow rapidly [5.36].

An important element of quality assurance in the monitoring of emissions and immissions in Germany was established in the form of so-called National Practice in the Field of Ambient Air Quality Monitoring [5.4]. Under this heading, the German Minister for the Environment, Nature Conservation, and Nuclear Safety has published several guidelines in the General Ministerial Paper (GMBl). These relate to the setting up and testing of measuring equipment for continuous monitoring of air quality [5.37], the selection of sites, the construction of automatic measurement stations for telemetric monitoring networks [5.38], and the establishment of reference, equivalence, and calibration methods [5.39].

5.2.2. User-Based Methods

The use of standardized and tested equipment and procedures for the measurement of pollution cannot of itself ensure an acceptable quality of measurement if adequate systems of quality management are not operated by the user. The series of Standards DIN–ISO 9000 (identical with the European series of Standards EN 29000) and EN 45000 play an important role in ensuring that the user does in fact operate correct quality assurance procedures. The first-named series, which describes both the fundamental principles and the setting up of quality management systems, includes the Standard DIN–ISO 9004 ("Handbook of Quality Assurance and Elements of a Quality Assurance System") [5.40].

According to the European Standard DIN–EN 45001 "General Criteria for the Management of Testing Laboratories" [5.41], laboratories intended for the measurement of pollution are also to be regarded as testing laboratories. This standard is currently being revised, and will adopt essential elements of the ISO Guide 25:1990 [5.42], which provides the most detailed recommendations for quality assurance management systems currently available.

In pollution measurement, good laboratory practice (GLP) is a basic requirement, which includes the exact documentation of all methods and procedures used. (This does not mean the strict formal restrictions of GLP prescribed by the German Chemical Act). The so-called standard operating procedures (SOPs), in which all details of the analytical procedures are described as fully as possible, are of central importance.

As well as the systematic repair and maintenance of all measuring equipment and rectification of faults, regular calibration and monitoring of functions are essential [5.43], [5.44]. There are various types of calibration operations [5.4], [5.39], which differ in complexity according to the tasks and type of application.

A further important element of quality assurance is interlaboratory testing [5.29], [5.45]–[5.47]. In pollution measurement technology, this is carried out with the aid of air sample distribution equipment [5.48] which enables standard gas mixtures and aerosols of widely different compositions and qualities to be produced and supplied. A number of participating investigators take samples repeatedly from the same gas/aerosol supply and analyze them chemically. This data are then statistically analyzed, enabling important conclusions to be drawn regarding measuring procedures and participants (accuracy, precision, repeatability, comparability).

Also, interlaboratory testing is an important element of accreditation by the Länder of institutes to carry out measurements in accordance with §§ 26 and 28 of the Antipollution Law (BImSchG). This is a special variation of an accreditation procedure based on the European Standards series EN 45000. It means that in the area of pollution measurements, which must be approved by the competent authorities, these measurements must largely be carried out by those agencies which are authorized by the highest authority in the Federal Land concerned. According to rules laid down by the Länder Committee for Pollution Protection (LAI) [5.4], the measuring institutes must prove their neutrality and independence, must have at their disposal the necessary specialist personnel and equipment, and must be able to fulfil certain quality requirements by means of interlaboratory testing methods of the kind outlined above. For the future, there are plans to use the work of private accreditation agencies for the national authorization procedures.

5.3. Measuring Methods

5.3.1. Measuring Methods for Inorganic Gases

A very wide range of manual and automatic measuring methods is available for inorganic gases [5.4], [5.49].

5.3.1.1. Manual Methods

In manual measuring methods for inorganic gases, the substance to be measured is generally absorbed in a solution or adsorbed on a solid material during sampling. In most cases, this is followed by a color reaction and photometric determination. Several manual methods are important reference methods. Because the personnel requirement is comparatively high, these are seldom used today for field measurements if alternative automatic methods are available. The most important methods are briefly outlined in Table 5.2.

Table 5.2. Manual measurement methods for inorganic gases

Substance	Method	Details	Comments*	Refs.
SO_2	TCM method	absorption in tetrachloromercurate solution (wash bottle); reaction with formaldehyde and *p*-rosaniline to give red-violet sulfonic acid; photometric determination	DL = 0.2 µg SO_2; reference method; s = 0.03 mg/m³ at 0.5 mg/m³	[5.50], [5,51], [5.54]
SO_2	silica gel method	removal of interfering substances by conc. H_3PO_4, adsorption on silica gel; thermal desorption in H_2 stream and reduction to H_2S; reaction to give molybdenum blue; photometric determination	DL = 0.3 µg SO_2; s = 0.03 mg/m³ to 0.5 mg/m³	[5.52], [5.53]
NO_2	Saltzman method	absorption in reagent solution to form a red azo dye (wash bottle); photometric determination	calibration with sodium nitrite; DL = 3 µg/m³; reference method	[5.55], [5.56], [5.57]
O_3	potassium iodide method	formation of iodine from aqueous potassium iodide solution (wash bottle); photometric determination	DL = 20 µg/m³; rel. $s = \pm$ 3.5% at 390 µg/m³; reference method	[5.58]
F^-	silver bead method; variant 1	sampling with dust removal; concentration of F^- on silver beads coated with sodium carbonate; elution and measurement with ion-sensitive lanthanum fluoride electrode	simultaneous detection of an unknown proportion of particulate fluoride pollutants	[5.61]
F^-	silver bead method; variant 2	sampling with heated membrane filter; concentration of F^- on silver beads coated with sodium carbonate; determination by electrochemical method (variant 1) or photometrically (alizarin-complexone method)	danger of low results due to partial sorption of gaseous fluoride pollutants on membrane filter; DL = 0.5 µg/m³	[5.62]
Cl^-	mercury rhodanide method	absorption in 0.1 M sodium hydroxide solution (wash bottle); reaction with mercury rhodanide and Fe(III) ions to form iron thiocyanato complex; photometric determination	DL = 9 µg/m³	[5.59], [5.49]
Cl_2	methyl orange method	decolorizing reaction with methyl orange solution (wash bottle); photometric determination	DL = 0.015 mg/m³	[5.64], [5.65]
NH_3	indophenol method	absorption in dilute H_2SO_4 (impinger/wash bottle); reaction with phenol and hypochlorite to form indophenol dye; photometric determination	DL = 3 µg/m³ (impinger); partial simultaneous detection of NH_4 compounds and amines	[5.66], [5.67]
NH_3	Nessler method	absorption in dilute H_2SO_4 (impinger/wash bottle); distillation and reaction with Nessler's reagent; photometric determination	DL = 2.5 µg/m³ (impinger); partial simultaneous detection of NH_4 compounds and amines	[5.68], [5.69]
H_2S	molybdenum blue method	sorption as silver sulfide on glass beads prepared with silver sulfate and potassium hydrogen sulfate (sorption tube); liberation as hydrogen sulfide and reaction to form molybdenum blue; photometric determination	DL = 0.4 µg/m³	[5.70], [5.71], [5.72]
H_2S	methylene blue method	absorption in cadmium hydroxide suspension to form CdS; liberation of H_2S; reaction to form methylene blue; photometric determination	DL = 0.3 µg/m³	[5.73], [5.74], [5.75]

* DL = detection limit; s = standard deviation.

For several decades, sulfur dioxide was regarded as the most important atmospheric pollutant. Consequently, this substance is included in most pollution measurement programs, and many methods for measuring pollution by sulfur dioxide exist. However, very few of these are of practical significance today. The most important manual methods are the tetrachloromercurate (TCM) method [5.50], [5.51] and the silica gel method [5.52], [5.53]. The TCM method is also the most widely used measuring method internationally (e.g., in EU Directives for limit values and guidance values for SO_2), and is also described in an ISO Standard [5.54]. It is therefore the fundamental reference method for practically all measurements of sulfur dioxide pollution carried out in Germany and Europe.

The Saltzman method [5.55]–[5.57] is the most important manual method for nitrogen dioxide, and is established in Germany as the reference method for measuring pollution by this substance [5.39].

The most important manual method for ozone pollution measurement is the potassium iodide method [5.58]. This is one of the two reference methods established in Germany for ozone [5.39].

Samples for measuring pollution by hydrogen fluoride and fluorides can be taken by impingers [5.59], sorption tubes [5.60]–[5.62], special filters, or denuders (diffusion separation tubes). Analytical determinations are mainly based on photometric, electrochemical, or ion chromatographic methods [5.49], [5.63].

Pollution by chlorine and chlorine compounds is seldom measured in practice, as chlorine itself occurs very rarely and in very few locations because of its high reactivity. The methyl orange method is often used [5.64], [5.65]. Spot sampling for HCl and chlorides can be carried out by using impingers containing dilute NaOH solution as the absorbent [5.59]. The solution obtained can be analyzed by ion chromatography, ion-selective electrodes, or the mercury rhodanide method [5.49].

Ammonia in the atmosphere is in chemical equilibrium with ammonium compounds, so that NH_3 and NH_4^+ compounds are often determined together. The most important manual methods are the indophenol method [5.66], [5.67] and the Nessler method [5.68], [5.69].

In the manual methods for measuring pollution by hydrogen sulfide, the H_2S is chemisorbed as the sample is taken. The sulfide formed is determined photometrically by the molybdenum blue [5.70]–[5.72] or methylene blue method [5.73]–[5.75], the latter being more specific.

A special sampling method, mainly used in conjunction with manual measurement, is based on the diffusion separation tube (denuder), which separates the gaseous phase from the particulate phase by means of the different diffusion rates. It is therefore mainly used for difficult separation problems involving systems of this type [5.76]. In the classical denuder technique, the air sample is drawn through a glass tube with a coating which depends on the substance or substances to be determined [5.77]–[5.80]. For example, in the determination of basic components such as ammonia and ammonium compounds, tubes coated with an acidic substance (e.g., oxalic acid or citric acid) are used. The gaseous substances are deposited and fixed on the tube wall by chemical reaction, while the particles mainly pass through the tube and can be collected on a filter.

Many variations of the denuder technique have been developed, some of which have been automated (annular gap denuders, wet denuders containing solutions of reagents, thermodesorption denuders) [5.81], [5.82]. Denuders are the preferred method for sampling systems such as ammonia/ammonium compounds, nitrogen oxides/nitric acid/nitrates, sulfur oxides/sulfuric acid/sulfates, or hydrogen halides/halides. The denuder technique has much extended the possibilities of taking differentiated samples. However, depending on the particular version of the method, it needs a comprehensive laboratory, and correct use of the technique requires great experience.

5.3.1.2. Automatic Measuring Methods

A large number of continuously measuring monitors are available for the substances sulfur dioxide, nitrogen oxides, carbon monoxide, and ozone. These are very widely used, especially in measurement networks, and the German Ministry for the Environment, Nature Conservation, and Nuclear Safety has declared that certain types of equipment, produced by various manufacturers, are suitable for pollution measurement [5.4], [5.83]. The main features of the individual methods are listed in Table 5.3.

All automatic measuring methods based on physicochemical principles must be calibrated

Table 5.3. Automatic measurement methods for inorganic gases

Substance	Measurement principle	Comments	Refs.
SO_2	conductometry; reaction of SO_2 with H_2O_2 in dilute H_2SO_4; measurement of conductivity increase	elimination of spurious reactions by selective filter ($KHSO_4/AgNO_3$)	[5.84], [5.85]
SO_2	UV fluorescence; excitation of SO_2 molecules by UV radiation (190–230 nm); measurement of the fluorescence radiation	spurious reactions (e.g., with hydrocarbons) must be eliminated by filters	[5.49], [5.86]
NO/NO_2	chemiluminescence; reaction of NO with O_3 to form NO_2; detection of chemiluminescence radiation by photomultiplier	NO_2 only indirectly measurable; use of converters to reduce NO_2 to NO; measurement of NO and NO_x ($=NO + NO_2$) in separate measurement channels	[5.87], [5.88], [5.89]
CO	nondispersive infrared absorption; measurement of IR absorption by substance-specific detector with reference cell	reference: a) cell with N_2; b) sample air after selective removal of CO by catalytic/chemical reactions; c) optical removal of the CO absorption (gas filter correlation)	[5.49], [5.90]
O_3	UV absorption; low-pressure Hg lamp as radiation source (253.7 nm); recording of the UV absorption according to the Lambert–Beer law; detector: vacuum photodiode, photosensitive valve	reference: air sample after selective removal of the ozone by chemical/catalytic reactions (e.g., Cu/MnO_2)	[5.91]
O_3	chemiluminescence; reaction of O_3 with ethene to form formaldehyde; detection of chemiluminescence by photomultiplier	good selectivity, ethylene required as reagent gas	[5.92]

using (manual) reference methods. This even applies to the measurement of ozone by UV absorption, although this in principle is itself a reference method under controlled conditions. Experience shows that knowledge of extinction coefficients at 253.7 nm coupled with the validity of the Lambert–Beer law does not ensure that measured values will be automatically correct. For example, soiling of the gas passages, malfunction of the internal ozone scrubber for the quantitative decomposition of the ozone in the reference measurement cycle, or electronic defects can lead to significant false readings.

5.3.1.3. Calibration of Automatic Measuring Methods

Automatic measuring equipment is mainly used in measurement stations, and these often form a part of a pollution measurement network (see Section 5.4).

As the equipment is operated under these conditions for long periods of time (sometimes several weeks) without direct personal supervision, it is essential to monitor their correct function by using an automated procedure on a regular basis. An essential requirement for this supervision is the availability of test gas generators, which supply blank samples and standard gases to the measuring equipment.

A blank sample gas is one in which the concentration of the substance to be measured and for which the equipment is being calibrated is present at only trace concentrations, below the detection limit of the measuring method. Blank sample gases must not contain substances that react with the component to be measured or interact in any other way. A test gas is a gas mixture which usually consists of a carrier gas (nitrogen, air) with one or more added components [5.4], preferably produced by dynamic test gas generators [5.93].

Nitrogen or synthetic air in pressurized gas cylinders can also be used as blank gases. Such blank gases are relatively easy to handle, but often have the disadvantage in pollution technology that their properties differ from those of the matrix air being analyzed (e.g., in moisture content). There is also the necessity for preparing and changing the gas cylinders. Therefore, blank gases are preferably prepared immediately before use from the air to be analyzed, e.g., by filtration through sorbents.

For many substances, standard gases can be obtained commercially in pressure cylinders. If the gas cylinders contain the standard gas in the concentration required, dilution is unnecessary. However, with such standard gases there are problems of high gas consumption and sometimes poor storage stability, which usually de-

Table 5.4. Production of test gases for calibration

Component		Atmosphere	Pressure cylinder (direct)	Pressure cylinder (with dilution)	Permeation	Gas-phase titration
SO_2	blank gas	x	x			
	test gas		x	x	x	
NO	blank gas	x	x			
	test gas		x	x		x
NO_2	blank gas	x	x			
	test gas		x	x	x	x
CO	blank gas	x	x			
	test gas		x	x		
O_3	blank gas	x	x			
	test gas	x				

creases with decreasing concentration. Alternatively, gas cylinders can be used in which concentrations are 50–100 times as high as that required in the standard gas. In this case, controlled dilution must be carried out with the base gas. A standard gas generator operating on this principle is more expensive, and the dilution stage is a potential source of error. Advantages include the generally better storage stability and lower rate of consumption of the bottled gas.

The standard gas mixture can also be prepared by permeation, or, in the case of nitrogen oxides, by gas-phase titration [5.94]. The most important methods used for preparing standard gas mixtures containing components measured in pollution measurement networks on a routine basis are indicated in Table 5.4.

5.3.2. Measurement and Analysis of Particulate Atmospheric Impurities

Particulate impurities in air can be divided into deposited dust and suspended particulate matter (SPM). Deposited dust consists of larger particles which descend to the ground by virtue of their size and density. The particle size range of such particles cannot be exactly defined and depends in a complex fashion on many exterior conditions. SPM is that proportion of atmospheric particles that are quasi-stable and quasi-homogeneously dispersed, so that they remain suspended in the atmosphere for a certain period.

5.3.2.1. Measurement of Suspended Particulate Matter (SPM)

A large number of methods of measuring SPM exist [5.4], [5.49], [5.94], but few of these are of any practical significance for pollution monitoring in Germany. As with the measurement of gaseous atmospheric pollution, continuous and discontinuous methods are used.

Only a few methods do not involve precipitation of SPM, e.g., optical methods (scattered light photometers) and acoustic methods (counting of dust particles by highly sensitive microphones) [5.49]. Both methods are of minor importance in routine pollution measurement at present.

SPM is generally first collected on glass-fiber or membrane filters. This is followed by a gravimetric or radiometric determination. Sampling methods can be divided into those that measure total SPM without particle-size fractionation, and fractionating methods (e.g., for measuring fine dust). The phenomenon of impaction (separation based on inertial forces) is generally exploited for the particle-size-selective measurement of SPM.

The advantages and disadvantages of fractionating measurement of airborne dust are the subject of national and international dispute. In Germany, all limit values and criteria are based on the measurement of total SPM. In contrast, in the United States, the so-called PM-10 method (particulate matter $\leq 10\ \mu m$) is widely used [5.95]. Here, only those particles with an aerodynamic diameter of up to 10 μm are measured (50 % of the total), these being respirable and able to enter the lungs. Introduction of the PM-10 method into the European Community as a reference model is planned. At present, the question of whether even smaller aerodynamic diameters (e.g., PM-2.5) should be monitored in assessing effects on human is being debated. Fractionating measurement of SPM is considerably more complicated than the measurement of total SPM, as the measuring equipment must be equipped with special sampling heads which require a high level of maintenance.

As the object of measurement, SPM, is not exactly defined, and variations in concentration, particle-size distribution, and composition can occur with time and location, there is no absolute measuring method. Different values of a given measured quantity can be obtained depending on the method used. The conversion of measured values obtained with one method to those obtained by another is only possible under closely defined conditions. Therefore, in Germany, a basic procedure for comparing nonfractionating methods as reference methods has been laid down [5.39], [5.96]. Details of the most important methods of measuring SPM are given in Table 5.5.

Recently, automatic filter changers have been developed that store a large number of filters and supply them in sequence to the sampling device. The exposed filters are stored in a magazine. The detection limits of the filter methods are usually in the range 5–10 µg/m³ SPM.

The black smoke method of measuring SPM, developed in the United Kingdom, is also of importance [5.4], [5.114], [5.115], and has been included in the EU Directive for SO_2 and SPM. A sampling device which can automatically take eight samples simultaneously is used to draw the air at a low rate (ca. 2 m³/d) through a filter paper. The darkening of the filter is then measured with a reflection photometer. This black smoke value is then converted into gravimetric units (µg/m³) by means of a calibration curve [5.116]–[5.118]. However, this calibration curve was determined some time ago with dust containing a considerably higher proportion of soot than that found in modern urban areas. The result is that SPM concentrations measured gravi-

Table 5.5. Measurement methods for SPM

Method	Measurement principle	Comments	Refs.
Small filter equipment	nonfractionating sampling; air flow rate 2.7–2.8 m³/h; filter diameter 50 mm; gravimetry; volume measurement by revolving vane anemometer	low noise; easy operation due to small size and low weight; control by time-control switch	[5.97], [5.98]
LIB equipment	nonfractionating sampling; air flow rate 15–16 m³/h; filter diameter 120 mm; gravimetry; volume measurement by gas meter	recovery of large amounts of dust, useful for analysis of dust components; control by time-control switch	[5.99], [5.100], [5.101]
High volume sampler	detection of particles <ca. 30 µm diameter; air flow rate ca. 100 m³/h; filter diameter 257 mm; gravimetry; volume measurement by flow meter	recovery of large amounts of dust, useful for analysis of dust components; German version with improved volume measurement (anemometer principle); high noise level	[5.102], [5.103], [5.104]
FH 62 I	continuous radiometric dust measurement equipment; nonfractionating sampling; air flow rate 1 or 3 m³/h; dust collection on filter band; dust measurement by attenuation of β-radiation (^{85}Kr) on passing through exposed filter (ionization chambers)	gravimetric calibration by known amounts of dust on single filters; operation with preliminary separation of PM-10 possible	[5.105], [5.107], [5.108], [5.111], [5.112]
BETA dust meter F 703	continuous radiometric dust measuring equipment; nonfractionating sampling; air flow rate 3 m³/h; dust collection on filter band; dust measurement by attenuation of β-radiation (^{14}C) on passing through exposed filter (Geiger counter)	gravimetric calibration by known amounts of dust on single filters; operation with preliminary separation of PM-10 possible	[5.106], [5.109], [5.110]
TEOM 1400	continuous dust measurement equipment; nonfractionating sampling; air flow rate 1 m³/h; dust collected on filter (which is part of a vibrating system with resonance frequency) in side stream (3 L/min); recording of frequency decrease with increase in dust load on the filter	relationship between frequency decrease and mass of dust must be established by calibration	[5.113]

metrically by other methods are around two to three times as high as those determined by the black smoke method. Thus the dust measured by the black smoke method is different from that measured by other methods.

5.3.2.2. Measurement of Heavy Metals and Metalloid Compounds in Suspended Particulate Matter

Metal compounds in SPM samples are today often determined on a routine basis [5.119].

The SPM is first collected on filters and is then usually chemically digested to bring the metal and metalloid compounds into aqueous solution before analysis by standard methods. The most important of these are atomic absorption spectrometry (AAS) and spectrometry with plasma excitation (ICP-OES) [5.4], [5.94], [5.119]–[5.121].

Atomic absorption spectrometry is a reliable and relatively simple routine method, and is now one of the most important standard methods in the field of pollution measurement [5.122], [5.123]. The advantages of AAS include its high specificity, speed (flame technique), and high sensitivity (graphite tube, hydride technique, and cold vapor technique). However, AAS is in principle a single-element method, and if several elements have to be determined, costs (lamps) and time requirements increase, and other techniques may then be advantageous.

Atomic emission spectrometry with plasma excitation (ICP-OES) is now as important as atomic absorption spectrometry as a standard routine method in the field of pollution measurement. The inductively coupled high frequency plasma (ICP) is produced in an ionized gas (argon) in a high-frequency field, and acts as the atomizing and excitation medium for the liquid or dissolved sample. The ICP-OES method is the method of choice especially if many elements are to be analyzed in many samples. Other advantages include its suitability for elements that are difficult to atomize and those with a high affinity for oxygen, high precision, low detection limits for a large number of elements (generally better than flame AAS, but not quite as good as graphite-tube AAS), and a high linear dynamic range. Disadvantages include the relatively long time required for analysis (compared with flame AAS), often many lines in the spectra, spectral interference, comparatively high capital and operating costs, relatively high detection limits for some elements (e.g., cadmium), and the necessity for a relatively large sample.

The ICP method has lately been increasingly used as an ion source for mass spectrometry (ICP–MS). This is a very powerful but very expensive method, with extremely low detection limits for many elements.

Other methods for determining metal compounds in SPM include X-ray fluorescence analysis (XRF), polarography, and neutron activation analysis.

The use of XRF has decreased to some extent, because of stricter requirements for lower detection limits for carcinogenic metals and metalloids and for insensitivity to disturbing influences. However, standardized measuring methods exist for several elements [5.122]. The use of polarography and neutron activation analysis requires experienced operators or the presence of a neutron source, so that these methods are mainly used at certain research institutes.

Although metal compounds as constituents of SPM have already been measured in some locations for over a decade, there are still some important unsolved questions. For example, in the conventional method of sampling by collecting SPM on filters, it is assumed that all the heavy metal compounds are collected on the filter. However, there are reports that compounds of some heavy metals (Cd, Cr, Ni) can pass through the filter, although the results are inconsistent [5.119].

A further problem is that analysis of metal compounds in SPM by conventional methods does not distinguish between the various oxidation states of the elements or between their different compounds. In many cases, a determination of the total metal present is sufficient. However, with certain metals, especially those with carcinogenic properties (As, Cr, and Ni), this further information would be desirable.

There are often large differences in carcinogenicity not only between elements but also between different compounds of the same element (e.g., only chromium compounds of oxidation state VI are carcinogenic). In such cases, specific determination of individual compounds (specification) would be desirable. In spite of thc importance of this problem, such specification in pollution measurement is only at the experimental stage [5.119], [5.124], [5.125].

5.3.2.3. Measurement of Deposited Dust

Two methods are used for measuring deposited dust [5.126]:

1) Collection in collecting vessels
2) Collection on adhesive surfaces

The most common method of measuring deposited dust is the Bergerhoff method [5.126]. The entire material deposited from the atmosphere (dry and wet deposition) is collected in vessels for 28–32 d at ca. 1.5–2.0 m above ground level (bulk deposition). The collection vessels are then taken to the laboratory and processed (filtration, evaporation of the water, drying, weighing). The result is quoted per unit area of the collecting surface of the vessel and per unit of exposure time, i.e., in grams per square meter per day. According to [5.126], the detection limit is $0.035\ g\,m^{-2}\,d^{-1}$.

Other methods of determining the deposited dust include the Liesegang–Löbner apparatus [5.126] and methods in which the deposited dust is collected on adhesive films [5.126], [5.127].

All measurements of deposited dust are relative values that depend on the apparatus used, as the recovery of dust is influenced by air flow behavior and other apparatus-dependent parameters. Measurements by different methods can differ by as much as 50% [5.49].

5.3.2.4. Measurement of Heavy Metals and Metalloids in Deposited Dust

Both the amount and the composition of deposited dust are important, especially the content of lead, cadmium, and other metal compounds. The analytical methods used are fundamentally the same as those used in the analysis of SPM (Section 5.3.2.2).

5.3.2.5. Measurement of Fibers

It is important to determine the fibers present as atmospheric pollution, as asbestos has been classified as a proven carcinogen [5.128], [5.129], [5.130] and mineral fibers as suspected carcinogens. Carcinogenic fibers have a diameter of $d \leq 3$ μm and a length $l \geq 5$ μm, where $l:d \geq 3$.

Measuring methods for fibrous materials depend on collecting the fibers on filters and counting them under a microscope.

All methods of measuring fibers in the atmosphere use the electron microscope [5.131]. The method is used for fibers of diameter $d < 3$ μm and length l 2.5–100 μm. Samples are taken on gold-coated porous filters and ashed on the filter by plasma treatment to remove organic substances. The fibers on a randomly selected part of the filter surface are then counted under the scanning electron microscope and classified by geometry and type of fiber. Under the usual conditions of measurement and counting recommended in the guideline (1 m^3 sampled air per cm^2 filter surface, 1 mm^2 filter surface inspected), the detection limit is 300 fibers per cubic meter. Energy-dispersive X-ray microanalysis (EDXA) can distinguish between asbestos, calcium sulfate, and other inorganic fibers by determining their elemental composition. The procedure is complex and requires great care to obtain reliable results.

5.3.2.6. Measurement of Soot

Soot emitted from diesel engines has become relevant since it has been classified as carcinogenic [5.132], [5.133].

Owing to the variable and complex composition of diesel soot and because some of its constituents are also emitted from other sources, there is no specific method for measuring it. However, existing methods used for workplace monitoring have been adapted to enable reliable figures for the concentrations to be expected in outdoor air to be obtained [5.134]–[5.136]. The fraction of very fine dust (e.g., of particle diameter <7 μm in accordance with the Johannesburg Convention [5.133], or PM-10 [5.95]) in a certain volume of air is separated by a preliminary separator and collected on a glass-fiber filter. After solvent extraction and thermodesorption of organic constituents, the carbon content is determined by combustion in a stream of oxygen and coulometric titration of the carbon dioxide formed. This coulometric method is now the reference method for measurement of soot.

Optical determination of the soot content of the dust can be carried out using an aethalometer, based on the light transmission of an exposed dust filter compared with a reference measurement on the unexposed filter. The concentration can be calculated from the light attenuation, using an empirically determined absorption coeffient for particulate elemental carbon and the volume of air passed through the filter. The aethalometer values show good correlation with the coulometric method, but the absorption coefficient is dependent on the specific site [5.137].

In the photoelectric aerosol sensor (PAS), the particles are exposed to UV radiation ($\lambda = 185$ nm) which causes photoionization of particles adhering to the soot. The charge on the aerosol is measured by an electrometer. The ionization of the aerosol is mainly due to adhering polycyclic aromatic hydrocarbons (PAH). The PAS data also correlate with the coulometric method, but the proportionality factor depends on the site, the age of the aerosol and meteorological conditions [5.137].

The black smoke method [5.4] also shows a stable relationship to the coulometric reference method for concentrations above ca. 5 µg/m^3, although there is an indication that the regression coefficients determined depend to some extent on the measuring location [5.137]. Because the detection limit of the black smoke method is comparatively high (4.2 µg/m^3, based on the concentration of elemental carbon) [5.137], this method is in any case unsuitable for low concentration ranges.

5.3.3. Measurement of Wet Deposition

Wet deposition (in rain, snow, fog, and dew) and dry deposition are the two routes which harmful substances can pass from the air to the ground, groundwater, or the surface of vegetation. The measurement of wet deposition has therefore been carried out for some decades as part of the study of environmental problems such as the acidification of soil and lakes and the destruction of forests. For example, the European EMEP measurement network [5.138] has measured the rate of deposition of sulfate and nitrate anions and some cations, and also the pH of the precipitation at ca. 90 measurement stations. There are also many measurement grids in the United States [5.139]. Unlike the concentration in the air and the directly correlated dry deposition, the fluctuations in the concentration of atmospheric pollution in rain are relatively small, so that a representative measurement of wet deposition can be performed with a wide-mesh grid. Wet deposition in Germany is traditionally measured by the wide-mesh grid of the Federal Environmental Office [5.140], rather than by the air measurement stations of the Länder.

5.3.3.1. Sampling

To completely separate the measurement of wet deposition by rain and snow (fog and dew represent special problems) from the measurement of total deposition (bulk deposition, Section 5.3.2.3) and dry deposition, rain collectors are used whose sampling orifices are covered during rain-free periods (wet only samplers [5.141]). Rain sensors, mostly depending on the principle of conductivity change, trigger the removal of the cover at the beginning of a period of rain and its replacement at the end. A review can be found in [5.142].

However, comparison of measurements obtained by various rain collectors [5.143] show that, other things being equal (e.g., evaporation), measured amounts of rain collected can vary because rain sensors differ in sensitivity, and this can considerably influence the consistency of the readings. With light rainfall, the boundary between wet and dry deposition becomes indistinct and is determined by the sensitivity of the sensor.

The collecting vessel consists of a funnel (collection area ca. 500 cm^2 or more [5.141]) and the collection vessel itself, which should be thermostatically controlled and kept in dark conditions to avoid artefacts. (Rain in collecting vessels exposed to light stimulates the growth of algae, and organic micropollutants may undergo photochemical degradation by, for example, H_2O_2 or other peroxides.) If the collected precipitation is to be analyzed for inorganic substances, the funnel and collection vessel should be composed of materials such as polyethylene or polypropylene. If organic constituents are also to be measured, borosilicate glass is preferable.

Artefacts should be avoided by daily analysis of samples for pollutants.

Totalizators are used to determine both the wet deposition and the approximate dry deposition (on a standardized model surface). The totalizator consists of a funnel and a sample receiver for measuring the wet deposition and a vessel (often a standardized glass jar for measuring dust deposits, see Section 5.3.2.3) for collecting the dry deposition. A rain sensor is used to switch from wet to dry collection and vice versa.

Snowfall cannot always be detected with certainty by the sensors of rain collection equipment. The best method for measuring snow is therefore still to collect it on an exposed plastic sheet.

The sampling of fog, clouds, and dew requires special equipment. Fog and cloud water can be collected actively by impactors [5.144] or passively by mesh screens [5.144]. The latter consist of plastic fibers from which the collected water from the fog or cloud runs into the collection vessel. Dew samples are collected on cooled glass surfaces [5.145], although it is doubtful whether this method corresponds exactly to the behavior of natural surfaces.

5.3.3.2. Analysis

Collected water can usually be analyzed for inorganic constituents without preliminary sample preparation, and samples need preliminary centrifugation or filtration only if there is visible cloudiness. Routine measurements are made of conductivity, pH, important anions (NO_3^-, SO_4^{2-}, Cl^-) and cations (Ca^{2+}, K^+, Mg^{2+}, Na^+, NH_4^+, etc.) [5.140], and also, for research purposes, unstable trace substances and intermediates such as H_2O_2 and HSO_3^-.

The analytical methods used are those generally available for aqueous solutions such as conductivity measurement [5.146], pH measurement by electrodes [5.146], atomic absorption for cations (see Section 5.3.2.2), and ion-exchange chromatography with conductometric detection for anions [5.147]. Wet chemical methods are also used, e.g., the colorimetric determination of nitrate [5.148], and nephelometric methods based on barium chloride to determine sulfate [5.149].

Organic components are extracted from rainwater (e.g., with dichloromethane), or volatile substances are removed by a stream of argon and adsorbed in Tenax tubes [5.150]. Gas chromatographic analysis is then carried out (see Section 4.3.6).

5.3.4. Optical Methods of Remote Measurement

Whereas the methods so far described enable atmospheric pollution to be measured at a single location, remote methods of optical measurement usually enable integrating measurements to be carried out with light beams up to several kilometers in length. They exploit the absorption properties of atmospheric gases in the UV, visible, or IR spectral regions, and are based on the Lambert–Beer law, which states that the measured extinction is proportional to the product of the path length and the concentration. If the wavelength of the radiation source and the detector in the measuring equipment can be varied, several components can be measured in parallel or in sequence with a single device.

The five most commonly used measuring systems—FTIR, DOAS, TDLAS, DIAL, and COSPEC—are discussed below. Reviews can be found in [5.151], [5.152].

5.3.4.1. Fourier Transform Infrared Spectroscopy

The Fourier transform infrared spectroscopy (FTIR) method is a true multicomponent system which simultaneously measures the spectra of all compounds present in the light beam that absorb in the IR range between ca. 700 and 3000 cm^{-1}. The heart of the equipment is a Michelson interferometer, which produces interferograms from which the IR spectra are obtained by Fourier analysis [5.151]. In most applications close to ground level, artificial light sources are used, which are either at a distance of as much as several hundred meters from the spectrometer (bimodal arrangement), or which direct the light beam into the spectrometer by means of reflecting mirrors. If white cells are used, measurements at almost point locations are also possible. A large number of organic components at concentrations of a few ppb can be measured in a light beam ca. 500 m in length. The FTIR method can thus be used as an "optical fence" for the effective monitoring of large-area sources or organic compounds such as petrochemical plants or waste dumps. If the light source and the spectrometer face towards each other and are mobile, gridlike surveys of large-area sources can be carried out [5.152].

The sun can be used instead of an artificial light source, especially for measurements from aircraft and balloons. Integrated column concentrations of trace gases in the atmosphere (e.g., ozone) can be determined in this way from ground level, as well as pollutants present in high smoke plumes [5.151], [5.152]. If hot exhaust gases (e.g., from aircraft engines) are observed, the FTIR system can be used to detect the emission spectra of the thermally excited molecules without using a light source.

5.3.4.2. Differential Optical Absorption Spectroscopy

Differential optical absorption spectroscopy (DOAS) is probably the most widely used method of pollution monitoring on account of its comparative ease of operation. There are several manufacturers of commercial equipment [5.151]. It operates in the UV and visible spectral regions, usually with a high-pressure xenon lamp as light source. The light source and spectrometer can either be in the bimodal arrangement, as used in the FTIR method, or the light beam can be directed into the spectrometer by a reflecting mirror. By using beam lengths of up to 10 km, some harmful substances (e.g., HONO), can be measured at concentrations of <100 ppt [5.152]. However, this requires very good atmospheric visibility, as fog and rain, for example, attenuate the light beam so much that the beam length must be limited to <100 m for practical purposes [5.153]. The light is collected by a telescope and focused onto the slit of a diffraction spectrometer. A photomultiplier acts as detector and measures the differential absorption coefficients, which are a measure of the difference between the absorption I measured at the wavelength and the broad band background I_0. Because the light source has a broad bandwidth, a wide spectral range is available, allowing several characteristic lines of the trace gas to be measured.

The sequential measurement of several components is possible, as is the use of several spectrometers simultaneously, although the number of components that can be measured is appreciably smaller than is possible with FTIR. However, the DOAS system has very low detection limits, enabling unstable trace substances such as NO_3 (see Section 5.3.4.7) to be measured. The DOAS system is used for measuring line sources (e.g., streets) and area sources (e.g., airports), but also for general air-pollution monitoring as a supplement or alternative to conventional measurement stations, e.g., in Switzerland and Sweden [5.152].

5.3.4.3. Long-Path-Length Absorption Spectroscopy with Tunable Diode Lasers

Tunable diode lasers can be tuned specifically to the absorption bands of atmospheric pollutants of interest and can provide monochromatic, coherent light beams in the visible to IR range for use in absorption spectrometry (TDLAS); sensitivities are in the ppt range [5.154]. By using long path lengths in low-pressure cells (a few millibar), it is possible to measure unstable trace substances that are important in air pollution chemistry (e.g., ˙OH radicals) [5.155]. In open absorption paths, the detection limits increase to several ppb—the concentration range that is also measurable by FTIR spectroscopy—because of pressure broadening of the absorption bands.

TDLAS has found wide application as a research tool [5.154]. However, the relatively high capital costs, the need to cool the laser to 20–110 K, the difficult operation of the equipment, and the uneven quality of the laser have prevented routine application in pollution monitoring up to now. A simplified mobile absorption spectrometer in which the laser is replaced by a different light source in the infrared range is commercially available [5.156] and is used for monitoring large-area sources. However, it cannot achieve the low detection limits of laser systems.

5.3.4.4. Differential Adsorption LIDAR

If a laser beam is directed into the atmosphere without a reflector, part of the radiation is scattered by particles and air molecules, whereby a very small fraction is reflected in the direction of the laser source and can be detected by an adjacent telescope. If pulsed laser light is used and the time taken for the reflected light to return is measured, the distance of the back-scattering particles can be determined. This technique is known as LIDAR (light detection and ranging). If, in addition, the laser beam is tuned to the wavelength of an absorption band in the visible, UV, or IR range and matched to a reference wavelength of the near background, it is possible to determine the spatial distribution of the pollution. This measuring technique is known as differential absorption LIDAR (DIAL) [5.151], [5.152]. The measuring system can be stationary or mobile (detection vehicle, aircraft), and has been used for large-area pollution measurements in urban regions, large-area sources such as waste dumps and refineries [5.156], loading facilities for petrochemical products [5.156], and for monitoring combustion gas plumes [5.156]. If the laser beam is periodically swung over a particular area or rotated in a circle, a two-dimensional picture of the pollution can be obtained in which point sources, for example, show up clearly as concentration peaks [5.152]. The remote monitoring of emissions from exhaust gas plumes [5.51] is hindered by the fact that only the concentration of a pollutant is measured, and not other important parameters such as volume flow rate, temperature, or the oxygen content of the combustion gas. So far, DIAL has not become widely used on a routine basis. Apart from equipment costs and the difficulty of operating the laser, another reason could be the fact that the first commercial system available (ARGOS) can only detect a limited range of substances (SO_2, NO_2, O_3) which are easily detectable by conventional techniques [5.157].

5.3.4.5. Correlation Spectroscopy and Gas Filter Correlation Spectroscopy

In correlation spectrometers, a spectral region of the undisturbed incident light is correlated with a prepared etched slit mask spectrum of the gas to be measured (COSPEC) [5.152] or with a reference cell which contains the gas to be measured (GASPEC) [5.151]. Both the slit mask spectrum and the reference cell rotate in front of the detector, so that the measured signal and the reference signal are detected alternately.

COSPEC passively measures the incident light, and determines the vertical concentration column of the components to be measured (SO_2 and NO_2). Both stationary and mobile operation (aircraft, detector vehicle) are possible, and commercial equipment is available.

The main applications are in studying atmospheric chemical reactions (e.g., conversion of NO to NO_2 in combustion gas plumes), and in the measurement of cross-border pollution from aircraft or detector vehicles [5.158].

5.3.4.6. Significance of Remote Measurement Methods

Remote measurement methods offer a number of advantages over conventional point-measurement technology:

1) Measurements at places and areas to which it is difficult or impossible to gain access (e.g., factory land, waste dumps, chimney stacks).
2) Continuous or quasi-continuous integrating measurement over an area (e.g., factory fence, factory land). This enables, for example, localized gas emissions from diffuse sources to be detected immediately, whereas single-point measuring equipment can often be at the "wrong" location. Moreover, integrating measurements provide a more representative picture than point measurements.
3) Measurement without physical contact (e.g., no spurious results caused by the sampling process itself), so that the measurement of highly reactive trace substances (e.g., $NO_3^{\cdot}$ radicals) is also possible.
4) Measurement of many pollutants simultaneously (FTIR), and measurement of substances that are very difficult to measure by conventional methods (FTIR, DOAS, TDLAS).

These advantages must be balanced against the disadvantages of high costs; operation difficulties, especially of the laser systems; high personnel requirement; the unacceptably high detection limits in many applications (especially FTIR and DIAL); and the difficulty in incorporating the remote measuring method into regulations for air pollution monitoring, which are not conceived for integrating measurements but for conventional point measurements. Some examples of this are listed below:

1) The incompleteness of the results obtained by some remote measurement methods in darkness or fog (e.g., COSPEC, DOAS) does not comply with the requirement for measurements to be representative of all times of the day (see Section 5.1.2).
2) Limit values often apply to a point or to a certain reference area (e.g., 1 km^2).
3) For routine monitoring in some countries (e.g., Germany, the United States, EC Guidelines), the equivalence of measuring methods to reference methods, which are always based on point measurements, must be proved. In the open areas measured by remote measuring methods, this is only approximately possible [5.151].
4) Existing systems of quality assurance, e.g., the regular use of test gases, cannot easily be used with remote measuring methods operating on open areas.

As a result of these difficulties, the remote measuring methods have so far only become established for special applications (e.g., monitoring diffuse sources) in spite of their obvious advantages. Nevertheless, the announcement in 1993 of a DOAS system as an approved method for general pollution monitoring [5.153] by the German Federal Administration constituted a development that, in association with the continuing improvements in remote measuring systems, should lead to wider practical use in the future.

5.3.4.7. Measurement of Unstable Trace Substances

The use of long light paths for remote measurement gives high sensitivity and lack of contact with the substance during "sampling", thereby enabling important unstable trace substances in the atmosphere to be measured (see

Section 5.3.4.6). Examples include $NO_3^•$ and HNO_2 (DOAS) [5.159]; $OH^•$ radical and H_2O_2 (TDLAS) [5.155], [5.160]; and HNO_2, N_2O_5, CH_2O, and organic intermediates (FTIR) [5.161]. Although remote measuring techniques do not involve contact with the substance, they can nevertheless lead to spurious results. Thus, in addition to $OH^•$ radicals being present at low concentrations (< 1 ppt), unless special countermeasures are taken, laser light at 308 nm causes photolysis of ozone followed by reaction with water vapor, producing additional $OH^•$ radicals in the measurement path [5.155].

Conventional point measurement techniques are also used for this group of substances, e.g., gas chromatography for peroxyacetyl nitrate [5.162] and fluorescence measurements on H_2O_2 [5.163].

5.3.5. Measurement of Organic Atmospheric Pollutants

5.3.5.1. Overview

The main problem with the measurement of organic atmospheric pollutants is the large number of substances of this type. Several hundred individual substances are listed in emission registers and air quality management plans for densely populated areas. These have a very wide range of toxicological, chemical, and physical properties. TA Luft alone lists ca. 120 organic pollutants and 15 organic carcinogens with varying degrees of relevance to air pollution [5.7].

The large number of substances suggests that the high cost of measuring the individual substances could be met by summation measurements, e.g., of parameters such as the total content of organically bonded carbon (total C) in the atmosphere. This was the first route taken in monitoring atmospheric pollution.

However, it subsequently became clear that because the effects and concentrations of the various organic compounds were so variable, measurements of total C were not very useful. For example, a benzene concentration of 100 μg/m³ in the atmosphere represents a considerable cancer risk [5.129], but the same concentration of *n*-hexane is considered harmless. This applies even more strongly to highly toxic compounds.

Total C measurements were therefore increasingly supplemented or replaced during the 1980s by single substance determinations of the relevant organic components.

This means that the analysis for organic compounds in the atmosphere is an exercise in "applied chromatography" (see Basic Principles of Chromatography, **B5**), as it is necessary to separate the complex mixture of organic compounds present. Highly sensitive detectors and mass spectrometers are available for a wide range of concentrations extending over ten orders of magnitude, i.e., from the mg/m³ range (e.g., tetrachloroethene in close proximity to emitters) down to the fg/m³ range (dioxins, furans). Chromatographic methods are therefore the methods of choice if the organic compounds are fairly stable both thermally and chemically. However, there are also special methods for a number of organic substances with reactive functional groups. Here, the physical properties or chemical reactions of the functional groups are utilized to detect the substances.

5.3.5.2. Summation Methods (Total-C Measurements)

Two main methods are used to determine the total C content of the atmosphere, and these have become standards [5.164], [5.165]. In the flame ionization detector (FID), the air sample is fed into a hot hydrogen flame, and the ionization current produced by the presence of organic compounds is measured. The sensitivity (response factor) of FID is not the same for all organic compounds. Whereas the response factors of hydrocarbons lie close to unity relative to the calibration standard propane, this figure can greatly decrease for compounds with several heteroatoms (e.g., oxo compounds, halogenated hydrocarbons). Compounds of low volatility, largely associated with the aerosol, are detected incompletely or not at all. Therefore, the FID method underestimates the content of organic substance not consisting of high- to medium-volatility hydrocarbons.

Apart from total C determination, FID is an important detector with a wide range of applications in gas chromatography. As FID can be operated by continuously feeding the air sample, it is suitable for use in automated stations.

Methane, which is nearly always present in the mg/m³ range, can be removed by installing a short separating column [5.166] or a cooled storage column [5.167]. This compound would otherwise mask differences in the concentrations of other hydrocarbons present at the level of a few hundred μg/m³.

Even if results are difficult to interpret from the point of view of air pollution control, they are always of some significance when linked to the problems of increased ozone concentrations and to the detection of unusual pollutants in areas occupied by the petrochemical industry.

The second method, which was mainly used in the past for random sampling over a wide area to determine total C content, was based on the adsorption of organic compounds on silica gel. The adsorbed substances are desorbed at elevated temperature in a stream of oxygen and converted to carbon dioxide by catalytic combustion. The amount of carbon dioxide is determined coulometrically [5.165]. The method has the relatively high detection limit of 0.3 mg C/m^3, and includes particulate organic substances such as soot and pollen. It is now rarely used.

Attempts have been made to develop methods of measuring total contents of reactive hydrocarbons of particular importance in photochemical ozone formation (e.g., alkenes), by observing the chemiluminescence of their reaction with ozone [5.168]. Another method utilizes the chemiluminescence produced in the reaction with atomic oxygen [5.169]. However, these have not found widespread practical application.

5.3.5.3. Chromatographic Methods

In the analysis of organic compounds in air gas chromatography (GC, see → Gas Chromatography, **B5**) and high pressure liquid chromatography (HPLC, see → Liquid Chromatography, **B5**) are widely used for separating the complex mixtures of substances. In gas chromatography high-resolution capillary columns with small cross-sections (ca. 0.2–0.3 mm) and lengths of 30–100 m are now used almost exclusively (HRGC).

After separation on the column, the compounds are detected individually by various detectors, including FID (see Section 4.3.5.2), electron capture detectors (ECD, specifically for electrophilic substituents such as halogens), photoionization detectors (PID, which are very sensitive to aromatic hydrocarbons and other π-electron systems), and thermoionic detectors (NPD, specifically for nitrogen and phosphorus compounds). The HPLC system incorporates special flow-through detectors, similar to the flow-through cells of a UV spectrometer.

An especially powerful method is to use a mass spectrometer as the detector. Knowledge of the mass spectrum of an organic compound often enables it to be identified with certainty, especially in unknown mixtures of substances. Qualitative knowledge of the retention time from chromatograms obtained with conventional detectors, is generally only useful for mixtures whose composition is largely known.

When high-resolution gas chromatography is combined with mass spectrometry (HRGC–MS), the specific detection of individual components using mass fragmentograms is combined with high detection sensitivity, so that difficult analyses such as the determination of polyhalogenated dibenzodioxins and dibenzofurans in the atmosphere then become possible.

Analysis is preceded by sampling, for which various techniques are used, as organic compounds can differ so widely with respect to volatility, polarity, and concentration range.

The choice of sampling method is mainly determined by volatility. It is therefore useful to divide the analytical techniques into those for volatile and those for low-volatility compounds, methods for the latter being subdivided into those based on concentration measurement and those based on deposition measurement.

5.3.5.3.1. Chromatographic Methods for Volatile Organic Compounds

The boiling points of volatile compounds range from methane (*bp* −161.5 °C) to the dichlorobenzenes (*bp* 170–180 °C). In spite of the high sensitivity of gas chromatographic detectors, direct measurement in the air is not possible without preliminary concentration (with the exception of methane), as detection limits in the range ng/m^3 to µg/m^3 must be achieved.

An overview of the usual methods for determining the active concentration and chromatographic analysis of organic compounds is given in Table 5.6.

The most common adsorbent used for sampling is Tenax-GC, an organic polymer which combines a high adsorption capacity for organic substances with a very low affinity for water vapor and is thermally stable up to ca. 400 °C [5.174]. Other solid adsorbents such as carbon molecular sieves [5.4], activated carbon, aluminum oxide, and silica gel are also used. With carbon molecular sieves, which have a higher adsorption power than that of Tenax, it is possible

Table 5.6. Widely used chromatographic methods of determining organic pollutants

Substance group	Concentration range	Sampling, treatment	Final analysis	Refs.
Hydrocarbons, C_1–C_9	µg/m³	gas mice (rapid sampling); gas-tight syringe; cold trapping before capillary column; controlled thermal desorption	GC/FID	[5.171]
Low-boiling hydrocarbons, volatile halogenated hydrocarbons	ng/m³–µg/m³	evacuated passivated stainless steel cylinders (also for measurement of pure air); samples fed via gas loops; cold trapping; thermal desorption	PID GC/FID/ECD	[5.175] [5.173]
Organic compounds, C_6–C_{30}, *bp* range 60–350 °C	µg/m³	adsorption on activated carbon; a) desorption with CS_2, b) desorption with solvents; vapor space (head space) analysis	capillary GC/FID	[5.176], [5.177], [5.178]
Organic compounds, *bp* range 20–300 °C	ng/m³–µg/m³	adsorption on organic polymers (e.g., Tenax) or carbon molecular sieve (Carbopack), thermal desorption with cold trapping before capillary column or solvent extraction	capillary GC/FID/ECD/MS	[5.170], [5.174]
Modification for low-boiling substances (< −120 °C)	ng/m³–µg/m³	adsorption on cooled polymers (e.g., Thermogradient tube); cooling to −120 °C; use of Carbopack	capillary GC/FID/ECD/MS	[5.174]
High-boiling organic substances partly bonded to particles (e.g., PAH, PCB, PCDD/PCDF); high sample volume	fg/m³–ng/m³	sampling on filters (e.g., small filter equipment) or high volume sampler with attached polyurethane cartridge for the gaseous fraction; solvent desorption of filter and sometimes of polyurethane; various purification and other treatment stages; sublimation of PAH	capillary GC–GCMS (PCDD/PCDF); capillary GC–FID or MS (PAH); HPLC fluorescence detector (PAH)	[5.194], [5.191], [5.188]
High-boiling organic compounds, e.g., CHC, PCDD, PCDF, PBDD, PBDF; low sample volume	fg/m³–ng/m³	adsorption on organic polymers (e.g., polyurethane foam cylinders) with filters (e.g., glass fibers) or inorganic adsorption (e.g., silica gel); extraction with solvents; various purification and other treatment steps (e.g., multicolumn chromatography); formation of derivatives of chlorophenols	HRGC/ECD (CHC) capillary GC-GCMS (PCDD, PCDF)	[5.170] [5.186]
High-boiling organic compounds bonded to particles, e.g., components of organic aerosols; deposition samples	ng/m³ (aerosol: ng/g–µg/g) (pg–ng)/m² d⁻¹	Collection of the aerosol on glass fiber filters (e.g., high- or low-volume samplers) or collection of dust on standard surfaces; extraction with solvents; various purification and other treatment steps	HRGC/MS HPLC (for PAH)	[5.191], [5.192]

to take samples of compounds with *bp* <40 °C. However, these adsorbents have the disadvantage that higher boiling substances (*bp* > ca. 100 °C) are difficult to desorb [5.174].

The solid adsorbent is contained in tubes that can be sealed at both ends, and through which a definite amount of air is drawn by a pump (ca. 0.1–1 L). Depending on the flow rate and quantity of air, sampling times can be varied from a few minutes to a few hours.

For analysis, the Tenax tubes are connected to the sampling system of a gas chromatograph, and the adsorbed compounds are thermally desorbed in an inert gas stream. Direct feeding to the capillary column is not possible, as the large gas volume would lead to loss of the separation effect of the capillary. Intermediate condensation in the gas chromatograph (cold trapping) is therefore carried out, followed by controlled thermal desorption to feed the capillary separation column (cryofocusing). Detection is performed by FID (hydrocarbons), ECD (halogenated hydrocarbons), or mass spectrometer.

This method is also suitable for the analysis of ground air [5.179], e.g., over abandoned polluted areas and waste dumps.

For the analysis of very volatile compounds (*bp* <40 °C) such as Freons, the process must be modified. These compounds can be collected by cooling the sampling tube, or by using a combination tube in which the Tenax GC tube is connected to another containing a more effective adsorbant (e.g., carbon molecular sieve or activated carbon) [5.174].

Desorption is often carried out with solvents such as toluene or carbon disulfide instead of thermally [5.176], [5.177]. This method also enables thermolabile halogenated hydrocarbons such as dibromoethane to be analyzed [5.180], but can only be used for *bp* $>80\,^{\circ}C$. Strongly polar compounds can also sometimes be analyzed.

In the United States, active sampling using specially passivated stainless steel canisters is more widely used than collection on solid adsorbents. This method has been standardized by the EPA [5.173]. Containers with a capacity of ca. 1–15 L are evacuated, taken to the test site, and filled with the air to be analyzed via a critical nozzle or an active pump in the sampling system. A disadvantage of using a pump is that it must be attached to the pressure side of the canister, so that there is a danger of contaminating the sample. However, some problems associated with the use of adsorbents are avoided, such as impurities leading to high blank figures and the decomposition of substances being measured. The volume flow rate during filling of the canister can be adjusted to such low values by means of flow regulators that sampling times of up to 24 h are possible. If the containers are at subatmospheric pressure before analysis, they are brought to atmospheric pressure with high-purity nitrogen, and a definite volume of air, e.g., 100–500 cm^3 is then drawn from the container into a gas sample loop by a pump. Moisture in the air can be removed by passing the sample through a dryer (risk of absorption of polar compounds). The substances being measured are then collected in a cold trap and released in a controlled manner as in the thermodesorption method described above, and gas chromatography is then performed.

If stainless steel containers with a blank value of zero are used, this method is also suitable for measuring background concentrations of Freons in the ppt range [5.173]. Terpenes and isoprene can also be measured in this way [5.181].

Sampling can also be carried out using bags made of Tedlar or Teflon. However, there is a considerable risk of contamination associated with this system, which is therefore mainly used for analysis at higher concentrations, e.g., in gases from waste dumps [5.172]. Polar compounds are partially adsorbed on the walls of the bag.

Special gas mice can also be used very successfully for sampling volatile C_2-C_9 hydrocarbons. These are sealed with septums [5.171]. Some of the collected air is removed with a gas-tight syringe, fed into a gas-sample loop, and cryofocused [5.171].

All the sampling techniques described above are particularly suitable for measuring hydrocarbons and halogenated hydrocarbons. For the analysis of strongly polar compounds such as alcohols or ketones, the suitability of the methods must be tested in each case.

Although gas chromatography is mainly used for manual and random pollution measurements, there have been many attempts to automate the method for use in pollution measurement stations. Suitable process chromatographs are available from several suppliers, (e.g., Chrompack, Airmotec, Siemens, Horiba, and Air Instruments). These have FID or PID detectors, and are mainly used for the automated analysis of volatile hydrocarbons such as benzene, toluene, and xylene (BTX aromatics). A process chromatograph for the analysis of vinyl chloride (without preconcentration, detection limit 10 $\mu g/m^3$) has been standardized in a German guideline [5.182]. Remote measuring methods for organic compounds are described in Section 5.3.4.

5.3.5.3.2. Chromatographic Methods for Nonvolatile Organic Compounds

Concentration Measurements. The difficulties in analyzing and sampling compounds of this class are due not only to their low concentrations (from a few ng/m^3 to the fg/m^3 range) but also to the fact that low-volatility organic compounds can exist in the atmosphere both bonded to particles and in gaseous form. In sampling, several cubic meters of air must be concentrated, so that if only the particles are collected by the sampling procedure, "blow-off" effects from the filters can be produced by the air flow being sampled [5.183].

This applies to the sampling of polycyclic aromatic hydrocarbons (PAH) with three or four rings, whereas PAHs with five or more rings (*bp* $>475\,^{\circ}C$) can be sampled satisfactorily by particle collection on filters [5.184].

The equipment used for sampling compounds with $bp \leq 500\,°C$ (e.g., polychlorinated biphenyls; hexachlorohexane; polyhalogenated dibenzodioxins and furans, PAH) therefore includes filters (usually of glass fiber) for collecting particles, followed by equipment to adsorb gases and other components that pass through the filter.

There are two types of sampling systems in current use. In the first of these, the glass fiber filter is integrated into the top of the sampling cartridge. After the filter there is an adsorption unit, usually a polyurethane foam plug [5.186] or an organic polymer such as XAD/2 (Amberlite) [5.185]. The advantages of this sampler are compact construction and the small air volume required, ranging from ca. 2 m^3 (e.g., for PCP) to ca. 70 m^3 for PXDD/F. Problems can arise if the filter head geometry and sampling rate are not optimized for sampling SPM, as this can lead to particle size discrimination.

This problem is avoided by the use of sampling devices in which the first stage is a conventional unit standardized for SPM measurements with the correct sampling rate (e.g., equipment in accordance with VDI 2463 or the high-volume sampler [5.187] described in Section 5.3.2.2). The air then passes through an adsorption unit containing polyurethane foam or some other adsorbent. For some purposes, the high air volume (ca. 1000 m^3) required by this technique and the noise produced by the sampling pump can be a disadvantage.

Before sampling, standards (e.g., ^{13}C-labeled congeners) can be added to determine whether losses occur during the sampling operation.

After sampling, both the filter and the adsorbent are extracted, and further standards are added for calibration purposes. Workup and chromatographic cleaning of the extracts are then carried out prior to analysis by high-resolution gas chromatography coupled with mass spectrometry (GC/MS; PXDD/F), or using detectors (ECD, FID) with spot checking by GC/MS. Labeled standards are added before or immediately after sampling for quantification purposes.

High pressure liquid chromatography (HPLC) can be used as an alternative to HRGC, especially for the analysis of PAH [5.188], with detection by fluorescence detectors [5.188] or UV detectors. PAH derivatives with nitro and oxo groups are determined by HRGC [5.190].

Examples of these procedures are listed in Table 5.6.

These analytical techniques can be used to determine both brominated and mixed halogenated PXDD/F. The analytical problem is not only the large number of possible congeners but also the lack of standard compounds for calibration. Therefore, only a few approximate analyses for PXDD/F in air have been undertaken so far [5.189].

Very little information on the accuracy and comparability in the measurement of pollution by PXDD/F have so far been published. In cooperative tests between four laboratories in 1989 [5.186], the experimental results (expressed as toxicity equivalents) showed standard deviations of 16% in the first series of measurements and 23% in the second.

Deposition Measurements. As ingestion in food accounts for most of the daily dose of PXDD/F or PCB absorbed by the general population, it is more important when studying the effects of low-volatility chlorinated compounds to measure their deposition on surfaces—and ultimately into the food chain—than to measure their concentration in the atmosphere.

As with deposition measurements in general, wet deposition (rain, fog, snow), dry deposition, and total (bulk) deposition must be measured. Bulk deposition is measured with standardized dust deposition measurement apparatus (e.g., Bergerhoff method, see Section 5.3.2.3) that has been simply modified (e.g., by shading from strong light). Some serious measurement problems (e.g., the resuspension of already precipitated particles, evaporation, and possible photolytic decomposition) have not been systematically researched, so that results are of an approximate character.

Bulk deposition of PCB and PAH has been determined in an analogous manner [5.193].

Wet deposition of PAH and chlorinated compounds such as PCB [5.195] or PCDD/F [5.196] has been investigated by using special rain and fog collectors, in connection with the transport of these compounds into relatively unpolluted areas.

5.3.5.4. Determination of Organic Compounds with Functional Groups

Organic compounds with functional groups (e.g., aldehydes and phenols) are difficult to analyze by chromatography because of their polarity, unless they are derivatized. However, the physical properties of the functional groups can often be utilized for detection by optical methods, such as infrared absorption or differential optical absorption spectrometry (Section 5.3.4). The chemical reactivity of functional groups can be used in reactions to form dyes, which are then determined photometrically, or derivatives are produced for chromatographic analysis.

Thus, aldehydes react readily and specifically with amines to form hydrazones. At concentrations of $>4\ \mu g/m^3$, formaldehyde can be determined by the color produced in the sulfite – *para*-rosaniline reaction followed by photometric measurement [5.197].

Reaction with 2,4-dinitrophenylhydrazine to form the hydrazone, followed by HPLC separation, is more sensitive [5.198], [5.199]. This method can also be used to determine higher aldehydes and ketones [5.199], and can be automated [5.199].

Similarly, phenol and the cresols react with *p*-nitroaniline in dilute NaOH to give azo dyes which can be determined photometrically [5.200]. In a method used for measuring pollution in Berlin in 1989, solid-phase extraction from the NaOH solution was followed by HPLC.

The alkylnitrosamine compounds present more serious problems of pollution measurement. There is in principle a gas chromatographic method based on a nitrosamine-specific detector (TEA) and concentration of the alkylnitrosamines on Tenax tubes followed by thermal desorption [5.201]. However, at the relevant low atmospheric concentrations (a few ng/m^3), there is considerable risk of the unwanted formation of nitrosamines from amines and NO_x on the concentration or separation columns.

Future developments in the analysis of organic atmospheric pollutants will mainly be directed towards inexpensive screening methods, so that the costly analysis of single substances can be concentrated on the most important cases and air pollution monitoring can be kept within economically acceptable limits.

5.3.6. Olfactometric Methods

Olfactometric methods are used in pollution monitoring to enable odors to be objectively assessed and to establish the extent of anticipated odor problems in licensing procedures. Thus, they are mainly used to judge the significance of existing or predicted odors.

In principle, there are three possible methods:

1) Measurement of the emission concentration (number of odor units) with an olfactometer (see Chap. 6), followed by diffusion modeling [5.202]
2) Measurements on single components of pollution (e.g., NH_3, see Section 5.3.1.1) or mixtures of substances (e.g., gas chromatography of gases from waste dumps, see Section 5.3.5), if these adequately characterize the odor event
3) Odor determination by inspections

Odors are often so complex that a direct measurement of the main odor carrier (method 2) is infeasible. Method (1) combines emission measurements with diffusion modeling and cannot strictly be regarded as pollution measurement. In the third method, the human nose is used as the detector. However, the direct relevance of this detector to the effect to be assessed must be balanced against its considerably lower precision compared with physicochemical methods of pollution measurement, as the sensitivity of the sense of smell of different people can differ by a factor of up to 100.

Inspections are carried out in accordance with precisely laid down measurement plans by a group of testers, who record at the locations and times specified whether odors are above the limit of odor detection (an odor event). From the number of odor events during the measurement time, the number of odor hours per annum can be determined, and these give a measure of the significance of the odor problem. Questionnaires can be used to obtain semiquantitative information on the intensity of the odor and its position on a scale of pleasantness or unpleasantness (e.g., "fragrant" or "nauseating"). This information is affected by subjective attitudes. Hence, the assessment method prescribed in administrative regulations is usually based on the more objective measurement of odor hours per annum, although this means that some of the information provided by the human nose is ignored. Reviews on olfactometric assessment methods can be found in [5.204], and standardized guidelines (VDI 3940, draft) are in preparation.

German Antipollution Laws do not contain scales of judgement for assessing the significance of odor problems, but this gap is being filled by administrative regulations prepared by some Länder [5.203].

5.3.7. Screening Measurements

Simplified measuring methods are used from time to time for preliminary investigations (screening). Examples include passive samplers, detector tubes, biological methods, and sensors.

In passive collectors [5.205], [5.206], the substance to be determined is collected by processes such as diffusion, permeation, or adsorption in simple collection equipment (test tubes, panels), and becomes concentrated on adsorption media such as impregnated filters and meshes. This contrasts with active sampling processes in which air is drawn through by a pump. After a certain exposure time, the collected substance is determined analytically, and its original concentration is calculated from based on the relevant physical laws (e.g., of diffusion), taking into account the collection time and the geometrical parameters of the collection equipment. The method is widely used in monitoring work environments (personal sampling) and air quality in buildings, and sometimes for epidemiological investigations. The method is less often used for classical measurement. The most widely used methods include surface active monitoring (SAM) [5.207] and Palmes passive collectors [5.208]. Passive collectors are very cheap compared with active measurement devices, a further advantage being that no energy source is required. Disadvantages include the sometimes poor detection limits, poor specificity, and poor reproducibility. The calculation of concentrations is especially problematical. For some compounds, valid relationships can only be established by on-site calibration.

Detector tubes are used for the sampling and approximate rapid analysis of gases [5.209]. A hand pump or automatic sampling device is used to draw a given volume of air through a glass tube containing a specific reagent supported on a carrier. The reagent changes color in a characteristic manner depending on the concentration of the substance tested in the air. Detector tubes are mainly used for workplace monitoring or as rapid measurement methods in accidents. They are not used for routine pollution measurements because the detection limits are normally too high and selectivity is poor. Detector tubes are available for many substances over a wide range of concentrations [5.210].

Two biological methods are routinely used in pollution monitoring. In the standard lichen exposure method [5.211], the death rate of the lichen is determined over an exposure time of 300 d. In another method, a species of rye grass is exposed for 13–15 d [5.212], and the growth of the grass is determined. Both methods enable the general effect of pollution to be determined.

Immunoassay methods are under investigation. In an early example, atrazine is determined in the gas phase using a cryogenic sampling technique [5.213].

Further development can be expected in the field of sensors which determine chemical states or parameters [5.214]. They often have the disadvantage of unacceptably high detection limits, poor selectivity and stability, drifting phenomena, and sensitivity to environmental conditions such as temperature. Sensors are already used in higher concentration ranges where great accuracy is not required (workplace monitoring, warning equipment, leak tracing).

5.4. Pollution Measurement Networks

A wide variety of pollution measurement grids are operated in Europe. These range from arrays of automatic computer-controlled measurement stations (measurement containers) to predetermined patterns that simply define measurement sites (e.g., Gauss–Krüger coordinates).

The tasks and concepts of measurement grids are described in Section 5.1.

5.4.1. Continuous Measurement Networks

Detailed regulations have been issued in the German Federal "guidelines for designing automated monitoring stations and selecting their locations in telemetric pollution measurement networks" [5.38]. These ensure comparability of the measurement grids operated in the various parts of Germany. While the Federal Government itself operates a measurement grid for monitoring background pollution [5.215], the regional measurement networks are mainly operated by the Länder [5.216]. These networks mainly cover investigation areas and regions of high population density, but also extend to rural and forest areas. In total, ca. 550 automatic container-type monitoring stations are currently in operation in Germany.

These systems of stationary stations are supplemented by mobile stations, which are operated for limited periods of time, usually between one month and one year, in changing locations in connection with inquiries of various types.

Atmospheric pollutants measured include sulfur dioxide, airborne dust, nitrogen monoxide, nitrogen dioxide, carbon monoxide, ozone, and sometimes hydrocarbons. Apparatus for the determination of single organic substances which is suitable for measurement grids is mostly at the development stage. Meteorological parameters are also recorded at certain stations (wind direction, wind velocity, air temperature, relative humidity, precipitation, and global radiation or radiation balance).

A monitoring instrument usually consists of three items:

1) The analyzer itself (e.g., for measuring sulfur dioxide).
2) A calibration unit (test-gas generator).
3) Electronic equipment for supervision and control. This supervises the entire monitoring instrument and contains a standard interface to the data collection equipment. All communications between the monitoring instrument and the computer unit take place via this interface.

Each monitoring instrument produces a signal representing the actual measured value (analog or digital) and a series of digital data. These so-called status signals indicate the current operating status (measurement of atmospheric parameters, maintenance, calibration, etc.), and provide information on fault conditions. In this way, important components of the monitoring instrument are themselves continuously monitored (flows of gases and solutions, temperatures in measuring chambers, lamp intensities, high voltage supplies, etc.).

The calibration unit provides a pollutant-free blank gas and one or two test gases of defined composition. Calibration cycles can be initiated manually or automatically at specified time intervals (e.g., every 25 h).

The monitoring stations are usually permanently connected to a computer via data lines, switched lines, or other data-transfer systems. The calculation of half-hour mean values from single values is performed locally in the monitoring stations or centrally in the measurement network computer, depending on the configuration of the system. At various time intervals (between once per minute and once every several hours), the measured values and status information for all monitoring stations are transferred to the central computer of the measurement network. Here, mean values over other time periods (1 h, 3 h, 24 h, etc.) can be calculated as required from

Table 5.7. Systematic air quality monitoring in North Rhine-Westphalia, Germany

Continuous measuring systems (TEMES/MILIS)	Partly automated measuring systems (LIB method)	Discontinuous measuring systems, multicomponent measurements	Special analytical techniques for special task force
Sulfur dioxide	SPM composition:	benzene and other hydrocarbons	wide range of inorganic and organic components with the aid of special instrumentation:
Nitrogen monoxide	lead		
Nitrogen dioxide	cadmium	halogenated hydrocarbons	
Carbon monoxide	nickel		
SPM	copper	deposition measurements (dust deposition and composition)	detector tubes
Ozone	iron		sensors
Hydrocarbons	arsenic		continuous measuring equipment
Wind direction	beryllium		
Wind velocity		soot	
Air temperature	benz[*a*]pyrene		gas chromatographs
Air pressure	benz[*e*]pyrene	aldehydes and ketones	mobile GC/MS coupling
Relative humidity	benz[*a*]anthracene		mobile electron scanning microscopy
Radiation balance	dibenz[*a, h*]anthracene	polychlorinated biphenyls	
Precipitation	benzo[*ghi*]perylene		
	coronene	polyhalogenated dibenzodioxins and dibenzofurans (PCDD/PCDF)	meteorological measuring equipment

the half-hour mean values. If limit values are exceeded, this is registered by the central computer or specially trained personnel, enabling critical air pollution situations to be recognized at any time. This is of great importance for smog monitoring (smog warning service) and the publication of up-to-date information on high ozone concentrations.

5.4.2. Measurement Networks for Random Sample Measurements

Partially automated measurement grids of dust samplers (e.g., small filtering devices) are used to determine SPM and its inorganic and organic constituents.

The most important routine discontinuous measurements are those of volatile organic substances (hydrocarbons and chlorinated hydrocarbons) and of dust deposition and its inorganic components. There are also special measurement programs for other substances such as dioxins, furans, and polychlorinated biphenyls.

A number of substances measured in this way are classified as carcinogenic, e.g., cadmium compounds, PAH, and benzene. Their monitoring is therefore of major importance.

A typical example of a structure for the systematic monitoring of air quality in North Rhine-Westphalia is given in Table 5.7 [5.217], [5.218].

6. Emission Measurements

6.1. Introduction

To protect the environment, air-polluting emissions must be limited using state-of-the-art technology. Here, the word emissions is used in the legal sense, in accordance with § 3 Section 3 of the Federal Antipollution Law (BImSchG) [6.1]. Emissions can pass into the atmosphere in exhaust gases or waste air from fixed sources, such as chimney stacks. However, they can also be produced diffusely, e.g., from openings in buildings or waste-storage areas. Emissions are usually expressed as mass concentrations, but also as mass flow rates or mass ratios (mass of emitted material per unit mass of substance produced or processed). Atmospheric impurities are changes to the natural composition of the air due to smoke, soot, dust, gases, aerosols, vapors, or odoriferous substances (odoriphores). Methods for determining emissions comprise both manual methods of measuring spot samples and continuous measurement techniques. Depending on the components measured and the measurement requirements, emission concentrations can be from the g/m^3 range down to the trace range (pg/m^3). As environmental protection standards have become stricter, new measurement methods have been developed with lower limits of detection. The demand for continuously operating measurement equipment has greatly increased. The methods described here therefore represent only an overview of the present state of measurement technology.

6.2. Objectives and Planning of Measurements

The definition of objectives must specify the quantities to be determined, the boundary conditions, and the required accuracy of the results. There can be many different reasons for determining emissions. Typical objectives include: monitoring compliance with emission limit values; determination of agreed guaranteed figures for production plants and exhaust gas purification plants (acceptance measurements); calibration of continuously recording emission measurement equipment; testing the functioning of continuously recording emission measurement equipment; emission control monitoring fitness for use; analysis of the causes of noncompliance with guaranteed values or emission limit values; determination of changes in the emissions from a plant after process changes, operational disturbances, or design modifications.

Pollution control regulations require that compliance of plants with emission limit values be monitored. In Germany, the main regulations are the Federal Antipollution Law (BImSchG) [6.1], TA Luft [6.2], 13th BImSchV [6.3], 17th BImSchV [6.4], and 2nd BImSchV [6.5] (see Chap. 4). In the licensing procedure, the requirements included in the individual regulations form part of the conditions and supplementary provisions under which a plant must be operated (requirements of the authorities). The rules concerning the monitoring of performance characteristics (proof of guarantee) of items of plant and gas purification installations are civil law agreements. The monitoring of compliance with emission limit values is already regulated by the Federal Antipollution Law. According to this, the authorities can order that, in the case of licensable plants [6.6], measurements should be carried out after bringing into service, after a significant change, or after certain periods of time. As well as random sample measurements (spot measurements), continuous measurement and recording of certain emissions may be required.

In measurement procedures, the measurement plan is of great importance. Measurements for determining plant emissions should be carried out such that the results correctly represent these emissions and are comparable to similar plants and operating conditions [6.2]. Before the measurement plan is produced, the objectives of the measurements must be exactly defined [6.7]. These objectives are usually based on licensing information, official regulations, or contractual documents (acceptance measurements). Alternatively, guidance can be obtained from relevant administrative regulations or VDI Guidelines. According to these, the necessary prior knowledge of the plant under investigation can be used, e.g., technical data about the plant, information about operational behavior, materials used, positions of the measurement points, etc. This is supplemented by knowledge about the measurement methods to be used, the personnel plan, and the time period. In the measurement plan, the following must be specified:

1) Where will the measurements be carried out?
2) What must be measured?
3) How are the measurements made?
4) What accuracy is required?
5) How will the results be determined?
6) Who will carry out the investigations?
7) When will the measurements by carried out?

VDI 2448 describes the planning of spot sampling measurements of stationary source emissions [6.7]. The measurement plan should be produced at an early stage. For example, the planning of acceptance measurements must be started at the planning stage of the plant. The type of measurements to be carried out, the arrangement of measuring ports, and the equipment for the measurement points must be considered and fixed. If this work is done at a later stage, the installation costs are often higher. A detailed plan is especially important if the equipping of measurement points will be impractical at a later stage, as in the case of coated exhaust gas ducting, or if the measurement points are no longer accessible or are unsuitable. In planning, the accessibility and the minimum requirements that apply to exhaust gas ducting must be considered. VDI 2066, Sheet 1 gives advice on the minimum length required for the exhaust gas ducting upstream and downstream of the measurement point [6.8]. For measurements at existing plants, the measurement plan must conform with plant conditions. The arrangement of measurement points then cannot always be freely chosen. In every case, it must be judged whether a measurement can be regarded as representative. If it is not, modifications to the measurement technique to improve its accuracy must be considered.

6.3. Requirements for Measurement Techniques and Measurement Stations

Emission measurements must be carried out using state-of-the-art techniques and equipment. TA Luft prescribes that emission measurements must be carried out in conformity with certain guidelines and standards of the Commission for Atmospheric Pollution in the VDI and DIN. Alternative or supplementary measurement methods are permitted, especially if they are accepted by the Federal Environmental Ministry with agreement with the highest Land authorities. The guidelines by the Commission for Atmospheric Pollution in the VDI and DIN describe the established state of the art in the measurement field.

6.3.1. Quality of Measurement Methods

The quality of the measurement method must meet the requirements of the measurement task. To specify measurement performance, quantities characteristic of the measurement method are used, such as detection limits and determination limits as defined in VDI 2449, Sheets 1 and 2 [6.9], [6.10]:

Detection limits and *determination limits* must lie significantly below the lower emission limit. In general, the detection limit should be less than one-tenth of the lower emission limit. Where there are no emission limits, but compliance with other constraints must be monitored, requirements are analogous.

The *measurement uncertainty* can be determined from the reproducibility found in duplicated determinations. The reproducibility of a measurement method is determined from a series of simultaneous determinations using two complete measurement procedures. The reproducibility can be based on the mean value of the sample of measured values or on classes of measured values. Thus, the standard deviation of the measurement process can be estimated from the number of pairs of measured values and the concentration of the substance.

The *selectivity* of the measurement method chosen must be sufficiently high with respect to other substances present in the exhaust gas, i.e., the cross-sensitivity must be low.

If continuously recording measurement methods are used, other parameters are important, e.g., linearity of the measurement signal, dead time, rise time, drifting of zero point and sensitivity, and setting-up times. The measurement method must cover the expected range of values. Similar considerations apply to the choice of methods for the measurement of reference quantities (O_2 content) and the state quantities temperature, humidity, and pressure.

6.3.2. Choice and Equipment of Measurement Stations

"Measurement stations should be of adequate size, easily accessible, and so designed and selected that technically problem-free emission measurements that correctly represent the emissions from a plant can be carried out" [6.2].

Measurement and sampling stations must be included at the planning stage of new plants. With existing plants, the requirements for measurement stations should be considered to enable the necessary modifications to be carried out and/or equipment to be installed during planned shut-downs. For all measurement stations, the following are fundamental requirements:

The association of the measurement station with the emission source must be indicated by unambiguous labeling. For this, the source numbers used in the description of the emissions or the company source numbers can be used. It should also be possible to a make an assignment to the relevant part of the plant or operation.

Flow into the measurement line should be as smooth and undisturbed as possible. Recommendations for the measurement of dust in flowing gases can be found in VDI 2066, Sheet 1 (dust measurements). The flow properties in the exhaust gas duct can be affected by merging gas flows, changes in direction, changes in cross-section, inserted items of equipment, and valves. To ensure a flow which is as undisturbed as possible, the measurement/sampling point should be in a straight section of the exhaust gas duct with a constant size and shape. VDI 2066 specifies uninterrupted inlet and outlet pipes whose length is each at least three times the hydraulic cross-section. To prevent sedimentation (e.g., when taking measurements in the unpurified exhaust gas prior to the dust filter), the measuring point for the determination of particulate emissions should if possible be located in a vertical part of the ducting. The sampling points should be so arranged that grid-based measurements can be made for determining both dust emissions and emissions in gaseous form.

The size of the measurement openings is determined by the type of measurement equipment and sampling probes. The number of openings depends on the size and shape of the cross-section under investigation. The arrangement, number, and dimensions of openings usually used for dust measurement has also been found to be suitable for measuring gaseous components where measurement grids are required. A good solution consists of large circular openings with closable gas-tight lids which can be modified to suit the measurement task. This possibility of modifying the lid at a later stage is especially important if it is not possible to modify the ducting under investigation (e.g., because of the presence of a rubber lining for protection against acid).

The operating platforms for sampling and measurement must be sufficiently large to accommodate the sampling probes, sampling equipment, and associated measurement apparatus. Access to measurement stations must be provided in accordance with the relevant regulations. There must be no extraneous emission sources (e.g., bursting disks, safety valves) in the operating area of the measurement station. The measurement station must at least be provided with protection from rain and weather if the operating platforms are in the open. Adequate, correctly sized, and fused electrical supply points must be installed at the measurement station. Depending on the type of measurement, compressed air and water supply points may also be required.

6.4. Requirements for Continuous Measurement and Associated Computer Systems

For a number of licensable plants, systems for continuously monitoring emissions are legally required. In addition, owners of industrial plants in their own interest may carry out continuous monitoring of emissions to show that emission limit values are being complied with, e.g., if local residents complain about plant emissions. The regulation on power-station emissions (13. BImSchV) states that power stations must be provided with equipment for continuously monitoring emissions of dust, carbon monoxide, nitrogen oxides, and sulfur oxides, and that the results must be continuously and automatically evaluated [6.3]. The regulation covering incineration plants for waste and similar combustible materials (17. BImSchV) requires the continuous determination, recording, and evaluation of emissions of carbon monoxide, total dust, organic substances (as total carbon), gaseous inorganic chlorine compounds (as hydrogen chloride), gaseous inorganic fluorine compounds (as hydrogen fluoride), sulfur dioxide and sulfur trioxide (expressed as sulfur dioxide), and nitrogen monoxide and nitrogen dioxide (as nitrogen dioxide). Additionally, other exhaust gas parameters such as oxygen content, exhaust gas volume flow rate, and temperature must be determined and recorded [6.4].

For other licensable plants, TA Luft prescribes the conditions under which quantitatively significant emissions of dust and gaseous atmospheric pollutants must be continuously automatically measured and evaluated. The measuring equipment is specified in publications by the Federal Ministry responsible for antipollution. Testing for suitability, installation, calibration, and maintenance of measurement equipment are described in binding guidelines by the Environment Minister in conjunction with the Länder so as to provide quality assurance and to ensure uniform federal practice in the monitoring of emissions.

6.4.1. Minimum Legal Requirements

A series of parameters, the so-called minimum requirements, ensure that continuously recording measurement equipment does not fall below a certain standard of performance [6.11]. These minimum requirements are from time to time reassessed in the light of the latest state of the art in measurement technology. They are subdivided into various general requirements, e.g., position of reference and zero points (live zero), recording range, and availability. Other requirements refer to measurement principles (e.g., photometry and radiometry), substance-specific properties such as dust content and smoke number, required degree of reproducibility, drift with time, and detection limits. Minimum requirements are also set for detection limits, availability, and the continuous measurement of reference values, e.g., oxygen content and waste gas volume flow rate. A separate guideline describes the requirements that apply to computers for the calculation of emissions [6.12].

There is a further guideline on the evaluation of continuous emission measurements on incineration plants (17. BImSchV) [6.13].

6.4.2. Testing the Suitability of Continuous Measurement Methods and Evaluation Computers

When testing for suitability, compliance with minimum requirements must be determined in laboratory tests and also in a long-term test lasting at least three months on an emitting plant. Standards of performance, but not measurement principles or any specific measurement equipment, are prescribed in the minimum requirements. Any kind of measuring equipment can be used, whatever its measurement principle, provided that it complies with the minimum requirements. Testing for suitability usually begins with the determination of characteristic performance figures in a laboratory test. After this, a long-term test under practical conditions is carried out on full-scale plants that require continuous monitoring of pollution emissions in accordance with legal and administrative regulations of the Federal Antipollution Law (13. and 17. BImSchV, TA Luft). The suitability test should be carried out with two similarly constructed measurement devices operating in parallel, with separate sampling, sample preparation, and data processing. The three-month test should be carried out under the severest known boundary conditions. If important minimum requirements are not complied with, the test is discontinued. The applicant can then carry out improvements, repairs, and modifications. Consideration must then be given to the question to what extent the changes to the measurement system make it necessary to retest any of the minimum requirements. On conclusion of the test for suitability, the testing institute produces a test report on the results. This is sent to the Federal Environmental Office and to the Länder Committee for Antipollution. If the Federal Ministry for the Environment, the relevant Land authorities, and the testing institutes agree that the overall judgment is positive, the certification of suitability is issued at the instigation of the Federal Ministry for the Environment in the general ministerial paper issued by the Federal Ministry [6.14], [6.15].

6.4.3. Testing and Calibration of Measurement Equipment

Quality assurance tests for continuously recording measurement methods can be divided into tests of internal equipment function, maintenance, annual functional verification, and calibration [6.16].

Internal equipment functions tested can include automatic changeover of mesh filters, automatic electrical balancing at given time intervals, or the regular functional checking of certain components (e.g., pumps, flowmeters).

Normally, the equipment must be maintained and tested at definite time intervals by the operator. This requires functional testing with calibration aids such as mesh filters and test gases.

For in-situ measurement equipment, the optical boundary surfaces must be cleaned, air filters must be cleaned or changed, and the zero point, the sensitivity, and data recording must be checked. In methods involving extractive sampling, the heater for the sampling system, the leak-tightness of the tubing, the sampling flow rate, and the condensate flow must be tested. The zero point and sensitivity must be tested and if necessary adjusted by using test gases.

For continuously recording measurement equipment, an annual functional test must be carried out on the complete measurement system, including sampling. In a general test of the state of maintenance, and, in the case of extractive sampling systems, a test of the leak-tightness of the measurement system, various testing activities are required. These can include: testing for soiling of the optical components or of the sampling system; determination of the changes with time of zero and reference points; determination of cross-sensitivity towards CO_2, CO, NO, NO_2, SO_2, water vapor, etc.; testing of the functioning of the equipment with test gases, test solutions, filters, or other equipment-specific standards with three or four values over the measured range. Also, specific equipment functions relevant to operational dependability are tested, e.g., constancy of sample stream offtake, duration of measurement cycle, metering and composition of reagents. The transfer of the measured figures to the data collection system must also be tested [6.17].

Continuously recording measuring equipment must be calibrated after a new plant has been put into service, after significant alterations, and at intervals of 3–5 a. Calibration is the determination of the relationship between the readings of the complete measuring equipment and the actual concentration of the substance present in the exhaust gas from the plant as determined by comparison measurements with a conventional analytical method [6.18], [6.19]. Checks on the maintenance and functioning of the system must be carried out beforehand. The figures determined simultaneously by the two methods are compared and statistically evaluated. To calibrate dust measurement devices, at least 12–15 comparative measurements are usually necessary [6.8], [6.20], these being carried out at a number of points arranged in a grid pattern covering the cross-section of the ducting. This gives the grid-based analytical function.

With gas emission measurement equipment, there are two main procedures for determining the analytical function. In the first of these, which is analogous to that used in dust measurement, comparative measurements are carried out as grid analyses using a standard method suitable for a bulk phase, e.g., the H_2O_2–thorine method for SO_2 determination (VDI 2462, Sheet 8). Sampling at the individual measurement points on the grid must be carried out at a rate proportional to the gas velocity and if possible, for various concentrations of the measured substance. This gives the grid-based analytical function. In the second method, a single point analysis is carried out. With continuously recording measurement methods using extractive sampling, ca 20 comparative measurements are carried out in the first stage, with the conventional method and the continuously recording measurement method being operated at adjacent sampling points. The statistical evaluation gives the single-point analytical function for sampling by both methods at the same place [6.18], [6.19]. The difference between the equipment-specific characteristic line and the single point and grid-based analytical function is shown schematically in Figure 6.1.

For many applications where the concentration distribution over the cross-section is approximately uniform, this single-point analysis is sufficient. It includes the entire exhaust gas matrix and the variation of concentration with time. It might be thought that this procedure would

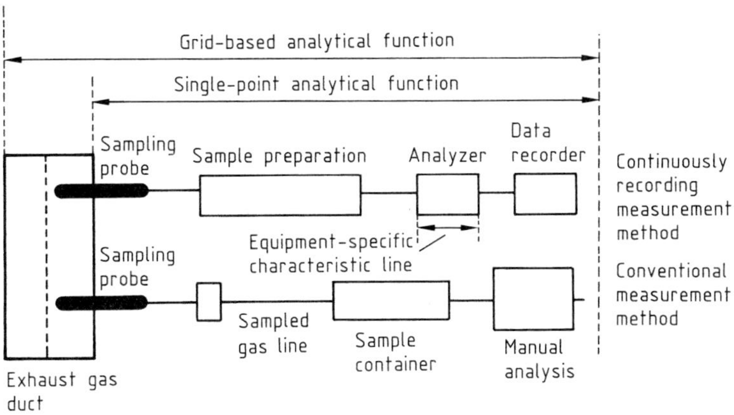

Figure 6.1. Calibration scheme for continuously recording emission measurement equipment

also be valid for linear sampling by in-situ measuring equipment. However, even when the single-point analytical function is used, the evaluation requires additional grid measurements to verify that the concentration distribution is approximately uniform. In all cases where the concentration distribution is not uniform, this single-point analytical function must be supplemented by additional measurements to give a grid-based analytical function. Therefore, in a second stage, a correction factor must be obtained by means of approximate grid-based measurements, enabling the representativity of the measurement point to be improved. This stepwise procedure is recommended for calibration with single-point sampling. It has the advantage that the cost of the grid-based measurement can be reduced, because instead of using a conventional analytical method to determine the correction factors, a second continuous emission measurement apparatus is used whose results have previously been compared with those from the equipment being calibrated, which also operates by extractive sampling [6.18]. For the measuring equipment to be calibrated, the single-point analytical function has already been determined by a conventional analytical method, taking into account the complete exhaust gas matrix. Thus, these results can be transferred to the second emission measurement equipment. Grid-based measurements are then carried out. The results of the grid-based measurements, in combination with the single-point analytical function of the equipment to be calibrated, lead to a grid-based analytical function. The arrangement of sampling probes for determining the equipment-specific factor for a second recording method and the procedure for extending the single point analytical function to a grid-based analytical function are shown schematically in Figure 6.2.

6.5. Physical State Parameters, Reference Value Measurements, and Evaluation Computers

The physical quantities temperature and humidity, and the reference values oxygen content and carbon dioxide content must be determined at the same measurement point as the measured

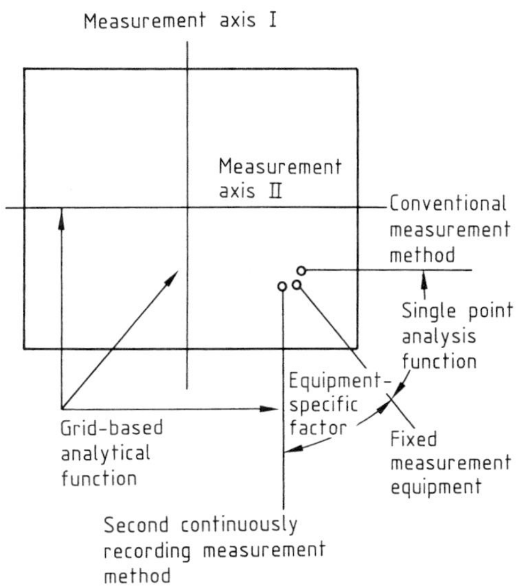

Figure 6.2. Arrangement of sampling probes to determine the equipment specific factor for a second continuously recording method, and a representation of the relationship between single point and grid-based analysis

components within the length of ducting under investigation. The temperature of the exhaust gas and sampled gas and the pressure of the exhaust gas must be known to enable the mass concentration of the measured components under standard conditions (273 K, 0.1013 MPa) to be calculated. Figures for the mass concentration based on dry exhaust gas are generally also required, so that the moisture content of the exhaust gases must also be determined. The standard density of a gas mixture is found from the standard densities of the individual components and the quantitative composition. The standard density of each component is found by dividing the molecular weight by the molar volume in the standard state (DIN 1306 and 1871) [6.21], [6.22]. The oxygen, carbon dioxide, water vapor, and nitrogen contents enable the density of the exhaust gas (used in dynamic pressure measurements) to be calculated precisely. The oxygen content is an important reference value from which the measured concentrations can be calculated according to the equation

$$E_B = \frac{21 - O_B}{21 - O_M} \cdot E_M$$

where E_M = measured emission, E_B = emission based on the oxygen content, O_M = measured oxygen content, O_B = reference oxygen content.

6.5.1. Temperature

The exhaust gas temperatures at which emission measurements are carried out are generally in the range 50–400 °C, which can conveniently be measured by resistance thermometers or preferably by thermocouples. The relevant regulations on temperature measurement should be referred to [6.23]–[6.26]. The furnace temperature is now of great importance in refuse incineration plants, and must be monitored as prescribed in 17th BImSchV. This requires that the minimum temperature after the final addition of combustion air must be maintained at least between 850 or 1200 °C. These temperatures must be determined continuously, recorded, and evaluated. In this temperature range, only thermocouples have become established for long term use. As there is significant radiation in this temperature range, the thermocouples must be calibrated. Gas extraction pyrometers can measure gas temperature without radiative contribution. The gas is sampled at a high rate by a pump and passed over a thermocouple. The thermocouple is covered with a protective tube to prevent errors due to thermal radiation. However, continuous long-term measurement is not possible at present [6.23].

Recently, recording measuring equipment based on acoustic pyrometry has been marketed. This relies on the fact that the speed of sound in gases depends mainly on the temperature of the gas and only slightly on its composition.

6.5.2. Moisture Content

Several methods are in principle suitable for determining the moisture content in off-gases, including psychrometric difference measurement, the lithium halide dew point method, and condensation and adsorption methods. In off-gases, the H_2O content usually lies between 2 and 30 vol%, corresponding to a water dew point temperature of 20–70 °C. The determination of water content has hitherto been carried out by manual analysis, as continuous measurement was not technically feasible.

The regulation that deals with waste incineration plants (17. BImSchV) prescribes continuous measurement, recording, and evaluation. Initial investigations into the use of continuously operating LiCl moisture measurement equipment were carried out in an R & D project intended to formulate the minimum requirements that would apply to moisture measuring equipment. The LiCl sensor was replaced by an LiBr sensor to give good service availability [6.27]. Legally binding minimum requirements for the measurement of moisture content have existed since 1992 [6.28]. A multicomponent (HCl, SO_2, NO, CO) photometer which uses the single-beam, dual-frequency method, produced by Perkin–Elmer, has been approved for measurements on waste-incineration plants. It can also measure the H_2O content in the range 0–40 vol% [6.29]. Another recommended measuring device manufactured by Ultrakust operates as a back-scattering psychrometer [6.30]. This is suitable for the exhaust gases from furnaces and from incinerators for process and domestic refuse. The lowest measurement range during the test for suitability is 300 hPa water vapor partial pressure.

Other measuring equipment operating on the principle of a difference measurement on "dry" and "wet" zirconium oxide probes is currently being investigated for suitability.

6.5.3. Carbon Dioxide Content

During the 1970s oxygen content replaced the formerly used reference value CO_2 for standardizing measured emission concentrations. There are therefore no minimum requirements and suitability tests for CO_2 measurement equipment. Investigations have been carried out to determine to what extent CO_2 measurement could be used as a substitute reference value [6.27]. In the combustion of a single substance, the two reference values CO_2 and O_2 are interchangeable without appreciable error. However, in the combustion of mixtures, the discrepancies are significant because of differences in the maximum CO_2 content associated with the fuel. In random sample emission measurements the CO_2 content is usually also measured to enable the density of the exhaust gas to be better characterized. In principle, the Orsat method is suitable, but the measuring methods usually used are based on instruments such as the IR absorption photometer. Electrochemical methods are also used.

6.5.4. Oxygen Content

All discontinuous and continuous measure methods include a mean-value calculation followed by conversion to STP, and this is usually followed by a further conversion to a definite oxygen content. Because errors in O_2 determinations have an immediate impact on the results of emission measurements, minimum requirements apply. A number of types of measurement equipment have been approved [6.11], [6.14], [6.15], [6.29]–[6.31]. Sampling can be performed by extraction or in situ. In extractive sampling, the exhaust gas together with the other pollutants is passed through a cooling system, so that measurement results based on dry exhaust gas and 273 K are immediately obtained. Suitable measurement principles include the magnetodynamic method using a torsion balance (manufacturers: Bühler Mess- und Regeltechnik, Mannesmann/H.u.B., MC Products Analysentechnik, Maihak, Rosemount), the equal pressure or variable-pressure measuring method with an auxiliary gas (manufacturers: Maihak, Siemens), and the electrochemical method with a zirconium dioxide solid-state electrolyte (manufacturers: ASEA BBC, Enotec, Westinghouse Controlmatic, Yokogawa). Using the last-named measuring principle, the O_2 content is determined either as "wet" O_2 content directly in the exhaust gas duct, or as "dry" O_2 content by extractive sampling followed by cooling and condensate removal.

6.5.5. Evaluation Computers

To comply with the requirement to evaluate measured results continuously and automatically and to document them, the measured values from the recording equipment (dust, SO_2, CO, NO, etc.) are converted into the appropriate physical quantities based on the regression curves determined in the calibration, and are averaged over the reference time (usually 30 min) [6.12]. After converting to STP and reference oxygen content (e.g., 11 % O_2 for refuse incineration equipment), the concentration figures are divided into 20 classes of equal range. A scheme for the production and evaluation of measured values is shown in Figure 6.3. These values are not classified in the true sense, but are counted in the computer memory. The emission limit value corresponds to the upper limit of Class 10. Values that exceed twice the limit value are counted in Classes 21 and 22.

The legally prescribed evaluation calculations should be carefully noted. For example, according to TA Luft and 13th BImSchV, emission values are considered to have been complied with if

1) Overall daily mean values do not exceed the emission limit value
2) 97% of all half-hour mean values do not exceed six-fifths of the emission limit value
3) Overall half-hour mean values do not exceed twice the emission limit value

For refuse incineration plants a different kind of evaluation is required [6.4], [6.13].

The limits of these two classes correspond to the tolerance range established in the calibration. In Class 23 is stored the number of half-hour mean values that exceed 1.2 times the limit value, up to the limit of the confidence range. The special classes are used to store the half-hour mean values which did not represent true measurement values for more than one-third of the measurement time because of breakdowns or maintenance work on the analyzers (Class 24). The classification of half-hour mean values is shown schematically in Figure 6.4.

The total number of these values, including those values that were integrated over less than two-thirds of the normal measurement interval owing to plant startup or

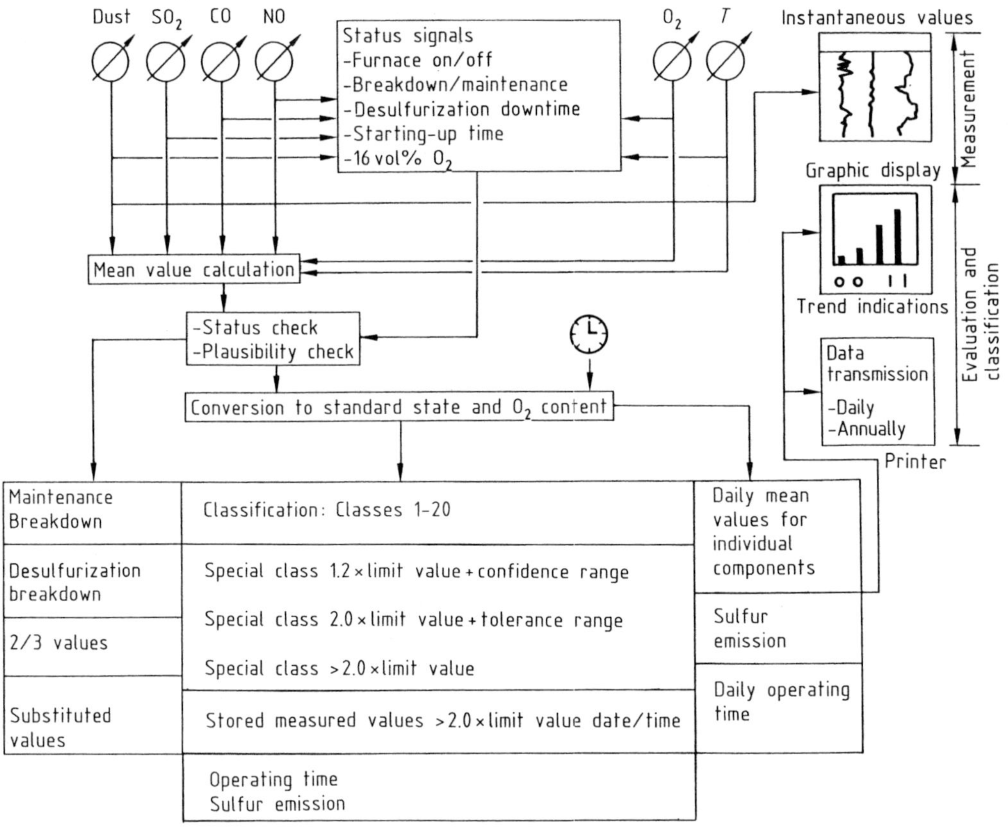

Figure 6.3. Collection and processing of measured values

shut-down are also counted in Class 26. In the absence of the reference value of oxygen due to a breakdown of the measurement equipment, the evaluation computer uses a substituted figure for O_2 which is established in the calibration. All half-hour mean values that were obtained using a substituted figure are classified in the normal way and also counted in Class 25. As well as counting the daily mean values in three classes (30, 31, and 32), the operating times of the plant during the day and the year are stored and reported (Classes 33 and 34). Class 35 is used to store all measured values where twice the limit value has been exceeded.

The time and date of each event is also recorded. Transmission of the data, i.e., the numbers of measurements counted in the 20 normal classes and special classes and the stored data takes place daily automatically at a fixed time. Furthermore, all the data is also transmitted at the end of each calendar year. The daily print-outs give continuous documentation of the emissions from the plant. The annual print-out provides information for the supervisory authority.

Evaluation computers tested for suitability include those manufactured by Durag, Gesytec, Industrie Electronic Schmitz, Maihak, NIS Ingenieurgesellschaft, Nukem, Sick, and RWE.

6.6. Measurement of Gas Velocity and Flow Rate

Normally, only the mass concentrations of the pollutants are measured for the control of emissions. However, in some cases, the emission mass flow rate of a pollutant must be measured discontinuously or continuously. It is determined as the product of the mass concentration and the exhaust gas volume flow rate. The mass flow rate can usually be calculated with sufficient accuracy from known plant parameters such as fuel consumption or steam production rate. If the operating parameters of the plant fluctuate widely or if the quality of the fuel is not known accurately, a direct determination of the volume flow rate of the exhaust gas must be carried out. Thus, for example, the regulation covering waste incineration plants usually requires continuously recording volume flow rate measurement [6.4]. The minimum requirements should be referred to (see Section 6.4.1).

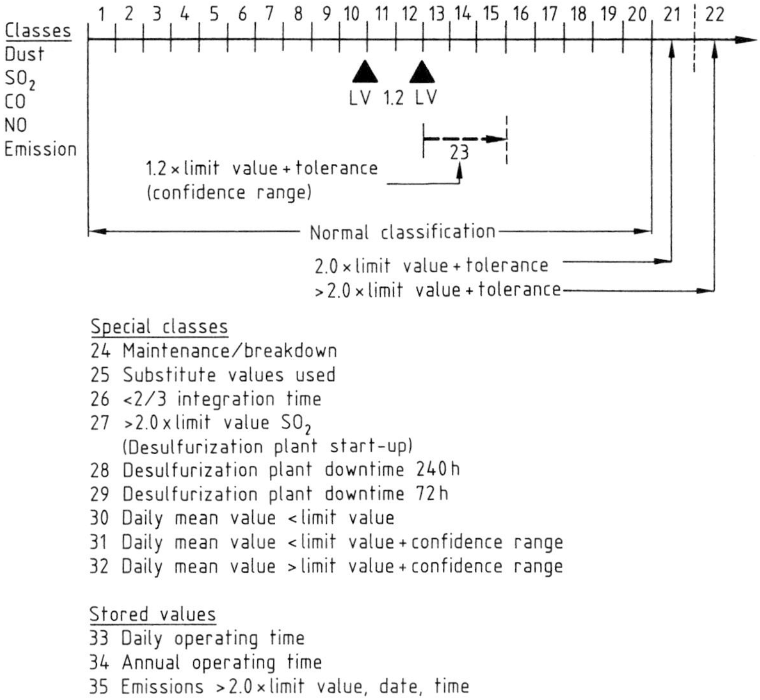

Figure 6.4. Classification scheme for half-hour mean values

Discontinuous Methods. To calculate the volume flow rate $\dot{V}$, the mean velocity $\bar{v}$ in the measured cross-section must be determined. The requirements that apply to the properties of the measured ducting (undisturbed inlet and exit) are described in the VDI Guideline 2066 [6.8]. The mean flow rate is determined by point velocity v_i measurements carried out in a grid arrangement covering the entire measured cross-section. For emission measurements, the arithmetic mean is obtained in accordance with the instructions given in VDI 2066, which also specifies the number of measurement points (see Section 6.7).

The value of $\bar{v}$ is given by

$$\bar{v} = \frac{1}{n} \sum_i v_i$$

The mean volume flow rate $\dot{V}$ is given by $\dot{V} = \bar{v}\ A$, where A is the measured cross-section.

Pitot tubes are used for manual measurement. The dynamic pressure is measured as the difference between the total pressures p_{tot} and the static pressure, and the local velocity is then given by

$$v_i = \sqrt{\frac{2}{\varrho}(p_{\text{tot}} - p_{\text{stat}})}$$

The Prandtl modification of the Pitot tube is widely used (VDI 2066). In the case of different Pitot tube designs, correction figures determined in fluid-dynamic experiments in a wind tunnel are sometimes used. The measurement range of the Prandtl tube is ca. 2–50 m/s. Anemometers are used for lower velocities; when used in stack gases special calibration conditions and the influence of gas temperature and gas properties on the reading must be considered.

Continuous Methods. Methods for the continuous measurement of gas velocity and volume flow rate include dynamic pressure difference methods with transducers, ultrasound measurements by the time-difference method, and determination of turbulence frequency by ultrasound measurement.

Dynamic pressure measurements usually require several openings on the upstream side which enable the mean total pressure over the cross-sectional area of the exhaust gas duct to be measured. Using the arrangement of measurement points specified in VDI 2066, with equal grid unit areas, the probe must be fitted at each of these points. For velocity measurement based on the determination of the turbulence frequency by its modulating effect on an ultrasound beam, precalibration must be carried out by the manufacturer for each user. In time difference measurement by ultrasound, the signals pass through the gas stream at an angle α in two directions. Each of the two transducers acts as both a transmitter and a receiver.

Equipment that operates according to these principles has now been officially approved. Note that all methods and equipment in the grid measurement system must be calibrated by carrying out simultaneous velocity measurements with a reference method, usually dynamic pressure measurement by the Prandtl method (see p. 499). Calibration is discussed in Section 6.4.3.

6.7. Measurement of Particles

Both manual and continuously recording methods can be used for measuring particulate materials. The manual methods and some of the recording methods are described in VDI 2066. When using recording methods, the minimum legal requirements (see Section 6.4) and the certification of suitability must be complied with. The manual methods are proven, but problems can occur with very low dust contents (<1 mg/m^3) and moist, supersaturated exhaust gas (e.g., after it has passed through water scrubbers). Although methods of continuous measurement are available, not all are equally suitable. For example, it is not possible to work with optical transmission equipment on supersaturated exhaust gases. Also, it is unsuitable for gases with low dust content inducts providing small optical path lengths. With exhaust gas saturated with water vapor, extractive sampling and reheating techniques are used.

The measurement principles used include the measurement of extinction, light scattering, and radiometry. For low dust contents, the more sensitive method based on scattered light measurement is used.

6.7.1. Sampling

The sampling of particles in a flowing exhaust gas must take place at the same velocity as that of the gas to prevent segregation or disturbance of the particle-size distribution due to the inertia of the particles, leading to falsification of the measured solids content. If the sampling rate is too high, the measured dust content will be too low, and vice versa. This mechanism depends on the particle size distribution. For particles of aerodynamic diameter <0.5 µm, the effect of this inertia is practically negligible. A schematic diagram of equal velocity sampling is given in Figure 6.5.

Another important aspect of sampling is that only part of the flowing gas in the measurement grid is removed. As the gas velocity and the dust flow can only be measured at certain points in the measurement cross-section, the measurement points chosen must be representative. The gas and dust flow rates must be determined as accurately as possible using a suitable number and position of measurement points within acceptable time scales and at acceptable labor cost.

For this measurement process, the total area is subdivided into smaller regions of equal area whose mid-points are used as the measurement points (grid measurement). It is assumed that at each point the gas velocity and the mass flow rate represent the mean for the small area. For rectangular cross-sections, the areas should be similar. The arrangement of measurement points in a rectangular cross-section is shown in Figure 6.6. With a circular cross-section, the measurement plane is divided into annular regions of

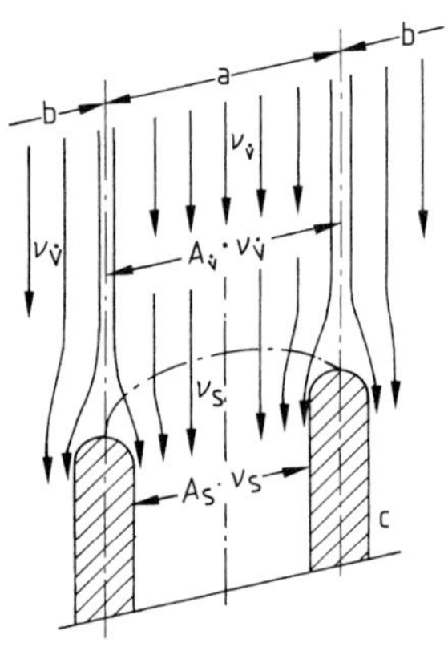

Figure 6.5. Scheme for equal velocity sampling
a) Sampled gas stream $\dot{v}$; b) Main gas stream $\dot{V}$; c) Sample probe
A = cross-sectional area; $A_{\dot{v}}$ = cross-sectional area of sampled gas stream = effective cross-sectional area of probe; v = velocity
Equal velocity condition: $v_{\dot{v}} = v_{\dot{V}}$; Equal flow rate condition: $A_{\dot{v}} \cdot v_{\dot{V}} = A_S \cdot v_S$

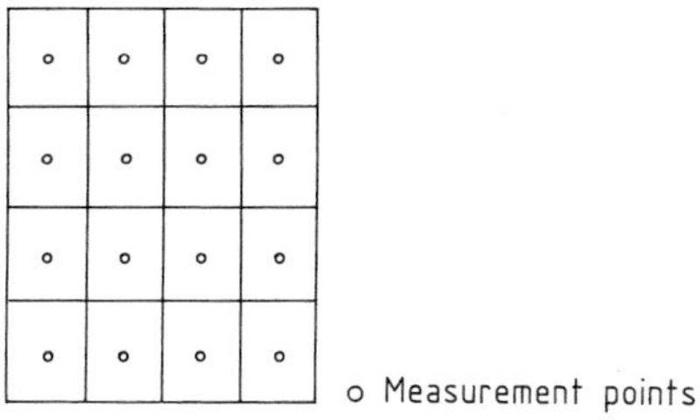

Figure 6.6. Arrangement of measurement points in a grid-based measurement in a rectangular cross-section

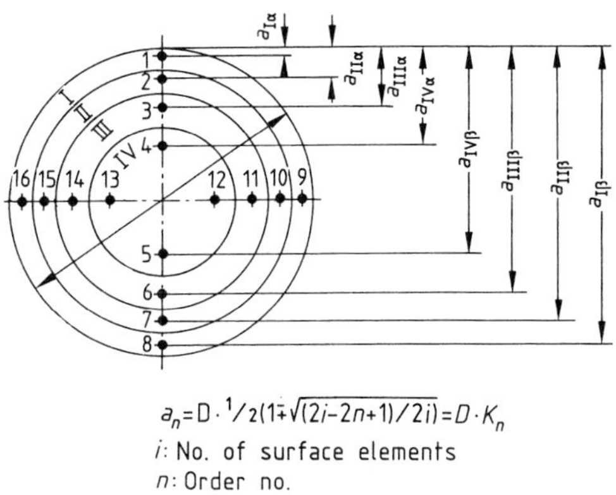

Figure 6.7. Arrangement of measurement points in a grid-based measurement in a circular section with two measurement axes

$$a_n = D\,{}^1\!/_2\left(1 \mp \sqrt{\frac{2i - 2h + 1}{2i}}\right) = D\,K_n$$

where i = number of surface elements, n = ordinal number

equal area. The typical measurement points for circular cross-sections are illustrated in Figure 6.7. Measurements are generally carried out on two perpendicular axes (diameters). For very uniform flow, a single measurement axis may suffice, although in exceptional cases more than two axes may be required.

The necessary number of equal rectangles or annular areas depends on the flow distribution and the area of the measured cross-section. Increased nonuniformity of gas and dust distribution means that more measurement points must be provided.

For a cross-sectional of 0.03–1 m², four measurement points are required, and for >1 m², four per square meter.

However, as a rule, 20 measurement points are sufficient for larger cross-sectional areas.

6.7.2. Discontinuous Methods

Particles can be measured discontinuously by various gravimetric methods, by optical methods in which the DIN smoke number is estimated visually, and by visual estimation of the gray value (Ringelmann method).

Gravimetric Determination. A probe removes a sample stream of dust-laden gas from the total volume flow, and the particles present in the sampled gas are collected on filters made of glass, quartz fiber, or quartz wadding. The dust content is found from its mass and the volume of sampled gas [6.8], [6.32]–[6.34]. In this method, particles are recovered under the conditions prevailing in the exhaust gas duct, and the filter is located immediately after the sampling probe.

These methods are used to determine particulate emissions, to investigate the efficiency of particle separators, and as reference methods for calibrating recording dust content measurement equipment.

A sampling arrangement for moist, saturated off-gases is described in VDI 2066, Sheet 3, in which the filter is located outside the duct. The sampling probe and the filter are heated so that the particles are collected at temperatures above the exhaust gas temperature (i.e., "quasi-dry").

A method for determining the dust content of the unpurified gas (up to 200 g/m³), i.e., before the dust filters, is described in VDI 2066, Sheet 3. A sampling flow rate of ca. 4 m³/h is used with a double filter cartridge with higher dust storage capacity.

The measurement method described in Sheet 2 is used in practice for dust contents of 1–1000 mg/m³. For concentrations of <20 mg/m³, a packed cartridge/flat filter combination is used. The arrangement described in Sheets 2 and 7 is illustrated in Figure 6.8.

This arrangement requires leak-proof suction equipment with a by-pass and a flow-control valve. If the suction equipment follows the gas meter, the volume flow rate can be controlled by bleeding in air.

The method of VDI 2066, Sheet 7 is suitable for measuring low dust contents (0.1–20 mg/m³). For the range 0.1–5 mg/m³, weighing must be very accurate. The system is designed for sampling rates up to 4 m³/h. A typical flat filter head is shown in Figure 6.9. The same method is used for measuring asbestos dust [6.35].

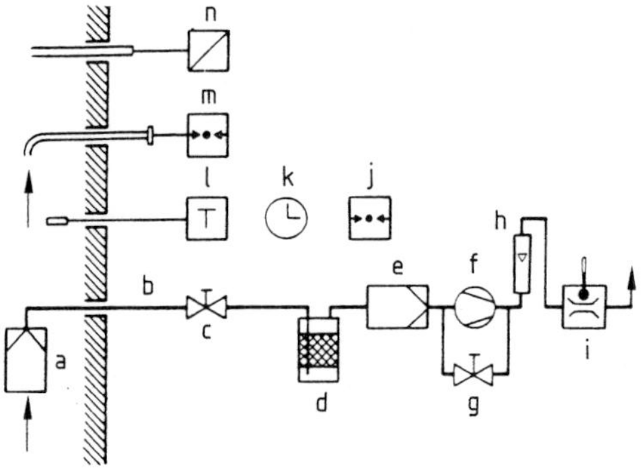

Figure 6.8. Sampling equipment (VDI 2066, Sheets 2 and 7)
a) Filter head with probe and diffuser; b) Gas sampling pipe; c) Shut-off valve; d) Drying tower (if necessary); e) Protective filter for suction equipment; f) Suction equipment (leak-proof); g) Bypass control valve; h) Flowmeter (rotameter); i) Gas meter with thermometer; j) Barometer; k) Timing instrument; l) Temperature sensor with display; m) Prandtl tube with micromanometer (alternatively anemometer); n) Gas measurement equipment (if necessary)

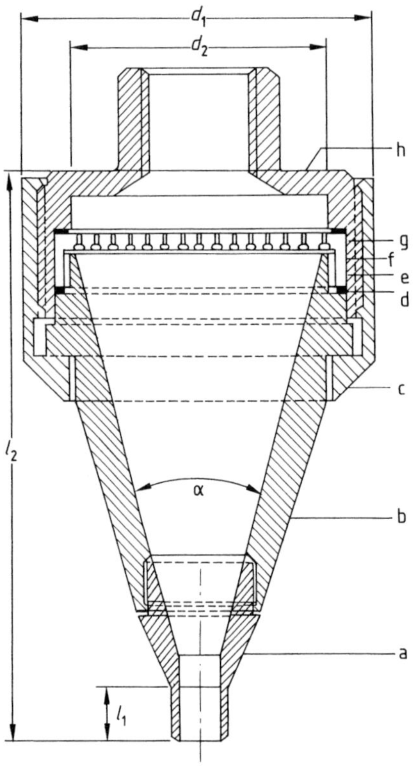

Figure 6.9. A typical flat filter head
a) Exchangeable sampling probe; b) Inlet cone; c) Union nut; d) Seal; e) Tension ring; f) Filter holder; g) Flat filter; h) End housing and connection piece
d_1 = diameter of housing (65 mm); d_2 = diameter of filter (50 mm); l_1 = length of inlet tube (10 mm), l_2 = length of flat filter head (100 mm); α = angle of inlet cone (30°)

The relative detection limits for the various dust measurement methods with the usual operating time of 30 min are 1–2 mg/m^3 with the 12 or 4 m^3/h filter head, 0.3–0.5 mg/m^3 with the 40 m^3/h filter head, and 0.1–0.2 mg/m^3 with the flat filter head.

Lower detection limits can be achieved by increasing the gas sampling rate or using longer running times. With a dust content of ca. 10 mg/m^3, duplicate determinations indicate approximate measurement uncertainties U (95%) of between ±(5–10)% with the 4–40 m^3/h equipment, and between ±(4–8)% with the flat filter head equipment.

Optical Methods. In the past, visual methods were prescribed for monitoring dust and smoke emissions. Official regulations at one time specified the Ringelmann method for checking the acceptability of exhaust gas plumes. However, now that dust contents are much lower, this requirement is now regarded as obsolete.

The Ringelmann scale is a series of gray values between white and black in five steps from 0% (white) to 100% (black). Estimation is carried out with a chart having four areas covered with black hatched lines that occlude 20, 40, 60, and 80% of the area, giving the impression to the eye of different gray values.

For combustion plants running on light heating oil, smoke emission limits must be complied with. Smoke emission is represented by the smoke number (DIN 51 402, Part 1), which is determined with a manual smoke pump. The value is determined from the change in the reflectivity of the measuring filter. An increase in the smoke number of 1 unit is defined as a decrease of 10% in the reflectivity of the measurement filter.

This method is mainly designed for monitoring small furnaces. For plants fired with light heating oil with a thermal capacity ≥5 MW, TA Luft requires continuous monitoring of the smoke number. These recording devices must be calibrated against a conventional measurement method. The DIN method using manual smoke pumps is unsuitable. A modified method is described in VDI 2066, Sheet 8 [6.36], based on DIN 51 402, Part 1. It is a photometric method developed from a technique for measuring low dust contents [6.34].

6.7.3. Continuous Methods

Where there is a legal requirement for equipment for continuously measuring emissions, the conditions described in Section 6.4 must be observed. Measurements carried out at the instigation of internal plant management are not affected.

6.7.3.1. Optical Methods

Optical monitoring of dust emissions is widely used, based on transmission/opacity, extinction, or scattered light.

In in-situ photometric dust measurement, the intensity of a beam of light passed through the off-gas duct is decreased as a result of absorption and scattering by the particles. The ratio of attenuated to emitted light is the optical transmission T. The quantity $1 - T$ is known as the opacity. The value $\log(1/T)$ is the extinction E.

The relationship between the length of the light beam L and the transmission T for constant dust properties in the exhaust gas is expressed by the Lambert–Beer law:

$$T = \exp(-E) = \exp(-\varepsilon L)$$

The extinction coefficient ε depends on the properties of the light used, the properties of the dust to be measured (e.g., shape, color, and particle size distribution), and on the dust content c. It has been found that, within certain limits, there is a linear relationship between dust content c and extinction coefficient ε if other parameters are constant. If ε' is the proportionality constant:

$$T = \exp(-\varepsilon' c L) \quad \text{or} \quad E = \varepsilon' c L$$

Depending on the application, the equipment can be designed for smoke density measurement or dust content measurement. In smoke density measurement equipment, only the transmission is measured. The object is to monitor the functioning of filter equipment and to determine the opacity of the exhaust gas, e.g., by determining its visibility threshold or by using the Ringelmann scale. There are several approved types of measuring equipment [6.11]. Manufacturers include Durag and Sick.

In dust content measurement equipment, the extinction is measured. An individual calibration by means of a separate determination of the dust content enables the measured extinction values to be calibrated, taking into account the effect of particle properties. Particle-size distributions, densities, colors, and refractive indices differ from plant to plant, but are comparatively constant for the same plant.

The usual arrangement for photometric dust content measurement apparatus in situ is illustrated in Figure 6.10. The measurement head is installed on one side of the exhaust gas duct and the reflector on the other. The light source and photoelectric detector are located close together in the measurement head. The reflector consists of an arrangement of triple mirrors (retroreflectors) which reverse the direction of the light beam. Thus, the light beam passes through the measurement path to the reflector and is returned. The reference light beam passes through a dust-free reference path inside the measurement head. A half-silvered mirror inserted alternately into the measurement and reference light paths causes the two light beams arrive at

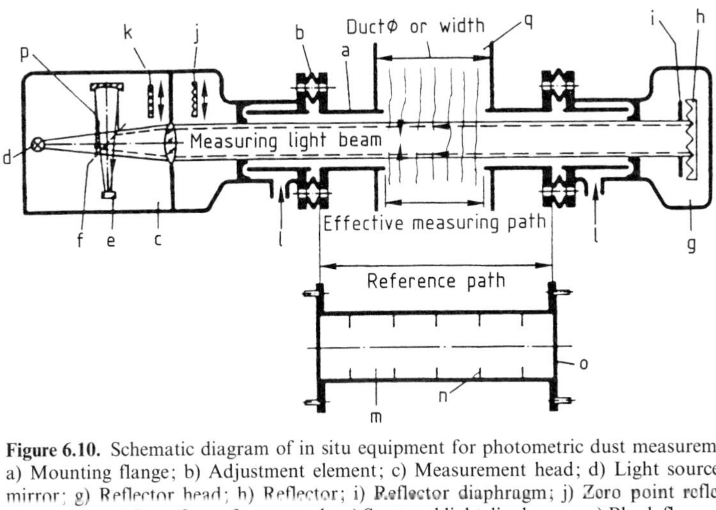

Figure 6.10. Schematic diagram of in situ equipment for photometric dust measurement
a) Mounting flange; b) Adjustment element; c) Measurement head; d) Light source; e) Light receiver; f) Half-silvered mirror; g) Reflector head; h) Reflector; i) Reflector diaphragm; j) Zero point reflector; k) Control filter; l) Purge air connection; m) Dust-free reference path; n) Scattered light diaphragm; o) Blank flange; p) Reference light beam; q) Exhaust gas duct

the photoelectric detector out of phase. The electrical signal produced by the detector is processed so that the final signal is proportional to the extinction.

Several types of measurement equipment have been approved [6.11]. Manufacturers include Durag, Mannesmann/H.u.B., and Sick.

Scattered light measurement exploits the fact that when a collimated light beam is passed through a dust-laden measured volume, part of the light is scattered in various directions. The intensity of the light scattered at a given angle to the original direction depends on the intensity, wavelength, and polarization of the incident light, on the angle itself, and also on the shape, particle-size distribution, and refractive index of the particles. The principle of scattered light measurement is illustrated in Figure 6.11. As with extinction measurement, a linear relationship exists between the dust content and scattered light intensity within certain limits. The sensitivity of light scattering measurement equipment is considerably higher than that of extinction measurement equipment. Depending on the method used, the detection limit can be as low as 0.1 mg/m^3. In practical applications, dust concentrations as high as 100 mg/m^3 have been determined. However, with both extractive sampling and in-situ equipment (small measurement volumes), care must be taken to ensure that sampling is representative.

The function scheme of the KTN/KTNR photometer manufactured by Sigrist is illustrated in Figure 6.12. This operates according to the double beam system, with forward scattering at 15° and automatic compensation.

The beam from a light source (a) passes through an optical path to the "flicker" mirror (b). In position (1), this deflects the incident light as the measuring beam (c) via an optical path into the measuring chamber (h). Part of the scattered light produced by the sample is received at an angle of 15° to the forward direction by a light detector (i)

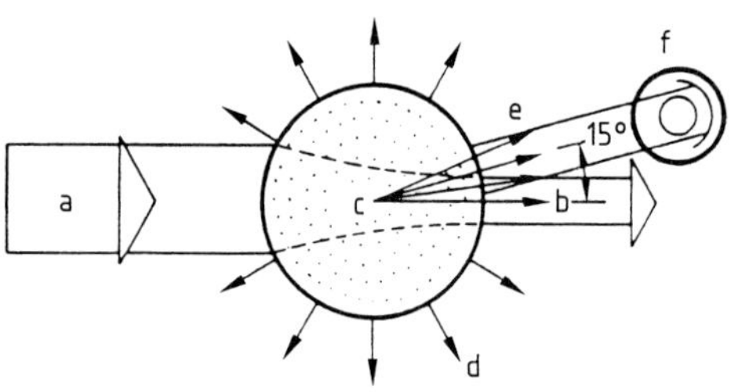

Figure 6.11. Principle of scattered light measurement
a) Incident light beam on sample; b) Light beam transmitted through sample; c) Sample; d) Scattered light; e) Measurement of selective scattered light beam; f) Photocell

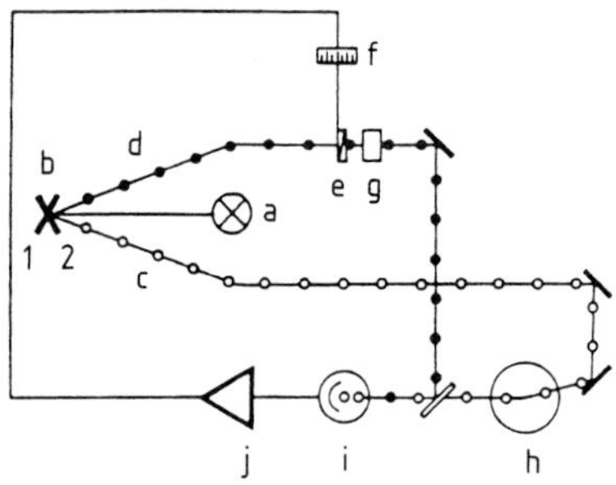

Figure 6.12. Functional scheme of a scattered light photometer
a) Light source; b) Flicker mirror; c) Measurement beam; d) Reference beam; e) Light attenuator; f) Display instrument; g) Reference standard; h) Measurement chamber; i) Light detector; j) Measurement amplifier
1) Measurement position of flicker mirror; 2) Reference position of flicker mirror

and measured. In position (2), the incident light is passed as a reference beam (d) through a light attenuator (e) and a reference standard (g) to the light detector. The signal currents produced by the light detector are compared in a measurement amplifier (j) and converted into a control signal which changes the reference beam by means of a light attenuator until its intensity corresponds to the intensity of the light which is scattered by the sample and falls on the light detector.

The measurement method and calibration are described in VDI 2066, Sheet 6 [6.37]. The equipment operates with an extractive sampling system. The sampling system can be heated to 180 °C, so that even water vapor saturated exhaust gases can be continuously measured [6.11], [6.14]. In another approved variation, the equipment is used to measure the smoke number [6.38].

In-situ scattered light photometers are produced by Durag and Sick for measuring both low dust contents and smoke numbers.

A schematic diagram of equipment for in-situ scattered light measurement is shown in Figure 6.13 [6.36]. This is suitable for measuring both dust contents and smoke numbers. Light is passed directly through the exhaust gas duct and received and absorbed on the opposite side by a light trap. The scattered light from a definite measured volume in the exhaust gas duct is received and measured at a certain angle (e.g., 90° scattering or back-scattering). Because of the geometry of the equipment, the measured volume lies near to the duct wall. Problems can therefore arise on installing this equipment in thick-walled ducts (e.g., brick-lined) or double-walled pipe. In-situ light scattering equipment is sensitive and

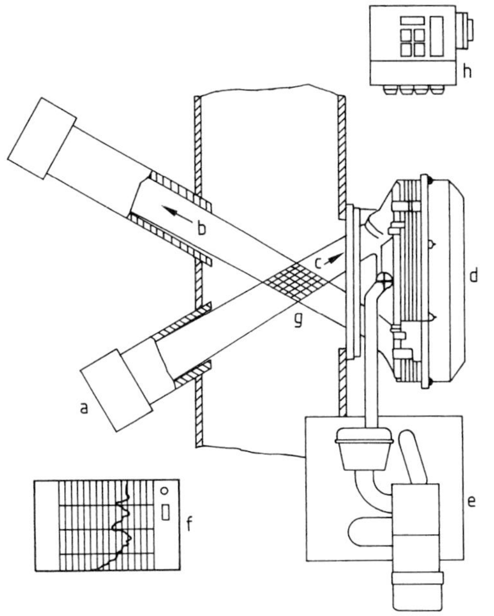

Figure 6.13. Schematic diagram of in-situ scattered light measurement equipment
a) Light trap; b) Direct light; c) Scattered light; d) Scattered light measurement; e) Purge air unit; f) Pen chart; g) Measurement volume; h) Display

suitable for measuring low dust contents (<10 mg/m^3) and low smoke numbers (<3 or <1). Unlike extinction equipment, it can be operated in narrow ducts. Various types of equipment have been approved [6.14], [6.15], [6.29], [6.38].

6.7.3.2. Radiometric Methods

Radiometric measurement equipment is manufactured by the companies Verewa (dust content, smoke number) and FAG Kugelfischer (dust content), and has been officially approved [6.31]. (Smoke number measurement equipment is based on reflection photometry rather than β-radiation.) In dust measurement based on β-radiation absorption, a sample gas stream is taken by a probe and passed through a strip of filter paper which collects the dust particles. The paper strip is moved along in a stepwise fashion. The quantity of dust deposited on the filter is measured by the attenuation of β-radiation (source: ^{14}C or ^{85}Kr) passed through the filter. The attenuated radiation is detected by a Geiger counter, and the attenuation is a measure of the mass of dust collected (Fig. 6.14).

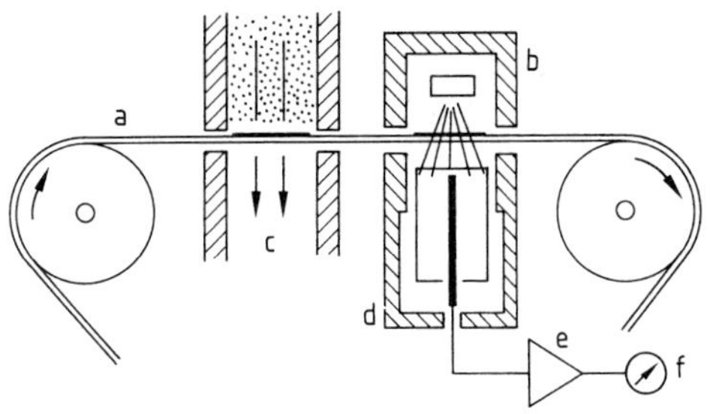

Figure 6.14. Principle of radiometric measurement of dust by β-radiation absorption
a) Filter band; b) β-Emitter; c) Partial stream; d) Detector; e) Amplifier; f) Display

As with the other methods, calibration is based on a conventional measurement method (see Section 6.7.2.1), carried out as a grid measurement over the duct cross-section.

In the Beta-Staubmeter F 904 equipment (Verewa), a sample stream is taken isokinetically from the gas flowing in the duct, and preheated dilution air is added to reduce the dew point of saturated exhaust gases. The gas is then filtered and the dust collected is measured using β-radiation. The result is the mean value of the dust content based on the cycle time used. The dust measuring equipment FH 62 E–NA (FAG) measures the actual dust content radiometrically. After the zero has been automatically set, a sample stream of gas is removed from the stack isokinetically. It is then heated, diluted with clean air, and aspirated through a measuring filter. The continuously increasing amount of dust on the filter causes the integral signal obtained to increase with time. An instantaneous dust concentration is obtained from the signals for rate of mass increase and flow rate.

In another continuous measurement method, the gas is extracted by a sampling probe which simultaneously measures the gas velocity. The dust-laden gas is trapped on a filter, which is changed at a time interval depending on the amount of dust collected. The dust is determined by measuring the differential pressure across the filter (LSP apparatus, Becker Verfahrenstechnik).

6.7.4. Particle Size of Dusts

Laboratory methods for determining the particle size of dusts include microscopy, sieving, air classification, sedimentation, and scattered light [6.39]. Dust quantities of 2–20 g are necessary. With low dust contents, relatively large volumes of the exhaust gas must be filtered. However, the particle size distribution can be changed by the sampling and collection processes (e.g., by agglomeration). Redispersion in the laboratory may result in the measured particle size differing from that in the gas [6.40]. It is therefore advantageous to measure the particle-size distribution in the gas. Various types of measurement equipment have been developed, e.g., light scattering, diffuser batteries, electrostatic aerosol analyzers, and aerosol centrifuges. Sampling must be carried out with a suction probe, as the equipment cannot be directly installed in the exhaust gas duct. A high proportion of particles of diameter >3 μm are lost during collection by deposition on the walls of the probe and on the inlet to the measuring equipment. The method is therefore limited to particles of diameter <3 μm, although particles in the range <1 to 100 μm can be expected in emissions from industrial and commercial plants. Aerosol measurement equipment generally cannot be used in hot and chemically aggressive exhaust gases. The sampled gas stream should not be strongly cooled because condensation of moisture and other substances can change the particle-size distribution. Cascade impactors are suitable for sampling a partial gas stream and separating the particles into mass fractions. They can be used directly in hot, chemically aggressive gases [6.41]. Gravimetric evaluation of cascade impactor measurements gives mass distributions as a function of aerodynamic diameter of the particles. However, this method does not replace the measurement of total dust content (VDI 2066, Sheets 2, 3, and 7). The cascade impactor method is described in detail in VDI 2066, Sheet 5 [6.42]. Cascade impactors can separate particles into several fractions between ca. 0.3 μm and 15 μm, depending on the design, the type of gas, and the particle properties. Larger particles (>10 μm) interfere with the fractionation process and must be removed by a preliminary separation process [6.43]. Low-pressure impactors are used for particles with a lower size range.

The impactor exploits differences in the inertia of particles to separate them into fractions. An impactor stage consists of a jet and an impact plate. The particles are accelerated in the jet and those having sufficient inertia strike the impact plate and are collected (see Fig. 6.15).

The particle fractionation depends on jet width D, jet length L, and the distance S between the impact plate and the jet. Cascade impactors consist of several impactors in series, whereby particles of decreasing inertia are separated in the successive stages, giving fractions of decreasing particle size. Particles not collected on the impact plates are trapped on a filter.

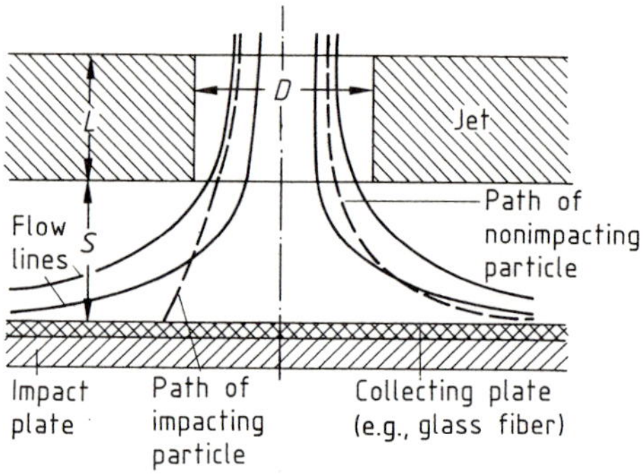

Figure 6.15. Principle of impaction of particles

Evaluation of impactor measurements is based on experimentally determined median values of the Stokes number St_{50} and the limiting aerodynamic particle diameter d_{ae50} of an impact stage. For an impactor with several circular jets per stage, a uniform value of $St_{50} = 0.26$ is used for all impactor stages for reasons of comparability [6.41].

The measurement setup is specified by VDI 2066, Sheet 7. The gas passes through the sampling probe, the cascade impactor, and then the angle bend. Without a preliminary separation stage, the cascade impactor described in VDI 2066, Sheet 5 can be used for dust contents of 1–2 mg/m^3. The impactor must not be overloaded. The maximum permissible load per stage is ca. 100 mg. If there is a preliminary separation stage, the dust content can be between 5 mg/m^3 and 25 g/m^3, depending on the fraction of coarse particles.

6.7.5. Heavy Metals as Dust Constituents

It is a legal requirement that all emissions of metals, metalloids, and their compounds should be measured [6.2]–[6.4]. Their emission limit values lie in the range 0.05–5 mg/m^3. These limits are defined as the sum of the concentrations of the substances in the form of dust, gas, and vapor. The duration of an individual measurement should normally be 30 min and should not exceed 2 h. The most important elements are antimony, arsenic, beryllium, lead, cadmium, chromium, cobalt, copper, manganese, nickel, palladium, platinum, mercury, rhodium, selenium, tellurium, thallium, vanadium, zinc, and tin. A proven measurement method is described in VDI 3868, Sheets 1 and 2 [6.44], [6.45]. It consists

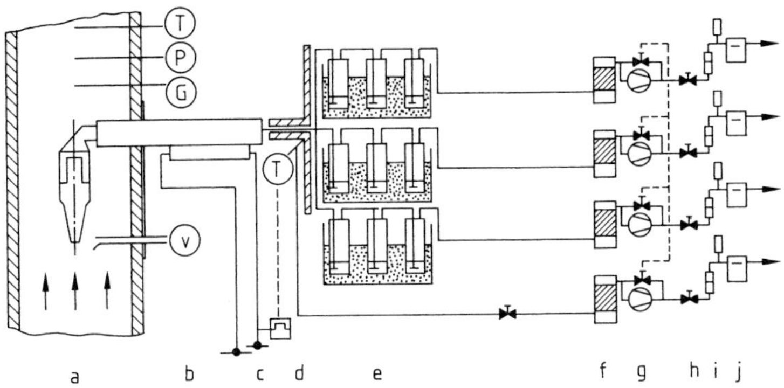

Figure 6.16. Equipment for the simultaneous collection of particulate materials and materials passing through the filter
a) System for retaining particles; b) Sample pipe with controlled heating; c) Temperature measurement and control for sampled gas stream; d) Adapter with insulation and heating if necessary, e) Absorption system; f) Condensate separator; g) Leak-proof pumping system with controllers; h) Isolation valve; i) Gas volume flow rate measurement; j) Gas meter with thermometer
T = temperature measurement; P = pressure measurement; G = gas analysis; v = gas velocity measurement

of a particle collection system (VDI 2066) and an absorption stage made up of a battery of wash bottles, as used for measuring emissions of gases such as SO_2 and HCl (Fig. 6.16). The exhaust gas is sampled isokinetically by a probe. The particles are retained by a filter system as specified in VDI 2066, Sheet 2 or Sheet 7. The materials passing through the filter are drawn through a heated sampling tube. A bypass stream is passed through one or more absorption systems in parallel, consisting of at least three gas-absorption bottles in series. The method enables the mass of the dust and of its constituents to be determined both per unit of time and per unit of volume. Two types of absorption solution are used:

1) Solution A: 3 volumes hydrochloric acid mixed with 1 volume nitric acid, 1 volume of this mixture being diluted with 9 volumes water
2) Solution B: 1 volume nitric acid mixed with 1 volume hydrogen peroxide, 2 volumes of this mixture being diluted with 13 volumes water

Mercury is determined with a solution of 20 g $KMnO_4$ in 1 L dilute sulfuric acid (VDI 3868, Sheet 2).

Under the usual sampling conditions, the detection limit for metals is usually <0.01 mg/m^3. The method described in Guideline VDI 3868, Sheet 2 for mercury is suitable for concentrations of >5 µg/m^3 [6.45]. Both methods are especially suitable for measurements on exhaust gases from thermal processes in which the constituent elements can accumulate on the smallest particles as well as being present in gaseous and vapor form.

The most important methods of analysis include X-ray fluorescence (XRF), atomic emission spectrometry with inductively coupled plasma source (ICP–AES), atomic absorption spectrometry (F-AAS and G-AAS), and instrumented neutron activation analysis (INAA). Analytical determination of the elements is described in VDI 2268 [6.46]–[6.49]. Because of the required detection sensitivity, atomic absorption is used preferentially for many components. Methods used include flame AAS, graphite-tube AAS, the hydride and cold vapor technique, and ICP–AES. In cases where even atomic absorption spectrometry does not enable all trace elements to be determined with sufficient sensitivity, INAA can be used if available.

6.8. Measurement of Gaseous Inorganic Emissions

6.8.1. Sampling

A measurement grid is necessary when sampling gaseous emissions if the concentration is not uniformly distributed over the measured cross-section [6.19]. There are two methods of determining the mean concentration. Either the local emission concentration c_i and the local velocity u_i (together with the oxygen content) are determined at each measuring point, or a partial mass m_i is determined by using a collection phase, with a sampling rate proportional to the velocity. However, this does not necessarily mean isokinetic sampling (as with dust measurement), but a sampling rate proportional to the velocity or the mass.

Generally, gaseous emissions are sampled at a velocity considerably lower than that of the exhaust gas. Also, the probe is not parallel to the flow direction as with dust measurement, but at an angle $<90°$ to the flow direction. The sampling volume rate is typically 30–120 L/h. The probe is made of, e.g., glass, quartz, titanium, stainless steel, or PTFE. The probes and associated pipework are usually heated to prevent condensation.

With recording analytical equipment that operates extractively, correct functioning and the results obtained depend on the processing of the sampled gas, which must be dust-free, dry, and free of corrosive materials before it enters the analyzer. Furthermore, the volume flow rate, pressure, and temperature must generally lie within limits imposed by the measurement method. If the dew point of the sampled gas is higher than the temperature of those parts of the analysis system with which it comes into contact, the gas must be cooled at the exit of the sampling equipment before it enters the analyzer to condense the moisture. The possibility as the substances being measured dissolving in the condensate must be considered.

6.8.2. Sulfur Oxides

When sulfur-containing fuels and wastes are burnt, sulfur dioxide is formed. Important emission sources are industrial furnaces, chemical plant, the cement industry, metal smelting, ceramics, the cotton industry, domestic heating, and traffic. Part of the SO_2 may be oxidized to SO_3. It is normally a legal requirement that the total sulfur oxides ($SO_2 + SO_3$) should be determined and expressed as SO_2.

6.8.2.1. Discontinuous Methods

At present, four manual analytical methods are available for determining SO_2, and these are also used as reference methods for the calibration of continuously measuring equipment. They are described in the Guideline VDI 2462, Sheets 1, 2, 3, and 8 [6.50]–[6.53] (Table 6.1).

Because of the low limits of detection and lack of cross-sensitivity towards nitrogen oxides, the method described in Sheet 8 is often used for emission measurements and as a reference method for the calibration of continuous measuring equipment. The gas is sampled through a heated probe, and is then passed through a heated silica wool filter to remove dust and through two wash bottles in series containing a 3% solution of hydrogen peroxide. The temperature of the heated probe and the silica wool filter is chosen so that condensate formation (acid dew point) is prevented. A temperature of 200–220 °C is adequate. The SO_2 in the exhaust gas is quantitatively oxidized to H_2SO_4. This method also determines SO_3 and is thus a summation method for sulfur oxides ($SO_2 + SO_3$) expressed as SO_2. A schematic diagram of the sampling apparatus is given in Figure 6.17. Analysis is by titration against a solution of barium perchlorate. A color change takes place in the presence of the added metal indicator thorine when the solubility product of barium sulfate is exceeded.

Table 6.1. Determination of SO_2 (VDI 2462)

Sheet	Brief title	Measurement principle	Remarks
1	Iodine–thiosulfate method	bubbling into I_2, oxidation to H_2SO_4, back-titration with sodium thiosulfate, detection limit 60 mg SO_2/m^3	cross-sensitivity towards hydrogen sulfide, nitrogen oxides, halogens; suitable for higher SO_2 concentrations
2	Hydrogen peroxide oxidation + titration	oxidation to H_2SO_4 by H_2O_2, titration with NaOH, detection limit 30 mg SO_2/m^3	cross-sensitivity towards components that can form acids or bases in the absorption solution; suitable for higher SO_2 concentrations
3	Hydrogen peroxide oxidation + gravimetric determination	oxidation to H_2SO_4 and precipitation of $BaSO_4$, detection limit 40 mg SO_2/m^3	cross-sensitivity towards other sulfur-containing components except SO_3; suitable for higher SO_2 concentrations
8	H_2O_2–thorine method	oxidation to H_2SO_4, titration with $Ba(ClO_4)_2$ with thorine as indicator, detection limit 0.72 mg SO_2/m^3	cross-sensitivity towards certain anions and metallic cations and other sulfur-containing components except SO_3; suitable for low SO_2 concentrations; no cross-sensitivity towards nitrogen oxides

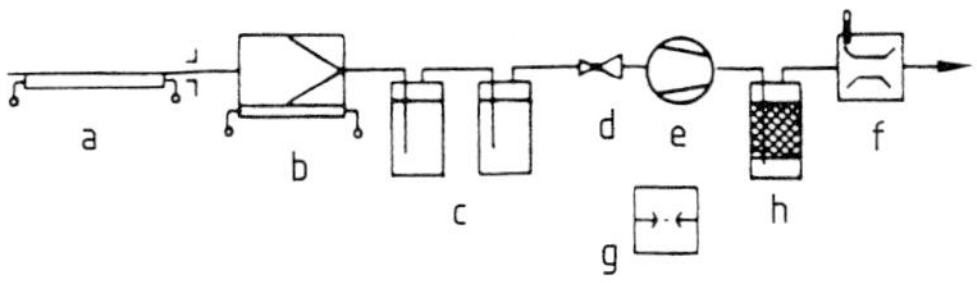

Figure 6.17. Schematic of sampling equipment for the measurement of sulfur oxides
a) Sampling probe, heated; b) Filter with silica wool packing, heated; c) Gas wash bottle, 250 mL, with grade G2 frit; d) Flow control valve; e) Suction pump; f) Gas meter with thermometer; g) Barometer; h) Drying tower (only used with dry gas meter)

The separate determination of SO_3, described in Sheet 7 [6.54], is also based on the thorine reaction. The SO_3 is absorbed in isopropanol, which does not absorb SO_2. The SO_3 in isopropanol solution is titrated with barium perchlorate with thorine as indicator, as described for SO_2. The relative detection limit is ca. 2 mg/m^3. Volatile sulfates that pass through the filter are also determined.

6.8.2.2. Continuous Methods

Several photometric measurement principles can be used for continuously measuring SO_2 emissions, with extractive and in-situ sampling. Conductivity methods are also used, with extractive sampling.

Photometry with Extractive Sampling. All heteroatomic molecules such as CO, SO_2, and NO have a characteristic absorption spectrum in the infrared; SO_2 and NO also absorb in the ultraviolet.

The simplest arrangement for an extractive absorption photometer is illustrated in Figure 6.18. The light from a radiation source passes through an optical filter to limit its spectral range to that required for the measured component, passes through an optical cell through which the gas being analyzed flows, and falls on a photodetector linked to an electronic signal processing device. Part of the light is absorbed by the pollutant molecules. The light attenuation is therefore a measure of the concentration of the pollutant. In this simple arrangement, very small changes in the radiation emission and the receiver sensitivity lead to unacceptably high zero point errors. Measures to avoid this problem are based either on a periodic zero point correction or a reference standard in the form of a second reference filter or a reference gas. This reference standard can either be inserted from time to time (and thus out of phase) into the radiation path, or it can be located in a parallel reference beam. The first arrangement is known as a single-beam photometer, and the second a double-beam photometer. The photometric analysis equipment can be caused to be sensitive to a selected component using either dispersive or nondispersive systems.

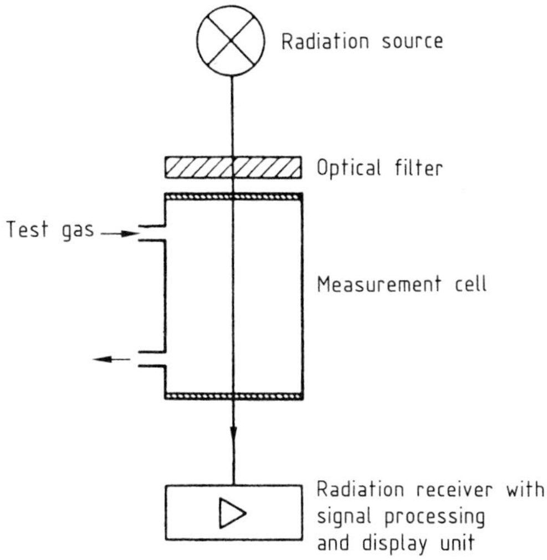

Figure 6.18. Absorption photometer (schematic)

Dispersive Systems. In dispersive systems, the light from a broad band radiation source is broken down into its spectral components by a spectrometer before the measurement is carried out. Prisms, gratings, or interference filters can be used to perform the dispersion.

Nondispersive Methods. In nondispersive methods, wavelength selection is carried out by the substances being measured, which are stored in the equipment. There are three methods of storage:

1) In the nondispersive infrared (NDIR) method, a gas-filled optical cell is used for storage (Fig. 6.19). The radiation is deflected to pass through the gas-filled chambers, and modulation by a rotating wheel produces periodic pressure fluctuations in the detector chambers. These are detected either by a membrane capacitor or a microflow detector which detects a flow

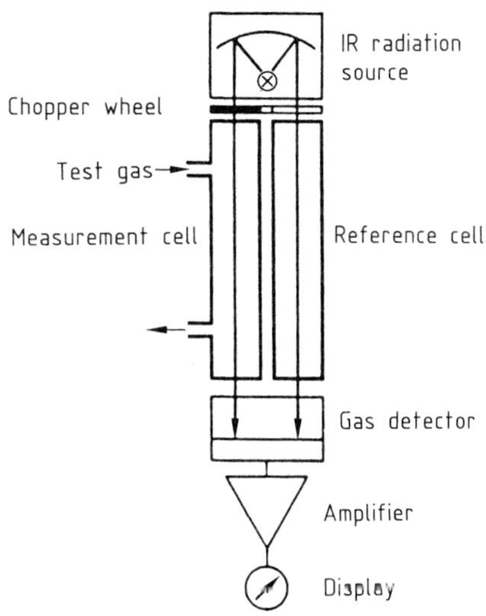

Figure 6.19. NDIR photometer (schematic)

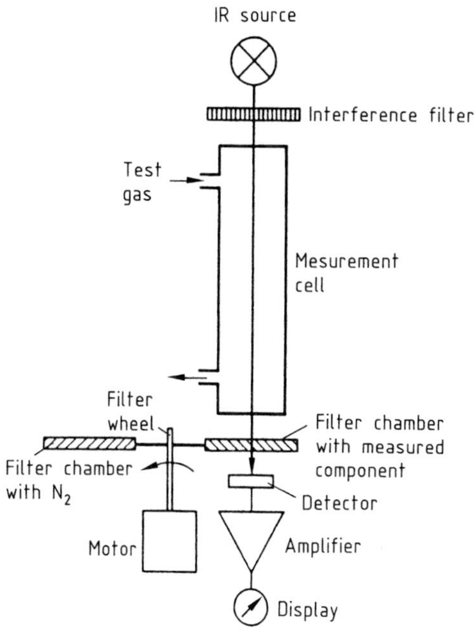

Figure 6.20. Gas filter correlation method (schematic)

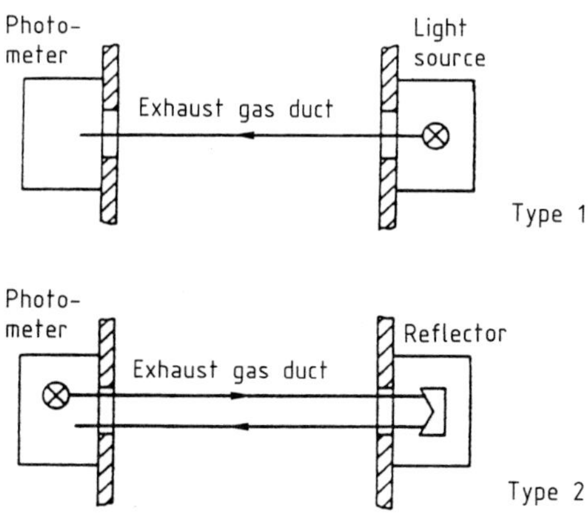

Figure 6.21. In-situ photometer (schematic)

between the two chambers caused by the pressure fluctuations, and converts this into an electrical signal.

2) In the gas filter correlation (GFC) method (Fig. 6.20), a gas-filled optical cell attached to a rotating wheel acts as the storage device. This is periodically brought into the radiation path, alternating with a nitrogen-filled gas filter.
3) In the nondispersive ultraviolet (NDUV) method, the component to be measured is stored in the radiation source. Gas-filled discharge lamps are used which emit spectral lines characteristic of the component being measured. UV photometers operate with 1 or 2 beams and with 1 or 2 photoelectric detectors. Some photometers use a combination of two of the optical selection methods described above.

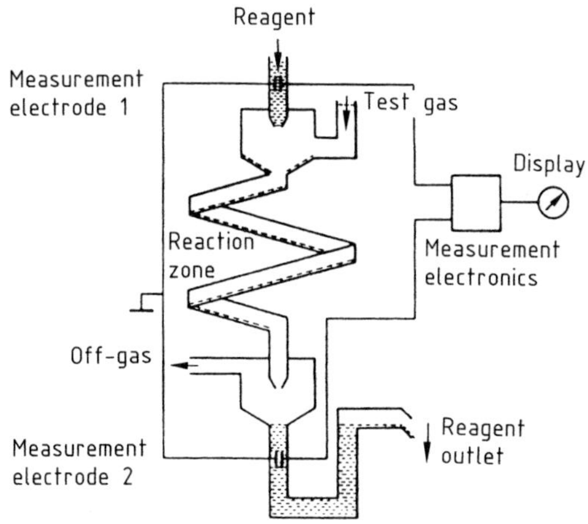

Figure 6.22. Conductometric measurement (schematic)

In-Situ Photometry. The absorption measurement path in in-situ photometers is located directly in the exhaust gas duct, so that the gas sample need not be transported to the optical cell by a sampling system. The photometer, consisting of a photometric detector, wavelength selection equipment, and electronics to calculate the result is situated outside the exhaust gas duct. In the UV range, wavelength selection is by diffraction grating. In the IR range, interference filters and gas-filled filter chambers (as with the GFC method) are also used.

Two arrangements are illustrated in Figure 6.21. In both cases, the windows on each side of the absorption path are protected from dust soiling by an air curtain, as in the photometric measurement of dust.

Conductometry. In conductometric gas analysis, the gas sample is passed into a liquid reagent, and the change in conductivity after reaction of the liquid with the gas is measured.

In continuous conductometry, the sampled gas and the reagent liquid are continuously fed into one end of a reaction tube. As the change in conductivity also depends on the ratio of the volume flow rates, both flow rates must be kept constant by a control device. The conductivity before addition of the gas phase and after its removal is measured by two pairs of electrodes (Fig. 6.22). The conductivity difference, which is proportional to the concentration of the measured component, is calculated by an electronic system. As conductivity is strongly temperature dependent, the effect of temperature must be compensated.

Details of officially approved equipment for measuring SO_2 have been published [6.55] (Table 6.2).

6.8.3. Nitrogen Oxides

Nitrogen oxides are formed in all thermal combustion processes in air and by oxidation of the combined nitrogen in fuel. Emission sources include furnaces burning solid, liquid, and gaseous fuels; internal combustion engines; power plants; cement works, metallurgical

Table 6.2 Equipment for continuous measurement of SO_2

Manufacturer	Type	Principle
Extractive sampling		
NDIR method		
Afriso	Infralyt 1210	double-beam equipment with two radiation sources
Mannesmann/H.u.B	URAS 3 G, 3 E	double-beam difference measurement method with one radiation source and double-layer detector
Maihak	UNOR 6 N So_2 UNOR 6 N - F UNOR 600/610	double-beam difference measurement method with one radiation source and double-layer detector single-beam analyzer with double-layer detector
Perkin–Elmer	MCS 100 HW	single-beam photometer, bifrequency method, two interference filters per measured component, sequential measurement
Siemens	Ultramat 32 Ultramat 5 Ultramat 21, 22	double-beam difference measurements method with one radiation source and double-layer detector single-beam analyzer with double-layer detector
NDUV method		
Maihak	Defor 3	double-beam difference measurement method with one radiation source
Rosemount	SO_2–UV Binos	double-beam difference measurement method with one radiation source
Conductivity		
Wösthoff	Mikrogas–SO_2	absorption in H_2O_2, change in electrical conductivity
In situ sampling		
Opsis	Opsis AR 600	single-beam photometer, absorption spectrum in certain UV wavelength range (diffraction grading)
Sick	GM 21, GM 30	single-beam photometer with reflector, absorption spectrum in certain UV wavelength range, spectral dispersion of the resulting light beam

plant; waste incinerators; and chemical plant. In general, the concentration of NO_2 is $<10\%$, but in a few cases (e.g., fertilizer production), it can be considerably higher.

Discontinuous Methods. Four manual analytical methods are described in VDI 2456. The phenoldisulfonic acid method and the titration method described in Sheets 1 and 2, respectively, are suitable for high concentrations in the g/m^3 range. The sodium salicylate method and the dimethylphenol method described in Sheets 8 and 9 are often used as reference methods for calibrating continuously measuring equipment on account of their low detection limits [6.56]–[6.59].

In the sodium salicylate method, total $NO + NO_2$ is determined. Sampling can be performed by a gas collection tube or an evacuated absorption vessel. When the gas collection tube is used, a volume equal to about ten times the volume of the tube is drawn through the sampling apparatus.

If an evacuated vessel is used, it is connected via a three-way tap between the solid filter and the pump. Its volume must be corrected for temperature and pressure. Oxidation to nitric acid is then carried out for 24 h. The nitric acid is reacted with sodium salicylate in a solution of H_2O_2 acidified with sulfuric acid to form nitrosalicylic acid. The anion is then measured photometrically. The relative detection limit, using 1 L gas collection tubes, is ca. 2 mg/m^3 (as NO_2). The total time between sampling and obtaining the result is 2 d.

In the dimethylphenol method, ozone is used as oxidizing agent. NO and NO_2 are oxidized by ozone in the gas phase to form dinitrogen pentoxide. The reaction takes place in 5–10 min. The N_2O_5 is absorbed in water, forming nitric acid. Samples are taken through a critical jet into an evacuated gas sample container. A schematic of the sampling and oxidation equipment is given in Figures 6.23 and 6.24.

The evacuated gas sample container is filled in the bypass to ca. 500 mbar with the sampled material, and, after temperature adjustment (ca. 5 min), is then filled with ozone until the pressures are equalized. Oxidation is complete after ca. 5 min. The nitric acid formed is reacted with 2,6-dimethylphenol (DMP) in a solution containing sulfuric and phosphoric acids to form 4-nitro-2,6-dimethylphenol. The 4-nitro-2,6-dimethylphenolate anion formed in alkaline solu-

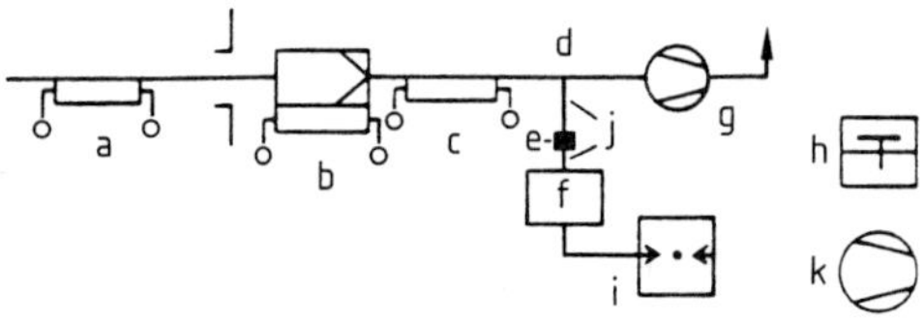

Figure 6.23. Sampling equipment for measurement of nitrogen oxide emissions (schematic)
a) Sampling probe, heated if necessary; b) Solid filter, heated if necessary; c) Test gas ducting, heated if necessary; d) T-piece of glass, heated if necessary; e) Critical jet, heated if necessary; f) Gas sample container with isolation valves; g) Suction pump; h) Thermometer; i) Pressure measurement (vacuum gauge); j) Tubing as short as possible; k) Vacuum pump

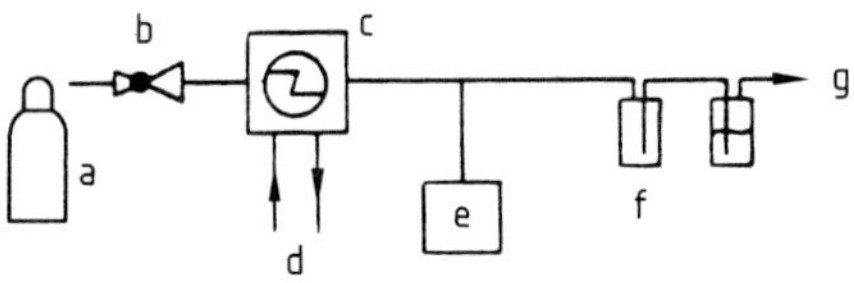

Figure 6.24. Example of oxidation equipment
a) Oxygen cylinder; b) Pressure reducer; c) Ozone generator; d) Water cooling (if necessary); e) Gas sample container with isolation valve; f) Wash bottles, the last of which containing; water g) Waste gas

tion is measured photometrically at ca. 430 nm. The relative detection limit is 15 mg/m^3, expressed as NO_2.

Ammonia at levels up to 150 ppm does not cause problems. If high concentrations of SO_2 are present in the exhaust gas, there is a risk of formation of other nitrogen-containing products (e.g., N_2O). It is then preferable, after pressure measurement, first to fill the evacuated gas sample container with ozone to ca. 250–300 mbar, then to add the gas being measured to ca. 700 mbar, and then to fill with ozone until the pressure is equalized.

As the time required to obtain an analytical result is short, this method is widely used for calibrating continuously measuring equipment in situ.

Continuous Methods. There are several methods of continuous measurement which use both extractive and in-situ sampling techniques. The measurement principles used are infrared and ultraviolet absorption, chemiluminescence reaction, and amperometric measurements with electrochemical cells. As a rule, only the component NO is selectively measured. However, equipment for selectively measuring NO_2 is also commercially available. By combining two analyzers in a single item of equipment, NO and NO_2 can be measured simultaneously. Various types of equipment measure NO and NO_2 sequentially. In the chemiluminescence method, a converter can be incorporated for reducing NO_2, so that total NO + NO_2 is measured [6.60]. Legal requirements are concerned with the determination of total nitrogen oxides (NO + NO_2) only. The manual method described above can be used to determine the concentration of NO_2 for calibration purposes if NO-selective equipment is used for continuous monitoring, provided that the proportion of NO_2 does not exceed 5–10%. Measurement by IR and UV absorption is described in Section 6.8.2.

The chemiluminescence method is based on the reaction of nitrogen monoxide with ozone:

$$NO + O_3 \rightarrow NO_2 + O_2$$

Some of the nitrogen dioxide molecules formed (ca. 10%) are in an excited state (NO_2^*) The excitation energy is partially released in the form of chemiluminescence

$$NO_2^* \rightarrow NO_2 + h\nu$$

At constant pressure, constant volume flow rate, and with a sufficient excess of ozone, measurement of the intensity of the radiation in the wavelength range 600–660 nm enables the concentration of nitrogen monoxide to be selectively measured. Commercial equipment operates at a pressure equal to or less than atmospheric pressure in the measurement chamber [6.61], [6.62]. Equipment legally approved for NO/NO_2 measurement is listed in Table 6.3.

Selective measurement of NO_2 can be carried out by the Perkin–Elmer multicomponent system MCS 100 CD, the Opsis multicomponent system AR 602 Z, the Rosemount NO_2-UV Binos, and the combined equipment Binos 1004 (selective for both NO and NO_2). In the chemiluminescence equipment, it is possible to determine NO_2 from the difference sum (NO + NO_2).

6.8.4. Carbon Monoxide

Carbon monoxide can be produced by the incomplete combustion of solid, liquid, and gaseous fuels and waste and is therefore likely to occur in the exhaust gases of all combustion plants. Other sources include internal combustion engines and gas turbines. Depending on the raw materials and the technological processes

Table 6.3 Equipment for continuous measurement of NO/NO_2

Manufacturer	Type	Principles
Extractive sampling		
NDIR method		
Maihak	UNOR 6 N-F UNOR 6 N - NO UNOR 600/610	double-beam difference measurement method with one radiation source and double-layer detector single beam analyzer with double-layer detector
Mannesmann/H.u.B.	URAS 3G NO URAS 3K NO URAS 10 P	double-beam difference measurement method with one radiation source and double-layer detector
Siemens	Ultramat 32 Ultramat 5 Ultramat 21, 22	double-beam difference measurement method with one radiation source and double-layer detector single-beam analyzer with double-layer detector
Perkin Elmer	Spectran 647 IR MCS 100 HW MCS 100 CD	single-beam photometer, gas filter correlation method with two miniature cells per measured component, sequential measurement
Rosemount	NO–IR Binos	double-beam difference measurement method with one radiation source and measurement and compensation chamber
NDUV method		
Mannesmann, H.u.B.	Radas 1 G Radas 1 G NO	gas filter correlation with moving filter cell
Chemiluminescence		
Rosemount	951	chemiluminescence
Eco Physics	CLS 700 El ht	chemiluminescence, two-channel equipment for NO and NO_2 (with converter), separate reaction chambers
Electrochemical cells		
AEG	NO_x-Monitor 4000	three gas diffusion cells with SO_2 absorption filter (cross-sensitivity)
MSI/Elektronik	MS/5600	three electrode micro fuel cells
In-situ sampling		
Opsis	Opsis AR 602 Z for SO_2, NO and NO_2	single-beam photometer measurement of the absorption spectrum in certain wavelength regions of the UV
Sick	GM 30	single-beam photometer with reflector, measurement of the absorption spectrum in certain wavelength regions of the UV

used, carbon monoxide emissions are produced by installations in the cement industry, the metal smelting industry, iron and nonferrous metal treatment plants, and ceramics.

Discontinuous Methods. The iodine pentoxide method described in the VDI 2459, Sheet 7 for the manual determination of carbon monoxide concentration is a reference method [6.63]. The principle is the oxidation of CO in the sampled gas to CO_2 with liberation of an equivalent quantity of iodine. The sample is passed over iodine pentoxide (I_2O_5) in a furnace at high temperature.

The iodine is absorbed in a solution of sodium hydroxide and oxidized with bromine to iodate. The iodate is reacted with potassium iodide to form iodine, which is titrated against a solution of sodium thiosulfate. Interfering substances are removed by various adsorption methods during sampling.

Either the exhaust gas is passed via the sample probe directly into the CO combustion furnace, or, more commonly, sampling is carried out with a gas collecting container (gas mouse), and the analysis is then carried out in the laboratory. The schematic design is shown in Figure 6.25. The working range of the method is between 10 mg/m^3 and 1.25 kg/m^3 CO. The relative detection limit is 2.5 mg/m^3 CO with a sample volume of 4 dm^3.

Continuous Methods. The continuous measurement of CO by NDIR absorption is described in Sheets 1–5 of VDI 2459, and the measurement principle in Sheet 6. These sheets are now partial obsolete, having been superseded by BMU (Bundesministerium für Umwelt) ap-

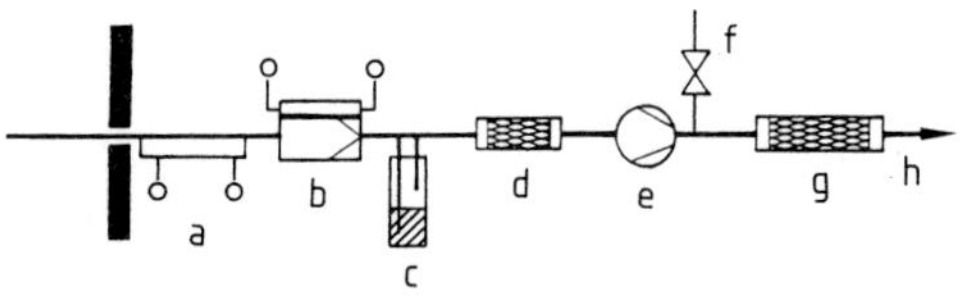

Figure 6.25. Arrangement of sampling equipment for CO with collecting vessel
a) Sampling probe, heated if necessary; b) Coarse dust filter, heated if necessary; c) Wash bottle with KOH solution; d) Absorption tube packed with soda–lime; e) Pump; f) Flow control valve for bypass; g) Absorption tube packed with IBr and activated carbon; h) Connection to gas sample container

proval of certain measurement equipment, including that for the determination of CO concentration by the IR absorption method. Two manufacturers use the gas filter correlation system; one type of equipment includes an electrochemical cell (Table 6.4).

6.8.5. Fluorine Compounds

Gaseous fluorine compounds can occur in the exhaust gases of combustion plants burning solid, liquid, and gaseous fuels and waste, chemical plants, metal treatment plants, and plants for processing rocks and earth.

Discontinuous Methods. Fluorine and gaseous inorganic fluorine compounds can be determined by absorption in sodium hydroxide solution followed by measurement with fluoride-sensitive electrodes or by photometry (method of VDI 2470, Sheet 1) [6.64]. The exhaust gas is sampled by a heated quartz probe, and is then filtered through a heated quartz wool filter and, if necessary, a heated fine filter or membrane filter to remove dust. It is then passed through two or three bottles filled with sodium hydroxide solution. After steam distillation, the fluoride content of the solution is determined photometrically by the Alizarin Complexone method or potentiometrically with a fluoride-ion-sensitive electrode. With the photometric method, cross-sensitivity effects rarely occur, as the very specific detection reaction is carried out after removal of interfering substances. Measurement with electrode chains gives high selectivity, provided fluoride complex formers such as Fe^{3+} or Al^{3+} are not present in the solution in large amounts. The detection limit is 0.05 mg F^-/m^3.

Continuous Methods. Two types of equipment have been officially approved for continuous potentiometric measurement. The gas is passed through a buffered electrolyte solution,

Table 6.4 Equipment for continuous measurement of CO

Manufacturer	Type	Principle
Extractive sampling		
NDIR method		
Afriso	Infralyt 1210	double-beam apparatus with two radiation sources
Maihak	UNOR 6 N–CO UNOR 6 N–F UNOR 600/610	double-beam difference measurement method with one radiation source and double-layer detector single-beam analyzer with double-layer detector
Mannesmann/H.u.B.	URAS 3 G/E URAS 3 K URAS 10 E	double-beam difference measurement method with one radiation source and double-layer detector
Perkin–Elmer	Spektran 647 IR MCS 100 HW MCS 100 CD	single-beam photometer gas filter correlation method two miniature cells per measured component, sequential measurement
Rosemount	CO-IR-Binos	double-beam difference measurement method with one radiation source and measurement and compensation chamber
Siemens	Ultramat 32 Ultramet 5 Ultramat 21/22	double-beam difference measurement method with one radiation source and double-layer detector single-beam analyzer with double-layer detector
Electrochemical cell		
MSI	MSI 5600	three-electrode micro fuel-cells
In-situ sampling		
Sick	GM 900/Modell 9200	single-beam photometer, gas filter correlation, cell wheel with four cells in the radiation source

and the ion concentration, changes by the substances being measured, is determined by an ion-sensitive electrode system.

In the Bran & Luebbe HF analyzer Sensimeter G, the gas is continuously extracted, passed the probe, the sampling line, the fine filter, and the absorption vessel where it is bubbled through a volume of buffer solution which is renewed every six minutes. The buffer solution is then transported to the ion-sensitive measurement cell. Measurement is thus only semicontinuous. To obtain an accurate result, the reagent solution must be carefully conditioned and its volume accurately measured, and the flow rate of the sampled gas must be constant during the concentration phase.

The Compur Ionotox HF equipment of Compur/Bayer for measuring HF emissions continuously measures the concentration of hydrogen fluoride and other inorganic, gaseous, hydrolyzable fluorine compounds. It is equipped with an electrochemical measurement cell containing an ion-sensitive electrode (fluoride electrode) and a reference electrode (silver/silver chloride). The gas being analyzed is passed continuously through an atomizer where it is mixed with the absorption solution. The latter is fed continuously into the gap between the electrodes whereby the electrode potential is measured. The interfering substances HCl, SO_2, H_2O, CO_2, CO, and NO lead to cross-sensitivity, so that the concentration of these substances must be limited.

6.8.6. Chlorine Compounds

Gaseous inorganic chlorine compounds are present in the exhaust gases from combustion plants burning solid fuels and waste materials, and also in the exhaust gases from rock and earth processing plants, the metallurgical industries, and the chemical industry.

Discontinuous Methods. Two methods for measuring chlorine are described in the VDI 3488, Sheets 1 and 2 [6.65], [6.66]. The methyl orange method is based on the bleaching effect of elemental chlorine on an acidified solution of methyl orange.

In the bromide–iodide method, bromine is liberated by chlorine from an acidified potassium bromide solution and determined iodometrically. These methods are generally unsuitable for exhaust gases from incineration plants because of their cross-sensitivity to nitrogen dioxide and sulfur dioxide.

A conventional method for measuring gaseous inorganic chlorine compounds is described in VDI 3480, Sheet 1 [6.67], with three alternative chloride determination methods: (1) Mohr titration, (2) potentiometric titration, and (3) photometric determination with mercury thiocyanate. The exhaust gas is sampled through a heated silica or borosilicate probe and a heated quartz wool filter, and is then passed into an absorption vessel containing distilled water. Methods (2) and (3) are suitable for HCl contents up to 100 mg/m^3. Methods (1) and (2) are suitable for higher concentrations. The relative detection limits are as follows: method 1: 20 mg/m^3, method 2: 2 mg/m^3, method 3: 2.5 mg/m^3.

Continuous Methods. Several proven types of equipment for the continuous measurement of gaseous inorganic chlorine compounds are commercially available. These make use of potentiometry, conductivity, and NDIR gas filter correlation. The Ecometer and Spectran 677 IR equipment are described in detail in VDI 3480, Sheets 2 and 3, in which the principle, design, and function of the equipment, the construction of the measurement station, and the calibration, evaluation, and parameters of the method are discussed [6.68], [6.69]. The Mikrogas conductivity equipment for measuring SO_2 is described in VDI 2462, Sheet 5. A summary of the commercially available equipment is given in Table 6.5.

The Sensimeter G for HCl operates like the HF equipment of the same name, with a timed collection phase in the measurement cycle. The other equipment described here operates continuously and is based on extractive sampling. In the conductometric method there is cross-sensitivity to HF, which must be calibrated for.

6.8.7. Hydrogen Sulfide

Hydrogen sulfide can occur in exhaust gases from coking plants, refineries, synthetic fiber plants, and cupola furnaces.

Table 6.5. Equipment for continuous measurement of inorganic chlorine components

Manufacturer	Type	Principle
Bran and Luebbe	Sensimeter G	ion-sensitive electrode, cyclic measurement
	Ecometer	ion-sensitive electrode, absorption path, continuous measurement
Perkin–Elmer	Spectran 677 IR MCS 100 HW	NDIR single-beam photometer, gas filter correlation
Wösthoff	Mikrogas HCl	conductivity, absorption in H_2O_2,
	Mikrogas HCl/SO_2	difference measurement (SO_2 + HCl)$-SO_2$ with two conductivity cells

Discontinuous Methods. Two methods for random sample measurement are described in VDI 3486, Sheets 1 and 2. These are manual analytical methods with potentiometric and iodometric titration [6.70], [6.71].

In the potentiometric titration, the test gas is first passed through a solution of hydrogen peroxide and sulfuric acid to remove any sulfur dioxide present, and the hydrogen sulfide is then absorbed in sodium hydroxide solution. The sulfide content of the solution is determined by potentiometric titration with silver nitrate. The relative detection limit is 0.5 mg H_2S/m^3. The method is suitable for determining hydrogen sulfide in exhaust gases. Because hydrogen cyanide and carbon disulfide interfere, the method cannot be used for coke oven gas. When determining hydrogen sulfide in combustion gases, the carbon dioxide content should be taken into account. (The CO_2 is absorbed in the alkaline wash bottle and must be subtracted from the sample volume.)

In the iodometric titration, hydrogen sulfide is reacted with cadmium acetate solution to form cadmium sulfide, which is determined iodometrically after removing the liquid phase and dissolution in hydrochloric acid. The relative detection limit is 1 mg H_2S/m^3. The method is generally applicable and is especialy suitable for gases with high concentrations of hydrogen sulfide. Collecting the cadmium sulfide precipitate by filtration and analysing the filter cake eliminates cross-sensitivity effects.

Continuous Methods. Hydrogen sulfide can be continuously measured and recorded by the Maihak Monocolor equipment. It is based on the colorimetric principle and operates semicontinuously, using a collecting phase and a dry reaction. Selectivity is achieved by means of chemically specific color indicators for hydrogen sulfide (lead acetate, silver nitrate, and silver sulfate), evaluated photometrically. The Monocolor equipment described in VDI 3486, Sheet 3 operates with a reagent paper strip impregnated with color indicators chemically selective for the components to be determined. The paper strip passes stepwise through an optical scanning device [6.72].

6.8.8. Ammonia

Emission sources of ammonia include coking plants and chemical fertilizer plants. In furnaces equipped with reduction catalysts and NH_3 addition (SCR) or with NH_3 addition alone (SNCR) to reduce sulfur emissions, unreacted NH_3 can be measured for monitoring purposes.

Discontinuous Measurement. Two methods are described for ambient air measurements in the μg/m³ range, namely the indophenol method and the Nessler method [6.73], [6.74]. In both, the sampled gas is passed through dilute sulfuric acid, trapping ammonia as ammonium sulfate. In the indophenol method, the sulfate is reacted with phenol and sodium hypochlorite to form a blue indophenol dye whose concentration is determined photometrically at 630 nm.

In the Nessler method, interfering substances (e.g., SO_2, NO, NO_2, and HCl) are removed by distillation of the ammonia from alkaline solution. If H_2S and formaldehyde are present, they are decomposed by potassium manganate(VII) before distillation. This is followed by reaction with the Nessler's reagent and photometric measurement at 450 nm.

Emission measurements are carried out in accordance with VDI 2461. Sampling is performed with heated quartz tubes leading to quartz or glass absorption vessels. The temperature of the sampling system is adjusted to the exhaust gas temperature, but should not exceed 330 °C. The dust is removed by a heated filter after the probe. The absorption solution is 0.1 N sulfuric acid. A flow rate of ca. 2 m³/h can be achieved by using two impingers in series. The rate is lower if wash bottles are used. Distillation, reaction with Nessler's reagent, and photometric determination at 450 nm are described in [6.74]. Determination with a sensitive electrode is also possible. For determinations in accordance with Sheet 2 [6.74], reproducibility is improved by buffering the reaction solution [6.75], enabling a detection limit of <0.1 mg/m³ to be attained.

Continuous Methods. Two types of measuring equipment that fulfil the minimum legal requirements are commercially available. Neither TA Luft, 13th BImSchV, nor 17th BImSchV prescribes the continuous measurement of ammonia in the exhaust gas. Official approval of the equipment is therefore unnecessary [6.15], [6.76]. The equipment Opsis AR 602 Z for NH_3 operates in situ as an NDUV single-beam photometer with radiation source and receiver. The absorption spectrum is determined in certain wavelength ranges of the UV (differential optical absorption spectroscopy). The smallest measurement range checked was 0–20 mg/m³. The detection limit is ca. 0.5 mg/m³.

The Mipan equipment of Siemens is based on absorption of electromagnetic radiation in the microwave range (microwave process gas analyzer). The method is suitable for gases with a permanent dipole moment. The equipment uses extractive sampling. The smallest measurement range checked was 0–15.4 mg/m^3, and the detection limit is ca. 0.2 mg/m^3.

6.9. Measurement of Gaseous Organic Compounds

TA Luft and 17th BImSchV regulations impose emission limits on ca. 150 organic substances. The regulations refer not only to individual compounds, but also groups of substances such as polychlorinated dibenzodioxins (PCDD), polychlorinated dibenzofurans (PCDF), and polychlorinated biphenyls (PCB). Almost all installations, including domestic heating, traffic, commerce and industry, can be sources of organic substances.

6.9.1. Sampling

The recommendations in Section 6.8.1. regarding velocity-proportional sampling, heating of the probe, construction materials, etc., also apply here. In addition, chemical changes to organic compounds during sampling due to heat, oxidation, nitration, chlorination, etc. must be prevented. As a rule, the substances must be collected at low temperature. Therefore, the sampled gas should be cooled during sampling by a heat exchanger or by adding purified air or nitrogen to reduce its temperature by dilution. An aqueous condensate is usually formed on cooling. Dust is removed from the sampled gas by a filter. For high-boiling substances such as PCDD/F, PAH, and PCB, the dust contains components that must be analyzed. Here, both gaseous and solid phases must be collected and analyzed to give the total figure. The relative proportion of gaseous and solid materials in the total sample depends on the exhaust gas temperature and sampling conditions. The gas can be collected in sample tubes, or in absorption apparatus with wash bottles or impingers. Solid adsorbents such as active carbon, silica gel, Florisil, resins, or various gas chromatographic media are also used. Depending on the detection limit or emission limit value, the collection apparatus must handle between a few dm^3 and 20 m^3 of sampled exhaust gas, and the sampling time can be up to several hours.

6.9.2. Measurement of Total-C

The TA Luft and 17th BImSchV Regulations state that volatile organic substances should be measured and recorded as total carbon.

Discontinuous Measurement. The silica gel method of VDI 3481, Sheet 2 [6.77] is a manual analytical method. The sampled gas is passed via a sampling probe (sometimes heated) through two adsorption tubes in series packed with silica gel.

The organic substances are adsorbed, though this may not be quantitative in some cases (i.e., CH_4, C_2H_6, C_2H_4, C_2H_2, C_3H_8, C_3H_6, C_3H_4, C_4H_{10}, C_4H_8, C_4H_6, C_5H_{12}, CH_2Cl_2, C_2H_3Cl, C_2H_5OH). Desorption is carried out in a stream of oxygen at elevated temperature followed by combustion to form CO_2, which is determined titrimetrically (acidimetric back-titration of barium hydroxide solution after precipitation of barium carbonate) or coulometrically (neutralization of a barium hydroxide solution). As this method does not adequately measure substances with very low boiling points, discrepancies with other methods can occur, depending on the composition of the exhaust gas. Additional gas chromatographic measurements are advisable. With water vapor concentrations of >20 g/m^3 (dew point ca. 20 °C) the adsorption may no longer proceed quantitatively, and the temperature of the sampled gas should not exceed 25 °C prior to the first sorption stage.

The relative detection limit for the titrimetric determination is 17 mg C/m^3, and for coulometric determination 2 mg C/m^3. When using the silica gel method, even at high temperatures and moisture contents, a condensate collecting vessel or a cooler with a condensate collecting vessel is fitted in front of the adsorption tubes [6.78]. The condensate must then also be analyzed, as it may contain some of the organic compounds. The concentration of CO_2 in the exhaust gas should be low, as it can dissolve in the condensate and lead to false results.

Continuous Methods. Continuous measurement of total-C concentration is usually carried out with flame ionization detectors (FID). These are described in VDI 3481, Sheets 1 and 2 [6.79], [6.80]. The limits of application of the FID process and a comparison with other methods for determining the carbon contents of exhaust gases are discussed in VDI 3481, Sheet 6 [6.81].

In the flame ionization detector, organically bonded carbon is ionized in a hydrogen flame. The ionic current produced in an electric field is amplified and displayed. It is proportional, over a wide range, to the number of carbon atoms introduced into the flame per unit time. The structure of the molecule (e.g., single and double bonds, number and nature of substituted atoms, chain length, or ring structure) considerably influences the oxidation behavior of the carbon and hence the magnitude of the detector signal. Thus, organic compounds with heteroatoms (e.g., N, O, S, or Cl) are generally detected with considerably lower sensitivity than hydrocarbons with the same number of carbon atoms per molecule.

The varying sensitivities to the various organic compounds are expressed as response factors. These are equipment-specific, depending, for example, on jet shape, electrodes, combustion space, and combustion air mixing, and must not be assumed to apply to other FID equipment.

A schematic diagram of an FID is given in Figure 6.26. The flame requires fuel gas (hydrogen or hydrogen-containing mixtures) and combustion air free of organic substances. The fuel gas and sample gas are mixed before they enter the flame. The combustion air is fed to the flame separately. The flow of combustion air is adjusted so that the flame burns steadily, while at the same time the dew point is kept as low as possible by dilution of the fuel gas, preventing undesired condensation in the exhaust gas duct of the detector. As the pressure and the volume flow rate at the FID affect the measured value, these must not be altered after calibration. Depending on the sensitivity (measurement range) and equipment type, quality requirements for the fuel gas, combustion air, test gas, and composition of the sampled gas (hydrogen/oxygen cross-sensitivity) vary.

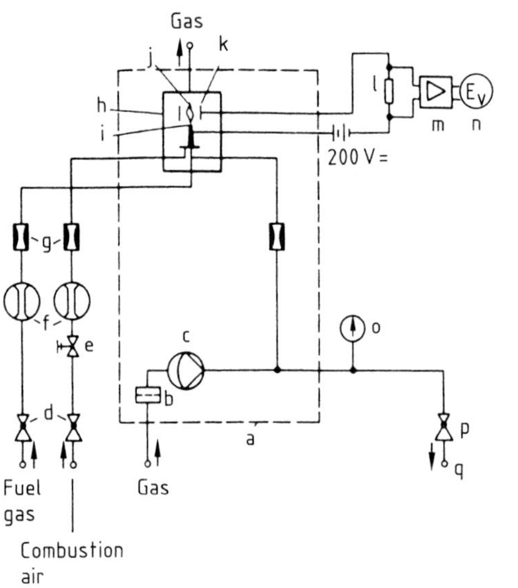

Figure 6.26. Flame ionization detector (schematic)
a) Heated housing; b) Fine dust filter; c) Pump for sampled gas; d) Pressure regulator; e) Flow control valve; f) Flowmeter; g) Capillary; h) Combustion chamber; i) Jet; j) Flame; k) Electrode; l) Resistor; m) Amplifier; n) Display of measured value; o) Manometer; p) Nonreturn valve; q) Bypass

An overview of equipment is given in Table 6.6. Monitoring limit values for total-C specified in 17th BImSchV (10/20 mg C/m^3) requires a measurement range of 0–30 mg/m^3.

One type of equipment operates on the principle of heat evolution. Catalytic oxidation of combustible components of the gas produces heat which is detected by temperature measurement. There is cross-sensitivity towards CO. The catalysts are poisoned by halogen, lead, and silicon compounds.

6.9.3. Chlorinated Hydrocarbons

The various chlorinated organic compounds that can occur in emissions can exist as gases,

Table 6.6 FID equipment

Manufacturer	Equipment name	Smallest measurement range tested, mg/m^3	Year of test report
Bayer	Compur FID	0–50	1989
Diagnostic	Compur Multi FID 100 E 17	0–25	1992
	Compur Multi FID 100 FE 17	0–25	1992
Bernath Atomic	BA 3001/BA 3004	0–160	1987/
	BA 3002 RC	0–160	1988
	BA 3002 RC	0–25	1991
J.U.M. Engineering	FID VE 7	0–25	1990
Mannesmann/ H.u.B.	FIDAS 2 T	0–645	1977/ 1982
	FIDAS 2 T	0–50	1987/ 1988
	FIDAS 3 E	0–50	1989
	FIDAS 3 E	0–25	1992
Siemens	FIDAMAT I/K	0–220	1977/ 1979
	FIDAMAT Ausf. M 52044-A 10	0–50	1990
Testa	TESTA FID 123	0–50	1992
Ados*	KM 2-CnHm-EM-ADOS	0–226	1977

* Heat evolution.

vapors, solids, materials adsorbed on solids, or as mixtures thereof. Therefore, different sampling and analysis methods are required, depending on the boiling points of the individual compounds and the exhaust gas temperature. Boiling ranges of halogenated compounds of Classes I–III (taken from Subparagraph 3.1.7 of TA Luft) are illustrated graphically in Figure 6.27 [6.82]. The boiling ranges of PCDD, PCDF, and PCB compounds are shown for comparison.

The high-boiling PCDDs and PCDFs have been thoroughly investigated owing to their environmental implications. Emission measurement methods are described in VDI 3499. A CEN Standard is in preparation, and investigations into PCBs are also in progress. The chlorobenzenes and chlorophenols in the middle boiling range have only been investigated individually from the point of view of emissions. This group of substances can be determined by the sampling and analysis methods described for high-boiling substances. However, there have been no systematic investigations of any significance.

6.9.3.1. Polychlorinated Dibenzodioxins and Polychlorinated Dibenzofurans (PCDD/F)

PCDDs and PCDFs are emitted at exhaust gas temperatures between 50 and 400 °C, depending on the design and exhaust gas purification technology of the plant. In this temperature range, the compounds exist in gaseous, solid, and adsorbed form, depending on boiling point. Because volume flow rates of exhaust gases from industrial plants are very high, analysis usually has to be based on extractive sampling, and, because of the presence of the particle phase, the sampling must be isokinetic, using a measurement grid for reasons of representativity. For limit values of ≤ 0.1 ng/m^3 as a toxicity equivalent, sampling volume rates of 10–20 m^3/h are necessary. To ensure mild collection of organic compounds, sampling must take place at a low temperature (if possible < 50 °C). At higher temperatures, the sample gas stream must be cooled by dilution with air or by cooling.

The dilution method is described in VDI 3499, Sheet 1 [6.83]. Cooling is carried out by

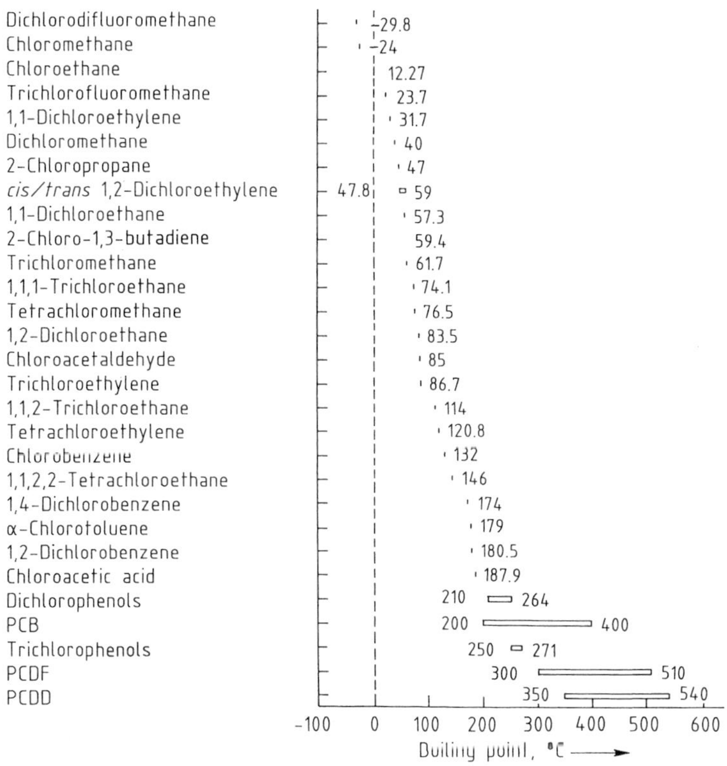

Figure 6.27. Boiling point ranges of organic compounds (taken from TA Luft)

adding dried air to the gas stream. This prevents the temperature from falling below the dew point of water and causing condensation. The design of the sampling equipment is illustrated in Figure 6.28.

The gas is sampled isokinetically by a probe heated to the temperature of the exhaust gas. The sampled gas stream is cooled to below 50 °C by adding dried, filtered air. The diluted exhaust gas is then passed through a filter which traps the solids and the condensed fraction of organic compounds, and the fraction adsorbed on the particles. The PCDDs and PCDFs are collected on the filter at ca. 40 °C almost quantitatively. A solid sorbent such as Florisil, XAD adsorber resin, Porapak, or PU foam, can be included to ensure complete filtration. Alternatively, the above system can be replaced by a flat filter with PU foam as the collecting phase, as is commonly used for the measurement of PCDD and PCDF as atmospheric pollutants.

The sampling system described in Sheets 2–4 involves condensation methods with direct cooling [6.84]–[6.86].

Sheet 2: Condensation–adsorption method with filter
Sheet 3: Condensation method, cooled sampling probe
Sheet 4: Condensation method with polyurethane foam for adsorption

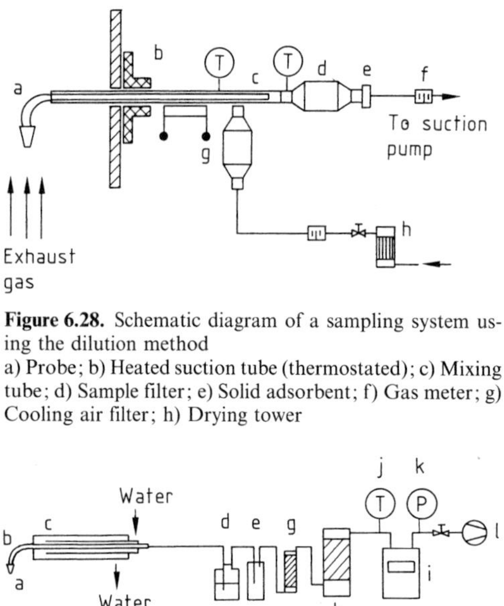

Figure 6.28. Schematic diagram of a sampling system using the dilution method
a) Probe; b) Heated suction tube (thermostated); c) Mixing tube; d) Sample filter; e) Solid adsorbent; f) Gas meter; g) Cooling air filter; h) Drying tower

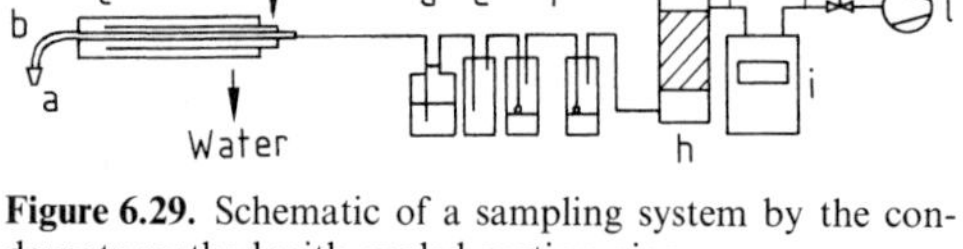

Figure 6.29. Schematic of a sampling system by the condensate method with cooled suction pipe
a) Probe; b) Bend; c) Cooled suction pipe; d) Condensation vessel; e) Droplet separator; f) Impinger with methoxyethanol; g) Solid adsorbent; h) Drying tower; i) Gas meter; j) Temperature at gas meter; k) Pressure at gas meter; l) Pump

In the condensation method of Sheet 3 the hot exhaust gas is cooled as it is withdrawn through the water-cooled probe, and the condensate formed is collected in a cooled separator. Filtration through a solid medium is thus avoided. The sampled gas is then passed through two parallel impinger lines filled with organic solvent, or through a solid adsorbent. The sampling apparatus is shown schematically in Figure 6.29.

In the method of Sheet 4, PU foam is used as an additional adsorbent. For inclusion of this sampling method in the CEN Guideline, a solid filter is planned prior to the last absorption or adsorption stage.

The first analytic step is extraction of the organic compounds from the collecting phase. For example, in the dilution method, a sampling filter, and, if necessary, a solid absorbent (XAD-2, Florisil, Porapak PS, and polyurethane foam) are used as collecting devices. For the solvent extraction stage, the whole glass fiber filter including loose dust particles is transferred to an extraction thimble. For isomer-specific measurement of the PCDD/F, ca. 5 ng of the following ^{13}C-labelled standards (one per isomer group) added before extraction:

^{13}C-2,3,7,8-tetraCDD/F, ^{13}C-1,2,3,7,8-pentaCDD/F, ^{13}C-1,2,3,7,8,9-hexaCDD, ^{13}C-1,2,3,4,7,8-hexaCDF, ^{13}C-1,2,3,4,6,7,8-heptaCDD/F, ^{13}C-1,2,3,4,6,7,8,9-octaCDD/F.

Extraction with toluene is then carried out in a Soxhlet apparatus for at least 20 h. The toluene extract is reduced to ca. 10 mL in a rotary evaporator at ca. 70 mbar and a bath temperature of ca. 45 °C. The adsorbents XAD-2 and Porapak PS are also extracted with toluene in a Soxhlet apparatus, for at least 20 h. Florisil is eluted with dichloromethane in a glass column. Before further treatment, the extract or eluate is reduced to ca. 5 mL under a controlled vacuum (70 mbar, 45 °C). The extracts can be combined for determination of total PCDD/F, or treated separately. Sample treatment is illustrated in Figure 6.30.

The ^{13}C-labelled standards are added to the fraction which experience shows is likely to have the higher PCDD/F content. This is either the filter (VDI 3499, Sheet 1) or the condensate (Sheet 2). The extracts are transferred to a column packed with 50 g aluminum oxide and 10 g sodium sulfate, which is eluted with benzene and then with hexane/dichloromethane (98:2). This eluate is collected and checked to ensure that at least 60 % of the ^{13}C-2,3,7,8-TCDD is present. The PCDD/F is then eluted with 200 mL hexane/dichloromethane (1:1). The eluate is reduced to ca. 2 mL under vacuum, transferred to a mixed column, and eluted with 250 mL hexane. This eluate is reduced to ca. 5 mL, and then chromatographed on a column of Bio-Beads (PCDD/F fraction 120–180 mL). The solvent is removed from this fraction under vacuum. The residue is taken up in a solution of 5 ng (or less) $^{13}C_6$-1,2,3,4-TCDD in 100 µL benzene, reduced if necessary in a stream of nitrogen to 30–50 µL, and analyzed by GC/MS. The quantitative evaluation is based on the added ^{13}C-labelled standard, identical response factors for all PCDD/F isomers being assumed for all degrees of chlorination. The analysis is carried out "isomer-specifically" for the 2,3,7,8-substituted PCDD/F. The capillary columns are charged with Silar C or an equivalent phase such as Sil 88, SP 2330, or SP 2331. Coupling with the mass spectrometer can be carried out either direct or open. Analysis is by mass fragmentography, with at least two masses being recorded both for the tetra- to octa-CDD/CDF and for the ^{13}C-

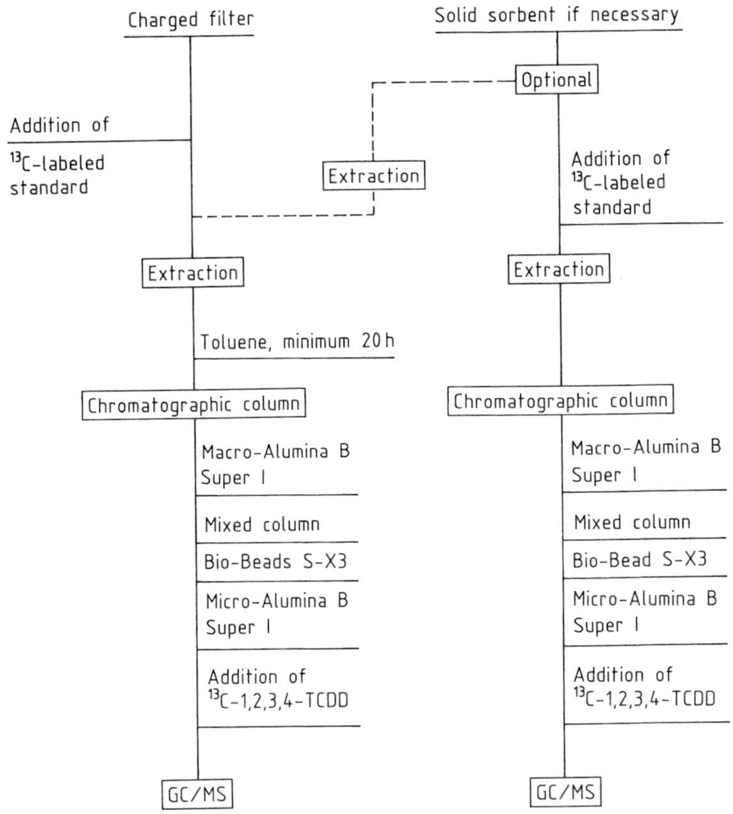

Figure 6.30. Scheme of sample preparation for PCDD/F

labelled standard. The absolute detection limit depends on the mass spectrometer used and the amount of material injected, and is normally ca. 0.001–0.1 ng per individual isomer and sample. For a gas sample volume of ca. 20 m^3, the detection limit is 0.00005–0.005 ng/m^3 for a single isomer.

6.9.3.2. Polychlorinated Biphenyls

Sampling of PCB can be performed by the same technique as used for PCDD/F. However, there is at present no mandatory guideline with proven parameters. The stages of the analysis are the same as those used in dioxin analysis, the choice of separation and extraction materials being based on the specific properties of PCB (polarity, solution properties). Before extraction of the solid filter (flat filter/PU foam, see VDI 3499, Sheet 1) or of the condensate (VDI 3499, Sheet 3) with toluene, internal standards (^{13}C-labelled for subsequent GC–MS analysis or PCB 28 and PCB 202 for subsequent GC–ECD analysis) are added.

After addition of 6 mL decane, the extract is reduced with the aid of a rotary evaporator and charged to the mixed column (see VDI 3499 and VDI 3498). It is eluted with 250 mL hexane, and the eluate is then reduced to 6 mL and charged to a alumina column. This is then eluted with 60 mL *n*-hexane, 90 mL toluene, and 250 mL *n*-hexane/dichloromethane (1:1) [6.87], [6.88]. The toluene contains the PCB compounds and the hexane–dichloromethane mixture the PCDD/F compounds. The fractions can be purified if necessary by a further cleanup with, e.g., Florisil or HPLC. A labelled standard is then added, e.g., for PCB, $^{13}C_{12}$-2,2′,4,5,5′-pentachlorobiphenyl, and, for PCDD/F, $^{13}C_6$-1,2,3,4-tetrachlorodibenzodioxin. It is then reduced to a final volume of 10–20 μL, and analyzed by GC–MS. The relative detection limit for the total content of PCB is 10–20 ng/m^3. For the individual components it is ca. 1–2 ng/m^3.

6.9.3.3. Chlorobenzenes and Chlorophenols

Sampling of chlorinated benzenes and phenols can in principle be carried out by the same methods as those for PCDD/F and PCB. There is no guideline with approved method parameters. Also, individual tests should be carried out to determine to what extent the more volatile mono- and dichlorocompounds are quantitatively recovered. Suitable collecting phases include solid filters, PU foam, methoxyethanol, and adsorber resins. Another possibility is absorption in cold solvents (e.g., methyldiglycol), as described in the draft version of VDI 2457, Sheet 2 [6.89].

Chlorobenzene. Toluene is used as an extraction medium for separating chlorobenzenes and chlorophenols from the collecting phase. The toluene extract is reduced to a few milliliter, and chromatographed on alumina. Elution is then carried out with hexane/dichloromethane (98:2), and this fraction is again reduced in volume. Analysis for chlorobenzenes is carried out, for example, by GC–ECD or GC–MS. The quantification of the chlorobenzenes is carried out externally using a mixture of di- to hexachlorobenzenes. Analysis can also be carried out with ^{13}C-labelled standards, using one standard for each degree of chlorination. The relative detection limit is ca. 1 ng/m^3 for each component.

Chlorophenols. The chlorophenols are determined by extracting the toluene extract with K_2CO_3 solution. Acetic anhydride is added, the samples are agitated, and the chlorophenol acetates are then extracted with hexane. The hexane extract is dried over Na_2SO_4, and analyzed by GC/ECD or GC/MS. Quantification of the chlorophenols is carried out externally using a mixture of di- to pentachlorophenols derivatized under the same conditions. Alternatively, ^{13}C-labelled standards can be used.

6.9.3.4. Volatile Chlorinated Hydrocarbons

Volatile chlorinated hydrocarbons (VCH) such as trichloroethane, trichloroethylene, tetrachloroethane, and tetrachloroethylene are used in industrial surface treatment, chemical cleaning, and solvent extraction. Limit values are imposed by 2nd BImSchV on emissions of the above VCHs and dichloromethane [6.5]. The use of trichloroethane is now prohibited in some cases.

Measurement of emissions is described in the VDI 2457, Sheets 2–4 of 1974–1976 for 1,1,1-trichloroethane, trichloroethene, and tetrachloroethene. They are absorbed in a liquid sorbent and determined by gas chromatography. However, in some cases the detection limit is relatively high.

Sampling and analysis are therefore now carried out by a method based on that used for ambient air pollution measurement [6.90], [6.91]. Small test tubes packed with activated carbon are used to collect the sample, and desorption is then carried out with solvents or by heating. The sampling itself is carried out with quartz probes provided with dust filters, if necessary heated to the temperature of the exhaust gas. With higher water vapor contents, a condensate trap follows the dust filter. The condensate is also analyzed for VCHs by gas chromatography. Another solid sorbent for pollutants is 2,6-divinylbenzene-*p*-phenylene oxide polymer (Tenax TA) [6.92]. The VCHs can be desorbed for gas chromatographic analysis and recondensed in a cooled container. Detection is by ECD or mass spectrometry. Another method is absorption in a cold solvent [6.89].

6.9.3.5. Chlorofluorocarbons

The regulation 2nd BImSchV referred to in Section 6.9.3.4 above also imposes limits on the emission of chlorofluorocarbons (CFCs) the only permitted CFCs being 1,1,2,2-tetrachloro-1,2-difluoroethane (R-112), 1,1,2-trichloro-1,2,2-trifluoroethane (R-113), and trichlorofluoromethane (R-11). Sampling is performed by the method described in Section 6.9.3.4, the substance being adsorbed on, e.g., activated carbon. With low-point components, gas collecting tubes can be used. Analysis is carried out by GC/ECD.

6.9.4. Hydrocarbons

TA Luft imposes emission limits on a number of hydrocarbons, depending on the substance group and mass flow rate. Total emission concentrations of 20–150 mg/m^3 are permitted, depending on the substance classification. Substances such as benzene, acrylonitrile, or polycyclic aromatic hydrocarbons, are controlled separately in Subparagraph 2.3 of TA Luft.

Polycyclic Aromatic Hydrocarbons. Polycyclic aromatic hydrocarbons (PAH) are present in many natural materials, for example, mineral oil. PAH structures exist in coal, but the compounds themselves are only formed after chemical transformation. In industrial processes, unreacted PAH residues and newly formed PAH are emitted during the incomplete combustion or pyrolysis of organic materials, especially fossil fuels. Some polycyclic aromatic hydrocarbons, especially tetracyclic to hexacyclic compounds, have a mutagenic and/or carcinogenic effect on human and animal organisms. A limit of 0.1 mg/m^3 is imposed on benz[*a*]pyrene and dibenz[*a*, *h*]anthracene by TA Luft. VDI 3873, Sheet 1 describes a method of measurement for use in industrial and commercial plants [6.93]. The sampling method used is the dilution process described in VDI 3499 (see Section 6.9.3.1). The four- to seven-ring PAH compounds have approximately the same boiling range as PCDD/F. They therefore occur in exhaust gases as gases, solids, and materials adsorbed on particles. The method of VDI 3873, Sheet 1 can deal with all states of aggregation. The four- to seven-ring compounds are quantitatively collected on a filter. The more volatile PAH compounds with two or three rings can be recovered by a downstream solid sorbent (e.g., Porapak PS). It is possible to collect PAH, PCDD/F, and PCB at the same time. The sampling times (collecting times) should not exceed ca. 2 h to ensure that the PAH profile remains the same, because some PAH compounds are very reactive. A detection limit of ca. 10 ng per component can be achieved. With a sample volume of 5 m^3 a relative detection limit of 2 ng/m^3 per component can be attained. The exposed filter is extracted with toluene for 30 min after adding the internal standard. The sample preparation scheme is illustrated in Figure 6.31. A detailed description can be found in VDI 3873, Sheet 1. Analysis is then carried out by capillary GC.

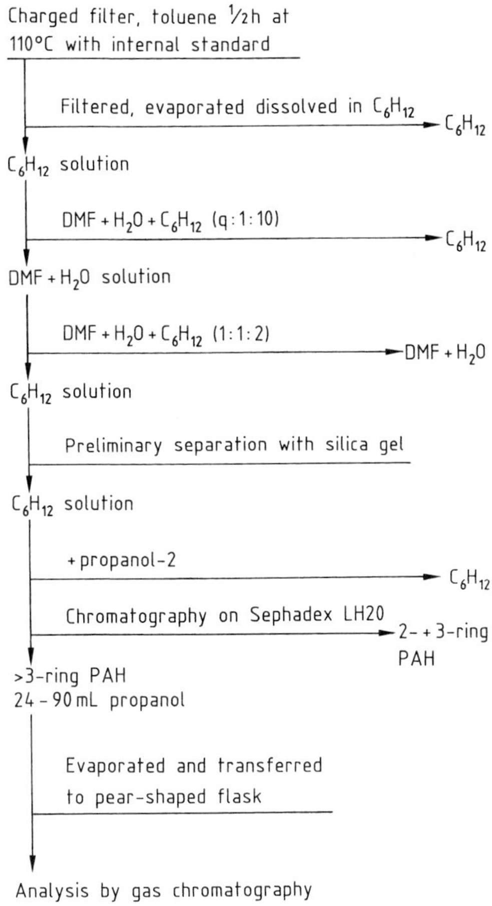

Figure 6.31. Scheme of sample preparation for PAH

Aromatic and Volatile Hydrocarbons. The sampling technique for aromatic hydrocarbons (benzene, toluene, xylene, and ethylbenzene: BTXE) is similar to that used for pollution measurement described in Section 6.9.3.4. The most important collecting phases used are activated carbon, Tenax, and cold solvents [6.90], [6.91]. Materials are desorbed from activated carbon mainly with solvents [6.94] or by thermal methods. Identification is by GC–FID or sometimes GC–MS. If condensate is formed during sampling, this must also be analyzed. With a sampling volume of 20 L, the detection limit is 5 µg/m^3 per component. The Guideline VDI 2457, Sheet 1, which describes the fundamental principles of sampling and determination of organic compounds by gas chromatography, should be consulted [6.95]. The method of absorption in nitrobenzene described in VDI 2457, Sheet 5 is less suitable because of the high detection limits and various cross-sensitivity effects [6.96].

Volatile organic compounds (VOC) such as aliphatic, cyclophatic, and aromatic hydrocarbons and solvents can be determined by the methods given for BTXE or VCH compounds. Activated carbon or other solid sorbents are mainly used.

1,3-Butadiene. Emissions of 1,3-butadiene, like those of acrylonitrile, must not exceed 5 mg/m^3 according to Subparagraph 2.3 of TA Luft. The Guideline VDI 3953 describes a measurement method by adsorption on activated carbon and desorption with CS_2 [6.104]. Gas chromatographic analysis is based on the GC–FID head space technique (vapor space analysis). The determination limit is 0.2 mg/m^3 with a sampled gas volume of 30 L. The limited storage time of the samples must be borne in mind.

6.9.5. Other Compounds

Phenol and Cresols. Phenol and the cresols can be collected on silica gel using the sampling arrangement described in Sections 6.9.3.4 and 6.9.4.2 for active carbon. Identification can be made by HPLC with a fluorescence detector. With a sampled gas volume of ca. 30 L, a detection limit of ca. 1 µg/m^3 is achieved.

Amines. Aliphatic amines such as methylamine, dimethylamine, ethylamine, propylamine, butylamine, and hexylamine can be quantitatively determined, especially in exhaust air, by the method described in VDI 2467 for ambient air pollution measurement [6.97]. The amines are dissolved as alkylammonium chlorides in dilute hydrochloric acid, and are then reacted to form dinitrophenylamines. Determination is by HPLC, with UV absorption analysis. Aromatic amines can be sampled with silica gel tubes and determined by HPLC–UV. Detection limits, e.g., in the case of 2-naphthylamine and toluidine, are ca. 5 µg/m^3.

Aldehydes. Short-chain aldehydes are formed by the incomplete combustion of hydrocarbons or alcohols, e.g., in internal combustion engines including diesel engines.

A method of measuring these emissions is described in the VDI 3862, Sheet 1 [6.98]. Aliphatic C_1–C_3 aldehydes are absorbed in solutions by the MBTH (3-methyl-2-benzothiazolinone hydrazone) method, and determined photometrically. The determination limit is 0.05 mg/m^3 if the volume of exhaust gas sampled is 25 L. If SO_2 is present at concentrations of >30 mg/m^3, the method is unsuitable because of interference effects.

In Sheet 2 (in preparation) of VDI 3862 the determination of long-chain and aromatic aldehydes and ketones is described [6.99]. These substances are reacted with 2,4-dinitrophenylhydrazine (DNPH method) to give the hydrazone and determined as individual components. Sampling is carried out with wash bottles containing acetonitrile to dissolve the hydrazine. The absorption solution is directly chromatographed without further treatment. Determination is by HPLC with UV absorption analysis.

Sheet 3 of the VDI 3862 (in preparation) describes sampling with the aid of DNPH dissolved in hydrochloric acid solution [6.100]. The hydrazones are extracted with carbon tetrachloride, the organic phase is chromatographed using HPLC, and analysis is performed by UV absorption. Depending on the substance, detection limits between 2 and 100 µg/m^3 have been quoted for a sampled volume of 50 L. A disadvantage is that the samples can be stored for a short time only.

Acrylonitrile. Three methods for the determination of acrylonitrile are described in VDI 3863, Sheets 1–3 [6.101]–[6.103].

Sheet 1: Sampling by a gas sample container. The detection limit with a flame ionization detector (FID) is 0.5 mg/m^3. With a phosphorus–nitrogen detector (PND), the limit is 0.05 mg/m^3. The samples can be stored only briefly. With high water vapor contents, the methods of Sheets 2 and 3 are used.

Sheet 2: Sampling in cold solvents, detection limit with 15 L sampled volume using FID 0.2 mg/m^3, or using PND 0.05 mg/m^3. Samples can be stored only briefly.

Sheet 3: Sampling on activated carbon, desorption with dimethylformamide (DMF), detection limit 12 µg/m^3 with 50 L sampled volume. GC with column switching and two FIDs.

All sampling equipment is fitted with a sampling probe and solid filter (heated if necessary).

Vinyl Chloride. The determination of vinyl chloride by gas chromatography of samples collected in gas sample containers is described in the Guideline VDI 3493, Sheet 1, of Nov. 1982 [6.105]. Analysis is by GC–FID, the determination limit being 0.3 mg/m^3. Subparagraph 2.3 of TA Luft imposes an emission limit value of 5 mg/m^3. By adapting the method of VDI 3494, Sheets 1 and 2, for determining vinyl chloride ambient air pollution by adsorption on activated carbon, it should be possible to achieve improved sensitivity [6.106], [6.107].

6.10. Measurement of Odor Emissions

Odor emissions can be produced in many industries, e.g., cocoa, coffee, and tobacco processing, fish production and processing, ruber production and processing, and the chemical industry. The aims of the measurement technique are to determine the odor intensity (based on determination of the odor threshold), to estimate nuisance potential, to estimate the effectiveness of techniques of odor emission reduction (crude/purified gas), and to establish data to enable spreading rates to be calculated and odor pollution to be forecast. Odors are determined using the human nose as a sensor (olfactometry). Olfactometry consists of the controlled presentation of odoriphores (odoriferous substances) and the assessment of the sensations which these produce in human subjects. An odor sensation can be caused even when the concentration of an odoriphore is less than the limit of detection by chemical or physical methods. The qualitative and quantitative assessment of mixtures with a large number of components is difficult. Even if analysis is possible, there may be no connection between this and the odor sensation.

The concentration of the odoriphore in the gas sample (single substance or mixture of substances) under investigation is determined by diluting with neutral air until the odor threshold is reached. One odor unit (OU) is the quantity of odoriphore which when uniformly distrubted in 1 m^3 of neutral air produces an odor sensation in accordance with the definition of the odor threshold. The unit of odoriphore concentration is one odor unit per unit volume (OU/m^3). The odor threshold is the concentration of a substance or mixture of substances that gives an odor sensation in 50% of a statistical sample of test subjects. The fundamental principles of odor threshold determination are described in the VDI 3881, Sheet 1 [6.108]. Olfactometers are used for measurement. A gas sample containing an odoriphore is diluted with various amounts of neutral air in a controlled manner and submitted to test subjects for odor testing.

Sampling can be dynamic or static. In dynamic sampling, a stream of exhaust gas or exhaust air is continuously removed. The olfactometer is attached by a T-piece to the sample line, so that a test volume can be fed into the olfactomer as required. The scheme for dynamic sampling is illustrated in Figure 6.32.

In static sampling, the sample of odoriferous substance is charged into a container, transported to the olfactomer, and connected to it for the

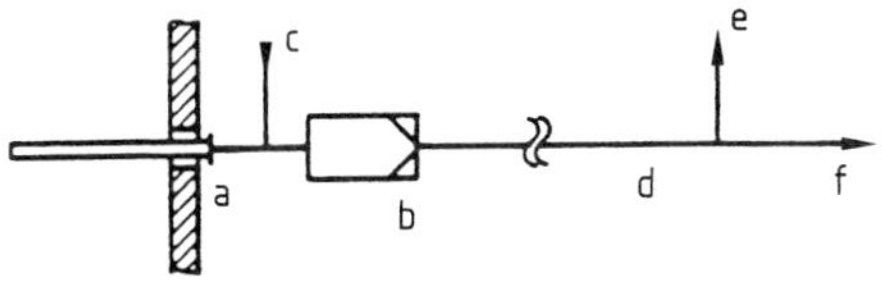

Figure 6.32. Dynamic sampling
a) Sampling probe; b) Dust filter (if necessary); c) Air addition for dilution (if necessary); d) Pipeline for gas samples; e) Excess to exhaust air; f) To the olfactometer

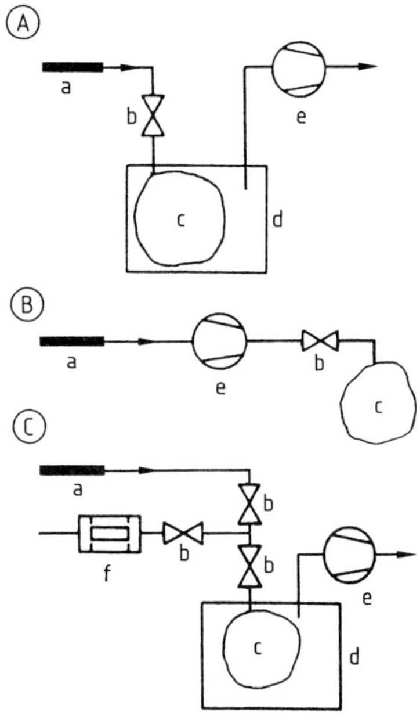

Figure 6.33. Various methods of static sampling
A) Bag filling by suction pump; B) Bag filling by pressure pump; C) Bag filling by suction pump with additional dilution with filtered air
a) Sampling probe; b) Valve; c) Bag; d) Bag container; e) Pump; f) Activated carbon filter

experiment. Possible methods of static sampling are shown schematically in Figure 6.33 [6.109]. If there is a possibility that water or other substances in the sample could condense inside the bag, the sample should be diluted. In this case, the sample bag is filled with a definite amount of odor-free air in an apparatus as shown in Figure 6.33. A definite amount of the test gas is then added. The dynamic sampling method is advantageous for constant emission condition because of the short time delay between sampling and analysis. The method is not suitable if concentrations change rapidly, as the sampled material changes during the course of an individual measurement, and there can then be a wide distributon of measured values.

With static sampling of varying concentrations, a representative mean value over the sampling time is obtained. Evaluation of the results is described in [6.108], [6.110]. Various types of olfactometer are described in [6.111]. The parameters of the method and an example of an odor measurement record are given in [6.112].

7.1. Separation of Solid and Liquid Particles from Gases [7.1.]–[7.8] (see also → Dust Separation, **B2**)

7.1.1. Introduction

Many industrial processes inevitably generate particles entrained by a carrier gas. These particles may be solid (dust) or liquid (droplets or mist). As a rule, the particles must be separated from the transporting gas, either because they are the desired product or because they are an undesirable emission. Product recovery by dust separation has been practiced long and intensively in process engineering. The importance of emission control has grown tremendously in recent years, because of rapid increases in industrial production, and better knowledge of potential hazards and harmful effects.

Dust collection methods used in practice can be classified as inertial and centrifugal separation (cyclones), wet collection (scrubbers), filtration (deep-bed filters, surface filters, and granular-bed separators), and electrostatic collection (precipitators). All of these collectors separate particles from gas by a common principle: forces acting on the particles cause them to enter regions from which the gas cannot transport them away. These regions may be the inner wall of a cyclone, the droplet surface in a scrubber, the fiber or grain surface in a filter, or the collecting electrode in a precipitator.

One difficulty in selecting and designing separation processes is that the size distribution of the particles to be collected generally extends over a wide range, from <0.1 μm to >100 μm. The knowledge about the dependence of the operational characteristics of each separator on particle size distribution is very important and is discussed widely.

7.1.2. Characterization of Particle Separation

To evaluate dust collection equipment, both the energy consumption (or pressure drop Δp) and the collection efficiency must be determined. The latter can be found by integration techniques (for the totality of particles) or differential methods (for each particle size fraction).

The overall collection efficiency E, also known as the total or cumulative collection efficiency, is given by

$$E = \frac{\dot{M}_c}{\dot{M}_e} = \frac{\dot{V}_e c_e - \dot{V}_f c_f}{\dot{V}_e c_e} = 1 - P \qquad (1)$$

where E is collection efficiency, $\dot{M}$ is mass flow rate of particles, $\dot{V}$ is volume flow rate of gas, c is particle concentration, and P is penetration. The subscript c stands for collected (coarse), e for entry, and f for penetrating (fine).

Note that if a collection device is rated in terms of overall collection efficiency, this quantity depends not only on the equipment design and operating conditions, but also on the type of dust and, especially, the particle-size distribution. The stated overall collection efficiency or overall penetration thus always refers to a particular case. A more useful and informative measure is the fractional collection efficiency $T(x)$, which measures collection as a function of particle size x:

$$T(x) = \frac{d\dot{M}_c(x)}{d\dot{M}_e(x)} = 1 - \frac{\dot{V}_f dc_f(x)}{\dot{V}_e dc_e(x)} \qquad (2)$$

With $\dot{V}_f = \dot{V}_e$ and particle size distribution q:

$$T(x) = 1 - \frac{P q_f(x)}{q_e(x)} = \frac{E q_c(x)}{q_e(x)} \qquad (3)$$

The cut size x_t is the particle size at which a given fractional collection efficiency T is achieved. A value of 50% is often used (x_{50}).

7.1.3. Gas Cyclones

7.1.3.1. Mode of Operation and Basic Designs

The simplest way to separate particles from gases is to allow them to settle out in so-called gravity settlers into zones of low gas velocity. Because this method is technically feasible only for coarse particles that are significantly larger than 100 μm, gravity settlers now have negligible importance and are not discussed here.

Collection by means of centrifugal forces is much more efficient than simple gravity separation, and this led to the development of centrifugal collectors. Because of their simple design, reliable operation, small space requirement, and low cost, gas cyclones are used widely in many branches of industry. They are frequently employed in material recovery from gas recycle systems, pneumatic conveying, and other areas. They are often used as first-stage collectors. Cyclones can be operated at pressures of 1 kPa

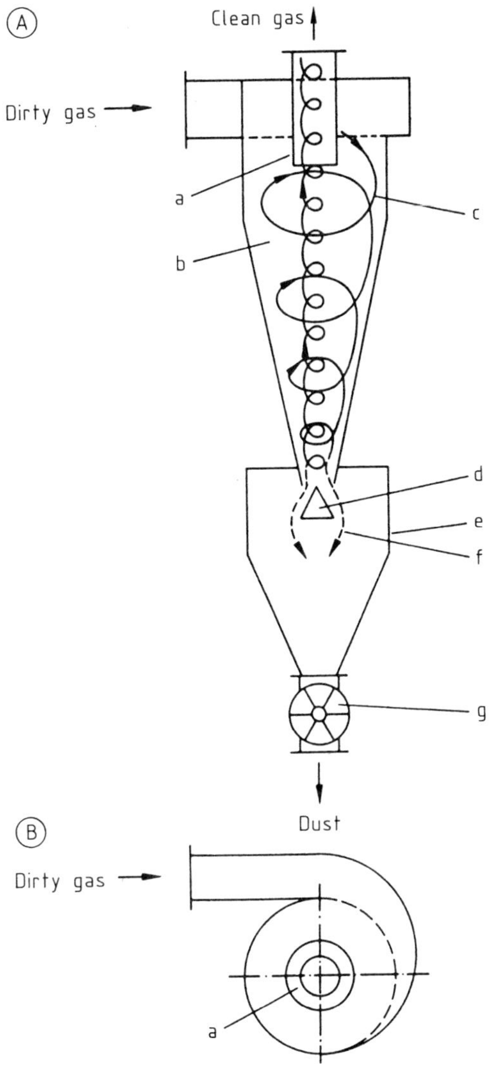

Figure 7.1. Schematic of a reverse-flow cyclone
A) Vertical section; B) Top view
a) Exit duct; b) Separating chamber; c) Gas flow pattern; d) Conical shield; e) Dust hopper; f) Dust; g) Lock

to 10 MPa and temperatures exceeding 1000 °C. The advantages of cyclones over other collectors are offset by the fact that they are less efficient collectors for particles below ca. 1–5 μm in size.

Figure 7.1 shows the most common type of cyclone with flow reversal. The gas path is designed to impart a twist to the particle-laden gas entering the rotationally symmetrical apparatus. A turbulent, three-dimensional, rotational flow is produced. In the separating chamber spiral, downward flow takes place at larger radii. Flow is then reversed, and the gas spirals upward at smaller radii, and reaches the exit duct. The particles entrained in the revolving gas stream experience centrifugal forces hundreds to thousands of times greater than the force of gravity. Thus, the larger particles migrate outward to the cyclone wall, where they collect as strands; they are carried down the tapered body and out of the separating chamber by boundary-layer flow. The conical shield, often built into the bottom of the tapered body, prevents already collected material from being reentrained. The particles that are not collected are carried inward by the gas flow and removed through the exit duct.

Cyclones vary in the form of the swirl-producing inlet (tangential, axial with swirl vanes), the shape of the separating chamber (cylindrical, conical), the design of the exit duct, and their dimensions. Diameters range from 0.02 m to 5 m and flow rates from 1 to 100 000 m^3/h.

7.1.3.2 Collection Efficiency and Pressure Drop

Important parameters in cyclone design and operation are the fractional collection efficiency $T(x)$ and the energy consumption, measured in terms of the pressure drop Δp.

Several models have been devised for the approximate calculation of collection efficiency at low dust loads. They describe the flow field in more or less simplified terms. The separation-zone model is now employed widely because of its simplicity and ease of use. However, this approach does not give the complete fractional efficiency curve; only a critical particle diameter x_{cr} can be calculated. According to this concept, all particles whose diameter x is less than x_{cr} pass through the device; larger particles are collected.

In theory, the fractional efficiency curve jumps from 0 to 1 at x_{cr}. It is more reasonable to assume that x_{cr} corresponds to the median diameter x_{50} of the real separation function $T(x)$, i.e., the 50 % value on the effective S-shaped fractional efficiency curve. Based on new experimental results the suggestion has been made that the actual cut diameter can be estimated as $x_{50} \approx 1.3 \cdot x_{cr}$.

In a more recent approach, particle motion is treated as a process in which diffusional particle transport is superimposed on a predetermined motion. Particle flux calculations are based on a modified flow field. The advantage of this model is that the complete fractional efficiency curve can be obtained.

At higher particle concentration ($>10\ g/m^3$), additional collection mechanisms contribute to particle separation. By analogy with pneumatic conveying a limiting dust load c_m is proposed. The excess particle mass cannot be transported by the gas and is accelerated to the wall immediately after entering the cyclone; the particles move down into the dust trap in the form of a strand. Along with particle collection due to excess loading, agglomeration of small particles with large ones can be assumed to occur at high particle concentration; this also improves overall collection efficiency.

The total pressure drop of a cyclone consists of pressure losses in the inlet, separating chamber, and exit duct; it is usually calculated as

$$\Delta p = (\xi_{\text{inlet}} + \xi_{\text{separator}} + \xi_{\text{exit}}) \cdot \frac{\varrho_f v_i^2}{2} \tag{3}$$

where ξ is the pressure drop coefficient, ϱ_f is the gas density, and v_i is the axial velocity in the exit duct. Experience has shown that the pressure drop in the exit duct accounts for as much as 90% of the total cyclone pressure drop. Typical values for Δp are in the range of 500–2000 Pa.

7.1.3.3. Operational Characteristics

The collection performance of a cyclone depends chiefly on its geometry, gas throughput, inlet gas concentration, and properties of the material to be separated. For example, the increase in gas viscosity when the temperature is raised decreases performance. If the exit-duct diameter or inlet cross section is diminished, the tangential velocity component at the exit-duct radius increases, so that a greater centrifugal force acts on the particles. As a result, the cut size x_{50} decreases and particle collection improves. Collection efficiency is also improved if the volume flow rate through the cyclone increases. Theoretically the cut size varies according to $\dot{V}^{-0.5}$.

When the tangential velocity in the cyclone increases, however, the pressure drop Δp also increases. To achieve good collection performance and acceptable energy consumption in cleaning gases with high flow rates, several small cyclones connected in parallel are preferable to a large cyclone because the reduction in cyclone diameter shifts the cut size toward smaller particle sizes. Figure 7.2 presents typical fractional efficiency curves for three cyclones with different diameters but the same pressure drop. If geometrically similar cyclones are assumed and Δp is constant, the result for cleaning a gas with a given flow rate $\dot{V}$ is that the cut size for n cyclones in parallel is proportional to $n^{-0.25}$.

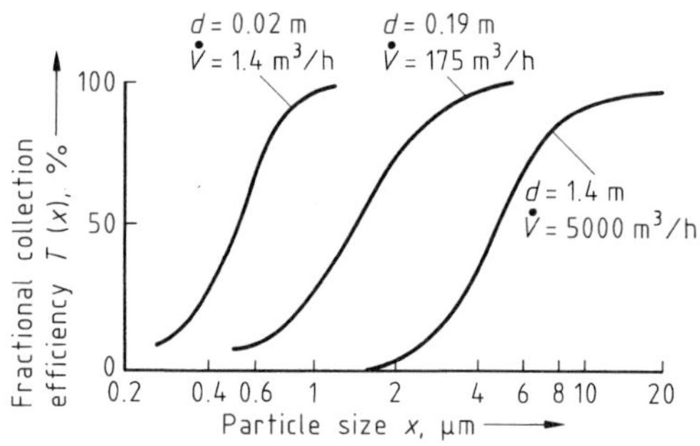

Figure 7.2. Fractional efficiency curves for cyclones of varying diameter d at a constant pressure drop
$\dot{V}$ = volume gas flow rate

7.1.4. Wet Scrubbers

7.1.4.1. Mode of Operation and Basic Designs

In wet collectors, both particulate and gaseous components can be separated from the carrier gas.

Even very fine particles ($<0.1\ \mu m$) can be collected in scrubbers. The particles are brought into contact with a scrubbing liquor and are removed from the gas stream in bound form as a slurry, which must be subjected to wastewater treatment or liquor recycling. In most scrubbers the scrubbing liquor is in droplet form. The first operation can, therefore, be regarded as the agglomeration of small solid particles and larger liquid particles. These agglomerates are readily collected from the gas stream. The transport step is decisive in separation. Diffusion and condensation can take place, but in general inertial forces appear to dominate. The wettability of the particles does not play a crucial role in binding; wettable particles penetrate the interior of the droplet, while nonwettable particles adhere to the surface.

Efficient dust collection in a scrubber requires separation of the dust-laden droplets in a separator that is either integrated into the scrubber or connected downstream. Because droplet sizes generally range from 1 µm to several hundred micrometers, commercial separators utilize

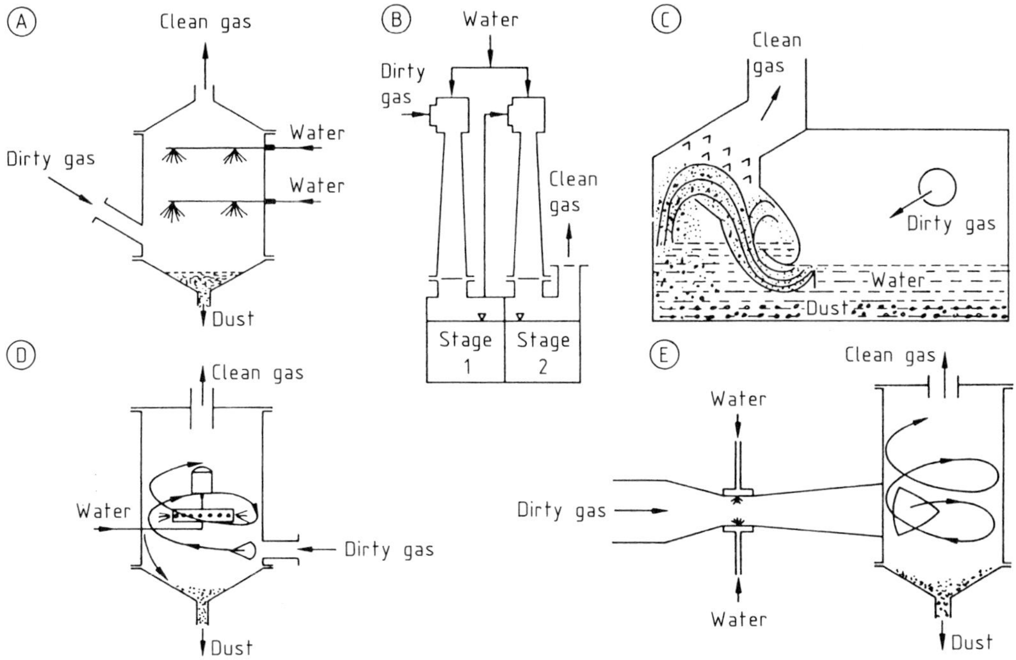

Figure 7.3. Basic types of scrubbers
A) Spray tower; B) Ejector venturi scrubber; C) Self-induced spray scrubber; D) Rotational scrubber; E) Venturi scrubber

inertial forces to separate droplets from the gas stream. Layers of wire mesh or fabric, stacks of plates, or centrifugal devices are employed.

Scrubbers are relatively easy to adapt to various operating conditions. They are employed mainly when products are to be recovered from a wet phase or when the danger of dust explosion or other hazards forbids dry collection. They have proved suitable chiefly for moderately large-volume gas flow rates (often $< 30000\ m^3/h$). Many scrubber designs are available. However, most designs fall into five basic types, shown in Figure 7.3. Typical operating data are listed in Table 7.1.

The spray tower (Fig. 7.3 A) is the oldest type of scrubber. The dirty gas flows upward through a large tube at a relatively low speed of ca. 1 m/s. The washing fluid is dispersed countercurrent to the gas by spray nozzles mounted at different levels. The dust-laden liquid leaves the bottom of the scrubber. Packings (e.g. glass beads) are sometimes used to improve collection efficiency.

The ejector venturi scrubber (Fig. 7.3 B) is basically a water-jet pump. Water is sprayed into the scrubber at a speed of ca. 25–35 m/s and provides a draft for moving the gas, which often attains a flow rate of 10–20 m/s. Two successive stages are often used because the collection efficiency of ejector venturi scrubbers is not much higher than that of spray towers.

In self-induced spray scrubbers (Fig. 7.3 C) the dust-laden gas is fed through the wash fluid.

Table 7.1. Operating data for scrubbers

	Packed-bed scrubber	Ejector venturi scrubber	Self-induced spray scrubber	Rotational scrubber	Venturi scrubber
Cut size, μm*	0.7–1.5	0.8–0.9	0.6–0.9	0.1–0.5	0.05–0.2
Relative velocity, m/s	1	10–25	8–20	25–70	40–150
Pressure drop, kPa	0.2–2.5		1.5–2.8	0.4–1.0	3–20
Water consumption, L/m^3	0.05–5	5–20**		1–3**	0.5–5
Energy consumption kW · h/1000 m^3	0.2–1.5	1.2–3	1–2	2–6	1.5–6

* Particle density $\varrho_p = 2.42\ g/cm^3$. ** For each stage.

The wash fluid is thus atomized, entrained, and mixed with the gas. This effect is reinforced by a redirecting vane. Entrained, dust-laden fluid collects inside the scrubber in the collection tank.

In rotational scrubbers (Fig. 7.3 D), the washing fluid is dispersed radially in the gas by rotating spray nozzles. The gas spirals upward and the dirty water is removed from the bottom of the scrubber.

Venturi scrubbers (Fig. 7.3 E) have the highest collection efficiency. The gas is accelerated up to 100 m/s in the venturi throat. The washing fluid enters the middle of the throat just in front of its narrowest point or is fed radially into the throat. The liquid is dispersed by the inflowing gas. The dust-laden water droplets are usually separated from the gas in a cyclone.

7.1.4.2. Collection Efficiency and Pressure Drop

Development and sizing of scrubbers are often purely empirical because the processes that occur in these complicated pieces of equipment are complex and difficult to simulate with theoretical models. However, the proposed models are quite helpful because they allow clear identification of important parameters and trends.

In most models, the scrubbing liquor is assumed to be in droplet form, and inertial forces are considered crucial for the transport of dust particles to the droplet. The calculation consists of three steps: calculation of single-droplet collection, calculation of the gas volume cleaned by the single droplet, and calculation of the change in dust concentration due to the action of all droplets. Important parameters are the Reynolds number and the inertial parameter (Stokes number). The effort required for the numerical computations is shortened substantially by fitting the equations to experimental values, so that the equation of motion for the droplets need not be solved. For the venturi scrubber, such an equation for the fractional collection efficiency $T(x)$ is given by

$$T(x) = 1 - \exp\left[-\frac{2\,\dot{V}_l d_l \varrho_l v}{55\,\dot{V}_g \mu} \cdot F(\psi, f) \right] \quad (4)$$

where $\dot{V}_l$ is the volume flow rate of scrubbing liquor, $\dot{V}_g$ is the volume flow rate of gas, d_l is droplet diameter, ϱ_l is droplet density, v is maximum gas velocity, μ is gas viscosity, and F a function of the inertial parameter ψ and an empirical fitting parameter f $(0.25 < f < 0.5)$. The following approximation is given for the function F in the range $1 \leq \psi \leq 4$.

$$F(\psi, f) \approx 0.312 \cdot \psi f^2 \quad (5)$$

Equation (4) can also be used, along with a pressure-drop equation, for optimization. In general, model calculations indicate that higher relative velocities and increased addition of scrubbing liquor lead to better collection efficiency; this is also observed experimentally.

Calculation of the pressure drop is of interest chiefly for venturi scrubbers. Empirical formulas and model calculations are available for this purpose. The empirical Equation (7) is based on the procedure for evaluating flow in pipes.

$$\Delta p = \left(C_1 + C_2 \frac{\dot{V}_l}{\dot{V}_g} 10^3 \right) \frac{\varrho_f v^2}{2} \quad (7)$$

C_1 and C_2 are empirical constants, with typical values $0 \leq C_1 \leq 0.5$ and $0 \leq C_2 \leq 1.7$; ϱ_f is gas density.

7.1.5. Filters

7.1.5.1. General

Filters constitute an important aid in modern particle collection technology, especially when highly efficient collection in the finest particle-size range is required. Many versions are available. Of all collectors, filters have the widest spectrum of applications and thus a large share of the market.

With regard to application, design, and mode of operation, filters for dust collection can be classified in different but overlapping groups. Three main classes are the so-called deep-bed filters, surface filters, and granular-bed filters.

7.1.5.2. Deep-Bed Filters

Deep-bed filters are used where dust loads are low, on the order of several milligrams per cubic meter of gas. Typical applications include air conditioning and ventilation, laboratory exhaust systems, clean rooms, and respiratory protection. The filters consist generally of a relatively loose fiber mat with a pore fraction over 90 % (often >99 %). As the gas flows through the fibrous layer, particle collection takes place in the interior of the layer, where the dust accumulates (deep-bed filtration). After becoming saturated with dust, these filters are usually discarded; some can be cleaned by washing or blowing. Typical gas velocities are 0.1 – 3 m/s.

7.1.5.3. Surface Filters

Mode of Operation and Basic Designs. Surface filters are used for high dust loads, on the order of several grams to hundreds of grams per cubic meter. This is the typical range for industrial dust removal and air-pollution control. The principal surface-filter media in current use are nonwovens and (needled) felts composed of randomly arranged glass, plastic, or ceramic fibers. Woven materials, chiefly constructed from glass-fiber yarns or metal wires, are also employed. Only at the beginning of filtration do these media trap particles in the interior of the fibrous layer. As more dust is deposited, it bridges the fibers and forms a dust layer (filter cake) on the surface.

The fractional efficiency curves in Figure 7.4 show the improvement in collection performance as more dust is collected and pressure drop increases. The new, clean filter gives a typical depth-filter curve. Since the cake-formation phase lasts only a short time for the dust loads usually encountered in practice, the interception effect inside the cake is dominant. This process is called surface filtration. Emission levels of <10 mg/m^3 can be achieved.

As the amount of dust deposited in and on the filter grows, the pressure drop also increases. After a certain time, or when a predetermined pressure-drop value is reached (ca. 1000 – 2500 Pa), the dust must be removed (cleaned off) to regenerate the filter.

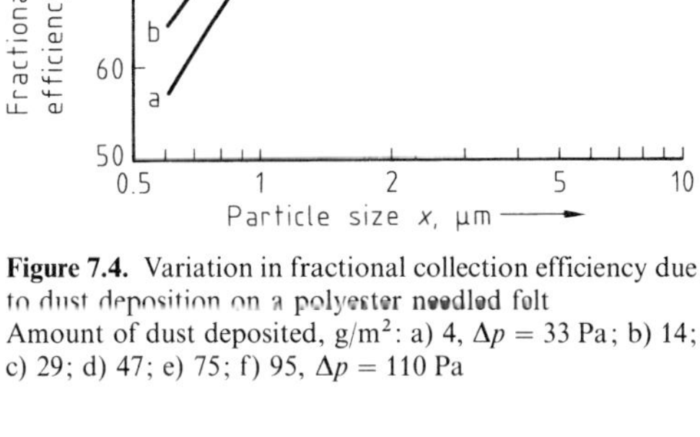

Figure 7.4. Variation in fractional collection efficiency due to dust deposition on a polyester needled felt
Amount of dust deposited, g/m^2: a) 4, $\Delta p = 33$ Pa; b) 14; c) 29; d) 47; e) 75; f) 95, $\Delta p = 110$ Pa

In newly developed commercial filter media, surface coating or treatment is used with the aim of collection on the surface right from the start and to prevent particles from penetrating into the medium. This approach yields high initial collection efficiencies, avoids the danger of plugging, and facilitates removal of the filter cake.

The three most important surface-filter designs are round bag, envelope, and cartridge. A filter unit contains the requisite number of filter elements connected in parallel. Designs differ primarily in the method of cleaning and sometimes in the way the dirty gas is introduced.

Round bag filters are used most widely. Typical bags are 100 – 300 mm in diameter and 1.5 – 10 m long. The classical design is a multichamber filter with shaker cleaning. Dirty gas is fed into the bags from below; it then flows through the filter medium and is led to a clean-gas duct at the top of the device. Particle collection takes place on the inside of the bags. For cleaning, the dirty-gas supply is cut off from one chamber at a time, and the dust is loosened by shaking or rapping the bag support. The dust removed from the bags falls into a dust hopper; purge air drawn into the bags from outside aids this process.

Pulse-jet filters (Fig. 7.5) became more widely used as progress was made in needle felts. The gas always flows from outside to inside, so the bags must be pulled over supporting frames. For cleaning, a jet of compressed air is admitted to the bag through its open top end. The compressed-air jet suddenly raises the pressure in the bag, which is rapidly inflated and gas flow is reversed. The dust is thus dislodged from the outside of the bag. Because single bags or single rows of bags must be removed from the filtration process for only a short time (ca. 0.1 – 0.3 s), the filter need not be divided into chambers, and cleaning can be done on-line.

Design Calculations. The main purpose of surface-filter design is to determine the required filtration area A, which is given by the ratio of the volume flow rate of dirty gas $\dot{V}$ and the filtration velocity or filter load v. The filtration velocity is selected on the basis of empirical values because of a lack of generally valid models.

The pressure drop Δp is assumed to be the sum of the contributions of the medium and the

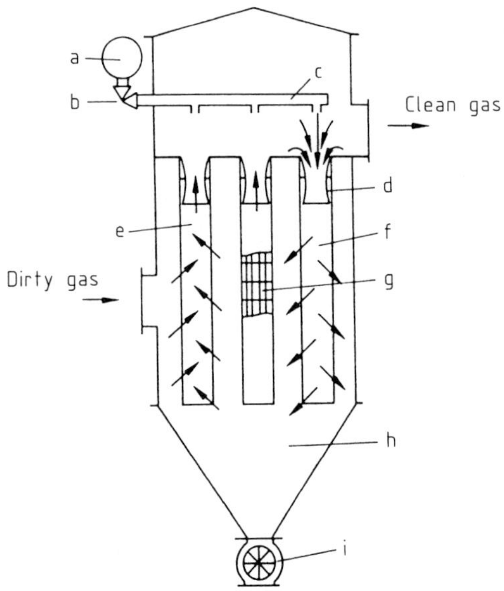

Figure 7.5. Round bag filter with reverse-pulse cleaning
a) Compressed-air tank; b) Right-angle diaphragm valve; c) Compressed-air duct with nozzles; d) Venturi nozzle; e) Filter bag in operation; f) Filter bag being cleaned; g) Retainer; h) Dust hopper; i) Air lock

filter cake. This leads to Equation (8), where K_1 is the residual resistance of the filter medium after cleaning, K_2 is the specific resistance of the filter cake, W is the mass of dust collected per unit filter area, and μ is the gas viscosity. Since W is proportional to filtration time, the second term describes the time dependence of Δp.

$$\Delta p = K_1 \mu v + K_2 \mu W v \qquad (8)$$

Important parameters in this equation are the resistances K_1 and K_2; as a rule, they are not constants but functions of several variables. Prediction of these parameters is still problematic because the functions are not sufficiently well known. Even when K_1 and K_2 can be determined experimentally, great care must be taken to ensure that conditions in the experimental device and the full-scale plant are comparable.

Operational Characteristics. Particle collection in a surface filter occurs mainly in the dust layer formed on the surface. Therefore, this highly efficient cake must be retained on the surface as long as possible.

Figure 7.6 shows that the pressure drop Δp increases as more dust is deposited, and periodic

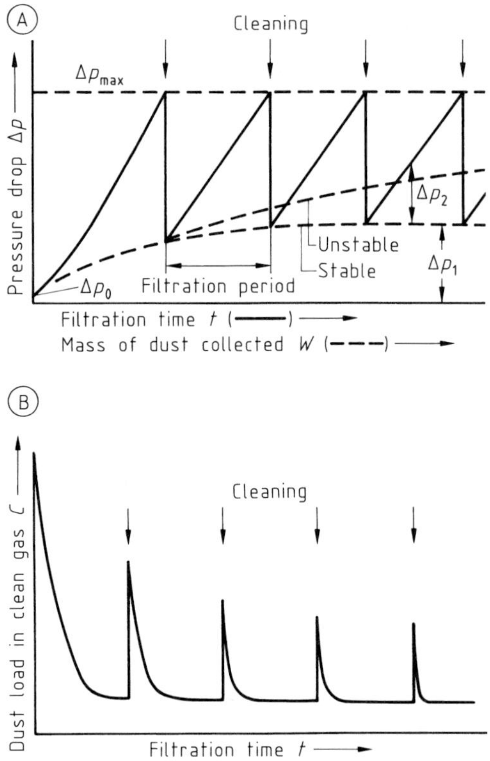

Figure 7.6. Pressure drop (A) and dust content (B) in clean gas versus filtration time at constant filtration velocity
The values Δp_1 and Δp_2 denote the pressure-drop contributions of the filtration medium and filter cake, respectively

cleaning is necessary. After cleaning, the pressure drop falls to a value that depends on the flow resistance of the filter medium (including residual dust); this value is higher than the initial pressure drop of the new, clean filter. For stable filter operation the residual pressure drop must reach an approximately constant value after an induction period. The filter medium, filtration velocity, and the method and intensity of cleaning determine whether stable filtration is attained for given gas and dust properties.

During and immediately after cleaning, the particle content of the clean gas increases. This occurs partly because the filter cake, which acts as a layer of filter aid, has been removed. Especially with reverse-pulse cleaning, some of the dust passes through the filter medium because after the pressure pulse inside the bag has decayed, the filter medium falls back against the cage and knocks remaining dust through the filter.

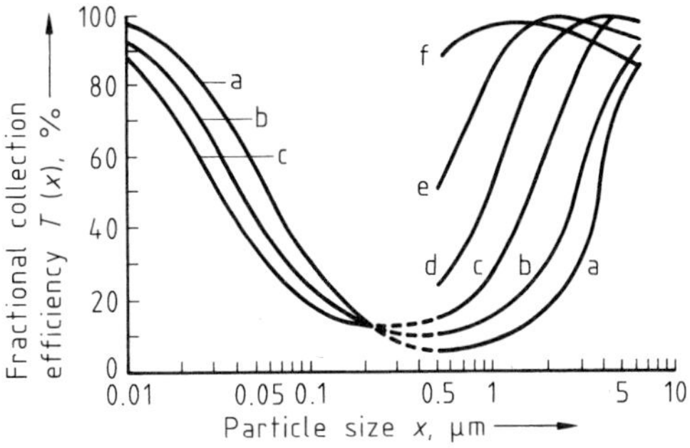

Figure 7.7. Fractional efficiency curves for a bed of glass beads at various filtration velocities
Bed height $H = 20$ mm; bead diameter $d_G = 0.58$ mm
Filtration velocity u_0, m/s: a) 0.05; b) 0.10; c) 0.25; d) 0.5; e) 1.0; f) 1.5

7.1.5.4. Granular-Bed Filters

In granular-bed filters, the dust-laden gas stream flows through a bed of gravel, sand, ceramics, activated carbon, etc. and is thereby cleaned. The granular-bed functions as a filter medium and can be regarded as a depth or surface filter. Besides pure particle collection, simultaneous removal of gaseous components can be performed in such filters. Granular-bed filters are used chiefly for the removal of hot, chemically aggressive, abrasive, or adhesive dust or when danger of fire exists.

Granular-beds can be either fixed or moving. Filters with fixed beds are operated batchwise; when dust deposition has raised the pressure drop to a predetermined limit, the dirty gas stream is cut off and the bed is regenerated. The dust is removed internally by reverse flushing and agitation or stirring, or externally by screening or other means. In filters with moving beds the granules creep through the filter chamber by gravity, either steadily or at intervals. Fresh granules are added continuously from above. Regeneration takes place in an external loop. This filtration method operates without interruption of the dirty-gas stream.

The design of granular-bed filters is mainly empirical (as in collection generally). Typical initial fractional collection efficiency curves (no dust load in the bed) are shown in Figure 7.7.

7.1.6. Electrical Precipitators

7.1.6.1. Mode of Operation and Basic Designs

An efficient way of separating particles from gases is based on the forces exerted on charged particles in an electric field. This principle is important when inertial forces are no longer effective, i.e., for fine particles with diameters < 1 μm. Because the required capital investment is rather large, the main application of these precipitators is in dust removal from large gas streams (up to several million cubic meters per hour). Emission levels significantly lower than 50 mg/m^3 can be achieved if the dust properties are favorable. Pressure drops are generally slight (< 500 Pa).

The collection principle used in electrical precipitators is depicted in Figure 7.8 and involves three successive steps: charging of particles, transport of particles to the collecting electrode, and removal of particles gathered on the collecting electrode.

The ions needed for particle charging are produced by corona discharge at the emission electrodes, which are made of thin wires or tapes with barbs. To obtain optimal collection efficiencies, high field strengths and high corona currents are required. The d.c. voltage is, therefore, adjusted to establish a condition as close as possible to electrical breakdown (ca. 60–80 kV), which depends on electrode geometry and gas properties.

Depending on particle size, the attachment of gas ions to particles occurs by one of two mechanisms: field charging or diffusion charging. Field charging dominates in the particle-size range above ca. 1 μm. The ions move toward the collecting electrodes along the electric field lines. As they move, the ions collide with particles and become attached to them. Diffusion charging results from random thermal motion of the ions

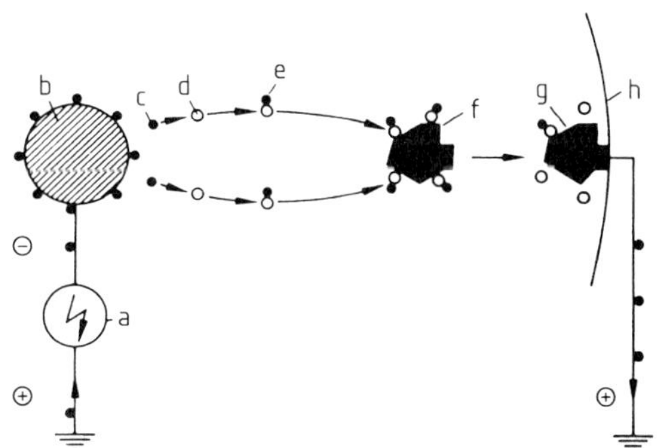

Figure 7.8. Schematic diagram of charging and collection process
a) High-voltage power supply; b) Emission electrode; c) Electron; d) Neutral molecule; e) Ionized molecule; f) Charged dust particle; g) Collected dust particle; h) Collecting electrode

(Brownian motion) and is important chiefly for particles smaller than 1 μm.

Charged particles are transported to the collecting electrode and must adhere to it. The electrical properties of the particles, particularly resistance, are important for the behavior of both the particle on impact and the resulting dust layer.

The favorable resistivity range for collection is $10^4 - 10^{11}$ Ωcm. Particles with a lower resistance give up their charge quickly when they collide with the collecting electrode; they change their polarity and are repelled from the electrode back into the gas stream. If the resistance is too high, the particles form a layer that continues to accumulate more charge, thus leading to a weakening of the electric field and discharge effects in the porous dust layer.

There are two main designs of precipitators. Wire-in tube precipitators are employed for small quantities of gas and especially for the collection of liquids. The wire emission electrodes are stretched along the central axis of parallel tubes. The inside walls of the tubes serve as collecting electrodes. If the tubes are set up vertically, the collected liquid runs down the walls as a film. If solid particles are collected, they are frequently rinsed out by flushing the tubes with water. The tubes are usually 0.2–0.3 m in diameter and 2–5 m long.

Parallel-plate precipitators contain many plane or profiled plates suspended at uniform intervals (200–600 mm) as collecting electrodes (Fig. 7.9). Gas flows horizontally in the channels between plates. Emission electrodes hang in the midplane between neighboring plates. Particle removal may be either dry or wet. In dry electrostatic precipitators, the dust layer collected on the plates is removed periodically by rapping the plates with hammers, whereby the dust falls into the hopper beneath the plates. The total plate length is divided into several zones for two purposes: dust reentrainment is reduced because the zones can be cleaned at different times; and separate zone-by-zone power controls allow the voltage to be adjusted in accordance with dust concentration.

In wet electrostatic precipitators, the collecting electrodes are cleaned by flushing with water. The dust flows down off the plates with the water; dust reentrainment is thus effectively prevented.

7.1.6.2. Design Calculations

Calculations for precipitator design are concerned with collection efficiency; the very small pressure drop is not important. The well-known Deutsch precipitator equation is still used as the basis for dimensioning precipitators and for evaluating and interpreting measurements. If turbulent flow in the precipitator is assumed and

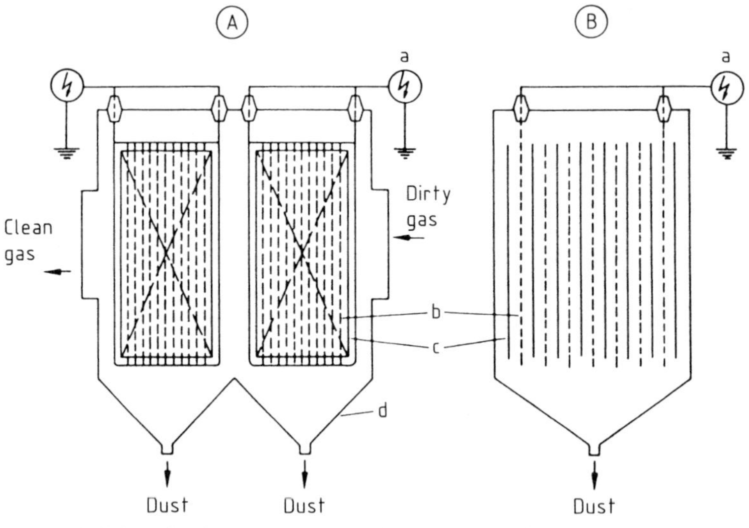

Figure 7.9. Schematic of a two-zone parallel-plate electrical precipitator (vertical sections A and B are shown at right-angles to each other)
a) High-voltage power supply; b) Emission electrode; c) Collecting electrode; d) Hopper

if reentrainment and reemission are neglected, the Deutsch equation for fractional efficiency is

$$T(x) = 1 - \exp\left[-\frac{2L}{av}w(x)\right] \tag{9}$$

where L is the collection length, a is the plate spacing (parallel plate precipitators) or the diameter of the tube (tube precipitators), v is the gas velocity, and $w(x)$ is the so-called migration velocity. For a parallel-plate device, Equation (9) can also be written as

$$T(x) = 1 - \exp\left[-\frac{A}{\dot{V}}w(x)\right] \tag{10}$$

where A is the surface area of the collecting electrodes and $\dot{V}$ is the volume gas flow rate.

The migration velocity $w(x)$ depends on the particle size and charge, the electric field, and the gas viscosity. It can be derived theoretically by solving the equation of particle motion. In practice migration velocity is calculated from the measured fractional efficiencies, by means of Equation (10); it often lies in the range of 10–30 cm/s. The parameter $w(x)$ is basically a transport coefficient that characterizes particle transport to the collecting electrode; it is thus vital for dimensioning electrical precipitators.

7.2. Waste-Gas Purification by Thermal or Catalytic Reactions

In thermal and catalytic waste-gas purification processes, gaseous air pollutants are converted by chemical reactions (oxidation, reduction, or decomposition) into harmless substances. These reactions can occur in flames (combustion [7.9]) or combustion chambers (e.g., thermal oxidation), or over catalysts (e.g., catalytic oxidation). In the following sections these processes are discussed with regard to their possible applications and the process and plant concepts.

7.2.1. Fundamentals

7.2.1.1. Types of Reaction

Air pollutants can be disposed of in the gas phase at the site of emission by thermal and catalytic reactions. By utilizing the reaction energy of the exothermic reactions, thermal recycling is possible but not the recycling of material. The latter can be achieved by absorption, adsorption, condensation, or membrane processes.

The disposal effect in the case of thermal and catalytic processes consists in the conversion of the air pollutants into harmless reaction products by chemical reactions. These can be divided into oxidation reactions, reduction reactions, and decomposition reactions.

Substances regarded as harmless here include species natural to the air such as CO_2, H_2O, and N_2. Some examples of reactions with their characteristic quantities and applications follow:

Oxidation reactions

Example	hydrocarbon combustion $C_mH_nO_p + (m + n/4 - p/2)O_2 \rightarrow mCO_2 + n/2\,H_2O$
Temperature	200–600 °C
Residence time	0.42–0.02 s
Space velocity	5000–50 000 h^{-1}
Application	catalytic afterburning, engine exhaust gas cleaning

Reduction reactions

Example 1	selective catalytic nitrogen oxide reduction (SCR) $4\,NH_3 + 4\,NO + O_2 \rightarrow 6\,H_2O + 4\,N_2$
Temperature	200–400 °C
Residence time	0.2–0.3 s
Space velocity	1000–5000 h^{-1}
Application	$DeNO_x$
Example 2	nonselective catalytic nitrogen oxide reduction (NSCR) $CO + NO \rightarrow 1/2\,N_2 + CO_2$ $C_mH_n + 2(m + n/4)\,NO \rightarrow (m + n/4)\,N_2 + n/2\,H_2O + m\,CO_2$
Temperature	200–600 °C
Residence time	0.42–0.02 s
Space velocity	5000–50 000 h^{-1}
Application	engine exhaust gas cleaning

Decomposition reactions

Example	ozone decomposition $2\,O_3 \rightarrow 3\,O_2$
Temperature	20–100 °C
Residence time	0.7–0.3 s
Space velocity	5000–10 000 h^{-1}
Application	water purification

Thus additional reactants (oxidizing agents, e.g., O_2, and reducing agents, e.g., NH_3) are required for conversion of air pollutants.

On the basis of the oxygen demand in oxidation processes, it is technologically reasonable to classify the gaseous emission streams as waste air streams (high O_2 content, differing little from the natural composition of the air and sufficient for safe burner operation) or waste gases (inert, low O_2 content, burners require additional fresh air for operation). In the following, however, in conformity with the definition of [7.10], carrier gases with solid, liquid, or gaseous emissions are referred to as waste gases.

7.2.1.2. Reaction Engineering

7.2.1.2.1. Homogeneous Gas-Phase Reactions

The course of homogeneous gas-phase reactions can be broken down into three process steps:

1) Preparation (e.g., atomization, swirling)
2) Mixing (diffusion, convection)
3) Reaction of the reactants

High- and low-temperature mechanisms have been formulated for the oxidation of hydrocarbons [7.11]–[7.18]:

1) High-temperature mechanisms:
   radical-chain mechanism at $T > 1000$ K
   hydrocarbon → intermediate products
   intermediate product → carbon monoxide
   carbon monoxide → carbon dioxide
   (the last step is rate-determining)
2) Low-temperature mechanisms:
   $T < 1000$ K, formation of partially oxidized intermediate products, such as aldehydes, ketones, alcohols, peroxides, hydroperoxides, and carboxylic acids

The rate of the overall process is determined by the design parameters temperature, time, and turbulence.

The exothermic oxidation and reduction reactions (e.g., combustion) used for waste-gas purification are in practice carried out at temperatures so high that the reaction step is not rate-determining. The main task in carrying out the reaction and designing the equipment is mixing the reactants [7.19].

Flame Combustion. The course of reaction in flames is characterized [7.20] by high temperatures (>1200 °C), high radical concentrations, and luminescence phenomena.

Flame combustion is used for waste gases containing pollutants in concentrations near the explosion limit. Combustion takes place in flares after addition of air or oxygen, without energy supply (autothermally) or with a small energy supply [7.21]. By using staged oxygen supply the combustion can be carried out such that reduction reactions occur [7.22].

Thermal Reactions. Thermal waste-gas purification (e.g., thermal postcombustion or thermal oxidation) is generally used for waste gases whose pollutant concentration is well below the flammability limit. If the explosion limit can be exceeded as a result of output variations, safety devices must be installed [7.23]. An auxiliary heater provides the required minimum operating temperature in the combustion chamber [7.11], [7.12], [7.19], [7.20]. The processes occurring there are influenced both by the reactivity of the pollutants [7.13]–[7.18] and by conditions depending on plant design, such as the combustion chamber geometry [7.15], nature of the auxiliary heating, and flow and temperature fields in the combustion chamber. The operating temperatures of thermal oxidizer plants for reasonable residence times (0.1–2 s) are between 700 and 1200 °C, so heat-resistant materials must be used. The thermal energy of the combustion flue gases can be utilized by heat exchange [7.24], [7.25].

7.2.1.2.2. Heterogeneous Gas-Phase Reactions, Catalytic Waste-Gas Purification

Kinetics. In catalytic waste-gas purification pollutant gases and vapors are converted to harmless substances by chemical reaction over a solid inorganic catalyst.

Most industrial catalysts are highly porous materials with a large surface for chemical reaction. Because the reactants are present in a different phase from the catalyst, physical transport processes occurring on the catalyst are also important.

The important individual steps of such heterogeneously catalyzed reactions are shown in Figure 7.10.

The use of solid catalysts in gas reactions (heterogeneous catalysis) raises the rate of the reaction by condensation of the reactants on the catalyst surface and, as a result, increased collision frequency, and by decreasing the activation energy (see Fig. 7.11).

Due to the increased reaction rate the conversion of air pollutants by chemical reaction is achieved at lower temperatures and with smaller reactors (shorter residence time) compared with thermal processes.

Requirements of Waste-Gas Purification Catalysts. The *activity* of a catalyst is its ability to

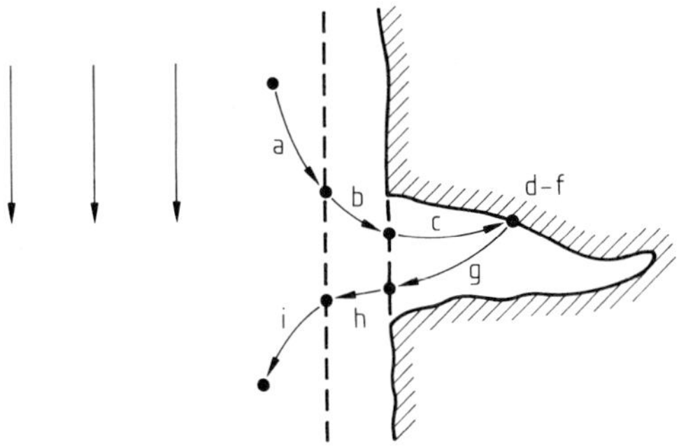

Figure 7.10. Individual steps of heterogeneous catalysis
a) Transport of the reactants through the gas stream to the boundary layer; b) Diffusion of the reactants through the boundary layer to the catalyst's external surface; c) Diffusion of the reactants through the pores to the catalyst's internal surface; d) Adsorption on the catalyst surface; e) Chemical reaction; f) Desorption of the products from the catalyst surface; g) Diffusion of the products through the pores to the catalyst's external surface; h) Diffusion of the products through the boundary layer; i) Transport of the products in the gas stream

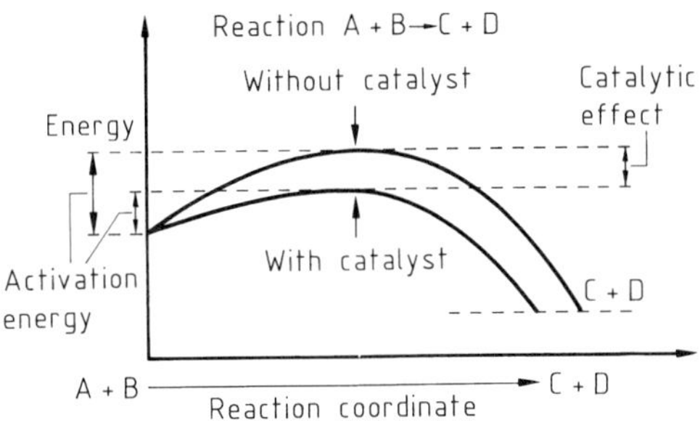

Figure 7.11. Simplified potential energy curves for homogeneous and heterogeneous exothermic reactions

increase the rate of a reaction in question under defined conditions. Activity is therefore not a fundamental property, but a property specific to a reaction and process parameters. The following are regarded as alternative measures of activity:

1) The percentage conversion at a given temperature and contact time and inlet concentration
2) The temperatures and contact times necessary to maintain certain percentage conversions or certain concentrations in the purified gas.

Only those reactions should occur on the catalyst that lead to the required products (*selectivity*). For example, during the catalytic total oxidation of hydrocarbons, partially oxidized intermediate products are converted with higher activity than the hydrocarbon. Intermediate products are therefore not emitted [7.26]. A catalyst should exhibit *stability towards deactivation*, i.e., its properties must remain unchanged for the longest possible lifetime. When selecting a catalyst, special attention must be paid to potential deactivation. It must be verified that

1) The catalyst is resistant to the substances present (often in traces) in the waste gas
2) The catalyst undergoes no structural changes in the operating temperature range
3) The catalyst is sufficiently resistant to abrasion and fracture by vibrations and stresses occurring in operation

Although the catalyst is not consumed in the chemical reaction, a change of properties during operation under industrial conditions must be expected. This leads to reversible or irreversible decrease in the catalytic activity (deactivation). Chemical and thermal factors can limit the effective life of catalysts.

Catalyst poisoning occurs when the waste gas contains substances that react with the catalyst support or the active component.

These undesirable reactions can lead to changes in chemical composition, reduction of surface area, loss of active components, or masking of active centers (e.g., coking).

The rate and frequency of structural changes (crystal growth, spinel formation, sintering) increase greatly with increasing temperature. These structural changes can lead to reduced internal surface area and decreased degree of dispersion of finely-divided active components.

For industrial catalysts, therefore, upper limits are specified for the temperature of use:

Metal oxide catalysts	
bulk catalysts	500 °C
supported catalysts	700 °C
Noble metal catalysts	
aluminum oxide supported	700 °C
stabilized aluminosilicate supported	1000 °C

Catalyst overheating during periods of intense heat of combustion release due to high pollutant concentrations can be avoided by introducing cold air or reducing the preheating.

Catalyst damage caused by mechanical influences include abrasion and fracture. Abrasion can be caused by vibration of the plant, pressure surges, loosening of the catalyst bed, and solid particles (abrasive dusts); and fracture by thermal stress owing to rapid temperature changes, and mechanical stress due to thermal expansions of the reactor.

Mechanical stress in reactors filled with honeycomb catalysts can be compensated by elastic internal seals.

The risk of catalyst beds being fluidized can be avoided by a vertical inlet flow from above. At the same time, however, the formation of a funnel-like bed surface by too high an inlet flow rate or by a swirling incident flow must be avoided.

The *pressure drop* in catalytic waste gas purification reactors can be influenced by selecting an appropriate geometry and type (e.g., honeycomb) for the catalyst as well as by the height of the catalyst bed.

Typical values of the specific pressure drops, per unit bed height, are 10 kPa/m for bulk catalysts and 1 kPa/m for honeycombs.

Reactor Design. Two methods can be used for reactor design.

Computing based on reaction engineering models [7.27], which are applicable if

1) The following kinetic data, necessary for determining the reaction rate, are known: activation energy, collision factor, reaction order, and kinetics of mass transfer.
2) There is no influence of interfering components.
3) Flow and temperature conditions are known. Simple reactor models are based on assumptions such as isothermal reactor, adiabatic reactor, or pseudohomogeneous reaction (i.e., no influence of mass-transfer processes).

Experimental simulation (*pilot-plant trials*) is more important in practice since:

1) Mixtures of pollutants are present, often with variable concentrations
2) Conditions of flow and temperature vary greatly
3) No kinetic data are available
4) The influence of further crude gas constituents is unknown
5) A pilot-plant trial is often necessary for detecting the presence of catalyst poisons

The results of these experiments are plotted in diagrams showing, for specific pollutants on specific catalysts, the dependence of the percentage conversion on inlet temperature, space velocity (under standard conditions), and inlet concentrations (Fig. 7.12).

These diagrams can be used as the basis for design according to the following scheme:

1) Determination of the percentage conversion required
2) Determination of the catalyst inlet temperature at the expected inlet concentration (taking into account the temperature rise due to the heat of reaction)
3) Determination of the permissible space velocity under normalized design conditions (SVN)
4) Calculation of the catalyst volume required

$$V_c = \frac{\Phi v, n}{\text{SVN}} = \frac{\text{volumetric flow rate (STP)}}{\text{space velocity (STP)}}$$

If mixtures of contaminants are present, the temperature and space velocity should be determined from the reaction behavior of the least reactive component.

During the catalytic afterburning of hydrocarbon mixtures the individual components are selectively oxidized, so the proportion of less re-

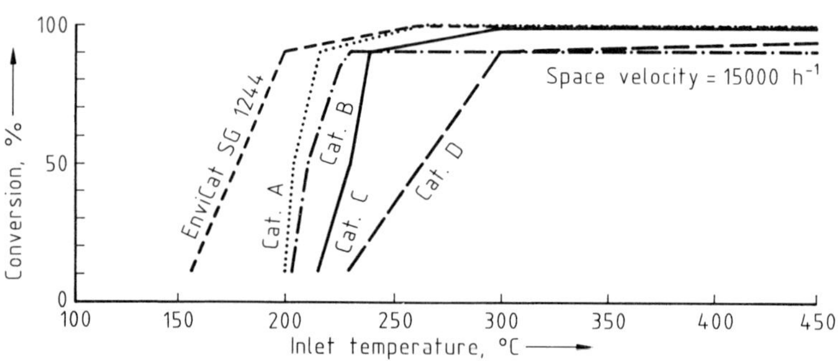

Figure 7.12. Curves of percentage conversion [7.26]
Activity comparison of various metal oxide catalysts (pollutant: 1000 ppm ethyl acetate) in air; catalyst aged for 168 h at 500 °C; space velocity = 15000 h^{-1})

active components increases with increasing temperature. The overall fractional conversions therefore exceed the value for the less reactive component.

When designing the catalytic waste-gas purification reactors, it must be ensured that the thoroughly mixed reactants are distributed homogeneously over the entire inlet flow cross section of the catalyst bed. This can be achieved by mixing blades, baffle plates, and straightening blades.

7.2.1.3. Process Selection and Plant Concepts

The process is selected primarily from knowledge of the appropriate type of reaction for pollutant conversion under the prevailing waste-gas conditions of temperature, pressure, composition (components and concentration) and required conversion.

The choice of process is also determined by economic considerations [7.28] such as investment costs; operating costs (energy supply, waste-heat utilization [7.24], [7.29], pressure drop); and the life of components such as catalysts, heat exchangers, and valves; formation of byproducts (e.g., HCl during the oxidation of chlorinated hydrocarbons), waste disposal (e.g., of catalysts).

The following process steps can be regarded as the basic operations of waste-gas purification processes:

1) Transfer of the waste gas (e.g., by suction)
2) Mechanical transport of the exhaust air (e.g., by ventilators, compressors)
3) Separation of interfering components (e.g., by filters, adsorbers, absorbers)
4) Heating of the waste gas by heat exchange
5) Supply of external energy (e.g., by burners, electrical resistance heating)
6) Mixing with oxidizing or reducing agents (e.g., oxygen, air, ammonia)
7) Pollutant conversion (e.g., by flame, thermal, or catalytic reaction)
8) Waste heat utilization (e.g., for heating steam or heat-transfer oil)
9) Separation of reaction products (e.g., HCl)
10) Discharge of purified gases (e.g., from the stack)

Economically, integration of waste-gas purification into the production process should be aimed for. In oxidation processes this can be achieved by waste-heat utilization [7.24], [7.27], [7.29], [7.40].

7.2.1.4. Operational Check [7.31], [7.32]

Waste-gas purification plants lead to reduced emission mass flow rates.

If waste gas mass flow rates are known, concentration measurements give information on the efficiency of the plant and are prescribed as emission values in permits.

In measuring emission concentrations, it must be taken into account for processes with pollutant conversion that the type of substance can change. Therefore, with thermal combustion processes, for example, the measurement of CO content in the purified gas is specified. Catalytic processes usually operate selectively (see Section 1.2.2.1).

The frequency of emission monitoring is prescribed for the specific plant and adapted to plant size and fluctuations in operating conditions.

For chemical conversions the reaction temperature largely determines the rate of the process. Temperatures (combustion chamber outlet, catalyst inlet/outlet) are therefore frequently used as control variables.

It is not useful to specify standard oxygen contents for oxidative waste-gas purification processes, since the oxygen concentration of the waste gas is predetermined by the emitting process and reduced as little as possible by an energy optimized conduct of the process (i.e. one with reduced fuel requirement) [7.19].

7.2.2. Thermal Processes for Waste-Gas Purification

7.2.2.1. Oxidation Processes

Thermal oxidation processes associated with flames are known as combustion. Complete combustion converts the carbon bound in the atmospheric pollutant or in auxiliary fuels to carbon dioxide. In the treatment of exhaust air streams containing sulfur or halogens, the formation of SO_2 and HCl or Cl_2 necessitates downstream separation. If chlorine-containing compounds are present, attention must be paid to the formation of polychlorinated compounds. In the presence of precursor compounds (e.g., chlorobenzene), polychlorinated dibenzodioxins and dibenzofurans can be formed under conditions such as local oxygen deficiency, low temperatures (300–600 °C), or inadequate residence time in the presence of oxygen.

Therefore in the design of combustion systems (intensive mixing) and waste-heat recovery (rapid cooling), these conditions must be avoided in order to avoid "de novo" synthesis from hydrocarbon radicals and organic or inorganic chlorine [7.19], [7.33], [7.34]. If organic nitrogen compounds or NH_3 are present, nitrogen oxides are formed [7.13], [7.16]. These can be avoided by control of the combustion or converted by waste gas aftertreatment (see Section 6.2.2.2).

The sequence of the following process descriptions follows the specific CO_2 formation to be expected, taking into account pollutant conversion, external energy supply, and waste-heat utilization. The amount of CO_2 formed is least if the combustible substances in the waste gas are used as replacement fuel in existing furnaces or stationary motors. With these processes, however, attention must be paid to nitrogen oxide formation. In separate waste-gas purification units, specific CO_2 formation decreases with increasing waste-gas preheating or with increased cooling of the purified gas by waste-heat utilization before discharge to the atmosphere.

7.2.2.1.1. Waste-Gas Disposal in Furnaces [7.24], [7.29]

Waste gases containing oxygen can be added to the combustion air of furnaces provided it is ensured, if necessary, by fitting (see Fig. 7.13), that

1) Deposits from the waste gas (e.g., cracking products) in the air supply system of the burner are avoided

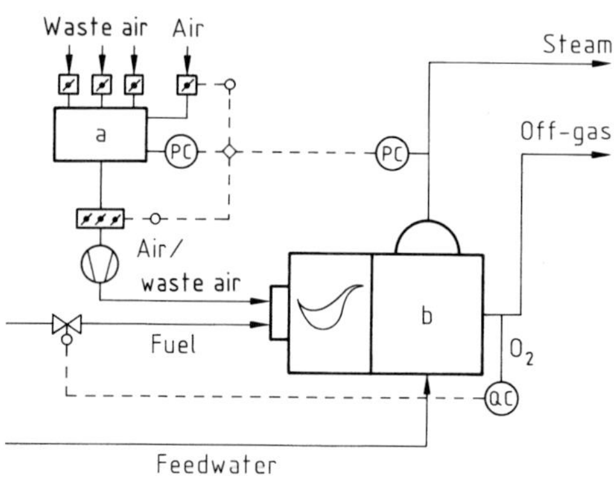

Figure 7.13. Waste gas thermal purification in a boiler furnace [7.24], [7.29]
a) Collection chamber; b) Steam boiler

2) The resulting waste-gas stream is always supplied to the heater
3) Fresh air is available for the generation of steam

Conditions (2) and (3) can be realized by using a vacuum-controlled waste-gas collecting chamber with a fresh-air valve [7.24].

With regard to condition (2) it must be checked whether there is simultaneity of waste gas production and furnace operation in the correct order of magnitude (approximately 800–1000 m^3/h waste gas for 1 t/h steam) [7.29].

7.2.2.1.2. Thermal Combustion

Waste-Gas Preheating by Regenerative Heat Exchangers [7.25], [7.35], [7.36]. Regenerative heat exchangers are characterized by:

1) The presence of a heat-storage material
2) Position- and time-dependent temperature profiles in the storage material for fixed beds (see Fig. 7.14)
3) An alternating flow direction (fixed bed) or moving storage materials (moving bed)

Advantages of these systems are:

4) High degrees of heat exchange

$$n = \Phi_E/\Phi_E \max = \frac{T_{hot, inlet} - T_{hot, outlet}}{T_{hot, inlet} - T_{cold, inlet}} > 90\,\%$$

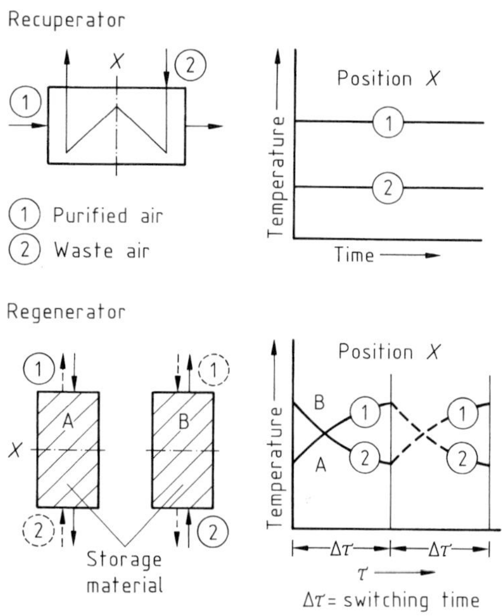

Figure 7.14. Difference between recuperators and regenerators [7.25]

5) Homogenization of the flow

The following features of this system must be taken into account:

1) Periodically changing mixture of waste gas and purified gas, since the waste gas path is the same as the purified gas path. Therefore, at high preheating temperatures the use of uneven-numbered tower layouts for carrying out a purging phase with recycle of the unpurified gas is advisable. By opening and closing the respective valves, the tower in purge mode is connected to a line leading to the suction side of the blower. After purging the crude gas from the void fraction of the bed, the hot purified gas from the burner chamber is connected to the purged tower. The bed is heated by heat transfer from the hot gases. Thus the cycle of modes is completed without giving untreated gas access to the purified gas duct.

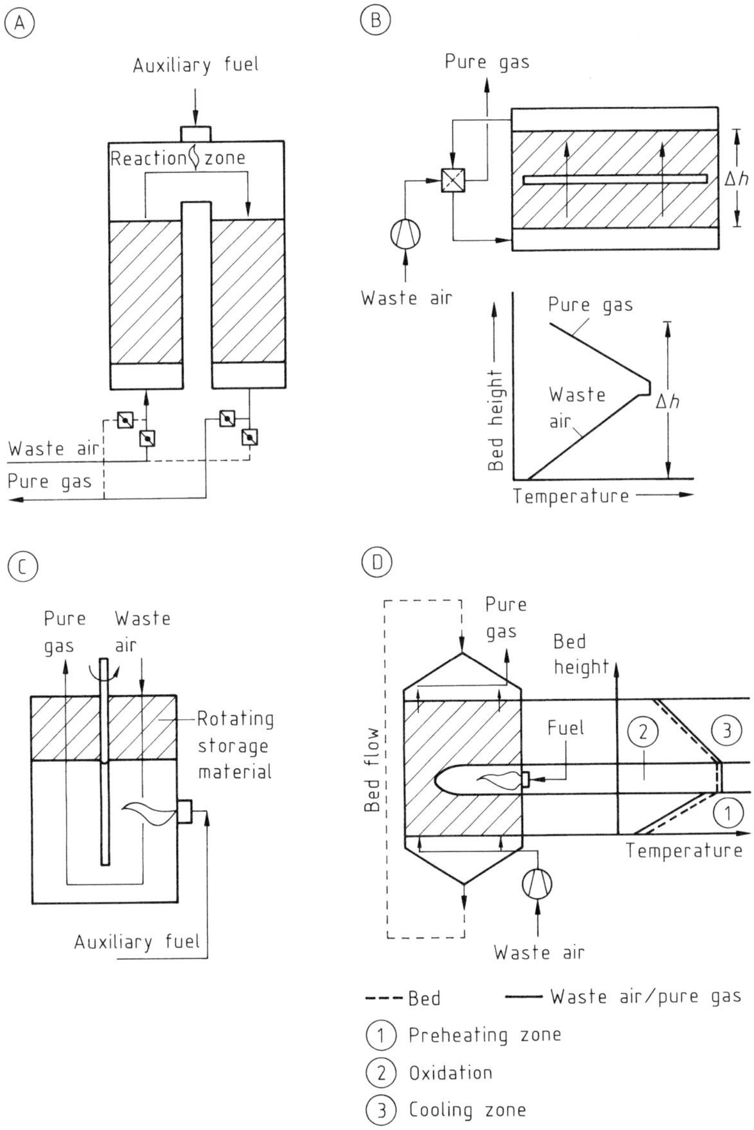

Figure 7.15. Schemes of various regenerative combustion systems [7.25]
A) Fixed-bed regenerator; B) Combu-Changer; C) Ljungstrom regenerator; D) Moving-bed regenerator

2) Large weight.
3) Mechanical stress on the moving parts (valves, rotors, seals) with cycle times of 30–200 s [7.25].
4) Fouling of the storage material by waste gas constituents [7.36].

The same applies to catalytic oxidation with heat exchanger designs of this type [7.52].

Figure 7.15 shows various designs of regenerative combustion systems.

The Combu-Changer and moving bed systems carry out the thermal oxidation in the bed of heat-storage material.

In the other systems, the combustion chamber and heat exchanger are separate. The dimensions are based on a given residence time in the combustion chamber (e.g., 0.5 s) [7.36]. Since the reactions also occur in the heat bed, the residence time in the total reaction volume is ca. 1 s [7.32]. Processes that mainly involve thermal oxidation without the action of a flame are known as regenerative thermal afterburning or regenerative thermal oxidation (thermo-reactor, etc.). At high concentration peaks, thermal energy must be withdrawn from the reaction zone by a bypass to avoid buildup of the temperature [7.25], [7.36]. Due to the relatively high combustion space temperature, the large excess of air, and small influence of flame, only small amounts of CO and NO_x are formed.

Typical process parameters and range of application [7.22]:

Temperature range	800–1000 °C
Residence time	ca. 1 s
Conversion	>99.5 %
Heat utilization, waste-gas preheating	90–97 %
Waste-gas volume flow rates (STP)	5000–10 000 m³/h

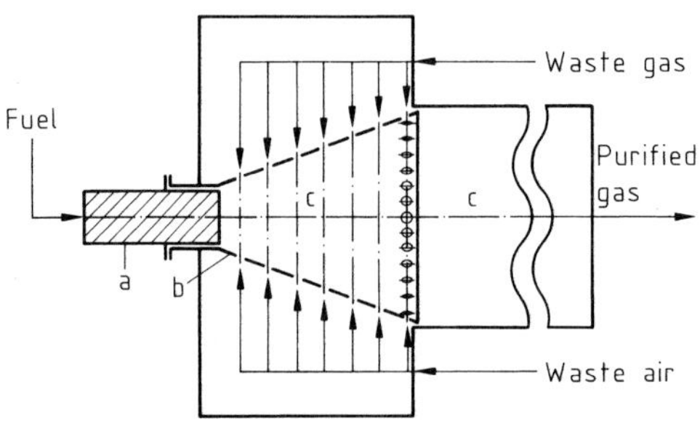

Figure 7.16. Flat burner [7.19]
a) Feeding and monitoring elements; b) Mixing cone; c) Mixing and reaction space

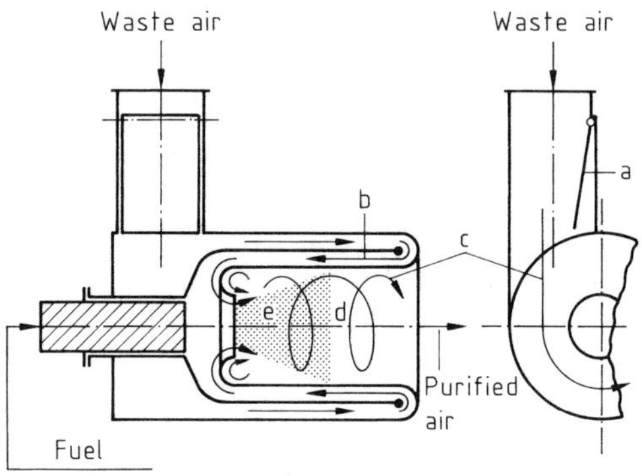

Figure 7.17. Swirl combustion chamber [7.19]
a) Swirl flap; b) Annular gap; c) Swirling flow; d) Combustion space; e) Swirling flame

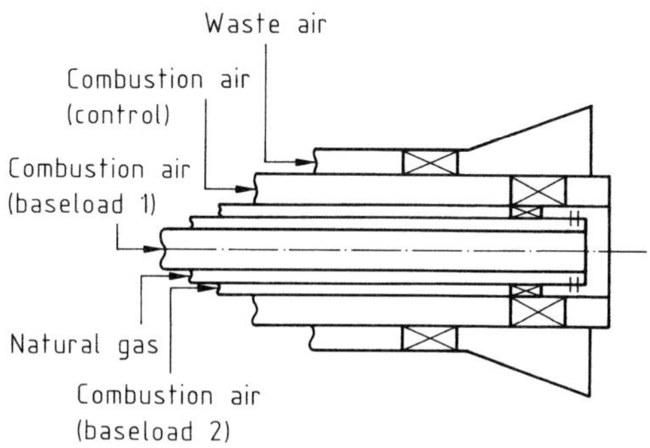

Figure 7.18. Lance burner [7.22]

Pollutant concentrations	≤10% of lower explosion limit
Suitable for continuous waste-gas emission	
Unsuitable for discontinuous waste-gas emission	

Waste-Gas Preheating by Recuperative Heat Exchangers. In these processes the combustion system consists of a burner and an afterburner chamber.

Burner designs are:

1) Flat burner (see Fig. 7.16)
2) Swirl combustion chamber (see Fig. 7.17)
3) Lance burner (see Fig. 7.18)

In the burner the auxiliary fuel (preferably natural gas) is burnt. The necessary oxygen can be provided by oxygen-containing waste gases (diffusion burner, types 1–3) or, if the oxygen content of the waste gas is insufficient, from combustion air fed to the burner together with the fuel (premixing burner, type 3). The added combustion air increases the waste gas mass flow rate, resulting in increased extraneous energy requirement [7.24]. Initial pollutant conversion occurs during the mixing of the waste gases with the hot flue gases of the burner, partly by contact with the flame.

The afterburner chamber provides additonal residence time at high temperatures for the oxidation reactions to proceed to completion, ensuring total combustion. In the swirl combustion chamber system, burner and afterburner chamber are combined in a single unit. The adjustability of the combustion system must ensure matching to different waste-gas mass flow rates.

For waste-gas preheating the following recuperative gas–gas heat exchangers are used (see Figs. 7.14, 7.19):

1) Circular-disk recuperator
   advantage: space-saving construction, low weight; suitable for compact units
   disadvantage: sensitivity to fouling, and thermal stress as a result of hindered differential expansions
2) Duct recuperator
   advantage: robust construction, easily maintained, good compensation of thermal expansions
   disadvantage: high weight

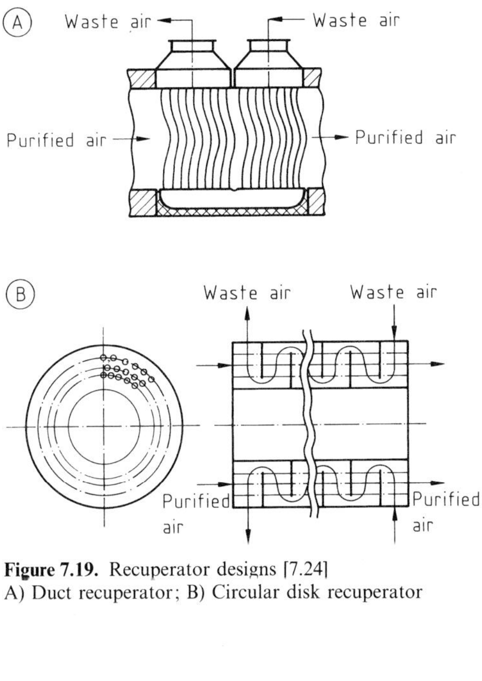

Figure 7.19. Recuperator designs [7.24]
A) Duct recuperator; B) Circular disk recuperator

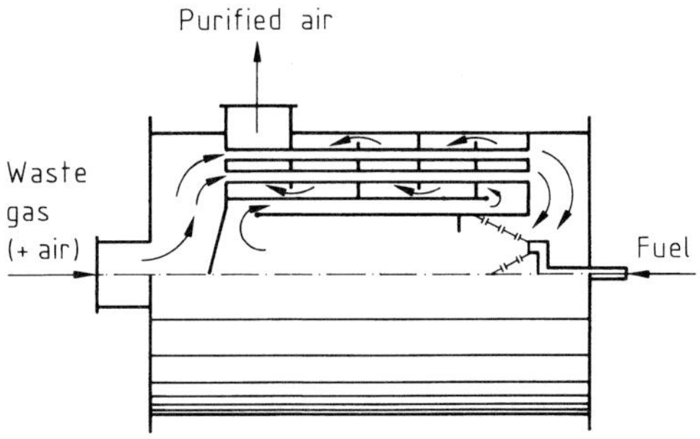

Figure 7.20. Compact thermal afterburning units [7.24]

Units can be classified on the basis of the structural arrangement of these components as compact units (see Fig. 7.20), which are often integrated into production plants [7.28], and units with separate arrangement of the components (see Fig. 7.20).

For units that are started up and shut down frequently, an all-steel construction with ceramic-fiber lining is preferable.

Waste-heat utilization can be carried out by downstream, steam generation, heat-transfer medium heating, or hot-air generation.

Typical process parameters and range of application [7.22]:

Temperature range	800–1200 °C
Residence time	0.3–2 s
Conversion	>99.5%
Heat utilization,	
waste-gas preheating	50–55%
waste-heat utilization	50–80%

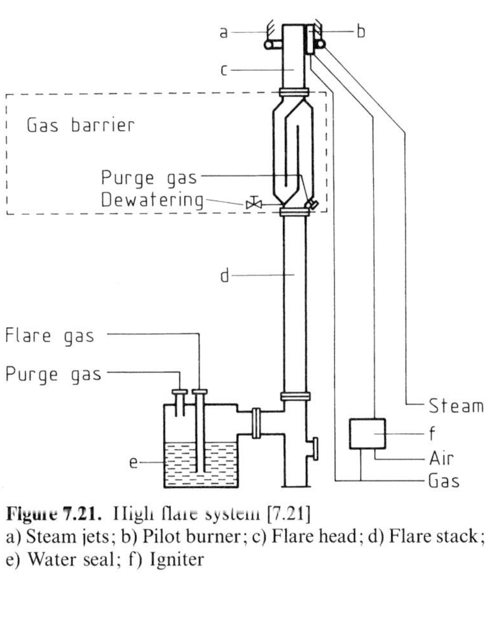

Figure 7.21. High flare system [7.21]
a) Steam jets; b) Pilot burner; c) Flare head; d) Flare stack; e) Water seal; f) Igniter

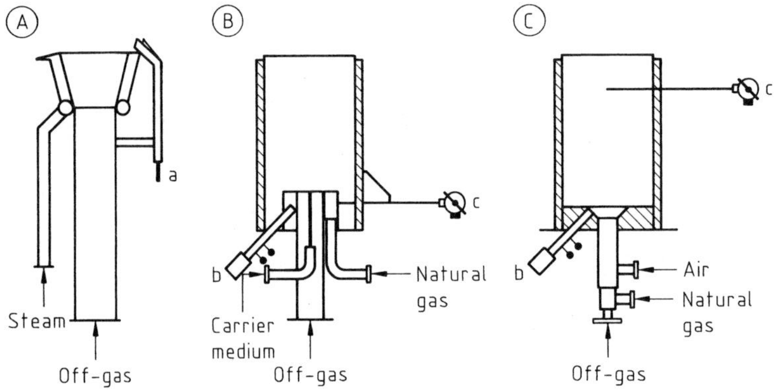

Figure 7.22. Types of flare [7.22]
A) Free-flame flare; B) Screen flare; C) Muffle flare
a) Pilot burner; b) Ignition burner; c) Thermoelement

Pollutant concentration, % of LEL
with safety engineering 0–100%
without safety engineering 0–50%
Volume flow rate (STP) 1000–50 000 m^3/h
Suitable for continuous waste-gas emission
Less suitable for discontinuous waste-gas emission

7.2.2.1.3. Combustion in Flares

Combustion in flares is used mainly in the chemical industry and in plants for oil and natural gas recovery and treatment (e.g. refineries), for waste gases originating from the following operating conditions of the associated production plants [7.21]:

1) Start-up and shut down
2) Disturbances with controllable operating conditions (e.g., safety valves)
3) Disturbances with uncontrollable operating conditions (e.g., power cut or cooling-water failure)

The construction of a high flare is shown in Fig. 7.21. The combustion processes occur in the flare stack tips which can be of three types (see Fig. 7.22) free flame flares, screen flares, and muffle flares.

There is no utilization of heat from flares.

Typical process parameters and range of application [7.22]:

Temperature range	>800 °C
Residence time	1–2 s
Conversion	>99.5%
Heat utilization	not provided

Pollutant concentration, % of LEL
with safety engineering 0–100%
without safety engineering 0–50%
Volume flow rate ≤1000 m^3/h

7.2.2.1.4. Safety Engineering in Thermal Waste-Gas Purification

The combustion chamber of waste-gas thermal purification plants constitutes a permanent ignition source. If the waste-gas composition can be explosive permanently (Zone 0), occasionally (Zone 1), or rarely (Zone 2), then three, two, or one independent protective measures, respectively, must be provided for introduction of the waste gas [7.23], [7.37]: (1) low-flashback inlet to burner, (2) explosion safeguard, (3) immersion safeguard.

7.2.2.2. Reduction Processes

7.2.2.2.1. Staged Combustion

Waste gases that contain organic nitrogen compounds or NO_x can be purified by combustion with staged oxygen supply. The nitrogen bound in the pollutants is emitted in molecular form (N_2). In the first step an atmosphere with reducing compounds (e.g., CO, H_2, NH_3) is generated by substoichiometric combustion of a fuel (oxygen deficiency). NO_x formation therefore stops, or that which is already present is reduced.

This effect can be intensified by cooling (steam addition) in the first step [7.22]. The unreacted combustibles emerging from this step are afterburnt in the second step with an excess of air ($\lambda > 2$).

For establishing substoichiometric conditions in oxygen-containing waste gases, these processes sometimes require considerable amounts of fuel, with which an energy release is associated. This is a stoichiometric requirement of the process and cannot be decreased by preheating the waste gas [7.22].

7.2.2.2.2. Thermal Reduction of Nitrogen Oxides

The reactions:

$$4\,NH_3 + 4\,NO + O_2 \rightarrow 6\,H_2O + 4\,N_2$$

$$CO(NH_2)_2 + 2\,NO + \tfrac{1}{2}\,O_2 \rightarrow 2\,N_2 + CO_2 + 2\,H_2O$$

occur as homogeneous gas-phase reductions in a temperature range 900–1000 °C. This temperature range can be established in furnace and waste-gas purification plants. Ammonia [7.22] or urea solution [7.38] is injected into this temperature zone by lances. By injecting additional chemicals, the temperature range of the urea process has been extended to 500–1200 °C, allowing NO_x conversions of >80 % to be attained [7.38].

7.2.3. Catalytic Waste-Gas Purification Processes

7.2.3.1. Catalytic Reduction of Nitrogen Oxides in Flue Gases and Process Off-Gases

Catalytic reduction is used for the removal of nitrogen oxides from flue gases and process off-gases (e.g., furnaces, nitric acid plant, glass melting, refuse incineration plants). The process is known as $DeNO_x$ or SCR (selective catalytic reduction).

In the SCR process, ammonia is added as a selective reducing agent (i.e., one that is effective even in presence of oxygen), and the nitrogen oxides are converted to molecular nitrogen:

$$4\,NH_3 + 4\,NO + O_2 \rightarrow 6\,H_2O + 4\,N_2$$

Process description: SCR for power station flue gases

Two process variants can be distinguished. In the *crude-gas circuit* (see Fig. 7.23).

The flue gases are taken off directly from the steam generator (after the boiler feedwater preheater, before the combustion air preheater) and, after addition of the reducing agent, passed over the catalyst.

The flue gas, now largely free of nitrogen oxides is introduced into the combustion air preheater to preheat the combustion air (heat

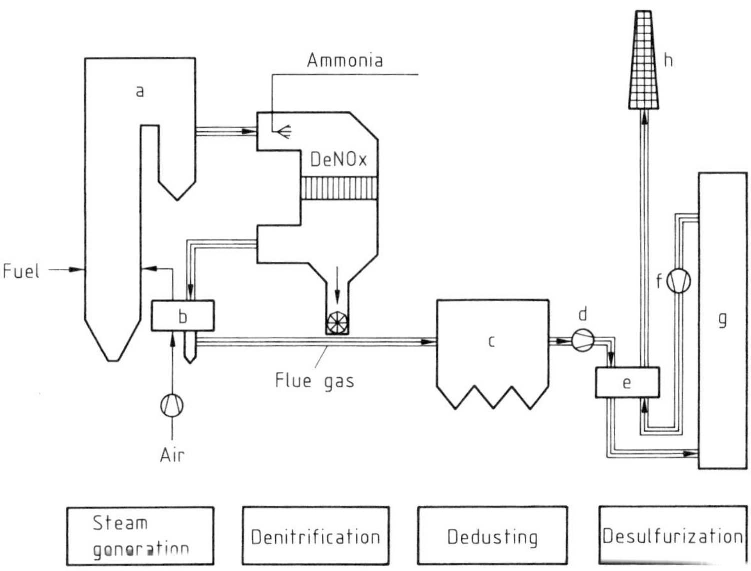

Figure 7.23. $DeNO_x$: crude-gas circuit [7.27]
a) Steam generator; b) Air preheater; c) Filter; d) Induced-draught blower; e) Gas heater; f) Blower; g) Washer; h) Stack

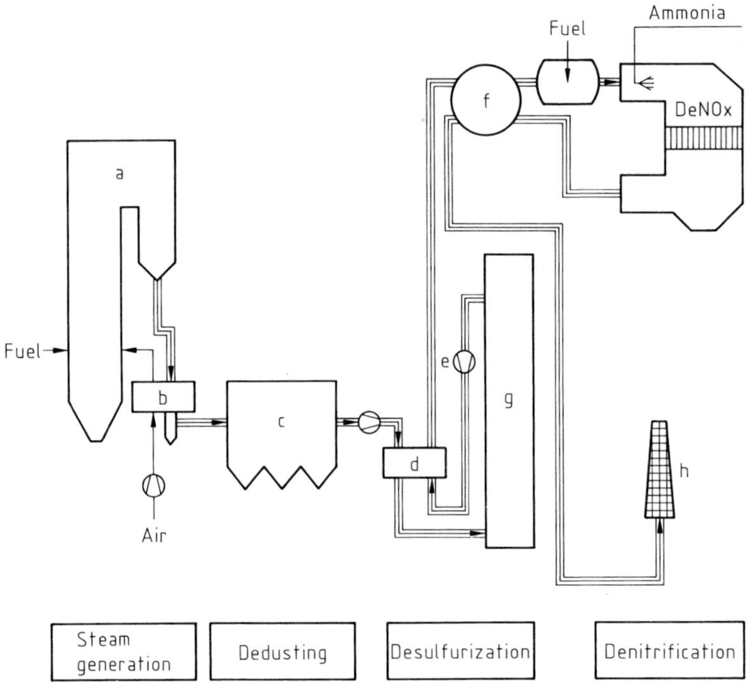

Figure 7.24. $DeNO_x$: pure-gas circuit [7.27]
a) Steam generator; b) Air preheater; c) Filter; d) Gas heater; e) Blower; f) Gas heater; g) Washer; h) Stack

recovery). Other off-gas purification stages (dedusting, desulfurization) follow.

Crude-gas preheating or addition of fuel prior to the catalyst is not necessary since the flue gas is removed from a point in the steam generator where it is sufficiently hot (350–450 °C).

The catalysts are either bulk honeycombs in the fixed-bed process (Ti, V, Fe, W, Ni, Co, Cu, Cr, U, Mo, Sn as oxides, Ag, Be, Mg, Zn, Al as metals); these can also be used for removing dioxins from flue gases in municipal waste incineration plants [7.39], or activated coke in the moving bed process [7.40].

Catalyst requirements are selectivity ($SO_2 \rightarrow SO_3$ conversion must be avoided), activity (low NH_3 consumption), and high abrasion resistance owing to the dust content of the flue gases.

In the *clean-gas circuit* (see Fig. 7.24), dust and sulfur oxides are removed by electric precipitators or flue-gas scrubbers before the crude gas/reducing agent mixture reaches the catalyst. The crude gas must be preheated to the required reaction temperature of 300–400 °C.

Honeycomb catalysts are used, similar to those of the crude-gas process, but with higher activity. Prior separation of interfering components means that demands on the selectivity are lower and a longer effective life can be expected.

Typical design parameters

Space velocity	1000–10 000 h^{-1}
Temperature	250–450 °C
NH_3 addition ratio of the stoichiometric ratio	≤100 %

7.2.3.2. Engine Exhaust Emission Control

Exhaust emissions from internal combustion engines contain pollutants of the following three classes: hydrocarbons (C_mH_n), carbon monoxide (CO), and nitrogen oxides (NO, NO_2).

By using special catalysts and a suitable air-to-fuel ratio in spark-ignited internal combustion engines the three classes of pollutants can be removed together in a single catalytic converter.

The main reactions, occurring in parallel, are [7.41]:

1) Oxidation reactions

$$C_mH_n + (m + n/4)\,O_2 \rightarrow m\,CO_2 + n/2\,H_2O$$

$$CO + 1/2\,O_2 \rightarrow CO_2$$

$$H_2 + 1/2\,O_2 \rightarrow H_2O$$

2) NO-reducing reactions (nonselective catalytic reduction)

$$CO + NO \rightarrow 1/2\,N_2 + CO_2$$

$$C_mH_n + 2(m + n/4)\,NO \rightarrow (m + n/4)\,N_2 + n/2\,H_2O + m\,CO_2$$

$$H_2 + NO \rightarrow 1/2\,N_2 + H_2O$$

The catalyst is integrated into the exhaust system of the engine.

Owing to the predetermined exhaust gas temperature, preheating or external energy supply is unnecessary. The required concentrations of oxidizing agents (O_2, NO) and reducing agents (C_nH_m, CO, H_2) are established by feedback to the air-to-fuel ratio in the engine [7.42] using an electronic lambda probe. In the case of block power stations driven by gas engines, the thermal energy of the purified gases is recovered. Noble-metal-coated honeycomb catalysts are used. The noble metals are mainly platinum, palladium, and rhodium. Coating and support are optimized for high thermal stability.

Typical design parameters

Start-up temperature	ca. 200 °C
Space velocity	up to ca. $10^5\ h^{-1}$
Pressure drop	<1 kPa

Exhaust emissions from diesel engines can lead to problems in industrial safety and environmental protection owing to their content of hydrocarbons and derivatives e.g., aldehydes) (HC), carbon monoxide (CO), nitrogen oxides (NO, NO_2), and particles (soot, polycyclic aromatics).

Oxidizable constituents (hydrocarbons, CO) can be removed from the exhaust gas by oxidation catalysts.

High catalyst selectivity is required since the oxidation of SO_2 to SO_3 and of NO to NO_2 must be avoided [7.43]. For retaining the particles various mechanical and electrostatic systems are known. The honeycomb filter has a catalytic coating and can be regenerated by burning off at ca. 400 °C.

Owing to the high oxygen content of diesel exhaust, the three-way catalysts cannot be used to decrease the NO_x content. The NO_x can be decreased by primary measures (e.g., exhaust gas recycling and filtration) or selective secondary measures (SCR: see Section 7.2.3.1).

7.2.3.3. Catalytic Purification of Claus Plant Off-Gases

Sulfreen Process [7.44], [7.45]. Typical Claus plant off-gases contain 1% H_2S and 0.5% SO_2.

In the Sulfreen process (Fig. 7.25) the reactions of the Claus process occur at fairly low temperatures (120–140 °C). Beds of modified alumina catalyst are used. The elemental sulfur formed by reaction of H_2S with SO_2 is adsorbed in the pores of the aluminum oxide, reversibly poisoning the catalyst. Cyclic purging with hot inert gas is used to regenerate the catalyst and recover sulfur.

Catalytic Afterburning [7.46]. Sulfur compounds (H_2S, COS, CS_2) remaining in the waste-

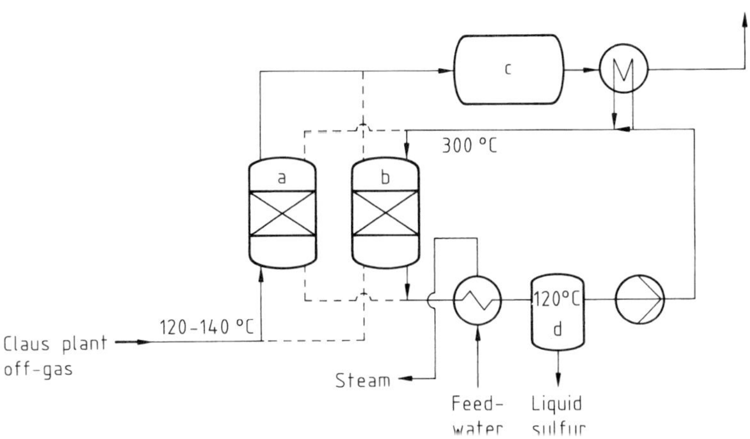

Figure 7.25. Sulfreen process [7.27]
a) Adsorption, reaction; b) Desorption; c) Afterburning; d) Condensation

Table 7.2. Combustion reactions and heating values of some air pollutants [7.39]

Compound	Combustion reaction	Heating value, kJ/kg	ΔT_{ad}, K, in air per 1 g/m^3
Toluene	$C_7H_8 + 9\,O_2 \rightarrow 7\,CO_2 + 4\,H_2O$	40935	30.6
Methanol	$CH_3OH + 1.5\,O_2 \rightarrow CO_2 + 2\,H_2O$	21119	15.8
Acetone	$C_3H_6O + 4\,O_2 \rightarrow 3\,CO_2 + 3\,H_2O$	29065	21.9
Ethyl acetate	$C_4H_8O_2 + 5\,O_2 \rightarrow 4\,CO_2 + 4\,H_2O$	23530	18
Carbon monoxide	$CO + 0.5\,O_2 \rightarrow CO_2$	10423	7.5
Ammonia	$NH_3 + 0.75\,O_2 \rightarrow N + 1.5\,H_2O$	18424	13.9
Hydrogen sulfide	$H_2S + 1.5\,O_2 \rightarrow SO_2 + H_2O$	15355	11

gas stream must be oxidized to SO_2 before the gas is discharged to the atmosphere. At high waste-gas volume flow rates the energy consumption is distinctly reduced, compared with thermal combustion methods, by using alumina-containing catalysts.

7.2.3.4. Catalytic Oxidation in Industrial Off-Gas Streams

Total Oxidation. Industrial and commercial exhaust air streams that contain oxidizable impurities can be purified by catalytic total oxidation (catalytic afterburning) [7.43], [7.47]–[7.50]. Table 7.2 shows some examples of the reactions. The use of catalytic afterburning is restricted by the presence of catalyst poisons and by the need to prevent catalyst overheating with highly variable pollutant loadings.

Noble metal and metal oxide catalysts can be used for total oxidation [7.49].

Example 1: catalytic afterburning unit with recuperative waste-gas preheating [7.51].

Styrene-containing exhaust air is emitted from a plant for producing glass fiber reinforced plastic moldings. After reducing the exhaust-air volume flow rate by enclosing the production machines, the catalytic afterburning unit was designed for a volumetric flow rate (STP) of 15000 m^3/h with a styrene loading (STP) of ca. 5 g/m^3. The heat exchanger allows the unit to operate autothermally above a loading (STP) of ca. 2.5 g/m^3. At higher loadings, heat recovery for heat-transfer fluid heating is possible.

Operating data:

Temperature	
Crude gas	30 °C
Catalyst inlet	320 °C
Catalyst outlet	470 °C
Stack	100 °C
Heat recovery (heat-transfer oil heater)	465 kW
Average natural gas consumption (for heating at low pollutant concentration)	6 m^3/h
Electrical load connected	50 kW
Catalyst volume	800 L
Catalyst type	honeycomb EnviCat KCO WK 4100 with 1 g Pt/L

The concentrations in the purified gas are less than 10 mg/m^3, and odor nuisance in the surroundings of the plant is hence avoided [7.27].

Example 2: unit with regenerative waste-gas preheating

Exhaust air streams from spray painting of metal or plastic components are characterized by:

1) The use of solvent mixtures (aromatics, acetates, glycols, alcohols, paraffins)
2) High volume flow rates
3) Low concentrations
4) Particulate loading (pigments)
5) High relative humidity

Total oxidation processes, optionally with adsorptive concentration, are used for exhaust air disposal. The design of catalytic units without concentration takes the high dilution into account as follows:

1) Low reaction temperature by using metal oxide catalysts
2) Low heat losses from the purified gas by using regenerative heat exchangers [7.52].

This is illustrated by a plant that disposes of the exhaust air from the robot spray painting of decorative electrical switch covers by using a

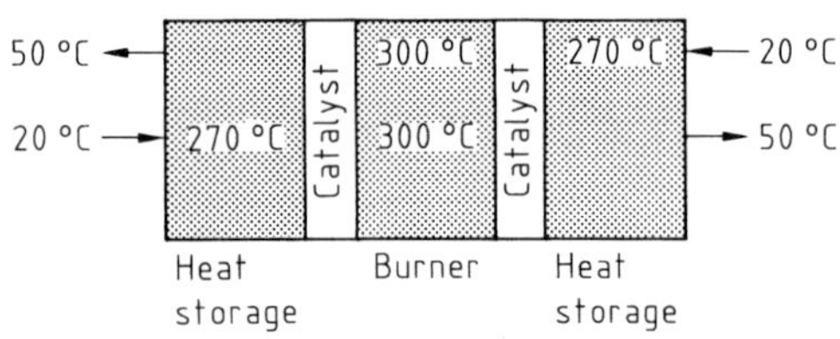

Figure 7.26. Process example of waste-gas catalytic purification with waste-gas regenerative preheating [7.26]

metal oxide catalyst EnviCat KCO SG 1244 (4 m^3, 2400 kg).

The overspray is retained by water spraying and cloth filters. The ceramic heat-storage bed and the catalyst bed are divided into four layers, through which the gas passes with alternating direction (reversing time: ca. 3 min). The heat exchanger designs described on p. 542 can usually be used.

The basic flow diagram and process parameters are shown in Figure 7.26. The heat exchanger and catalyst beds are arranged above each other. The flow direction can be reversed by a lateral shutter system.

The pollutant concentrations in the purified gas, measured as half-hourly averages, are 20–30 mg/m^3, including the concentration peaks that occur on reversal of the flow direction owing to the flushing-out of the intergranular volume.

Combination of Thermal and Catalytic Afterburning [7.27]. To the operator of thermal afterburning units it often seems desirable to reduce the combustion chamber temperatures, to reduce energy consumption (natural gas) and the fatigue or wear of components such as burners or exhaust air heat exchangers.

However, this also leads to formation of intermediate products of hydrocarbon oxidation. Relatively high combustion chamber temperatures (> 750 °C) are necessary to ensure that the aldehydes and carbon monoxide formed as intermediate products in the combustion chamber are converted to carbon dioxide within the predetermined residence time.

If conditions of flow, temperature, or residence time are unfavorable (e.g., during capacity changes), the emission of secondary pollutants (e.g., CO) can be detected in the flue gas of thermal afterburning units.

Increasing the combustion chamber temperature is a suitable countermeasure, but this often runs up against equipment limitations. Moreover, it must be investigated how far the temperature can be raised without significantly increasing the NO_x emissions.

The use of waste-gas purification catalysts enables secondary pollutant emission to be avoided. A lower combustion chamber temperature that ensures oxidation of the primary pollutants (solvents) with reliable observance of the emission limits can then be used.

The object of combining homogeneous and heterogeneous pollutant combustion (thermal and catalytic afterburning) is the conversion of hydrocarbons in the gas phase, and of carbon monoxide and unburnt hydrocarbons on the catalyst.

By dividing the reaction stream into two reaction zones as well as by matching the operating temperature to the thermal ignition temperature of the hydrocarbons, the catalyst can be designed for a higher space velocity than in the case of a single-stage catalytic combustion unit. As a result of the lower concentrations at the entrance to the catalyst stage, smaller temperature differences between inflow and outflow are to be expected.

Owing to the relatively high inlet temperatures for waste-gas purification catalysts, a high thermal stability is demanded of the catalyst. In addition, particularly when refitting old plants, only a relatively small pressure drop is available for flow through the catalyst stage. Therefore, honeycomb catalysts with a large surface area for incident flow are preferred.

For example, in an adhesive tape processing plant a thermal afterburning unit has been retrofitted with a catalyst stage, which allows the exhaust air flow of 30000 m^3/h (STP) and contaminated with ca. 10 g/m^3 solvent, to be purified at a combustion chamber temperature of ca. 500 °C [7.27]. The installed catalyst volume is 1 m^3. For hydrocarbons and CO the concentrations measured in the purified gas at the stack are distinctly below the emission limits.

Droplets of high-boiling hydrocarbon compounds (aerosols), which are present in the waste gas and on contact with the catalyst surface could lead to coking or the masking of active centers, are vaporized in the combustion chamber. These compounds therefore burn in the vapor state without residue.

Selective Oxidation of Ammonia [7.26]. In the catalytic afterburning of ammonia, in contrast to the total oxidation of organic compounds, the question of product formation also plays a part. In the oxidation of carbonaceous substances, carbon dioxide is the only stable end product. Partially oxidized intermediate products are easily converted with heat evolution into CO_2 and can only be obtained under special catalytic conditions (low temperature, high throughput of material, corresponding to short residence time in the catalytic reactor). In the desired oxidation of the ammonia to nitrogen, the gases N_2O, NO, and NO_2 are formed as byproducts, depending on the catalyst and reaction conditions:

$$2\,NH_3 + 2\,O_2 \rightarrow N_2O + 3\,H_2O$$

$$4\,NH_3 + 5\,O_2 \rightarrow 4\,NO + 6\,H_2O$$

$$4\,NH_3 + 7\,O_2 \rightarrow 4\,NO_2 + 6\,H_2O.$$

The formation of the more highly oxidized byproducts involves a greater oxygen consumption. In the oxidation of air – ammonia gas mixtures with ammonia contents that lead to gas heating in the catalytic reactor, the formation of byproducts cannot be absolutely excluded owing to the large excess of oxygen.

The use of noble metal catalysts offers several advantages for the catalytic afterburning of ammonia. Noteworthy is the low operating temperature required, which permits an energy-saving operation and reduced thermal stress of the plant components. The somewhat lower selectivity for N_2O can be improved by optimizing the operating temperature or eliminated by process engineering (reduction of the oxygen concentration). The noble metal catalysts can also bring about the simultaneous afterburning of organic compounds, if present, without forming organo-nitrogen substances (including hydrogen cyanide). A process example is the application for ammonia-containing drier exhaust air.

During the drying of webs coated with mineral powder, vapors contaminated with ammonia are released from the binder.

A design is carried out, based on the reaction behavior of a bed of the platinum-containing catalyst EnviCat KCO SG 1137, for the following exhaust gas conditions:

Volume flow rate (STP)	6500 – 22 000 m^3/h
NH_3 mass flow rate	ca. 13 kg/h
Water vapor concentration (STP)	100 – 400 g/m^3
Temperature	> 190 °C.

The process parameters fall in the temperature range of high NH_3 fractional conversion and of low formation of NO, NO_2 and N_2O.

7.3. Absorption

7.3.1. Introduction

Absorption is a process for removing some components of a gas stream by contact with a liquid phase (solvent). The desired selectivity with respect to the component(s) to be removed (solute) depends either on purely physical interactions (physical absorption) or on strong chemical bonding (chemical scrubbing or stripping). So-called physicochemical scrubbing occupies an intermediate position. Here, dissolution takes place in the absorbing liquid due to physical interaction accompanied by a reversible chemical reaction.

There are three major industrial applications of absorption processes. In one large group applications, undesired components such as H_2S or organic sulfur compounds, and, in some cases, water vapor, are removed from natural gas, refinery gases, synthesis gases, or process gases. Another major application is the removal of gaseous pollutants such as hydrogen halides or volatile organic solvents from waste air. The third group includes the removal of SO_2 and NO_x from the flue gases of power stations or domestic waste incinerators.

This section treats the industrial use of equipment for absorption processes. The discussion is limited to processes in which only absorption takes place. Apparatus principally intended for wet scrubbing of solids and only incidentally performing an additional absorption function, such as venturi or rotary scrubbers, is described in Section 7.1.

7.3.2. Basic Principles of Absorption Processes

The most important element of an absorption plant is the absorber (Fig. 7.27), in which the gas stream being treated and the solvent are brought into contact. The purified gas and the

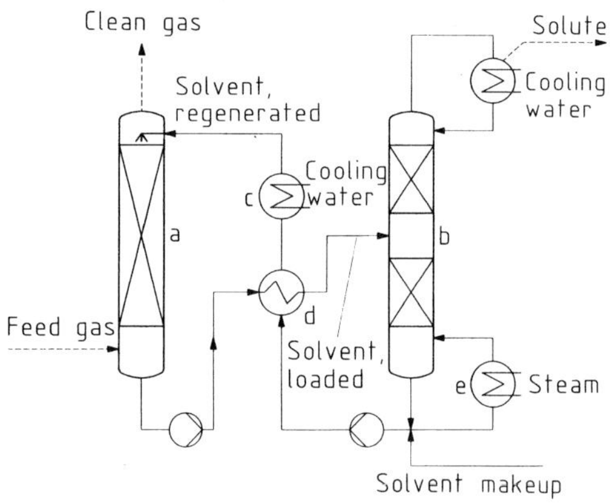

Figure 7.27. Flowsheet of an absorption process
a) Absorber; b) Desorber; c) Cooler; d) Recuperator; e) Boiler

loaded solvent then leave the absorber. In some cases (e.g., absorption of HCl and NO_x with water; absorption of NH_3 with sulfuric acid) the loaded solvent is itself a usable product, so that a solvent regeneration stage is not required.

However, in most cases it is necessary to regenerate the solvent for closed-loop recycling to the absorption stage. This takes place in the desorption stage, which can be a distillation step, as shown in Figure 7.27. Alternatively, steam or inert-gas stripping is sometimes used. For thermodynamic reasons, desorption should be operated at high temperature and low pressure. The efficiency of the recuperator (d) is thus important for the economics of the process. In physicochemical scrubbing, a desorption process can be considered if the chemical bonding can be broken by pressure reduction or temperature increase, but if the chemical bond is very strong the solvent cannot be regenerated. In most cases where this happens, the charged absorbing liquid is itself a usable product.

The absorption and desorption stages can be designed according to established methods (see → 4. Distillation and Rectification, **B 3**) based on knowledge of the equilibrium between the partial pressure of the gaseous component in the gas phase and its concentration in the liquid phase. Details of these solution equilibria can be found in the comprehensive reference books [7.52], [7.54]. For newer processes, such as the absorption of organic substances in high-boiling solvents, special literature should be consulted [7.55].

The operating costs of a combined absorption/desorption plant mainly depend on the recirculating flow of solvent. As heat recovery in the recuperator is incomplete, the recycled solvent must be further cooled. To maintain thermal balance, it is therefore necessary to supply energy to the desorption stage. This occurs at the bottom of the desorption column if distillative treatment is used, or by a superheater before the inlet to the desorption column in the case of stripping. Note that desorption (with the exception of inert gas stripping) requires a certain minimum supply of steam, depending on the substance being desorbed, and this requires additional energy input to the process.

The heat of absorption is liberated in the absorption process; it is usually slightly higher than the enthalpy of condensation of the absorbed gas. However, in the case of strongly nonideal equilibria, the heat of absorption may be very high (e.g., 2100 kJ/kg for HCl). This fact, together with the possibility of heat and mass transfer between the gas and liquid streams due to different inlet conditions, must be taken account of in the design, especially of the absorption stage.

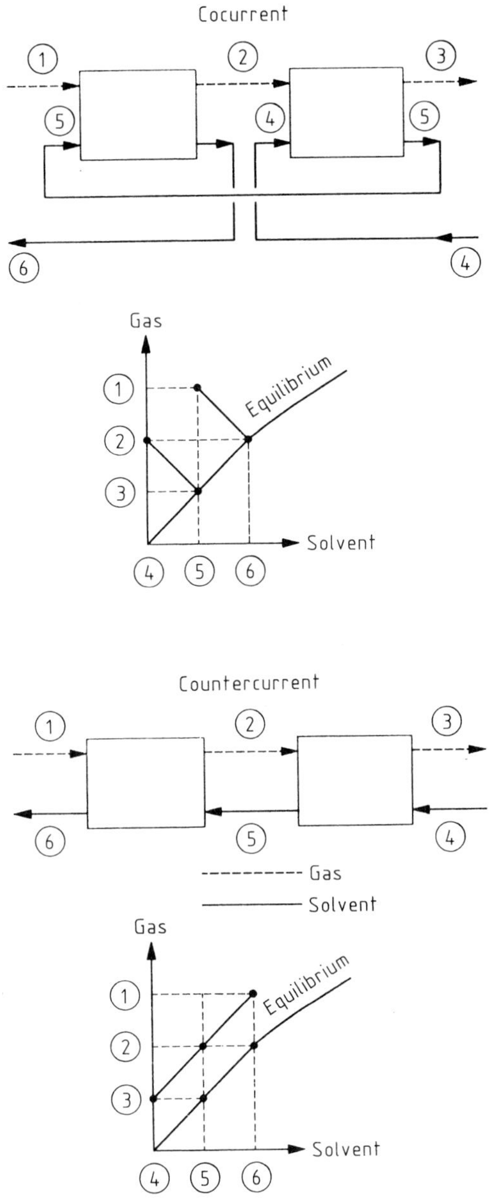

Figure 7.28. Internal flow in multistage co- and countercurrent processes
Representation of concentrations in McCabe–Thiele diagrams

7.3.3. Operating Principles of Absorption Apparatus

The thermodynamic design of an absorption apparatus usually leads to a requirement for a certain number of theoretical plates needed to produce the required quality of the purified gas. This can only be achieved by use of the countercurrent principle, whereby the two process streams pass through the apparatus in opposite directions. However, in this context it is immaterial whether the flow in the individual stages is countercurrent or cocurrent (see Fig. 7.28). The concentration change represented in the distribution diagram shows that the same degree of separation can be produced in both cases with the same number of theoretical stages. The essential difference is that in cocurrent flow, one stream (5) in a unit flows in the opposite direction to the principal flow direction. Therefore individual stages always represent separate units of apparatus. With countercurrent flow, the required number of theoretical stages can be achieved in a single column. In special cases, a semicontinuous mode of operation is used in which an absorption process takes place in each stage, and after a certain period of time the solvent is pumped in countercurrent flow to the next stage, with fresh solvent being added to the first stage.

The removal of large amounts of heat of absorption requires modifications of the apparatus (Fig. 7.29). In design (A) the heat removal takes place in the apparatus. Good results have been obtained using film absorbers or tray columns operating on this principle (see Section 7.3.4). In the alternative version (B) the liquid is pumped through a recirculating system containing a cooler which removes the heat of absorption. If the heat of absorption is very large and the solution equilibrium is favorable, as in HCl absorption, water can be evaporated in the apparatus. The water vapor leaves the top of the apparatus and is condensed in a downstream condensation stage.

For certain purposes, it can be advantageous or necessary to link together different designs of apparatus or completely different processes (see Fig. 7.30). For high heats of absorption, as in the removal of ammonia from waste air, the stage into which the gas is fed (Fig. 7.30 A) should be operated in cocurrent flow with heat removal, and final purification should be carried out in countercurrent flow apparatus [7.56]. The flow

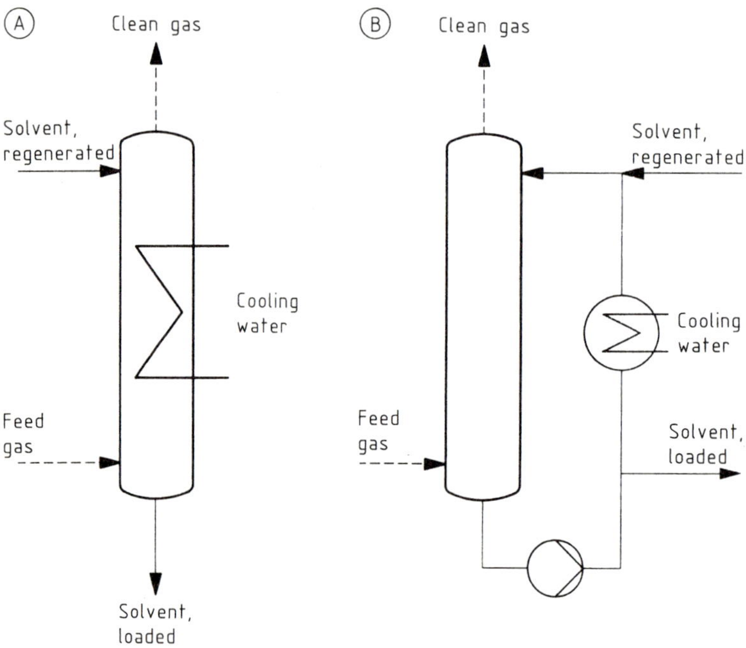

Figure 7.29. Absorption equipment with removal of heat of absorption
A) Removal takes place in the apparatus;
B) Removal through recycling system

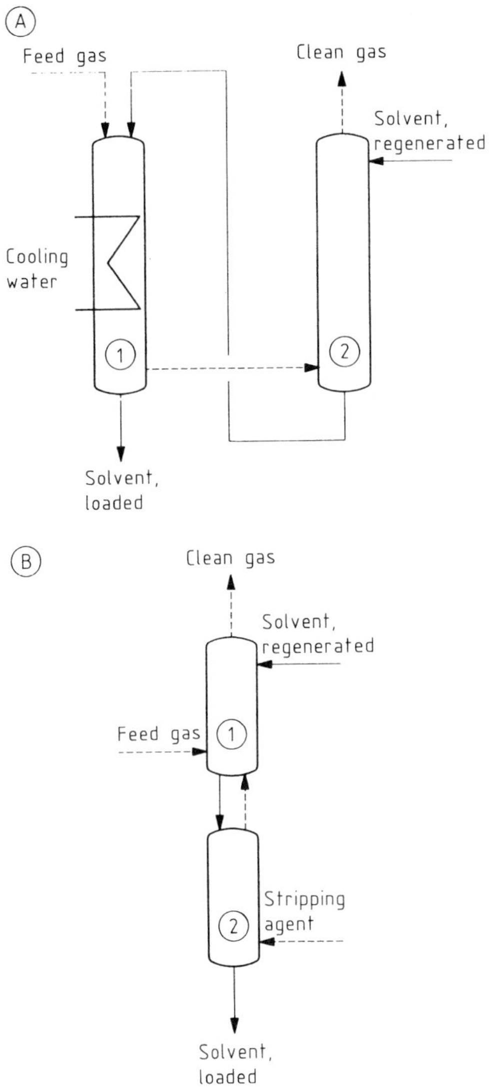

Figure 7.30. Combination of different process stages
A) Without stripping stage
B) Including stripping stage

sheet of Figure 7.30 B is used in, e.g., the absorption of $HCl-SO_2$ mixtures, where SO_2 has fairly low but nevertheless unacceptably high solubility in aqueous HCl. A stripping stage is then included in which SO_2 is removed by a stripping agent to give the required purity [7.57].

7.3.4. Absorption Apparatus

The apparatus described below are characterized by the largest possible interface between the gas and the liquid. In some types of apparatus, one phase is dispersed in the other. In another design, the liquid is spread over a surface to form a very thin film. This method has the advantage that only a small amount of energy is required to produce the interface, and the pressure drop is relatively low. Furthermore, the risk of entrainment of fine droplets is low, in contrast to processes in which one phase is dispersed in the other. In this case, droplet separation or demisting equipment must be included to prevent entrainment of the liquid phase in the gas stream.

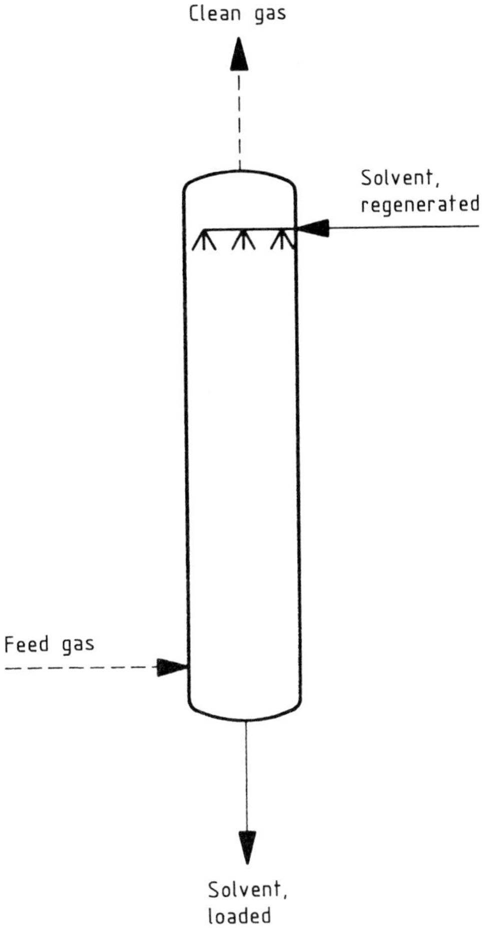

Figure 7.31. Spray absorber

Spray Absorbers. In spray absorbers (Fig. 7.31) the solvent is broken down into droplets by nozzles arranged in one or more planes. The momentum of the liquid is not high enough for hydraulic interactions to take place between the phases. Also, the absence of inter-

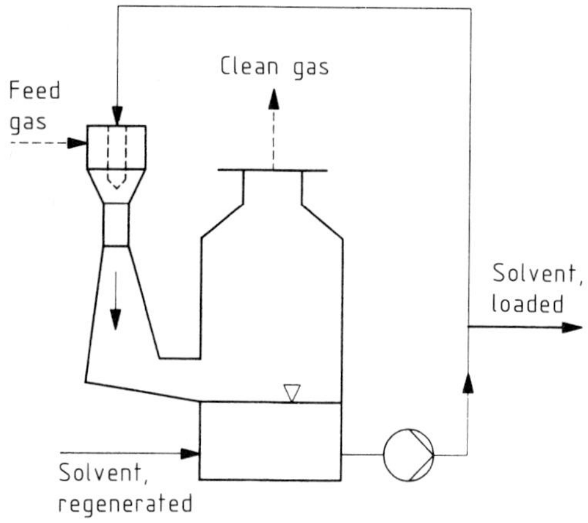

Figure 7.32. Jet scrubber

nals means that the pressure drop across the apparatus is almost zero and soiling is negligible. Spray absorbers are preferred for applications which require approximately one theoretical stage. These are mainly chemical absorption processes.

Jet scrubbers [7.58] represent a further development of spray absorbers (see Fig. 7.32). The gas and liquid pass through the jet tube in cocurrent flow. The absorbing liquid enters the interaction space at high velocity. The gas stream is accelerated by the momentum of the liquid, and the pressure increase produced can in some cases overcome the flow resistance of the apparatus including the associated pipework. The momentum transfer also causes the liquid phase to break up into droplets, so that intensive mass and heat transfer occur. The downstream separation space is also used as receiving vessel for the pump. Generally, a demister is included at the gas outlet.

Jet scrubbers can be used in a multistage arrangement if several theoretical stages are necessary. This also gives the possibility of semicontinuous operation using a timed system for pumping the liquid from stage to stage. As the energy for dispersing the liquid is contained in the liquid itself as it enters the apparatus, the separating power is largely independent of fluctuations in the gas flow rate. If heat must be removed, a cooler can be included in the liquid pumping circuit.

Packed Columns. In packed columns the interface is produced by forming a film on the sur-

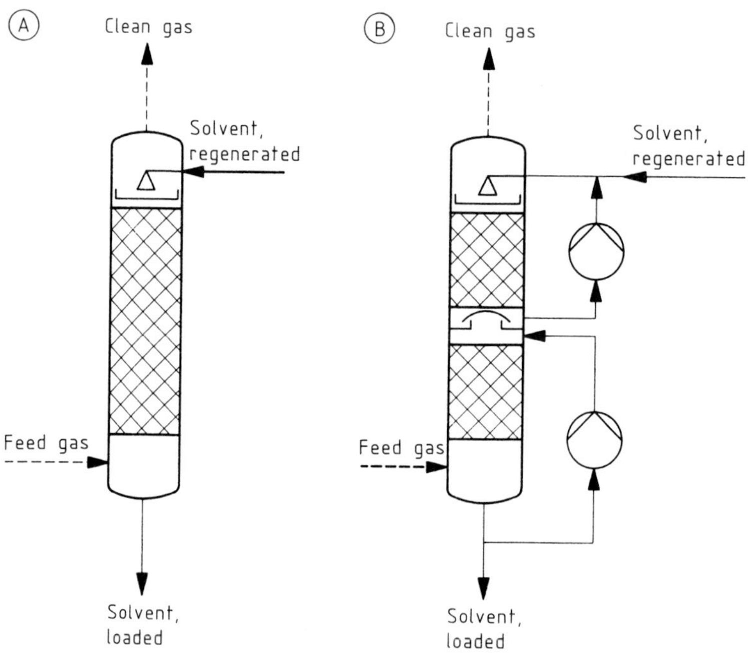

Figure 7.33. Packed column absorbers
A) For continuous countercurrent flow;
B) With packings in series

face of internals (see Fig. 7.33 A). This enables a continuous countercurrent flow of gas and liquid to be produced. Random packings like Berl saddles, which provide a large surface for the liquid to cover, have been widely used. More recent packings [7.595] are provided with holes and interior structures which not only give film formation but to some extent break the liquid up into droplets.

For some time, packings first used in distillation technology have been widely used for absorption [7.60]. These consist of parallel toothed strips of various types of material. The teeth are inclined with respect to the axis of the column, and neighboring strips incline in opposite directions. Regular packings produce a considerably lower pressure drop per theoretical stage compared with random packings. Also, the minimum liquid loads are lower than with random packings. The separation effect only slightly depends of the colum diameter. Both types of design require distribution devices that uniformly distribute the liquid over the cross section.

Because of the low liquid content, packed columns are not very suitable if fluctuations occur in the inlet gas velocities or in the gas concentrations. If it is not desired to use a different apparatus such as jet scrubbers or tray columns, another alternative is to use several packings in series, with recirculation (see Fig. 7.33 B). Fresh solvent is fed to the top of the column, and the product is withdrawn from the base. This design is relatively costly and should only be considered in special cases.

Tray columns used in absorption processes (see Fig. 7.34) are in principle identical to those used in distillation. A fundamental disadvantage is that typically the pressure drop in tray columns is high compared to packed columns. However, the following requirements or operating conditions can make the use of a tray column worthy of consideration:

1) In absorption processes where the loaded solvent is a useful product, high concentrations are often required. If the concentration of the component to be absorbed from the feed gas stream is low, this may mean that for mass balance reasons the necessary minimum loading of packed columns can no longer be maintained. In contrast the liquid holdup in tray columns can be reduced almost to zero.
2) With fluctuating gas flow rates or gas concentrations, a certain volume of liquid must be stored in the apparatus. This can be achieved by the use of bubble caps with chimneys. However, there is competion from other equipment, e.g., jet scrubbers.
3) A feature of the design is that contact between the gas and the liquid is concentrated on the trays. It is therefore possible to remove the heat of absorption liberated here (e.g., during the absorption of NO_x) by using tray columns fitted with cooling coils [7.61]. In this way, an absorption temperature that ensures optimum absorption can be maintained.

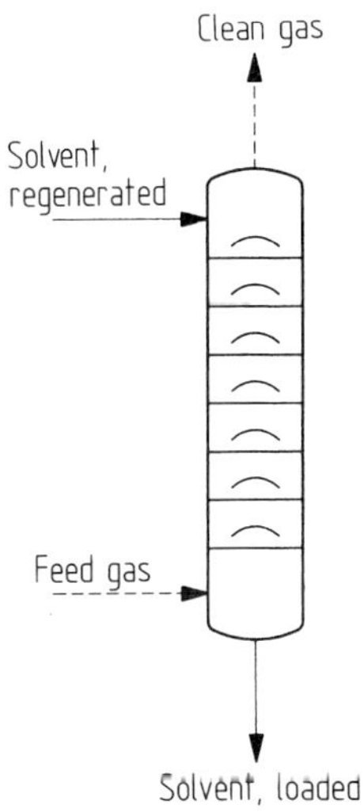

Figure 7.34. Tray column absorbers

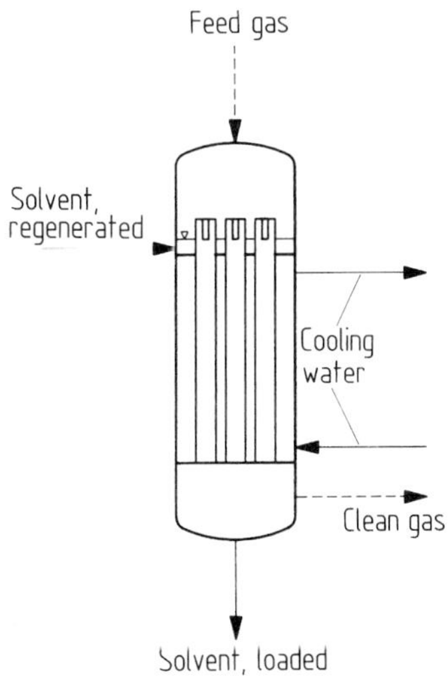

Figure 7.35. Film absorber

Tray columns are now commercially available in all the usual construction materials, including highly corrosion-resistant versions made of graphite [7.62] and glass [7.63].

Film absorbers (Fig. 7.35) are used where the heat of absorption must be removed in the apparatus. The design is illustrated in Figure 7.35. In principle, they are heat exchangers through which the gas and liquid pass in cocurrent flow. Distributors must be included to ensure that each tube is fed with the same amount of liquid. The liquid forms a film on the inner surface of the tube, while the gas flows through the remaining interior space. The heat of absorption is removed via the outer surface of the tubes.

Film absorbers represent one theoretical stage at the most. They are therefore only used where absorption must take place at low temperature to achieve high product concentrations. In all cases, a conventional final purification stage for the gas stream must be included (e.g., a packed column).

7.3.5. Selection of Equipment

The various types of absorption equipment described above have preferred areas of use, depending on their characteristics. These characteristics, subdivided into process and operational parameters, are reviewed below. However, an unequivocal choice of equipment is not always possible.

Process Parameters. One important process parameter is the efficiency of separation required, expressed in terms of the number of theoretical stages. For demanding applications, packed columns or tray columns are used. Jet scrubbers are usually designed as only two or three stages, and spray absorbers in practice represent only one theoretical stage.

A further important consideration affecting the choice of apparatus is the question of heat removal. If it is possible for process reasons to remove heat by evaporation of a liquid, tray or packed columns can be used. If the temperature increase in the column leads to no phase change of the liquid, it is possible to use jet scrubbers or packed columns with exterior recirculation of the liquid through coolers. Film absorbers are also used. Tray columns having cooling devices in the trays are preferentially used in NO_x absorption.

If for reasons of mass balance the liquid flow rate is very low compared to an appreciably higher gas flow rate, the use of tray columns should be considered first. If packed columns, preferably with ordered packing, are considered, uniform distribution of the liquid over the cross section of the column is essential.

Operational Parameters. Intermittent operation with a variable total gas flow rate or variable concentrations of the substances to be absorbed should be considered first. The apparatus should be able to store a certain amount of liquid, and, the efficiency should vary as little as possible with the gas flow rate. Tray columns are highly suitable in this case. Jet scrubbers and packed columns with an external liquid circulation system can also be used, although practical problems with large number of stages must be borne in mind.

A further important operational parameter is the pressure drop across the apparatus. Jet scrubbers are a special case because they can create a pressure increase in the gas stream. The pressure drop across spray absorbers and packed columns is usually low, whereas tray columns always give rise to a high pressure drop.

7.4. Adsorption

Research into the use of porous solids for removing vapors from gas mixtures began with the constructon of the first industrial plants for the production of activated prowdered carbon in 1900 to meet the demands of the sugar refining industry. The physicist J. Dewar, who worked with activated carbon, silica gel, and aluminum oxide gel, was awarded a patent on the subject in 1905. The first process for the adsorptive removal of organic solvents from waste air and recovery by steam desorption was patented in 1916 for the Friedrich Bayer dyeing company. The use of adsorption for purifying town gas increased the importance of this gas separation technique in the 1920s. Since the 1960s, the development of adsorption catalysis and the introduction of new adsorbents have opened up further applications [7.64]–[7.66].

7.4.1. Fundamentals

In the application area considered here, adsorption means the adherence of molecules of gas or vapor (the adsorptive) to the inner surface of a highly porous solid (the adsorbent). The latter contains atoms or groups of atoms with free valencies (active centers) to which foreign molecules can become attached, and which are known as the adsorpt when in the adsorbed state. The complete boundary surface phase is known as the adsorbate (see Fig. 7.36). The reverse of adsorption, namely desorption, yields the desorpt and the desorption medium, which together are known as the desorbate.

In addition to reversible physical sorption with purely physical bonding between adsorbent and adsorptive (van der Waals forces), there is also the phenomenon of chemisorption in which stronger interaction energies cause the adsorptive to undergo chemical change, which can lead to either reversible or irreversible attachment.

The adsorption capacity of an adsorbent for a given adsorptive at equilibrium at constant temperature depends on the size and structure of the inner surface of the adsorbent and is usually represented as a function of the concentration in the carrier gas, i.e., an adsorption isotherm $x = f(p/p_s)_T$. Figure 7.37 shows the equilibrium loading x as a function of the concentration in the carrier gas, expressed in terms of the ratio of partial pressure p to the saturation pressure p_s at constant temperature T. Industrially attainable breakthrough or additional loadings are considerably lower. Theoretical adsorption isotherms can be derived from various models [7.67], [7.68]:

The Freundlich model

$$x = k \cdot p_i^n$$

applies to low partial pressures of the adsorptive.

The Langmuir model

$$x = \frac{x_{max} \cdot b \cdot p_i}{1 + b \cdot p_i}$$

applies to homogeneous adsorbent surface and monomolecular covering.

The Brunauer–Emmert–Teller (BET) model

$$\frac{p_i}{V(p_s - p_i)} = \frac{1}{V_{MC} \cdot C} + \frac{C-1}{V_{MC} \cdot C} \cdot \frac{p_i}{p_s}$$

applies to homogeneous surface, multilayer adsorption and capillary condensation.

The Dubinin–Raduskevic model

$$\log V = \log V_s - 2.3 \cdot \frac{RT}{\varepsilon_0 \beta} \cdot \left(\log \frac{p_s}{p_i}\right)^n$$

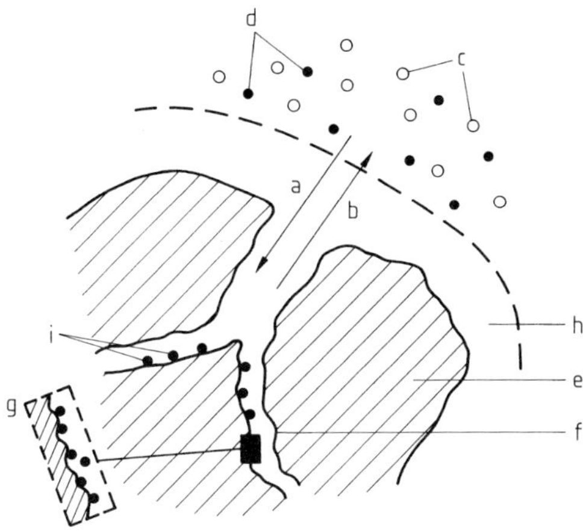

Figure 7.36. Adsorption and desorption processes at the boundary layer
a) Adsorption, exothermic; b) Desorption, endothermic; c) Carrier gas; d) Adsorptive; e) Adsorbent; f) Inner surface (sum of pore walls); g) Adsorbat; h) Boundary surface phase; i) Adsorpt

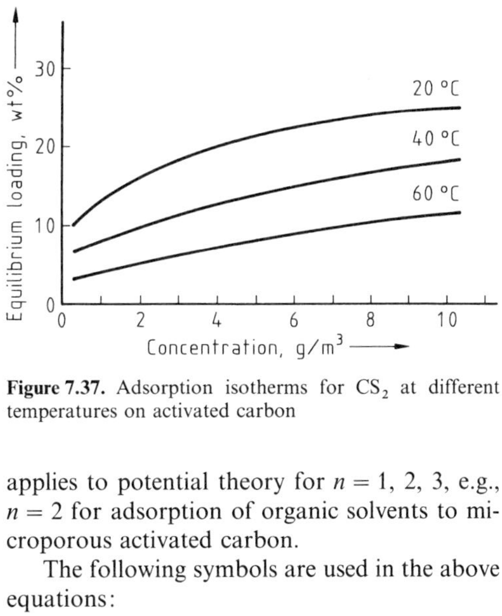

Figure 7.37. Adsorption isotherms for CS_2 at different temperatures on activated carbon

applies to potential theory for $n = 1, 2, 3$, e.g., $n = 2$ for adsorption of organic solvents to microporous activated carbon.

The following symbols are used in the above equations:

- k, n specific constants of the adsorptive,
- x equilibrium loading of the adsorbed substance, g/100 g
- x_{max} saturation value of the isotherm for monomolecular covering, g/100 g
- b adsorptive coefficient,
- C constant
- p_i partial pressure of the adsorptive, Pa
- p_s saturation vapor pressure of the adsorptive, Pa
- V adsorbed volume, ml

V_{MC} adsorbed volume for monomolecular covering, ml
V_s adsorbed volume at saturation pressure, ml
T temperature, K
R gas constant, $J K^{-1} mol^{-1}$
ε_0 characteristic adsorption energy, J/kg
β affinity coefficient

At adsorption temperatures below the critical temperature of the adsorptive, the pores of the adsorbent can become filled with liquid adsorpt. This capillary condensation increases the adsorption capacity of the adsorbent. Assuming that the pores are cylindrical, it can be represented quantitatively by the Kelvin equation, which indicates that the degree of saturation is inversely proportional to the pore radius:

$$\ln \frac{p}{p_s} = -\frac{2 V_M \sigma \cos \theta}{r R T}$$

The following additional symbols are used in the Kelvin equation:

p/p_s relative equilibrium pressure
V_M molar volume, m^3/mol
σ surface tension adsorpt/adsorbent, N/m
θ contact angle adsorpt/adsorbent
r minimum pore radius, nm

As the transition from the fluid to the adsorbed phase is associated with release of energy, adsorption is an exothermic process. The enthalpy of adsorption liberated is a measure of the binding energy of the adsorpt, and this decreases as the quantity adsorbed increases. The energy required for solvent recovery is ca. 1.5 times the evaporation energy, resulting in effective temperature increases of 10–20 °C. The energy liberated during the complete adsorption process is known as the integral enthalpy of adsorption. The differential enthalpy of adsorption represents thermal processes at different stages of the adsorption process.

Adsorption is usually carried out in a fixed bed, and the adsorption zone (mass-transfer zone) migrates through the adsorber in the same direction as the gas flow (Fig. 7.38). Below this zone the adsorbent is fully charged, and above it the adsorbent contains only the residual loading from the previous adsorption cycle. Within the adsorption zone, there is equilibrium between the adsorbed phase (adsorpt) and the adsorptive concentration in the gas phase. In adsorbents with wide pores, these complex processes are controlled less by matter transfer than is the case with adsorbents with narrow pores.

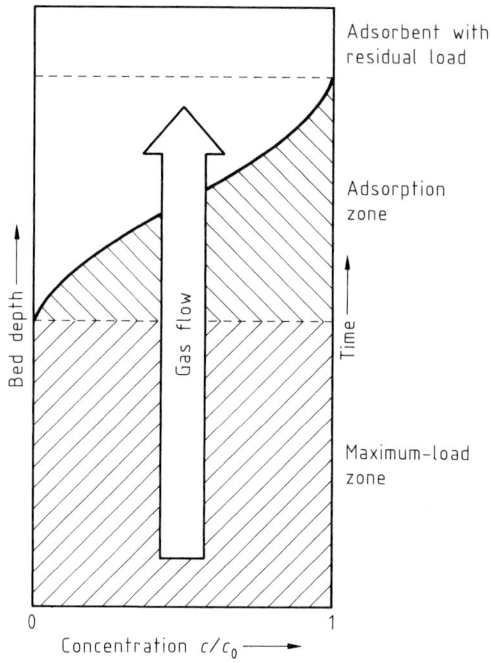

Figure 7.38. Loading profile in a fixed-bed adsorber

In practical applications, the waste gases to be purified often consist of a carrier gas containing mixtures of substances from which one or more components must be selectively removed. Separation can be achieved by utilizing the following effects:

1) Variation in the affinity of the adsorbent for the various gaseous components, this being apparent from the adsorption isotherms (equilibrium effect)
2) Differences in the critical molecular diameters of the adsorptives, enabling separation to be carried out if the effective pore diameter of the adsorbent (molecular sieve) is greater than that of the molecules of the component to be selectively adsorbed and smaller than those of the carrier gas (steric effect)
3) Different rates of diffusion of the adsorptives through the boundary layer and the pore structure to the active centers (kinetic effect)

Industrial adsorbents are regenerated when the breakthrough of the adsorptive occurs. Regeneration involves driving off the adsorpt as a gaseous or liquid phase (desorption), and drying and/or cooling treatment to restore the adsorbent to the optimum condition for reuse. As it is not possible to drive off the adsorpt completely at acceptable cost, a residual loading is left behind, reducing the capacity of the adsorbent. If side reactions occurring during adsorption or desorption change the chemical structure of the adsorpt such that the residual loading increases, it is still often possible to reactivate the adsorbent in a separate plant. The process conditions for this treatment correspond to those of activation, i.e., the adsorpt is removed by oxidation. For activated carbons, this involves a selective gas treatment of the adsorbents with water vapor or CO_2 at 700–900 °C without appreciably changing the carbon structure. Oxidic adsorbents are regenerated with hot air.

7.4.2. Industrial Adsorbents

The adsorbents used in gas technology are not usually finely divided, but are highly porous granules or fibers with a large inner surface on which the foreign molecules become concentrated. Depending on the raw material and method of manufacture, they have different adsorption properties, which can be classified according to the following criteria:

The extent of the inner surface formed by all the pore walls is the main factor determining the adsorption capacity. It may be determined by the method of DIN 66131 (BET method) [7.69], which is based on the adsorption isotherm for nitrogen at its boiling point and the space required by nitrogen molecules for monomolecular occupancy.

The pore structure is important for the diffusion of the adsorptive to the active centers of the inner surface. Unlike molecular sieves, in which the inner cavities are linked by pores of uniform diameter, other adsorbents have a wide spectrum of pore size (Fig. 7.39). The pore radii (assuming a cylindrical shape) determined by different methods depending on the pore diamater (macropores: mercury porosimetry; mesopores and micropores: Kelvin equation) can extend over several orders of magnitude, and are classified according to the IUPAC Standard into

macropores	$d > 50$ nm
mesopores	$50 > d > 2$ nm
micropores	$2 > d > 0.4$ nm
submicropores	$d < 0.4$ nm.

The porosity is the pore volume as a fraction of the adsorbent volume, or per gram of adsorbent.

The adsorption characteristic of an adsorbent is revealed by the isotherms obtained with different adsorptives. It yields important information on the concentration dependence of the adsorption properties and the establishment of equilibrium.

The hydrophobicity or hydrophilicity of an adsorbent often determines whether it can be used in waste-gas purification, because water vapor is usually present at a considerably higher concentration than the substances to be removed.

The selectivity of an adsorbent can be a result of chemical effects (hydrophobicity, hydrophilicity) or steric effects (similarly shaped pore open-

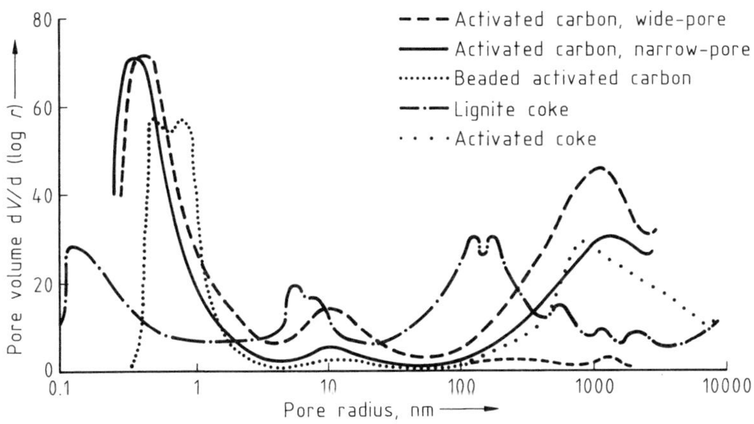

Figure 7.39. Differential pore distribution of carbonaceous adsorbents

ings through which only molecules smaller than a particular diameter are able to penetrate).

Catalytic properties resulting from the chemical constitution of the inner surface may enable certain tasks to be performed, but can also lead to undesired side reactions and irreversible formation of chemical bonds (oxidation, polymerization, hydrolysis).

The industrial adsorbents can be divided into the categories shown in Table 7.3 in accordance with the raw materials and treatment methods used in their production.

The activated carbons (→Carbon, treated in the A series) are hydrophobic adsorbents with a wide range of specific surface areas, pore structures, and pore volumes, over 90 % of which consist of carbon with a graphitic crystal structure. They are produced in the form of shaped, powdered, and fibrous activated carbon from raw materials of vegetable origin by low temperature carbonization (primary carbon), grinding, shaping, and activation, and play a major role in the adsorptive purification of waste gas.

Activated coke is an adsorbent produced from finely hard coal preoxidized with air, shaped using pitch as a binding agent, and activated by treatment with gas (CO_2, H_2O). The inner specific surface area, pore volume, and porosity so produced do not reach the figures typical of activated carbons.

Coke produced by the low temperature carbonization of lignite from the Rhine region in a hearth type furnace contains up to 87 % carbon. Its adsorptive and catalytic properties enable the gaseous impurities Hg, SO_2, HCl, HF, Cl_2, NO_x, dioxins and furans to be efficiently removed. This cheap material is not regenerated and is used as a crushed or powdered adsorbent to remove small quantities of impurities.

Carbon molecular sieves have micropores with diameters of 0.3–1.0 nm. These very narrow pores mean that these adsorbents cannot be used for emission reduction.

Activated aluminum oxide (>99 % Al_2O_3), an oxidic hydrophilic adsorbent, is mainly used for removing sulfur compounds.

Molecular sieve zeolites are crystalline aluminum silicates with an $SiO_2-Al_2O_3$ lattice. They contain cavities linked by pores of uniform diameter, so that their adsorption capacity combined with their steric separating effect can be used in gas purification. If a high proportion of Al_2O_3 is replaced by SiO_2, this gives a hydrophobic zeolite (trade name: Wessalith DAY) [7.70] suitable for selectively adsorbing organic components and mercury from waste gases with a

Table 7.3. Industrial adsorbents

Adsorbent	Specific surface, $m^2 g^{-1}$	Specific heat, $J kg^{-1} K^{-1}$	Compacted apparent density, $kg m^{-3}$	Typical applications
Activated carbons				
fine-pored	1000–1200	850	400–500	waste-air treatment, deodorization, separation of low-boiling substances
medium-pored	1200–1400	850	350–450	solvent recovery, adsorption of medium-boiling substances
wide-pored	1000–1500	850	300–400	adsorption of high-boiling substances
Activated coke	<400	850	500–600	separation of dioxines, furanes, SO_x and NO_x
Carbon molecular sieves	<100	1000	620	nitrogen and oxygen removal from air, methane adsorption from biogas
Aluminum oxide	100–400	920	700–800	desulfurization of gases
Silica gel				
fine-pored	250–350	920	700–800	deodorization, gas dehydration
wide-pored	600–850	920	400–800	adsorption of hydrocarbons
Molecular sieve zeolites				
hydrophilic	500–1000	900	600–800	gas dehydration, fractionation of hydrocarbons
hydrophobic	500–700	900	400–500	waste-air treatment, solvent recovery, solvent upgrading
Polymer adsorbents (styrene–divinylbenzene resins)	1000–1500	350	300–400	solvent recovery, gasoline recovery

Table 7.4. Reactions of impregnated activated carbon with contaminants of a gas stream

Contaminant	Impregnating agent	Active substance	Conversion
Ammonia	copper	$CuSO_4 \cdot 5\,H_2O$	$CuSO_4 \cdot 5\,H_2O + 4\,NH_3 \rightarrow [Cu(NH_3)_4]SO_4 + 5\,H_2O$
Arsenic hydride	silver, zinc	$Ag, Ag_2O, ZnO, Na_2ZnO_2$	$2\,AsH_3 + 3\,O_2 \xrightarrow{\text{cat.}} As_2O_3 + 3\,H_2O$
Hydrogen cyanide	copper	CuO, Cu_2O	$2\,HCN + Cu_2O \rightarrow 2\,CuCN + H_2O$
	zinc	ZnO, Na_2ZnO_2	$2\,HCN + ZnO \rightarrow Zn(CN)_2 + H_2O$
Phosgene	copper	CuO, Cu_2O	$COCl_2 + CuO \rightarrow CuCl_2 + CO_2$
	zinc	ZnO, Na_2ZnO_2	$COCl_2 + ZnO \rightarrow ZnCl_2 + CO_2$
Mercury	sulfur	S_8	$Hg + S \rightarrow HgS$

high moisture content, and is therefore known as "white activated carbon". As it is a mineral, it can be regenerated at high temperatures with air.

Spheroidal adsorber resins made of nonpolar macroporous polymers are produced by polymerization of styrene in the presence of a cross-linking agent, and have the advantage of ease of regeneration at low temperatures when used for the purification of solvent-containing waste air. They also have a low affinity for water vapor.

In silica gels, the SiO_2 particles form a continuous crystalline structure containing pores. Silica gels with small pores are used as drying agents, while the wide-pore types are suitable for removing hydrocarbons from waste gases.

The properties of some adsorbents can be modified by impregnation with organic or inorganic substances (Table 7.4) to give additional separation effects (e.g., the adsorption of H_2S by activated carbon impregnated with potassium iodide).

The following are the principal criteria used when selecting an adsorbent for waste-gas purification:

1) High selectivity with fast kinetics for the component to be removed
2) High adsorption capacity even at low concentrations
3) Ease of desorption
4) Stability towards temperature change and chemical attack
5) Hydrophobicity and attrition resistance

Relevant properties of the substances to be removed include:

1) Molecular mass
2) Ease of desorption
3) Boiling point or boiling range
4) Solubility in water
5) Chemical structure and stability
6) Flammability
7) Critical temperature
8) Surface tension
9) Corrosion attack
10) Molecular structure

7.4.3. Design Criteria for Adsorbers

To enable the gas that is to be purified to pass through the adsorbent as effectively as possible, the latter must be arranged as a fixed bed, moving bed, or other arrangement in the adsorber. The dimensions of the equipment are determined by the maximum permissible flow rate of the waste gas and bed depth of the adsorbent. For adsorbents in powder form, fluidized beds or entrained flow systems give intensive contact with the adsorptive. Most adsorbers are of the fixed-bed type, with a gas velocity of ca. 0.5 m/s in the empty cross section, which should be well below the velocity at which the granules in the upper region of the bed will be in motion. With most activated carbons, this velocity is 0.8 m/s for the carrier gas air at ca. 20 °C. In some cases, the adsorption and/or reaction rates make the residence time decisive, which can lead to larger adsorber cross sections or/and bed depth.

Flow resistance figures for various granular adsorbents under normal gas velocities are shown in Figure 7.40. They depend on the properties of the fluid phase (density, viscosity, velocity) and of the adsorbent bed (grain size, grain shape, surface area, voids fraction). The calculation method often quoted in the literature [7.71] also contains an experimentally determined resistance number. Uniform flow through the bed not only gives optimum utilization of its adsorption capacity, but tends to give efficient heat removal, preventing hot spots which can lead to spontaneous ignition of carbonaceous adsorbents. Equipment for monitoring CO in the impure and purified gas serves as a warning device. Bed temperature measurements provide inadequate warning owing to the poor heat-transfer properties.

For a given diameter, the bed depth of the adsorber depends on the amount of adsorption medium charged. This is usually determined

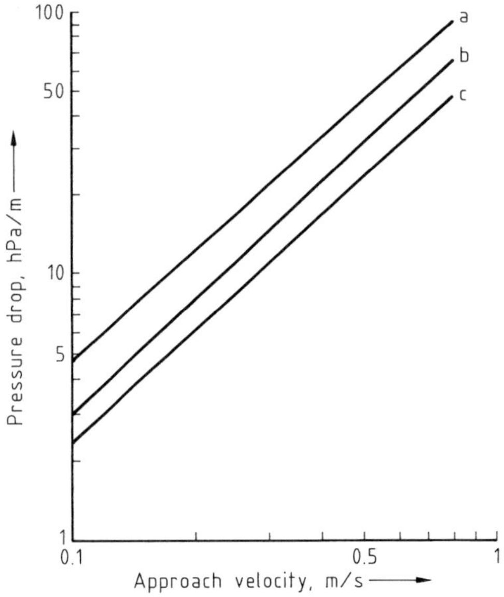

Figure 7.40. Pressure drop for various adsorbents ($p = 1$ bar abs.; $T = 20\,°C$)
a) Activated carbon cylinders, 3 mm; b) Activated carbon cylinders, 4 mm; c) Zeolite cylinders, 4 mm

from the equilibrium loading, after subtraction of factors that reduce the capacity such as residual loading and irreversible damage. This lowers the operating capacity of the adsorbent to a fraction of the total loading at equilibrium. In solvent recovery by activated carbons, this fraction is ca. 30 %. Moreover, a minimum bed depth must be provided to accommodate movement of the adsorption zone. This also ensures minimum pressure drop required for uniform gas distribution.

The effective bed depth can be derived from the profile of the adsorption zone [7.67], [7.68]. As the adsorption zone usually migrates through the bed at constant velocity in the same direction as the gas flow, breakthrough curves determined at the same flow rate in experimental adsorbers can be transferred to large-scale industrial plants. In breakthrough curves, the ratio of the concentration of the adsorptive at the exit of the adsorber to that at the inlet (c/c_0) is plotted as a function of adsorption time t. As shown in Figure 7.41, for bed depths h_1 and h_2, the half-value times t_{h_1} and t_{h_2} (where $c/c_0 = 0.5$) are determined. This gives the migration rate u of the adsorption zone:

$$u = \frac{h_2 - h_1}{t_{h_2} - t_{h_1}}$$

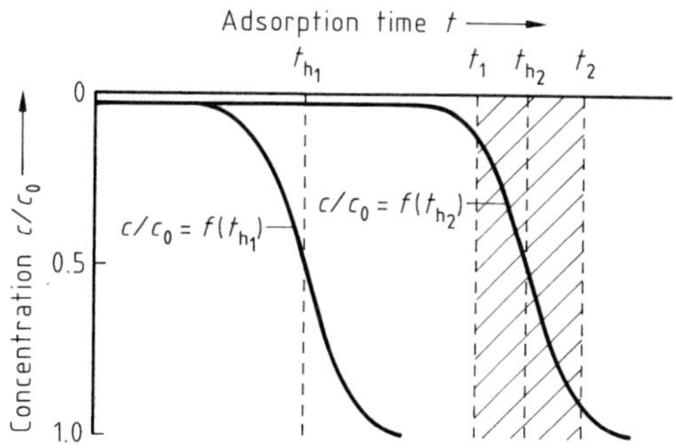

Figure 7.41. Determination of effective bed depth from breakthrough curves

Multiplication by the time interval $t_2 - t_1$ gives the length h_z of the adsorption zone:

$$h_z = u(t_2 - t_1)$$

Assuming a symmetrical profile of the adsorption zone, the effective height h_{eff} of the bed is given by $h_{\text{eff}} = h - 0.5\,h_z$. The amount of adsorbate $x(t)$ taken up by the adsorbent from the carrier gas $\dot{V}$ up to time t is given by

$$x(t) = \frac{\dot{V}}{m} \int_0^t (c_0 - c_t)\,\mathrm{d}t$$

Other methods are described in the literature [7.72], [7.73] for dimensioning an adsorber under given conditions from mass and energy balances.

The design of the adsorber calculated from adsorption criteria must also be suitable for the regeneration conditions, and this must be investigated separately.

7.4.4. Adsorber Designs

Specific types of adsorber have been developed for the various tasks of waste-gas purification [7.74]. These can be divided into fixed-bed and moving-bed equipment, and special designs (Fig. 7.42).

Because of the poor mechanical properties of most adsorbents, the fixed-bed design (Fig. 7.42 A) is usually used in industrial plants. This consists of a random packing on a carrier grid covered with perforated sheet, woven steel wire, or a perforated ceramic plate. Vertical or horizontal adsorbers may be preferred depending on the type of waste gas to be treated and the space available. In the vertical designs, the adsorbent is sometimes contained in a ring-shaped

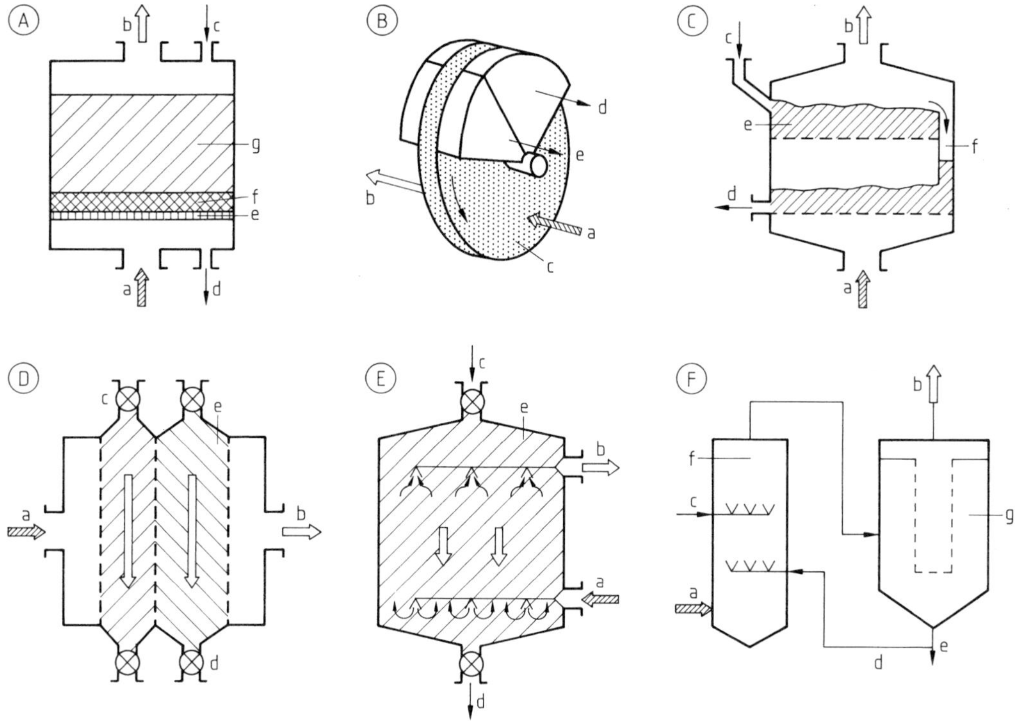

Figure 7.42. Adsorber designs
A) Fixed-bed adsorber
a) Crude gas; b) Clean gas; c) Desorption medium; d) Desorption medium and desorpt; e) Carrier grid; f) Intermediate layer; g) Adsorbent packing
B) Rotating adsorber
a) Crude gas; b) Clean gas; c) Adsorbent; d) Desorption segment; e) Cooling segment
C) Fluidized-bed adsorber
a) Crude gas; b) Clean gas; c) Adsorbent, regenerated; d) Adsorbent, loaded; e) Fluidized bed; f) Down pipe
D) Moving-bed adsorber
a) Crude gas; b) Clean gas; c) Adsorbent, unloaded; d) Adsorbent, loaded; e) Moving bed
E) Countercurrent adsorber
a) Crude gas; b) Clean gas; c) Adsorbent, unloaded; d) Adsorbent, loaded; e) Moving ded
F) Entrained-flow adsorber
a) Crude gas; b) Clean gas; c) Adsorbent, unloaded; d) Adsorbent, partially loaded; e) Adsorbent, transfered outward; f) Tubular reactor; g) Woven filter

shaft to save floor space. Even for longer service life of the adsorbent, the apparatus should include equipment for rapid charging and emptying. If desorption with water vapor is used, the adsorber must be designed as a pressure vessel.

Moving-bed adsorbers (Fig. 7.42 D) are almost always designed vertically to save floor space and to facilitate the demands made on the mechanical charging and discharging equipment. Virgin or regenerated adsorbent is fed from above while the same amount of exhausted material is removed from below. Locks are provided if necessary at both feed and discharge points. Slotted walls or screens are provided at the gas inlet and outlet sides to guide the adsorbent bed. The bed can be subdivided into several layers arranged in series, containing different adsorbents with different functions which can be moved at different rates. The mechanism for removing the adsorbents must ensure that they move downwards as a single mass, including the edge regions. In another design, a circular moving bed is used through which the gas flows from the inside to the outside or vice versa. In the design illustrated in Figure 7.42 E, the waste gas is routed through the adsorber in countercurrent flow to the moving bed.

The fluidized-bed technique (Fig. 7.42 C) requires hard adsorbent granules with high abrasion resistance, which has considerably restricted the use of this method. The apparatus contains a perforated sheet or nozzle tray on which the adsorbent is maintained in a fluidized state by the waste gas entering from below. The adsorbent passes from the inlet to the outlet perpendicular to the gas flow. The purification effect can be increased by using several fluidized beds in series. Like scrubbing columns, they are equipped with inlet and outlet weirs and are connected by down tubes. The comparatively small amount of adsorbent makes the removal of some gaseous components difficult, and the additional loading capacity of the adsorbent is lower than that obtainable with fixed beds. An advantage is the continuous operation of the adsorption and regeneration stages.

Continuous operation is also achieved by rotating adsorbers (Fig. 7.42 B). The rotor has a large specific surface area formed by granular activated carbon or ceramic trays which carries a layer of finely divided activated carbon or fibrous carbon. The rotor rotates round a horizontal or vertical axis inside a housing, passing through the adsorption, desorption, and cooling zones successively. The gas can pass through the rotor axially or radially. Because desorption is carried out with hot air at a flow rate that is only a fraction of that of the waste gas, the adsorber is not suitable for the final purification stage, but only for a preliminary concentration.

In the entrained-flow system (Fig. 7.42 F), the adsorbent powder is carried by the flow of waste gas. The main function of the apparatus is to provide a mixing space, while the adsorption of the impurities mainly takes place on a woven filter downstream.

Some special designs of adsorbers include adsorbent mats made of activated carbon fibers, filter elements in which activated carbon granules are bonded together, or powdered activated carbon bonded to highly porous carriers such as polyurethane foam.

7.4.5. Plant Concepts

Apart from those plants intended purely for filtration, in which no regeneration is provided for or possible because of the long service life of the adsorbent (several months), the concept of an adsorption plant for waste-gas purification is very much influenced by the regeneration method. Regeneration can be achieved in various ways:

1) Steam desorption
2) Hot gas desorption
3) Pressure-swing desorption
4) Solvent extraction

Steam desorption, which exploits the temperature dependence of the adsorption equilibrium and the high heat of condensation of the steam, is the most widely used process for solvent recovery (Fig. 7.43). The adsorber is heated with slightly superheated steam at ca. 150 °C, which also acts as a purging medium. The condition and consumption of steam (3 – 5 kg/kg solvent) depend mainly on the type of adsorptive and the

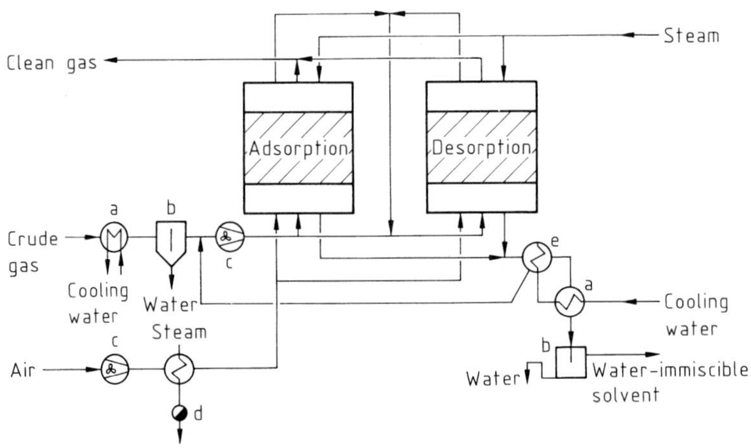

Figure 7.43. Flow sheet for activated carbon plant with steam desorption
a) Cooler; b) Separator; c) Fan; d) Steam trap; e) Condenser

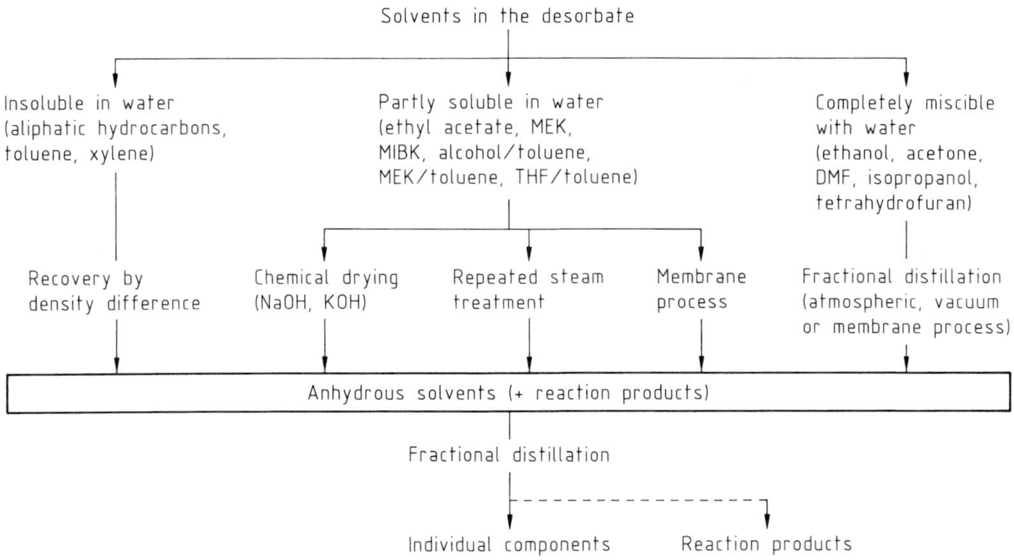

Figure 7.44. Methods of processing desorbates

desired residual loading of the adsorbent, according to the required purity of the waste gas. The desorbate produced, consisting of solvent vapors, steam, and condensate, is condensed, cooled, and separated. If the solvent is immiscible with water, the separation equipment used is based on density difference. As a rule of thumb the higher the mutual solubility of solvent and water, the more complex separation process, and a rectification column may even be required. An overview of methods used for processing the desorbate is given in Figure 7.44. After desorption, the adsorber is dried and cooled with purified waste gas or clean air to the temperature of the adsorption step. As the cooling gas contains small quantities of solvent it is passed through the operating adsorber to prevent emission peaks. This alternating adsorption/desorption system means that there must be at least two adsorbers to ensure continuity of the waste-gas purification process.

In special cases, an inert gas or air is used to heat the adsorber directly. This can be carried out in three ways (Fig. 7.45). The closed-loop system (Fig. 7.45 A) is advantageous if enough of the desorbed substances can be condensed to ensure that their residual partial pressures do not cause an unacceptably high residual loading. Otherwise, if oxygen-free waste gases are being purified or a noncombustible adsorbent is used, regeneration can be carried out in an open loop (Fig. 7.45 B). Here a certain amount of the purified waste gas is fed through the adsorber being cooled and that being heated, and, after partial condensation of the desorbed substances, passed into the impure gas and thus into the adsorber during the adsorption cycle. If the desorbed substances cannot be adequately condensed, or if they consist of a mixture of substances whose recovery is not feasible, a recirculation system cannot be used, and the desorpt is then treated by thermal or catalytic combustion. In this case, the adsorption plant functions as a concentration plant, using preheated air for desorption (Fig. 7.45 C).

Desorption by pressure reduction (pressure-swing process) requires a pressure difference between the adsorption and the desorption step so that desorption has to be carried out under vacuum combined with a small purge gas stream which carries the desorpt out of the adsorber.

In a few special cases, e.g., sulfur-loaded carbon treated with CS_2, desorption is carried out by solvent extraction, after which the solvent is desorbed by steam.

Plant sizes range from compact plants with throughputs of a few hundred m^3/h to large plants in which waste gas can be purified at a rate of several hundred thousand m^3/h.

7.4.6. Adsorption Processes

The availability of several adsorbents with various physical, chemical, and mechanical

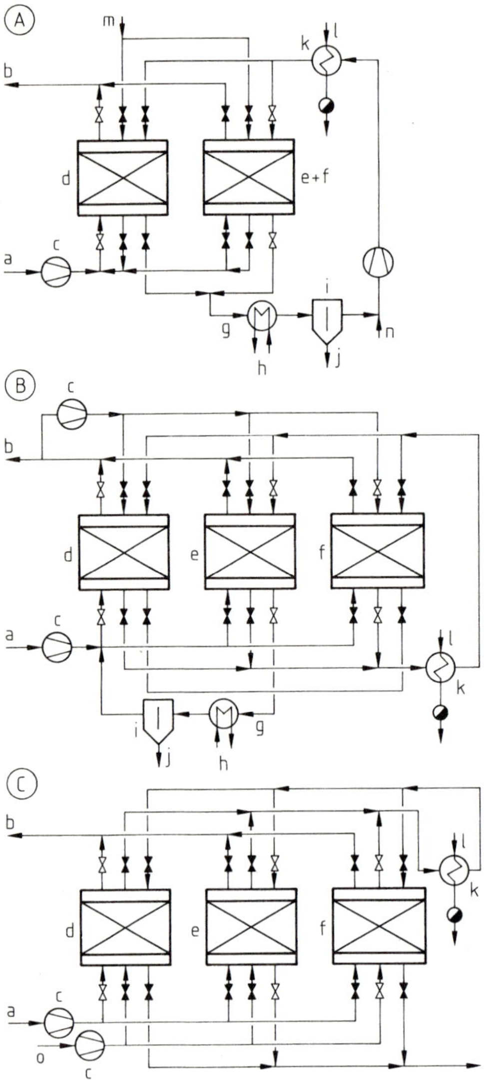

Figure 7.45. Plant design concepts with hot-gas desorption
A) Closed-loop system
B) Open-loop system
C) Plant using preheated air
a) Crude gas; b) Clean gas; c) Fan; d) Adsorption (adsorber); e) Desorption (adsorber); f) Cooling (adsorber); g) Condenser/cooler; h) Cooling, water; i) Separator; j) Desorpt; k) Heater; l) Heating steam; m) Purge gas, inert; n) Inert gas, makeup; o) Air
▷◁ = Open valve; ▶◀ = Closed valve

properties in association with the wide range of waste-gas purification tasks has led to the development of a variety of industrial techniques.

In the oldest application — the recovery of organic solvents from waste air — the most widely used processes are based on fixed beds of activated carbon (see Fig. 7.42). The waste air is extracted from hooded production areas conditioned by cooling, partial condensation of the water vapor and heating and then fed at ca. 40 °C to the adsorber on loading duty. At an approach velocity of ca. 0.5 m/s it passes upwards through the activated carbon bed, which has a height of 0.8–1.5 m. The granular adsorbent takes up the solvent vapors, and the purified waste air passes out of the adsorber. After the solvent breakthrough limit has been reached, the adsorption process is interrupted, after the freshly regenerated adsorber has taken over the purification of the waste air. The solvent is then desorbed by passing superheated steam downwards through the activated carbon. The steam heats the bed to >100 °C and carries away the desorpt. The desorbate is condensed, cooled, and passed to the gravity separator. This is a settling vessel in which the mutually insoluble organic and aqueous phases separate as a result of the density difference. Approximately 4 kg steam/kg solvent is required for this process step, which accounts for most of the operating costs. The subsequent drying and cooling steps are carried out with purified waste air or clean air, which are then cleaned by passing them through the adsorber on operating duty. If the recovered solvent is miscible with water, fractional distillation may be required.

If the time required for regeneration is appreciably shorter than that for adsorption, higher loading of the activated carbon can be achieved. In this case, the regenerated adsorber can be connected after the operating adsorber (two-stage adsorption). The waste air passes through the two vessels connected in series, the first adsorber being charged beyond breakthrough.

In activated carbon plants using hot gas desorption, the operation must be performed by purging the loaded adsorbent with an inert gas in a closed circuit, consisting of a cooler-condenser and separator, a heater, and a drying stage for the regeneration gas.

With the hydrophobic zeolites used for removing solvents from waste air, dry desorption with hot air can be used, preferably in a closed recirculating system, provided that partial condensation of the solvent is sufficient.

Polymer adsorber resins loaded with solvents can be regenerated at temperatures as low as 80 °C, either by steam or hot gas [7.75]. They are typically employed in moving-bed applications, and permit continuous operation of fluidized-bed adsorbers in combination with moving-bed regenerators (e.g., Polyad process) [7.75].

The Kontisorbon process for continuous adsorption and desorption (Fig. 7.46) [7.76] uses attrition-resistant activated carbon beads of diameter 0.7 mm in a multistage fluidized bed, maintained in the fluidized state by waste air at a velocity of 0.8–1.2 m/s. The carbon is regenerated by feeding it into a heat exchanger with vertical tubes underneath the adsorber. Indirect steam heating in combination with nitrogen purging drives off the solvent, which is partially recovered as condensate. The remaining solvent is adsorbed from the nitrogen purge cycle by the activated carbon above the desorber which has passed through the fluidized bed. Cooling of the activated carbon after desorption is also indirect. It is then pneumatically returned by air lift to the upper perforated tray of the fluidized bed.

Rotary adsorbers also enable the process to be operated continuously. The rotor is packed with granular activated carbon or consists of a ceramic honeycomb structure coated with fine-grained activated carbon or fibrous carbon. It rotates round its horizontal axis in a housing at 1–2 rpm passing through an adsorption zone, a desorption zone, and a cooling zone in succession (Fig. 7.42 B). The waste air passes through the adsorption segment at 1–2 m/s, the pressure drop being only 2–3 mbar [7.77]. Desorption can be carried out with steam, but preferably by hot air at a flow rate of 2–20% of the waste gas flow rate. This passes out of the plant as higher concentrated secondary waste air and requires purification.

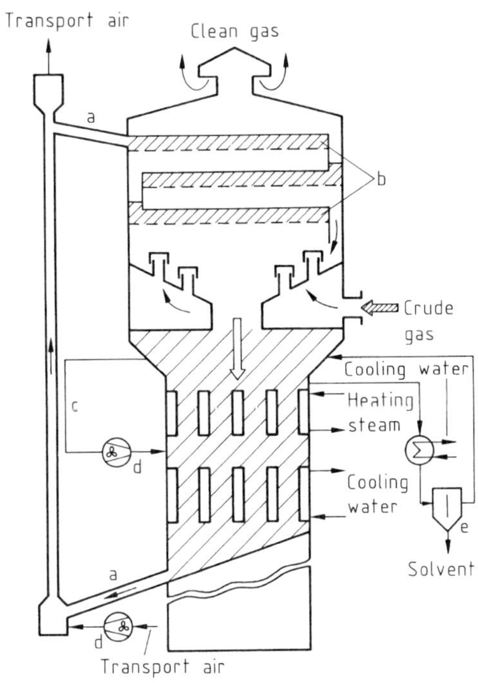

Figure 7.46. Kontisorbon process
a) Adsorbent, regenerated; b) Fluidized bed; c) Inert gas; d) Fan; e) Separator

The plant concept to clean waste air from e.g. gasoline tanks with high concentration of organic vapors involves a scrubber or a membrane unit combined with a downstream pressure-swing adsorption plant.

Sulfur compounds are removed from waste gases using either activated carbons or activated aluminum oxide as the adsorbent or adsorption catalyst. In the Sulfosorbon process [7.78] for purifying waste air containing H_2S and CS_2 from the viscose process, each adsorber is usually packed with two layers of activated carbon having two different effects. The lower part of the bed consists of wide-pore carbon impregnated with potassium iodine, which catalyzes oxidation of the H_2S and adsorbs the elemental sulfur formed. Oxidation to sulfuric acid is largely prevented. The CS_2 is removed mainly in the upper bed of narrow-pore activated carbon. As soon as the CS_2 concentration in the purified waste air exceeds the limit value, a regenerated adsorber takes over the purification process, and after purging with an inert gas, desorption of the CS_2 by steam at 110–130 °C commences. The organic and aqueous phases are separated in a gravity separator. Regeneration of the lower, sulfur-loaded bed is required only in intervals of several months and comprises the following process steps: water washing to remove sulfuric acid and/or ammonium sulfate formed as side products, extraction of the elemental sulfur with CS_2, separation of sulfur and CS_2 by distillation, desorption of CS_2 with steam, and drying and cooling the adsorber with air.

The Thiocarb process [7.79] developed for the same purpose uses a single bed of wide-pore activated carbon with a very low metal content (to suppress the catalytic formation of H_2SO_4). Any acid which is formed despite this is neutralized by adding NH_3. The regeneration process is as described above.

The Sulfacid process [7.80] can be used with SO_2-containing, dust-free process waste gases using activated carbon as an adsorption catalyst to purify the gas in a fixed bed, provided the O_2

content is sufficient to convert SO_2 and water to sulfuric acid. Washing with water yields dilute H_2SO_4 (10–15%), which can be concentrated to ca. 30% in combination with the waste-gas cooling in the venturi scrubber. For SO_2 concentrations of 0.1–1.0 vol% the removal efficiency is 90–95%.

Desulfurization and removal of nitrogen oxides from combustion waste gases is carried out in the BF–Uhde process [7.81] by adsorption catalysis at 120–150 °C on 6 mm activated coke in a two-stage moving-bed reactor with transverse gas flow (Fig. 7.47). As well as desulfurization, the first passage of the gas also removes HCl and HF. Ammonia is then added to convert the nitrogen oxides to N_2 and H_2O by selective catalysis in the second passage of the gas, through the upper part of the reactor. Ammonium sulfate is formed in side reactions. Regeneration is carried out by indirectly heating the dedusted activated coke to ca. 450 °C, and a gas with a high SO_2 content is formed by reaction of the sulfuric acid with the carbon to give a SO_2 rich gas which can be converted to sulfuric acid, elemental sulfur or liquid SO_2.

Both activated carbon and activated aluminum oxide catalyze the Claus reaction

$$2\,H_2S + SO_2 \rightleftharpoons 3\,S + 2\,H_2O$$

at ca. 130 °C and readily adsorb the sulfur formed. This is the basis of the Sulfreen process [7.82] for desulfurizing Claus waste gases. In the single-stage process (Fig. 7.48), a fixed bed of activated aluminum oxide gives removal efficiencies of 80–93%, depending on the inlet concentration of H_2S and SO_2. During desorption by heated clean gas at ca. 300 °C, most of the sulfur driven off can be condensed from the recirculating gas stream. Clean gas is also used for cooling to the adsorption temperature. The two-stage variant (the Carbosulfreen process) includes an activated carbon bed in which H_2S reacts with O_2, and the elemental sulfur formed is adsorbed. Preliminary catalysis by titanium dioxide allows COS and CS_2 to be hydrolyzed to H_2S (Hydrosulfreen process).

Another important application of adsorption is the removal of heavy metals (in particular mercury vapor) as well as dioxins and furans from the flue gases from incineration plants for munic-

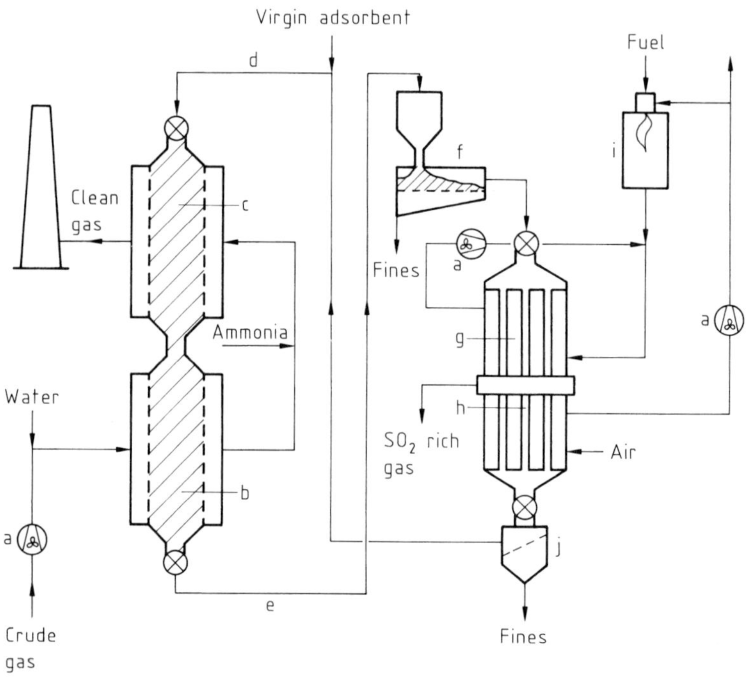

Figure 7.47. BF–Uhde process
a) Fan; b) Desulfurization; c) NO_x removal; d) Adsorbent, regenerated; e) Adsorbent, loaded; f) Air classifier; g) Desorber; h) Cooler; i) Combustion chamber; j) Sieve

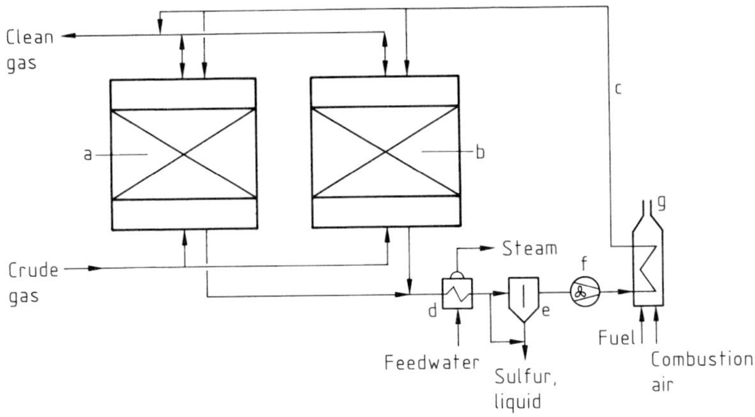

Figure 7.48. Sulfreen process
a) Adsorption (adsorber); b) Regeneration (adsorber); c) Regeneration gas; d) Condenser/cooler; e) Separator; f) Fan; g) Heater

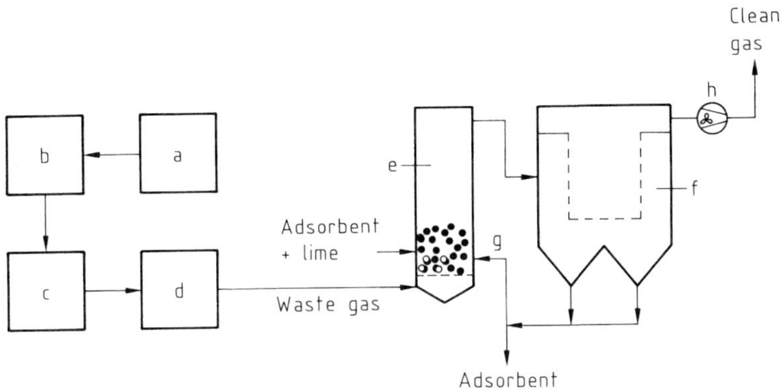

Figure 7.49. Circulating fluidized bed for removal of mercury, dioxins, and furans as tail end cleaning stage.
a) Waste incineration plant; b) Electrostatic precipitator; c) Scrubbers for $HgCl_2$, HCl, HF, SO_2; d) DENOX; e) Fluidized bed; f) Fabric filter; g) Recirculation of adsorbent; h) Fan

ipal waste, hazardous waste, and sewage sludge. Adsorbents used include activated carbon, activated coke, hearth-furnace coke, and hydrophobic zeolites. The efficiency of removal of mercury by activated carbon or zeolites can be considerably improved by impregnation with sulfur or sulfuric acid, and the same effect is produced by SO_2-containing flue gases. The long residence times required are a disadvantage, requiring the use of high-volume vertical adsorbers [7.83]. The exhausted adsorbent is removed discontinuously or continuously, and is often used as a fuel in combustion plants. The pollutants are recovered by means of dedusting filters or washing processes (Hg, $HgCl_2$, SO_2, HCl, and HF) or are incinerated (dioxins and furans). A further alternative is thermal regeneration, which can be more cost-effective than combustion in the case of expensive adsorbents [7.84]. If the space for large fixed-bed filters is not available, powdered hearth-furnace coke can be added to the wash suspension for desulfurization. High removal efficiencies have been obtained using powdered hearth furnace coke and lime in a pilot plant consisting of a fluidized-bed reactor and fabric filter in an hazardous-waste incineration plant (Fig. 7.49) [7.85]. The results correspond to those obtained on a full-scale plant for treating flue gases from municipal waste incineration operating on almost the same principle [7.86].

Powdered carbonaceous adsorbents injected in entrained-flow systems and separated downstream by fabric filters are gaining increased significance in removing heavy metals, dioxins and furans from waste gases and hazardous waste incineration plants, steel production plants, recycling plants for Pb- and Zn-containing filter dust, sintering plants and other processes. They collect those substances from the off-gas streams at high selectivity in spite of their low concentrations [7.87]–[7.89].

Natural zeolites boast the advantage of greater safety in operation. This means that the safety equipment normally required, addition of limestone, extinguishing facilities, measuring and control devices etc. can be omitted. There is no self-ignition hazard [7.90].

7.4.7. Recycling and Disposal of Spent Adsorbents

Depending on the type of adsorbent, the following alternatives are available for recycling or disposal of spent adsorbents:

1) Reactivation of carbon-based and oxidic adsorbents with recovery of the adsorbed substances, if possible
2) Hydrogenation of carbon-based adsorbents and polymer adsorber resins, together with other organic wastes such as spent oils, paint sludge, plastics, and solvents to produce base stocks for the oil industry
3) Co-burning of carbon-based adsorbents in the combustion plant producing the flue gas
4) Co-burning of carbon-based adsorbents and polymer adsorber resins in hazardous-waste incineration plants;
5) Disposal of oxidic adsorbents

7.5. Biological Processes

7.5.1. Overview and Classification

In biological methods for purifying waste gas, the hazardous and/or odorous constituents are reacted or degraded biologically. This does not take place in the waste gas itself, but after the materials to be treated have been sorbed by solids or liquids in the biological purification equipment. This equipment can be classified according to the process used for conversion or reaction as follows:

1) Biofilters containing organic and inert materials
2) Bioscrubbers containing fixed or suspended microorganisms
3) Combined processes in which biofilter and bioscrubber units are arranged in series.

The selection of the process to be used depends on several criteria which must be considered in each individual case [7.91].

7.5.2. Process Engineering Principles

To correctly operate and optimize a process, biological, physical, and chemical effects have to be considered. The main process engineering task is to establish and maintain optimum living conditions for the microorganisms.

7.5.2.1. Biological Conditions

Microorganisms are so-called destruents, which break down organic materials such as carbohydrates, fats, and proteins into their basic components (CO_2, CH_4, NO_3^-, PO_4^{3-}, etc.). These components are then used by the microorganisms to form their own biomass. The required energy is derived from the biochemical oxidation (i.e., the degradation of the organic substances). Microbial degradation is therefore associated with an increase in the biomass. If the required carbon is derived from an organic substance, the term "heterotrophic habit" is used. Most gaseous substances are available to microorganisms, although their biodegradability properties can range from easy to very difficult, depending on their chemical structure.

Inorganic gaseous compounds such as ammonia and hydrogen sulfide can also be used by microorganisms as sources of energy or as nutrients. The carbon required for growth is derived from atmospheric carbon dioxide (carbon autotrophism) [7.92].

There are ca. 3500 naturally occurring microorganisms which, together with specialized lower life forms, make it possible to break down even difficultly available substances. The microorganisms are either sessile (i.e., attached to organic filter materials or inert packings) or suspended freely in a scrubbing liquid. In both cases, the gaseous substances must be absorbed in a liquid to make them available for degradation.

The following conditions must be fulfilled to enable the equipment to function:

1) The components of the waste gas must be biologically degradable
2) The waste gas must not contain substances toxic to microorganisms
3) The components of the waste gas must be soluble in water

4) The products formed by the microorganisms must not have a detrimental effect on the biological degradation process.

The conditions in the medium must be optimized for the microorganisms. In biological processes, the following factors affect growth:

Oxygen Demand. In most waste gases not derived from purely chemical processes, sufficient oxygen is present to enable processes based on fixed microorganisms to be used. In bioscrubbers which use suspended microorganisms, adequate amounts of oxygen must be added at the regeneration stage.

Water Demand. Living microorganisms contain 75–85 % water, which is needed to transport nutrients into and metabolic products out of the cells. Thus they have a water demand which can be met by the water in which they are suspended or by water-containing solid substrates. The actual quantity of water is not as important as its availability in the substrate or medium. A measure of this is the water activity a_w, which is the ratio of the vapor pressure of an aqueous solution to the vapor pressure of the gas dissolved in water. The water activity of pure water is $a_w = 1.0$. The water requirement of bacteria is approximately $a_w = 0.91$, of molds $a_w = 0.80$, and of osmophilic yeasts $a_w = 0.60$ [7.92].

Nutrient Demand. The approximate composition of a bacterial cell on a dry basis is 50% carbon, 14% nitrogen, 2–6% phosphorus, and 1% sulfur and other elements.

Carbon, nitrogen, and phosphorus are essential for cell growth and producing energy. As a rule, a well-balanced nutrient contains these elements in the approximate ratios (C:N:P) 100:5:1 (bioscrubbers) or 200:10:1 (biofilters).

Temperature. Three temperature ranges, in which different types of microorganisms are active, are of interest for biological waste-gas purification processes:

Psychrophilic 10–20 °C
Mesophilic 20–37 °C
Thermophilic 50–65 °C

The activity of the microorganisms increases with temperature, but the solubility of gaseous substances in water decreases. It is important for the reliable operation of a biological waste-gas purification plant to keep its temperature within the appropriate range as consistently as possible to prevent depressions due to population changes. In practice, most installations are operated in the mesophilic range.

The temperature range 37–50 °C should be avoided if possible, as a continuous fluctuation between mesophilic and thermophilic microorganisms can occur, reducing the efficiency of the installation.

pH Value. The hydrogen ion concentration in the medium affects the growth and metabolism of the microorganisms. Whereas most heterotrophic bacteria prefer the neutral range, fungi prefer the acid range. The growth and metabolism of microorganisms also cause changes in the pH of the medium. Such changes can be prevented by controlled additions of H_2SO_4 and NaOH (buffering), or by operating multistage equipment at various pH ranges.

7.5.2.2. Physicochemical Conditions

Since microorganisms can exist only in or on moist media, they can only metabolize the components of the gas to be treated if these are absorbed in the liquid phase.

Biological waste-gas purification is a biochemical absorption, the result of which is not exclusively determined by the physical solution equilibrium because of the reactive conversion of the dissolved gas [7.93]. Henry's solubility coefficient plays an important role, and is affected by the temperature and pH of the washing water [7.94].

As the component dissolved in the wash liquid are biologically degraded by the microorganisms present in a suspended state or fixed in the biofilm, thermodynamic equilibrium cannot be established. The resulting fall in concentration influences diffusion and accelerates absorption. This process depends mainly on the biological reaction rate.

However, with increasing biological activity, the absorption of pollutants depends less and less on the reaction rate and increasingly on the diffusion rate [7.95].

There are thus two areas in which the absorption process and hence the effectiveness of biological waste-gas purification processes can be influenced:

1) Diffusion
For a constant concentration in the waste gas, the concentration difference can be increased by increasing the water supply or by addition-

al biological purification of the wash water.
The solubility can be increased by solubilizing agents (e.g., high-boiling oils) [7.96].
The specific exchange surface (phase boundary) in the absorber must be as large as possible.

2) Kinetics
The reaction conditions (temperature and pH) must be optimized.
The residence time can be increased by additional aeration stages.
The biological activity (cell concentration) can be influenced by adding nutrients at apropriate times.

As the microorganisms in a biofilter settle on the moist filter material, two types of transfer occur: adsorption on a solid substance (van der Waals forces) and biochemical absorption. The ratio of the two types of sorption in the filter bed depends on the moisture content of the materials. In purely dry zones only adsorption takes place up to the point of saturation, but no biological degradation can occur.

Biofilters are operated in a moist condition without water circulation, so that the absorption capacity and hence the efficiency of the equipment depends on a homogeneous and uniform flow through the filter bed. The flow behavior of the waste gas through the filter bed is thus determined by the void volume, the gas volume, and the associated relative pressures. The void volume of the material should be distributed uniformly over the filter cross section to ensure uniform flow rates and residence times at all points [7.97], [7.98].

Other important factors are the water retention properties of the filter material which determine the water content during operation. With increasing water content, the free void volume in the filter bed decreases and the gas flow rate and the pressure drop increase. The residence time and efficiency of the biofilter therefore both decrease.

7.5.3. Biofilters [7.99]

Designs. The classical biofilter is the flat filter (see Fig. 7.50). The conditioned crude gas is compressed in a chamber under a tray provided with holes or slits. The distance between the bottom of the pressure chamber and the grid above should be at least 0.7 m to ensure uniform distribution of the gas flow. The air then flows

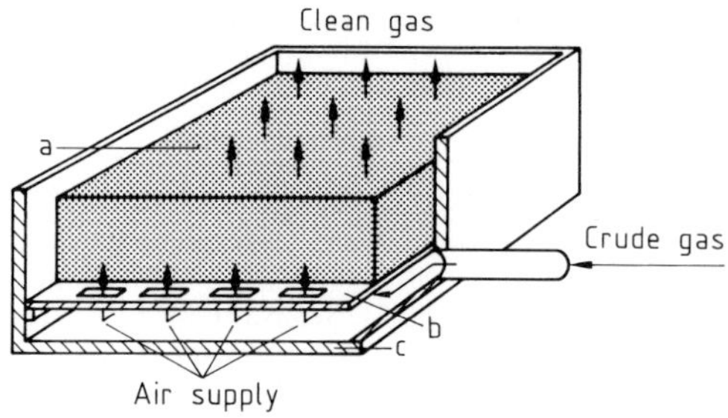

Figure 7.50. Schematic of a flat biofilter
a) Filter material; b) Slotted floor; c) Bottom seal

through the filter bed whose height depends on the material chosen, and should be at least 1 m in a settled state. Flat filters are used where enough floor space is available, and where there are large flow rates with relatively small concentrations of pollutant gas or odoriferous material (e.g., slaughterhouses, composting plants, sewage works). If floor space is restricted, multistage filters of equal area can be stacked one above the other. These can be operated either in parallel or, with multistage operation, in series. In highly loaded waste gases with a low flow rate, e.g., from pharmaceutical and chemical plants especially paint manufacture and processing and in printing works, open or closed container plants (compact plants) can be used.

Design Parameters. Because biofilter plants are used to treat complex (odoriferous) gas mixtures, it is seldom possible to perform theoretical calculations. Plant design usually involves comparison with other plants and/or the results of pilot plant experiments. The most important design criterion is the volumetric filter loading, i.e., the amount of waste gas in m^3/h that can be passed through each m^3 of filter material while maintaining a given efficiency.

The bed volume determines the residence time of the waste gases in the filter bed, and depends on the structure and type of filter material. The following materials are used at present:

1) Composted garbage (20–40 mm) is sieved out of composted plants and organic domestic waste. This material is biologically highly active, but gives problems with flow obstruction by settled material and uncontrolled gas breakthrough. It is therefore mainly used in combination with inert material (clay, Styropor, and activated carbon) in compact equipment. The bed height is 1.0 m.
2) Coarse material (40 mm, 80 mm, 120 mm) sieved from composted plant material with a bed height of 1.0–1.5 m (depending on the humus fraction). The material is biologically very active, but has to be sieved annually.

3) Mixtures of brushwood/fibrous peat, brushwood/coconut fiber, or brushwood/heather are very suitable for flat filters, but can be difficult to procure. The bed height is 1.0 m.
4) Bark mulch from spruce and fir can be used if the material is homogeneously shredded to 50–150 mm. The bed height is ca. 1.2 m. Pine bark is unsuitable because it is friable, swells with moisture, and hence obstructs the flow.
5) Cracked root wood pieces, 60–180 mm long, can be used if the bed height is at least 1.5–1.8 m and is provided with a covering layer of ca. 0.2 m bark mulch for pressure equalization and protection from rain.

The function of a biofilter can be optimized by ensuring that the material has the following properties [7.100]:

1) Uniform structure (uniform flow and low pressure drop)
2) Adequate void volume (low pressure drop, low specific filter resistance, low energy consumption, good drainage and oxygen supply)
3) High portion of organic material (good additional supply of nutrients, long service life of the filter)
4) Large surface of the carrier material (large specific surface area for occupation by microorganisms)
5) No extreme pH fluctuations in the filter material (good buffer for variations in the waste gas)
6) Good water retention properties (constant moisture)
7) Low rate of decay (long service life, long maintenance intervals)
8) Low odor (low dispersion range of the odor, no unpleasant odor quality)

The balance of these properties is an important condition for optimum functioning of the filter.

The density of the filter material in the wet state is $\leq 1\ t/m^3$, depending on the type of material and its condition.

Design Details. Biological methods have only limited ability to smooth out peaks in the concentration of impurities in the waste gas. It is therefore preferable to operate the equipment continuously or to add buffers such as activated coke, to the system.

Biofilter materials become clogged by dust and fats from the waste gas. Also, certain operating ranges of temperature and moisture must be maintained. The crude gas must therefore be conditioned upstream of the biofilter (see Fig. 7.51). To prevent breakthrough of crude gas at the edge of the filter, a grid floor covering should be provided, extending to 1.0–1.5 m from the edge. At a bed height of ≥1.5 m, another edge screen 0.5 m wide should be added.

On concrete surfaces where biologically active material is present under pressure, e.g., on prefabricated slatted floors used in agricultural

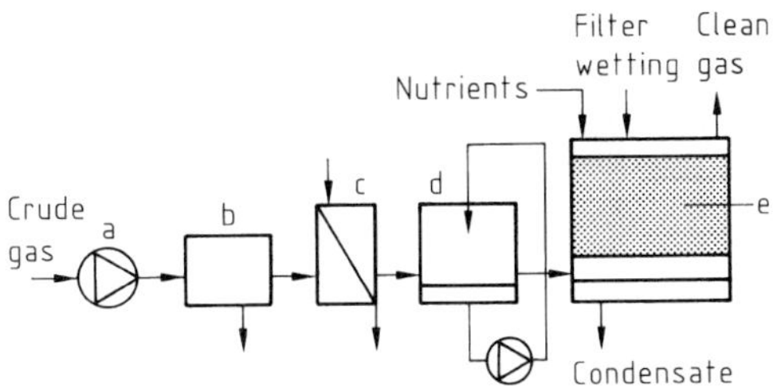

Figure 7.51. Preconditioning for a biofilter
a) Fan; b) Dust separator; c) Temperature controller; d) Scrubber; e) Filter bed

equipment, biological corrosion can take place. These surfaces should be sealed with liquid resin. Corrosion has so far not been found on the concrete walls or other equipment under the grid.

It is not necessary to provide a cover on flat filters, but a duct under the grid is required for seepage water. This water can be supplied to the scrubber for humidifying the crude gas, or can be sprinkled on the filter. Thus, the biofilter operates without producing waste water.

Maintenance and Repair. The filter bed must be kept free of plant growth, and must be regularly checked for settled material and gas breakthroughs. Fine filter materials has to be loosened and sieved at least once a year. In the case of cracked root packings, it is sufficient to loosen the top layer. As degradation of organic substances takes place in the filter material, an addition of ca. 5–10% of the material volume must be made annually. Complete replacement of the material is only necessary if the degradation process has proceeded to such an extent that the required air flow rate can no longer be achieved. To enable replacement or sieving of the filter material, the biofilter equipment has to be designed and constructed such that the grid is accessible for vehicles (wheel loader) or the filter can be maintained with a mobile gripper. The biofilter should be divided into several segments so that the entire production need not be shut down for maintenance work.

Process Monitoring. Biological waste-gas purification processes require a start-up period of 4–8 weeks, depending on the composition and concentration of the waste gas. It is only necessary to inoculate the filter if it is filled with an inert material. If waste-gas compositions match those of the associated sewage plants, sewage sludge can be sprinkled on the biofilter to increase its efficiency and to reactivate the biological activity rapidly after a long shutdown. The water content and pH of the material in the filter must be continuously monitored and controlled.

Additions of nutrients are only necessary if there is a deficiency in carbon, nitrogen, and phosphorus in both the waste gas and the filter material. Such a deficiency can occur after a long shutdown period or production changeovers.

Another important operating parameter is the continuous monitoring of the pressure drop in the filter bed. This indicates whether the filter bed has settled or whether surface densification (e.g., due to heavy rain) has occurred. Permeability properties are tested with smoke cartridges or fog generators.

7.5.4. Bioscrubbers [7.93]

Designs. Biological gas scrubbing can be carried out in bioscrubbers containing suspended microorganisms (see Fig. 7.52). They include an absorber in which the constituents of the waste gas are dissolved in the scrubbing liquid. This is collected in the absorber, and fed to an aerated activating tank. The regenerated scrubbing liquid is then recirculated to the absorber. The degradation efficiency depends on the active sludge mass formed from the microorganisms suspended in the scrubbing liquid and the dissolved pollutants. The limiting factor for this process is the distribution coefficient: for sparingly water-soluble substances in the waste gas extremely large amounts of water–sludge mixture have to be available and kept in circulation.

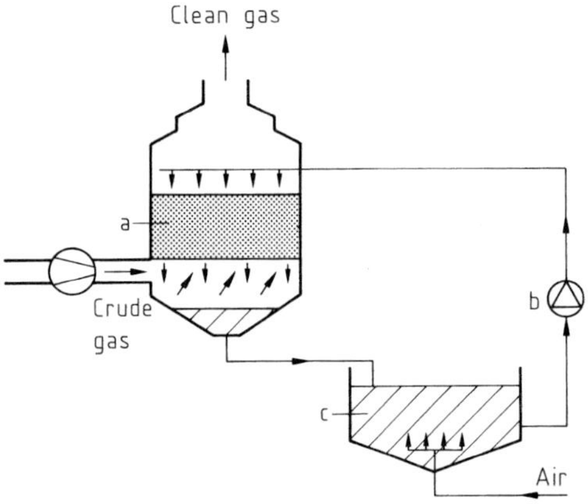

Figure 7.52. Bioscrubber with suspended microorganisms
a) Absorber; b) Activated sludge circulation; c) Activating tank

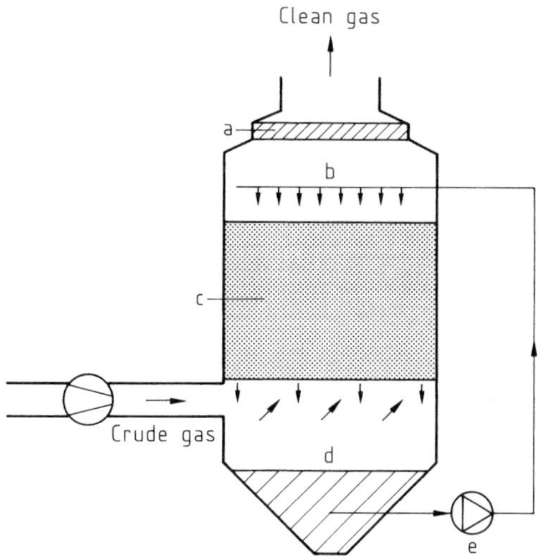

Figure 7.53. Trickling-bed reactor with fixed microorganisms
a) Demister; b) Nozzles; c) Packing material with biological coating; d) Pump sump; e) Water circulation

For economic operation, the distribution coefficient should be higher than the limit value of 0.0005 g/m^3 gas to 1 g/m^3 liquid. The addition of a high-boiling oil (e.g., a methylsilicone fluid) as a 10–30% dispersion to the water–sludge mixture can increase the absorption rate by a factor of 100–1000 [7.96], [7.103], [7.104].

In a trickling-bed reactor, absorption and biological degradation take place in the same part of the equipment (see Fig. 7.53). The scrubbing liquid is circulated over an inert packing material and collected in a pump sump. A biological film forms on this material, which can consist of loose filling material or a structured packing. The oxygen needed for biological degradation and the constituents of the waste gas to be degraded are absorbed by the water phase and transported to the biofilm where the reaction takes place. The degradation capacity depends on the specific surface of the packing, i.e., of the biofilm, and the residence time in the trickle bed.

Because mass transfer from the gas to the liquid phase takes place at the same time as the reaction on the biofilm of the absorber, the trickling-bed reactor can be used if the distribution coefficient (g/m^3 gas to g/m^3 liquid) of the components to be degraded is higher than 0.0005 [7.100].

Design Parameters. There are more design criteria to be considered with bioscrubbers and trickling-bed reactors than with biofilters. A correctly dimensioned and operated biofilter has a large buffering capacity which a bioscrubber cannot have, as the latter has to give the same purification efficiency with only a fraction of the biologically active volume. Whereas in biofilters only small parts can become overloaded or sour and respond to toxic substances, disturbances in bioscrubbers have an immediate effect on the entire biosystem.

Therefore, it has to be determined at a preplanning stage whether certain components of the waste gas that are harmful to the microorganisms or can clog the packing need to be removed by dedusters or prescrubbers before they reach the bioscrubbers. It must also be decided whether the composition of the waste gas makes a two-stage operation (acid/neutral) necessary.

The reaction products and cell mass become concentrated in the circulating scrubbing liquid. Part of the scrubbing liquid therefore has to be removed at regular intervals and replaced by fresh water or solution. Any solubilizing agents used have to be separated before the spent liquid is disposed of.

Thus, bioscrubbers are waste-gas purification devices producing wastewater (according to the German Water Management Act they are not wastewater purification devices). As a rule, the mixture of water and suspended material which is continuously removed from the installation does not contain heavy metals and other hazardous substances, and can therefore be disposed of in domestic sewage plants without problems. Ranges of characteristic parameters for bioscrubbers can be found in [7.93], [7.96], [7.101], [7.102].

If the necessary design parameters are unknown because the mixture of substances to be treated cannot be precisely determined, as in the case of emissions of odorous substances, pilot plant investigations have to be carried out. If the waste-gas load cannot be reproduced, reference substances may be used to test the purifying efficiency of the pilot plant.

Design Details. Materials that come into contact with intensive biological activity have to be protected from (biological) corrosion. Metals inside a scrubber, including stainless steels and seawater-resistant aluminum, have to be coated with plastic.

Packings in absorbers and trickling filters have to be designed and constructed such that a biofilm can form while a low pressure drop is guaranteed and the packing does not become clogged with activated sludge. A good impingement force of water will prevent clogging and flow obstructions, and there large inspection openings are necessary to enable an easy addition and removal of the packing material.

The scrubbing liquid distribution system above the packing material needs to be designed to operate without blockages. The water circulation pumps have to be suitable for transporting activated sludge, and need to run slowly so that the sludge flocks formed in the activating tank are not broken down. Displacement or slow-running plunger pumps are therefore recommended.

Process Monitoring. All the parameters relevant to operation of the process, such as temperature, pH of the scrubbing liquid, pressures in the scrubbers, oxygen and nutrient concentrations, and characteristics of the spent scrubbing liquid have to be determined continuously or at regular intervals for process control. Maintenance at regular intervals is therefore unnecessary, as the packed bed need only be cleaned when the pressure drop indicates flow obstruction. When biological scrubbing equipment is in the initial start-up phase, e.g., after inoculation with activated sludge from an associated sewage treatment plant, natural selection of the microorganisms (adaptation) takes place for 4–6 weeks. Significant changes or fluctuations in the crude gas composition may require re-adaptation, which can also take several weeks.

During plant shutdowns when no pollutants are added via the waste gas, nutrients have to be added to the scrubbing liquid by metering devices. The biological oxygen demand must be continuously supplied by aeration of the activating tank or by passing indoor air through the absorber.

Process Optimization. The purification efficiencies of biological plants can be improved by connecting a trickling-bed reactor with a biofilter to form a two-stage process. The first stage acts as a neutralizer, reacting with acid or basic components of the waste gas. The biofilter is then used without special control techniques to reduce the residual pollution.

Further developments have led to the biosorber principle [7.105], a biofilter and a bioscrubber combined in a single unit. This consists of a packing of inert material on the surface of which nutrient media specially suitable for the application are grown.

In the biomembrane principle [7.106], strongly hydrophobic waste gas components diffuse under pressure through a silicone tube located in a mixture of activated sludge, water, and, if necessary, solubilizing agent. The tube is coated on the outside by a biofilm, so that the biological reaction takes place in the biofilm and the waste gas components need not be dissolved in a liquid. Waste gas components that are not picked up by the biofilm can then be treated in the activating tanks.

Measurement and Assessment. Determination of the gaseous components of a waste gas before and after biological purification is carried out by a wide range of methods, including gas chromatography, and spectrometry. To determine odoriferous substances, olfactometry is used for measuring emissions [7.107], and field inspection with panelists is used for ambient air pollution [7.108]. The following criteria should be taken into account:

1) Determination of the odor concentration in the waste gas is by itself not a criterion for odor nuisance. The odor intensity [7.109] and the position on the hedonic scale [7.110] also need to be determined.
2) The odor of the crude gas is changed by the biological treatment. This "biological smell", which is undetectable a few meters away from the equipment, also has an odor concentration, so that the calculated efficiency gives a false impression of the purification performance of the equipment.
3) Biofilter material, e.g., composted domestic waste, can have its own specific odor.

Assessment of the ability of a biological process to remove odor emissions therefore has to be carried out in accordance with the VDI Guidelines [7.93], [7.99], which state that the aim of the purification process is achieved when the characteristic odor of the crude gas is undetectable after purification in the biofilter or bioscrubber.

7.6. Flue-Gas Purification in the Incineration of Hazardous Waste

In the incineration of hazardous waste, the methods and equipment described in Sections 7.1–7.5 are used together with additional technical and operational procedures to enable high standards of environmental protection to be met. As a result of their composition, condition, or quantity, such materials can endanger health, the atmosphere, or water supplies, or they can constitute an explosive hazard. Typical substances include:

1) Solvent mixtures
2) Industrial sludges
3) Paint residues
4) Plastic waste
5) Inorganic waste (salts)
6) Halogenated hydrocarbons
7) Petrochemical residues
8) Halogenated acids/alkalis
9) Pharmaceutical waste
10) Acid sludge
11) Waste oil
12) Pesticides
13) Herbicides

If hazardous wastes cannot be treated or reused and direct dumping is impossible, they must be incinerated to destroy their organic constituents and produce inert combustion products that can be reused or safely dumped. By far the largest quantities of hazardous waste are incinerated in rotary furnaces. This enables the incineration of solids, liquids, and pastes and the thermal treatment of substances that produce molten slags during the incineration process. Solids and pastes are added by crane loaders and liquids through burners and lances, or barrels are charged to the furnace intact.

A simplified flow diagram of a complete plant is shown in Figure 7.54 and a section through an actual plant in Figure 7.55.

In the incineration process, the elemental constituents of the waste are retained, but their chemical and physical forms are changed. To prevent unacceptable contamination of the environment by hazardous substances, either by their entry into the atmosphere or as solids, the operation of hazardous-waste incineration plants must include comprehensive measures to reduce emissions.

Typical concentrations of pollutants in crude gas produced by "publicly" operated incineration plants used to treat waste from trade, business, and industry are:

Dust	1000–10 000 mg/m^3
Inorganic chlorine compounds (as HCl)	3000–10 000 mg/m^3
Inorganic fluorine compounds (as HF)	50–550 mg/m^3
Sulfur compounds ($SO_2 + SO_3$)	1500–5000 mg/m^3
Nitrogen oxides (as NO_2)	100–300 mg/m^3
Water vapor (H_2O)	6–20 vol%

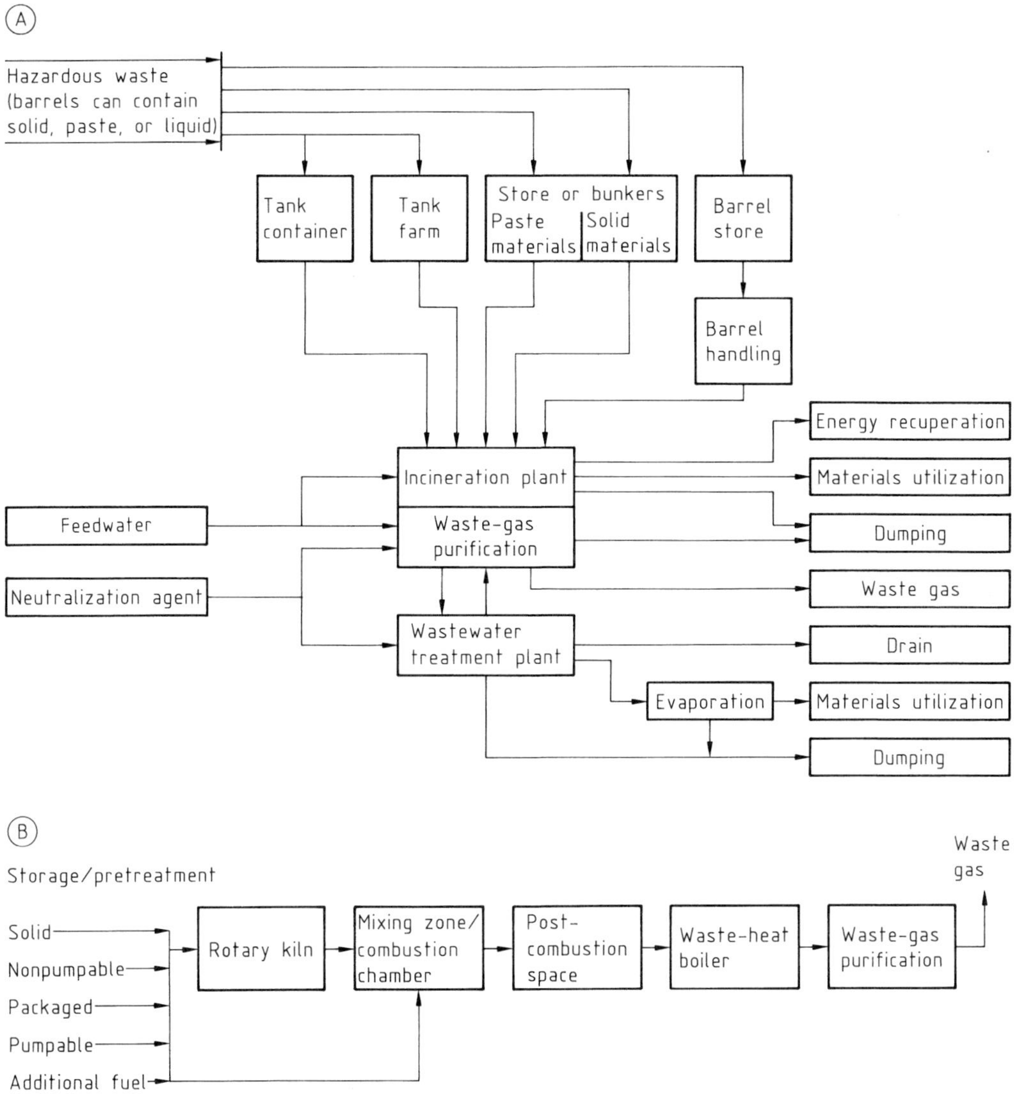

Figure 7.54. A) Simplified flow diagram of a hazardous-waste incineration plant
B) Flow diagram of a rotary furnace

In plants operated by the chemical industry for the incineration of production residues with high contents of reactive substances, a much wider range of concentrations can occur in the waste gas. Table 7.5 lists the legally required emission limit values for existing and newly constructed plants in Germany. The emission limit values in an EC Guideline proposed in 1993 and given legal force in the 1994 plan are also given. These are to be incorportaed in National Law within two years.

The emission limits for conventional air pollutants have been reduced considerably, and a new limit value for the emission of polychlorinated dibenzodioxins and dibenzofurans (PCDD/PCDF) has been introduced. Whereas emissions of conventional atmospheric pollutants are essentially a function of the type, amount, and composition of the waste materials incinerated, dioxins and furans and other organic hydrocarbons are formed by complex decomposition and side reactions in the combustion and cooling processes.

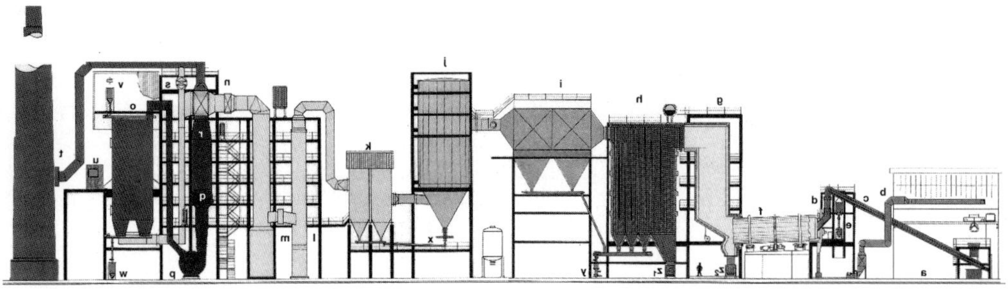

Figure 7.55. Hazardous-waste incineration plant
With kind permission of the L. & C. Steinmüller GmbH, Gummersbach, Federal Republic of Germany

Table 7.5. Waste incineration emission limit values simplified synopsis of EG Guideline/17th BImSchV, Germany (limit values in mg/m^3 based on STP and 11% O_2)

Component	EG Guideline	17. BImSchV
Carbon monoxide		
Daily mean value	50	50
Hourly mean value		100
90% of all measurements (peak concentration) in 24 h		150
95% of all 10 min mean values in 24 h	150	
Daily mean values (DMV)		
Total dust	10	10
Organic substances (total-C)	10	10
Hydrogen chloride	10	10
Hydrogen fluoride	1	1
Sulfur dioxide	50	50
Nitrogen dioxide	(none)	200
Half-hour mean values (HMV)		
Total dust	30	30
Organic substances (total-C)	20	20
Hydrogen chloride	60	60
Hydrogen fluoride	4	4
Sulfur dioxide	200	200
Nitrogen dioxide	(none)	400

According to current knowledge, most of the dioxins and furans found in the flue gas are formed from unchlorinated organic material in the presence of a chlorine source. In the temperature range 450–250 °C, catalytic formation of chlorine from metal chlorides in the presence of oxygen is the essential initiation reaction. The mechanism of formation assumes the presence of chlorine and oxygen, and this is unavoidable in the flue gases produced by incineration of hazardous waste. Dioxins are formed in a solid state reaction, and are mainly bonded to particles.

The primary emission control measures are meticulous planning of the incineration process and control of the firing process. The incineration plan determines how the waste materials are metered and added. If possible, they are size reduced and mixed to ensure that the furnace feed has consistent properties, including calorific value, content of components such as halogens, sulfur, nitrogen, heavy metals and organic compounds, and other properties relevant to the incineration process such as ease of ignition and burning rate. Drums to be added intact to the incinerator for reasons of operational safety must be of a size and energy content suitable for the firing equipment. Owing to the heterogeneous nature of the fuel, the usual systems for controlling the combustion air are unsuitable. The amount of combustion air is often fixed and the waste material is added in a controlled manner to give the required temperatur, controlling the CO content at the same time. Emission control measures in the incineration process affect almost exclusively the carbon monoxide level and the total content of organic compounds.

The special characteristics of the rotary furnace, such as the frequent change of fuel for the burners and sludge lances and feeding of drums lead to large, brief fluctuations in the flue gas composition. High air pollutant concentrations that change according to the feed and the combustion conditions must be dealt with by secondary emission control measures. For a flue-gas cleaning system for a hazardous-waste incineration plant today, not only must the capacity to remove pollutants be high, but also the following requirements must be met:

1) The process must be designed so that no wastewater is produced
2) Residues from the various processes used in waste-gas purification must be reused as far as economically possible
3) Residues that cannot be reused must be safely dumped

The costs of dumping depend on the nature of the waste material. For example, the disposal of residues containing heavy metals, dioxins, and soluble salts in an underground dump is approximately five times as expensive as disposal in an open hazardous-waste dump supervised in accordance with modern technical regulations. Therefore, on cost grounds, every operator endeavors to concentrate potentially environmentally hazardous substances as far as possible, so that only small amounts of residues will need expensive treatment. Finally, flue-gas purification must function with a high degree of safety and at low operating cost.

Flue-gas purification equipment for hazardous-waste incineration plants usually incorporates a dry dedusting stage and several wet scrubbing stages. A design of such elements including spray tower scrubbers is shown in Figure 7.56. The combustion exhaust gases are cooled in a steam-generating heat exchanger to ca. 250 °C. The design concept requires

1) Preliminary dedusting of the waste gases by a double-field electrostatic filter
2) An initial scrubbing stage to cool the combustion gas to the cooling limit temperature (quench), with removal of HF, HCl, and heavy metals
3) A second scrubbing stage to remove SO_2, producing the required waste-gas purity
4) Wastewater treatment including gypsum production and external evaporation of the $CaCl_2$ solution produced

The first scrubbing stage, also termed the prescrubber, is operated at pH <2 to remove HCl, HF, and heavy metals. At this pH, the chloride content of the wastewater can be high, which is indispensible or advantageous for the external evaporation step. In the first scrubbing stage, only a limited amount of SO_2 is removed. The sulfite formed is oxidized to sulfate by aeration of the sump. The second scrubbing stage, also termed the main scrubber, predominantly removes SO_2. In the spray tower, the suspension used for the scrubbing is introduced via nozzles at different three levels. The droplet content on the pure gas side should be as low as possible. The droplets are removed by an internal and an external demister. Sulfur dioxide scrubbers are operated at pH 6.5–7; the absorption medium is milk of lime. The salts (mainly gypsum) formed by reaction with SO_2 are removed in the wastewater treatment stage. To improve the ease of water removal from the insoluble salts and to ensure high purity of the separated sulfate (>99% $CaSO_4$ in the solid), the sulfite formed during the absorption is oxidized to sulfate by passing air into the sump of the scrubber. Fine crystals must be avoided as far as possible in the sulfate crystallization. Therefore, a solids con-

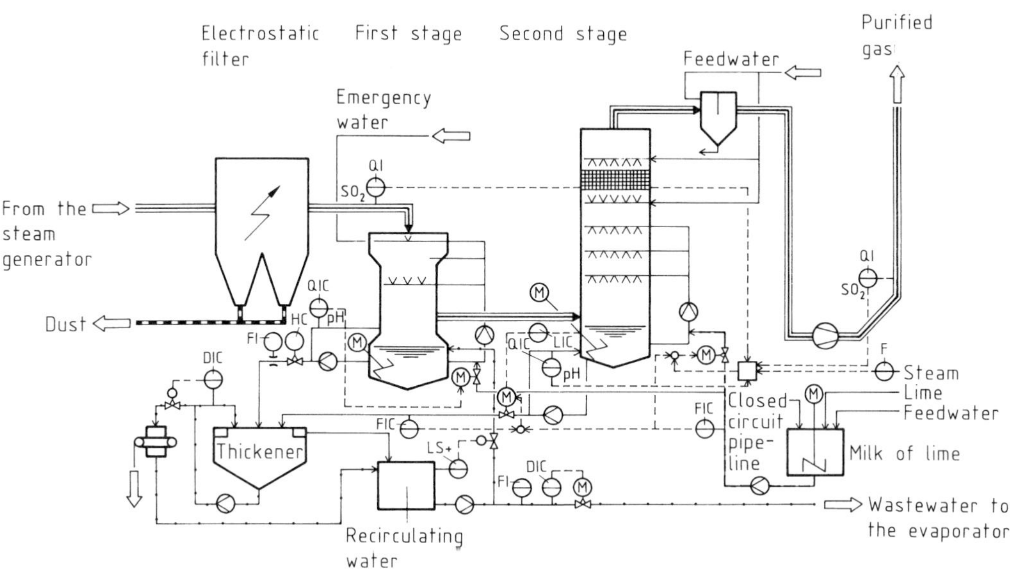

Figure 7.56. Dedusting and wet scrubbing stages as elements of flue-gas purification system

tent of ca. 8% is maintained in the wash liquor suspension to ensure that enough seed crystals are produced during the crystallization and spontaneous nucleation is avoided. The $CaSO_4$ formed is removed from the scrubber overflow by a thickener and band filter. The clear fraction is returned to the scrubbers, and a small percentage is taken directly to wastewater treatment.

Special technical measures must be taken to deal with fluctuations in the concentration of the waste-gas constituents. By using rapidly responding control systems and by directly feeding the absorption medium into the scrubber inlet pipes, an adequate quantity of absorption medium is provided at all operating points. The rate of addition of the suspension is controlled in accordance with the concentrations measured in the crude gas after the electrostatic filter. Both scrubbers include normal methods for measuring and controlling pH, density of the suspension, and flow rates by cascade controllers. Temporary supersaturation in the first scrubbing stage can lead to incrustation and scaling. This is prevented by spraying the top of the scrubber with fresh water. This mechanical and chemical rinsing also provides the water required for the process. Hydrogen chloride peaks not eliminated in the prescrubber are reduced to permissible levels in the main scrubber. Any significant chloride loading in the main washer is undesirable because it has an unfavorable effect on the equilibrium conditions for SO_2 removal. Therefore, an external demister is included between the prescrubber and the main scrubber. As well as the cascade control system in the main scrubber, additional equipment is necessary in the liquid circulation sytem to maintain the liquid balance and achieve the necessary oxidation and crystallization of the insoluble solids simultaneously with the removal of SO_2.

The spray tower scrubber can be used as an example of the problems to be solved when transferring a process technology to full industrial scale. In other full scale plants, spray scrubbers, rotary scrubbers, venturi scrubbers, or packed tower scrubbers are used. Sodium hydroxide solution can be used as an alternative to a milk of lime as absorption medium. Sodium salts are soluble in water, which avoids the danger of incrustations in pipework and equipment. Instead of the electrostatic filter, a cloth filter can also be used. The choice depends on the material to be incinerated. Cloth filters, which are usually operated at $\leq 200\,^{\circ}C$, are unsuitable, if there is a danger of clogging by fine dust produced in the incineration process, or if the dust is hygroscopic and forms coatings on the filter that cannot be removed. Electrostatic filters are unsuitable if the dust cannot be electrostatically precipitated, if the electrical resistance of the dust layer is too high, or if the dust has such a low density (e.g., in the incineration of chlorosilane) that it is not removed on trapping the electrostatic elements. Therefore, in some areas of the chemical industry dry removal of dust is infeasible. For wet-dust removal, single-stage or multistage venturi or rotary scrubbers can be used (see Section 7.1.4). These operate on the principle of contacting the dust-laden waste gas with a relatively finely dispersed liquid. The dust particles come into contact with the liquid droplets, are wetted, and are collected along with the liquid. The dust must therefore be wettable.

The liquid from the scrubbers goes to a water-treatment stage which includes the precipitation of heavy metals, flocculation, and solids separation. The heavy metal sludge also contains dioxins bound to the dust collected by the scrubbers. The clarified liquid is evaporated externally (if the dry product can be sold) or internally by a spray dryer that forms part of the main waste-gas stream. Internal evaporation produces a relatively large amount of a dust–salt mixture containing dioxin and heavy metals. This residue is usually disposed of by underground dumping.

In the process described above, mercury is removed in the first, acid scrubbing stage. The residual concentration in the purified gas depends on the vapor pressure of mercury. If mercury concentration peaks occur or if mercury is present in the scrubber in metallic form (e.g., caused by dissolved sulfur dioxide), this can lead to the breakthrough of mercury, which can then exceed emission limit values. This system also does not include any process to limit the emission of nitrogen oxides. If large amounts of aerosols are present, there is an increase in the dust content and hence also in the dioxin content of the purified gas.

Thus, the process technology for waste-gas purification must be improved for new plants. There are many proposals based on laboratory or pilot plants. Emissions of nitrogen oxides can be reduced by the selective catalytic reduction (SCR) process as used in power station technology (see pp. 145–146). Only the low-dust variant is relevant. The waste gas must be reheated to ca. 250–300 °C before it enters the catalyst, by heat transfer from the hot to the cold waste gas or by heating with burners. Experiments have shown that when catalysts are used to reduce emissions of nitrogen oxides, organic trace substances such as dioxin can also be catalytically oxidized. Reaction efficiencies (removal efficiencies) of ca. 90 % for PCDD and PCDF have been found in semitechnical experiments. The oxidation catalyst would then perform two of the additional emission-control tasks required. A particular advantage is that no additional residue is produced in this process.

In many hazardous-waste incineration plants, nitrogen oxide concentrations in the crude gas are <300 mg/m^3. To achieve the legally required value of 200 mg/m^3, a reduction of 30 % is necessary. This can be achieved either by the selective noncatalytic reduction (SNCR) process in which ammonia or aqueous ammonia solution is added to the waste gas at 850–1000 °C. As the reaction does not proceed to completion, ammonium chloride can be formed as an aerosol. Furthermore it can be assumed that some of the ammonia will go into solution in the first scrubbing stage and must then be removed in a stripping operation.

To remove aerosols, a wet electrostatic filter can be used as final purification stage. The purified gas leaving the last scrubbing stage is cooled until droplets form and agglomerate, and the droplets are removed by an electric field. Decreasing the dust content of the purified gas automatically reduces dioxin emissions. If cooling before the wet electrostatic filter gives a temperature of up to 20 °C below the cooling limit temperature, the mercury content is also reduced because of the lower vapor pressure.

Multifunctional separations are achieved by activated carbon filters (see Section 7.4), which have undergone rapid development in recent years. They have been tested in full-scale experiments, and are already used in one hazardous-waste incineration plant in Germany. Activated carbon filters, operated as fixed or moving beds (Fig. 7.57), adsorb heavy metals and organic trace substances simultaneously. They also adsorb any hydrogen chloride and sulfur dioxide that have passed through the earlier stages. If they are used on the purified-gas side, the consumption of activated carbon is very low. Entrainment reactors can also be used, in which a mixture of activated calcium carbonate with a proportion of activated carbon is blown into the waste gas and then removed on a cloth filter.

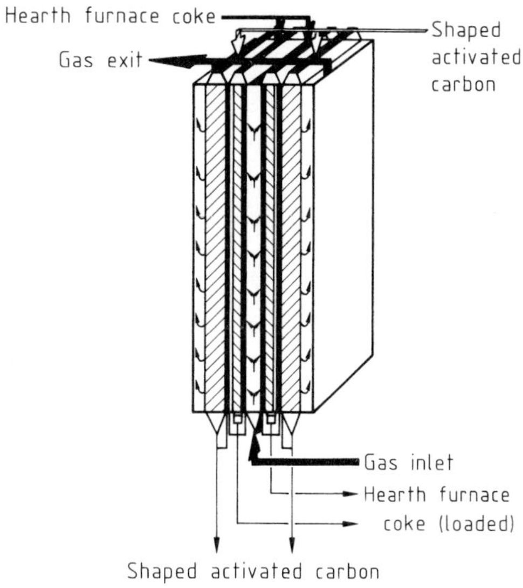

Figure 7.57. Hearth furnace coke adsorber–shaped activated carbon catalyst (countercurrent flow)

Using this as the final stage waste-gas purification gives results as good as those obtained with moving or fixed bed activated carbon filters. Examples of analysis figures for the purified gas in full-scale plants are given in Table 7.6. Technical-grade hydrochloric acid can be produced from the chloride solution generated in the first scrubbing stage. For this, the first scrubbing operation is carried out in three stages, in which dust and heavy metals are removed in the first stage, the required concentration of HCl is obtained in the second stage, and the third stage performs the final purification. The hydrochloric acid is removed and distilled. The distillation process would replace an external evaporation of chlorides, which are generally difficult to market.

All these processes for additional waste-gas purification and reuse of residues can lead to complex design and process technology (Fig. 7.58). Plants of this type have not yet been realized. In Germany the first plants of this type will be put into operation in 1996 and 1997.

It was formerly intended that hazardous-waste incineration plants should be designed for the treatment of all types of hazardous waste using all techniques available. These requirements now appear to be changing. It is apparent that the cost of incineration is increasing rapidly, and cheaper methods for disposing of hazardous waste will be sought.

Table 7.6. Comparison of emission limit values according to EC Regulations and 17th BImSchV (Germany) with initial results from a hazardous waste incineration plant with a carbon filter with an activated coal–lime cloth filter. Values are in mg/m^3 at STP based on 11% O_2

	Emission limit (mean values)		Daily mean values Dec. 91–Feb. 92
	17. BImSchV		
	Daily	Half-hourly	
Total dust	10	30	<1 [a]
$\sum C_{org}$	10	20	<2
CO	50	100 [c]	25
HCl	10	60	<2 [a]
HF	1	4	<0.5 [a]
So_2 (SO_2, SO_3)	50	200	8
NO_x (NO, NO_2)	200	400	120
Heavy metals			
	∑0.05 Cd, Tl		<0.007 [b]
	0.05 Hg		<0.002 [a]
	∑0.5 Sb, As, Pb, Cr, Co, Cu, Mn, Ni V, Sn		<0.2 [b]
Dioxins and furans as ∑TE [d]	0.0000001		0.000000028 (=0.028 ng)

[a] Below detection limit. [b] Sometimes below detection limit. [c] Hourly. [d] TE = Toxic equivalent values.

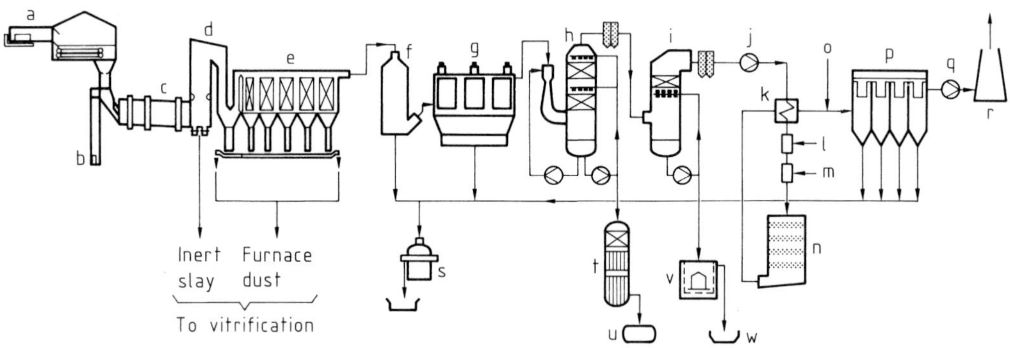

Figure 7.58. Possible plant design for hazardous-waste incineration plant
a) Material supplied in containers; b) Charging of drums; c) Rotary furnace; d) Post-combustion equipment; e) Waste-heat boiler; f) Evaporative cooler; g) Electrostatic scrubber; h) HCl scrubber; i) SO_2 scrubber; j) Suction pump 1; k) Heat exchanger; l) Burner; m) Aqueous ammonia addition; n) Combination catalyst; o) Addition of activated coke–lime mixture; p) Filter/adsorber; q) Suction pump; r) Stack; s) Fly ash vitrification; t) Hydrochloric acid recovery; u) Container for hydrochloric acid; v) Gypsum dewatering; w) Gypsum silo

The positive results obtained using the entrainment method and the fact that the SNCR method for removing nitrogen oxides has been very effective have led to a reassessment of the situation. It is likely that in future hazardous waste will be more carefully sorted to match the material to the incineration method used. Wastes that do not require high-quality gas purification will then be treated in the appropriate plants using simpler techniques like dry or semidry absorption methods.

8. Photochemical Degradation

8.1. Introduction

In terms of mass and energy turnover, photochemical degradation is one of the largest chemical reactions in nature. It is ultimately an oxidation process involving numerous physicochemical intermediates. It takes place in the troposphere and makes an essential contribution to global recycling because of its mineralizing effects, leading to the formation of carbon dioxide and water. In simple terms, photochemical degradation starts with the formation of reactive species, usually radicals, when the molecules of certain atmospheric trace substances are cleaved by sunlight. These species then react with other, mineralizable trace substances.

These in turn can be of natural and anthropogenic origin: organic and inorganic components of sea spray, exudates from forests, products of biotic transformation processes, agriculture, industrial emissions, etc. [8.1]–[8.3].

The most important single process which brings about natural purification of the atmosphere is the formation of hydroxyl radicals, which oxidize hydrocarbons and their derivatives, e.g.,

$$O_2 + NO_2 \xrightarrow{h\nu} NO + O_3$$

$$O_3 + H_2O \xrightarrow{h\nu} O_2 + 2\,^{\bullet}OH$$

$$2\,^{\bullet}OH + CO \longrightarrow H_2O + CO_2$$

Here, as in most other cases, it is indirect photochemical degradation that is taking place. Direct photochemical degradation—the absorption of a light quantum followed immediately, or after rearrangement, by oxidation—is comparatively rare and plays only a minor role in natural purification of the atmosphere and in special cases of industrial air purification. The term photochemical degradation is often used in the literature for reasons of simplicity when indirect photochemical degradation is really meant.

The hydroxyl radical is the most important oxidizing agent formed photochemically, although it is present at a concentration of only 0.5×10^6 molecules/cm^{-3}. However, it reacts rapidly and fairly unselectively, is continuously formed, and is involved in the first stage of most oxidative transformations in the troposphere. In contrast the activation energy for direct oxidations by molecular oxygen in the troposphere is so high that measurable effects are not accomplished in reasonable periods of time.

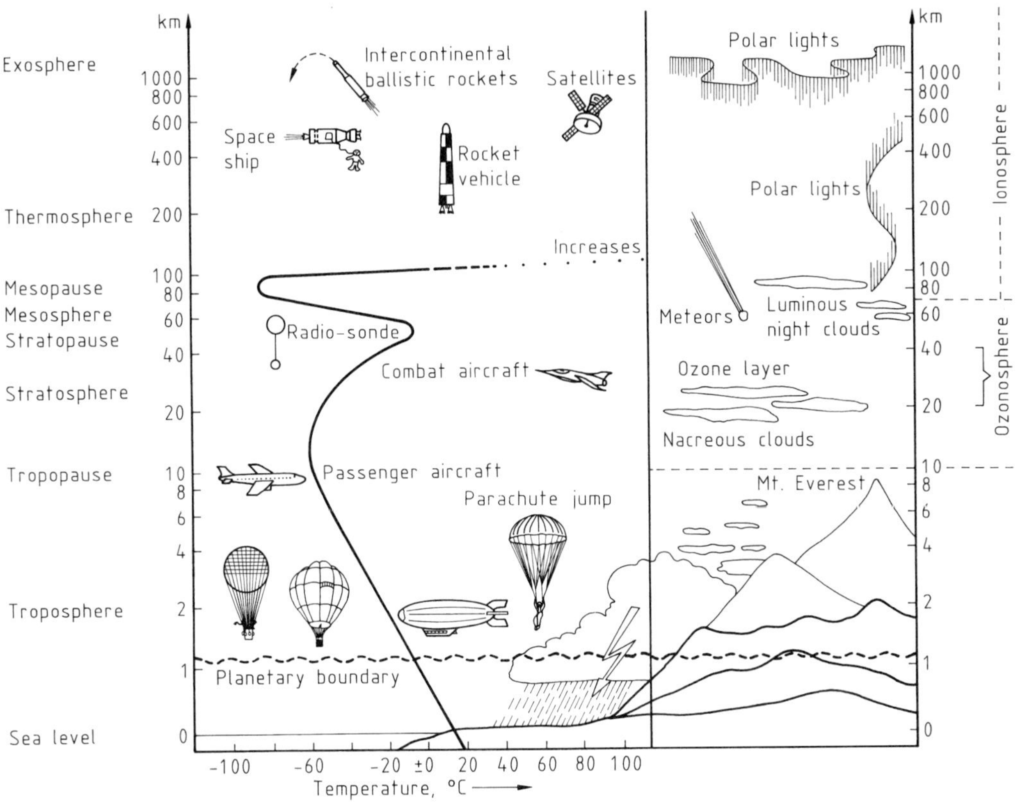

Figure 8.1. Atmospheric strata with temperature profile
(Origin: Funds of Chemical Industry "Ecological Compartment Air", 1987)

For determining the lifetimes of chemical substances in the environment or of their residence times in individual environmental compartments, atmospheric conditions and processes are of great importance, especially in the case of emissions into the atmosphere of homogeneous gases or substances bound to particles. This also holds for deposition and other immissions from the air. Interactions with energy fluxes in the atmosphere and with atmospheric cycles of trace substances are particularly important [8.4]–[8.9].

The role of the atmosphere has many aspects: air acts as a container, vehicle, reactor, reactant, and filter, these functions sometimes being mutually interdependent. Here, the decisive feature is the characteristic temperature profile with increasing elevation (Fig. 8.1) [8.2], [8.10], [8.11].

From ground level, the temperature at first decreases linearly by ca. 6.5 °C/km. This zone, the troposphere, extends to an altidue of ca. 10 km (9 km at the poles, 15 km at the equator). It is formed by rising masses of air heated by the earth's surface and then subsequently cooled. Above the boundary zone (the tropopause), the stratosphere begins, and temperature now increases with altitude as a result of solar radiation; short-wave sunlight is directly absorbed. The absorbing substance is ozone in the lower stratosphere (altitude ca. 25 km) where the O_3 concentration is ca. 10^{12} molecules/cm³. The ozone concentration decreases rapidly both above and below this zone. The presence of ozone in the lower parts of the troposphere is a result of other formation mechanisms (see Section 8.3).

The inversion of the temperature gradient above the tropopause is essential for maintaining the stability of the stratosphere. It causes the formation of a zone which acts as a barrier layer that prevents rapid ascent of trace gases from the earth's surface into the stratosphere. The vertical diffusion coefficient is at a minimum at this point, and penetration through this layer takes several years.

The composition, i.e., the nature and amount of the principal reactants in the zones of the troposphere close to the earth's surface, determines the processes that lead to degradation (Fig. 8.2).

Apart from concentration, the various reaction rate constants of the trace substances are decisive (see Section 8.5). The only source of energy is the sun. The energy flux for vertical radiation has a mean value of 1378 W/m², corresponding to a total annual energy intake of 1.56×10^{18} kW · h, with the radiation maximum at 480 nm (yellow-green). Of this, ca. 70% is absorbed at the earth's surface on a cloudless day in Central Europe (Fig. 8.3).

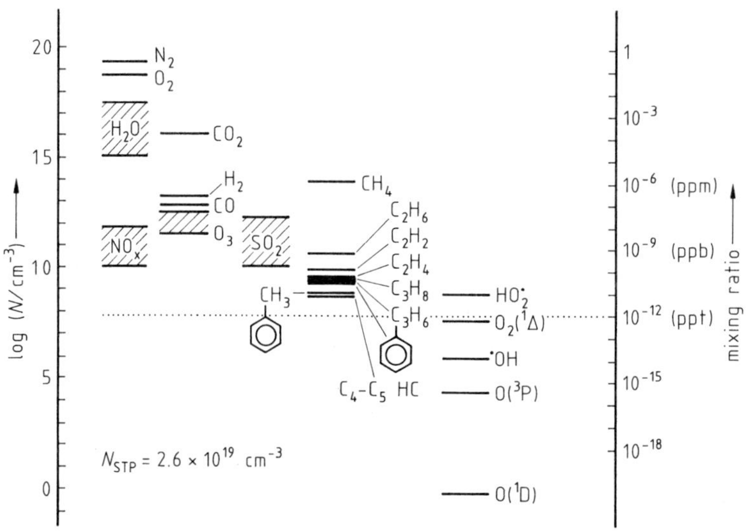

Figure 8.2. Characteristic concentrations (in molecules/cm³) and volume mixing ratios (= mole fraction) of tropospheric main and selected trace gases
The hatched areas indicate common variabilities
HC = hydrocarbons

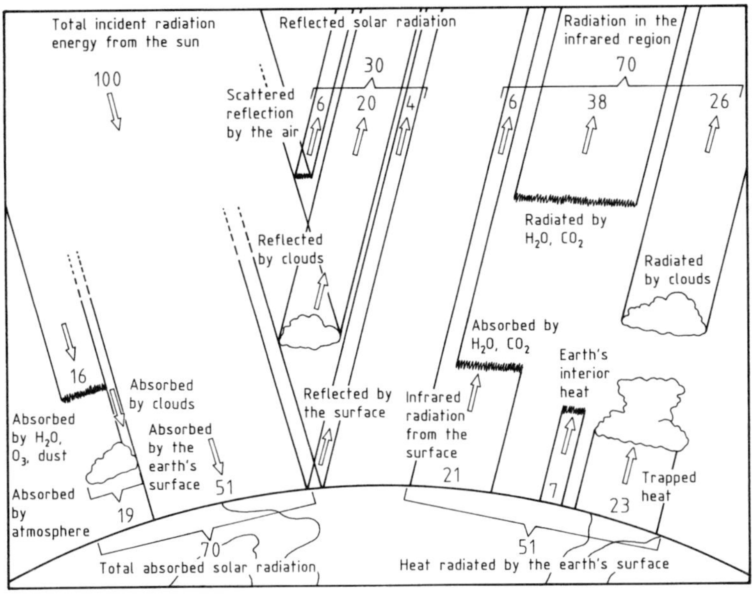

Figure 8.3. Percentage energy balance of the earth's surface
(Origin: Funds of Chemical Industry "Ecological Compartment Air", 1987)

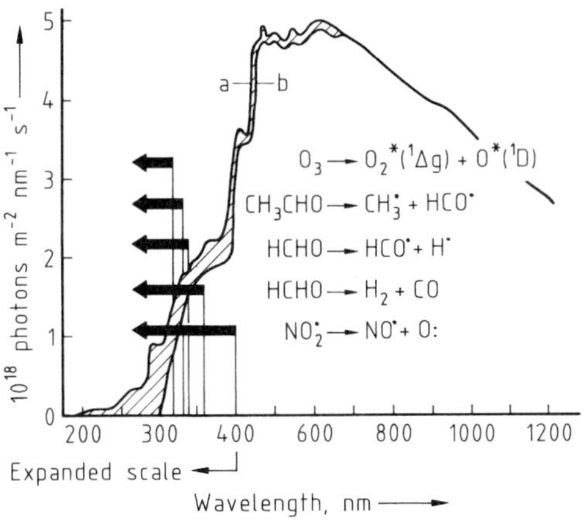

Figure 8.4. Spectral distribution of sunlight and wavelength limits for some important photolytic reactions
a) Incident sunlight reaching the outer atmosphere; b) Incident sunlight reaching the earth's surface
Hatched area: portion of light absorbed in higher atmospheric strata

In the lower troposphere the fraction of the radiation absorbed is 17%. This corresponds to a daily amount comparable to annual global energy consumption by humans today. After subtraction of the dissipated light from the incident light, a proportion remains for tropospheric photochemistry [8.12]. Figure 8.4 shows the spectral distribution of sunlight and the limiting wavelengths for important photolysis reactions.

The formation of photooxidants and their reaction with pollutants entering the troposphere primarily depends on the photodissociation of O_2 and O_3. The fundamental reactions were formulated as early as 1930, albeit for the stratosphere [8.13].

8.2. Natural and Anthropogenic Emissions

Photoinduced degradation of emissions that enter the atmosphere from the earth's surface is achieved by physical processes, particularly those occurring in the zones of the troposphere near ground level, and by trace gases that are already present there, especially carbon monoxide, methane, and nitrogen oxides. Via formation of photooxidants, they determine the rate of degradation of other trace substances emitted into the atmosphere. Several billion tonnes of such substances are emitted from the earth's surface annually. They are chemically transformed in the troposphere, mainly by mineralization to form oxidation products, and return as such to the earth's surface [8.14], [8.15].

The concentration of a trace substance is determined by its rates of emission and degradation. The most important of such substances (Table 8.1), originating from various sources, are

Table 8.1. Global annual emissions*

	Emission rate, 10^9 kg/a	Main sources, 10^9 kg/a	
Carbon dioxide	830000	respiration, biological degradation	370000–520000
	(700000–1000000)	oceans	300000–450000
		combustion of fossil fuels and forest clearance	20000–30000
Carbon monoxide	3400	combustion of biomass	400–1600
	(1700–5800)	combustion of fossil fuels	500–800
		oxidation of hydrocarbons in the atmosphere	800–2200
Hydrocarbons except CH_4 (calculated as C)	1000	trees (terpenes, isoprenes)	600–1300
	(640–1400)	motor vehicles	30–50
		solvents	10–20
Methane	500	swamps and geothermal activity	100–300
	(350–1600)	rice paddies	50–300
		animal husbandry (ruminants)	100–200
		termites	100–150
Sulfur compounds (calculated as SO_2)	400	combustion of coal and mineral oil	160–240
	(290–500)	oxidation of sulfur compounds from oceans, swamps, etc.	80–200
		volcanoes	10–20
Nitrogen oxides excluding N_2O (calculated as NO_2)	160	combustion of fossil fuels	40–80
	(70–320)	combustion of biomass	10–80
		lightning	7–70
		atmospheric oxidation of ammonia	3–30
Ammonia	100	biological degradation, soil	up to 1200
	(30–1200)	animals	20–30
		wastewater treatment	10
Dinitrogen oxide	50	soil	up to 600
	(10–600)	oceans	6–15
Hydrogen	50	hydrocarbon oxidation	20–50
	(40–130)	industry and traffic	10–60

* This table lists the probable total annual quantities of emissions of various substances. The spread of values quoted in the literature is given in parentheses. Because the estimates of the main emission sources show such wide divergence, only the ranges of these estimates is shown.

measured and calculated [8.2], [8.4], [8.16]–[8.22].

In many countries of the world (Table 8.2), certain emissions are continuously measured [8.16]: CO_2, CH_4, and N_2O because of the greenhouse effect, and CFCs because of their effect on the ozone layer.

By 1978, over 1300 trace substances had been identified in the atmosphere [8.20]. Substances of low vapor pressure enter the atmosphere bound to dusts [8.6], [8.23]–[8.25], and take part in photochemical degradation like other aerosols (see Sections 8.2 and 8.5.2). Horizontal transport processes occur in this way, and often removal of transformation products from the troposphere by precipitation. Smog, which can have a variety of compositions, modes of formation, and reaction paths is also included here, especially in connection with reaction cycles, atmospheric lifetime, and quality assessment.

So-called London smog differs from Los Angeles smog, also referred to as photosmog. London smog was formed when the smoke from domestic fires typical of the technological state of the art in the early 1900s combined with fog. Owing to the use of sulfur-containing fuels, the smog was acidic, and catalytic oxidation by heavy metal containing dusts converted sulfur dioxide to the trioxide. This type of smog occurred particularly under conditions of atmospheric inversion typical of the winter. Today, improved combustion technology and low-sulfur fuels have led to its virtual disappearance worldwide.

Los Angeles smog is also formed in inversion conditions but is caused by intense solar radiation acting on emissions from motor vehicles. Volatile organic compounds (VOCs) behave similarly, with the exception of methane. Systems of this general type are referred to as photosmog

Table 8.2. Main emissions by countries (10^3 t/a)

	CO_2			CH_4	CFCs	N_2O
	From energy use	From industrial processes	Total			
Canada	435 400	25 700	461 100	3 700	14	92
United States	5 035 000	31 950	5 066 950	29 000	90	428
Japan	1 078 900	46 600	1 125 500	1 338	111	47
Germany	956 900	25 000	981 900	6 060	15	220
Italy	418 500	92 078	510 578	3 646	25	42
Spain	236 200	25 764	261 964	1 400	11	
Sweden	56 300	3 250	59 550	490	2	11
United Kingdom	608 400	5 981	614 381	4 844	31	109
OECD	10 438 660	372 880	10 811 540	63 840	350	
World	21 613 300	733 850	22 347 150	274 930	480	

[8.2], [8.26]. It consists of photooxidants—products of the oxidation of VOCs by nitrogen oxides. Generally, these are organic nitrates and peroxo compounds such as hydrogen peroxide, methyl hydroperoxide, peroxyacetic acid, peroxynitric acid, and peroxyacetyl nitrate. The most important component of photosmog is ozone, which far exceeds all other compounds in concentration [8.27]. The formation and reactivity of photosmog depend very much on the time of day, and follow complex mechanisms (see Section 8.3.2).

Information about the earlier condition and behavior of the atmosphere is provided by investigations of sample cores from boreholes drilled in the ice of polar regions, especially Greenland and the Antarctic, and also from cores obtained from the seabed. Contained in the occluded gas bubbles are trace substances from earlier times, including materials deposited as solid particles from the atmosphere, e.g., from volcanic eruptions or meteorite impacts. The mightiest continental ice shield is found in the South Polar regions, with a thickness profil corresponding to a time span of over 100 000 a. The composition of the troposphere can thus be traced back over this time period. The carbon-containing compounds, especially methane, carbon monoxide, and carbon dioxide, are the most meaningful as they are partly of anthropogenic origin and participate in the most important tropospheric reaction cycles. They have been thoroughly, and in some cases continually investigated, as shown by the example of emission data for methane:

Sources (10^{12} g/a)

1) Fossil deposits

As 18–24% of the methane does not contain any radioactive ^{14}C, the amount from fossil deposits is $(100 \pm 20) \times 10^{12}$ g/a:

from coal mines	25 ± 5
from methane hydrates	~5
from losses of natural gas	70 ± 15

Thus, 6–9% of natural gas production is lost as CH_4 emissions

2) Biogenic sources

Ruminants	~80
Dumps	50 ± 20
Ocean and termites	~30
Wet regions and rice paddies	215 ± 50
Combustion of biomass	30 ± 15
Total sources	505 ± 105

Sinks

OH oxidation	420 ± 80
Uptake by the soil	30 ± 15
Stratospheric degradation by Cl and O (^{1}D)	10 ± 5
Subtotal	460 ± 100
Annual increase	45 ± 5

The main sources of methane are in the undeveloped countries [8.28], and the main sources of nitrogen oxides in the industrialized countries [8.29].

Emissions of NO_x (10^6 t/a, calculated as NO_2) can be summarized as follows:

Ground sources	124
Combustion of fossil fuels	69
Emissions from the soil	18
Combustion of biomass	37
Atmospheric sources	30
NH_3 oxidation	10.2
Lightning	16.4
Aircraft exhaust gases	1.0
NO_x from the stratosphere	2.0
Total	154 ± 76

Emissions in the form of aerosols make a large contribution to the input to the troposphere. In total, $1-2 \times 10^6$ t are added annually, of which 10% is of anthropogenic origin [8.6] (Fig. 8.5). The inorganic material in aerosols alone is considered to have a total surface area of

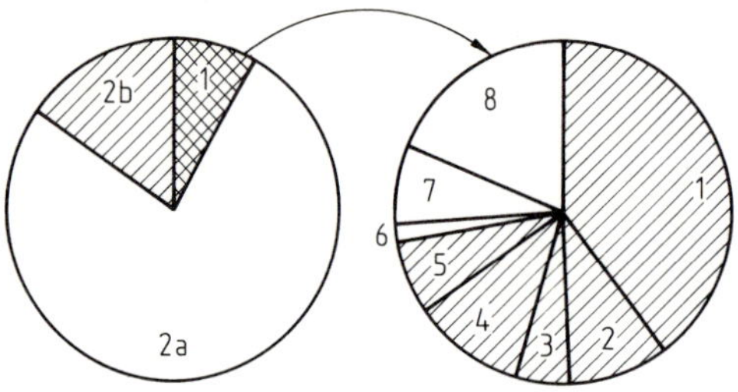

Figure 8.5. Source strength of the inorganic fraction of the atmospheric aerosol
The anthropogenic portion dominating in industrialized countries is split into various contributions in the right-hand diagram. The hatched areas denote light-absorbing aerosol materials with semiconductor properties.

2–10 times that of the earth [8.30] and a residence time of 1–14 d [8.31].

Trends are of particular interest, especially those that influence anthropogenic activities or are initiated or directly caused by them. Of primary concern today is the increase in CO_2 emissions. The carbon dioxide content of the atmosphere has increased continuously since 1700, and currently by ca. 1 ppm/a. The volume concentration today is 355 ppm, the total amount added annually being 22×10^9 t [8.16]. As with CO_2, the global concentrations of CH_4, H_2, CO, and O_3 are also increasing, but only in the northern hemisphere (Table 8.3) [8.15].

Table 8.3. Concentration trends of chemically reactive trace substances [X]*

X	[X], ppbv	Increase above preindustrialized period, %	Annual increase, %
CO	NH: 100–150		NH: 1 [8.32]
	SH: 50–70		SH: ≈0 [8.32]
CH_4	1700	130	1 [8.32]
H_2	512		0.6 [8.32]
O_3	NH: 40–45**	160	NH: 1 [8.32]
	NH: 20–30***		SH: ≈0
Nitrate		100 [8.33]	
Sulfate		100 [8.33]	
H_2O_2	0.2	50 [8.34]	
˙OH	ca. $(2-3) \times 10^{-5}$?		?

* NH = Northern hemisphere; SH = Southern hemisphere. ** Free troposphere. *** Near ground level.

8.3. Reaction Cycles in the Atmosphere
(see also Chap. 2)

The atmosphere contains a total of ca. 5×10^{18} kg air, of which 3×10^{18} kg make up the troposphere. The reactions above and below the tropopause differ, and there is almost no mass transfer between troposphere and stratosphere. Emissions take part in chemical transformations in the troposphere, and eventually return to the earth's surface, partly by transport reactions, but mainly as mineralized substances.

The atmosphere obtains its energy from solar radiation, which amounts to 3.7×10^{23} kW, or 3.2×10^{27} kW · h/a. Of this a light flux of 1.6×10^{18} kW · h/a is incident on the stratosphere. After passing through the ozone layer, where light is absorbed in the wavelength range below 300 nm, 8×10^{17} kW · h/a passes into the troposphere and to the earth's surface.

8.3.1. The Troposphere

Material cycles in the troposphere are partly mass-transfer cycles with a high energy requirement and partly chemical transformation processes involving reactive intermediates formed photochemically. These cycles are partly interlinked, especially the water cycle and the carbon cycle.

The amount of water entering the troposphere annually is 4.6×10^{17} kg, of which 4.0×10^{17} evaporates from the oceans and 0.6×10^{17} from the land surfaces. This corresponds to an energy of 3×10^{17} kW · h, which is released as heat of condensation on cloud formation and precipitation. The largest part of the carbon cycle is accounted for by CO_2, with a concentration in air of 0.034 vol% at sea level and a turnover of 8.3×10^{14} kg/a. Of this, 3.7×10^{14} kg CO_2 takes part in a direct partial cycle with the ocean.

The degradation of organic compounds, as well as other oxidation processes there, in the troposphere is initiated by only a small number of photochemical primary reactions. Very few reactions act as radical sources, as only light of wavelength ≥ 300 nm reaches the lower atmosphere [8.15].

The fundamental reactions are:

1) Photolysis of nitrogen dioxide at 400 nm

$$NO_2^{\cdot} \xrightarrow{h\nu} NO^{\cdot} + O:$$

followed by the formation of ozone

$$O: + O_2 \xrightarrow{M} O_3$$

where M is a collision partner, an inert molecule that absorbs energy without itself being chemically changes.

2) Photolysis of ozone at 300 nm with formation of excited oxygen atoms

$$O_3 \xrightarrow{h\nu} O_2^*(^1\Delta g) + O^*(^1D)$$

followed by formation of the hydroxyl radical

$$O^*(^1D) + H_2O \rightarrow 2\,^{\cdot}OH$$

3) Oxidation of carbon compounds, ultimately of CO

$$HO^{\cdot} + CO \rightarrow H^{\cdot} + CO_2$$

followed by formation of the peroxyhydroxyl radical and regeneration of nitrogen dioxide and hydroxyl

$$H^{\cdot} + O_2 \rightarrow HO_2^{\cdot}$$
$$HO_2^{\cdot} + NO^{\cdot} \rightarrow {}^{\cdot}OH + NO_2^{\cdot}$$

This regeneration ensures that the oxidation rate in the troposphere is maintained.

Together with VOCs complex reaction mechanisms result.

Radical reactions in the system $CH_4 + CO + \text{air} + NO_x + h\nu$ are listed in the following

Radical source

$O_3 + H_2O + h\nu \rightarrow 2\,^{\cdot}OH + O_2$

Radical transformation: $HO^{\cdot} \rightarrow HO_2^{\cdot}/CH_3O_2$

$^{\cdot}OH + CO + O_2 \rightarrow CO_2 + HO_2^{\cdot}$
$^{\cdot}OH + CH_4 + O_2 \rightarrow CH_3O_2^{\cdot} + H_2O$
$2\,CH_3O_2^{\cdot} + O_2 \rightarrow 2\,HO_2 + 2\,HCHO$
$^{\cdot}OH + O_3 \rightarrow HO_2^{\cdot} + O_2$

Radical regeneration: $HO_2^{\cdot}/CH_3O_2 \rightarrow {}^{\cdot}OH$

$HO_2^{\cdot} + O_3 \rightarrow {}^{\cdot}OH + 2\,O_2$
$HO_2^{\cdot} + NO \rightarrow {}^{\cdot}OH + NO_2$
$CH_3O_2 + 2\,NO + O_2 \rightarrow {}^{\cdot}OH + 2\,NO_2 + HCHO$
(via $CH_3O^{\cdot} + O_2$)

Radical loss

$^{\cdot}OH + NO_2 \rightarrow HNO_3$
$HO_2^{\cdot} + HO_2^{\cdot} \rightarrow H_2O_2 + O_2$
$HO_2^{\cdot} + CH_3O_2^{\cdot} \rightarrow CH_3OOH + O_2$
$2\,CH_3O_2^{\cdot} \rightarrow CH_3OH + HCHO$

Ozone formation

$NO_2 + O_2 + h\nu \rightarrow NO + O_3$

Some intermediates react with one another, with already formed photooxidants, or with direct photolysis products of certain VOCs, e.g.,

$$H_2CO \xrightarrow{h\nu} H^{\cdot} + HCO^{\cdot}$$
$$HCO^{\cdot} + O_2 \longrightarrow HO_2^{\cdot} + CO$$

These radical-loss reactions lead to oxidation products which are washed out by precipitation.

Oxidation of VOCs during the day proceeds via radical chains (Fig. 8.6 A), and at night via the $NO_3^{\cdot}$ radical (Fig. 8.6 B).

The continuous regeneration of oxidizing species consumed in oxidation, especially the $^{\cdot}OH$ radical, depends basically on the concentration of the nitrogen oxides. With increasing emissions of organic compounds, emissions of NO_x ought also to increase to ensure their degradation [8.14]:

$HO_2^{\cdot} + NO^{\cdot} \rightarrow {}^{\cdot}OH + NO_2^{\cdot}$
$RO_2^{\cdot} + NO^{\cdot} \rightarrow RO^{\cdot} + NO_2^{\cdot}$
$HO_2^{\cdot} + NO_3^{\cdot} \rightarrow {}^{\cdot}OH + NO_2^{\cdot} + O_2$
$RO_2^{\cdot} + NO_3^{\cdot} \rightarrow RO^{\cdot} + NO_2^{\cdot} + O_2$

In this context, tropospheric ozone plays an ambivalent role. Depending on its concentration and that of other trace substances, especially nitrogen oxides, it is an essential provider of excited oxygen atoms, which form hydroxyl radicals and hence cause degradation of VOCs. However, ozone formed in the same way in photosmog causes formation of the nitrogen-containing components of the photooxidants. At lower NO_x concentrations (i.e., < 10 ppt), ozone is consumed, and at higher NO_x concentrations (i.e., up to 10 ppb), O_3 and $^{\cdot}OH$ are continuously regenerated. Above 10 ppb, $^{\cdot}OH$ radicals are captured by NO_2 radicals and the chain is broken.

Ozone-forming and ozone-decomposing reactions are summarized in the following:

Ozone formation

$NO_2 + O_2 + h\nu \rightarrow O_3 + NO$
$O_3 + NO \rightarrow O_2 + NO_2$

$[O_3] = cJ[NO_2]/[NO]$, photochemical equilibrium, J = photolysis frequency

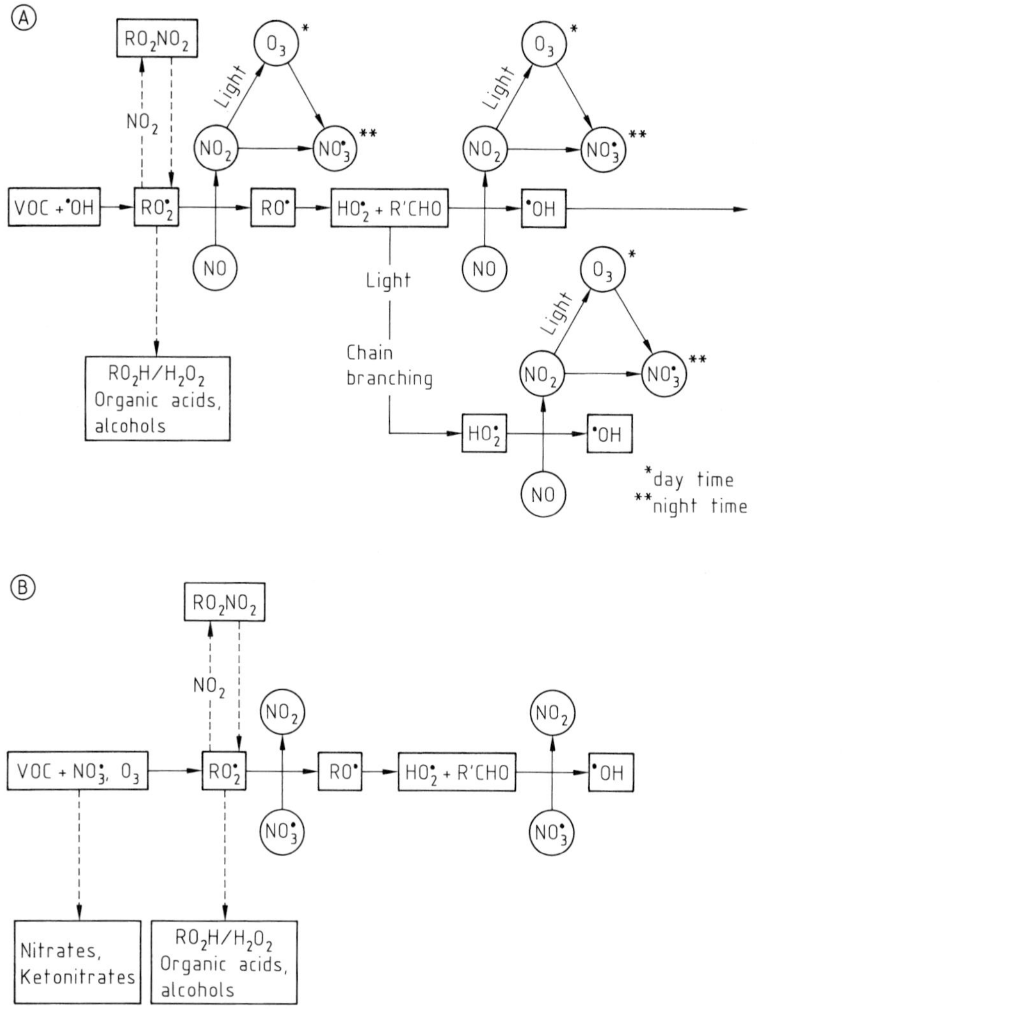

Figure 8.6. A) Day time chemistry: VOC oxidation by $^{\bullet}OH/^{\bullet}O_2H$ radical chain; B) Night time chemistry
The atmospheric oxygen that takes part in the reaction is present in excess and is not shown in the reaction scheme

$RO_2^{\bullet} + NO \rightarrow RO^{\bullet} + NO_2$
$HO_2^{\bullet} + NO \rightarrow {}^{\bullet}OH + NO_2$

These two reactions cause the $[NO_x]/[NO]$ ratio to increase without consuming O_3

Critical competitive reactions which retard the oxidation of NO:

$HO_2^{\bullet} + HO_2^{\bullet}$
$HO_2^{\bullet} + RO_2^{\bullet}$
$RO_2^{\bullet} + R'O_2^{\bullet}$
$HO_2^{\bullet} + O_3$
$RO_2^{\bullet} + O_3$
$RO_2^{\bullet} + NO \rightarrow RONO_2$

Destruction of radicals on surfaces and in raindrops

$RO_2 + NO_2 \rightarrow RO_2NO_2$ (only transitory, as thermal decomposition occurs; stable alkyl nitrates can be formed on surfaces)

		Ozone loss
$O_3 + h\nu(\lambda < 310\ \text{nm}) \rightarrow O(^1D) + O_2$ $O(^1D) + H_2O \rightarrow 2\ {}^{\bullet}OH$		30%
${}^{\bullet}OH + O_3 \rightarrow HO_2^{\bullet} + O_2$ $HO_2^{\bullet} + O_3 \rightarrow {}^{\bullet}OH + 2\ O_2$	} HO_x cycle	30%
Overall: $2\ O_3 \rightarrow 3\ O_2$		
$RO_2^{\bullet} + O_3 \rightarrow RO^{\bullet} + 2\ O_2$	(with low NO_x)	
Alkenes + $O_3 \rightarrow$ products		<2%
$NO_2 + O_3 \rightarrow NO_3^{\bullet} + O_2$	(night time)	<1%
Dry deposition		ca. 25%

In the case of photochemical degradation in the troposphere above densely populated areas, the lifetimes and residence times of photooxidants are important evaluation criteria together with the emission rates of nitrogen oxides and VOCs [8.14]. With large-area emission sources such as dense motor traffic and combustion in general, the chemistry and physics of the photosmog form the decisive basis for technical and administrative action.

Some lifetimes τ (298 K, 1 bar) for the reaction $RO_2NO_2 \rightarrow RO_2 + NO_2$ are:

Ethyl peroxynitrate ($C_2H_5O_2NO_2$)	0.3 s
Acetyl peroxynitrate ($CH_3C(O)O_2NO_2$)	2174 s (0.6 h)
Acetonyl peroxynitrate ($CH_3C(O)CH_2O_2NO_2$)	0.4 s

The mean ozone volume concentration in the free troposphere of the northern hemisphere is 40 ppb (80 µg/m³), which is three times that of 100 a ago, and is mainly due to anthropogenic emissions of nitrogen compounds. However, these are needed to degrade the increasing emissions of organic compounds, including methane and carbon monoxide, by continuously forming oxidizing species in the troposphere.

The ozone in the troposphere takes part in a complex series of reactions, and is to a degree indispensable to maintain the purity of the lower atmospheric strata. However, these chemical processes and their significance are entirely different from those of the ozone of the stratosphere. If an interaction between the two strata does exist, the possible role of ozone must again be considered. Although this substance is formed differently in the two strata, has different effects, and is not involved in mass-transfer processes, there is an indirect relationship between the ozone in the troposphere and that in the stratosphere (see Section 8.3.2).

A further contribution to atmospheric photochemical degradation is made by halogen chemistry. Halogenated alkanes and volatile halogenated organic compounds in general are mostly degraded fairly rapidly, similar to chlorine-free organic compounds, though some are degraded relatively slowly [8.5], [8.14], [8.35]. These processes are directly relevant to questions raised by some trace substances which are not degraded particularly rapidly in the troposphere, have high vapor pressures (e.g., fluorinated hydrocarbons; see Section 8.5), and as a result can therefore pass into the stratosphere. Degradable halogen is converted to hydrogen halide, and is washed out of the troposphere by precipitation.

A direct contribution to photochemical degradation in the lower strata of the atmosphere above the oceans and coastal regions (extending as far as 100 km inland, depending on the wind) is made by entrained sea spray. An aerosol is formed and a photochemical reaction, mainly with NaCl, then forms Cl atoms, which degrade organic trace substances in the troposphere more rapidly than ·OH radicals [8.6].

Of the large mass cycles, that of carbon and its compounds forms the foundation for living matter. It is dominated by carbon dioxide [8.2], which at a concentration of 0.034 vol% in the atmosphere, amounts to 2.6×10^{12} t, with an annual turnover of 830×10^9 t and a mean residence time of 3.3 a.

The turnovers of photochemical partial cycles are comparatively small.

Entry of carbon compounds into the atmosphere (in 10^9 t) is as follows:

CO_2 from respiration of biomass	220	
CO_2 from putrefaction of biomass	220	
(formation on biomass by photosynthesis)		440 (removal)
CO_2 from the sea	365	
Returned to the sea (equilibrium)		365 (removal)
CO_2 from forest clearance	7	
CO_2 from combustion processes	18	
CO_2, total emission	830	
CO, total emission	3.4	
CH_4, total emission	0.5	
C (in VOCs), total emission	1.0	

Worldwide efforts to reduce CO_2 emission are intended to

1) Reduce the greenhouse effect
2) Save energy
3) Conserve resources of carbon-containing fossil fuels, partly to retain them as chemical raw material

Since, the majority of anthropogenic CO_2 emissions are produced by combustion processes, which simultaneously produce NO_x, a decrease in CO_2 would also be associated with a decrease in NO_x and hence in tropospheric ozone. Decreases in the concentration of hydrocarbons and nitrogen oxides would immediately lead to a noticeable reduction in the concentration of ozone, as its lifetime in the lower atmosphere is short.

8.3.2. The Stratosphere

The most important trace gas in the stratosphere is ozone. It is contained in a layer 25–30 km thick in the lower stratosphere at a latitude-dependent concentration of 10^{12} cm^{-3}. This accounts for 90% of the total atmospheric ozone [8.10], [8.11], [8.32], [8.36], [8.37]. Ozone is mainly produced above the equatorial regions and is transported towards the polar regions. Its existence depends on a delicate equilibrium between formation and degradation, and it is essential that its concentration in the stratosphere should be maintained for the benefit of climatic and living conditions on the earth's surface. Stratospheric photochemical degradation of volatile substances emitted into the troposphere but are degraded so slowly there that they migrate to the stratosphere would lower the stratospheric ozone concentration and lead to impaired conditions at the earth's surface.

Ozone is formed in the stratosphere exclusively by the photolysis of oxygen.

$$O_2 \xrightarrow[\lambda \leq 250\ nm]{h\nu} 2\,O$$

$$O + O_2 \xrightarrow{M} O_3$$

The inert collision partner M is provided from the surrounding supply of O_2 or N_2 molecules. Photolytic degradation then follows, but caused by light of a longer wavelength

$$O_3 \xrightarrow[\lambda \geq 300\ nm]{h\nu} O_2 + O$$

$$O + O_3 \longrightarrow 2\,O_2$$

A chemical degradation process also exists

$$Y + O_3 \rightarrow O_2 + YO$$

$$YO + O \rightarrow O_2 + Y$$

The net reaction is thus

$$O + O_3 \rightarrow 2\,O_2$$

In the natural stratosphere, $Y = {}^{\bullet}OH$ and $YO = HO_2^{\bullet}$. The natural ozone concentration is formed in this way.

The exhaust gases from supersonic aircraft, rockets, and the undecomposed components of emissions from the earth's surface ascending through the troposphere can provide further reactive species that can take part in ozone degradation and affect the above-mentioned equilibrium concentration:

$$Y = NO, Cl^{\bullet}, Br^{\bullet}$$

$$YO = NO_2, ClO^{\bullet}, BrO^{\bullet}$$

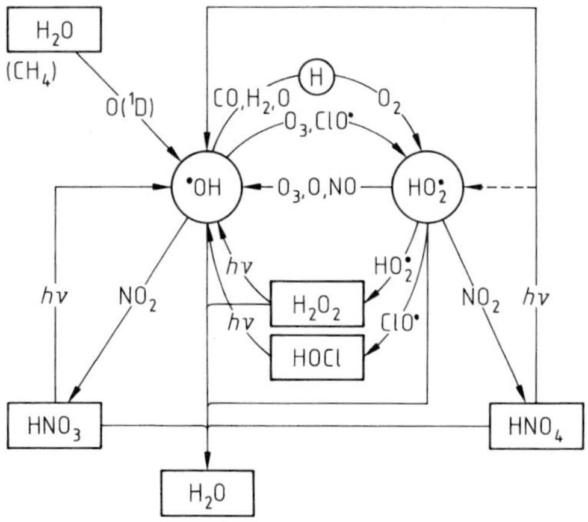

Figure 8.7. HO_x chemistry of the stratosphere

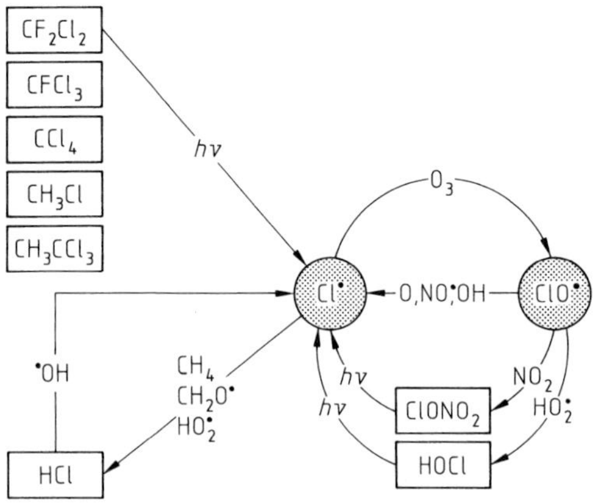

Figure 8.8. Scheme of the chlorine-induced cycle causing degradation of ozone in the global stratosphere

In this system, the transformation of $^{\bullet}OH$ into $HO_2^{\bullet}$ and back again takes place very rapidly (Fig. 8.7).

H_2O_2 and HOCl are formed and act as temporary sinks, while nitric acid and water act as permanent sinks. If these degradation reactions consume ozone more rapidly than it is regenerated or replaced by transport mechanisms, which is regularly the case in the region of the antarctic polar cap, this leads to a transitory but periodically recurring decrease in the concentration of stratospheric ozone (Fig. 8.8).

This can lead to a 50% depletion of the normal ozone layer over an area the size of Antarctica — the so-called ozone hole (Fig. 8.9).

The chlorine compounds HOCl, $ClONO_2$, HCl, and elemental chlorine are formed from ascending CFCs and chlorine injections from the boosters of rocket carriers [8.38].

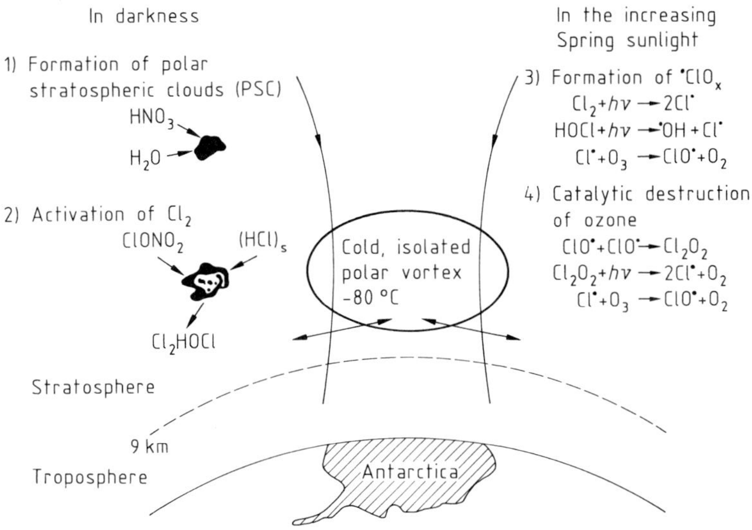

Figure 8.9. Meteorology and chemistry of antarctic ozone depletion

The ozone hole starts to form at the end of the antarctic winter by mid September and disappears again by mid November. It is preceded by a rapid cooling of the stratosphere to −80 °C in June and July. This causes formation of polar stratospheric clouds (PSCs), one type of which consists of nitric acid, hydrochloric acid, and water, and the other exclusively of water. The latter type is larger and leads to precipitation and hence dehydration of the polar stratosphere. In a reaction that takes place at the surfaces of the PSCs, $ClONO_2$ forms Cl_2, HOCl, and HNO_3. This amounts to condensation of the nitrogen species into sinks but to a redistribution of the chlorine species from less active sinks into more reactive species (see Figs. 8.7 and 8.8).

The degradation of emissions that have entered the stratosphere also leads to the degradation of the stationary reaction partner ozone. The direct result is an increase in the UV-B radiation at the earth's surface, and an indirect result is increased permeability of the stratosphere towards other emissions ascending from the troposphere. This would be enabled by cooling of the stratosphere due to the loss of the ozone.

8.4. Interaction with Surface Layers of Aquatic and Terrestrial Environments

Under the influence of sunlight, active oxidizing agents are formed in water [8.39] with the following concentrations (mol/L):

Hydroperoxyl ($HO_2^{\cdot}$)	10^{-9}–10^{-8}
Peroxyalkyl ($ROO^{\cdot}$)	10^{-11}–10^{-10}
Singlet molecular oxygen	10^{-14}–10^{-13}
Hydrated electron	10^{-17}–10^{-15}
Hydroxyl ($^{\cdot}OH$)	10^{-18}–10^{-19}

Of these, the hydroxyl radical, with 10^6 radicals per cubic centimeter down to a depth of ca. 1 m [8.40], is the most significant for photochemical degradation in water as well as in air. The concentration is comparable to that in air, but can be lower in pure water or pure natural water bodies [8.41]. In natural surface water, the most important source of $^{\cdot}OH$ is the photolysis of nitrate.

Dissolved or dispersed organic compounds or inorganic compounds in lower oxidation states are also oxidized in the aqueous phase by indirect photochemical degradation. These processes are similar to those that take place in the mixture of hydrogen peroxide and iron(II) salts known as Fenton's reagent, in which intermediate stages also involve radicals (see → Water, **B8**). Reactions of this type can already take place with constituents of natural surface water. For example, in seawater the bromide ion is oxidized to hypobromite, and in the layers near the surface of fresh water the humic matter present at a concentration of 1–10 mg/L is oxidized to form oxidizing radicals such as $RO^{\cdot}$, $RO_2^{\cdot}$, and $HO_2^{\cdot}$.

Most rate constants for the reaction of organic compounds with $^{\cdot}OH$ radicals in water have values of the order of k_{OH}

$\approx 10^9$ L mol^{-1} s^{-1} [8.42]. The rate constants of other reactions show a similar scatter around the mean figure [8.40], e.g., K_{OH} values in air (see Section 8.5.1).

k_{OH} (L mol^{-1} s^{-1}) values for some compounds in water follow:

Urea	2.1×10^6
Acetonitrile	3.5×10^6
Chloroform	8.5×10^6
Ethylene oxide	4.0×10^7
Methanol	8.9×10^8
Thiophene	3.3×10^9
Phenol	1.6×10^{10}
Aniline	1.7×10^{10}

The dependence on solar radiation is more pronounced than in the case of air (Table 8.4) [8.41].

In water droplets, fog, dew, etc., or on the earth's surface, the situation is similar, and the substances in these environmental compartments do not have a high vapor pressure. In the vapor-pressure range from 10^{-7} to 1 Pa they are referred to as semivolatile organic compounds (SOCs) [8.43]. For the photochemical degradation of these substances, not only the formation but also the distribution of the photooxidants is of direct relevance [8.44]. In the interaction between biotic and abiotic degradation steps of the aquatic systems of rivers, lakes, and wet ground, and mass transfer into the troposphere at ground level, the photochemical degradation of organic substances in water makes a contribution to the total mineralization [8.7], [8.9], [8.40], [8.43], [8.44].

8.5. Measurement and Computation of Degradability

To derive ecological relationships from the properties of chemical substances, their biotic and abiotic interactions must be considered. The behavior of a substance in the environment is determined and assessed by drawing on the results of ecotoxicological investigations (see → Ecology and Ecotoxicology). However, only since the late 1970s has it been possible to answer a fundamental aspect of this question, i.e., how long a substance exists in the atmospheric environment. This came with the development of reliable measurement techniques and equipment.

With regard to the biotic degradation pathways in the carbon cycle (see Chap. 2), the reaction chains, mechanisms, and rates in the

Table 8.4. Lifetime of $^{\bullet}$OH in surface (fresh) water

Substance	τ (sun)[a], h	τ (average)[b], h
Aniline, phenol	14, 5	6.37×10^3
Methanol	320	10.43×10^4
Chloroform	33 000	12.46×10^6

[a] Optimum conditions (intense sunlight, at water surface) $[^{\bullet}OH] \approx 10^{-15}$ mol/L. [b] With annual average values $\langle[^{\bullet}OH]\rangle \approx 3 \times 10^{-18}$ mol/L.
The time for (a) is hours of intense sunlight, for (b) it is "real" time.

water and soil environmental compartments are well known in many cases, and provide a working knowledge for science, economy, administration, and legislation. An abiotic process in the atmosphere was only confirmed with the development of laboratory methods that supplemented earlier field tests of mainly meteorological interest. Besides benefitting from progress in other fields, especially the various types of laser spectroscopy, a number of techniques have been developed specifically for investigating photodegradation.

In the environmental compartment troposphere, degradability (or its converse, chemical persistence) is characterized by abiotic processes which are photochemically induced and then proceed in oxidative steps. Degradability is determined by

1) The concentration of the oxidizing agent in the troposphere, sometimes in a particular locality and at a particular time
2) The rate constants of the reaction with a reaction partner which is volatile and homogeneous in the gas phase or is bonded to particles as an aerosol

There is now much data on both factors. A number of experimental methods have been developed for determining rate constants, and figures for K_{OH} are now available for 600 compounds in air, as are rate constants for other oxidizing agents in other media and in heterogeneous phases. Rate constants for photochemical degradations are available in spreadsheets [8.46]–[8.48] and data banks [8.49]. Values for K_{OH} can also be derived from bond energy increments of individual structural groups and classes of substances [8.50]–[8.52].

The chemical lifetime is found from the rate constants of bimolecular reactions. The degradation rate of substance X

$$-\frac{dc_{(X)}}{dt} = k\,c_X\,c_Y$$

and the rate constant k are both expressed as cm^3 per molecule per second (cm^3/s). It depends on temperature and pressure. It is simplified by incorporating c_Y, the time-averaged concentration of the oxidizing agent, into the constant, and a new monomolecular rate constant K is derived:

$$K = k\,c_Y$$

The degradation rate is then given by

$$-\frac{dc_X}{dt} = K\,c_X$$

Sometimes, the less obvious term "chemical half-life" ($\tau_{1/2} = 0.69\,\tau_X$) is used, i.e., the time for the concentration of the starting material to reach half its initial value. However, this is dependent on the physicochemical environment of a substance, and is thus not a property inherently characteristic of a substance like the half-life for radioactive decay, which is unaffected by outside influences. The residence time in the atmosphere is usually smaller than the chemical lifetime because in some cases part of the undecomposed material can be removed by transport processes.

The main primary oxidizing agents in the troposphere are ˙OH, $HO_2^{\bullet}$, O_3, and $NO_3^{\bullet}$. The ˙OH radical is present at rather low concentration, but reacts so much more rapidly than the other oxidizing agents that it has been found possible to deduce the chemical lifetime of a substance in the troposphere solely from its rate constants [8.43].

The limits of measurability are:

K_{OH} values	Corresponding chemical lifetime
10^{-10} cm^3/s for rapid reactions	1 h
10^{-16} cm^3/s for slow reactions	130 a

To measure rate constants of the photochemical degradation of a substance in the atmosphere, a vapor pressure of >1 Pa is required in the laboratory to give a sufficiently high concentration of the substance in the photoreactor [8.42], [8.52]. Nonvolatile compounds are first adsorbed on an inorganic carrier, and the adsorbate is then investigated as an aerosol in a smog chamber [8.6], [8.25]. This method has the advantage that it simulates a real dust-laden atmosphere, and any catalytic acceleration of the photochemical degradation by traces of heavy metals in the carriers can be measured. This also applies to the use of TiO_2 as a carrier for the test aerosol [8.25].

If only the rate constant K_{OH} is to be determined, this can be carried out in aqueous solution with H_2O_2 as the source of ˙OH radicals [8.42] at a stationary concentration of $c_{OH} = 5\times10^{-14}$ mol/L or 4×10^{7} cm^{-3}. For SOCs, which usually are only sparingly soluble in water, 1,1,2-trichlorotrifluoroethane can be used as solvent [8.45], with H_2O_2 as ˙OH source.

A guideline for the standardization and international harmonization of test methods and standards for photochemical degradation in air [8.52] and water [8.53] has been produced by the OECD countries. Methods for the derivation of rate constants from structural increments have also been produced.

8.5.1. Lifetime in the Homogeneous Gas Phase

The various important naturally-occurring oxidizing agents can be present in widely different concentrations. The effects of the intensity and duration of the solar radiation can also differ widely (Table 8.5) [8.43].

The $NO_3^{\bullet}$ radical is rapidly photolyzed and is therefore mainly active at night. It forms nitric acid by hydrogen abstraction and undergoes addition to double bonds.

In the medium lifetime range (ca. 10 d) the fluctuation in ˙OH concentration caused by irregular radiation is sufficient to cause significant differences in rate of degradation (Table 8.6) [8.14], [8.40].

A relationship between K_{OH} values and chemical structure is only observed with homologues [8.6], [8.46]. The first member reacts much more slowly than the rest (Table 8.7).

Photochemical degradation by ozone also shows no dependence on chemical structure [8.6], [8.54]. Alkenes and other substances with olefinic double bonds are significantly degraded by ozone, but other compounds containing multiple bonds and aromatics are degraded rather slowly (Table 8.8).

The rates of degradation of chlorinated hydrocarbons also do not show any correlation with structure [8.55]. They include very slowly degraded substances with lifetimes of several years and typical ones with lifetimes of around 14 d (Table 8.9).

Table 8.5. Concentration of oxidizing agents

Molecules/cm^3	Day	Day/night	Night
˙OH	1×10^{6}	0.5×10^{6}	
O_3	1×10^{12}		
$NO_3^{\bullet}$	instable	1.5×10^{8}	2.5×10^{8}

Table 8.6. Chemical lifetimes of some volatile organic compounds in the troposphere*

Compound	Chemical formula	*k* values for ˙OH, 10^{-12} cm^3/s	Lifetime, d	
			Tropics	Global
Ethane	C_2H_6	0.29	20	61
Methanol	CH_3OH	0.80	6.5	15
Acetylene	C_2H_2	0.90	5.5	13
Benzene	C_6H_6	1.31	4.5	12
Propane	C_3H_8	1.59	3.6	11
n-Butane	C_4H_{10}	2.4	2.1	7.5
Isobutane	C_4H_{10}	2.3	2.2	8
Dimethyl ether	CH_3OCH_3	2.6	1.9	7
Ethyl alcohol	C_2H_5OH	3.0	1.8	5.9
n-Pentane	*n*-C_5H_{12}	3.7	1.5	4.8
Toluene	$C_6H_5CH_3$	6.19	0.9	2.8
Ethylbenzene	$C_6H_5C_2H_5$	7.73	0.75	2.5
Ethene	C_2H_4	7.85	0.7	1.9
Formaldehyde	HCHO	10	0.5	0.9
o-Xylene	$C_6H_4(CH_3)_2$	14	0.35	0.7
Acetaldehyde	CH_3CHO	16	0.3	0.6
Propene	C_3H_6	29	0.2	0.4

* ˙OH concentration: 2×10^6 cm^{-3} at 24 h mean in tropical regions, 6.5×10^5 cm^{-3} as global mean; O_3 concentration: 30 ppbv.

Table 8.7. Rate constants *k* (in 10^{-12} $cm^3 s^{-1}$) for the reactions of ˙OH with various organics at room temperature

Compound	Ref. [8.6]	Ref. [8.46]	Compound	Ref. [8.6]	Ref. [8.46]
Propane	1.3	1.18	2,2-Dimethylhexane	4.9	
n-Butane	2.5	2.53	2,2,4-Trimethylpentane	3.4	3.66
n-Pentane	4.1	4.04	2,3,5-Trimethylhexane	7.9	
n-Hexane	5.7	5.58	Cycloheptane	11.7	13.1
n-Heptane	7.3	7.2	Cyclooctane	14.1	
n-Octane	8.8	8.72	1,1,3-Trimethylcyclohexane	8.7	
n-Nonane	10.3	10.0	Adamantane	21.9	22.7
n-Decane	12.4	11.2	Propene	28	26.3
n-Undecane	13.3		Acrylonitrile	4.3	4.8
n-Dodecane	13.9		Propionitrile	0.8	0.19
n-Tridecane	15.5		Butyronitrile	1.1	
Isobutane	2.5	2.37	Benzene	1.3	1.28
Isopentane	4.0	3.9	Toluene	5.6	6.19
3-Methylpentane	5.7	5.6	*o*-Xylene	10.3	14.7
4-Methyloctane	9.7		*p*-Xylene	11.6	15.2
2-Methyloctane	10.1		Chlorobenzene	1.0	0.7
2,3-Dimethylbutane	5.8	6.2	Vinyl chloride	5.6	6.6
2,3-Dimethylpentane	6.0	5.2	1,1-Dichloroethene	7.0	
2,4-Dimethylhexane	7.8		1,2-Dichloroethene	2.2	
2,2-Dimethylbutane	2.4	2.59	Trichloroethene	2.2	2.36
2,2-Dimethylpentane	3.4		Carbon tetrachloride	<0.06	
3,3-Dimethylpentane	3.7				

The large number of K_{OH} values known show that degradability rate constants of organic compounds can have a range of six orders of magnitude. Most organic compounds have a photochemical lifetime within a range of two orders of magnitude, i.e., a lifetime of 1 – 100 d. Some are degraded within hours, and a few require more than a year (Fig. 8.10). If the lifetime exceeds 10 a, ascent from the troposphere into the stratosphere has to be taken into consideration [8.14].

Standardized graphical and tabular presentations are usually based on a concentration of 0.5×10^6 ˙OH radicals/cm^3 as a basis for K_{OH} values and lifetimes.

8.5.2. Lifetimes of Adsorbates Aerosols

A portion of natural and anthropogenic emissions reach the troposphere bound to aerosol particles. Experimental investigations in an aerosol/smog chamber with a volume of 2.4 m^3

Table 8.8. Rate constants, k (in 10^{-17} $cm^3 s^{-1}$), for the reaction of ozone with various organics at room temperature

Compound	Ref. [8.6]	Ref. [8.54]
Propene	1.1	1.13
Isobutene	1.3	1.21
1-Pentene	0.95	1.1
1-Hexene	1.1	1.17
1-Heptene	1.15	0.8–1.7
1-Octene	1.25	0.8
1-Nonene	1.45	
Vinyl chloride	0.014	0.017–0.2
Acrylonitrile	0.002	
Propionitrile	<0.001	
Butyronitrile	<0.001	
o-Xylene	<0.007	7×10^{-5}
p-Xylene	<0.01	4×10^{-5}

Table 8.9. Tropospheric concentrations and lifetimes of some chlorinated hydrocarbons

Compound	Troposheric concentrations, pptv	Tropospheric lifetimes, a
C_2HCl_3	16	0.04
C_2Cl_4	40	0.4
$CHCl_3$	15	1
CH_3Cl	600	2–3
CH_3CCl_3	110	5–10
CCl_4	120	60–100

[8.6], [8.25] have shown that the degradation rate of nonvolatile substances transported into the atmosphere in adsorbed form is comparable to that of volatile substances in a homogeneous phase.

The aerosols have a multilayer structure. Surrounding a water-insoluble core of, for example, desert dust, fly ash, volcanic ash, is a salt-containing aqueous envelope onto which the organic trace substances are adsorbed, with the hydrophilic ends of the molecules directed inwards. The hydrophobic groups directed outwards hinder the penetration of hydrophilic gases such as acids and tend to keep the aerosol in suspension [8.57].

Most of the global total of $(1-2) \times 10^9$ t of aerosol is formed by sodium chloride from sea spray. Some of the chlorine is transformed photolytically to chlorine-containing reactive species. Hence a photo-induced degradation of emissions in the troposphere also takes place in this way. The reaction can be measured in a smog chamber [8.6], [8.25]. The k_{Cl} rate constants found are in agreement for various Cl sources [8.56]. The degradation process is about two orders of magnitude more rapid than that via $OH^{\bullet}$ radicals (Table 8.10).

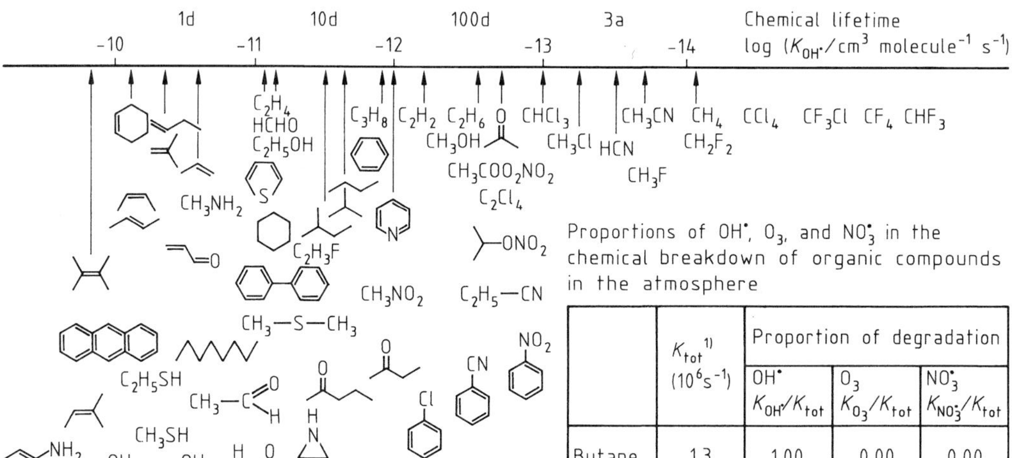

	K_{tot} [1) ($10^6 s^{-1}$)	Proportion of degradation		
		$OH^{\bullet}$ $K_{OH^{\bullet}}/K_{tot}$	O_3 K_{O_3}/K_{tot}	$NO_3^{\bullet}$ $K_{NO_3^{\bullet}}/K_{tot}$
Butane	1.3	1.00	0.00	0.00
Propene	27	0.56	0.41	0.03
Toluene	3.1	1.00	0.00	0.00
Dimethyl sulfide	123	0.02	0.00	0.98

1) $K_{tot} = K_{OH^{\bullet}} + K_{O_3} + K_{NO_3^{\bullet}}$

Figure 8.10. Chemical lifetimes of organic compounds in the atmosphere

Table 8.10. Rate constants k (in 10^{-10} $cm^3 s^{-1}$) for the reaction of chlorine atoms with various organics

Compound	Source*			Ref. [8.56]
	I [8.6]	II [8.6]	III [8.6]	
Propene	1.09		1.02	1.34
n-Butane**	1.97	1.97	1.97	1.97
n-Pentane	2.65	2.50	2.35	2.52
n-Hexane	3.15	3.00	3.20	3.03
n-Octane	4.25	3.65		
n-Nonane	4.40			
Isobutane	1.35		1.30	1.37
Isopentane	2.10		1.75	2.03
3-Methylpentane	2.70		2.80	
2,2-Dimethylbutane	1.60	1.75		
2,3-Dimethylpentane	3.00			
2,2-Dimethylhexane	3.30			
2,2,4-Trimethylpentane	2.25	2.25		
2,2,3,3-Tetramethylbutane		1.55		
Propene	3			2.44
Vinyl chloride	2.05	1.55	1.90	1.97
1,1-Dichloroethene	2.10			
Benzene	0.05			0.15
Toluene	0.74	0.66		0.589

* Source of Cl: NaCl aerosol in photosmog (I); photolysis of Cl_2 (II), photodegradation of chlorobenzene, vinyl chloride, trichloroethene and 1,1-dichloroethene (III).
** Reference compound.

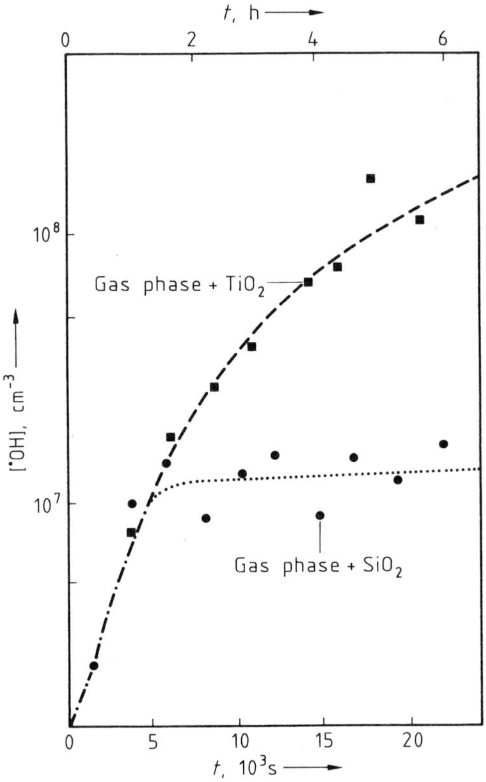

Figure 8.11. Typical formation of ˙OH in an aerosol smog chamber in the presence of SiO_2 or TiO_2 aerosol
In the absence of aerosol the formation of ˙OH is similar to the curve observed in the presence of SiO_2

A further decrease in the lifetime is observed if the airborne dusts of the aerosols contain heavy metals, have semiconducting properties, or can react in redox systems. With pure titanium dioxide as the carrier particle, the ˙OH concentration is increased by a factor of ten, and the degradation rate is correspondingly increased (Fig. 8.11) [8.6], [8.25].

This reaction principle is also used in aqueous suspensions [8.58] (see → Water, **B8**). Multiphase systems are important in addition to airborne aerosols in the earth–air boundary region (see Section 8.3).

8.5.3. Assessment and Evaluation

Photochemical degradation is of major and direct importance because it eliminates substances that enter the atmosphere as emissions from natural and anthropogenic sources or from products used in the open. Some of these substances are considered to affect the climate, others to be detrimental to the filtering effects of the ozone layer on the UV-B fraction of sunlight, and many others to be pollutants of soils and bodies of water on the earth's surface, from where they can return to the atmosphere.

The question of the whereabouts of a substance in a given environment [8.40], [8.59] is largely tantamount to enquiring about its compatibility with conditions prevailing in that particular environmental compartment; holistic considerations contribute greatly [8.60]–[8.63].

Moreover, it is processes much more than states which furnish the determinants. Already the environmental concentration of a substance is merely the difference between two rates, that of input and that of output. In that sense, cross sections of an environment are better expressed by mass and energy fluxes than by properties of any particular substance or a group of substances.

Data on individual substances are much less suitable for describing the status of the environmental compartment "air". Instead they express the oxidative sensitivity of a substance in sunlight as a first step towards mineralization [8.64].

9. References

References for Chapter 1

[1.1] Der Bundesminister für Umwelt, Naturschutz und Reaktorsicherheit (BMU) (eds.): *Beschluß der Bundesregierung vom 7. November 1990 zur Reduzierung der CO_2-Emissionen in der Bundesrepublik Deutschland bis zum Jahr 2005,* Druck-Service E. Böhm, Hansen 1991.

[1.2] Der Bundesminister für Umwelt, Naturschutz und Reaktorsicherheit (BMU) (eds.): *Klimaschutz in Deutschland, Nationalbericht der Bundesregierung für die Bundesrepublik Deutschland im Vorgriff auf Artikel 12 des Rahmenübereinkommens der Vereinten Nationen über Klimaänderungen,* 1993.

[1.3] Dritter Bericht der Enquete-Kommission "Schutz der Erdatmosphäre" – "Klimaschutz durch umweltgerechte Landwirtschaft und Erhalt der Wälder", Drucksache 12/8350, 1994.

[1.4] Der Bundesminister für Umwelt, Naturschutz und Reaktorsicherheit (BMU) (eds.): *Bericht der Bundesregierung an den Deutschen Bundestag,* Fünfter Immissionsschutzbericht der Bundesregierung, 1993.

[1.5] Umweltbundesamt (UBA) (eds.): *Daten zur Umwelt 1990/91,* Erich Schmidt Verlag, Berlin 1992.

[1.6] Bericht der Enquete-Kommission "Schutz des Menschen und der Umwelt – Bewertungskriterien und Perspektiven für umweltverträgliche Stoffkreisläufe in der Industriegesellschaft", Drucksache 12/8260, 1994.

[1.7] P. Davids, M. Lange. *Die TA Luft 1986 – Technischer Kommentar,* VDI-Verlag, Düsseldorf 1986; *Die Großfeuerungsanlagen – Technischer Kommentar,* VDI-Verlag, Düsseldorf 1984.

[1.8] Verband der Chemischen Industrie e.V.: *Chemiewirtschaft in Zahlen,* V. edition, 1993, M. Erhardt KG, Frankfurt.

[1.9] Der Bundesminister für Umwelt, Naturschutz und Reaktorsicherheit (BMU) (eds.): *Zweiter Bericht der Bundesregierung an den Deutschen Bundestag über Maßnahmen zum Schutz der Ozonschicht,* 1992.

[1.10] Der Bundesminister für Umwelt, Naturschutz und Reaktorsicherheit (BMU) (eds.): *Tagungsband Alternativen zu FCKW und Halonen,* Berlin, Feb. 24–26, 1992.

[1.11] VDI 3455: Anlagen zur Serienlackierung von Automobilkarossen, Beuth-Verlag, Berlin 1993.

[1.12] O. Lenz, Verband der Chemischen Industrie e.V. (VCI), personal communication, 1994.

References for Chapter 2

[2.1] J. v. Iribarne, H. R. Cho: *Atmospheric Physics,* D. Reidel Publ., Dortrecht 1980.

[2.2] P. Fabian: *Atmosphäre und Umwelt,* Springer Verlag, Berlin 1984.

[2.3] C. D. Schönwiese, B. Dieckmann: *Der Treibhauseffekt,* Rowohlt-Verlag, Hamburg 1989.

[2.4] R. P. Hosker, Jr. in D. Randerson (ed.): "Flow and Diffusion near Obstacles, Atmospheric Science and Power Production," *Tech. Inf. Cent. U.S. Dep. Energy* DOE/TIC-27601 (DE 84005177).

[2.5] A. Knaps, G. Manier, R. Röckle: *Untersuchung der Strömungsverhältnisse im Bereich von Industrieanlagen zur Beurteilung von Störfallemissionen,* Bundesministerium für Forschung und Technologie, no. RGB 8606, Darmstadt 1989.

[2.6] L. F. Richardson: "The Supply of Energy from and to Atmospheric Eddies," *Proc. R. Soc. London A* **97** (1920) 354–373.

[2.7] A. M. Obukhow: "Turbulence in an Atmosphere with Non-uniform Temperature," *Boundary Layer Meteor.* **2** (1946) 7–29.

[2.8] G. Manier: "Vergleich zwischen Ausbreitungsklassen und Temperaturgradienten," *Meteorol. Rundsch.* **28** (1975) 6–11.

[2.9] H. E. Cramer, F. A. Record, H. C. Vaughan: "The Study of the Diffusion of Gases in the Lower Atmoshere," M.I.T. Dept. of Meteorology, *Final Report under Contract,* no. AF 19 (604) 1058.
J. S. Hay, F. Pasquill: "Diffusion from a Continuous Source in Relation to Spectrum and the Scale of Turbulence. Atmospheric Diffusion and Air Pollution," in F. N. Frenkiel, P. A. Sheppard (eds.), *Adv. Geophys.* **6** (1959) 345.
K. Vogt, H. Geiss, G. Polster: "New Sets of Diffusion Parameters Resulting from Tracer Experiments with 50 and 100 m Release Hight," *Proc. 9th Int. Meet. Air Poll. Modelling Appl.* No. 103, *Nato Comm.* (1978) 221–239.

[2.10] S. Blinda, A. Hopf, G. Manier, K. Marburg: Entwicklung und Verifizierung eines Rechenmodells zur Simulation der Ausbreitung von Kfz-Emissionen im Bereich komplexer Gebäudekonfigurationen. Im Auftrage des Ministers für Umwelt, Raumordnung und Landwirtschaft des Landes Nordrhein-Westfalen, Institut für Meteorologie, Darmstadt 1993.

[2.11] Technische Anleitung zur Reinhaltung der Luft (TA-Luft), Erste Allgemeine Verwaltungsvorschrift zum Bundesimmissionsschutzgesetz, C. Heymanns Verlag, Köln 1980.

[2.12] Zweite Allgemeine Verwaltungsvorschrift zur Störfall-Verordnung (2. Störfall VwV), April 27, 1982, Gemeinsames Ministerialblatt A 33, no. 14, pp. 205–216.

[2.13] VDI 3783, Blatt 2: Umweltmeteorologie – Ausbreitung von störfallbedingten Freisetzungen – schwerer Gase – Sicherheitsanalyse, Beuth Verlag, Berlin 1990.

[2.14] VDI 3783, Blatt 1: Umweltmeteorologie – Ausbreitung von störfallbedingten Freisetzungen – Sicherheitsanalyse, Beuth Verlag, Berlin 1987.

[2.15] VDI 3782, Blatt 4: Umweltmeteorologie – Ausbreitung von Geruchsstoffen in der Atmosphäre, Beuth Verlag, Berlin 1994.

[2.16] H. Schirmer, W. Kuttler, J. Löbel, K. Weber: *Lufthygiene und Klima. Ein Handbuch zur Stadt- und Regionalplanung,* VDI-Verlag, Düsseldorf 1993.

[2.17] VDI 3958, Blatt 1: Umweltsimulation, Grundlagen und Methoden, Beuth Verlag, Berlin 1994.

References for Chapter 3

[3.1] H. Behrendt in: N. H. Seemayer, W. Hadnagy (eds.) *Environmental Hygiene* II, 1991, 253–256.

[3.2] R. Doll, R. Peto, *JNCI J. Natl. Cancer Inst.* **66** (1981) 1191–1308.

[3.3] M. Gough, J. *Am. Chem. Soc.* **23** (1989) 925–930.

[3.4] T. Eikmann in W. Bachmann (ed.): *Das Grüne Gehirn,* Schulz-Verlag, Starnberg 1991, E2, pp. 1–57.

[3.5] D. Sheppard, *Am. Rev. Respir. Dis.* **123** (1981) 486–491.

[3.6] G. Ericsson, P. Canner, *Scand. J. Work Environ. Health* **9** Suppl. 3 (1983) 1 52.

[3.7] D. S. Dockery, *J. Air Pollut. Control Assoc.* **32** (1982) 937–942.
[3.8] A. Cohen, A. Nelson, C. J. Bromberg, *Am. J. Public Health* **64** (1974) 442–449.
[3.9] S. Tani, *Nippon Koshu Eisei Zasshi* **22** (1975) 431–438.
[3.10] H. E. Wichmann, D. Sugiri, G. Herold, E. Knülle, *Zentralbl. Bakteriol Mikrobiol. Hyg. Abt. 1 Orig.* **B 185** (1988) 509–519.
[3.11] H. E. Wichmann, C. Spix, G. Mücke: *Bericht für den Minister für Arbeit, Gesundheit und Soziales,* Medizinisches Institut für Umwelthygiene, Düsseldorf 1987.
[3.12] Technische Anleitung zur Reinhaltung der Luft – TA Luft vom 27.02.1986, Erste Allgemeine Verwaltungsvorschrift zum Bundes-Immissionsschutzgesetz der Bundesrepublik Deutschland.
[3.13] VDI-Richtlinie Schwefeldioxid, no. 2310, Blatt 11, Verein Deutscher Ingenieure, Düsseldorf 1984.
[3.14] WHO, Air Quality Guidelines for Europe, Europ. Series no. 23, Copenhagen 1987.
[3.15] H.-W. Schlipköter, N. Manolovic, R. Dolgner, *Luftverunreinigung*, Medizinisches Institut für Umwelthygiene, Jahresbericht 1985, Düsseldorf 1985, 1–6.
[3.16] D. B. Menzel, in T. Schneider, L. Grant (eds.): *Air Pollution by Nitrogen Oxides*, Elsevier Scientific Publ., Amsterdam 1982, pp. 417–426.
[3.17] WHO, Environ. Health Criteria no. 4, Geneva 1977.
[3.18] VDI-Richtlinie Stickstoffoxid, no. 2310, Blatt 12, Verein Deutscher Ingenieure, Düsseldorf 1985.
[3.19] EPA – U.S. Environmental Protection Agency, Air Quality Criteria for Ozone and Other Photochemical Oxidants, 1986.
[3.20] VDI-Richtlinie Ozon (und photochemische Oxidantien), no. 2310, Blatt 15, Verein Deutscher Ingenieure, Düsseldorf 1987.
[3.21] WHO, Environmental Health Criteria 13, Geneva 1979.
[3.22] W. S. Aronow, M. W. Isbell, *Ann. Intern. Med.* **79** (1973) 392–395.
[3.23] W. S. Aronow, J. Cassidy, *Ann. Intern. Med.* **83** (1975) 496–499.
[3.24] W. S. Aronow, E. A. Stemmer, M. W. Isbell, *Circulation* **49** (1974) 415–417.
[3.25] VDI-Richtlinie Kohlenmonoxid, no. 2310, Blatt 13, Verein Deutscher Ingenieure, Düsseldorf (Entwurf) 1992.
[3.26] VDI-Richtlinie Schwebstaub, no. 2310, Blatt 19, Verein Deutscher Ingenieure, Düsseldorf 1992.
[3.27] J. W. B. Douglas, R. W. Waller, *Br. J. Prev. Soc. Med.* **32** (1966) 1–8.
[3.28] ZEBS – Zentrale Erfassungs- und Bewertungsstelle für Umweltchemikalien des Bundesgesundheitsamtes, *ZEBS-Hefte* **1** (1984).
[3.29] WHO, Environmental Health Criteria no. 3, Geneva 1977.
[3.30] G. Winneke, A. Brockhaus, U. Ewers, U. Kauer, M. Neut, *Neurotoxicol. Teratol.* **12** (1990) 553–559.
[3.31] U. Ewers, H.-W. Schlipköter in E. Merian (ed.): *Metalle in der Umwelt*, Verlag Chemie, Weinheim 1984, pp. 351–374.
[3.32] Verein Deutscher Ingenieure, *VDI Ber.* **203** (1974).
[3.33] DFG – Deutsche Forschungsgemeinschaft, VCH-Verlagsgesellschaft, Weinheim 1992.
[3.34] LAI – Länderausschuß für Immissionsschutz: *Krebsrisiko durch Luftverunreinigungen*, Düsseldorf 1992.
[3.35] T. Eikmann, M. Kramer, H. Goebel, *Zbl. Hyg.* **193** (1992) 41–52.
[3.36] T. Eikmann in H. Wichmann, H.-W. Schlipköter, G. Füllgraf (eds.): *Handbuch der Umweltmedizin*, ecomed-Verlag, Landsberg 1992.
[3.37] F. Pott, *Staub Reinhalt. Luft* **45** (1985) 369–379.
[3.38] BGA – UBA – Bundesgesundheitsamt – Umweltbundesamt, Bundesgesundheitsbl., Sonderheft 1993.
[3.39] H. Beck, *Bundesgesundheitsbl.* **33** (1990) 99–104.
[3.40] J. Abel, *VDI Ber.* **634** (1987) 487–501.
[3.41] Y. P. Drogan et al., *Carcinogenesis* **13** (1992) 1389–1395.
[3.42] J. A. Stöckhardt: "Untersuchungen über die schädlichen Einwirkungen des Hütten- und Steinkohlenrauches auf das Wachstum der Pflanzen, insbesondere der Fichte und Tanne," *Forstl. Jahrb.*, 1871, p. 218 ff.
[3.43] K. Garber: *Luftverunreinigungen und ihre Wirkungen*, Borntraeger, Berlin-Nikolassee 1967.
[3.44] R. Guderian: *Air Pollution Phytotoxicity of Acidic Gases and its Significance in Air Pollution Control, Ecological Studies,* vol. 22, Springer Verlag, Berlin 1977.
[3.45] Umweltbundesamt Berlin (eds.): *Medizinische, biologische und ökologische Grundlagen zur Bewertung schädlicher Luftverunreinigungen,* Sachverständigenanhörung Berlin, Feb. 20–24, 1978, Aug. 1978, pp. 33–99.
[3.46] R. Guderian, K. Reidl: "Höhere Pflanzen als Indikatoren für Immissionsbelastungen im terrestrischen Bereich," *Dechema-Beih.* (Bonn) **26** (1982) 6–22.
[3.47] Biologische Meßverfahren zur Ermittlung und Beurteilung der Wirkung von Luftverunreinigungen. Grundlagen, VDI-Richtlinie 3957, Blatt 1, Vorentwurf, Stand July 1994.
[3.48] G. Scholl, D. Schwela: "Kriterien zur Auswahl von Akzeptoren und Meßgrößen des Wirkungskatasters," *Schriftenr. Landesanst. Immissionsschutz Landes Nordrhein-Westfalen, Essen* **40** (1977) 49–52.
[3.49] H. Schönbeck: "Eine Methode zur Erfassung der biologischen Wirksamkeit von Luftverunreinigungen durch transplantierte Flechten, *Staub Reinh. Luft* **29** (1969) 14–18.
[3.50] G. Scholl: "Die Immissionsrate von Fluor in Pflanzen als Maßstab für eine Immissionsbegrenzung," *VDI-Ber.* **164** (1971) 39–45.
[3.51] Bioindikation. Wirkungsbezogene Erhebungsverfahren für den Immissionsschutz, *VDI-Ber.* **609** (1987) 377–393.
[3.52] Bioindikation: Ein wirksames Instrument der Umweltkontrolle, *VDI-Ber.* **901** (1992) vols. 1 and 2.
[3.53] Zielsetzung und Bedeutung der Richtlinie Maximale Immissions-Werte, VDI 2310, Blatt 1 (10/88).
[3.54] A. C. Posthumus, *VDI-Ber.* **314** (1978) 225–230.
[3.55] A. S. Heagle, D. E. Bady, W. W. Heck: "An Open-Top-Field Chamber to Assess the Impact of Air Pollution on Plants," *J. Environ. Qual.* **2** (1973) 265–268.
[3.56] VDI-Fachdokumentation Reinhaltung der Luft: *Sauerstoffhaltige Schwefelverbindungen. Bibliographie 1974–1978*, VDI-Verlag, Düsseldorf 1978, pp. 286 ff.
[3.57] J. Ziegler, H. Ziegler, *VDI-Ber.* **314** (1978) 219–223.
[3.58] Maximale Immissions-Werte zum Schutz der Vegetation. Maximale Immissions-Werte für Schwefel-

dioxid, VDI-Richtlinie 2310, Blatt 2 (Entwurf) Stand Aug. 1978.
[3.59] Maximale Immissions-Werte zum Schutz der Vegetation. Maximale Immissions-Konzentrationen für Ozon. VDI-Richtlinie 2310, Blatt 6, April 1989.
[3.60] Maximale Immissions-Werte zum Schutz der Vegetation. Maximale Immissions-Konzentrationen für Fluorwasserstoff. VDI-Richtlinie 2310, Blatt 3, Dezember 1989.
[3.61] D. F. Adams et al.: "Relationship of Atmospheric Fluoride Levels and Injury Indexes on Gladiolus and Ponderosa Pine," *J. Agric. Food Chem.* **4** (1956) 64–66.
[3.62] H.-J. Crecelius, W. Forwerg, R. Zahn, *Staub Reinh. Luft* **36** (1976) 201–205.
[3.63] E. Brennan, J. A. Leone, R. H. Daines, *For. Sci.* **12** (1966) 286–390.
[3.64] H. van Haut, B. Prinz, *Staub Reinh. Luft* **39** (1979) 408–414.
[3.65] L. Steubing, *Staub Reinh. Luft* **39** (1979) 161–164.
[3.66] G. H. Krause, *Staub Reinh. Luft* **39** (1979) 165–167.
[3.67] Deutsche Gesellschaft für Mineralölwissenschaft und Kohlechemie, Forschungsbericht 174-1, Hamburg 1978.
[3.68] D. T. Tingey, G. E. Taylor, jr.: "Variation in Plant Response to Ozone: A Conceptional Model of Physiological Events," in D. P. Orurod (ed.): *Effects of Gaseous Air Pollutants in Agriculture and Horticulture*, Butterworth Scientific 1982, p. 111 ff.
[3.69] W. A. Feder: "Plant Response to Chronic Exposures to Low Levels of Oxidant Type Air Pollution," *Environ. Pollut.* **1** (1970) 73–79.
[3.70] Umweltbundesamt Berlin (eds.): Berichte 5/83, Luftqualitätskriterien für photochemische Oxidantien, Erich Schmidt Verlag, Berlin 1983.
[3.71] WHO Regional Office for Europe: "The Effects of Ozone and Other Photochemical Oxidants on Vegetation," in: *Air Quality Guidelines for Europe*, WHO Regional Publications, Eur. Ser. No. 23, Copenhagen 1987.
[3.72] Minister für Arbeit, Gesundheit und Soziales des Landes Nordrhein-Westfalen (eds.): *Umweltprobleme durch Schwermetalle im Raum Stolberg*, Düsseldorf 1975.
[3.73] Minister für Arbeit, Gesundheitheit und Soziales des Landes Nordrhein-Westfalen (eds.): *Umweltbelastung im Raum Datteln, Analyse und Problemlösung*, Düsseldorf 1977.
[3.74] Verein Deutscher Ingenieure, Kommission Reinhaltung der Luft (eds.): *Säurehaltige Niederschläge – Entstehung und Wirkungen auf terrestrische Ökosysteme*, Düsseldorf 1983.

References for Chapter 4

General References for Section 4.1–4.3

[4.1] P. Davids, M. Lange, Die Großfeuerungsanlagenverordnung, Komm., 1984.
H. Engelhardt, Bundesimmissionsschutzgesetz, Losebl.-Komm., 1993.
[4.2] G. Feldhaus, Bundesimmissionsschutzrecht, Losebl.-Komm., 1995.
K. Hansmann, TA Luft, Komm. 1987.
H. D. Jarass, BImSchG, Komm., 1995
[4.3] Landmann, Rohmer, vol. III, *Umweltrecht*, Losebl.-Komm. 1995.
C. Offermann-Clas, *Luftreinhaltung in der Bundesrepublik Deutschland*, 1984.
H. Schmatz, M. Nöthlichs, *Sicherheitstechnik*, vol. VIII, Losebl.-Komm. 1995.
D. Sellner, *Immissionsschutz und Industrieanlagen*, 1988.
[4.4] R. Stich, K. W. Porger, *Immissionsschutzrecht des Bundes und der Länder*, Losebl.-Komm. 1995.
C. H. Ule, W. H. Laubinger, *Bundesimmissionsschutzgesetz*, Losebl.-Komm., 1995.

General References for Sections 4.4–4.5

[4.5] H. Reihlen: *Technische Normen – Freiheit und Bindung*, DIN-Mitteilung 70, 1991, no. 10, D. 527–532.
[4.6] DIN-Normenheft 10: Grundlagen der Normungsarbeit des DIN, Beuth-Verlag, Berlin 1991.
[4.7] Richtlinie des Rates vom 21. Dez. 1988 zur Angleichung der Rechts- und Verwaltungsvorschriften der Mitgliedsstaaten über Bauprodukte (89/106/EWG) A Bl. no. 40, Feb. 12, 1989, p. 12.
[4.8] Amtsblatt der Europäischen Gemeinschaften: "Richtlinie des Rates vom 28. März 1983 über ein Informationsverfahren auf dem Gebiet der Normen und technischen Vorschriften," 83/109/EWG.
[4.9] C. Mohr: *ISO and CEN*, DIN-Mitteilung 70, 1991, no. 6, pp. 319–320.
[4.10] Europäische Normen für 1992 – Ein Leitfaden des DIN, DIN Deutsches Institut für Normung e.V., Berlin.
[4.11] H. Pohle: *Chemische Industrie – Umweltschutz, Arbeitsschutz, Anlagensicherheit. Rechtliche und Technische Normen, Umsetzung in die Praxis*, VCH Verlagsgesellschaft, Weinheim 1991.
[4.12] R. Kleinert: *Marktzugang durch Normenkonformität*, DIN-Mitteilung 71, 1992, no. 1, pp. 17–23.
[4.13] K. Grefen: "Harmonisierung technischer Regeln im EG-Binnenmarkt – Beitrag der Kommission Reinhaltung der Luft im VDI und DIN," *Staub – Reinhalt. Luft* **51** (1991) 199–205.
[4.14] C. A. Aronds: "International Efforts for Standardization of Measuring Methods in Air Pollution Control – Requirements and Aims," *Tagung Kommission Reinhaltung der Luft im VDI und DIN über Aktuelle Aufgaben der Meßtechnik in der Luftreinhaltung*, Heidelberg 1990, pp. 3–10.
[4.15] Richtlinie (EWG) des Rates über die Verbrennung gefährlicher Abfälle, Bundesrats-Drucksache 290/92, Ratsdok. 5761/92, Heger-Verlag, Bonn.
[4.16] Bauproduktenrichtlinie 89/106/EWG: Richtlinie des Rates vom 21.12.1988 zur Angleichung der Rechts- und Verwaltungsvorschriften der Mitgliedsstaaten über Bauprodukte, geändert durch die Richtlinie 93/68/EWG des Rates vom 22. Juli 1993.
[4.17] G. Feldhaus: *Grundlagen des Umweltschutzes – Bericht über die Arbeit des NAGUS*, DIN-Mitteilung 73. 1994, no. 7, pp. 456–460.

References for Chapter 5

[5.1] R. C. West, M. J. Astle: *Handbook of Chemistry and Physics*, CRC Press, Palm Beach 1979.
[5.2] T. E. Graedel: *Chemical Compounds in the Atmosphere*, Academic Press, New York 1978.
[5.3] L. S. Lochran, R. A. Pielke, E. Kovács, *J. Air Waste Manage. Assoc.* **42** (1992) 1567–1572.

[5.4] E. Lahmann: *Determination and Evaluation of Ambient Air Quality – Manual of Ambient Air Quality Control in Germany,* Federal Minister for the Environment, Nature Conservation and Nuclear Safety (eds.), Berlin 1992.
[5.5] D. L. Fox, H. E. Jeffries, *Anal. Chem.* **51** (1979) 22R–34R.
[5.6] Vierte Allgemeine Verwaltungsvorschrift zum Bundes-Immissionsschutzgesetz (Ermittlung von Immissionen in Untersuchungsgebieten) vom 26.11.1993. Gemeinsames Ministerialblatt 1993, 827–840.
[5.7] Technische Anleitung zur Reinhaltung der Luft (TA Luft) vom 27.2.1986. Gemeinsames Ministerialblatt 37 (1986) 95–144.
[5.8] "Musterentwurf einer neuen Smogverordnung," *Neue Z. Verwaltungsrecht* **7** (1988) 138–140.
[5.9] Richtlinie des Rates vom 21.9.1992 über die Luftverschmutzung durch Ozon (92/72/EWG), Amtsblatt der Europäischen Gemeinschaften L 297/1–7.
[5.10] Richtlinie des Rates vom 7.3.1985 über Luftqualitätsnormen für NO_2 (85/203/EWG). Amtsblatt der Europäischen Gemeinschaften 28 (1985) L 87, 1–7.
[5.11] Richtlinie des Rates vom 15.7.1980 über Grenzwerte und Leitwerte der Luftqualität für Schwefeldioxid und Schwebstaub (80/779/EWG), Amtsblatt der Europäischen Gemeinschaften, L 229/30–48.
[5.12] Bundes-Immissionsschutzgesetz vom 14.5.1990. Bundesgesetzblatt 1990, part I, pp. 880–901.
[5.13] P. Bruckmann, H.-U. Pfeffer: "Umweltschutz in Städten," *VDI Ber.* **952** (1992) 151–166.
[5.14] D. H. F. Atkins, I. Quirino: *A Survey of Nitrogen Dioxide in Paris,* Report EUR 13369 EN, Commission of the European Communities (eds.), Brussels 1992, pp. 1–59.
[5.15] P. Bruckmann, R. Beier, S. Krautscheid, *Staub Reinhalt. Luft* **43** (1983) 404–410.
[5.16] J. C. Frohne, J. Reis, *Erdöl Kohle Erdgas, Petrochem.* **34** (1981) 209–213.
[5.17] M. Buck, S. Krautscheid, *Umwelt* **1988**, no. 5, 188–190.
[5.18] M. Buck, R. Beier, *Clean Air* (*Melbourne*) **22** (1988) no. 2, 64–67.
[5.19] R. Beier, *Staub Reinhalt. Luft* **50** (1990) 265–270.
[5.20] R. Beier, M. Buck, H.-U. Pfeffer "Man and His Ecosystem," in L. J. Brasser, W. C. Mulder (eds.): *Proceedings of the 86th World Clean Air Congress,* vol. 3, Elsevier Science Publ., Amsterdam 1989, pp. 55–60.
[5.21] H. Stratmann, M. Buck, U. Hölzel, D. Rosin, *Schriftenr. LIB Landesanst. Immissions- Bodennutzungsschutz d. Landes Nordrhein-Westfalen* **1** (1965) 25–49.
[5.22] J. P. Butler, G. B. Post, P. J. Lioy, J. M. Waldmann, A. Greenberg, *J. Air Waste Manage. Assoc.* **43** (1993) 970–977.
[5.23] N. D. van Egmond, D. Onderdelinden, *Atmos. Environ.* **15** (1981) 1035–1046.
[5.24] R. Beier et al.: "A Study of Network Design and Measurement Methods in the Member States for the EC Air Quality Directive for Nitrogen Dioxide (85/203/ECC)", Commission of the European Communities (eds.), *EUR-Report,* Brussels 1987.
[5.25] R. Beier, *Schriftenr. Landesanst. Immissionsschutz d. Landes Nordrhein-Westfalen* **55** (1982) 7–13.
[5.26] ISO/CD 11222.2, Air Quality–Comparison of Mean Values Determined from Air Quality Measurements with Prescribed Values.
[5.27] B. Schneider, *Atmos. Environ.* **21** (1987) 1275–1283.
[5.28] W. A. McClemey, K. D. Oliver, J. D. Plell, *Environ. Sci. Technol.* **23** (1989) 1273–1379.
[5.29] M. Buck, *Staub Reinhalt. Luft* **49** (1989) 337–342.
[5.30] Kommission Reinhaltung der Luft im VDI und DIN (eds.): *VDI-Handbuch Reinhaltung der Luft,* VDI-Verlag, Düsseldorf (loose-leaf book).
[5.31] DIN ISO 5725, Genauigkeit (Richtigkeit und Präzision) von Meßverfahren und Meßergebnissen.
[5.32] DIN 32645 (05/94).
[5.33] DIN ISO 6879 (01/84).
[5.34] VDI-Richtlinie 2449, Bl. 1 (02/95), in [5.30].
[5.35] H.-U. Pfeffer, M. Buck, *Proceedings of the 11th Triennial World Congress of the International Measurement Confederation (IMEKO),* Houston, Texas, ACTA IMEKO 1988, vol. 4, 527–536.
[5.36] K. Grefen, *Staub Reinhalt. Luft* **51** (1991) 199–205.
[5.37] Gemeinsames Ministerialblatt 1981, p. 355.
[5.38] Gemeinsames Ministerialblatt 1983, p. 78.
[5.39] Gemeinsames Ministerialblatt 1988, p. 191.
[5.40] DIN ISO 9004: Qualitätssicherung und Elemente eines Qualitätssicherungssystems – Leitfaden, 1990.
[5.41] DIN EN 45001: Allgemeine Kriterien zum Betreiben von Prüflaboratorien, 1990.
[5.42] ISO-Guide 25: General Requirements for the Competence of Calibration and Testing Laboratories, 1990.
[5.43] H.-U. Pfeffer, H. Dobrick, *Staub Reinhalt. Luft* **47** (1987) 28–33.
[5.44] H.-U. Pfeffer: *Proceedings of the International Workshop on Harmonisation of the Technical Implementation of the Three EEC Air Quality Directives 80/779 (SO_2, SPM), 82/884 (Lead), 85/203 (NO_2),* Nov. 22–24, 1988, Lyon, Editions GAP, La Ravoire 1990.
[5.45] M. Buck: *Staub Reinhalt. Luft* **41** (1981) 365–369.
[5.46] H.-U. Pfeffer: "Qualitätssicherung in automatischen Immissionsnetzen,", *LIS-Ber.* **52** (1984) 1–73.
[5.47] D. Dahmann, J.-C. Frohne, H. Manns, *Staub Reinhalt. Luft.* **45** (185) 359–362.
[5.48] H. Manns, H. Gies, G. Nitz: "Anlage zur Qualitätssicherung von Immissionsmessungen," *LIS-Ber.* **27** (1982) 1–139.
[5.49] E. Lahmann: *Luftverunreinigung–Luftreinhaltung,* Verlag Paul Paray, Berlin 1990.
[5.50] P. W. West, G. C. Gaeke, *Anal. Chem.* **28** (1956) 1816–1819.
[5.51] VDI-Richtlinie 2451, Bl. 3 (11/94, draft), in [5.30].
[5.52] H. Stratmann, *Mikrochim. Acta* **1954**, 668–678.
[5.53] VDI-Richtlinie 2451, Bl. 1 (08/68), in [5.30].
[5.54] ISO 6767, 1990.
[5.55] B. E. Saltzman, *Anal. Chem.* **26** (1954) 1949–1955.
[5.56] VDI-Richtlinie 2453, Bl. 1 (10/90) in [5.30].
[5.57] ISO 6768, 1985.
[5.58] VDI-Richtlinie 2468, Bl. 1 (05/78) in [5.30].
[5.59] VDI-Richtlinie 2452, Bl. 1 (03/78) in [5.30].
[5.60] M. Buck, H. Stratmann, *Brennst. Chem.* **46** (1965) 231–235.
[5.61] VDI-Richtlinie 2452, Bl. 2 (02/75) in [5.30].
[5.62] VDI-Richtlinie 2452, Bl. 3 (07/87) in [5.30].
[5.63] K. Svoboda, H. Ixfeld, *Staub Reinhalt. Luft* **31** (1971) 1–8.
[5.64] H. Kettner, W. Forwerg, *Atmos. Environ.* **3** (1969) 215–220.
[5.65] VDI-Richtlinie 2458, Bl. 1 (12/73) in [5.30].
[5.66] S. Häntzsch, E. Lahmann, *Schriftenr. Ver. Wasser-, Boden- Lufthyg.* **33** (1970) 35–39.

[5.67] VDI-Richtlinie 2461, Bl. 1 (03/74) in [5.30].
[5.68] M. Buck, H. Stratmann, *Z. Anal. Chem.* **213** (1965) 241–250.
[5.69] VDI-Richtlinie 2461, Bl. 2 (05/76) in [30].
[5.70] M. Buck, H. Gies, *Staub Reinhalt. Luft* **26** (1966) 379–384.
[5.71] M. Buck, H. Stratmann, *Staub Reinhalt. Luft* **24** (1964) 241–250.
[5.72] VDI-Richtlinie 2454, Bl. 1 (03/82) in [5.30].
[5.73] K.-E. Prescher, E. Lahmann, *Gesund. Ing.* **87** (1966) 351–353.
[5.74] VDI-Richtlinie 2454, Bl. 2 (03/82) in [5.30].
[5.75] ISO DIS 6769, 1983.
[5.76] D. Klockow, *Fresenius Z. Anal. Chem.* **326** (1987) 5–24.
[5.77] M. Ferm, *Atmos. Environ.* **13** (1979) 1385–1393.
[5.78] M. Ferm, *Atmos. Environ.* **20** (1986) 1193–1201.
[5.79] J. Slanina et al., *Int. J. Environ. Anal. Chem.* **9** (1981) 59–70.
[5.80] N. A. Dimmock, G. B. Marshall, *Anal. Chim. Acta* **185** (1986) 159–169.
[5.81] M. P. Keuken, A. Wayers-IJpelaan, J. J. Mols, R. P. Otjes, J. Slanina, *Atmos. Environ.* **23** (1989) 2177–2185.
[5.82] J. Slanince, G. P. Wyers, *Fresenius J. Anal. Chem.* **350** (1994) 467–473.
[5.83] Gemeinsames Ministerialblatt: 1990, 242–244; 1991, 1046; 1992, 1142–1143.
[5.84] VDI-Richtlinie 2451, Bl. 6 (07/87, draft) in [5.30].
[5.85] M. Buck: *Schriftenr. LIB Landesanst. Immissions-Bodennutzungsschutz d. Landes Nordrhein-Westfalen* **6** (1967) 28–31.
[5.86] H. Okabe, P. L. Splitstone, J. J. Ball, *J. Air Pollut. Acontrol. Assoc.* **23** (1973) 514–516.
[5.87] A. Sabadeli, A. Frontijn, R. J. Ronco, *Anal. Chem.* **42** (1970) 575–579.
[5.88] ISO 7996, 1985.
[5.89] J. Sigsby, F. M. Black, T. A. Bellar, D. L. Klosterman, *Environ. Sci. Technol.* **7** (1973) 51–54.
[5.90] VDI-Richtlinie 2455, Bl. 1 (08/70), Bl. 2 (10/70) in [5.30].
[5.91] VDI-Richtlinie 2468, Bl. 6 (07/79) in [5.30].
[5.92] VDI-Richtlinie 2468, Bl. 4 (05/78) in [5.30].
[5.93] VDI-Richtlinie 3490, Bl. 1–17 in [5.30].
[5.94] J. P. Lodge, Jr. (ed.): *Methods of Air Sampling and Analysis*, 3rd ed., Lewis Publishers, Chelsea, Michigan, 1989.
[5.95] DIN EN 481, 1993.
ISO TR 7708, 1992.
[5.96] VDI-Richtlinie 2463, Bl. 8 (08/82) in [5.30].
[5.97] D. Bake, L. Laskus, *Staub Reinhalt. Luft* **38** (1978) 21–24.
[5.98] VDI-Richtlinie 2463, Bl. 7 (08/82) in [30].
[5.99] E. Herpertz, *Staub Reinhalt. Luft* **29** (1969) 408–413.
[5.100] VDI-Richtlinie 2463, Bl. 4 (12/76) in [5.30].
[5.101] VDI-Richtlinie 2463, Bl. 9 (02/87) in [5.30].
[5.102] G. A. Jutze, K. E. Foster, *J. Air Pollut. Control. Assoc.* **17** (1967) 17–25.
[5.103] US Environmental Protection Agency: Quality Assurance Handbook for Air Pollution Measurement Systems, vol. II: Ambient Air Specific Methods, Research Triangle Park, N.C., 1977.
[5.104] VDI-Richtlinie 2463, Bl. 2 (07/77, draft).
[5.105] VDI-Richtlinie 2463, Bl. 5 (12/87) in [5.30].
[5.106] VDI-Richtlinie 2463, Bl. 6 (11/87) in [5.30].
[5.107] H. Manns, H. Gies, W. Stramplat, "Schwebstaubmeßgerät FH 62 I für die automatisierte Immissionsmessung," *LIS-Ber.* **11** (1980) 1–26.
[5.108] H.-U. Pfeffer, H. W. Lohse, *Staub Reinhalt. Luft* **44** (1984) 67–71.
[5.109] H. Dresia, R. Mucha, *Staub Reinhalt. Luft* **34** (1974) 125–128.
[5.110] H. Dresia, F. Spohr, *Staub Reinhalt. Luft* **38** (1978) 431–435.
[5.111] H. G. Wirweitzky, *VDI-Ber.* **1059** (1993) 533–569.
[5.112] H.-U. Pfeffer, M. Buck, *J. Aerosol Sci.* **17** (1986) 290–292.
[5.113] H. Patashnick, E. G. Rupprecht, *J. Air Waste Manage. Assoc.* **41** (1991) 1079–1083.
[5.114] Organisation for Economic Cooperation and Development (OECD): *Methods of Measuring Air Pollution*, Paris 1964.
[5.115] *Instruction Manual for the National Survey of Smoke and Sulfur Dioxide*, Warren Spring Laboratory, Stevenage, Hertfordshire, 1966.
[5.116] D. L. R. Balley, P. Clayton; Report LR 325 (AP), Warren Spring Laboratory, Stevenage, Hertfordshire, 1980.
[5.117] L. Laskus, D. Bake, L. Armbruster, *Staub Reinhalt. Luft* **40** (1980) 18–26, 101–105.
[5.118] G. E. Pashel, D. R. Egner, *Atmos. Environ.* **15** (1981) 919–927.
[5.119] P. Bruckmann, H.-U. Pfeffer, *VDI-Ber.* **888** (1991) 377–396.
[5.120] *Nachr. Chem. Tech. Lab.* **37** (1989) Special issue "Spektroskopie".
[5.121] "Analytische Chemie 1991," *Nachr. Chem. Tech. Lab.* **40** (1992) 146–154.
[5.122] VDI-Richtlinie 2267, Bl. 3, 4, 6, 7, 11, 12, in [5.30].
[5.123] VDI-Richtlinie 2268, Bl. 1 to 4 in [5.30].
[5.124] B. Neidhardt et al., *VDI-Ber.* **838** (1990) 485–498.
[5.125] E. S. Rabano, N. T. Castillo, K. J. Torre, P. A. Solomon, *J. Air Pollut. Control. Assoc.* **39** (1989) 76–80.
[5.126] VDI-Richtlinie 2119, Bl. 1 to 4 in [5.30].
[5.127] E. Schultz, *Staub Reinhalt. Luft* **43** (1983) 482–489.
[5.128] WHO Regional Office for Europe (eds.): Air Quality Guidelines for Europe, WHO Regional Publ. European Series No. 23, Copenhagen 1987.
[5.129] "Krebsrisiko durch Luftverunreinigungen," Studie des Länderausschusses für Immissionsschutz, Ministerium für Umwelt, Raumordnung und Landwirtschaft Nordrhein-Westfalen (eds.); Düsseldorf 1992.
[5.130] "Faserige Stäube," *VDI-Ber.* **475** (1982).
[5.131] VDI-Richtlinie 3492, Bl. 1 (08/91) in [5.30].
[5.132] F. Pott, *VDI-Ber.* **888** (1991) 211–244.
[5.133] Technische Regeln für Gefahrstoffe, BArbBl. 2/92, 57.
[5.134] G. Elbers, S. Muratyan, *VDI-Ber.* **888** (1991) 143–170.
[5.135] M. Buck, G. Elbers, *Erdöl Kohle Erdgas Petrochem.* **45** (1992) 219–223.
[5.136] G. Elbers, S. Lutz, *VDI-Ber.* **1059** (1993) 337–356.
[5.137] A. Petzold, R. Niessner, *Mikrochim. Acta* **117** (1995) 215–237.
[5.138] Norwegian Institut for Air Research (NILU), "Annual Summary Report from the Chemical Co-ordinating Centre", *EMEP-CCC-Report 6/88*, 1988.
[5.139] V. Mohnen, J. Wilson, J. Kadlecek in H. W. Georgii, W. Jaeschke (eds.): *Chemistry of the Unpolluted*

and Polluted Troposphere. D. Reidel Publ., Dordrecht 1982, pp. 395–423.
[5.140] W. Fricke, S. Beilke, *Environ. Pollut.* **75** (1992) 121–127.
[5.141] VDI 3870 Bl. 1 (draft (07/85)): Messen von Regeninhaltsstoffen. Kriterien für Aufbau, Aufstellung und Betrieb von Regensammlern; in [5.30].
[5.142] P. Winkler, *Staub Reinhalt. Luft* **4** (1985) 256–260.
[5.143] P. Winkler: "Vergleichende Untersuchungen von Niederschlagssensoren und Sammlern im Hinblick auf eine Standardisierung," *VDI Ber.* **608** (1987) 483–492.
[5.144] H. W. Georgii, G. Schmitt, *Staub Reinhalt. Luft* **45** (1985) 260–264.
[5.145] F. G. Römer, B. H. TE Winkel, L. H. J. M. Janssen in G. Restelli, G. Angeletti (eds.): *Physico-chemical Behaviour of Atmospheric Pollutants,* Kluwer Acad. Publ., Dordrecht 1990, pp. 444–451.
[5.146] U. Bartels, J. Block, *J. Plant Nutr. Soil Sci.* **148** (1985) 689–698.
[5.147] M. Small, T. S. Stevens, W. C. Baumann, *Anal. Chem.* **47** (1975) 1801–1809.
[5.148] H. P. Orland (ed.): *Standard Methods for the Examination of Water and Wastewater,* 12th ed., American Public Health Assoc., New York 1975.
[5.149] M. C. Rand, A. E. Greenberg, M. J. Taras, M. A. Franson (eds.): *Standard Methods for the Examination of Water and Wastewater,* 14th ed., American Public Health Assoc., New York 1975.
[5.150] T. Class, K. Ballschmiter, *Fresenius Z. Anal. Chem.* **325** (1986) 1–7.
[5.151] W. B. Grant, R. H. Kagann, W. A. McClenny, *J. Air Waste Manage. Assoc.* **42** (1992) 18–30.
[5.152] K. Weber, V. Klein, W. Diehl: "Optische Fernmeßverfahren zur Bestimmung gasförmiger Luftschadstoffe in der Troposphäre," *VDI-Ber.* **838** (1990) 201–246.
[5.153] Eignungsprüfung Mehrkomponentenmeßsysteme AR 500 Komponente SO_2 im Auftrag der Fa. Opsis AB. Prüfbericht nr. 33-01/93. UMEG GmbH (ed.), Karlsruhe 1993.
[5.154] K. Josek, W. Junkermann, P. Werle, F. Slémr: "Erfahrungen an Absorptionsspektrometern mit abstimmbaren Diodenlasern für die Mehrkomponentenmessung von Luftschadstoffen," *VDI Ber.* **838** (1990) 181–200.
[5.155] A. Hofzumahans, H.-P. Dorn, U. Platt: "Tropospheric OH Radical Measurement Techniques: Recent Developments," in G. Restelli, G. Angeletti (eds.): *Physico-chemical Behaviour of Atmospheric Pollutants,* Kluwer Acad. Publ., Dordrecht 1990, pp. 103–108.
[5.156] P. T. Woods, R. H. Partridge, M. J. Milton, A. J. Davenport, *Bull. Int. Org. Legal Metrology* **131** (1993) 11–25.
[5.157] C. Weitkamp et al.: "Das mobile, ortsauflösende Schadgasfernmeß-System ARGOS" in C. Werner, V. Klein, K. Weber (eds.): *Laser in der Umweltmeßtechnik in Remote Sensing,* Springer Verlag, Berlin 1992, pp. 197–200.
[5.158] S. Beilke et al.: "Air Sampling Flights at Low Altitudes along the Border between the FRG and its Neighbours," in G. Angeletti, G. Restelli (eds.): *Physico-Chemical Behaviour of Atmospheric Pollutants,* D. Reidel Publ., Dordrecht 1987, pp. 685–703.
[5.159] U. Platt, D. Perner, A. M. Winer, G. W. Harris, J. N. Pitts, *Geophys. Res. Lett.* **7** (1980) 89–92.
[5.160] F. Slemr, G. W. Harris, D. H. Hastic, G. I. Hackay, H. I. Schiff, *J. Geophys. Res.* **91** (1986) 5371.
[5.161] E. C. Tuazon et al., *Environ. Sci. Technol.* **18** (1984) 981–984.
[5.162] P. Bruckmann, W. Mülder, *Schriftenr. Landesanstalt Immissionsschutz d. Landes Nordrhein-Westfalen, Essen* **47** (1979) 30–41.
[5.163] T. E. Kleindienst et al., *Environ. Sci. Technol.* **22** (1988) 53–61.
[5.164] VDI-Richtlinie 3483 Bl. 1 (12/79): Messen der Summe organischer Stoffe mit einem Flammen-Ionisations-Detektor (FID). Grundlagen, in [5.30].
[5.165] VDI-Richtlinie 3495 Bl. 1 (09/80): Bestimmung des durch Adsorption an Kieselgel erfaßbaren organisch gebundenen Kohlenstoffs in der Luft, in [5.30].
[5.166] VDI-Richtlinie 3483 Bl. 4 (11/81): Messen der Summe organischer Stoffe und von Methan mit dem Flammen-Ionisations-Detektor (FID) Bendix 8202, in [5.30].
[5.167] VDI-Richtlinie 3483 Bl. 2 (11/81): Messen der Summe organischer Stoffe ohne Methan mit dem Flammen-Ionisations-Detektor (FID) Siemens U 100, in [5.30].
[5.168] K. H. Becker, U. Schurath, A. Wiese, *Int. Conf. Photochem. Oxid. Pollut. Contr. Proc.* **1** (1977) 31–40.
[5.169] F. M. Black, L. E. High, A. Fontijn, *Environ. Sci. Technol.* **11** (1977) 597–601.
[5.170] P. Bruckmann et al., *Chemosphere* **17** (1988) 2363–2380.
[5.171] W. Schneider, J. C. Frohne, H. Bruderreck, *J. Chromatogr.* **155** (1978) 311–327.
[5.172] H. Kern, N. Kirshen: "Halogenierte organische Verbindungen in der Umwelt," *VDI-Ber.* **745** (1989) 335–344.
[5.173] W. McClenny, J. Pleil, G. Evans, K. Oliver, M. Holdren, W. Winbery, *J. Air Waste Manage. Assoc.* **41** (1991) 1308–1318.
[5.174] VDI-Richtlinie 3482 Bl. 6 (07/88): Gaschromatographische Bestimmung organischer Verbindungen – Probenahme durch Anreicherung. Thermische Desorption, in [5.30].
[5.175] J. Rudolph, A. Khedim, *Int. J. Environ. Anal. Chem.* **20** (1985) 265–282.
[5.176] K. Grob, G. Grob, *J. Chromatogr.* **62** (1971) 1–13.
[5.177] VDI-Richtlinie 3482 Bl. 4 (11/84): Gaschromatographische Bestimmung organischer Verbindungen mit Kapillarsäulen. Probenahme durch Anreichung an Aktivkohle-Desorption mit Lösemittel, in [5.30].
[5.178] VDI-Richtlinie 3494 Bl. 1 (12/94): Messen von Vinylchlorid-Konzentrationen. Gaschromatographische Bestimmung. Manuelle und automatische Dampfraumanalyse, in [5.30].
[5.179] H. P. König: "Halogenierte organische Verbindungen in der Umwelt," *VDI-Ber.* **745** (1989) 321–334.
[5.180] M. Zingg, C. Schlatter: "Halogenierte organische Verbindungen in der Umwelt," *VDI-Ber.* **745** (1989) 245–255.
[5.181] M. A. K. Khalil, R. A. Rasmussen, *J. Air Waste Manage Assoc.* **42** (1992) 810–813.

[5.182] VDI-Richtlinie 3494 Bl. 3 (05/85): Messen von Vinylchlorid-Konzentrationen; Automatisches Gas-Chromatisches Verfahren (A. I. R. Instruments Model 755 GC), in [5.30].
[5.183] J. König et al., *Atmos. Environ.* **14** (1980) 609–613.
[5.184] R. Westerholen, U. Stenberg, T. Alsberg, *Atmos. Environ.* **22** (1988) 1005–1010.
[5.185] C. Rappe, L. O. Kjeller, P. Bruckmann, K. H. Hackhe, *Chemosphere* **17** (1988) 3–20.
[5.186] P. Bruckmann et al., *Chemosphere* **27** (1993) 707–720.
[5.187] G. T. Hunt, B. E. Maisel, *J. Air Waste Manage Assoc.* **42** (1992) 672–680.
[5.188] L. Huber, J. Emmert, A. Gratzfeld-Hüsgen, W. Dulson, *Staub Reinhalt. Luft* **47** (1987) 22–27.
[5.189] Umweltforschungsplan des Bundesministers für Umwelt Naturschutz und Reaktorsicherheit, Forschungsbericht 10403364/01: "Weiterführende Untersuchung zur Bildung von polybromierten Dioxinen und Furanen bei der thermischen Belastung flammengeschützter Kunststoffe und Textilien," Feb. 1991.
[5.190] B. Zielinska, J. Arey, R. Atkinson, A. M. Winer, *Atmos. Environ.* **23** (1989) 223–229.
[5.191] M. Oehme, S. Man, A. Mikalsen, P. Kirschmer, *Chemosphere* **15** (1986) 607–617.
[5.192] H. Knoeppel et al., in: *Proceedings of the First European Symposium on Physico-Chemical Behaviour of Atmospheric Pollutants,* Ispra (1980) 25–40.
[5.193] A. R. Fernandez et al., *Chemosphere* **25** (1992) 1311–1316.
[5.194] W. N. Billings, T. F. Bidleman, *Environ. Sci. Technol.* **14** (1980) 679.
[5.195] D. A. Kurtz (ed.): *Long Range Transport of Pesticides,* Lewis Publishers, Chelsea, Michigan, 1990.
[5.196] C. Tashiro et al., *Chemosphere* **18** (1989) 777–782.
[5.197] VDI-Richtlinie 3484 Bl. 1 (01/79): Messen von Aldehyden. Bestimmen der Formaldehyd-Konzentration nach dem Sulfit-Pararosanilin-Verfahren, in [5.30].
[5.198] H. I. Schiff et al., *Atmos. Environ.* **22** (1988) 1931–1939.
[5.199] P. Kirschmer, P. Eynck: "Meßverfahren mit automatisierter Probenahme zur Bestimmung von Aldehyden in der Luft," *LIS-Ber.* **92** (1989) 1–34.
[5.200] VDI-Richtlinie 3485 Bl. 1 (12/88): Messen von Phenolen, *p*-Nitroanilin-Verfahren, in [5.30].
[5.201] B. Spiegelhalder, R. Preussmann, *IARC Sci. Publ.* **84** (1984) 411–414.
[5.202] W. Medrow in: *Gerüche,* Schriftenreihe der VDI-Kommission Reinhalt. Luft, vol. 12, Düsseldorf 1990.
[5.203] Verwaltungsvorschrift des Sächsischen Staatsministeriums für Umwelt und Landesentwicklung zur Feststellung und Beurteilung von Geruchsimmissionen, Geruchsimmissions-Richtlinie vom 16.3.1993, Sächs. Amtsblatt Nr. 17 vom 22.4.1993.
[5.204] V. Thiele in: *Special Meßtechnik im Umweltschutz,* VDI Verlag, Düsseldorf 1987, pp. M101–M111.
[5.205] R. H. Brown, *Pure Appl. Chem.* **65**, no. 8 (1993) 1859–1874.
[5.206] B. Seifert, *VDI-Ber.* **608** (1987) 181–190.
[5.207] J. Alexander, R. Träm, M. Drüeke, K.-J. Rumpel, *Staub-Reinhalt. Luft* **51** (1991) 5–6.
[5.208] E. D. Palmes, A. F. Gunnison, J. Dimattio, C. Tomczyk, *Am. Ind. Hyg. Assoc. J.* **37** (1976) 570–577.
[5.209] K. Leichnitz: *Prüfröhrchen-Meßtechnik,* Ecomed-Verlagsgesellschaft, Landsberg/Lech 1981.
[5.210] *Dräger-Röhrchen Handbuch,* 9th ed., Drägerwerk Aktiengesellschaft, Lübeck 1994.
[5.211] VDI-Richtlinie 3799, Bl. 2 (10/91), in [5.30].
[5.212] VDI-Richtlinie 3792, Bl. 1 (07/78), in [5.30].
[5.213] R. Niessner, *VDI-Ber.* **1059** (1993) 33–47.
[5.214] M. Birkle, *VDI-Ber.* **1059** (1993) 49–65.
[5.215] W. Fricke, R. Sartorius, *UWSF – Z. Umweltchem. Ökotox.* **3** (1992) 83–84.
[5.216] H.-U. Pfeffer, *Staub – Reinhalt. Luft* **42** (1982) 233–236.
[5.217] K. Ellermann, A. Borowiak, H.-U. Pfeffer: *LIMES – Jahresbericht 1993,* Reihe B, Landesumweltamt Nordrhein-Westfalen (eds.), Essen 1994.
[5.218] M. Buck, H.-U. Pfeffer, *Clean Air* **22** (1988) 7–12.

References for Chapter 6

[6.1] Gesetz zum Schutz vor schädlichen Umwelteinwirkungen durch Luftverunreinigungen, Geräusche, Erschütterungen und ähnliche Vorgänge (Bundes-Immissionsschutzgesetz-BImSchG) vom 15. März 1974, Bundesgesetzblatt, Teil 1, no. 27, Jahrgang 1974, pp. 721 ff., last version 11. Mai 1990, BGBl. I, pp. 881 ff.
[6.2] Erste Allgemeine Verwaltungsvorschrift zum Bundesimmissionsschutzgesetz (Technische Anleitung zur Reinhaltung der Luft – TA Luft) vom 27.2.1986, Gemeinsames Ministerialblatt, Ausgabe A, 37. Jahrgang, Feb. 28, 1986, pp. 95 ff.
[6.3] Dreizehnte Verordnung zur Durchführung des Bundes-Immissionsschutzgesetzes über Großfeuerungsanlagen – 13. BImSchV vom 22. Juni 1983, Bundesgesetzblatt I, pp. 719 ff.
[6.4] Siebzehnte Verordnung zur Durchführung des Bundes-Immissionsschutzgesetzes über Verbrennungsanlagen für Abfälle und ähnliche brennbare Stoffe – 17. BImSchV vom 23.11.1990, Bundesgesetzblatt 1990, no. 64, pp. 2545 ff.
[6.5] Zweite Verordnung zur Durchführung des Bundes-Immissionsschutzgesetzes (Verordnung zur Emissionsbegrenzung von leichtflüchtigen Halogenkohlenwasserstoffen – 2. BImSchV) vom 10. Dezember 1990 BGBl. I, pp. 2694 ff.
[6.6] Vierte Verordnung zur Durchführung des Bundes-Immissionsschutzgesetzes über genehmigungsbedürftige Anlagen – 4. BImSchV in der Fassung vom 24. 7. 1985 (BGBl. I, p. 1586) last version 28. Aug. (BGBl. I, pp. 1838, 2044).
[6.7] Richtlinie VDI 2448, Blatt 1, April 1992, Planung von stichprobenartigen Emissionsmessungen an geführten Quellen.
[6.8] Richtlinie VDI 2066, Blatt 1, Oct. 1975, Messen von Partikeln, Staubmessungen in strömenden Gasen, Gravimetrische Bestimmung der Staubbeladung – Übersicht.
[6.9] Richtlinie VDI 2449, Blatt 1, Entwurf, Dec. 1991, Prüfkriterien von Meßverfahren, Ermittlung von Verfahrenskenngrößen für die Messung gasförmiger Schadstoffe (Immission).
[6.10] Richtlinie VDI 2449, Blatt 2, Jan. 1987, Grundlagen zur Kennzeichnung vollständiger Meßverfahren, Begriffsbestimmungen.

[6.11] RdSchr. vom 1.3.90, Bundeseinheitliche Praxis bei der Überwachung der Emissionen – Richtlinien über die Eignungsprüfung, den Einbau, die Kalibrierung und die Wartung von Meßeinrichtungen für kontinuierliche Emissionsmessungen, Gemeinsames Ministerialblatt, 41. Jahrgang 1990, pp. 226ff.

[6.12] RdSchr. d. BMU vom 26.7.1988 – IGI 2-556134/4, Bundeseinheitliche Praxis bei der Überwachung der Emissionen, Richtlinie über die Auswertung kontinuierlicher Emissionsmessungen, GMBl. 1988, pp. 426ff.

[6.13] RdSchr. d. BMU vom 16.10.1992 – IGI 3-51134/3, Bundeseinheitliche Praxis bei der Überwachung der Emissionen, Richtlinien über die Auswertung kontinuierlicher Emissionsmessungen nach der Verordnung über Verbrennungsanlagen für Abfälle und ähnliche brennbare Stoffe (17. BImSchV), GMBl. 1992, pp. 1137ff.

[6.14] RdSchr. d. BMU vom 28.10.1992 – IGI 3-51134/2, Bundeseinheitliche Praxis bei der Überwachung der Emissionen und der Immissionen, Eignung von Meßeinrichtungen zur kontinuierlichen Überwachung von Emissionen/Bezugsgrößen, GMBl. 1992, pp. 1140ff.

[6.15] RdSchr. d. BMU vom 1.7.1992 – IGI 3-51134/2, Bundesheinheitliche Praxis bei der Überwachung der Emissionen und der Immissionen. Eignung von Meßeinrichtungen zur kontinuierlichen Überwachung von Emissionen, Bezugsgrößen (Abgasvolumenstrom, Sauerstoff), Eignung von elektronischen Systemen zur Auswertung kontinuierlicher Emissionsmessungen, GMBl. 1992, pp. 294ff.

[6.16] K. Lützke: "Leitlinien zur Messung und Bewertung von Emissionen," *VDI-Ber.* **608**, Aktuelle Aufgaben der Meßtechnik in der Luftreinhaltung, Kolloquium Heidelberg, Sept., 17–19, 1986, VDI-Kommission Reinhaltung der Luft, VDI-Verlag, Düsseldorf 1987.

[6.17] Luftreinhaltung, Leitfaden zur kontinuierlichen Emissionsüberwachung, Vorschriften und Verfahren der Emissionsmeßtechnik unter Berücksichtigung der TA Luft 86 und Datenblätter eignungsgeprüfter Meßgeräte, 4th ed., *Ber. Umweltbundesamt* (*Ger.*) **11/90** (1990).

[6.18] Richtlinie VDI 3950, Blatt 1, Entwurf Januar 1991, Kalibrierung automatischer Emissionsmeßeinrichtungen.

[6.19] K. Lützke, H.-D. Burk: "Kalibrierung automatischer Emissions-Meßeinrichtungen – Das Konzept der Richtlinie VDI 3950", *VDI-Ber.* **1059**. *Aktuelle Aufgaben der Meßtechnik in der Luftreinhaltung*, Tagung Heidelberg, June 2–4, 1993, Kommission Reinhaltung der Luft im VDI und DIN, VDI-Verlag, Düsseldorf 1993.

[6.20] Richtlinie VDI 2066, Blatt 4, Jan. 1989, Staubmessung in strömenden Gasen, Bestimmungen der Staubbeladung durch kontinuierliches Messen der optischen Transmission.

[6.21] DIN 1306, Dichte (Begriffe), Dec. 1971.

[6.22] DIN 1871, Gasförmige Brennstoffe und sonstige Gase, Teil 1 (1981), Teil 2 (1990).

[6.23] Richtlinie VDE/VDI 3511, Feb. 1967, Technische Temperaturmessung.

[6.24] Richtlinie VDE/VDI 3512, Blatt 2, Sept. 1972, Meßanordnungen für Temperaturmessungen.

[6.25] Richtlinie VDE/VDI 3511, Blatt 5, Entwurf, June 1993, Technische Temperaturmessungen, Einbau von Thermometern.

[6.26] Richtlinie VDE/VDI 3511, Blatt 1, Entwurf, Nov. 1991, Technische Temperaturmessung, Grundlagen und Übersicht über besondere Temperaturmeßverfahren.

[6.27] K. Lützke, R. Wilkes: "Erprobung von Meßverfahren zur Durchführung der Großfeuerungsanlagen-Verordnung, Umweltforschungsplan des Bundesministers für Umwelt, Naturschutz und Reaktorsicherheit", *Luftreinhaltung Forschungsbericht 88-10402164.*

[6.28] RdSchr. d. BMU vom 1.4.1992, Bundeseinheitliche Praxis bei der Überwachung der Emissionen, Richtlinien über die Eignungsprüfung, den Einbau, die Kalibrierung und die Wartung von Meßeinrichtungen für kontinuierliche Emissionsmessungen, Mindestanforderungen bei der Eignungsprüfung, Messen von Bezugsgrößen, Feuchtegehalt, GMBl. 1992, pp. 377ff.

[6.29] RdSchr. d. BMU vom 30.10.1991 – IGI 3-51134/4, Bundeseinheitliche Praxis bei der Überwachung der Emissionen und Immissionen, Eignung von Meßeinrichtungen zur kontinuierlichen Überwachung von Emissionen, GMBl. 1991, pp. 1045ff.

[6.30] RdSchr. d. BMU vom 7.6.1991 – IGI 2-51134/4, Bundeseinheitliche Praxis bei der Überwachung der Emissionen und der Immissionen, Eignung von Meßeinrichtungen zur kontinuierlichen Messung von Bezugsgrößen, GMBl. 1991, pp. 526 ff.

[6.31] RdSchr. d. BMU vom 1.6.1990 – IGI 2, Bundeseinheitliche Praxis bei der Überwachung der Emissionen und der Immissionen, Eignung von Meßeinrichtungen zur kontinuierlichen Überwachung von Emissionen, Eignung von Meßeinrichtungen zur kontinuierlichen Messung von Bezugsgrößen, GMBl. 1990, pp. 400ff.

[6.32] Richtlinie VDI 2066, Blatt 2, Aug. 1993, Messen von Partikeln, Manuelle Staubmessung in strömenden Gasen, gravimetrische Bestimmung der Staubbeladung, Filterkopfgeräte (4 m^3/h, 12 m^3/h).

[6.33] Richtlinie VDI 2066, Blatt 3, 1993, Messen von Partikeln, Manuelle Staubmessung in strömenden Gasen, Gravimetrische Bestimmung der Staubbeladung, Filterkopfgerät (40 m^3/h).

[6.34] Richtlinie VDI 2066, Blatt 7, Aug. 1993, Messen von Partikeln, Manuelle Staubmessung in strömenden Gasen, Gravimetrische Bestimmung der Staubbeladung, Planfilterkopfgeräte.

[6.35] Richtlinie VDI 3861, Blatt 1, Dec. 1989, Messen faserförmiger Partikeln, Manuelle Asbest-Staubmessung im strömenden Reingas, IR-spektrographische Bestimmung der Asbeststaub-Massenkonzentration.

[6.36] Richtlinie VDI 2066, Blatt 8, Entwurf, 1993, Staubmessungen in strömenden Gasen, Messung der Rußzahl an Feuerungsanlagen für Heizöl EL.

[6.37] Richtlinie VDI 2066, Blatt 6, Jan. 1989, Staubmessungen in strömenden Gasen, Bestimmung der Staubbelastung durch kontinuierliches Messen des Streulichtes mit dem Photometer KTN.

[6.38] RdSchr. d. BMU vom 7.11.1990 – IGI 2-40038-2, Bundeseinheitliche Praxis bei der Überwachung der Emissionen und Immissionen, Eignung von Meßeinrichtungen zur kontinuierlichen Überwachung von Emissionen.

[6.39] Richtlinie VDI 2031, Oct. 1962, Feinheitsbestimmungen an technischen Stäuben.

[6.40] W. Muhr: "Feinstaub – Bedeutung und Messung", *Schriftenreihe RW-TÜV,* 13. Heft.

[6.41] K. Lützke, W. Muhr: "Erprobung von Emissionsmeßverfahren zur Feststellung von Korngrößenfraktionen", *RWTÜV Forschungsbericht* 81-10402121.

[6.42] Richtlinie VDI 2066, Blatt 5, 1993, Staubmessung in strömenden Gasen, Fraktionierende Staubmessung nach dem Impaktionsverfahren – Kaskadenimpaktor.

[6.43] K. Lützke: "Mit Kaskadenimpaktoren – Feinstaubmessungen an Industrieanlagen", *Ind. Anz.* **104** (1982) no. 4, 22–25; no. 19, 24–26.

[6.44] Richtlinie VDI 3868, Blatt 1, Entwurf, Oct. 1992, Messen der Gesamtemission von Metallen, Halbmetallen und ihren Verbindungen. Manuelle Messung in strömenden, emittierten Gasen, Probenahmesystem für partikelgebundene und filtergängige Stoffe.

[6.45] Richtlinie VDI 3868, Blatt 2, Vorentwurf, Feb. 1992, Bestimmung der Gesamtemission von Metallen, Halbmetallen und ihren Verbindungen, Messen von Quecksilber, Atomabsorptionsspektrometrie mit Kaltdampftechnik.

[6.46] Richtlinie VDI 2268, Blatt 1, Stoffbestimmung an Partikeln: Bestimmung der Elemente Ba, Ca, Cd, Co, Cr, Cu, Ni, Pb, Sr, Zn in emittierten Stäuben mittels atomspektrometischer Methoden.

[6.47] Richtlinie VDI 2268, Blatt 2, Stoffbestimmung an Partikeln; Bestimmung der Elemente Arsen, Antimon und Selen in emittierten Stäuben mittels Atomabsorptionsspektrometrie nach Abtrennung über ihre flüchtigen Hydride.

[6.48] Richtlinie VDI 2268, Blatt 3, Stoffbestimmung an Partikeln; Bestimmung des Thalliums in emittierten Stäuben mittels Atomabsorptionsspektrometrie.

[6.49] Richtlinie VDI 2268, Blatt 4, Stoffbestimmung an Partikeln; Bestimmung der Elemente Arsen, Antimon und Selen in emittierten Stäuben mittels Graphitrohr-Atomabsorptionsspektrometrie.

[6.50] Richtlinie VDI 2462, Blatt 1, Feb. 1974, Messen gasförmiger Emissionen, Messen der Schwefeldioxid-Konzentration. Jod-Thiosulfat-Verfahren.

[6.51] Richtlinie VDI 2462, Blatt 2, Feb. 1974, Messung gasförmiger Emissionen, Messen der Schwefeldioxidkonzentration, Wasserstoffperoxid-Verfahren, Titrimetrische Bestimmung.

[6.52] Richtlinie VDI 2462, Blatt 3, Feb. 1974, Messung gasförmiger Emissionen, Messen der Schwefeldioxidkonzentration, Wasserstoffperoxid-Verfahren, gravimetrische Bestimmung.

[6.53] Richtlinie VDI 2462, Blatt 8, March 1985, Messen gasförmiger Emissionen, Messen der Schwefeldioxid-Konzentration, H_2O_2Thorin-Methode.

[6.54] Richtlinie VDI 2462, Blatt 7, March 1985, Messen gasförmiger Emissionen, Messen der Schwefeltrioxid-Konzentration, 2-Propanol-Verfahren.

[6.55] RdSchr. d. BMU vom 26.3.1991, Bundeseinheitliche Praxis bei der Überwachung der Emissionen, Eignung von Meßeinrichtungen zur kontinuierlichen Überwachung von Emissionen, GMBl. 1991, pp. 470ff.

[6.56] Richtlinie VDI 2456, Blatt 1, Dec. 1973, Messung gasförmiger Emissionen, Messen der Summe von Stickstoffmonoxid und Stickstoffdioxid, Phenoldisulfonsäure-Verfahren.

[6.57] Richtlinie VDI 2456, Blatt 2, Dec. 1973, Messung gasförmiger Emissionen, Messen der Summe von Stickstoffmonoxid und Stickstoffdioxid, Titrations-Verfahren.

[6.58] Richtlinie VDI 2456, Blatt 8, Jan. 1986, Messung gasförmiger Emissionen, Analytische Bestimmung der Summe von Stickstoffmonoxid und Stickstoffdioxid, Natriumsalicylat-Verfahren.

[6.59] Richtlinie VDI 2456, Blatt 10, Nov. 1990, Messen gasförmiger Emissionen, Analytische Bestimmung der Summe von Stickstoffmonoxid und Stickstoffdioxid, Dimethylphenol-Verfahren.

[6.60] Richtlinie VDI 2456, Blatt 6, May 1978, Messen gasförmiger Emissionen, Messen der Summe von Stickstoffmonoxid und Stickstoffdioxid als Stickstoffmonoxid unter Einsatz eines Konverters.

[6.61] Richtlinie VDI 2456, Blatt 7, April 1981, Messen gasförmiger Emissionen, Messen von Stickstoffmonoxid-Gehalten, Chemiluminiszenz-Analysatoren (Atmosphärendruckgeräte).

[6.62] Richtlinie VDI 2456, Blatt 5, May 1978, Messen gasförmiger Emissionen, Messen von Stickstoffmonoxid-Gehalten, Chemiluminiszenz-Analysator, Thermo Electron Modell 10.

[6.63] Richtlinie VDI 2459, Blatt 7, Entwurf, Jan. 1990, Messen gasförmiger Emissionen, Messen der Kohlenmonoxid-Konzentration, Jodpentoxid-Verfahren.

[6.64] Richtlinie VDI 2470, Blatt 1, Oct. 1975, Messen gasförmiger Emissionen, Messen gasförmiger Fluor-Verbindungen, Absorptions-Verfahren.

[6.65] Richtlinie VDI 3488, Blatt 1, Dec. 1979, Messen gasförmiger Emissionen, Messen der Chlorkonzentration, Methylorange-Verfahren.

[6.66] Richtlinie VDI 3488, Blatt 2, Nov. 1980, Messen gasförmiger Emissionen, Messen der Chlorkonzentration, Bromid-Jodid-Verfahren.

[6.67] Richtlinie VDI 3480, Blatt 1, July 1984, Messen gasförmiger Emissionen, Messen von Chlorwasserstoff, Messen der Chlorwasserstoff-Konzentration von Abgas mit geringem Gehalt an chloridhaltigen Partikeln.

[6.68] Richtlinie VDI 3480, Blatt 2, Jan. 1992, Messen gasförmiger Emissionen, Messen von Chlorwasserstoff, kontinuierliches selektives Messen von Chlorwasserstoff mit dem SPECTRAN 677 IR.

[6.69] Richtlinie VDI 3480, Blatt 3, Jan. 1992, Messen gasförmiger Emissionen, kontinuierliches Messen von gasförmigen anorganischen Chlorverbindungen mit dem ECOMETER.

[6.70] Richtlinie VDI 3486, Blatt 1, April 1979, Messen gasförmiger Emissionen, Messen der Schwefelwasserstoff-Konzentration, Potentiometrisches Titrations-Verfahren.

[6.71] Richtlinie VDI 3486, Blatt 2, April 1979, Messen gasförmiger Emissionen, Messen der Schwefelwasserstoff-Konzentration, Jodometrisches Titrations-Verfahren.

[6.72] Richtlinie VDI 3486, Blatt 3, Nov. 1980, Messen gasförmiger Emissionen, Messen der Schwefelwasserstoff-Konzentration, Colorimetrisches Verfahren (Monocolor-Analysator).

[6.73] Richtlinie VDI 2461, Blatt 1, March 1974, Messung gasförmiger Immissionen, Messen der Ammoniak-Konzentration, Indophenol-Verfahren.

[6.74] Richtlinie VDI 2461, Blatt 2, May 1976, Messung gasförmiger Immissionen, Messen der Ammoniak-Konzentration, NESSLER-Verfahren.

[6.75] Bericht Nr.: 936/800010, Modellhafte Eignungsprüfung der Mehrkomponenten-Emissionsmeßeinrichtung OPSIS AR für NH_3, Institut für Umweltschutz und Energietechnik, Köln.

[6.76] RdSchr. d. BMU v. 2.6.1993 – IGI 3-51134/2 – Bundeseinheitliche Praxis bei der Überwachung der Emissionen und der Immissionen, Eignung von Meßeinrichtungen zur kontinuierlichen Überwachung von Emissionen, GMBl. 1993, pp. 467ff.

[6.77] Richtlinie VDI 3481, Blatt 2, April 1980, Messen gasförmiger Emissionen; Bestimmung des durch Adsorption am Kieselgel erfaßbaren organisch gebundenen Kohlenstoffs.

[6.78] Richtlinie VDI 3481, Blatt 4, Vorentwurf (Arbeitspapier), Bestimmung des durch Adsorption an Kieselgel erfaßbaren organisch gebundenen Kohlenstoffs in Abgasen mit höherem Wassergehalt.

[6.79] Richtlinie VDI 3481, Blatt 1, Aug. 1975, Messung gasförmiger Emissionen, Messen der Kohlenwasserstoff-Konzentration, Flammen-Ionisations-Detektor (FID).

[6.80] Richtlinie VDI 3481, Blatt 3, Entwurf, Sept. 1992, Messen gasförmiger organischer Verbindungen, insbesondere von Lösemitteln, mit dem Flammen-Ionisations-Detektor (FID).

[6.81] Richtlinie VDI 3481, Blatt 6, Entwurf, Sept. 1992, Messung gasförmiger Emissionen, Auswahl und Anwendung von C-Summenverfahren.

[6.82] K. Lützke: "Emissionsmessung leicht- und schwerflüchtiger halogenierter Verbindungen; Halogenierte organische Verbindungen in der Umwelt: Herkunft, Messung, Wirkung, Abhilfe-Maßnahmen," *VDI-Ber.* **745**, Tagungsbericht der VDI-Kommission Reinhaltung der Luft; Kolloquium Mannheim, April 25–27, 1989, VDI-Verlag, Düsseldorf 1989.

[6.83] Richtlinie VDI 3499, Blatt 1, 1993, Messen von Emissionen – Messen von Reststoffen, Messen von polychlorierten Dibenzodioxinen und -furanen im Rein- und Rohgas von Feuerungsanlagen mit der Verdünnungsmethode, Bestimmung in Filterstaub, Kesselasche und in Schlacken.

[6.84] Richtlinie VDI 3499, Blatt 2, Entwurf, 1993, Messen von Emissionen, Messen von polychlorierten Dibenzo-*p*-dioxinen und Dibenzofuranen, Filter/Kühler-Methode.

[6.85] Richtlinie VDI 3499, Blatt 3, Entwurf, 1993, Messen von Emissionen, Messen von polychlorierten Dibenzo-*p*-dioxinen und Dibenzofuranen an industriellen und gewerblichen Anlagen, Kondensationsmethode – Gekühltes Absaugerohr.

[6.86] Richtlinie VDI 3499, Blatt 4, Entwurf, 1993, Messen von Emissionen, Messen von polychlorierten Dibenzodioxinen und Dibenzofuranen in Emissionen von Verbrennungsanlagen und bei anderen Verbrennungsprozessen, Polyurethan-Adsorptions-Methode.

[6.87] P. Kirschmer, W. Mülder, P. Eynk: "Comparison of Sample Preparation and Extraction Procedure for the Analysis of Polychlorinated Dibenzo-*p*-dioxins in Deposited Particulate Matter and Balkdeposition," *Chemosphere* **24** (1992) no. 5, 575–580.

[6.88] Richtlinie VDI 2464, Blatt 1, Vorentwurf, 1992, Messen von Immissionen, Messen von Innenraumluft, Messen von polychlorierten Biphenylen.

[6.89] Richtlinie BDI 2457, Blatt 2, Vorentwurf, Messen gasförmiger Emissionen, Gaschromatographische Bestimmung organischer Verbindungen, Probenahme durch Absorption in tiefkaltem Lösemittel, (2-(2-Methoxylethoxy)ethanol, Methyldiglykol).

[6.90] Richtlinie VDI 3482, Blatt 4, Nov. 1984, Messen gasförmiger Immissionen, Gaschromatographische Bestimmung organischer Verbindungen mit Kapillarsäulen, Probenahme durch Anreicherung an Aktivkohle – Desorption mit Lösemittel.

[6.91] Richtlinie VDI 3482, Blatt 6, July 1988, Messen gasförmiger Immissionen, Gaschromatographische Bestimmung organischer Verbindungen – Probenahme durch Anreicherung – Thermische Desorption.

[6.92] Richtlinie VDI 3864, Blatt 2, Vorentwurf, April 1993, Messen gasförmiger Immissionen, Gaschromatographische Bestimmung von leichtflüchtigen halogenierten Kohlenwasserstoffen, Probenahme durch Adsorption, Thermische Desorption.

[6.93] Richtlinie VDI 3873, Blatt 1, Nov. 1992, Messen von Emissionen, Messen von polycyclischen aromatischen Kohlenwasserstoffen (PAH) an stationären industriellen Anlagen – Verdünnungsmethode (RWTÜV-Verfahren) – Gaschromatographische Bestimmung.

[6.94] Richtlinie VDI 3482, Blatt 5, Nov. 1984, Messen gasförmiger Immissionen, Gaschromatographische Bestimmung von aromatischen Kohlenwasserstoffen, Probenahme durch Anreicherung an Aktivkohle – Desorption mit Lösemittel.

[6.95] Richtlinie VDI 2457, Blatt 1, Entwurf, March 1991, Messen gasförmiger Emissionen, Gaschromatographische Bestimmung organischer Verbindungen – Grundlagen.

[6.96] Richtlinie VDI 2457, Blatt 5, June 1981, Messen gasförmiger Emissionen, Gaschromatographische Bestimmung von Benzol, Toluol und Xylol.

[6.97] Richtlinie VDI 2467, Blatt 2, Aug. 1991, Messen gasförmiger Immissionen, Messen der Konzentration primärer und sekundärer aliphatischer Amine mit der Hochleistungs-Flüssigkeits-Chromatographie (HPLC).

[6.98] Richtlinie VDI 3862, Blatt 1, Dec. 1990, Messen gasförmiger Emissionen, Messen aliphatischer Aldehyde (C_1 bis C_3) nach dem MBTH-Verfahren.

[6.99] Richtlinie VDI 3862, Blatt 2, Entwurf, Messen gasförmiger Emissionen, Messen aliphatischer und aromatischer Aldehyde und Ketone nach dem DNPH-Verfahren, Acetonitril-Verfahren.

[6.100] Richtlinie VDI 3862, Blatt 3, Entwurf, Messen gasförmiger Emissionen, Messen aliphatischer und aromatischer Aldehyde und Ketone nach dem DNPH-Verfahren, Tetrachlorkohlenstoff-Methode.

[6.101] Richtlinie VDI 3863, Blatt 1, April 1987, Messen gasförmiger Emissionen, Messen von Acrylnitril, Gaschromatographisches Verfahren, Probenahme mit Gassammelgefäßen.

[6.102] Richtlinie VDI 3863, Blatt 2, Feb. 1991, Messen gasförmiger Emissionen, Messen von Acrylnitril, Gaschromatographisches Verfahren, Probenahme durch Absorption in tiefkalten Lösemitteln.

[6.103] Richtlinie VDI 3863, Blatt 3, Entwurf, Oct. 1988, Messen gasförmiger Emissionen, Messen von Acrylnitril, Adsorption an Aktivkohle, Desorption durch Dimethylformamid (DMF).

[6.104] Richtlinie VDI 3953, Blatt 1, Entwurf, April 1991, Messen gasförmiger Emissionen, Messen von 1,3-Butadien, Gaschromatographisches Verfahren, Probenahme durch Adsorption an Aktivkohle, Dampfraumanalyse.

[6.105] Richtlinie VDI 3493, Blatt 1, Nov. 1982, Messen gasförmiger Emissionen, Messen von Vinylchlorid, Gaschromatographisches Verfahren, Probenahme mit Gassammelgefäßen.
[6.106] Richtlinie VDI 3494, Blatt 1, Entwurf, May 1988, Messen gasförmiger Immissionen, Messen von Vinylchlorid-Konzentrationen, Gaschromatographische Bestimmung, Manuelle und automatische Dampfraumanalyse.
[6.107] Richtlinie VDI 3494, Blatt 2, April 1986, Messen gasförmiger Immissionen, Messen von Vinylchlorid-Konzentrationen, Gaschromatographische Bestimmung mit der Trennsäulenschalteinrichtung für Live-Chromatographie.
[6.108] Richtlinie VDI 3881, Blatt 1, May 1986, Olfaktometrie, Geruchsschwellenbestimmung – Grundlagen.
[6.109] Richtlinie VDI 3881, Blatt 2, Jan. 1987, Olfaktometrie, Geruchsschwellenbestimmung – Probenahme.
[6.110] Richtlinie VDI 3882, Blatt 1, Oct. 1992, Olfaktometrie, Bestimmung der Geruchsintensität.
[6.111] Richtlinie VDI 3881, Blatt 3, Nov. 1986, Olfaktometrie, Geruchsschwellenbestimmung, Olfaktometer mit Verdünnung nach dem Gasstrahlprinzip.
[6.112] Richtlinie VDI 3881, Blatt 4, Entwurf, Dec. 1989, Olfaktometrie, Geruchsschwellenbestimmung, Anwendungsvorschriften und Verfahrenskenngrößen.

References for Chapter 7

General references for Section 7.1

[7.1] F. Baum: *Luftreinhaltung in der Praxis*, Oldenbourg Verlag, München 1988.
[7.2] J. Böhm: *Electrostatic Precipitators*, Elsevier, Amsterdam 1982.
[7.3] R. C. Brown: *Air Filtration*, Pergamon Press, Oxford 1993.
[7.4] F. A. L. Dullien: *Introduction to Industrial Gas Cleaning*, Academic Press, San Diego 1989.
[7.5] W. Licht: *Air Pollution Control Engineering*, Marcel Dekker, New York 1988.
[7.6] F. Löffler: *Staubabscheiden*, Thieme-Verlag, Stuttgart 1988.
[7.7] F. Löffler, H. Dietrich, W. Flatt: *Dust Collection with Bag Filters and Envelope Filters*, Vieweg, Braunschweig 1988.
[7.8] A. Ogawa: *Separation of Particles from Air and Gases*, vol II, CRC Press, Boca Raton 1984.
[7.9] VDI-Richtlinie 2442, Abgasreinigung durch thermische Verbrennung, VDI-Handbuch Reinhaltung der Luft, VDI-Verlag, Düsseldorf 1985.
[7.10] Erste allgemeine Verwaltungsvorschrift zum BImSchG (Technische Anleitung zur Reinhaltung der Luft, TA-Luft), Feb. 27, 1986.
[7.11] H. Meier zu Köcker: "Abgasreinigung durch thermische Verbrennung," part I, "Thermische Stabilität im Wärmetauscherrohr," *Staub Reinhalt. Luft* **38** (1978) 101–105.
[7.12] H. Meier zu Köcker: "Abgasreinigung durch thermische Verbrennung," part II, "Thermische Abluftverbrennung mit indirekter Beheizung," *Staub Reinhalt. Luft* **38** (1978) 167–206.
[7.13] H. Meier zu Köcker, F. Weger: "Abgasreinigung durch thermische Verbrennung," part V, "Reaktionsverhalten stickstoffhaltiger Schadstoffe," *Staub Reinhalt. Luft* **40** (1980) 245–249.
[7.14] H. Meier zu Köcker, G. Müller: "Abgasreinigung durch thermische Verbrennung," part VI, "Reaktionsverhalten halogenhaltiger Schadstoffe," *Staub Reinhalt. Luft* **40** (1980) 280–283.
[7.15] H. Meier zu Köcker, F. Weger, R. Vögtlin: "Abgasreinigung durch thermische Verbrennung – Reaktionen im Röhrenwärmetauscher einer TNV-Anlage," *Staub Reinhalt. Luft* **42** (1982) 287–292.
[7.16] H. Meier zu Köcker: "Verbrennung N-haltiger Verbindungen," *VDI-Ber.* **525** (1985) 333–346.
[7.17] J. K. Lesmana, E. Schütt, H. Meier zu Köcker: "Abgasreinigung durch thermische Verbrennung – Reaktionsmodell der Carbonsäureverbrennung," *Staub Reinhalt. Luft* **42** (1982) no. 5, 208–214.
[7.18] H. Meier zu Köcker, J. Lesmana: "*Stoffspezifische Einflüsse auf den Reaktionsverlauf der thermischen Nachverbrennung*," VDI Bericht Nr 423, VDI Verlag, Düsseldorf, 1981, pp. 61–65.
[7.19] O. Carlowitz: "Grundlagen der thermischen Abgasreinigung," *Tech. Mitt.* **82** (1989) no. 5, 325–332.
[7.20] H. Meier zu Köcker, J. Dühr: "Abgasreinigung durch thermische Verbrennung," part III, "Thermische Abluftverbrennung mit Flammenbeheizung," *Staub Reinhalt. Luft* **38** (1978) 370–374.
[7.21] D. Siegel: "Anwendung von Fackeln und Ergebnisse von Emissionsmessungen," *VDI-Ber.* **525** (1985) 365–380.
[7.22] H. Seifert, R. Becker, G. Hemmer: "Verfahrenstechnische Lösungen bei der thermischen Abgasreinigung," *VDI-Ber.* **1034** (1993) 39–68.
[7.23] W. Hünig: "Sicherheitstechnische Maßnahmen bei der thermischen Abluftreinigung," *Tech. Mitt.* **82** (1989) no. 5, 333–343.
[7.24] O. Carlowitz: "Wirtschaftlichkeitsbetrachtungen bei der Gestaltung von thermischen Abgasreinigungsanlagen," *Tech. Mitt.* **82** (1989) no. 5, 344–351.
[7.25] St. Jauns, O. Carlowitz, M. Jäger: "Realisierung hoher Abluftvorwärmtemperaturen in thermischen Nachverbrennungsanalgen mit Hilfe von Festbettregeneratoren," *Gas Wärme Int.* **40** (1991) no. 1, 59–65.
[7.26] T. Schmidt, M. Hoffmeister, H. Falke: "Entfernung organischer Lösemittel und Ammoniak aus Abluft durch Adsorption und katalytische Oxidation," *VDI-Ber.* **1034** (1993) 139–175.
[7.27] T. Schmidt: "Anwendungsbeispiele für die katalytische Abgasreinigung," *Tech. Mitt.* **82** (1989) no. 5, 360–369.
[7.28] W. Golibrzuch: "Thermische Verfahren der Abgasreinigung – Einsatzbereiche, Anwendungsbeispiele, Wirtschaftlichkeitsbetrachtungen," *VDI-Ber.* **1034** (1993) 89–108.
[7.29] O. Carlowitz: "Anlagen zur thermischen Abgasreinigung – Ansatzpunkte zur Umsetzung der Wärmenutzungsverordnung," *Staub Reinhalt. Luft* **52** (1992) 325–330.
[7.30] F. H. Ackermann: "Wirtschaftlicher Einsatz von thermischen Abluftreinigungsanlagen," *VDI-Ber.* **525** (1985) 275–298.
[7.31] W. Bitter: "Einführung in die TA-Luft im Hinblick auf den Einsatz von Verbrennungsverfahren zur Emissionsminderung," *Tech. Mitt.* **82** (1989) no. 5, 320–324.

[7.32] J. Guggenberger, A. Brandl: "Meßtechnische Überwachung von Verfahren zur thermischen und katalytischen Reinigung von Abgasen," *VDI-Ber.* **525** (1985) 91–119.
[7.33] J. M. Freidel: "Giftresistente Katalysatoren in der Abgasreinigung," *VDI-Ber.* **1034** (1993) 177–186.
[7.34] H. Müller, K. Deller: "Katalytische Reinigung CKW-haltiger Abgase aus dem VCM-Herstellungsverfahren," *VDI-Ber.* **1034** (1993) Anhang.
[7.35] K. H. Reitze: "Der "Thermo-Reaktor": Ein neuartiges System der thermischen Abgasreinigung," *VDI-Ber.* **525** (1985) 323–332.
[7.36] O. Ebbecke, K. Noll: "Abgasreinigung mit regenerativer thermischer Oxidation – Leistungsbereiche, Betriebsergebnisse," *VDI-Ber.* **1034** (1993) 69–88.
[7.37] Braunschweiger Flammenfilter Armaturen- und Apparatebau GmbH, company brochure, Braunschweig 1990.
[7.38] T. von Harpe: "Nichtkatalytische Reduktion von NO_x mittels Harnstoff – Erste Ergebnisse," in: *Aktuelle Probleme der Abluftreinigung,* VDI-Kommission Reinhaltung der Luft, April 12/13, 1989.
[7.39] G. Mayer-Schwining, R. Knoche, G. Schaub: "Abscheidung von Dioxinen, Furanen und Schwermetallen bei der thermischen Abfallverwertung," *VDI-Ber.* **1034** (1993) 107–122.
[7.40] E. Richter: "Die katalytische NO_x-Reduktion – eine Übersicht," *VDI-Ber.* **730** (1989) 121–156.
[7.41] E. Koberstein: "Katalytische Motorabgasreinigung," *VDI-Ber.* **525** (1985) 217–246.
[7.42] Engelhard Kali-Chemie AutoCat GmbH, company brochure, Hannover 1985.
[7.43] B. Engler, E. Koberstein: "Katalysatoren in der Abgasreinigung," *VDI-Ber.* **730** (1989) 97–120.
[7.44] E. Ruhl: "Katalytische Reinigung von Claus-Anlagen-Abgasen," *VDI-Ber.* **730** (1989) 255–274.
[7.45] R. Lell: "Katalytische Reduktion von SO_2-Emissionen mit Wertstoffgewinnung," *VDI-Ber.* **1034** (1993) 203–226.
[7.46] R. Kettner, T. Lübcke: "Betriebserfahrungen mit der KNV an Claus-Anlagen," *VDI-Ber.* **730** (1989) 255–274.
[7.47] K. Kirchner, A. Kaiczik: "Grundlagen der katalytischen Nachverbrennung," *VDI-Ber.* **525** (1985) 119–145.
[7.48] T. Schmidt: "Katalytische Verbrennung chlorierter Kohlenwasserstoffe," *VDI-Ber.* **730** (1989) 201–237.
[7.49] J. J. Spivey: "Complete Catalytic Oxidation of Volatile Organics," *Ind. Eng. Chem. prod. Res. Dev.* **26** (1987) 2165–2180.
[7.50] C. Herion, R. Meißner: "Katalytische Abgasreinigung in der chemischen Industrie," *VDI-Ber.* **1034** (1993) 123–138.
[7.51] EnviCat, Solvay Catalysts GmbH, company brochure, Hannover 1992.
[7.52] U. Tanger: "Energiesparendes KNV-Verfahren durch neuartige Prozeßführung mit regenerativen Wärmeaustauschern," *VDI-Ber.* **730** (1989) 189–200.
[7.53] *Landolt-Börnstein*, **IV**, **4c**.
[7.54] D. Tiegs et al.: "Activity Coefficients at Infinite Dilution," *DECHEMA Chem. Data Ser.* **IX** (1986).
[7.55] M. Schiller, J. Gmehling, *J. Chem. Eng. Data* **37** (1992) 37.
[7.56] G.-G. Börger, *Chem. Ing. Tech.* **64** (1992) 905.
[7.57] R. Schmidt, D. Wolf: "Absorption von Chlor und Chlorwasserstoff aus Abgasen," *VDI Ber.* **253** (1976) 77–84.
[7.58] K. H. Schaber, *Chem. Ing. Tech.* **59** (1987) 376.
[7.59] R. Billet, J. Mackowiak, *Chem. Eng. Technol.* **11** (1988) 213.
[7.60] G. K. Chen, *Chem. Eng.* **91** 1984, March 5, 40.
[7.61] J. Roiron: "How to Solve the Problem of NO_x Pollution from Nitric Acid Plants," The British Sulphur Corp., London 1986.
[7.62] Sigri GmbH, company brochure, Meitingen.
[7.63] QVF Process Systems Ltd., Catalogue, Sect. 6, "Column Components," Stone, United Kingdom, 1992.
[7.64] H. Krill, H. Menig: "Aktivkohlen im Umweltschutz," *Umwelt Aktuell* **1** (1990) 46–55.
[7.65] H. Krill: "Adsorptive Abgasreinigung – eine Bestandsaufnahme," *VDI-Ber.* **1034** (1993) 339–371.
[7.66] M. Bank: *Basiswissen Umwelttechnik,* 1st ed., Vogel Verlag, Würzburg 1993.
[7.67] W. Kast: *Adsorption aus der Gasphase – Grundlagen und Verfahren,* VCH, Weinheim 1988.
[7.68] H. Jüntgen: "Physikalisch-chemische und verfahrenstechnische Grundlagen von Adsorptionsverfahren," Haus der Technik, no. 404 (1977) 5–22.
[7.69] DIN 66131, Bestimmung der spezifischen Oberfläche von Feststoffen durch Gasadsorption nach Brunauer, Emmet und Teller, Beuth Verlag, Berlin.
[7.70] W. Otten, E. Gail, T. Frey: "Einsatzmöglichkeiten hydrophober Zeolithe in der Adsorptionstechnik," *Chem. Ing. Tech.* **64** (1992) no. 10, 915–925.
[7.71] VDI-Richtlinie 3674, Abgasreinigung durch Adsorption – Prozeß- und Abgasreinigung.
[7.72] A. Mersmann, G.-G. Börger, S. Scholl: "Abtrennung und Rückgewinnung von gasförmigen Stoffen durch Adsorption," *Chem. Ing. Tech.* **63** (1991) no. 9, 892–903.
[7.73] W. Otten, W. Kast: "Entwicklung von Auslegungsdiagrammen für Festbettadsorber," *Chem. Ing. Tech.* **60** (1988) no. 11, 929–933.
[7.74] H. Menig: *Emissionsminderung und Recycling – Grundlagen und Technologien,* 3rd ed., ecomed, Landsberg/Lech 1995.
[7.75] R. Backes, H. Plinke: "Mikroporöse Polymere zur Reinigung lösemittelhaltiger Abluft im Fließbettverfahren," *VDI Ber.* **1034** (1993) 473–486.
[7.76] H. W. Bräuer: "Adsorptive Abluft- und Abgasreinigung mit der Festbett- und Wirbelschichttechnik," *VDI Ber.* **1034** (1993) 445–460.
[7.77] G. Konrad, G. Eigenberger: "Rotoradsorber zur Abluftreinigung und Lösemittelrückgewinnung," *Chem. Ing. Tech.* **66** (1994) no. 3, 321–331.
[7.78] K. Storp: "Abluft- und Abgasentschwefelung durch Adsorption und Katalyse an Aktivkohlen," *Dechema Monogr.* **64** (1970) 91–94.
[7.79] H. W. Bräuer, F. Fischer: "Die Rückgewinnung von Schwefelkohlenstoff bei gleichzeitiger Entfernung von Schwefelwasserstoff," *Dechema Monogr.* **59** (1968) 173–189.
[7.80] Lurgi AG, Sulfacid-Process, Desulfurization of SO_2-laden Waste Gases with Direct Production of Dilute Sulfuric acid, company brochure, no. 1657 e/2.94/10, Frankfurt/Main 1994.
[7.81] K. Schneiders, J. Zimmermann: "Erste Betriebserfahrungen mit dem BF/Uhde-Verfahren im Kraftwerk Arzberg," *VGB Kraftwerkstech.* **69** (1989) no. 1, 66–69.

[7.82] R. Lell: "Katalytische Reduktion von SO_2-Emissionen mit Wertstoffgewinnung," *VDI Ber.* **1034** (1993) 187–202.
[7.83] K.-A. Theis, W. Dolkemeyer, H. Kreusing: "Braunkohlenkoks – Anwendungsgebiete in der Umwelttechnik," *ET – Z. Energiewirtsch. Recht Tech.* **4** (1989) 2–6.
[7.84] E. Richter: "Die Abscheidung von polychlorierten Dioxinen und Furanen aus Abgasen mit Aktivkoksverfahren," *Chem. Ing. Tech.* **64** (1992) no. 2, 125–136.
[7.85] M. Hefni-Omar, E. Jury: "Abscheidung von Schadstoffen aus den Rauchgasen von Müllverbrennungsanlagen in einem zirkulierenden Wirbelschichtreaktor," *Chem. Ing. Tech.* **64** (1992) 442–446.
[7.86] P. Janositz: "Kat knackt Dioxin," *VDI Nachr.* (1992) no. 10, 26.
[7.87] M. Eicken, W. Esser-Schmittmann, J. Lambertz, G. Ritter: "Braunkohlenkoks zur Rauchgasreinigung und Reststoffminimierung von Abfallverbrennungsanlagen," *BWK-TÜ-UT-special* Nr. 10/90, R 37–49.
[7.88] J. Gottschalk: "Flugstromverfahren – ein Verfahren der Kohlenstoffadsorptionstechnik zur Minimierung der Restemissionen, "*Abfallwirtschaftsjournal* **4** (1992) Nr. 12, 997–1001.
[7.89] E. Richter: "Aktivkokse als Adsorbentien/Katalysatoren in der Rauchgasreinigung," *Chemie-Technik* **19** (1990) Nr. 4, 122–126.
[7.90] G. Mayer-Schwinning, H. Herden, H. W. Bräuer: "Zeolithe zur Dioxin/Furan- und Schwermetallabscheidung," *Staub Reinhaltung der Luft* **55** (1995) 183–188.
[7.91] G. Feldhaus: *Feststellen und Beurteilen des Standes der Technik bei der Richtlinienarbeit,* no. 10, VDI-Kommission Reinhaltung der Luft, Düsseldorf 1980.
[7.92] D. Bardtke: "Mikrobiologische Voraussetzungen für die biologische Abluftreinigung," *Kontakt Stud.* **212** (1990) 1–12.
[7.93] VDI 3478 E, Biologische Abgasreinigung – Biowäscher und Rieselbettreaktoren, VDI-Verlag, Düsseldorf 1994.
[7.94] W. J. Moore, D. O. Hummel: *Physikalische Chemie,* De Gruyter, Berlin – New York 1973.
[7.95] K. Kirchner, P. Krämer: "Zur Absorption und Oxidation von Schadstoffen mittels Bakterien-Suspensionen," *VT Verfahrenstech.* **17** (1983) no. 12, 675–681.
[7.96] E. Schippert: "Biowäschertechnologie," *VDI Ber.* **1104** (1994) 39–56.
[7.97] F. Sabo et al.: "Entwicklung von Hochleistungsbiofiltern – Optimierung der Materialdurchströmung," *Entsorgungspraxis* **9** (1992) 576–578.
[7.98] F. Sabo: "Verfahrenstechnische Grundlagen der biologischen Abluftreinigung," *Kontakt Stud.* **212** (1990) 13–34.
[7.99] VDI 3477, Biologische Abgasreinigung – Biofilter, VDI-Verlag, Düsseldorf 1991.
[7.100] R. M. M. Diks, S. P. P. O. Ottengraf: "Verification Studies of an Simplified Model for the Removal of Dichloromethane from Waste Gases Using a Biological Trickling Filter," *Bioprocess Eng.* **6** (1991) part I: pp. 93–99, part II: pp. 131–140.
[7.101] C. A. Gossen: "Abgasreinigung mit fixierten Bakterien im Rieselbett," Ph.D. thesis, TU München 1990.
[7.102] R. M. M. Diks: "The Removal of Dichloromethane from Waste Gases in a Biological Trickling Filter," Dissertation, Eindhoven University of Technologie, Eindhoven 1992.
[7.103] E. Schippert: "Abscheidung von Lösemitteln im Biowäscher," *VDI Ber.* **1034** (1993) 577–610.
[7.104] S. Fröhlich: "Entscheidungshilfen und Kriterien für den Einsatz von Biowäschern zur Abluftreinigung und Geruchsminderung," *VDI Ber.* **1104** (1994) 81–93.
[7.105] P. Kunz et al.: "Biosorber – ein neues Prinzip für die biologische Abgasreinigung," *VDI Ber.* **1104** (1994) 239–250.
[7.106] M. Reiser et al.: "Kombination aus Biowäscher und Biomembranverfahren zur Reinigung von Abluft mit hydrophilen und hydrophoben Inhaltsstoffen," *VDI Ber.* **1104** (1994) 103–112.
[7.107] VDI 3881, Blatt 1–4, Olfaktometrie, VDI-Verlag, Düsseldorf 1986.
[7.108] VDI 3940, Bestimmung der Geruchsstoffimmission durch Begehung, VDI-Verlag, Düsseldorf 1993.
[7.109] VDI 3882, Blatt 1, Bestimmung der Geruchsintensität, VDI-Verlag, Düsseldorf 1992.
[7.110] VDI 3882, Blatt 2, Bestimmung der hedonischen Geruchswirkung, VDI-Verlag, Düsseldorf 1994.

References for Chapter 8

[8.1] B. J. Finlayson-Pitts, J. N. Pitts, Jr.: *Atmospheric Chemistry: Fundamentals and Techniques,* John Wiley & Sons, New York 1986.
[8.2] Umweltbereich Luft, Fonds der Chemischen Industrie, Frankfurt 1987.
[8.3] P. Warneck: *Chemistry of the Natural Atmosphere,* Academic Press, San Diego 1988.
[8.4] R. Zellner, K. J. Niemitz in *Dechema Monogr.* **104** (1987) 5.
[8.5] P. Ott, D. Helf, K. Kirchner in Ref. [8.4], p. 91.
[8.6] C. Zetsch in Ref. [8.4], p. 187.
[8.7] W. Simmler: "Air – an Environmental Compartment," *Proceedings of the GDCh-Hauptversammlung*, Berlin 1987.
[8.8] W. Simmler in Ref. [8.4], p. 1.
[8.9] W. Simmler: "Abiotic Degradation," *Chemie-Report 10/91,* Verband der Chemischen Industrie, Frankfurt 1991, p. 3.
[8.10] R. Zellner, *UWSF – Z. Umweltchem. Ökotox.* **3** (1991) no. 1, 52.
[8.11] R. Zellner, *Chem. Unserer Zeit* **27** (1993) no. 5, 230.
[8.12] I. M. Campbell: *Energy and the Atmosphere,* John Wiley & Sons, New York 1977.
[8.13] S. Chapman, *Mem. R. Meteor. Soc.* **3** (1930) 103.
[8.14] K. H. Becker, *UWSF – Z. Umweltchem. Ökotox.* **3** (1991) no. 1, 48.
[8.15] K. H. Becker, *UWSF – Z. Umweltchem. Ökotox.* **4** (1992) no. 1, 25.
[8.16] Environmental Indicators, OECD Core Set, Organisation for Economic Cooperation and Development, Publications Service, Paris 1994.
[8.17] DFG Deutsche Forschungsgemeinschaft: *Atmosphärische Spurenstoffe,* VCH Verlagsgesellschaft, Weinheim 1987.
[8.18] J. D. Nriaga (ed.): "*Gaseous Pollutants:* Characterization and Cycling," *Advances in Environmental Science and Technology,* vol. 24, John Wiley & Sons, New York 1992.

[8.19] P. Fabian: *Atmosphäre und Umwelt*, 4th ed., Springer Verlag, Berlin 1992.

[8.20] T. E. Graedel: *Chemical Compounds in the Atmosphere*, Academic Press, New York 1978.

[8.21] K. H. Becker in: *Atmosphärische Spurenstoffe und ihr physikalisch-chemisches Verhalten*, Springer Verlag, Berlin 1985.

[8.22] O. Hutzinger, M. Blumich, M. Berg, K. v. d. Olie, *Chemosphere* **14** (1985) 581.

[8.23] G. W. Israel in Ref. [8.4], p. 147.

[8.24] K. R. Spurny in Ref. [8.4], p. 161.

[8.25] C. Zetsch, *UWSF – Z. Umweltchem. Ökotox.* **3** (1991) no. 1, 59.

[8.26] "Photochemical Smog," *OECD Report*, Organisation for Economic Cooperation and Development, Publications Service, Paris 1982.

[8.27] K. H. Becker, W. Fricke, J. Löbel, U. Schurath in: *Air Pollution by Photochemical Oxidants*, Springer Verlag, Berlin 1985.

[8.28] X. J. Crutzen, *Nature (London)* **350** (1991) 380.

[8.29] D. H. Ehalt, J. W. Drummond in: *Tropospheric Ozone*, Reidel Publ., Dordrecht 1988.

[8.30] H. Güsten in: *Chemistry of Multiphase Atmospheric Systems*, Springer Verlag, Heidelberg 1986, p. 567.

[8.31] R. Jaenicke, *Ber. Bunsen-Ges. Phys. Chem.* **82** (1978) 1198.

[8.32] "Scientific Assessment of Stratospheric Ozone," *World Meteorological Organization Global Ozone Research and Monitoring Project*, Report 20, vol. 1, WMO, Geneva 1990.

[8.33] P. A. Mayewski et al., *Nature (London)* **346** (1990) 554.

[8.34] A. Sigg, A. Neftel, *Nature (London)* **351** (1991) 557.

[8.35] K. Bächmann, G. Fuchs in Ref. [8.4] p. 79.

[8.36] "WMO and the Ozone Issue," *World Meteorological Organization*, Geneva 1992.

[8.37] "Scientific Assessment of Stratospheric Ozone," *World Meteorological Organization* Report 25, WMO, Geneva 1992.

[8.38] M. J. Prather, R. Watson, *Nature (London)* **344** (1990) 729.

[8.39] T. Mill: Reprint Extended Abstract, Division of Environmental Chemistry, Am. Chem. Soc., Boston, April 1990.

[8.40] W. Klöpffer in G. Rippen (ed.): *Handbuch der Umweltchemikalien*, part 2, suppl. 26, no. 10/1994, ecomed-Verlag, Landsberg 1991.

[8.41] R. G. Zepp, J. Hoigné, H. Bader, *Environ. Sci. Technol.* **21** (1987) 443.

[8.42] W. Klöpffer, E.-G. Kohl, *Ecotoxicol. Environ. Saf.* **22** (1991) 67.

[8.43] W. Klöpffer, *Sci. Total Environ.* **123/124** (1992) 145.

[8.44] J. Hoigné: "Bildung und Bedeutung von Photooxidantien in der wässrigen Phase," *Proc. Int. Symp. Verteilung Wirkung Photooxidantien Alpenraum*, 1988.

[8.45] W. Klöpffer, E. G. Kohl, *Ecotoxicol. Environ. Saf.* **26** (1993) 346.

[8.46] R. Atkinson, *Chem. Rev.* **86** (1986) 69.

[8.47] R. Atkinson, *J. Phys. Chem. Ref. Data Monograph 1*, Am. Inst. Physics, New York 1989.

[8.48] W. Klöpffer: Contract R-67.233, Batelle Europe to VCI Frankfurt and UBA Berlin, Batelle Institute, Frankfurt 1989/90.

[8.49] W. Klöpffer, B. Daniel: "Reaktionskonstanten zum abiotischen Abbau," *texts 57*, Umweltbundesamt, Berlin 1991.

[8.50] R. Atkinson, *Environ. Sci. Technol.* **21** (1987) 305.

[8.51] W. L. Dilling, S. J. Gonsior, G. V. Boggs, C. G. Mendoza, *Environ. Sci. Technol.* **22** (1988) 1447.

[8.52] OECD-Test "Guideline on Photochemical Oxidative Degradation in the Atmosphere," Umweltbundesamt Berlin, draft, 2nd ed. 1987, final draft 1991.

[8.53] OECD-Test "Guideline on Photochemical Transformation of Chemicals in Water," Umweltbundesamt Berlin, draft for OECD workshop, Berlin 1987.

[8.54] R. Atkinson, W. P. L. Carter, *Chem. Rev.* **84**, (1984) 437.

[8.55] K. Bächmann, G. Fuchs in Ref. [8.4] p. 79.

[8.56] R. Atkinson, S. M. Aschmann, *Int. J. Chem. Kinet.* **17** (1985) 33.

[8.57] C. J. Weschler, T. E. Graedel: "Theoretical Limitations on Heterogeneous Catalysis by Transition Metals in Aqueous Atmospheric Aerosols," in "Heterogeneous Atmospheric Chemistry," *Geophys. Monogr. Am. Geophys. Union* **26** (1982) 196.

[8.58] V. Brezova, et. al., *J. Photochem. Photobiol. A* **59** (1991) 385.

[8.59] W. Klöpffer in G. Rippen (ed.): *Handbuch der Umweltchemikalien*, part 1, ecomed-Verlag, Landsberg 1991.

[8.60] W. Klöpffer: "Environmental Hazard Assessment of Chemicals and Products," part I, *ESP Environ. Sci. Pollut. Res.* **1** (1993) 47.

[8.61] W. Klöpffer: "Environmental Hazard Assessment of Chemicals and Products," part II, *ESP Environ. Sci. Pollut. Res.* **1** (1994) 108.

[8.62] W. Klöpffer: "Environmental Hazard Assessment of Chemicals and Products," part III, *ESP Environ. Sci. Pollut. Res.* **1** (1994) 179.

[8.63] W. Klöpffer: "Environmental Hazard Assessment of Chemicals and Products," part IV, *ESP Environ. Sci. Pollut. Res.* **1** (1994) 272.

[8.64] F. Korte, et al. *Lehrbuch der ökologischen Chemie*, Thieme-Verlag, Stuttgart 1987.

Soil

STEPHEN NORTCLIFF, University of Reading, Department of Soil Science, Reading, United Kingdom (Chap. 1)

HERWIG HULPKE, Bayer AG, WV Umweltschutz, Leverkusen, Federal Republic of Germany (Chap. 2)

CLAUS G. BANNICK, KONSTANTIN TERYTZE, Umweltbundesamt, Berlin, Federal Republic of Germany (Section 3.1)

GERHARD KNOOP, Bayer AG, WV Umweltschutz, Leverkusen, Federal Republic of Germany (Section 3.2, Section 6.1 in part)

KARL AUERSWALD, Technische Universität München, Lehrstuhl für Bodenkunde, Freising, Federal Republic of Germany (Chap. 4)

NORBERT LITZ, Umweltbundesamt, Berlin, Federal Republic of Germany (Section 5.1)

ROBERT MAYER, Universität Gh Kassel, Fachbereich Stadtplanung/Landschaftsplanung, Kassel, Federal Republic of Germany (Section 5.2)

ALEXANDER STOY, Fachhochschule Kiel, Fachbereich Landbau, Rendsburg, Federal Republic of Germany (Section 5.3.1)

KASSEM ALEF, Universität Bayreuth, Lehrstuhl für Ökologische Chemie, Bayreuth, Federal Republic of Germany (Section 5.3.2)

HELMUT KERNDORFF, Umweltbundesamt, Berlin, Federal Republic of Germany (Section 5.4)

GERD CRÖSSMANN, Landwirtschaftliche Untersuchungs- und Forschungsanstalt, Münster, Federal Republic of Germany (Section 6.1 in part)

THOMAS EIKMANN, Universität Gießen, Hygiene-Institut, Gießen, Federal Republic of Germany (Section 6.2)

VOLKER FRANZIUS, Umweltbundesamt, Berlin, Federal Republic of Germany (Section 7.1)

NORBERT MÖHLENBRUCH, Rheinbraun AG, Abt. Forstwirtschaft, Erftstadt, Federal Republic of Germany (Section 7.2)

PETER-CHRISTOPH STORM, Umweltbundesamt, Berlin, and Universität Tübingen, Federal Republic of Germany (Chap. 8)

The entire topic was coordinate by HERWIG HULPKE

3-527-20137-8/95/$5.00 + .50

1. Definition

Soil is normally considered as the fine earth which covers land surfaces as a result of the in situ weathering of rock materials or the accumulation of mineral matter transported by water, wind, or ice. The distinctive feature of soil is that to this weathered mineral material is added organic material. This organic material may be both living and dead. The dead organic matter will include little altered and freshly added dead plant roots and leaf and other plant litter, dead fauna, and organic material in various stages of decomposition from little modified relatively fresh materials to the complex decomposed material called humus. It is this mixture of mineral and organic material which gives the soils their distinctive characteristics. Across the surface of the earth there are many different types of soil which reflect, at least in part, varying combinations of mineral and organic matter and their differing responses—both separately and often in complex association—to different environmental conditions. Indeed soil (and the soil constituents), together with the plant life it supports, the rock on which it lies, and the climate it experiences, forms a finely balanced system.

Depending upon the context, the word "soil" may have very different meanings. A simple definition of soil is the material that plants grow in and which provides them with physical support, water, and nutrients. There are other more particular uses of the term soil. To the engineer, soil is the finely divided and relatively loose "rock" material at the earth's surface, and this overburden is often considered an inconvenience because it must be removed. The geologist calls this layer the regolith and geological investigations frequently begin below it. The hydrologist looks on the soil as if it were a large reservoir storing water to supply streams and rivers. The most widely held view of the soil, however, is as a medium for plant growth and the provision of food and fiber directly or through the intermediate stage of animals.

Soil is important as a medium for plant growth and for the support of much animal and human activity. The soil acts as a reservoir for nutrients and water providing the plants' needs for these requirements throughout their growth. The soil may also provide an environment for the breakdown and immobilization of materials added to the surface (in addition to the aforementioned plant and animal remains) such as fertilizers and pesticides and waste products such as sewage sludge, animal wastes and slurries, and composted refuse materials. The soil is a complex dynamic system in which the interactions of the biological, chemical, and physical environments results in the transformation of materials, possibly rendering initially harmful materials less dangerous and immobilizing others as a result of the interactions between these added materials and the organic and inorganic soil constituents. This immobilization may enable breakdown of the potentially dangerous materials to less dangerous forms. These interactions and transformations may be long term, over decades, medium term over months or years, short term between individual events such as rainstorms, or almost instantaneous.

The soil usually consist of a vertical sequence of layers or horizons differentiated by physical, chemical, or biological characteristics. This sequence of layers is normally described as the soil profile. Surface horizons are usually characterized by accumulations of organic materials as a result of the addition of plant and animal residues (it is, however, becoming increasingly apparent that organic additions in deeper layers from roots may have a significant role to play in the overall additions of organic material and in the operation of the soil processes). Surface horizons with a mixture of mineral and organic materials are usually identified as *A horizons* and may be further described by the amount and nature of the organic matter present within the horizon. Where there is a substantial proportion of organic material in the surface layer it may be called *O horizon*. The A and O horizons are generally considered the biologically most active part of the soil and are also subject to considerable changes as a result of the weather, being frequently exposed to wetting and drying cycles and to a wide range of temperatures, which results in high rates of activity in physical and chemical processes. Both organic and mineral materials are transformed in these surface layers and frequently some of the byproducts of these transformations are released and removed to lower layers within the soil (eluviation). Below the A horizons usually *B horizons* are encountered where the balance shifts from predominantly biological processes as in the A horizons to predominantly chemical and physical processes. Pedologists normally consider this zone to be that of peak activity of the pedological or soil-forming processes, and it is for this reason that many of the currently used soil classifications pay particular attention to the characteristics of the B horizon in the allocation of soils to classes. The B horizons are frequently subject to the maximum intensity of weathering processes and transformations and may also exhibit the accumulation of materials translocated from surface layers (illuviation) such as clays, organic matter, and iron and aluminum hydrous oxides. The description of B horizons frequently emphasises the degree of weathering and the nature of the accumulated materials. The *C horizon*, below the B, is the relatively unaltered material from which the A and B horizons may have developed.

The importance of the incorporation and breakdown of organic matter and the production of complex organic materials, as well as the weathering of mineral materials to produce secondary products (such as clay minerals) and to release minerals which may be taken up by plants or leached out of the system, provides the distinctive nature of the soil. In particular the large surface area associated with soil materials and their ability both to retain and exchange ions (both those added in fertilizers and those released during weathering and biological decomposition) and to absorb water are important properties which distinguish soil from rock materials.

Soil is essential for many human activities. It is also a basic part of the natural environment. The development of humans and society since prehistoric times has been closely linked with an increasing ability to manage the soil to human benefit. This progress has been achieved by adjusting the balance between the soil and its natural environment. These adjustments have not always resulted in positive responses and benefits, and there are records throughout history describing soil destruction as a result of mismanagement and misuse of the soil, often as a result of the failure to understand the nature and complexity of the soil–environment relationships. Future use of the soil must take account of these relationships even in a society where we are increasingly able to manipulate our environment. As yet we know too little neither of the nature of the soil–environment relationships nor of the changes in these relationships and the response of the soil as the environmental conditions are changed. It is essential that this potential for change is acknowledged in any attempts to manipulate the soil–environment system.

2. Ecological and Political Aspects of Soil Protection

Soils are complex ecosystems that are strongly cross-linked to the other environmental compartments (water and air). Because of these linkages far reaching interactions with the other environmental compartments exist, through which soils have an important ecological mediator function. With respect to material transport processes, soils can be both sources and sinks.

In significant contrast to the other environmental compartments (air, water systems) humans took possession of soil from early times. The concept of "owned land", which implies "possession of soil" underlines this fact. In some languages (e.g., German) land (Land) and soil (Boden) are often used as synonyms. Soil therefore represents the most important resource for human life in both its material and cultural aspects. Because soil is possessed as property, it has long been assumed that special care of the soil by the state could be dispensed with, since the soil's owners would already take care of it. This assumption may well be true in many cases for specific forms of use, such as agricultural use. But less care is felt when the utilization of the soil does not serve a single aim but several parallel or even conflicting purposes. Such multiple uses are typically the result of an increasing density of human population.

Soil is utilized by exploiting its functions. The great importance of soils for the existence of humankind is clearly suggested by the term "culture", which is derived from the Latin word "colere" (to till).

The exploitation of certain soil functions does not only create benefits but can also cause damage. There are many examples for such conflicts. Agricultural use as opposed to planned use for housing or industrialization is a typical conflict of this type.

In the course of this, the use of the soil for nature conservation purposes or as a groundwater reservoir for the production of drinking water can be restricted or even endangered. Often, however, with cross-linked and complex ecosystems the conflicting interactions are more complicated and often neither transparent nor easy to understand. Nevertheless, when exploiting the soil it must always be borne in mind that the benefit from the use of one function can at the same time be detrimental to the use of another function. Soil protection is therefore chiefly a task whereby the utilization of soil functions has to be prudently selected. The properly targeted management of these uses must then be carried out in such a way that no unreasonable and unaccepted conflicts of the various intended uses will occur. At the same time the kind of use and its extent must also be taken into account. The understanding of soil protection as the management of competing claims for use by human beings is undoubtedly an anthropocentric view. But criticism raised against this view frequently fails to understand that this approach to action implies a large number of optional actions that can extend from the utilization of one or a few soil functions to the use of all conceivable or available functions. In the limiting case the utilization of any soil function can be entirely dispensed with. The use can therefore admit the range from "no functions" to "all." As a result of this, and despite the anthropocentric approach, even extensive ecological requirements are taken into account. A system designed for the protection of soil functions consequently offers starting points for practicable as well as ecologically demanding measures.

With regard to its utilization and exploitation, the soil is the memory and archive of these actions. In this respect, pollutions arising from earlier uses (contaminated sites) play an important part in soil protection. As a result of the exploitation of soils in the past, restrictions often arise today on the utilization of some or all soil functions. Thus a soil that functions as a reservoir for groundwater regeneration and was contaminated in the past cannot be used as a water protection zone if through this contamination, owing to displacement processes, concentrations of contaminants appear in the groundwater making it unacceptable for drinking water catchment. The problem of exploitations in the past (contaminated sites) for soil protection is that the actual results of this exploitation often cannot be undone. Care must then be taken that this exploitation in the past has no harmful effects for present-day or future uses. The selection of appropriate uses therefore plays an important part. The precautionary principle, generally important in environmental protection, cannot apply here since, understandably, actions in the past cannot be cancelled by means of precautionary measures.

In the case of contaminated sites, the protection of soil functions can therefore not be managed by applying the precautionary principle. Instead, it must be orientated to the reduction or elimination of risks to soil functions intended for future utilization.

3. Function and Utilization of Soil

3.1. Soil Ecosystems

3.1.1. Introduction

Pedogenesis—the formation of soil—involves interactions between the lithosphere, hydrosphere, atmosphere, and biosphere (Fig. 3.1) [3.1]. At the earth's surface, the lithosphere becomes the pedosphere under the influence of weathering forces and organisms.

Thus soil is a complex, multiphase mixture of minerals, water, air, and organisms, together with products of their transformation and degradation. Depending on the climatic conditions and the nature of the underlying rocks, widely differing types of soil can be formed, whereby the rate of formation depends on how fast the rocks are weathered in the lower layers and on how rapidly biomass is degraded in the upper layers.

Lithosphere. The lithosphere is the rock mantle of the continents and the earth's surface beneath the oceans. The soil—the upper, inhabited part of the earth's crust—is formed from rocks by weathering, mineralization, decomposition, humification, texture formation, and displacement, under certain conditions (e.g., climate, leaf fall from vegetation). Weathering is the chemical, physical, and biological transformation of inorganic material such as basalt, granite, gneiss, carbonate, loess, and sand at the earth's surface by size reduction, degradation, transformation, and dissolution. With progressing soil development, the petrogenic starting materials are increasingly replaced by pedogenic materials such as clay minerals, which are more stable at the earth's surface.

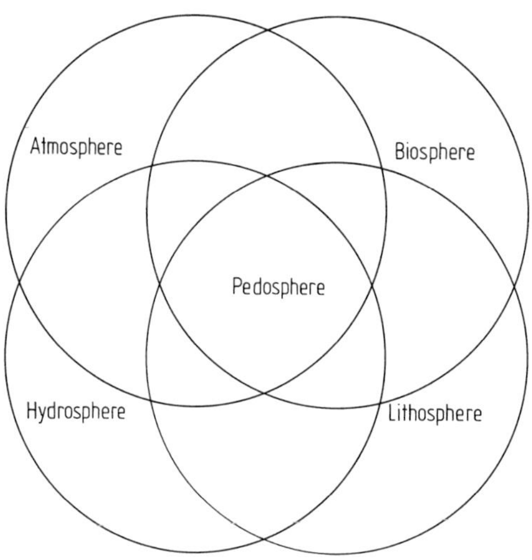

Figure 3.1. The soil in terms of the interaction of various spheres [3.1]

Nine groups of minerals can be distinguished on the basis of chemical composition [3.2]:

Class I	Elements: metals (Ag, Au, ...), metalloids (As, Sb, ...), nonmetals (C, Se, ...)
Class II	Sulfides: galena (PbS), sphalerite (ZnS)
Class III	Halides: halite (NaCl), sylvinite (KCl)
Class IV	Oxides, hydroxides: corundum (Al_2O_3), quartz (SiO_2), hematite (Fe_2O_3)
Class V	Carbonates, nitrates, borates: calcite ($CaCO_3$)
Class VI	Sulfates, chromates, molybdenates, tungstates: baryte ($BaSO_4$)
Class VII	Phosphates, arsenates, vanadates: apatite $[Ca_5[(F,Cl;OH)/(PO_4)_3]$
Class VIII	Silicates: quartz (SiO_2)
Class IX	Organic compounds

As a result of the various soil-forming factors, characteristic horizontal layers, known as horizons, are formed, which differ in their biological, chemical, and physical properties. At the soil surface, the horizons contain larger proportions of organic material (O/A horizons), whereas the mineral component increases with increasing depth.

The horizons together make up the solum. The vertical distribution of the horizons in a two-dimensional cross section through the soil (soil profile) is the main criterion for classification of soils.

The smallest, undividable unit of a soil body is known as a pedon. Neighboring peda of the same soil type form a polypedon. The properties of the peda of a polypedon vary only slightly (e.g., horizon depth). On a larger scale, polypeda may be collected together as pedocomplexes.

Atmosphere and Hydrosphere. The spacial arrangement of the voids, the pore system, and the solid soil components determins the soil structure. The structures are classified in two main categories: primary structures and secondary structures (aggregates). In primary structures, aggregation is not apparent, due to uniform binding forces between the primary particles. Only in coherent structures are the primary particles permanently attached, forming a nonorganized mass, as in humus and clay soils. In secondary structures, primary particles are bound to one another far more strongly than to neighboring groups, thus forming secondary particles.

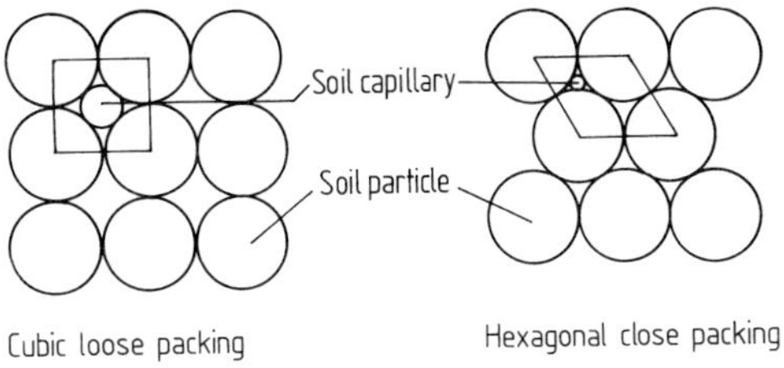

Figure 3.2. Packing of spherical particles [3.3]

Aggregates such as crumbs (1 – 10 mm) are formed by aggregation of individual grains. Segregates form by shrinkage from a coherent bulk. Interparticle binding can be by pure adhesion, or by cementation by mineral precipitates, inorganic colloids, roots, fungal hyphae, or bacterial slime. The soil solid phase can be separated into various particle sizes by classification processes such as sieving and sedimentation. The mass composition of the solid phase as a function of particle size is known as granularity or texture. Particles >2 mm make up the coarse fragment, and those <2 mm, the fine earth, which is subdivided into sand (63 – 2000 μm), silt (2 – 63 μm), and clay (<2 μm) fractions.

Soil pores are also classified according to size: large pores (>50 μm), medium pores (0.2 – 50 μm), and fine pores (<0.2 μm). Large pores bind soil water only weakly and are easily drained. Medium pores can be penetrated by root hairs and microorganisms and store capillary water available to plants. Fine pores bind water so strongly that it is no longer available to plants. The soil contains a pore system formed by mineral particles generated by weathering. The two main types of particle packing (see Fig. 3.2), cubic loose and hexagonal close packed, influence various soil processes.

Biosphere. The totality of the ecosystems of the earth's crust, in close interaction with the hydro-, litho-, and atmospheres, makes up the biosphere. The total organic fraction of the solid phase of the soil is divided into biomass, which includes all living organisms (soil flora and fauna), and the necromass, which contains all dead organisms and their transformation products. The latter category is derived from both underground (e.g., roots) and overground sources (e.g., leaf fall). The degradation products of the necromass are the humic substances which, depending on their solubility, are divided into the alkali-insoluble humins and the alkali-soluble humic acid precursors and humic acids (see Section 3.1.2.1).

Within their biotope, the soil organisms form a community known as an edaphon. The edaphon is divided according to the size and way of life of the organisms. In the division according to morphological size [3.4] the edaphon is divided into microflora (bacteria, fungi, algae), microfauna (protozoa), mesofauna (nematoda, rotifera, tardigradae, mites, and collembola), macrofauna (annelids, snails, insects and their larvae), and megafauna (vertebrates); see Figure 3.3.

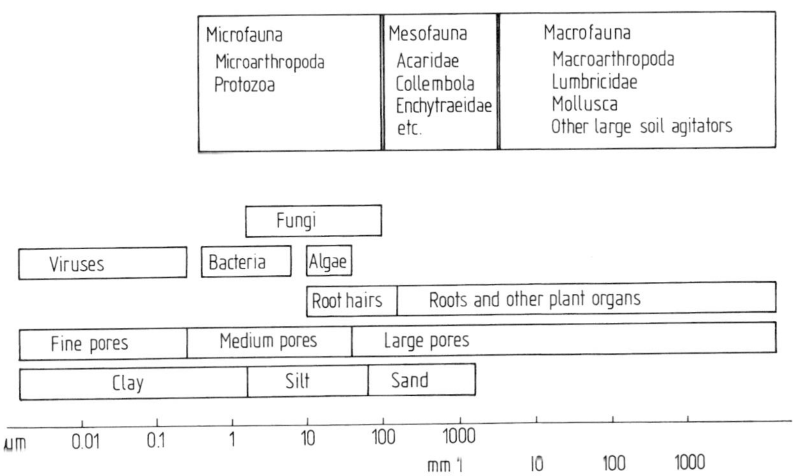

Figure 3.3. Classification of soil organisms by size [3.4]

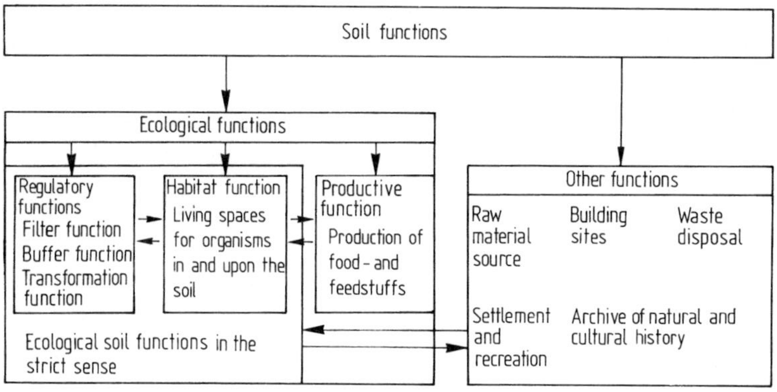

Figure 3.4. Soil functions

The particular living conditions in the soil are coupled with typical modes of behavior of the organisms. Thus the group of soil burrowers (fodient edaphon) is formed mainly of organisms from the mega- and macrofauna (annelids, insects and their larvae) which generate their own living environment by burrowing and agitation of the soil. Serpentine edaphon move through the pores of the soil (rhizopods, annelids, collembola). They are restricted to open, air-containing pores and are found mainly in the upper soil layers. The soil swimmers (natant edaphon) are adapted to life in the water of the soil. They include some microflora and the majority of the mesofauna (rotifera, nematodes, tardigradae). Bacteria and fungi, which rapidly cover the walls of pores or enter soil voids belong to the sessile edaphon [3.5].

The soil flora and fauna introduce organic material into the soil system. Almost 100% of the elemental composition is accounted for by C, O, H, N, and S. The central importance of the soil is due to the fact that fundamental ecological processes interact there. Thus the soil is an essential basis for all life processes. Soil functions are summarized in Figure 3.4.

From the environmental viewpoint, the most important functions are the regulating and productive functions, known together as ecological soil functions. Other soil functions include raw material resources, land for building, recreational areas, and waste disposal.

The filter function is the mechanical retention of particules and colloids by the soil-pore system.

The buffer function includes the binding of dissolved substances by absorption and desorption, the inclusion and release of substances (including water) by pore spaces, and the in- and excorporation of substances by soil organisms. The habitat function is the ability of the soil to support colonies of organisms within and upon the soil. This includes the function of the soil as substrate for higher plants.

The productive function differs from the habitat function in that here the economic exploitation of the biomass is of primary importance.

The other soil functions are mainly or exclusively of anthropogenic nature. Here, the ecological functions are of minor importance compared to the anthropogenic exploitation or are indirectly exploited, for example, in the degradation or sorption of pollutants. In contrast to the other soil functions (raw material deposits, building, recreation, waste disposal), the regulatory and habitat functions can be defined solely on the basis of the natural pedological and ecological properties of the soil, regardless of the use to which it is put. The productive function can be distinguished on the basis of both ecological aspects (habitat for cultivated plants) and socio-economic criteria, since it is directly interlinked with agriculture and forestry.

3.1.2. Ecological Soil Functions

3.1.2.1. Regulating Functions

The regulating function of the soil depends on the regulation of natural material and energy cycles. It includes all material and nonmaterial introductions to the soil and the process therein which they initiate.

Filter Function. Filtration is defined as mechanical retention of material >100 nm in size [3.6]. Filtration occurs during the flow of aqueous suspension, through the pore systems.

The behavior of the soil water (hydraulic potential ϕ) is determined by the gravitation potential z (downwards) and the matrix potential ψ (upwards). At equilibrium, these two potentials are equal, i.e., $\phi = (-\psi) + (+z) = 0$. Addition of water (e.g., by precipitation) increases the matrix potential and displaces the equilibrium. This infiltration (→ Ecology and Ecotoxicology, Section 2.1.2) is one of the most important effects counteracting erosion.

Decreased filter performance can arise through pore blockage due, for example, to compaction or clogging with mud. Pore blockage can also have deleterious effects on the transformation and habitat functions (stagnant water, air deficiency).

Buffer Function. The buffer function encompasses the following major areas:

1) Buffering of water balance
2) Buffering of acid inputs
3) Buffering of nutrients and pollutants

The buffering of water balance is largely a result of the ability of soils to store water in pore spaces by means of capillary forces. Since the pressure of a liquid (tension) is lower than atmospheric pressure, the liquid rises in a capillary until hydrostatic equilibrium is attained. The capillary pressure and hence the height to which the liquid rises increase with decreasing capillary diameter.

The water-storage capacity of a soil is characterized by the amount of water that can be retained for a given value of capillary pressure (pF characteristic). The pF value is defined as the logarithm of the negative capillary force $(-\psi)$, or the pressure in centimeters water column. Of particular importance for the habitat and productive functions is water available to plants, with pF = 1.8–4.2 (exploitable field capacity). For each soil type, at a given pF value, the quantity of water that can be retained against the force of gravity can be determined. The exploitable field capacity (in mm/dm) for some substrates is given in the following [3.7]:

Fine sand	13
Silt	25
Sandy loam	19
Clay	15

Buffering of acid inputs is achieved by various soil buffer systems, depending on the pH range:

Neutral (6.2–8.0)	carbon dioxide/ carbonate buffer
Acidic (4.2–6.2)	exchange buffer
Highly acidic (2.8–4.2)	aluminum buffer

Buffer systems consist of a weak acid or base and the corresponding salts, and maintain constant pH within a certain range. All soils are exposed to acids, both of natural (e.g., absorption of nutrient ions by plants with release of protons) and anthropogenic origin (e.g., fertilization, atmogenic inputs).

Carbonate-containing soils can buffer protons by formation of hydrogencarbonate

$$CaCO_3 + H^+ \rightleftarrows Ca^{2+} + HCO_3^-$$

After exhaustion of the carbonate buffer, buffering is performed by the exchange buffer system, whereby all cations of the mineral and organic ion exchangers are replaced by protons. Below pH 4.2 in the aluminum buffer system, destruction of the crystal lattice structure and release of aluminum ions occurs

$$AlO(OH) \cdot H_2O + 3H^+ \rightarrow Al^{3+} + 3H_2O$$

The buffering of nutrients and pollutants affects most ecological soil functions. It is made possible by the presence of organic and inorganic exchangers, whereby the following adsorption phenomena can occur:

1) Van der Waals forces: relevant for organic pollutants, humus
2) Hydrogen bonds: relevant for organic pollutants, mineral surfaces
3) Electron donor–acceptor complexes: relevant for organic pollutants, humus
4) Hydrophobic bonding: relevant for organic pollutants, humus
5) Ion bonding: relevant for organic pollutants, inorganic pollutants and nutrients, humus, clay minerals
6) Covalent bonding: relevant for organic and inorganic pollutants

For nutrients and inorganic pollutants, ion formation is the most important process. Ion exchangers have positive or negative excess charge, thus binding anions or cations. Sorption of this type is reversible and an ion-concentration-dependent equilibrium is established in the soil solution. The sorption strength depends on the distance between the bonding partners, which is largely determined by their hydration shells,

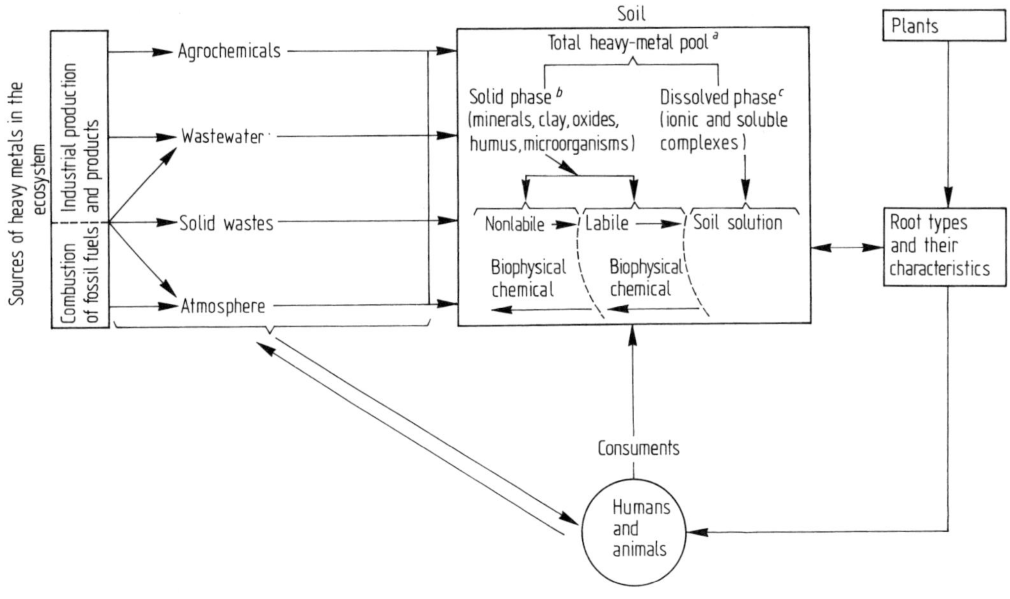

Figure 3.5. Heavy-metal cycle in the ecosystem [3.8]
[a] Can be used as indicator for accumulation. [b] Potentially biorelevant. [c] Momentary biorelevant increase of metal concentration in the dissolved phase as indicator for biotoxicity and bioacumulation.

decreasing with increasing distance according to the Coulomb law $P \sim Q_1 Q_2 / r^2$.

In addition, the strength of sorption increases with increasing ion charge, which is taken into account in the Hofmeister ion series of decreasing hydration shell and increasing exchange potential:

$$Li^+ < Na^+ < K^+ < Rb^+ < Cs^+$$

$$Mg^{2+} < Ca^{2+} < Sr^{2+} < Ba^{2+}$$

The input of nutrients and inorganic pollutants into the soil, their dissolution in soil water, buffering by adsorption and desorption, uptake by plants, and loss of eluates can be depicted in nutrient and pollutant cycles (→ Ecology and Ecotoxicology, Section 2.1.4). The ecological cycle for heavy metals is shown in Figure 3.5.

Heavy metals can be bound in the soil in the following ways (R represents a solid/dissolved inorganic/organic compound, e.g., iron oxide, clay mineral, humic substance):

Solid state

Unspecifically adsorbed, e.g., $R\langle^{O}_{OCd^{2+}\,(aq.)}$

Specifically adsorbed, e.g., $R-OCu^+$

Metal salts, e.g., $PbCO_3$, $Pb_3(PO_4)_2$, PbS

Oxides, e.g., As_2O_3

Bound in silicates e.g., tin in Hornblende

Dissolved

Aquo complexes, e.g., $Cd^{2+}_{(aq.)}$, $Pb^{2+}_{(aq.)}$

Inorganic complexes, e.g., AsO_4^{3+}, $Zn(OH)_3^-$, $Pb(CO_3)_2^{2-}$, $CdCl^+$

Organic complexes and chelates, e.g., $CH_3Hg^+_{(aq.)}$, $R-Cu-R$

The pollutant content of the soil is characterized by background values, which are representative values for the general background content of a substance or group of substances in the soil. When statistically determined they can be used to give the local, regional, or nationwide pollution of the soil. Table 3.1 lists values based on the evaluation of over 16000 soil samples collected nationwide in Germany [3.9]–[3.11]. The samples were collected by a uniform method, and the 50th and 90th percentiles of the heavy-metal content determined for three types of land usage. The shown values are representative for the rural area.

Soil has to fulfill a buffer function to protect humans, plants, and groundwater from soil pollutants. Soil pollutants and the corresponding pathways are listed in Table 3.2.

Table 3.1. Heavy-metal contents in rural areas for arable land (90th percentile)

Substrate	As	Cd	Cr	Cu	Hg	Ni	Pb	Sb	Zn
Sand	3.0	<0.3		3.0	0.03	<3.0	13	<0.3	14
Sandstone	7.0	0.9		15	0.11	30	75	0.6	43
Loess	9.6	<0.3	149	25	0.16	44	51	1.1	89
Boulder clay	5.4	<0.3		14	0.08	19	32	0.9	76
Marshes									
Claystone	10	1.1		27	0.14	74	61	0.6	121
Limestone									
Basalt	5.0	0.8		71	0.1	339	49		168

Table 3.2. Priorities of relevant soil pollutants

Relevant pollutants in the pathway		
Soil–human	Soil–plant	Soil–groundwater
Arsenic	arsenic	heavy metals
Lead	lead	crop-protection agents
Cadmium	cadmium	volatile chlorinated hydrocarbons
Chromium	chromium	
Nickel	copper	
Mercury	mercury	
Vanadium	thallium	
Zinc	zinc	
Benzo[*a*]pyrene	aldrin/dieldrin	
Benzene	benzo[*a*]pyrene	
DDT (total)	endrin	
Dieldrin	hexachlorobenzene	
PCB (total)	pentachlorophenole	
PCDD/PCDF*	DDT (total)	
Phthalates	nitrate	
1,1-Dichloroethane	PCDD/PCDF*	
Dichloromethane		
Tetrachloroethylene		
Toluene		
Trichloromethane		
Xylene		
Vinylchloride		
Cyanides		

* PCDD = polychlorinated dibenzodioxins;
PCDF = polychlorinated dibenzofurans.

Transformation Function. The behavior of chemical compounds in the pedosphere depends on the properties of the substance, the location, and in some cases is affected by anthropogenic activities. The transformation function provides for self-cleaning of the soil. Degradation and transformation of substance in soil can occur abiotically (e.g., reactions on mineral surfaces, photochemical reactions) or biotically. One of the most important forces of the self-cleaning capacity of soils are the microorganisms, which provide the enzymes necessary for biotic transformation processes. Microorganisms use substances that enter the soil as nutrients or energy sources, whose exploitability increases with decreasing persistence of the compounds. After a substance has entered the soil, the microflora require a latent period (lag phase) for the development of the specific species spectrum required for its metabolization. This time period depends on the xenobiotic properties of the substance. If the substance is introduced more often, the lag phase becomes shorter and the persistence decreases. In addition, all processes that stimulate microbial activity in the soil are suitable for increasing its metabolic potency. These include application of organic and inorganic nutrients and cultivation of the soil.

The degradation and transformation of postmortal plant and animal material to humic substance is also attributed to the transformation function. Humic substances are naturally occurring biogenic substances with characteristic yellow to black color, high particle density, and persistence. Humic substances have been found in materials that have undergone coalification, such as peat, liquites, and lignites. The humic substances can be classified according to their solubility. Humic substance fractions and their solubility, as determined by various authors, are as follows:

1) Classification according to Scheffer, Ulrich [3.12]:

Crenic acids	acids, alkalis, water, alcohol
Apocrenic acids	acids, alkalis, water, alcohol
Hymatomelanic acids	alkalis, alcohol, acid (partially)
Gray humic acids	alkalis
Brown humic acids	alkalis

2) Classification according to Flaig [3.13]:

Fulvic acids	acids, alkalis
Hymatomelanic acids	alkalis, alcohol
Humic acids	alkalis
Humines	insoluble

3) Classification according to Swift [3.14]:

Fulvic Acids	acids and alkalis
Humic acids	alkalis
Humines	insoluble

In contrast, Ziechmann [3.15] classifies the humic acid fractions according to their reactivity and their ability to undergo further humification with uptake of oxygen:

Starting materials (nonhumic)	organic substances undergoing humification lightly colored to colorless
Humic acid precursors	reactive fraction, gradually transforming into humic acid light brown to brown
Humic acids	fraction containing carboxylic acid groups, phenolic OH group base-soluble, precipitate in acid intense brown color
Humines	inactive fraction insoluble intense brown color

The principle route for the formation of humic substances is shown in Figure 3.6. In reality, the formation mechanism presented here may be far more complex. This is due firstly to the wide variety of substance groups that can be involved in the humification process, and secondly to the activities of the micro- and macrofauna, which are responsible not only for producing these substances but also for mineralization, which competes with humification. Substances produced by microorganisms during the degradation of organic matter, as well as partially degraded, but considerably modified, substances, are involved in the humification process. In contrast to this incorporation of degraded biopolymers in a preexisting humic substance system, it is now being discussed whether such degraded biopolymers should themselves be designated as humic acids.

Degradation and transformation of nutrients (→ Ecology and Ecotoxicology, Section 2.1.4) is another important transformation process in soils and also involves microorganisms. Of major importance for the cleansing of the soil is the degradation and transformation of pollutants. The microbial degradation of various groups of chemical substances with ultimate mineralization is a first-order reaction. This has been shown, among others, for some chlorinated hydrocarbons, triazine derivatives, organic phosphorus compounds, carbamates, thiocarbamates, thiurams, urea derivatives. Under comparable conditions, the rate of microbial degradation is given approximately by

$$c = e^{-kt}$$

where c is the concentration of the compounds, t is time, and k is the mineralization rate.

Plotting log c against t gives a straight line whose gradient is the mineralization rate k, which allows predictions of the persistence of the compound under given soil and other environmental conditions to be made. Strongly sorbent, dry, low-nutrient soils with relatively low microbial activity exhibit small gradients, whereas soils in which the nutrient supply, air and water balance and hence the microbial activity are optimal give a steep gradient.

Continuation of the microbial transformation process in the soil is essential for plant growth and hence for the existence of animals and humans. In this context, the question of whether the increasing production of chemicals in the last few decades, which has led to intentional or accidental introduction of intermediate, final, and waste products into the soil, can seriously affect the biological functions of the soil, is becoming of increasing importance.

To investigate this, it is necessary to study the behavior of numerous chemical substances in the

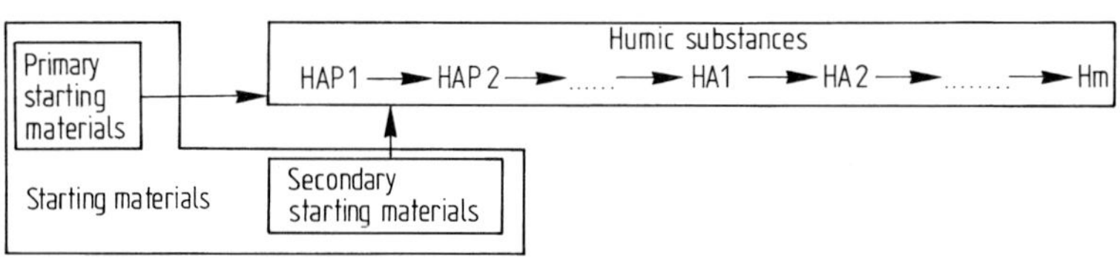

Figure 3.6. Formation of humic substances [3.15]
HAP = humic acid precursors; HA = humic acid; Hm = humines

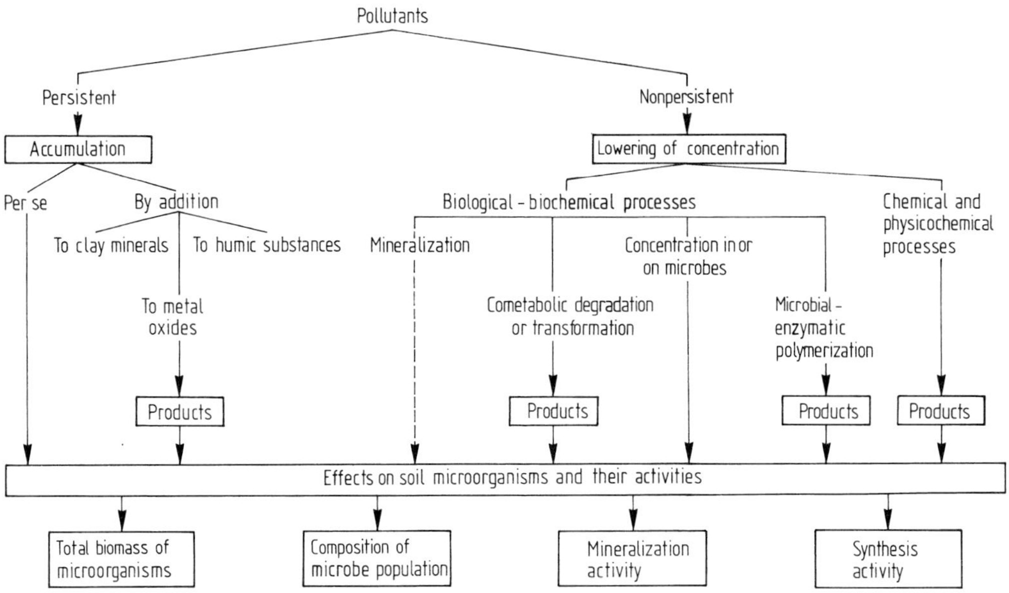

Figure 3.7. Behavior of pollutants in the soil and their interaction with microorganisms [3.16]

soil and to elucidate their effects on soil organisms. Depending on their properties, chemical pollutants can persist in the soil or be degraded. Thus a distinction can be made between persistent and nonpersistent pollutants. Both groups can affect soil microorganisms and their activity in various ways (Fig. 3.7). Persistent pollutants accumulate in the soil, acting on the soil microflora either alone or as complexes with clay minerals, humic acids, and oxides. Nonpersistent pollutants undergo a decrease in concentration due to chemical or physicochemical, but mainly biological/biochemical processes. If such processes result in mineralization of the pollutant, a detrimental effect on soil microorganisms is unlikely. However, since complete mineralization is generally the exception, it can be expected in many cases that intermediate degradation products or the final products of transformation reactions could adversely affect the soil microorganisms and their activity. Such effects can change the condition of the total biomass as well as the species composition of the microorganism population and its mineralization and synthesis activity.

3.1.2.2. Habitat Function for Organisms in and upon the Soil

The organic material of the soil consists of 85% postmortal substance, 10% plant roots, 4% bacteria and fungi, and 1% soil fauna.

Figure 3.8 shows the soil microorganisms and their important ecological functions.

Bacteria and fungi account for 90% of degradation, while soil animals are responsible for only ca. 10%. However, soil animals are highly effective in accelerating microbial degradation and can increase the availability of nutrients by up to 50%. Field studies have shown that decreasing their numbers decreases the degradation performance of the soil by up to 40%. In general, single-cell organisms, worms, and anthropods each make up one-third of the soil biomass. The number of organisms per square meter in the upper 30 cm of the soil is as follows [3.17]:

Bacteria	60×10^{12}
Fungi	10^9
Algae	10^6
Single-cell organisms	500×10^6
Threadworms	10×10^6
Mites	150 000
Collembola	100 000
White ringworms	25 000
Earthworms	200
Snails	50
Spiders	50
Isopods	50
Millipedes	150
Centipedes	50
Beetles	100

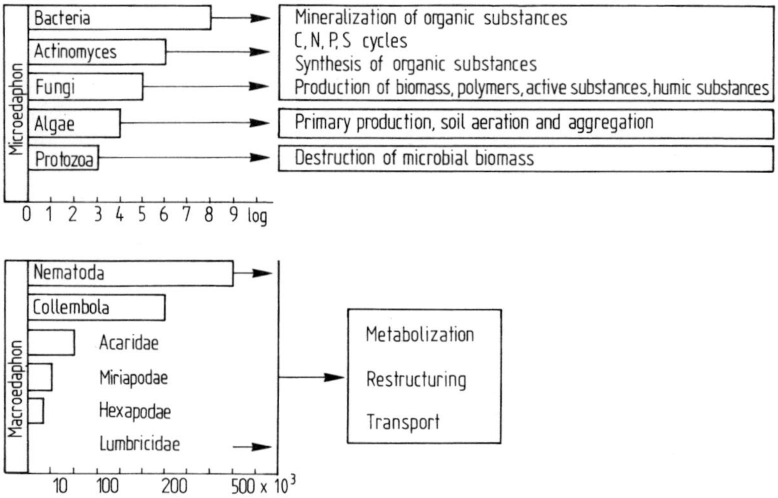

Figure 3.8. Soil microorganisms and their ecological functions

Fly larvae	200
Vertebrates	0.001

Respiration of an organism is a measure of its metabolic activity and, therefore, its material turnover. Protozoa account for 70% of respiration—considerably higher than their biomass fraction of about one-third. It is not general knowledge that these soil-inhabiting microfauna produce approximately as much biomass as earthworms. In the food chain, protozoa represent an ecologically important link between bacteria and small metazoans (e.g., rotifera and nematoda) and thus exhibit pronounced catalytic activity towards bacterial processes and remineralization processes in the soil.

Taxonomically, soil protozoans can be divided into three main groups: small flagellates of size 3–10 μm; 2 μm flat naked amoebas with diameter 10–300 μm and testate amoebas of size 20–200 μm; and ciliates (20–300 μm).

Their occurrence is mainly limited to the upper 10 cm of the soil. Due to their small size or flat, flexible form, flagellates and naked amoebas can enter even the smallest soil pores (≥ 2 μm). In terms of ecological function, biomass (Table 3.3), and distribution, the three groups of protozoans (flagellates, amoebas, and ciliates) can be regarded as being of equal importance.

The broad-spectrum activity of pollutants in soils, in particular their effects on entire colonies of organisms, has remained largely unregarded, partly due to the lack of suitable test systems. Classical laboratory tests (e.g., LC_{50} values) are unsuitable for predicting the true toxicity in open country. To make more accurate predictions for both existing and new substances, current laboratory tests, which are restricted to a few selected animal species, generally on standardized substrates, must be extended to further soil animal species, and additional tests on organisms, or-

Table 3.3. Distribution of "naked" protozoa in different soils and in the rhizosphere (numbers given: individuals/g dry matter [3.18] and, below, relative biomass fraction (%) calculated according to [3.19])

	Arable soil		Forest soil, subtropical	
	Uncultivated soil	Rhizosphere	Soil (total)	Rhizosphere
Flagellates	570–14000	1600–65000	22000–3127000	4000–8231000
	4–16	8–15	19–58	12–16
Naked amoebas	10500–50000	18000–181000	14800–823000	1100–30750000
	79–56	43–92	5–39	3–61
Ciliates	30–280	0–1750	10–132600	280–132600
	17–29	0–41	3–76	23–84

gans, single cells, and cell compartments are also required. Novel test methods based on enzymes as biomarkers avoid the need for animal tests and can also be used at sublethal concentrations. Although such tests provide important ecological data on the lethal and sublethal action of pollutants, they are not suitable for predicting effects on species colonies in ecosystems. This information can only be obtained by means of field experiments designed to take into account temporal, spatial, and genetic population variations in the ecosystem and the varying bioavailability of pollutants from species to species.

3.1.3. Productive Function

The productive function is the ability of the soil to act as substrate for cultivated plants (food and feed plants, renewable resources). Effects on the productive function are generally assessed in economic terms, rather than ecologically, since cultivation in agriculture, forestry, and horticulture is intended to produce a profit. Nevertheless, a number of ecologic soil functions are affected by cultivation.

3.1.3.1. Agriculture

Plants obtain their nutrient elements, but also pollutants, mainly from the soil. The supply of nutrients is generally typical for the type of soil and region. However, in modern agriculture external measures are applied to increase the harvest: inorganic and organic crop-protection agents, organic and mineral fertilizers, and utilization of organic and mineral wastes, all of which can lead to pollution of the soil and can damage soil functions.

Crop-Protection Agents. In Germany ca. 30 000 t/a of crop-protection agents (CPA) are used. About 250 approved agents are available. The entry of CPA into the soil as a result of agriculture is a potential influence on soil function. The CPAs are designed to alter the balance of nature in favor of the cultivated plants.

The behavior of CPA in soil depends on their properties, the way in which they are used, and on local conditions. The soil determines important properties of the location, especially the hydrological properties.

The amount of CPA applied, the frequency of application, and the time at which it is applied are relevant to movement of the agents in the soil and their potential entry into water systems. The manner in which a CPA is used affects its degradation, which is faster at the higher temperatures of spring and summer and leads to a more rapid decrease in the amount of CPA present. In autumn and winter degradation is generally slower. However, general statements valid for all CPAs cannot be made.

Fertilization and Waste Utilization. Fertilization and the utilization of waste materials can lead to the introduction of both organic and inorganic pollutants into the soil. In contrast to the application of crop-protection agents, this is an unintentional contamination. The aim of fertilization, which can include utilization of waste, is to improve conditions for the growth of cultivated plants, either by direct action on the plants themselves or by effecting the soil. The macronutrients nitrogen, phosphorus, and potassium are the most important for the nutrition of cultivated plants. Thus in the sustainable development of agriculture, both nutrient and pollutant aspects must be taken into account.

Nutrient Aspect. The high productivity in Germany and other countries of Central Europe, compared to average world harvests, is only attainable by intensive application of fertilizers. The concept of location-specific soil fertility formulated by Klapp [3.20], who defined this as the soil's own level of natural, persistent performance which distinguishes it from other soils, is not applied in conventional agriculture. According to Klapp the most important aspect is the sustainability: the long-term ability of the soil to provide all requirements for plants to flourish, that is, to provide constant or only slowly decreasing harvests without fertilization. In contrast, in modern agriculture, fertilization is regarded as guaranteeing the nutrient requirement of the plants, taking into account the nutrients already available in the soil. The natural productivity of the soil is thus completely disregarded. In modern agriculture, the target of fertilization to an extent appropriate to the requirements of the plants is not met. If a balance is determined for nitrogen, the result is a considerable fertilizer excess. The starting point for sustainable use of the soil should be a nutrient supply that is appropriate to the needs of the plants and which avoids an excess in the soil and release into other environmental compartments. Furthermore, in intensive agriculture, the proportion of nutrients introduced into organic materials should be restricted to 60% of the total nitrogen requirement. The reference points for fertilization planning are the exploitable components in the following plant population, which increase with increasing frequency of utilization and the amount of organic material utilized (see transformation function, pp. 623–627). Residual material remaining on the field after harvesting must also be included in the balance, and manure should be used in preference to external waste materials.

Pollutant Aspect. Mineral fertilizers, organic fertilizers, and wastes all introduce inorganic and organic pollutants into the soil and can cause adverse effects there and in other environmental compartments. Mineral fertilizers can contain various amounts of pollutants, depending on the origin of their raw materials and the production process used.

The main pollutants entering soils via fertilization are Cd and Cr, and to a lesser extent Pb, Ni, and As. The Mn, Cu, and Zn contents of fertilizers are, with the exception of the high Cu and Zn content of liquid pig manure, low, lying in the nutrient element/trace nutrient element range. Most of the heavy metal load comes from phosphate rock and P fertilizers derived therefrom (super and triple phosphate, partially digested and soft phosphate rock), Thomas phosphates, and phosphate-containing multinutrient fertilizers. P fertilizers derived from phosphate rock can exhibit high Cd contents, depending on the origin of the phosphate rock. Thomas phosphates have high Cr contents, resulting from the production process. The specific quantity of heavy metals introduced is strongly influenced by the various land-use and fertilization systems. According to BOYSEN [3.21], the Cd entry is up to 4.1 g/ha in conventional farming, up to 4.8 g/ha for meadowland, but only 0.9 g/ha for ecological agriculture. Data on organic pollutants in fertilizers are not yet available in sufficient quantity. In terms of quantity, the most important organic fertilizers are liquid manure, sewage sludge, and compost.

In Germany, only sewage sludge is subject to federal regulations which set limit values for heavy metals and some organic pollutants in sewage sludge and soil. In addition, the methods of application are specified, allowing theoretical permissible pollutant loads to be calculated. For compost, numerous quality marks exist, whereby the pollutant concentration is specified. They do not, however, set limits on quantity. If an average application of 10 $t\,ha^{-1}\,a^{-1}$ is assumed, a cadmium load of 10–40 g/ha can be calculated [3.22].

However, depending on the region and the spectrum of discharge or compost starting materials, these theoretical pollutant loads are not always fully exploited.

The inputs can be compared to theoretical outputs, as exemplified below for the Cd and Hg balance. For the average German wheat harvest of 5 $t\,ha^{-1}\,a^{-1}$ and a tolerable Cd content of 0.2 mg Cd/kg dry matter wheat (marketability criterion), a potential output of ca. 1 $g\,ha^{-1}\,a^{-1}$ results. The Hg output is 0.3 $g\,ha^{-1}\,a^{-1}$ (0.06 mg Hg/kg dry matter).

If the values specified e.g., by the Drinking Water Regulations (Trinkwasserverordnung), are taken as the tolerable pollution of the leachate, then for the average leachate quantity in Germany of 200 mm/a (2×10^6 $L\,ha^{-1}\,a^{-1}$), a potential Cd output of 10 $g\,ha^{-1}\,a^{-1}$ and an Hg output of 2 $g\,ha^{-1}\,a^{-1}$ result. Thus the potential outputs are already met by the potential inputs from the utilization of waste materials alone.

3.1.3.2. Forestry

After agriculture, forestry is the second largest use of land in Germany, in terms of area. In forestry, atmospheric nutrients and pollutants are of major importance.

The problem of pollution in forestry is exemplified by a comparison between Cd inputs and outputs (in $g\,ha^{-1}\,a^{-1}$) for a forest soil (spruce trees, ca. 120 years old) in a mountainous region of northwest Germany [3.23]:

Total input	6
Precipitation	2.7
Interception	3.3
Total output	15.3
Leachate	13.5
Crop (trunk wood without bark)	1.8

Thus the mean annual output is 2.5 times higher than the atmospheric input by precipitation and dry deposition. It is questionable whether the limit values of the German drinking water regulations are suitable quality standards for assessing outputs in the leachate. For a limit value of 0.005 mg Cd/L water, a leachate flow of 424 mm/a, and given the requirement that the leachate water quality must comply with the drinking water regulations, a tolerable Cd output of 21.2 $g\,ha^{-1}\,a^{-1}$ can be calculated. This is considerably higher than the actual leachate output of 13.5 $g ha^{-1} a^{-1}$. In analogy to the specified quality target that the output should not exceed the input, according to this model calculation, an additional Cd input of 9.3 $g ha^{-1} a^{-1}$ would be possible without exceeding limit values specified for drinking water.

The fact that for the material balance for cadmium in this example, the output is more than twice as high as the input shows that this soil has already lost its function as a filter medium for the formation of new groundwater. The mobilization of cadmium accumulated in the forest soil as a result of acidification demonstrates the problems that arise when the filter capacity of the soil is exhausted. The filtration capability of the soil can only be regarded as undisturbed if the output (at the background level) is lower than the current input. If changes in soil properties lead to release of accumulated materials, then pollution of the groundwater or of the habitat for organisms cannot be excluded.

3.1.4. Other Soil Functions

3.1.4.1. Utilization of Wastes

The exploitation of waste is possible in cases where it is both safe and useful. Benefits to the soil occur if at least one soil function is improved without permanent detriment to other natural soil functions or other environmental compartments. In the use of organic wastes in agriculture, the benefits can be assessed in terms of the fertilizer effect. Utilization of waste should be safe with regard to nutrients and pollutants. An important part of the recycling and waste business in Europe is the material utilization of waste. Although in Germany, material and energy recycling are of equal importance, there is increasing pressure to use wastes in soil-related applications.

In Germany, the distinction between mineral and organic wastes is defined in the regulation TA Siedlungsabfall, which excludes wastes containing > 5 % TOC (organic wastes) from dumps to minimize the pollution of the leachate associated with the degradation of organic materials.

Soils should be included under mineral wastes because the organic matter (up to 30 %) they contain differs strongly from organic wastes in type and constitution. They therefore represent a special case. Organic wastes are used in agriculture, gardening, and landscaping, whereby a distinction can be made between repeated application, mainly in agriculture, and one-time use, mainly in recultivation and landscaping. Mineral wastes are mainly used in landscaping and in recultivation and building projects; smaller amounts are utilized in agriculture. An exception is soil material, which is used in considerable amounts for improving soil functions.

An important criterion for assessing the harmlessness of such measures are the background concentrations of pollutants (see Section 3.1.2.1), which should not be exceeded by the input pollutant load of organic wastes. In the utilization of organic wastes, the pollutant load is the most important factor, since they remain almost entirely on the surface after mineralization of the organic component, contributing to an increase in the pollutant potential.

In the case of mineral wastes, there is generally no loss of mass after use in the soil, so that a different pollution assessment concept must be used.

The pollutant concentrations in the waste should not exceed the background concentrations in the location of use. If the values are compatible, application is possible, since a deterioration in soil quality with respect to pollutant concentration will not occur. For the application of mineral wastes that exceed the background values, eluate investigations must be undertaken, which indicate whether risks to groundwater quality are present.

The use of soil material in the topsoil region is subject to strict requirements for the soil properties. This is due to the requirements of plants for water retention capacity, aeration, rootability, warmability, and nutrient storage capacity. For this reason, the soil should have similar composition and properties to that at the site of use. Poorer soil materials should not be applied to soils of higher quality.

3.1.4.2. Raw Material Deposits

Soils that act as raw material deposits include sands for building and clays for ceramic production, but the largest areas are taken up by lignite mining. Other areas that generally require recultivation are the slagheaps from salt, ore, and coal mining; dumps; and military training areas. All of these lack a plant-supporting upper layer and have lost the natural soil functions.

Table 3.4. Breakdown of areas under mining law [3.24]

	Lignite mining districts, ha		
	Rhenish	Central German	Lausatia
Total area used	24 500	46 900	48 999
Area subjected to recultivation	15 600	22 000	24 655
Operated area	8 900	24 900	24 344

Table 3.4 compares the three main lignite mining areas in Germany.

Especially in the former German Democratic Republic, there are a large number of areas still to be recultivated. In the Land Brandenburg alone, as of 1994, the total area affected by open-cast mining of lignite was 50 000 ha. About half of this has so far been rendered reusable. The recultivation of devastated and mined areas is intended to restore or improve the natural soil function. This includes building up a plant-bearing soil layer in sufficient quality and depth. Other targets are:

1) Increasing the content of organic substances in the topsoil
2) Introduction of plant nutrients
3) Regulation of the pH value
4) Stimulation of soil life

At the same time, deterioration of soil function due to pollution input, addition of inappropriate nutrients, and compaction must be avoided.

3.2. Land Utilization and Soil Destruction (Statistics)

3.2.1. Introduction

The consumption and destruction of soils, i.e., the destruction of the surface soil or of the topsoil [3.25], is one of the greatest problems of the world. The soil is the basis for food production and will remain so in future [3.26]. While in the highly developed and industrialized countries problems of heavy-metal accumulations, soil acidification, and of contaminated sites (see Section 5.4) are discussed [3.27], [3.28], the soil erosion (Section 4.2), soil salination, and desertification in other parts of the earth progress unabatedly.

Many thousands of hectares of cultivated land are lost each year through erosion, salination, and desertification. Thus, in 1975 the soil of half of the Indian subcontinent, 150×10^6 ha, was seriously affected. In the United States in 1975, erosion had made nearly 40×10^6 ha unusable for agricultural cultivation [3.26]. In Argentina in 1980, 22×10^6 ha, nearly 16 % of the cultivated land, had been destroyed by wind erosion. In the 1970s – 1990s, according to estimates, the farmers lost 480×10^9 t of topsoil, which equals the entire topsoil of India [3.29]. These few examples should prove the seriousness of the problem of soil destruction.

3.2.2. Land Utilization

Table 3.5 shows a breakdown according to continents with regard to the land area and its use [3.30].

In the following, the land utilization and land use are considered in more detail for the Federal Republic of Germany, as an example of an industrialized country. Figure 3.9 shows the results of the land survey of 1989 according to the type of actual use in the Federal Republic of Germany. Settlement areas, which are composed of residential and office buildings, industrial plant and equipment, administrative and educational establishments, as well as sports fields and airfields, cemeteries and parks, in 1989 occupied about 12 % [3.31]. In contrast, in 1950, the proportion of residential, industrial, and traffic land was 7.2 %. This proportion of land has more than doubled in the last 40 years [3.32]. Figure 3.10 [3.33] shows the daily change in the years 1981 to 1985.

The settlement area in Germany increases by 100 ha daily [3.34].

Table 3.5. Land area and use (figures in 10^6 ha; source [3.39])

	Cropland	Pasture	Forest, woods
World	1475	3152	4082
Africa	184	782	695
North and Central America	274	362	660
South America	139	451	928
Asia	455	644	528
Europe	140	85	155
Former Soviet Union	232	373	929
Oceania	49	455	158

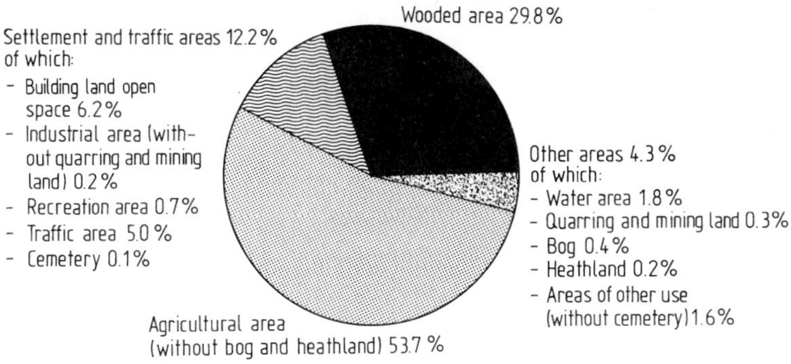

Figure 3.9. 'Land use in the Federal Republic of Germany (source: [3.31])

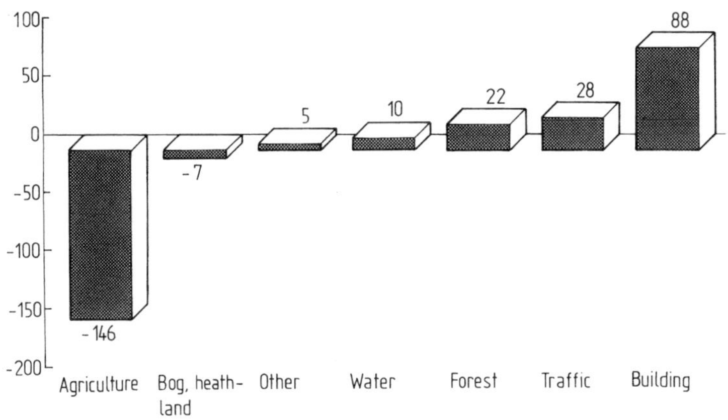

Figure 3.10. Daily change in land use from 1981 to 1985 in hectares (source: [3.33])

3.2.3. Soil Destruction

Proceeding on another scale than the sealing of land mentioned above are the destruction of soil by erosion, salination, desertification, and destruction of forests.

In the mostly extremely labile ecosystems in the nonindustrialized countries, where 80% of the world population growth occurs, industrial methods of agriculture lead much more quickly to the destruction of the valuable topsoil. In the subtropics and tropics, this leads rapidly, because of erosion, to the expansion of dry areas and desertification after a short phase of meager yields [3.35].

Salination. The agricultural lands rendered usable by artificial irrigation are threatened, after some time of use, by salination. Table 3.6 shows the large proportion of soil areas degraded by salination [3.36], [3.37].

Table 3.6. Soil area degraded by salination together with irrigated area under cultivation (source [3.36])

Country	Irrigated area under cultivation, 10^6 ha	Soil area degraded by salination	
		10^6 ha	% of irrigated area
India	55.6	20.0	36
China	47.1	7.0	15
United States	19.3	5.2	27
Pakistan	16.0	3.2	20
Territory of the former USSR	20.8	2.5	12
Total	158.8	37.9	24

Table 3.7. Estimated annual soil erosion in the catchment areas of great rivers (source [3.46])

River	Mouth	Size of the catchment area, 1000 km^2	Average annual suspended load, 10^6 t	Estimated annual soil erosion, t/ha
Niger	Gulf of Guinea	1114	5	0
Congo	Atlantic Ocean	4014	65	3
Nile	Mediterranean Sea	2978	111	8
Amazon	Atlantic Ocean	5776	363	13
Mekong	South China Sea	795	170	43
Irrawaddy	Bay of Bengal	430	299	139
Ganges	Bay of Bengal	1076	1455	270
Huang (Yellow River)	Yellow Sea	668	1600	479

Erosion. Erosion by wind or water plays the most important part in soil destruction. Worldwide, 25000×10^6 t of soil are washed away by water erosion every year [3.38]. The extent of erosion is shown in Table 3.7. The Huang (Yellow River), for example, washes away 479 t of soil annually from one hectare of land.

Chinese soil scientists estimate that in their country since the 1960s almost 10 % of the arable land has been lost through erosion and desertification [3.39]. In the United States in the last 200 years about one-third of the surface soil has been destroyed by erosion [3.40].

At the 7th Conference of the International Soil Conservation Organization in 1992 in Sydney, the following figures were mentioned from the project for the worldwide estimation of soil degradation: Between the 57th degree of latitude South and the 72nd degree of latitude North, 164×10^6 ha, i.e., 15% of the earth's surface, show more or less strong signs of degradation. In addition, 295×10^6 ha, corresponding approximately to the size of India, have to be classed as highly degraded. About 124×10^6 ha of these are in African and 106×10^6 ha in Asian countries [3.36].

Desertification. About 25% of the earth's surface is subject to the spread of deserts and every sixth human being is affected [3.41]. Every year, about 6×10^6 ha of pasture land become desert [3.42]. The deserts have in this way expanded by 120×10^6 ha since 1970. This corresponds to a greater land area than is cultivated in China today [3.43].

In the late 1970s the United Nations estimated the risk that nondesert land will become desert as follows:

50 % desertification of nondesert land in Africa
32 % desertification in Asia
19 % desertification in South America [3.44].

The Sahara alone expanded in the 50 years before 1977 by 650000 km^2 [3.41].

Forests. The felling of trees in forests leads indirectly to the destruction of soil. Through the felling of trees the soil is exposed to erosion by wind or water.

More than 5×10^6 ha of the tropical rain forest are wiped out each year to bring land into cultivation. Since 1970, farmers and ranchers in South America have converted more than 20×10^6 ha of wet tropical forests into cattle pasture [3.42].

Table 3.8. Estimates of world forest resources 1978 and 2000 (source [3.43])

	Closed forest, 10^6 ha		Growing stand, 10^6 m^3 overbark	
	1978	2000	1978	2000
Former Soviet Union	785	775	79	77
Europe	140	150	15	13
North America	470	464	58	55
Japan, Australia, New Zealand	69	68	4	4
Subtotal	1464	1457	156	149
South America	550	329	94	54
Africa	188	150	39	31
Asia and Pacific developing countries	361	181	38	19
Subtotal (developing countries)	1099	660	171	104
Total (world)	2563	2117	327	253

The worldwide forest lands for the years 1978 and 2000 (estimated) are given in Table 3.8 [3.43], according to which, in South America, Africa, Asia, and the Pacific area, the forest lands will decrease by almost one-half. The growing timber stand will decrease by about 23%.

3.2.4. Forecast

The degeneration of soils is difficult to determine quantitatively and the estimates are based on the numbers for past years. According to MEADOWS [3.45], urbanization in the less industrialized regions will increase (Fig. 3.11). In the "Third World", the town population has doubled every 20 years up to now. But the population in the towns is increasing in the industrialized countries also (Table 3.9) [3.46]. This development will lead to a further sealing of the soil.

According to a forecast of the Global 2000 Report on the changes of the vegetation and land resources of the earth, the desert areas will increase greatly while the closed forest area will decrease by 17% (Table 3.10) [3.44]. Since ever more land will have to be irrigated, the salinated areas will greatly increase in the future. If the forecast of BROWN and WOLF proves to be correct, 32% less soil will be available for each human being in the year 2000 than in the early 1980s [3.47].

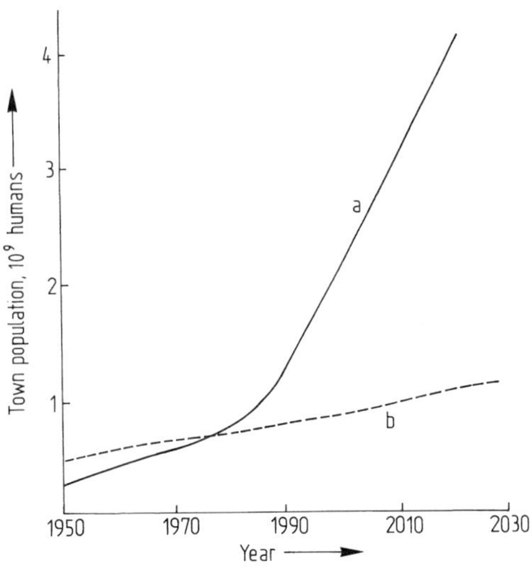

Figure 3.11. Urbanization (source: [3.45])
a) Less industrialized region; b) More industrialized region

Table 3.9. Cities in the study by region population in 1960 and 1980 and population growth 1960–1980 (source [3.44])

Cities by region	Population 1960	Population 1980	Population 1960–1980
Australia			+ 47.4%
Perth	475 398	898 918	+ 89.1%
Melbourne	1 984 815	2 722 817	+ 37.2%
Sydney	2 289 747	3 204 696	+ 40.0%
United States			+ 19.8%
Phoenix	663 510	1 509 052	+127.4%
Houston	1 430 394	2 905 353	+103.1%
Los Angeles	6 032 771	7 477 503	+ 23.8%
New York	16 834 500	17 925 200	+ 6.5%
Western Europe			+ 6.2%
Frankfurt	670 048	631 287	− 5.8%
Amsterdam	869 600	716 900	− 17.6%
Munich	1 046 000	1 298 941	+ 24.2%
Stockholm	1 271 014	1 528 200	+ 20.2%
Copenhagen	1 607 526	1 739 860	+ 8.2%
London	7 992 400	6 713 200	− 16.0%
Paris	8 400 000	10 094 000	+ 20.2%
Asia			+ 31.4%
Singapore	1 646 000	2 413 945	+ 46.7%
Hong Kong	3 129 648	4 986 560	+ 59.3%
Tokyo	9 683 802	11 597 211	+ 19.8%

Table 3.10. Projected changes in global vegetation and land resources 1975–2000 (source [3.43])

	1975, 10^6 ha	2000, 10^6 ha	Change, 10^6 ha	Change, %
Deserts	792	1284	+492	+62
Closed forest	2563	2117	−446	−17
Irrigated area	223	273	+ 50	+22
Irrigated area, damaged by salination and related problems	111.5	114.6	+ 3.1	+ 3
Arable land	1477	1539	+ 62	+ 4

4. Physical Damage to Soils

Soils play a central role in ecosystems, e.g., for plant growth or the storage of rainwater, since soils exchange matter and energy with the atmosphere, the biosphere, and the hydrosphere. This exchange depends on their physical structure, e.g., on their depth and the structure of their pore system. Physical damage therefore impairs all functions of soils in the ecosystem.

4.1. Compaction [4.1]–[4.3]

Process. Natural compaction is caused by the soil's own weight, which leads to higher bulk densities at greater depths. Menisci of the water in the soil pores excert forces on the pore wall. These forces increase strongly with the drying-out of the soil and therefore stabilize aggregates. The force excerted by the water menisci is 1.5 kPa when the soil has been dried to the point where plants begin to wilt irreversibly. Further natural compaction is caused by the burden of the vegetation cover, especially under woodland; and by the intercalation of illuvial material (clay, iron oxides, carbonates) in the course of pedogenesis (luvizol, podzol, gleyzol). With increasing compaction of the soil, its shear strength, its bearing capacity and the capillary rise height increase. Soils loosened by cultivation are therefore reconsolidated, e.g., by rolling, after sowing.

The compactibility of soils decreases with increasing content of sand, stones, and organic matter; grain roughness; increasing aggregation (from blocky via prismatic and angular to subangular aggregates); rooting; and preconsolidation.

Damage. Only above a certain limit does compaction harm plant growth. In what follows, this compaction is called *excessive compaction*. When a compaction becomes excessive depends, apart from the soil properties, greatly on the weather and thus changes in the course of the year and from year to year [4.4]. In dry seasons a fairly high soil density is advantageous, because of the faster capillary water supply. In wet seasons, on the other hand, a low density is advantageous, because the soil then contains more coarse pores, in which the excess water is drained away. Since the weather changes, it is also impossible to quote an optimum soil density. If the bulk density deviates from those shown in Figure 4.1, however, on average reduced yields have to be reckoned with under Central-European climatic conditions.

Excessive compaction harms plants in all crop stages, but especially in the early stages. Compacted soils warm up more slowly (increased thermal conductivity, increased evaporation) to the temperatures necessary for germination and N mineralization, and the young plants develop only a weak root system. In the later stages, root development in compacted soils is more shallow and less dense. This lowers the nutrient uptake and impedes root respiration. The susceptibility to root diseases increases and in leguminous plants fewer root nodules are formed. Even with sufficient water and nutrient

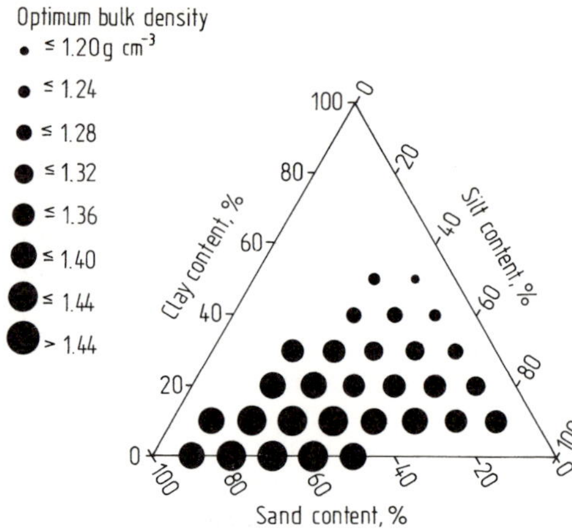

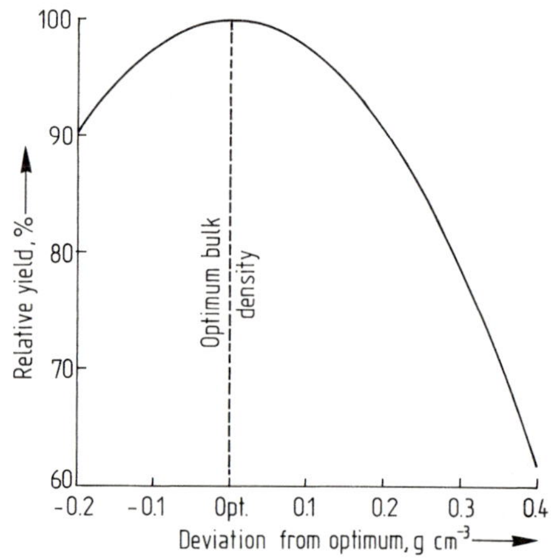

Figure 4.1. Optimum bulk density for soils of different texture (A); change in yield of grain, potatoes, and sugar beet on deviation from the optimum bulk density (according to [4.21], modified) (B)

supply, plants react to compaction with a modified hormone balance, which reduces their growth [4.5].

The damage due to compaction is difficult to quantify, since large differences occur over the years and even within one field. Estimates with grain crops assume mean yield reductions of 5–15%. Other crops, especially root crops, react more strongly than grains. In addition to the losses of yield, the harvest is more difficult (clods during the potato harvest). The products have a lower quality (more clinging dirt in sugar beet, deformations in carrots). The tractive force requirement and machine wear during soil cultivation rise. (If compaction causes only 5% more slip, the tire wear already increases by 20 %.) The soil compaction can persist many years, especially if the subsoil has been compacted. Up to 50 years persistence has been reported [4.6].

Excessive compaction does not occur only in mechanized agriculture but in all soil uses, also during forestry [4.6] and as the result of cattle tracks on pastures [4.7]. In modern agriculture the loads and the number of machine passes have risen, increasing the compaction risk. The loads rose because the yields increased through the intensification of agriculture and an increased stocking of pastures with cattle became possible. In addition, to reduce the working costs the operating widths were increased and larger tractors, which can move the greater loads and are more convenient, were used. Particularly high loads on the soil occur in lime and slurry spreading and in the harvesting of crops of high water content (sugar beet, potatoes, green maize). The number of passes increased, because a well-devised system of agriculture necessitates, among other things, intensive tillage, fertilization in many small doses and the regular application of pesticides and growth regulators. In particular the fungicides are frequently sprayed after rainstorms because of the high risk of infection at that time. But at that time also the bearing capacity of the soil is low. In grain cultivation, 10–15 passes are customary today and in the case of more intensive crops (especially, e.g., hops) considerably more. Five passes, which are necessary for the sowing of barley, already cover 91 % of the field with tire tracks [4.2]. Compaction is a problem particularly in perennial crops, such as hops and wine, and in fruit plantations, since the same tire tracks are always used and very many passes are necessary.

Loosening. Freezing and thawing, swelling and shrinking, soil animals (the edaphon), and roots loosen the soil naturally. Roots displace the soil mainly to the side (radial pressure: 0.5–1 MPa) and because of that compact it in addition, but create the vertical pores necessary for gas exchange and percolation. Earthworms also mainly displace and compact the soil. Because of their small axial force (0.1 MPa) they require zones of weakness (cracks), into which they penetrate in order to widen them. If the soil strength is higher than the radial forces that the earthworms can exert (0.2 MPa) [4.8], this strategy fails and the earthworms then loosen by ingesting compacted soil and transporting it to the soil's surface. Since this is exhausting, they loosen only slowly and only in cases when displacement is no longer possible but at the same time their other living conditions (e.g., availability of food) are advantageous.

Frequency and intensity of frost (up to 200 MPa at −20 °C) and swelling and shrinking (up to 2 MPa in Central Europe) decrease with increasing soil depth. In Central Europe the soil at a depth of 0.4 m freezes on average only 10 times per year. Since freezing and thawing or shrinkage must occur many times before a compaction is loosened, the compaction in the subsoil persists for a long time. It is therefore especially problematic there.

In the surface soil, which is loosened considerably faster by natural processes, humans also intervene with tillage. In doing so they create above all coarse pores between the clods, but cannot loosen the clods themselves. Natural processes are necessary for that. Cultivation of the topsoil loosens but also compacts it, since moldboard ploughing runs of the subsoil, the loosening destabilizes the soil and the forces necessary for the loosening are in turn supported by the soil.

Protection Measures. Excessive compaction can be avoided by larger supporting contact areas of agricultural machines (dual wheels, cage wheels, wide-section and ultra-low pressure tires, more axles, tracked vehicles). The contact area can be raised further by reducing the tire inflation pressure, the use of radial instead of cross-ply tires and a small tire profile. Since as a result of larger contact surfaces the compacted fraction of the surface increases, an improvement is achieved only if the contact pressure is decreased by more than 50 % [4.2]. The critical contact pressure is considered to be 60–80 kPa.

Above all, heavy loads should be avoided, since with heavy loads the compaction depends more on the load than on the contact pressure and penetrates deep into the soil. The critical wheel load is considered to be 30 kN [4.2]. Faster passing compacts less deep (e.g., at 1 km/h, 28 cm deep; at 12 km/h, 16 cm deep). However, the compaction at the soil surface rises as a result, since the shearing of the soil increases.

Because the optimum density to obtain high yields depends on the weather, a well settled soil with rapid capillary water supply is advantageous, if large vertical pores, e.g., earthworm channels, regulate excess water and gas exchange and a mulch cover simultaneously reduces evaporation losses. These conditions are ideally present in the case of the no-till cultivation system. In the no-till system tillage is replaced by chemical weeding with herbicides. The crop residues accumulate at the settled soil surface where they supply food for earthworms over a long period. Under this system the soils are also more stable to compaction because loads are well supported in the settled soil and the compaction of vertical coarse pores is more difficult than that of horizontal ones. Despite these advantages, the no-till system has hitherto been successful only in isolated cases because, among other reasons, the lack of working-in of crop residues causes problems with plant diseases which survive in the crop residues.

4.2. Erosion [4.9], [4.10]

Soil erosion is the loss of soil through the action of wind (wind erosion) or water (water erosion). The water can originate from rain, dew, or snowmelts. Transportable soil particles are initially detached from the assemblage and then transported away by water or wind.

Wind Erosion. Small aggregates and in particular loose sand particles are caught and dragged along by the wind and roll or saltate along the soil surface. On impact these particles loosen further particles that initially are still bound and accelerate them. As a result these particles also can be caught and transported by the wind. The amount detached and transported increases in the wind direction. The detached fines (clay, organic material) are transported very far as aerosols, while the sand, in rolling and jumping motion, is usually soon redeposited. Consequently a loss of fines, which are important for the soil's productivity, occurs while sand accumulates.

Particularly susceptible to erosion are fine sandy soils, because they are only weakly aggregated, and peaty soils which, because of their low specific gravity, can easily be blown away when they dry out at the surface. The high wind speeds necessary occur particularly in spring and autumn storms. In these seasons the fields are often inadequately covered with vegetation. Large fields without fairly high structures (hedges, ridges) at the edges favor high wind speeds at the soil surface.

Water Erosion. Transportable particles are detached by striking raindrops, air entrapment by rapid wetting, dispersion, and shear forces of the runoff. (When water is imbibed from both sides of a capillary pore, it compresses the entrapped air. Air pressure may be strong enough to disrupt the aggregate.) The detached particles are transported by lateral splash and above all by the water running off. Sheet erosion occurs if the detachment by the runoff is small. With increasing detachment the runoff cuts its way in. If this takes place at many points simultaneously, *rill erosion* occurs (depth usually <10 cm). If the runoff concentrates, the rills become fewer and deeper. It then comes to the point of *ephemeral gully erosion*. In this case the channels can still be regularly refilled by the tillage. Still deeper cutting-in (>30 cm) leads to *gully erosion*. *Tunnel erosion* occurs when a readily dispersible subsoil is washed out below a relatively stable topsoil, e.g., one interlaced with roots.

Water erosion is especially high for soils rich in silt and poor in organic matter, which have a low stability in water and easily form transportable particles. The erosivity of the rain increases approximately as the square of the rain intensity. For this reason summer thunderstorms with their high intensities (>10 mm/h) are particularly erosive. In early summer they impact—above all in the case of row crops such as maize, beet, potatoes, and soya—on a surface which still has only a sparse vegetation cover, and can cause large soil losses. With increasing slope and size of the *drainage* area (dependent on slope length and slope shape) and with decreasing roughness of the soil surface, the transportation force of the runoff rises. In the absence of cover, snowmelts likewise can lead to erosion, since the frost destabilizes the soil.

Extent. Natural ("geological") erosion is at a maximum in semi-arid to arid regions because of the scattered vegetation cover there. Rare but in some cases violent rains or strong winds can therefore trigger off considerable erosion. High erosion losses are natural in the mountains also, owing to the steep slopes and frequently unfavorable vegetation conditions. But erosion is principally a problem of farming, because this at times leaves soils bare, and wind and water can then attack. Wind erosion occurs above all near coasts, where high wind speeds and sandy soils are common. Leaving aside a few areas, it is less important than water erosion. The amounts of soil lost through wind in the United States are assumed to be less than 1/10 of those lost through water [4.11]. During farm use in moderate latitudes, soil erosion by water is frequently of the order of 10 t $ha^{-1} a^{-1}$, but soil losses of up to 200 t $ha^{-1} a^{-1}$ have also been observed over fairly long periods [4.12]. Probably the highest soil losses, of the order 1000 t $ha^{-1} a^{-1}$ [4.13], occur in South-East Asia (see Section 3.2.3) [4.14]. High soil losses are reported for other tropical regions also (Brazil, West and East Africa, and India) [4.10].

Damage. In wind erosion, saltating grains of sand, above all, damage the plants by exposing, covering, or by injuring the cuticle. In water erosion, rainfall flows unproductively away and cultivation is made difficult by uneven crop development, and by gullies and channels. In both forms of erosion, nutrients and pesticides are lacking at the eroded sites and overdosed at the deposition sites.

Aside from these short-term forms of damage, cumulative, irreversible, long-term damage occurs because the soil depth and the water- and nutrient-holding capacity decrease. This damage is particularly marked where unfavorable parent material, such as gravel, hard rock, or acid subsoil, is covered with a fertile but shallow material such as loess.

To the on-site damage to the eroded areas must be added the off-site damage at the points where the sediment and runoff accumulate. Examples are the soiling of roads and houses, the siltation of waters, and the input of nutrients, pesticides, and heavy metals into waters and other neighboring ecosystems.

In some cases erosion can also be advantageous. Thus the soil fertility rises if an impoverished or acidified topsoil is eroded and exposes more favorable material. In addition the input of soil into waters can reduce the concentration of dissolved nutrients by sorption if high nutrient concentrations are caused by point sources.

Erosion Protection [4.15]. A permanently covered soil surface offers the best protection against erosion. 30–50 % covering is usually sufficient, since a vegetation cover reduces erosion more than proportionate. The covering protects from drop impact, reduces wind speed and runoff velocity, raises the soil moisture at the surface, and promotes vertical pore formation by earthworms. The cover need not be present permanently, but only during the seasons of erosive rain or winds. In conventional farming the soil is least covered directly after sowing. In minimum or conservation tillage a sufficient cover can also be guaranteed after sowing.

Besides that, runoff and wind can be slowed down by a rough surface, which is obtained by cultivating transversely to the slope or to the main wind direction. A small-scale, varied land-use system in which areas with different degrees of erosion protection adjoin each other prevents that too much runoff or eroding saltation material is collected and that high wind velocities arise. This mutual protection of different crops is deliberately made use of in strip cropping, in which the individual crops are grown in strips of any length and a width of usually 10–50 m, running transversely to the slope or to the main wind direction. The runoff can be slowed down further by the laying-out of terraces, and the wind by wind breaks.

Models. Because of the great variability of erosion in time and space, many prediction models have been developed to estimate its extent and plan protection measures. Far more and better supported models exist for water erosion than for wind erosion (Table 4.1). The oldest, best known, and most used model of water erosion is the Universal Soil Loss Equation (USLE) [4.16], on which many more recent models are based. These models differ in the processes considered (wind, water erosion; sheet, gully erosion), in the time scale (single-event, long-term models), in the spatial scale (field, drainage area), and in the modeling technology (stochastic, deterministic). Table 4.1 assembles the most common models with their main applications. Many have been extended beyond this.

4.3. Building Over and Sealing [4.17], [4.18]

Building over and earth sealing extensively damage all soil functions; infiltration and gas exchange, for example, are impeded, temperature buffering is impaired, and plant growth is scarcely possible any more. Even a minor amount of sealing may have consequences. Thus a consolidation of arable land with asphalting of only 6% of the surface by field paths led to a twelvefold increase in the peak runoff [4.19]. Despite the higher runoff, the infiltration is in some cases greater under sealed areas (e.g., asphaltic surfacings) because of the greatly reduced evapo-

Table 4.1. Main areas of application for five common erosion models (special adaptations were not considered)[a]

Areas of application	Model				
	USLE/RUSLE[b] [4.16], [4.22]	CREAMS[c] [4.23]	ANSWERS[d] [4.24]	EPIC[e] [4.25]	WEPP[f] [4.26]
Time scale					
Long-term	•			•	•
Annual	o	•		o	•
Single event	o	•	•	o	•
Spatial scale					
Field	•	•		•	•
Drainage area	o		•		•
Applicability in the field	•				•
Soil use					
Agriculture	•	•	•	•	•
Rangeland	•				•
Forest	•				•
Construction sites	•				•
Erosion control structures	•		•		•
Types of erosion					
Gully erosion		•	•		
Ephemeral gullies				•	•
Sheet and rill	•	•	•	•	•
Snowmelt	o	•		•	•
Irrigation					•
Wind erosion				•	
Sedimentation		•	•	•	•
Nutrient loss		•		•	
Pesticide loss		•			
Water movement (runoff, infiltration, percolation)		•	•	•	•
Weather generator				•	
Soil temperature				•	
Plant growth				•	o
Economic evaluation				•	

[a] • = applicable; o = restricted applicability. [b] USLE = Universal Soil Loss Equation; RUSLE = Revised Universal Soil Loss Equation. [c] CREAMS = Chemicals, Runoff, and Erosion from Agricultural Management Systems. [d] ANSWERS = Areal Non-point Source Watershed Environment Response Simulation. [e] EPIC = Erosion Productivity Impact Calculator. [f] WEPP = Water Erosion Prediction Project.

ration. This impairs the microclimate in sealed areas (higher temperatures and lower atmospheric humidities by day, lower temperatures by night). Moreover, additional damage to the neighboring soils often emanates from sealed surfaces such as roads, owing to thawing salt, for example.

In contrast to compaction and erosion, sealing affects limited areas only, but there frequently leads to a complete loss of soil. In some industrialized countries more soil is being lost annually by development than by erosion on arable areas, although the area affected by development is only 0.2% of that affected by erosion.

Whereas compaction and erosion have natural causes also and are merely increased by humans, they alone are responsible for building over. Nevertheless, just this kind of damage is increasing more and more, especially in industrialized countries. There, the settlement area rises by ca. 1% per year, since the area needs of the individual increase and there is a trend to styles of residential construction that use up much soil area. In order to limit the damage or partly to cancel it, the following measures are necessary in case of existing or planned building over:

1) More compact styles of construction with smaller consumption of soil area per unit of useful area
2) Land recycling, i.e., redevelopment of formerly developed areas instead of allocation of new building land
3) Replacement of coverings that greatly impede mass transfer by more permeable coverings (Fig. 4.2)
4) Removal of sealing where it is unnecessary (e.g., central strip between the track grooves of roadways)

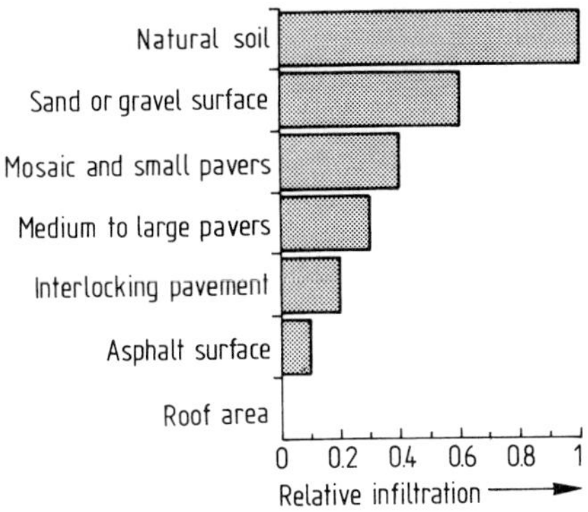

Figure 4.2. Relative infiltration of precipitation with different kinds of sealing (according to [4.17], modified)

As a rule these measures are also perceived as an esthetic gain (e.g., [4.20]).

5. Effects of Anthropogenic Substances on Soil

5.1. Persistent Pollutants [5.1]–[5.4]

The growing production and use of organic chemicals, their worldwide dissemination, and ignorance about the behavior of substances in the environment are some of the reasons why persistent organic substances can be detected in the environment, and not only in industrial conurbations. About 10×10^6 chemical substances are registered with the CAS. Day after day about 1500 new substances are registered, that is, 500 000 new substances are developed or registered every year. But of these 10×10^6 substances only about 100 000 are produced and traded industrially at present, and of those, according to information from the German Association of the Chemical Industry (Verband der Chemischen Industrie, VCI) about 5000 substances are produced in amounts of more than 10 t and 1080 in amounts of more than 1000 t/a [5.5]. By comparison with the number of substances produced and traded, the number of persistent substances known up to now is small.

5.1.1. Definition and Criteria

Persistent compounds are durable substances that are stable in the biosphere under natural climatic and soil conditions and also towards soil flora and fauna (the edaphon). Substances can be called persistent whose total degradation in the biosphere does not occur in one vegetation period. To be classified as nonpersistent half of the amount of a substance must be degraded within 1.5 months. Such criteria are also drawn up for plant-protection agents [5.6]. Ottow defines persistence as the time required for 95 % degradation [5.7]. Persistent compounds are frequently xenobiotic and are among the environmental chemicals.

5.1.2. Sources and Entry Paths

Chloroorganic compounds in particular represent the bulk of persistent compounds in the environment. At present, about 3×10^6 t of chloroorganic compounds are produced annually in Germany. Mainly low-boiling chlorinated hydrocarbons are emitted into the environment in noteworthy amounts, i.e., between 60 and 90 %. Pollutants are released not only from production sites but also during the consumption and application of products [5.8].

Pollutants can enter the soil via the *air path* (dry deposition); the *water path* (sewage water, precipitation, surface water, irrigation water, i.e., wet deposition); and via *organic solids*, such as sewage sludges, composts, organic residual substances, or agricultural inputs (soil-improvement agents, fertilizers, manure, plant-protection agents).

Air Path. The importance of the air path for the entry of organic compounds into soils is recognized more and more. For 1986 the total emission of volatile organic substances into the atmosphere originating from the Federal Republic of Germany was estimated at 2.45×10^6 t. According to a description of Schleyer et al. for the Federal Republic of Germany before 1990 (area 248 798 km^2), 7000 t/a benzene, 250 000 t/a toluene, and 250 000 t/a nonvolatile chlorinated hydrocarbons were released and gave calculated mean concentrations in the rainwater of 336 µg/L, 1200 µg/L, and 1200 µg/L, respectively [5.9].

Organic substances can enter the soil in gas form, dissolved, or particle-bound. The relative importance of individual kinds of deposition is determined by the substance properties, the particle supply, the kind of emission, and meteorological parameters. Substances with low vapor pressure, e.g., polychlorinated aromatic biphenyls (PCBs) are present in the air mainly in the gas phase and hardly at all in the dissolved aqueous phase [5.10]. Similar results have been obtained for polycyclic aromatic hydrocarbons (PAHs) and other chloroorganic compounds [5.11]. Dry deposition means the deposition from the atmosphere onto the earth's surface. Substances can be deposited from the gaseous or particle-bound phase.

Wet Deposition. Studies on wet deposition (i.e., entries as a result of rain, snow, mist and dew) have been performed in the case of low-boiling chlorinated hydrocarbons and aromatics [5.12]–[5.15]. The behavior of PAHs and PCBs in rainwater has also been studied [5.14], [5.16]–[5.18]. A survey can be found in [5.19]. CAPEL et al. [5.20] determined depositions of *n*-alkanes in Switzerland at concentrations of 150–170 $\mu g m^{-2} a^{-1}$ by mist and at 400–4000 $\mu g m^{-2} a^{-1}$ by rain and depositions of PAHs at 90–900 $\mu g m^{-2} a^{-1}$ by mist and at 100–600 $\mu g m^{-2} a^{-1}$ by rain. In the United States, PAH depositions of 350 $\mu g m^{-2} a^{-1}$ for urban locations and 100 $\mu g m^{-2} a^{-1}$ for rural areas have been determined.

Surface Waters. The pollution of soils by organic pollutants through infiltration of contaminated surface water (caused mainly by sewage waters) tends to decrease. However, the pollutants concentrations in surface waters are distinctly higher than the concentrations in rainwater, sometimes by a factor 10 or more [5.21]. When flooded and irrigated by surface water, soils come into contact with its constituents. Areas of flooding therefore often constitute endangered areas [5.22].

Sewage Water and Sewage Sludge. Some substances, e.g., surfactants, enter the soil environment exclusively via sewage water and sewage sludge. It is estimated, for example, that large amounts of linear alkyl benzene sulfonates remain in the sewage sludge [5.24]–[5.25]. Leakages in sewage water systems can also contribute significantly to soil pollution.

Within the framework of measures for groundwater enrichment, old shutdown irrigation fields are increasingly being reactivated and rainwater percolating areas (infiltration ponds) used. Studies on substance retaining powers of the upper active soil layers and on the contamination of the underground and groundwater showed that an accumulation of PAHs, PCBs, and dibenzodioxins/dibenzofurans in particular occurs in the mud zone of infiltration ponds [5.26]. In the sediment of such a pond, 15 μg/kg toluene, <10 μg/kg benzene, and <1 μg/kg trichloroethylene and tetrachloroethylene were determined. The PCB contents were up to 260 μg/kg and the PAH contents between 1 and 2 mg/kg d.m. (dry matter). Hexachloro- and pentachlorobenzene also were detected in traces.

The revision of the German TA Siedlungsabfall (Technical Instructions for Estate Waste) will lead to an increasing consideration of *organic solids*, such as sewage sludge, composts, organic residues, etc. In Germany, ca. 30 % of all municipal sewage sludges are spread in agriculture. The concentration of some pollutants in sewage sludges is therefore subject to limit values: thus, e.g., the dibenzodioxins and dibenzofurans to 100 ng TEQ (Toxicity Equivalents of the BGA)/kg d.m.; the PCB congeners 28, 52, 101, 138, 153, and 180 each to 0.2 mg/kg d.m.; and AOX (adsorbable organic chloro compounds) to 500 mg/kg d.m. The pollutant load of sewage sludges—on which limit values are imposed in the German regulation on sewage sludge (Klärschlammverordnung) [5.27]—is continuously decreasing since the 1970s. The extreme and mean values of different pollutants in more than eightynine different German municipal sewage sludges in 1989 are listed in Table 5.1 [5.28].

In the case of PAHs the proportions by mass were between 1 and 21.81 mg/kg d.m. for individual PAHs, with the exception of fluoranthene (0.26–9.38 mg/kg d.m.). Benzo[*a*]pyrene reached values between 0.07 and 5.16 mg/kg d.m. [5.28]. Contamination of the sewage sludges by surfactants (linear alkyl benzene sulfonates and nonylphenol ethoxylate) in some

Table 5.1. Extreme and mean values of various organic pollutants in sewage sludges (data in mg/kg d.m.) [5.28]

Compound	Minimum	Maximum	Mean value
Fluoranthene	0.26	9.38	1.495
Benzo[*b*]fluoranthene	0.13	4.07	0.665
Benzo[*k*]fluoranthene	0.05	1.97	0.36
Benzo[*a*]pyrene	0.07	5.16	0.57
Indano[1,2,3-*cd*]-pyrene	0.03	3.07	0.36
Benzo[*g,h*]perylene	0.03	2.50	0.36
Di(2-ethylhexyl) phthalate	15	314	110.9
AOX	42	4400	360.9
PAHs (total)	1	21.81	5.508

cases reaches the g/kg range and can strongly influence the behavior of other pollutants. The mineral oil hydrocarbons also reach similar orders of magnitude (g/kg d.m.).

5.1.3. Processes in Terrestrial Soils

Distribution. The processes in terrestrial ecosystems determine the distribution of chemicals in the soil. The distribution via the *air path* is influenced by climatic/meteorological factors (main wind direction, air temperature, frequency and intensity of precipitation, terrain topography, distance from emitter). The spread via the *water path* by seepage and infiltration is influenced by the intensity and frequency of the water input, the permeability of the soil, the properties of chemicals and the chemical properties of the soil. If an ecosystem is regarded as a compartment, the input and output of substances must be considered. Both processes are time-dependent. Between the two processes fixation, reaction, or mineralization occurs.

Sorption. Organic pollutants are subject in the soil to the same processes as, e.g., pesticides. After the pollutant enters the soil, it may be sorbed on solids (soil matrix) or remain in the gas phase or solution phase. Substances that have a higher density than air move downwards in the soil, while substances that are lighter than air can also be transported upwards.

The *bioavailability* is closely connected to the effect thresholds. The toxic threshold value is below the no observed effect level (NOEL, → Toxicology, Section 4.1). For estimating inputs over area and time and their effect on the ecosystem, the concept of critical load and critical level [5.29] is used. This means the pollutant loads that, for certain organisms in the soil and plants, still just cause no effects.

Organic pollutants are bound in the soil as a function of their water solubility, molecular structure, and functional groups. The soil is basically a multiphase system, with solid, liquid, and gas phase. The liquid phase can be broken down spatially into a hydrophobic and a hydrophilic phase and therefore considerably changed in its properties by dissolved organic substances, organic complexes, and suspended particles [5.30]. Contaminated soil water can contain miscible or immiscible organic phases of cosolvents that can considerably influence the sorption behavior of contaminants [5.31]–[5.34]. The soil sorption coefficients of chemicals decrease semilogarithmically with increasing concentration of the cosolvent in water. In addition to the type of bonding (van der Waals, hydrophobic, hydrogen-, charge-transfer, complex, and ionic bonding, induced ion–dipole or dipole–dipole interactions, see also [5.35]), the soil properties (contents of humus, clay, oxides, and hydroxides, the pH value, redox potential, and charge [5.36]), and the properties of the chemicals (water solubility, molecular size and structure, lipophilicity, acidity, polarity, vapor pressure, [5.37]) are also important. The sorbents contained in the soil have very different properties. The organic matter is an important sorbent for all kinds of organic compounds, e.g., cationic or anionic, polar or nonpolar, because it can interact with the functional groups contained in these substances. The degree of decomposition of the organic matter has to be taken into account. Thus, the sorption capacity increases in the sequence mor, moder, and mull. Oxides and hydroxides can be further important sorbents, especially for anionic organic compounds. Clay minerals, because of their negative surface charge, are capable of binding cationic compounds as well as negatively charged compounds, such as basic nitrophenols and anilines after protonation [5.38]. The bonding relationships are mainly determined in batch experiments. However, more realistic predictions on sorption are obtained in leaching tests. Sorption constants are usually smaller in leaching tests since the penetration into the soil particle is less. The sorption is also time dependent, organic compounds are initially bound at the bonding sites at the surface of clay minerals and humus. Transposition and incorporation into the sorbents can occur gradually. Only in this way can it be explained that bound residues can scarcely be desorbed from the soil matrix with aqueous solution [5.39], [5.40].

Elimination here means the sum of the chemical, physical, and biological degradation processes in the soil, including volatilization and loss because of uptake of the substances by plants and seepage into the groundwater (nonextractable substance residues are often considered being part of the eliminated portion). The first degradation step is often a *hydrolysis*, which can proceed only in the presence of water. The rate of hydrolysis is influenced by the pH value and by the presence of metal cations. Hydrolysis is especially important in inputs via the air (precipitation) and surface waters. In the soil, *reductive and oxidative processes* can lead to chemical alteration of substances. Oxidative reactions occur with "reactive oxygen species" or molecular oxygen (autoxidations) in soil and bodies of water. Abiotic reductions possibly occur in the degradation of pentachloronitrobenzene under anaerobic conditions to pentachloroaniline [5.41].

The *physical degradation* is based on photochemical and photophysical processes, which are mainly restricted to the soil surface. Dechlorination reactions of chlorohydrocarbons are typical photoreactions. The photomineralization of adsorbed organic pollutants is based on the absorption of UV light of fairly long wavelengths (>290 nm), as is assumed for pentachlorophenol (PCP) [5.42]. Dibenzodioxins and -furans were 90% dechlorinated within a week in soils [5.43].

Organic compounds are mainly degraded biologically (metabolic or cometabolic). Chemical and physical degradation plays a part with substances that are degradable with difficulty, for example, highly chlorinated dibenzodioxin congeners (octa- and heptachlorodibenzodioxins).

The persistence of a substance depends on its physicochemical properties and the soil properties, including the soil's microbiological potential. According to Ottow [5.7], the degradation of near-natural compounds is mainly metabolic and that of xenobiotics mainly cometabolic. Persistent substances usually have unphysiological, chemically stable groups. These include, e.g., $C-Cl$, $C-SO_3H$, $C-NO_2$ groups, trisubstituted N atoms, quaternary carbon atoms, ether bridges, polychlorinated compounds, chlorine substituents in benzene rings in the *ortho* or *meta* positions, pyridines, and long-chain hydrocarbons. Furthermore, these substances are persistent because of their low water solubility and bioavailability and large molecular size. They are mainly cometabolic and frequently only to very stable metabolites ("dead-end" metabolites).

In soils that contain various persistent organic pollutants, the degradation may be reduced. In saturated, anaerobic soils, other microorganisms are usually necessary for the degradation of a substance than under aerobic conditions. Many chloroorganic substances, for instance, are degraded more easily by reductive dechlorination under anaerobic than under aerobic conditions [5.44]. Alternately moist and dry conditions therefore increase the degradation efficiency of soils. The oxygen supply of the soil influences the activity of aerobic microorganisms that not only use oxygen as an energy donor but bring about hydroxylations of aliphatics and aromatic compounds by means of microbial oxygenases [5.45]–[5.47]. The degradation rate usually decreases with increasing soil depth, but a heterogeneous distribution of the organism density can lead to quite different degradation rates [5.48], as has been shown in the cases of methanol, phenol, and *tert*-butanol. The degradation is in general distinctly slower in winter because of lower temperatures [5.23].

Volatility. The transfer of substances from the soil into the atmosphere is known as volatilization or volatility. For aqueous solutions Henry's law constant describes the ratio of the concentration of a substance in the gas phase to its concentration in the solution phase. The volatility of highly volatile adsorbed substances is liquid-phase controlled and that of low-volatility substances is gas-phase controlled. In addition, the volatility is strongly influenced by the water solubility of the substance, the vapor pressure, the temperature, the soil moisture, and the humus content of the soil. Substances can as a rule be more easily released from moist soils, because they are more easily desorbed from the aqueous phase. The soil water can moreover make an important contribution to the evaporation of substances at the soil surface that are delivered subsequently from deeper soil layers. Diffusion plays a certain part in the subsequent release at the soil surface and therefore in the volatility of chemicals. Even low-volatility compounds, such as dibenzodioxins/dibenzofurans, with extremely persistent properties can be volatilized into the atmosphere [5.49].

Uptake by Plants. The uptake of organic pollutants in plants is determined by their water solubility, molecular mass, and lipophilicity, as well as their concentration in the soil and the soil properties. The uptake can occur via the root with the transpiration stream; as a result of uptake via the leaves (dry and wet deposition); and by dust particles, degassing and spraying of the soil by the impact of rain.

5.1.4. Persistent Organic Compounds

The persistent organic substances also include organochlorine compounds, such as hexachlorocyclohexane (HCH) hexachlorobenzene (HCB), DDT, and 1,1-dichloroethenylidenebis-(4-chlorobenzene) (DDE), that have been used as pesticides (→ Insect Control, **A 14**, pp. 278–284). These substances and those dealt with below have in some cases been detected worldwide in various environmental media (water, sediment, soil, organisms, and the atmosphere). Their contents in environmental media decrease only slightly, even over fairly long periods.

5.1.4.1. Chlorinated Aliphatic Hydrocarbons

Some haloalkanes (used as organic solvents) behave as persistent compounds. The low-boiling chlorinated hydrocarbons occur ubiquitously, they are partly of biological origin. Algae in particular are capable of forming halogen compounds. It is assumed that ca. 10000 t/a of organohalogen compounds are emitted into the atmosphere worldwide from macroalgae [5.50]. In 1987 about 250000 t/a of low-boiling chlorinated hydrocarbons of industrial origin were consumed in the Federal Republic of Germany and at the same time partly emitted into the environment [5.51]. Relative to the average rainfall of 837 mm/a, this would yield a hypothetical mean concentration of 1.2 g per milliliter rain water, i.e., at uniform deposition ca. 1 g per square meter soil surface [5.9].

The most important low-boiling chlorinated hydrocarbons are dichloromethane, 1,1,1-trichloroethane, trichloroethylene, and tetrachloroethylene. Some of the low-boiling chlorinated hydrocarbons, e.g., tetrachloroethylene, have a carcinogenic potential. They reach the environment via the air and via the sewage water path, from contaminated sites, and by accidents. Some of the low-boiling chlorinated hydrocarbons are quite stable in the atmosphere (e.g., 1,1,1-trichloroethane has a tropospheric half-life of 63 a [5.52]) and are photooxidized (e.g., tetrachloroethylene to trichloroacetyl chloride). After hydrolysis, the water-soluble, low-volatility trichloroacetic acid [5.53] is formed in amounts of ca. 6 $g\,d^{-1}\,km^{-2}$, which yields an estimated deposition of 4 $g\,ha^{-1}\,a^{-1}$. SCHLEYER et al. [5.9] estimate depositions of 23 $g\,ha^{-1}\,a^{-1}$ (in a spruce stand) and 6 $g\,ha^{-1}\,a^{-1}$ (in open land) (Table 5.2) [5.9]. Trichloroacetic acid is also formed in the photochemical degradation of trichloroethylene. Carbon tetrachloride is especially stable in the atmosphere (half-life 60 a). Many low-boiling chlorinated hydrocarbons accumulate in soils relative to the atmosphere (see Table 5.3, e.g., $CHCl_3$, CH_3CCl_3, C_2Cl_4), but their outgassing as well as their metabolites also have to be reckoned with [5.54].

Table 5.2. Mean concentrations of selected organic parameters in rainwater calculated from four measuring stations and deposition rates determined from them; basis of calculation is a mean rainfall quantity in Germany of 837 mm/a [5.9]

Substance	Spruce stand		Open land	
	Mean concentration, ng/L	Deposition rate, $g\,ha^{-1}\,a^{-1}$	Mean concentration, ng/L	Deposition rate, $g\,ha^{-1}\,a^{-1}$
AOX	29000	243	4700	39
Chloroform	56	0.5	14	0.1
Trichloroacetic acid	2800	23	730	6.1
Toluene	47	0.4	48	0.4
γ-HCH (lindane)	36	0.3	39	0.3
Dibutyl phthalate	310	2.6	265	2.2
Di(2-ethylhexyl) phthalate	790	6.6	580	4.9

Table 5.3. Contents of various low-boiling chlorinated hydrocarbons (LBCHs) in soil atmosphere and in the atmosphere of a forest in the Black Forest in 1989 (means of six measurements) [5.15]

LBCH	Soil atmosphere, µg/m³	Atmosphere, µg/m³
$CHCl_3$	41.8	0.167
CH_3CCl_3	2.27	1.3
CCl_4	0.5	0.66
$CHClCCl_2$	0.23	0.58
C_2Cl_4	2.98	1.85

The persistent low-boiling chlorinated hydrocarbons, the half-lives of which in soils are more than 5 months, include in particular trichloroethylene, dichloromethane, trichloromethane, carbon tetrachloride, and vinyl chloride. Under *water-saturated conditions* the dominant degradation processes are reductive dehalogenation (e.g., with tetrachloroethylene), dehydrohalogenation (e.g., hexachloroethane to 1,1,1-tetrachloroethane), and hydrolytic dehydrohalogenation (thiolysis of the C–Cl bond in dichloromethane). In *anaerobic media* mainly higher halogenated alkanes are dehalogenated. Reductive dehalogenation depends on the redox potential of the reactants and requires an anaerobic medium [5.55]. *cis*-1,2-Dichloroethylene and vinyl chloride frequently accumulate as degradation products in groundwater, because of the slow transformation of vinyl chloride to ethylene. 1,1,1-Trichloroethane proves to be extremely persistent in water-saturated soils (half-life in the groundwater 2.8–19 a [5.56]). The assumption of slow degradation in water-saturated soil is moreover supported for tri- and dichloromethane [5.57], [5.58].

Under *aerobic conditions* haloalkanes are oxidatively dehalogenated by monooxygenases. In the case of C_1 haloalkanes, the corresponding aldehydes and carboxylic acids are formed successively. Mono- and dihaloalkanes in particular are degraded more rapidly under aerobic than under anaerobic conditions. The degradation of carbon tetrachloride is quite slow, even under aerobic conditions. During degradation by the various oxygenases the metabolite trichloroethylene epoxide, as a cell poison, inhibits the further elimination of trichloroethylene and trichloromethane [5.59].

The degradation should not lead to toxic "dead-end" metabolites such as *cis*-1,2-dichloroethylene or vinyl chloride. In the degradation of dichloromethane, which is regarded as easily degradable, the presence of 1,2-dichloroethane causes a competitive inhibition of the dichloromethane-degrading enzymes [5.60]. Dichloromethane is degraded more strongly (faster) in presence of acetate than in its absence [5.61]. The presence of formate also had a stimulating effect [5.62]. Formate can thus be advantageously used in in-situ remediations. Low-boiling chlorinated hydrocarbons are weakly bound in the soil, because of their high water solubility (up to 20 g/L) and high vapor pressure. The main sorbent is the organic matter of the soil. The Freundlich sorption constants K (K_{OC} relative to the humus content) vary from <1 to 300. The low binding by the soil also explains the large number of nearly 3000 cases of groundwater pollution by chlorinated hydrocarbons reported to the German water authorities in 1990 [5.63]. Remediation measures that enhance the bioavailability of chlorinated hydrocarbons by increasing the desorption from the soil matrix, e.g., by surface active substances, should only be used if no additional risk arises to soil and groundwater.

5.1.4.2. Aromatics and Chlorinated Aromatics

Aromatic hydrocarbons are among the most widespread organic compounds [5.64]. Thus, e.g., in the Federal Republic of Germany in 1985 alone, ca. 76.6×10^6 t of crude oil were distilled and subsequently processed to energy carriers and fuels. The motor vehicle fuels produced contained ca. 680000 t of benzene, 2.4×10^6 t of toluene, and ca. 3.1×10^6 t of xylenes. Simultaneously, 1.6×10^6 t of pure benzene, 360000 t of toluene, and ca. 470000 t of xylenes were produced for further chemical processing [5.65]. According to an estimate of the Umweltbundesamt (UBA, Federal Environmental Agency) for 1982, the total emisson of benzene from motor vehicles, heating plants, and industrial production was ca. 60×10^6 t/a, but distinctly smaller amounts can be assumed today. At the end of the 1980s, the 70000 t/a of benzene and 250000 t/a of toluene were emitted into the environment by motor vehicle traffic. Despite the enormous amounts introduced into the environment, the benzene, toluene, xylene (BTX) aromatics can only be detected in small concentrations (0.1–1 μg/L) in water bodies. The concentrations in municipal sewage water are greater by a factor ten and industrial wastewaters can be loaded with 1–10 g/L. More than 90% of the BTX is eliminated in sewage treatment plants. Annually, 3.5 t benzene are said to be introduced to the environment via sewage sludges and ca. 80 t into bodies of water [5.66]. The benzene contents of the air in urban zones are between 5000–50000 ng/m^3, in rural zones 500–5000 ng/m^3, and in remote areas <2000 ng/m^3. In the case of toluene, the pollution is 5000–10000 ng/m^3 for urban zones, 1000–7000 ng/m^3 for rural areas, and ca. 750 ng/m^3 for remote areas. The concentration of the various xylene isomers varies from 800–70000 ng/m^3 and from 200–1200 ng/m^3, no data being available for areas remote from sources [5.9], [5.67]. The ethylbenzene contents are also relatively high. In precipitation waters the pollution due to aromatics is higher by a factor of ca. 1000, depending on the distance from emitters. Thus, e.g., <220 ng/L toluene, <260 ng/L xylene (*m*-,*p*-), and up to 72 ng/L ethylbenzene are found in rainwaters [5.68]. Rain loadings of a similar order were determined in the Rhine–Main area and in the hills of Hesse, the concentrations being distinctly higher under spruce stands than in open land. This is a result of the effect of trees in "mopping up" air pollution. In both cases the loads were higher in winter than in summer.

No representative data are available for background concentrations in soils for BTX, but data exist for higher-chlorinated benzenes and phenols [5.22]: an average of 0.1 µg/kg d.m. was measured for pentachlorobenzene and 1–2 µg/kg d.m. for hexachlorobenzene.

In Switzerland the background values vary between 0.15 and 0.5 µg/kg d.m. for hexachlorobenzene [5.69]. According to studies of [5.70] and [5.71], the pentachloro- and hexachlorobenzene contents in the humus cover of forest sites (6.5–11.5 µg/kg d.m.) are distinctly higher than those of the mineral horizons (1–2 µg/kg d.m.) or of sites on open land (< 5 µg/kg [5.72]). In forest soils near Linz [5.73], the hexachloro- and pentachlorobenzene contents were 0.38–0.4 µg/kg d.m. and in the humus cover 0.147 µg/kg d.m.

In the soils themselves, benzene, toluene, and xylene are only moderately bound. K_{OC} values are < 100 for benzene, < 500 for toluene, and < 1400 for xylene (Table 5.4). The aromatics are bound mainly to the organic matter. The aerobic degradation of BTX in the soil is very rapid (half-lives 2–16 d [5.74]). In water-saturated soils, on the other hand, these substances prove to be more stable; apparently volatilization often makes the main contribution to elimination [5.75]. The biradical oxygen molecule oxidizes BTX aromatics to phenols. Subsequently the ring opens. This process is inhibited under anaerobic conditions and only occurs at a sufficient rate if the ring is substituted by carboxyl groups. Thus, a strong benzene contamination in anaerobic underground was not degraded even after 4 a [5.76]. Strong persistence and small degradation under anaerobic aquifer conditions was also established for aromatics [5.77], [5.54]. This behavior is strictly related to the lack of molecular oxygen. After dissolved oxygen was supplied, a rapid degradation occurred [5.78]. On the other hand, a degradation could occur when ferulic acid (4-hydroxy-3-methoxycinnamic acid) [5.79] was present or iron reduction was possible.

Table 5.4. Sorption of aromatics and haloaromatics to organic matter [5.33], [5.53]

Substance	K_{OC}	Substance	K_{OC}
Benzene	60–100	Phenol	< 100
Chlorobenzene	210–390	Chlorophenol	100–300
Dichlorobenzene	230–970	Dichlorophenol	300–700
Trichlorobenzene	4300	Trichlorophenol	100–570
Tetrachlorobenzene	85 000	Tetrachlorophenol	150–750
Pentachlorobenzene	11 400	Pentachlorophenol	300–950
Hexachlorobenzene	3900–11 800	Toluene	300–510
		p-Xylene	400–1400
Ethylbenzene	300–600		

Before aromatics are illuviated into the soil, they can evaporate into the air. Owing to the high vapor pressure of BTX, high volatilization rates are to be expected. In the atmosphere, benzene, for example, undergoes a photochemical degradation by OH radicals, NO_x, and ozone (→Air, Chap. 8).

The volatility of aromatics is used especially in the remediation of contaminated areas. Volatility depends greatly on the porosity and the moisture of the soil. The relative humidity (rH) at a certain water content of the soil determines the composition of the vapor phase of BTX, etc. very strongly. A stepwise increase of the relative humidity from 0–20% produced a reduction of the binding of *p*-xylene to kaolinite [5.37]. At 67 rH only 50% of the xylene remained bound. Compounds such as BTX with Henry's law constants of $> 10^{-3}$ bar mol^{-1} m^{-3} can be removed from the soil and groundwater by stripping [5.80], they can also be reduced relatively rapidly in bioreactors with 100% turnover. Under field conditions the remediation measure was distinctly slower.

Chlorinated Aromatics. Chloro-substituted aromatics behave in the environment, especially in terrestrial ecosystems, distinctly differently from the pure aromatics. The sorption in the soil increases with increasing degree of chlorination. The solubility of the BTX and chlorinated aromatics in the soil can be raised by addition of vegetable oils or by biosurfactants formed by microorganisms, so that a more efficient remediation is possible via the water phase [5.81]. But the degradation can also be reduced if the toxic threshold concentration for microorganisms is exceeded. It has been reported that biopolymer-bound trichlorophenol was stable to degradation, and a negative effect of the surfactant on the microorganisms also being possible [5.82].

5.1.4.3. Nitroaromatics and Aromatic Amines

Nitroaromatics play a notable role in regard to the amount of inputs from the atmosphere.

They are formed from BTX after photochemical reaction with NO_2 and OH radicals in the troposphere. Through rainfall nitroaromatics enter soil, leachates, and spring waters in low concentrations over a wide area [5.9]. As intermediate and end products in the production of plant-protection agents, pharmaceuticals, plastics, dyes, and explosives, they reach the environment in the course of their processing and use. Most important are the various nitrophenols, nitrotoluenes, nitroanilines, and pyridines, as well as chlorine-derivatives of these compounds. It is estimated for Germany that ca. 20 000 t of nitrophenols are emitted annually into the atmosphere or formed there [5.83]. Aromatic amines such as chloroanilines are determined in surface water in concentrations of 1–5 μg/L but also distinctly higher values are found [5.84].

Because of their considerable toxicity to microorganisms and humans and their mutagenic potential, which is in some cases considerable, nitroaromatics and aromatic amines are described in more detail here. In the the soil aromatic amines and nitrocompounds get into contact with the clay minerals and the organic matter via their nitro and amino groups. Anilines react as bases and are often present in protonated form, depending on the pH value. With decreasing pH value, increasing sorption to the mineral matter occurs because protonated amino groups are more strongly bound than the free bases. Anilines can form bonds with exchangeable protons of silicate surfaces or positively charged hydroxide surfaces. They react with hydrated exchangeable cations of silicates and can be indirectly bound via hydrogen bridges. They can also be coordinately bound by means of electron transfer [5.85]. In soils, aromatic amines are oxidized by free radicals, e.g., at the amino group. The reaction is catalyzed by clay minerals. Aromatic amines can form complex compounds with metals. At fairly high pH, pyridines and other compounds occur as neutral species and can then also bind to the organic matter. Ionic bonding to the soil matrix occurs especially when the pH of the soil is near the pK_a of the amine [5.86]. Pyridines can also be hydrophobically bound [5.87]. With few exceptions, the binding of aromatic amines and nitro compounds to the organic matter is only moderately strong (see Table 5.5). Pyridines [5.87], [5.89] and also nitro compounds [5.90] are therefore repeatedly found in mg/L quantities in the groundwater. Particularly in acid soils nitrobenzenes are reduced to anilines, by sulfides, especially by Na_2S in anaerobic soils [5.36].

The nitrotoluene concentrations at a blasting site of a former explosives factory were in the mg/kg range and led to the closing of a drinking water well [5.91]. As follows from subsequent analyses, the concentrations clearly decreased (Table 5.6). Under anaerobic conditions a distinctly slower degradation occurs. A trinitrotoluene contamination of groundwater near Grand Island, Nebraska was also reported [5.90].

Table 5.5. Degradation sorption and volatilization in % of pyridine (-derivatives) in soil suspensions after 30 d [5.88]

Pyridine (-derivatives)	Position	Bound	Degraded	Volatilized	Loss*
Pyridine		15.6	6.3	16	100
-OH	2,3-	2.4	<1		4
-COOH	2,3-	3.0	81		100
-CCl	2,3-	15.1	<1	50	79
-NH_2	2,3-	14.3	<1		4
-CH_3	2,6-	7.7	<1	23	29

* Difference indicates the recovery rate.

Table 5.6. Concentrations of nitrotoluenes in the surface soil of a blasting site near Torgau in μg/kg d.m. in 1992 as a function of time (data by the Regional Environmental Agency, Torgau)

	January	February	March	April	May	August	October
2,4-DNT[a]	390	18	n.d.*	96	n.d.*	<5	n.d.*
2,6-DNT[a]	360	8	<5	26	<5	7	n.d.*
TNT[b]	2860	156	31	260	10	8	n.d.*
2,4-Dinitro-6-aminotoluene	1220	38	33	560	<5	7	n.d.*
2,4-DN-4-AT[c]	2240	82	46	920	<5	8	n.d.*
Cyclonite	2700	78	56	210	8	24	n.d.*

[a] DNT = dinitrotoluene. [b] TNT = trinitrotoluene. [c] DN-4-AT = dinitro-4-aminotoluene. * n.d. = not detected.

The degradation of 2-nitroaniline and 4-nitrophenol was studied [5.92]. The bacterium *Nocardia* sp. can degrade these substances to 4-hydroxy-2-nitroaniline and nitrocatechol. The splitting-off of a nitro or amino group associated with a hydroxylation has been observed with 2-nitrophenol, 4-chloroaniline, and 4-chloro-2-nitrophenol. An extremely slow degradation of chloroanilines by reductive dehalogenation by microorganisms under anaerobic aquifer conditions was observed [5.93]. Under methanogenic conditions the degradation rates were slightly greater ($K < 0.01-1.36$ after 8 months) than under sulfate-reducing conditions ($K < 0.01$). The only exception was provided by 2,3,4-trichloroaniline, which was completely transformed after 10 months. Under some circumstances, pyridines are rapidly degraded; thus, e.g., pyridine itself is eliminated within 7 d [5.86]. However, also very persistent pyridine derivatives exist (Table 5.5), that frequently are also rather volatile. The differences between aerobic and anaerobic degradation are small. The persistence of the pyridine derivatives decreases in the following order: pyridinecarboxylic acid < pyridine = monohydroxypyridine < methylpyridines < aminopyridines = chloropyridines [5.88].

5.1.4.4. Polychlorinated Dibenzodioxins (PCDDs) and Dibenzofurans (PCDFs)

Dibenzodioxins and dibenzofurans are classes of aromatic ethers, with totals of 75 and 135 congeners, respectively. There are congeners substituted with 1 to 8 chlorine atoms (mono- to octachlorinated), which can also differ in the position of the chlorine atoms in the organic rings. Chlorine can be replaced by bromine, some substances can contain chlorine and bromine, so that finally 5020 different compounds of this type are possible. The toxicity of the dioxins and furans is rated by quoting a toxicity equivalent (TEQ) according to the BGA [5.94].

Chlorinated and/or brominated dibenzodioxins and dibenzofurans reach the environment mainly through thermal reactions (motor traffic, refuse incineration plants, industrial incinerators and blast furnaces, domestic heating systems, as byproducts of syntheses), etc. [5.95]. Also flame retardants that consist of bromine-containing phenols, benzenes, and biphenylene contain dibenzodioxins and -furans. Precursor compounds for chlorinated dibenzodioxins and dibenzofurans are chlorinated phenols, benzenes, polychlorinated biphenyls, and other aromatics in combination with oxygen and chlorine. Sulfur-containing octachlorodibenzodioxin (OCDD) analogues reach the environment via tire abrasion. Since dioxins and furans reach the environment mainly particle-bound or gaseous, degradation in the atmosphere is important. The hepta- and octachlorinated congeners especially are strongly bound and carried in the soil via dust particles or dust precipitation. The lower-chlorinated congeners seemingly undergo a more rapid photochemical degradation as a result of their higher vapor pressure. The annual PCDD/PCDF inputs to the environment of Germany, Sweden, Switzerland, and Norway have been estimated [5.96], [5.97]; inputs of ca. 0.928 kg (Germany), 0.14–0.82 kg (Sweden), 0.285–0.37 kg (Switzerland) and 0.430 kg (Norway) were determined. This calculation disregarded the large number of industrial blast furnaces and incineration plants with considerable PCDD/PCDF emissions. Only after these plants have been re-equipped can large reductions of the input be expected. At the end of the 1980s, 48 g of PCDD/PCDF were said to be introduced into soils annually through sewage sludges used in agriculture and amounts of the same order through liquid manure and composts (2.6–47 and 32–80 g TEQ, respectively) [5.98]. Today the inputs are presumably distinctly less because the loads of sewage sludges and composts are no longer as high as in the 1970s or 1980s. Dioxin contents in surface waters of Hesse (the Main river) were 0.1–0.2 pg/L TEQ, in the corresponding suspended material 8–20 ng/kg d.m. TEQ, and in the sediments 9–19 ng/kg d.m. TEQ [5.99].

The immission air loads of rural regions are 0.025 to 0.07 pg TEQ/m^3, with a deposition of 5–20 pg TEQ m^{-2} d^{-1}. In urban regions the load is 0.07–0.35 pg TEQ/m^3 and 10–85 pg TEQ m^{-2} d^{-1}. In the neighborhood of emitters, immission values of 0.35–1.6 pg TEQ/m^3 and deposition values of $\leq$1000 pg TEQ m^{-2} d^{-1} are reached [5.100]. Loads in the rainwater (Dübendorf, Switzerland) were 0.03–0.1 and in mist 0.3–3 ng/L [5.20].

Viewed over time, the inputs of dioxins and furans to soils distinctly rose from 1920 and reached their maximum in the 1970s [5.101]. Studies in England on lake sediments show increases of dioxin since 1940 [5.102] (Fig. 5.1). Core drillings of lake sediment in Berlin indicate inputs falling overall by now.

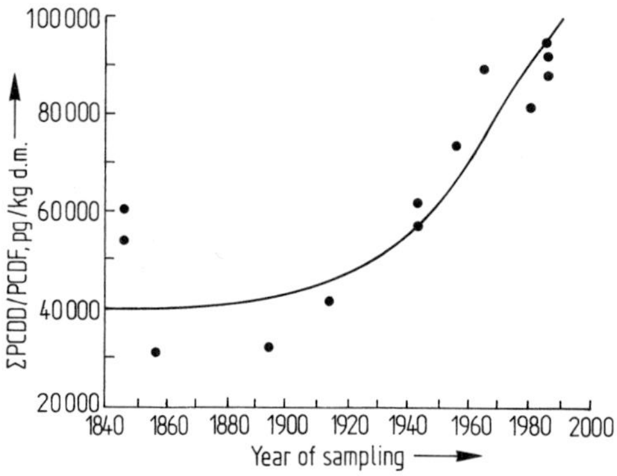

Figure 5.1. Development of PCDD and PCDF contents in the surface soil of Rothamsted, United Kingdom [5.101]

The background load for soils was quoted as 0.8 ng TEQ/kg d.m. for Baden-Württemberg [5.103]. The contents in Bavaria are of a similar order [5.104]. Loads in urban areas and areas with greater industrial use are 3.4 and 8.4 ng TEQ/kg d.m., respectively [5.52]. On the other hand, the dioxin loads in North Rhine-Westphalia in rural areas were 0–10 ng/TEQ/kg d.m. (50 % < 5 ng TEQ/kg d.m.), and in the conurbations 10–15 ng TEQ/kg d.m. [5.22]. The dioxin load in forest soils reaches values of 50–80 ng TEQ/kg d.m. in the litter cover. Table 5.7 gives a compilation. The soils (0 to 5 cm) in the surroundings of metal works have distinctly higher loads (e.g., in Linz, Maulach, and Rastatt with up to 332 ng TEQ/kg d.m., 7926 ng TEQ/kg d.m., and 29039 ng TEQ/kg d.m., respectively [5.73]. In the grassland young growth near such plants, 20–40 ng TEQ/kg d.m. were determined [5.105]. Correspondingly, increased dioxin loads of the milk were found.

Because of the low water solubility and high lipophilicity of dioxins and furans, they are sorbed to the organic matter (Table 5.8). The K_{OC} values increase with increasing degree of chlorination, in this case from 10000 to 57×10^6. Because of this strong sorption the higher-chlorinated congeners in particular have a low bioavailability and are nearly immobile in the soil. This is evident also in the congener distribution in soil, where the hexachloro- (HCDD) and octachlorodibenzodioxins (OCDD) dominate. The origin of the dioxins and furans can in some cases be deduced from the congener patterns. High proportions of octa- and hexachlorinated dioxins point to inputs from pentachlorophenol (PCP) sources.

In a sandy-loamy silt, increased mobility of low-chlorinated dioxins (down to 40 cm depth) was observed over 12 months [5.108]. Deeper displacement of 2,3,7,8-tetrachloro-*p*-dioxin (2,3,7,8-TCDD) down to a depth of 5.40 m was determined at a contaminated site [5.109]. After the addition of lauryl sulfate and at pH values of 8 instead of 7, the K_{OC} fell by a factor 10. Such behavior of a substance can be influenced by cosolvents (e.g., oil) [5.110], [5.111].

Near the soil's surface (0.06–0.13 mm) the higher-chlorinated dioxins and furans in particular are subject to a photolysis at 250–360 nm [5.112], [5.113] and volatilization, that can be enhanced by solvents. Photodegradation and volatilization on plant surfaces can also contribute to the elimination of dioxins [5.114]. The degradation of dioxins and furans proceeds extremely slowly. The half-lives vary from 10–14 years [5.115] to several decades. In other investigations scarcely any losses, but only small lateral displacements were found after 17 years [5.116], [5.117], [5.118].

The dichlorobenzofurans are less persistent than the dichlorobenzodioxins. While the degradation routes of the PCDDs have not yet been elucidated, microorganisms exist in polluted waters that can degrade dibenzofurans [5.119]. The organisms capable of degradation include *Pseudomonas* spp. [5.120]. The degradation of dibenzofuran leads after dehalogenation in an anaerobic environment finally to branched and unbranched paraffins [5.121]. Diphenyl ether, 2-methoxydiphenyl, and 2,3-dihydroxydibenzofuran are formed as intermediates. FORTNAGEL et al. [5.122] assume a primary degradation by *Pseudomonas*-sp *HH69*, in which 1-hydroxybenzo-*p*-dioxin is formed via an unstable dihydrodiol and after ring cleawage is degraded to 2,2,3-trihydroxybiphenyl which is transformed to catechol after a second cleavage. The inability of the microorganisms to degrade TCDD is a result of TCDD's high toxicity to microorganisms [5.123], its low ability to permeate membranes, its strong sorption, and its low water solubility or bioavailability.

Studies on the degradation of dioxins showed that succinate- and biphenyl-adapted microorganisms degrade monochlorinated dioxins because succinate and biphenyl bring about the enzyme induction required for this [5.124]. The relative conversion rate compared to biphenylene is > 72 % for monochlorodibenzodioxin and 8.9 % for 1,2,4-trichlorodibenzodiox-

Table 5.7. Range of the dioxin loads of soils in Germany, compiled from data of the different states (situation at May 15, 1991; TEQ-calculation according to BGA [5.100])

Loaded area		Mean (ng/kg TEQ)	Range of the means	Minimum value	Maximum value	Number of samples	Sampling depth, cm
Soils in rural areas (background)	forest soil	5	1.4–14	0.04	46	33	[a]
	forest litter	35	7–56	0.3	140	41	
	grassland	3	0.6–7	0.04	10.7	95	1 state 0–5 4 states 0–10
	arable land	2	0.2–5	0.01	7.7	178	0–30
	other	5 (1 state)	2–7	0.03	35	34	0–2
	soils treated with sewage sludge	5	3–7	0.025	21	46	grassland 0–10 arable land 0–30
Variously polluted soils in conurbations	forest	2 (1 state)		0.8 (1 state)	7.4	48	1 state 0–5 1 state 0–15
	forest litter	10			75.1	29	
	arable land	2.8			32	98	0–30
	grassland	3.8		0.5	100	223	1 state 0–5 2 states 0–10
	residential areas	9 (1 state)			30	10	
	parks and children's playgrounds			5	1594	126	1 state 0–10 1 state 0–5 & 0–40
	kindergartens	12		2.5	27.3	86	1 state 0–30 1 state 0–5 & 0–40
	other	32 (1 state)		8.6 (1 state)	73.0 (1 state)		
Soils in the area around defined emitters	refuse incineration plants	43		0.01	202	118	[b]
	landfills			0.8	188 (1 state)	5	
	old sites			3	200000		1 state 0–10 1 state 0–5 & 0–40
	chemical industry	8	5–13	0.05	1658	57	1 state 0–10 1 state 0–5 & 0–40
	metal processing			0.07	4958	262	[c]
	road traffic	10	10–20		200	30	1 state 0–5 1 state 0–30
	other therm. processes	65	1–160	0.02	800	47	[d]
	other chem. processes	7		0.05	89	27	1 state 0–10 1 state 0–10

[a] 1 state 0–4 cm; 2 states 0–5 cm; 1 state 0–12 cm; 1 state 0–15 cm (horizon-dependent).
[b] 1 state 0–10 cm; 1 state 0–5 and 0–30 cm; 1 state 0–40 cm.
[c] 1 state 0–5 cm; 2 states 0–10 cm; 1 state in addition 10–30 cm; 2 states 0–30 cm.
[d] 1 state 0–5 and 0–40 cm; 3 states 0–10 cm; 1 state 0–25 cm; 1 state 0–30 cm.

Table 5.8. Sorption constants, water solubility (WS) and Henry's law constant of various dioxins

p-Dioxin	WS, µg/L	$\log k_d$	$\log K_{OC}$	$\log K_{OW}$	Henry's law constant
Dibenzo-	842		3.91*	4.3**	12.29
2-Chloro-	278	3.3	4.60	5.00	14.82
2,3-Dichloro-	14.9		5.19	5.60	6.61
1,2,4-Trichloro-	8.41		5.93	6.35	3.84
1,2,3,4-Tetrachloro-	0.63	4.65	6.18	6.60	3.77
2,3,7,8-Tetrachloro-	0.2	4.48	6.38	6.80	1.63
1,2,3,4,7-Pentachloro-	0.118		6.97	7.40	0.264
1,2,3,4,7-6-Hexachloro	0.004		7.37	7.80	4.52
1,2,3,4,6,7,8-Heptachloro	0.002		7.57	8.00	0.133
Octachloro-	0.0004		7.76	8.20	0.683

* Calculated from $\log K_{OC} = 0.989 \log K_{OW} - 0.346$ [5.106]. ** $\log K_{OW}$ of [5.107].

ins. The degradation product is 1,2-dihydroxydibenzo-*p*-dioxin. In the case of more highly chlorinated dioxins a steric hindrance of the enzymes is to be expected. It is assumed that in the presence of other carbon sources (e.g., naphthalene, benzene, phenol) the microorganisms utilize these first. It can furthermore be assumed that the concentration range of the dioxins between enzyme induction and toxicity is very small. Furans are as a rule taken up in distinctly greater amounts than dioxins by soil-activating organisms or organisms living in the sediment [5.125]. The ovaries, liver, and intestine of *Fundulus heteroditus* (a species of fish) which take up higher amounts of dioxins provide an exception.

The estimation of the loading of the food chain by dioxins and furans plays a part in the determination of limit values. Over 90% of the PCDD/PCDF uptake of humans occurs via food [5.126]. The dioxin uptake by dairy cattle, sheep, etc. as a result of polluted soils can amount to 10% or more of the amount available in the soil. On average, 20–56% of the dioxins absorbed by dairy cattle enter the milk [5.127]. Hexachlorodibenzodioxin is more rapidly metabolized than OCDD. Compared with other dioxins/furans, the 2,3,7,8-substituted tetra- and hexachlorinated dibenzodioxins and -furans are absorbed particularly in the milk and not mainly in adipose tissue of dairy cattle. The consumption of eggs is another source of PCDD/PCDF uptake by humans.

In rats 2,3,7,8-TCDD was 100% bioavailable after transpulmonary uptake. The transfer factors via the air–plant path were estimated for leaf vegetable species to be 0.2 ng I-TEQ/kg d.m. (International Toxicity Equivalent Factor according to NATO-CCMS, 1988) [5.100], [5.128]. Deviations: green cabbage 0.4 ng I-TEQ/kg d.m.; pasture land 0.1–0.2 ng I-TEQ/kg d.m. A value of 0.01 ng I-TEQ/kg d.m. was assumed for the soil–plant transfer path. Such a definition often overlooks the fact that the transfer factors frequently vary just as strongly between the plant as between the plant species. The transfer of PCDD/PCDF via the water–sediment–fish path, is relatively high (30% transfer). Thus the congener profile found in eels, for example, is reflected in that of the sediments.

Because of the enormous toxicity, the BGA has defined use-dependent limit values for soils [5.94], that do not, however, take account of the soil properties.

5.1.4.5. Polycyclic Aromatic Hydrocarbons (PAHs) (→ Hydrocarbons, **A 13**, pp. 265–274)

Polycyclic Aromatic hydrocarbons are formed during combustion processes (domestic heating, road traffic, fires, refineries, coking plants, waste incineration plants, etc.). They include a whole series of carcinogenic compounds. In all, there are assumed to be several hundred such PAHs. They enter the environment via the air, sewage water, and sewage sludges. Usually 18 PAHs in all are considered. On lands used for agriculture the contents of benzo[*a*]pyrene (BaP) were mainly <0.2 mg/kg d.m. and of fluoranthene <0.3 mg/kg d.m. soil [5.72]. In urban backyards and allotments, 10–20% of the soils have contents of >1 mg BaP and >2 mg fluoranthene. In Bavaria, the concentration of six PAHs (fluoranthene; benzo[*b*]- and benzo[*k*]fluoranthene; benzo[*a*]pyrene, indeno[1,2,3-*c,d*]pyrene; and benzo[*g,h*]perylene) was determined in various soils [5.129]. The mean of the benzo[*a*]pyrene was a factor of ten higher than the reference value of 0.05 mg/kg d.m., and two soils out of twelve were above the B value (limit value for soil investigation [5.129a]) of 1.0 mg/kg d.m. Soils in urban center areas are more heavily polluted. These areas consist in some cases of rubble and burnt building material and contaminated sites, such as gasworks. The total PAH load of soils frequently amounts to 10 times the BaP value.

With the exception of locally concentrated inputs, it is assumed today that PAH inputs into the soil occur mainly via the air path. Loads of PAHs (21 compounds) in Dübendorf (Switzerland) were 0.1–0.6 µg/L in the rain and up to 60 µg/L in the mist, with deposition rates of 10–60 and 90 $ng\,cm^{-2}a^{-1}$ [5.20]. The annual increases in the soil of Rothamsted (United Kingdom), at 4.56 mg/m^2, were higher by a factor 2–3 [5.130] (see Fig. 5.2). Since the 1890s the

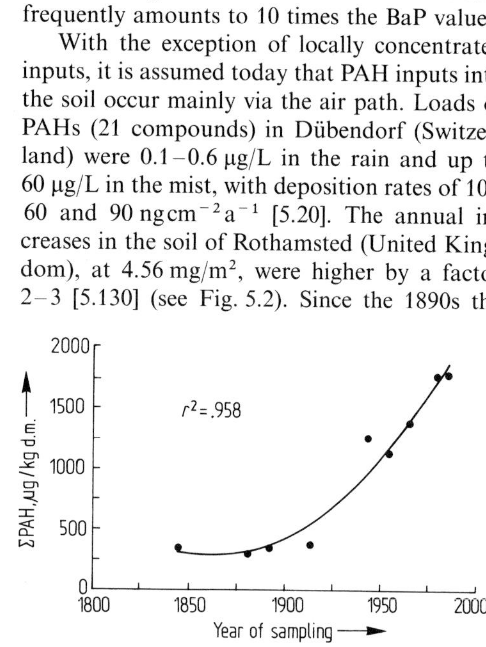

Figure 5.2. Development of PAH contents in the surface soil of Rothamsted, United Kingdom [5.130]

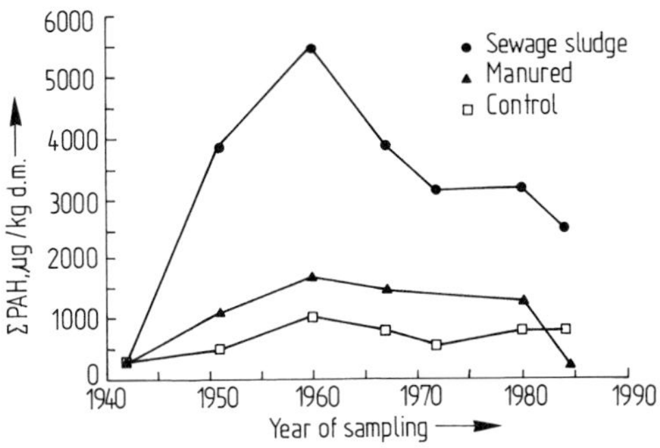

Figure 5.3. PAH pollution of experimental soils fertilized by sewage sludge and of a soil treated with farmyard manure [5.130]

concentration of PAH components in soil increased by the following factors: for benzo[*b*]-fluoranthene, benzo[*k*]fluoranthene, and benzo-[*a*]pyrene >15; for pyrene, benzo[*a*]fluoranthene, benzo[*e*]pyrene, and indenopyrene 10–15; for chrysene, benzo[*g,h,i*]pyrene, acenaphthylene, acenaphthene, 5–10; and for phenanthrene, anthracene, fluoranthene, and naphthalene, <5. Long-term sewage sludge application from 1940 to 1960 caused a remarkable increase of PAHs (see Fig. 5.3) compared to manured areas or the control area [5.130]. Since 1960 the PAHs contents decrease gradually due to increasing quality of the sewage sludge.

The strength of binding of the PAHs in the soil differs, depending on their water solubility. The K_{OC} values increase with the number of condensed benzene rings, and specifically from ca. log 2.8 to log 7 [5.131]. Because of the chemical structure of this class of compound mainly hydrophobic bonding can be assumed.

The degradation proceeds very slowly in some cases [5.132]. The half-lives in an arable soil vary for less condensed PAHs from 14–48 d and for more highly condensed PAHs from 603 to 2030 d, depending on the stability of the compound and its molecular mass (Table 5.9). The results of this study show that laboratory experiments permit conclusions on the behavior under natural conditions only to a very limited degree. In smoked soils the sorption is weaker and the degradation more rapid because the sorption is not yet at equilibrium. Bioavailable PAHs can be more rapidly degraded. In two soils treated in 1968 with sewage sludge rich in heavy metals the half-lives of PAHs were 2–9 a [5.133], i.e., degradation was impeded. The biological degradation is important in the case of naphthalene, which is degraded via the dihydrodiol to dihydroxynaphthalene and a hydroxybenzopyruvate derivative and finally via salicylic acid to catechol [5.47]. The degradation of phenanthrene is basically similar, although *o*-phthalic acid is split off in addition. Aside from *Pseudomonas* spp., *Alcaligenes denitrificans* also is capable of degrading PAHs [5.134]. The more highly condensed PAHs (>4 rings) are considered to be extremely resistant to enzymatic degradation.

Table 5.9. Half-lives (d) of various PAHs in soils from microcosm studies (losses in % in brackets) [5.123]

Compound	RA 1[a]		RA 2[b]		CF[c]		RS[d]		Mean	SD
	Soil	R^2 [e]	Soil	R^2 [e]	Soil	R^2 [e]	Soil	R^2 [e]		
Naphthalene	33 (100)	0.72	16 (100)	0.86	14 (95)	0.67*	48 (91)	0.54*	28	16.0
Acenaphthene/fluorene	74 (86)	0.68	70 (84)	0.88	73 (100)	0.90*	44 (96)	0.94	65	14.3
Phenanthrene	108 (85)	0.88	193 (50)	0.8*	112 (56)	0.63*	83 (94)	0.55*	124	47.8
Anthracene	48 (61)	0.88*	210 (22)	0.34	105 (90)	0.54*	200 (59)	0.98	141	77.9
Fluoranthene	110 (78)	0.88	184 (35)	0.69	143 (60)	0.93	110 (73)	0.6	137	35.1
Pyrene	285 (36)	0.69	167 (49)	0.78	320 (20)	0.57*	127 (57)	0.90*	225	92.4
Benzo[*b*]chrysene	199 (66)	0.68	106 (76)	0.79	313 (34)	0.42*	240 (67)	0.94*	215	86.3
Benzo[*b*]fluoranthene	135 (90)	0.65	113 (44)	0.62	279 (8)	0.22*	282 (34)	0.93*	202	90.8
Benzo[*k*]fluoranthene	350 (53)	0.93	143 (57)	0.76*	351 (8)	0.23	359 (30)	0.97*	301	105.2
Benzo[*a*]pyrene	120 (85)	0.62	270 (18)	0.9	197 (44)	0.43*	258 (48)	0.92*	211	68.7
Benzo[*g,h,i*]pyrene	460 (15)	0.75*	365 (20)	0.7	460 (16)	0.83	535 (25)	0.73*	455	69.6
Coronene	2030 (16)	0.65	603 (14)	0.79	753 (17)	0.5	669 (7)	0.41	1014	680.3
PAHs (total)	146	0.90	165	0.85*	192	0.79*	177	0.82	170	19.4

[a] RA 1 = rural arable soil, no preceding sewage sludge application. [b] RA 2 = rural arable soil. [c] CF = rural coniferous forest. [d] RS = roadside soil. [e] R^2 = linear regression. $^{+}$ = significance.

According to studies the half-lives of >4-ring PAHs exceed 300 d [5.135]. PAHs are not degraded under anaerobic conditions [5.93]. In addition to photolytic degradation and chemical oxidation, volatilization also plays a certain part during the elimination in and on soils. Simple PAHs, such as naphthalene and methylnaphthalene, volatilize in proportions of 20–30 %, and are also degraded abiotically to 2–20 %. Neither is true of more highly condensed compounds.

Analyses of grass samples since 1900 prove that the highest PAH values occurred between 1931 and 1935, whereas distinct reductions occurred after 1955.

The "light" PAHs stay largely in the gas phase and can be found in the leaves or parts of fruit, whereas the heavier ones remain in the soil [5.136] or root parts. They can therefore be found mainly on root surfaces, such as those of carrots [5.133]. Eleven years after the application of sewage sludge 30 μg ∑PAH/kg d.m. were found in potatoes. The light PAHs were dominant in sugar beets, beets, and carrots [5.137]. There is little indication that the contents of PAHs were raised by the sewage sludge in comparison with the control.

The mobility of PAHs in soils can be distinctly raised by increased sorption on the dissolved organic matter. It has been that naphthalene and other PAHs can be displaced in the order in which they are bound to dissolved organic matter (DOM) [5.30]. From 100 mg/L DOM, hydrophobic substances with K_{OW} values $>10^{4.5}$ can be transported at the same rate as the dissolved organic matter. Under aquifer conditions, the mobility of the mobile phase of a PAH-containing coal tar in sandy material amounted to barely 7 cm in 54 d ca. 14 % of the naphthalene, 0.2 % of the fluoranthene and 0.06 % of the benzo[*a*]anthracene having been transformed in 91 d [5.138]. The transport is attributed essentially to molecular diffusion.

5.1.4.6. Polychlorinated Biphenyls (PCBs)

Polychlorinated biphenyls can occur in 209 congeners. About 60 individual substances with chlorine contents between 21 and 68 % have been detected. PCBs have a very low water solubility and are chemically very stable. They are therefore detected ubiquitously. About 2×10^6 t of PCBs have been produced worldwide since 1929, but they can also occur as byproducts in organic syntheses and in the pyrolysis of chlorobenzene. The coplanar PCBs 77, 120, and 169 have proved to be more toxic than the nonplanar *ortho*-substituted PCBs. PCBs are present in used oil, condensers, adhesive and sealing compounds, chipboards, ceiling panels, flame retardants, hydraulic fluids, films (as plasticizers), etc., but an extensive ban has existed on their use for open systems, in the Federal Republic of Germany since 1972 and EC-wide since 1985.

Wide-area pollution occurs mainly via the air path. In conurbations, 180 $g\,km^{-2}\,a^{-1}$ of wet and dry deposition of PCBs is assumed. Low-chlorinated congeners distributed more by the gas phase, and high-chlorinated ones are mainly distributed particle-bound. According to CAPEL et al., PCBs (31 compounds) can be discharged at 7–22 μg/L in the mist and 0.002–0.003 μg/L in the rain, which results in a deposition of ca. 110–330 $g\,km^{-2}\,a^{-1}$ [5.20].

Municipal sewage sludges contain 0.1–1.0 mg/kg d.m. [5.28] and river sediments 0.01–1.0 mg/kg d.m. of PCBs. The 90 % percentile values are about 4.8 μg/kg d.m. for rural areas (grassland) and ca. 15.5 μg/kg d.m. of PCBs (13 congeners) for conurbations. The three coplanar PCBs usually amount to 10 % of the whole. The PCB contents were raised by a factor 10–15 especially in the flooded area of the Rhine [5.22]. In forest soils, the mean PCB content in the humus cover was ca. 85 μg/kg d.m. The contents in the subsoil were as high as in the surface soil [5.99].

PCBs are bound in soils as a function of their water solubility and lipophilicity, i.e., sorption increases from the singly chlorinated to the polychlorinated biphenylene. The K_{OC} values rise from 10^3 to 8×10^5 (see Table 5.10). At low humus contents, the content of clay and allophanes is also important in PCB sorption [5.199].

The degradation of PCBs in soils under aerobic conditions proceeds slowly, and under anaerobic conditions very slowly [5.141]. Anaerobic degradation occurs as a reductive dehalogenation. After input over many years, the hepta- and octachlorinated compounds accumulate, because of extremely small degradation and more powerful sorption than the low-chlorinated compounds [5.73]. After 300 d no reduction of PCBs was found in soils treated with sewage sludge [5.142]. In other field experiments only after application of 50 t sewage sludge per hectare was a five- to seventeenfold enrichment observed, a value of 0.05 mg/kg d.m. not being exceeded [5.143]. Mono- and dichlorinated PCBs degrade

Table 5.10. Physicochemical data of PCBs

Biphenyl	IUPAC No.	WS, mg/L	K_{OC}*	k_d**
Biphenyl		7.5	1400	
2-Chloro-	6	0.062	4460–7940	
4,4′-Dichloro-	15	0.062	20 000–100 000	
2,4-Dichloro-	20		151 000	7–20
2,4,4-Trichloro-	28	0.113	166 000	
2,2,5,5′-Tetrachloro-	52	0.113	257 000	21
2,2,4,4′-Tetrachloro-		0.085	17 000	
2,2′,3′,4,5-Pentachloro-	87	0.099	490 000	740
2,2′,4,5,5′-Pentachloro-	101	0.029	425 000	810
2,2′,3,4,4′,5-Hexachloro-	138		70 800	2420
2,2′,4,4′,5,5′-Hexachloro-	153	0.001	724 000	2500
2,2′,3,4,4′,5,5-Heptachloro-	180		794 000	4640
2,2,3,4,4,4′,5,6-Heptachloro-	183	0.00047	620 000	3620
2,2′,3,3′,4,5,5′,6-Octachloro-	201			8700

* [5.139], low clay contents. ** [5.140], low humus contents.

microbially, e.g., by *Achromobacter* or *Pseudomonas* spp. to benzoic acid via the dihydrodiol and *meta* cleavage of the rings.

With increasing chlorination, the degradation proceeds more slowly, especially if a chlorine atom occupies a position *ortho* to the second ring [5.144]. The volatilization of low-chlorinated PCBs is regarded as a possibility of elimination, especially because of their great persistence [5.145]. Aroclor 1254 was released from the soil by 40–50 % within 2–4 months. Hexa- and heptachlorobiphenyls were released only half as well as trichlorobiphenyl.

In case of high PCB pollution and long-term action, PCBs can reach even deep soil layers and the groundwater (distance from ground elevation: 1–1.2 m). The migration depends on the solubility of the PCB [5.146]. If used oils or an oil phase are present in the underground, PCBs can also move laterally. The greater solubility of PCBs in the oil phase than in the dissolved organic matter causes a tenfold enrichment in the oil phase [5.147].

For many applications (e.g., risk assessment) it is necessary to estimate the behavior of PCBs and other pollutants in soil. For landscape planning, for groundwater protection, for sewage sludge application on farmland, etc. a site-specific prediction system to evaluate the behavior of such pollutants is important for the administration to protect the environment. Results from field trials with more than ten different soil types are in good agreement with estimates obtained for a qualitative evaluation [5.148].

PCBs are absorbed by plants only in small amounts. Studies with soils treated with sewage sludge containing PCBs showed that carrots in particular, because of their lipoid-containing roots, accumulated PCBs in their skin [5.149]. The bioconcentration factors (→ Ecology and Ecotoxicology, Section 2.4.2.7) were: in the roots, 0.02–0.04; in the skin, 0.02–0.1; and into the foliage, <0.002–0.0008. The low-chlorinated congeners in particular are absorbed. Small uptakes were also determined in various grasses, grains, and vegetable species [5.143]. The bioconcentration factors were between 0.2 and 0.02. The action of PCBs in other useful plants was also studied [5.150]. At PCB concentrations above 10 000 mg/kg d.m., growth inhibition was established in soya and beets. No metabolization and only minor xylene transport was found [5.151]. Only 20–25 % of PCBs absorbed by rats via the soil was not excreted. Different soil properties scarcely affected the excretion.

Within an animal food chain (oak leaves, earthworm, great tit's egg) an accumulation (biomagnification) of 1 ng/g to 170 ng/g of PCB 153 occurred [5.152].

5.2. Acid Precipitation

5.2.1. Definitions

Used in a strict sense, the term acid precipitation is defined as increased hydrogen ion content of precipitation water due to anthropogenic sources [5.153]. From a chemical point of view this term must be extended, and other rain components included that are subject to acid–base reactions after deposition. As a consequence of these reactions, hydrogen ions may be generated when rain water comes into contact with vegetation and soil surfaces. The term "potential acidity" therefore takes into account all acidic components carried in atmospheric deposition fluxes.

Meteorology distinguishes different forms of atmospheric deposition such as rain, snow, dry particulates, aerosols, gases. Assessment of deposition fluxes and separation of single fluxes are very difficult to accomplish for methodological reasons [5.154]. Nevertheless, in order to estimate effects of acid precipitation on soils and ecosystems, assessment of all forms of atmospheric deposition is indispensable.

With regard to the impact of acid deposition on soils, it is also necessary to define the terms acidity and acidification for soils, soil water, groundwater, and surface water. *Soil acidity* is usually expressed as the hydrogen ion activity (pH) measured in an aqueous suspension of a soil sample. The pH is a key parameter for most soil-chemical reactions and for the conditions in which soil organisms live and plant roots are growing. However, pH is an intensity parameter which gives no information on the extent to which acid–base reactions can be maintained in a given soil over time. Therefore, when soil changes in time are considered soil acidity must be described using capacity parameters (such as mass of acids and bases available in the surface soil layers).

According to [5.155], *soil acidification* is related to (1) a decrease in acid neutralizing capacity (ANC, common units mol H^+ per kilogram soil or per square meter of a defined soil horizon) of the soil by loss of base cations such as Ca^{2+}, Mg^{2+}, and K^+, and/or (2) an increase in base neutralizing capacity by retention and accumulation of H^+ and acid cations such as Al [Al^{3+}, $Al(OH)^{2+}$, $Al(OH)_2^+$] and heavy-metal ions. These changes are accompanied with a decrease in soil pH when the acid strength of the acidic groups of the solid phase in the soil increases. Acidification may take place in one or more soil horizons, depending on the balance between acid input (or production) and acid output (or consumption).

In soil water, groundwater, and surface water chemistry, acid–base conditions are usually described in terms of alkalinity. Alkalinity can be defined as the sum of

$$2\,[CO_3^{2-}] + [HCO_3^-] + [OH^-] - [H^+] \\ - 3\,[Al^{3+}] - 2\,[Al(OH)^{2+}] + [Al(OH)_2^+]$$

where brackets denote molar concentrations of each ion species [5.156]. Alkalinity defined in this way may take positive or negative values. Water acidification is then defined as a loss in alkalinity.

Thus, the choice of appropriate parameters to characterize acidity of precipitation in relation to its effects on soils and ecosystems depends on the time scale in which the impact of acid deposition is considered. Processes occurring in the soil can be classified into different categories, *direct* and *long-term processes*. The first category of processes is directly and instantaneously related to acid intensity. Examples for such processes are: dissolution of solid mineral phase, ion exchange on mineral particles and humic substances, formation and destruction of ionic complexes and, as a consequence, toxic effects on soil organisms and plant roots. These processes instantaneously follow changes in thermodynamic conditions at any given point of the soil or ecosystem. Here acidity influences soil-chemical processes which may be connected to life processes on the organism or ecosystem level, but must not necessarily have any visible effects on soil components (soil minerals, aggregates, humus). For these processes acidity is appropriately described by the pH of solution. The presence of cation acids and the redox potential may be additional conditions to be considered.

The second category of acidity-related soil processes becomes important in long-term considerations. As a result of this type of processes, changes in composition and structure of soil minerals and soil organic matter, in nutrient availability, and soil biology are observed, followed by changes on the ecosystem level (changes in tree population, plant, and animal communities). The underlying processes—acid–base and redox reactions—are the same as those described in the first category and are occurring naturally in soils. Acid impact often results in an acceleration and intensification of natural processes such as weathering and alteration of minerals, leaching of nutrient cations, aluminum, iron, and manganese, depletion of the exchange complex, formation of secondary minerals, redistribution of elements in the soil profile, build-up of organic matter at the soil surface, retreat of roots from deeper soil layers, changes in the edaphon (soil animals) and in the herb layer. All these changes tend to be irreversible at least within a human life span. In contrast to the first category, these processes result in visible changes in the structure and composition of soil itself. In order to characterize acid deposition with respect to the described effects, capacity parameters such as deposition rates of acidic substances (expressed, e.g., in $kmol\,m^{-2}a^{-1}$) are appropriate.

Table 5.11. Acidic components of atmospheric deposition and their main sources

Substances	Anthropogenic sources	Natural sources
(1) CO_2 and H_2CO_3	combustion of fossil fuels	atmospheric CO_2, respiration, mineralization of organic matter
(2) Strong mineral acids and their precursors		
SO_2 and H_2SO_4	combustion of fossil fuels, incineration, smelting of ores	volcanic exhalation
NO_x and HNO_3	combustion of gasoline, agriculture (livestock)	electric discharge (lightning), microbial denitrification in soils, forest and grassland fires
HCl_{gas}	incineration of refuse, chemical industry	
(3) NH_4^+, when transformed into ammonia or when taken up by organisms to form organic N-compounds	agriculture, sewage treatment, chemical industry	decomposition of organic N-compounds (proteins)
(4) H_2S	chemical industry	biogenic emissions from marine algae and sulfate-reducing bacteria
(5) Organic acids	chemical industry	
(6) Cation acids (as Al^{3+}, heavy metals)	soil and industrial dust particles when dissolved in acid precipitation	soil particles when dissolved in acid precipitation

5.2.2. Sources of Acidity in Atmospheric Deposition

Most acidic components of atmospheric deposition may originate both from natural and from anthropogenic sources. In order to justify control measures for soil protection and to set priorities, it is necessary to quantify the contribution of single sources of acidity, natural as well as anthropogenic.

Table 5.11 shows substances which form protons in cloud droplets and precipitation water, or which carry acidity into soil. The substances are listed together with their most important sources. In industrialized countries, carbonic acid (1) and strong mineral acids (2) are generally the most important contributors to precipitation acidity, the acid intensity (pH of precipitations; see below) being determined mainly by (2). In class (2), man-made sources by far exceed natural sources, at least in the industrialized countries of the northern hemisphere [5.153]. Ammonium (3) plays an important role in areas with intensive agriculture mainly with a high density of livestock. Hydrosulfuric acid (4) and its oxidation products may be important contributors to acidity in bogs, swamps, and on mangrove coasts. Organic and cation acids [substances (5) and (6)] are usually of minor importance in precipitation chemistry.

As pointed out above, when impact of acid deposition on soils is studied, alkaline components of precipitation must be taken into account. The most important basic components are listed in Table 5.12, together with their sources.

5.2.3. Source and Sink Processes for Acidity in the Atmosphere and in the Soil Environment

For the effect of acid deposition on soils it is unimportant whether the hydrogen ions are already present in the cloud droplet or in precipitation, or are generated by chemical reaction of a pollutant at vegetation surfaces or in the soil.

Table 5.12. Components contributing basicity to atmospheric deposition and their main sources

Substances	Anthropogenic sources	Natural sources
(1) HCO_3^-	see (1), Table 5.11	see (1), Table 5.11
(2) NO_3^-, when taken up by organisms to form organic N-compounds	see NO_x in (2), Table 5.11	see NO_x in (2), Table 5.11
(3) NH_3, when transformed to ammonium NH_4^+ (see Table 5.11)	agriculture (livestock)	decomposition of organic matter (proteins)
(4) dust particles from limestone and other carbonates	cement industry, agriculture	soil dust

Table 5.13. Hydrogen ions generation processes in atmosphere and in soil

Process	Conditions
(1) Formation of carbonic acid from carbon dioxide originating	
in atmosphere: from atmospheric CO_2	
in soil: from soil respiration	
$CO_2 + H_2O \rightarrow H_2CO_3 \rightarrow H^+ + HCO_3^-$	pH >5.0
(2) Formation of strong mineral acids	
$SO_2 + 1/2\ O_2 \rightarrow SO_3 \rightarrow H_2SO_4 \rightarrow 2\ H^+ + SO_4^{2-}$	independent on pH; oxidizing conditions with an oxygen source (free atmosphere)
$NO_2 + H_2O \rightarrow HNO_3 \rightarrow H^+ + NO_3^-$	independent on pH
$HCl_{gas} \rightarrow H^+ + Cl^-$	independent on pH
$H_2S + 3/2\ O_2 \rightarrow H_2O + SO_2$ (see Table 5.11)	oxidizing conditions
nitrification $N_{org} \rightarrow HNO_3$ (see Table 5.11) $NH_4^+ \rightarrow HNO_3$ (see Table 5.11)	oxidizing conditions
(4) Formation of organic acids	presence of organic matter and soil microorganisms
(5) Hydrolysis/protonation of weak metal hydroxides, e.g., $Al^{3+} + H_2O \rightarrow Al(OH)^{2+} + H^+$	pH >4
(6) Formation of biomass by uptake of nutrient cations (most important: ammonium)	

Chemical reactions not only depend on presence and amount of reaction agents, but also on thermodynamic conditions at the site of reaction. Therefore, in order to understand the effects these conditions must be specified. The most important variable in acid–base reactions is the pH at the site of reaction, but other factors such as redox potential or temperature must also be taken into account.

Source processes are listed in Table 5.13 together with the range of physicochemical conditions—in a generalized form—under which they take place. Under these conditions atmospheric substances undergo chemical transformation (indicated by an arrow), generating hydrogen ions in the atmosphere (cloud droplets, precipitation water), at the contact of rain with surfaces (vegetation, soil), or in soil water.

Finally, Table 5.14 gives a list of *sink processes* which consume hydrogen ions, thus neutralizing acidity.

Table 5.14. Hydrogen ion consumption processes in atmosphere and in soil

Process	Conditions
(1) Transformation of bicarbonate into carbonic acid	
$HCO_3^- + H^+ \rightarrow H_2CO_3$	pH <5
(2) Dissolution of carbonates	
$CaCO_3 + H^+ \rightarrow Ca^{2+} + HCO_3^-$	
(3) Weathering of silicates	
(4) H^+ exchange on cell walls of leaf surface and on the exchange complex of the soil against (base) cations	
(5) Weathering of clay minerals	
(6) Formation of biomass by uptake of nutrient anions (most important: nitrate)	
(7) Transformation of ammonia into ammonium	
$NH_3 + H^+ \rightarrow NH_4^+$	pH >6.5

5.2.4. Direct Effects of Acid Deposition on Soils

Traditionally, direct effects are studied separately for different environment compartments, such as vegetation cover, soils, subsoils, groundwater bodies, lakes, buildings, etc. In landscapes subject to acidic deposition, chemical reactions occurring at any point are interrelated, mostly by fluxes of water and dissolved matter, but also by transfer in the solid and gaseous phase, or by organisms. The ecosystem approach helps to find the linkage between individual processes, and to close the gap of understanding between chemical characterization of acid rain and damage to the environment.

When precipitation reaches the soil, acids or acidifying components usually undergo chemical reactions with the soil matrix which is not in a (chemical) equilibrium with rain water. The kind of reaction and the effect of acid input depend very much on type of soil and underlying bedrock. It has been suggested to group soil types according to criteria which reflect the type of buffering system in effect, and its buffering capacity [5.157].

In soils on calcareous bedrocks that are not totally decalcified and contain rather unweathered lime particles in the surface horizons, acidity is readily neutralized by weathering of carbonates, resulting in the release and leaching of Ca^{2+} with seepage water to the subsoil, together with bicarbonate [reaction (2) in Table 5.14]. This reaction is usually fast enough to buffer even high atmospheric input of acids. Only if limestone is distributed very disproportionate over the soil volume, even this soil type may acidify in the fine fraction of upper soil layers.

In soils free of limestone the weathering of primary silicates leads to the release of base cations such as Na^+, K^+, Mg^{2+} and Ca^{2+}, which then buffer the protons [reaction (3) in Table 5.14]. The consequence is again cation loss by leaching, soil is depleted from essential plant nutrients. The rate of silicate weathering depends on contents, type, and grain size of silicates in the mineral fraction of soil. The rate is estimated to range between 0.2 (soils on sandy sediments, sandstone, quartzitic bedrocks) and 2 (soils on basaltic rock) $kmol\,ha^{-1}a^{-1}$ for a soil depth of 1 m [5.158]. These figures represent the maximum proton consumption rate. In comparison, the acid load from atmospheric deposition is 0.5–6 $kmol\,ha^{-1}a^{-1}$ [5.159].

In cases where acid deposition exceeds silicate weathering rate, or when silicate content in mineral soil is low or zero (sandy soils, soils on acid loess loam, quartzitic bedrocks), protons are neutralized by displacement of exchangeably bound Ca^{2+} and Mg^{2+} from the surfaces of clay minerals, iron and manganese oxides, and humic matter [reaction (4) in Table 5.14]. As a consequence, soil suffers further leaching losses of plant nutrients. The total ANC of the exchange complex in the rooting zone of vegetation ranges between 7 (acid sandy soils) and 700 (well-buffered clay soils). Comparing these figures with total acid input during a 10-year period of 5–60 kmol/ha (see above) it is obvious that acid input may lead to a complete depletion of the exchange complex on less buffered soils within a short period of time. This depletion is reversible after a decrease in acid load from atmospheric input, but at low weathering rates (see above) complete recovery may require decades or centuries.

When the ANC decreases because of leaching of buffering substances to deeper soil layers, protons begin to attack the crystal lattice of clay minerals at pH <4.2 [reaction (5) in Table 5.14]. The consumption of protons is then balanced by release of cation acids (mainly monomeric or polymeric Al ions) from the clay minerals which, in turn, tend to displace base cations from exchange sites. Saturation of the exchange complex with Al ions is followed by an increase of Al and heavy metal ions in soil solution. This process is ecologically harmful because Al and most heavy metal ions are toxic in their ionic form for plant roots and soil microorganisms, and acidification of subsoil and groundwater is promoted with leaching of these cation acids to deeper soil layers.

When the exchange sites are saturated with Al ions, the concentration of Al ions in soil solution is mainly determined by the concentration of sulfate, which is usually the dominant anion in acid soils. Under these conditions aluminum hydroxosulfates are formed and accumulate in acidified soil horizons [5.160]. These sulfates have to be considered as a quite labile pool of cation acids, which is readily mobilized under changing chemical conditions in the soil (lowering of pH), and then may be carried to deeper soil layers or groundwater with seepage water.

5.2.5. Effects Mediated by Organisms (Indirect Effects)

Indirect effects, are mediated by organisms, mainly by plant roots, soil animals, and microorganisms. In such cases, atmospheric input of acidity initiates processes in organisms which, as a consequence, react by producing or transmitting acidity to the soil. This gives rise to further soil acidification with reactions described in Section 5.2.4.

5.2.5.1. Soil Acidification by Canopy Leaching

When acid precipitation hits vegetation, hydrogen ions may be buffered on leaf surfaces by exchange with Ca^{2+} or Mg^{2+} bound to acidic groups in the cell walls [reaction (4) in Table 5.14]

[5.161], [5.162]. The basicity used up for neutralization of atmospheric acidity must be replenished from the exchangeable pool of Ca^{2+} and Mg^{2+} in the soil. Consequently, soil is acidified by this process, its base neutralizing capacity decreases. Such effects are predominantly observed on slightly acid soils in the main rooting zone of the trees (B horizon). Surface soil (A horizon) and organic surface layer (O horizon), which receive base cations from canopy leaching and litterfall, remain in a state of high base saturation. This shows how acid deposition reaches soil via root surfaces of the trees. Buffering in the canopy does not prevent the soil being acidified in the rhizosphere within very short time.

5.2.5.2. Acidification by Uptake of Nitrogen Compounds

Atmospheric deposition of nitrogen compounds occurs mainly in the form of nitrate and ammonia. The plants usually take up these compounds wholly or partially, nitrogen being one of the most important plant nutrients. The uptake of nitrate by organisms and transformation into organic molecules results in the equivalent consumption of protons [Table 5.12, reaction (2)].

In contrast, if ammonium (NH_4^+) is taken up to form organic nitrogen compounds, an equivalent amount of protons is generated [Table 5.11, reaction (3)]. If nitrate and ammonium are taken up in equivalent amounts, consumption and production of protons are balanced and no net acidification takes place. But when ammonium is deposited in excess, as in the vicinity of intensive agriculture (livestock), and is taken up by plants or transformed to nitrate and leached, this results in a considerable acidification of the soil horizon in which ammonium uptake is located.

5.2.5.3. Acidification by Accumulation of Organic Matter

The emission of nitrogen oxides (from industrial processes and combustion of oil and gasoline) and of ammonia (mainly from intensive agriculture/livestock) leads to considerable deposition of nitrogen compounds on large areas ($20 - > 50$ kg N $ha^{-1} a^{-1}$). In many forests on poor soils, nitrogen would represent a minimum factor in the nutritional supply of the trees without this input. The consequence of nitrogen deposition is an enhancement of tree growth and accumulation of biomass and litter at the soil surface. During the buildup phase of this pool of biomass, these processes result in a depletion of the mineral soil from nutrients (Ca, Mg, K), and a decrease of buffering capacity in the root zone, connected with a temporary or permanent acidification of the root zone.

Another effect of soil acidification is the inhibition of mineralization of litter and other forms of dead organic matter. In forests, an organic layer is built up at the soil surface, in which the activities of soil animals are reduced, and the microorganism population shifts from a dominance of bacteria to one of fungi. The mineral horizons underlying the organic surface layer are subject to intensified podzolization, with leaching of base cations, low-molecular organic matter and metal ions (Al, Fe, Mn) in complex form. In shallow soils, this can result in groundwater and surface water acidification.

5.2.5.4. Acidification by Humus Disintegration

Forests on well buffered, noncalcareous soils, e.g., on basaltic rocks under deciduous forests in favorable climatic conditions, usually show a soil profile with a deeply developed A horizon very rich in humus. The amounts of nitrogen stored in the organic fraction of such soils may be considerable. Under the conditions of high acid input, these stores of organic matter tend to disintegrate, i.e., the buildup of new humic substances is delayed (possibly through inhibition by Al ions) compared with the mineralization through accelerated degradation of high polymeric humic substances [5.163]. As a consequence of humus disintegration, the organically bound stores of nitrogen are transformed into nitrate (nitrification by soil bacteria). If the nitrates thus formed are not taken up by roots or soil organisms, but are leached with seepage water, a considerable acidification of the soil according to reaction (2), Table 5.13, results. This effect may be estimated by keeping in mind that a loss of 1000 kg N by nitrification and leaching would be equivalent to 70 kmol H^+ per hectare.

5.2.6. Consequences of Soil Acidification

The most important consequences of soil acidification for structure and function of soil itself, for the ecosystem of which soil is part, and for adjacent systems such as groundwater, lakes, and rivers are summarized below:

1) Loss of plant nutrients by enhanced leaching of basic cations which function as buffering against acidity—as secondary effects disturbance in the nutrition of trees and other plants, destabilization of forest ecosystems
2) Mobilization of cation acids, mainly Al and heavy metal ions from native soil minerals and/or atmospheric pollutants accumulated in the upper soil layers—as secondary effects toxicity to fine roots of trees and other forest vegetation, to soil animals and microorganisms, destabilization of forest ecosystems
3) Acidification of subsoil, underlying bedrock, groundwater and surface water, and concentration increase of undesired or toxic ions (nitrate, Al and heavy-metal ions) in the water

5.2.7. Field Observations of Soil Acidification

Since the beginning of industrialization of Western Europe, and more intensively from the 1950s onward, soil acidification has increased progressively. This is indicated by a number of studies, refering to different interrelated acidification parameters (see Sections 5.2.1–5.2.6).

Decrease of Soil pH. The earliest widespread pH measurements in forest soils was made in the late 1920s. These data were reexamined in 1987 [5.163 a]. Comparison with new measurements suggests that strong soil acidification has occurred in areas close to industrial zones and on mountain ridges with high acid atmospheric deposition. Soil acidification in soils under new planted spruce stands is attributed to ecosystem-internal acidification. pH measurements were repeated in a forest soil inventory in Bavaria after 20–30 a [5.164]. Depending on the initial acidity status, a decrease in soil pH of 0.3 to 1.8 units was observed on almost 50 % of 1000 cases. A recent review of soil pH changes found in a number of European countries is given in [5.165]. Berden et al. 1987 critically reviewed pH and base saturation data from Scandinavian and Central European forest soil samples since the 1940s, resampled after 20 to 50 years [5.166]. The drop in pH after this time span decreases, as expected, with decreasing pH at the time of first sampling. Similar results are reported for North Rhine-Westphalia, which suggest acid deposition as cause for soil acidification in the 1960s–1980s [5.167]. A good correlation was found between acidification parameters in soils and exposition of the site resulting in very different atmospheric deposition rates. Further evidence for soil pH decrease under forest vegetation is given for the Berlin region [5.168], mountain areas of North Rhine-Westphalia [5.169] Hesse [5.170], and the Hamburg region [5.171].

Decrease in Exchangeable Bases (Ca and Mg) and Base Saturation. In forest soils in North Rhine-Westphalia the exchangeable Ca decreased by about 50 % since the 1950s [5.172]. In 1987 an average of 50 % decrease in base cation stores was found in Swedish forest soils since the 1970s to 1950s [5.173]. Similar results were found for the Harz [5.174], the Oberpfalz [5.175], [5.176], and South Sweden [5.177]. A large number of field studies showed increased soil acidification around the trunks of beech trees, where large amounts of stemflow with high acid load infiltrates the soil [5.178]–[5.181]. Beech stands in remote areas of Southern Europe that are very little influenced by acid deposition do not show this pattern [5.182].

Increase of Protons and Cation Acids at the Exchange Complex [5.183], [5.184]. For south Swedish forest soils an increase in the pool of exchangeable Al of 100 % was shown [5.177].

The appearance of cationic acids, mainly ionic species of Al, in groundwaters and surface waters as a consequence of soil acidification was reported in many studies, for example, in forest-covered Central German mountain areas such as Kaufunger Wald [5.185], Harz [5.174], [5.186], Hesse [5.187], Fichtelgebirge [5.188]. For a review, see [5.189], [5.190]. Distinction between naturally occurring acidification of ground- and surface water and acidification caused by atmospheric deposition is usually made by using sulfate concentration as indicator (mobile anion concept, see [5.156]). When sulfate becomes the dominant anion in solution, the cause for acidification is attributed to atmospheric input, provided the underlying bedrock does not contain sulfate or sulfide minerals. In the mountains of the Harz (Lower Saxony), rivers and creeks of different acidification types exist in close vicinity, depending on acid load from atmospheric deposition and types of soil and bedrock [5.191], [5.192].

5.3. Fertilizers and Agricultural Chemicals

5.3.1. Fertilizers

5.3.1.1. Development of Fertilization

The development of fertilization in the 20th century is marked by an increased production of manures and a large rise in the consumption of inorganic fertilizers. From the beginning of the century until the mid-1980s, the use of nutrients in Germany in agriculture developed as follows: Inorganic fertilizers increased from a few kg/ha to an average of 130 kg N, 65 kg P_2O_5 and 90 kg K_2O per hectare of agricultural areas. In the 1980s the annual consumption of manures of animal origin was 100 kg N, 60 kg P_2O_5, and 110 kg K_2O per hectare [5.193]. In Germany, for example, from 1985 to 1993 the annual consumption of inorganic fertilizers (excluding former German Democratic Republic) fell to 108 kg/ha N, 35 kg/ha P_2O_5, and 48 kg/ha K_2O [5.194]. Similar results are reported for The Netherlands and the United Kingdom, whereas a much lower reduction is found in southern Europe. The major causes of this decline in Germany were the reorganization of the agricultural market ordinance (reduction of selling prices to world market level, payment of land grants), structural change, and the increasingly selective use of manures. These and other causes (limitation of the amounts of fertilizers by legislation, smaller nutrient losses and greater utilization of the fertilizers, higher fertilizer prices in combination with decreasing selling prices of agricultural products, increase of ecological farming with abandonment of the use of fast-acting inorganic fertilizers, declaration of water protection and conservation areas) will have the effect that the trend, of a declining use of inorganic fertilizers with simultaneously more selective use of the manures will also continue, in weakened form, in future.

Because of the use of composts and of secondary raw material fertilizers (e.g., sewage sludge), an additional input of nutrient into soils will occur in future. While the proponents consider this a possibility of closing the nutrient cycle of nature to the greatest possible extent and of dispensing with the feeding of inorganic fertilizers, critics fear a contamination of the soil by heavy metals and organic pollutants.

5.3.1.2. Legal Framework of Fertilization

In the following, the legislation in Germany is given as an example. It is anticipated that these laws will also be adapted by the EC in the future. In Germany, various laws and regulations greatly limit the use of fertilizers by restrictions and even prohibitions on their application. The fact that in 1994 almost all laws and regulations affecting fertilization were amended or reenacted (at least in draft form) reflects the increased attention of society to the application of substances to soils. The essential contents and objectives of the laws and regulations are described in brief below:

The fertilizer law (Düngemittelgesetz, DümiG of November 15, 1977, last amended by law on September 27, 1994) defines fertilizers and thereby lays down what substances may be applied to agricultural lands. Beyond that the law governs the application of fertilizers and defines the expression "fertilization according to good technical practice" as follows: "The fertilization must be directed, as regards type, quantity and time, to the needs of the plants and of the soil, taking into consideration the nutrients and organic matter available in the soil as well as the site and cultivation conditions; the nutrient needs of the plants in consideration of their productivity under the prevailing site and cultivation conditions and to the standards of quality of the produce".

The fertilizer regulations, which exist in draft, are intended to regulate fertilization according to "good technical practice". Aside from restrictions and prohibitive rules on the application of substances, it is intended to achieve and guarantee a selective fertilization that avoids any overprovision of nutrients by obliging farmers to make soil analyses and written records and checks of fertilizer use.

The liquid manure regulations of various federal states (Bundesländer) regulate the use of the most important manure in Germany. In particular, restrictions, both quantitative and in terms of time of the year, on the application of liquid manure are intended to ensure a high utilization of the nutrients it contains and to minimize the risk of an environmental load.

The recycling and waste law (→ Waste) can stipulate prohibitive rules and restrictions concerning contaminants for the application of manures and secondary raw material fertilizers.

The German soil protection law (Bundesbodenschutzgesetz, BBodSchG), which is in draft form, is intended to protect the soil from damages. The landowner is forced to practice precautionary soil protection, which naturally applies also to the application of materials such as fertilizers.

5.3.1.3. Forms of Fertilizers (→ Fertilizers, **A 10**)

The fertilizers applied can be divided into inorganic and organic fertilizers.

Inorganic fertilizers contain inorganic nutrients or release them after a reaction. Since the nutrients are frequently water-soluble salts or salt mixtures, they are usually available to plants immediately after application or their period of availability can at least be calculated. However, adverse effects can occur when these fertilizers are not properly used; in particular, too high a supply of readily soluble nutrients can lead to increased nutrient loads of the soils and nutrient losses, in addition to reductions in crop yield and quality.

Organic Fertilizers. Two types of organic fertilizers exist, (1) *commercial organic fertilizers*, which are produced by the processing of peat (→ Peat, **A 19**, pp. 27–28) or of vegetable and animal materials (→ Fertilizers, **A 10**, pp. 369–371) and are important in the backyard sector at most, and the *organic manures* whose origin is animal (liquid manure, slurry, solid manure) (→ Fertilizers, **A 10**, 371–374). The nutrients of organic fertilizers are organically combined to different degrees, depending on the type of fertilizer and must the mineralized in the soil in order to be taken up by the plants. Since mineralization processes are highly dependent on temperature and soil moisture, the nutrients are not always available specifically and the amounts taken up by the plants cannot be calculated. The feeding of crops via organic fertilizers therefore makes great demands on the farmer, especially because the nutrient contents of organic fertilizers can vary within wide limits and the precision of application is not optimum.

With the amendment to the German Fertilizer Law of September 27, 1994, *sewage sludge and other wastes* also are for the first time assigned to the fertilizer sector as "secondary raw material fertilizers."

5.3.1.4. Effects of Fertilizers in the Soil

Fertilization pursues several goals. On the one hand, the real task of fertilization—feeding the crop plants with the aim of improving the harvest in quantity and quality—must be carried out; and on the other the soil's fertility must not be threatened. Fertilization can affect various parameters of the soil and therefore the soil's fertility. Strongly influenced are chemical characteristics of the soils (nutrient concentration, soil reaction, pollutants and heavy-metals content), followed by biological parameters (humus content, soil edaphon). Soil-physical parameters are relatively little affected by fertilization.

5.3.1.4.1. Nutrient Contents and Nutrient Losses (→ Fertilizers, **A 10**, pp. 332–341)

Fertilization should supply the soil with plant nutrients without enriching it with nutrients beyond a tolerable extent. The tolerability in this connection corresponds less to the concentration above which toxicity symptoms appear than to the contents above which increased nutrient discharges occur from the soil to groundwater, surface waters, and the air. If soil pollution by nutrient enrichment can be spoken of at all, it is by possible environmental loads as a result of losses from the soils. In addition, disproportionately high concentrations of certain nutrients in particular impair the composition of the plants via the nutrient uptake and reduce yield and quality.

Nitrogen. More than 90% of the nitrogen present in soils is combined in the organic matter. Depending on the depth of topsoil and humus content, between 2000 and 10000 kg nitrogen per hectare are stored under Central European conditions. This nitrogen is mineralized by microbial conversion processes and released as ammonium (NH_4^+); at an average annual mineralization rate of 2% about 40–200 kg of nitrogen per hectare and year are supplied. The NH_4^+ formed can be bound by the exchange complex in the soil, but in aerated soils usually undergoes nitrification by bacteria to nitrate (NO_3^-). Nitrate is not adsorbed in the soil, is thus freely mobile, and is the most important nutrient form of N for plants. High nitrate concentrations in the soil solution are advantageous if they are provided specifically in the period of uptake by the plants and can be largely utilized by these plants.

In the absence of uptake by plants or if the nitrate supply exceeds the uptake capacity of the plants, NO_3^- is displaced from the root space as a result of the movement of water in and out of the soil. A surface runoff or discharge via drainage water contributes to the eutrophication of surface waters and the seas; increased contents of nutrients (N in particular) bring about increased algae growth which results in decreasing oxygen contents in the water. A far greater part in public discussion is played by the washing-out of nitrate from the soil and its translocation into the groundwater, which can greatly impair the drinking water quality. Both during the nitrification and especially during the denitrification in the soil nitrous oxide as an intermediate product of these processes can escape and contribute to the greenhouse effect [5.195].

Before the influence of fertilization on the nitrogen content and nitrogen dynamics in the soil as well as on potential nitrogen losses is described, it has to be pointed out that the agricultural use itself already decisively interfers with the natural nitrogen cycle. Because of the complexity of the effects of individual variables, the influence of fertilization on nitrogen loads in soil will not be exactly quantified in the following.

In long-term nitrogen fertilization the amount of combined nitrogen in the soil is influenced both by the level of fertilization and by the form of fertilizer chosen. Compared with similar, but unfertilized soils, soils exploited agriculturally have higher humus contents with greater depths of topsoil and higher nitrogen contents. The extent of the accumulation of organic matter and therefore also of organically combined nitrogen as a result of long-term fertilization is very strongly influenced by temperature and rainfall [5.196]. Not only the productivity but particularly the reliability of yield of soils, which is independent of current fertilization, can be improved by raising the N contents and the additional N supply. With the exception of crops that react to higher N supplies from the soil with a decrease especially in processing quality (brewing barley, sugar beet, etc.), the positive aspect of the feeding of crop plants predominates. Moreover, N fertilization positively affects the soil biota and the microbial biomass in the soil [5.197]. The nitrogen-fixing microorganisms whose activity falls with increasing nitrogen supply are an exception [5.198].

In addition to the level of N compounds supplied, the type of fertilizer especially influences the amount of nitrogen stored in the soil. Since organic nitrogen fertilizers already contain the nitrogen in the storage form found in the soil, they can raise the N contents in the soil more rapidly and—via higher reproduction rates of the humus in the soil—also more effectively than inorganic nitrogen fertilizers [5.199]. Long-term fertilization with liquid manure improves the humus and nitrogen contents by about the same order of magnitude as an equivalent dosage of farmyard manure [5.200]. The chain of action following inorganic fertilization—more plant growth bringing about a greater additional supply of organic matter—also can have a positive influence on the nitrogen concentration in the soil [5.201], [5.202]. A combination of farmyard manure and inorganic fertilization can increase the N content more than farmyard manure alone [5.196]. Hydrolyzable organic N compounds give the most important contribution to the higher total N contents [5.203].

With increasing additional nitrogen supply from the organic matter or with increased direct N fertilization, there is an increased risk of nitrogen losses from the soil which exceed an unavoidable basic emission [5.204], [5.205]. The factor usually responsible for increased nitrate discharges is a discrepancy between the amount and the period of time of the NO_3^- supply in the soil and the amount extracted by the plants during the same period of time. In this connection, the form of the applied nitrogen plays a minor part: thus, even with ecologically cultivated lands, high nitrate translocations are observed under certain conditions [5.206]. The limit of the N fertilization level above which environmental loads occur depends on many factors (soil, weather, rotation of crops, etc.) and cannot be defined in general. At present the nitrate content in the soil is determined after the harvest of a crop at the start of the translocation period in the autumn and used as a criterion for a balanced NO_3^- supply during the period of uptake by plants [5.207]. Nitrogen balances also can give clues to a possible risk potential from a N fertilization measure. However, they do not permit direct conclusions on the nitrate discharge [5.208], [5.209]. Beyond that, computer simulation models of the nitrogen cycle are being developed that shall reproduce the processes of N dynamics in the soil [5.210] and thus can provide concrete informations for an N fertilization that accords with requirements and is environmentally acceptable. Even though, the models are not yet suitable for practical use at the present time [5.211] because of the large number of input parameters, they will become important in future.

The influence of N fertilization on losses in the form of nitrous oxide cannot be estimated at present. All fertilization measures that trigger nitrification and denitrification processes in the soil contribute to nitrous oxide formation in the soil. Even if N_2O is released in any conversion of organic matter (also in the absence of fertilization) [5.212], it can be assumed that these losses will be greater after N fertilization [5.213]. Organic and ammonia-containing fertilizers increase N_2O production at a higher extent than nitrate fertilizers [5.214]; in all, the N_2O emission level probably depends less on the form of fertilizer than on its amount.

Phosphates. Fertilizer phosphates are absorbed in the soil, especially by iron oxides, and thereby fixed. Because of their low mobility, phosphates accumulate mainly in the topsoil. With rising P fertilization the translocation into the subsoil increases, however, losses from the root zone can be observed only under special conditions (peat soils, [5.215]). Phosphates from organic fertilizers are more mobile in the soil and, compared with phosphates from inorganic fertilizers, can be translocated into deeper soil layers. However, the phosphate contents in groundwaters near the surface is not increased [5.193], [5.216]. The removal of phosphates from the surface soil by erosion and drainage water can more likely be of ecological relevance.

Phosphate fertilization raises the P contents in the topsoil, as is shown by comparison with nonfertilized forest sites. If the P concentrations necessary for plant nutrition are not substantially exceeded no harmful effects in the soil from P fertilization are to be expected.

Potassium. Fertilization with potassium can raise the potassium contents, depending on the buffering capacity of the soil. A disproportionately high K supply can hinder the magnesium uptake by plants and so induce a Mg deficiency [5.217]. In isolated cases, too high potassium concentrations which, however, occur in practice only as the result of a long-term excessive potash fertilization, can impair the quality of crop products. Avoidable losses from the soil as a result of fertilization measures can occur on light soils [5.205], [5.218], but no ecological relevance is known.

5.3.1.4.2. Soil Reaction

Agricultural production is possible only on soils with a soil reaction that is optimum for crop plants. This optimum reaction depends on the kind of plant, the nutrient contents, the kind of soil, and the soil struture. From the plant nutrition standpoint, the optimum pH values are considered to be 6–7. In this range most microorganisms [5.197] and the soil edaphon also show their highest activity. As a result of the natural acidification of the soil by translocation of alkali metal and alkaline-earth metal ions from the soil, which is further reinforced by agricultural production, it is usually necessary to raise the current pH value.

Fertilization measures can decisively influence the extent of soil acidification. Ammonium-containing fertilizers, which are nitrified in the soil and in the course of this release H^+, have a physiologically acidifying effect. Since inorganic N fertilization is carried out worldwide mainly using NH_4^+ containing fertilizer, N fertilization contributes considerably to the lowering of the pH value and involves an increased lime requirement. Phosphate-containing fertilizers that are obtained by solubilizing apatites or are combined with Ca in the course of production can promote the soil reaction. Organic manures show no short-term effect on the soil reaction but in the long term can lead to an increase of the pH value by replacement of cations [5.219].

Through liming, the natural soil acidification and the lowering of the pH value as a result of fertilization measures and acidic inputs from the air can be compensated. This is true not only for agricultural but also, e.g., for forestry lands.

5.3.1.4.3. Humus Content

Humus influences the chemical, physical, and biological properties of soils and is thus an important factor in soil fertility. The humus content is essentially determined by quantities typical of the site, such as nature of soil and climate as well as the conditions for conversion of the organic matter and the possibilities of formation of durable humus materials. Within the limits set hereby, the fertilization can influence the humus content.

Obviously, the humus content of soils can best be influenced by organic fertilizers. Long-term application of solid or liquid manure improves the humus content for a long time afterward. Similar effects are produced by compost and sewage sludge. The humification rates in this connection decrease with increasing supply of organic matter and increasing humus content of the soil [5.229]. Inorganic fertilization can improve the humus content to a limited extent by way of an increase in the vegetation cover and the resulting crop residues (straw, foliage, green manure) or the humus content can at least be maintained with regular working-in [5.221]. Only in crop rotations with high proportions of root crops, crop and root residues even in increased amounts cannot prevent a decrease of the humus content and fertilization in organic form is necessary [5.222].

Very close relations exist between the organic matter content of the soil and its nitrogen pool and nitrogen supply capacity. It is today very questionable from the ecological standpoint to improve the humus content typical of the site by intensive fertilization measures, because with this the risk of N losses detrimental to the environment increases.

5.3.1.4.4. Heavy Metals and Organic Pollutants

Even though heavy metals occur naturally in all soils, a possible accumulation in the soil as a result of fertilization measures must be avoided. The ecological relevance of an accumulation in that connection consists not so much in increased emissions but rather in an increased uptake of heavy metals by plants with the risk of an impact on the health of humans and animals. Similar effects are to be exptected from an application of toxic organic pollutants.

Inorganic fertilizers and farm-produced manures usually contain no organic pollutants and only occasionally small amounts of heavy metals that are not hazardous to the soil. Phosphate fertilizers, which are made from crude phosphates of organogenic origin, can contain considerable amounts of cadmium. By avoiding the use of highly contaminated crude phosphates, the Cd content of P fertilizers has been reduced and currently is usually not more than 40 g/t P_2O_5; hence only an extremely small Cd enrichment occurs in the soil [5.223]. In the EU countries limit values have been drawn up for further reducing the undesirable Cd intake. The level of Cd intake can in addition be influenced by the form of fertilizer chosen [5.224].

Farm-produced manures also can contain heavy metals; high Cu and Zn contents in pig manure originating from the pig feed considerably exceed uptake by plants and these metals can accumulate in the soil over the long term. The risk of a possible load has been reduced as a result of the restriction of fertilization by liquid manure (see Section 5.3.1.2) and to the restriction of copper feed supplements by the German law on feed materials (Futtermittelgesetz).

With the parity of legal treatment between sewage sludge and farm-produced manures, a possible load of soils by "fertilization" with sewage sludge could occur in the future. The present stage of the discussion is given below:

1) If disposal areas are limited, there is a high sociopolitical pressure to use sewage sludge alternatively on agricultural lands
2) From the scientific standpoint, the pollutant tolerance levels of the soil and the permissible pollutant contents are controversal subjects of discussion and in some cases called in question
3) In agricultural practice the reservation against the use of sewage sludge prevail, despite considerable financial incentives

Sewage sludges have a relatively small fertilizing and soil-improving effect and are not necessary as fertilizers from the plant nutrition standpoint. Possible soil loads from the application of sewage sludge have to be justified not by agriculture but by society and the political world in their desire for waste disposal using agricultural areas.

5.3.2. Pesticides (→Crop Protection; →Acaricides; →Fungicides, Agricultural; →Insect Control; →Molluscicides; →Nematocides; →Rodenticides; →Weed Control, all treated in the A series)

5.3.2.1. Introduction

Pesticides are in general formulations of active substances and formulation aids. *Active substances* are chemically defined substances or mixtures of substances and are the actual carriers of the plant-protecting action. The *formulation aids* (e.g., solvents and extenders, deposit builders, emulsifiers, wetting agents, stabilizers, and warning dyes) provide the mixtures with the physical properties which make them suitable for application as sprays, dusting agent, etc. in solid, liquid, or gaseous form. Pesticides arrive, either directly or via the crop plants, onto and into the soil or in groundwater and surface water.

Pesticides are used in agriculture, forestry, industry, and in urban areas. They are frequently used for medical and military purposes. Agriculture forms the largest sales outlet for pesticides. Thousands of tonnes are applied worldwide each year, including 110000 t in the Federal Republic of Germany alone in 1981 [5.225]–[5.227], the use of herbicides and insecticides being predominant internationally. Various pesticides and their chemical structures are given below. They have been selected mainly from chemical considerations and less from their economic importance or biological action (I = insecticide, H = herbicide, F = fungicide, A = acaricide).

Alicyclic insecticides

Endosulfan

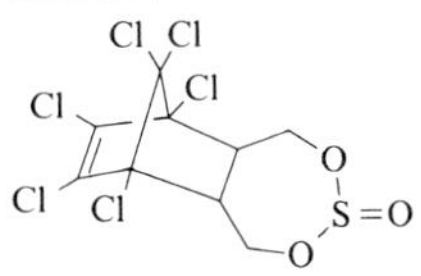

1

Aldrin

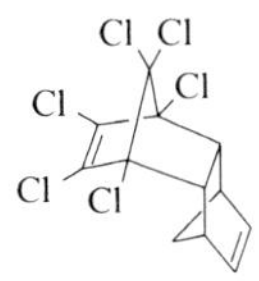

2

Dieldrin

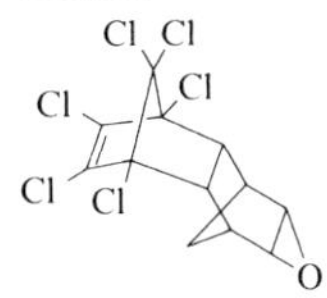

3

Heptachlor

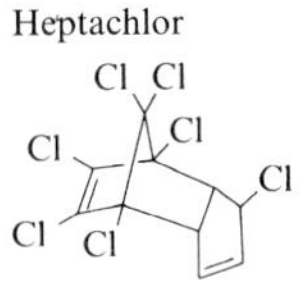

4

cis-Chlordane

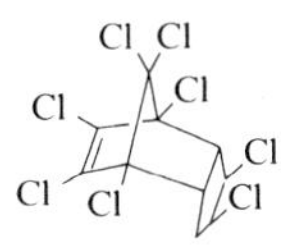

5

Lindane

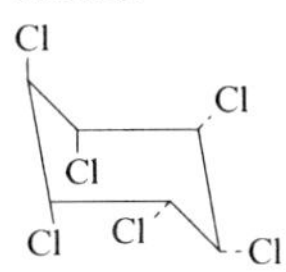

6

Anilides

Propanil (H)

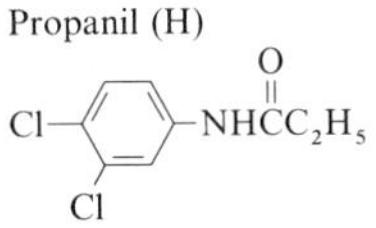

7

Metolachlor (H)

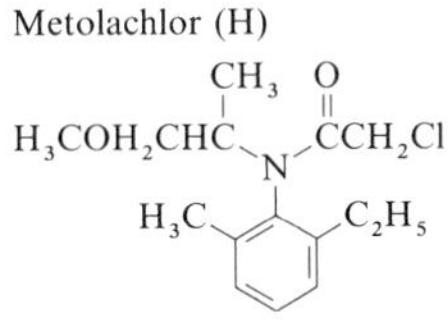

8

Homocyclic herbicides

2,3,6-TBA

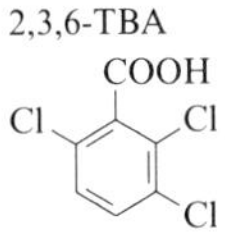

9

Dinoseb

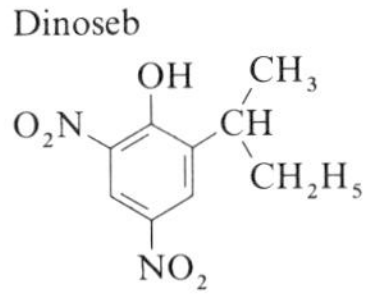

10

DNOC

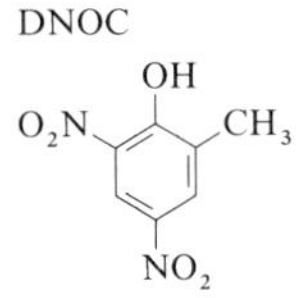

11

Pentachlorophenol

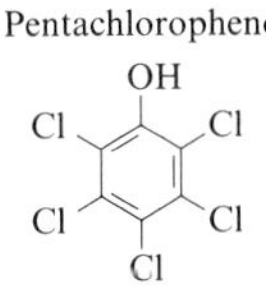

12

Nitralin

13

Trifluralin

14

Heterocyclic herbicides

Paraquat

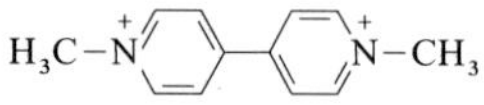

15

Prometryn

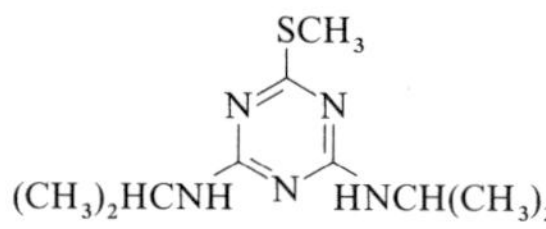

16

Atrazine

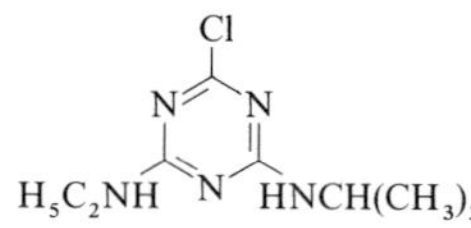

17

Simazine

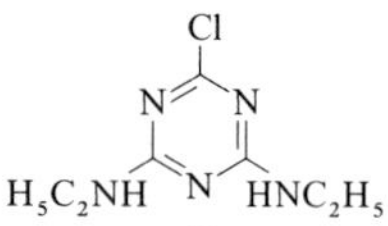

18

Picloram

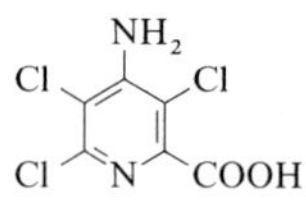

19

Methyl and phenyl carbamates

Triadimenol

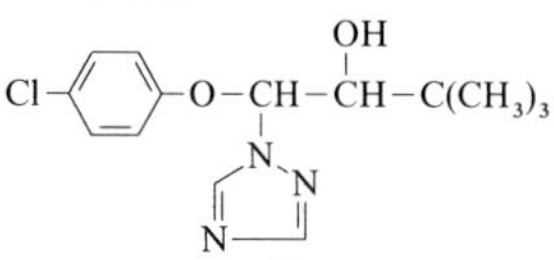

20

Carbaryl (I)

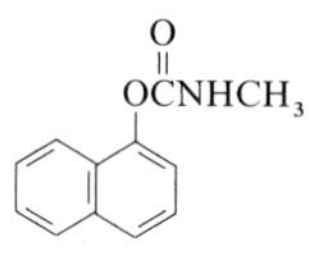

21

Carbofuran (I)

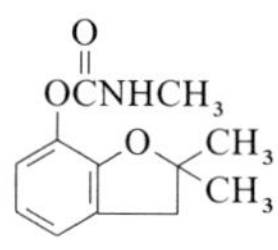

22

Aldicarb (I)

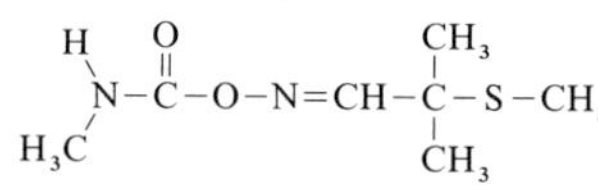

23

Thiophanatemethyl (F)

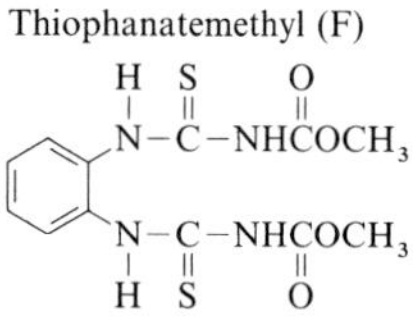

24

Carbendazim (F)

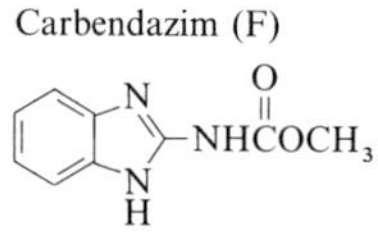

25

Captan (F)

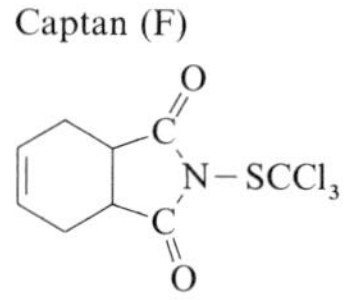

26

Organophosphorus compounds

Malathion (I)

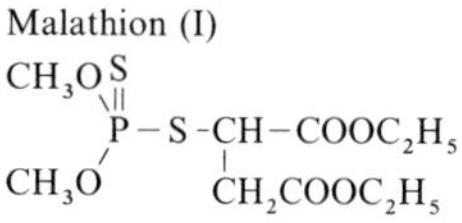

27

Diazinon (I)

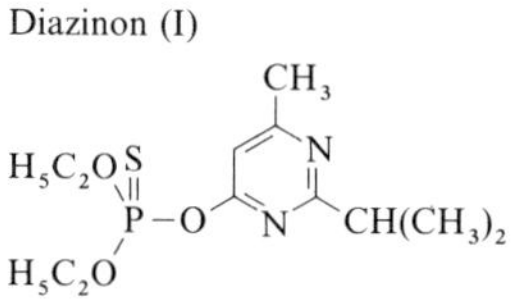

28

Derivatives of phenoxyacetic acid

2,4-D

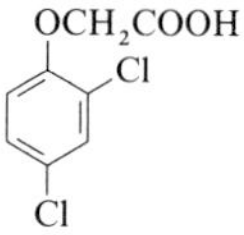

29

MCPA

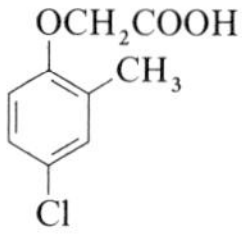

30

2,4,5-T

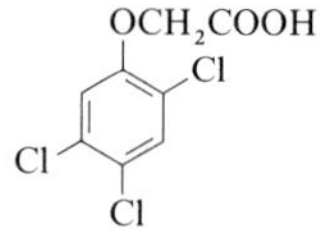

31

Substituted ureas

Monuron

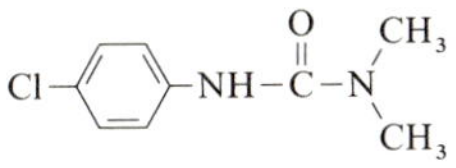

32

Diuron

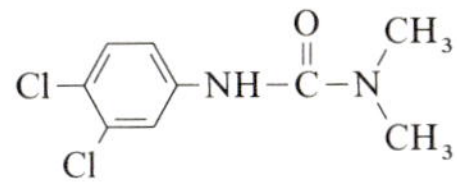

33

Monolinuron

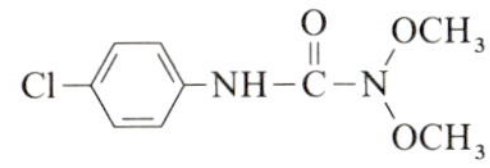

34

Methabenzthiazuron

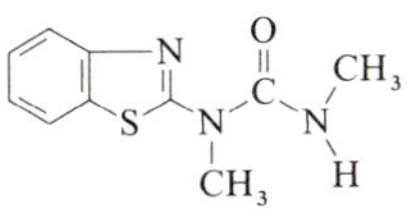

35

Linuron

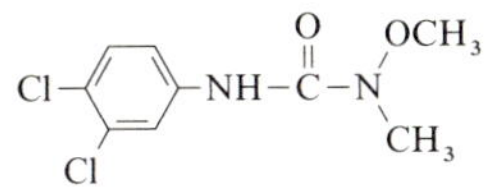

36

The utilization and the toxicity of pesticides is decisively influenced by their behavior in the environment. They can be adsorbed/desorbed, vaporized, leached out, biodegraded and changed photochemically [5.225], [5.227], [5.228]. These processes are clearly influenced by abiotic and biotic factors. Important factors are physicochemical properties of the pesticides such as solubility, volatility, mobility, and degradability [5.229]–[5.231], but also the soil characteristics and the boundary conditions.

Mobility means the translocation of the pesticides in the soil. The mobility is strongly dependent on the water solubility and the sorbability of the pesticides. The water solubility of the individual pesticides varies within wide limits. While triazoles, phenoxy fatty acids, and the benzoic acids have a high water solubility of more than 500 mg/L, chlorinated hydrocarbons such as DDT, aldrin (**2**), and heptachlor (**4**) are only slightly soluble (< 1 mg/L, Table 5.15). The mobility of pesticides is determined under laboratory conditions using percolation tests with soil columns. The organic matter and the clay content of a soil and the pH value have to be taken into account as important factors in this connection. The herbicides 2,4-(dichlorophenoxy)acetic acid [2,4-D, (**29**)] and 2,4,5-(trichlorophenoxy)-acetic acid [2,4,5-T, (**31**)], for example, percolate as rapidly in a low-humus subsoil (pH = 6.6, 0.26 % C_{org}, 19 % clay) as tritium-labeled water, because at pH values above 4 they are not bound to clay (Table 5.15, pp. 314–320 in [5.225]). In the cases of monolinuron (**34**) and chloridazon, a translocation in humus-containing soil (pH = 6.4, 3.2 % C_{org}, 8 % clay) could not be detected. Generally, the adsorption on soils with a very low content of organic matter is small. With high content of organic matter, an increase of adsorption is usually found. It can be assumed that both very low and very high contents of soil organic matter retard the degradation of active substances and reduce their effect. In the case of numerous active substances, 60–95 % of the substances used can be immobilized by adsorption.

Transformation. Pesticides can be biologically transformed (cometabolized) in the soil under aerobic or anaerobic conditions and/or degraded to CO_2 and H_2O (mineralized). In the course of the degradation process, metabolites, which can be harmless or toxic, can be formed.

In the presence of microorganisms pesticides undergo oxidations, hydrolyses, and aerobic and anaerobic ring cleavage [5.229], [5.231]. The stability of the pesticides in the soil is described by their half-life ($t_{1/2}$). The half-lives are regarded as ecologically relevant band-widths of stability. If deviations from an average band-width occur, this is in most cases due to soil properties or soil/environment conditions that lead to particularly low or particularly high stability of active substances [5.227]. Degradation rates usually mean the disappearance of the pesticides within a certain period of time. The majority of the test systems previously proposed and used for pesticide degradation in the soil assume that the active substances are retained in the Ap horizon and are also degraded there. However, because of the detection of certain pesticides in the groundwater [5.232], [5.233] the reactions and effects in the subsoil were investigated to an increased extent [5.234].

The *biodegradability* of the active substances is influenced by biological factors such as the quantity of the microbial biomass, the physio-

Table 5.15. Properties and mode of action of crop-protection agents in soil at doses used in practice and favorable conditions of degradation (11–16 °C, loose, loamy arable soils, pH 5.5–6.5, 2–4% humus, [5.225])[a]

Common name (action)[b]	Chemical name	Prohibited, permitted, (FRG)[c]	Solubility	Volatility	Bonding by			Degradation		Mobility	Formula
					Humus	Clay Fe-Ox	pH influence	Aerobic	Anaerobic		
Cation-active											
Bipyridyl derivatives											
Deiquat (H)	1,1′ethylene-2,2′-bipyridylium	+	4	1	5	4–5	+	3		1–2	
Paraquat (H)	1,1′-dimethyl-4,4′-bipyridylium	+	4	1	5	4–5	+	1		1	**(15)**
Weakly basic diazines											
Chloridazone (H)	5-amino-4-chloro-2-phenyl-(2*H*)-pyridazinone	+	3	1	3	2	(–)	4–5		2	
Bentazone (H)	3-isopropyl-2,1,3-benzothiadiazin-4-one-2,2-dioxide	(+)	4	1				4–5			
s-Triazines											
Prometryn (H)	4,6-diisopropylamino-2-methylthio-*s*-triazine		4	1	3	2	(–)	4		2	**(16)**
Atrazine (H)	4-ethylamino-2-chloro-6-isopropylamino-*s*-triazine	(+)	2–3	1	2	1	–	3–4	3–4	3	**(17)**
Simazine (H)	chlorobis(ethylamino)triazine	(+)	2	1	2	2	(–)	3–4		3	**(18)**
Propazine (H)	2-chloro-4,6-diisopropylamino-*s*-triazine	(+)	2	1	2		(–)	3–4		3	
Anilazine (F)	2,4-dichloro-*N*-(6-chlorophenyl)-1,3,5-triazine-2-amine	+	2	1			(–)	5			
Terbuthylazine (H)	4-ethylamino-6-*tert*-butylamino-2-chloro-*s*-triazine	+	2	1	2–3	2	(–)	3–4			
s-Triazoles											
Amitrole (H)	3-amino-1,2,4-triazole	(+)	4	1	2	2	(–)	4		4	
Propiconazol (F)	1-[2-(2,4-dichlorophenyl)-4-propyl-1,3-dioxolan-2-yl(methyl)]-1*H*-1,2,4-triazole	+	3	1			(–)				
Triadimenol (F)	1-(4-chlorophenoxy)-3,3-dimethyl-(1*H*-1,2,3-triazole-2-yl)-butane-2-ol	+	3	1	(1)		(–)	(1)			**(20)**
Triazinones											
Metribuzin (H)	4-amino-6-*tert*-butyl-3-methylthio-1,2,4-triazine	(+)	4	1	2	(2)	(–)	4			
Metamitron (H)	4-amino-3-methyl-6-phenyl-1,2,4-triazine	+	4	1			(–)	5			
Acid											
Phenoxy fatty acids											
2,4-D (H)	2,4-(dichlorophenoxy)acetic acid	+	4	1	1–2	1	–	4		4	**(29)**
MCPA (H)	4-(4-chloro-*o*-tolyloxy)acetic acid	+	3	1	2	1	–	4		4	**(30)**
Mecoprop (H) CMPP	2-(4-chloro-2-methylphenoxy)propionic acid	+	4	1	2	1	–	4		3	
2,4,5-T (H)	2,4,5-(trichlorophenoxy)acetic acid	+	3	1	2–3	1	–	4	3	3	**(31)**
Fluazifop-butyl (H)	butyl-2-[4(*S*-trifluoromethyl-2-pyridyloxy)-phenoxyl] propionate	(+)	1–2	1	2		(–)	5			
Dichlorprop (H)	2-(2,4-dichlorophenoxy)propionic acid	(+)	3–4	1			(–)	4–5			

Table 5.15. (continued)

Common name (action)[b]	Chemical name	Prohibited, permitted, (FRG)[c]	Solubility	Volatility	Bonding by: Humus	Bonding by: Clay Fe-Ox	Bonding by: pH influence	Degradation: Aerobic	Degradation: Anaerobic	Mobility	Formula
Benzoic acid											
Dicamba (H)	3,5-dichloro-2-methoxybenzoic acid	(+)	4	1	2	1	–	5		4	
Chlorambene (H)	3-amino-2,5-dichlorobenzoic acid	(+)	4	2			(–)	2		4	
2,3,6-TBA (H)	2,3,6-trichlorobenzoic acid		4	2	2	1	–	3		4	**(9)**
Picolinic acid derivatives											
Picloram (H)	3-amino-3,5,6-trichloropicolinic acid	(+)	3–4	1	2	1	–	2–3		3	**(19)**
Phenols, benzenes											
Dinoseb (H)	4,6-dinitro-2-(1-methylpropyl)phenol	(+)	4	2	2		(–)	5		4	**(10)**
DNOC (H)	4,6-dinitro-2-methylphenol	(+)	3	1	2–3		–	4		3	**(11)**
PCP (I)	pentachlorophenol	86	2	1	4	2	–	3–4	4	1	**(12)**
Quintozene (F)	pentachloronitrobenzene (PCNB)	+	1	2			0	2–3	4	1	
Hexachlorobenzene (F)	hexachlorobenezene	77	1	1	4	2	0	2	1	1	
Apolar											
Halogenated hydrocarbons											
Endosulfan (I)	6,7,8,9,10,10-hexachloro-1,5,5*a*,6,9,9*a*-hexahydro-6,9-methano-2,4,3-benzodioxathiepin 3-oxide	+	1	2	4		0	3	(2)	1	**(1)**
Aldrin	1,2,3,4,10,10-hexachloro-1,4,4*a*,5,8,8*a*-hexahydro-1,4-*endo,exo*-5,8-dimethanonaphthalene	81	1	1	(5)	(3)	0	3	(2)	1	**(2)**
Dieldrin (I)	3,4,5,6,9,9-hexachloro-1*a*, 2,2*a*,3,6,6*a*,7,7*a*-octahydro2,7:3,6-dimethanonaphth[2,3*b*]oxirene	71	1	1	5	3	0	1–2	1–2	1	**(3)**
Heptachlor (I)	1,4,5,6,7,8,8-heptachloro-3*a*,4,7,7*a*-tetrahydro-4,7-methanoindene	86	1	2	5	3	0	2–3	2–3	1	**(4)**
Chlordane (I)	1,2,4,5,6,7,8,8-octachloro-2,3,3*a*,4,7,7*a*-hexahydro-4,7-methano-1*H*-indene	71	1	2	5		0	3		1	**(5)**
Lindane (γ-HCH) (I)	γ-1,2,3,4,5,6-hexachlorocyclohexane	(+)	2	1	3–4	3	0	2–3	5	1	**(6)**
DDT (I)	1,1′-(2,2,2-trichloroethylidene)bis(4-chlorobenzene)	77	1	1	5	4	0	1–2	3	1	
Dichloropropene (N)	1,3-dichloropropene	(+)	2	2	2	2	0	5	4	4	
Methyl bromide (I, N)	methyl bromide	(+)	4	3			0				
Phosphate esters											
Malathion (I)	*O,O*-dimethyl-*S*-(1,2-dicarbethoxy)ethyl phosphorodithioate	+	3	2	2–3		0	3	3–4	2	**(27)**
Parathion (I, A)	*O,O*-diethyl-*O*-(4-nitrophenyl)-thionophosphate	+	2	2	4–5	2–3	0	4–5	4	2	
Phorate (I, A)	*O,O*-diethyl-*S*-[(ethylthio)-methyl] phosphodithioate		2–3	2	4	2	0	4		2	
Diazinon (I)	*O,O*-diethyl-*O*-[(2-isopropyl-4-methyl)pyrimidin-6-yl)]-phosphorothioate		2	2	3	2	0	4	4	2–3	**(28)**

Table 5.15. (continued)

Common name (action)[b]	Chemical name	Prohibited, permitted, (FRG)[c]	Solubility	Volatility	Bonding by			Degradation		Mobility	Formula
					Humus	Clay Fe-Ox	pH influence	Aerobic	Anaerobic		
Carbamates											
a) Methylcarbamates											
Carbaryl (I)	1-naphthyl methylcarbamate	86	2–3	2	2–3		0	3–4	4	2–3	**(21)**
Carbofuran (I, A, N)	2,3-dihydro-2,2-dimethyl-7-benzofuranyl methylcarbamate	(+)	3	1	3		0	4	3–4	2–3	**(22)**
Propoxur (I)	2-(1-methylethoxy)phenyl methylcarbamate	(+)	4	2	2		0	(4)			
Phenmedipham (H)	3-[(methyloxycarbonyl)-amino]-phenyl-*N*-(3-methylphenyl)carbamate	+	2	1	3–4		0	4			
b) Phenylcarbamates											
Propham (H)	isopropyl *N*-phenylcarbamate	+	2	1	2	1–2	0	4–5		3	
Chlorpropham (H)	isopropyl *N*-(3-chlorophenyl)carbamate	+	3	1	3	1–2	0	4		2	
c) Thiocarbamates											
Diallate (H)	*S*-2,3-dichloroallyl *N,N*-diisopropylthiocarbamate		2	1	4		0	4–5		2	
Triallate (H)	2,3,3-trichloroallyl diisopropylthiocarbamate	+	2	1	4	2	0	3–4		2	
Butylate (H)	*S*-ethyl *N,N*-diisobutylthiocarbamate	+	2	2	3		0	4		2	
EPTC (H)	*S*-ethyl *N,N*-dipropylthiocarbamate	+	3	2	2		0	4		2	
d) Dithiocarbamates											
Maneb (F)	manganese(II)-1,1-ethylene-bis(dithiocarbamate)	+	1	1			0	5	5		
Metam-Na (F, N, H)	*N*-methyldithiocarbamate	(+)	4	1	1–2		0	4		3	
Zineb (F)	zinc-ethylene-bis(dithiocarbamate)	+	2	1	3–4		0	4–5		2	
Thiram (F)	tetramethylthiuram disulfide	+	3	1	3		0	4–5			
Propanil (H)	*N*-(3,4-dichlorophenyl)-propionamide		3	1–2	2	1–2	0	4			(7)
Mancozeb (F)	ethylene-bis(dithiocarbamate)	+	1	1			0	4–5			
e) Oxicarbamates											
Aldicarb (I, N)	2-methyl-2-(methylthio)propanal *O*-[(methylamino)carbonyl]oxime	(+)	4	1	2	1	0	3–4	4	4	**(23)**
Acetanilides											
Propachlor (H)	2-chloro-*N*-isopropyl-*N*-phenylacetamide	(+)	4	1	2	1–2	0	4		3	
Carboxin (F)	5,6-dihydro-2-methyl-1,4-oxathiin-3-carboxanilide	+	3	1	5	3	0	4–5		2	
Metolachlor (H)	chloroacetanilide	+	4	1	2–3		0	4			**(8)**
Metazachlor (H)	*N*-(2,6-dimethylphenyl)-*N*-(1-pyrazolylmethyl)chloroacetanilide	(+)	3	1	2–3		0	4–5			
Urea derivatives											
Fenuron (H)	1,1-dimethyl-3-phenylurea		4	1	2–3	2	0	3		3	
Monuron (H)	3-(4-chlorophenyl)-1,1-dimethylurea	+	3	1	3	2	0	4		3	**(32)**
Diuron (H)	3-(3,4-dichlorophenyl)-1-1-dimethylurea	+	4	1	3	2	0	3		2	**(33)**
Monolinuron (H)	3-(4-chlorophenyl)-1-methoxy-1-methylurea	+	3–4	1	3	2	0	4		2	**(34)**

Table 5.15. (continued)

Common name (action)[b]	Chemical name	Prohibited, permitted, (FRG)[c]	Solubility	Volatility	Bonding by			Degradation		Mobility	Formula
					Humus	Clay Fe-Ox	pH influence	Aerobic	Anaerobic		
Methabenzthiazuron (H)	1-(benzthiazol-2-yl)-1,3-dimethylurea	+	3	1	3	2	0	3		2	**(35)**
Metoxuron (H)	3-(3-chloro-4-methoxyphenyl)-1,1-dimethylurea	+	4	1	2	(2)	0	4–5		2	
Linuron (H)	3-(3,4-dichlorophenyl)-1-methoxy-1-methylurea		3	1	3–4	2	0	4		2	**(36)**
Chlortoluron (H)	3-(3-chloro-4-methylphenyl)-1,1-dimethylurea	+	3–4	1	2–3		0	4–5			
Isoproturon (H)	*N*-(4-isopropylphenyl)-*N'*,*N'*-dimethylurea	+	3	1	2		0	5			
Dinitroanilines											
Nitralin (H)	4-(methylsulfonyl)-2,6-dinitro-*N*,*N*-dipropylaniline	+	1	1	3–4	(3)	0	3–4		1	**(13)**
Trifluralin (H)	2,6-dinitro-*N*,*N*-dipropyl-4-trifluoromethylaniline	+	1	1	4–5	(3)	0	3–4		1	**(14)**
Pendimethalin (H)	*N*-(3-phenyl)-3,4-dimethyl-2,6-dinitroaniline	+	1	1			0	4			
Benzimidazoles											
Benomyl (F)	methyl-1-(butylcarbamoyl)-2-benzimidazolcarbamate	+	2	1	5	3	0	4		1	
Thiophanatemethyl (F)	1,2-bis[(3-methoxycarbonyl)-2-thioureido]-benzene	+	2	1			0	4–5		1	**(24)**
Carbendazim (F)	methyl(1-*H*-benzimidazol-2-yl)carbamate	+	2	1	4		–	3–4		1	**(25)**
Prochloraz (F)	*N*-propyl-*N*-2,4,6-trichlorphenoxy-ethylimidazole-carboxamide	+	2	1							
Phthalimide derivatives											
Captan (F)	*N*-trichloromethylthio-cyclohex-4-ene-1,2-dicarboximide	+	1	1	1–2		0	4–5		2	**(26)**
Captafol (F)	*N*-(1,1,2,2,2-tetrachloroethylthio)-3,6,7,8-tetra-hydrophthalimide	+	1–2	1	1–2		0	4–5			
Aromatic cyanides											
Dichlobenil (H)	2,6-dichlorobenzocyanide	+	2	2	2		0	3		2	
Ioxynil (H)	3,5-diiodo-4-hydroxybenzonitrile	+	2	1	2		0	5			
Bromoxynil (H)	3,5-dibromo-4-hydroxybenzonitrile	+	3	1	2		0	4–5			
Amino acid derivatives											
Glyphosate (H)	*N*-(phosphonomethyl)glycine	+	4	1	3			3–4			
Metalaxyl (F)	methyl-D,L-*N*-2,6-dimethylphenyl-*N*-(2'-methoxyacetyl)alanate	(+)	4	1	3			4			
Aldehyde oligomers											
Metaldehyde (M)	2,4,6,8-tetramethyl-1,2,3,5,7-tetraoxacyclooctane	+	3	2	2						
Morpholine derivatives											
Fenpropimorph (F)	*cis*-4-[5-(4-*tert*-butylphenyl)-2-methylpropyl]-2,6-dimethylmorpholine	(+)	2	1							

Table 5.15. (continued)

Common name (action)[b]	Chemical name	Prohibited, permitted, (FRG)[c]	Solubility	Volatility	Bonding by			Degradation		Mobility	Formula
					Humus	Clay Fe-Ox	pH influence	Aerobic	Anaerobic		
Quarternary ammonium compounds											
Chlormequat (H)	chlorocholine chloride	+	4	1							
Uracil derivatives											
Bromacil (H)	5-bromo-3-*sec*-butyl-6-methyluracil	+	4	1				3			
Benzofuran derivatives											
Ethofumesat (H)	(2-ethoxy-2,3-dihydro-3,3-dimethylbenzofuran-5-yl) methanesulfonate	+	3	1	2			3–4			

[a] Ranking: **0** practically not; **1** very slight, very weak; **2** slight, weak; **3** medium to increased; **4** high, strong; **5** very high, very strong; ( ) doubtful data; solubility (mg/L H_2O by 20 °C): **1** <1; **2** 1–50; **3** 50–500; **4** >500; strength of binding: Freundlich sorption coefficient of humus and clay: K_{OC}: **1** 1–100; **2** 100–300; **3** 300–1000; **4** 1000–10000; **5** >10000; K_{Clay}: **1** 0.5–50; **2** 50–150; **3** 150–500; **4** 500–5000; **5** >5000; influence of bonding increases with (+) or decreasing (–) pH; 0 no effect; degradation (>75 %): **1** >3 years; **2** 1–3 years; **3** 18 weeks to 1 year; **4** <18 weeks; **5** <6 weeks; volatility (vapor pressure in hPa at 20–25 °C): **1** $<10^{-3}$; **2** 10^{-3}–100; **3** 100–5×10^3; **4** $<5\times10^3$ (data for pure substance). Mobility (annual depth of translation by 1500 mm rain, at 25 °C in loamy soil with ca. 2% humus and pH 6 in the Ap horizon: **1** <10; **2** 5–20; **3** 15–40; **4** >35 cm. [b] (A) acaricide; (F) fungicide; (H) herbicide; (I) insecticide; (N) nematocide; (R) rodenticide; (M) molluscicide. [c] Since 19.. prohibited, + permitted, ( ) prohibited in water protection zones since September 1, 1988 (Pflanzenschutzanwendungsverordnung).

logical state (activities), and adaptability of the microorganisms.

Factors such as temperature, salinity, water content, oxygen availability and the supply of organic substances determine the composition of the microbial populations at the site and have therefore to be considered during the application of the pesticides.

On the soil surface, pesticides are *photochemically degraded* and *chemically converted.* Pentachlorophenol and DDT can be degraded photochemically, e.g., in presence of UV radiation to CO_2 and hydrochloric acid. Chemical conversions of the pesticides also occur by hydrolysis, oxidation, and isomerization.

Table 5.15 shows that the pesticides differ greatly in their biological and chemical degradability. While the phenoxy fatty acids can be degraded in arable soils, mainly in their topsoil, in about 18 weeks, pentachlorobenzene and the majority of the chlorinated hydrocarbons remain detectable for many years.

Behavior in the Environment. The applied amount, the formulation, and the type of application are very important for the action and the environmental behavior of pesticides. The interplay of the various factors finally determines the actual behavior of the pesticides in the environment.

Composition and properties of pesticides and their behavior in soils at doses used in practice and favorable degradation conditions are listed in Table 5.15 [5.225].

Legal Aspects, Test Systems. The use of pesticides causes a contamination of the soils; the magnitude of the enrichment depends on the persistence of an active substance, the amount and frequency of application, and also on the sorption, degradation, and translocation conditions of the soil [5.228], [5.230]. The formation of pesticide residues in agricultural soils as well as the contamination of groundwater and surface water cannot be excluded. According to the German law on crop protection (Pflanzenschutzgesetz), pesticides, if applied properly, must have no impact on humans, animals, or groundwater and not affect the natural economy in a way that does not comply with the present state of scientific knowledge. In order to guarantee these requirements, pesticides are subjected before use to extensive studies, so that sufficient information is available on their behavior (physicochemical properties, adsorption/desorption, degradability, etc.) and possible side-effects on plants, selected animals, and microorganisms under standardized laboratory conditions. Test systems for determining the behavior of pesticides, such as the degradability or the percolation behavior in soil columns have been introduced in the industrialized countries and by international organization (OECD, EU, WHO, FAO). Test systems such as those for acute and chronic toxicity, mutagenicity, carcinogenicity, and bioaccumulation are used for estimating the risk potential [5.235]–[5.241]. The information gained is only to a limited extent applicable to the ecosystem. According to experience with aquatic organisms, the informative value of tests can certainly be increased by compilation of test batteries, but the uncertainties remain considerable.

In the soil area the action of the pesticides on the whole microbial population can be estimated. With the exception of the oxidation of ammonia to nitrite (nitrification) the microbial indicators used hitherto, such as respiration, enzyme activities, or bacterial counts, have not been very sensitive. This was actually to be expected, because the ecological relevance of the result of a measurement generally increases dramatically with multi-species test system, usually at the expense of sensitivity [5.235]. Investigations of the action of pesticides under field conditions are certainly desirable. At present this seems difficult for logistical and financial reasons and because of the low reproducibility of the results.

Almost all known tests investigate the action of environmental chemicals under aerobic conditions. Anaerobic test systems are hardly available [5.236], [5.242]. It is to be expected, though, that aerobic degradation will differ greatly from anaerobic degradation, with regard both to the composition of the microbial populations and to degradation pathways. An action of the pesticides differing according to O_2 availability is therefore to be expected. ALEF et al. [5.243] have developed the arginine mineralization as a simple and reproducible method for estimating the action of environmental chemicals on anaerobic microbial populations. Both the NOEL and the EC_{50} values can be determined from the dose–response relationship (Fig. 5.4; the greatest inhibition of the arginine mineralization by the environmental chemicals was usually found on the third day of incubation). The new method can be used for determining the acute and chronic effects of environmental chemicals [5.243].

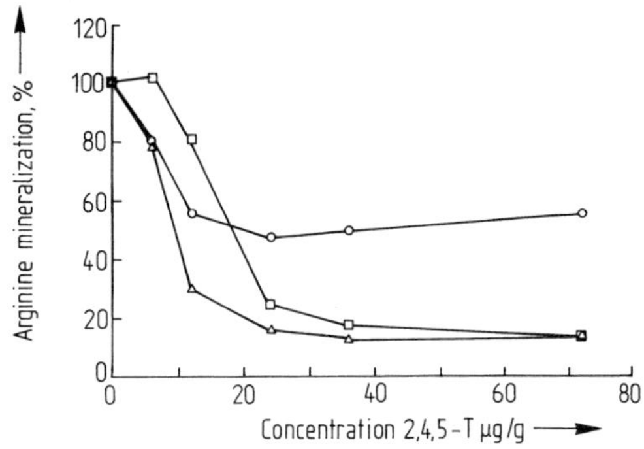

Figure 5.4. Action of 2,4,5-T on the anaerobic arginine mineralization; data in % of the control values [5.243] Spheres = 2 days; triangles = 3 days; squares = 4 days

To obtain registration for a pesticide, generally studies carried out under defined "worst-case" conditions must be submitted investigating the effect on soil microorganisms and on selected organisms. The results obtained give information on the behavior of pesticides in the environment; they should be reasonably comprehensive so that the authorities can make a decision taking into account the intended use. The objective of the investigations can never be a complete picture of the behavior of the pesticides in all environmental compartments.

5.3.2.2. Side-Effects

Pesticides can interfere with various processes in the cell. Their mechanisms of action differ [5.244]. While photosynthesis is the main point of attack for herbicides, fungicides attack nuclear division, nucleic acid and protein synthesis, respiration, biogenetic pathways, and the membrane function. Insecticides influence respiration and the membrane structure.

To be able to use pesticides profitably, active substances are required whose selective action on crop plants is as different as possible from that on target organisms. A very high specificity of action, though, is not always achievable. Herbicides, for example, can influence respiration and biogenetic pathways aside from photosynthesis, and also inhibit germination. Undesirable effects on soil organisms, plants, animals, and humans can therefore be expected. Pesticides can reach humans via the food chain or drinking water.

Side-Effects on Soil Organisms. The action of pesticides on soil organisms is to be regarded as negative if the soil biota is so extensively disturbed that essential functions of the ecosystem such as decomposition and humification, aggregation, nutrient immobilization and mobilization, loosening, and bioturbation are lastingly impaired. In this connection, negative effects can be called "critical" that last for more than one month [5.227], [5.245]. From the standpoint of species preservation, action of pesticides should preferably also not lead to any complete change of species or at least guarantee a rapid recolonization after the specific use of soil is finished. The potential regenerability of the soil biota has in any case to be preserved.

The effects of herbicides, fungicides, and insecticides are not completely specific. Their action depends on the concentration used, on the mode of application, and on the properties of the site. Fungicides such as benomyl and thiabendazole, for example, can have a negative effect on earthworms. Other pesticides can influence bacteria and stimulate ammonification.

Herbicides such as paraquat, linuron, and 2,4-D can stimulate bacterial activities (e.g., respiration), but on the other hand injure algae [5.246]–[5.248]. Insecticides can in some cases also act as nematocides and then be used as such. Carbamates and phosphate esters are in some cases toxic to earthworms and their activity even at a normal dose [5.246].

Different site conditions not only modify the expected impact of pesticides but can also lead to different side-effects, a topic which has been little studied up to now. It is impossible to consider all environmental factors in the laboratory test; notwithstanding that, basic information on the possible ecotoxicological action of pesticides is indispensable in practice.

Various groups have proposed uniform procedures or tests of estimating the action of pesticides on soil microorganisms (respiration, substrate-induced respiration, nitrification, N mineralization, dehydrogenase activity, [5.245]). A series of data is available in the literature, on the effect of pesticides on the composition of microbial populations (bacterial counts, fungal counts, or plate counts). This information should be regarded with caution, especially because no uniform procedure is given and comparability of the data is therefore scarcely possible. Furthermore, reliable information on the action of pesticides on the interactions between the various organisms in the soil is scarcely obtainable. There is a great need for research in this field.

Uptake, Accumulation, and Side-Effect on Plants. In principle, the uptake and accumulation of pesticides can occur in all parts of plants. This depends on the plant species, on the physicochemical properties of the substances and on the site (pH value, content of organic substances in the soil, soil structure).

In *soil application*, pesticides are absorbed with the soil water by the plant roots and transported in the xylem. With high vapor pressure, the active substances can also reach the roots in the soil airspace.

In *leaf application*, the pesticides can penetrate into the leaf by diffusion. The uptake is considerably easier for lipophilic than for hydrophilic substances. Both in soil and in leaf application the active substances can accumulate in the plant leaf. However, pesticides can also remain at the site of application, and not be transported in plants. Aside from the positive effects of pesticides, negative effects can also occur. The side-effects can extend as far as to serious damage [5.249]. Under extreme weather conditions, damage, e.g., to crop plants cannot be ruled out. After the use of urea derivatives, for example, an increased attack by fungi has occasionally been observed in grain cultivation. The selectivity of the active substances is of great importance and is therefore tested during the registration process.

A consequence of the use of pesticides in agriculture is that occasional residues in foods cannot be prevented. In order to be able to estimate the risk potential, both the acute and the chronic action on animals (e.g., rats, guinea-pigs) are tested during the registration process. In these studies, the NOEL value (→ Toxicology, p. 239) for the experimental animal is determined. In application to humans this maximum inactive dose is still for safety's sake reduced by a factor of at least 100. From this, there is obtained the highest still acceptable daily dose in milligrams per kilogram food (ADI; → Foods, 3. Food Additives, p. 563; → Insect Control, **A 14**, p. 315). To prevent that the maximum quantity regulation is exceeded when using pesticides, the concept of "waiting time" is introduced. This means how many days before the harvest of the crop, at the latest, a pesticide may be used to make sure that the maximum amount is not exceeded. As a result of the maximum quantity regulation, the marketing of foods with excessive pesticide residues is forbidden [5.250]. The strict conditions cannot, however, completely prevent human foods from containing pesticide residues [5.251].

Contamination of Groundwater and Surface Waters. Because of their mobility, pesticides can be translocated after their application into the groundwater and surface water. Despite high dilution in the surface water, however, remobilization, can occur. The pollution of surface water by polluted groundwater depends on the length and bonding strength of the filter section, the groundwater flow rate, and the degradation rate of the pesticides in the groundwater. With long-term and intensive use of mobile active substances on the soil, pollution of the groundwater must be reckoned with.

In order to protect water bodies from contamination by pesticides, national and international regulations have come into force in which the maximum permitted concentrations have been stipulated. The EC drinking water regulations, for example, provide that drinking water may contain no more than 0.5 μg/L of pesticides in total and 0.1 μg/L of a single active substance.

5.4. Effects of Abandoned Waste Disposal Sites

5.4.1. Introduction

The genesis of contaminated sites is different from site to site. A distinction can be made, however, between those sites in which the chemical nature, quantity, number and distribution of the contaminants are largely production-specific and those in which they are waste- and deposition-specific.

The first group includes contaminated plant sites (*contaminated land*). Aside from different localities and extents, these sites also differ in the arrangements of manufacturing plants, handling facilities for water-endangering substances, intermediate storage facilities, waste deposits, etc., depending on the type of plant (gasworks, oil refinery, a factory for manufacturing explosives, weapons, plant-protection agents, dyes, etc.). Old sites are contaminated from specific manufacturing operations or production residues. At present no statistically supported results on the extent of this contamination exist. In addition, usually no data are available on the persistence and mobility of specific contaminants in soil and groundwater and on their possible ecotoxicity or human toxicity.

The situation is usually different in the case of abandoned waste disposals (*abandoned sites*). By contrast with the contaminated land, from which dangers often arise only from changes in ground use (e.g., excavation of foundation pits) or by direct contact with "easily accessable" hazardous substances (for example by children at play), dangers originating from old deposits can be attributed almost entirely to their emissions. Especially important are the leachates, loaded with considerable quantities of pollutants that reach the soil and subsequently the groundwater and can then endanger its use as drinking water. In Germany alone account must be taken of more than 100 000 old deposits. This kind of contamination of soil and groundwater therefore represents a problem also with regard to the area concerned, which is steadily increasing because the pollutant plumes continue to expand in the groundwater. This section describes the qualitative and quantitative effects of leachate emissions from abandoned waste disposals on groundwater.

5.4.2. Material Contents in Groundwater Affected by Abandoned Waste Disposal Sites

Inorganic Substances. The natural substance contents of groundwater are changed considerably by the leachates of abandoned sites [5.252]. Table 5.16 assembles the mean, median, and other percentile values calculated for contaminated groundwaters from unconsolidated rocks, together with the maximum values of the various inorganic parameters.

Organic Substances. As regards the number of substances the possible contamination of groundwater in the outflow from abandoned sites by organic compounds is far greater than that by inorganic compounds. About 1200 organic groundwater contaminants have up to now been detected by extensive GC–MS analytical studies in Germany and the United States [5.253] in the outflow from abandoned disposal sites. Only a relatively small number of these contaminants, however, are detected fairly frequently and at the same time have concentrations distinctly above 1 μg/L (United States) or above 0.1 μg/L (West Germany), respectively. The majority of the identified organic contaminants (>1000) have a detection frequency (DF, see Section 5.4.3.2) of less than 0.1 %; i.e., in 1000

Table 5.16. Inorganic constituents of groundwaters contaminated by abandoned waste disposal sites and their statistical parameters

Parameters	DL [1]	Number of measured values [2], n	n >DL	DF [3], %	Mean A [4]	Mean B [5]	Median 50% percentile	75% percentile	Maximum
Main cations									
Calcium, mg/L	5.0	318	317	99.7	177.	175.	156.	229.	785.
Magnesium, mg/L	2.0	321	320	99.7	39.3	38.1	22.0	42.3	436.
Sodium, mg/L	5.0	338	337	99.7	140.	128.	45.6	130.	3600.
Potassium, mg/L	1.0	217	217	100.	34.4	32.9	12.0	36.0	350.
Ammonium, mg/L	0.02	276	231	83.7	14.4	11.0	0.41	6.2	945.
Iron, mg/L	0.01	335	330	98.5	10.3	9.6	2.0	11.0	240.
Manganese, mg/L	0.01	329	311	94.5	1.6	1.5	0.43	1.2	33.1
Main anions									
Hydrogencarbonate, mg/L	18.0	233	233	100.	527.	519.	408.	661.	2458.
Chloride, mg/L	1.0	360	360	100.	218.	202.	74.2	168.	6020.
Sulfate, mg/L	1.0	352	350	99.4	218.	194.	122.	223.	8560.
Nitrate, mg/L	0.5	347	244	70.3	58.7	25.6	2.5	24.3	11 500.
Trace elements									
Aluminum, µg/L	40.0	168	135	80.4	947.	596.	80.0	573.	59 600.
Lead, µg/L	0.05	334	129	38.6	7.6	6.3	<0.05	2.1	450.
Cadmium µg/L	0.05	335	126	37.6	40.3	1.5	<0.05	0.3	13 000.
Chromium (total), µg/L	10.0	279	179	64.2	39.5	21.2	<10.	<10.	5 120.
Copper, µg/L	10.0	275	189	68.7	26.5	24.5	<10.	20.0	577.
Nickel, µg/L	10.0	249	187	75.1	116.	22.9	13.0	24.0	23 200.
Mercury, µg/L	0.2	232	39	16.8	<0.2	<0.2	<0.2	<0.2	3.4
Strontium, µg/L	40.0	205	201	98.0	604.	580.	399.	764.	5 580.
Zinc, µg/L	10.0	328	276	84.1	921.	410.	64.5	220.	168 000.
Trace anions									
Arsenic [arsenates(III)/(V)], µg/L	0.2	253	172	68.0	61.0	45.4	1.3	6.0	4 000.
Nitrite, mg/L	0.1	285	107	37.5	0.77	0.49	<0.1	0.3	79.0
Selenium [selenates(I)/(VI)], µg/L	0.2	112	20	17.9	0.22	0.10	<0.2	<0.2	14.0
Boron (borates, total), µg/L	20.0	259	253	97.7	949.	832.	181.	530.	31 200.
Fluoride, mg/L	0.5	97	57	58.8	<0.5	<0.5	<0.5	0.50	1.7
Phosphate (total), mg/L	0.1	207	113	54.6	2.3	1.8	0.1	0.60	115.

[1] DL = Detection limit. [2] In some cases the mean values of one well are used, therefore n is the number of measured wells. [3] DF = Analytical detection frequency (percentage from $n >$ DL to n_{total}). [4] Arithmetic mean of all measured values including maximum value. [5] Arithmetic mean rejecting maximum value.

contaminated groundwater samples polluted by abandoned waste deposits they are detected on average less than once.

A bias for substances or classes of substances detected in the contaminated groundwater due to the analytical method used can be almost ruled out because the sample preparation (substances extractable under acid or basic/neutral conditions, highly volatile substances, and pesticides) and the analytical method (essentially GC–MS) were aimed at the detection of the largest number of groundwater contaminants possible [5.254], [5.255].

Table 5.17 lists, in the order of decreasing *detection frequency*, the important organic groundwater contaminants determined in Germany (old states) with various statistical parameters of their concentrations. Of the 19 contaminants with detection frequencies ≥10%, 12 (63%) are volatile halogenated hydrocarbons. Of these, 5 are alkane derivatives, 4 are alkene derivatives, and 3 are chlorosubstituted

Table 5.17. Organic constituents of groundwater contaminated by abandoned waste disposal sites and their statistical parameters

Parameters	DL[1], µg/L	Number of measured values[2], n	n > DL	DF[3] %	Nonlogarithmic values: Mean µg/L	Nonlogarithmic values: Median 50% percentile, µg/L	Nonlogarithmic values: 75% percentile, µg/L	Nonlogarithmic values: Maximum µg/L	Mean of logarithmic values: log	Mean of logarithmic values: µg/L
Tetrachloroethylene	0.1	277	195	70.4	56.1	1.4	3.7	6500.	0.62	4.2
Trichloroethylene	0.1	277	154	55.6	1010.	2.3	11.0	128000.	1.06	11.5
cis-1,2-Dichloroethylene	4.	153	46	30.1	22100.	166.	1150.	411000.	2.37	234.
Benzene	1.0	127	38	29.1	141.	13.5	38.0	1795.	1.24	17.4
1,1,1-Trichloroethane	0.1	206	47	22.8	16.5	1.0	7.4	270.	0.25	1.8
m/*p*-Xylene	0.1	92	21	22.8	39.9	2.8	4.9	447.	0.53	3.4
Trichloromethane	0.1	236	52	22.0	76.2	1.4	10.1	2800.	0.32	2.1
1,2-Dichloroethane	5.0	16	3	18.8	107.	<5.	<5.	210.	1.86	72.4
Vinyl chloride	1.0	136	24	17.7	1690.	99.5	1950.	12000.	1.93	85.1
Toluene	0.1	127	21	16.5	73.2	3.5	11.0	911.	0.61	4.1
Dichloromethane	10.0	114	17	14.9	38100.	438.	16800.	499000.	2.94	871.
Tetrachloromethane	0.1	201	29	14.4	1.2	0.2	0.3	23.0	0.48	3.0
4-Methylphenol (*p*-Cresol)	1.0	124	17	13.7	42.0	8.6	20.7	283.	1.08	12.0
Chlorobenzene	0.1	93	12	12.9	52.9	2.3	6.5	388.	0.67	4.7
2-Methylphenol (*o*-Cresol)	0.5	124	16	12.9	1.0	5.9	9.2	63.3	0.66	4.6
1,2-Dichlorobenzene	0.1	90	11	12.2	1.4	0.9	2.1	6.6	0.08	0.83
1,4-Dichlorobenzene	0.1	90	11	12.2	31.9	2.2	37.5	265.	0.53	3.4
Naphthalene	0.1	124	15	12.1	2.2	1.1	2.1	12.6	0.16	1.5
Ethylbenzene	0.1	124	14	11.3	32.2	4.4	62.5	160.	0.82	6.6
o-Xylene	0.1	127	12	9.5	13.8	4.7	19.6	69.0	0.77	5.9
2,4,6-Trichlorophenol	0.1	124	11	8.9	3.2	0.6	1.5	24.1	−0.15	0.71
3,5-Dimethylphenol	0.1	124	10	8.1	16.2	6.9	35.6	61.0	0.60	4.0
Phenol	0.1	124	10	8.1	2.2	1.5	3.6	5.6	0.24	1.7
1,1,2-Trichloroethane	0.5	140	11	7.9	36.0	3.0	70.0	190.0	0.88	7.6
1,3-Dichlorobenzene	0.1	90	7	7.8	11.5	1.1	38.2	74.0	0.23	1.7
trans-1,2-Dichloroethylene	5.0	134	10	7.5	57.1	50.0	95.0	135.	1.61	40.7
Cumene	0.1	90	5	5.6	2.4	3.1	4.5	4.7	0.20	1.6
1,1-Dichloroethane	10.0	130	7	5.4	52.7	53.0	90.0	110.	1.64	43.7
Acenaphthene	0.1	124	6	4.8	6.3	1.3	17.1	32.0	0.19	1.6
2,4-Dichlorophenol	0.1	124	6	4.8	3.5	0.9	9.6	17.2	−0.07	0.85
3-Chlorophenol	0.1	124	6	4.8	12.7	12.3	20.8	22.8	1.02	10.5
1,2,4-Trimethylbenzene	0.1	90	4	4.4	1.9	1.5	2.6	3.5	0.22	1.7
2-Ethyltoluene	0.1	90	4	4.4	0.6	0.6	0.8	1.0	−0.21	0.62
2,4,5-Trichlorophenol	0.1	127	5	3.9	7.1	1.3	16.6	31.9	0.23	1.7
Mesitylene	0.1	90	3	3.3	1.7	<0.1	<0.1	4.0	−0.02	0.95
Phenanthrene	0.1	124	4	3.2	1.5	0.6	2.6	4.4	−0.48	0.33
Tribromomethane	1.0	130	4	3.1	3.0	2.5	4.5	6.0	0.39	2.5

[1] Detection limit. [2] In some cases mean values of one well are used, therefore n is the number of measured wells. [3] Analytical detection frequency (percentage from n > DL to n_{total}).

benzenes. Among the 7 halogen-free contaminants, benzene and its alkyl derivatives (4 compounds) are most important; in addition 2 phenols and naphthalene have detection frequencies ≥10%.

Highly volatile halogenated hydrocarbons also show the highest *mean concentrations* in groundwaters, both when evaluated from the concentrations themselves and when evaluated from the logarithms of the concentrations. Dichloromethane (38 mg/L and 871 µg/L, respectively), *cis*-1,2-dichloroethylene (22 mg/L and 234 µg/L, respectively), vinyl chloride (ca. 1.7 mg/L and 85 µg/L, respectively) and trichloroethylene (ca. 1 mg/L and 11.5 µg/L, respectively) should be mentioned in the first place, so confirming quantitatively the dominance of these compounds as groundwater contaminants from abandoned waste disposal sites [5.256].

5.4.3. Contamination Criteria and their Quantification

The influence of leachates from abandoned sites on the groundwater can be characterized and quantified by comparing for relevant substances the statistical parameters of data records of contaminated and noncontaminated groundwaters calculated from the concentrations determined. In the case of inorganic substances, the geogenic fraction must be taken into account and in the case of a few organic substances the ubiquitous contamination by these compounds.

5.4.3.1. Contamination Factor

The contamination factor (*CF*) of a substance represents the ratio of the observed concentration in the contaminated area to the concentration in the uncontaminated area of a site [5.257]. The contamination factor indicates how many times the concentration of a substance in the groundwater at the contaminated area is increased compared with that in the unaffected area. It is therefore a measure of the intensity of the effect of a substance on groundwater. It is furthermore also an indirect measure of the mobility and persistence of a contaminant on the emission path from the waste site into the groundwater, because only substances that are highly water soluble, have a low sorbability on C_{org} and clay minerals (i.e., a low geoaccumulability), and that are not or not rapidly degradable can ocur in significantly enhanced concentrations in groundwaters.

As a measure of the strength of influence of the individual *inorganic parameters*, the ratio is formed of the mean, rejecting the maximum value, from the data record of the contaminated groundwater (see Table 5.16) to the mean from the data record of groundwaters that are unpolluted or affected as little as possible (see [5.252]). In principle a contamination factor can be calculated for other statistical parameters also (median, percentile values, maximum value). A subscript is used (as in CF_{mean} in this case) for differentiating the contamination factors. For all inorganic parameters studied, with the exceptions of selenium and nitrate, the contamination factors CF_{mean} are >2 (Table 5.18). Values >20 are reached by ammonium (CF_{mean} 64.7), arsenic (CF_{mean} 37.8), chromium (CF_{mean} 36.6), and boron (CF_{mean} 23.9). Values >10 are found for phosphate (CF_{mean} 12.9), aluminum (CF_{mean} 11.9), cadmium (CF_{mean} 11.5), and potassium (CF_{mean} 11.3). This verifies the strongly increased concentration of these substances in groundwaters as compared with their natural concentrations.

Table 5.18. Inorganic groundwater contaminants from abandoned waste disposal sites listed in the order of decreasing contamination factors

Substance	Mean*, contaminated**	Mean, uncontaminated [5.252]	Contamination factor (CF_{mean})
Ammonium, mg/L	11.0	0.17	64.7
Nitrite, mg/L	0.49	<0.01	>49.0
Arsenic, µg/L	45.4	1.2	37.8
Chromium, µg/L	21.2	0.58	36.6
Boron, µg/L	832.	34.8	23.9
Phosphate, mg/L	1.8	0.14	12.9
Aluminum, µg/L	596.	50.	11.9
Cadmium, µg/L	1.5	0.13	11.5
Potassium, mg/L	32.9	2.9	11.3
Manganese, mg/L	1.5	0.18	8.3
Nickel, µg/L	22.9	2.8	8.2
Sodium, mg/L	128.	17.5	7.3
Chloride, mg/L	202.	33.4	6.0
Iron, mg/L	9.6	2.0	4.8
Lead, µg/L	6.3	1.4	4.5
Zinc, µg/L	410.	117.	3.5
Copper, µg/L	24.5	7.0	3.5
Sulfate, mg/L	194.	59.0	3.3
Magnesium, mg/L	38.1	14.3	2.7
Strontium, µg/L	580.	240.	2.4
Hydrogencarbonate, mg/L	519.	238.	2.2
Calcium, mg/L	175.	83.2	2.1
Nitrate, mg/L	25.6	13.3	1.9
Selenium, µg/L	0.10	0.55	0.2
Fluoride, mg/L	<0.5	<0.1	
Mercury, µg/L	<0.2	<0.1	

* Arithmetic mean, rejecting maximum value.
** see Table 5.16

5.4.3.2. Emission Detection Frequency

The emission detection frequency of a substance is by definition the percentage of detections of a substance that has entered the groundwater with the leachate from the abandoned site [for example, if the boron contents of 259 contaminated groundwater samples (different wells) is determined and the concentration in 222 samples is higher than the background, the emission detection frequency is 85.7 %]. It is important in this connection whether an emitted substance (contaminant) can occur naturally in the groundwater or not. Naturally occurring are almost all inorganic contaminants, but not the purely anthropogenic organic substances. To recognize an increase of contents of the natural inorganic constituents of water, their geogenic content must be known.

In the case of the anthropogenic organic contaminants, the natural geogenic "background" concentration is zero; i.e., analytical detection frequency and emission detection frequency are usually identical. In the case of the detection frequency, it is not important—as in the case of the contamination factor—how large the increase is but only that the value is increased.

Inorganic Substances. In the case of inorganic substances, the arithmetic mean value from the data record of the "least affected" groundwaters from unconsolidated rock aquifers is by definition the concentration for the average geogenic background [5.252]. The number of measured values above this geogenic background value is determined from the data record of the groundwaters affected by abandoned waste disposal sites and the percentage detection frequency calculated. If the background value is below the detection limit, the corresponding detection limit is used instead (fluoride, nitrite, mercury). The results are listed in Table 5.19. Hydrogencarbonate shows the highest emission detection frequency (85.8 %), followed by boron (85.7 %), sodium (83.4 %), chloride (79.2 %), and magnesium (76.0 %). A lower detection frequency is found for lead (38.3 %), nitrite (35.4 %), aluminum (33.0 %), selenium (22.3 %), cadmium (14.9 %), and mercury (14.7 %).

Comparison of the detection frequencies (Table 5.19) with the contamination factors (Table 5.18) shows that a high contamination factor does not necessarily also imply a high detection frequency. Thus aluminum and cadmium have relatively high contamination factors of 11.9 and 11.5, respectively, but comparatively small detection frequencies (33.9 % and 14.9 %, respectively). The reverse situation also occurs (e.g., HCO_3^-, Mg). For inorganic substances high contamination factors in combination with high detection frequencies can be regarded as typical of emissions of abandoned sites into the groundwater. Contamination factors >10 and detection frequencies >50 % are found for arsenic (37.8, 61.3 %), ammonium (64.7, 53.6 %), boron (23.9, 85.7 %), potassium (11.3, 74.2 %), and chromium (36.6, 56.3 %).

Table 5.19. Inorganic groundwater contaminants from abandoned waste disposal sites listed in the order of decreasing detection frequencies

Substance	"Background" (mean unaffected [5.252])	Number of samples affected*	Number of samples with value > "background"	Detection frequency, %**
Hydrogen-carbonate	238. mg/L	233	200	85.8
Boron	34.8 µg/L	259	222	85.7
Sodium	17.5 mg/L	338	282	83.4
Chloride	33.4 mg/L	360	285	79.2
Magnesium	14.3 mg/L	321	244	76.0
Calcium	83.2 mg/L	318	237	74.5
Potassium	2.9 mg/L	217	161	74.2
Strontium	240. µg/L	205	152	74.1
Sulfate	59. mg/L	352	227	64.5
Nickel	2.8 µg/L	249	160	64.3
Manganese	0.18 mg/L	329	206	62.6
Arsenic	1.2 µg/L	253	155	61.3
Chromium	0.58 µg/L	279	157	56.3
Ammonium	0.17 mg/L	276	148	53.6
Fluoride	<0.1 mg/L	97	51	52.6
Copper	7.0 µg/L	275	144	52.4
Nitrate	13.3 mg/L	347	176	50.7
Phosphate	0.14 mg/L	207	86	41.5
Zinc	117. µg/L	328	135	41.2
Iron	2.0 mg/L	335	136	40.6
Lead	1.4 µg/L	334	128	38.3
Nitrite	<0.01 mg/L	285	101	35.4
Aluminum	50. µg/L	168	57	33.9
Selenium	0.55 µg/L	112	25	22.3
Cadmium	0.13 µg/L	335	50	14.9
Mercury	<0.1 µg/L	232	34	14.7

* see Table 5.16. ** Emission detection frequency.

Organic Substances. For some organic substances the analytical detection limit is so low that even ubiquitous loadings not attributable to old deposit emissions are recorded. Thus the analytical detection limit for trichloroethylene, tetrachloroethylene and trichloromethane is 0.1 µg/L, whereas the ubiquitous loadings are in the order of up to 0.4 µg/L [5.256]. Thus, an analytical detection of these substances in the concentration range of the ubiquitous loading cannot be interpreted as the detection of emission from an abandoned site.

For determining the emission detection frequency, the so-called "relevance limit" was created. This is the concentration above which a detection of an organic substance can be attributed with considerable certainty to the emission from an old deposit. In nearly all cases the relevance limit corresponds to the detection limit. Only for a few volatile chlorinated hydrocarbons was the relevance limit set at 1 µg/L.

Table 5.20. Organic groundwater contaminants from abandoned waste disposal sites listed in the order of decreasing emission detection frequencies

Substance	Number of analyses	Relevance limit, µg/L	Number of measured values > relevance limit	Detection frequency, %
Tetrachloroethylene	277	1.0	128	46.21
Trichloroethylene	277	1.0	113	40.79
cis-1,2-Dichloroethylene	153	4.0	46	30.07
Benzene	127	1.0	37	29.13
p/*m*-Xylene	92	0.1	21	22.83
1,1,1-Trichloroethane	206	0.1	47	22.82
Trichloromethane	236	0.1	52	22.03
1,2-Dichloroethane	16	5.0	3	18.75
Vinyl chloride	136	1.0	24	17.65
Toluene	127	0.1	21	16.54
Dichloromethane	114	10.0	17	14.91
Tetrachloromethane	201	0.1	29	14.43
4-Methylphenol (*p*-Cresol)	124	1.0	17	13.71
2-Methylphenol (*o*-Cresol)	124	0.5	16	12.90
Chlorobenzene	93	0.1	12	12.90
1,2-Dichlorobenzene	90	0.1	11	12.22
1,4-Dichlorobenzene	90	0.1	11	12.22
Naphthalene	124	0.1	15	12.10
Ethylbenzene	124	0.1	14	11.29
o-Xylene	127	0.1	12	9.45
2,4,6-Trichlorophenol	124	0.1	11	8.87
3,5-Dimethylphenol	124	0.1	10	8.06
Phenol	124	0.1	10	8.06
1,1,2-Trichloroethane	140	0.5	11	7.86
1,3-Dichlorobenzene	90	0.1	7	7.78
trans-1,2-Dichloroethylene	134	5.0	10	7.46
Cumene	90	0.1		5.56
1,1-Dichloroethane	130	10.0	7	5.38
2,4-Dichlorophenol	124	0.1	6	4.84
3-Chlorophenol	124	0.1	6	4.84
Acenaphthene	124	0.1	6	4.84
2-Ethyltoluene	90	0.1	4	4.44
1,2,4-Trimethylbenzene	90	0.1	4	4.44
2,4,5-Trichlorophenol	127	0.1	5	3.94
Mesitylene	90	0.1	3	3.33
Phenanthrene	124	0.1	4	3.23
Tribromomethane	130	1.0	4	3.08
1,1,2,2-Tetrachloroethane	124	1.0	3	2.42
2,6-Dichlorophenol	124	0.1	3	2.42
1,2,4-Trichlorobenzene	90	0.1	2	2.22
2-Methylnaphthalene	90	0.1	2	2.22
3-/4-Ethyltoluene	90	0.1	2	2.22
Fluoranthene	123	0.1	2	1.63
2-Chlorophenol	124	0.1	2	1.61
1,2,3-Trichlorobenzol	90	0.1	1	1.11
1,3,5-Trichlorobenzol	90	0.1	1	1.11
1-Methylnaphthalene	90	0.1	1	1.11
2-Chlorotoluene	90	0.1	1	1.11
Pentachlorobenzene	93	0.1	1	1.08
Fluorene	124	0.1	1	0.81
Pentachlorophenol	124	0.1	1	0.81
Pyrene	124	0.1	1	0.81
Dibromochloromethane	125	1.0	1	0.80
Acenaphthylene	124	0.1	0	0.00
Anthracene	124	0.1	0	0.00
Benz[*a*]anthracene	124	0.1	0	0.00
Benzo[*a*]pyrene	124	0.1	0	0.00
Benzo[*b*]fluoranthene	124	0.1	0	0.00
Chrysene	124	0.1	0	0.00
Hexachlorbenzene	93	0.1	0	0.00

Since in the case of tetrachloroethylene, for example, of 195 measured values above the analytical detection limit (0.1 µg/L), 67 measured values (24.2%) fall below the relevance limit of 1 µg/L, the emission detection frequency is only 46.2% (Table 5.20), compared with the analytical detection frequency of 70.4% (Table 5.17).

Table 5.20 shows, in the order of decreasing emission detection frequencies, the organic substances that have been detected in groundwaters affected by old deposits. Those of prime importance are tetrachloroethylene (DF 46.21%) and trichloroethylene (DF 40.79%). Altogether only 7 substances have a detection frequency >20% and 19 substances a detection frequency >10%. "Only" 49 substances have a detection frequency of >1%.

5.4.3.3. Mean Emission Concentration

A suitable measure of the quantitative influence of a substance is its mean emission concentration. This is the difference between the mean concentration from affected groundwaters and the mean concentration from unaffected groundwaters of a given origin. In the case of organic substances, the small ubiquitous concentration

Table 5.21. Inorganic groundwater contaminants from abandoned waste disposal sites listed in the order of decreasing mean emission concentrations

Substance	Mean* affected, μg/L	Mean unaffected [5.252], μg/L	Mean emission concentration, μg/L
Hydrogencarbonate	519 000	238 000	281 000
Chloride	202 000	33 400	168 600
Sulfate	194 000	59 000	135 000
Sodium	128 000	17 500	110 500
Calcium	175 000	83 200	91 800
Potassium	32 900	2 900	30 000
Magnesium	38 100	14 300	23 800
Nitrate	25 600	13 300	12 300
Ammonium	11 000	170	10 830
Iron	9 600	2 000	7 600
Phosphate	1 800	140	1 660
Manganese	1500	180	1 320
Boron	832	34.8	797
Aluminum	596	50.0	546
Nitrite	490	< 10.	480
Strontium	580	240	340
Zinc	410	117	293
Arsenic	45.4	1.2	44.2
Copper	24.5	7.0	17.5
Chromium	21.2	0.58	20.6
Nickel	22.9	2.8	20.1
Lead	6.3	1.4	4.9
Cadmium	1.5	0.13	1.4
Selenium	0.10	0.55	0.
Fluoride	< 500.	< 100	
Mercury	< 0.2	< 0.1	

* Mean rejecting maximum value (cf. Table 5.16).

has a negligible influence on the difference, so that here the mean emission concentration is equal to the mean concentration in the affected groundwaters.

Inorganic Substances. The mean emission concentrations of inorganic substances from abandoned sites are listed in Table 5.21. For better comparison, all concentrations are listed in the unit μg/L and sequenced in the order of decreasing mean emission concentration. Hydrogencarbonate has the highest mean emission concentration (281 mg/L), followed by chloride (169 mg/L), sulfate (135 mg/L), sodium (111 mg/L), and calcium (92 mg/L). In the old deposit (which functions as a "bioreactor") CO_2 is formed in significant amounts and converted to HCO_3^-. Thus, hydrogencarbonate must be a principal emitted species, just like Na^+ and Cl^-, which are emitted together (both originating from NaCl). Since construction waste rubble makes up a considerable part of the volume of waste, SO_4^{2-} and Ca^{2+} are also among the substances that are emitted most frequently and in high quantities from abandoned waste disposal sites.

Organic Substances. To calculate the mean emission concentration of organic substances, the logarithms of the individual concentrations are first taken, because a few measured values are so high that the arithmetic mean values of concentrations before taking their logarithms are meaningless and make comparison impossible. Since logarithms cannot be taken of the concentrations below the detection limit (exact concentrations cannot be determined, but the substance is present in small amounts), only the measured values above the detection limit are included in the mean emission concentration of the organic substances. This procedure is reasonable also because the detection frequencies of most organic substances are very small (see Table 5.20), which means that a high proportion of measured values are below the detection limit. If these measured values were to be included in the calculation, most mean concentrations would fall below the detection limit.

Table 5.22 assembles the organic substances in the order of decreasing mean emission concentrations. The list is headed by dichloromethane and a number of other chlorinated and highly volatile compounds, a fact which again confirms the predominance of this group of organic substances in emissions from abandoned sites.

Table 5.22. Organic groundwater contaminants from abandoned waste disposal sites listed in the order of decreasing mean emission concentrations

Substance	Number of analyses	Number of measured values > relevance limit	Mean of log. concentrations*	Mean emission concentrations, μg/L
Dichloromethane	114	17	2.94	870.96
cis-1,2-Dichloroethylene	153	46	2.37	123.42
Vinyl chloride	136	24	1.93	85.11
1,2-Dichloroethane	16	3	1.86	72.44
1,1-Dichloroethane	134	10	1.61	40.74
trans-1,2-Dichloroethylene	130	7	1.64	43.65
Benzene	127	37	1.24	17.38

Table 5.22. (Continued)

Substance	Number of analyses	Number of measured values > relevance limit	Mean of log. concentrations*	Mean emission concentrations, µg/L
4-Methylphenol (*p*-Cresol)	124	17	1.08	12.02
Trichloroethylene	277	113	1.06	11.48
3-Chlorophenol	124	6	1.02	10.47
1,1,2-Trichloroethane	140	11	0.88	7.59
Ethylbenzene	124	14	0.82	6.61
o-Xylene	127	12	0.77	5.89
Chlorobenzene	93	12	0.67	4.68
2-Methylphenol (*o*-Cresol)	124	16	0.66	4.57
1,1,2,2-Tetrachloroethane	124	3	0.63	4.27
Tetrachloroethylene	277	128	0.62	4.17
Toluene	127	21	0.61	4.07
3,5-Dimethylphenol	124	10	0.60	3.98
p-/*m*-Xylene	92	21	0.53	3.39
1,4-Dichlorobenzene	90	11	0.53	3.39
Tetrachloromethane	201	29	0.48	3.02
Tribromomethane	130	4	0.39	2.45
Trichloromethane	236	52	0.32	2.10
1,1,1-Trichlorethane	206	47	0.25	1.78
Phenol	124	10	0.24	1.74
1,3-Dichlorobenzene	90	7	0.23	1.70
2,4,5-Trichlorophenol	127	5	0.23	1.70
Pseudocumene	90	4	0.22	1.66
Cumene	90	5	0.20	1.58
Acenaphthene	124	6	0.19	1.55
Naphthalene	124	15	0.16	1.45
Mesitylene	90	3	−0.02	0.95

Substance	Number of analyses	Number of measured values > relevance limit	Mean of log. concentrations*	Mean emission concentrations, µg/L
2,4-Dichlorophenol	124	6	−0.07	0.85
1,2-Dichlorobenzene	90	11	−0.08	0.83
2,6-Dichlorophenol	124	3	−0.10	0.79
2,4,6-Trichlorophenol	124	11	−0.15	0.71
2-Ethyltoluene	90	4	−0.21	0.62
Phenanthrene	124	4	−0.48	0.33
1,2,4-Trichlorobenzene	90	2	< −1	<0.1
2-Methylnaphthalene	90	2	< −1	<0.1
3-/4-Ethyltoluene	90	2	< −1	<0.1
Fluoranthene	123	2	< −1	<0.1
2-Chlorophenol	124	2	< −1	<0.1
1,2,3-Trichlorobenzol	90	1	< −1	<0.1
1,3,5-Trichlorobenzol	90	1	< −1	<0.1
1-Methylnaphthalene	90	1	< −1	<0.1
2-Chlorotoluene	90	1	< −1	<0.1
Pentachlorobenzene	93	1	< −1	<0.1
Fluorene	124	1	< −1	<0.1
Pentachlorophenol	124	1	< −1	<0.1
Pyrene	124	1	< −1	<0.1
Dibromochloromethane	125	1	< −1	<0.1
Acenaphthylene	124	0	< −1	<0.1
Anthracene	124	0	< −1	<0.1
Benz[*a*]anthracene	124	0	< −1	<0.1
Benzo[*a*]pyrene	124	0	< −1	<0.1
Benzo[*b*]fluoranthene	124	0	< −1	<0.1
Chrysene	124	0	< −1	<0.1
Hexachlorobenzene	93	0	< −1	<1

* Only measured values > relevance limit.

6. Measurement and Evaluation of Soil Pollution

6.1. Sampling and Investigation of Soil Pollutants

6.1.1. Introduction

Sampling and investigation of samples are integral parts of procedures for the hazard and risk assessment of soil pollution. Soil pollution here means the presence and usually uneven spatial distribution (both horizontally and vertically) of certain substances in soil. These substances either do not occur naturally in these soils or their natural inventory has been enriched to an undesirably high degree by inputs via various exposure paths. The first substance group includes high- and low-volatility halogenated hydrocarbons, aromatic and polycyclic aromatic hydrocarbons, petroleum hydrocarbons, but

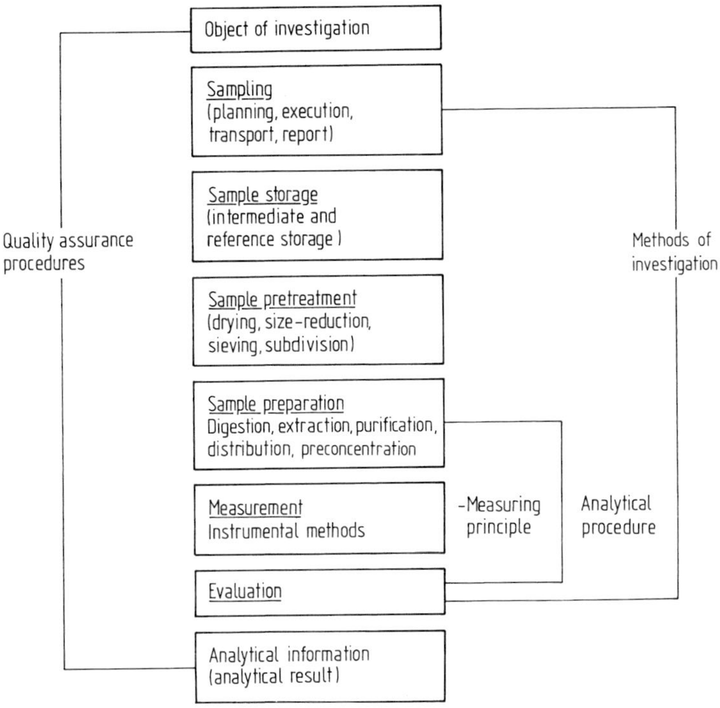

Figure 6.1. Individual steps of object-related procedures for detection and identification of soil contaminants

also organically bound residues of pesticides. Nonessential heavy metals such as lead, cadmium, mercury, and thallium are to be assigned to the second group. Procedures for the detection and recording of soil pollution are combinations of various methods and techniques. The most important individual steps are given in Figure 6.1.

Sampling has a central importance in this process (see also, → Sample Preparation for Trace Analysis, **B5**). All other individual steps of investigation processes depend on planning and execution of sampling. Compared with the endeavour and successes in the field of chemical and instrumental analysis, the sampling, but also the sample pretreatment for the preparation of a representative laboratory sample, have up to now been underestimated.

For international harmonization of test standards the ISO has undertaken the task of collecting the national standards for soil investigations, and—after thorough tests, improvements, and adaptations to the current state of analysis of introducing them into international ISO standard. The Technical Committee "Soil Quality" (ISO TC 190–Soil Quality) has already submitted draft standards for numerous test methods, including methods for sampling and sample pretreatment, to which reference is repeatedly made in the following [6.1]. The German *Methodenhandbuch Bodenschutz I* ("Methods Book for Soil Protection I"), submitted in 1994, first as a draft and without claiming completeness, is a valuable reference compendium, in which national and international standards and rules in the field of soil investigations have been collated in systematically abridged version [6.2].

6.1.2. Sampling

Sampling is the first and most important step of soil investigation procedures. Not only the quality of the test data but also decisions that can have far-reaching consequences at many levels (economic, health, and environmental policy) depend on the well-designed planning and careful execution of sampling. The soil sample once taken is seldom reproducible with respect to its chemical and physical composition, i.e., a second sample from the same sampling point may not be identical.

The basis for the development of an object-related sampling strategy is the definition of the objective of the investigation, i.e., the future use of an area, taking into account all site-specific features that should be protected. If knowledge about the nature, position, and extent of soil pollution is lacking or inadequate, a preliminary investigation is required. In the preliminary investigation sampling points are rather randomly selected and smaller in number: both selection and number of sampling points must suffice to delimitate spatially the area under suspicion. Information about previous uses of the land or soil and about material inputs and their paths is very helpful in this phase. If the suspicion of soil pollution is confirmed, the sampling for the detailed investigation has to be planned.

For this, criteria according to ISO/CD 10381-1 [6.3], which are listed in brief below, have to be observed:

1) *Definition of the sampling site*

The sampling site is delimitated and defined, taking into account the nature of the relevant substances and their lateral spread and vertical distribution. This information has been obtained in the preliminary investigation.

2) *Selection of sampling patterns and sampling points*

Sampling patterns can be based on

Known "hot spots" and substance distribution
Distribution along a line
Strip-wise distribution
No specific estimation of substance distribution

The design of patterns includes the setting up of single sampling points to complex patterns developed on the basis of statistical and computer-aided methods.

The following patterns can be taken [6.3], [6.4]:

Systematic sampling (regular grids), e.g. sampling on a circular grid or triangular grid; sampling along a linear source
Nonsystematic (irregular) sampling, e.g. random sampling; stratified random sampling, unaligned random sampling

The location and the number of sampling points in the site depend on the aim of investigation and the results of preliminary studies.

3) *Selection of the test parameters*

Depending on the objective, the test parameters are selected in accordance with the results from the preliminary investigation.

4) *Single or composite samples*

Single samples are necessary during the preliminary investigation to characterize the heterogeneity of the soil, to delimit the contaminated area, and to determine the vertical substance distribution. Single samples are taken from one sampling point. Composite samples are individual samples from different sampling points but from the same sampling depth. It has to be determined for each individual case which single samples shall be combined to give composite samples.

5) *Number of samples*

The number of samples of one site follows from the results of preliminary investigation and is linked with the objective of the investigation. Because of the general complexity of the sampling, it is advisable to take more single or composite samples than are required for the laboratory investigations. The reserve samples can be investigated subsequently on request.

6) *Sampling from disturbed or undisturbed soil stratification*

For most laboratory tests for soil pollutants, sampling from disturbed stratification is sufficient. For certain soil-physical and soil-biological parameters, a sampling from undisturbed stratification is necessary. For this, the arrangement of soil prospecting (trial pits) can also be useful.

7) *Sampling equipment*

Depending on the problem, various equipment or techniques be used, such as hand or power auger, percussion drilling systems, tube borers, but also mobile hydraulic drilling systems. For sampling from undisturbed stratification, metal cylinders and cutting frames (e.g., for sampling of forest soils) are used. ISO Standard ISO/CD 10381-2 [6.5] contains an assortment of the techniques concerned. Cross contaminations of the samples by abraded metal and by carry-over of soil material from sample to sample, must be avoided.

8) *Sampling depth*

The sampling depth depends on the type of soil use, but can also relate to the particular route of effect for special kinds of acceptors (e.g., children's playgrounds). In cultivated soils, samples are usually taken from the top soil, i.e., to a depth of 20–30 cm for arable soils and to a depth of max. 10 cm for pasture soils. In uncultivated soils, samples are usually taken to a soil depth of max. 10 cm. For refilled sites and in analogous cases, sampling by soil layers is appropriate.

9) *Sample quantity*

The sample quantity per sampling point and layer is determined by the kind and size of the laboratory investigation, the condition of the soil material (e.g., mineral grains, sand, pebble, or foreign materials), and required reserve samples. For chemical analysis 500 g of fine soil should be obtained.

10) *Use of containers*

The containers must ensure that the sample material is not contaminated by foreign substances and that no losses or changes of material occur by adsorption, volatilization, or degradation processes. For soils containing organic pollutants, glass vessels with tight closures should be used. For preserving soil samples after sampling on the spot, cooled transport containers are useful. They should be used for samples whose constituents to be tested are volatile or degradable. The soil samples can also be temporarily stored cool in these containers for a limited time. The sample should be clearly and unmistakably labeled on both the container and its lid.

11) *Safety precautions*

The personal protection of the sampling staff also has to be included in the sampling strategy [6.6].

12) *Sampling report*

The sampling report should contain the following information [6.3], [6.7]:

Description of site and surrounding area, time of sampling

Cartographic sketch of the area and the sampling points

Execution of the sampling in all details

Deviations from plan and their cause

Delivery to the laboratory

The quality of the sampling and of the reporting is closely related to the sampling staff. It has to be ensured that the sampling, as a decisive quality parameter of the laboratory data, does not put that data in question. Qualification and advanced training of the sampling staff are therefore of utmost importance.

6.1.3. Sample Pretreatment

After their entry into the laboratory or after appropriate temporary storage, soil samples are usually pretreated for analysis in various ways, depending on the analytical problem.

Like sampling, the sample pretreatment also is a step of the analytical method, to which little attention often is paid, yet which nevertheless can contribute considerably to the total error of a method. Especially, contamination or losses of materials to be tested must be avoided or minimized. Contamination can arise, e.g., during mechanical size-reduction by grinding devices as well as by the carry-over of soil material. Material losses or changes often occur through inappropriate drying procedures, the effects of light, prolonged periods of storage, or change of storage conditions.

It is advisable to pretreat the whole field sample. The sample is dried sometimes after preceding manual or mechanical preliminary size-reduction with sieves. A mild thermal drying is carried out at up to 40 °C. Freeze-drying is to be preferred for tests on organic compounds (see Section 6.1.8.2) as well as on volatile inorganic substances (e.g., mercury). For this, though, the often large initial sample has first to be reduced, because the capacity of the freeze-drying units is often limited.

After drying, the whole sample is mechanically size-reduced. For sample material from the freeze drying size reduction is sometimes not necessary.

Afterwards, sample division is advisable if reserve samples have to be formed. This is performed in mechanical sample-dividers, because manual sample division is not without errors.

To prepare the laboratory sample ready for analysis, the size-reduced soil material is sieved to obtain different grain-size fractions. In general, fractions with a grain size < 1 mm are used for the laboratory sample. Other grain fractions are, however, also customary. Sieving to definite grain sizes is performed to obtain samples of the same chemical and physical composition, and also for improving the extraction yield in the analysis.

During the storage of dried soil samples, the soil reabsorbs water, depending on the room temperature and humidity. The water content of so-called air-dried samples is usually about 3–5%. In individual cases a further drying can be required. An ISO draft standard exists for the handling of samples [6.8]. The laboratory is in charge of the correct care during sample pretreatment. Documentation of the laboratory operations is mandatory for reasons of quality assurance.

6.1.4. Sample Preparation

The sample preparation is usually the first individual step of an analytical method and an important part of combined analytical methods. Its objective is

1) To separate the elements/compounds to be tested from interfering matrix constituents and to convert them into a form accessible to the measurement (e.g., aqueous solution)
2) To render low substance concentrations (ng/g or lower) detectable or determinable by preconcentration steps

In the case of instrumental direct methods (e.g., X-ray fluorescence), this preparation is omitted; at the most, melt digestion is used. The relatively low power of detection of direct methods is disadvantageous [6.9]. Wet-chemical combined methods, on the other hand, enable the use of various measuring systems as well as simple calibration by aqueous standard solutions (→ Trace Analysis, **B5**, pp. 97–98).

Systematic sources of error can originate from contamination of equipment, vessels, and reagents (blank values), as well as from the working environment (laboratory air). Material losses occur through volatilization and adsorption.

The digestion techniques available today are increasingly powerful, because of their degree of automation, ensuring optimal decomposition conditions. Consequently they are more economical than previous techniques, e.g., because of higher sample throughput. In principle, the following modes of processing can be used:

1) Digestion
2) Extraction
3) Elution

In *inorganic analysis* the terms digestion and extraction are frequently used synonymously. According to the recent ISO methods, *digestion* means the total digestion of soils, but also of geological samples with the aid of strong acids (mixtures). The digestion yield should amount to 100%.

The term *extraction* means that relatively weak acids, acid mixtures, complexing agents, neutral salt solutions, etc. are applied. Extraction is used when relatively low material yields can be tolerated or a certain element fraction or form of element binding (e.g., plant-available compounds, exchangeable cations) has to be determined selectively. Environmentally relevant material mobilities in soils can be characterized approximately by various extractants. Figure 6.2 shows the corresponding relationships. Extraction is usually used for the detection of organic substances in soils.

The *elution* of soil contaminants is frequently carried out from the standpoint of groundwater management. However, elution with water, which is still practiced today, is very controversial, since the significance of the results can fluctuate greatly owing to considerable influences from the soil matrix.

For sample preparation for the detection of *inorganic substances* in soils, high-performance digestion techniques are available today that enable an efficient, low-cost, environmentally compatible, and largely low-error operation:

1) Wet digestion (extraction) by means of open systems with convective heat transfer (heating block, reflux); closed high-pressure digestion techniques [6.10], [6.11]; or microwave technique [6.12], [6.13]
2) Dry decomposition [6.14] by means of combustion systems such as pressureless decomposition in open systems [flask combustion (e.g., Schöniger digestion), Wickbold combustion with oxyhydrogen gas flame], cold-plasma ashing and dry ashing or melt digestions

Of the wet-chemical digestion techniques, the microwave technique is becoming increasingly important. It is time-saving, environmentally

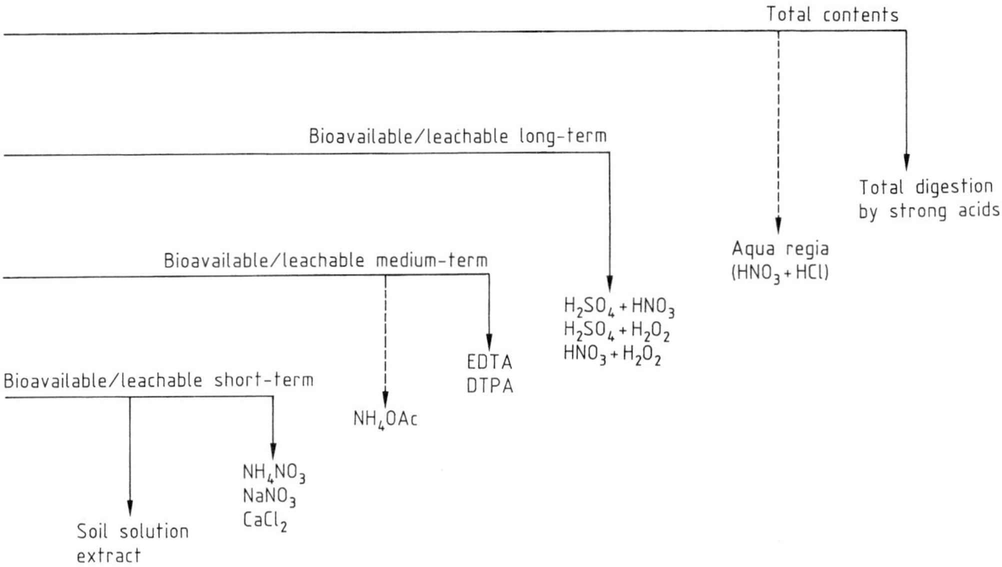

Figure 6.2. Relations between the state of mobility of heavy metals in soils and state-related digestion or extraction solutions [6.37]
DTPA = diethylenetriaminepentaacetic acid; EDTA = ethylenediaminetetraacetic acid

compatible and, with regard to the reproducibility of the material yield, often more favorable than, e.g., pressure digestion techniques.

For the determination of *organic compounds* in soils, there are standard procedures for the preparation of measuring solutions. The separation processes are based on partitioning processes between phases, which obey physicochemical laws. For organic soil pollutants the following technical processes are available in routine analysis [6.15], [6.16]:

1) Head-space processes (→Gas Chromatography, **B5**, pp. 211–212) for highly volatile halogenated hydrocarbons and aromatic hydrocarbons (benzene, toluene, xylenes) that are also suitable for difficultly vaporizable and polar compounds after converting them to suitable derivatives
2) Soxhlet extraction for low-volatility halogenated hydrocarbons
3) Extraction with supercritical fluids (SFE) as a relatively new process route for the detection of PAHs, PCBs, and pesticides in soils

In individual cases, preconcentration steps are sensible and necessary in order to determine the true substance concentration (e.g., preconcentration by ion exchange or in cartridges).

6.1.5. Measurement

Inorganic Substances. Inorganic soil pollutants, which are found matrix-free and without interfering accompanying substances in the measuring solution, can be determined by different instrumental methods. Of these, atomic absorption spectrometry (AAS, →Atomic Spectroscopy, **B5**, pp. 602–616) and its various special techniques are used most widely:

1) Flame technique for alkali metals and heavy metals (F–AAS)
2) Graphite furnace technique (GF–AAS) for determining elements in the lower ng/g range
3) Hydride technique for hydride-forming elements such as As, Se, or Sb
4) Cold vapor technique for mercury

All of the techniques are used widely in soil and environmental analysis, depending on the kind of element and its concentration and on the desired sensitivities of detection. Correction of the nonspecific background by deuterium or Zeeman compensation leads to improvement of the data in the trace and ultratrace range [6.17].

Atomic emission spectroscopy with inductively coupled plasma impulse (AES–ICP, →Atomic Spectroscopy, **B5**, pp. 623–628) is already widely used in the routine analysis of inorganic substances in soils because of its great technical advantages [6.17]. The important advantages include high measurement frequencies and the simultaneous determination of several elements. Disadvantages are at present relatively low sensitivities of detection. However, measurement of most relevant elements (with the exception of cadmium, mercury, and thallium) is possible because of their relatively high geogenic background contents. With AES–ICP, survey analyses for a wide range of elements in unknown samples also can be carried out.

The flow injection analysis (FIA) is applied with the mentioned AAS systems to improve analytical and economical aspects by automation of, e.g., sample introduction, dilution or preconcentration, or addition of internal standards [6.17].

For element measurement other techniques exist, not or scarcely inferior in their specificity and sensitivity of measurement to the AAS methods, but less widely used in routine analysis for the detection of soil pollutants. They include electrochemical methods (e.g., polarography, voltammetry) and molecular spectroscopic methods (e.g., UV/VIS spectroscopy). Multielement determinations also are possible with X-ray fluorescence spectroscopy after melt digestion.

Organic Compounds. The determination of organic compounds in polluted soil samples is performed according to the following process steps [6.15]:

1) Chromatographic separation
2) Detection
3) Evaluation

The *vapor-phase separation* of organic substances is today carried out mainly in capillary columns, usually with high resolution and sensitivity. Substance-specific separating columns are used, e.g., for the analysis of PAHs and PCBs. Aside from the "miniaturization" of separating columns, further advances in the separation of substances are emerging in connection with the development and adaptation of microchips and chemosensors [6.16].

Many substance- and structure-specific detectors are suitable for the detection of organic compounds. Examples are flame-ionization detectors (FID), thermionic detectors (TID), or electron-capture detectors (ECD) (→ Gas Chromatography, **B5**, pp. 213–218). The coupling of gas chromatography with mass spectrometry (GC–MS) offers an improved sensitivity of detection as well as a reliable identification of individual compounds or substances.

High-performance liquid chromatography (HPLC) has developed into one of the important and efficient separation techniques for environmental analysis [6.18]. This technique enables the determination of difficultly vaporizable, thermally labile, or ionogenic substances in extracts or eluates. Columns both for general uses (e.g., in reversed-phase chromatography) and for special applications (e.g., for the determination of PAHs) have a high degree of standardization with good reproducibility. Improvements in detection are possible with the photodiode array detector, which permits simultaneous measurement at different wavelengths. Coupling techniques with the mass spectrometer (HPLC–MS) having the same effects as those of GC–MS are awaited.

6.1.6. Evaluation

Today, instrumental measuring systems are generally equipped with microprocessors and computers. The equipment is used for controlling apparatus and operating parameters as well as the entire procedure from the input of the sample up to the documentation of the measurement signals. Editing, processing, and documentation of the data—as far as to adoption in dispatchable test certificates and in data banks—become more and more computerized. Interfaces allow the individual measuring systems to be included in networks such as the laboratory information and management system (LIMS). This makes it possible to record all information on sampling and sample pretreatment.

6.1.7. Quality Assurance

Analytical data are basically subject to error. Systematic and random errors result from the various process steps, from sampling to measurement. While errors during the individual analytical steps can more easily be identified, minimized, or eliminated, little is yet known about sources of error in sampling and sample pretreatment and their proportion in the total error, or the analyst may often not be conscious of them.

The inhomogeneity of contaminated soil samples is unquestionably the most frequent cause of a sometimes high systematic error.

Therefore, as already explained in Section 6.1.2, the sampling on the spot has to be carried out very carefully. In all later steps—from the immediate storage in containers on the spot, via transport, to the conclusion of the sample pretreatment—the minimization or elimination of errors is more easy if the sampling standard operation procedures (SOP) are exact and obeyed. The documentation of the sampling in all its details is very important here.

Features of the procedure such as selectivity, accuracy, and precision as well as limits of detection and determination are indispensable for the statistical evaluation of analytical procedures [6.19]. Often they have still to be worked out.

Quality assurance systems should comprise the following main issues:

- Sample and laboratory management
- Quality assurance manual according to ISO 45001–3
- Participation in round robin tests
- Staff qualification
- Maintenance of equipment
- Sample handling
- Methodology
- Statistic control
- Recording and reporting of data

The realization of these requirements is the basis for the competence and reliability of the laboratory carrying out the test. It is therefore recommendable to entrust with the investigation of soils only such test laboratories as meet all conditions mentioned [6.20].

6.1.8. Analytical Procedures for Selected Inorganic and Organic Soil Pollutants

From the multiplicity of analytical procedures for determining soil pollutants in this section detailed procedures are described for the elements cadmium and thallium as well as for PAHs and petroleum hydrocarbons. These are analytical procedures that are already far advanced in their degree of standardization—thus the extraction of trace elements soluble in aqua regia (ISO/DIS 11466), the measurement of cadmium and other elements by F–AAS and GF–AAS (ISO/DIS 13877).

First drafts for national or international standardization are available for the extraction of mobile trace elements from soils with ammonium nitrate solution (E DIN 19730) and for petroleum hydrocarbons (ISO TR 11046).

6.1.8.1. Analysis of Cadmium and Thallium

Both heavy metals occur ubiquitously in soils, although cadmium is more widespread than thallium, which has a more local importance. They are nonessential trace substances which, however, have toxicological relevance because of their mobility behavior in soil–plant transfer and their enrichment capacity in food and forage plants. Previously only the so-called total content after extraction with aqua regia or equivalent acid mixtures was used as the basis for evaluation, this is today supplemented by the determination of that part of the total content which is mobile, i.e., short- or medium-term available by plants or translocatable by percolate water into deeper soil layers (leaching).

6.1.8.1.1. Sample Preparation

According to ISO 11466 *cadmium* is extracted from air-dried fine soil (grain size <0.1 mm) with aqua regia under reflux for a mean duration of ca. 2 h. The supernatant suspension (filtrate) in the measuring flask after sedimentation of the undissolved fraction is ready for measurement. A detailed description is given in [6.21].

Owing to the formation of volatile thallium compounds, HNO_3 and H_2O_2 are used as an extraction mixture for the determination of *thallium* [6.22]. Pressure digestion systems should be used, with closed vessels of PTFE (polytetrafluoroethylene), PFA (perfluoroalkoxy resin) or other suitable materials and with thermal heating devices or with microwave heating systems. Undissolved substances must immediately be removed from the extraction solution, since otherwise deficiency of the true concentration is possible.

The extraction with ammonium nitrate detects nearly all environmentally relevant trace elements, including cadmium and thallium, in one solution. The mobile element fractions, i.e., those available or leachable in a short time, are determined. Extraction is carried out by shaking overhead at room temperature. After the filtration, the extract is stabilized with HNO_3 [6.23].

6.1.8.1.2. Measurement

In instrumental analysis, the relevant heavy metals are today determined with, among others,

F–AAS, electrothermal AAS (ET–AAS), and AES–ICP, which are part of the basic equipment of laboratories working in environmental analysis. Other measuring systems will consequently not be discussed.

Flame AAS and electrothermal AAS are described in detail in ISO 11047. They differ above all in the sensitivity of detection for the elements named therein. Thus with expected cadmium contents >2 mg/kg soil, the use of F–AAS is recommendable and with contents <2 mg/kg soil that of ET–AAS.

Because the thallium contents in soils usually are very low, ET–AAS (graphite tube furnace) and appropriate background compensation (e.g., Zeeman method) are recommended for the measurement [6.22].

The same instrumental systems can be used for the measurement of the heavy metal contents from the ammonium nitrate solution. Since the elements often occur in very low concentrations in these extracts, measurement with ET–AAS is advisable. At higher assumed contents, F–AAS also can be used. Efforts to determine cadmium concentrations also with AES–ICP have at present not yet reached the routine stage, because here the sensitivity is still unsatisfactory.

Taking into account instrumental equipment, extraction agents, and laboratory staff, the following limits of determination specific to different procedures can be assumed (Table 6.1) [6.24].

Information is at present still being compiled on procedural characteristics, such as precision and accuracy.

6.1.8.2. Analysis of Organic Soil Pollutants

In this chapter the analytical methods for the determination of two typical organic soil pollutants, PAHs and petroleum hydrocarbons are presented.

6.1.8.2.1. Analysis of Polycyclic Aromatic Hydrocarbons

Polycyclic aromatic hydrocarbons occur ubiquitously in the environment (see Section 5.1.4.5). While previously their main anthropogenic source originated from coal combustion and coal processing, today they are mainly emitted by scattered sources, such as motor vehicle traffic, or domestic fuel.

Table 6.1. Limits of determination of different AAS methods with respect to cadmium and thallium [6.24]

Element	F–AAS, mg/L	GF–AAS, µg/L	AES–ICP, µg/L
Cadmium	0.02	0.1	10
Thallium	0.2	2	100

The EPA has stipulated 16 PAHs—some of which are suspected of being carcinogenic—in its "priority pollutant list" for monitoring and restriction (see below):

Naphthalene	($C_{10}H_8$)
Acenaphthylene	($C_{12}H_{10}$)
Acenaphthene	($C_{12}H_{10}$)
Fluorene	($C_{13}H_{10}$)
Phenanthrene	($C_{14}H_{10}$)
Anthracene	($C_{14}H_{10}$)
Fluoranthene	($C_{16}H_{10}$)
Pyrene	($C_{16}H_{10}$)
Benz[*a*]anthracene	($C_{18}H_{12}$)
Chrysene	($C_{18}H_{12}$)
Benzo[*k*]fluoranthene	($C_{20}H_{12}$)
Benzo[*b*]fluoranthene	($C_{20}H_{12}$)
Benzo[*a*]pyrene	($C_{20}H_{12}$)
Indeno[1,2,3-*cd*]pyrene	($C_{22}H_{12}$)
Dibenzo[*a,h*]anthracene	($C_{22}H_{14}$)
Benzo[*g,h,i*]perylene	($C_{22}H_{12}$)

After extraction with an organic solvent PAHs are determined by capillary gas chromatography with mass-selective detector (quadrupole or ion trap). In addition HPLC with fluorescence detection (FLD) is becoming increasingly established [6.2], [6.25]–[6.28].

The ISO has published in the Technical Committee 190 "Soil Quality" a draft (ISO/CD 13877) on the analysis of the PAHs [6.1].

Samples from lightly loaded and heavily loaded soils are prepared differently. *Lightly loaded soils* are extracted with acetone. The extracts are purified by mixing with cyclohexane, extraction of acetone with water, and subsequently purification via an aluminum oxide column.

The *heavily loaded soils* are extracted with toluene in a Soxhlet apparatus, and the PAHs measured, without further purification, with an HPLC–FLD or diode array detector (DAD) (→ Liquid Chromatography, **B5**, pp. 248–249).

In the case of lightly loaded soils the purification is labor-intensive, while in the second method, only one Soxhlet extraction taking 4–8 h is required.

A HPLC method [6.29] has been developed to improve the efficiency of the extraction process, its reproducibility, practicability, profitability, industrial safety, and environmental compatibility. In this method an ultrasonic extraction is performed with tretrahydrofuran or acetonitrile, and a purification carried out on hydrophilized silica gel [6.29].

Samples for gas chromatography (GC) are subjected to a Soxhlet extraction with toluene. Purification is carried out on silica gel or aluminum oxide.

Because of the sample preparation by air drying which is generally used, the measurements for low-molecular PAHs are unreliable. This applies particularly to the four easily volatile components of the EPA list, naphthalene, acenaphthylene, acenaphthene, and fluorene. To avoid these possible losses, the mild freeze-drying method should be used preferably [6.30]. In this technique the samples are dried at ca. $-40\,^{\circ}C$ under vacuum.

Other laboratories use a cold acetone–cyclohexane mixture for well homogenized, field-fresh samples. The measurements are made with HPLC–FLD or –DAD, or GC–MS. This method was developed and validated in collaboration with state and private laboratories for the analysis of soil, waste, and contaminated site samples [6.31].

Immunoassay test methods have recently been offered as screening processes [6.32].

The use of SFE is also customary for soil samples (see Section 6.1.4). In this, supercritical carbon dioxide is passed through the sample at high pressure. The carbon dioxide is subsequently depressurized and collected in a solvent receiver. The advantage of this method is the low solvent requirement and short extraction times. An obstacle to routine application is the different extraction yields under the same process conditions but different soil samples. An optimized method must be developed here for each type of soil [6.33]–[6.35].

6.1.8.2.2. Petroleum Hydrocarbons

The risk of pollution exists anywhere where petroleum hydrocarbons are processed, stored, reloaded, transported, or used [6.36].

In some countries standards and draft standards exist for the determination of petroleum hydrocarbons in soils [6.37], [6.38].

In the extraction of petroleum hydrocarbons since 1981, carbon tetrachloride (which had usually been used up to that time) has been dispensed with for reasons of industrial safety in favor of 1,1,2-trichlorofluoroethane. However, the use of chlorofluorocarbons (CFCs) is controversial because of their ozone-depleting effects in the atmosphere. Whether CFCs will be available in the future for analytical purposes is therefore uncertain.

For quantitative determination of petroleum hydrocarbons an *infrared spectrum* is recorded from wave number $3125\ cm^{-1}$ to $2800\ cm^{-1}$. The CH_2 absorption band, the CH_3 band at $2925\ cm^{-1}$, the CH_3 band at $2958\ cm^{-1}$, and the aromatic CH absorption at $3030\ cm^{-1}$ are used for quantification.

An alternative to this is offered by a *GC–FID method*, whereby the total area of the peaks from decane ($C_{10}H_{22}$) to *n*-tetracontane ($C_{40}H_{82}$) is evaluated as petroleum hydrocarbons content.

The ISO TC 190 "Soil Quality" has submitted a draft technical report [ISO/TR 11 046: 1994 (E)] [6.38]. The procedure recommends a shaking method or a Soxhlet extraction with 1,1,2-trichlorofluoroethane. The "clean-up" step uses a column filled with magnesium silicate or aluminum oxide.

The validity of the aforementioned GC method has been checked by an interlaboratory comparison. According to this the difficulty of the procedure is in the external calibration with a petroleum standard whose petroleum hydrocarbon distribution does not in every case agree with that of the soil sample. Sample injection is the largest source of error in the procedure. Split or splitless injection (→Gas Chromatography, **B5**, pp. 200–204) techniques should not be used in any case. The results determined with the GC method should be supported by means of IR [6.39].

With the draft method 3560/8440 [6.40] the EPA has proposed an extraction with *supercritical carbon dioxide*. The sample is dried with magnesium sulfate before the extraction. The SFE is carried out at $80\,^{\circ}C$ and 34 MPa. Tetrachloroethylene is used as collection solvent. The "clean-up" is carried out with a silica gel column or a solid-phase cartridge. The SFE method is equivalent to a Soxhlet extraction in respect of reproducibility and recovery.

Accelerated solvent extraction (ASE) has recently come onto the market as a new development. Conventional solvents are used at up to $200\,^{\circ}C$. In order to maintain the solvent in the liquid state above its atmospheric boiling point, the extraction is carried out at 0.3–20 MPa in pressure-tight tubes [6.41], [6.42].

Outlook. In soil analysis, the development of digestion procedures, extraction procedures, and methods of measurement is aimed at efficiency, as much automation as possible, and at little solvent consumption.

Soil sampling remains the most relevant area for accuracy because it is subject to relatively large errors. Inaccuracy of the analytical values caused by the sampling can only to a limited extent be compensated by improving the analytical procedure [6.43].

6.2. Evaluation Concept

6.2.1. Uptake Paths to Humans

The hazard potential of pollutant-containing soils to humans is normally assessed from a number of different basic conditions. In the foreground—aside from the investigation of the nature and concentration of the pollutants in the soil—are principally the uptake paths (possibilities of transfer) by which pollutants can enter the human body. In addition the present use of the soil areas is important, because conclusions can be drawn from this on predominant uptake paths or special exposure situations. The relevance of the pollutant contents of the soil can only be assessed by considering all these parameters. On the other hand, the evaluation of pollutant concentrations on the basis of clinical-toxicological data alone without reference to the soil environment must be rejected as inadmissible [6.44].

6.2.1.1. Inhalation Uptake

Nonvolatile Pollutants. The uptake of particle-bound, nonvolatile pollutants is always assumed when fairly large parts or even the whole of a contaminated site have no vegetation or other kind of cover (e.g., asphalt). The lack of vegetation is observed relatively frequently on foundry heaps in which high concentrations of phytotoxic metals are present [6.44]. Here dust can be blown up and subsequently inhaled by human beings.

The uptake by inhalation depends on the concentration, exposure time, substance-dependent toxic effect, and especially on the particle size. The fine dust (particles with an aerodynamic diameter of 1 – 7 μm) which can pass into the alveoli must be regarded as especially biologically relevant. These dust particles reach as far as the pulmonary vesicles. From there easily absorbable toxic agents can pass directly into the bloodstream.

It can be presumed from literature data that hazardous effects on humans as a result of the blowing-up of dusts are encountered only very rarely [6.45]. Effects on humans have to be assumed chiefly when there is an active release of dusts from contaminated material, e.g., during activities on sports grounds with surface made of blast-furnace ash or on leisure grounds in the vicinity of contaminated sites. In dry weather, these activities can result in severe dust formation. This leads to a potential hazard not only for the participants themselves, but also for spectators or neighbors [6.46].

Volatile Pollutants. The uptake of volatile pollutants from gas emanating from the soil is considerably more important than the inhalation of particles. In the case of volatile pollutants, a genuine exposure of residents and users of contaminated sites and people living nearby is frequently demonstrable [6.47].

The exposure to gases or vapors from contaminated sites is possible when, for example, a building is constructed directly on the contaminated soil. Taking into account that residents stay up to 24 h a day in the polluted area, a considerable uptake of pollutants is certainly possible. But even people living near contaminated sites can be at considerable risk as a result of the migration of gases and vapors beyond the immediate area of the contaminated site. However, the concentrations of the volatile pollutants are usually less in the neighborhood than directly on the contaminated site itself. The possible accumulation of gas in indoor air is particularly relevant. The occurence of odor nuisances and/or the detection of methane and carbon dioxide in cellars of buildings on contaminated sites is always a indication of this risk [6.48].

6.2.1.2. Oral Uptake

The risk of oral uptake of pollutants from the soil appears to be greatest for young children. Many authors report an oral uptake, above all of heavy metals by hand-to-mouth activity, but give no information on the amount of dust or soil thereby ingested. Estimates in the literature are in a range between 200 mg and 10000 mg of soil per day. Generally it is assumed that children at play take up 200 mg to 1000 mg of soil per day. Taking this amount as the basis for children aged three to five years, the uptake of pollutants via the soil plays an important part for this group when compared with the other contamination paths and must definitely be taken into account in the estimates of effect.

The oral exposure for adults originates in most cases from the uptake of contaminated edible plants from backyards or garden plots. The uptake of heavy metals by plants has been proved in a large number of studies. As a result, the agricultural use of lands in the neighborhood of heavy-metal emitters has been restricted. Aside from the uptake of pollutants present in the soil via the roots, additional contamination of the plants can occur by deposition of pollutant-containing dust.

The direct oral uptake of pollutants by adults from the soil of backyards and garden plots can be rated as negligible. The danger to small children is certainly considerably less than that of playgrounds, because the use of backyards and garden plots, especially for the cultivation of fruit and vegetables on patches, usually allows no play activities that could lead to noteworthy contamination.

Another exposure path for oral uptake of pollutants can be contamination of groundwater, which frequently occurs at contaminated sites. However, the problem is not the public supply with drinking water of adequate quality, but rather the possible use of contaminated groundwater from domestic wells [6.46].

6.2.1.3. Uptake via the Skin

The contamination of the skin has to be rated as rather unimportant. The theoretically possible uptake of fat-soluble substances or solutions by direct contact with the skin requires very high pollutant concentrations, such as are not normally found on the surface of contaminated sites or other contaminated soils. The absorption of gas or vapor via the skin must be rated as almost impossible even under extreme exposure conditions and can therefore be ruled out in this context [6.45].

6.2.2. Relevant Soil Uses

To enable an evaluation of soil contamination with regard to the health hazard for persons concerned and for the establishment of tolerable soil values, it is reasonable to reduce the huge number of possible uses of the soil to a small number of use scenarios. Starting points could be on the one hand particular user groups (playing children or young people, sportsmen and sportswomen, garden plot holders, etc.) and on the other the complex uptake paths of nonvolatile or semivolatile substances (predominantly oral uptake, mixed inhalation and oral uptake, etc.). It does not appear sensible to consider all possible transfer modes from the soil to humans in detail but rather to use simple models and scenarios as a basis for a general assessment. Table 6.2 shows use scenarios according to the assessment system of Eikmann and Kloke with parameters characterizing use and protected subjects [6.49].

Table 6.2. Use criteria for various types of soil use (objective: protection of humans)

Type of use	User group/ protected subject	Uptake path	Soil areas	Soil depth
Children's playgrounds	small children and escorts	oral	sand pits vegetation-free environment	35 cm
Backyards and garden plots	children adults	oral by inhalation	patches low-vegetation areas	35 cm
Sports and recreation grounds	sportsmen and sportswomen young people	by inhalation	gravel surfaces vegetation-free areas	10 cm
Parks and leisure grounds	adults children	oral by inhalation	unpaved, low-vegetation areas	10 cm
Industrial and commercial areas	adults of working age	by inhalation water path	unpaved, vegetation-free areas	10 cm

6.2.3. Evaluation Criteria

In the development of regulative values or values for the soil, there is always a need to establish conventions, for example, for the comparability of measurement methods and of measurement results. Although these conventions are certainly based on scientifically supported knowledge, they are nonetheless to be regarded within a certain tolerance range as "arbitrary".

The calculation of tolerable soil values depends to a considerable degree on the particular conventions chosen. The influence of these conventions on the quantification of the uptake is clearly greater than that of toxicologically derived criteria (body loads). Moreover the tolerable soil values calculated as a function of the chosen assumptions on exposure are frequently far above or far below the usual background concentrations of these substances in soils.

In the assessment methodology, owing to the considerable uncertainties and difficulties which have been described, there is a trend of describing the exposure—especially of persons on contaminated soils—by characteristic simplified individual scenarios for types of use and protected property and of establishing a quantitative assessment on this basis or to orient oneself according to the main uptake path. Thus the base values (BW I) stated by Eikmann and Kloke for (harmful) substances in soils are derived essentially from the corresponding background concentrations in the soils in Germany. The tolerance (test) values (BW II) and intervention values (BW III) are defined as a function of the toxic properties of the substances involved and the specific exposure situation (inhalation) by application of appropriate factors [6.49].

Other evaluation models (e.g., the EPA model or the Holland List) attempt to avoid these methodical difficulties. Here, either the risk is estimated by using evaluation models referring to individual cases without defining soil values; or for defining acceptable soil values, only difficultly understandable correction factors or so-called reliability scores are applied. In addition, all estimation parameters used in these models (e.g., according to information from the authors of the Holland List) have uncertainties that frequently can amount to a factor of 10 or more [6.50].

The repeated demand for "strictly toxicologically derived" soil values, taking into account the various kinds of soil use, cannot be satisfied owing to the paramount influence of the conventions chosen for quantifying the (inhalation) uptake of pollutants. When different evaluation models are applied in the context of an individual-case assessment according to the classical derivation model, this can result in the definition of different soil concentrations for a certain pollutant. However, different intervention or remediation target values, depending on the particular contaminated site, are certainly unacceptable. In particular the frequently existing necessity for immediate action—in relation to a further possible use of contaminated soil areas—requires at least the definition of (standardized) intervention values. In addition, different remediation target values for the same kind of uses will certainly not be accepted by the public because of the incomprehensibly higher remediation costs in single cases.

In the preparation of balances for the transfer of pollutants from the environment to humans the pollutant contaminations of the various environmental media (atmospheric and indoor air, drinking water, and food) are considered. No attention is usually paid to the direct exposure from the soil, because it must be rated unimportant in relation to the total uptake. The indirect uptake from the soil via food plants and animals is normally taken into account by recording the contamination of the food [6.51].

The direct intake of pollutants from the soil—which is frequently considered the main problem of contaminated sites—is a very specific pollution problem for only part of the population, and therefore also cannot be considered in a general balance. If tolerable body doses are reached or exceeded as a result of the general environmental pollution with particular pollutants, such as heavy metals, in theory also a reduction of the soil contamination can be considered to lower the metals content in the environment. However, because of the practically nonexistent discharge from soil and the nondegradability of metals, a reduction cannot normally be achieved here, by contrast, e.g., with air (by reduction of emissions). An inclusion of the soil in such balances as an equivalent medium appears therefore to be unjustified and leads to an assessment dead-end [6.52].

Extensive contamination of the soil by heavy metals (caused by immissions) frequently occurs, especially in the conurbations, and, in the use of garden plots or backyards, results in (officially recommended) restrictions of the cultivation of various useful plants. This is one of several possibilities of reacting to such wide-ranging soil contamination. On the other hand, an extensive remediation of the soils of whole regions to achieve tolerable heavy metal concentrations must certainly be regarded as difficult, if not quite impossible. For the purposes of general soil protection (above all also for the persistent organic compounds), any further input must rather be kept as small as possible.

7. Remediation of Soil and Land

7.1. Remediation Methods

7.1.1. Introduction

The effect of point sources of pollution on soil and groundwater is especially clear from the example of the contaminated sites. In view of the importance and relevance of the contaminated sites problem, industry and commerce have equipped themselves with strategies and techniques for reducing or removing the soil contamination. The methods for soil purification described in this chapter relate exclusively to applications to contaminated sites. The application of these processes to the treatment of extended soil contaminations, caused, for example, by airborne immissions or agricultural measures, is excluded for economic reasons.

As of December 31, 1993 about 139000 suspect contaminated sites were registered in Germany alone (Table 7.1), divided into about 86000 abandoned waste disposal sites and about 53000 closed industrial sites. In addition there are about 4700 sites suspected of contamination due to armaments as well as about 10000 properties used by the military that are being examined for suspect contaminated sites. The degree of registration in the new Federal States (former German Democratic Republic), at 85%, is relatively high by comparison with the nationwide registration ratio of about 65%.

Present forecasts of the number of suspect contaminated sites to be expected in future must clearly be revised upwards. Final registration results of more than 240000 suspect sites may be realistic for Germany.

Like the number of suspect contaminated sites, the number of actual contaminated sites has been unknown up to now, because the risk assessments which follow the registration have not yet been started or completed for the whole area of Germany. On a cautious forecast, at least 10% of the total number of registered suspect contaminated sites can be assumed, so that overall a gigantic remediation requirement can be expected. As a rule, suitable remediation measures for individual cases can be put in concrete terms only after a final risk assessment.

Table 7.1. Number of registered and estimated suspect contaminated sites in Germany

	Registered suspect contaminated sites		Total suspect contaminated sites	
	Old waste disposal sites	Old industrial sites	Registered	Estimated
Baden-Württemberg	6500	460[a]	6960	35000[b]
Bavaria	3820[c]	1119	4939	no data
Berlin	746	4242	13565	15000
Bremen	100	4189[d]	4289[d]	4290[d]
Hamburg	392[e]	212[e]	604[f]	2600[e]
Hesse	3400[g]	no data	3400	13400[h]
Mecklenburg-West Pomerania	4749	7209	11958	no data
Lower Saxony	7488	no data	7488	57550[i]
North Rhine-Westphalia	14043[j]	4153	18196	no data
Rhineland-Palatinate	14760	no data[k]	14760	no data
Saarland	1700	no data[l]	1700	4250[m]
Saxony	8045	10597	18642	22000
Saxony-Anhalt	7100	7853	14953[n]	17000
Schleswig-Holstein	2822	3871	6693	no data
Thuringia	4618	969	5587[o]	12000
Old Federal States	55025	14004	69029	161678
New Federal States[p]	30008	39685	69693	83248
Old and new Federal States	85033	53689	138722	244926

[a] Previously evaluated old industrial sites. [b] Extrapolation based on 16 pilot communities. [c] In addition 4000 former municipal landfills from the survey at the start of the 1970s. [d] Old industrial sites identified during a study typical in the line of business. [e] In addition, 2000 old industrial sites are reckoned on. [f] In addition, 1060 probably unproblematic site changes were registered. [g] Of 6000 registered old waste disposal sites, ca. 3400 are to be classified as suspect contaminated sites. [h] Including estimated old industrial sites. [i] Old industrial sites not yet registered; estimated ratio of old waste disposal sites to old industrial sites ca. 1 to 7. [j] Registration as of Dec. 31, 1992. [k] Registration starts 1993. [l] In the area of the city unit of Saarbrücken, 2113 areas suspected of contamination were identified. [m] Including 2500 estimated old industrial sites. [n] In addition, 38 extensive soil contaminations registered. [o] In addition, 13702 applications for exemption. [p] Including Berlin.

Aside from the remediation of contaminated sites by structural containment measures, which is used preferentially in the case of abandoned waste disposal sites that endanger the environment, decontamination processes have been implemented. These methods have been used extensively for the remediation of contaminated industrial sites that endanger the environment.

7.1.2. Processes for Soil Treatment

Processes for the treatment of contaminated soils and groundwaters can in principle be applied in three ways: (1) at the natural location (in situ); (2) after excavation or pumping-off (on-site); or (3) after transportation, in central treatment plants (off-site). While methods and processes for groundwater treatment and soil remediation (on-site/off-site) already exist on a technical scale, processes for the in-situ decontamination of contaminated soils are mainly at the stage of large-scale demonstration or research and development. Specific soil treatment processes are also already used successfully in practice to a fairly large extent.

Possible on-site/off-site soil treatment processes are thermal, physicochemical and biological processes, which are as a rule carried out in combination with supplementary treatment processes, such as waste-air and wastewater treatment. The purpose of these soil remediation processes is to eliminate the pollutants present in the soil or to reduce them so far that the treated soil can without environmental risks be reused on the spot as back-fill material or used elsewhere in order to avoid landfill disposal. The pollutant concentrates arising from soil remediation processes must be treated further or appropriately disposed of. In the following, the most important on-site/off-site soil treatment processes are described in principle first, followed by some of the in-situ soil treatment processes that have been used in practice.

7.1.2.1. On-Site/Off-Site Processes [7.1]

7.1.2.1.1. Thermal Processes

Thermal processes are suitable for the treatment of highly concentrated organic pollutants in soils. This technology is based on the oxidation of the pollutants or their transfer to the gas phase with subsequent thermal decomposition at high temperature and subsequent multistage waste-gas purification.

According to present information in Germany about 20 different designs or processes exist for thermal soil treatment. Aside from a few thermal treatment processes for hazardous waste, that are also licensed for soil treatment, at present only three thermal soil treatment plants work in large-scale continuous operation in Germany. The reason for this is that the licensing process takes many years and in actual individual cases has not yet been completed. Table 7.2 contains a survey of operating large-scale treatment plants.

For Germany it is expected that sixteen stationary plants for the thermal treatment of soils or fine-grain residues from other treatment processes can be assumed, which will go into operation in 1994 or later [7.2]. Proof has been provided in The Netherlands of the suitability of thermal soil treatment with more than 1×10^6 t of purified soils, and also in Germany with several experimental or pilot plants.

Direct Heating Process of Züblin. Under a research and development project supported by the Federal Ministry for Education, Science, Re-

Table 7.2. Operating large-scale plants for the thermal treatment of contaminated soils

Process	Operator	Brief description	Capacity t/h	Location	Characteristics*
Boran	Boran Bodenreinigung, Berlin	fluidized-bed soil treatment plant, forced circulation, 900–1200 °C	10	plant in Berlin, start-up 1995	s
Broerius	Broerius Bodensanierung, Borna	direct-heated rotary kiln, treatment temperatures on an average of 520–600 °C, storage capacity 135000 t	9	plant in Deutzen, start-up December 1995	s
Hochtief	Hochtief Umwelt GmbH, Essen	indirect-heated rotary kiln with afterburning	7	plant in Herne, start-up July 1992	s

* s = stationary.

search, and Technology (Bundesministerium für Bildung, Wissenschaft, Forschung und Technologie, BMBF), an experimental plant has been developed by the Züblin Company (Stuttgart). This plant has a throughput rate of ca. 400 kg/h and was used successfully at several locations for the thermal treatment of soil. High purification capacities were achieved, even with varying pollutant spectra and soil structures, by direct heating at 1200°C and residence times of 30–45 min. A pilot plant with a throughput rate of 5 t/h is planned for a treatment station in Bischofswerda in the new Federal States.

Terra-Dest Process of Ruhrkohle Umwelt. In contrast to the above directly heated plant, the plant of Ruhrkohle Umwelt GmbH, Bottrop, which has been operated since May 1988, also under a BMBF project, involves an *indirectly heated system* which was used especially for sites with specific coke-oven contamination. The plant, with a rated output of 7 t/h, has been shut down after carrying out the research and development project. A development based on this plant is offered by Ruhrkohle Umwelttechnik of Bottrop under the name Terra-Dest process.

Pyrolysis Process of Hochtief Umwelt. A thermal soil treatment plant of Hochtief Umwelt GmbH, Essen, has been located since June 1992 in Herne/Westphalia. The plant, based on the pyrolysis principle, has a rated capacity of 7 t/h, which with three-shift operation corresponds to a throughput rate of 35 000 t/a. In the soil treatment section the plant consists essentially of an indirectly heated rotary dryer (100–120°C) and an evaporation stage in which the soil is indirectly heated to 600°C. After 30–45 min treatment time, the soil is cooled to temperatures <100°C. Figure 7.1 shows the pyrolysis drum of this first large-scale plant in Germany.

Terra-Therm Process of Ruhrkohle Umwelt. Further plants for the thermal treatment of contaminated soils are planned by Ruhrkohle Umwelt GmbH. This technology involves a directly heated rotating cylindrical drum, based on the process of the Dutch company Ecotechniek. In this so-called Terra-Therm process, the treated contaminated soil is heated in the rotary kiln to 350–600°C, depending on the type of pollutant, so that the organic chemical contaminants volatilize from the soil. The volatilized pollutants are oxidized in a thermal gas purification at

Figure 7.1. Pyrolysis drum of the thermal soil treatment plant of Hochtief Umwelt GmbH in Herne

1200°C. According to the present plan it can be assumed that a first plant of this design will start operation on site with an output of 30–40 t/h. Figure 7.2 shows the process diagram of this plant.

Terravac Process of Harbauer Umwelttechnik. For the treatment of highly contaminated fine-grain fractions, especially from wet-mechanical soil cleaning (soil washing), the company Harbauer Umwelttechnik (Berlin) has developed a vacuum distillation plant (Terravac process), which works in the low-temperature range and was brought into operation in June 1993 in Marktredwitz (Germany). The fine-grain suspension coming from the soil washing is dewatered and dried and then fed via vacuum locks to the vacuum distillation plant. The material is treated at 10 hPa and an average temperature of 310°C. After the thermal treatment the fine material is mixed again with the material from the soil washing. Figure 7.3 shows the vacuum distillation plant under construction as a component of the soil washing plant erected on the site of the former Chemische Fabrik Marktredwitz (CFM).

Boran Process of Boran Bodenreinigung. An innovative thermal soil treatment plant started operation in 1995 in the Berlin Westhafen. The fluidized-bed soil treatment plant, erected under the name Boran, was developed for the thermal treatment of contaminated soils and pollutant

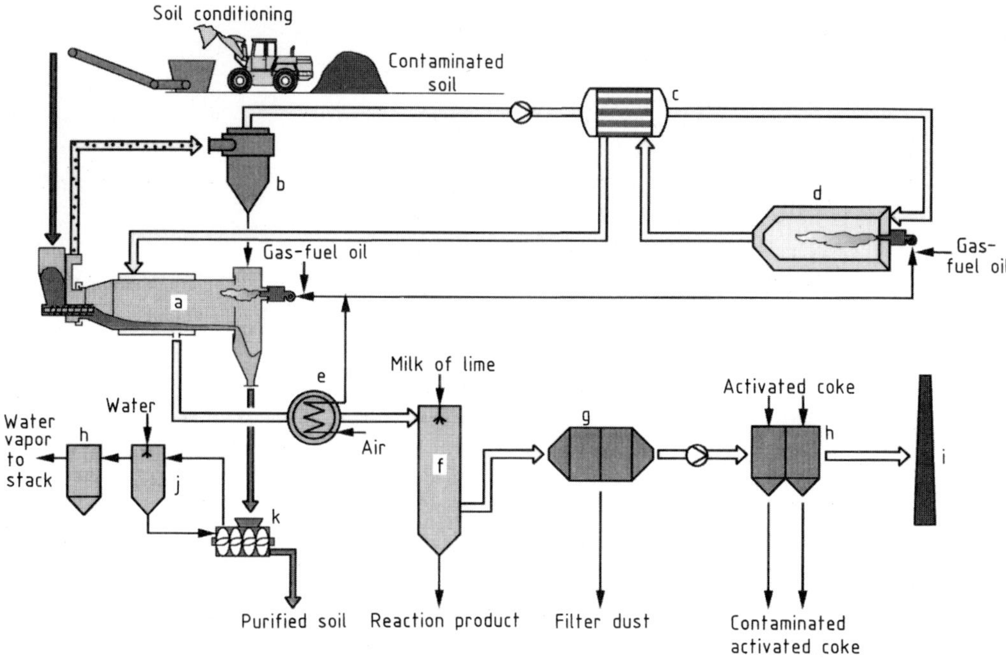

Figure 7.2. Terra-Therm process scheme of the thermal soil-cleaning plant of Ruhrkohle Umwelt GmbH, Bottrop (based on a process of Ecotechniek, The Netherlands)
a) Rotary kiln, directly heated; b) Cyclone; c) Process gas preheater; d) Thermal gas cleaning; e) Air preheater; f) Spray absorber; g) Electrostatic precipitator; h) Activated coke filter; i) Stack; j) Wet washer; k) Mixer

Figure 7.3. Vacuum distillation plant (Terravac Process) of Harbauer Umwelttechnik Berlin under construction at Marktredwitz

concentrates from soil washing plants. The plant has a throughput of 10 t/h. The soil is cleaned above 800°C in a rotating fluidized bed. The thermal treatment section consist of fluidized-bed reactor, hot cyclone, afterburning chamber designed for 1200°C, and air preheater for raising the temperature of the combustion and fluidizing air to 500°C. This section is followed by two high-performance cyclones and the flue-gas cleaning. The multistage flue-gas cleaning plant consists of quenches, cloth filters for first dust separation, spray absorbers for removing the acid gases, post-reaction section for blowing in calcium hydroxide with added activated carbon, a cloth filter for removing the reaction products and an activated coke adsorber as final purification stage and activated carbon filter. The emission values are below the limit values of the 17th BImSchV (Federal Technical Instructions on Air Quality Control). Figure 7.4 shows the plant under construction in March 1994.

Figure 7.4. Boran fluidized-bed soil treatment plant under construction at Berlin Westhafen

Direct Heating Process of Broerius. The first thermal soil treatment plant in the new Federal States is operated by the company Broerius Bodensanierung und Umwelttechnik GmbH, Borna since 1 December 1994 on a former coal briquet factory site in Deutzen. The treatment plant has been licensed under the Federal Mining Act and the Federal Technical Instructions on Air Quality Control. The direct heated rotary kiln is operated on average treatment temperatures between 520 and 600 °C. The treatment capacity is about 9 t/h, an extension to 15 t/h is planned. The treatment facility includes a roofed storage for contaminated soil with a capacity of 135000 t. The same direct heating process has been operated since 1989 by Broerius in The Netherlands.

Thermal soil treatment processes are suitable for almost all soil types and pollutant groups. As a rule, they have only limited suitability for the treatment of heavy-metal contamination. At high treatment temperatures, however, heavy-metal compounds become largely immobilized in the soil matrix. High degrees of cleaning, usually >90%, are achievable for the pollutant groups hydrocarbons/petroleum products, polycyclic aromatic hydrocarbons (PAHs), volatile chlorinated hydrocarbons, aromatic hydrocarbons, polychlorinated biphenyls (PCBs), halogenated hydrocarbons, cyanides, organic nitro compounds, pesticides, as well as dibenzodioxins and dibenzofurans, according to treatment temperatures (ca. 350–1200°C) and residence times in the decontamination process (15–60 min). Depending on the pollutant spectrum and other boundary conditions, the costs for thermal soil cleaning have hitherto been above 300 DM/t. However, against the current background of scarce resources in public budgets, an increased trend to safety procedures is anticipated. This trend is leading to a low occupancy of existing soil cleaning plants and therefore to an extreme fall in price for these and the following on-site/off-site soil cleaning techniques in the range around 200 DM/t.

7.1.2.1.2. Soil Washing Processes

In contrast to the thermal soil treatment processes, physicochemical processes—wet-mechanical and chemical pollutant extraction (soil washing processes) in particular—have already been used successfully on the large scale in the practice of contaminated site remediation in numerous plants. The variety of concepts and plants, with 40 different processes, is twice as high as for thermal soil treatment. Table 7.3 contains a survey of plant concepts, plans, and plants in operation for soil washing.

In Germany 11 stationary soil washing processes were in operation up to September 1993 [7.2]. Thirty five further plants will start operation in 1994 or later, so that in future more than 45 soil washing plants can be assumed. Some large-scale plants are described in the following.

Soil Washing Center of Harbauer Umwelttechnik. The soil washing plant developed by Harbauer Umwelttechnik in Berlin under a BMBF project has successfully cleaned at its location in Berlin not only the highly contaminated soils present, but also other soils with different degrees of contamination from other old sites in Berlin. After further development the washing process, which optionally also uses surface-active substances, will be usable for the treatment of fine-grain soils down to a grain size of ca. 15 μm. The plant was shut down in 1991. On the basis of the experience available from this plant, a soil washing center was erected in Berlin on a neighboring site. This latter plant started its experimental run in July 1993 and officially entered continuous operation in December 1993. The plant has a throughput rate of 25–30 t/h.

CFM Soil Cleaning Plant of Harbauer. Also in mid-1993, the CFM soil cleaning plant of Harbauer at the site of the former Chemical Factory Marktredwitz (production of pesticides, mercury and mercury products, etc.) started up. The cleaning plant, with a throughput of 25 t/h, comprises a combination of soil washing process and

Table 7.3. Plants and processes for the extractive treatment of contaminated soils

Process	Operator, concept	Brief description	Capacity, t/h	Location, form of operation	Characteristics*
AB-Umwelttechnik	AB-Umwelttechnik, Lägerdorf	washing process with closed water circuit, residue disposal in cement works	40	plants since May 1989 in Lägerdorf, Munich, and Coswig; Hanau planned	m
ASRA	Krupczik Umwelttechnik, Hamburg	hydrogenative separation process for treatment of sludges, dredged material and sands, overall disposal plan	ca. 10	plant since Sep. 1987 in "Stellinger Moor" sewage treatment plant (Hamburg)	m
Baresel ÖkoChem	Baresel, Stuttgart	chemicobiological earth washing process using surfactant	ca. 16	plant in Dresden, 1992	m
Baur	Inst. Dr. Baur, Fernwald	treatment of heavy-metal contaminated soils by mineral acids	1 m^3/h	R & D project	m
B + B suspension-stripping process	Bilfinger + Berger, Mannheim	combined suspension and stripping process for removing chlorinated hydrocarbons from cohesive soils	20	plant 1993, Karlsruhe	m
Biological-physical soil purification	Contracon, Cuxhaven	washing process by means of free-fall mixers, active agents, water and bacteria	max. 18	pilot plant 1987, several mobile plants, including Heidenheim, Hamburg, and Gütersloh, 4 stationary plants planned	m, s
Soil washing process, Züblin system	Ed. Züblin AG, Stuttgart	2 washing circuits with mechanical fluidized bed (water, detergents), countercurrent rinsing, thermal disposal of residue	up to 20	pilot plant experiments, plant concept 1989	m
BOWA 20	WU-Walter Umwelttechnik, Augsburg	discontinuous intensive washer with hydromechanical agitator, extractant water with closed water circuit	20	plant since 1992, further plant planned	m
Bremer Vulkan	Bremer Vulkan	washing process and fine-particle extraction with organic solvents	18	pilot plant, large plant planned	r
CBBR	Possehl Umweltschutz, Hamburg	chemicobiological soil treatment process, compressed-air fluidized bed	ca. 12	8 plants in operation at various sites	m
Contamex	Contamex Industrieanlagen, Bremen-Stuhr	physicochemical washing process for track chippings and soils with integrated wet bioreactor	30	Kriebitzsch plant in operation May 1994 (as development of Trebbin plant)	m
DJS	Dr. Joachim Schilling, Apolda	physicobiological treatment in 3-chamber horizontal silo system, surfactant and nutrient addition		plant (30 000 t/a) in Apolda, Wormstedt	r
DSU	Duisburger Schlackenaufbereitung	soil washing by intensive attrition with extractants, washing cylinder	30	pilot plant	s
Dywinex	D & W AG, Munich	separation with water or extractant, variable container construction	ca. 5	1989 Hamburg trial operation, 1992 large-scale operation	m
Eco Tech	CB-Chemie und Biotechnologie, Verl	physicochemical washing with integrated water and valuable-product treatment (solid cycle process)	8	plants 1992	s, m
EUR	EUR, Emmerich	soil washing process for treating oil-contaminated soils and chippings		plant (> 40 000 t/a) is operating in Emmerich	s
Fermenta-Tec	Peter Neumann, Eckernförde	physicochemical and biological treatment in drum mixers	150–200 t/d	in operation since 1990 at several sites	m
Franken	Franken Umwelttechnologie, Breitscheid	physicochemical soil washing plant, use of additives	> 20	plant 1993	m
Countercurrent extraction	Rethmann, Selm/Weßling, Altenb.	countercurrent extraction principle, use of readily vaporizable extractants	5	pilot plant, Essen, large plant planned	m

Table 7.3. (Continued)

Process	Operator, concept	Brief description	Capacity, t/h	Location, form of operation	Characteristics*
GESU CP4 extraction	Gesellschaft für saubere Umwelt, Berlin	agitators and extractors, flow energy, use of wash liquids	3 30	pilot plant 1987 large plant planned	m r
GHU/AKW	GHU Berlin/AKW Hirschau	wet-cleaning process by means of washing cylinder, multihydrocyclone, container method of construction	30	large plant planned	m
Hafemeister	Hafemeister, Berlin/Hochtief	physical separation and chemical treatment in 2 handling stages	5–15 m^3/h	pilot plant February 1989 plant Berlin/Reinickendorf June 1992, shut-down 1995	m sm
Harbauer	Harbauer, Berlin	wet-extraction process, energy input by means of vibratory washing screw	20 25–30	plant 1986–1991 in Berlin (Pintsch site), next plant in Vienna plant Berlin-Gradestraße in operation July 1993	sm s
		combined soil washing and distillation plant (low-temperature–vacuum distillation – Terravac)	25	plant in Marktredwitz in operation June 1993	s
Heijmans/ Preussag	Preussag, Kiel	physicochemical process (attrition wash), container method of construction/stationary plant	10–30 12	2 plants since 1984 in The Netherlands new plant 1992, Kiel	r/ sm
High-pressure wash process	NORDAC Group/ Afu Berlin	high-pressure water-jet process, 250 bar, water as extraction medium	15–40	plant since 1986 in Berlin	r
	NORDAC Group, Hamburg	Oecotec high-pressure soil washing plant 2000, new development, 2nd and 3rd generation	45	plant since April 1989 in Düsseldorf/Hamburg/ Stuttgart/Munich, Thale (sale to GFA Thale), 2nd plant NORDAC waste disposal center February 1991	r
Huber system	Gebr. Huber, Munich	combined plant for mechanical and biological working-up of petroleum contaminations	20	pilot plant since September 1991, Munich	m/r
In-situ soil washing process	Arge soil remediation (Keller Grundbau, WUE, S + I, Hamburg)	in-situ high-speed water jet soil washing and on-site treatment of soil and water, also usable below buildings	15	experimental plant 1988, use 1991 in Hamburg planned	m
In-situ high-pressure washing process	Philipp Holzmann AG, Düsseldorf	high-pressure soil washing process for in-situ and on-site use, extraction medium water, 500 bar, microbiological mud treatment	ca. 12	plant in use for the first time in Bremen since 1989	m
ISA steam extraction	RWTH Aachen, Bronnenberg und Drescher	treatment of orangic- and heavy-metals-contaminated soils by steam extraction (R & D project)		experimental plant since March 1993	r
KRC	KRC, Würzburg	soil washing process by means of high-pressure spray washer and high-pressure nozzling (fine-grain washing)		–	
LGA	Landesgewerbeanstalt, Bayern	countercurrent reactors with extraction solution, especially for heavy metals		laboratory tests and plant concept 1988 (see SMEA)	
LRS Terrastream	Lurgi-Renner, Regensburg	wet-mechanical soil treatment with downstream thermal stage (steam distillation)	0.5 40	pilot plant in Regensburg large plant since July 1992 in Regensburg	m m
Lurgi (Deconterra)	Lurgi (L.U.B.) Frankfurt	wet-mechanical treatment by means of attrition-washing cylinder, extractant water	1 20	demonstration plant 2 large plants, shut down	r

Table 7.3. (Continued)

Process	Operator, concept	Brief description	Capacity, t/h	Location, form of operation	Characteristics*
MCT Teufert's process	MCT Umwelttechnologie, Berlin	mechanicochemico-thermal soil washing process, countercurrent drum mixer with surfactants and saturated steam	15	plant planned in Berlin, 1992 large plant in Bavaria planned	m s
Prof. Müller's process	ROM, Hamburg	acid digestion of aquatic sediments with combined hydroxide–carbonate precipitation for heavy-metal removal	260 kg/h with 30% solids	experimental plant, Hamburg 1991	m
Preussag Noell Wassertechnik	Preussag Noell, Darmstadt	physicochemical process: heavy-metal extraction with acids, attrition wash with surfactants	1	pilot plant	m
San-O-Clean	SAN, Bremen	multistage washing process, extractant San-O-Clean, preferentially for oil-contaminated soils	10–15	laboratory experiments and plant since 1985 at various sites	m
Schauenburg	Schauenburg, Mühlheim	plant for treatment of oil-contaminated sands, plan for further development		pilot plant in Mühlheim	m
SEG	SEG, Lübeck	decontamination of oil-contaminated soils, chippings, and building rubble; washing with surfactants	10	pilot plant 1989, Kornwestheim	m
SMEA	Bauer and Mourik, Schrobenhausen	decontamination of heavy-metal-contaminated soils with nitric acid	0.1	pilot plant in Kornstein (LGA license) 1990	
TerraCon	Arge W & F/Tereg/Eggers, Hamburg	drum washer with detergent surfactants, especially for soils contaminated with petroleum hydrocarbons	120 t/d	plant in Hamburg 1989, large-scale operation 1992	m
Weiss	L. Weiss, Crailsheim	extraction of contaminants by means of high-pressure jet tube	10–25	large plant with collaborator	r

* m = mobile; sm = semimobile; r = relocatable; s = stationary.

low-temperature vacuum distillation. In combination with the Terravac process already mentioned, in particular fine-grain soils highly contaminated with mercury or fine-grain residues from the soil washing plant will be treated.

High-Pressure Washing Process of NORDAC. Since 1986 the high-pressure washing process of Klöckner-Oecotec, Duisburg has been used by Afu of Berlin at various Berlin sites. This process has been developed in The Netherlands. The centerpiece of the washing plant consists of a high-pressure steel jet tube, in which water under high pressure up to ca. 5000 N/m^2 impinges on the pollutant-laden soil and brings about a separation of pollutant and soil particles. This process is usable, like the process previously mentioned, for a broad pollutant spectrum. A development of Klöckner-Oecotec based on this principle started up in April 1989 on the former site of the Mannesmann Röhrenwerke in Düsseldorf-Lierenfeld. The Oecotec high-pressure soil washing plant 2000, which consists of 50 standard containers and is equipped among other things with a 3-stage high-pressure steel tube, uses steam injections to partially dissolve the pollutants before the treatment in the high-pressure steel tube. This plant was subsequently transferred to contaminated sites in Hamburg, Stuttgart/Bad Cannstatt, Munich and Thale. In 1992 the plant was sold to the GFA company in Thale. Figure 7.5 shows this plant at the site of use in Düsseldorf-Lierenfeld.

Since 1991 a further development of this process has been in continuous operation on the site of the North-German Contaminated Sites Remediation Center (Norddeutsches Altlastensanierungszentrum, NORDAC) in Hamburg. Since then more than 300 000 t of contaminated material has been purified with the NORDAC soil washing plant. In 1993 the NORDAC Group took over all the soil washing activities of Klöckner-Oecotec and Afu.

Suspension-Strip Process of Bilfinger + Berger. The B + B suspension-strip process has been developed by Bilfinger + Berger Bau AG (Mannheim) for the removal of chlorinated hy-

Figure 7.5. Oecotec high-pressure soil washing plant 2000 at the first site of application in Düsseldorf-Lierenfeld

drocarbons from clayey-silty soils. In this process the soils to be treated are completely suspended in water. The dissolved pollutants are then stripped by air blown into the suspension, and adsorbed from the gas stream on activated carbon. In addition to chlorinated hydrocarbons, aromatics and low-boiling petroleum fractions can also be removed from the soil. The mobile plant has a troughput of 20 t/h and, with 9 containers, has a space requirement of only ca. 800 m^2.

Extraction with Supercritical Fluids (University of Dortmund). Aside from the use of water or organic solvents as extractants, the use of supercritical fluids has been investigated in research work at the University of Dortmund. In this high-pressure extraction process, carbon dioxide in particular is used as extractant, the extraction conditions (temperatures around 100°C and pressures to 40 MPa) being varied according to the pollutant spectrum and type of soil. Experiments have been carried out successfully on the pilot plant scale on soils contaminated with aromatics, polycyclic aromatics, and chlorinated hydrocarbons.

Soil washing plants are suitable principally for the treatment of coarse-grain soils, such as gravel and sand. They can be used with limitations for loamy and clayey soils as well as soils with a high content of organic constituents. Decisive for the mass balance of the washing process is the cut between the grain size of the purified material and the grain size of the material remaining in the wash liquid or in the residual sludge. In highly automated soil washing plants this cut is in the region of 0.015 mm. Depending on the extraction medium (water, surfactants, alkalis, acids) soil washing processes can treat a wide variety of pollutants successfully. The range of pollutants that can be dealt with corresponds to that of thermal treatment processes except for dibenzodioxins and dibenzofurans. The degrees of purification for petroleum hydrocarbons, PAHs, chlorinated hydrocarbons, heavy metals, and cyanides are also above 90 %. The costs have hitherto been on average between 200 and 350 DM/t, depending especially on pollutant spectrum and treatment type. According to the actual fall in price the treatment costs are falling below 200 DM/t.

7.1.2.1.3. Biological Processes

Biological soil remediation processes are aimed at the degradation of organic compounds, especially by activation of bacteria already present in the soil or else by adding specially cultivated bacteria and optimization of their environmental conditions. At present a large number of processes are on the market both for on-site/off-site as well as for in-situ treatment. Up to September 1993, thirty three stationary biological treatment plants were in operation in Germany [7.2]. A further 45 plants will start to operate in 1994 or later.

Of the pollutant spectrum to be encountered in contaminated soils of old sites, many organic pollutants are biologically easily degradable, mainly aerobically. This applies especially to aliphatic and aromatic hydrocarbons and with certain limitations to polycyclic aromatic hydrocarbons also, whereas chlorinated aromatic hydrocarbons are biologically degradable only with difficulty.

The biological treatment in on-site/off-site operation occurs in heap processes or also recently in reactor processes.

Heap Process. In the heap process, contaminated soils are packed closely on an underground seal to degrading beds. In the case of enclosed heaps, special measures for underground sealing can be dispensed with. According to the process, different process steps are used, such as soil preparation, homogenization, nutrient and bacteria addition, leaching water recycling, aeration and deaeration (optionally with waste air purification), as well as periodically repeated treatment processes.

Figure 7.6. Terraferm heap process of Umweltschutz Nord for the biological treatment of soils contaminated with petroleum hydrocarbons and polycyclic aromatic hydrocarbons in Frankfurt/Main

Experience with biological treatment in such dynamic soil heaps is available chiefly for petroleum-contaminated heaps. The degradation processes in soil heaps extend over several months. In the case of petroleum hydrocarbons, the treatment is finished after 4 months on average. The equivalent throughput rates correspond in order of magnitude with those of soil washing processes and in the case of soils contaminated with petroleum hydrocarbons can be in the range of 20 t/h.

Aside from soils contaminated with aliphatic hydrocarbonss (e.g., petroleum, gasoline), soils contaminated with aromatic hydrocarbons (e.g., benzene, toluene, xylene) and polycyclic aromatic hydrocarbons (e.g., naphthalene, anthracene, benzo[*a*]pyrene) from gasworks sites have also recently been biologically treated in heaps. Figure 7.6 shows an example of the biological treatment of soils contaminated with petroleum hydrocarbons and polycyclic aromatic hydrocarbons according to the Terraferm process in roofed heaps by Umweltschutz Nord (site Frankfurt).

Reactor Process. A variant of the biological on-site/off-site treatment in heaps is the biological treatment in reactors, in which the degradation processes can be shortened to a minimum period of time by optimum adjustment of the aforementioned boundary conditions. The first large-volume reactor, with a length of 45 m and a capacity of 240 m^3, was put into operation at the beginning of June 1989 by Umweltschutz Nord on their plant site in Ganderkesee.

For the treatment of volatile chlorinated hydrocarbons in soils, a newly developed biological reactor was used within the framework of a large-scale pilot project on a pilot site in Baden-Württemberg. Figure 7.7 shows the biological soil purification process according to the Terranox process of Umweltschutz Nord.

Biological heap processes are principally suitable for the treatment of gravel, medium sand, fine sand, and silty sand. For fine-grain soils such as clayey or loamy soils, heap processes are applicable only with limitations. With a continuous intermixing of the soil during treatment as, e.g., in the Terranox reactor process, fine-grain soils also are accessible to biological treatment.

Petroleum and aromatic hydrocarbons in particular are suitable for biological treatment in degrading heaps. Polycyclic aromatic hydrocarbons, volatile chlorinated hydrocarbons, other halogenated compounds, cyanides, organic nitro compounds, and pesticides can also be treated in this way with limitations. According to experience up to now, PCBs, heavy metals, dibenzodioxins, and dibenzofurans are unsuitable.

Biological reactor processes are basically suitable for a broad pollutant spectrum. Similar to heap processes they also are unsuitable for the treatment of soils contaminated with heavy metals and, according to experience up to now,

Figure 7.7. Terranox Process of Umweltschutz Nord for the biological treatment of soils contaminated with chlorinated hydrocarbons (pilot site)

for the large-scale treatment of dibenzodioxins and dibenzofurans. Depending on contamination and amount of treatment, the specific treatment costs for heap processes are on average between 100 and 250 DM/t. Biological treatment in reactors costs more than 300 DM/t.

7.1.2.2. In-Situ Processes

In-situ soil treatment processes have the advantage that the soil compartment requiring treatment remains in its natural position, and an excavation and subsequent back filling is therefore unnecessary. Since in in-situ processes cost-intensive investments for plant technology also are generally unnecessary, these processes are generally regarded as particularly inexpensive by comparison with on-site/off-site processes. This is certainly correct for suitable soil conditions and specific pollutants. In general, however, the cost situation must be evaluated differently in each individual case taking into account the limited applicability and the actual state of development.

In the following a description is given of some in-situ soil treatment processes that are already used successfully in the practice of soil remediation or can be regarded as promising.

7.1.2.2.1. Soil Vapor Extraction Processes

Soil vapor extraction has been used very successfully for a long time and on a large scale for the remediation of soils that are contaminated with volatile compounds (e.g., volatile halogenated hydrocarbons, aromatic hydrocarbons, gasoline hydrocarbons). As a result of the subatmospheric pressure generated by high-pressure ventilators, bypass compressors, or vacuum pumps, not only is the contaminated soil atmosphere present exhausted but increased evaporation is brought about by the air streams sweeping past near the pollutant molecules. The exhausted soil vapor is purified by activated carbon. Soil vapor extraction processes are usable preferentially in the unsaturated soil zone as in-situ processes. For decontamination of the saturated soil zone, soil vapor extraction is also used in combination with compressed air injection. In the course of this stripping air is introduced to the groundwater layer and desorbs the pollutants present in the water. The air is then captured via the soil vapor extraction system and led off to the activated carbon purification unit.

The advantage of the soil vapor extraction process is that high purification capacities can be achieved with relatively simple design. It is suitable preferentially for volatile chlorinated hydrocarbons.

Figure 7.8. Compact plant for soil vapor extraction (HUT company) on a plant site

Figure 7.8 shows a soil vapor extraction arrangement for the remediation of underground contamination at a plant site. The compact system, consisting of an exhaustion unit with water separator, electrical control unit, and activated carbon filter, has only a small space requirement. Soil vapor extraction processes have been part of the state of the art since the mid-1980s and since then have been used successfully for the aforementioned pollutants.

7.1.2.2.2. Soil Washing Processes

Soil washing processes also are used as in-situ processes. Keller Grundbau (Hamburg) has developed on the basis of the Soilcrete process (used in civil engineering) an in-situ soil washing process, with which soils can be washed to a depth of 35 m, using a high-speed water jet. The contaminated wash water and the extracted mud are purified aboveground. The purified soil constituents are refilled into the erosion cavities in parallel with the in-situ wash. The process was first tested in collaboration with S + I Schlammentsorgung and WUE Umwelt Engineering, Hamburg, on the site of a former disinfectant factory.

Another innovative in-situ soil washing process was used on a large scale by Philipp Holzmann (Düsseldorf) in the remediation of a former gasworks site in Bremen. In this case the soil to be cleaned was delimited from the remaining soil area by steel encasing tubes. Amounts of water up to 300 L/min were injected at up to 50 MPa with a rotating high-pressure injection lance. As a result of this high-pressure process, an extensive separation of the impurities from the soil particles took place already in situ. The purification of the soil–water–pollutant mixture took place in a multistage on-site cleaning plant.

7.1.2.2.3. Biological Processes

The possible uses of biological in-situ processes are at present still relatively restricted by comparison with on-site processes. This is due to the generally unfavorable in-situ boundary conditions, such as inhomogeneous pollutant distribution, deficient hydraulic permeabilities of the undergroud, inadequate moisture and oxygen conditions, and unfavorable conditions of temperature and pH.

A prerequisite for the biological degradation of pollutants in undisturbed soil is an active microflora whose degradation activity can be increased by external influences, for example, by artificial oxygen sources, such as free oxygen, ozone, hydrogen peroxide, or nitrate. Special nutrient additions (e.g., nitrogen, phosphorus) can furthermore contribute to accelerated microbial degradation.

In addition, the soil must have a suitable permeability in order to enable the access of sources of oxygen and nutrient to the microflora. Biological in-situ remediation is regarded as reasonable at soil permeability characteristics of $10^{-3} > k_f > 10^{-4}$ m/s.

At low soil permeabilities (e.g., in soils with a high fine-grain fraction or high degree of compaction), the permeability can be raised by external intervention. By use of the frac technique (→ Resources of Oil and Gas, **A 23**, p. 183) [7.3], the pore volume is enlarged in situ, which results in improved purging circuits for oxygen and nutrient transport.

In the sphere of biological in-situ treatment, successes are definitely recognizable. Thus the example of the biological treatment of contaminated soils in in-situ operation by Messer Griesheim (Krefeld) shows that the hydrocarbons typical of gasworks sites can be degraded to a considerable extent by the Biox-S-process.

Questions about the duration of remediation and the achieveable degradation output cannot in general be finally answered. From previous findings, however, the remediation duration, at several years, is considerably longer than in the case of on-site/off-site biological treatment.

7.1.2.2.4. Thermal Processes

In principle, thermal processes for the in-situ treatment of contaminated soils should also be possible. Various techniques to this end have been tried experimentally in the past. A promising technique using in-situ heating by high-frequency energy has been outlined [7.4]. In this process the pollutants present in the soil, such as halogenated hydrocarbons and petroleum hydrocarbons, are heated to temperatures of ca. 90–160°C in a powerful high-frequency field and vaporized together with the pore water. Through the effect of the steam stripping, the pollutants reach the soil surface, where they are collected by a steam trap and concentrated with the aid of a gas purification system. The special advantage of this technique, which has yet to be tested in practice, lies especially in the relatively short treatment duration by comparison with biological in-situ remediation. Another advantage can be seen in the fact that this process can be used in combination with the in-situ processes already mentioned.

7.1.3. Outlook

In 1995 in Germany a total capacity of 1.5×10^6 t/a is available by 69 operational stationary soil treatment facilities. These facilities include 50 biological, 16 soil washing and 3 thermal soil treatment facilities.

The development of on-site/off-site soil treatment processes goes in the direction of integrated soil remediation stations, where soil treatment with several different component processes is carried out to achieve efficient soil purification. Also in the case of in-situ remediation it is advantageous to use a combination of different techniques corresponding to the particular purification objective. A broad application spectrum is predicted for treatment in soil remediation stations. In particular, inexpensive and effective possibilities of treatment are opening up for small batches of soils—which arise to a considerable extent in contaminated sites practice—and for which the use of mobile treatment plants would be too expensive.

7.2. Recultivation

7.2.1. Definition

The term *recultivation* has become established as a collective name for all human activities for the restoration of a landscape. Following definition is given in [7.5]: "Recultivation denotes all the initiatives and measures that serve for the restoration of the scenic and economic structure of those parts of the cultivated land whose natural productivity have been considerably reduced or destroyed by human interventions, primarily by the mining of raw materials." In this definition the economic, ecological, and structural factors are regarded as of equal rank.

In recent scientific discussion the term *renaturation* has been introduced for the treatment and development of abandoned mining areas, preference here being given to free succession without significant additional shaping of the landscape. As a rule renaturation areas are small. In recultivation, by contrast, various, extensive landscape elements are combined with one another. At the same time the main characteristics of the cultivated land—that is its settlement and use by human beings—are deliberately incorporated.

Successional processes are integrated also in the modern notion of recultivation. If a certain nature conservation goal has to be followed, priority is given to the taking over of geological and ecological conditions exposed by mining or to the conscious shaping with initial substrata that are not immediately cultivable. A landscape made largely uniform by agriculture and forestry, especially in Central Europe, can require this.

7.2.2. Demands on Recultivation

The mining of raw materials is usually accompanied by a radical change of the earth's surface, and ecological equilibria that have become established over thousands of years and have been more or less modified by human use are destroyed. For this reason the ultimate goal of recultivation is considered the restoration of a balanced ecological state and fitting of the recultivated area into the surrounding landscape. Negative examples from many countries of the world show that this cannot be achieved without a target for soil development or establishing an ecologically compatible chemical and physical state. Those examples include the tertiary sand waste heaps of the East German lignite open mines, the erosion-threatened and nutrient-free dumps of lateritic soils in Asia, and the heavy-metals contents of waste materials from mines worldwide.

7.2.2.1. Recultivation Goals

The recultivation goals for mined areas have to be as various as the settlement, industrial, and countryside areas concerned themselves. However, some basic principles can be generalized:

1) Recultivation, in contrast to renaturation, requires such a large area of land that its organization becomes the subject of district planning and regional planning procedures. Recultivation thus becomes a task of regional policy and is therefore exposed to political processes even in the goal setting.
2) Interventions in the economic area necessitate interference with property. Depending on the initial situation, decisive courses are often set here. Thus, in West Germany, for example, clear preferences exist for the reestablishment of agricultural or subsequent commercial uses.
3) The nature of the intervention to be compensated for, especially its technical or commercial layout, creates far-reaching conditions that exclude particular recultivation goals and favor others. If, for example, the landscape is changed by the creation of inner and outer dumps, only woods can be planted because of the resulting slopes, especially also with regard to the erosion dangers. Microclimatic changes due to the intervention also occur, but their influence is weaker. Furthermore, the leaving of final voids will inevitably lead to formation of water bodies. Altogether, changes of the groundwater conditions are far-reaching for the shaping of the landscape in certain regions of the world.
4) Important factors for the formulation of the recultivation goal are the geological conditions and thereby the possibility of selecting adequate soils. The lignite commission mining plans in West Germany, which were drawn up for the Rhenish lignite district, can serve as examples for the drawing up of recultivation goals.

In the Rhineland, in accordance with landscape use before coal mining, attention is centered on *agricultural* reutilization. With this, the careful choice of initial substrate and the development of soil fertility become very demanding tasks. The application of the tillable cultivated layer, mostly consisting of loess or loam, is followed by seven years of interim farming, which leads, via the first sowing and maintenance of alfalfa, after five years to the main crops barley, wheat, rye and recently also vegetables.

In the winning of other raw materials, agricultural recultivation after termination of extraction is not an attainable goal anyway, since no soil substrata at all are available originally: examples are the hard coal heaps which even after applying native soil, merely allow planting with trees or grasses, but on the other hand are well suitable for the restoration of recreation areas.

In other climates, especially in the tropics, agricultural recultivation is started best with agroforestry, especially in order to remove the initial food bottlenecks and solve the erosion problems.

For *reafforestation*, a much wider range of different substrata can be used because of the greater ranges of site requirements of the wood plants.

For the *moderate climatic regions of West Germany*, determined by loam–gravel soils, the possibilities of forestry can be described with the following types of management objective.

1) Common oak – small-leaved lime – hornbeam mixed forests with transitions to sessile oak forest on dry locations, 40% of recultivated area
2) Beech forests with strong inclusion of valuable broad-leaved trees (e.g., ash, red oak, walnut, lime, maple, and common elder) and a minor amount of European larches, 30%
3) Deciduous forests with a high proportion of valuable broad-leaved trees distributed in groups or small stands, 10%
4) Coniferous woods, principally Douglas fir, but also pine and fir, 10%
5) Meadows, succession areas, hedges, bodies of water, 10%

Things are quite different in the drier and sandy *East German recultivation areas*. In East Germany the following types definitely predominate:

1) Pine–sessile oak forest
2) Pine–common oak forest
3) Birch–heath succession stages

Water bodies can be laid out as a further important building block of recultivation landscapes. These bodies of water can be economically important as retention areas, but in arid regions they take over the important function of water reservoirs. Water bodies are regarded in Central Europe as undesirable for water management owing to the exposure of the groundwater strata and the risk of entry of substances. As recreation areas and above all as nature conservation areas, however, they are outstandingly important. If recultivation areas are declared as national preserves in West and East Germany this is based mostly on the wealth of species on the bodies of water.

Owing to the enormous decline of species in the cultivated land it must be assumed that at least 10% of the recultivated area are used for the protection of species in highly developed settlement and economic areas. In other areas, e.g. in East Germany, that have a considerably smaller demand for agricultural use, the proportion of areas used for the protection of endangered species amounts to more than 50% of the total area. By this means recultivation comes ever closer to its goal of being a compensation for an intervention in the ecosystem.

7.2.2.2. Initial Substrata

A key role in recultivation planning attaches to the analysis and selection of the geological strata exposed during mining. Materials from great depths often have on the one hand advantages in nutrient content and grain size structure, but on the other hand outright poverty or absence of biogenic contents. As a rule, a topsoil application is impossible for technical reasons. The main problem in this connection is that because bulldozers and lorries must be used for the application of topsoil, the possible positive effects are nullified again by soil compaction.

The displacement process inevitably entails the destruction of the live substratum "soil" and its specific bedding. The recultivation cannot start to develop a pedosystem and finally an ecosystem from spent soils but merely from introduced substrata.

For Europe three initial materials in particular play an important part: loess and loess loam; river gravel; Pleistocene sands and marl.

Loess is a yellow-brown to gray-brown, limey, eolian loose sediment from the dry-cold climatic phases of the Pleistocene. Over a period of about 20000 years, the upper layer of this silt-rich material has developed through pedogenetic processes (clay formation, clay displacement, deliming) as a rule to para-brown earths, similigleys (including their transitional phases) and, if groundwater influence is present, to gleys.

This loess loam grades downward into lime-rich, loose loess. Until into the present century, this loose loess was extracted as an agricultural ferilizer in small underground "marl pits". Below this "marl" there follows a layer, of Riss date, of carbonate-poor, partly clayey or sandy-gravelly loam.

This stratigraphic sequence found in the Rhineland is of variable thickness, because during the deposit of wind-borne loess erosion phenomena also naturally occurred, and may occasionally have removed a whole bed again.

The terrace sediments lying under the loess beds are melt water deposits (sands, gravels, and boulder flint) from the Pleistocene. Also found are oligotrophic, weakly sorbing clays from the Tertiary, consisting mainly of kaolinite.

7.2.3. Soil-Physical Properties

7.2.3.1. Silvicultural Recultivation

The investigations carried out by the State Geological Office of North Rhine-Westphalia led in 1967 to the laying-down of directions on the application of tillable soil material for silvicultural recultivation in the lignite districts of the Rhineland. After a certain trial period, these directions were revised in 1973 and finally once more in 1984 in order also to support the use of pure loesses instead of sand–gravel–loess mixtures, though at quite small thickness.

The essential elements of the directions are: the rough heaps must be so shaped that no water damage or damming occurs. Their top 1–2 m should consist of sufficiently permeable sandy–gravelly material. Tillable soil material should be applied about 3–5 m thick on the rough heap. Possible tillable material is a mixture of sandy–gravelly beds of the high and middle terraces and loess or other cohesive types of soil called forest gravel. Also usable are highly loamified loesses unsuitable for agricultural recultivation as well as loess beds affected by groundwater and dam water. On level or gently inclined surfaces the tillable material should include 25–66% of loess or loess loam. On slopes (i.e., areas of an inclination exceeding 1:3) the loess fraction should be limited to 20–25%. Instead of a 4 m thick mixture of loam and gravel, 2 m of loess or loess loam also can be dumped. Processing the surface thus applied with earthworking machines is not required.

In 1984, STRATMANN carried out profitability studies on recultivation areas in the Rhenish lignite district for the Niedersächsische Forstliche Versuchsanstalt (Lower Saxony Forestry Research Institute) [7.6]. One of the purposes of the investigation was to assess in detail the relationships between tree growth and soil. The soils included were mixtures of gravels and loams: i.e., substrata that corresponded to the instructions of the State Supreme Office of Mining for the Rhineland, even if these came from times when the directives for the deliberate blending of such mixtures were not yet laid down. However, among others locations with pure loams, sands, and silts having more or less small additions of fly-ash and coal were also investigated.

According to the results, the prospects for future oak cultivation on forest gravel must be judged favorable. In accordance with the specific site requirements of the two oak species, the sessile oak should be placed on the drier sites and the common oak on the wetter. The hornbeam or the small-leaved lime can be mixed in instead of the beech for the clearing of the stems.

In the final assessment it is stated that sandy forest gravels can lead to a considerable reduction of the yield of forest stands. Soil compaction significantly restricts the growth. The effects of soil compaction were investigated in detail in the following years. On spreading the forest gravel, more or less strongly marked ridges arise which, according to forestry opinion of the 1970s, had to be removed by leveling. The standards of the final operation schedules (especially the required slope inclinations) made in many areas an enormous amount of leveling necessary; and the resulting compactions, which in some cases were high, could not be removed satisfactorily, even by subsequent ripping.

The compactness of the forest gravel was subjected to a detailed investigation. Finally the decision was made to preferably dump the material such that small ridge heights are obtained and leveling is therefore omitted [7.7]. The tipped substrata then have the highest possible pore volumes which gives the best prerequisites for good rooting, a sufficient water storage and subsequent water supply, as well as an assured gas exchange for plant growth.

The basis of these considerations was that each excavation process destroys the soil structure, and as a result of the deposition of both loess and boulder gravel, a coherent structure that tends to instability or *compaction* is finally formed. This relates especially to compaction of the coarse pores, whereas the distribution of the fine pores is little changed.

The total pore volume fluctuates between 40 and 25%, forest gravels of high loam content show relatively high pore volumes. The influence of the leveling operation was investigated and established in detail [7.8]. It is interesting that merely a single aftertreatment shows a positive effect, while several operations are again accompanied by negative compaction effects (Table 7.4).

It remains to be seen whether the better starting conditions on unleveled forest soils persist: future research will have to keep an eye especially on the biological regeneration and structuring of the soils. Previous investigations show that there is apparently a connection between amount of initial compactness and earthworm settlement.

In forest gravel investigations the compactness ranges from 1.5–2 g/cm^3, and therefore approaches that of pure main terrace gravels. Forest gravels richer in loess, with their densities of 1.7–1.85 g/cm^3, correspond to the old forest sites on similigleys.

Recultivation soils must be generally classified as "dense" to "very dense", an increase of

Table 7.4. Effects of various forms of tillage before afforestation on soil structure, void ratios per measuring depth

	Surface			
	Leveled, ripped and harrowed	Leveled, ripped and smoothed	Leveled, ripped	Small spreader, not treated
Measuring points	1–8	1–6	1–6	1–6
Depth	Void ratio %			
0.3 m	30.1 ± 2.9	28.0 ± 1.9	29.2 ± 1.0	35.2 ± 1.0
0.6 m	27.7 ± 2.6	27.8 ± 1.2	29.0 ± 2.3	33.5 ± 3.5
0.9 m	28.1 ± 2.5	28.3 ± 1.5	30.5 ± 2.4	35.3 ± 1.1
1.2 m	28.5 ± 3.0	30.1 ± 1.5	32.5 ± 1.8	35.0 ± 3.5

Table 7.5. Most frequent values for consolidation, pore volume and void ratio of the soils (105°C dry)

Soil type	Consolidation, g/cm^3	Pore volume, %	Void ratio (or ε)*
Sandy	1.67–1.19	37–55	0.58–1.22
Loamy	1.96–1.19	26–55	0.25–1.22
Silty	1.53–1.19	42–55	0.72–1.22
Clayey	1.32–0.92	50–65	1.00–1.85
Organic	0.48–0.12	60–90	1.50–9.00

* ε = porosity.

loess and loam fractions enhances pore space and loosening. For typical physical soil characteristics see Table 7.5.

A peculiarity shown in the investigations on forest gravel was a strong tendency to a natural compaction of the materials when the admixture of loams and loesses in the forest gravel was such that the proportion of cohesive components taken up was between 15 and 18 % [7.8]. In these cases a complete filling of the pore space can be assumed due to the composition and grain size of the material, so that compaction occurs even without fairly strong external action.

The investigations in the Rhineland show that with loess, but in the end also with forest gravel with the appropriate admixture of cohesive components, the moisture of the material to be excavated is of great importance. It is generally accepted that wet areas should not be excavated. Wet areas are also often encountered where as a result of preliminary work soil materials are stockpiled and are then wetted for a fairly long period. The materials scarcely release this moisture even after digging.

The erosion risk on the numerous slopes leads to the conclusion that the forest gravels should contain a maximum of 25 % of cohesive components. However, investigations in open pits that have limey loess available showed, that on increasing the lime content of the admixed loess the stability of the forest gravel increases considerably even on the slopes, so that 30–40 % admixtures of loess in the forest gravel are acceptable without considerable erosion problems occurring.

7.2.3.2. Agricultural Recultivation

The greatest demands on the choice of soil naturally arise for agricultural recultivation. If loess materials are available, it is clearly preferable here to feed them into the agricultural recultivation. The following procedure has been successful [7.9]. The quality of the loess to be spread is analyzed by sampling the mining slope and hand drilling in the zone immediately in front of the working boundary. During the building up of the rough heaps and the spreading of the loess, technical aids are used (e.g., rotary lasers) in order to produce the most accurate possible surface even with the high-capacity equipment and to minimize the leveling expenditure. Bulldozers with extremely low specific soil pressure are used. In the loess leveling of the arable land, surface roughnesses can be allowed to remain, provided they do not prevent subsequent treatment with agricultural machines. The loess must not be leveled immediately after spreading, but only after complete drying out. To obtain a small ridge height during the spreading of the loess, high-capacity equipment must be operated at reduced power, if required. Wherever possible the dumping height of the topmost spreader must be so reduced that a horizontal position of the discharge boom is maintained. By this means a more precise filling is made possible.

The questions of the aeration and rooting of the soil are most important, especially in the technical design of the measures. The comparison of yield between old and new land allows a positive assessment of the loess recultivations (Table 7.6). The variation of the basic soil-physical data in loess recultivations has been pointed out [7.10] (see below):

Soil type and grain size distribution	lU (medium-loamy silt) with 14–17% clay, 70–80 % silt, 2–5 % sand
Humus content	0.5 to >1.0 %; average increase ca. 0.03 %/a
Bulk density	1.55–1.75 g/cm^3
Total pore volume	35–52 vol %

Table 7.6. Comparison of crop yields on recultivated areas and neighboring old land in Neuss county (t/ha)

Crop type	1990		1991	
	Recultivated area	Old land	Recultivated area	Old land
Wheat	6.1	7.1	8.5	8.5
Rye	5.4	6.1	6.7	6.0
Barley	6.3	5.1	6.9	6.8
Rape	4.1	3.1	4.2	4.0
Sugar beet	71.4	58.6	66.0	58.4

Table 7.7. Use of various substrates

	Loess			Marl	Pleistocene river gravel	Pleistocene sands	Tertiary sands	Clay	Marl–sand mixture	Loess–gravel mixture
	Para-brown earth	Similig-ley–gley	Carbonate-containing underlying loess							
Agriculture	x	■	x	■	o	■	o	o	■	o
Forestry	x	x	x	o	o	x	o	o	x	x
Protection of species	x	x	x	x	x	x	■	x	x	x
Recreation areas	x	■	x	■	o	x	o	o	x	x

x = suitable; ■ = of limited suitability; o = unsuitable.

Proportion of air-conveying voids (>30 μm)	4 to >12 vol%
Utilizable water capacity relative to 1 m root space	high, 160 to >200 L/m^2
Consistency limits	
yield point	26.0–29.0 wt%
plastic limit	18.0–22.0 wt%
plasticity index	5.5–8.0 wt%
shrinkage limit	5.0–10.0 wt%

The suitability of different starting materials for different recultivation goals is shown in Table 7.7.

7.2.4. Soil-Chemical Properties

7.2.4.1. Silvicultural Recultivation

Investigations on nutrients content at various problem sites and on nutrient uptake by their vegetation have yielded fundamental information.

Thus it has been found for oaks and beeches on comparable difficult stands that as regards soil nutrient supply, only a few nutrients such as phosphorus and magnesium showed a deficiency. In addition, the pH values were too low in many cases, because at the problem sites the forest gravels were usually very poor in loam and loess. The leaf nutrient values found there were sufficient, except for nitrogen in the case of the oak and potassium and magnesium in the case of the beech. This finding runs through all other investigations, which have shown that the leaf nutrient contents from the dumping of forest gravels are sufficient.

If, on the other hand, the nutrient contents of the soils analyzed at the problem sites are considered, it becomes clear that in some cases a deficiency in the nutrients phosphorus and magnesium and in many areas also nitrogen occurred. With the other nutrients, the supply could not be called particularly good, but it could not in the end be held responsible for causing the harm. In the case of some soils it was also shown that the supply of calcium is inadequate and a strong acidification was found as far as into the aluminum buffer range.

Friedrich has demonstrated a relationship between the nutrient conditions and the different proportions of loess in the forest gravel [7.9]. Thus, for the forest gravel group with ≥50% loess and the soils with 25–33% loess neutral to weakly acidic reactions are found. At unlimited exchange capacity calcium and magnesium almost exclusively make up the exchangeable cations. The reserve of phosphorus in the loess-rich forest gravels can also be regarded as high, but naturally in young forest gravels the proportions of carbon and nitrogen in the dry material are low. The forest gravels with 20 or even 10% loess appear distinctly worse. The pH values of these gravels are in the acid range (Table 7.8).

Especially in forest gravels with less than 10% loess, aluminum and in some cases iron and hydrogen can have a high to predominant proportion of the cation exchange. The exchange capacities fall in each case by 30% relative to the preceeding soil groups (10–20% loess, 25–33% loess, etc.) that for forest gravels with small proportions of loess nutrient problems must occur.

In the bad forest gravels, possible heavy-metals concentrations have been suggested as the cause of poor growth. Here, though, the analyses

Table 7.8. Chemical state of soils of forest gravel with varying proportions [7.9]

Surface profile no.	Proportion of loess, %	Depth, 10^{-1} m	pH ($CaCl_2$)	AK_E	%C/d.m.	%N/d.m.	C:N ratio	P, mg/100 g
62	50–66	0–4	7.35	145.692	0.37	0.03	12.23	38.0
61	25–33	0–4	7.8	271.761	0.51	0.017	29.82	26.0
		4–8	7.7	137.350	0.64	0.021	30.24	18.0
21	20	0–4	5.6	31.752	0.098	0.01	9.80	10.0
72	20	0–4	7.7	118.008	0.14	0.01	14.30	14.5
		4–8	7.4	58.129	0.076	0.008	9.50	11.0
45	20	0–4	5.8	61.104	0.44	0.036	12.28	22.0
11	20	0–4	5.3	24.611	0.22	0.02	13.06	11.0
22	10	0–4	5.4	26.008	0.077	0.009	8.56	12.0
65	10	0–4	3.75	9.500	0.14	0.011	12.45	11.0
41	10	0–4	4.95	106.509	0.05	0.006	7.67	8.0
31	10	0–4	3.55	23.843	0.14	0.012	11.58	13.0

AK_E = effective exchange capacity; d.m. = dry matter.

have shown that the heavy metals copper, silver, chromium, and zinc are present only in very small concentrations and thus cannot be the cause for the impairment of plant growth.

7.2.4.2. Agricultural Recultivation

DUMBECK has assembled the most important chemical parameters for initial soils in the Rhineland (see below) [7.12].

Carbonate content	4.5–11.5%
Nitrogen	total N = 0.02–0.05%
Potassium	total K_2O = 300–700 mg/100 g soil calcium lactate soluble K_2O = 6–12 mg/100 g soil
Phosphorus	calcium lactate soluble P_2O_5 = 3–6 mg/100 g soil
Magnesium	$CaCl_2$ soluble Mg = 12 mg/100 g soil
Cation exchange capacity	8–10 mval/100 g soil

It is significant that, because of the admixture of carbonate-containing loesses, the initial pH values (7.5–8.0) are in the weakly alkaline range. The initial soils often show excellent nutrient supply. However, it can be observed in the course of recultivation when intermixing the soils that the nutrient contents in the recultivation substrata are decidedly low for agricultural use. Here, appropriately increased amounts of fertilizer are supplied, which exceed the uptake by plants and therefore lead to an enrichment in the soil. Since the humus contents are low, the nitrogen mineralization is also lower than on old sites. A 15–30% nitrogen supply compensates the nitrogen release resulting from the lower mineralization.

Great problems have occurred in the recultivation of poor tertiary material, especially sands, in the mining districts of Central and East Germany. The content of marcasite and pyrite led to an enrichment of the soil with acids, which can only be contained by extensive liming using power station ashes or lime fertilizers. Above all, the tertiary sands having a total sulfur content >2% also must be viewed critically for improvement measures, since often with them only a superficial remediation is possible; and, with deep rooting of plant species, this can lead to considerable growth reductions.

In the discussion of improving agricultural or silvicultural use on poor substrata, the recycle of recycling substances also has been considered recently. It is being discussed above all in the case of recultivation, to what extent sewage sludge, wood chips, composts, and other recycling products can be used. The previous studies show that for difficult recultivation situations a distinct improvement can be seen, especially in the medium term.

8. Legal Aspects of Soil Protection

8.1. Foundations of Soil Protection Law

8.1.1. Land Law

Following Roman legal tradition, the open air and flowing water are not things in possession in the legal sense and not accessible to action under (private) law. In contrast, the partition of the land, is considered an original law of the earth. The taking of land is regulated by land law: this allocates the land to persons in law as bearers of rights and duties and governs legal relationships with regard to the land. It is expected that the user of the land will also protect, maintain, and preserve it. Until recently the land has been regarded not only as immovable and nonincreasable but also as indestructible. Since the 1970s, however, concern about human's natural fundamental requirements of life has developed on the one hand from the recognition that natural resources are limited, and on the other from a growing awareness both of the increasing anthropogenic pollution (or other harmful change of the natural condition) of soil, water, and air and of their dwindling renewability. The protection and maintenance of human's natural fundamental requirements of life, i.e., care of the environment, has developed into a public task for public benefit going far beyond the interests of the individual for private benefit protected by neighbor law. Within the framework of care of the environment, care of the soil has been added as a new terrestrial primary law. The soil is no longer indestructible. The supposed identity of interests between soil use and soil care appears no longer adequate to maintain the soil as a part of the natural fundamental requirements of life for present and future generations.

8.1.2. Soil Protection Law

Land law is therefore supplemented by a soil care law or soil protection law [8.1], that

1) As a collective concept gathers together the legal provisions that serve the protection, maintenance, and development of the soil,
2) Can be described as a legal system of measures for intervention incentives, and planning for the realization of certain soil-protection goals, including the necessary processes, organization and sanctions—and therefore as a part of environmental law, and
3) As a result of its primary purpose of protection, differs from regulations for land use in the narrow sense, without it being possible to avoid the overlapping of the two spheres in the face of the subject of regulation named.

8.2. Foundations of Soil Protection Policy

8.2.1. Soil Protection in Germany

The environmental program of the government of the Federal Republic of Germany of 1971 [8.2] already designated the soil in its "triad of goals" as an environmental medium, alongside water and air, that had to be protected from harmful human actions. The soil protection concept of the German Federal Government of 1985 [8.3] described soil protection as a cross-sectional task of environmental protection and emphasized two central approaches for enhanced soil protection:

1) Minimization of the qualitatively or quantitatively problematic entry of substances and
2) A "trend reversal" in land use [8.4]

8.2.2. International Soil Protection

Already in 1972 in the regional framework of the Council of Europe a "European Soil Charter", which formulates a Europe-wide soil protection policy in twelve articles, was adopted as a recommendation for the Member States [8.5].

In 1981, the protection of the soil was recognized worldwide as a problem in environmental law by the Montevideo Environmental Law Program of the United Nations [8.6]. The program describes as goals of the legal measures in the soil protection subsection at national, regional and global levels

1) To prevent the impairment as a result of human actions of the sustained productivity of the soil, resulting in erosion, desertification, salination, deforestation, overuse, contamination, or excessive land use, and
2) If the need arises, to remediate polluted soils

Aside from the application of appropriate worldwide basic principles, guidelines, and recommendations at all levels of control ("World Soil Charter", "World Charter for Nature"), a stronger emphasis on soil protection in the laws at national level is encouraged.

The Conference of the United Nations for Environment and Development of 1992 in Rio de Janeiro worked out in the action program "Agenda 21" among other things instructions for an integrated approach to the planning and management of land resources (Chap. 10 of the Agenda) and the combating of desertification (Chap. 12 of the Agenda), which need national enforcement [8.7]. In the following sections the soil protection law in Germany is described as an example.

8.3. Constitutional Soil Protection

Soil protection is part of the national task of care of the environment, even if this national task was not expressly enshrined in the German "Grundgesetz" until 1994. With regard to intervention measures, soil protection experiences constitutional barriers. The measures must be suitable, necessary, and proportionate.

The soil protection legislator has correctly to value the realities underlying his decisions, to which end scientifically based data are drawn upon, especially for the definition of soil-protecting environmental standards (soil values) to quantify soil damage. This does not exclude precautionary measures if there is not (yet) sufficient scientific knowledge. In addition the remaining constitutional standards of competence and of guaranty have to be observed, above all the warranty of property under the Grundgesetz (Art. 14 Grundgesetz). The social obligation of land ownership is of increased importance because of the nonincreasability of the land. Where constitutions of the German "Bundesländer" (federal states) contain state objectives in favor of care of the environment, these also include soil protection within the scope of the federal state's spheres of jurisdiction.

8.4. Soil Protection under Administrative Law

The protection of the soil is found in a large number of legal regulations from all stages of legislation, without being concentrated at the level of the German federal law in a single standard work. Where a federal state has its own soil protection law, this is applied, provided regulations under federal or state law do not contain provisions to the same effect or to the contrary [8.8].

The soil protection standards are mainly administrative law. Environmental private law (e.g., environmental liability, Umwelthaftungsgesetz) and environmental criminal law, on the other hand play only a supporting role in soil protection owing to their essentially sanctioning character. Soil protection law is accordingly principally administrative law and as such a part of environmental administrative law, which in turn is part of special administrative law. If no special regulations exist under the law of a particular federal state, soil protection follows the provisions of the remaining environmental administrative law.

8.5. Soil Protection Goals

8.5.1. Soil

The soil, at the point of intersection of atmosphere, hydrosphere, biosphere and lithosphere, has not yet received an authoritative definition in law. The meaning of the term soil has in each case to be obtained from the context of legal rules in which it is used; the regulatory purpose of the rules can circumscribe the connotation. However, as a starting point for a legal description of the soil as a value to be protected it can be defined as the topmost layer of the earth's surface, inhabited by animals, plants, and microorganisms, on the firm ground and below the water surface of waters above ground. Beyond that, the protected zone of standards significant for the soil can also cover the uninhabited underground, as far as it is accessible to anthropogenic influences.

8.5.2. Protection of Soil Functions

The soil is protected legally as a part of human's fundamental requirements of life for the sake of its functions, especially as a habitat for biomass, as a filter, as a stock of raw materials, and as building land. The legal protection has to guarantee that the soil can provide its varied functions in the required quality and quantity for a long time. The reference to the soil as part of the natural fundamental requirements of life shows the ecological approach to soil protection, which not only views the soil sectorally and as a medium comparable with the environmental media water and air, but also sees it together with these media and the sunlight as the basis of all life [8.9]. Futhermore the reference makes clear that soil protection is not an end in itself, but rather, like care of the environment as a whole, in the end always occurs for human's sake.

8.5.3 Impairments of the Soil

The protection of the soil is directed against impairments of the soil quality and quantity that are caused in the soil by material and nonmaterial anthropogenic intrusions in the soil.

Material intrusions in the soil can be brought about by the directly- or indirectly-caused entry of substances (soil change by substance entry). Soil impairment by substance entry means the effects of chemical elements or chemical compounds that are capable of detrimentally changing the natural condition of the soil.

Nonmaterial intrusions in the soil include intrusions through an alteration of the soil's use, whereby the efficiency of the soil or of the ecosystem or of the landscape can be impaired considerably or for a long time (soil consumption). In addition they include other non- material intrusions into the soil, which, without being soil consumption in the sense just named, can bring about corresponding impairments.

8.5.4. Soil Protection by Provision and Aftercare

Protection of the soil signifies not only the warding-off of impairments of the soil. Protection of the soil also comprises maintenance and development in the sense of shaping, improvement and restoration (remediation of contaminated sites) and provision that impairments do no occur at all and that the soil can perform its diverse functions for a long time. Soil protection accordingly means both precautionary care and aftercare.

8.5.5. Direct and Indirect Soil Protection

Soil protection can occur directly as the main or subsidiary goal of regulation, if the soil is identified explicitly as the value to be protected. Direct soil protection can also be spoken of when the legal system uses terms such as environment, natural fundamental requirements of life, natural resources, ecosystem, or nature and landscape, because these notions include the soil as protected property. Soil protection can also occur indirectly, however, by other goods and values being protected, the soil being co-protected without itself expressly being protected property.

8.5.6. Relative and Absolute Soil Protection

In view of the variety of the soil's functions and of the multitude of demands that are made upon it, the legal system has to provide for an appropriate solution of possible conflicts of goal. When weighing different interests, consideration is accorded to soil protection (relative soil protection). Soil protection can also be given priority, either generally or in the individual case (absolute soil protection), for instance, by regulations on sewage sludge application.

8.6. Soil Protection Measures

8.6.1. Types of Measure

The set of instruments under administrative law, consisting of soil protection measures for intervention incentives and planning (triad of measures) is used to realize soil protection goals. The measures interfering with the legal rights of the individual are orders and prohibitions or restrictions directed to the carrying out, toleration, or neglect of particular actions. They are supplemented by instruments that, by means of specific incentives such as the grant of tax allowances or of user advantages, indirectly guide human behavior to the realization of goals serving soil protection. With the aid of the incentive measures the environmental administration influences corporate behavior. This involves the erection and maintenance of supply and waste-disposal equipment, public consultation and information, the public financial support of activities appropriate to soil protection, and public compensation. Within the triad of measures the measures of environmental planning have a special function, because they above all are capable of realizing the precautionary principle.

The measures can be restricted spatially to certain areas (protected areas). They can distinguish chronologically between old soil contaminated sites (Federal Waste Act, remediation of contaminated sites) and new soil impairments.

8.6.2. Classification of Soil Protection Measures

The standards significant for soil protection can be classified according to the type of soil intrusion into protective standards against impairments that are brought about by material intrusions and those that are brought about by nonmaterial intrusions.

8.6.3. Measures against Material Soil Impairments

The soil receives direct or indirect protection as a result of environmental law from impairment(s) of the soil brought about by directly or indirectly caused entry of substances in the environmental care areas of the protection of waters (Wasserhaushaltsgesetz, Water Resources Policy Act), avoidance and disposal of wastes (Abfallgesetz, Federal Waste Act), immission protection (air pollution control; Bundesimmissionsschutzgesetz, Benzinbleigesetz, etc.), radiation protection (Atomgesetz, Atomic Act), protection from harmful materials (ChemG) and genetically engineered organisms (Gentechnikgesetz, Genetic Engineering Act), and nature conservation (nature protection and landscape preservation, soil protection; Bundesnaturschutzgesetz). These are supplemented by the Building Regulations Law (Landesbauordnungen) and the General Police and Regulatory Law (Landespolizeigesetze) but also by the Regional Planning Law and Federal Planning Laws (Raumordnungsgesetz, Landesplanungsgesetze) so far as this can control entries of substances by appropriate allocation of soil uses. The main emphasis of the regulations that are intended to protect the soil from the direct entry of substances is in protection from harmful substances and organisms modified by gene technology, avoidance and disposal of wastes, and protection of waters. The immission protection regulations, in particular, give protection from the entry of substances via the air path by measures that operate at the source of emissions. In the area of general environmental care, the environmental impact assessment prescribes for particular projects the determination, description, and evaluation of their effects, also on the soil, and requires consideration of the result of the evaluation during the licensing of the project.

8.6.4. Measures against Nonmaterial Soil Impairments

Standards governing nonmaterial soil impairments are stipulated in laws of nature protection and landscape preservation, and of soil protection (Bundesnaturschutzgesetz, Bundeswaldgesetz, Landesbodenschutzgesetze). This core area is supplemented by the standards controlling soil protection of environmental planning and regional planning (Raumordnungsgesetz, Landesplanungsgesetz), of zoning (Baugesetzbuch), and of individual technical plans. For projects subject to an environmental impact assessment the observance of soil protection is assured. This is supplemented by direct and indirect soil-protection standards of agricultural law (e.g., land consolidation, Flurbereinigungsgesetz) and of the Federal Mining Law (Bundesberggesetz).

8.7. Development Trends

8.7.1. Further Development of German Soil Protection Law since 1985

At first sight, the soil protection law in force in Germany has an appearance of strong fragmentation. But this alone must not lead to the conclusion that the soil is unprotected or only very incompletely protected. The legal translation of the ecopolitical demands for a reduction of the direct and indirect entry of substances into the soil and for a reduction of soil consumption certainly signals weaknesses of standardization of the law in force. But this is up to now combated less by a special soil protection law than by a strict application of the soil protection law in force and its development. Regulations for soil protection are now being considered, like those enacted in Baden-Württemberg and Saxony in 1991.

8.7.2. Draft of a German Soil Protection Act 1993

A draft bill for an amending act for the protection of the soil is being discussed at present (ministerial draft for a soil protection law). It provides in its Article 1 for an "Act for the Protection from Harmful Changes to the Soil and for the Remediation of Contaminated Sites (Bundes-Bodenschutzgesetz, BBodSchG)"; Articles 2 to 7 contain amendments to pertinent environmental laws to ensure the dovetailing of the soil protection law with the remaining environmental law.

The draft of the Federal Soil Protection Act aims to protect the soil from harmful changes and to make provisions against the build-up of harmful soil changes. Aside from its precautionary function, comprising the defence against hazards and the reduction of detrimental effects on the soil, the law also stipulates soil protection by aftercare for the removal of harmful soil changes and contaminated sites.

The purpose of the Act is highlighted by two principles, which present soil protection as soil function protection:

1) Preservation or restoration of the natural functions of the soil as a fundamental requirement of life and as living space for humans, animals, plants, and soil organisms, as part of the ecosystem, as a degradation, equilibration, and synthesis medium for substances based on its filtering, buffering, and matter conversion properties (§ 2 no. 1–3, E-BBodSchG)
2) Preservation or restoration of the soil in its functions as a raw material deposit, as a site for agricultural and silvicultural use, as an area for settlement and recreation, as a site for economic uses, traffic, supply, and waste disposal as well as an archive of natural history and the history of civilization (§ 2 no. 4–8, E-BBodSchG)

The Act defines *harmful soil changes* as changes of the physical, chemical, or biological condition of the soil that, depending on their nature, extent, or duration, are capable of impairing the soil functions and thereby of bringing about dangers, considerable disadvantages, or considerable nuisances for the individual or the general public (§ 3 par. 1, E-BBodSchG). Here the ecological functions of the soil are included as protected property of the general public. The law also takes into account whether the possibility of a sustained impairment of the natural functions of the soil exist which cannot be reconciled with the protective and user interests of the individual or with the common good (§ 3 par. 2, E-BBodSchG).

Obligations to provide against harmful soil changes (§ 6, E-BBodSchG), for accident prevention, and for the elimination of harmful soil changes (§ 7, E-BBodSchG) support the objective of the law. Special regulations are made for the registration and remediation of contaminated sites (§§ 17–26, E-BBodSchG). The obligation of remediation of contaminated sites (§ 18, E-BBodSchG) is put in concrete form by rules and orders on soil-dependent pollution limits and test values as well as requirements for removal of pollutants (§ 21, E-BBodSchG). The federal states can also prepare soil protection plans for the remediation of areas in which harmful soil changes occur, or are to be expected (§§ 28–30, E-BBodSchG).

9. References

References for Chapter 3

[3.1] C. Bosch in D. Rosenkranz, G. Bachmann, G. Einsele, H.-M. Harreß (eds.): *Bodenschutz*, Erich Schmidt Verlag, Berlin 1988, p. 1480.

[3.2] H. Strunz: *Mineralogische Tabellen 3*, 7th ed., Akademische Verlagsgesellschaft Egeestu. Portig., Leipzig 1970.

[3.3] C. Ahl et al.: *Aspekte und Grundlagen der Bodenkunde*, Eigenverlag, Göttingen 1994.

[3.4] V. Wolters: "Bodenökologische Aspekte der Freisetzung gentechnisch veränderter Organismen," *UBA-Texte* **20/93** (1993) 182–196.

[3.5] N. Knauer, D. Trautz: "Böden als Lebensraum von Organismen," in H. P. Blume (ed.): *Handbuch des Bodenschutzes: Bodenökologie und Bodenbelastung; vorbeugende und abwehrende Schutzmaßnahmen*, Ecomed Verlag, Landsberg/Lech 1990, pp. 59–70.

[3.6] H.-P. Blume: *Handbuch des Bodenschutzes*, ecomed, 2nd ed., Landsberg 1990.

[3.7] AG Bodenkunde: *Bodenkundliche Kartieranleitung*, 3rd ed., E. Schweitzerbart'sche Verlagsbuchhandlung, Stuttgart 1982.

[3.8] S. K. Gupta, F. X. Stadelmann, F. X. Häni, A. Rudaz: "Interrelation of Cd-Ion Concentrations and the Growth and Activities of Microorganisms in two Growth Media." *Toxicol. Environ. Chem.* **8** (1984) 173–184.

[3.9] R. Hindel, W. Fleige: "Schwermetalle in Böden der Bundesrepublik Deutschland – geogene und anthropogene Anteile," *UBA-Texte* **10/91** (1991).

[3.10] M. Gruppe, H. Kuntze: "Bodenuntersuchungen im Bereich des UBA-Meßnetzes als Bestandsaufnahme zur Beweissicherung von Umweltveränderungen," *UBA-Texte* **60/93** (1993).

[3.11] R. Hindel et al.: "Kartiertechnisches Konzept zur flächenhaften Erfassung von Schwermetallgehalten in Böden," *UBA-Texte*, in press.

[3.12] F. Scheffer, B. Ulrich: *Humus und Humusdüngung*, Enke Verlag, Stuttgart 1960.

[3.13] W. Fleig, H. Beutelspacher, E. Rietz in J. E. Gieseking (ed.): *Soil Components*, vol. 1, Springer Verlag, New York 1975, pp. 1–211.

[3.14] R. S. Swift in G. R. Aiken, D. U. McKnight, R. L. Wershaw, P. McCarthy (eds.): *Humic substances in Soil, Sediment and Water*, J. Wiley & Sons, New York 1985, pp. 387–408.

[3.15] W. Ziechmann: *Huminstoffe*, Verlag Chemie, Weinheim 1980.

[3.16] Z. Filip: "Bodenorganismen und ihre Aktivitäten als Anzeiger ökologisch bedeutender Auswirkungen anthropogener Bodenbelastungen," in: *Tagungsbericht-Symposium mit osteuropäischen Staaten*

"Untersuchungsmethoden, Bewertungsmaßstäbe und staatliche Regelungen für den Bodenschutz", BMU, 1993, pp. 47–56.

[3.17] Verband der Chemischen Industrie (ed.): *Chemie und Umwelt: Boden*, VCI, 1987, 28 pp.

[3.18] J. D. Stout, S. S. Bamforth, J. D. Lousier: "Protozoa," in: *Methods of Soil Analysis 2—Chemical and Microbiological Properties*, American Society of Agronomy, Madison, Wis., 1982.

[3.19] E. Aescht, W. Foissner: "Effects of Mineral and Organic Fertilizers on the Microfauna in a High-Altitude Reafforestation Trial," *Biol. Fertil. Soils* (1992) 17–24.

[3.20] E. Klapp: *Lehrbuch des Acker- und Pflanzenbaus*, Paul Parey, Hamburg 1958.

[3.21] P. Boysen: "Schwermetalle und andere Schadstoffe in Düngemitteln," *UBA-Texte* **55/92** (1992).

[3.22] C. Bannick, *Mitt. Dtsch. Bodenk. Ges.* **73** (1994) 23–26.

[3.23] R. Schulz: "Vergleichende Betrachtung des Schwermetallhaushaltes verschiedener Waldökosysteme Norddeutschlands," *Ber. Forschungszentr. Waldökosysteme/Waldsterben der Universität Göttingen* **32** (1987) 1–27.

[3.24] M. Abo-Radi, H. Meyer-Steinbrenner, *Wasser Boden* **46** (1994) 58–65.

[3.25] DIN 4047, part 3, Landwirtschaftlicher Wasserbau, Begriffe, Beuth Verlag, Berlin 1985.

[3.26] FAO, Guidelines for Control of Soil Degradation, Rome 1983.

[3.27] D. Rosenkranz, G. Einsele, H.-M. Harreß (eds.): *Bodenschutz*, supplement, Erich Schmidt Verlag, Berlin 1988.

[3.28] Der Rat der Sachverständigen für Umweltfragen: Altlasten, Sondergutachten, Stuttgart, Dec. 1989.

[3.29] L. Brown: *State of the World*, W. W. Norton, New York - London 1991.

[3.30] *Environment Statistics in Europe and North America*, part 1, United Nations, New York 1987.

[3.31] Statistisches Bundesamt, Land- und Forstwirtschaft, Fischerei (eds.): *Reihe 5.1: Bodenfläche nach Art der tatsächlichen Nutzung*, Metzler Poeschel, Stuttgart 1989, p. 9.

[3.32] BINE, Projekt Info-Service, *Bodenschutz in der Bauplanung*, no. 8, Karlsruhe 1992.

[3.33] Bundesanstalt für Landeskunde und Raumordnung, zitiert in VCI (ed.): *Chemie und Umwelt: Boden*, 1986, p. 7.

[3.34] E. Cordsen: "Bodenüberformung und -versiegelung," in H.-P. Blume (ed.): *Handbuch des Bodenschutzes*, ecomed, Landsberg 1990, p. 118.

[3.35] Bundesminister für Forschung und Technologie: *Unsere Erde im Wandel*, Bonn 1993.

[3.36] L. Brown: *State of the World 1990*, W. W. Norton, New York - London 1990.

[3.37] E. Fleischhauer: "Menschen schützen ihr Land," *Umwelt* **1** (1993) 25–27.

[3.38] FAO, Protect and Produce – Putting the Pieces Together, Rome 1992, p. 5.

[3.39] R. W. Arnold, Iszabocs, V. O. Targulian (eds.): *Global Soil Change*, Laxenburg 1990, p. 78.

[3.40] *Entsorgungspraxis* **3** (1993) 169.

[3.41] FAO, The Conservation and Rehabilitation of African Land, an International Scheme, ARC/90/4, p. 9.

[3.42] L. Brown: *Zur Lage der Welt 1987/88*, World Watch Institute, Washington, D.C., S. Fischer Verlag, Frankfurt/Main 1987, p. 93.

[3.43] G. O. Barney (Study director); prepared by the Council of Environmental Quality and the Department of the State, *The Global 2000 Report to the President of the U.S. Entering the 21st Century*, vol. 1: The Summary Report, Pergamon Press, New York 1980.

[3.44] B. Roberts: "The Health of the World's Land: A Perspective," *7th ISCO Conference Sydney, Proc.*, vol. 1, 1992.

[3.45] D. Meadows, D. Meadows, J. Randers: *Die neuen Grenzen des Wachstums*, DVA, Stuttgart 1992, p. 77.

[3.46] J. R. Kenworthy, P. Newman: "Cities and Transport Energy-lessons from a Global Survey," *Ekistics* **344/345** (1990) 258–268.

[3.47] L. Brown, E. Wolf: *Soil Erosion: Quiet Crisis in the World Economy*, World Watch Institute, Washington, D.C., 1984.

References for Chapter 4

[4.1] R. Horn, *Soil Technol.* **6** (1993) 47–75.

[4.2] B. D. Soane: "Compaction by Agricultural Vehicles: A Review," *Scott. Inst. Agric. Eng. Tech. Rep.* **5** (1983) 95 pp.

[4.3] W. B. Voorhees, *Adv. Soil Sci.* **19** (1992) 73–95.

[4.4] J. Eriksson, I. Håkansson, B. Danfors: "The Effect of Soil Compaction on Soil Structure and Crop Yields," *Swed. Inst. Agric. Eng. Bull.* **354** (1974) 101 pp.

[4.5] J. B. Passioura, *Aust. J. Soil Res.* **29** (1991) 717–728.

[4.6] E. L. Greacen, R. Sands, *Aust. J. Soil Res.* **18** (1980) 163–189.

[4.7] R. Horn, *Z. Kulturtech. Flurbereinig.* **26** (1985) 42–51.

[4.8] B. M. McKenzie, A. R. Dexter, *Biol. Fert. Soils* **5** (1988) 328–332.

[4.9] R. F. Follett, B. A. Stewart: *Soil Erosion and Crop Productivity*, Soil Sci. Soc. Am., Madison, Wis., 1985.

[4.10] R. Lal: *Soil Erosion in the Tropics – Principles and Management*, McGraw-Hill, New York 1990.

[4.11] Committee on Conservation: *Assessing the National Resources Inventory*, vols. 1 and 2, National Academy Press, Washington 1986, 114 pp., 314 pp.

[4.12] U. Schwertmann, F. Schmidt in M. DeBoodt, D. Gabriels (eds.): *Assessment of Erosion*, J. Wiley, Chichester 1980, pp. 403–406.

[4.13] K. Auerswald, *Z. Geomorphol.* N.F. *Suppl.* **83** (1991) 155–160.

[4.14] H. E. Dregne, *J. Soil Water Conserv.* **47** (1992) 8–13.

[4.15] U. Schwertmann, R. J. Rickson, K. Auerswald: "Soil Erosion Protection Measures in Europe," *Soil Technol. Ser.* **1** (1989) 216pp.

[4.16] W. H. Wischmeier, D. D. Smith: "Predicting Rainfall Erosion Losses - A Guide to Conservation Planning," *USDA Agric. Handbook*, Washington 1978, pp. 537.

[4.17] H.-P. Blume: *Handbuch des Bodenschutzes*, ecomed, 2nd ed., Landsberg 1990.

[4.18] D. Rosenkranz, G. Einsele, H.-M. Harreß: *Bodenschutz*, Erich Schmidt Verlag, Berlin 1990.

[4.19] G. Luft, G. Morgenschweis, *Z. Kulturtech. Flurbereinig.* **25** (1984) 138–148.

[4.20] R. Hoisl, W. Nohl, S. Zekorn, G. Zöllner: *Landschaftsästhetik in der Flurbereinigung, Materialien zur Flurbereinigung 11*, Bayer. Staatsministerium für Ernährung, Landwirtschaft und Forsten, München 1987, 161 pp.

[4.21] H. Petelkau, *Agrartechnik* **41** (1991) 54–57.
[4.22] K. G. Renard, V. A. Ferreira, *J. Environm. Qual.* **22** (1993) 458–466.
[4.23] W. G. Knisel: "CREAMS: A Field-Scale Model for Chemicals, Runoff, and Erosion from Agricultural Management Systems," *Conserv. Res. Rep.* (U.S. Dep. Agric.) **26** (1980) 643 pp.
[4.24] D. B. Beasley, L. B. Huggins: *ANSWERS User Manual*, Purdue University Chicago, Agric. Eng. Dep., 1981.
[4.25] A. N. Sharpley, J. R. Williams: "EPIC-Erosion Productivity Impact Calculator: I. Model Documentation," *Tech. Bull. U.S. Dep. Agric.* 1768 (1990) 235 pp.
[4.26] L. J. Lane, M. A. Nearing (eds.): *USDA - Water Erosion Prediction Project: Hillslope Profile Model Documentation*, National Soil Erosion Research Laboratory, report 2, West Lafayette, 1990.

References for Chapter 5

[5.1] F. Scheffer, P. Schachtschabel: *Lehrbuch der Bodenkunde*, 13th ed., Enke Verlag, Stuttgart 1992.
[5.2] H.-P. Blume: *Handbuch Bodenschutz*, ecomed, Landsberg 1990.
[5.3] F. Korte: *Lehrbuch der Ökologischen Chemie*, 2nd ed., Thieme Verlag, Stuttgart 1987.
[5.4] U. Gisi: *Bodenökologie*, Thieme Verlag, Stuttgart 1990.
[5.5] B. O. Wagner, W. Mücke, H.-R. Schenk: *Umweltmonitoring – Umweltkonzentrationen organischer Chemikalien/Literatur-Recherche und -Auswertung. Angewandter Umweltschutz*, ecomed, Landsberg 1989.
[5.6] W. Koch, K. Hurle: *Grundlagen der Unkrautbekämpfung*, Ulmer Verlag, Stuttgart 1978.
[5.7] J. C. Ottow: "Umwandlung von Xenobiotika durch Bodenmikroorganismen und ihre potentielle Steuerung," in: "Biotechnologische In-situ-Sanierung kontaminierter Standorte," *Schriftenr. Ver. Wasser Boden Lufthyg.* **80** (1988) 61–92.
[5.8] T. M. Sack, D. H. Steele: "A Survey of Household Products for Volatile Organic Compounds," *Atmos. Environ.* **26 A** (1992) no. 6, 1063–1070.
[5.9] R. Schleyer, I. Renner, D. Mühlhausen: "Beeinflussung der Grundwasserqualität durch luftgetragene organische Schadstoffe," *WaBoLu-Ber.* **1991**, no. 5, 1–96.
[5.10] J. S. Eisenreich, B. Looney, J. D. Thoronton: "Airborne Organic Contaminants in the Great Lake Ecosystem," *Environ. Sci. Technol.* **15** (1981) 30–33.
[5.11] T. F. Bidleman: "Atmospheric Processes," *Environ. Sci. Technol.* **22** (1988) 361.
[5.12] V. Neumayr: "Verteilungs- und Transportmechanismen von chlorierten Kohlenwasserstoffen in der Umwelt," *WaBoLu-Ber.* **1981**, no. 3, 24–40.
[5.13] G. Milde, P. Friesel, V. Neumayr: *Verhalten flüchtiger halogenierter Stoffe in Böden auch im Hinblick auf Grundwassergefährdung*, Final Report to the Ministry of Science and Technology of the Federal Republic of Germany, no. 037210 (1983).
[5.14] J. Förster: "Roof Runoff: A Source of Pollutants in Urban Storm Drainage Systems?," *Proc. Int. Conf. Urban Storm Drainage*, 5th, Osaka 1990, p. 469.
[5.15] H. Frank, W. Frank, D. Thiel: "C_1- and C_2-Halocarbons in Soil-Air of Forests," *Atmos. Environ.* **23** (1989) 1333–1335.
[5.16] H. Hellmann: "Leichtflüchtige Chlorkohlenwasserstoffe in den Gewässern der Bundesrepublik Deutschland – Auftreten und Bilanz," *Gesund.Ing.* **105** (1984) no. 5, 169–278.
[5.17] H. D. Winkeler, U. Puttins, K. Levsen: "Organische Verbindungen im Regenwasser," *Vom Wasser* **70** (1988) 107–117.
[5.18] J. Daub, T. Striebel: "Niederschlagsbedingte Schadstoffe im Straßenabfluß," in: *Schriftenreihe des Instituts für Siedlungswasserwirtschaft der Universität Karlsruhe*, no. 58: Schadstoffe im Regenabfluß aus städtischen Gebieten (1990).
[5.19] R. Rippen et al.: "Do Airborne Nitrophenols Contribute to Forest Decline?" *Environ. Technol. Lett.* **8** (1987) 475–482.
[5.20] P. D. Capel, T. Lichtensteiger, P. H. Brunner: "The Use of Sludge-only Landfills as Historical Records of Persistent Organic Chemicals and Heavy Metals in Sewage Sludge," *Water Research* **23** (1989) 525–527.
[5.21] U. Pagga: "Biologischer Abbau von Stoffen bei geringen Konzentrationen," *Z. Wasser Abwasser Forsch.* **20** (1987) 101–107.
[5.22] NRW-Meßprogramm: "Chloraromaten – Herkunft und Transfer, Abschlußbericht 1991," Ministerium für Raumordnung und Landwirtschaft des Landes Nordrhein-Westfalen, Düsseldorf 1991.
[5.23] N. Litz, M. Thiele, H. W. Döring, H.-P. Blume: "The Behaviour of Linear Alkylbenzenolsulfonates in Different Soils: A Comparison between Field and Laboratory Studies," *Ecotoxicol. Environ. Saf.* **14** (1987) 103–106.
[5.24] R. Leschber, N. Litz: "Klärschlammverwertung und -beseitigung im Licht der Klärschlammbelastung mit organischen Schadstoffen. Verfahrenstechnik der mechanischen, thermischen, chemischen und biologischen Abwasserbehandlung," *2nd GVC-Congress*, Würzburg, Oct. 19–21, 1992, VDI-Gesellschaft Verfahrenstechnik und Chemieingenieurwesen, pp. 477–482.
[5.25] N. Litz: "Abschätzung des Verhaltens von Umweltchemikalien in Böden," 2. Internationale Tagung und Ausstellung über Umweltinformation und Umweltkommunikation, Bayreuth, Sep. 14–18, 1993, *ECOINFORMA* **2** (1992) 323–335.
[5.26] D. Pernak, R. Leschber: "Tiefenverteilung persistenter Schadstoffe im Sediment einer Regenwasserversickerungsanlage," *Vom Wasser* **79** (1992) 321–331.
[5.27] Klärschlammverordnung (AbfKlärV) of April 28, 1992, Bundesgesetzblatt, part 1, pp. 912–928.
[5.28] R. Leschber, R. Mergler-Völkl, U. Zimmermann, N. Litz: *Untersuchung von relevanten organischen Schadstoffen im Klärschlamm*, Forschungsbericht 10303521 an das Umweltbundesamt, 1992.
[5.29] K. R. Bull: "The Critical Loads/Levels Approach to Gaseous Pollutant Emission Control," *Environ. Pollut. Ser. A* **69** (1991) 105–123.
[5.30] A. T. Kan, M. B. Tomson: "Groundwater Transport of Hydrophobic Organic Compounds in the Presence of Dissolved Organic Matter," *Environ. Toxicol. Chem.* **9** (1990) 253–263.
[5.31] R. L. Wershaw: "Application of a Membrane Model to the Sorptive Interactions of Humic Substances," *EHP Environ. Health Perspect.* **83** (1989) 191–203.

[5.32] P. Nkedi-Kizza, P. Rao, A. Hornsby: "Influence of Organic Solvents on Sorption of Hydrophobic Organic Chemicals by Soils," *Environ. Sci. Technol.* **19** (1985) 975–979.

[5.33] B. T. Walton et al.: "Soil Sorption of Volatile and Semivolatile Organic Compounds in a Mixture," *J. Environ. Qual.* **21** (1992) 552–558.

[5.34] K. B. Woodburn, P. S. C. Rao, M. Fukui, P. Nkedi-Kizza: "Solvophobic Approach for Predicting Sorption of Hydrophobic Organic Chemicals on Synthetic Sorbents and Soils," *J. Contam. Hydrogeology* **1986**, 227–241.

[5.35] B. Oepen, W. Kördel, W. Klein: "Sorption of Nonpolar and Polar Compounds to Soils: Processes, Measurements and Experience with the Applicability of the Modified OECD-Guideline 106," *Chemosphere* **22** (1991) 285–304.

[5.36] Y. S. Yu, G. W. Bailey: "Reduction of Nitrobenzene by Four Sulfide Minerals: Kinetics, Products and Solubility," *J. Environ. Qual.* **21** (1992) 86–94.

[5.37] K. D. Pennell, R. D. Rhue, A. G. Hornsby: "Competitive Adsorption of Para-xylene and Water Vapors on Calcium, Sodium, and Lithium-saturated Kaolinite," *J. Environ. Qual.* **21** (1992) 419–426.

[5.38] J. D. Ortego, M. Kowalska, D. L. Cocke: "Interactions of Montmorillonite with Organic Compounds – Adsorptive and Catalytic Properties," *Chemosphere* **22** (1991) no. 8, 769–798.

[5.39] I. Scheunert, M. Mansour, F. Andreux: "Binding of Organic Pollutants to Soil Organic Matter," *Int. J. Environ. Anal. Chem.* **46** (1991) 189–199.

[5.40] I. Scheunert, C. Meer-Beek, F. Korte: "Distribution and Biodegradation of ^{14}C-Residues Bound in Various Soilfractions after Treatment of the Soil with Modern ^{14}C-Chemicals," *Proceedings of the final Research Co-ordination Meeting on Isotopic Tracer-aided Studies of Unextractable or Bound Pesticide Residues in Soil, Plant, and Food,* International Atomic Energy Agency, Vienna 1986, pp. 31–40.

[5.41] N. Kamal, I. Scheunert, F. Korte: "Mass Balance of Pentachloronitrobenzene and Metabolite in a Closed Aerated Soil-plant- or Soil-system," *Bull. Environ. Contam. Toxicol.* **31** (1983) 559–565.

[5.42] D. G. Crosby, A. S. Wong: "Environmental Degradation of 2,3,7,8-Tetrachlorodibenzo-*p*-dioxin (TCDD)," *Science (Washington, D.C. 1883)* **195** (1977) 1337–1338.

[5.43] I. Gebefügi, R. Baumann, F. Korte: "Photochemical Degradation of 2,3,7,8-TCDD under Simulated Environmental Conditions," *Naturwissenschaften* **64** (1977) 486.

[5.44] M. L. Rochkind-Dubinski, G. S. Sayler, J. W. Blackburn: "Microbial Decomposition of Chlorinated Aromatic Compounds," *Microbiol. Ser.* **18** (1987) 314 pp.

[5.45] R. C. Bayly, M. G. Barbour: "The Degradation of Aromatic Compounds by the Meta and Gentisate Pathways: Biochemistry and Regulation," in D. T. Gibson (ed.): "Microbial Degradation of Organic Compounds," *Microbiol. Ser.* **13** (1984) 253–294.

[5.46] L. Britton: "Microbial Degradation of Aliphatic Hydrocarbons," in D. T. Gibson (ed.): "Microbial Degradation of Organic Compounds," *Microbiol. Ser.* **13** (1984) 89–130.

[5.47] D. T. Gibson, V. Subramanian: "Microbial Degradation of Aromatic Hydrocarbons," in D. T. Gibson (ed.): "Microbial Degradation of Organic Compounds," *Microbiol. Ser.* **13** (1984) 181–252.

[5.48] G. Hickman, J. Novak, M. Morris, M. Rebhuhn: "Effects of Site Variations on Subsurface Biodegradation Potential," *J. Water Pollut. Control. Fed.* **61** (1989) 1564–1575.

[5.49] A. L. Young: "Long-term Studies on the Persistence and Movement of TCDD in a Natural Ecosystem," *Environ. Sci. Res.* **26** (1983) 173–190.

[5.50] P. Geschwend, S. Wu: "On the Constancy of Sediment-water Partition Coefficients of Hydrophobic Organic Compounds," *Environ. Sci. Technol.* **19** (1985) 90–96.

[5.51] K. E. Quentin et al.: *Halogenorganische Verbindungen in Wässern*, Forschungsbericht 102 04 323 im Auftrag des Umweltbundesamtes, Berlin 1987.

[5.52] K. Ballschmiter: "Transport und Verbleib organischer Verbindungen im globalen Rahmen," *Angew. Chem.* **104** (1992) 501–528.

[5.53] H. Seip, J. Alstad, G. Carlberg, K. Martinsen: "Measurement of Mobility of Organic Compounds in Soils," *Sci. Total Environ.* **50** (1986) 87–101.

[5.54] B. H. Wilson, G. B. Smith, J. F. Rees, *Environ. Sci. Technol.* **20** (1986) 997.

[5.55] T. Vogel, C. Criddle, P. McCartey: "Transformations of Halogenated Aliphatic Compounds," *Environ. Sci. Technol.* **21** (1987) 722–736.

[5.56] T. Vogel, P. McCartey: "Biotransformation of Tetrachloroethylene to Trichloroethylene, Dichloroethylene, Vinyl Chloride, and Carbon Dioxide under Methanogenic Conditions," *Appl. Environ. Microbiol.* **49** (1985) 1080–1083.

[5.57] B. Zoeteman et al.: "Persistent Organic Pollutants in River Water and Groundwater of The Netherlands," *Chemosphere* **9** (1981) 231–249.

[5.58] M. Nerger, R. Mergler-Völkl: "Biologischer Abbau von leichtflüchtigen Chlorkohlenwasserstoffen in Grund- und Abwasser," *Z. Wasser Abwasser Forschg.* **21** (1988) 16–19.

[5.59] H. Scholz-Muramatsu, H.-C. Flemming: "Unter welchen Milieubedingungen erfolgt ein mikrobieller Abbau leichtflüchtiger chlorierter Kohlenwasserstoffe (LCKW)?" in R. Wagner (ed.) *Wasserkalender 1991,* Erich Schmidt Verlag, Berlin 1991, pp. 135–158.

[5.60] V. Schneider, H. Scholz-Muramatsu, S. Gaiser: "Hemmung der biologischen Elimination von Dichlormethan durch kontaminiertes Grundwasser," *GWF Gas Wasserfach. Wasser/Abwasser* **129** (1988) 755–758.

[5.61] A. B. Zehnder, G. Schraa: "Biologischer Abbau xenobiotischer Verbindungen bei niedrigen Konzentrationen und das Problem der Restkonzentration," *GWF Gas Wasserfach. Wasser/Abwasser* **129** (1988) 369–373.

[5.62] L. Semprini, D. Grbic-Galic, P. L. McCartey, P. V. Roberts: "Introduction," in P. V. Roberts et al. (eds.) "In-situ Aquifer Restoration of Chlorinated Aliphatics by Methanotrophic Bacteria," *Tech. Rep. Dep. Civil Eng. Univ. Stanford* **310** (1989) 65–90.

[5.63] B. Toussaint, H.-R. Langguth: "Das Reinigungsvermögen des Untergrundes – Vorstellungen, Wünsche und Wirklichkeit," *Wiss. Umwelt* **1991**, 1–2.

[5.64] World Health Organization: "Chlorobenzenes other than Hexachlorobenzene," *Environ. Health Crit.* **128** (1991).

[5.65] BUA-Report, (Beratergremium für Umweltrelevante Altstoffe) *24: Benzene*, VCH Verlagsgesellschaft, Weinheim 1992.

[5.66] ATV (Abwassertechnische Vereinigung e.V.): "Zusammenfassender Bericht über einige ökotoxikologisch gefährliche Stoffe," *Dokumentation und Schriftenreihe aus Wissenschaft und Praxis 27*, St. Augustin 1991.

[5.67] R. Schleyer, G. Milde: "Zur Bewertung luftbürtiger Grundwasserqualitätsbeeinflussungen," *VDI Ber.* **837** (1990) 663–684.

[5.68] M. P. Ligocki, C. Leuenberger, J. F. Pankow: "Organic Compounds in Rain – II. Gas Scavenging of Neutral Organic Compounds," *Atmosp. Environ.* **19** (1985) 1609–1617.

[5.69] Bundesamt für Umweltschutz: "Polychlorierte organische Verbindungen in Rheinfelden (Schweiz)," *Schriftenreihe Umweltschutz* **18** (1983).

[5.70] R. Seibold, W. Arnold, R. Schmid: "Persistente Chlorkohlenwasserstoffe in Böden, Klärschlämmen und Flußsedimenten," *Hohenheimer Arb.* **1986**, 65–76.

[5.71] A. Reischl, M. Reissinger, O. Hutzinger: "Organische Luftschadstoffe und ihre Bedeutung für die terrestrische Vegetation," *UWSF Z. Umweltchem. Ökotox.* **2** (1989) 32–41.

[5.72] W. König, A. Hembrock-Heger, M. Wilkens: "Persistente organische Chemikalien im Boden – Eintragspfade und Vorkommen," *UWSF Z. Umweltchem. Ökotox.* **3** (1991) no. 1, 33–36.

[5.73] P. Weiss, A. Riss: "Bodenuntersuchungen von Grünland- und Waldböden im Raum Linz auf Schwermetalle, polyzyklische aromatische Kohlenwasserstoffe und chlorierte Kohlenwasserstoffe," in: "Schadstoffe im Raum Linz," *Monographie*, vol. 20, Bundesministerium für Umwelt, Jugend und Familie, Wien 1992.

[5.74] T. A. Anderson, J. J. Beauchamp, B. T. Walton: "Organic Chemicals in the Environment," *J. Environ. Qual.* **20** (1991) 420–424.

[5.75] Y. Jin, G. O'Connor: "Behavior of Toluene Added to Sludge-Amended Soils," *J. Environm. Qual.* **19** (1990) 573–579.

[5.76] E. Merian, M. Zander: "Volatile Aromatics," in O. Hutzinger (ed.) *Handbook Environ. Chem.* **3** (1982) 117–161.

[5.77] M. Reinhard, N. L. Goodman, J. F. Barker, *Environ. Sci. Technol.* **18** (1984) 953.

[5.78] C. Y. Chiang et al.: "Aerobic Biodegradation of Benzene, Toluene, and Xylene in a Sandy Aquifer – Data Analysis and Computer Modeling," *Ground Water* **27** (1989) no. 6, 823–834.

[5.79] D. Grbic-Galic, T. M. Vogel, *Appl. Environ. Microbiol.* **52** (1986) 200.

[5.80] J. P. van der Hock, L. G. C. M. Urlings, C. M. Grobben: "Biological Removal of Polycyclic Aromatic Hydrocarbons, Benzene, Toluene, Ethylbenzene, Xylene, and Phenolic Compounds from Heavily Contaminated Ground Water and Soil," *Environ. Technol. Lett.* **10** (1989) 185–194.

[5.81] D. M. Falatko, J. T. Novak: "Effects of Biologically Produced Surfactants on the Mobility and Biodegradation of Petroleum Hydrocarbons," *Water Environ. Res.* **64** (1992) no. 2, 163–169.

[5.82] K. G. Robinson, W. S. Farmer, J. T. Novak: "Availability of Sorbed Toluene in Soils for Biodegradation by Acclimated Bacteria," *Water Res.* **24** (1990) 345–350.

[5.83] R. Herterich, R. Herrmann: "Comparing the Distribution of Nitrated Phenols in Atmosphere of German Hill Sites," *Environ. Technol.* **11** (1990) 961–972.

[5.84] B. Scholz, N. Palauschek: "The Determination of Substituted Aromatic Amines in Water and Sediment Samples," *Fresenius Z. Anal. Chem.* **331** (1988) 282–289.

[5.85] J. Zachara, J. Lawrence, J. Sauer: "Sorption on Aniline on Selected Alfisols from the Eastern Coal Region," *Soil Sci.* **138** (1984) 209–219.

[5.86] G. Sims, L. E. O'Loughlin: "Degradation of Pyridines in the Environment," *CRC Crit. Rev. Environ. Control.* **19** (1989) 309–340.

[5.87] J. A. Leenheer, H. A. Stuber: "Migration Through Soil of Organic Solutes in Oil-shale Process Water," *Environ. Sci. Technol.* **15** (1981) 578.

[5.88] G. K. Sims, E. Sommers: "Biodegradation of Pyridine Derivates in Soil Suspension," *Environ. Toxciol. Chem.* **5** (1985) 503–509.

[5.89] D. H. Stuermer, D. J. Ng, C. J. Morris: "Organic Contaminants in Groundwater Near an Underground Coal Gasification in Northeastern Wyoming," *Environ. Sci. Technol.* **16** (1982) 582.

[5.90] R. F. Spalting, J. W. Fulton: "Groundwater Munitions Residues and Nitrate Near Grand Island, Nebraska, USA," *J. Contam. Hydrol.* **2** (1988) 139–153.

[5.91] A. Oberholz: *Tödliche Gefahr aus der Tiefe. Bittere Erkenntnisse zu Kriegs- und Rüstungsaltlasten*, Kommunal-Verlag, Düsseldorf 1991.

[5.92] B. Ehring, J. Prassler, P. R. Wallnöfer, W. Ziegler: "Mikrobielle Transformation von *o*-Nitroanilin und *p*-Nitrophenol," *UWSF Z. Umweltchem. Ökotox.* **4** (1992) 81–85.

[5.93] E. P. Kuhn, J. M. Suflita: "Microbial Degradation of Nitrogen, Oxygen, and Sulfur Heterocyclic Compounds under Anaerobic Conditions: Studies with Aquifer Samples," *Environ. Toxicol. Chem.* **8** (1989) 1149–1158.

[5.94] W. Rothard: "Aktuelles zur Dioxinproblematik – Ableitung von Dioxinrichtwerten für die Bodensanierung," *Bundesgesundheitsbl.* **4** (1991) 155–158.

[5.95] Environmental Health Criteria 88: *Polychlorinated Dibenzopara-dioxins and Dibenzofurans*, World Health Organization, Geneva 1989.

[5.96] O. Hutzinger, H. Fiedler: "From Source to Exposure: Some Open Questions," *Chemosphere* **27** (1993) 121–129.

[5.97] H. Kaune, H. Fiedler, O. Hutzinger: *Dioxine und Furane – Quellen, Einträge in die Umwelt und Aufnahme durch den Menschen*, Im Auftrag der Hamburger Umweltbehörde, Bayreuth 1990.

[5.98] H. Fiedler, O. Hutzinger: "Sources and Sinks of Dioxins: Germany," *Chemosphere* **25** (1992) no. 10, 1487–1491.

[5.99] M. Büchen et al.: *Dioxine und Furane in der hessischen Umwelt – Meßergebnisse aus Hessen*, Hessische Landesanstalt für Umwelt, Umweltplanung, Arbeits- und Umweltschutz, no. 126, Wiesbaden 1991.

[5.100] First VGA/UBA-Evaluation of the Internationale Dioxin Symposium and of the Public Expert-Hearing of 09.–13. Nov. 1992, Berlin, 28. Jan. 1993.

[5.101] L. O. Kjeller, C. Rae, C. K. C. Jones, A. E. Johnston: "Evidence for Increases in the Environmental Burden of Polychlorinated Dibenzo-*p*-Dioxins and Furans (PCDD/PCDFs) over the last Century," *Chemosphere* (1990) in K. Meyer: Bodenverschmutzung in der Schweiz. Themenbereich des Nationalen Forschungsprogrammes "Boden," Liebenfeld-Bern, 1991.

[5.102] R. M. Smith et al.: "Measurement of PCDFs and PCDDs in Air Samples and Lake Sediments at Several Locations in Upstate New York," *Chemosphere* **25** (1992) nos. 1-2, 95-98.

[5.103] Ministerium für Umwelt Baden-Württemberg (eds.): *Umweltdaten 89/90,* Stuttgart 1990.

[5.104] Bayerisches Geologisches Landesamt: "Dioxine in Böden Bayerns," *GLA Fachberichte* **7** (1991) 1-96.

[5.105] A. Riss, H. Hagenmaier, W. Rotard: "Wirkungen von Dioxinimmissionen auf Boden, Grünlandaufwuchs und Kuhmilch - Fallstudie anhand einer Metallrückgewinnungsanlage in Österreich," in: *VDI Ber.* **837** (1990) 831-848.

[5.106] S. Karickoff: "Semiempirical Estimation of Sorption of Hydrophobic Pollutants on Natural Sediments and Soils," *Chemosphere* **10** (1981) 833-346.

[5.107] W. Y. Shiu et al.: "Physical-Chemical Properties of Chlorinated Dibenzo-*p*-dioxins," *Environ. Sci. Technol.* **22** (1988) 651-658.

[5.108] S. Kapila et al.: "Field and Laboratory Studies on the Movement of Tetrachlorodibenzo-*p*-dioxin in Soil," *Chemosphere* **18** (1989) nos. 1-6, 1297-1304.

[5.109] H.-J. Jürgens, R. Roth: "Case Study and Proposed Decontamination Steps of the Soil and Groundwater Beneath a Closed Herbicide Plant in Germany," *Chemosphere* **18** (1989) nos. 1-6, 1163-1169.

[5.110] R. K. Puri et al.: "Studies of Parameter Affecting Translocation of Tetrachlorodibenzo-*p*-dioxin in Soil," *Chemosphere* **18** (1989) 1291.

[5.111] D. Dougherty, L. Peters, M. Overcash, R. Carbonelle: "Transport of 2,3,7,8-Tetrachlorodibenodioxin in Soil Containing Organic Solvents," *Organohalogen Compounds* **3** (1991) 315-318.

[5.112] R. Orth, C. Ritchie, F. Hileman: "Measurement of the Photoinduced Loss of Vapor Phase TCDD," *Chemosphere* **18** (1989) 1275-1282.

[5.113] G. C. Miller et al.: "Photolysis of Octachlorodibenzo-*p*-dioxin on Soils: Production of 2,3,7,8-TCDD," *Chemosphere* **18** (1989) nos. 1-6, 1265-1274.

[5.114] J. K. McCrady, S. P. Maggard: "Uptake and Photodegradation of 2,3,7,8-Tetrachlorodibenzo-*p*-dioxin Sorbed to Grass Foliage," *Environ. Sci. Technol.* **27** (1993) 343-350.

[5.115] S. Cerlesi, A. Di Domenico, S. Ratti: "2,3,7,8-Tetrachlorodibenzo-*p*-dioxin (TCDD) Persistence in the Seveso (Milan, Italy) Soil," *Ecotoxicol. Environ. Saf.* **18** (1989) 149-164.

[5.116] A. Yanders, S. Kapila, R. Schreiber in C. Rappe, G. Choudhary, L. Keith (eds.): *Chlorinated Dioxins and Dibenzofuran in Perspective,* Lewis Publ., Chelsea, Michigan, 1986, pp. 237-239.

[5.117] R. A. Freeman, J. M. Schroy, *Chemosphere* **18** (1989) 1305.

[5.118] A. L. Young, L. G. Cockerham, C. E. Thalken: "A Long-term Study of Ecosystem Contamination with 2,3,7,8-Tetrachlorodibenzodioxin," *Chemosphere* **16** (1987) 1791-1815.

[5.119] E. Arvin, B. Jensen, J. Aamand, C. Jorgensen: "The Potential of Free Living Ground Water Bacteria to Degrade Aromatic Hydrocarbons and Heterocyclic Compounds," *Water Sci. Technol.* **20** (1988) 109.

[5.120] G. Springer, H. G. Rast: "Biologischer Abbau mehrfach halogenierter mono- und polyzyklischer Aromaten," *GWF Gas Wasserfach*: Wasser/Abwasser **129** (1988) 70.

[5.121] D. Frahne et al.: "Zum biologischen Abbau von Dibenzofuran im Hinblick auf doxinähnliche Verbindungen," *Fat Sci. Technol.* **93** (1991) no. 9, 355-360.

[5.122] P. Fortnagel, R.-M. Wittich, H. Harms, S. Schmidt: "New Bacterial Degradation of the Biaryl Structure," *Naturwissenschaften* **76** (1989) 523-524.

[5.123] M. F. Arthur, J. I. Frea: "2,3,7,8-Tetrachlorodibenzo-*p*-dioxin: Aspects of its Important Properties and its Potential Biodegradation in Soils," *J. Environ. Qual.* **18** (1989) no. 1, 1-11.

[5.124] G. Klecka, D. Gibson: "Metabolism of Dibenzo-*p*-dioxin and Chlorinated Dibenzo-*p*-dioxins by a Beijerinckia Species," *Appl. Environ. Microbiol.* **39** (1980) 288-296.

[5.125] K. R. Cooper, A. Christini, P. Bergqvist, C. Rappe: "Bioavailability and Bioconcentration of Polychlorinated Dioxins (PCDD) and Furans (PCDF) to Organisms Inhabiting a Heavily Contaminated Estuarine Ecosystem," *Chemosphere* **25** (1992) nos. 1-2, 25-28.

[5.126] G. F. Fries: "Assessment of Potential Residues in Food Derived from Animals Exposed to TCDD-Contaminated Soil," *Chemosphere* **16** (1987) 2123-2128.

[5.127] M. S. McLachlan, H. Thoma, M. Reissinger, O. Hutzinger: "PCDD/F in Agricultural Food Chain Part 1: PCDD/F Mass Balance of a Lactating Cow," *Chemosphere* **20** (1990) nos. 7-9, 1013-1020.

[5.128] C. S. Nessel et al.: "Transpulmonary Uptake and Bioavailibility of 2,3,7,8-TCDD from Respirable Soil Particles," *Chemosphere* **25** (1992) nos. 1-2, 29-23.

[5.129] W. Feige, R. Schmalzried: *Stoffbestand von Gartenböden,* Bayerisches Staatsministerium für Landesentwicklung und Umweltfragen, München 1991.

[5.129a] VROM: Leidraad Bodembescherming, Staatsuitgeverij, Den Haag 1983/1990.

[5.130] K. C. Jones et al.: "Increase in the Polynuclear Aromatic Hydrocarbon Content of an Agricultural Soil over the Last Century," *Environ. Sci. Technol.* **23** (1989) no. 1, 95-101.

[5.131] A. Sablijic: "On the Prediction of Soil Sorption Coefficients of Organic Pollutants from Molecular Structure: Application of Molecular Topology Model," *Environ. Sci. Technol.* **21** (1987) 358-365.

[5.132] S. R. Wild, K. C. Jones: "Biological and Abiotic Losses of Polynuclear Aromatic Hydrocarbons (PAHs) from Soils Freshly Amended with Sewage Sludge". *Environ. Toxciol. Chem.* **12** (1993) 5-12.

[5.133] S. R. Wild, M. L. Berrow, K. C. Jones: "The Persistence of Polynuclear Aromatic Hydrocarbons (PAHs) in Sewage Sludge Amended Agricultural Soils," *Environ. Pollut. Ser. A* **72** (1991) 141-157.

[5.134] W. D. Weissenfels, M. Beyer, J. Klein: "Degradation of Phenantrene, Fluorene, and Fluoranthene by Pure Bacterial Cultures," *Appl. Microbiol. Biotechnol.* **32** (1990) 479–484.

[5.135] K. S. Park et al.: "Fate of PAH Compounds in Two Soil Types: Influence of Volatilization, Abiotic Loss and Biological Activity", *Environ. Toxicol. Chem.* **9** (1990) 187–195.

[5.136] S. R. Wild, K. C. Jones, A. E. Johnston: "The Polynuclear Aromatic Hydrocarbon (PAH) Content of Herbage from a Long-term Grassland Experiment," *Atmos. Environ.* **26 A** (1992) no. 7, 1299–1307.

[5.137] S. R. Wild, M. L. Berrow, S. P. McGrath, K. C. Jones: "Polynuclear Aromatic Hydrocarbons in Crops from Long-termed Field Experiments Amended with Sewage Sludge," *Environ. Pollut. Ser. A* **76** (1992) 25–32.

[5.138] W. Pyka, C. Schüth, P. Grathwohl: "Dissolution and Transport of Coaltar Constituents and their Impact on Groundwater Quality," *Int. Symp. Environ. Contam. Central Eastern Europe,* Budapest 1992, Oct. 12–16, 1992.

[5.139] A. Cortes, J. Riego, A. B. Paya-Perez, B. Larsen: "Soil Sorption of Co-Planar and Non-Planar PCB's," *Toxicol. Environ. Chem.* **31–32** (1991) 79–86.

[5.140] A. B. Paya-Perez, F. Pelusio: "Sorption of 3,4,5-Trichloroaniline in a Soil Profile," *Sci. Total Environ.* **123/124** (1992) 519–529.

[5.141] J. F. Brown et al.: "Polychlorinated Biphenyl Dechlorination in Aquatic Sediments," *Science (Washington, D.C., 1883)* **236** (1987) 709–712.

[5.142] A. Marcomini et al.: "Behaviour of Aromatic Surfactants and PCB's in Sludge-treated Soil and Landfills," *J. Environ. Qual.* **18** (1989) 523–528.

[5.143] W. Kampe: "Organische Stoffe in Böden und Pflanzen nach langjährigen, intensiven Klärschlammanwendungen," *Korresp. Abwasser* **43** (1987) no. 8, 820–827.

[5.144] S. W. Hooper, C. A. Pettigrew, G. S. Sayler: "Ecological Fate Effects and Prospects for the Elimination of Environmental Polychlorinated Biphenyls," *Environ. Toxicol. Chem.* **9** (1990) 655–667.

[5.145] D. Pal, J. Weber, M. Overcash: "Fate of Polychlorinated Biphenyls (PCB) in Soil-plant System," *CSIR Publ. Rev.* **74** (1980) 45–98.

[5.146] S. Ketterer, J. Michel, W. Holzwarth: "Verhalten von polychlorierten Biphenylen (PCB) in der wassergesättigten Bodenzone eines kontaminierten Standortes", Bayreuth, Sep. 14–18, 1992 *ECO-INFORMA* **2** (1992).

[5.147] S. Boyd, S. Sun, *Environ. Qual.* **24** (1990) 1920–1923.

[5.148] G. A. O'Connor, R. L. Chaney, J. A. Ryan: "Bioavailability of Plants of Sludge-Borne Toxic Organics," *Rev. Environ. Toxicol.* **121** (1991) 129–155.

[5.149] N. Litz: "Verhalten von Umweltchemikalien in Böden – Bewertungsverfahren für die Umweltplanung," *Z. Pflanzenernähr. Bodenkd.* **156** (1993) 203–208.

[5.150] G. F. Fries, G. S. Marrow: "Influence of Soil Properties on the Uptake of Hexachlorobiphenyls by Rats," *Chemosphere* **24** (1992) 109–113.

[5.151] Q. Ye et al.: "Studies on the Transport and Transformation of PCBs in Plants," *Chemosphere* **25** (1992) 1475–1479.

[5.152] S. Winter, B. Streit: "Organochlorine Compounds in a Three Step Terrestrial Food Chain," *Chemosphere* **24** (1992) no. 12, 1765–1774.

[5.153] VDI (ed.): *Acidic precipitation – Formation and Impact on Terrestrial Ecosystems,* VDI Verlag, Düsseldorf 1983, 1987, 281 p.

[5.154] H.-W. Georgii (ed.): *Mechanisms and Effects of Pollutant-Transfer into Forests,* Kluwer Academic Publ., Dordrecht 1989, p. 361.

[5.155] B. Ulrich, *Z. Pflanzenernähr. Bodenkd.* **149** (1986) 703–717.

[5.156] J. O. Reuss, D. W. Johnson: *Acid Deposition and the Acidification of Soils and Waters,* vol. 8, Springer Publ., New York 1986., pp. 247–256.

[5.157] B. Ulrich, *Z. Pflanzenernähr. Bodenkd.* **144** (1981) 289–305.

[5.158] B. Ulrich, *Forstarchiv* **58** (1987) 232–239.

[5.159] R. Mayer, *VDI Ber.* **837** (1990) 295–303.

[5.160] J. Prenzel in B. Ulrich, J. Pankrath (eds.): *Effects of Accumulation of Air Pollutants in Forest Ecosystems,* D. Reidel Publ. Comp., Dordrecht 1983, pp. 157–170.

[5.161] B. Ulrich in B. Ulrich, J. Pankrath (eds.): *Effects of Accumulation of Air Pollutants in Forest Ecosystems,* D. Reidel Publ. Comp., Dordrecht 1983, pp. 1–20.

[5.162] W. Flückiger in G. Glatzel (ed.): *Möglichkeiten und Grenzen der Sanierung immissionsgeschädigter Waldökosysteme,* Institut für Forstökologie der Hochschule für Bodenkultur, Wien 1987, pp. 65–81.

[5.163] B. Ulrich, *Experentia* **41** (1985) 578–584.

[5.163 a] B. Ulrich, H. Meyer, *Ber. Forschungszentrum Waldökosysteme Göttingen* **6** (1978) 133 pp.

[5.164] O. Wittmann, K. D. Fetzer: "Aktuelle Bodenversauerung in Bayern," *Bayer. Staatsmin. Landesentw. Umwelt. Materialien* **20** (1982) 71 pp.

[5.165] J. Nilsson (ed.): *Critical Loads for Sulfur and Nitrogen,* Nordisk Ministerrad, Miljo Rapport 1986.

[5.166] M. Berdén, S. I. Nilsson, K. Rosen, G. Tyler: "Soil Acidification: Extent, Causes, Consequences," *Natl. Swed. Environ. Protection Board Rep.* **3292** (1987) 164 pp.

[5.167] E. v. Zezschwitz, *Allg. Forst. Jagdztg.* **158** (1987) 136–147.

[5.168] H.-P. Blume: "Alarmierende Versauerung Berliner Forsten," *Berliner Naturschutzblätter* **75** (1981) 713–715.

[5.169] H. Butzke: "Versauern unsere Wälder? Erste Ergebnisse der Überprüfung 20 Jahre alter pH-Messungen in Waldböden Nordrhein-Westfalens," *Forst Holzwirt* **36** (1981) 542–548.

[5.170] R. Riebeling, Ch. Schaefer: "Jahres- und Langzeitentwicklung der pH-Werte von Waldböden in hessischen Fichtenbeständen," *Forst Holzwirt* **39** (1984) 177–178.

[5.171] N. Rastin, B. Ulrich: "Bodenkundliche Standortscharakterisierung zur Beurteilung des Stabilitätszustands von Waldökosystemen in Hamburg," *Ber. Forschungszentrum Waldökosysteme Göttingen* **10** (1985) 1–91.

[5.172] E. v. Zezschwitz, *Geol. Jahrb. Reihe F.* **20** (1985) 3–41.

[5.173] U. Falkengren-Grerup, *Environ. Pollut. Ser. A* **43** (1987) 79–90.

[5.174] M. Hauhs, *Ber. Forschungszentr. Waldökosysteme Göttingen* **17** (1985) 1–206.

[5.175] R. Grimm, K. E. Rehfuess, *Allg. Forst. Jagdztg.* **157** (1986) 205–213.

[5.176] K. Rost-Siebert, *Texte Umweltbundesamt* 1, Deutsches Umweltbundesamt, Berlin 1986.

[5.177] L. Hallbäcken, *Z. Pflanzenernähr. Bodenkd.* **155** (1992) 51–60.

[5.178] H. Koenies, Ph. D. Thesis, Universität Kassel 1982.

[5.179] G. Glatzel, M. Kazda, L. Lindebuer, *Düsseldorfer Geobot. Kolloq.* **3** (1986) 15–32.

[5.180] R. Wittig, H. Neite, *Vegetatio* **64** (1985) 113–119.

[5.181] A. Schulte, M. Spiteller, *Forst Holzwirt* **42** (1987) 150–154.

[5.182] H. Jochheim, H. Schäfer, *Z. Pflanzenernähr. Bodenkd.* **151** (1988) 81–85.

[5.183] W. Hetsch, K. Meiwes, B. Ulrich, *Mitt. Forstl. Bundesversuchsanst. Wien* **140** (1981) 223–230.

[5.184] E. E. Hildebrand, *Forstwiss. Centralbl.* **105** (1986) 60–76.

[5.185] J. Puhe, B. Ulrich, *Arch. Hydrobiol.* **102** (1985) 331–342.

[5.186] H. Heinrichs et al., *Neues Jahrb. Mineral. Abh.* **156** (1986) 23–62.

[5.187] M. Brechtel, *Forst Holzwirt* **43** (1988) 298–302.

[5.188] R. Hantschel, *Bayreuther Bodenkd. Ber.* **3** (1987) pp. 219.

[5.189] S. Linkersdörfer, P. Benecke: *Auswirkungen von sauren Depositionen auf die Grundwasserqualität in bewaldeten Gebieten. Eine Literaturstudie,* Materialien des Umweltbundesamtes no. 4, Erich Schmidt Verlag, Berlin 1987, pp. 170.

[5.190] M. Hauhs, R. F. Wright (undated): "Acid Deposition: Reversibility of Soil and Water Acidification," *Air Pollution Research Report* 11, Commission of the European Communities, Luxemburg, pp. 42.

[5.191] M. Schmidt, *Allgem. Forst Z.* **30/31** (1990) 790–792.

[5.192] V. Malessa: "Depositionsbedingte Tiefengradienten der Bodenversauerung in der Sösemulde (Westharz)," *Ber. Forschungszentr. Waldökosysteme Göttingen* **98** (1993) 223 pp.

[5.193] H. Vetter, G. Steffens: *Wirtschaftseigene Düngung,* DLG-Verlag, Frankfurt/Main 1986.

[5.194] Bundesministerium für Ernährung, Landwirtschaft und Forsten (ed.): *Statistisches Jahrbuch über Ernährung, Landwirtschaft und Forsten,* Landwirtschaftsverlag Münster, Münster 1994.

[5.195] Deutscher Bundestag, ENQUETE-KOMMISSION: "Vorsorge zum Schutz der Erdatmosphäre," 3. Bericht zum Thema Schutz der Erde, Drucksache 11/8030, 1990.
J. Jung, K. Isermann, G. Henjes: "Der Einfluß von cadmiumhaltigen Düngerphosphaten auf die Cadmiumanreichungen von Kulturböden und Nutzpflanzen," *Landwirtsch. Forsch.* **32** (1979) 262–282.

[5.196] J. Baier, W. Werner: "Einfluß langjähriger Stallmistdüngung auf chemische Parameter der Bodenfruchtbarkeit unter verschiedenen Klimabedingungen," *VDLUFA-Schriftenr.* **33** (1991) 177–182.

[5.197] M. Bosch, A. Amberger: "Einfluß langjähriger Düngung mit verschiedenen N-Formen auf pH-Wert, Humusfraktionen, biologische Aktivität und Stickstoffdynamik einer Ackerbraunerde," *Z. Pflanzenernaehr. Bodenkd.* **146** (1983) 714–724.

[5.198] W. Kolb: "Nitrogenaseaktivität und Populationsdichte von luftstickstoffbindenden Bakterien in Abhängigkeit von Umweltfaktoren", Ph. D. Thesis, Universität Hoheneim 1984.
H. Kuhlmann: "Ursachen und Ausmaß der N-, P-, K- und Mg-Ernährung der Pflanzen aus dem Unterboden," Habilitationsschrift, Universität Hannover 1988.

[5.199] A. Klansink, G. Steffens: "Der Einfluß unterschiedlicher organischer Düngung auf N-Bedarf, Ertrag und Qualität von Ackerkulturen sowie auf den N- und Humusgehalt im Boden," *VDLUFA-Schriftenr.* **37** (1993) 269–272.

[5.200] W. Heymann: "Humusgehalt verbessert," *Neue Landwirtschaft* **7** (1993) 40–42.

[5.201] M. Körschens, *Archiv Acker Pflanzenbau Bodenkd.* **22** (1978) no. 3, 175.

[5.202] T. Beck, *VDLUFA-Schrifenr.* **16** (1986) 31–46.

[5.203] H. W. Schwerer, W. Werner, U. Ortseifen: "N-Haushalt und N-Nachlieferungsvermögen einer Lößparabraunerde nach langjähriger Strohdüngung in Kombination mit unterschiedlichen N-Düngerformen," *VDLUFA-Schriftenr.* **32** (1990) 199–203.

[5.204] P. Foerster: "Einfluß hoher Güllegaben und üblicher Mineraldünger auf die Stoffbelastung (NO_3, NH_4, P und SO_4) im Boden und Grundwasser in Sandböden Nordwestdeutschlands," *Z. Acker Pflanzenbau* **137** /1973) 270–286.

[5.205] P. Boysen: "Nährstoffauswaschungen aus gedüngten und ungedüngten Böden in Abhängigkeit von Standorteigenschaften und Nutzung der Moränen- und Sandergebiete Schleswig-Holsteins," Ph. D. Thesis, Universität Kiel, 1977.

[5.206] I. Hess, H. Franken: "Über die Reduzierung von Nitratverlusten nach Leguminosen nach Anbau von Cruziferen," *Mitt. Dtsch. Bodenkdl. Gesellschaft* **57** (1988) 55–60.

[5.207] J. Bischoff, K.-W. Krüger: "Einfluß von N-Düngermenge und Applikationszeitpunkt auf die Restmengen an Nitrat-N im Boden nach dem Anbau von Kartoffeln," *VDLUFA-Schriftenr.* **38** (1994) 125–128.

[5.208] T. Appel, M. Schmücker, U. Schultheiss: "Möglichkeiten und Grenzen schlagbezogener N-Bilanzen zur Reduzierung der Nitratbelastung in Wasserschutzgebieten," *VDLUFA-Schriftenr.* **37** (1993) 137–140.

[5.209] O. Günther-Borstel, M. Finck, H. Hanus: "N-Dynamik im Boden und N-Aufnahme von Winterweizen, Wintergerste und Winterraps unter besonderer Berücksichtigung von mineralischer Stickstoff- und Gülledüngung – Versuch einer Bilanzierung," *VDLUFA-Schriftenr.* **37** (1993) 125–128.

[5.210] K.-C. Kersebaum: "Die Simulation der Stickstoffdynamik von Ackerböden," Ph. D. Thesis, Universität Hannover 1989.

[5.211] H. Hersemann: "Stickstoffsimulationsmodelle zur Unterstützung einer umweltgerechten Landbewirtschaftung und Düngebedarfsermittlung," *VDLUFA-Schriftenr.* **38** (1994) 223–226.

[5.212] Council for Agricultural Science and Technology: "Effect of Increased Nitrogen Fixation on Stratospheric Ozone," *CAST-Report* **53** (1976).

[5.213] W. Seiler, R. Conrad: "Mikrobielle Bindung von N_2O (Distickstoffoxid) aus Mineraldüngern – ein Umweltproblem," *Forum Mikrobiologie* **6** (1981) 322–328.
[5.214] M. J. Eichner: "Nitrous Oxide Emissions from Fertilized Soils, a Summary for Available Data," *J. Environ. Qual.* **19** (1990) no. 2, 272–280.
[5.215] H. Munk: "Zur vertikalen Wanderung mineralischer Phosphate bei starker Phosphatdüngung," *Landwirtsch. Forsch. Sonderh.* **27** (1972) 192–199.
[5.216] H. Kuntze: "Düngung und Gewässergüte," *Landwirtsch. Forsch. Sonderh.* **35** (1979) 14–24.
[5.217] B. Seggewiss: "Magnesiumdynamik im wurzelnahen Boden und Magnesiumverfügbarkeit unter dem Einfluß von Kalium," Ph. D. Thesis, Universität Göttingen 1986.
[5.218] P. Foerster, H. Neumann: "Stoffbelastung kleiner Fließgewässer in landwirtschaftlich genutzten Gebieten Norddeutschlands," *Mitt. Niedersächs. Wasseruntersuchungsamt* 7, Hildesheim 1981.
[5.219] E. Rübesam: "Die Wirkung der Stallmistdüngung dargestellt an Dauerversuchen in Thyrow und Müncheberg," *Tagungs-Ber. Dtsch. Akad. Landwirtschafts. Wiss.*, 1962, 23–37.
[5.220] F. Asmus: "Einfluß organischer Dünger auf Ertrag, Humusgehalt des Bodens und Humusreproduktion," *Landwirtsch.* **4** (1993) 127–139.
[5.221] H. Sturm, A. Buchner, W. Zerulla: *Gezielter Düngen*, DLG-Verlag, Frankfurt/Main 1994.
[5.222] F. Asmus: "Zeit, Menge und Art bestimmen den Erfolg," *Neue Landwirtschaft* **4** (1994) 47–51.
[5.223] J. Jungk, K. Isermann, G. Henjes: "Einfluß von cadmiumhaltigen Düngerphosphaten auf die Cadmiumanreicherung von Kulturböden und Nutzpflanzen," *Landwirtsch. Forsch. Sonderh.* **32** (1979) 262.
[5.224] A. Finck: "Phosphat-Düngung – Aktuelle Fragen aus agronomischen und ökologischen Perspektiven," *VDLUFA-Schriftenr.* **37** (1993) 133–136.
[5.225] H.-P. Blume: "Pflanzenschutzmittel (Pestizide)," in H.-P. Blume (ed.): *Handbuch des Bodenschutzes*, ecomed, Landsberg 1990, pp. 311–340.
[5.226] C. M. Himel, H. Loats, G. W. Bailey: "Pesticide Sources to the Soil and Principles of Spray Physics," in H. H. Cheng (ed.): *Pesticides in the Soil Environment: Processes, Impacts, and Modeling*, Soil Science Society of America, Madison, Wis., 1990, pp. 7–50.
[5.227] K.-H. Domsch: *Pestizide im Boden, mikrobieller Abbau und Nebenwirkungen auf Mikroorganismen*, VCH Verlagsgesellschaft, Weinheim 1992.
[5.228] H. H. Cheng: "Pesticides in the Soil Environment – An Overview," in H. H. Cheng (ed.): *Pesticides in the Soil Environment: Processes, Impacts, and Modeling*, Soil Science Society of America, Madison, Wis., 1990, pp. 1–5.
[5.229] J.-M. Bollag, S.-Y. Liu: "Biological Transformation Processes of Pesticides," in H. H. Cheng (ed.): *Pesticides in the Soil Environment: Processes, Impacts, and Modeling*, Soil Science Society of America, Madison, Wis., 1990, pp. 169–212.
[5.230] B. L. Sawhney, K. Brown (eds.): *Reactions and Movement of Organic Chemicals in Soils*, Soil Science Society of America, American Society of Agronomy, Madison, Wis. 1989.
[5.231] R. W. Wallnöfer, G. Engelhardt: "Microbial Degradation of Pesticides," *Chem. Plant Prod.* **2** (1989) 1–115.
[5.232] J. M. Dougherty, G. R. Lanza: "Anaerobic Subsurface Soil Microcosms: Methods to Monitor Effects of Organic Pollutants on Indigenous Microbial Activity," *Toxic. Assess.* **4** (1989) 85–104.
[5.233] R. J. Hance: "Herbicide Behaviour in the Soil, with Particular Reference to the Potential for Ground Water Contamination," *Prog. Pestic. Biochem. Toxicol.* **6** (1987) 223–247.
[5.234] G. T. Hickmann, J. T. Novak: "Relationship Between Subsurface Biodegradation Rates and Microbial Density," *Environ. Sci. Technol.* **23** (1989) 525–532.
[5.235] K. Alef: "Ökotoxikologische Verfahren," in K. Alef (ed.): *Biologische Bodensanierung*, VCH Verlagsgemeinschaft, Weinheim 1994, pp. 171–186.
[5.236] K. Alef: "Anaerobic Microbial Activities in Soil," in K. Alef, P. Nannipieri (eds.): *Methods in Applied Soil Microbiology and Biochemistry*, Academic Press, London 1995, pp. 271–310.
[5.237] Chemikaliengesetz (1982) Gesetz zum Schutz vor gefährlichen Stoffen, BGBl. I, novelliert 14.03.1990.
[5.238] Pflanzenschutzgesetz – PflSchG (Gesetz zum Schutz der Kulturpflanzen), Sep. 15, 1986 (BGBl. I, S. 1505), in C. H. Beck (ed.): *Umwelt-Recht*, Verlag C. H. Beck, München 1987, Chap. 9.2.
[5.239] P. Rudolph, R. Boje: *Ökotoxikologie. Grundlagen für die ökotoxikologische Bewertung von Umweltchemikalien nach dem Chemikaliengesetz*, ecomed, Landsberg 1986.
[5.240] OECD-Guidelines for Testing of Chemicals, Paris 1984.
[5.241] C. A. M. Van Gestel, N. M. Van Straalen: "Ecotoxicological Test System for Terrestrial Invertebrates," in M. H. Donker, H. Eijsackers, F. Heimbach (eds): *Ecotoxicology of Soil Organisms*, Lewis Publishers, Boca Raton 1994.
[5.242] K. Alef: *Methodenhandbuch Bodenmikrobiologie*, ecomed, Landsberg 1991, pp. 135–150.
[5.243] K. Alef, J. C. Forster, A. Hauck, G. Bergmann: "Arginin-Mineralisation zur Beurteilung der Wirkung von Chemikalien in Umweltproben," in K. Alef, H. Fiedler, O. Hutzinger (eds.): *Umweltmonitoring und Bioindikation*, Ecoinforma Press, Universität Bayreuth, Bayreuth 1994, pp. 325–334.
[5.244] J. R. Corbett: *The Biochemical Mode of Action of Pesticides*, Academic Press, London 1974.
[5.245] L. Somerville, M. P. Greaves: *Pesticide Effects on Soil Microflora*, Tylor & Francis, London 1987.
[5.246] K. E. Lee: *Earthworms*, Academic Press, Sydney 1985.
[5.247] G. Simon-Sylvester, J. C. Fourner: "Effects of Pesticides on the Soil Microflora," *Adv. Agron.* **31** (1979) 1–92.
[5.248] B. Weischer, J. Müller: "Nebenwirkungen von Pflanzenschutzmitteln auf Nematoden und ihre Antagonisten," in: *Pflanzenschutzmittel und Boden*, Paul Parey, Berlin 1985, pp. 159–175.
[5.249] B. Hock E. F. Elstner: *Pflanzentoxikologie*, Wissenschaftverlag, Mannheim 1984.
[5.250] E. Grigo: "Pflanzenschutz," in J. Oehmichen (ed.): *Pflanzenproduktion, 2. Produktionstechnik*, Paul Parey, Hamburg 1986, pp. 131–219.

[5.251] H. W. Berg, G. F. Diehl, H. Frank: *Rückstände und Verunreinigungen in Lebensmitteln,* D. Steinkopff Verlag, Darmstadt 1978.

[5.252] R. Schleyer, H. Kerndorff: *Die Grundwasserqualität westdeutscher Trinkwasserressourcen,* VCH Verlagsgesellschaft, Weinheim 1992.

[5.253] R. H. Plumb, Jr.: "The Importance of Volatile Organic Compounds as a Disposal Site Monitoring Parameter," in S. Lesage, R. E. Jackson (eds.): *Groundwater Contamination and Analysis at Hazardous Waste Sites,* Marcel Dekker, New York 1992.

[5.254] R. H. Plumb, Jr.: "Disposal Site Monitoring Data: Observation and Strategy Implications," in B. Hitchon, M. Trudell (eds.): "Hazardous Wastes in Groundwater: A Soluble Dilemma," *Proceedings of the Second Canadian/American Conference on Hydrology,* Banff, AB, Canada 1985, pp. 66–67.

[5.255] R. H. Plumb, Jr., A. M. Pitchford: "Volatile Organic Scans: Implications for Groundwater Monitoring," in: *Proceedings of the National Water Well Association/American Petroleum Institute Conference on Petroleum Hydrocarbons and Organic Chemicals in Groundwater,* Housten, TX, Nov. 13–15, 1985, pp. 1–15.

[5.256] H. Kerndorff, R. Schleyer, H. H. Dieter: "Bewertung der Grundwassergefährdung von Altablagerungen – Standardisierte Methoden und Maßstäbe," *WaBoLu-Hefte 1/1993.*

[5.257] H. Kerndorff et al.: "Erfassung grundwassergefährdender Altablagerungen – Ergebnisse hydrogeochemischer Untersuchungen," *WaBoLu-Hefte* 5/**1985**.

References for Chapter 6

[6.1] ISO Standard Compendium: Environment – Soil Quality – General Aspects, Chemical and Physical Methods of Analysis: Biological Methods of Analysis, 1st ed., 1994.

[6.2] P. Dominik, A. Paetz: *Methodenhandbuch Bodenschutz I,* Forschungsbericht UFOPLAN no. 10705001/07, Umweltbundesamt, Berlin 1994.

[6.3] ISO/CD 10381-1/4: Soil Quality - Sampling, Part 1: Guidance on the Design of Sampling Programmes 1994; Part 4: Guidance on the Investigation of Natural, Near Natural, and Cultivated Sites, 1994.

[6.4] N. Nothbaum, R. W. Scholz, Th. W. May: *Probenplanung und Datenanalyse bei kontaminierten Böden,* Erich Schmidt Verlag, Berlin 1994.

[6.5] ISO/CD 10381-2: Soil Quality – Sampling, Part 2: Guidance on Sampling Techniques, 1994.

[6.6] ISO/CD 10381-3: Soil Quality – Sampling, Part 3: Guidance on Safety, 1994.

[6.7] Bodenschutzzentrum Nordrhein-Westfalen: *Mindestdatensatz Bodenuntersuchungen* (Abschlußbericht des AK Mindestdatensatz Bodenuntersuchungen), Landesumweltamt, Essen 1991.

[6.8] ISO 11464: Soil Quality – Pretreatment of Samples for Physico-chemical Analysis, 1994.

[6.9] P. Tschöpel: "Fehlerquellen beim Aufschluß," in M. Stoeppler (ed.): *Probenahme und Aufschluß,* Springer Labormanual, 111–120, Springer Verlag, Berlin 1994.

[6.10] O. Fränzle: *Methodik für Schadstoffuntersuchungen in Böden* (draft), Geograph. Institut der Universität Kiel, 1994.

[6.11] E. Jackwerth, M. Würfels: "Der Druckaufschluß – apparative Möglichkeiten, Probleme und Anwendungen," in M. Stoeppler (ed.): *Probenahme und Aufschluß,* Springer Labormanual, 121–138, Springer Verlag, Berlin 1994.

[6.12] L. Dunemann: "Mikrowellenaufschluß," in M. Stoeppler (ed.): *Probenahme und Aufschluß,* Springer Labormanual, 139–150, Springer Verlag, Berlin 1994.

[6.13] G. Knoop, B. Pehl, D. Schlösser: "Vergleichende Untersuchungen zum Feststoffaufschluß mit der Mikrowellentechnik," *Korresp. Abwasser* **41** (1994) 1836–1839.

[6.14] L. Dunemann: "Aufschlußmethoden für die Schwermetallanalytik (Marktübersicht)," *Nachr. Chem. Tech. Lab.* **39** (1991) no. 10, M 2–M 16.

[6.15] H. Hein: "Aktuelle chromatographische Umweltanalytik," *CLB Chem. Labor Betr.* **43** (1992) 7–15, 65–75.

[6.16] W. Engewald: "Gaschromatographie (Marktübersicht)," *Nachr. Chem. Tech. Lab.* **42** (1994) no. 4, 356–366; **43** (1995) no. 3, M 1–M 6.

[6.17] G. Schwedt: *Taschenatlas der Analytik,* Thieme Verlag, Stuttgart 1992.

[6.18] K. Unger: "HPLC-Quo vadis?" *Labor 2000,* **1994**, 90–93.

[6.19] W. Funk, V. Dammann, G. Donnevert: *Qualitätssicherung in der Analytischen Chemie,* VCH Verlagsgesellschaft, Weinheim 1992.

[6.20] L. Viereck, U. Ewers, M. Fliegner: "Erarbeitung und Zusammenstellung von Hinweisen zur Entnahme und Untersuchung von Bodenproben," report (unpublished), 1993.

[6.21] VDLUFA: "Aufschluß mit Königswasser," in: *Umweltanalytik,* Methodenbuch vol. VII, Kennziffer 2.1.2 (draft), VDLUFA-Verlag, Darmstadt 1995.

[6.22] VDLUFA: "Bestimmung von Thallium," in: *Umweltanalytik,* Methodenbuch vol. VII, Kennziffer 2.2.4 (draft), VDLUFA-Verlag, Darmstadt 1995.

[6.23] DIN, Extraktion von Spurenelementen aus Böden mit Ammoniumnitratlösung, E DIN 19730 (2nd draft), 1995.

[6.24] P. Dominik, A. Paetz: *Methodenbuch Bodenschutz I,* Forschungsbericht UFOPLAN no. 10705001/07, Umweltbundesamt, Berlin 1994.

[6.25] W. Jennings: *Analytical Gas Chromatography,* Academic Press, New York 1987.

[6.26] K. J. Hyver, P. Sandra: *High Resolution Gas Chromatography,* Hewlett-Packard, Palo Alto 1989.

[6.27] H. Engelhardt: *Practice of High Performance Liquid Chromatography,* Springer Verlag, Heidelberg 1986.

[6.28] J. W. Dolan, L. R. Snyder: *Troubleshooting LC Systems,* Humana Press, London 1989.

[6.29] Merkblätter no. 1, Bestimmung von polyzyklischen aromatischen Kohlenwasserstoffen in Bodenproben, Landesumweltamt Nordrhein-Westfalen, Essen 1994.

[6.30] J. Bortlisz, L. Velten: "Herstellung und Bestimmung der Trockenmasse von Klärschlämmen, Sedimenten und Böden mit Hilfe der Gefriertrocknung," *Abwassertech. Abfalltech. Recycl.* **1991**, no. 4, 35–37.

[6.31] I. Blankenhorn, D. Meijer, R. J. van Delft: "Interlaboratory Comparison of Methods Used for Analysing Polycyclic Aromatic Hydrocarbons (PAHs) in Soil Samples," *Fresenius J. Anal. Chem.* **343** (1992) 497–504.

[6.32] C. Herziger-Möhlmann, A. Manns: "Erkundung und Bewertung von Bodenkontaminationen mit immunochemischen Meßverfahren," *Draegerheft* **357** (1994) 2–6.

[6.33] S. B. Hawthorne: "Analytical Scale Supercritical Fluid Extraction," *Anal. Chem.* **62** (1990) no. 11, 633.

[6.34] F. Höfler: "Überkritische Fluidextraktion, Teil I," *LaborPraxis* **16** (1992) 350.

[6.35] F. Höfler: "Überkritische Fluidextraktion," *LaborPraxis* **16** (1992) 506.

[6.36] E.-G. Zeschmann: *Beurteilung und Sanierung ölverunreinigter Standorte*, Expert Verlag, Böblingen 1987.

[6.37] EPA Method 9071 A/8440, Total Recoverable Petroleum Hydrocarbons by Infrared Spektrophotometry, Revision Draft, 1992.

[6.38] ISO/TR 11046, Determination of Mineral Oil Content-Method by Infrared Spectrometry and Gas Chromatography Method, 1994.

[6.39] H. Fischer et al.: "Bestimmung von Mineralöl-Kohlenwasserstoffen im Boden," *UWSF-Z Umweltchem. Ökotox.* **6** (1994) no. 4, 189–195.

[6.40] EPA Method SW 846/3560, Supercritical Fluid Extraction, Revision Draft 1, 1992.

[6.41] F. Höfler, J. Ezzell, B. Richter: "Beschleunigte Lösemittelextraktion (ASE)," *LaborPraxis* **19** (1995) nos. 3, 4.

[6.42] EPA M 3545, Proposal, Accelarated Solvent Extraction, 1994.

[6.43] M. Stoeppler: *Probenahme und Aufschluß*, Springer Verlag, Heidelberg 1994.

[6.44] S. Michels, T. Eikmann, *VDI Ber.* **837** (1990) 509.

[6.45] H. J. Einbrodt et al., *Wiss. Umwelt* **2** (1985) 149–157.

[6.46] T. Eikmann in V. Franzius, R. Stegmann, K. Wolf (eds.): *Handbuch Altlastensanierung*, Section 1.3.2.2, R. v. Decker's Verlag, Heidelberg 1988, p. 1.

[6.47] T. Eikmann, S. Michels: "Wirkung von Luftverunreinigungen auf Böden," *VDI-Ber.* **837** (1990) 489.

[6.48] H. Heil et al., *Vom Wasser* **72** (1989) 321–348.

[6.49] T. Eikmann, A. Kloke in D. Rosenkranz, G. Einsele, H.-M. Harreß (eds.): *Handbuch Bodenschutz 3590*, 2nd ed., Erich Schmidt Verlag, Berlin 1993.

[6.50] T. Eikmann, *Altlasten-Spektrum* **1** (1993) 31–38.

[6.51] B. Hudec et al., *Altlasten-Spektrum* **3** (1993) 164–168.

[6.52] S. Eikmann, U. Lieser, T. Eikmann, *Altlasten-Spektrum* **2** (1993) 75–84.

References for Chapter 7

[7.1] P. Doetsch, P. Dreschmann: "Verfahrensdokumente wichtiger Dekontaminationsverfahren als Leitfaden für die Sanierungsuntersuchung bzw. -planung bei Altlasten," in V. Franzius, R. Stegmann, K. Wolf, E. Brandt (eds.): *Handbuch der Altlasten-Sanierung*, R. V. Decker's Verlag, G. Schenck, Heidelberg, 1993, chaps. 5.4.1.0.0 ff.

[7.2] G. Kielburger, H.-J. Schmitz: "Bodenbehandlungszentren: Die Jagd nach dem Boden hat begonnen," *TerraTech* **3** (1993) 46–57.

[7.3] W. Neemann, F. Burkant: "In-situ Sanierung unter Einsatz des Frac-Verfahrens," in V. Franzius (ed.): "Sanierung kontaminierter Standorte 1993," *Abfallwirtsch. Forsch. Prax.* **60** (1993) 321–327.

[7.4] ARBES GmbH, Bodensanierung mit Hochfrequenzenergie, company brochure, Berlin 1994.

[7.5] G. Schmitz: "Rekultivierung," in *Taschenwörterbuch der Umweltplanung*, vol. 1614, F. Malz, München 1974, p. 463.

[7.6] I. Stratmann: "Ertragskundliche Untersuchungen auf Rekultivierungsflächen im rheinischen Braunkohlegebiet," *Braunkohle (Düsseldorf)* **37** (1985) 484–491.

[7.7] E. Friedrich: "Forststandörtliche Eignung des 'Forstkieses' aus Löß und Rheinschotter im Hinblick auf Lagerungsdichte, Mischungsverhältnis, Nährstoffe," Untersuchung im Auftrag der Rheinbraun AG, Köln 1985 (unpublished).

[7.8] K. Winter: "Bodenmechanische und technische Einflüsse auf die Qualität von Neulandflächen," *Braunkohle (Düsseldorf)* **42** (1990) no. 10, 15–23.

[7.9] R. Starke: "Wiedernutzbarmachung im rheinischen Braunkohlerevier," *Braunkohle (Düsseldorf)* **42** (1990) no. 10, 4–10.

[7.10] G. Dumbeck: "Bodenkundliche Aspekte bei der landwirtschaftlichen Rekultivierung im rheinischen Braunkohlerevier," *Braunkohle (Düsseldorf)* **44** (1992) no. 9, 8–11.

References for Chapter 8

[8.1] M. Kloepfer: *Umweltrecht*, Verlag C. H. Beck, München 1989, p. 814 ff.
P.-C. Storm: "Bodenschutzrecht," in O. Kimminich, H. v. Lersner, P.-C. Storm (eds.): *Handwörterbuch des Umweltrechts*, 2nd ed., vol. I, Erich Schmidt Verlag, Berlin 1986.
W. Burhenne: *Umweltrecht*, loose leafe collection, 7 vols., Erich Schmidt Verlag, Berlin.
M. Kloepfer: *Umweltschutz*, Verlag C. H. Beck, München 1981 ff.
P.-C. Storm: *Umwelt-Recht, Wichtige Gesetze und Verordnungen zum Schutz der Umwelt*, Textausgabe, 8th ed., Deutscher Taschenbuch Verlag, München 1994.

[8.2] Umweltprogramm der Bundesregierung (Environmental Program of the Federal Government), 1971 (Bundestags-Drucksache 6/2710).

[8.3] Bundesinnenministerium (eds.): *Bodenschutzkonzeption der Bundesregierung* (Soil Protection Concept of the Federal Government) (Bundestags-Drucksache 10/2977), Kohlhammer Verlag, Stuttgart 1985.
Bericht der Bundesregierung, Maßnahmen zum Bodenschutz, 1988 (Bundestags-Drucksache 11/1625).

[8.4] H. Ebersbach: *Rechtliche Aspekte des Landverbrauchs am ökologisch falschen Platz*, Erich Schmidt Verlag, Berlin 1985 (UBA-Berichte 1/85).

[8.5] D. Rosenkranz, G. Einsele, H.-M. Haress (eds.): *Bodenschutz*, Ergänzbares Handbuch, Erich Schmidt Verlag, Berlin 1988.

[8.6] P.-C. Storm: "Umweltrechtsprogramm von Montevideo," in O. Kimminich, H. v. Lersner, P.-C. Storm (eds.): *Handwörterbuch des Umweltrechts*, 2nd ed., vol. II, Erich Schmidt Verlag, Berlin 1994.

[8.7] Bundesumweltministerium (eds.): *Bericht der Bundesregierung über die Konferenz der Vereinten Nationen für Umwelt und Entwicklung Rio de Janeiro, 1992*, Bonn 1992, pp. 30.

[8.8] Bodenschutzgesetz (BodSchG, Soil Protection Act) § 3, Baden-Württemberg, 24. June 1991 (GBl. S. 434).

[8.9] In Ref. [8.3], p. 21.